MATERIALS PROCESSING HANDBOOK

MATERIALS PROCESSING HANDBOOK

EDITED BY

JOANNA R. GROZA
JAMES F. SHACKELFORD
ENRIQUE J. LAVERNIA
MICHAEL T. POWERS

CRC Press is an imprint of the
Taylor & Francis Group, an informa business

Cover: Si nano-bridges between two vertical Si walls synthesized using metal-catalyzed chemical vapor deposition. The work was done at Quantum Science Research group of Hewlett Packard Laboratories. -M. Saif Islam, University of California, Davis, CA

CRC Press
Taylor & Francis Group
6000 Broken Sound Parkway NW, Suite 300
Boca Raton, FL 33487-2742

First issued in paperback 2019

ISBN-13: 978-0-8493-3216-6 (hbk)
ISBN-13: 978-0-367-38930-7 (pbk)

Library of Congress Cataloging-in-Publication Data

Materials processing handbook / editors: Joanna R. Groza ... [et al.].
p. cm.
Includes bibliographical references and index.
ISBN-13: 978-0-8493-3216-6 (0-8493-3216-8 : alk. paper)
1. Manufacturing processes--Handbooks, manuals, etc. 2. Materials science--Handbooks, manuals, etc. I. Groza, Joanna R.

TS183.M3827 2007
670--dc22 2006030545

Visit the Taylor & Francis Web site at
http://www.taylorandfrancis.com

and the CRC Press Web site at
http://www.crcpress.com

Contents

SECTION II Deposition Processes

SECTION III Dislocation-Based Processes

SECTION IV Microstructure Change Processes

SECTION V Macroprocesses

SECTION VI Multiscale Processes

Preface

By the dawn of the 21st century, the field of materials science and engineering has evolved into a science of its own, embracing the well-established disciplines of physical metallurgy and ceramic/glass engineering, along with new and emerging developments in electronic, optical, and magnetic materials, as well as semiconductors, polymers, composites, bio- and nano-materials. Despite the enormous diversity in modern day advanced engineering materials, they are tied together by unifying concepts and first principles in areas such as thermodynamics of equilibria, statistical mechanics, phase transformations, matter and energy transport, as well as fundamental material structure from the atomic to macroscopic level. In the traditional representation of materials science as a tetrahedron, *processing* plays a central and critical role: processing generates the *microstructure* of a material, which in turn imparts the desired *properties* and *performance.* With the impetus created by the rapid pace of contemporary technological innovation, the field of materials processing has grown exponentially in both popularity and importance. The explicit dependence of the ultimate properties of a material on the specific processing steps employed in its fabrication, places materials processing in a decisive position not only for the production and application of conventional engineering materials, but for the future of new and novel materials as well.

Traditionally, materials processing has been considered part of materials technology or engineering and as such, was deemed the practical complement of materials science, with a high degree of associated empiricism. The evolution and complexity of new materials, such as cutting edge semiconductors, smart materials, high T_c superconductors, and materials based on spintronics, has enthused contemporary materials processing beyond this stage. However, in contrast to the rigor and unity of materials property or structure treatments in the literature, materials processing has been somewhat neglected. First, few materials curricula provide in depth coverage of materials processing. Second, when processing is addressed in handbooks or textbooks, it is primarily from a technological or practical engineering point of view, with a conspicuous dearth of materials science fundamentals. Our intent is that the *Materials Processing Handbook* will fill these gaps.

This handbook is intended to provide broad coverage of a number of materials processes associated with a myriad of solid materials, including ceramics, polymers, metals, composites, and semiconductors. Our goal is to present the fundamentals of a particular materials process by emphasizing the integral processing–structure–property relationship. Principles of thermodynamics, phase transformations, mechanisms, and kinetics of energy and mass transport are defined for each process category. Simulation and modeling of materials processes are an important part of the chapter presentations. Traditional, as well as novel processes are covered and the scale of the materials structures and associated processing spans from the nanometer level to macroscopic. Several challenges have been recognized with this approach. First, some materials processes have minimal or no associated microstructural change (e.g., the production of raw materials, chemical synthesis, machining processes, etc.) and as such, will not be covered in this work. The mechanics and design aspects of process development have not been emphasized here. Although some treatment of processing equipment is provided, it is done so only to enhance the understanding of a specific materials process. Since materials professionals and practitioners occasionally require quick insights or know-how to help them solve a demanding process problem, this handbook balances an emphasis on fundamentals with practical examples, case studies, and applications for each of the materials processes covered.

The thirty-one chapters covered in this handbook are organized into six sections by the type of materials change (phase, structure, or shape), with the sections roughly corresponding to increasing lengths of scale. Each chapter within a section describes the principles, processing techniques, and means for controlling microstructural evolution to achieve the final desired properties and performance. The first section, Small Scale Processes, addresses process events that occur at atomic or nanoscale dimensions. The second section deals specifically with deposition processes, while the third section focuses on processes that involve dislocations and plastic deformation. The fourth section is devoted to phase transitions, shape, and chemistry changes that modify the microstructure and hence, the properties of materials. The fifth section addresses processes that occur at the macroscopic scale. Finally, the sixth section, Multiscale Processes, considers the basics of process integration, that is, combinations of any of the above processes occurring over a range of length scales.

Since this handbook is intended to be a wide-ranging "one-stop" reference in materials processing for a variety of advanced engineering materials, it is our hoped that engineers, scientists, and students will be provided with an appreciation of the fundamental principles behind each of the processes presented. World class experts in materials processing have been brought together to convey the principles and applications contained in this compilation. Their creative transfer of knowledge spans the gamut from traditional to emerging industries, from conventional to novel materials, across length scales, from simulation and modeling to real materials processes. It is our aspiration that this handbook will foster an understanding of the technical challenges associated with these processes, that it will help practitioners avoid processing inconsistencies, which may be counterproductive and costly, that it will aid in the selection of a particular process for an intended application, and that it provides inspiration to researchers, designers, and inventors.

About the Editors

Joanna R. Groza is a professor at the University of California at Davis in the Department of Chemical Engineering and Materials Science. She holds a master of science with honors in metallurgical engineering, a Ph.D. in physical metallurgy and is a registered professional engineer in California. She worked in industry in materials processing and powder metallurgy, gaining experience with traditional processing and nonconventional bonding and sintering techniques.

Her research focuses on field assisted sintering of difficult to sinter powders, including nanocrystalline particles, simultaneous sintering and reactions, phase transformations, processing and microstructural characterization of various materials, net shape manufacturing. She is active in several professional organizations and is a Board of Review member for *Metallurgical and Materials Transactions.* She authored (or coauthored) more than 100 technical papers, invited lectures, three book chapters on nanomaterials processing and properties, and a couple of chapters in the ASM Handbook, vol. 7 (*Powder Metallurgy*). She has served on numerous panels and grant reviews for NSF, ARO, LLNL, DOE, and has been involved in a number of NSF/NIST workshops on materials processing.

Enrique J. Lavernia became the dean of the College of Engineering at UC Davis, on September 1, 2002. He came to UC Davis from the Department of Chemical Engineering and Materials Science at UC Irvine, where he was the chair of the department.

Dean Lavernia received his M.S. and Ph.D. degrees from the Massachusetts Institute of Technology and his bachelor's degree from Brown University. After completing a postdoctoral research program at MIT, he joined the faculty at UCI. He has held prestigious fellowships from the Ford Foundation; from the Alexander von Humboldt Foundation, from the Iketani Science and Technology Foundation of Tokyo, and from Rockwell International. Dean Lavernia has coauthored over 350 journal papers, over 190 conference papers, and one book, *Spray Atomization and Deposition.* He is currently the principal editor of the international journal, *Materials Science and Engineering A* and associate editor of the *Journal of Metastable and Nanostructured Materials.*

Awards Received	Dates
Fellow, American Society of Mechanical Engineers (ASME)	2006
Best Paper Award, International Thermal Spray Conference Seattle, WA	2006
Elected as an Honorary Member of the Materials Research Society of India (MRSI)	2006
Brown University, *Outstanding Engineering Alumnus Medal*	2005
Named *Highly Cited Researcher*, ISI, Citation Impact for Research in Materials	2002
Chancellor's Professor, UCI	2002
Fellow, *American Association for the Advancement of Science*	2000
1999 Marcus A. Grossmann Award for "Best Paper," *Metallurgical and Materials Transactions* (with Weidong Cai — Ph.D. Student)	1999
Chemical and Biochemical Engineering and Materials Science Teacher of the Year, UCI	1998
1998 Fellow, Board of Trustees of ASM International	1998
Marion Howe Medal for "Best Paper," *Metallurgical and Materials Transaction* (with D. Lawrynowicz and B. Li – Ph.D. Students)	1998
Silver Medal of the Materials Science Division of ASM International	1996
Best Paper Award, *Journal of Thermal Spray Technology* with X. Liang (Ph.D. Student) And J. Wolfenstine (Faculty, Colleague)	1995
Fellowship, *Ford Foundation*	1995
Alexander Von Humboldt Fellowship from Germany	1995
Fellowship from the *Iketani Science and Technology Foundation*, Tokyo, Japan	1993
ASM International *1993 Bradley Stoughton Award for Young Teachers*	1993
Ranked 21st in the World by Science Watch, ISI, *Citation Impact for Research in Materials* [1]	1990–1994
Young Investigator Award, Office of Naval Research (ONR)	1990–1993
Presidential Young Investigator, National Science Foundation (NSF)	1989–1994

Some of Dean Lavernia's research accomplishments include the study and development of spray atomization and coinjection for the manufacture of metal matrix composites (MMCs); the analysis of fundamentals of microstructure evolution during spray deposition; the modeling and analysis of ceramic particle-molten droplet interactions; the development of elevated temperature spray facilities for reactive materials processing; the development of physical models to explain microstructure phenomena in spray processes; and the study and development of the spray atomization, reaction, and deposition technique. His more recent work has been in the area of nanostructured materials (NMs), which have the potential of using atomic-level structural control to revolutionize traditional material, since their engineering properties can be tailored to specific requirements. His research has demonstrated that the production of fully dense, bulk nanostructured materials is possible by means of a novel approach that uses mechanical alloying/milling under liquid nitrogen. This technique yields thermally stable, nanostructured powder materials. His recent work has demonstrated that the mechanical properties of bulk NMs at room temperature can be greatly improved over the properties of their conventional, coarse-grained counterparts.

James F. Shackelford earned B.S. and M.S. degrees in ceramic engineering from the University of Washington and a Ph.D. in materials science and engineering from the University of California, Berkeley. He is currently a professor in the Department of Chemical Engineering and Materials Science at the University of California at Davis. For many years, he served as the associate dean for undergraduate studies in the College of Engineering and currently serves as the director of the Integrated Studies Honors Program, an invitational, residential program for first-year students from a wide spectrum of majors. Dr. Shackelford also serves as an associate director for education for the National Science Foundation-funded

Center for Biophotonics Science and Technology (CBST). He teaches and conducts research in the areas of Materials Science, the structure of materials, nondestructive testing, and biomaterials. A member of the American Ceramic Society and ASM International, he was named a Fellow of the American Ceramic Society in 1992 and received the Outstanding Educator Award of the American Ceramic Society in 1996. In 2003, he received a Distinguished Teaching Award from the Academic Senate of the University of California, Davis. He has published over 100 archived papers and books including *Introduction to Materials Science for Engineers* now in its 6th Edition and *The CRC Materials Science and Engineering Handbook* now in its 3rd Edition. The *Introduction to Materials Science for Engineers* has now been translated into Chinese, German, Korean, and Spanish.

Michael T. Powers is currently a materials scientist in the Electronic Measurements Group of Agilent Technologies (formerly part of the Hewlett-Packard Company) at their facility in Santa Rosa, California, where he has worked for the past 15 years. His technological expertise is in the field of joining process technology, including glass-to-metal sealing, active metal brazing, reactive joining, and lead-free soldering. Mike earned his B.S. (1989) and M.S. (1992) from the University of California, Berkeley, both in materials science and engineering.

Powers holds eight U.S. patents, with two pending, in the areas of chemical engineering and high-frequency connector design. He has published a number of technical papers on joining of advanced engineering materials and materials characterization, and has served as editor for various conference proceedings. In 1995, he codeveloped a nonozone depleting solvent chemistry to replace CFC-113 for critical cleaning of components used in the electronics industry. This solvent technology, called Synergy CCS™, received the 1996 U.S. Department of Energy Pollution Prevention Award and the 1997 Excellence in Technology Transfer Award from the U.S. Federal Laboratory Consortium. Mike is a member of the American Welding Society, TMS, and the American Society for Metals International, which recently elected him to Fellowship. He is a lecturer at the University of California, Davis in the Department of Chemical Engineering and Materials Science, where he teaches a senior level capstone projects course called Materials in Engineering Design.

Contributors

John Ågren
Department of Material Science and Engineering
Royal Institute of Technology
Stockholm, Sweden

Pulickel M. Ajayan
Nanocenter
Rensselaer Polytechnic Institute
Troy, NY

C. Y. Barlow
University of Cambridge, UK

Matthew R. Barnett
GTP/Centre for Material and Fibre Innovation
Deakin University
Geelong, Victoria, Australia

Carol Barry
University of Massachusetts
Lowell, MA

Michel Belett
Ecole des Mines de Paris
Paris, France

Peter Bjeletich
Intel Corporation
Santa Clara, CA

H. P. Buchkremer
Forschungszentrum Jülich
Institute for Materials and Processes in Energy Systems, IWV-1
Juelich, Germany

Indrajit Charit
Department of Materials Science and Engineering
University of Missouri
Rolla, MO

Julie Chen
University of Massachusetts
Lowell, MA

J. A. Christodoulou
U.S. Office of Naval Research
Arlington, VA

Alexis G. Clare
New York State College of Ceramics
Alfred, NY

J. P. Colinge
University of California
Davis, CA

Florian H. Dalla Torre
Laboratory of Metal Physics and Technology
ETH
Zurich, Switzerland

Jean-Pierre Delplanque
University of California
Davis, CA

Jürgen Eckert
IFW Dresden Institut für Komplexe Materialien (IKM)
Darmstadt, Germany

Uwe Erb
Departement of Materials Science and Engineering
University of Toronto
Toronto, Canada

Hans J. Fecht
Department of Materials
Institute of Nanotechnology
University of Ulm
Ulm, Germany

Roberto Fornari
Institute of Crystal Growth
Berlin, Germany

Randall M. German
Center for Advanced Vehicular System
Mississippi State University
Mississippi State, MS

Marty E. Glicksman
Material Science Engineering Department
Rensselaer Polytechnic Institute
Troy, NY

Joanna R. Groza
Chemical Engineering and Materials Science Department
University of California
Davis, CA

B. Q. Han
Chemical Engineering and Materials Science Department
University of California
Davis, CA

N. Hansen
Risø National Laboratory
Roskilde, Denmark

Peter D. Hodgson
GTP/Centre for Material and Fibre Innovation
Deakin University
Geelong, Victoria, Australia

M. Saif Islam
University of California
Davis, CA

G. D. Janaki Ram
Department of Mechanical and Aerospace Engineering
Utah State University
Logan, UT

Chad Johns
University of California
Davis, CA

Samuel Johnson
University of California
Davis, CA

John J. Jonas
Materials Engineering
McGill University
Montreal, Canada

Bernard H. Kear
Rutgers University
New Brunswick, NJ

O. M. Knio
Department of Mechanical Engineering
Johns Hopkins University
Baltimore, MD

Sergei O. Kucheyev
Lawrence Livermore National Laboratory
Livermore, CA

Tonya L. Kuhl
Department of CHMS
University of California
Davis, CA

Andreas A. Kündig
Laboratory of Metal Physics and Technology
ETH
Zurich, Switzerland

Wilfried Kurz
Swiss Federal Institute of Technology
Lausanne, Switzerland

E. J. Lavernia
College of Engineering
University of California
Davis, CA

Jörg F. Löffler
Laboratory of Metal Physics and Technology
ETH
Zurich, Switzerland

A. Lupulescu
Department of Materials Science and Engineering
Union College, Schenectady
Troy, NY

James W. Mayer
Arizona State University
Tempe, AZ

Joey Mead
University of Massachusetts
Lowell, MA

N. H. Menzler
Forschungszentrum Jülich
Institute for Materials and Processes in Energy Systems, IWV-1
Juelich, Germany

Rajiv S. Mishra
Department of Materials Science and Engineering
University of Missouri
Rolla, MO

Amit Misra
Los Alamos National Laboratory
Los Alamos, NM

Nathan W. Moore
Department of CHMS
University of California
Davis, CA

Amiya K. Mukherjee
University of California
Davis, CA

Michael Nastasi
Los Alamos National Laboratory
Los Alamos, NM

A. P. Newbery
Department of Chemical Engineering and Materials Science
University of California
Davis, CA

T. Nguyentat
Aerojet
Rancho Cordova, CA

M. Powers
Agilent Technologies, Inc.
Santa Rosa, CA

Subhash H. Risbud
University of California
Davis, CA

M. P. S. Saran
Department of Materials
University of Oxford
Oxford, England

Daniel Schmidt
University of Massachusetts
Lowell, CA

Sergio Scudino
IFW Dresden Institut für Komplexe Materialien (IKM)
Darmstadt, Germany

S. Sen
Department of Chemical Engineering and Materials Science
University of California
Davis, CA

Leon L. Shaw
University of Connecticut
Storrs, CT

Brent E. Stucker
Department of Mechanical and Aerospace Engineering
Utah State University
Logan, UT

C. Suryanarayana
Department of Mechanical, Material and Aerospace Engineering
University of Central Florida
Orlando, FL

Brian G. Thomas
University of Illinois
Urbana-Champaign, IL

R. I. Todd
Department of Materials
University of Oxford
Oxford, England

Rohit Trivedi
Ames Laboratory
Iowa State University
Ames, IA

Robert Vajtai
Nanocenter
Rensselaer Polytechnic Institute
Troy, NY

Ke Gang Wang
Material Science Engineering Department
Rensselaer Polytechnic Institute
Troy, NY

T. P. Weihs
Reactive Nanotechnologies
Hunt Valley, MD

G. Wilde
University of Münster
Institute of Materials Physics and Forschungszentrum Karlsruhe
Institute of Nanotechnology
Karlsruhe, Germany

Yizhang Zhou
University of California
Davis, CA

Yuntian T. Zhu
Los Alamos National Laboratory
Los Alamos, NM

I

Small-Scale (Atomic/Cluster/Nanoscale) Processes

1

Controlled Processes for Growth of Carbon Nanotube Structures

Robert Vajtai and
Pulickel M. Ajayan
Rensselaer Polytechnic Institute

Abstract

Carbon nanotubes (CNTs), which are fullerenes[1] elongated to extremely high aspects, have been studied extensively since their discovery[2] in 1991, mainly because of the extraordinary physical properties they exhibit in electronic, mechanical and thermal processes. For different applications, different properties and structures are important. Perfect CNTs have high Young modulus, but others may have lots of defects, which nevertheless provide the possibility for covalent or noncovalent functionality. Individual nanotubes make use of quantum effects; and organized structures contain millions of nanotubes to harness their synergy. Above all, the helicity in nanotubes is the most revealing feature to have emerged out of the first experimental[3] and theoretical work.[4–6] This structural feature has great importance, as electrical properties of nanotubes change drastically as a function of helicity and tube diameter.

In this chapter, we briefly summarize the history and the most important achievements of the last several years of CNT growth, concentrating on chemical vapor deposition (CVD), and more narrowly on experiments where the catalyst is fed via vapor during the process. We demonstrate our state-of-the-art method of tailored nanotube growth, both for the single- and multi-walled ones. To demonstrate the possible applications of CNTs and their structures we describe high-efficiency filters, nano-sized brushes and springs, devices based on field-emission, low-voltage gas breakdown sensors, and applications based on the enhanced properties of different composite materials that consist of nanotubes.

1.1 Basic Properties and Electrical Applications of Carbon Nanotubes

A single-walled nanotube (SWNT) may be considered as a specific, one-dimensional (1D) giant molecule composed purely of sp^2 hybridized carbon, while properties of multi-walled nanotubes (MWNTs) are closer to those of graphite. To prepare closed-shell graphite structures, one needs to insert topological defects into the hexagonal structure of graphene sheets. The extraordinary physical and chemical properties[7] and their possible applications can be attributed to the one-dimensionality and helicity of the nanotube structure. In addition to the extraordinary properties of individual CNTs their collective behavior is also important in most of the systems; it is critical how they interact with each other and also with their environment, for example, a polymer matrix. Several possible applications of CNTs are shown in Figure 1.1. For example, the first case displayed (top left) uses an individual MWNT contacted by metal electrodes. These nanotube structures may be used as electrical interconnects or as active semiconducting devices but a priori knowing and defining the specific characters of individual nanotubes have been difficult tasks. Electrical testing of this and similar structures showed that current density in CNTs can reach 10^9 to 10^{10} A/cm^2 (top right) without much damage.[8] The MWNTs used in this experiment were prepared by forming the structure in a carbon arc-discharge; the diameters of the two samples displayed were 8.6 and 15.3 nm. Two- and four-terminal resistances were measured in air at an ambient temperature of 250°C. The measurements were continued for 334 h and the nanotubes sustained these high currents without getting destroyed (high electro-migration resistance).

On the other hand, when one increases the current density in a specific shell of a MWNT or in a specific SWNT in a bundle, that shell or nanotube can be selectively destroyed (middle left schematics), tailoring the remaining part of the MWNT or SWNT bundle to be purely semiconductive or metallic.[9] An interesting approach is when nanotubes are organized vertically,[10] when CNTs may be used as interconnects or they may perform as active device elements (see the middle right schematics in Figure 1.1). The last two applications cited in Figure 1.1 (bottom) apply to a larger number of CNTs. One interesting possibility is to prepare nanotube filaments for conventional light bulbs (figure at bottom left).[11] The mechanical properties and the resistance to oxidation of long CNTs are far better than those of any metal. Shorter MWNTs may be organized into pure nanotube yarns, and textiles can be fabricated out of that yarn. One possible application of the CNT textile is in the preparation of planar light emitting devices,[12] which have various interesting applications (figure at bottom right). Recent success also demonstrates that extremely thin transparent films can be fabricated from SWNT mats, with possible applications in light emitting devices (transparent electrodes).[13]

1.2 Different Methods of Carbon Nanotube Production

CNT production methods can be classified based on the type of nanotubes that are produced, i.e. MWNT or SWNT production. First, both MWNTs[14] and SWNTs[15,16] were produced via electric arc-discharge

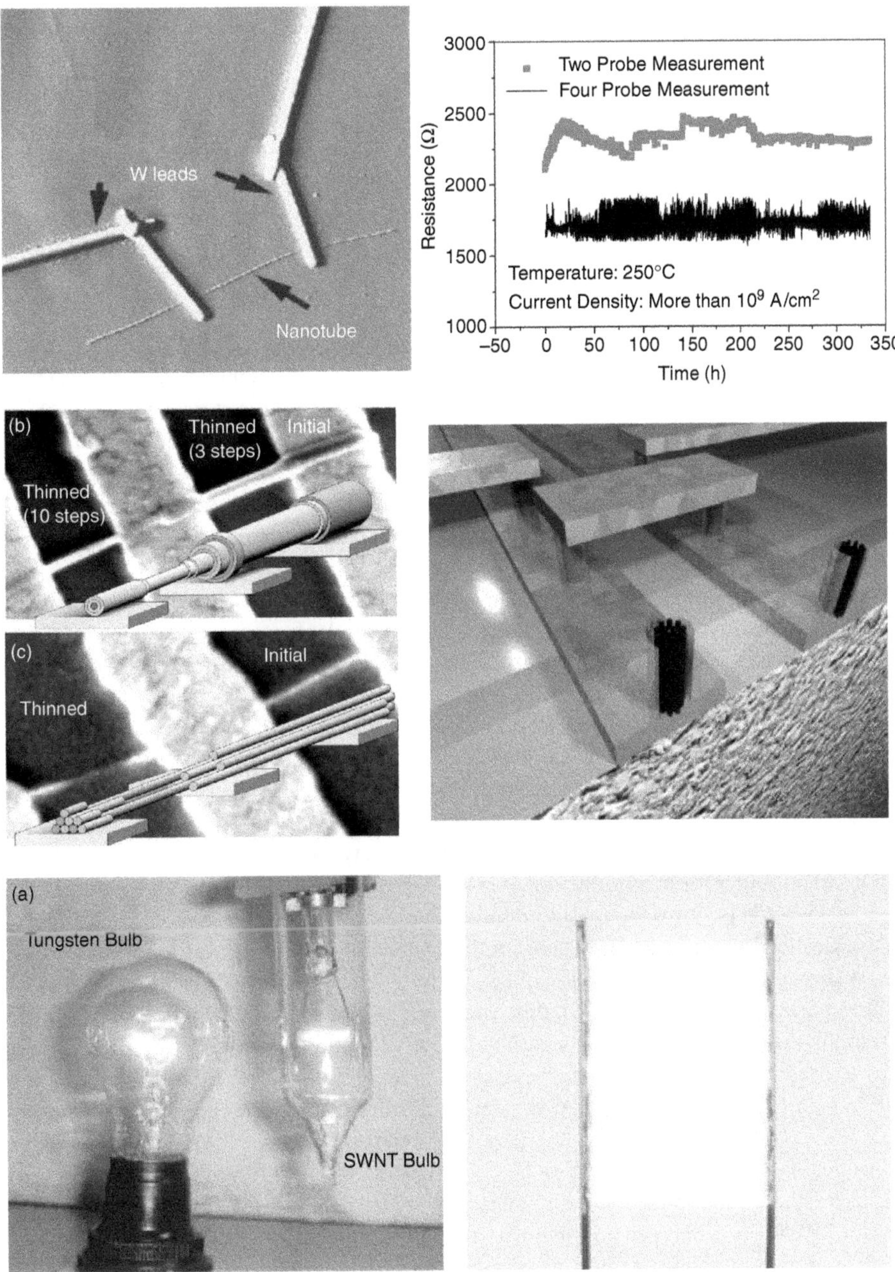

FIGURE 1.1 Select applications of CNT structures. (top) High current density measurements carried out on multi-walled CNTs at elevated temperature. (From Wei, B.Q., Vajtai, R., and Ajayan, P.M., *Appl. Phys. Lett.* 79, 1172, 2001. With permission from American Institute of Physics.) (middle left) Removing of the outer shell(s) of a MWNT or a specific SWNTs from a bundle to maintain an exclusively semiconductive or metallic structure. (From Collins, P.G., Arnold, M.S., Avouris, Ph., *Science*, 292, 706, 2001. With permission from AAAS.) (middle right) Vertical interconnects (via) prepared from CNT bundles. (From Graham, A.P., Duesberg, G.S., Seidel, R.V., Liebau, M., Unger, E., Pamler, W., Kreupl, F., and Hoenlein, W., *Small*, 1, 382, 2005. With permission from John Wiley & Sons, Inc.) (bottom left) Light bulb made with SWNT filament compared with a conventional tungsten bulb. (From Wei, J.Q., Zhu, H.W., Wu, D.H., and Wei, B.Q., *Appl. Phys. Lett.*, 84, 4969, 2004. With permission from American Institute of Physics). (bottom right) Photograph showing a free-standing MWNT sheet used as a planar incandescent light source that emits polarized radiation. (From Zhang, M., Fang, S.L., Zakhidov, A.A., Lee, S.B., Aliev, A.E., Williams, C.D., Atkinson, K.R., and Baughman, R.H., *Science*, 309, 1215, 2005. With permission from AAAS.)

carried out in an inert gas atmosphere between carbon and catalyst-containing carbon electrodes, respectively. Nowadays, CNTs and related materials are produced by a wide variety of processes,[17] such as high-temperature arc methods,[18–20] laser vaporization of graphite targets,[21,22] as well as techniques using chemical vapor deposition. The electric arc and laser methods are basically difficult to scale up; however even these techniques are used routinely to produce gram quantities of nanotubes. Nanotube samples produced by these diverse methods are now commercially available in smaller quantities in their "as grown" or purified (from other carbon and metallic impurities)[23–27] form. In this chapter, due to the relatively high importance assigned to CVD methods in recent years, we will focus on this method of production.

1.2.1 CVD as a Versatile Tool for CNT Growth

CVD is a versatile and powerful tool in modern chemistry, chemical engineering, materials science and nanotechnology. More details on CVD may be found in Chapter 8. It is at present the most common method for CNT production and a well-known method[28] of carbon fiber production. In contrast to other methods, CVD can be scaled up, and already exist industrial-scale methods that use this technique for the production of nanofibers[29] that are dimensionally similar to the nanotubes. Furthermore, CVD can be tailored, that is, it can also be used to create oriented nanotube arrays on flat and 3D substrates, and so on.

The basic CVD technique for CNT growth is simple; a hydrocarbon gas feed line supplies the carbon feedstock and another inert gas (He or Ar) line provides the outlet for a carrier gas. The reaction occurs inside a chamber heated to higher temperatures. The other advantage of the CVD method is that it may work without templates but also on substrates, and is able to directly deposit CNTs onto predefined locations. Nanotubes have been deposited on Si, SiO_2, SiC,[30,31] MgO,[32,33] Al_2O_3,[34,35] zeolites,[36–38] quartz,[38,39] and glass.[40] Catalytic metals Fe, Co, Ni, and Mo or a combination of these are either sprayed into the chamber or deposited on substrates to promote growth.[32,37,39–42] The main parameters to control the growth process are the hydrocarbon source, the flow rates of the feedstock gases, the reaction temperature, the catalyst and the substrate. At lower temperatures acetylene (C_2H_2) or ethylene (C_2H_4) may be used as the carbon source for the production of MWNTs[36] while at higher temperatures CH_4,[43] or CO[44] may used be for SWNTs. Introducing sulfur-containing organics[13] or water vapor[39] facilitates SWNT production; alcohols[38,45] have also been used for the production of SWNTs. As the diameter of the CNTs produced is strongly dependent on the size of the catalyst particles initially distributed,[46–48]the synthesis of SWNTs requires a very well-controlled, thin, precoated catalytic metal film. Thick catalyst metal films usually result in the growth of MWNTs or carbon fibers.[49] Another advantage of the CVD method is the possibility to produce aligned CNT layers. Several examples of well-controlled growth of CNT layers are displayed in Figure 1.2 and Figure 1.3. At present, both MWNT and SWNT layers are being produced with high levels of alignment. Figure 1.2 shows SEM images of aligned MWNT films (top two SEM images) produced on a laser ablated catalyst layer,[50,51] on catalyst particle layers applied on substrates by stamping techniques,[52] and by the pyrolysis of ferrocene–xylene mixture precursors.[53] Figure 1.3 presents examples for MWNT growth based on a template-modification method to control catalyst particle size,[54] (left panel), and the application of multilayer catalyst films for the production of aligned SWNTs.[39]

Recent results point to the flexibility of CVD techniques. By tailoring the catalyst particles on substrates and controlling the CVD conditions, one hopes to achieve precise control of nanotube architecture. With further control, the ultimate goal of the nanotechnology community would be to provide nanotube growth with a predefined number of tubes, with a given diameter, helicity, and length along predefined locations.

1.2.2 CVD with Vapor-Phase Catalyst Delivery

In spite of a lot of work on catalyst and catalyst/substrate optimization for achieving better control on the CVD process, precise control and transformation of the catalyst film (into catalyst particles) at elevated temperatures before and during the growth has remained a challenge. An alternative approach where CVD growth of nanotubes is stimulated by exposing the substrate to a ferrocene–xylene vapor mixture at about

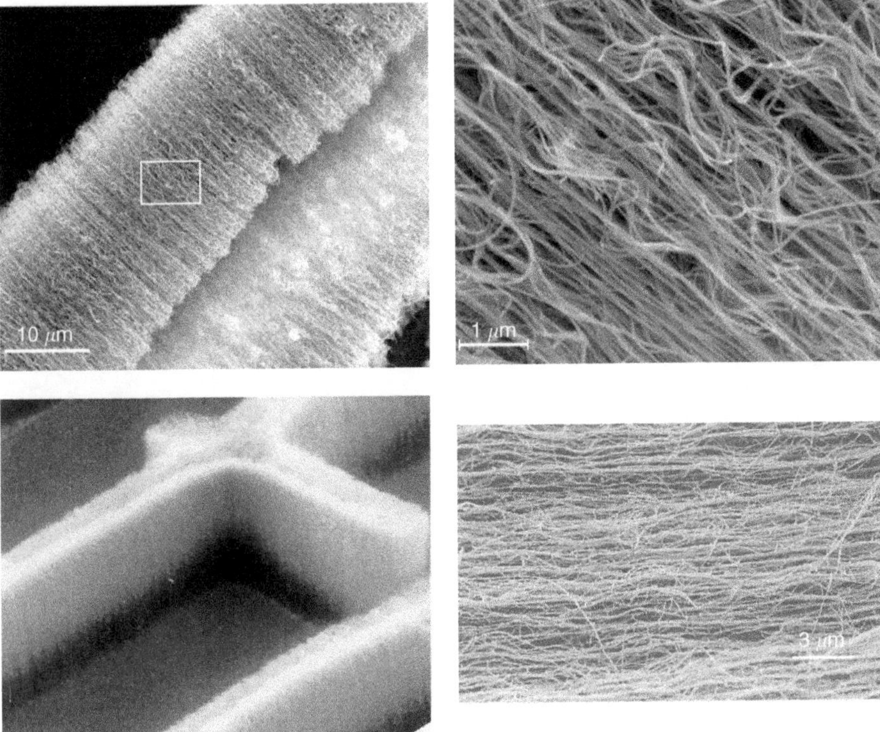

FIGURE 1.2 Aligned CNT structures achieved via CVD growth. (top) Aligned MWNT structure grown on laser pretreated catalyst layer. (From Terrones, M., Grobert, N., Zhang, J.P., Terrones, H., Kordatos, K., Hsu, W.K., Hare, J.P., Townsend, P.D., Prassides, K., Cheetham, A.K., Kroto, H.W., and Walton, D.R.M., *Chem. Phys. Lett.* 285, 299, 1998. With permission from Elsevier B. V.) (bottom left) Effective catalyst patterning by use of a micro-stamp. (From Kind, H., Bonard, J.M., Emmenegger, C., Nilsson, L.O., Hernadi, K., Maillard-Schaller, E., Schlapbach, L., Forro, L., and Kern, K., *Adv. Mater.* 11, 1285, 1999. With permission from John Wiley & Sons, Inc.) (bottom right) Aligned nanotubes produced by pyrolysis of Ferrocene/Xylene. (From Andrews, R., Jacques, D., Rao, A.M., Derbyshire, F., Qian, D., Fan, X., Dickey, E.C., and Chen, J., *Chem. Phys. Lett.* 303, 467, 1999. With permission from Elsevier B. V.)

800°C,[53–57] simplifies the process dramatically. In these experiments, nanotubes are grown on thermally oxidized silicon wafers using xylene (C_8H_{10}) as carbon source and ferrocene ($Fe[C_5H_5]_2$) provides the iron catalyst. One such example is shown in Figure 1.2 (bottom right). To carry out the CVD process ferrocene is dissolved in xylene at concentrations of $\sim$0.01 g ml^{-1}, preheated at about 150°C, coevaporated, and fed into the CVD chamber that is preheated gradually to the desired temperature of 800 to 900°C. This approach provides the catalyst particles (Fe nanoparticles formed from the pyrolysis of ferrocene) directly to the growth zone and avoids coalescence and any other substrate effect on the catalyst form.

1.2.3 Selective Growth with the Catalyst Delivery Method

While the vapor-phase catalyst delivery method eases the problem of controlling the particle size of catalysts, it lacks the ability to control the positioning of the catalyst and hence the patterning of nanotubes selectively on the substrates. However, akin to the patterning of the catalyst film on the substrate by photolithography, the positioning of CNT by the vapor-phase catalyst growth method can be controlled by differentiating the substrates[58] on which the nanotubes are grown; that is the growth of CNTs is observed to be strongly substrate-dependent. It was clearly demonstrated that CNTs can grow on the SiO_2 substrates, with no observable growth on the Si substrate. The nanotubes grown on the SiO_2, as shown in scanning

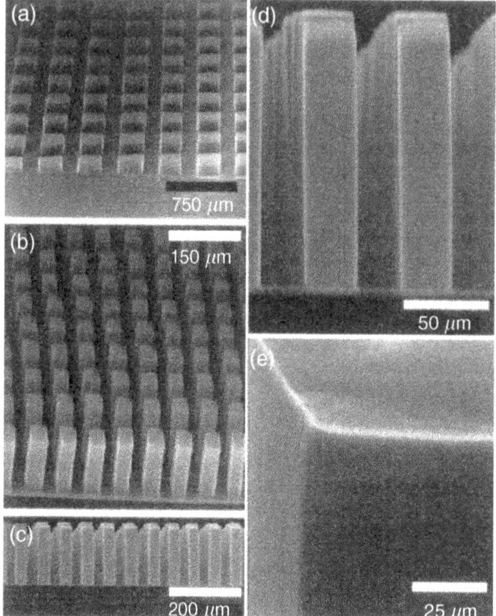

FIGURE 1.3 Patterned multi-walled and single-walled CNT structures achieved via CVD growth. (left panel) Self-oriented MWNT structures grown on porous silicon substrate. (From Fan, S.S., Chapline, M.G., Franklin, N.R., Tombler, T.W., Cassell, A.M., and Dai, H.J., *Science*, 283, 512, 1999. With permission from AAAS.) (right panel) Aligned single-walled CNT structures. (From Hata, K., Futaba, D.N., Mizuno, K., Namai, T., Yumura, M., Iijima, S., *Science*, 306, 1362, 2004. With permission from AAAS.)

electron microscope (SEM) characterization (Figure 1.2), are well-aligned. TEM analysis showed evidence for MCNTs with diameters of 30 to 50 nm and lengths of several hundreds of microns.

1.2.4 Features of Vapor-Phase Catalyst Delivery CVD

1.2.4.1 Selective Growth on Si and SiO_2 Substrates

The template has a curious role (Si *vs.* SiO_2) in the CVD process; and the reason for growth selectivity in the cases of Si and SiO_2 were analyzed in detail.[59] A close view of the surface of the samples after the CVD growth shows that many particles (metallic, originating from the ferrocene precursor) are formed on the silicon surface, but these apparently do not aid in CNT growth. On the other hand, a dense film of nanoparticles (Fe) is observed on the silicon oxide surface where aligned nanotubes grow very well. Particle sizes are observed to be around 20 to 40 nm in the silicon oxide region but larger in the Si region. Cross-sectional TEM was carried out to gather more detailed information on the Fe-containing particles after nanotube growth on both the silicon and silicon oxide templates. TEM images of the cross-section of the substrates with the corresponding electron diffraction patterns are shown in Figure 1.4 (top). On the SiO_2 substrate the nanotubes as well as the Fe-containing particles can be easily identified. The most important feature here is the presence of the irregularly shaped particles 20 to 40 nm in diameter on the top surface of the oxide area as well as inside the nanotubes (particles have dark contrast). On the surface of the Si template no trace of nanotubes is found; however larger, submicron-size particles are observed beneath the surface. Electron beam diffraction results indicate that these irregular-shaped nano-sized particles on the top of silicon oxide surfaces are pure gamma iron (fcc Fe), but on the silicon surface, diffraction patterns suggest the formation of noncatalytic iron silicide ($FeSi_2$), and iron silicate (Fe_2SiO_4).

On the silicon oxide template, carbon from the gas phase can dissolve into Fe particles formed during the decomposition of ferrocene. The Fe particles may easily become saturated or supersaturated with carbon,

FIGURE 1.4 Selective growth of multi-walled CNT structures on Si/SiO_2 substrate. (top) TEM images of the cross-section of the substrates. (a) SiO_2 area after CVD growth and removing of CNTs. (b) Si area without any nanotube growth but precipitate of submicron-size particles near the surface. (c) Enlarged picture from the nanotube/SiO_2 interface in (a) showing the formation of gamma iron particles on silicon oxide surface and the growth of nanotubes from the particles formed. (d) Enlarged area from (b) showing the formation of iron silicide and iron silicate crystals during CVD processing. (Reprinted from Jung, Y.J., Wei, B.Q., Vajtai, R., and Ajayan, P.M., *Nano Lett.*, 3, 561, 2003. Copyright (2003) American Chemical Society.) (bottom) SEM images showing 2D arrays of pillars, each made of eight stacks of aligned nanotube layers. The substrate made of SiO_2 has been patterned using Au patterns, and the nanotubes grow selectively in the SiO_2 exposed areas. The higher magnification SEM image shows the interfaces (position indicated by arrows) between the separate stacks of nanotubes in a single pillar. Interestingly, each subsequent stack forms at the base, pushing the rest of the stack up, with the first stack ending at the top at the end of the growth sequence. (Reprinted with permission from Li, X.S., Cao, A.Y., Jung, Y.J., Vajtai, R., and Ajayan, P.M., *Nano Lett.*, 5, 1997, 2005. Copyright [2005] American Chemical Society.)

and the precipitation of carbon from the surface of the Fe particle leads to the formation of tubular carbon structures of sp^2 bonding. Nanotube formation is possible, as the iron particle is chemically stable and has the appropriate size on the silicon oxide during the entire CVD growth. At the same time, on the silicon surface, a chemical reaction occurs between the silicon template, iron, and residual oxygen, forming compounds that become catalytically inactive for growing nanotubes.

1.2.4.2 Usage of Different Substrates

As we demonstrated in the previous section, the template plays a crucial role in the tailored growth of CNTs by the catalyst delivery method. In addition to the above mentioned, that is, initially discovered and most used Si and SiO_2 selectivity, a wider range of the substrates can be used for fulfilling different goals. To obtain the different growth features we used MgO substrates with different crystal orientation,[60] SiC planar and 3D templates,[61] and gold plated substrates of different metal film thicknesses.[61,62] As can be expected from the nature of Si and SiO_2 selectivity, stable ceramics are good candidates for being template materials, as most of them interact weakly with the catalyst. As a second order approach we found differences between the growth activity on differently oriented MgO crystal facets, namely, the (111) crystal face is more active than the (100) ones. The most important difference between the two facets from this point of view is that (100) is a neutral plane while (111) is oxygen terminated (charged). Some structures derived from this approach using these templates will be discussed in Section 1.3.

1.2.4.3 Extraordinary Structures Resulting from Multilayered Growth

Using the technique described above, it is possible to build interesting structures, for example multilayers. In our study, we were able to grow new layers of the nanotube forests, however, surprisingly the new layer did not grow on the top of the old one, but it formed as a new layer below it.[63] At the bottom part of Figure 1.4 we display SEM images of pillars where the CVD was carried out eight times, and therefore eight layers of the CNT forest can be distinguished. The top layer was the first to grow (as it was identified by its thickness) and the layers below it grew in the order of the numbers denoted. This extraordinary growth mechanism is unforeseen in any vapor-phase film growth on substrates. What we observed was that during growth, each layer, consisting of uniformly aligned arrays of hundreds of microns-long MWNTs, and nucleates grows from the buried original substrate plane (silicon oxide) even after the substrate gets completely covered by continuous and multiple layers of nanotubes deposited during previous growth sequences. For this to happen, it is imperative that the hydrocarbon and the catalyst metal precursors diffuse through several hundreds of microns of porous nanotube films and start growing on top of the substrate, underneath the bottom of existing multilayered nanotubes. It also means that, every time a fresh layer is nucleated and grown from the bottom, the rest of the layers in the stack get lifted up from the substrate, moving upwards with the newly growing nanotube layers. When the new layer lifts up the older one an interaction based on mainly on van der Waals' forces builds up between the adjacent layers, and accordingly, the whole structure stays intact. At the same time, these layers may be easily peeled off from each other, showing that the individual nanotubes are not continuously growing from one layer to the next. Recently there have been several reports to corroborate this phenomenon.[64,65]

1.2.4.4 Multi-walled vs. Single-walled Nanotube Growth

The ultimate goal of controlling nanotube growth is the placement of nanotubes on predefined locations with a given orientation and at the same time controlling the structural, dimensional and molecular properties of the nanotubes; that is, single- or multi-walled type, length and diameter, and chirality. Some of these features can already be controlled, for example, density, orientation, and the size of the nanotube bundles.[66,67] Major control parameters are temperature and the catalyst. For SWNT production one normally needs to use higher temperature ($\sim$1000 to 1100°C) and carbon sources that have lower C/H ratio, for example, methane *vs.* acetylene. The catalyst particles need to be on the nanometer size-scale to be able to produce SWNTs having diameter ranges between 0.7 and 2 nm. To keep the catalyst size small at the high temperature, we need to use very thin layers of metal films[66]; or use catalyst phase embedded structures such as copolymers or polymer composites that would prevent the agglomeration problem.[67]

By modifying the parameters of the floating catalyst method one may obtain SWNTs. By simply raising the reaction temperature to 850°C a mixture of SWNTs and MWNTs is usually produced[57]; similar structures have been achieved by other groups, too. On the other hand, when we used *n*-hexane as a carbon source, and additives such as thiophene in small concentrations, the reaction product constituted very long (in the range of 10 cm) strands of nanotubes, consisting of continuous oriented SWNTs. This level of control that researchers have obtained for CVD growth of nanotubes is spectacular in recent years, allowing for the growth of various kinds of nanotube structures, organized in a range of dimensionalities.

1.3 Nanotubes and Nanotube Architectures Produced via Vapor-Phase Catalyst Delivery Growth

1.3.1 Structures Grown on Planar Substrates

To exploit of the selective growth feature of the floating catalyst CVD we used patterned substrates; namely, silicon wafers capped with thin oxide layer were processed with standard photolithography and the resulting Si and SiO_2 micro patterns were used for deposition of the organized MWNT structures.[68] Figure 1.5 shows examples of aligned nanotubes selectively placed on predefined locations. The SEM images show that highly aligned nanotubes grow readily on the SiO_2 islands in a direction normal to the substrate surface, and the selectivity is retained down to several micrometers. No nanotube growth is observed on pristine Si surfaces or on the native oxide layer. The nanotubes are well oriented and packed with uniform density and the height of these blocks can be precisely controlled in the range of 10 to 100 μm by adjusting the deposition time. Only the in-plane dimensions of silica templates determine the number of nanotubes in each block (e.g., pillar) and the lateral separation between the blocks are again tailored during lithography. The adhesion of the as-grown nanotubes to the substrate is not strong and can be removed quite easily; however the samples remain intact during the treatment and applications, without any further processing. The growth features displayed in Figure 1.5 show excellent control over the placement of nanotubes at desired locations and their inheritance of the underlying template pattern's shape and separation. In addition to the shapes shown, one can build other shapes simply by designing the SiO_2 substrate pattern accordingly. This approach can be used to build porous nanotube films with control over pore size, shape, and separation, too. Porous architectures are obtained by using a template with a silica film (on which nanotubes grow) with holes of different shapes etched at different locations; literally a negative pattern of the one used to make free-standing nanotube blocks.

1.3.2 Three-Dimensional (3D) Nanotube Structures

In the above cases, we showed directionally oriented nanotube architectures that are perpendicular to the planar substrate surface, by keeping the thickness of silicon oxide patterns small, normally below 100 nm. By using significantly thicker (e.g., 5 to 8 μm) silica islands, we are able to form nanotube blocks oriented in multiple directions, including those in the plane of the macroscopic substrate surface.[69] We can also realize nanotube growth in mutually orthogonal directions by using templates consisting of deep-etched trenches, separating several micrometers-tall SiO_2 towers or lines. In Figure 1.6, vertically and horizontally aligned nanotube arrays are displayed. These nanotube structures were grown in a single step of the CVD process. Here we were also able to create structures that are more complex, based on silica structures machined with different techniques known in the fabrication of MEMs, for example, see the 3D nanotube structures "nanotube daisies." For this the nanotubes were grown with oblique inclinations on truncated cone-shaped silica features. The inverse structures, grown on truncated thick oxide layers,[68] could be useful as thin membranes for electromechanical applications.

Another impressive structure made of nanotubes grown on SiO_2 in a direction normal to the surface, shown also in Figure 1.6 (bottom), displays two layers of nanotubes, growing in two opposite directions (up and down) from a suspended SiO_2 layer with bottom and top surfaces exposed.[68] The suspended silicon oxide was machined on the silicon pillar by under-etching the silica disk (40 to 50 μm into the

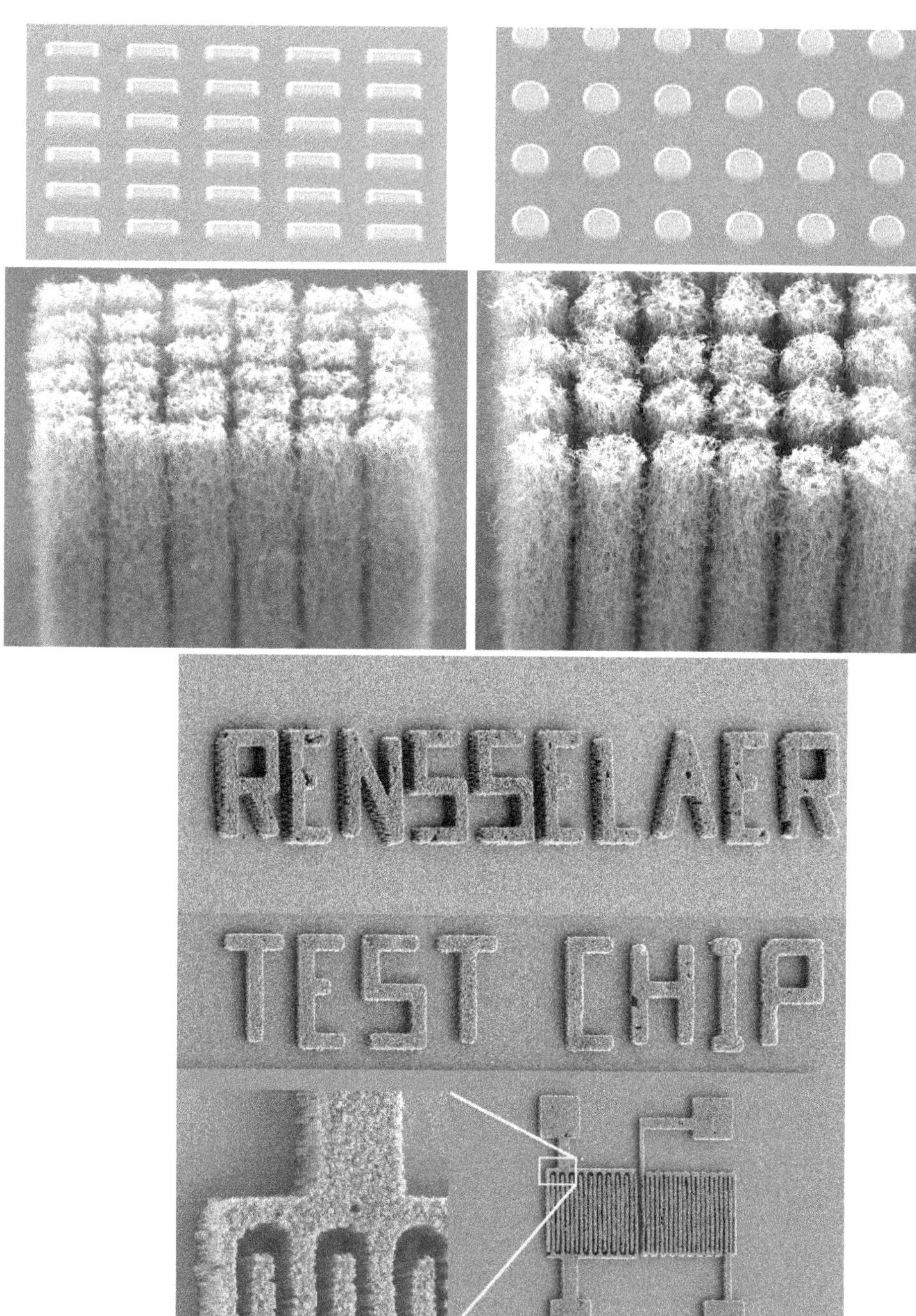

FIGURE 1.5 Patterned growth of multi-walled CNTs on planar substrates. Different Si/SiO_2 patterns were prepared by conventional photolithography, and the selective CVD growth resulted in CNT structures only on the SiO_2 areas, on the Si substrate there is no CNT growth. (Reprinted with permission from Wei, B.Q., Vajtai, R., Jung, Y., Ward, J., Zhang, R., Ramanath, G., and Ajayan, P.M., *Chem. Mater.* 15, 1598, 2003. Copyright [2003] American Chemical Society.)

FIGURE 1.6 3D structures achieved by growth of nanotubes. (Reprinted with permission from Wei, B.Q., Vajtai, R., Jung, Y., Ward, J., Zhang, R., Ramanath, G., and Ajayan, P.M., *Chem. Mater.* 15, 1598, 2003. Copyright (2003) American Chemical Society. Wei, B.Q., Vajtai, R., Jung, Y., Ward, J., Zhang, R., Ramanath, G., and Ajayan, P.M., *Nature*, 416, 495, 2002. With permission from Nature Publishing Group.) (top) Nanotube bundles oriented in both vertical and horizontal orientations grown simultaneously on a template with deep, etched trenches that separate 8.5 μm-high SiO_2 islands on Si. The cross-section of the specimen shows the orthogonal configuration of arrays. The length of the nanotubes in both vertical and horizontal directions is 60 μm. (middle) CNT "daisies." Repeating patterns containing mutually orthogonal nanotube arrays produced on deep (about 5 μm) silica features (circular cross-section) machined on silicon substrates. Growth in the vertical direction occurs from the top silica surface (seen as arrays emanating from the center of each pattern); growth on the sides occurs as horizontal arrays (sideways growth seen on each pattern). (bottom) Schematics and SEM micrograph showing simultaneous multilayer and multidirectional growth of oriented nanotubes from a 2 μm-thick SiO_2 layer suspended on deep-etched Si pillars; the three directions of aligned CNT growth are marked by arrows.

silicon substrate). In the schematic, the disk-shaped transparent layer represents the silica membrane, and the darker region shows the silicon pillar supporting it. By increasing the thickness of the suspended SiO_2 layer by several micrometers, one can obtain multilayers as well as orthogonally oriented nanotube architectures pointing in the radial direction in one step.

1.3.3 Directed Growth of Complex Structures

Further control on the size and orientation of larger CNT structures can be achieved by applying the above mentioned approach in combination with conventional metallization methods.[62] To achieve alignment only in selected directions, one needs to create 3D silica surfaces where some of the exposed surfaces can be capped with sputtered gold layers (over several select sides), and direct the CNT growth only on the preselected directions (Figure 1.6). The SEM pictures in Figure 1.7 show three different structures, namely, one totally covered by nanotubes, a nanotube bridging two adjacent blocks, and a shorter nanotube structure where the nanotubes only partially bridge the gap. These different configurations are useful for different applications, for example, from a brush contact to a field-emission device.

1.3.4 Two-Dimensional (2D) Structures of SWNTs

By simple CVD methods 2D SWNT networks can also be easily fabricated, in additional to vertically aligned arrays. SEM investigations showed that high-density SWNT networks may form structures of various shapes on nanoscale-patterned silicon oxide.[66] By using Fe catalyst deposited on the top and side walls of the pillars we observe that the yield of suspended SWNT can be significantly high (see Figure 1.8, top). Closer observation of nanotube networks indicates that many nanotubes grow on the bottom of the substrate as well as the sidewalls of the patterned structures. The nanotube directions are controlled by the locations of the pillars, and it results in a highly organized SWNT architecture following the predesigned geometry of the patterns. In the case of line patterns, nanotubes preferentially grow normal to the topography of the substrate surfaces, regardless of the direction of the gas flow. TEM investigation of nanotubes on pillars showed that small bundles consisting of few SWNTs having 1 to 1.3 nm diameters are produced.

Other transition metals such as cobalt may also be used to create high-density SWNT networks. Compared with Fe catalyst, CVD for nanotube growth using Co catalyst can be done at a lower deposition temperature, 800°C, yielding approximately the same density of the SWNTs compared with Fe catalyst at 900°C. The substrate once again has strong effect on nanotube growth. Here SWNT networks grown under the same CVD parameters on SiO_2 pillars have higher density compared with the ones grown on Si pillars. An interesting result from these SWNT structures is their behavior under the ion (Ga) irradiation using a focus ion beam (FIB).[70] Figure 1.8 (bottom) shows how consecutive scans of the ion beam remove carbon atoms from the nanotubes resulting in shorter effective length and straighter geometry. With consecutive scans some nanotubes can be selectively eliminated from the sample, for example, those in the vicinity of the substrate compared to those bridging the pillars and this allows one to perform a postgrowth processing (irradiation) to configure the nanotube networks into structures of practical use (e.g., interconnects).

1.4 Applications of Larger Assembled Nanotube Structures

1.4.1 Macroscopic Nanotube Filters

Macroscopic structures made of aligned CNTs can be synthesized, with a certain amount of control, not only on flat or patterned substrates, but also on curved substrates, and could be easily removed from the substrates. We have reported the fabrication of free-standing monolithic uniform macroscopic hollow cylinders having radially aligned CNTs, with diameters and lengths up to several centimeters.[71] These cylindrical membranes were used as filters to demonstrate their utility in two important settings: the

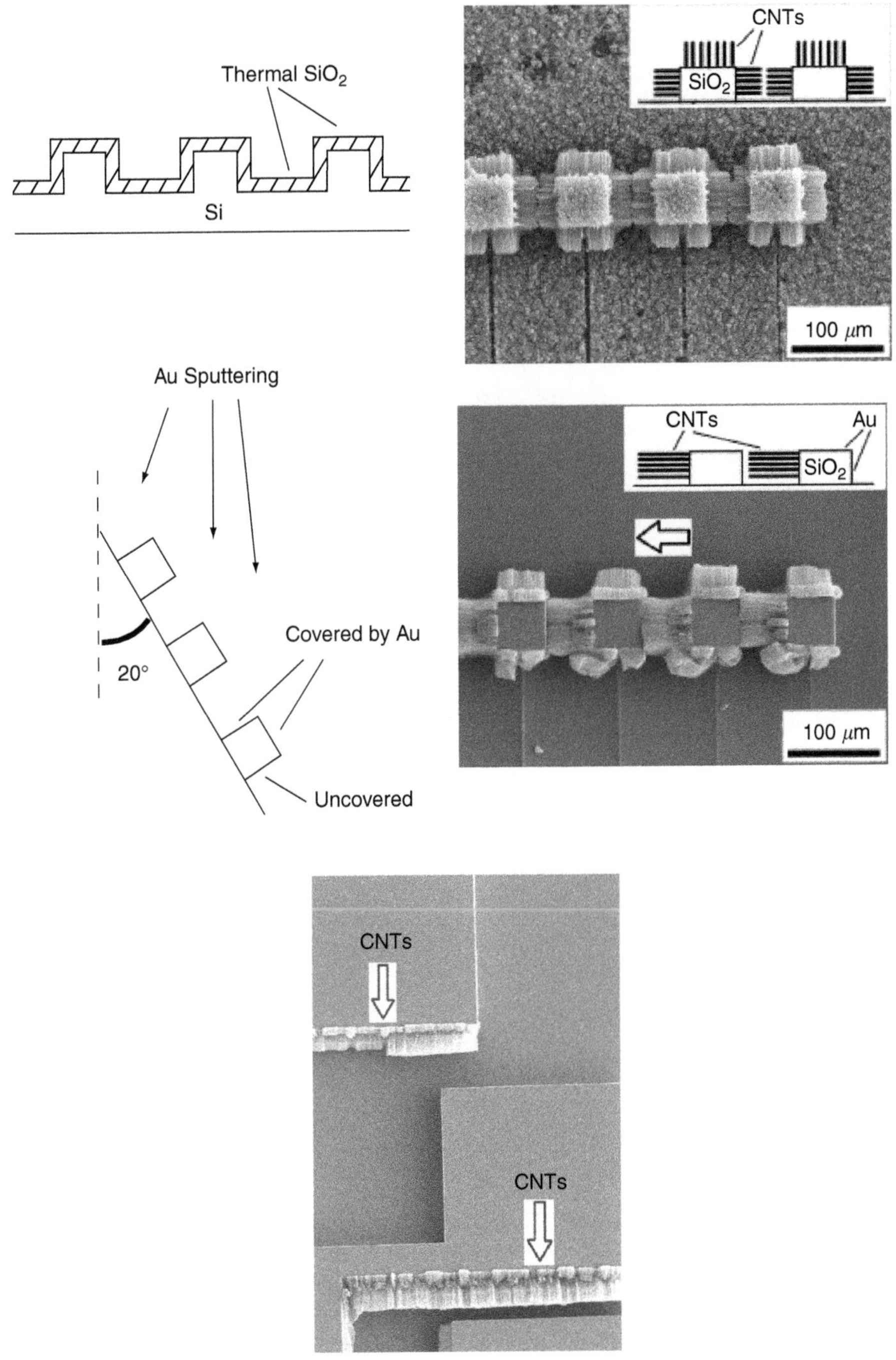

FIGURE 1.7 Tailoring CNT structures by blocking the growth into the unwanted directions. (top left) Schematic illustrations of substrates with thermally oxidized SiO_2 relief pattern and their near-vertical placement during Au sputtering to selectively cover the top and several side surfaces. (top right and bottom) Star-like (3D) growth of aligned multi-walled CNTs grown into the perpendicular directions of each surfaces uncoated, and total absence of CNT growth on the Au coated surfaces. (From Cao, A.Y., Baskaran, R., Frederick, M.J., Turner, K., Ajayan, P.M., and Ramanath, G., *Adv. Mater.* 15, 1105, 2003. With permission from John Wiley & Sons, Inc.)

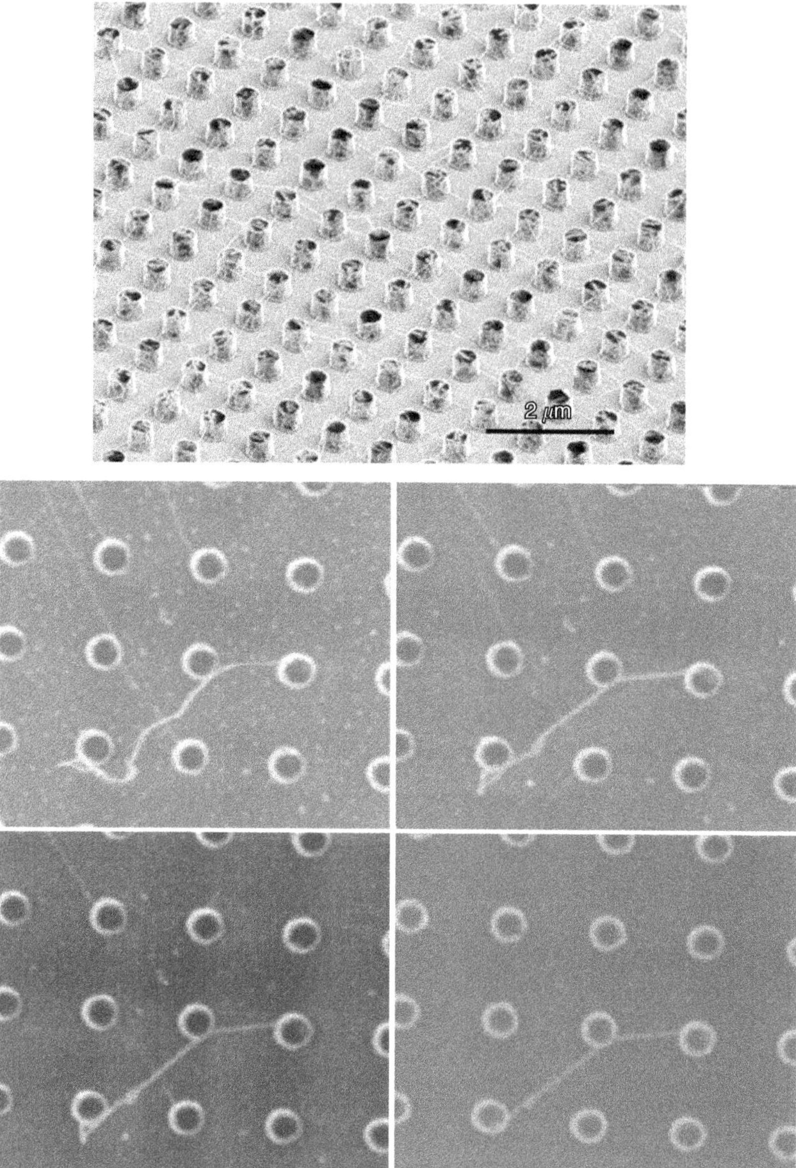

FIGURE 1.8 Growth and modification of SWNTs on Si or SiO_2 pillars. (Reprinted with permission from Jung, Y.J., Homma, Y., Ogino, T., Kobayashi, Y., Takagi, D., Wei, B.Q., Vajtai, R., and Ajayan, P.M., *J. Phys. Chem. B.* 107, 6859, 2003 and Jung, Y.J., Homma, Y., Vajtai, R., Kobayashi, Y., Ogino, T., and Ajayan, P.M., *Nano Lett.*, 4, 1109, 2004. Copyright (2003) and (2004) American Chemical Society.) (top) SEM images of single-walled CNT networks formed on submicrometer-scale patterned silicon oxide substrates using a Fe thin-film catalyst at 900°C produced by methane CVD. (bottom) Series of FIB images showing the sequential straightening of suspended SWNTs on the patterned pillars and selective removal of SWNTs on the substrate; 0, 4, 12, and 16 FIB scans, respectively.

elimination of multiple components of heavy hydrocarbons from petroleum — a crucial step in postdistillation of crude oil — with a single-step filtering process (Figure 1.9), and the filtration of bacterial contaminants such as *Escherichia coli* (Figure 1.10) or nanometer-sized poliovirus (~25 nm) from water. These nanotube based macroscale filters could be cleaned for repeated filtration through ultrasonication and autoclaving. The exceptional thermal and mechanical stability of nanotubes, and the high surface

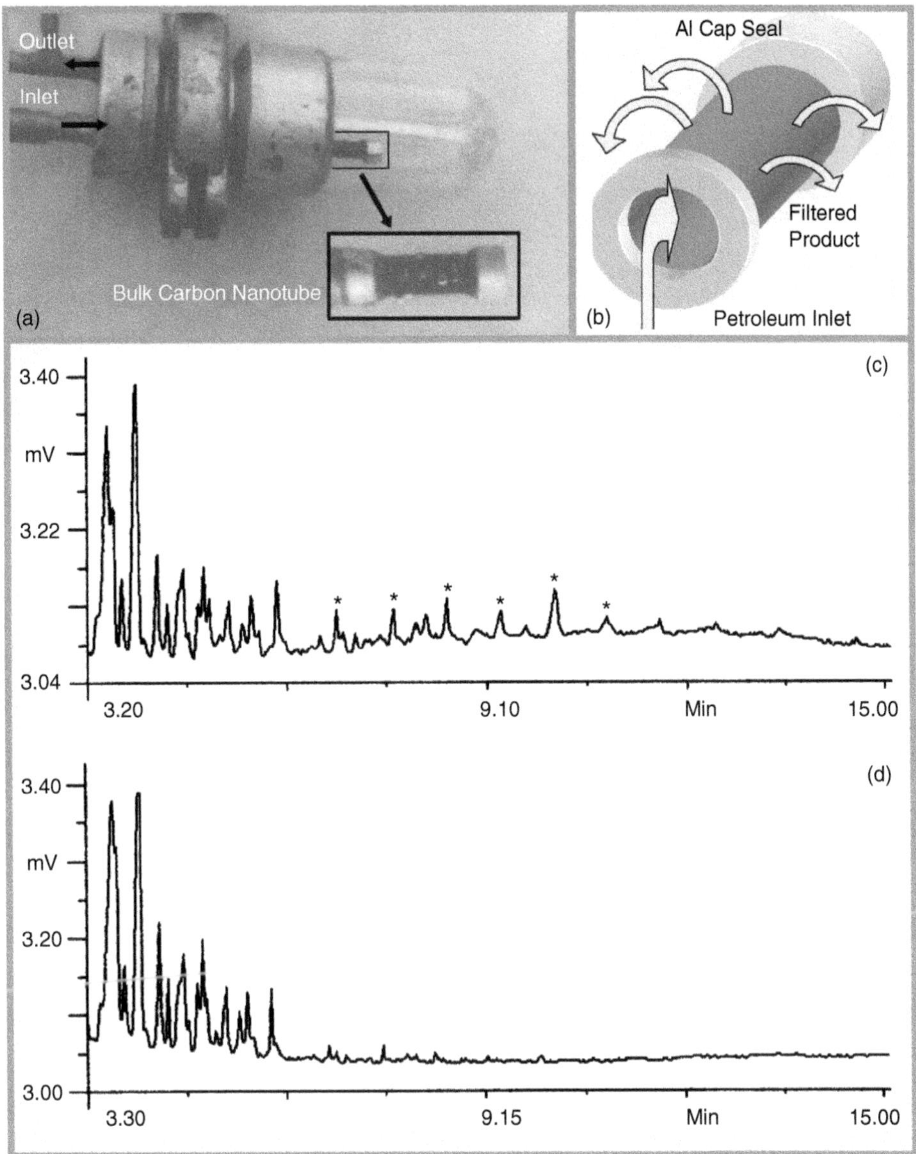

FIGURE 1.9 Macroscopic nanotube filter structure and removal of heavy petroleum components. (a) Photograph of the arrangement used for heavy hydrocarbon separation. The inset shows the bulk tube mounted as a filter. The tube is closed at the right end with an aluminum cap, and the other end is kept open serving as an inlet port for injection of petroleum. (b) Schematics of the petroleum dynamics through the bulk tubes. (c) Gas chromatography (GC) spectrum of the unfiltered products. The asterisks show the heavier hydrocarbon components in the unfiltered sample. (d) GC spectrum of the sample after it was passed through the nanotube filter, showing the absence of heavier hydrocarbon peaks. (From Srivastava, A., Srivastava, O.N., Talapatra, S., Vajtai, R., and Ajayan, P.M., *Nat. Mater.* 3, 610, 2004. With permission from Nature Publishing Group.)

area, easy, and cost-effective fabrication of the nanotube membranes, suggest that they could compete with commercially used ceramic- and polymer-based separation membranes.

A major advantage of using nanotube filters over conventional membrane filters will be their ability to be cleaned repeatedly after each filtration process to regain their full filtering efficiency. A simple process of ultrasonication and autoclaving (~121°C for 30 min) has been found to be sufficient for cleaning

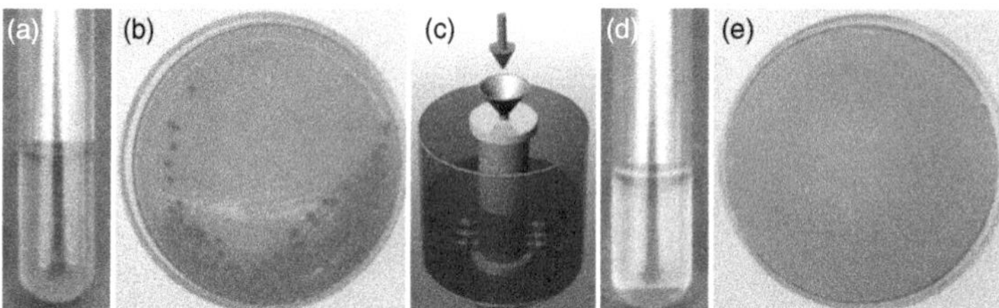

FIGURE 1.10 Removal of bacteria using a nanotube filter. (a) Unfiltered water containing *E. coli* bacteria; the turbid and light-pink color is suggestive of the presence of bacteria colonies scraped from the surface of MacConey agar, which contains Phenol red as an indicator. (b) Colonies of *E. coli* bacteria (marked by arrows) grown by the culture of the polluted water. (c) Assembly for the filtration experiment. The nanotube filter with the bottom end-capped is placed inside a container, and the liquid flows through the macrotube (shown by the vertical arrow). The horizontal arrows show the flow direction of the filtered liquid. (d) Water filtered through the nanotube macrofilter. The product obtained is relatively clear compared with the original bacterial suspension in (a), indicating the absence of the bacteria (as well as coloring particles) in the filtrate. (e) Filtrate after culture, showing the absence of the bacterial colonies. (From Srivastava, A., Srivastava, O.N., Talapatra, S., Vajtai, R., and Ajayan, P.M., *Nat. Mater.*, 3, 610, 2004. With permission from Nature Publishing Group.)

these filters; cleaning can also be achieved by purging for the reuse of these filters. In conventional cellulose nitrate and acetate membrane filters used in water filtration for strong bacterial adsorption on the membrane surface affects their physical properties preventing their reusability as efficient filters[72]; most of the typical filters used for virus filtration are not reusable.[73] Because of the high thermal stability of nanotubes, nanotube filters can be operated at temperatures of ~400°C, which is several times higher than the operating temperatures of the conventional polymer membrane filters (~52°C). The nanotube filters, owing to their high mechanical and thermal stability, may also compete well with commercially available ceramic filters; furthermore, these filters may be tailored to specific needs by controlling the nanotube density in the walls and the surface character by chemical functionalization.

1.4.2 Sensors Made of Nanotube Films

The low voltage needed for electron emission from the sharp nanotube tips makes possible an interesting application of the aligned nanotube arrays as electrodes in ionization sensors.[74] The CNTs possess advantages over the currently used ionization sensors in terms of size, simple operation, and not being affected by external conditions like temperature and humidity. The setup used is similar to that of field emission with the nanotubes as anode and an Al sheet as cathode, separated by a vacuum at a pressure of 10^{-4} torr with a spacing of about 150 μm. The gas that needs to be analyzed is then allowed to flow into the chamber. The voltage and current are monitored by an ammeter and voltmeter, to detect when the voltage is increased, meaning the gas is ionized within a small portion of the tips of the nanotubes. This cloud of ions then gains energy from the field and generates more electron-hole pairs. Later, more electron-hole pairs are formed till this process eventually leads to an avalanche breakdown between the electrodes. The voltage at which the breakdown occurs is unique for different gases and is referred to as the "breakdown voltage," which is a fingerprint of the given gas. The measurement of current is also important, as it is connected to the concentration of that gas. Proper measurement of the breakdown voltage can identify the gas present in the chamber. Some examples of the gases detected are helium, ammonia, argon and oxygen, at a constant gas concentration of 4×10^{-2} mol/l as shown in Figure 1.11. The values of the breakdown voltage for individual gases remain the same even at different concentrations of the gas.

1.4.3 Nanotube Composites for Superior Damping

CNT-based nanocomposites have been investigated owing to their applications related to specific strength.[75] Recently, there has been significant interest in developing nanotube–polymer composites for

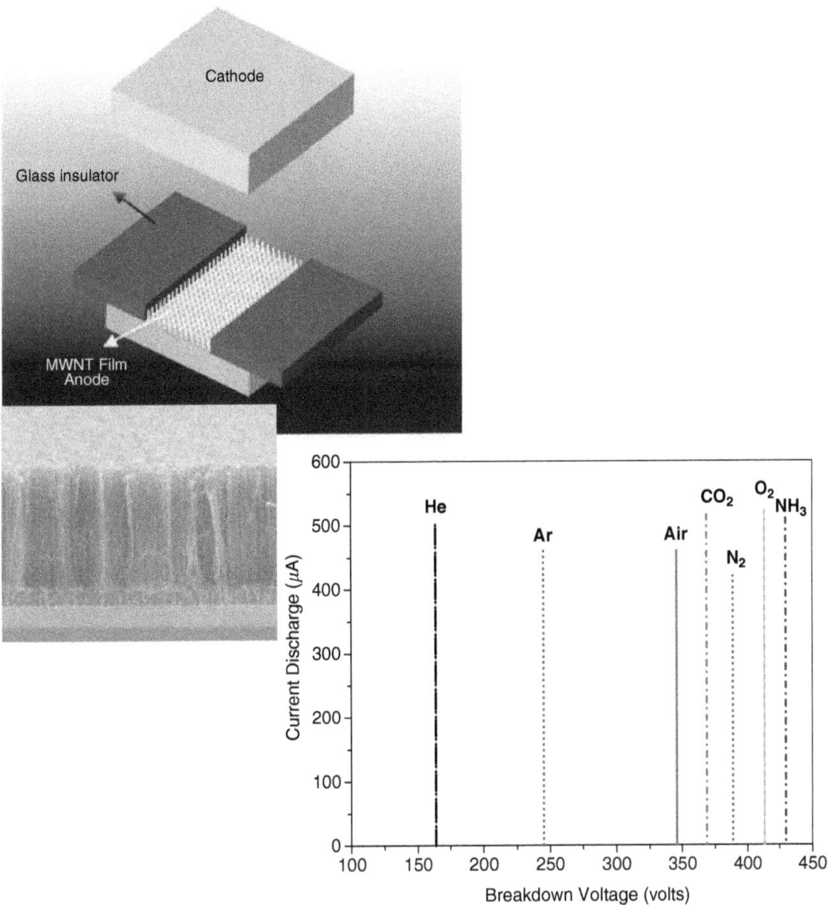

FIGURE 1.11 Nanotube ionization sensor. (From Modi, A., Koratkar, N., Lass, E., Wei, B.Q., and Ajayan, P.M., *Nature*, 424, 171, 2003. With permission from Nature Publishing Group.) Schematics of the nanotube sensor device and an exploded view of the sensor showing MWNTs as the anode on an SEM micrograph of a CVD-grown, vertically aligned MWNT film used as the anode. I–V curves for NH_3, CO_2, N_2, O_2, He, Ar and air, showing distinct breakdown voltages. Ammonia displays the highest breakdown voltage, and helium the lowest.

applications requiring unique combinations of properties.[76] Nanotube–polymer composites have shown promise in applications such as ultrafast all-optical switches,[77] and as biocatalytic films.[78] The potential of these nanocomposites can be exploited by overruling one of the main limitations, the lack of control over the orientation and dispersion of nanotubes in the polymer matrix, as well as the difficulty in tailoring the nanotube–polymer interface.[79] One approach to simultaneously control the nanotube alignment and dispersion in a polymer composite is to infiltrate monomers into the prealigned arrays of nanotubes, followed by *in situ* polymerization.[80] The resulting composite films can have good distribution, dispersion, and alignment of nanotubes in a polymer matrix, and they also provide reinforcement in the out-of-plane direction. An important application for nanotube composites is based on the unique mechanical properties of the interfaces, that is, damping behavior.[81] We have found that by using nanotube-epoxy films as inter-layers within laminated composite systems or nanotube layers used to reinforce the interfaces between composite plies, can enhance laminate stiffness as well as the mechanical damping properties (Figure 1.12). The experiments conducted using a composite beam with an embedded nanotube film sublayer indicate up to 200% increase in the inherent damping level and 30% increase in the baseline bending stiffness with minimal increase in structural weight. SEM characterization of the nanotube film

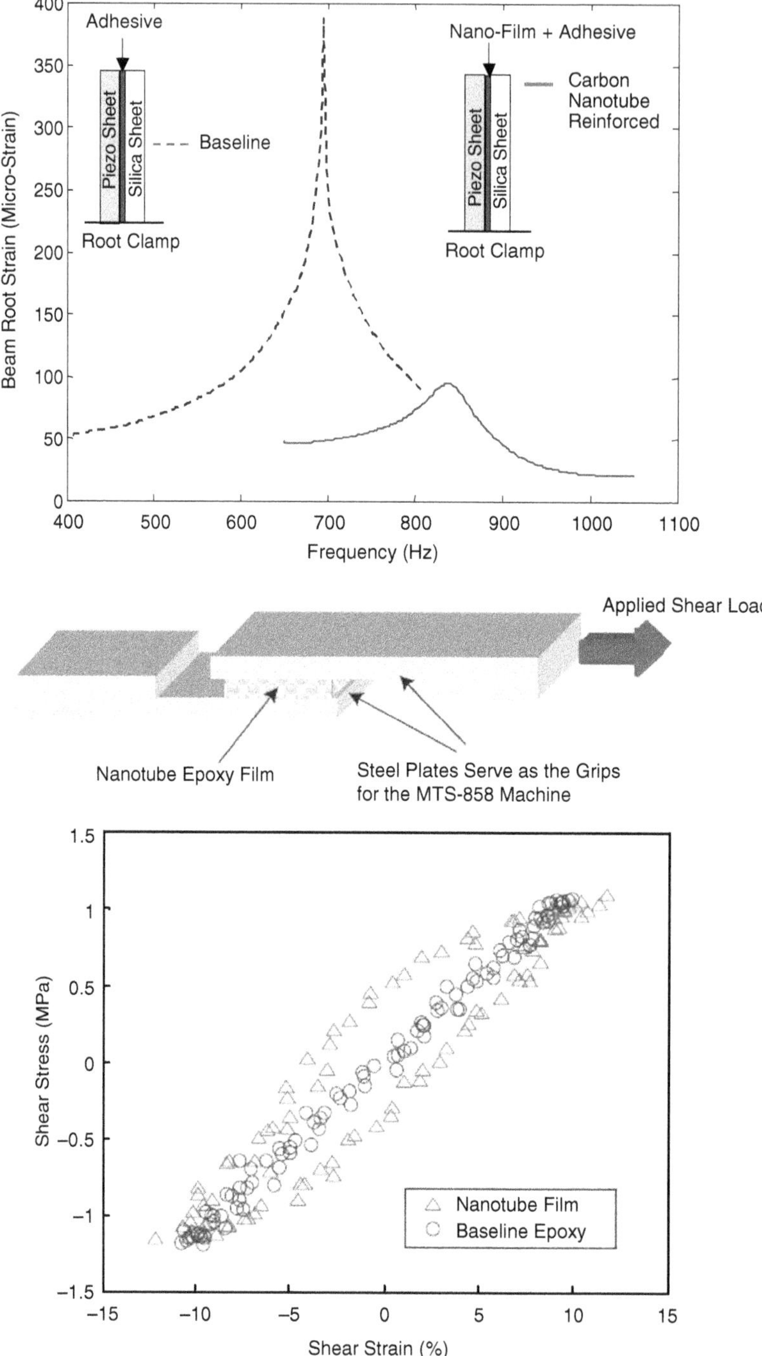

FIGURE 1.12 Applying CNTs for mechanical damping. (top) Comparison of the dynamic response of the baseline beam and nanotube reinforced sandwich beam for a frequency-sweep test at 50 Vrms. (From Koratkar, N., Wei, B.Q., and Ajayan, P.M., *Adv. Mater.*, 14, 997, 2002. With permission from John Wiley & Sons, Inc.) (middle) Schematic of viscoelastic shear-mode testing of nanotube thin film. (From Suhr, J., Koratkar, N., Keblinski, P., and Ajayan, P.M., *Nat. Mater.*, 4, 134, 2005. With permission from Nature Publishing Group.) (bottom) Shear stress versus shear strain response for nanotube film and baseline epoxy at a test frequency of 10 Hz (both films have identical dimensions: 20 mm × 12 mm × 0.05 mm). The large hysteresis observed in the response of the nanotube film is indicative of energy dissipation.

reveals a fascinating network of densely packed, highly interlinked MWNTs.[82] This inter-tube connectivity results in strong interactions between adjacent nanotube clusters as they are shear related to each other, causing energy dissipation within the nanotube film. The cross-links between nanotubes also provide improvement in load transfer within the network, resulting in enhanced stiffness properties. Our new study further developed the insight into the role of CNTs in the damping composite materials.[83] Nanotube fillers are minimally intrusive but their inter-tube or interfacial (with the polymer) sliding interaction in composites can be useful for structural damping, without sacrificing other mechanical properties. The measurements reveal strong viscoelastic behavior with up to 1400% increase in loss factor (damping ratio) of the baseline epoxy. Based on the interfacial shear stress ($\sim$0.5 MPa) at which the loss modulus increases sharply for the system, it has been concluded that the damping is related to frictional energy dissipation during interfacial sliding at the large, spatially distributed, nanotube–nanotube interfaces.

There are multiple advantages of using a nanotube-epoxy film as a damper compared to pure polymer materials. First, traditional viscoelastic polymers degrade in performance at higher temperatures. The operating range for commercial damping films varies in the range of 0 to 100°C. CNTs can withstand high temperature without significant degradation and consequently, for high-temperature applications nanotube films would offer superior performance and reliability. Second, the stiffness of traditionally used polymers is much lower than the composite structures with nanotubes. Also, the integration of viscoelastic damping films within composite systems poses significant technical challenges.[82,84] When damping materials are cocured with the host composite, the material experiences the temperature cycle of the cure. Commercially available damping materials have maximum recommended temperatures below that of the composite cure cycles. CNTs are stable at temperatures well beyond composite cure cycles and do not exhibit degradation in structure and properties due to resin penetration; for these reasons CNT films could potentially be perfectly integrated within composite and heterogeneous systems.

1.4.4 Nanotube Reinforced Polymer Skins

As we mentioned in Section 1.4.3, CNT–polymer composites have been extensively researched for many applications requiring the combination of unique electronic, optical, and mechanical properties of CNTs, and polymer materials. Another promising area of applications for nanotube–polymer composites are those involving their use in flexible electronics. We developed a new approach for the preparation of CNT flexible electronic devices[85]; namely, a direct and effective method of incorporating aligned MWNT structures into flexible polydimethylsiloxane (PDMS) polymer matrix. These structures could be used as strain gauges, tactile and gas sensors, and also as field-emission devices. In particular, these structures were used as flexible field-emitters as they conduct electrical signals through the structures and can sustain harsh mechanical environment and are also advantageous due to the suppression of mutual screening effect in the devices resulting in impressive field-emission behavior.

Figure 1.13 depicts the schematic for designing and building prepatterned, vertically aligned MWNT architectures into a PDMS matrix. First, the patterning of a SiO_2 and Si substrate was performed using a conventional photolithography process and then aligned MWNTs were selectively grown on these patterned SiO_2 regions using the method described before. Later, a PDMS prepolymer solution was poured over the nanotube structure and the excessive PDMS solution was removed. Finally, the PDMS was cured and then self-standing nanotube–PDMS composite films were carefully peeled off from the Si substrate. The SEM image of the MWNT structure after the infiltration of PDMS polymer is also displayed in the figure. The nanotube–PDMS composite structures were found to be resilient against large physical deformations. A patterned NT-PDMS matrix of cylindrical shape with diameter 500 μm was used to test the field-emission properties. SEM investigation showed that the vertically aligned MWNTs were completely surrounded by PDMS. The nanotube ends on the back of the sample were coated with Ti or Au and mounted on a Ti or Au coated cathode. The structures exhibited excellent field emission at a very modest vacuum (5×10^{-4} torr). Figure 1.13 (bottom) shows the emission current as a function

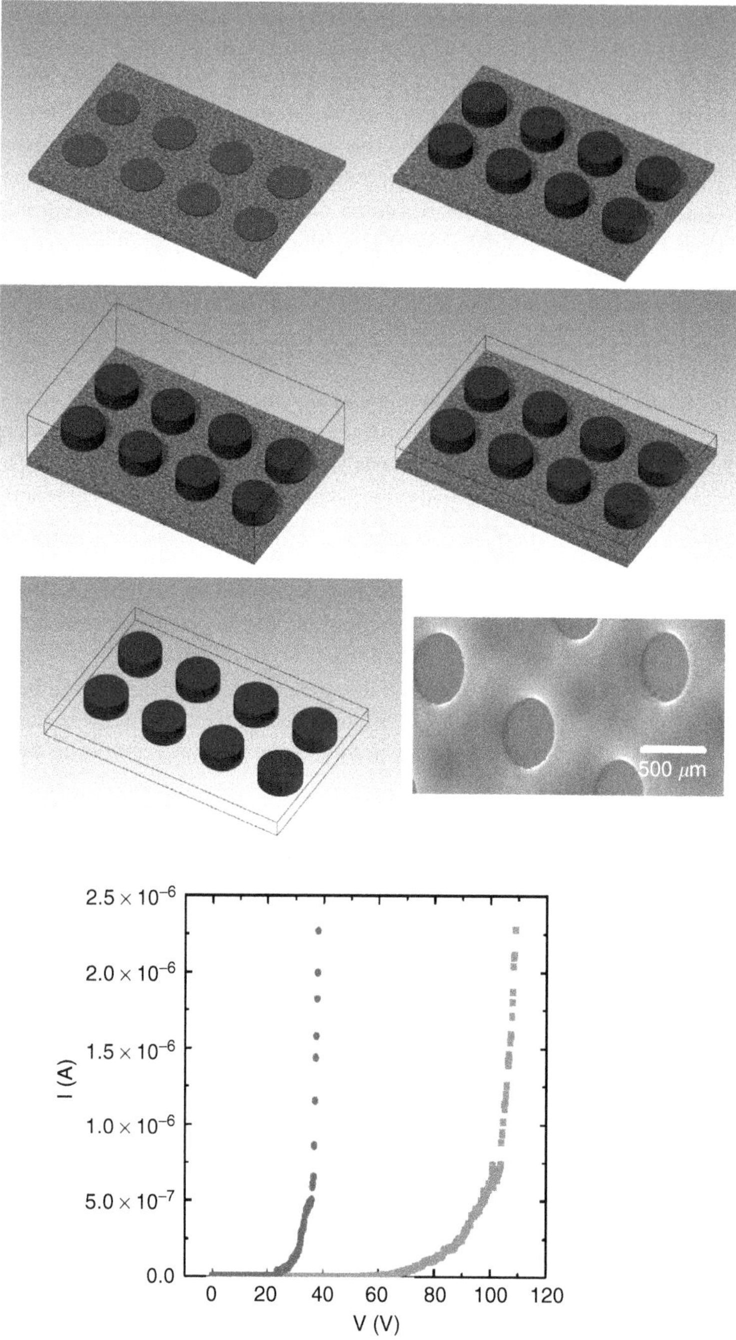

FIGURE 1.13 Nanotube-PDMS structure for field-emitter. Schematics are illustrating fabrication steps of aligned MWNT architectures/PDMS composite: preparing micron scale patterns of Si/SiO_2 structures by photolithography, selective growth of aligned MWNTs on the patterned SiO_2 regions using a CVD of Ferrocene/Xylene, PDMS infiltration into arrays of organized MWNT structure, removing excessive PDMS and curing of nanotubes/PDMS composite, peeling cured MWNT architectures/PDMS composite film from the silicon substrate. The graph displays the emission current — applied voltage characteristics for two field-emission devices fabricated from the flexible nanotube–polymer composite layer. (Reprinted with permission from Jung, Y.J., Kar, S., Talapatra, S., Soldano, C., Vishwanathan, G., Li, X.S., Yao, Z.L., Ou, F.S., Avadhanula, A., Vajtai, R., Curran, S., Nalamasu, O., and Ajayan, P.M., *Nano Lett.*, 6, 413, 2006. Copyright [2006] American Chemical Society.)

FIGURE 1.14 Nanotube brush preparation via masking the unwanted areas of the SiC fiber with gold and the resulting structure. (top left) Illustration of partial masking of SiC fibers to grow nanotubes only on the top part of the fiber; (top right) SEM images of an as-grown brush (resembling a dust sweeper) consisting of nanotube bristles and a fiber handle and an image to demonstrate the structure of the tip of the nanotube brush. The bristles have a length of 60 μm, and a span of over 300 μm along the handle. (middle) A nanobrush attached to a small electrical motor. (bottom) A dump of nanoparticles formed by a sweep brush and 10-μm-wide trenches cleaned by sweeping the brush over the surface. Inset: Dispersed nanoparticles inside trenches before brushing. (From Cao, A.Y., Veedu, V.P., Li, X., Yao, Z., Ghasemi-Nejhad, M.N., and Ajayan, P.M., *Nat. Mater.*, 4, 540, 2005. With permission from Nature Publishing Group.)

of the voltage for two different samples. The emitted current follows the well-known Fowler–Nordheim mechanism, in which the current density is approximately related to the square of the effective field multiplied by the exponential function of the effective field. CNTs that have high aspect ratio and also high curvature at the tips have high effective field; the field enhancement factor (β) may have values of several thousands, for example $\sim 8 \times 10^3$ and $\sim 1.91 \times 10^3$ for sample 1 and 2, respectively, better than what has been reported for free-standing nanotube arrays.

1.4.5 Nanotube Brushes

We have successfully constructed multifunctional brushes consisting of CNT bristles grafted on fiber handles (Figure 1.14). We also demonstrated several unique tasks for the application of these brushes, such as cleaning of nanoparticles from narrow spaces, coating of the inside of holes, selective chemical adsorption, and as movable electromechanical brush contacts and switches.[61] The nanotubes were grown on select areas of SiC micro-fibers by masking the rest of the fiber with Au thin film coating. The different positioning of the fibers resulted in different configurations of the brush heads; for example, vertical placement of the SiC fibers usually yielded three nanotube prongs. The formation of this three-pronged morphology is due to the self-organized growth of dense nanotube arrays as they emerge outwards from the cylindrical surface, having circular cross-section, as the circumference surrounding the nanotube front surface is enlarged as the front moves away from the fiber–nanotube interface. Two- and one-pronged structures were obtained by laying the fibers down on a flat surface during CVD, to block the growth of the nanotube in several directions. These brushes were used to effectively remove debris from a microchannel array, which is normally a challenging task. The flexibility of the brushes ensures that the nanotube structure does not scratch the surface, while the porosity of the nanotube film near the end of the bristle structure makes it possible to "absorb" the debris of the nanoparticles. Various other applications such as selective cleaning (with functionalized brushes), painting and applying in micro-capillaries and cavities, and compliant brush contacts were also demonstrated with these brushes.

1.4.6 Nanotube Springs

Considering the porous membrane-like constitution of aligned CNT films grown by CVD, they can be considered as foams or cell-like structures in several aspects. We showed that a uniform CNT film loses its homogeneity and forms cell-like structures when different solvents are poured into and dried out from the nanotube forest.[86] Beyond forming these cell-like structures, the free-standing film of vertically aligned CNTs also exhibits foam-like behavior under compression. The nanotube film acts as a collective spring and nanotubes in the film form zigzag buckles that can later unfold to their original length when the load is removed.[87] Compared with other low-density flexible foams, the nanotube films have higher compressive strength, recovery rate, and sag factor.

We tested compression cycles on these nanotube spring-like films, repeatedly for thousands of cycles. MWNT films squeezed to 15% of their thickness retrieved their original size by the end of the loading–unloading cycle. The porosity of the nanotube films is high, CNTs occupy only ~20% of the macroscopic volume of the film, allowing a large volume reduction (up to 85% seen in the experiment) when the film is compressed. In Figure 1.15, the schematic describes the loading–unloading cycle as well as SEM images of the nanotube film with different loads. Nearly perfect thickness recovery was achieved in hundreds of loading–unloading cycles before any (very small) reduction in the thickness was recognized. The nanotube film did not fracture, tear, or collapse under compression. Regarding the dynamics, we investigated the speed of the recovery of the nanotube film thickness, which was much higher than the recovery speed of conventional flexible foams and spongy structures, especially those made of polymers with viscoelasticity. Aligned SWNT films are expected to have better performance, and the compressive strength of the films also could be tailored by controlling the wavelength of buckles. This kind of nanotube structures may have many applications, such as in flexible electromechanical systems, interconnects, actuators, as well as they may be used in mechanical damping application.

1.5 Conclusions

In this chapter, we presented a brief summary of the ongoing work focusing on the ability to grow architectures of nanotubes using CVD techniques. The progress in this area has been substantial and we can today build 2D as well 3D architectures of both MWNT and SWNT organized microscale features on various platforms in different geometries. These structures, once assembled on effective substrates, can be transferred to other substrates including flexible polymer films. The nanotube architectures will be the building blocks for many different applications for a broad range of areas that include electronics, MEMS, chemical industry, and structural materials. The challenge that remains is in controlling the molecular

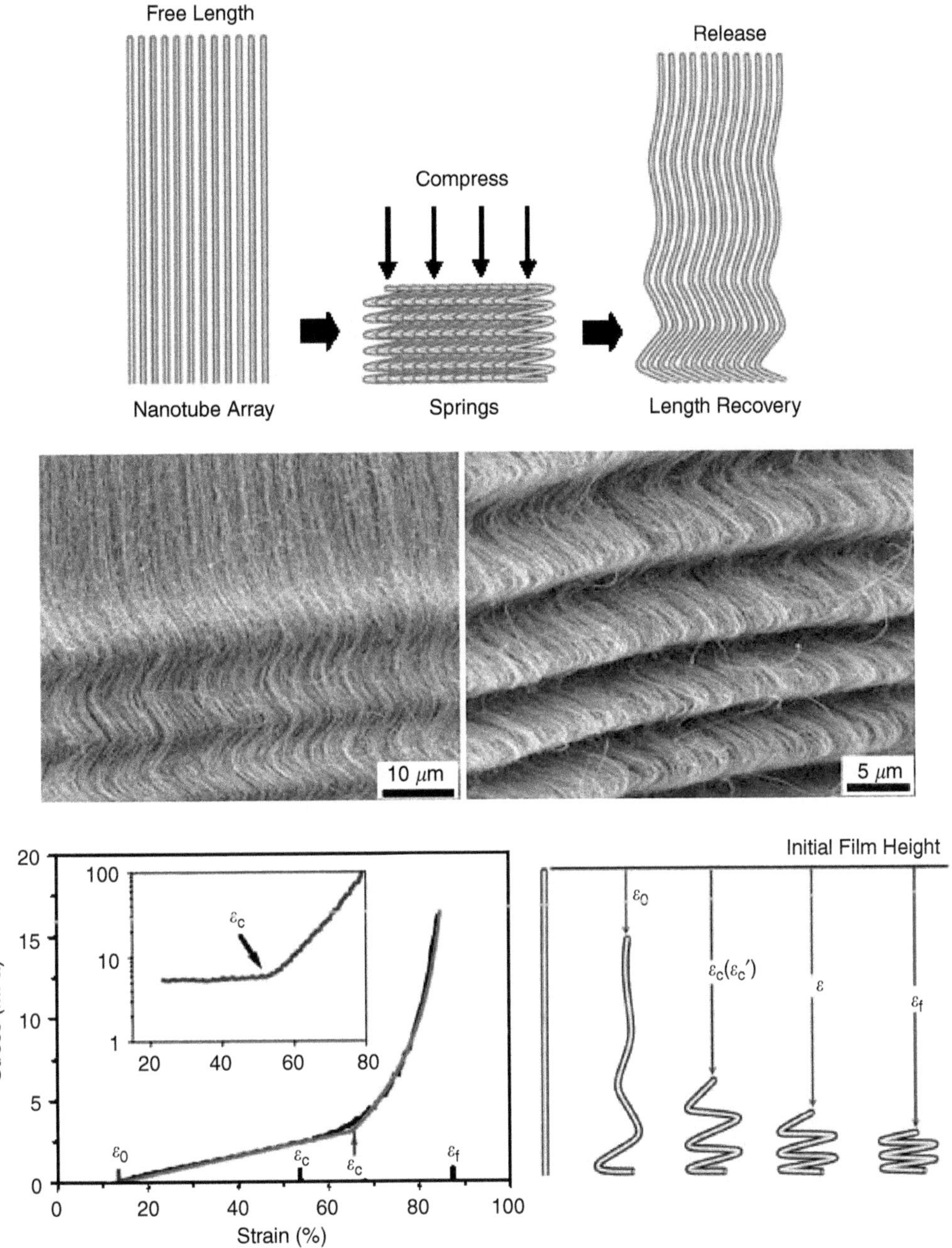

FIGURE 1.15 Compression testing of aligned CNT films. A schematic illustration shows a nanotube array compressed to folded springs and then regaining the free length upon the release of compressive load. The image also shows modeling of nanotube behavior under large strain compression, namely buckling under compression with certain permanent initial strain, with specific critical strain where the buckled nanotube folds begin to collapse from the bottom side, and with final strain where all folds are fully collapsed. Experimental data (black curve) and the model (red curve) fit to this data. (From Cao, A.Y., Dickrell, P.L., Sawyer, W.G., Ghasemi-Nejhad, M.N., and Ajayan, P.M., *Science*, 310, 1307, 2005. With permission from AAAS.)

architecture (chirality) of the individual nanotubes that are grown and in obtaining precise control of dimensions and density of individual units in macroscale assembled structures. Rapid progress is being made in these areas and it seems that soon, some of these structures will appear as prototypes in working devices and ultimately become miniaturized parts of our high tech daily life.

Acknowledgments

The authors acknowledge funding from NSF-NSEC at RPI for direct assembly of nanostructures (Nanoscale Science and Engineering Initiative of the National Science Foundation under NSF award number DMR-0117792) and the Focus Center New York for Electronic Interconnects at Rensselaer Polytechnic Institute.

References

[1] Kroto, H.W. et al., C_{60}: Buckminsterfullerene, *Nature*, 318, 162, 1985.
[2] Iijima, S., Helical microtubules of graphitic carbon, *Nature*, 354, 56, 1991.
[3] Ajiki, H. and Ando, T., Electronic states of carbon nanotubes, *J. Phys. Soc. Jpn.*, 62, 1255, 1993.
[4] Hamada, N., Sawada, S., and Oshiyama, A., New one-dimensional conductors: Graphitic microtubules, *Phys. Rev. Lett.*, 68, 1579, 1992.
[5] Mintmire, J.W., Dunlap, B.I., and White, C.T., Are fullerene tubules metallic? *Phys. Rev. Lett.*, 68, 631, 1992.
[6] Saito, R. et al., Formation of general fullerenes by their projection on a honeycomb lattice, *Phys. Rev. B*, 46, 1804, 1992.
[7] Ajayan, P.M., Nanotubes from carbon, *Chem. Rev.*, 99, 1787, 1999.
[8] Wei, B.Q., Vajtai, R., and Ajayan, P.M., Reliability and current carrying capacity of carbon nanotubes, *Appl. Phys. Lett.*, 79, 1172, 2001.
[9] Collins, P.G., Arnold, M.S., and Avouris, Ph., Engineering carbon nanotubes and nanotube circuits using electrical breakdown, *Science*, 292, 706, 2001.
[10] Graham, A.P., Duesberg, G.S., Seidel, R.V., Liebau, M., Unger, E., Pamler, W., Kreupl, F., and Hoenlein, W., Carbon nanotubes for microelectronics? *Small*, 1, 382, 2005.
[11] Wei, J.Q., Zhu, H.W., Wu, D.H., and Wei, B.Q., Carbon nanotube filaments in household light bulbs, *Appl. Phys. Lett.*, 84, 4969, 2004.
[12] Zhang, M., Fang, S.L., Zakhidov, A.A., Lee, S.B., Aliev, A.E., Williams, C.D., and Atkinson, K.R., Baughman, R.H., Strong, transparent, multifunctional, carbon nanotube sheets, *Science*, 309, 1215, 2005.
[13] Wu, Z.C. et al., Transparent, conductive carbon nanotube films, *Science*, 305, 1273, 2004.
[14] Ebbesen, T.W. and Ajayan, P.M., Large-scale synthesis of carbon nanotubes, *Nature*, 358, 220, 1992.
[15] Iijima, S. and Ichihashi, T., Single-shell carbon nanotubes of 1-nm diameter, *Nature*, 363, 603, 1993.
[16] Bethune, D.S. et al., Cobalt-catalyzed growth of carbon nanotubes with single-atomic-layerwalls, *Nature*, 363, 605, 1993.
[17] Terrones, M., Carbon nanotubes: Synthesis and properties, electronic devices and other emerging applications, *Int. Mater. Rev.*, 49, 325, 2004.
[18] Lambert, J.M. et al., Improving conditions towards isolating single-shell carbon nanotubes, *Chem. Phys. Lett.*, 226, 364, 1994.
[19] Ajayan, P.M. et al., Growth morphologies during cobalt-catalyzed single-shell carbon nanotube synthesis, *Chem. Phys. Lett.*, 215, 509, 1993.
[20] Ando, Y. et al., Mass production of single-wall carbon nanotubes by the arc plasma jet method, *Chem. Phys. Lett.*, 323, 580, 2000.

[21] Guo, T. et al., Self-assembly of tubular fullerenes, *J. Phys. Chem.*, 99, 10694, 1995.

[22] Eklund, P.C. et al., Large-scale production of single-walled carbon nanotubes using ultrafast pulses from a free electron laser, *Nano Lett.*, 2, 561, 2002.

[23] Rinzler, A.G. et al., Large-scale purification of single-wall carbon nanotubes: Process, product, and characterization, *Appl. Phys. A*, 67, 29, 1998.

[24] Chiang, I.W. et al., Purification and characterization of single-wall carbon nanotubes (SWNTs) obtained from the gas-phase decomposition of CO (HiPco process), *J. Phys. Chem. B*, 105, 8297, 2001.

[25] Chiang, I.W. et al., Purification and characterization of single-wall carbon nanotubes, *J. Phys. Chem. B*, 105, 1157, 2001.

[26] Banerjee, S., Hemraj-Benny, T., and Wong, S.S., Covalent surface chemistry of single-walled carbon nanotubes, *Adv. Mater.*, 17, 17, 2005.

[27] Hirsch, A. and Vostrowsky, O., Functionalization of carbon nanotubes, *Top. Curr. Chem.*, 245, 193, 2005.

[28] Baker, R.T.K., Catalytic growth of carbon filaments, *Carbon*, 27, 315, 1989.

[29] See, e.g., http://www.apsci.com/ppi-about.html

[30] Ding, D. et al., Synthesis of carbon nanostructures on nanocrystalline Ni-Ni3P catalyst supported by SiC whiskers, *Carbon*, 41, 579, 2003.

[31] Murakami, T. et al., Raman study of SWNTs grown by CCVD method on SiC, *Thin Solid Films*, 464, 319, 2004.

[32] Colomer, J.F.C. et al., Large-scale synthesis of single-wall carbon nanotubes by catalytic chemical vapor deposition (CCVD) method, *Chem. Phys. Lett.*, 317, 83, 2000.

[33] Ago, H. et al., Growth of double-wall carbon nanotubes with diameter-controlled iron oxide nanoparticles supported on MgO, *Chem. Phys. Lett.*, 391, 308, 2004.

[34] Cheung, C.L. et al., Diameter-controlled synthesis of carbon nanotubes, *J. Phys. Chem. B*, 106, 2429, 2002.

[35] Hongo, H. et al., Chemical vapor deposition of single-wall carbon nanotubes on iron-film-coated sapphire substrates, *Chem. Phys. Lett.*, 361, 349, 2002.

[36] Fonseca, A. et al., Synthesis of single- and multi-wall carbon nanotubes over supported catalysts, *Appl. Phys. A*, 67, 11, 1998.

[37] Willems, I. et al., Control of the outer diameter of thin carbon nanotubes synthesized by catalytic decomposition of hydrocarbons, *Chem. Phys. Lett.*, 317, 71, 2000.

[38] Maruyama, S. et al., Generation of single-walled carbon nanotubes from alcohol and generation mechanism by molecular dynamics simulations, *J. Nanosci. Nanotechnol.*, 4, 360, 2004.

[39] Hata, K., Futaba, D.N., Mizuno, K., Namai, T., Yumura, M., and Iijima, S., Water-assisted highly efficient synthesis of impurity-free single-walled carbon nanotubes, *Science*, 306, 1362, 2004.

[40] Ren, Z.F. et al., Synthesis of large arrays of well-aligned carbon nanotubes on glass, *Science*, 282, 1105, 1998.

[41] Kitiyanan, B. et al., Controlled production of single-wall carbon nanotubes by catalytic decomposition of CO on bimetallic Co-Mo catalysts, *Chem. Phys. Lett.*, 317, 497, 2000.

[42] Ng, H.T. et al., Growth of carbon nanotubes: A combinatorial method to study the effect of catalyst and underlayers, *J. Phys. Chem. B*, 107, 8484, 2003.

[43] Franklin, N.R. et al., Patterned growth of single-walled carbon nanotubes on full 4-inch wafers, *Appl. Phys. Lett.*, 79, 4571, 2001.

[44] Nikolaev, P. et al., Gas-phase catalytic growth of single-walled carbon nanotubes from carbon monoxide, *Chem. Phys. Lett.*, 313, 91, 1999.

[45] Maruyama, S. et al., Growth process of vertically aligned single-walled carbon nanotubes, *Chem. Phys. Lett.*, 403, 320, 2005.

[46] Kobayashi, Y. et al., CVD growth of single-walled carbon nanotubes using size-controlled nanoparticle catalyst, *Thin Solid Films*, 464–465, 286, 2004.

[47] Moisala, A., Nasibulin, A.G., and Kauppinen, E.I., The role of metal nanoparticles in the catalytic production of single-walled carbon nanotubes — A review, *J. Phys. Condens. Matter.*, 15, S3011, 2003.

[48] Shibuta, Y. and Maruyama, S., Molecular dynamics simulation of formation process of single-walled carbon nanotubes by CCVD method, *Chem. Phys. Lett.*,382, 381, 2003.

[49] Teo, K.B. et al., Catalytic synthesis of carbon nanotubes and nanofibers, in *Encyclopedia of Nanoscience and Nanotechnology*, Ed. H.S. Nalwa, American Scientific Publishers, 2003.

[50] Terrones, M. et al., Controlled production of aligned-nanotube bundles, *Nature*, 388, 52, 1997.

[51] Terrones, M., Grobert, N., Zhang, J.P., Terrones, H., Kordatos, K., Hsu, W.K., Hare, J.P., Townsend, P.D., Prassides, K., Cheetham, A.K., Kroto, H.W., and Walton, D.R.M., Preparation of aligned carbon nanotubes catalysed by laser-etched cobalt thin films, *Chem. Phys. Lett.*, 285, 299, 1998.

[52] Kind, H., Bonard, J.M., Emmenegger, C., Nilsson, L.O., Hernadi, K., Maillard-Schaller, E., Schlapbach, L., Forro, L., and Kern, K., Patterned films of nanotubes using microcontact printing of catalysts, *Adv. Mater.*, 11, 1285, 1999.

[53] Andrews, R., Jacques, D., Rao, A.M., Derbyshire, F., Qian, D., Fan, X., Dickey, E.C., and Chen, J., Continuous production of aligned carbon nanotubes: A step closer to commercial realization, *Chem. Phys. Lett.*, 303, 467, 1999.

[54] Fan, S.S., Chapline, M.G., Franklin, N.R., Tombler, T.W., Cassell, A.M., and Dai, H.J., Self-oriented regular arrays of carbon nanotubes and their field emission properties, *Science*, 283, 512, 1999.

[55] Cao, A.Y. et al., Vertical aligned carbon nanotubes grown on Au film and reduction of threshold field in field emission, *Chem. Phys. Lett.*, 335, 150, 2001.

[56] Zhang, X.F. et al., Rapid growth of well-aligned carbon nanotube arrays, *Chem. Phys. Lett.*, 362, 285, 2002.

[57] Cao, A.Y. et al., Grapevine-like growth of single walled carbon nanotubes among vertically aligned multiwalled nanotube arrays, *Appl. Phys. Lett.*, 79, 1252, 2001.

[58] Zhang, Z.J. et al., Substrate-site selective growth of aligned carbon nanotubes, *Appl. Phys. Lett.*, 77, 3764, 2000.

[59] Jung, Y.J., Wei, B.Q., Vajtai, R., and Ajayan, P.M., Mechanism of selective growth of carbon nanotubes on SiO_2/Si patterns, *Nano Lett.*, 3, 561, 2003.

[60] Wei, B.Q. et al., Carbon nanotube-magnesium oxide cube networks, *J. Nanosci. Nanotechnol.*, 1, 35, 2001.

[61] Cao, A.Y., Veedu, V.P., Li, X., Yao, Z., Ghasemi-Nejhad, M.N., and Ajayan, P.M., Multifunctional brushes made from carbon nanotubes, *Nat. Mater.*, 4, 540, 2005.

[62] Cao, A.Y., Baskaran, R., Frederick, M.J., Turner, K., Ajayan, P.M., and Ramanath, G., Direction-selective and length-tunable in-plane growth of carbon nanotubes, *Adv. Mater.*, 15, 1105, 2003.

[63] Li, X.S., Cao, A.Y., Jung, Y.J., Vajtai, R., and Ajayan, P.M., Bottom-up growth of carbon nanotube multilayers: Unprecedented growth, *Nano Lett.*, 5, 1997, 2005.

[64] Zhu, L.B. et al., Aligned carbon nanotube stacks by water-assisted selective etching, *Nano Lett.*, 5, 2641, 2005.

[65] Pinault, M. et al., Evidence of sequential lift in growth of aligned multiwalled carbon nanotube multilayers, *Nano Lett.*, 5, 2394, 2005.

[66] Jung, Y.J., Homma, Y., Ogino, T., Kobayashi, Y., Takagi, D., Wei, B.Q., Vajtai, R., and Ajayan, P.M., High-density, large-area single-walled carbon nanotube networks on nanoscale patterned substrates, *J. Phys. Chem. B*, 107, 6859, 2003.

[67] Lastella, S. et al., Density control of single-walled carbon nanotubes using patterned iron nanoparticle catalysts derived from phase-separated thin films of a polyferrocene block copolymer, *J. Mater. Chem.*, 14, 1791, 2004.

[68] Wei, B.Q., Vajtai, R., Jung, Y., Ward, J., Zhang, R., Ramanath, G., and Ajayan, P.M., Assembly of highly-organized carbon nanotube architectures by chemical vapor deposition, *Chem. Mater.*, 15, 1598, 2003.

[69] Wei, B.Q., Vajtai, R., Jung, Y., Ward, J., Zhang, R., Ramanath, G., and Ajayan, P.M., Organized assembly of carbon nanotubes, *Nature*, 416, 495, 2002.
[70] Jung, Y.J., Homma, Y., Vajtai, R., Kobayashi, Y., Ogino, T., and Ajayan, P.M., Straightening suspended single walled carbon nanotubes by ion irradiation, *Nano Lett.*, 4, 1109, 2004.
[71] Srivastava, A., Srivastava, O.N., Talapatra, S., Vajtai, R., and Ajayan, P.M., Carbon nanotube filters, *Nat. Mater.*, 3, 610, 2004.
[72] Chwickshank, R. et al., *Medical Microbiology*, Churchill, London, 1975.
[73] http://www.millipore.com/publications.nsf/docs/DS1180EN00
[74] Modi, A., Koratkar, N., Lass, E., Wei, B.Q., and Ajayan, P.M., Miniaturized gas ionization sensors using carbon nanotubes, *Nature*, 424, 171, 2003.
[75] Ajayan, P.M. et al., Single-walled carbon nanotube–polymer composites: Strength and weakness, *Adv. Mater.*, 12, 750, 2000.
[76] Raravikar, N.R. et al., Embedded carbon-nanotube-stiffened polymer surfaces, *Chem. Mater.*, 17, 974, 2005.
[77] Chen, Y.-C. et al., Ultrafast optical switching properties of single-wall carbon nanotube polymer composites at 1.55 μm, *Appl. Phys. Lett.*, 81, 975, 2002.
[78] Rege, K. et al., Enzyme-polymer-single walled carbon nanotube composites as biocatalytic films, *Nano Lett.*, 3, 829, 2003.
[79] Wagner, H.D., Nanotube-polymer adhesion: A mechanics approach, *Chem. Phys. Lett.*, 361, 57, 2002.
[80] Philip, B. et al., Carbon nanotube/PMMA composite thin films for gas-sensing applications, *Smart Mater. Struct.*, 12, 935, 2003.
[81] Koratkar, N., Wei, B.Q., and Ajayan, P.M., Carbon nanotube films for damping applications, *Adv. Mater.*, 14, 997, 2002.
[82] Koratkar, N., Wei, B.Q., and Ajayan, P.M., Multifunctional structural reinforcement featuring carbon nanotube films, *Compos. Sci. Technol.*, 63, 1525, 2003.
[83] Suhr, J., Koratkar, N., Keblinski, P., and Ajayan, P.M., Temperature-activated interfacial friction damping in carbon nanotube polymer composites, *Nat. Mater.*, 4, 134, 2005.
[84] Biggerstaff, J.M. and Kosmatka, J.B., Damping performance of cocured graphite/epoxy composite laminates with embedded damping materials, *J. Compos. Mater.*, 33, 1457, 1999.
[85] Jung, Y.J., Kar, S., Talapatra, S., Soldano, C., Vishwanathan, G., Li, X.S., Yao, Z.L., Ou, F.S., Avadhanula, A., Vajtai, R., Curran, S., Nalamasu, O., and Ajayan, P.M., Aligned carbon nanotube-polymer hybrid architectures for diverse flexible electronic applications, *Nano Lett.*, 6, 413, 2006.
[86] Chakrapani, N. et al., Capillarity-driven assembly of two-dimensional cellular carbon nanotube foams, *Proc. Natl. Acad. Sci. USA*, 101, 4009, 2004.
[87] Cao, A.Y., Dickrell, P.L., Sawyer, W.G., Ghasemi-Nejhad, M.N., and Ajayan, P.M., Super-compressible foam-like carbon nanotube films, *Science*, 310, 1307, 2005.

2

Controlled Self-Assembly

Nathan W. Moore and
Tonya L. Kuhl
University of California at Davis

Abstract

Self-assembly is the synthesis of either microscopic or macroscopic materials assembled from the bottom-up into larger structures. Controlled self-assembly differs in that the desired properties of the assembled materials are "programmed" into their structure through a systematic selection of the constituent components. This rapidly growing field — which traditionally has applications as broad as solid materials fabrication, lubrication, biotechnology, and more — is also at the forefront of modern nanoscience, yet its theoretical foundation can be explained by classical thermodynamics and kinetics. In this chapter we review the principles governing controlled self-assembly and provide examples of applying these principles with monodisperse colloidal microspheres. Specifically, we show how colloidal crystals and fractals can be fabricated by tuning the inherent particle–particle interactions and from controlling the flow of a supporting liquid or gas phase. Also presented is the production of materials with novel, programmed microstructure through symmetry reduction via surface patterning, microsphere templating, and cross-linking with biospecific polymer linkers. Finally, we present challenges for the future of controlled self-assembly.

2.1 Introduction

If one knows what atoms are present in a material and how these atoms are arranged, then often the properties of the material may be understood.[1] Capitalizing on this approach is self-assembly, a bottom-up approach where one simply adds components together and they spontaneously order into some defined (hierarchical) and reproducible structure with the desired material properties. Although achieving materials through design is a far reaching goal, even with simple building blocks of spheres and rods quite complex structures can be realized, for example, structures analogous to the various crystal unit cells of different salts.[2] One crucial difference is, whereas the physical properties of materials are usually directly related to the arrangement and type of bonds that make up the material, self-assembly relies on the accurate and controlled application of intermolecular forces. These much weaker "bonds" allow the building blocks to selectively assemble under equilibrium (or close to equilibrium) conditions into regularly patterned structures — typically in a fluid phase at room temperature. This emphasis on softer, fluid phase materials is a new frontier for material science, which has primarily concentrated on rigid, condensed states of matter.

At the micron-scale, colloidal particles can be self-assembled into complex materials, such as gels or solids, driven solely by weak, nonspecific interactions between the particles. For single-component repulsive particles, substantial work has been done to characterize their phase behavior as a function of volume fraction and temperature.[3,4] Interactions between colloids can be modulated by the addition of free macromolecules such as polymers, or by the attachment of polymers to the particle surfaces to promote steric stabilization. Moreover, particle surfaces can also be modified with immobilized molecules, such as specific lock-and-key biomolecules.[5–7] In the latter case, complimentary molecules allow for assembly of particles through attractive interactions. High affinity biomolecular interactions, such as biotin–avidin[8] and antigen–antibody recognition, have been shown to make suspensions of strongly attractive particles leading to disordered, noncrystalline structures. Weakly attractive interactions may allow more equilibrium structures to be formed.

At the nanoscale, these same repulsive and attractive interactions can be harnessed for programmed self-assembly.[9] Controlled or programmed self-assembly seeks to teach the building blocks how to arrange themselves into desired configurations to yield micro- to macroscopic functional materials with distinct physical properties and controlled size and composition. As opposed to bulk material properties, the particle or building block interface dictates the physical and chemical interactions of interest between the particle and its surroundings. Programmed self-assembly frequently relies on symmetry breaking at surfaces to align or orient sub-blocks, so that subsequent complimentary recognition between groups on the blocks allows them to bind — locking the desired structure into place. This methodology offers great flexibility in composition, as the building blocks may be any type of material with complimentary recognition tailored via organic components, which act as macromolecular glue. Most existing approaches utilize the strong affinity of thiols to gold or silver,[10] DNA linking,[11–13] or hydrogen bonding.[14,15] For example, polymers functionalized with molecular recognition groups have recently been used to mediate the formation of spherical or related assemblies of gold nanoparticles.[14] While these approaches have shown remarkable capabilities in assembling nanoparticles into functional nanostructures, the ability to control the size and shape of the aggregates has not been fully realized. One of the greatest challenges for materials science is the creation of supramolecular materials in which the constituent units are nanostructures with a high degree of molecular ordering. Learning how to create supramolecular units from nanoscale objects and elucidating rules for mediating their macroscopic organization into functional materials offer fascinating prospects for technology. Such synthetic nanostructures, analogous to folded proteins in definition of chemical sectors, shape, and topography, will be interesting building blocks for materials because they must pack in ways that fill space efficiently. The replicate structures produced using this methodology may then be integrated into a device or used as constituents of a bulk material.

In this chapter, we address the combining of organic macromolecules that have been tailored to recognize each other in specific ways, with inorganic components that carry the desired electronic, magnetic, or photonic properties, to construct new composite materials based on the principles of programmed self-assembly. We cover the basic principles of self-assembly, give examples, and address some current challenges in implementing self-assembly for material design.

2.2 Theory of Self-Assembly

It is a fundamental goal of materials science to understand how the microstructures of component materials imbue observable properties to bulk materials. While in controlled self-assembly the product is often not a bulk material but a microstructure built from nanostructures or even atoms, it is still true that the manner in which a product's properties evolve can only be understood by examining the fundamental physics that govern the way its components interact. In this section, we first explain how such constituents typically interact and how one can tune their properties to build materials with prescribed properties from the bottom up.

The theoretical basis of particle self-assembly is rooted in the symbiotic fields of colloidal and interfacial science — a study that has traditionally dealt with the "bulk" behavior of materials sized between 10 nm and a few microns. In recent years, this field has been renamed "nanoscience" to emphasize that as particles approach the sizes of atoms, their surface area to volume ratio increases dramatically; consequently, it is their interfacial — and not their bulk — properties that dominate their interaction. As an example, in these size regimes the movement of particles through fluids can depend less on inertial forces, such as gravitational acceleration, than on viscous drag. Likewise, smaller masses diffuse more easily through Brownian motion and thus present additional kinetic constraints than in bulk material processing techniques. Thus, their kinetics of assembly requires considerations not present in bulk materials. Once in contact, whether two particles "stick" together depends on the molecular interactions between their two surfaces. It is these sorts of interactions that we attempt to clarify through the theories in this section and the many examples in the next section.

2.2.1 Thermodynamics of Self-Assembly

By definition, self-assembly does not require work or energy input into the system. Rather, individual components spontaneously assemble into larger aggregates. One commonly encountered class of materials that self-assemble into a range of structures is amphiphilic molecules, a typical soft condensed matter. Depending on concentration and temperature a wide range of structures can be formed from spatially arranged molecules, for example, micelles, spherical vesicles, and inverted hexagonal rods of varying length. In each case, the system is at equilibrium. Similar phenomena have been observed with monodisperse spherical particles, where close-packed structures such as hexagonally close-packed (hcp) or face-centered cubic (fcc) crystals self-assemble when their volume fraction exceeds about 50% for spheres with short-ranged isotropic interactions.[3]

For a closed system at equilibrium with a fixed temperature and pressure, the Gibbs free energy is at a minimum and the chemical potential, μ, of all phases is equal. The local curvature of the Gibbs free energy of solution with respect to composition determines whether an unstable composition will achieve stability by small, widely spaced composition fluctuations or large isolated fluctuations in composition. The latter is an important process called nucleation in materials science (typically solids) or aggregation in soft condensed matter (e.g., amphiphilic molecules).

For self-assembling systems, the chemical potential of all identical particles in different aggregates is the same at equilibrium so

$$\mu = \mu_N^0 + \frac{kT}{N}\log\left(\frac{X_N}{N}\right) = \text{constant} \tag{2.1}$$

where μ is the chemical potential of a particle in an aggregate composed of N particles, μ_N^0 is the standard part of the chemical potential per particle in aggregates of N particles, k is the Boltzmann constant, T is the absolute temperature, and X_N is the activity (typically using the concentration is sufficiently accurate) of particles in aggregates of N particles.[16] For the simple case of particle monomers (isolated particles) or dimers we have

$$\mu = \mu_1^0 + kT\log(X_1) = \mu_2^0 + \frac{kT}{2}\log\left(\frac{X_2}{2}\right)\ldots = \mu_N^0 + \frac{kT}{N}\log\left(\frac{X_N}{N}\right) \tag{2.2}$$

If we are only concerned with the case of particles being associated into either a bulk aggregate or dispersed state, the situation can be simplified to

$$\underset{N\ \text{particles}}{\mu_1^0} \qquad Np \underset{k_N}{\overset{\Delta G^0,\ k_1}{\rightleftharpoons}} p_N \qquad \underset{\text{Aggregate of } N \text{ particles}}{\mu_N^0} \tag{2.3}$$

Using the law of mass action the equilibrium constant, K, is given by the ratio of the reaction rates of association and dissociation where

$$\text{rate of association} = k_1 X_1^N \tag{2.4}$$

$$\text{rate of dissociation} = k_N \frac{X_N}{N} \tag{2.5}$$

Combined with Equation 2.2 we find

$$K = \frac{k_1}{k_N} = \exp\left[\frac{-N(\mu_N^0 - \mu_1^0)}{kT}\right] \tag{2.6}$$

Provided one has knowledge of the aggregation number and chemical potentials or there is change in the Gibbs free energy upon aggregation, such an analysis enables one to predict the aggregation of particles into larger structures, for example, surfactants aggregating into micelles above the critical micelle concentration[16] and the aggregation of hard spheres with increasing volume fraction.[4]

In the following discussion, we shall limit ourselves to the aggregation of two particles in solution and predict whether or not particles will self-assemble into a dimer (aggregate) or stay dispersed based on the interaction energy.

2.2.2 Particle–Particle Interactions

If one can reproducibly couple particles together into aggregates of a prescribed structure, a large range of novel hierarchical materials based on defined composition and properties may be produced. Toward this objective, knowledge of the relevant forces operating between particles can be used to manipulate their assembly at various stages. For simplicity, we will restrict ourselves to spherical objects because many different materials are commonly available in a wide range of spherical sizes.

Considering the interaction of two spheres across a fluid medium, the most pertinent physical forces include:

1. Van der Waals forces — due to the polarizability of the particles.
2. Electrostatic forces — due to charged groups on the particle.
3. Steric forces — due to the local concentration and configurational entropy of the molecular groups (frequently polymers) attached to the particle surface.
4. Hydrophobic forces — strong attraction between inert, nonpolar surfaces in water. The nonpolar surfaces may be particles themselves or coatings on the particle surface.
5. Hydrogen bonding — due to spatial ordering of positively charged hydrogen atoms covalently bound to electronegative atoms. Hydrogen bonding is particularly important in water and between complementary DNA base pair interactions.

These forces can combine with complementary geometry, for example, a lock to a key, to yield the high specificity associated with biological recognition interactions between ligands and receptors or

TABLE 2.1 Sphere–Sphere Interaction Energies

Interaction	Comments	Energy	Parameters
DLVO = van der Waals + Electrostatic			
van der Waals	Always present. Always attractive between two particles of the same material interacting across any medium	$E = \frac{-AR}{6D}$, $A = \frac{3}{4}kT\left(\frac{\varepsilon_1 - \varepsilon_2}{\varepsilon_1 + \varepsilon_2}\right)^2 + \frac{3h\upsilon_e}{16\sqrt{2}}\frac{(n_1^2 - n_2^2)^2}{(n_1^2 + n_2^2)^{3/2}}$	A = Hamaker Constant n_i = refractive index of material i (subscripts 1 & 2 refer to the particles and medium, respectively) ε = permittivity h = Plank's constant ν_e = absorption frequency (typically ~$3.0 \times 10^{15}\ s^{-1}$)
Electrostatic	Approximate for charged spheres Constant potential, ψ_0	$E = 2\pi R\varepsilon\varepsilon_0\psi_0^2 \ln(1 + \exp(-\kappa D))$	For $\kappa > 10$. The Debye screening length of a 1:1 electrolyte is $\kappa^{-1}[\text{Å}] \approx 3.04/\sqrt{M}$ at room temperature and M is electrolyte molarity ε_0 = permittivity of vacuum.
	Constant charge, σ	$E = \pi R\left(\frac{\sigma^2}{\varepsilon\varepsilon_0\kappa^2}\right)\exp(-\kappa D)$	$\sigma \cong \varepsilon\varepsilon_0\kappa\psi_0$ Valid for $\psi_0 < 25$ mV.
Polymeric			
Steric — Polymer mushroom	Always repulsive. For nonlaterally overlapping grafted polymer chains.	$E = 72\pi\Gamma kTR_g R\exp(-D/R_g)$	Γ = surface density R_g = chain dimension or radius of gyration $R_g = a \cdot \eta^\nu$ a = monomer size η = # of mers per chain ν = 0.6 for a good solvent. $L = \Gamma^{1/2}R_g^{5/3}$ (brush extension)
Steric — Polymer brush	Always repulsive. For strongly overlapping grafted polymer chains.	$E = 64\pi\Gamma^{1/2}L^2kTR\exp(-D/L)$	
Polymer bridging	Attractive (nonspecific) polymer bridging between two parallel plates.	$E \cong -\alpha R\exp(-D/R_l)$	R_l = characteristic length of tails and/or loops of adsorbed polymer chains α depends on the adsorption energy of the polymer to the surfaces[17,18]
Depletion			
Polymeric or sphere–sphere	Attractive — due to presence of smaller, nonadsorbing particles (or polymers) in solution with larger particles. By aggregating into larger particles there is a free volume gain for the smaller particles. Considerable when particles differ in size by a factor of 10 or more.	$E = -(\pi\Pi/12)(2(R+d)^3 - 3(R+d)^2D + D^3)$ $E_{max} = -(\pi\Pi/6)(R+d)^3$ at $D = 0$	d = depletion layer thickness (distance over which the concentration of smaller particles at the surface of the large particles goes from zero to the bulk concentration. For polymers $d \approx R_g$. The osmotic pressure is $\Pi \approx \rho kT$, (ρ = concentration of mers or smaller particles between the larger particles). $E \approx \Pi \times d$ for flat surfaces.

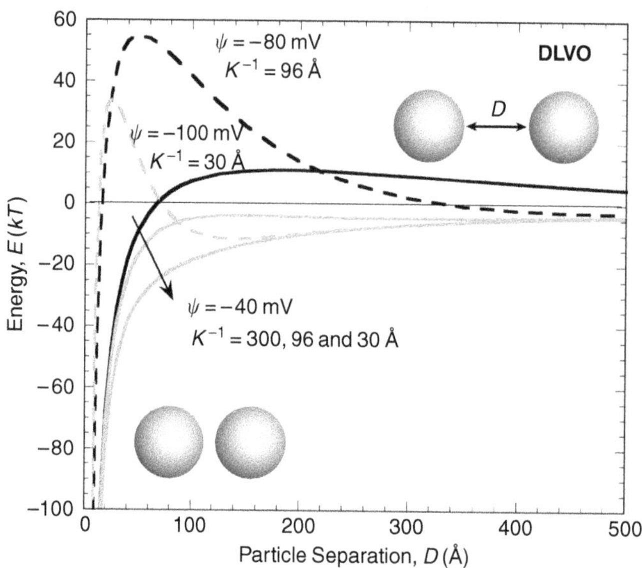

FIGURE 2.1 Calculated DLVO interaction between two 100-nm TiO_2 spheres in water. Parameters used in the calculations were $A = 2.0 \times 10^{-19}$J, $\psi = -40$ to -100 mV, in 0.1, 1, or 10 mM monovalent electrolyte ($K^{-1} = 300$, 96, or 30 Å). Positive interaction energies are repulsive, whereas negative values are attractive. Because of the large refractive index difference between TiO_2 ($n_1 \approx 2.4$) and water ($n_2 = 1.33$) the van der Waals attraction between the particles is significant. The electrostatic interaction can be tuned by varying the amount of electrolyte (salt) in the water. A typical surface potential of TiO_2 is $\psi = -40$ mV. (From Li, B.K. and Logan, B.E., *Colloids Surf., B Biointerfaces*, 36, 81, 2004.)

complementarity between DNA strands. Table 2.1 provides a convenient summary of the governing equations that can be used to calculate the magnitude of these interactions between spherical particles as a function of the intersphere distance (D).[16] In Table 2.1 the interactions are presented as energies (E) rather than as forces so that these interactions can be compared to the available thermal energy in the medium, kT, which as we discuss later has some bearing on the kinetics of these interactions. The interaction forces between two spherical particles can be computed as $F = \mathrm{d}E/\mathrm{d}D$. Note that with the exception of depletion interactions, all these interaction energies (and hence interaction forces) depend linearly on R, which is the properly normalized effective interaction radius between two spheres with radii R_1 and R_2, namely, $R = R_1R_2/(R_1 + R_2)$. Thus, the sizes of the particles play an important role in the strength of their interaction. For materials interacting at plane–parallel interfaces, the interfacial energy per unit area can be estimated as $W = F/2\pi R$, the well-known Derjaguin approximation.

To illustrate the use of Table 2.1, the sum of the van der Waals and electrostatic interactions (DLVO) acting between two titanium oxide particles (TiO_2) in water is shown in Figure 2.1. The van der Waals attraction between any two particles of the same material is always attractive in any medium. In this example, the van der Waals attraction is very large due to the significant difference in the refractive index of TiO_2 ($n_1 \approx 2.4$) and water ($n_2 = 1.33$). In comparison, the electrostatic interactions between the two particles can be tuned by changing the concentration of the electrolyte in solution (monovalent in this case). Typically, TiO_2 would have a surface potential $\psi = -40$ mV in 1-mM monovalent electrolyte.[19] A range of surface potentials (−40, −80, −100 mV) is shown in Figure 2.1 to demonstrate the wide variety of interaction profiles that can be encountered — from purely attractive to repulsive with primary and secondary minima.

When the interaction potential is combined with the laws of mass action, the equilibrium behavior of a particle dispersion can be predicted. Unfortunately, thermodynamics does not tell us about the rates of processes, which we address in detail in the next section.

2.2.3 Kinetics of Self-Assembly

There are several reasons to study the kinetics of self-assembly. First, completely equilibrated systems rarely exist in nature, as evidenced by the observation that even the most regular crystals have imperfections, grain boundaries, or other irregularities. An understanding of the kinetics can be used to predict the range of structures that will evolve in a given time frame. The timescale of self-assembly can range from picoseconds for molecules to milliseconds for colloidal particles exhibiting purely attractive potentials. One of the longest-living suspensions in historical record that has not yet self-assembled is a gold sol prepared by Michael Faraday in 1856, presently on display at the Royal Institution of Great Britain.

At the simplest level, the difference between particle systems that self-assemble in a given time frame and those that do not is the height of the energy potential barrier (Figure 2.1) compared to the available thermal energy, kT. A significant amount of assembly is possible provided: (1) the energy barrier is sufficiently small (less than a few kT); and (2) the potential well is sufficiently deep compared to kT so that assembly is favored over disassembly.

In general, predicting the rate of aggregate formation becomes increasingly complex as the number of particles per aggregate increases. Dimers can only be formed from monomers, and trimers can only be readily formed from one dimer plus one monomer. However, tetramers can be formed either by uniting two dimers, or a monomer and trimer, and pentamers can be formed either by uniting a tetramer with a monomer, or a dimer with a trimer, and so on. Thus, any useful kinetic description must utilize a matrix method to track the many possible reactions that can lead to a wide distribution of aggregate sizes. One such method is the Smoluchowski coagulation equation, which relates the rates of change in the concentrations of aggregates C_i with i number of particles as

$$\frac{\mathrm{d}C_i}{\mathrm{d}t} = \frac{1}{2}\sum_{l+m=i} K_{lm}C_l(t)C_m(t) - C_i(t)\sum_{j=1}^{\infty} K_{ij}C_j(t) \tag{2.7}$$

where K_{lm} is a matrix of second-order rate constants between aggregates with l and m numbers of particles $(l+m=i)$.[20] These rate constants must usually be determined experimentally except for a few very special cases.

2.2.3.1 Rapid Aggregation

As one simplified example we will consider the special case of dimers forming from a dilute suspension of identical monomers so that we can neglect the formation of aggregates with $i > 2$. We also consider the interparticle repulsion to be negligible so that aggregation is diffusion-limited, the so-called "rapid" aggregation regime wherein upon colliding particles bind irreversibly. To estimate the reaction rate we use Fick's law, which estimates the frequency of collisions between any two particles as

$$J_f = 8\pi \mathscr{D} R C_1 \tag{2.8}$$

where $\mathscr{D}$ is the particle diffusion coefficient, R is the particle radius as before, and C_1 is the monomer concentration. If all particles undergo Brownian motion, the diffusion coefficient $\mathscr{D}$ can be estimated using the Einstein relation and the initial rate of dimer formation is

$$R_f = 4kTC_1^2/3\eta \tag{2.9}$$

where η is the solvent viscosity. The time required for half of all particles to couple is:

$$t_{1/2} = 3\eta/4kTC_1 \tag{2.10}$$

At room temperature, aqueous sols at typical industrial concentrations have $t_{1/2}$ in the range of milliseconds or less; consequently, stabilizing particles by modifying their pair interaction potential is often of great practical importance.

2.2.3.2 Slow Aggregation

Although it is useful as a starting point for understanding the kinetics of aggregation, the case of rapid aggregation is a rare scenario. More commonly, particles repel each other at some distance, such as if the particles all carry a net charge or are decorated with polymer. Then not all collisions will result in particle binding, but only those for which there is sufficient kinetic energy to overcome the energy barrier. If the adhesive minimum is sufficiently weak compared to kT, there is also the possibility that bound particles will dissociate. For slow aggregation the rate constants in K_{lm} must be determined experimentally. However, it is useful to consider the stability ratio:

$$W = \frac{\text{Frequency of collisions between particles}}{\text{Frequency of collisions that result in coagulation}} \tag{2.11}$$

The stability ratio characterizes the propensity of particles to self-assemble (coagulate). It can be calculated numerically as:

$$W = 2R \int_{2R}^{\infty} \exp(E(D)/kT) \frac{\mathrm{d}D}{D^2} \tag{2.12}$$

A value of $W > 10^5$ can be achieved with an energy barrier of even 15 kT, and $W > 10^{10}$ (very stable) is not entirely impossible. Although it is difficult to correlate the reaction rate to the energy barrier numerically, this feature of the interaction potential has been found to be most influential in determining the rate of coupling. The rate of slow coagulation can be estimated using Equation 2.9 and Equation 2.12 as:

$$R_s = R_f / W \tag{2.13}$$

which accounts for the limitations of both diffusion and the presence of any repulsive potential. For particles whose surfaces interact principally by DLVO forces — that is, via repulsive electrostatic and attractive van der Waals forces — the stability ratio for many systems has been correlated to:

$$\log_{10} W = -k_1 \log_{10} C - k_2 \tag{2.14}$$

where k_1 and k_2 are constants, $k_1 = 2.1 \times 10^7 RZ^2/y_d^2$, the dimensionless surface potential is $y_d = ze\varphi_0/kT$, R is the particle radius as before, e is the fundamental electron charge, z is the charge valency, φ_0 is the surface potential, and $Z = \tanh(ze\varphi_0/4kT)$.

If a sufficient amount of salt is added to a suspension, electrostatic interactions can be screened to nullify any long-range repulsion and yield rapid aggregation. The concentration of salt at which rapid aggregation becomes kinetically significant is termed the critical coagulation concentration (c.c.c.), which can be roughly estimated for aqueous dispersions of particles with high surface potentials at 25°C as:

$$\text{c.c.c.}(\text{mol/L}) \cong 87 \times 10^{-40}/z^6 A^2 \tag{2.15}$$

where z is the valency of the electrolyte and A is the Hamaker constant for the particle surface material in water. Experimentally one can estimate the propensity for aggregation by measuring a particle suspension's zeta potential, which is typically ± 25 to 50 mV or smaller in magnitude for rapid aggregation. Because Equation 2.15 predicts approximate values only, if one can compare to known data it is better to estimate the critical coagulation concentration from:

$$\text{c.c.c.} \propto \varphi_0^4/z^2 \tag{2.16}$$

In contrast, a suspension of particles or aggregates can be stabilized (protected from aggregation) by keeping the salt concentration less than the critical coagulation concentration. Alternately, for ionizable particles in aqueous suspensions, the critical coagulation concentration can be modified by shifting the pH farther away from the material's isoelectric point, thereby increasing the magnitude of φ_0.

Another way of stabilizing particles or aggregates is to introduce an adsorbent molecule that coats the suspended material. The adsorbent molecules should be sufficiently large as to create a steric repulsion that extends beyond the range of the attractive van der Waals force (typically > 5 nm), cover a large fraction of the particle surface, and be in sufficient concentration in the bulk solvent as to disfavor desorption. Common molecules used for particle stabilization include proteins, polyelectrolytes, diblock copolymers, and amphiphilic surfactants (see also polymer forces in Table 2.1).

One other barrier to aggregation is the existence of a repulsive force between particles that originates from having to squeeze out a thin film of solvent from between approaching particles. This hydrodynamic force has been well studied both theoretically and experimentally.[21,22] The simplest approach is to cast this effect as a change in the diffusivity of particles undergoing Brownian motion, and then to calculate W numerically, for which it is found that the rate of assembly is roughly half of what is predicted by Equation 2.8.

Increasing the temperature of a self-assembly process generally increases the rate of assembly due to increased Brownian motion. Notable exceptions include cross-linking via coupling molecules that lose adhesive strength with increased temperature, such as with DNA base pairing or other ligands that bind to complimentary molecules via hydrogen bonds.

In summary, the kinetics of aggregation depends intimately on the interaction profile between particles, which is determined by the chemical composition of the particle surface, the electrostatic environment, and solvent properties. The following section will explore ways to tune these interactions to favor assembly.

2.3 Examples of Self-Assembly

In this section, we explore both traditional and nontraditional approaches to self-assembly. The traditional approach, often termed, "controlled self-assembly" is to modulate the DLVO, polymeric, and depletion forces listed in Table 2.1. However, one often desires to aggregate materials whose interfacial properties are not easily modified. Thus, we also survey recent works that demonstrate "programmed self-assembly," which is the art of tuning individual particle interactions by functionalizing particle surfaces with molecules that are encoded to interact specifically with complimentary molecules. With programmed self-assembly, a greater variety of aggregate structures can be formed in much the same way that atoms with specific electron orbitals interact uniquely with complimentary atoms to form complex atomic clusters or crystal structures.

2.3.1 Controlled Self-Assembly

In this section, we explore three approaches to *controlled* self-assembly: colloidal crystals made via strong repulsion between particles, fractal aggregates made from strongly attracting particles, and assembly aided by fluid flows.

2.3.1.1 Strong Repulsion: Colloidal Crystals

Many solid materials are crystals with atoms ordered on the angstrom length scale. Colloidal crystals, which have subunits sized between nanometers and microns, can exhibit order on a micron to millimeter length scale (Figure 2.2). The best-known example of a colloidal crystal is opal. Opals are produced in nature by the very slow precipitation and alignment of small silica spheres over geologic time scales. Their beautiful color patterns are due to interference and diffraction of light passing through colloidal silica spheres in the microstructure — hydrated silica, $SiO_2 \cdot n(H_2O)$. Regions that contain spheres of approximately equal size and have a regular structure diffract light at various wavelengths, creating color patterns. Similarly, monodisperse colloids can be used to rapidly fabricate periodic lattices in solution

FIGURE 2.2 Cross-sectional electron micrograph of a planar opal made from 1-μm SiO_2 spheres using an evaporation method. (From Norris, D.J. et al., *Adv. Mater.*, 16, 1393, 2004. With permission.)

for optical devices. For example, such particles and media with large dielectric contrasts and particular crystal symmetries can be used to create photonic band gaps (frequency ranges that will not propagate light because of multiple Bragg reflections).[23] A significant hurdle remaining is the creation of macro-structures with sufficient order to possess a full photonic band gap.

Pusey and van Megen were the first to investigate the phase behavior and crystallization of monodisperse, hard sphere colloidal suspensions.[3] At about 50% volume fraction, hard spheres with strong, short-range repulsion order and arrange themselves into a periodic lattice to increase their entropy. Whereas one usually associates an increase in entropy with disorder, at high concentrations particles gain entropy by arranging themselves into a lattice, thereby increasing their available space. This gain more than offsets the loss in translational free energy.[24] Theory predicts that for a colloidal crystal of hard spheres the fcc structure is the most energetically stable. However, the energy difference between the fcc and hcp is extremely small.[25] As a result, hcp structures are typically observed and the crystals are prone to defects such as vacancies, grain boundaries, and staking faults.[26] Near 60% volume fraction there is a glass transition, where crystallization is kinetically suppressed.[3]

Although one might anticipate that weak attraction between colloids would also lead to phase transitions and crystallization as with small molecules, this is not the case. The condition of equilibrium is fairly stringent in that particles must be allowed to adhere and de-adhere to obtain the minimum energy configuration. If mutual attraction is strong enough to bring two particles together (slightly greater than kT), it is ordinarily more than strong enough to bring pairs or larger groups together.[27] As the attractive energy between particles scales with their radii (Table 2.1), irreversible attraction occurs as the colloids assemble in the dense phase and fractal growth occurs.

2.3.1.2 Strong Attraction: Fractals

As described earlier, particles exhibiting strong self-attraction relative to the available thermal energy (kT) exhibit fractal-like growth. Materials in nature that exhibit fractal properties include bacterial colonies, tumor growths, condensing particulates in smoke and dust, and large composite snowflakes.[27,28] Fractal aggregates can also be produced artificially out of a variety of materials. They are used industrially to thicken liquids, increase fluid viscosity, and to form gel networks through solidification or cross-linking. An example of an artificially produced fractal aggregate of 30-nm CdS@SiO_2 particles is shown in Figure 2.3.[29]

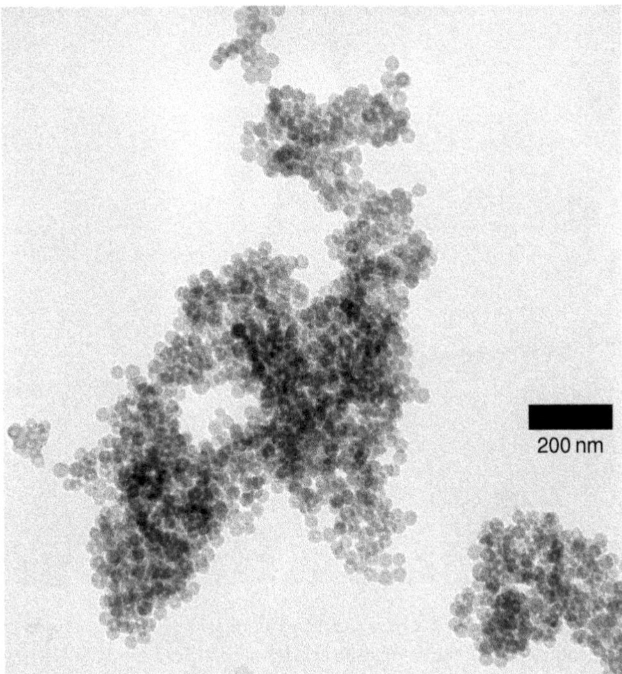

FIGURE 2.3 TEM of fractal structure comprised of 30-nm $CdS@SiO_2$ particles made strongly attractive by functionalizing their exteriors with biotin on short spacers of poly(ethylene glycol) and introducing the protein avidin to cross-link the particles. (From Costanzo, P.J., Patten, T.E., and Seery, T.A.P. et al., *Chem. Mater.*, 16, 1775, 2004. With permission.)

Fractal growth has been well studied both theoretically and experimentally.[28,30–31] A useful review of characterizing fractals made from polydisperse components is provided by Witten and Pincus.[27] For fractals made of monodisperse components, the compactness of the aggregate is characterized by its fractal dimension, D_f, a scaling exponent that describes how its mass is radially distributed, that is, $M \propto r^{D_f}$, where r is the magnitude of the radial vector circumscribing the fractal. The fractal dimension can range from three for completely dense materials (i.e., crystals) to one for completely linear aggregates. Most self-assembled aggregates have a fractal dimension between 1.7 and 2.1, the latter being more common for slow homoaggregation processes where the energy barrier is large compared to kT.[32]

Materials that prefer to aggregate into crystalline structures can be imbued with fractal tendencies by modifying their surfaces to produce stronger interparticle attractions. Particle surfaces can be made more attractive by changing the particle environment (e.g., pH, ionic concentration, or temperature as described previously), adsorbing macromolecules at low concentrations, through direct chemical modification (e.g., covalent coupling), or by grafting a variety of cross-linking agents (e.g., DNA, fibronectin, actin filaments, or synthetic polymers).

While we have talked about aggregation that is diffusion-limited (forming crystals) and reaction-limited (forming fractals), it should be noted that there is a third category of assembly termed "ballistic aggregation" wherein particles collide along straight-line paths. Ballistic aggregation can produce materials with highly fractal or porous microstructures, such as from particles settling in gravitational or electromagnetic fields, and from deposition processes such as thermal/electrochemical vapor deposition, ion sputtering (covered in Chapter 8), the drying of particle suspensions over a substrate, and aerosol impaction.

2.3.1.3 Assembly Driven by Fluid Flows

Hydrodynamic forces external to particles can be used to rapidly assemble aggregates. Controlled flow fields can maneuver particles toward impaction zones or areas of fluid stagnation where particles collide or

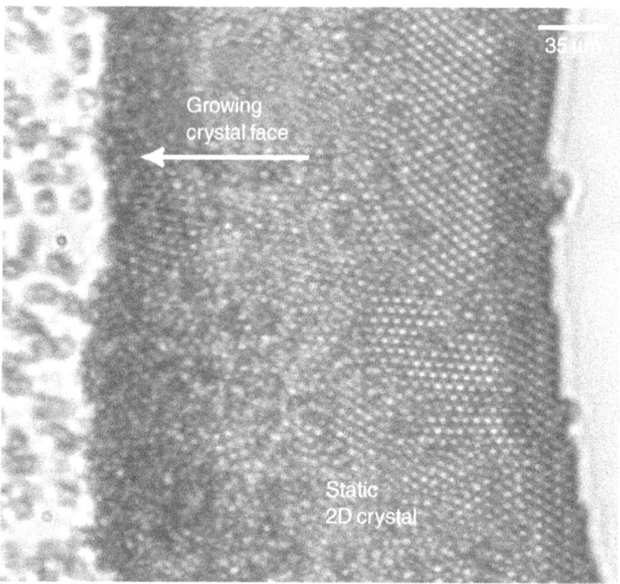

FIGURE 2.4 6-μm SiO_2 particles forming a hexagonal two-dimensional crystal atop an atomically smooth mica substrate. Laplace pressure induced from the drying solvent draws particles from far away toward the growing crystal face. Optical micrograph. (From Moore, N.W. and Kuhl, T.L., unpublished data, 2003.)

settle, respectively. By effectively increasing the frequency of particle collision, the assembly can be made less diffusion-limited and more reaction-limited. Whether crystals or fractals are formed still depends on the features of the energy interaction profile as described previously.

When the fluid motion is induced by drying, the bulk of the liquid flow occurs near the liquid–air interface. When enough liquid has evaporated that the tops of particles become exposed to air, the surface tension of the fluid–air interface induces exceptionally strong lateral attraction between particles that can range hundreds of particle diameters away. Such flows are termed "Laplace-driven flows" due to the decreased pressure between particles relative to the pressure in the surrounding fluid. This pressure differential is minimized through particle–particle contact. The result is rapid coagulation of the particles into a two-dimensional fractal or crystal, as shown in Figure 2.4.[33] In this example, the particles initially travel at speeds up to 35 μm/s but slow down near the growing crystal face due to the repulsive pressure induced by squeezing out fluid from between the particle and the preformed crystal, an effect that diminishes linearly with decreased speed.[34] This fundamental technique has been modified in a variety of ways to self-assemble three-dimensional crystals as well, for example, Reference 35 and Reference 36.

Laplace-driven assembly is not limited to batch processes. An example includes the assembly of nanoparticles via aerosol drying. In the work by Borra et al.,[37] aerosols of oppositely charged nanoparticles are formed by electrospraying and then mixed. Their drying results in rapid aggregation in flight. *In situ* spectral monitoring can permit tight control over the resultant cluster size by changing the flow characteristics of the aerosol feed streams.

2.3.2 Programmed Self-Assembly

It is often desired to assemble materials that may not necessarily possess surface properties that are conducive toward self-assembly by DLVO forces alone. A broad range of assembly phenomena can be adequately described by DLVO theory, which identifies two forces that dominate the interactions between colloidal particles: a van der Waals force and an electrostatic force, which between like media are attractive and repulsive, respectively. These forces are typically long range compared to steric, structural, fluctuation, and other forces that decay at a rate comparable to the sizes of molecules. Consequently, these latter forces are often considered unimportant in understanding colloidal self-assembly, a science whose very objects of study occupy lengths of tens of nanometers and larger. However, one can magnify the effects of these lesser forces by introducing macromolecules that extend between particles, thereby increasing their effective range.

Of the three short-ranged forces above, the most easily modified is the steric force, which for macromolecules manifests a long-range repulsion between particles they decorate (Table 2.1).[16] Macromolecules can be attached to particle surfaces via physisorption or covalently grafted (chemisorption). In either case, if the macromolecule has a high affinity for the particle surface and has sufficient length, it can bridge two nearby particles, creating an attractive "bridging force." Bridging forces can operate at long range, up to almost the full extension of the tethering macromolecule.[17,18,38] Long adsorbed polymers may bridge particles multiple times along their lengths. If a tether is sufficiently long, its long-range attraction can overpower electrostatic repulsion and cause two inherently repulsive particles to unite.

To create a bridging force between particles, the polymer bridges need nothing more than an affinity to adsorb on an opposing particle. However, we will show that much more control over the aggregation can be attained if one covalently anchors polymers that would prefer to desorb from surfaces if they weren't covalently anchored. If the free end of such a polymer is functionalized so that it can mate with a receptor (i.e., protein conjugate) on an opposing particle, then while it is bridged the polymer will exert a spring-like tug between the two particles owing to its loss of entropy, an effect that can be measured as an attractive bridging force.[38] The magnitude of the bridging force depends on the effective spring constant of the tether and the strength of the ligand–receptor bond. The frequency with which these cross-bridges form depends on the dynamics of the polymer chain, but in most practical situations can be assumed to be much faster than the adhesion event, so that the ligand and receptor are often in chemical equilibrium.[39] With this assumption as a starting point, the bridging force between two particles can be shown to scale linearly with the interaction radius (R) along with the nonspecific forces.[40] Thus, particle size and shape significantly influence how particles self-assemble.

When a ligand and receptor bind with lock-and-key specificity, as occurs with many biological interactions, the bridging is called site-specific cross-linking. By making cross-linking site-specific, nonspecific interactions can be minimized, the aggregates made more stable, and specific/complimentary interactions can be favored. For example, particles can be made immunogenic by coating them with suitable polymers for safe use *in vivo* while retaining the ability to tune how they bind to target cells.[41] A wide range of tethers has been used for these applications, including the immunogenic poly(ethylene glycol) (PEG) and other synthetic polymers, and biologically harvested materials such as DNA strands, actin filaments, and fibronectin.[9,42–47] Likewise, wide ranges of chemical functionalities have been imparted to many of these tethers to achieve the desired specificity with receptor surfaces.

A careful choice of parameters can tune the interaction between tethered particles to meet the demands of a particular application. For example, the interaction can be made more attractive to favor aggregating fractal structures or weaker to favor crystallization. The height of the energy barrier can be raised to slow the kinetics of assembly, that is, to stabilize the suspension, or lessened to promote aggregation. We now review two specific examples of site-specific cross-linking to demonstrate how tuning the interaction can lead to the creation of novel materials.

2.3.2.1 Assembly with DNA Cross-Linking

Mirkin et al.[12,48,49] have demonstrated a way to make aggregation reversible by using DNA as a cross-linker (Figure 2.5). DNA oligomers grafted to particle surfaces by chemisorption serve both to sterically stabilize a suspension of gold particles and provide an avenue for site-specific cross-linking. However, to control the reaction the DNA strands that are installed are chosen to *not* be complimentary. Consequently, the particles stay suspended and aggregation does not proceed until a third DNA oligomer is added to the solvent. Each end of this third portion binds specifically to one kind of DNA-functionalized particle in the solution. With its addition those two particle types aggregate and precipitate, while other particle types remain suspended. Raising the temperature diminishes the strength of the hydrogen bonding that holds the complementary DNA base pairs together, and above 42°C the particles separate without

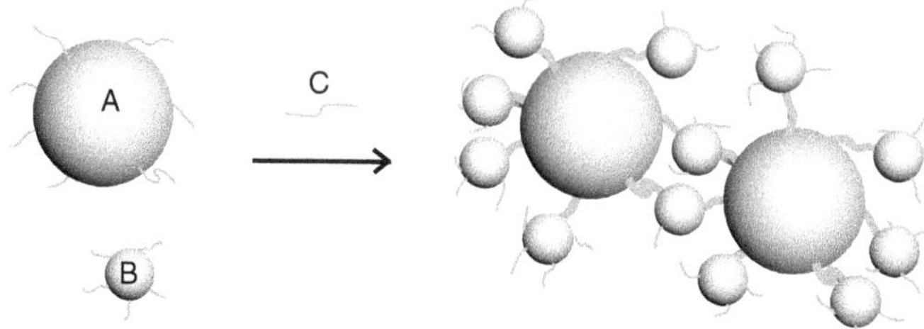

FIGURE 2.5 Self-assembly of colloidal gold particles by specific cross-linking with DNA. (a) 31-nm Au particle functionalized by adsorption with 3′ $HS(CH_2)_3O(O')OPO$-ATG-CTC-AAC-TCT; (b) 8-nm Au with 3′ TAG-GAC-TTA-CGC-$OP(O)(O')O(CH_2)_6SH$; (c) Nucleotide fragment with left and right ends complimentary to A and B, respectively, that completes the cross-linking: 5′ TAC-GAG-TTG-AGA-ATC-CTG-AAT-GCG. (From Mucic, R.C. et al., *J. Am. Chem. Soc.*, 120, 12674, 1998. With permission.)

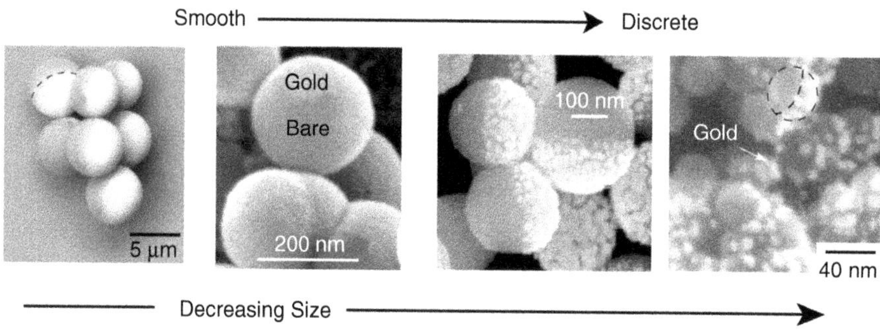

FIGURE 2.6 Spheres made asymmetric by evaporation of Ag (bright hemispheres). Spheres are polystyrene (left image only) and SiO_2. Aggregation occurred during drying in preparation for this SEM imaging. (From Barber, S.M. et al., *J. Phys. Chem. A.*, 110, 4538, 2006. With permission.)

incurring damage. Thus, the aggregation is reversible and repeatable, though it should be noted that resuspending precipitated particles generally requires agitation.

2.3.2.2 Assembly of Bilateral, Difunctional Aggregates

The work in Kuhl's Laboratory[50] expands on the above approach by using asymmetric particles as building blocks. Using asymmetric particles permits the self-assembly of asymmetric aggregates. Such asymmetry allows the development of multifunctional particles for use as biological triggers, biosensors, and separation media, as well as the bottom-up self-assembly of novel structures with unique mechanical properties.

To produce asymmetric particles, a suspension of uniform nanospheres are deposited onto a flat substrate and dried. A combination of DLVO and Laplace-driven fluid flows self-assemble the spheres into a two-dimensional crystal. After drying, the spheres are coated with a noble metal via thermal vaporization.[51] As the spheres are packed into a two-dimensional crystal, the backsides of the spheres are shadowed from the metal vapors, and upon resuspension the spheres have a metal coating on only one side (Figure 2.6).

A variety of subsequent treatments are employed to independently modify the chemical reactivity of the metal and the exposed particle surface on each hemisphere. In the simplest case, SH-PEG-NH_2 conjugates are adsorbed to the metal surface via thiol anchors. The PEG imparts mobility to the amine ligand, which can be coupled to other particles exhibiting carboxylic groups. A variety of structures can be made using this technique, including "satellite" structures (Figure 2.7a) and "dumbbells" (Figure 2.7b).

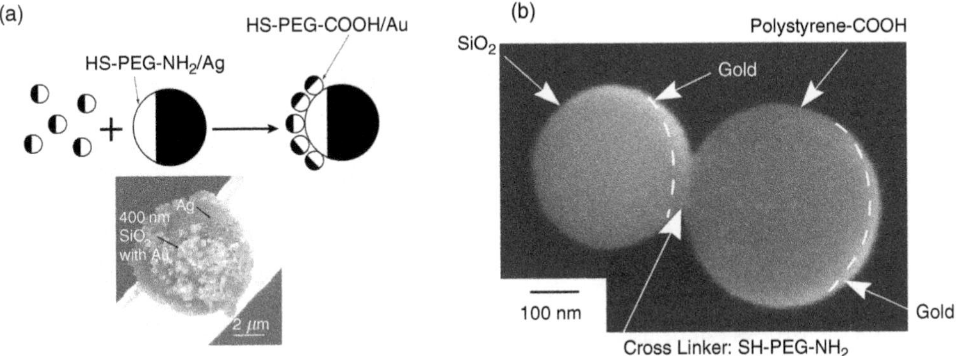

FIGURE 2.7 SEM images of "half-raspberries" and "dumbbells." (a) The large particle is 5-μm polystyrene plated with Ag and functionalized with HS-PEG-NH$_2$; small particles are 400-nm SiO_2 plated with Au that have been functionalized with HS-PEG-COOH. (b) Left particle is 400-nm SiO_2 plated with Au and functionalized with HS-PEG-NH$_2$; right is 500-nm COOH-functionalized polystyrene plated with Au. The reaction between the PEG-NH$_2$ and COOH-functionalized surfaces covalently cross-links the particles. (From Barber, S.M. et al., *J. Phys. Chem. A.*, 110, 4538, 2006. With permission.)

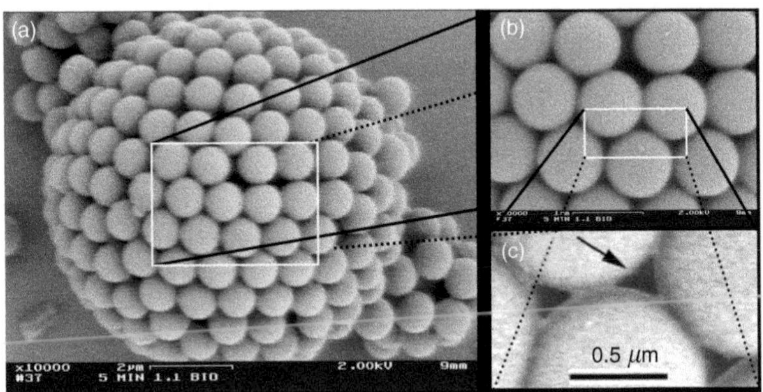

FIGURE 2.8 0.9 μm polystyrene spheres assembled into a 10-μm "colloidosome" via templating at a water–oil interface. A–C are successive magnifications of the SEM images of the dried samples. Arrows point to pores whose size depends on the size of the component spheres and the amount of sintering (5 min at 105°C in this case). (From Dinsmore et al., *Science*, 298, 1006, 2002. With permission.)

2.3.2.3 Templating

As particle interactions are becoming increasingly understood, controlled self-assembly is becoming increasingly sophisticated. One developing technique is the use of interfacial templates to guide self-assembly. Templates are always an interface between phases that preferentially house one or more kinds of particles due to a minimization of the Gibbs free energy upon the particle(s) settling at the interface. Examples of templates include patterned surfaces immersed in a medium, the boundaries between two immiscible liquid phases, and particles themselves. The following are a few recent examples of the use of templating in controlled self-assembly.

Dinsmore et al.[52] have used water in oil immersions to form hollow microparticles they term "colloidosomes" (Figure 2.8). The suspension of a dispersed water phase in oil creates a vast amount of interfacial area between oil and water phases that are each spherical in shape. At this interface, 0.7 μm poly(methyl methacrylate) (PMMA) spheres spontaneously aggregate. The PMMA spheres are locked together either by light sintering or by the introduction of a cross-linking agent, such as the polyelectroylte, poly-L-lysine. Inorganic nanoparticles can also be organized using this method.[53–55]

Similarly, Caruso et al.[56] have used particles as templates upon which to aggregate other particles and ultimately to form hollow microspheres. In short, the linear cationic polymer poly(diallydimethylammonium chloride) is adsorbed to a suspension of 640-nm poly(styrene) (PS) latex particles. SiO_2 particles of 25-nm are then introduced and spontaneously assemble around the coated PS particle surfaces. Exposure to a suitable solvent erodes the PS core, leaving a hollow raspberry-like assembly of the smaller SiO_2 particles. When the templating particle is the same size as particles aggregating at its surface, a variety of unique crystal structures can be grown provided the aggregate number is kept under 20.[55]

Monodisperse particles can be used as masks to template surfaces in so-called "particle lithography." Microspheres can be arranged into two-dimensional lattices on crystalline substrates, for example, through drying of a concentrated suspension.[51] The tightly packed spheres will shield the substrate from line-of-site physisorption, such as from vapor deposition or sputtering.[57] Subsequent removal of the microsphere templates reveals regularly spaced triangular islands of modified substrate. If two monolayers of microspheres are used to mask a substrate, the patches of substrate modified will appear circular and less frequently spaced. Three layers of microspheres mask the substrate completely.[58] Polymer particles patterned onto surfaces can also be melted to form hexagonal lattices[59] or used as optical lenses in projection lithography.[60] The transpose perspective of this technique can be viewed as a method to pattern the microspheres themselves. That is, surfaces can be used to pattern particles by controlling the orientation and position of particles subjected to various treatments.[57,58]

2.4 Challenges for Self-Assembled Materials

Incorporating self-assembled materials into industrial processes requires not only a working knowledge of the surface interactions that govern their assembly but also of the ways they interact with their environment. Some of the challenges in working with nanoscale materials include scale-up, characterization, separation, and asymmetric assembly.

2.4.1 Scale-Up

Producing large quantities of aggregates with engineered properties requires tight control of process variables and consistency in the properties of the starting materials. Although so far we have only discussed particles of uniform size, obtaining such uniformity adds considerable cost to their purchase and it may be preferred to assemble cheaper, more polydisperse starting materials. Consequently, the complexities of modeling the kinetics and energetics of these nanoscale processes typically mandates that their optimization be performed through trial and error. Due to their high surface activity, self-assembled aggregates are more likely to accumulate in or foul process equipment, or themselves be modified by the vessels or equipment they travel through. To retain the benefits of small-scale handling in large-scale production, scaled-up processes may consist of many parallel, small-scale processes.

2.4.2 Characterization

It is true both for marketability and practicability that one should aim to sell only materials that one can characterize. However, establishing an effective protocol for evaluating the properties of self-assembled materials can be a significant challenge. Problematically, aggregates may appear indistinguishable from their component building blocks if imaged with insufficient resolution. In contrast to bulk materials, self-assembled materials are relatively fragile, and each handling step may change their shape or properties. For example, if aggregates are dried onto a substrate in preparation for imaging in an electron microscope, aggregates and individual particles will tend to cluster due to capillary forces as solvent evaporates, and it may be impossible to tell which aggregates had self-assembled prior to drying. Thus, the best assessments may be *in situ* monitoring during the final stage of processing (such as just prior to incorporation into another material). Owing to their large surface area to volume ratio, many of the spectral techniques used for probing molecular structure can also be used to characterize the composition of self-assembled materials. Also useful are techniques that measure aggregate size (dynamic light scattering, acoustosizing), surface area or porosity (gas adsorption, pycnometer), and surface charge (zeta potential). In all cases, using multiple techniques to evaluate the properties of self-assembled materials can be extremely helpful.

2.4.3 Separation

It may be desired to separate self-assembled aggregates to purify them as a product or intermediate or to recycle unaggregated components. Traditional separation techniques that operate by establishing a gradient of fluid pressure or electromagnetic field may become unsuitable if the aggregates are sufficiently small. For example, filters with nanoscale pores have enormous operating costs for the quantity of material they are able to separate due to the high pressures required to squeeze fluid through the pores. Batch separations based on density gradients can be cheap, low energy alternatives. Examples include centrifugation, fractionation columns, and on a larger scale, pond sedimentation. In all cases, fractionation of different sizes can be achieved by sampling the suspension at multiple depths. However, because the settling time is proportional to $1/R^2$, these techniques are often impractical on the nanoscale. A variety of other techniques have been employed to separate self-assembled aggregates, including electroosmosis and magneto-, and hydrodynamic processes. Perhaps the most promising of these are shear-driven flows that separate aggregates in microfluidic channels as narrow as just a few particle diameters.[61]

2.4.4 Asymmetric Assembly

As we increase our understanding of the forces that govern self-assembly, it has been an expanding area of research to build nanoscale materials with programmed functionalities, from iconic "micro machines" to drug-delivery devices, miniaturized sensors, and electrophotonic devices.[62] These and most other useful tools are like a hammer — asymmetric — and are assembled from asymmetric parts. A variety of recent materials research has aimed toward developing methods for producing asymmetric particles that can self-assemble into novel materials from the bottom up,[63,64] most of which employ a substrate to orient the building blocks. Examples of these techniques include biphasic adsorption,[65] layer-by-layer deposition,[66,67] photolithography,[68] microcontact printing,[69] chemical surface modification,[70] chemical evaporation deposition,[71] laser photochemical deposition,[72] metallic dewetting,[73] thermal evaporation[51] and combinations of the above techniques, for example, Reference 50.

Acknowledgments

We gratefully thank everyone who contributed figures and images for this chapter, and Pieter Stroeve for his review. This work was supported by NSF NER DMII-0404457.

References

[1] Carter, G.F. and Paul, D.E., *Materials Science and Engineering*, ASM International, USA, 1991, 351.

[2] Yoshimura, S. and Hachisu, S., Order formation in binary-mixtures of monodisperse lattices 1. observation of ordered structures, *Prog. Colloid Polym. Sci.*, 68, 59, 1983.

[3] Pusey, P.N. and van Megen, W., Phase-behavior of concentrated suspensions of nearly hard colloidal spheres, *Nature*, 320, 340, 1986.

[4] Yoshida, H. et al., Transitions between ordered and disordered phases and their coexistence in dilute ionic colloidal dispersions, *Langmuir*, 15, 2684, 1999.

[5] Bruchez, M. et al., Semiconductor nanocrystals as fluorescent biological labels, *Science*, 281, 2013, 1998.

[6] Chan, W.C.W. and Nie, S.M., Quantum dot bioconjugates for ultrasensitive nonisotopic detection, *Science*, 281, 2016, 1998.

[7] Parak, W.J. et al., Conjugation of DNA to silanized colloidal semiconductor nanocrystalline quantum dots, *Chem. Mater.*, 14, 2113, 2002.

[8] Connolly, S. and Fitzmaurice, D., Programmed assembly of gold nanocrystals in aqueous solution, *Adv. Mater.*, 11, 1202, 1999.

[9] Mann, S. et al., Biologically programmed nanoparticle assembly, *Adv. Mater.*, 12, 147, 2002.

[10] Novak, J.P. and Feldheim, D.L., Assembly of phenylacetylene-bridged silver and gold nanoparticle arrays, *J. Am. Chem. Soc.*, 122, 3979, 2000.

[11] Coffer, J.L. et al., Dictation of the shape of mesoscale semiconductor nanoparticle assemblies by plasmid DNA, *Appl. Phys. Lett.*, 69, 3851, 1996.

[12] Mirkin, C.A. et al., A DNA-based method for rationally assembling nanoparticles into macroscopic materials, *Nature*, 382, 607, 1996.

[13] Alivisatos, A.P. et al., Organization of "nanocrystal molecules" using DNA, *Nature*, 382, 609, 1996.

[14] Boal, A.K. et al., Bricks and mortar self-assembly of nanoparticles, *Tetrahedron*, 58, 765, 2002.

[15] Shenhar, R. and Rotello, V.M., Nanoparticles: Scaffolds and building blocks, *Acc. Chem. Res.*, 36, 549, 2003.

[16] Israelachvili, J., *Intermolecular and Surface Forces*, 2nd ed., Academic Press, San Diego, 1992.

[17] Ji, H. et al., Polymer bridging between two parallel plates, *Macromolecules*, 23, 698, 1990.

[18] Marla, K.T. and Meredith, J.C., Simulation of interaction forces between nanoparticles in the presence of Lennard-Jones polymers: Freely adsorbing homopolymer modifiers, *Langmuir*, 21, 487, 2005.

[19] Li, B.K. and Logan, B.E., Bacterial adhesion to glass and metal-oxide surfaces, *Colloids Surf., B-Biointerfaces*, 36, 81, 2004.

[20] Hunter, R.J., *Foundations of Colloid Science*, Vol. 1, Oxford University Press, Oxford, 1987.

[21] Honig, E.P., Roebersen, G.J., and Wiersema, P.H., Efffect of hydrodynamics interaction on coagulation rate of hydrophobic colloids, *J. Colloid Interface Sci.*, 36, 97, 1971.

[22] Lichtenbelt, J.W. Th., Pathmamanoharan, C., and Wiersema, P.H., Rapid coagulation of polystyrene latex in a stopped-flow spectrophotometer, *J. Colloid Interface Sci.*, 49, 281, 1974.

[23] Colvin, V.L., From opals to optics: Colloidal photonic crystals, *MRS Bull.*, 637, August 2001.

[24] Gast, A.P. and Russel, W.B., Simple ordering in complex fluids, *Phys. Today*, 51, 24, 1998.

[25] Norris, D.J. et al., Opaline photonic crystals: How does self-assembly work?, *Adv. Mater.*, 16, 1393, 2004.

[26] Glotzer, S.C., Solomon, M.J., and Kotov, N.A., Self-assembly: From nanoscale to microscale colloids, *AIChE J.*, 50, 2978, 2004.

[27] Witten, T.A. and Pincus, P.A., *Structured Fluids: Polymers, Colloids, Surfactants*, Oxford University Press, Oxford, 2004.

[28] Sander, L. M. Diffusion-limited aggregation: A kinetic critical phenomenon? *Contemp. Phys.*, 41, 203, 2000.

[29] Costanzo, P.J., Patten, T.E., and Seery, T.A.P., Protein-ligand mediated aggregation of nanoparticles: A study of synthesis and assembly mechanism, *Chem. Mater.*, 16, 1775, 2004.

[30] Lin., M.Y. et al., Universality in colloid aggregation, *Nature*, 339, 360, 1989.

[31] Kolb, M., Botet, R., and Jullien, R., Scaling of kinetically growing clusters, *Phys. Rev. Lett.*, 51, 1123, 1983.

[32] Kim, A. Y. et al., Linear chains and chain-like fractals from electrostatic heteroaggregation, *J. Colloid Interface Sci.*, 260, 149, 2003.

[33] Moore, N.W. and Kuhl, T.L., unpublished data, 2003.

[34] Chan, D.Y.C. and Horn, R.G., The drainage of thin liquid films between solid surfaces, *J. Chem. Phys.*, 83, 5311, 1985.

[35] Goldenberg, L.M. et al., Ordered arrays of large latex particles organized by vertical deposition, *Mater. Sci. Eng. C*, 22, 405, 2002.
[36] Xia, Y. et al., Monodispersed colloidal spheres: Old materials with new applications, *Adv. Mater.*, 12, 693, 2000.
[37] Borra, J.-P. et al., A new production process of powders with defined properties by electrohydrodynamic atomization of liquids and post-production electrical mixing, *J. Electrostat.*, 40 & 41, 633, 1997.
[38] Jeppesen, C. et al., Impact of polymer tether length on multiple ligand-receptor bond formation, *Science*, 293, 465, 2001.
[39] Moreira, A.G. et al., Irreversible vs. reversible bridging: When is kinetics relevant for adhesion?, *Europhys. Lett.*, 62, 876, 2003.
[40] Moore, N.W. and Kuhl, T.L., The role of flexible tethers in multiple ligand-receptor bond formation between curved surfaces, *Biophys. J.*, 91, 1675, 2006.
[41] Harris, J.M. and Zalipsky, S., Eds., *Poly(ethylene glycol): Chemistry and Biological Applications*, American Chemical Society, Washington, D.C., 1997.
[42] Lasic, D. and Martin, F., Eds., *Stealth Liposomes*, CRC Press, Boca Raton, FL, 1995.
[43] Photos, P.J. et al., Polymer vesicles in vivo: Correlations with PEG molecular weight, *J. Controlled Release*, 90, 323, 2003.
[44] Goubault, C. et al., Self-assembled magnetic nanowires made irreversible by polymer bridging, *Langmuir*, 21, 3726, 2005.
[45] Liu, A.P. and Fletcher, D.A., 2005, Photopatterning of actin filament structures, *Nano Lett.*, 5, 625, 2005.
[46] Li, Y. et al., Controlled assembly of dendrimer-like DNA, *Nat. Mater.*, 3, 38, 2004.
[47] Meadows, P.Y. and Walker, G.C., Force microscopy studies of fibronectin adsorption and subsequent cellular adhesion to substrates with well-defined surface chemistries, *Langmuir*, 21, 4096, 2005.
[48] Storhoff, J.J. and Mirkin, C.A., Programmed materials synthesis with DNA, *Chem. Rev.*, 99, 1849, 1999.
[49] Mucic, R.C. et al., DNA-directed synthesis of binary nanoparticle network materials. *J. Am. Chem. Soc.*, 120, 12674, 1998.
[50] Barber, S.M. et al., Bilateral, difunctional nanosphere aggregates and their assembly mediated by polymer chains, *J. Phys. Chem. A.*, 110, 4538, 2006.
[51] Takei, H. and Shimizu, N., Gradient sensitive microscopic probes prepared by gold evaporation and chemisorption on latex spheres, *Langmuir*, 13, 1865, 1997.
[52] Dinsmore, A.D. et al., Colloidosomes: Selectively permeable capsules composed of colloidal particles, *Science*, 298, 1006, 2002.
[53] Lin., Y. et al., Nanoparticle assembly and transport at liquid-liquid interfaces, *Science*, 299, 226, 2002.
[54] Kwon Y.K., Kang, J.P., and Kim, T.H., Synthesis of hollow TiO2 nanoparticles prepared by colloidal templating, *Abstracts of Papers of The American Chemical Society*, 228, U347, 149, 2004.
[55] Manoharan, V.N. and Pine, D.J., Dense packing and symmetry in small clusters of microspheres, *Science*, 301, 483, 2003.
[56] Caruso, F. et al., Nanoengineering of inorganic and hybrid hollow spheres by colloidal templating, *Science*, 282, 1111, 1998.
[57] Haynes, C. L. and Van Duyne, R.P., Nanosphere lithography: A versatile nanofabrication tool for studies of size-dependent nanoparticle optics, *J. Phys. Chem. B*, 105, 5599, 2001.
[58] Zhang, G., Wang, D., and Möhwald, H., Patterning microsphere surfaces by templating colloidal crystals, *Nano Lett.*, 5, 143, 2005.
[59] Mezzenga, R. et al., Templating organic semi-conductors via self-assembly of polymer colloids, *Science*, 299, 1872, 2003.

[60] Wu, M.H. and Whitesides, G.M, Fabrication of arrays of two-dimensional micropatterns using microspheres as lenses for projection photolithography, *Appl. Phys. Lett.*, 78, 2273, 2001.
[61] Clicq, D. et al. Shear-driven flow approaches to LC and macromolecular separations, *Anal. Chem.*, 76, 430A, 2004.
[62] Fudouzi, H. and Xia, Y. Colloidal crystals with tunable colors and their use as photonic papers, *Langmuir*, 19, 9653, 2003.
[63] Zhang, Z. and Glotzer, S.C., Self-assembly of patchy particles, *Nano Lett.*, 4, 1407, 2004.
[64] Zhang, Z. et al., Tethered nano building blocks: toward a conceptual framework for nanoparticle self-assembly, *Nano Lett.*, 3, 1341, 2003.
[65] Crisp, T.M. and Kotov, M.A., Preparation of nanoparticle coatings on surfaces of complex geometry, *Nano Lett.*, 3, 173, 2003.
[66] Petit, P. et al., Dissymmetric silica nanospheres: a first step to difunctionalized nanomaterials, *J. Mater. Chem.*, 10, 253, 2000.
[67] Nakahama, K., Kawaguchi, H., and Fujimoto, K., A novel preparation of nonsymmetricaly microspheres using the Langmuir-Blodgett technique, *Langmuir*, 16, 7882, 2000.
[68] Bao, Z. et al., Toward controllable self-assembly of microstructures: selective functionalization and fabrication of patterned spheres, *Chem. Mater.*, 14, 24, 2006.
[69] Cayre, O., Paumov, V.N., and Velev, O.D., Fabrication of asymmetrically coated colloid particles by microcontact printing techniques, *J. Mater. Chem.*, 13, 2445, 2003.
[70] Bauer, L.A., Reich, D.H., and Meyer, G.J., Selective functionalization of two-component magnetic nanowires, *Langmuir*, 19, 7043, 2003.
[71] Choi, J.Y. et al., Patterned fluorescent particles as nanoprobes for the investigation of molecular interactions, *Nano Lett.*, 3, 995, 2003.
[72] Hugonnot, E.A. et al., "Smart" surface dissymmetrization of microparticles driven by laser photochemical deposition, *Langmuir*, 19, 226, 2003.
[73] Lu, Y. et al., Asymmetric dimers can be formed by dewetting half-shells of gold deposited on the surfaces of spherical oxide colloids, *J. Am. Chem. Soc.*, 125, 12724, 2003.

3

Ion-Beam Processing

Sergei O. Kucheyev
Lawrence Livermore National Laboratory

Abstract

Ion beams are widely used for material modification. Here, we briefly review the fundamental physical and chemical processes occurring in solids exposed to ion bombardment. We also discuss the major current applications of ion beams for materials processing and some recent trends in the research in this field. Emphasis is given to the current understanding of the formation and evolution of radiation damage due to the profound effects of ion-beam-produced defects on material properties and their role in materials processing.

3.1 Introduction

Bombardment with energetic ions has been widely used for material processing since the middle of the twentieth century. Prior to about 1960, most of the research efforts in this field focused on understanding the effects of radiation damage produced by fission fragments in nuclear reactors. The rapid development of semiconductor device technology, where ion bombardment is a powerful processing tool, has been the major driving force of ion-beam research since the 1960s.

In this chapter, we give an overview of the fundamental physical and chemical processes that take place in various classes of solids under ion bombardment. In particular, we discuss the formation and evolution of radiation damage because ion-beam-produced defects can greatly affect material properties and often limit technological applications of ion bombardment. We also briefly discuss the major current applications of ion beams and the research areas that have received the most interest in recent years. Many excellent books and detailed reviews emphasizing different aspects of ion-beam processing of various materials have been written in the past four decades. Throughout this chapter, the reader will be referred to such specialized reviews and books for additional details.

3.2 Ballistic Processes

In this section, we briefly discuss the general phenomena related to ballistic interactions of energetic ions with solids. These ballistic phenomena include atomic collisions and displacements, electronic energy loss, surface sputtering, atomic mixing, spatial separation of vacancies and interstitials in collision cascades, and ion-beam-induced stoichiometric changes in compound materials (due to preferential surface sputtering and spatial separation of elements of different masses within collision cascades).

When an energetic ion impinges on the surface of a solid (target), it undergoes a succession of *ballistic collisions* with the stationary atoms of the target. Penetrating into the target, the ion gradually loses its energy in such collisions and comes to rest. These collisions can be conveniently classified as *nuclear* (or elastic) and *electronic* (or inelastic) energy loss processes. In nuclear energy loss processes, interaction occurs between (screened by electrons) the nuclei of the impinging ion and target atoms. In electronic collisions, the kinetic energy of the ion is transferred to the target electrons. Nuclear energy loss is the dominant mechanism for ions of keV energies typically used for ion implantation, while the contribution of electronic energy loss increases with increasing ion velocity and becomes dominant for energies above several MeV for almost all ion species-target combinations.

The target atoms displaced by the primary ion and receiving energy above the *threshold displacement energy* will create further displacements in the target. A *collision cascade* is thus formed, which consists of *Frenkel pairs*: vacancies and interstitials. The threshold displacement energy is typically ~20 eV. Its value depends on the material and on the direction in the lattice for crystalline solids. Figure 3.1 shows a schematic of a collision cascade and a typical depth profile of atomic displacements (averaged over many cascades). It is seen that ion implantation produces Gaussian-like (unimodal) depth profiles of atomic displacements. The shape of the depth profiles of implanted atoms (not shown in Figure 3.1) resembles that of the profiles of atomic displacements. The thickness of the layer modified by ion bombardment, the total number of atomic displacements generated, and the shape (i.e., effective width and skewness) of the depth profiles of atomic displacements and implanted atoms depend on ion mass and energy. This is schematically illustrated in Figure 3.1 for the two limiting cases of low-energy (keV) and high-energy (MeV) ions.

In the first approximation, the unimodal depth distributions of implanted species and atomic displacements can be described by the first moments (R_p and R_{p_d} for the distributions of ion ranges and atomic displacements, respectively) and by the second central moments (ΔR_p and ΔR_{p_d} for ion ranges and atomic displacements, respectively). Because $R_p > R_{p_d}$ (always), the depth distribution of implanted species is shifted slightly deeper in the bulk of the solid with respect to the distribution of atomic displacements. However, $\Delta R_p > \Delta R_{p_d}$ in the case of light ions, and $\Delta R_p < \Delta R_{p_d}$ for heavy ions.

In addition to atomic displacements, ballistic collisions of energetic ions with target atoms result in *surface sputtering* (i.e., ejection of target atoms from the surface) and *atomic mixing* of elements across an interface. A number of material analysis and device processing tools use the phenomenon of ion sputtering for a controlled removal of the near-surface layer. Another important ballistic phenomenon that occurs in dense collision cascades in compound solids (i.e., solids made of more than one element) is local *stoichiometric imbalance.* Indeed, in collision cascades in a compound solid, an excess concentration of the heavier element exists at shallow depth, while the region at greater depth is enriched with atoms of the lighter element. Calculations show that such stoichiometric disturbances are greatest when the mass ratio of the constituent elements of the solid is high, and when the ion mass is large.[1] In addition to local material stoichiometric imbalance within dense collision cascades, *preferential sputtering* (i.e., differences in surface sputtering yields for different elements in the solid) may significantly change the composition of the near-surface region of a compound solid when high ion doses are involved.

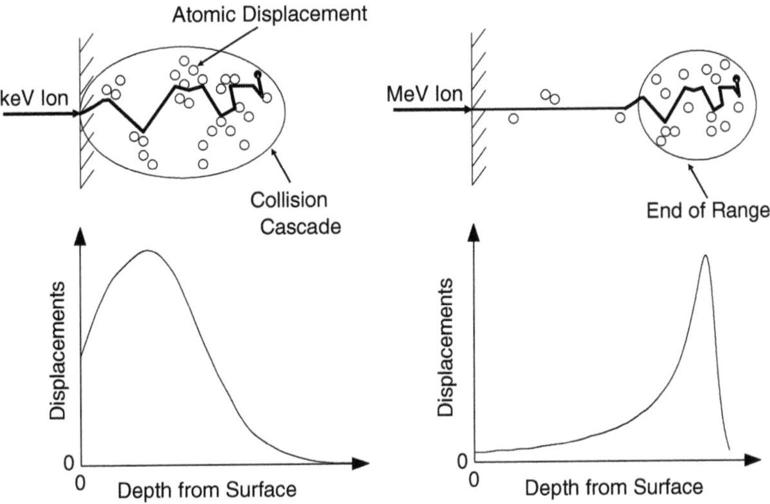

FIGURE 3.1 Schematic of collision cascades and typical depth profiles of atomic displacements created by keV ions (left) and MeV ions (right).

Note that, in this chapter, we are using the term "ion dose" (in ions/cm^2), which is well accepted in the literature on ion implantation into semiconductors and metals. However, it should be noted that, in the literature on irradiation effects in ceramics and polymers, ion dose is often referred to as "integrated ion flux" or "ion fluence."

Finally, another important ballistic process is worth mentioning. Vacancies and interstitials are spatially separated in a collision cascade, with an interstitial excess at the ion end-of-range and a vacancy excess closer to the surface.[2] The effect of such a *spatial separation of vacancies and interstitials* for each collision cascade becomes more pronounced with increasing ion dose. As a result, this effect is important in controlling damage accumulation for irradiation regimes that require high ion doses.

Parameters of the ballistic processes discussed above (i.e., ion ranges, atomic displacements, surface sputtering, atomic mixing, stoichiometric imbalance, and spatial separation of vacancies and interstitials) can be readily calculated by several methods. The currently most common approach in the ion-beam community to obtain such information is calculations with the TRIM code,[3] a Monte Carlo computer simulation program. Other methods often used to calculate the profiles of implanted species and atoms displaced involve analytical calculations such as developed by Lindhard, Scharf, and Schiott (often called the LSS theory) and an approach based on the Boltzmann transport equation.[4] The physical processes underlying these ballistic phenomena are fairly well understood. Their detailed discussion can be found, for example in References 5 to References 7, and will not be reproduced here.

3.3 Non-Ballistic Processes

3.3.1 Limitations of the Ballistic Approach

The theoretical calculations of ion ranges, atomic displacements, sputtering, mixing, and stoichiometric changes discussed in the previous section, however, take into account only ballistic processes and completely neglect more complex phenomena including (i) diffusion processes (such as dynamic annealing), (i) high-density cascade processes, and (iii) effects of electronic excitation. Such nonballistic processes are highly dependent on irradiation conditions such as ion mass, energy, dose, substrate temperature, and beam flux. In contrast to our ability to calculate collisional processes, material modification as a result of nonballistic processes is usually very difficult to predict. For example, the total amount and the depth distribution of lattice disorder experimentally observed in crystalline solids after ion bombardment in most cases depart from predictions based only on collisional processes. In addition, the ion sputtering yield (i.e., the number of target atoms ejected from the surface per incident ion) of crystalline solids, measured experimentally, typically depends on the crystallographic orientation of the surface. Sputtering in nonmetallic solids is also known to be affected by electronic energy loss for ion irradiation regimes with large electronic excitation, which will be discussed in Section 3.3.6. In most cases, experimental studies are essential to understand the effects of implant conditions on material properties.

Recent research efforts in the field of ion-beam processing have been largely devoted to understanding such nonballistic processes. Most such studies, in fact, focus on understanding the following fundamental aspects: (i) the degree of dynamic annealing, (ii) the nucleation of an amorphous phase, (iii) the nature of ion-beam-produced defects, (iv) the effect of electronic energy loss, (v) the influence of the density of collision cascades, (vi) chemical effects of implanted species, (vii) material decomposition during bombardment, (viii) mobility and lifetime of point defects, and (ix) thermal stability of radiation damage. We discuss these fundamental processes below. In addition, more applied studies often focus on the effects of implantation damage on technologically important material properties such as various electrical, magnetic, mechanical, and optical characteristics, which will be briefly discussed in Section 3.4.

3.3.2 High-Density Cascade Effects

Given the current understanding of ion-beam processes in solids, one can think of two limiting cases based on the characteristics of the collision cascades generated by energetic ions penetrating through a solid. These two cases are bombardment by light and heavy ions relative to the masses of the host atoms of the material under bombardment. In the case of light ions, collision cascades are dilute and consist mostly of simple point defects such as vacancies and interstitials. In the case of heavy ions, where the nuclear energy loss rate is large, each ion generates a dense collision cascade which, upon very fast quenching (over ~1 to 10 ps), can result in the formation of amorphous zones in dense regions of the collision cascade (which are often called *subcascades*).[8,9] Such amorphous zones are believed to form due to collective nonlinear processes when the damage level (and the energy deposited) in the volume of the collision cascade exceeds some threshold value. It should also be noted that the amorphous zones formed in dense regions of collision cascades or subcascades are often called *amorphous pockets* in more recent literature on molecular dynamics simulations of ion-beam-defect processes.

The formation of amorphous zones has been experimentally observed in many nonmetallic crystals.[8,9] However, under certain ion irradiation conditions (i.e., ion mass, energy, substrate temperature, and beam flux) for most crystalline materials, amorphous zones are not observed after ion bombardment. Under these irradiation conditions, amorphous zones (if they form) appear to be unstable and anneal via, presumably, direct thermal and ion-beam-induced processes.

The concept that collective nonlinear processes in dense collision cascades result in the formation of amorphous zones has been successfully used to explain the fact that, for ion bombardment conditions with suppressed defect mobility, lattice disorder often accumulates faster than predicted by ballistic calculations.[8,9] The same concept has also been used to explain the dependence of the sputtering yield on the density of collision cascades. In particular, studies of the so-called *molecular effect* (i.e., bombardment with atomic and cluster ions with the same velocity per each atom comprising the molecular ion) have been very instrumental in understanding the influence of the density of collision cascades on the sputtering yield and the formation of stable lattice disorder. For a more detailed discussion of high-density cascade effects in solids, the reader is referred to comprehensive reviews in Reference 8 and Reference 9. A more recent discussion of cascade density effects for irradiation conditions with efficient defect mobility can be found in Reference 10 and Reference 12.

3.3.3 Dynamic Annealing

By *dynamic annealing* we mean migration and interaction of ion-beam-generated defects *during* ion irradiation. Ion-generated simple point defects, which survive after the thermalization of collision cascades,

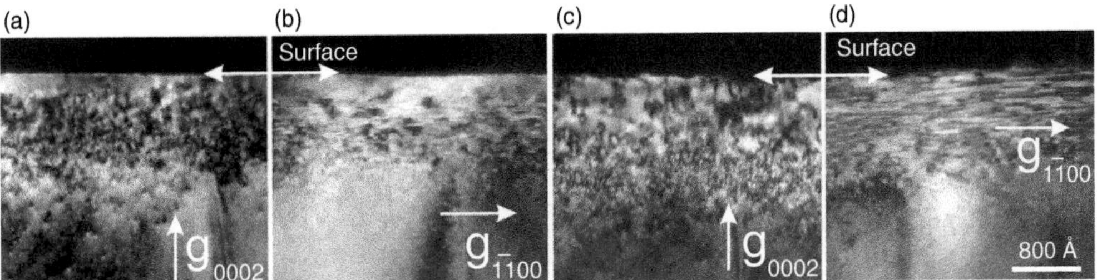

FIGURE 3.2 Dark-field cross-sectional transmission electron microscopy images [(a),(c) **g** = 0002* and (b),(d) **g** = $1\bar{1}00^*$] of the GaN epilayers bombarded at 550°C with 300 keV Au ions with a beam flux of 3.1×10^{12} cm^{-2} s^{-1} to doses of 6×10^{14} cm^{-2} [(a) and (b)] and 4×10^{15} cm^{-2} [(c) and (d)]. (After Kucheyev, S. O., Williams, J. S., Zou J., Jagadish, C. and Li, G., *Appl. Phys. Lett.*, 78, 1373, 2001. With permission.)

in many crystalline materials for common ion-irradiation conditions, can migrate through the lattice and experience annihilation and cluster formation. Numerous defect interaction processes are possible. For example, vacancies and interstitials can annihilate via both direct and indirect recombination processes. In an indirect annihilation process, recombination of a vacancy (interstitial) occurs via trapping at an interstitial (vacancy) complex. Dynamic annealing processes may also result in the formation of antisite defects in compound solids. Migration energies of point defects in semiconductors can also depend on the defect charge state.

However, defect annihilation is not perfect in most cases, and point defect clusters (such as complexes of vacancies and interstitials as well as defect-impurity complexes) will form and grow with increasing ion dose. For conditions with large dynamic annealing (as for most metals and ionic solids at room temperature and above), energetically favorable extended defects may form during ion irradiation. Ion-produced extended defects can consist of a regular array of planar defects (like for wurtzite group-III nitrides and ZnO) or dislocation tangles as in metals and other semiconductors (such as Si and GaAs).[6] An example of such *energetically favorable defects* formed as a result of efficient dynamic annealing processes is given in Figure 3.2,[13] which shows the presence of some point defect clusters (Figure 3.2a and Figure 3.2c) and planar defects (Figure 3.2b and Figure 3.2d) in the near-surface region of GaN damaged by ion bombardment (up to ~1500 Å from the GaN surface). A similar band of planar defects has been observed in GaN bombarded under a wide range of irradiation conditions.[11,14]

Figure 3.3a gives an example of how dramatically the experimental depth profiles of lattice disorder can deviate from the predictions of ballistic calculations. It is seen from Figure 3.3a that the depth profiles of relative gross lattice disorder in GaN irradiated with 300 keV Au ions significantly deviate from the Gaussian-like (unimodal) shape expected based on only ballistic processes (see Figure 3.1). Such deviations include (i) damage saturation in the crystal bulk close to the region of the maximum nuclear energy loss and (ii) preferential surface disordering with the formation of a surface amorphous layer that grows layer by layer from the surface with increasing ion doses. These features result from complex dynamic annealing processes.[10,11,14,15] This is also an example of a profound *influence of the sample surface or interfaces* on dynamic annealing. Indeed, the sample surface or an interface often represents an effective sink for migrating point defects. The influence of the surface on the evolution of point defects generated in the crystal bulk is determined by the effective mobility of defects and by the distance between the surface and the region where these defects are generated.

As alluded to earlier, for regimes with strong dynamic annealing, ion-beam-defect processes strongly depend on ion irradiation conditions. As an example of large effects of ion species on the level of pre-amorphous disorder in crystals, Figure 3.3b (the left axis) shows the ion mass dependence of the ion dose required to produce 30% relative disorder, $\Phi_{0.3}$, in GaN at room temperature for different ion species of keV energies. It is seen from Figure 3.3b (the left axis) that $\Phi_{0.3}$ generally decreases with increasing ion mass. This qualitative trend is expected as the number of ion-beam-generated atomic displacements increases with ion mass.

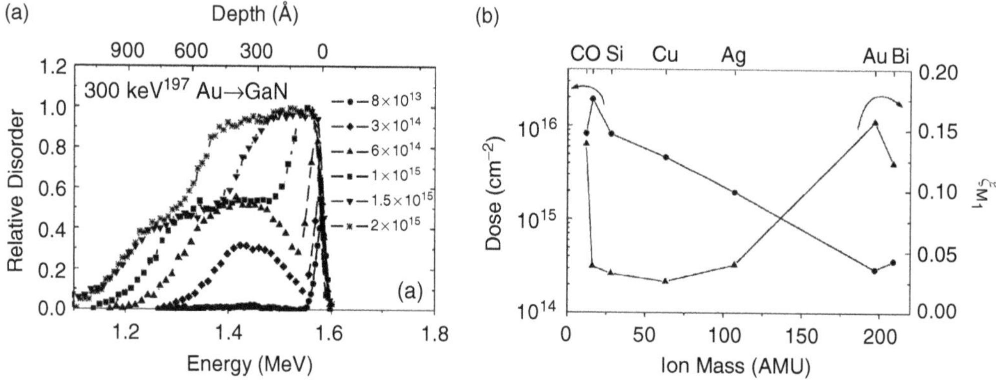

FIGURE 3.3 (a) Depth profiles of relative disorder (measured by Rutherford backscattering/channeling (RBS/C) spectrometry) in GaN films bombarded at 20°C with 300 keV Au ions with a beam flux of 4.4×10^{12} cm^{-2} s^{-1}. Implantation doses (in cm^{-2}) are indicated. (After Kucheyev, S. O. Williams J. S., Jagadish C., Zou J., Li G., and Titov A. I., *Phys. Rev. B*, 64, 035202, 2001. With permission.) (b) Left axis: the ion mass dependence of the ion dose necessary to produce relative disorder of 0.3 (as measured by RBS/C) by implantation of GaN at 20°C with different ion species. (b) Right axis: the dependence of ξ_{M_1} (ξ_{M_1} = the ratio of the level of lattice disorder measured experimentally to the damage level predicted based on ballistic calculations) on ion mass for GaN at 20°C. (See Kucheyev S. O., Williams J. S., Jagadish C., Zou J., Li G., and Titov A. I., *Phys. Rev. B*, 64, 035202, 2001 for the other details of implant conditions.)

However, two deviations from the trend expected are clearly seen in Figure 3.3b (the left axis) for the cases of ^{12}C and ^{209}Bi ions. In particular, $\Phi_{0.3}$ for ^{12}C ions is about two times smaller than $\Phi_{0.3}$ for ^{16}O ions despite the fact that ^{16}O ions produce ~1.6 times more vacancies than ^{12}C ions in the maximum of the nuclear energy loss profile. This is a somewhat extreme example of *chemical effects of implanted species*, discussed in more detail below. In addition to ^{12}C, irradiation with ^{209}Bi ions produced less stable lattice damage than bombardment with ^{197}Au ions, although ^{209}Bi ions generate a larger number of atomic displacements than ^{197}Au ions of the particular energies used. This result is due to a lower beam flux value in the case of Bi ions compared with the beam flux value of Au ions,[11] which illustrates the importance of *beam flux* on the formation of stable defects under ion bombardment.

A further insight into the effect of ion mass on implantation-produced lattice disorder is given in Figure 3.3b (the right axis), which shows the ion mass dependence of ξ_{M_1}, the ratio of the level of lattice disorder of 0.3 (measured experimentally) to the damage level predicted based on ballistic calculations. In the first approximation, the parameter ξ_{M_1} reflects the effectiveness of the production of stable lattice disorder (N^{def}) by ion bombardment under particular implant conditions: $N^{def} = \xi_{M_1} \Phi N_{vac}^{max}$ for relatively low levels of lattice disorder, where Φ is ion dose, and N_{vac}^{max} is the number of lattice vacancies in the maximum of the nuclear energy loss profile. If postimplantation stable lattice damage were the same as the one predicted based on ballistic calculations (such as by TRIM [3]), ξ_{M_1} would be equal to unity and independent of ion mass.

In contrast to such expectations, Figure 3.3b (the right axis) shows a rather complex dependence of ξ_{M_1} on ion mass. First of all, for all ion species used, ξ_{M_1} is significantly below unity. This is a direct consequence of strong dynamic annealing processes when a large fraction of ion-beam-generated point defects experiences annihilation. It is also seen from Figure 3.3b (the right axis) that, with increasing ion mass from ^{12}C to ^{63}Cu, ξ_{M_1} decreases. For ions heavier than ^{63}Cu, ξ_{M_1} shows an increase with increasing ion mass. Such a complex behavior is due to a combination of the following factors: (i) strong dynamic annealing processes, (ii) changes in defect clustering efficiency, (iii) collective energy spike effects, (iv) variations in the effective displacement energy, and (v) chemical effects of implanted species.[11]

The chemical enhancement of damage buildup mentioned above is a common phenomenon for both nonmetallic and metallic crystalline solids [6,11]. Indeed, chemical effects often dominate the damage buildup behavior in implantation regimes with strong dynamic annealing. Such strong dynamic annealing, which occurs in some materials (i.e., metals and some semiconductors and insulators such as GaN, AlN, ZnO, etc.) even during heavy-ion bombardment at cryogenic temperatures, is typical even for covalently bonded semiconductors (such as Si, Ge, SiC, and GaAs) during ion irradiation at elevated temperatures. The scale of chemical effects depends on the particular ion species used. A chemical enhancement of damage buildup can be due to (i) the trapping of ion-beam-generated mobile point defects by implanted impurity atoms, (ii) second phase formation and associated lattice distortion, and (iii) changes in energy barriers for defect migration and interactions.[11] Additional studies are currently needed to ascertain the contributions to the chemical effect in crystalline solids from each of the above three processes.

3.3.4 Amorphization

Ion bombardment of initially crystalline solids under certain conditions can result in a complete disordering of the long-range order in the crystal lattice; that is, amorphization. Numerous, both theoretical and experimental, studies have focused on understanding the physical and chemical mechanisms of ion-beam-induced amorphization in different materials and for different ion irradiation conditions. In Section 3.3.2, we have already discussed one type of amorphization that occurs locally in dense regions of individual collision cascades. In this case, with increasing ion dose, the amorphization of the entire near-surface layer proceeds via the accumulation (and overlap) of completely amorphous zones produced by individual ions. This case is sometimes called *heterogeneous amorphization.*

Amorphization of the entire near-surface layer can also occur when stable amorphous zones are not formed in collision cascades generated in previously intact (undamaged) regions of the crystal lattice. In such cases, with further irradiation, when the concentration of ion-beam-produced stable lattice defects accumulates, the free energy of the defective material may rise above that of the amorphous phase, resulting in a catastrophic collapse of the defective crystalline lattice into an amorphous phase.[16] This case is sometimes called *homogeneous amorphization.* Damage accumulation can often be described by the so-called *defect-overlap model* developed in the 1970s. This model takes into account a spatial overlap of regions with an incompletely disordered crystal structure. It is assumed that a spatial overlap of such incompletely disordered regions is required for complete lattice amorphization. Quantitative analysis of the experimental damage–dose curves with this model provides information about the number of overlaps needed for complete amorphization and the effective (lateral) size of amorphous zones. A detailed discussion of the defect-overlap model and its variations can be found, for example, in Reference 17 and Reference 18.

It is often overlooked that ion-beam-induced amorphization occurs only for certain materials under a relatively narrow range of ion irradiation conditions. Indeed, many crystalline solids (such as all pure metals and most ionic crystals) remain crystalline (though heavily damaged) even after bombardment with keV heavy ions (which generate dense collision cascades) to very high doses. Moreover, even the materials that exhibit little dynamic annealing at room temperature for irradiation with keV heavy ions and readily amorphize for moderate ion doses of $\sim 10^{14}$ to 10^{15} cm^{-2} will exhibit pronounced dynamic annealing effects and will not amorphize for irradiation at elevated temperatures or with light ions even for very large ion doses.

Over the past several decades, many studies have focused on understanding the influence of material parameters on susceptibility to ion-beam-induced amorphization. [15,19–27] However, most research efforts have been concentrated on material susceptibility to amorphization at room temperature, without taking into account the often dramatic influence of ion irradiation conditions such as substrate temperature, cascade density, the generation rate of atomic displacements (determined by ion mass, energy, and beam flux), the proximity of defect sinks such as surfaces and interfaces, chemical effects of implanted species for high-dose regimes, and possible ion-beam-induced stoichiometric imbalance.

The following criteria for material susceptibility to amorphization induced by ballistic ion-beam processes have been proposed: (i) bond type,[19,20,27] (ii) the ratio of crystallization temperature to the melting point,[20] (iii) structural connectivity,[23] (iv) enthalpy or free energy difference between crystalline and amorphous phases,[24] (v) an empirical combination of several of these parameters,[22,25] and (vi) bond energy.[15] However, the structural connectivity criterion[23] fails to explain the large difference in amorphization susceptibility experimentally observed in many systems with the same crystal structure.[15,27] In addition, a quantitative analysis based on (i) the ratio of the crystallization temperature to the melting point[20] or (ii) an enthalpy or free energy difference[24] is difficult because enthalpies, free energies, and crystallization temperatures for many materials are currently unknown. Finally, based on the bond type criterion,[20,21] ion-beam-induced amorphization should occur in solids with bond ionicities (e.g., as defined, by Phillips[28]) ≤ 0.47, whereas solids with metallic bonds or with ionicities of chemical bonds above 0.60 remain crystalline even after high-dose ion bombardment. Materials with bond ionicities between 0.47 and 0.60 typically show varying structural stability under ion bombardment.[20,21] Unfortunately, many materials fall into the last category of solids with ionicities between 0.47 and 0.60 which may or may not exhibit ion-beam-induced amorphization.

It should also be noted that the physical mechanisms underlying the empirical bond type criterion[20] are not well understood. It has been suggested that an increase in electrostatic energy associated with substitutional disorder as ionicity increases[20] and the rotational rigidity of covalent bonds[19] may underlie the ionicity criterion for amorphization. Large bond ionicity can also facilitate defect annihilation by electrostatically lowering energy barriers to defect interaction processes.[15]

In addition, the level of dynamic annealing and the susceptibility to amorphization for some systems scale with the energy of the chemical bonds in the solid. This is at least the case for continuous alloys of III–V semiconductors such as AlGaN, InGaN, and AlGaAs.[15] This experimental finding can be explained by noting that the buildup of radiation damage is associated with the formation of lattice defects involving broken and reconstructed bonds. It is expected that dynamic annealing processes, including defect annihilation, will be more efficient in a system with a larger energy gain due to the recovery of broken, distorted, and nonstoichiometric bonds, which are ballistically generated by the ion beam. However, although the efficiency of dynamic annealing in these alloys scales with the energy of chemical bonds (or the melting point, which is typically proportional to the bond energy), variations in other parameters can also be responsible for changes in the damage buildup behavior. For example, activation energies for various defect migration and interaction processes can also dramatically affect damage accumulation. In addition, possible segregation of some elements (such as In and Al atoms in these alloys) during ion bombardment could influence the buildup of stable lattice disorder. Hence, although the bond energy gives a clear trend in the efficiency of dynamic annealing in a number of semiconductor systems such as AlGaN, InGaN, and AlGaAs, a better understanding of the physical mechanisms controlling dynamic annealing in these semiconductors will need to await more detailed data on defect migration and interaction processes.

3.3.5 Swelling and Porosity

A combination of ballistic, diffusion-related, and chemical processes can result in spectacular macroscopic effects in some solids exposed to ion bombardment—swelling and porosity of implanted layers. Figure 3.4 illustrates ion-beam-induced swelling and porosity in GaSb and GaN.[29] Such effects of ion-beam-induced porosity have been observed in the following semiconductors: Ge,[30,31] InSb,[32] GaSb,[33] GaN,[34,35] and AlGaN.[15] Numerous other materials have been found to exhibit swelling (and associated mechanical stresses) upon irradiation with energetic ions. However, for all the materials studied (except for Ge, InSb, GaSb, GaN, and AlGaN discussed above) the volume expansion is only up to several percent because a porous structure is not formed.

Although studied in some detail (particularly in the case of Ge), ion-beam-induced formation and evolution of porous structures are not well understood. One of the possible explanations for this effect is energetically favorable agglomeration of vacancy-like defects, generated by an ion beam in an amorphous matrix, resulting in the formation of voids.[30,31,36] In addition, in the case of GaN and AlGaN, the porosity has been attributed to ion-beam-induced material decomposition with the formation of N_2 gas bubbles in a Ga-rich matrix.[15,34,35]

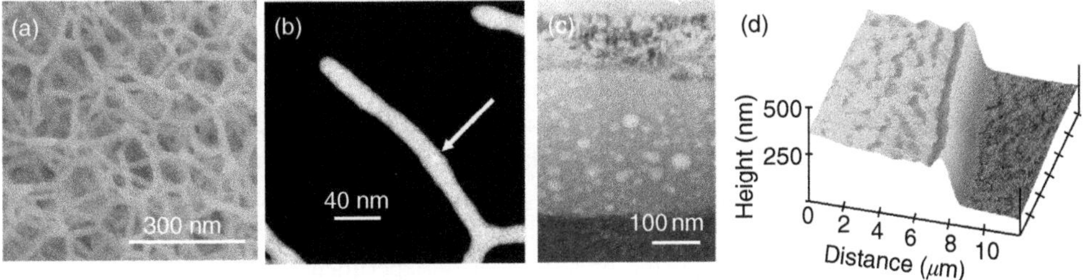

FIGURE 3.4 (a) Scanning electron microscopy and (b) dark-field XTEM images of GaSb irradiated at room temperature with multiple energy ^{69}Ga ions to total doses of (a) 5×10^{15} and (b) 1×10^{16} cm^{-2}. (c) Bright-field XTEM and (d) tapping-mode AFM images of GaN bombarded with 2 MeV Au ions at −196°C to a dose of 10^{16} cm^{-2}. The AFM image in (d) illustrates the border between implanted (on the left) and unimplanted (on the right) areas of a GaN sample. (After Kucheyev S. O., Williams J. S., Jagadish C., Craig V. S. J., and Li G., *Appl. Phys. Lett.*, 77, 1455, 2000; Kucheyev S. O., Williams J. S., Zou J., Jagadish C., and Li G. *Appl. Phys. Lett.*, 77, 3577, 2000; Kluth S. M., Fitz Gerald J. D., and Ridgway M. C., *Appl. Phys. Lett.*, 86, 131920, 2005.)

3.3.6 Effects of Electronic Excitation

For the ion energies typically used for ion implantation (i.e., <1 MeV), lattice defects in most radiolysis-resistant solids are produced as a result of nuclear (or elastic) collisions only, while electronic energy loss processes have a negligible role in defect formation. Indeed, in the case of irradiation with keV ions, the level of electronic excitation is ≤ 1 keV/nm. Such low electronic energy losses typically do not result in the formation of lattice defects in radiolysis-resistant materials like metals and most semiconductors, and electronic excitation is dissipated as heat.

The level of electronic excitation gradually increases with increasing ion energy and atomic number. Electronic energy losses can reach some tens of keV/nm in the case of heavy ions with energies of several hundred MeV. In this so-called *swift heavy ion* (SHI) bombardment regime, intense ultrafast excitation of core and valence electrons occurs along ion paths. The dissipation of such intense electronic excitation can often result in the formation of stable lattice defects and spurious surface sputtering.

There have been numerous previous studies of SHI-induced effects in different solids. It has been found that SHI bombardment of various polymers and insulators results in the formation of *latent tracks*. Such tracks are continuous or discontinuous cylindrical damaged zones created along the paths of rapidly moving ions. Numerous experiments have shown that track formation processes are nonlinear. Indeed, tracks are created only when the level of electronic excitation exceeds a certain threshold, typically corresponding to electronic energy losses of ~1 to 30 keV/nm, depending on the material. Excellent reviews on the formation of latent tracks in polymers and insulators can be found.[37,38] More recently, swift-ion-induced lattice damage has also been studied in metals and a wide range of semiconductors.[39,40]

As energy loss processes of energetic ions are relatively well understood and are similar for different crystals, the variations in the track formation behavior of different materials are mostly determined by processes of energy dissipation. Several models have been developed to explain the transfer of the energy of electronic excitation into atomic motion, resulting in track formation in crystalline solids. The main physical models are the thermal spike,[41] Coulomb explosion,[37,42] and material instability at high levels of electronic excitation (i.e., lattice relaxation model).[43]

In the *thermal spike* model, excited electrons transfer their energy to atoms via the electron-phonon coupling. The amount of heat deposited into the thermal spike volume can be large, and the spike temperature can significantly exceed the melting point. Hence, a pseudo-liquid ("molten") region is formed around the ion trajectory. Subsequent very rapid quenching of such a hot region results in the formation of an amorphous track if the cooling rate is too fast for epitaxial crystallization to occur. Imperfect re-crystallization will result in the formation of a track consisting of a damaged but not amorphous material. An alternative mechanism of track formation, related to thermal spikes, is *plastic deformation* due to the high pressure on the material surrounding the ion path. Indeed, heating of the track core material, after electrons transfer their energy to the lattice, results in thermal expansion and an associated stress field. Within the *Coulomb explosion* model,[37,42] a large density of positive charge in the track core results in repulsion between atoms, leading to atomic motion and track formation. Finally, according to the *lattice relaxation* model,[43] intense electronic excitation weakens the covalent bonds and causes a repulsive force between atoms, resulting in collective atomic rearrangement and track formation.

Track formation in insulators has generally been explained within the Coulomb explosion approach,[37,38,42] which has been supported by a number of experiments.[44] Tracks in semiconductors (which typically have chemical bonds with relatively low ionicities[28]) have been attributed to the formation of thermal spikes.[41] It should be noted, however, that there has been no direct experimental evidence to support the formation of thermal spikes along paths of SHIs in semiconductors. Miotello and Kelly[45] have given a detailed discussion of difficulties associated with using the thermal spike approach to describe the formation of latent tracks in solids under SHI bombardment. They have pointed out that, due to large kinetic energies of inner-shell electrons excited by SHI,[39] a large part of the energy deposited by the fast moving particle is carried away from the ion track volume and dissipated over larger distances, not confined by the ion trajectory. Hence, the existence of thermal spikes along the trajectories of SHIs is not obvious. It is likely that, in different solids, different physical mechanisms are responsible for track formation.

There is also synergy between physical processes occurring in solids during two seemingly different types of excitation: irradiation with SHI and with *slow, highly charged ions* (SHCI). Irradiation with SHCI (discussed in detail in, for example, Reference 46) also results in ultrafast intense local excitation of the electronic subsystem and is associated with high temperatures and pressures. In both cases, the relaxation of intense excitation can result in various nonequilibrium phenomena. These phenomena include the formation of stable lattice defects, changes in material density, phase transformations, and material ablation.[46] It is also interesting to note that, for both cases, the same models (i.e., the thermal spike, Coulomb explosion, and material instability at high levels of electronic excitation) have typically been used to explain defect formation.

However, there are significant differences in electronic excitation and relaxation processes during SHI and SHCI irradiation. In particular, the proximity of the sample surface for SHCI in contrast to the deeper penetration of SHIs with confinement of SHI-induced tracks by the surrounding matrix could play a role. Hence, although similar physical phenomena occur in solids under intense electronic excitation with SHI and SHCI, the transfer of experimental and theoretical approaches between these research fields is not necessarily straightforward.

Another spectacular SHI-induced phenomenon is worth mentioning — radiation-induced *anisotropic plastic flow* in amorphous solids. This effect, sometimes also called the *ion hammering* effect, has been known since the early 1980s.[47] The plastic flow occurs in the direction perpendicular to the ion beam as if the sample were hammered. Changes in material density are typically negligible during the plastic flow. The ion hammering effect has been attributed to thermal stresses associated with thermal spikes generated along ion tracks.[47] This phenomenon has also recently been explored to modify the shape of nanostructures.[48]

Finally, a comment should be made on the so-called *radiolytic processes* that can lead to the formation of atomic displacements (lattice defects) even when the level of electronic excitation is low compared to the cases of bombardment with SHIs and SHCIs. In radiolytic processes, the energy released during electronic relaxation (such as recombination of an electron-hole pair in an insulator) is transferred to atomic motion, resulting in a scission/cleavage of chemical bonds. Radiolytic processes have been observed in many insulators and polymers exposed to ionizing radiation. However, these effects are typically negligible in all metals and most semiconductors (including wide band-gap semiconductors such as GaN, InN, SiC, ZnO, etc.). Radiolytic effects are commonly studied not for cases of bombardment with heavy ions of keV energies (used for ion implantation) but for irradiation with relatively low energy electron beams and photons (with energies above the fundamental absorption gap of the solid), which do not produce ballistic displacements. This is partly due to the fact that, during keV ion bombardment of insulators and polymers that are known to have radiolytic effects, such processes are often negligible as they are overshadowed by much more efficient collision-related processes of defect production. A detailed discussion of various radiolytic processes can be found, for example, in Reference 49.

3.4 Applications

In this section, we very briefly discuss only the major current applications of ion beams for processing different classes of materials. Our attention will primarily focus on ion-beam processing of semiconductors as most technological applications of ion implantation are related to semiconductor technology.

3.4.1 Semiconductors

In the fabrication of semiconductor devices, ion bombardment represents a very attractive processing tool for several technological steps, including selective-area doping, electrical isolation, dry etching, quantum-well intermixing, ion slicing, nanofabrication, and gettering. *Doping* is the most common application of ion implantation in semiconductor technology. In this process, electrical, optical, and magnetic properties of the host material are modified by implanting impurities. Advantages of ion implantation over the other doping approaches include an excellent control of the depth profile and concentration of dopants and the possibility of selective-area doping in cases when diffusion-based approaches cannot be used. The major disadvantage of ion implantation is related to adverse effects of ion-beam-produced lattice defects. A postimplantation thermal annealing step is typically required to restore lattice crystallinity and to activate the dopants (i.e., to promote the short-range diffusion of dopants into the lattice locations desired, which are typically substitutional positions).

Ion irradiation under appropriate conditions can render semiconductors highly resistive and, hence, can be used for selective-area *electrical isolation* of closely spaced devices. It is generally believed that irradiation-induced degradation of carrier mobility as well as the trapping of carriers at deep centers associated with irradiation-produced damage (*defect isolation*) or with implanted species (*chemical isolation*) is the mechanism responsible for electrical isolation of semiconductors. Additional information on ion-irradiation-induced electrical isolation of different semiconductors can be found elsewhere.[13,50,51]

As another process based on ion irradiation, *dry etching* is used for a controlled layer-by-layer removal of the material. The most common approach is reactive ion etching (RIE), when the sample surface is bombarded with ions, and the removal or sputtering of the material is assisted by specific chemical reactions between ions and the target atoms. The *chemical component* of the RIE process is related to the formation of volatile products of the bombarding ions with target atoms, while the *physical component* of RIE is due to collisional ion sputtering processes. Numerous studies focused on finding appropriate processing parameters for dry etching of various semiconductors have been reported. For additional details, the reader is referred to, for example, Reference 52.

Ion irradiation can also be used for so-called *quantum well intermixing*. This process involves irradiation of quantum well structures with energetic light ions that create very dilute collision cascades and generate mostly point defects. Ion irradiation is followed by a thermal annealing step when ion-beam-generated defects diffuse and cause the mixing of atoms through the barrier–quantum well interface. Such mixing results in changes to the potential profile of the quantum well from the initially approximately square shape to a more rounded shape, causing an increase in the optical gap (i.e., in emission and absorption energies) of the quantum-well-based optoelectronic devices. Additional information on the use of ion irradiation for quantum well intermixing can be found, for example, Reference 53.

Another interesting application of ion bombardment is based on the formation of open volume defects (small cavities) outside the active part of the device. Most research in this area has been done on Si — the major material of modern microelectronics. Such open volume defects can be introduced in Si by high-dose irradiation with Si ions, which does not change the material stoichiometry but introduces excess of vacancies near the surface due to the spatial separation of vacancies and interstitials within

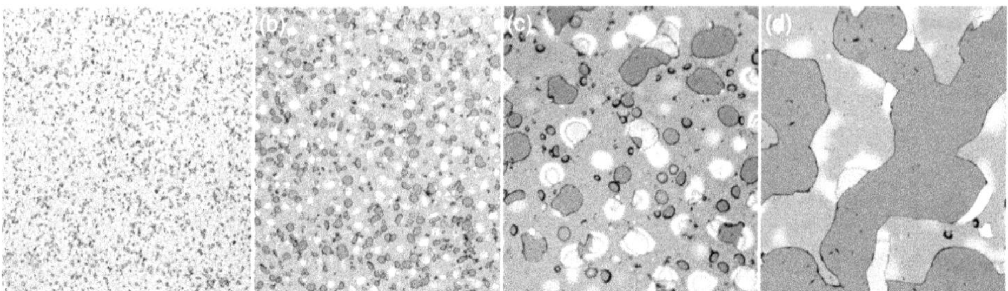

FIGURE 3.5 Typical optical micrographs of GaN samples implanted at 20°C with 20 keV H ions to a dose of 3.3×10^{17} cm^{-2} (a), 50 keV H ions to a dose of 4.4×10^{17} cm^{-2} (b), 100 keV H ions to a dose of 5×10^{17} cm^{-2} (c), and 150 keV H ions to a dose of 5.9×10^{17} cm^{-2} (d). After implantation, samples were annealed at 900°C for 5 min in a nitrogen ambient. The horizontal field width of each image is 250 μm. (After Kucheyev S. O., Williams J. S., Jagadish C., Zou J., and Li G., *J. Appl. Phys.*, 91, 3928, 2002. With permission.)

collision cascades (a ballistic effect discussed in Section 3.2). Open volume defects can also be introduced by high-dose irradiation with H ions with subsequent thermal annealing to form H_2 gas bubbles and then to release H_2 from the bubbles during an additional annealing step, leaving empty cavities. These open volume defects can be used to selectively remove metal impurities; that is, for *gettering*. Additional information about this application can be found in Reference 54.

Ion slicing (or ion-cut) is a more "exotic" application of ion irradiation. It involves high-dose H ion implantation in combination with wafer bonding and postimplantation annealing. It provides a method to transfer thin surface layers onto other substrates. Ion slicing has been studied in a number of semiconductors such as Si, Ge, GaAs, InP, SiC, C (diamond), and GaN.[55–58] It has been found that different semiconductors respond differently to H-ion implantation.[56–58]

Studies of ion-cut are usually made in two steps: (i) investigation of surface blistering and (ii) studies of wafer bonding and layer splitting during postimplantation annealing. It has been found that both processes of blistering and layer splitting have the same activation energy.[57] Moreover, implantation and annealing conditions resulting in the formation of surface blisters and layer splitting (after wafer bonding) are similar.[57,58] An example of ion-induced blistering is shown in Figure 3.5, which shows typical optical micrographs of GaN implanted at room temperature with 20, 50, 100, or 150 keV H ions to doses resulting in $\sim$30 *at.*% of H atoms in the maximum of the ion distribution profile and subsequently annealed at 900°C for 5 min. Figure 3.5 reveals that, with increasing ion energy from 20 to 100 keV, blister size also increases. The dependence of blister size on ion energy shown in Figure 3.5 is consistent with extensive studies of blistering in metals.[59]

Finally, there have been numerous, more recent reports on using ion bombardment for the fabrication of various *nanostructures*. These include (i) high-dose implantation of impurities (typically to doses as high as $\sim 10^{17}$ to 10^{18} cm^{-2}) followed by a subsequent annealing step resulting in the precipitation of nanoparticles and (ii) the formation of surface nanostructures during ion sputtering, the process that relies on the peculiarities of ion sputtering at oblique angles of incidence of the ion beam. For more details on the application of ion implantation for the fabrication of nanostructures, the reader is referred to Reference 60 and to the constantly growing periodic literature on this topic.

3.4.2 Inorganic insulators and polymers

The effects of ion bombardment on various properties of polymers have attracted significant research efforts for the past several decades. Most investigations have focused on electrical, optical, and mechanical properties as many polymers exhibit a very large increase in electrical conductivity, optical density, and hardness as a result of irradiation.[61,62] Such irradiation-induced changes are often due to the effects of electronic (rather than nuclear) energy loss processes of energetic ions. As an example, Figure 3.6 shows

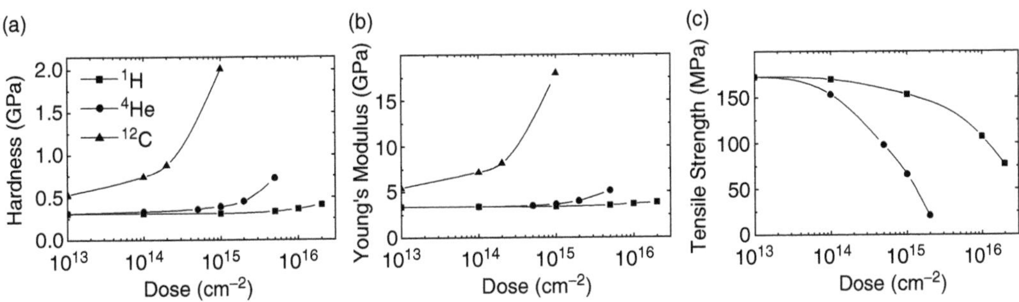

FIGURE 3.6 Ion dose dependencies of nanoindentation hardness (a), Young's modulus (b), and tensile strength (c) after irradiation of polyimide films with MeV ^{1}H, ^{4}He, and ^{12}C ions. (See Kucheyev S. O., Felter T. E., Anthamatten M., and Bradby J. E., *Appl. Phys. Lett.*, 85, 733, 2004 for more details.)

ion-dose dependencies of hardness, H, (Figure 3.6a), Young's modulus, E, (Figure 3.6b), and tensile strength, T, (Figure 3.6c) of a polyimide (Kapton H) bombarded with ^{1}H, ^{4}He, and ^{12}C ions. It is seen from Figure 3.6 that, for all the three ion species, bombardment results in an increase in both H and E, with a corresponding decrease in T. Figure 3.6 also shows that, with increasing ion mass (and, hence, the stopping power), the changes in mechanical properties are observed for lower ion doses. These changes depend close-to-linearly on ion dose and superlinearly on electronic energy loss. Additional details on the modification of mechanical properties by ion irradiation can be found in Reference 61 and Reference 63.

Other applications of ion irradiation of polymers and insulators include improvement of adhesion of films on various substrates (sometimes called the *ion stitching* effect, attributed to the energy deposition by irradiation into the electronic subsystem of the solid, leading to changes in local electron distribution and atomic configuration at film–substrate interfaces),[64] ion tracking with the formation of nanopores with a large aspect ratio used for filters and membranes,[37,38] modification of optical properties (such as the luminescence, refractive index, and reflectance),[65] and the fabrication of field emitters.[66] For ceramic insulators, such as yttria-stabilized zirconia (YSZ), processing with neutrons or various ions may result in increases in thermal conductivity, a property important for YSZ use as electrolytes in solid oxide fuel cells.[67]

3.4.3 Metals

Ion-beam processing of metals has received less attention than processing of other materials. This is primarily due to the fact that significantly larger ion doses are required to modify the properties of metals than semiconductors, for instance. Nevertheless, numerous studies have been reported on the use of ion beams for modification of mechanical, oxidation and corrosion, magnetic, adhesive, and catalytic properties of metals. A detailed discussion of such studies can be found, for example, in Reference 6.

3.5 Concluding Remarks

The goal of this chapter has been to give an introduction into a wide range of fundamental phenomena occurring in solids under ion bombardment. We have also discussed the current major applications of ion beams for processing various classes of materials. No attempt, though, has been made to give a literature review of this very broad research field. Such a literature overview is beyond the scope and space limit of this chapter.

A final comment should also be made on what has not been even mentioned in the chapter. Due to space limitations, we have entirely omitted the following fields that have received significant attention in the ion beam community in the past two decades: molecular dynamics and kinetic Monte-Carlo simulations of ion-beam processes, the synthesis of new material phases by high-dose ion implantation, the development of single ion implantation, focused ion-beam processing, ion lithography, and the peculiarities of ion-beam processing of astronomically relevant materials and materials for immobilization of high-level nuclear waste.

Acknowledgments

The author is grateful to Prof. B. D. Wirth for reading the manuscript and making helpful comments. This work was performed under the auspices of the U.S. DOE by the University of California, LLNL under Contract No. W-7405-Eng-48.

References

[1] Christel, L. A. and Gibbons, J. F., Stoichiometric disturbances in ion-implanted compound semiconductors 1, *J. Appl. Phys.*, 52, 5050, 1981.

[2] Mazzone, A. M., Defect distribution in ion-implanted silicon — A Monte-Carlo simulation, *Phys. Status Solidi A*, 95, 149, 1986.

[3] Ziegler, J. F., Biersack, J. P., and Littmark, U., *The Stopping and Range of Ions in Solids*, Pergamon, New York, 1985.

[4] Gibbons, J. F. and Christel, L. A., An application of the Boltzmann transport-equation to ion range and damage distribution in multilayered targets, *J. Appl. Phys.*, 51, 6176, 1980.

[5] Kumakhov, M. A. and Komarov, F. F., *Energy Loss and Ion Ranges in Solids*, Gordon and Breach, New York, 1981.

[6] Williams, J. S., Materials modification with ion beams, *Rep. Prog. Phys.*, 49, 491, 1986.

[7] Nastasi, M., Mayer, J. W., and Hirvonen, J. K., *Ion-Solid Interactions: Fundamentals and Applications*, Cambridge University, Cambridge, 1996.

[8] Thompson, D. A., High density cascade effects, *Radiat. Eff.*, 56, 105, 1981.

[9] Davies, J. A., High energy collision cascades and spike in *Ion Implantation and Beam Processing*, Williams, J. S. and Poate, J. M., Eds., Academic Press, Sydney, 1984.

[10] Kucheyev, S. O. et al., Damage buildup in GaN under ion bombardment, *Phys. Rev. B*, 62, 7510, 2000.

[11] Kucheyev, S. O. et al., Effect of ion species on the accumulation of ion-beam damage in GaN, *Phys. Rev. B*, 64, 035202, 2001.

[12] Kucheyev, S. O. et al., Effect of the density of collision cascades on implantation damage in GaN, *Appl. Phys. Lett.*, 78, 2694, 2001.

[13] Kucheyev, S. O. et al., Disordering and anomalous surface erosion of GaN during ion bombardment at elevated temperatures, *Appl. Phys. Lett.*, 78, 1373, 2001.

[14] Kucheyev, S. O., Williams, J. S., and Pearton, S. J., Ion implantation into GaN, *Mater. Sci. Eng., R*, 33, 51, 2001.

[15] Kucheyev, S. O. et al., Dynamic annealing in III-nitrides under ion bombardment, *J. Appl. Phys.*, 95, 3048, 2004.

[16] Swanson, M. L., Parsons, J. R., and Hoelke, C. W., Damaged regions in neutron-irradiated and ion-bombarded Ge and Si, *Radiat. Eff.*, 9, 249, 1971.

[17] Weber, W. J., Models and mechanisms of irradiation-induced amorphization in ceramics, *Nucl. Instrum. Methods B*, 166-167, 98, 2000.

[18] Titov, A. I. et al., Damage buildup in Si under bombardment with MeV heavy atomic and molecular ions, *J. Appl. Phys.*, 90, 3867, 2001.

[19] Pavlov, P. V. et al., Phase transformations at bombardment of Al and Fe polycrystalline films with B, C, N, P, and As ions, *Phys. Status Solidi A*, 19, 373, 1973.

[20] Naguib, H. M. and Kelly, R., Criteria for bombardment-induced structural changes in non-metallic solids, *Radiat. Eff.*, 25, 1, 1975.

[21] Matzke, Hj., Radiation damage in crystalline insulators, oxides and ceramic nuclear fuels, *Radiat. Eff.*, 64, 3, 1982.
[22] Eby, R. K., Ewing, R. C., and Birtcher, R. C., The amorphization of complex silicates by ion-beam irradiation, *J. Mater. Res.*, 7, 3080, 1992.
[23] Hobbs, L. W. et al., Structural freedom, topological disorder, and the irradiation-induced amorphization of ceramic structures, *Nucl. Instrum. Methods*, 116, 18, 1996.
[24] Lam, N. Q., Okamoto, P. R., and Li, M., Disorder-induced amorphization, *J. Nucl. Mater.*, 251, 89, 1997.
[25] Wang, S. X. et al., Ion beam-induced amorphization in MgO–Al_2O_3–SiO_2. II. Empirical model, *J. Non-Cryst. Solids*, 238, 214, 1998.
[26] Williams, J. S., Ion implantation of semiconductors, *Mater. Sci. Eng. A*, 253, 8, 1998.
[27] Trachenko, K. et al., How the nature of the chemical bond governs resistance to amorphization by radiation damage, *Phys. Rev. B*, 71, 184104, 2005.
[28] Phillips, J. C., *Bonds and Bands in Semiconductors*, Academic, New York, 1968.
[29] Kluth, S. M., Fitz Gerald, J. D., and Ridgway, M. C., Ion-irradiation-induced porosity in GaSb, *Appl. Phys. Lett.*, 86, 131920, 2005.
[30] Appleton, B. R. et al., Characterization of damage in ion implanted Ge, *Appl. Phys. Lett.*, 41, 711, 1982.
[31] Wang, L. M. and Birtcher, R. C., Amorphization, morphological instability and crystallization of krypton ion irradiated germanium, *Phil. Mag. A*, 64, 1209, 1991, and references therein.
[32] Destefanis, G. L. and Gailliard, J. P., Very efficient void formation in ion implanted InSb, *Appl. Phys. Lett.*, 36, 40, 1980.
[33] Callec, R. et al., Anomalous behavior of ion-implanted GaSb, *Appl. Phys. Lett.*, 59, 1872, 1991.
[34] Kucheyev, S. O. et al., Ion-beam-induced porosity of GaN, *Appl. Phys. Lett.*, 77, 1455, 2000.
[35] Kucheyev, S. O. et al., Ion-beam-induced dissociation and bubble formation in GaN, *Appl. Phys. Lett.*, 77, 3577, 2000.
[36] Chaki, T. K. and Li, J. C. M., Radiation damage in an amorphous Lennard-Jones solids. A computer simulation, *Phil. Mag. B*, 51, 557, 1985.
[37] Fleischer, R. L., Price, P. B., and Walker, R. M., *Nuclear Tracks in Solids*, University of California Press, Berkeley, 1975.
[38] Fischer, B. E. and Spohr, R., Production and use of nuclear tracks: Imprinting structure on solids, *Rev. Mod. Phys.*, 55, 907, 1983.
[39] Dammak, H. et al., Tracks in metals by MeV fullerenes, *Phys. Rev. Lett.*, 74, 1135, 1995.
[40] Kucheyev, S. O. et al., Lattice damage produced in GaN by swift heavy ions, *J. Appl. Phys.*, 95, 5360, 2004; and references therein.
[41] Toulemonde, M., Dufour, C., and Paumier, E., Transient thermal process after a high-energy heavy-ion irradiation of amorphous metals and semiconductors, *Phys. Rev. B*, 46, 14362, 1992.
[42] Fleischer, R. L. et al., Criterion for registration in dielectric track detectors, *Phys. Rev.*, 156, 353, 1967.
[43] Stampfli, P. and Bennemann, K. H., Theory for the instability of the diamond structure of Si, Ge, and C induced by a dense electron-hole plasma, *Phys. Rev. B*, 42, 7163, 1990.
[44] Schiwietz, G. et al., Energy dissipation of fast heavy ions in matter, *Nucl. Instrum. Methods B*, 175-177, 1, 2001; and references therein.
[45] Miotello, A. and Kelly, R., Revisiting the thermal-spike concept in ion-surface interactions, *Nucl. Instrum. Methods B*, 122, 458, 1997.
[46] Schenkel, T. et al., Interaction of slow, very highly charged ions with surfaces, *Prog. Surf. Sci.*, 61, 23, 1999.
[47] Hedler, A., Klaumunzer, S., and Wesch, W., Boundary effects on the plastic flow of amorphous layers during high-energy heavy-ion irradiation, *Phys. Rev. B*, 72, 054108, 2005; and references therein.

[48] van Dillen, T. et al., Ion beam-induced anisotropic plastic deformation of silicon microstructures, *Appl. Phys. Lett.*, 84, 3591, 2004; and references therein.

[49] Hayes, W. and Stoneham, A. M., *Defects and Defect Processes in Nonmetallic Solids*, Dover, New York, 2004; Itoh, N. and Stoneham, M. A., *Materials Modification by Electronic Excitation*, Cambridge University Press, Cambridge, 2001.

[50] Pearton, S. J., Ion implantation for isolation of III–V semiconductors, *Mater. Sci. Rep.*, 4, 313, 1990.

[51] Kucheyev, S. O. et al., Implant isolation of ZnO, *J. Appl. Phys.*, 93, 2972, 2003.

[52] Pearton, S. J. and Norton, D. R., Dry etching of electronic oxides, polymers, and semiconductors, *Plasma Processes Polym.*, 2, 16, 2005.

[53] Charbonneau, S. et al., Photonic integrated circuits fabricated using ion implantation, *IEEE J. Selected Topics Quantum Electron.*, 4, 772, 1998.

[54] Williams, J. S. et al., Ion irradiation-induced disordering of semiconductors: Defect structures and applications, *Phil. Mag.*, 85, 677, 2005.

[55] Kucheyev, S. O. et al., Blistering of H-implanted GaN, *J. Appl. Phys.*, 91, 3928, 2002.

[56] Bruel, M., Application of hydrogen ion beams to Silicon On Insulator material technology, *Nucl. Instrum. Methods B*, 108, 313, 1996.

[57] Tong, Q.-Y. et al., Layer splitting process in hydrogen-implanted Si, Ge, SiC, and diamond substrates, *Appl. Phys. Lett.*, 70, 1390, 1997; Tong, Q.-Y. et al., A "smarter-cut" approach to low temperature silicon layer transfer, *Appl. Phys. Lett.*, 72, 49, 1998.

[58] Weldon, M. K. et al., On the mechanism of the hydrogen-induced exfoliation of silicon, *J. Vac. Sci. Technol. B*, 15, 1065, 1997.

[59] McCracken, G. M., The behaviour of surfaces under ion bombardment, *Rep. Prog. Phys.*, 38, 241, 1975; Scherzer, B. M. U., Development of surface topography due to gas ion implantation, in *Sputtering by Particle Bombardment II*, Behrisch, R., Ed., Springer-Verlag, Berlin, 1983, p. 271.

[60] Roorda, S., Bernas, H., and Meldrum, A., Eds., *Nanostructuring Materials with Energetic Beams*, Materials Research Society Proceedings, Warrendale, 777, 2003.

[61] Lee, E. H., Ion beam modification of polyimides, in *Polyimides: Fundamentals and Applications*, Ghosh, M. K. and Mittal, K. L., Eds., Marcel Dekker, NY, 1996, p. 471.

[62] Davenas, J. and Boiteux, G., Ion beam modified polyimide, *Adv. Mater.*, 2, 521, 1990.

[63] Kucheyev, S. O. et al., Deformation behavior of ion-irradiated polyimide, *Appl. Phys. Lett.*, 85, 733, 2004.

[64] Baglin, J. E. E., Ion beam enhancement of metal–indulator adhesion, *Nucl. Instrum. Methods B*, 65, 119, 1992.

[65] Townsend, P. D., Optical effects of ion implantation, *Rep. Prog. Phys.*, 50, 501, 1987.

[66] Bernhardt, A. F. et al., Arrays of field emission cathode structures with sub-300 nm gates, *J. Vac. Sci. Technol. B*, 18, 1212, 2000.

[67] Johnsen, J. N. et al., In-Situ Thermal Anneal of Proton and Xe^{++} Irradiated YSZ, unpublished data, 2005.

4

Spinodal Decomposition

Subhash H. Risbud
University of California at Davis

Abstract

A brief overview of phase transformations by spinodal decomposition inside a miscibility gap is presented. Essential basic concepts of unmixing of solutions are summarized followed by a concise thermodynamic description of the connection between free-energy vs. composition curves and the location of the spinodal region on a binary phase diagram. The kinetics of spinodally decomposed structures and their evolution with time is outlined, and finally properties and applications of nanostructures arising from spinodal mechanisms are listed.

4.1 Introduction

The transformation of a homogeneous solution of chemical species into unmixed states gives rise to nano- or microstructures with two distinct phases (for a binary two-component system). These phases maybe stable or metastable and the boundaries of existence of the phase-separated material are commonly shown on a phase diagram (temperature-composition sketch at one atmosphere pressure) as regions of solid–solid, liquid–liquid immiscibility. If the melt is supercooled without crystallization and is then transformed into glassy structures, a metastable immiscibility region is superimposed on the stable phase equilibrium curves. Immiscibility domes on phase diagrams concisely capture both the thermodynamic and kinetic aspects of the unmixed states, the latter becoming important when the phase(s) predicted by stable phase equilibria cannot form due to kinetic reasons. These can be rapid cooling rates, high pressure conditions, or intrinsically high viscosities of the melt that hinder matter transport. Figure 4.1 is a schematic drawing of the free energy–temperature relationships between stable and metastable phase assemblages and demonstrates the progressively decreasing free energy as the stable liquid transforms from above the liquidus temperature to solid, liquid, and immiscible mixtures. The free-energy curves

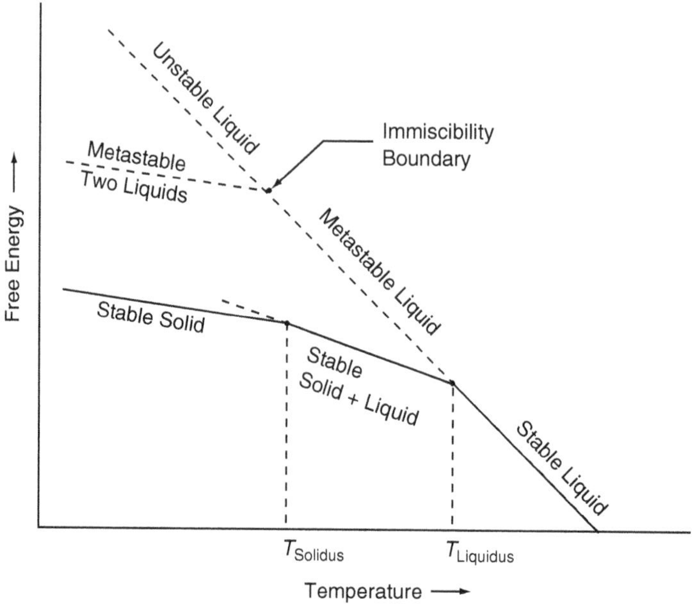

FIGURE 4.1 Schematic of the free energy–temperature relationships between stable and metastable phases.

can also be used to illustrate (Figure 4.2) the relative positions of free energy for stable and metastable states at a temperature T below the eutectic point where the diagram shows an $\alpha + \beta$ mixture being stable. When processing conditions give rise to phase separation into metastable states, the relationships between free-energy composition and the resulting metastable phase diagram can be shown as in Figure 4.3. The dotted lines in Figure 4.3 are classic representations of metastable liquid–liquid immiscibility at sub-liquidus temperatures and show two distinct regions: the outer dome-shaped curve defining the overall temperature-composition space in which immiscibility occurs and the inner dome marked "spinodal." The spinodal is a region of the immiscibility dome (also called the miscibility gap) where phase separation does not occur by the classical nucleation-growth mechanisms but is spontaneously driven by thermodynamics. The essential concepts of immiscibility that arise owing to the mechanism of spinodal decomposition are summarized in this chapter.

4.2 Thermodynamics of Spinodal Transformations

Let us consider the temperature-composition (top) and free-energy-composition (bottom) sketches shown in Figure 4.3 for a hypothetical two-component phase diagram. A homogeneous molten liquid of composition marked X_s is cooled fast enough to avoid solid phase crystallization, and instantaneously (rapidly) reaches a temperature T_1 that is inside the inner dome of the immiscibility region (marked by points S_1 and S_2 in the bottom sketch of Figure 4.3). Let us examine the thermodynamics of this situation with respect to phase stability in the presence of small composition fluctuations around the original (homogeneous) composition X_s. Figure 4.4 shows a magnified view of the free-energy vs. composition curve at the temperature T_1 and the composition boundaries of the inner spinodal dome (S_1 and S_2). On this sketch the free energy of the homogenous solution X_s is given by the quantity G_h marked at the top of the curve. As small composition fluctuations always exist in a material, we can easily visualize the free energy of a slightly phase-separated material with compositions slightly more than X_s (shown as X_+) and slightly less than X_s (shown as X_-); this value is G_u which is lower than G_h. A similar reasoning can be applied to any small composition fluctuation within the spinodal region to show that starting with a homogeneous solution inside the spinodal dome, a spontaneous decrease in free energy results upon

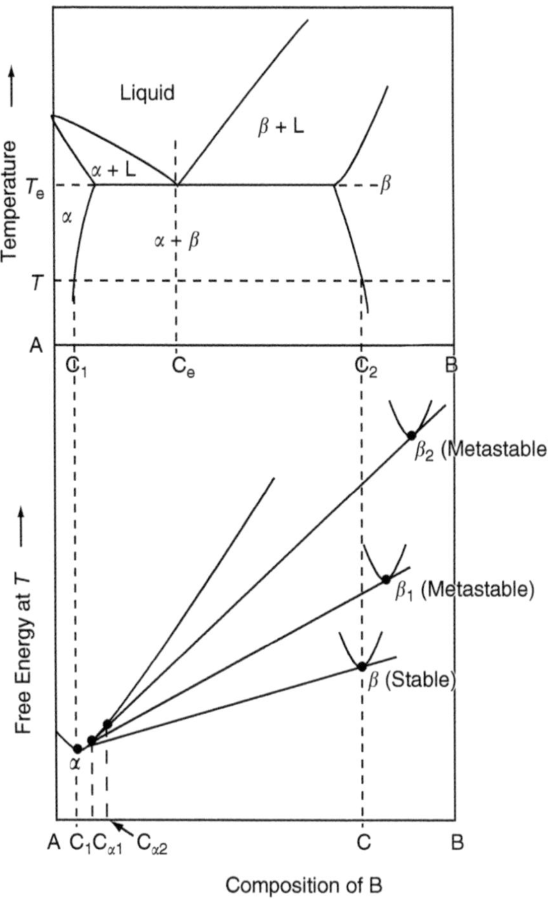

FIGURE 4.2 Free energy curves to illustrate the relative positions of free energy for stable and metastable states at a temperature T below the eutectic point.

unmixing, even if the unmixed compositions are only slightly apart. Thus all compositions inside the spinodal dome are inherently thermodynamically unstable as homogeneous solutions and will spontaneously phase separate without the need for nucleation. On the other hand, note that when the starting composition is outside the spinodal dome but still inside the immiscibility boundary (e.g., the composition marked X_N in Figure 4.4), small composition fluctuations actually increase the free energy so that there is no spontaneous thermodynamic driving force for the homogeneous solution to break up into two phases. This is the case of nucleation and growth-based phase separation that involves nucleation.

Table 4.1 encapsulates the key differences between spinodal decomposition and nucleation growth as mechanisms by which a homogeneous solution transforms into an immiscible solution. Spinodal decomposition is characterized by continuous composition changes that start out as tiny composition fluctuations that amplify until the two-phase metastable equilibrium is achieved at the composition points L_1 and L_2 at temperature T_1 (Figure 4.4). Once again the argument for spinodal decomposition is thermodynamic spontaneity without the need for nucleation. Also, because the composition changes at the beginning of spinodal decomposition are quite small, the interfaces between the two phases are diffuse but sharpen gradually as the phase-separation proceeds.

Morphologically, it is commonly observed that the second-phase structure that evolves by spinodal decomposition is not spherical (unlike the almost spherical nuclei in nucleation growth) but has worm- or tweed-like interconnectivity inherited from the composition fluctuations that are small in degree but large in spatial extent. It is important to note, however, that morphology alone is not sufficient proof of a

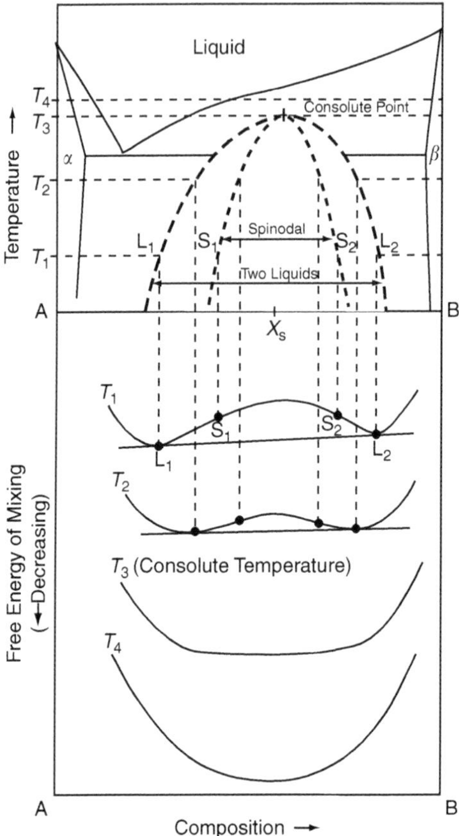

FIGURE 4.3 Metastable phase diagram (dotted lines) and corresponding free-energy-composition curves at different temperatures.

spinodal decomposition mechanism. Indeed, interconnected structures have been shown to occur by the coalescence of nuclei that may give the appearance of a spinodal structure. Confirming and documenting the continuous composition variations during the evolution of the spinodal decomposition structures represents the most unambiguous proof of this mechanism. It is useful to realize, from an applications point of view, that "quenching in" the structure is possible at any point in the evolution of the composition segregation from X_s to L_1 and L_2. Indeed, if the quenching rate is fast enough, one could completely avoid spinodal decomposition and obtain a homogeneous supercooled material — a point that stresses the importance of considering the kinetics of this process.

4.3 Kinetics of Spinodal Decomposition

The kinetics of spinodal transformations is best described by considering the amplification of very small composition fluctuations (around the composition X_s in Figure 4.4 for example). It is common to think of these initially small composition waves as fluctuations that are small in degree but large in spatial extent (i.e., they are spread all over the material in a random fashion). Based on the classic Hillert–Cahn treatment the free-energy change is calculated between the original homogeneous solution and one that is just slightly apart with a small concentration gradient, as shown in Figure 4.5. This initial stage composition fluctuation can be mathematically expressed by the equation:

$$X - X_s = A\cos(\beta X) \tag{4.1}$$

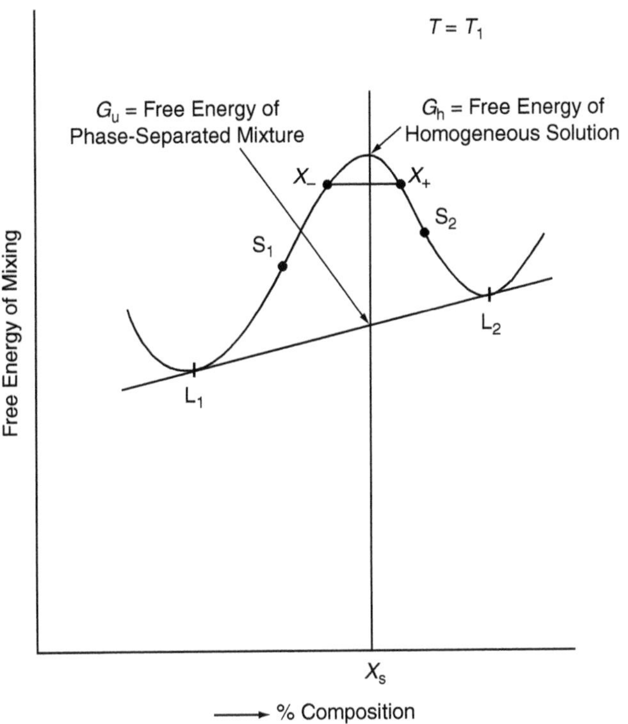

FIGURE 4.4 Comparison of free energies of homogeneous solid solution and phase-separated mixture at and around the composition X_s (see Figure 4.3).

TABLE 4.1 Comparison of the Essential Features of Spinodal Decomposition with Nucleation and Growth Mechanism of Phase Separation within a Miscibility Gap

Classical nucleation	Spinodal decomposition
The Composition of the second phase remains unaltered in time–equilibrium composition.	A continuous change of the composition occurs until the equilibrium values are attained.
The interface between the phases is always sharp during the growth.	The interface is initially very diffuse but sharpens gradually.
A tendency towards a random distribution of the particle position in the matrix.	Regularity of the second-phase distribution both in size and position, characterized by geometric spacing.
A tendency of the second phase to separate into spherical particles with little connectivity.	A tendency of the second phase to separate as a nonspherical structure with high connectivity.

where A is the amplitude of the fluctuating wave, and the wave number $\beta = (2\pi/\lambda)$ for a wave characterized by the wavelength λ.

By substituting the above simple equation for the initial composition fluctuations in the free energy equation it is easy to obtain an expression for the change in free energy for the system with fluctuations. From this one can identify a critical wavelength, λ_c, given by the expression:

$$\lambda_c = \left\{ \frac{(-8\pi^2\kappa)}{(d^2G/dX^2)} \right\}^{1/2} \tag{4.2}$$

where κ is a positive materials constant called the gradient-energy coefficient.

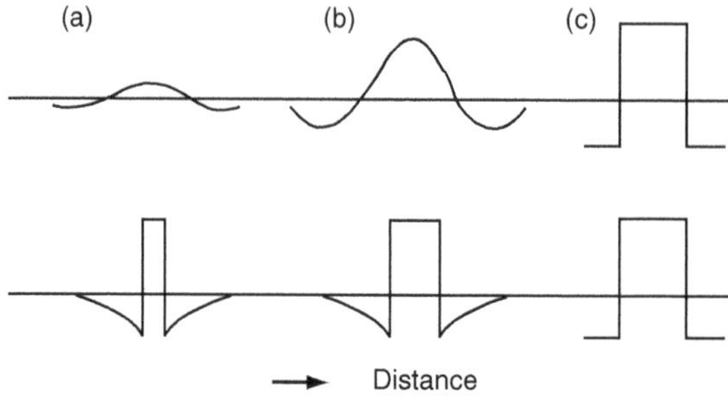

FIGURE 4.5 Schematic composition fluctuations in the (a) early stages, (b) intermediate stages, and (c) final stages of phase separation by the spinodal decomposition mechanism (top sketches) and nucleation-growth mechanism (bottom sketches).

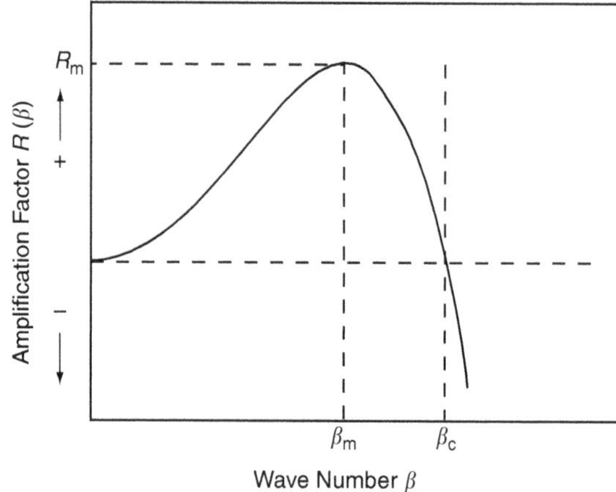

FIGURE 4.6 Amplification factor $R(\beta)$ plotted as a function of the wave number β for a set of sinusoidal waves that are assumed to initiate the spinodal decomposition process. Waves with the β corresponding to the maximum $R(\beta)$ are predicted to grow most rapidly.

Thus, the homogeneous solution is unstable at all waves longer than this critical wavelength and the waves will amplify to trigger the spinodal decomposition process. Using the approximation of sinusoidal waves, Cahn [1] has developed an amplification factor $R(\beta)$ that can be plotted as a function of wave number or wavelength to predict the wavelength that grows most rapidly with time (Figure 4.6).

4.4 Properties and Applications

Interconnectivity is a unique feature of nanostructures that arise from spinodal decomposition, and the mechanical, chemical, and electrical properties that result from this feature have been exploited for many applications. To cite one of the best examples, a family of commercial products classified as VYCOR are based on a spinodally decomposed structure processed within a miscibility gap in sodium borosilicate glass compositions, a pioneering invention at Corning Glass Works 35 to 40 years ago. The compositions for VYCOR glasses are based on sodium borosilicate melts that first undergo spinodal decomposition into two continuous but highly interconnected phases. The predominant phase is silica-rich whereas the sodium and boron oxides are contained in the other, sodium borate-rich, phase. As the sodium borate phase is soluble in certain acid solutions, it is dissolved or leached out. This leaves a residue of a porous skeleton that is approximately 96% silica. The porous "thirsty" glass is directly usable as a molecular sieve, or a template in pharmaceutical applications or as a catalyst support. VYCOR products however are

TABLE 4.2 Properties and Applications of a Typical VYCOR Created by a Process that Involves Spinodal Decomposition

VYCOR (corning 7) properties
Approx. Specific Gravity (dry): 1.5 g/cm^3
Void Space: 28% of vol.
Internal Surface Area: 250 m^2/g
Avg. Pore Diam (std): 4 nm
Appearance: opalescent
Loss Tangent at 250°C, 100 Hz: 0.007
Dielectric Constant at 250°C, 100 Hz: 3.1
Electrical Impedance: 500 Ω
Examples of applications
• Electrode junctions
• Diffusion barrier
• Electroanalytica
• Battery studies

TABLE 4.3 Applications of Toughmet Spinodal alloys made by Brush Wellman Company

Mobile equipment	**Drilling & mining equipment**
• Linkage bushings	• Centralizers
• Thrust washers	• Rifles nuts
• Wheel bearings	• Piston rings
Pumps	• Guide bushings
• Cylinder barrels	• Spindle bearings
• Wear plates	**Aircraft**
• Slippers	• Landing gear bearings
• Shaft bushings	• Guide plates
• Seals	• Shaft bearings
• Stators	• Spherical bearings
• Rotors	**Metal working**
• Fasteners	• Wear plates
Trucks, automobiles, and motorcycles	• Forming rolls/punches
• Wrist pin bushings	• Die inserts
• Thrust washers	• Guide/Slides
• Valve guides	• Postbushings
• Roller bearing cages	
• Cam roller pins	

made by sintering the porous skeleton by heat treatment to a dense silica-rich glass. Table 4.2 lists some applications for VYCOR and related products.

Many other uses exist for spinodally decomposed nanoporous glasses and alloys including emerging applications based on glass-polymer composites (e.g., see Reference 2). (For more details on spinodal decomposition to form amorphous/nanocrystalline structures in oxide glasses see Chapter 6.)

Turning to examples of applications in metallic alloy systems, one finds nanoscale structures created by spinodal decomposition in Cu–Ni–Sn alloys. These spinodal alloys exhibit high performance characteristics including high strength, resistance to elevated temperatures, and very good tribological properties. These desirable properties are attained while retaining the customary qualities of copper alloys such as high thermal conductivity and resistance to general corrosion and hydrogen embrittlement. Nicomet bronze spinodal alloys, for instance, made by Anchor Bronze and Metals, Inc., find use in wear-resistance applications as they wear at less than half the rate of beryllium copper and aluminum bronze alloys. Unlike in precipitation hardened copper alloys, Nicomet is a single-phase alloy that is strong due to the spinodal structure, an ordered arrangement of nickel and tin atoms in wave-like configurations. The tin and nickel content of Nicomet is an attribute in marine applications as well as the corrosive environments encountered in oil and gas well drilling.

Table 4.3 summarizes a range of applications of the spinodal alloys "Toughmet" made by Brush Wellman Company. These alloys based on Cu–Ni–Sn compositions show a diverse set of technological applications.

References

[1] R. W. Cahn, Spinodal decomposition, *Trans. Metall. Soc. AIME*, 242, 166, 1968.

[2] Yu. S. Lipatov, Hybrid binders in polymer composites. Peculiarities of structure and properties conditioned by microphase separation. *Pure Appl. Chem.*, 57, 1691, 1985.

5

Ostwald Ripening in Materials Processing

K. G. Wang
Florida Institute of Technology

M. E. Glicksman
University of Florida

Abstract

Ostwald ripening is a ubiquitous phenomena occurring in the fields of chemistry, physics, materials science, geology, and economics. In this chapter we describe Ostwald ripening and its applications in metallic, ceramic, polymer, semiconductor, and nanostructured systems. Ostwald ripening is discussed in various processes such as aging, epitaxial growth, sintering, and nanostructure fabrication. The advance over the past four decades in theoretical and computational studies of Ostwald ripening is described, starting with the seminal studies by Lifshitz, Slyozov, and Wagner, extensions of mean-field theory, and, finally, recent approaches using stochastic theory. Two classes of computer simulation, viz., multiparticle diffusion, and phase-field simulation, have been used extensively to simulate the phenomenon of Ostwald ripening. Finally, statistical predictions, including particle-size distributions and the maximum expected radii in ripened populations of particles are predicted from theory and simulations. These approaches are shown to agree well with experimental results obtained from measurements performed on δ' (Al_3Li) precipitates in binary Al–Li alloys.

5.1 Introduction: Ostwald Ripening

Ostwald ripening[1] refers to the coarsening of phase particles in solid, liquid, or gaseous media. Phase coarsening — that is, the growth of the average size particle — occurs at the expense of small particles within a system, which shrink and finally disappear. Over a century ago, in 1900, Wilhelm Ostwald reported that the solubility of small HgO particles depends on their radii.[1] The eponymous phenomena of size-dependent solubility and the resultant phase coarsening have collectively borne his name. Fundamentally, the explanation for Ostwald ripening is straightforward: the free energies of smaller particles with their larger geometric curvatures are higher than those of larger particles with their smaller curvatures. Stated otherwise, smaller particles express a larger surface-to-volume ratio than do larger particles, so they have a larger amount of surface energy per unit volume. Expressed as the chemical potential, the free energy per unit mass is larger for small particles than it is for large ones.

Ostwald ripening, in fact, also appears in many two-phase materials as a kinetic relaxation phenomenon, occurring at elevated temperatures during late-stage microstructural evolution. It is driven by a decrease, or dissipation, in the total stored interfacial energy, and consequently, the average particle tends to grow during Ostwald ripening by absorbing solute atoms from the surrounding matrix phase. These solute atoms are released into the matrix at the expense of smaller particles in the system that tend to dissolve. Over time, a "competitive diffusion" process develops, resulting in an increase in the average size of the particle population, with a concomitant decrease in their number density.

In a typical phase separation process, nucleation, growth, and, eventually, phase coarsening, all usually follow a temperature quench from an initially homogeneous phase that results in a two-phase microstructure. Specifically, a typical precipitation sequence occurring within a supersaturated solution, or matrix, initially involves spinodal decomposition of the supersaturated matrix, or, more commonly, nucleation of a second phase, followed by its growth. In the late stage of phase separation, the onset of competitive coarsening occurs among the precipitate particles, that is, Ostwald ripening sets in. The nucleation and growth processes consume most of the available free energy, leaving only about one part in 10^3 of the initial free energy to reside among the interfaces created between the second phase and the matrix. It is this tiny residual of the initial driving free energy that is gradually consumed in Ostwald ripening. Figure 5.1 shows, in a schematic manner, Ostwald ripening. Indeed, in addition to the geometric changes occurring within such a ripening system, the physical and mechanical properties of the resulting two-phase materials, such as hardness and toughness, will often depend sensitively on the material's average particle size and particle-size distribution (PSD) function. Ostwald ripening is such a widespread phenomenon because it arises during many different types of materials processes, from the solidification and aging of alloys and rocks, to the interaction of droplets in clouds and crystallites in snow.[2]

5.2 Phenomenology of Ostwald Ripening

Ostwald ripening occurs in a truly vast array of diverse phenomena. More specifically, Ostwald ripening processes appear widely in the fields of physics, chemistry, geology, earth science, chemical engineering, and materials science. We now describe engineering applications of Ostwald ripening to metallic, ceramic, polymer, semiconductor, and nanostructured materials.

5.2.1 Ostwald Ripening in Processing Metals

Binary or multicomponent alloys freezing from a single-phase liquid state are quenched to room temperature to produce a supersaturated crystalline matrix. If such a supersaturated solid solution alloy is subsequently annealed at a sufficiently high temperature, precipitation hardening or aging precipitates

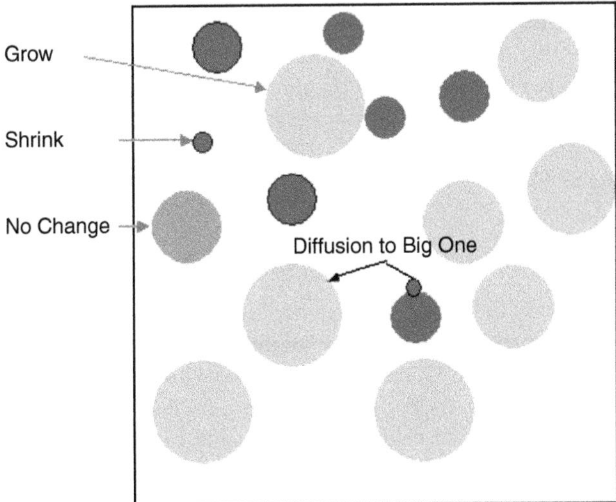

FIGURE 5.1 Coarsening of particles in an active two-phase medium via matter transport from smaller to the large particles. Bigger particles tend to grow, smaller particles tend to shrink, and some "average" sized particle might not be changing size. As the average size for the population continues to increase by Ostwald ripening, eventually every particle tends to fall below this average and shrink. Thus, competitive diffusion, at least in principle, will reduce the structure in an infinite time to a single particle representing the thermodynamic end-state of complete phase separation.

nucleate and grow by diffusion of solute from the matrix toward the population of growing precipitates. After a sufficiently long annealing time the initially large supersaturation of solute decreases enough for interfacial curvatures to play an important role: Ostwald ripening starts. Among the earliest quantitative studies are those by Ardell and Nicholson,[3] who studied Ostwald ripening of Ni–Al alloys in 1966. Baldan[4] recently reviewed the applications of Ostwald ripening concepts to nickel-base superalloys.

5.2.2 Ostwald Ripening in Sintering Ceramics

Liquid-phase sintering (LPS)[5] is usually considered to occur in three stages, (1) rearrangement; (2) solution-precipitation; and (3) Ostwald ripening. In the third stage of sintering, further densification becomes slow because of the enlarged diffusion distances encountered in the coarsened structure. Consequently, this stage takes most of the time required for completion of LPS. Microstructural coarsening by Ostwald ripening thus dominates the final stages of LPS. For the resulting particulate solid produced through LPS, three common types of grain structures are observed: (1) Where the volume fraction of liquid present during LPS is moderate ($\sim>5\%$), grains with rounded shapes are observed. (2) For higher liquid content, the grain shapes after LPS become spheroidal. For low volume fraction of liquid ($\sim<2$–5%), grains undergo considerable changes in shape. They develop a polyhedral network morphology in which the contacts between neighboring grains are nearly flat. (3) When the liquid content remains high during LPS,[6] grains may adopt a prismatic shape (instead of the spheroidal shape), and they may also develop considerable solid–solid contacts between neighboring crystallites. Al_2O_3 grains, for example, develop in a complex manner during LPS in the presence of a small amount of melt. A variety of grain types have been observed ranging from equiaxed to elongated with curved sides, to platelet and platelike grains with flat, faceted sides.[7]

5.2.3 Ostwald Ripening in Processing Polymers

Far fewer quantitative applications of Ostwald ripening to polymer–polymer melts have been carried out than for other phase-separating systems, such as metallic alloys. Crist and Nesarikar[8] performed experimental studies of phase coarsening in artificially blended binary mixtures of polyethylene and hydrogenated polybutadiene. These investigators found that Ostwald ripening is the dominant kinetic process when both solubility and diffusivity of the minority component in the matrix phase are large. Haas and Torkelson[9] studied two-dimensional coarsening and phase separation in thin polymer solution films, and recently, Cheng and Nauman[10] performed experimental and computational studies of Ostwald ripening in binary polymer blends. The latter authors also obtained a self-similar (affine) PSD.

5.2.4 Ostwald Ripening in Processing Semiconductors

The homoepitaxial system Si/Si(100) is of particular interest here due to its strong technological relevance (for more details on epitaxy, see Chapter 9). For example, Bartelt et al.[11] studied Ostwald ripening occurring among two-dimensional Si islands deposited homoepitaxially on Si(001) substrate using low-energy electron microscopy (LEEM). By studying the behavior of individual Si islands compared to their surroundings, these investigators quantified the step-edge attachment and terrace-diffusion processes that are kinetically responsible for the ripening process. By comparing the time dependence of specific configurations of islands from their LEEM studies to computer simulations, Bartelt et al. discovered correlations in the rate of change of an island's area with the sizes of its neighboring islands. The clear implications of this study point to the fact that the chemical potential of the surrounding adatom "sea" is nonuniform, whereas classical theories of Ostwald ripening assume uniformity of all locales as a "mean field." Additional works have been published on islands of germanium or compound semiconductors such as GaAs residing on a semiconductor surface. Zinke-Allmang[12] published a comprehensive review on this semiconductor application.

5.2.5 Ostwald Ripening in Processing Nanostructured Materials

Recently, Yang and Zeng[13] reported a so-called "one pot" processing method to prepare hollow anatase TiO_2 nanospheres via Ostwald ripening under hydrothermal conditions. When the reaction time is kept short, the TiO_2 spheres develop solid cores. Solid TiO_2 spheres are in turn comprised of numerous smaller crystallites. Compared to the size of those crystallites forming the outer surface, the crystallites located in the inner cores are even smaller. Thus, those titania crystallites located within the cores are more easily dissolved than the surface crystallites. A hollowing-out effect is observed with increase of reaction time. Larger crystallites grow at the expense of the smaller ones. Thus, Ostwald ripening provides the underlying selective mass-transport mechanism operative in this process. The diameters of these ceramic nanocrystallites fall in the range of 30 to 50 nm. Liu and Zeng[14] also extended their method to the fabrication of homogeneous core-shell semiconductors.

5.3 Experimental Studies of Ostwald Ripening

There are several advanced experimental methods developed to study and quantify the phenomenon of Ostwald ripening and late-stage phase coarsening. The most popular experimental method used is transmission electron microscopy (TEM). There are also a considerable number of experimental studies of Ostwald ripening that rely on small-angle x-ray scattering (SAXS) and on scanning electron microscopy (SEM). Only a few studies have been carried out, however, using small-angle neutron scattering (SANS), three-dimensional atom-probe microscope (3DAP), and local electrode atom-probe (LEAP) microscope. Some experimental work used a combination of different observational methods. In the following sections, we briefly describe the main experimental methods that have proven useful in the study of Ostwald ripening and phase coarsening.

Recently, Baldan[4] also reviewed the status of experimental studies to quantify phase coarsening in nickel-base superalloys, especially those based on the binary system Ni–Al. Indeed, because of the considerable engineering significance of these materials in gas turbines and jet engines, a great deal of experimental data currently exist on a variety of interesting high-temperature alloy systems. In Ni–Al superalloys, specifically, there exist elastic interactions between the γ' precipitate particles and the γ-phase matrix.

5.3.1 Transmission Electron Microscopy

To the current authors' knowledge two-phase binary metallic alloys based on Al–Li provide a nearly ideal binary alloy system for kinetic ripening studies, because the δ' particles in this alloy system exhibit an extremely small lattice misfit and interfacial strain with the solid solution matrix phase. The low interfacial strain contributes a negligible amount of strain-induced free energy to the coarsening process. The near lack of any strain energy allows the particles to configure themselves as near perfect spheres. Gu et al.[15] first studied phase coarsening kinetics in binary Al–Li alloys, employing quantitative TEM. Then Mahalingam et al.[16] studied coarsening of δ'–Al_3Li precipitates in binary Al–Li alloys using TEM to obtain additional microscopic details. The measurements performed by Mahalingam et al. allowed accurate estimates of the steady-state PSDs and phase coarsening rates for Al–Li microstructures having different precipitate volume fractions.

Much earlier, in 1966, Ardell and Nicholson[3] used TEM in their important study of the coarsening of γ' precipitates in Ni–Al alloys. These investigators experimentally determined the PSD and the kinetic coarsening rate constant that allows estimation of the average particle size with aging time. In addition, there are many experimental groups that have studied different binary systems using TEM. More recently, studies have been reported on Ostwald ripening in multicomponent systems. For example, H. A. Calderon et al.[17] reported on Ostwald ripening in Fe–Ni–Al–Mo ferritic alloys.

5.3.2 Scanning Electron Microscopy

Redmond et al.[18] used SEM in their study of spontaneous morphological reformation of evaporated silver nanoparticles on indium-tin oxide deposited on flat graphite surfaces in pure water. Over a period of several hours at 23°C, large faceted 100 to 300 nm crystallites grew as small 20 nm particles dissolved. The kinetic mechanism operative in this process is electrochemical Ostwald ripening — a mechanism driven by the slight particle-size dependence of the Ag standard electrode potential as facilitated by the high electrochemical exchange rate of polycrystalline silver electrodes.

The technique of ion-beam synthesis has been successfully used to fabricate a variety of compound layers on Si, such as insulators, metals, and barrier silicides (see Chapter 10). During the dopant implantation step a Gaussian-like concentration distribution is built up in the near-surface region of the substrate, which upon annealing, transforms into a flat-topped profile. The redistribution of the implanted species depends on the Ostwald ripening of the subsequent precipitation process. Weber and Skorupa[19] studied the Ostwald ripening in such an implanted semiconductor system by taking a series of SEM-micrographs. These investigators experimentally determined the spatial correlations among precipitates, such as the radial distribution function, and next-neighbor pair correlations.

5.3.3 Three-Dimensional Atom-Probe and Local Electrode Atom-Probe Microscopes

TEM and SEM produce two-dimensional planar images. Recently, however, the study of three-dimensional microstructures has become important. 3DAP and LEAP microscopes are now in use in the detailed study of three-dimensional microstructures.

Sudbrack[20] studied phase coarsening in a Ni-5.2Al-14.2Cr at.% alloy aged at 600°C, for different times using 3DAP microscopy. She obtained in the quasi-steady-state coarsening regime the temporal kinetic exponents of the Al and Cr supersaturations in matrix, respectively. She established the composition "trajectory" within the γ-matrix. By using the experimentally determined coarsening rate constants for the average precipitate radius, and the mean matrix supersaturation, Sudbrack determined the interfacial free energy and the effective diffusion coefficients of Al and Cr independently.

5.3.4 Small-Angle X-Ray Scattering and Small-Angle Neutron Scattering

SAXS and SANS diffraction techniques are best considered in Fourier reciprocal space, rather than in real space as is common in electron microscopy. The intensity of scattered x-rays or neutrons is directly proportional to the structure function, which in turn is defined as the Fourier transform of the pair correlation function. Therefore, SAXS and SANS directly measure the effects of spatial correlation among second-phase particles, and, as such, are useful analysis techniques for testing theories and simulations of Ostwald ripening in which significant microstructural correlations occur.

SAXS, in particular, is a powerful experimental tool to study the kinetics of phase separation in optically opaque materials such as the metals. Several distinct processes may be distinguished during phase separation: (1) nucleation and growth, (2) spinodal decomposition, and (3) phase coarsening or Ostwald ripening. Structural parameters, including the dimensions of second-phase particles, the mean-squared electron density fluctuation, the volume fraction or number density of the particles, and the interfacial area per unit volume can all be determined through means of SAXS experiments. Evaluation of these parameters allows each of the different physical processes mentioned above to be distinguished qualitatively by study of the development of the overall phase-separation process as a function of time.

Möller et al.[21] presented the procedure to calculate a set of valuable size parameters, the average intersection length, the correlation length, the correlation surface, the correlation volume, the mean electron distance, and the electronic radius of gyration from the scattering intensity function, $I(s)$, where s is the wavenumber. Walter et al.,[22] Damaschun et al.,[23] and Sjöberg [24] showed that these characteristic length scales are associated with certain ratios of various moments of the PSD without specification of the detailed form of the PSD. Möller et al.[21] specified the size distribution given by Lifshitz and Slyozov,[25] and by Wagner[26] as well as the Gaussian size distribution for diffusion-limited or reaction-limited ripening. These investigators calculated the higher moments of different size distribution functions and compared them with the SAXS data. This approach has been applied to a specimen of photochromic glass containing a population of fine silver-halide particles undergoing Ostwald ripening. The applicability of this method is restricted to homogeneous spherical precipitates with miniscule volume fractions.

Che and Hoyt[27] extended the phase coarsening model of Marder[28] and derived a form for the scaled structure function based on it. Che et al.[29] studied phase coarsening in Al–Li alloys using SAXS experiments. The scaled structure function and the peak breadth as a function of equilibrium volume fraction were measured in several Al–Li alloy samples.

Abis et al.[30] studied Ostwald ripening in Al–Li alloy using SANS. These investigators measured the pair distribution function $g(r)$, estimated the first peaks of $g(r)$ corresponding to the nearest-neighbor distance, or the center-to-center interparticle spacing. Abis et al. also studied in detail the spatial correlations developed during phase coarsening. They obtained the scaling of the structure function, and confirmed that the cube of the average radius increases linearly with time, allowing an estimate to be made of the kinetic coarsening rate constant. Finally, these investigators measured the PSD function for different aging times.

5.4 Theoretical and Computational Studies of Ostwald Ripening

Surprisingly, for almost 50 years following Ostwald's original discovery at the turn of the twentieth century, only qualitative explanations of Ostwald ripening existed. Greenwood[31] and Asimow[32] made early attempts to develop quantitative theories, but neither succeeded because their analyses were based upon an unrealistic solution to the diffusion field in the matrix. The first successful quantitative descriptions of Ostwald ripening was published by Lifshitz and Slyozov,[25] and by Wagner.[26] This theory is often referred to simply as "LSW theory." Subsequent to LSW theory, many efforts arose from physicists, chemists, and materials scientists to modify and extend LSW theory to the practical case of systems with a nonzero volume fraction of precipitates. As the computational power of digital computers increased rapidly over the past two decades, many research groups throughout the world attempted numerical simulations of

Ostwald ripening. In the following sections, we describe the worldwide theoretical and computational efforts expended to understand and interpret Ostwald ripening generally, and make specific predictions in many types of systems.

Baldan[33] has reviewed the development of phase-coarsening theories during the second half of the twentieth century. In just the past few years, however, the sophistication and predictive capability of the theory and numerical simulation of phase coarsening kinetics have increased. This large body of research now helps underpin our ability to describe the kinetic complexities of evolving two-phase microstructures. Recently, Wang et al.[34] briefly reviewed the remarkable advances that have occurred over the past 5 years.

5.4.1 Lifshitz, Slyozov, and Wagner Theory

In the formulation of Lifshitz, Slyozov, and Wagner (LSW) theory,[25,26] three main equations are posed. The first equation is the so-called "continuity equation" in size space that governs the PSD, which is the function $F(R, t)$, usually written as

$$\frac{\partial F(R,t)}{\partial t} + \frac{\partial}{\partial R}\left[\frac{\mathrm{d}R}{\mathrm{d}t}F(R,t)\right] = 0, \tag{5.1}$$

where R and t are the particle radius and coarsening time, respectively. The second equation comprising LSW theory is the mean-field kinetic equation describing the growth rate of particles. For Ostwald ripening controlled by volume diffusion[25] the growth rate of a particle is related to the radius of the particle itself and the system's critical particle. The critical particle is a particle of such a size and curvature that it is provisionally stable with respect to the matrix, and consequently has a zero growth rate. The mean-field LSW kinetic law is

$$\frac{\mathrm{d}R}{\mathrm{d}t} = \frac{1}{R}\left(\frac{1}{R^\star} - \frac{1}{R}\right), \tag{5.2}$$

where $R^\star$ is the critical radius. The time and radii are appropriately made dimensionless.[25] The growth rate of particles with a critical radius is zero. Equation 5.2 is a deterministic rate law that states that the growth of a spherical particle is proportional to the difference in the mean curvatures between the critical particle and the particle of interest. For the interface-controlled mechanism, the growth rate of particle is

$$\frac{\mathrm{d}R}{\mathrm{d}t} = \frac{1}{R^\star} - \frac{1}{R}. \tag{5.3}$$

Finally, the third equation needed in formulating the LSW theory is that of global mass conservation for the coarsening system. LSW theory retains it validity only in the extreme limit of a vanishing (zero) volume fraction. Taken literally, LSW kinetics predict the growth rate and population dynamics of a single particle interacting with an infinite collection of neighbors set at an infinite distance away. The important prediction from LSW theory that the cube of the average length scale of particles increases linearly with time has, however, been shown to remain robust by numerous experiments, even for cases where the assumption of a zero volume fraction of particles is seriously violated. Furthermore, LSW theory predicts that coarsening systems enter a long-time limit of phase separation characterized by a self-similar, or affine set of properties, including a re-normalized PSD, wherein the microstructure changes continuously by just a scale factor. In different words, the metric features of an affine coarsening microstructure change by a mere magnification factor as the annealing time increases. The fact that the average length scale of an LSW system increases as the cube-root of annealing time implies that initially the scale changes very rapidly, but then slows down as the scale of phase separation increases. Such extreme sublinear kinetics is the hallmark of three-dimensional Ostwald ripening.

LSW theory ignores the effect of the volume fraction of the precipitate particles and their interactions by assuming that neighboring particles are so far away from the particle of interest that the particle just senses the "mean field" induced collectively by all its distant neighbors. However, in realistic systems, such as alloys, a finite, nonzero volume fraction of particles, $V_V \neq 0$, is distributed through the microstructure. Such distributions maintain most of the particles in close proximity to each other, so both local and many-body interactions will arise among them. Numerous attempts have been made to improve upon LSW theory by extending its applicability to the more realistic situation of systems with nonzero volume fractions of precipitates. Needless to say, as the departure from the ideal LSW limit of zero volume fraction increases, the system behaves more and more differently from the kinetic predictions based on LSW theory. Simply stated, as the volume fraction of precipitates in a microstructure increases, the kinetic laws, Equation 5.2 and Equation 5.3, become less and less accurate, and the mean-field formulation on which the LSW theory depends becomes questionable.

The theoretical PSD derived from LSW theory for the case of diffusion-limited ripening is

$$G_{\mathrm{LSW}}(\rho) = \begin{cases} \frac{4}{9}\rho^2 \left(\frac{3}{3+\rho}\right)^{7/3} \left(\frac{3}{3-2\rho}\right)^{11/3} \exp\left[\frac{-\rho}{(1.5-\rho)}\right], & \left(0 \leq \rho < \frac{3}{2}\right), \\ 0 & \left(\rho \geq \frac{3}{2}\right), \end{cases} \tag{5.4}$$

where ρ is the particle's re-normalized radius, $R/R^{\star}$. Note that the variable ρ is a *time-independent* ratio of the time-dependent particle size, R, divided by the time-dependent critical size, $R^{\star}$. In essence, the variable ρ compares the instantaneous particle size to the entire population's increasing (ripening) size scale. Thus ρ — which is both a local variable and a global variable — causes the PSDs expressed in terms of it to have special affine (scale-independent) features. The corresponding theoretical PSD for interface-controlled ripening, where some interfacial reaction, rather than volume diffusion through the matrix, becomes rate limiting, is

$$G_{\mathrm{ICR}}(\rho) = \begin{cases} \left[\frac{24\rho}{(2-\rho)^5}\right] \exp\left[\frac{-3\rho}{(2-\rho)}\right], & (\rho < 2), \\ 0 & (\rho \geq 2), \end{cases} \tag{5.5}$$

where the subscript ICR means-connotes interface-controlled ripening.

5.4.2 Mean Field Theory

Ardell,[35] for example, was among the first to modify LSW theory for a nonzero volume fraction of the dispersed phase. He considered the influence of nearest neighbors on the growth rate of particles. Ardell's theory showed that the PSD broadened in response to these local interactions, and that the coarsening rate itself increased with increasing volume fraction. His detailed theoretical results, however, seem to overestimate the influence of the volume fraction and, consequently, disagree quantitatively with more recent theories and computer simulations.[36] Variations on Ardell's basic method were implemented later by Tsumuraya and Miyata[37] by using a series of different kinetic coarsening interaction laws, referred to as "T–M models." Specifically, each T–M model defines some appropriate radius of "influence" surrounding each particle. Tsumuraya and Miyata proposed and explored six mean-field interactions, and used them to predict the growth rates of the interacting particles and the affine form of their PSDs. Two of the six T–M models predicted similar broadening of the PSDs. All of the T–M models, however, employed heuristic extensions of the basic mean-field LSW approach. Ardell's original model, and the subsequent six T–M models, all belong to the same universality class, and, therefore, all share some common approximations and, consequently, exhibit similar traits as to their kinetic predictions for late-stage microstructure evolution.

Brailsford and Wynblatt[38] were the first investigators of Ostwald ripening to employ "effective medium" theory. They obtained the theoretical growth rates of particles and a broadened PSD relative to that given by LSW theory. Brailsford and Wynblatt established an implicit relationship between the coarsening rate and volume fraction. Marsh and Glicksman[39] then introduced the concept of a statistical "field cell" acting around each size-class of the particles undergoing coarsening. Marsh and Glicksman obtained coarsening rate constants that are in good agreement with a variety of data derived from LPS experiments, particularly where the range of volume fractions of the dispersed phase lies between $0.3 \leq V_V \leq 0.6$. However, the PSD derived by Marsh and Glicksman was not compared with other results. All of the theoretical models mentioned above employed growth rate equations based on Laplace's equation as the quasi-static approximation for the time-dependent diffusion field. These models also rely on finding self-consistent global microstructural constraints that provide a so-called "cut-off" distance to terminate mass diffusion at some appropriate distance from a particle centered in an otherwise infinite Laplacian field. This diffusive cut-off adjusts the precise position of the mean field potential provided by the population of coarsening particles relative to the particle centers. The success of the statistical field cell approach depends sensitively on the global constraints that are used.

A very different approach to the modification of LSW theory was posed by Marqusee and Ross.[40] By contrast with earlier work, these investigators limited the spatial extent of the diffusion field by taking into account the collective diffusion "screening" provided by an "active" two-phase medium. This active medium is actually the transmitting matrix plus the distribution of coarsening particles that they considered to be a globally neutral collection of active diffusion point sources and sinks. Instead of using Laplace's equation as the quasi-static approximation, Marqusee and Ross showed that Poisson's equation is actually more appropriate for deriving a suitable kinetic expression using the growth rates for an "effective medium." They found the maximum particle radius expected in effective media at different volume fractions, the relationship between the coarsening rate and the volume fraction, and the affine, or self-similar, PSDs. Following their interesting approach, Fradkov et al.[41] and then Mandyam et al.[42] studied coarsening kinetics in finite clusters using Poisson's equation to approximate multiparticle diffusion with interactions included up to dipolar order.

Recently, Wang et al.[43,44] developed diffusion screening theory for application to Ostwald ripening. In diffusion screening theory, the range of diffusion interactions occurring among particles is described by employing a "diffusion screening length," R_{D}. The diffusion screening length — a collective property of the particle population — sets the range over which interactions occur, and beyond which they effectively cease. The diffusion screening length may be related directly to the ratio of moments of the PSD and to the square-root of the system's volume fraction as follows:

$$R_{\mathrm{D}} = \sqrt{\frac{\langle R^3 \rangle}{3\langle R \rangle V_V}}. \tag{5.6}$$

The growth rate of a spherical particle to first-order in a screened active medium is written as

$$\frac{\mathrm{d}R}{\mathrm{d}t} = \frac{1}{R}\left(\frac{1}{R^{\star}} - \frac{1}{R}\right)\left[1 + \frac{R}{R_{\mathrm{D}}}\right]. \tag{5.7}$$

Comparing Equation 5.2 with Equation 5.7 we found that the interactions among particles increase the LSW growth rate by the factor $(1 + R/R_{\mathrm{D}})$. The continuity equation and conservation law in LSW theory remain valid in diffusion screening theory. In addition, it was discovered that the relative coarsening rate $K(V_V)/K_{\mathrm{LSW}}$ depends approximately on the volume fraction as

$$\frac{K(V_V)}{K(0)} \cong 6.41\left[\frac{2 - (1 - \sqrt{3V_V})(1 - (1/\sqrt{3V_V}) + \sqrt{(1/3V_V) + (1/\sqrt{3V_V} + 1)}}{[1 - (1/\sqrt{3V_V}) + \sqrt{(1/3V_V) + (1/\sqrt{3V_V}) + 1}]^3}\right]. \tag{5.8}$$

Figure 5.2 shows that for volume fractions that are smaller than $V_V = 0.1$, our result derived from diffusion screening theory is in good agreement with other theoretical and simulation results. However, with further increases of the volume fraction, our result becomes the lower bound, which agrees with

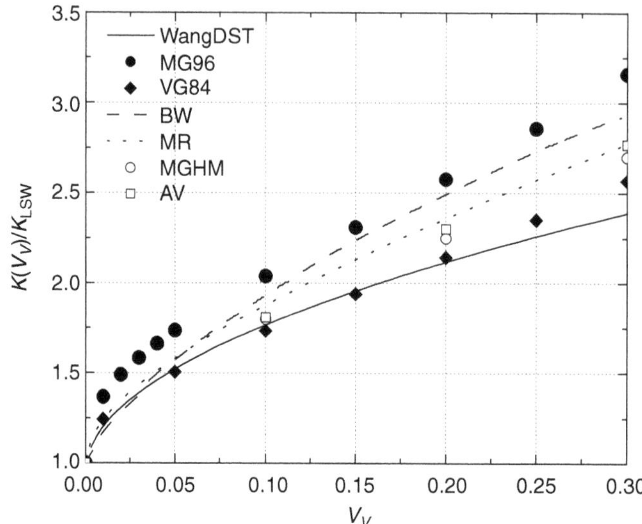

FIGURE 5.2 Coarsening rates, $K(V_V)$, normalized by the LSW coarsening rate, K_{LSW}, vs. volume fraction, V_V. The result from diffusion screening theory[33] (solid line) is compared here with results from other phase-coarsening theories: Brailsford and Wynblatt[38] (dashed line), Marqusee and Ross[40] (dotted line), and Marsh and Glicksman[39] (filled circles), and Voorhees and Glicksman[45] (filled diamonds). Simulation results (open symbols) of Akaiwa and Voorhees'[57] and Mandyam et al.[42] are also included. The solid line is a plot of Equation 5.8.

Voorhees and Glicksman'[45] earlier simulation results, whereas Marsh's statistical field cell result provides the upper bound.

The approximate relationship found between the maximum size particle's normalized radius, ρ_{max}, and the system's volume fraction is

$$\rho_{max} = 1 - \frac{1}{\sqrt{3V_V}} + \sqrt{\frac{1}{3V_V} + \frac{1}{\sqrt{3V_V}} + 1}. \tag{5.9}$$

Ardell, using his coarsening interaction model, predicted a relationship between the maximum radius of a coarsening particle to the microstructure's volume fraction,[35] as did Marqusee and Ross.[40] A comparison of the maximum radius of a coarsening particle predicted from diffusion screening theory with those estimated earlier by Ardell and then by Marqusee and Ross is shown in Figure 5.3 as functions of the microstructure's volume fraction. As $V_V \to 0$, the appropriate asymptotic limit is approached, namely, $\rho_{max} \to 1.5$, thus recapturing the classical infinitely dilute limit obtained from LSW theory.[25] Approximations used in diffusion screening theory and those employed by Ardell and by Marqusee and Ross result in the detailed differences shown in Figure 5.3. The important issue exposed here from the standpoint of improving our understanding of the physics of microstructural evolution is that theoretical predictions, such as estimating ρ_{max}, suggest the need for additional careful experimental studies to test them.

Figure 5.4 shows the PSD derived from diffusion screening theory for various values of V_V, over the range 0 to 0.3, all of which are obtained by numerical solution of the continuity equation. As a comparison, the PSD derived from LSW theory is also plotted in Figure 5.4.

The common finding exhibited by all of the mean-field theories mentioned above is that a nonzero volume fraction of particles does not alter the temporal coarsening exponent in the kinetic law from that predicted using LSW mean-field theory. The robustness of that temporal exponent devolves from fundamental scaling arguments that are independent of certain detailed kinetic assumptions. However, changing the volume fraction alters both the coarsening rate constant (by changing the kinetic coefficient of the growth law) and the resultant PSD. Despite this rough qualitative agreement occurring between

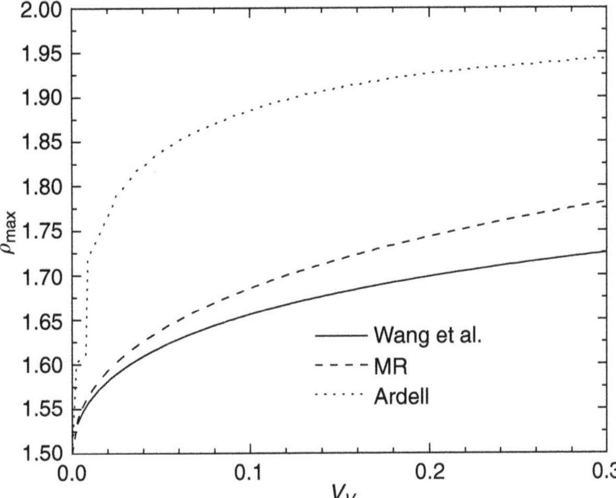

FIGURE 5.3 The relationship between the normalized maximum radius, ρ_{max}, and volume fraction, V_V. The solid, dotted, and dashed lines represent work from Wang et al.,[43] Ardell,[35] and Marqusee and Ross (MR),[40] respectively.

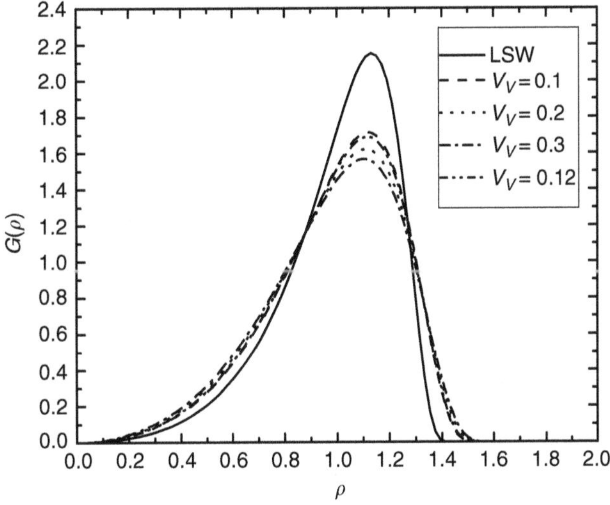

FIGURE 5.4 Particle size distributions derived from our diffusion screening theory,[43] $G(\rho)$, versus normalized radius, ρ, for several volume fractions, $V_V = 0.1, 0.12, 0.2, 0.3$. The $G(\rho)$ function from LSW theory ($V_V = 0$) is shown for comparison.

mean-field and effective media models, quantitative coarsening rate constant and significant details of the form of the PSDs actually differ markedly from theory to theory. The key to improving this situation still resides with finding the best physical approximations for the particle interactions, and eliminating as many heuristic aspects of the model as is possible.

5.4.3 Stochastic Theory

LSW theory as well as other mean-field theories predict, in general, that particles with identical size should have the same growth rates, regardless of their location and environment within the microstructure (see again Equation 5.2). Experimentalists,[11] however, repeatedly show the presence of fluctuations, or correlations, indicating that particles of the same size can exhibit very different growth rates. Rogers et al.[46] found that some particles that are larger than their nearest neighbors shrank in their experiments, and suggested that the growth rates of individual particles depended not only on their size, but also on the details of their immediate environment, or "locale."

Wang et al.[47,48] first developed a stochastic analytic model of coarsening that employed a Fokker–Planck equation (FPE) to estimate the theoretical PSD. Recently, Wang et al.[49] applied the technique known as "multiplicative noise" to describe the observed fluctuations in the mean growth rate of particles during diffusion-limited coarsening. Wang et al.[49] derived the following expression for the "locale noise" of the observed particle volume fluxes,

$$\zeta\left(\frac{\rho}{\rho_0}\right) = \frac{\rho}{2\rho_0}\left(1 + \frac{\rho}{\rho_0}\right)\eta. \tag{5.10}$$

Here η represents a Gaussian random variable with mean-value zero and unit width, and $\zeta(\rho/\rho_0)$ is the Gaussian multiplicative noise. ρ_0 is the re-normalized diffusion screening length.[49] All the simulation data are scattered between bands predicted using Equation 5.10, showing that multiplicative noise provides a reasonable basis for correlating the observed simulation noise.[49]

This approach established a stochastic equation to describe the growth rates of particles at low-to-moderate volume fractions.[49] The resulting stochastic expression for a particle's growth rate becomes

$$\frac{\mathrm{d}R}{\mathrm{d}t} = \frac{1}{R}\left(\frac{1}{R^\star} - \frac{1}{R}\right)\left(1 + \frac{R}{R_\mathrm{D}}\right) + \frac{1}{2R_\mathrm{D}}\left(\frac{1}{R} + \frac{1}{R_\mathrm{D}}\right)\eta. \tag{5.11}$$

Equation 5.11 is an ordinary stochastic differential equation, and provides the kinetic law for particles interacting in a "noisy" microstructure. Moreover, Wang et al. also developed the FPE for calculating the PSD associated with Equation 5.11, namely,

$$\frac{\partial F(u,\tau')}{\partial \tau'} = -\frac{\partial}{\partial u} D_1(u) F(u,\tau') + \frac{\partial^2}{\partial u^2} D_2(u) F(u,\tau'), \tag{5.12}$$

where the interested reader can find the functions $D_1(u)$ and $D_2(u)$ in Reference 50. $F(u,\tau')\mathrm{d}u$ is the number of particles per unit volume, the scaled sizes of which are between u and $u + \mathrm{d}u$ at time τ'. In the case of LSW theory, the volume fraction vanishes, so $\rho_0 \rightarrow \infty$, the interaction noise also vanishes, and Equation 5.12 reduces to the well-known LSW continuity equation. Wang et al.[50,51] found, using the asymptotic solution of this FPE, that fluctuations present in the microstructure are effective in broadening the PSD (see Figure 5.5). Figure 5.5 shows clearly how the interaction and fluctuation effects from the various microstructural locales influence the scaled PSD.

5.4.4 Multiparticle Diffusion Simulation

There are two different simulation methods that are applied popularly: (1) multiparticle diffusion simulation (monitoring the particles in the system) and (2) phase-field simulation (monitoring the field variables in the system).

The earliest attempt to simulate multiparticle diffusion of discrete particles by using numerical methods was made in 1973 by Weins and Cahn.[52] These investigators were severely limited at that time — by the extant computational capability — to describing just a few interacting particles arranged in several configurations to demonstrate basic coarsening interactions. Their work was followed by an investigation by Voorhees and Glicksman,[53] who, with greater computing power, studied the behavior of several hundred particles randomly distributed in a periodic, three-dimensional, unit cell, thus simulating extended phase coarsening. Later, Beenaker[54] enhanced further the capability of multiparticle diffusion simulation

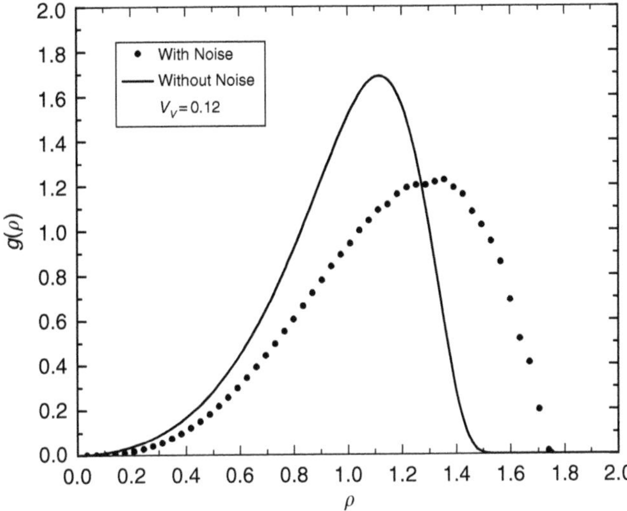

FIGURE 5.5 PSDs are obtained with, and without, stochastic interactions, respectively, during Ostwald ripening at $V_V = 0.12$.

procedures and was able to increase the total number of interacting particles during simulation to several thousand. Other investigators,[49–51,55–57] continued to improve upon the accuracy and statistical basis of large-scale simulations of late-stage phase coarsening.

A two-phase coarsening system may be simulated by placing n spherical particles of the dispersoid phase in a cubic box. The contiguous spaces between the particles are filled by the matrix phase, throughout which the dispersoid population is embedded. Particles are located by specifying the positions of their centers with three random coordinates, and by their radii chosen initially from a relatively narrow Gaussian distribution. Particle overlaps are precluded.

Wang et al., using this approach, carried out simulations of late-stage coarsening for various values of the volume fractions of the dispersoid phase from a lower limit of $V_V = 10^{-10}$ to an upper limit of $V_V = 0.3$.[34] The PSDs observed for different volume fractions are published in References 43, 55, and 56 which show that the height of a PSD is gradually reduced and its width commensurately broadened with increasing volume fraction. For more details, the reader is referred to References 43, 51, and 55.

5.4.5 Phase-Field Simulation

Phase-field modeling, was originally developed in the 1970s by a number of physicists[58] for the purposes of modeling phase transition and modeling system dynamics with diffuse interfaces.[59] Later, Kobayashi[60] demonstrated qualitatively that phase-field methods appeared to be useful for simulating dynamical phenomena in multiphase systems that exhibit interfaces that were relatively thin on small scales, but spatially complex on larger scales. Crystalline dendrites growing from a supercooled or supersaturated melt were archetypical of "complex" systems that appeared suitable for phase-field analysis. Over the past decade, phase-field modeling and simulation were rapidly developed and then applied more broadly toward studies of phase separation[61] in alloys. Chen and his coworker[62] used phase-field methods for two-dimensional and three-dimensional simulations of phase coarsening.[63,64]

We describe in this section, just briefly, the application of phase-field approaches for the simulation of coarsening in two-phase microstructures in a binary alloy. A two-phase microstructure in a binary system can be described by a concentration field, $c(\vec{\mathbf{r}}, t)$, which represents the spatial compositional distribution, and a structural order-parameter field that distinguishes the matrix phase from precipitates, $\eta_i(\mathbf{r}, \mathbf{t}), (i = 1, 2, \ldots, p)$. The free energy density for the binary system can be generally written as follows:[65]

$$f(c, \eta_i) = f_{\text{bulk}} + f_{\text{int}}, \tag{5.13}$$

where f_{bulk} is the bulk free energy density of the system. The interfacial energy density between two phases is introduced through the gradient of other spatially dependent field variables.[59] The isotropic interfacial energy, neglecting the effects of interfacial anisotropy, may be written as

$$f_{\text{int}} = \frac{1}{2}\kappa_c(\nabla c)^2 + \sum_{i=1}^{p} \frac{1}{2}\beta_i(\nabla \eta_i)^2, \tag{5.14}$$

where κ_c and β_i are the gradient energy coefficients. These coefficients weigh the energy cost of encountering gradients in concentration or in the phase indicator, η_i, within the system. The total energy of the system is expressed by integrating f_{int} over the total volume of the system as follows

$$F(c, \eta_i) = \int_{\Omega} f(c, \eta_i)\text{d}v, \tag{5.15}$$

where Ω represents the domain of the binary system. The temporal evolution of the field variables is obtained through solving the Cahn–Hilliard or Allen–Cahn equations.[59,66,67,68] The evolution equations with thermal noise added, become

$$\frac{\partial c(\vec{\mathbf{r}}, t)}{\partial t} = M\nabla^2 \frac{\delta F}{\delta c(\vec{\mathbf{r}}, t)} + \xi(\vec{\mathbf{r}}, t), \tag{5.16}$$

and

$$\frac{\partial \eta_i(\vec{\mathbf{r}}, t)}{\partial t} = -L\frac{\delta F}{\delta \eta_i(\vec{\mathbf{r}}, t)} + \zeta_i(\vec{\mathbf{r}}, t); \qquad i = 1, \ldots, p, \tag{5.17}$$

where the functions $\xi(\vec{\mathbf{r}}, t)$ and $\zeta_i(\vec{\mathbf{r}}, t)$ are thermodynamic noise terms that are Gaussian distributed and which satisfy the usual correlation conditions from the fluctuation-dissipation theorem.[69] The coefficients M and L appearing in Equation 5.16 and Equation 5.17 denote mobility constants that are related to atom or interface mobility.

In the application of phase-field simulations to a specific phase-separating binary system, one needs first to construct the bulk free energy density, f_{bulk}, in Equation 5.13 and to determine the phase gradient coefficients β_i in Equation 5.14. Chen et al.,[62,63] using phase-field modeling, simulated two-dimensional and three-dimensional phase coarsening. In their study of Al–Li alloys,[63] they specified f_{bulk} as follows:

$$f_{\text{bulk}} = \frac{A_1}{2}(c - C_1)^2 + \frac{A_2}{2}(C_2 - c)\sum_{i=1}^{3}\eta_i^2 - \frac{A_3}{3}\prod_{i=1}^{3}\eta_i + \frac{A_4}{24}\sum_{i=1}^{3}\eta_i^4 + \frac{A_5}{24}\sum_{i\neq j}^{3}\eta_i^2\eta_j^2, \tag{5.18}$$

where, values of the coefficients C_1, C_2 are 0.0571, 0.2192, and those for A_1, A_2, A_3, A_4, and A_5 are 125.12, 44.74, 21.21, 22.14, and 22.14, respectively.[63] They choose $p = 3$, $\kappa_c = 0$, $\beta_i = 0.75$ for the determination of interfacial energy density term.

5.5 Comparison of Theory, Simulation, and Experiment

The binary alloy Al–Li provides a nearly ideal system for studies of coarsening, primarily because this system develops microstructures containing spherical precipitates of δ' phase (Al_3Li) with extremely small strains developed at their interfaces. Mahalingam et al.[16] used quantitative TEM to study the coarsening behavior of δ' precipitates in a series of binary Al–Li alloys. These authors carried out

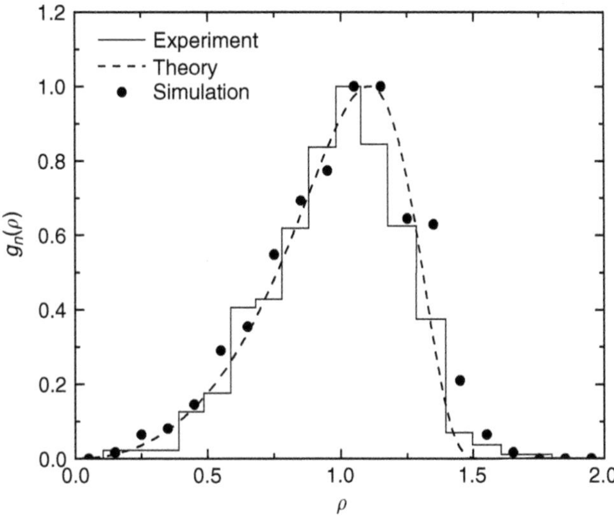

FIGURE 5.6 Particle size distributions (normalized), $g_n(\rho)$, from diffusion screening theory and multiparticle diffusion simulations,[43] compared with the experimental PSD of Mahalingam et al.[16] for an Al–Li alloy at a dispersoid volume fraction of $V_V = 0.12$. The normalization used is to set the peak value, $g_n^{\max} = 1$.

careful experiments and obtained the microstructure's steady-state PSD. Figure 5.6 shows the PSDs for an Al–Li alloy with a volume fraction of δ' set at $V_V = 0.12$. The PSDs are derived from diffusion screening theory, multiparticle diffusion simulations, and from Mahalingam et al.'s[16] experimental data. Figure 5.6 shows that the PSDs predicted from Ostwald ripening theory are in good agreement with the experimental data.

Moreover, the PSD derived from multiparticle diffusion simulations appears to be in even closer agreement with the experimental results of Mahalingam et al.[16] at a δ' volume fraction $V_V = 0.12$. The reason for this closer correspondence with experiment, as compared with diffusion screening theory, is that in these simulations *all* interactions among particles are included throughout the system. However, in analytic diffusion screening theory, one approximates these interactions and neglects the higher order interactions. It is further demonstrated in Figure 5.6, that the tail of the PSD at large particle radii derived from computer simulations mimics closely the experimental data.

Wang et al.[43] found that at a volume fraction $V_V = 0.12$ of δ' precipitates, the maximum normalized radius, $\rho_{\max} \approx 1.67$. The corresponding maximum normalized radius found from multiparticle diffusion simulations is $\rho_{\max} \approx 1.74$. The experimental result for the maximum precipitate radius reported by Mahalingam et al. shows that $\rho_{\max} \approx 1.80$.[16] These correspondences represent the first instance, to the authors' knowledge, in which three-way agreement was successfully achieved among the values for the maximum normalized particle radius derived independently from theory, simulation, and experiment. Figure 5.3 shows that at a dispersoid volume fraction of $V_V = 0.12$, Marqusee and Ross's theory predicts $\rho_{\max} \approx 1.70$, whereas Ardell's theory predicts $\rho_{\max} \approx 1.90$. LSW theory, by comparison, which ignores all interactions, predicts the much smaller value $\rho_{\max} = 1.5$.

In an attempt to calculate and compare the relative kinetic coarsening rates from diffusion screening theory and from simulation data, Wang et al. obtained the following results for the ratios of the kinetic rate constants in Al–Li alloys at a δ' precipitate volume fraction of 0.12; from diffusion screening theory Wang et al. found that $K(0.12)/K_{\text{LSW}} = 1.81$, whereas from multiparticle diffusion simulations Wang et al. found that $K(0.12)/K_{\text{LSW}} = 1.83$.[43] It is interesting that both these values are markedly different from the experimental result[16] $K(0.12)/K_{\text{LSW}} = 3.72$. The reason for this disparity in the estimated coarsening rate constants might lie in the particular values selected for the particle–matrix interfacial energy and the interdiffusion coefficient for the matrix. Both these parameters have a wide range of

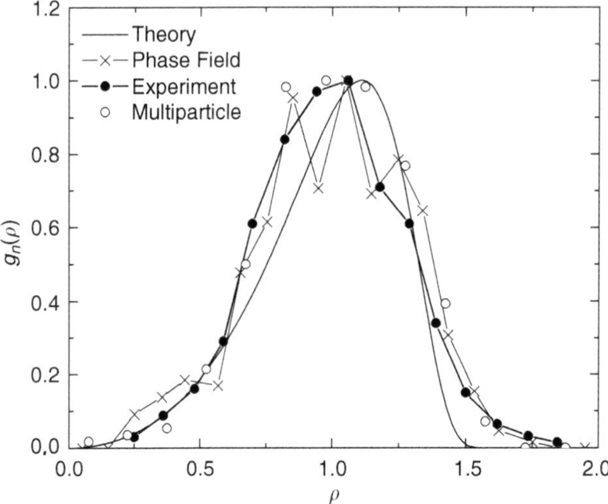

FIGURE 5.7 Particle size distributions (normalized), $g_n(\rho)$, from diffusion screening theory,[43] multiparticle diffusion simulation,[43] and phase-field simulation at a dispersoid volume fraction of $V_V = 0.2$,[63] compared with the experimental PSD of Gu et al.[15] for Al–Li alloy at a dispersoid volume fraction of $V_V \approx 0.2$. The normalization used is to set the peak value, $g_n^{max} = 1$.

uncertainty. This result strongly suggests that one needs more extensive and accurate thermophysical data for experimental alloys used to test the accuracy of estimated kinetic coarsening rate constants.

Recently, Vaithyanathan and Chen carried out three-dimensional phase-field simulations for Al–Li alloys at a volume fraction of $V_V \approx 0.2$.[63] Guo et al. performed TEM experiments for this alloy at a δ' volume fraction of $V_V = 0.2$.[15] Wang et al.[33] calculated the PSD at $V_V = 0.2$ from diffusion screening theory and multiparticle diffusion simulation. All of these experimental, theoretical and simulation results are compared in Figure 5.7. Data from phase-field simulation are scattered because the simulation box size used in three-dimensional phase-field simulation is severely limited by present day computational capability. Thus, at present, phase-field simulations of Ostwald ripening phenomena continue to yield relatively poor statistics compared to multiparticle diffusion simulation that can be performed for very large-scale systems with corresponding good statistics.

5.6 Conclusions

1. Although Ostwald ripening is a classical capillarity effect discovered over a century ago, its understanding and applications remain in use to create novel materials such as nanostructures.[13,14] Ostwald ripening phenomena are still commonly found in a range of different processes.[18] At small-to-moderate volume fractions of the precipitates phase ($0 < V_V < 0.3$), the diffusion screening length created among interacting particles is proportional to $V_V^{-1/2}$ — a theoretical prediction, which has now been tested by simulation and placed on a firm theoretical footing. This result may now be verified through direct experiments.
2. Diffusion screening theory itself is classified as a mean-field theory, and thus leads to a set of deterministic results for the *average* behavior of the particle or precipitate population. Diffusion screening theory provides a physically based approach to Ostwald ripening, and the mathematical physics to implement it are straightforward. Diffusion screening theory is capable of predicting subtle changes in the effective diffusion length scale, the maximum particle radius, the PSD, and the coarsening rate constant variation with precipitate volume fraction. A few predictions derived

from diffusion screening theory are found to be in agreement with selected experimental results. A major challenge that remains is to develop a physics-based theory for cases involving high volume fractions of the precipitate.

3. If local microstructural details are important and need to be included in the interaction physics, then fluctuations from mean-field behavior must be considered. Multiparticle simulations and experimental observation show convincingly that such fluctuations indeed occur during the Ostwald ripening of particles at any sensible nonzero volume fraction above about $V_V = 10^{-4}$. Although Wang et al.[47,53] initially developed a stochastic theory of Ostwald ripening in which the observed spectrum of fluctuations is approximated, the theory remains in need of additional theoretical work covering more systems.
4. Multiparticle diffusion simulations are capable of determining the maximum precipitate radius within the interacting population of particles, their steady-state PSD and spatial correlations, the fluctuation spectra of the growth rates, the diffusion screening length, and the coarsening rate constants. Phase-field simulations, a somewhat newer computational approach also allow determination of the maximum precipitate radius, the PSD and spatial correlations, and the coarsening rate constants. Phase-field results lack fluctuation spectra of the growth rates and an associated diffusion screening length.
5. Al–Li alloys provide a nearly ideal experimental solid-state test system for the study of late-stage phase coarsening, because these alloys develop nearly spherical δ' precipitates that develop negligible strains magnitudes at the particle–matrix interfaces. The lack of interfacial strains and dislocations allows easy comparison between theoretical and computational predictions and those observed experimentally. Although the theoretical and experimental PSDs are in relatively good agreement, large disparities remain between the experimentally observed coarsening rate constant and those predicted from theory. Perhaps some of the critical materials parameters for the experimental test system need to be further evaluated so that comparisons with theoretical and experimental results are made more meaningful.
6. Multiparticle diffusion simulation might prove to be easier to extend to much larger systems ($>10^4$ particles) for the study of phase coarsening in three-dimensions than to attempt an ambitious scale-up of phase-field simulations. However, phase-field simulations provide the distinct advantage over multiparticle simulations of being capable of dealing with the geometrically complicated nonspherical shapes of precipitates and particles that undergo shape accommodation and topological changes in microstructures at higher volume fractions ($V_V \geq 0.5$). The authors also suggest that new experiments be designed to derive precise microstructural data that can be compared quantitatively with the continually improving predictions derived from theory and simulation. New experimental results will always remain of fundamental importance to making further progress in the field of Ostwald ripening phenomena. New experimental data are essential to establishing more reliable quantitative comparisons with theory and simulation. Finally, direct linkage of the effects of Ostwald ripening on the kinetics of microstructure evolution to the properties of multiphase materials constitutes an important next step[70] in this field, and more effort expended in this direction may be expected in the years ahead.

Acknowledgments

The author, K.G.W. thanks Professor H.C. Zeng, National University of Singapore, Singapore, for his discussion on applications of Ostwald ripening to the synthetic architecture of nanomaterials, and Professor David N. Seidman, Northwestern University, USA, for his experimental studies of three-dimensional phase coarsening. The authors are also especially pleased to acknowledge partial financial support received from the National Aeronautics and Space Administration, Marshall Space Flight Center, under Grant NAG-8-1468, and M.E.G. for financial support derived from the John Tod Horton Distinguished Professorship in Materials Engineering, Rensselaer Polytechnic Institute.

References

[1] Ostwald, W.Z., Über die vermeintliche Isomerie des roten und gelben Quecksilberoxyd und die Oberflächenspannung fester Körper, *Z. Phys. Chem.*, **34**, 495, 1900.

[2] Rogers, R.R. and Yau, M.K.A., *A Short Course in Cloud Physics*, Pergamon Press, Oxford, 1989.

[3] Ardell, A.J. and Nicholson, R.B., The coarsening of γ' in Ni-Al alloys, *J. Phys. Chem. Solids*, **27**, 1793, 1966.

[4] Baldan, A., Progress in Ostwald ripening theories and their application to γ'-precipitates in nickel-base superalloys, Part II Nickel-base superalloys, *J. Mater. Sci.*, **37**, 2379, 2002.

[5] Rahaman, M.N., *Ceramic Processing and Sintering*, Marcel Dekker, Inc., New York, 1995.

[6] Kingery, W.D., Niki, E., and Narasimhan, M.D., Sintering of oxide and carbide-metal compositions in presence of a liquid phase, *J. Am. Ceram. Soc.*, **44**, 29, 1961.

[7] Song, H. and Coble, R.L., Origin and growth-kinetics of platelike abnormal grains in liquid-phase-sintering alumina, *J. Am. Ceram. Soc.*, **73**, 2077, 1990.

[8] Crist, B. and Nesarikar, A.R., Coarsening in polyethylene-copolymer blends, *Macromolecules*, **28**, 890, 1995.

[9] Haas, C.K. and Torkelson, J.M., Two-dimensional coarsening and phase separation in thin polymer solution films, *Phys. Rev. E*, **55**, 3191, 1997.

[10] Cheng, M.H. and Nauman, E.B., Phase ripening in particulate binary polymer blends, *J. Polym. Sci., Part B: Polym. Phys.*, **42**, 603, 2004.

[11] Bartelt, N.C., Theis, W., and Tromp, R.M., Ostwald ripening of two-dimensional islands on Si (001), *Phys. Rev. B*, **54**, 11741, 1996.

[12] Zinke-Allmang, M., Phase separation on solid surface: Nucleation, coarsening and coalescence kinetics, *Thin Solid Films*, **346**, 1, 1999.

[13] Yang, H.G. and Zeng, H.C., Preparation of hollow anatase TiO_2 nanospheres via Ostwald ripening, *J. Phys. Chem. B*, **108**, 3492, 2004.

[14] Liu, B. and Zeng, H.C., Symmetric and asymmetric Ostwald ripening in the fabrication of homogeneous core-shell semiconductors, *Small*, **1**, 566, 2005.

[15] Gu, B.P., Liedl, G.L., Kulwicki, J.H., and Sanders, T.H. Jr., Coarsening of δ' (Al_3Li) precipitates in Al-2.8 Li-0.3 Mn Alloy, *Mater. Sci. Eng.*, **70**, 217, 1985.

[16] Mahalingam, K. et al., Coarsening of δ' Al-3Li precipitates in binary Al-Li alloys, *Acta Metall.*, **35**, 483, 1987.

[17] Calderon, H.A., Fine, M.E., and Weertman, J.R., Coarsening and morphology of β' particles in Fe-Ni-Al-Mo ferritic alloys, *Metall. Trans. A*, **19**, 1135, 1988.

[18] Redmond, P.L., Hallock, A.J., and Brus, L.E., Electrochemical Ostwald ripening of colloidal Ag particles on conductive substrates, *Nano Lett.*, **5**, 131, 2005.

[19] Weber, R. and Skorupa, W., Experimental investigation of Ostwald ripening in an implanted system, *Nucl. Instrum. Methods Phys. Res. B*, **149**, 97, 1999.

[20] Sudbrack, C.K., Decomposition Behavior in Model Ni-Al-Cr-X superalloys: Temporal Evolution and Compositional Pathways on a Nanoscale, Ph.D. thesis, Northwestern University, 2004.

[21] Möller, J., Kranold, R., Schmelzer, J., and Lembke, U., Small-angle X-ray scattering size parameters and higher moments of the particle-size distribution function in the asymptotic stage of Ostwald ripening, *J. Appl. Crystallogr.*, **28**, 553, 1995.

[22] Walter, G., Kranold, R., Gerber, T., Baldrian, J., and Steinhart, M., Particle size distribution from small-angle X-ray scattering data, *J. Appl. Crystallogr.*, **18**, 205, 1985.

[23] Damaschun, G.J., Müller, J., Pürschuel, H.-V., and Sommer, G., Computation of shape of colloid particles from low-angle X-ray scattering, *Monatsh. Chem.*, **100**, 1701, 1969.

[24] Sjöberg, B., Small-angle X-ray investigation of equilibria between copper(II) and glycyl-L-histidylglycine in water solution — Method for analyzing polydispersed systems, *J. Appl. Crystallogr.*, **7**, 192, 1974.

[25] Lifshitz, I.M. and Slyozov, V.V., The kinetics of precipitation from supersaturated solid solution, *J. Phys. Chem. Solids*, **19**, 35, 1961.

[26] Wagner, C., Theorie der Älterrung von Niederschlagen durch Umlösen, *Z. Elektrochem.*, **65**, 581, 1961.

[27] Che, D.Z. and Hoyt, J.J., Spatial correlations during Ostwald ripening: A simplified approach, *Acta Metall. Mater.*, **43**, 2551, 1995.

[28] Marder, M., Correlation and Ostwald ripening, *Phys. Rev. A*, **36**, 858, 1987.

[29] Che, D.Z., Spooner, S., and Hoyt, J.J., Experimental and theoretical investigation of the scaled structure function in Al-Li alloys, *Acta Mater.*, **45**, 1167, 1997.

[30] Abis, S. et al., Late stage of δ' precipitation in an Al-Li alloy by small-angle neutron scattering, *Phys. Rev. B*, **42**, 2275, 1990.

[31] Greenwood, G.W., The growth of dispersed precipitates in solutions, *Acta Metall.*, **4**, 243, 1956.

[32] Asimow, R., Clustering kinetics in binary alloys, *Acta Metall.*, **11**, 72, 1963.

[33] Baldan, A., Progress in Ostwald ripening theories and their applications to nickel-base superalloys, *J. Mater. Sci.*, **37**, 2171, 2002.

[34] Wang, K.G., Glicksman, M.E., and Rajan, K., Length scales in phase coarsening: Theory, simulation, and experiment, *Comput. Mater. Sci.*, **34**, 235, 2005.

[35] Ardell, A.J., The effect of volume fraction on particle coarsening: Theoretical considerations, *Acta Metall.*, **20**, 61, 1972.

[36] Yao, J.H., Elder, K.R., Guo, H., and Grant, M., Theory and simulation of Ostwald ripening, *Phys. Rev. B*, **47**, 14110, 1993.

[37] Tsumuraya, K. and Miyata, Y., Coarsening models incorporating both diffusion geometry and volume fraction of particles, *Acta Metall.*, **31**, 437, 1983.

[38] Brailsford, A.D. and Wynblatt, P., The dependence of Ostwald ripening kinetics on particle volume fraction, *Acta Metall.*, **27**, 489, 1979.

[39] Marsh, S.P. and Glicksman, M.E., Kinetics of phase coarsening in dense systems, *Acta Mater.*, **44**, 3761, 1996.

[40] Marqusee, J.A. and Ross, J., Theory of Ostwald ripening: Competitive growth and its dependence on volume fraction, *J. Chem. Phys.*, **80**, 536, 1984.

[41] Fradkov, V.E., Glicksman, M.E., and Marsh, S.P., Coarsening kinetics in finite clusters, *Phys. Rev. E*, **53**, 3925, 1996.

[42] Mandyam, H. et al., Statistical simulations of diffusional coarsening in finite clusters, *Phys. Rev. E*, **58**, 2119, 1998.

[43] Wang, K.G., Glicksman, M.E., and Rajan, K., Modeling and simulation for phase coarsening: A comparison with experiment, *Phys. Rev. E*, **69**, 061507, 2004.

[44] Glicksman, M.E., Wang, K.G., and Marsh, S.P., Diffusional interactions among crystallites, *J. Cryst. Growth*, **230**, 318, 2001.

[45] Voorhees, P.W. and Glicksman, M.E., Ostwald ripening during liquid-phase sintering: Effect of volume fraction on coarsening kinetics, *Metall. Trans. A*, **15**, 1081, 1984.

[46] Rogers, J.R. et al., Coarsening of three-dimensional droplets by two-dimensional diffusion: Part I experiment, *J. Electron. Mater.*, **23**, 999, 1994.

[47] Wang, K.G. et al., Kinetics of particle coarsening process, *Z. Phys. B*, **94**, 353, 1994.

[48] Wang, K.G., Nonequilibrium statistical mechanical formulation for grain growth, *Z. Phys. B*, **99**, 593, 1996.

[49] Wang, K.G., Glicksman, M.E., and Crawford, P., Multiplicative noise in microstructure evolution, in *Modeling and Numerical Simulation of Materials Behavior and Evolution*, ed. A. Zavaliangos et al., Materials Research Society, **731**, 227, 2002.

[50] Wang, K.G. and Glicksman, M.E., Noise of microstructural environments in late-stage phase coarsening, *Phys. Rev. E*, **68**, 051501, 2003.

[51] Glicksman, M.E., Wang, K.G., and Crawford, P., Stochastic effects in microstructure, *J. Mater. Res.*, **5**, 231, 2002.

[52] Weins, J. and Cahn, J.W., The effect of size and distribution of second phase particles and voids on sintering, in *Sintering and Related Phenomena*, ed. G.C. Kuczynski, Plenum, New York, 151, 1973.

[53] Voorhees, P.W. and Glicksman, M.E., Solution to the multi-particle diffusion problem with applications to Ostwald ripening-II. Computer simulations, *Acta Metall.*, **32**, 2013, 1984.

[54] Beenakker, C.W.J., Numerical simulation of diffusion-controlled droplet growth: Dynamical correlation effects, *Phys. Rev. A*, **33**, 4482, 1986.

[55] Glicksman, M.E., Wang, K.G., and Crawford, P., Simulations of microstructural evolution, in *Computational Modeling of Materials, Minerals and Metals Processing*, ed. Mark Cross, J.W. Evans, and C. Bailey, The Minerals, Metals & Materials Society, 703, 2001.

[56] Wang, K.G., Glicksman, M.E., and Rajan, K., Computational modeling for high speed screening of polymer microstructures, *Macromol. Rapid Commun.*, **25**, 377, 2004.

[57] Akaiwa, N. and Voorhees, P.W., Late-stage phase separation: Dynamics, spatial correlations, and structure functions, *Phys. Rev. E*, **49**, 3860, 1994.

[58] Hohenberg, P.C. and Halperin, B.I., Theory of dynamic critical phenomena, *Rev. Mod. Phys.*, **49**, 435, 1977.

[59] Emmerich, H., *The Diffuse Interface Approach in Materials Science*, Springer-Verlag, Berlin, Heidelberg, 2003.

[60] Kobayashi, R., Modeling and numerical simulations of dendritic crystal-growth, *Phys. D*, **63**, 410, 1993.

[61] Bray, A.J., Theory of phase-ordering kinetics, *Adv. Phys.*, **43**, 357, 1994.

[62] Fan, D., Chen, S.P., Chen, L.-Q., and Voorhees, P.W., Phase-field simulation of 2-D Ostwald ripening in the high volume fraction regime, *Acta Mater.*, **50**, 1895, 2002.

[63] Vaithyanathan, V. and Chen, L.Q., Coarsening kinetics of δ'-Al_3Li precipitates: Phase-field simulation in 2D and 3D, *Scripta Mater.*, **42**, 967, 2000.

[64] Zhu, J.Z., Wang, T., Ardell, A.J., Liu, Z.K., Zhou, S.H., and Chen, L.Q., Three-dimensional phase-field simulations of coarsening kinetics of γ' particles in binary Ni-Al alloys, *Acta Mater.*, **52**, 2837, 2004.

[65] Hudson, J.B., *Surface Science*, Butterworth-Heinemann, Boston, 1992.

[66] Kupper, T. and Masbaum, N., Simulation of particle growth and Ostwald Ripening via the Cahn-Hilliard Equation, *Acta Mater.*, **42**, 1847, 1994.

[67] Cahn, J.W. and Hilliard, J.E., Free energy of a nonuniform system: I. interfacial free energy, *J. Chem. Phys.*, **28**, 258, 1958.

[68] Allen, S.M. and Cahn, J.W., A microscopic theory for antiphase boundary motion and its application to antiphase domain coarsening, *Acta Metall.*, **27**, 1084, 1979.

[69] Elder, K., Langevin simulations of non-equilibrium phenomena, *Comput. Phys.*, **7**, 27, 1993.

[70] Wang, K.G., Guo, Z., Sha, W., Glicksman, M.E., and Rajan, K., Property predictions using microstructural modeling, *Acta Mater.*, **53**, 3395, 2005.

6

Crystallization of Amorphous Material

Jürgen Eckert and
Sergio Scudino
IFW Dresden, Institut für Komplexe Materialien (IKM)

Abstract

Alternatively to the direct synthesis by solidification of the melt, nanostructured glassy-matrix composites can be prepared by controlled devitrification (crystallization) of amorphous solids. This two-stage technique (i.e., glass formation and subsequent crystallization) can be usefully employed when a material with specific properties, such as microstructure or thermal stability, is required. This is a very crucial aspect since the mechanical properties of composite materials are strongly dependent on their microstructure. For a given alloy composition the microstructure of the composite can be controlled by choosing the proper annealing conditions (i.e., temperature, time, and heating rate). On the other hand, if the crystallization mechanism is known, in order to obtain a material with the desired properties several features of the alloy can be tuned by appropriately varying the chemical composition. In this chapter, the recent progress in the study of the devitrification of glassy materials is described for multicomponent Zr-based alloys. The basic concepts of the crystallization process are given and the effect of important variables, that is, composition, impurities, and sample preparation, which can influence the crystallization behavior of metallic glasses, are described for selected materials. The purpose is to give some guidelines in order to tune and control the thermal stability as well as the microstructure evolution upon heating, which is a necessary prerequisite for any possible commercial application of such glassy-matrix composites.

6.1 Introduction

In 1959, Duwez et al.[1] developed a rapid solidification technique capable of achieving cooling rates of the order of 10^7 K/s. By applying such a high cooling rate to the $Au_{75}Si_{25}$ melt, they were able to bypass the process of nucleation and growth of the stable crystalline phase and to produce a frozen liquid where the atoms have lost the ability to modify their respective configurations, a metallic glass. Soon after, Cohen and Turnbull suggested that the formation of the glassy phase in $Au_{75}Si_{25}$ was connected with the existence of a deep eutectic near this composition.[2] The presence of a deep eutectic gives the melt the opportunity to cool to a temperature at which its viscosity is quite high and, consequently, the atomic mobility in the melt has been reduced, thus hindering nucleation. This idea provided a logical basis to look for other glass-forming systems and led to the discovery of glass formation in a number of alloys, including Pd–Si,[3] Nb–Ni,[4] Fe–C–P alloys,[5] and Zr–TM (TM = Ni, Pd, Cu, Co)[6] alloys. Later, by systematically investigating Pd–M–P (M = Ni, Co, and Fe) alloys, Chen was able to produce glassy samples with thickness of about 1 mm.[7] If one arbitrarily defines the millimeter scale as "bulk," this can be considered the first example of bulk metallic glass (BMG). In 1984, Kui et al.[8] successfully prepared the first centimeter-thick glass in the Pd–Ni–P system by using boron oxide flux to purify the melt and to eliminate heterogeneous nucleation. For more details on amorphization see Chapter 17.

A great impulse in the development of BMGs was given by Inoue and coworkers. In the late 1980s, they observed a supercooled liquid (SCL) region in Al–La–Ni alloys, that is, a distinct temperature span between the crystallization temperature T_x and the glass transition T_g ($\Delta T_x = T_x - T_g$), of about 70 K.[9] This wide interval implies that the SCL can exist in a large temperature range without crystallization. They proposed that such a high resistance to crystallization at temperatures above the glass transition might reflect a lower critical cooling rate and, consequently, a high glass-forming ability (GFA). Therefore, assuming a link between high ΔT_x values and high GFA, they looked for alloys with large ΔT_x. Indeed, they found a number of Mg- and Zr-based alloys with large values of ΔT_x, which can be cast into fully glassy rods with thickness of several millimeters.[10–12] In particular, the $Zr_{65}Al_{7.5}Ni_{10}Cu_{17.5}$ alloy displays a SCL region of 127 K and can be cast into rods with diameter up to 16 mm, indicating a particularly low critical cooling rate.[12] The increased thermal stability against crystallization compared with binary Zr–TM (TM = Cu, Ni) alloys[13,14] was interpreted in terms of higher packing density of the amorphous phase due to the addition of Al. The intermediate atomic size of Al with respect to Zr and the TM should be appropriate to fill up the vacant sites in the disordered amorphous structure, increasing the packing density and, as a result, enhancing the thermal stability of the SCL.[10] Following this idea, Peker and Johnson[15] developed a new family of glassy alloys based on Zr–Ti–Cu–Ni–Be with critical cooling rates as low as 1–100 K s^{-1},[16] sufficient to obtain high levels of undercooling and allowing to produce fully glassy rods with diameters of several centimeter with a SCL region of about 135 K.[17] These new glassy alloys are sufficiently stable against crystallization to explore the highly undercooled liquid up to more than 100 K above T_g and to provide insight into the kinetics and thermodynamics of the SCL state.[18–20]

Besides their thermodynamic properties, multicomponent Zr-based metallic glasses show a positive combination of high thermal stability against crystallization with remarkable mechanical properties at room temperature, namely, yield stress up to 2 GPa, microplasticity up to 1%, a low Young's modulus, and high fracture toughness comparable with that of Al alloys[21–23] as well as improved electrochemical behavior and corrosion resistance.[24] In addition, the easy forming and shaping capability at elevated temperatures above the glass transition temperature[25] opens up the possibility to shape the material into parts with complex three-dimensional geometry and promises applications in the field of near-net-shape fabrication of structural components.

The plastic deformation of metallic glasses at room temperature is inhomogeneous and occurs via highly localized shear bands.[26,27] Shear bands normally appear in a distorted region to accommodate the applied strain, forming thin regions in which very large strains are concentrated. Although the local plastic strain in a shear band can be quite high,[28] only a few shear bands become active prior to occurrence of catastrophic failure. Improvements in the mechanical properties of single-phase glassy alloys can be achieved by the presence of second-phase particles embedded in the glassy matrix.[29–31] The particles act as barriers to the propagation of the shear bands promoting homogeneous deformation by multiple shearing events, thus increasing the yield and the fracture stress.[32,33] However, the effectiveness of the property improvement strongly depends on the volume fraction as well as on the grain size of the phase(s) formed.[22,33–35] Therefore, the commercial application of such composite materials requires the ability to

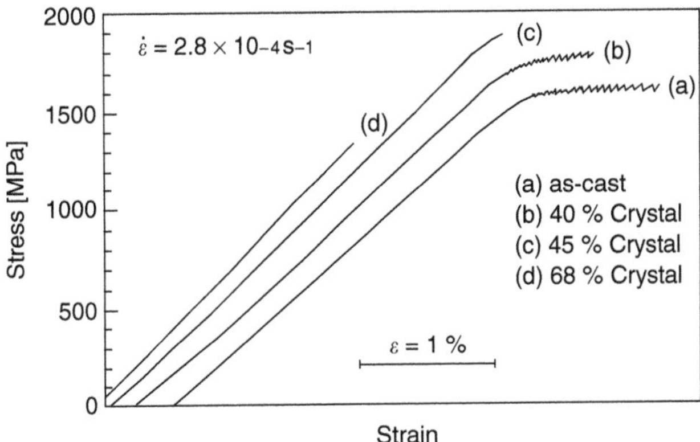

FIGURE 6.1 Compressive stress–strain curves of (a) as-prepared amorphous and partially crystallized $Zr_{57}Ti_5Cu_{20}Al_{10}Ni_8$ with (b) 40%, (c) 45%, and (d) 68 vol.% nanocrystals. (After Leonhard, A., Xing, L.Q., Heilmaier, M., Gebert, A., Eckert, J., and Schultz, L., *Nanostruct. Mater.*, 10, 805–817, 1998. With permission from Elsevier Science Ltd.)

produce a controlled microstructure. In fact, it has been reported that the microstructure of nanosized particles embedded in a glassy matrix exhibits high strength and good ductility when the volume fraction is less than about 40 to 50 vol.%,[22,33–35] and that particles of micrometer size have a strong effect on crack initiation leading to early fracture.[36] As an example, typical compressive stress–strain curves at room temperature for as-cast fully amorphous and partially crystallized $Zr_{57}Ti_5Cu_{20}Ni_8Al_{10}$ samples prepared by controlled annealing of the glass are presented in Figure 6.1.[22] Even though the ductility of the material decreases with increasing volume fraction of precipitates and finally disappears when a critical volume fraction is reached, the dispersion of nanocrystals in the amorphous matrix can lead to a distinct strength increase.[22] For samples up to about 50 vol.% of nanocrystals, the fracture surface exhibits a well-defined vein pattern, indicating that the deformation mechanism is governed by the glassy phase and not by the nanocrystals.[22] When the volume fraction of the nanocrystalline precipitates increases to more than about 50 vol.%, the nature of the brittle intermetallic phases is likely to dominate the mechanical behavior, leading to a marked decrease in the ductility of the sample.[22]

Materials such as glass–matrix or nanostructured composites can be produced by introducing metal or ceramic solid phases as reinforcements into the glass-forming melt during solidification [32,37–39] or from the melts directly during solidification.[40–44] For example, He et al. recently developed Ti-based nanostructure–dendrite composites produced by copper mold casting, which show an ultimate compression stress of more than 2 GPa and a ductility of about 14.5%.[40] As against to such direct processing routes, glass–matrix composites consisting of nanosized particles embedded in a glassy matrix can be produced by controlled devitrification (crystallization) of metallic glasses.[45–48] This technique has been used for long time for conventional glasses[49] to produce composite materials with a wide variety of microstructures and advantageous properties.

The basic principle for the production of glass–matrix composites by crystallization of a glassy precursor is to control the crystallization kinetics by optimizing the annealing conditions (annealing temperature and time, heating rate, etc.) and chemical composition to obtain a glassy phase that partially or completely transforms into a nanocrystalline material with the desired microstructure.[50,51] To do so, detailed knowledge of the crystallization mechanism and of the possible factors affecting the crystallization process is of extreme importance.

In this chapter, the recent progress in the study of the devitrification of glassy materials is described for multicomponent Zr-based alloys as a typical example for such materials. The basic concepts of the crystallization process are given and the effect of important variables, that is, composition, impurities, and sample preparation, which can influence the crystallization behavior of metallic glasses, will be described for selected materials. The purpose is to give some guidelines to tune and control the thermal stability as well as the microstructure evolution upon heating, which is a necessary prerequisite for any possible commercial application of such glassy–matrix composites.

6.2 Formation and Crystallization of Glassy Materials

Generally a glassy phase can be formed through two ways, that is, through liquid-to-solid and through solid-to-solid transformation (see also Chapter 13 and Chapter 17). The theory of glass formation has been described in several review articles or books, such as in References 50, 52–56. Therefore, only the essential elements are reviewed here. The liquid-to-solid transformation, for example, rapid solidification, consists of cooling a liquid so quickly that crystallization has not enough time to occur. On continued cooling, as the temperature is lowered, the viscosity of the liquid increases and the atomic mobility decreases. At low cooling rate the liquid is able to crystallize, but crystallization can be avoided if the cooling is sufficiently rapid, resulting in a progressive freezing of the liquid configuration. That is, the viscosity becomes so high and the atomic mobility so slow that the liquid cannot change its structure rapidly enough to stay in internal equilibrium and, ultimately, the final product is a solid with a liquid-like atomic arrangement. In contrast, the solid-to-solid transformation, for example, interdiffusion in thin film diffusion couples[57,58] or mechanical alloying (MA) of elemental powder mixtures for the formation of an amorphous phase,[59–61] involves an increase in the energy of the starting crystalline material by the addition of some externally provided energy, and the storage of this energy in the crystal up to a point at which it becomes unstable with respect to the amorphous state. The highly energized material then lowers its energy by transforming into a different atomic structural arrangement, that is, the glass.[62]

Regardless of the processing route used for their production, amorphous materials are not in a state of internal equilibrium and, when heated to a sufficiently high temperature, they tend to a more stable condition. Upon annealing below the glass transition temperature, the glass initially relaxes toward a state corresponding to the ideal frozen liquid with lower energy.[50] The structure evolves to one with higher density, which could be considered characteristic of glass formation at a slower cooling rate[50] and finally, above the glass transition temperature, the glass crystallizes.

Structural relaxation. Structural relaxation involves complex phenomena that occur during annealing of glassy materials at temperatures below the glass transition. On heating, starting from a metastable condition with respect to the corresponding crystalline state, the glassy structure reduces its free energy by undergoing small rearrangements through short-range ordering (SRO).[50,52] As a consequence, several changes in the physical properties of the material, for example, density, hardness, and Young's modulus, can be observed.[63,64] Such properties may change upon thermal relaxation in an irreversible or reversible way as the annealing process is cycled. Egami proposed a distinction between these two different types of relaxation behavior:[65] topological short-range ordering (TSRO) and chemical short-range ordering (CSRO). TSRO describes the density of packing of atoms. It involves the rearrangement of atomic sites in the amorphous structure and thus requires considerable atomic mobility. On relaxation annealing, TSRO changes through the progressive, irreversible removal of free volume, and this causes irreversible changes in properties such as the density. Differently, CSRO involves the rearrangement of chemically different atomic species on fixed sites without appreciable variation of density. CSRO is assumed to change reversibly with temperature. This accounts for the reversible changes in particular properties, such as elastic properties.[50]

Crystallization. Following structural relaxation, crystallization concludes the path of the metastable glass toward the stable crystalline state. Crystallization studies on metallic glasses are of primary importance not only to analyze their thermal stability against crystallization but also to investigate the fundamental aspect of the processes of nucleation and growth.[66] As the crystallization process upon annealing of a glass is much slower than during solidification of liquids, it is relatively easier to investigate crystallization in glasses than in liquids. Crystallization of glasses/SCLs is known to generally proceed by a nucleation and growth mechanism[67] in a similar way as for liquids below their *liquidus* temperature. The first step that initiates crystallization is nucleation, which selects the starting phase for solidification, which can be stable or metastable, depending on its thermodynamic and kinetic characteristics. Following nucleation, growth completes the crystallization.

A metallic glass can be considered as a metastable deeply undercooled liquid[50] and, consequently, has the tendency to crystallize. The driving force for crystallization is the free energy difference between the metastable glass and the more stable crystalline or quasicrystalline (QC) phase(s).[66] The dimensions and morphologies of the crystallization products strongly depend on the transformation mechanism, which is closely related to the chemical composition of the amorphous phase and to the thermodynamic properties of the corresponding crystalline phase.[68] According to Köster and Herold[67] the crystallization reactions that may occur during devitrification can be classified depending on the composition change involved: polymorphous, eutectic, and primary crystallization. In *polymorphous* crystallization the glass transforms to a single phase with different structure but with same composition as the glass. This reaction can only occur in concentration ranges near to those of pure elements or stable compounds and needs only single jumps of atoms across the crystallization front. Crystallization of a single phase accompanied by a composition change is referred to as *primary* crystallization. During this reaction, a concentration gradient occurs at the interface between the precipitate and the matrix until the reaction reaches the metastable equilibrium. Finally, a crystallization in which two phases simultaneously and cooperatively precipitate from the glass is termed *eutectic* crystallization. This reaction has the largest driving force and can occur in the whole concentration range between two stable crystalline phases. There is no difference in the overall concentration across the reaction front and diffusion takes place parallel to the reaction front. Therefore, this type of reaction is slower compared to a polymorphous crystallization, which does not show any separation between the components.

The possible crystallization reactions are illustrated by the hypothetical free energy diagram in Figure 6.2, where the free energy curves for the metastable glassy and β phases are plotted as a function of composition, together with those of the stable crystalline phases α and γ. Examples of polymorphous transformations are the reactions 1 and 5 in which the glass crystallizes to the α and β phases, respectively, without change in composition. The devitrification products of the polymorphous reaction can subsequently decompose to the equilibrium mixture of α and γ (reactions 1′ and 5′). Alternatively, the glass can reduce its free energy to a point on the common tangent between α and β or α and γ (reactions 3 and 6, respectively) via a eutectic mechanism with the simultaneous precipitation of two crystalline phases. Reaction 2 illustrates an example of primary transformation. The glass crystallizes by the precipitation of the phase α with composition C_α, which is different from the composition of the parent glassy phase. As a consequence, solute partitioning takes place and the residual glassy matrix of changed composition (C_0) may subsequently transform by one of the mechanisms described above (e.g., reaction 4).

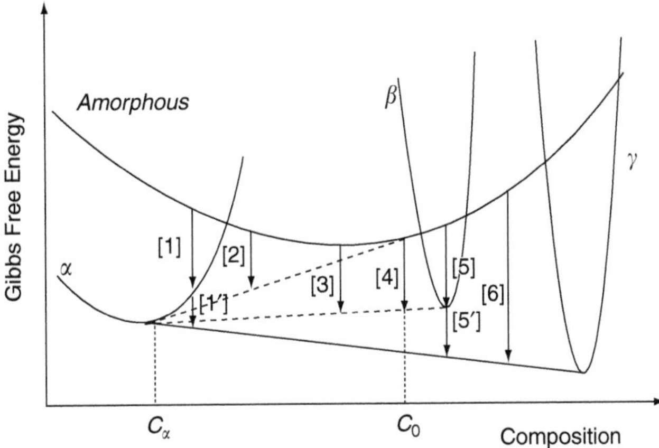

FIGURE 6.2 Schematic free energy diagram as a function of composition for the metastable glass and β phases, and for the stable crystalline α and γ phases, displaying the possible crystallization reactions.

Metallic glasses can be used as precursors for nanocrystalline materials, perhaps the most attractive microstructure from the point of view of functional properties. The crystallization product made from the glass can have a very fine and uniform microstructure.[50] The prerequisites for the formation of a fine-grained material by crystallization are low growth rates combined with very high nucleation rates.[69] Among the different types of crystallization modes, primary crystallization reactions are expected to give the finest microstructure.[69] Primary transformations are inherently more complex than polymorphous reactions. While polymorphous reactions involve only local topological rearrangement, in which atomic movements are of the order of the interatomic distances, primary transformations, which require composition changes, are characterized by long-range atomic transport over distances that are at least as large as the critical nuclei of the product phase.[70] In primary crystallization the composition of the emerging nuclei differs from the parent amorphous matrix and, as a result, the composition of the matrix changes during the transformation. At the beginning of the transformation, for small particles the interface reaction is most likely the rate-controlling step, as the diffusion distances are very short.[71] Once the particles have grown to a certain size, the surrounding matrix, whose composition is enriched in the atoms rejected from the growing particles, approaches saturation and the associated reduction in the driving force makes diffusion to be the rate-controlling step, slowing down grain growth.[71]

Amorphous phase separation A large GFA is generally related to deep eutectics, indicating a strong negative enthalpy of mixing of the constituents.[67] However, there is some evidence for the existence of systems which show, at least in some concentration range, a different behavior, that is, a positive enthalpy of mixing, leading to a miscibility gap that may be due to singularities in the structure of the amorphous state.[67] Therefore, a concentration range exists where the free energy of a mixture of two coexisting glassy phases is lower than that of a single glassy phase. Glassy materials in this concentration range not only have the possibility of transforming by one of the crystallization reactions mentioned above, but may also decompose into two amorphous phases, that is, may undergo amorphous phase separation.[67]

Amorphous phase separation occurring by nucleation and growth processes, or even by spinodal decomposition without any nucleation process, are well-known for oxide glasses.[72,73] There, extremely fine-grained glass ceramics (partially crystallized glasses) can be produced, which is generally assumed to be a result of amorphous phase separation followed by subsequent crystallization. As illustrated in Figure 6.3, for an alloy that exhibits a positive enthalpy of mixing, a concentration range between *a* and *b* exists, where phase separation into two glassy phases can occur by nucleation and growth or in a narrower composition range between *c* and *d* by spinodal decomposition[69] (See also Chapter 4). The spinodal is the boundary between the metastable and the unstable state of the phase. In the metastable state, a finite fluctuation, that is, a nucleus of the new phase, is required for any phase transformation.[69] In contrast,

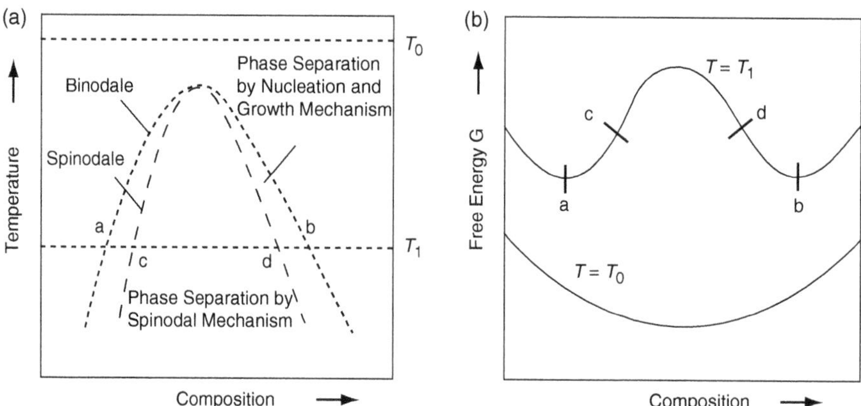

FIGURE 6.3 Spinodal decomposition: (a) schematic phase diagram showing the phase boundary and the spinodal of a two-liquid immiscibility region, (b) free energy vs. composition diagrams for the temperatures given in (a). (From Köster, U., *Mater. Sci. Forum*, 235–238, 377, 1997. With permission.)

beyond the spinodal, a single phase is unstable with respect to infinitesimal fluctuations in composition and begins to separate into two related phases with different composition without nucleation.[69] In binary systems, the condition for metastability is given by $\partial^2 G/\partial c^2 < 0$, where G is the molar free enthalpy and c is the mole fraction.[69] The driving force for increasing the composition fluctuations is the gain in free energy upon decomposition. Outside this concentration range decomposition into two amorphous phases will proceed by nucleation and growth processes.[69]

Amorphous phase separation is expected to have a strong potential for increasing the nucleation rate, and to decrease the growth rate of crystals.[74] Therefore, it promotes nanocrystal formation. For example, an increase in the nucleation rate may be due to an increase in the diffusivity of the respective element(s), a decrease in the crystal/glass interfacial energy, a local increase in the thermodynamic driving force, or heterogeneous nucleation of crystals at the amorphous/amorphous interface.[74] On the other hand, phase separation is expected to reduce the crystal growth rates due to composition shifts induced by the phase separation and by interference between the growing crystals and the amorphous regions.[69]

Amorphous phase separation has been reported in Pd–Au–Si,[75] Pd–Ni–P,[76] or Ti–Zr–Be systems,[77] either prior to crystallization or even in the as-quenched state. Subsequent reinvestigation of the Ti–Zr–Be alloy ruled out the possibility of phase separation.[78] Based on small-angle scattering results, phase separation in the amorphous phase has been projected to operate in Zr–Ti–Cu–Ni–Be metallic glasses.[79,80] It was proposed that the occurrence of phase separation in the undercooled liquid state is thermodynamically caused by a miscibility gap that opens up between two thermodynamically favored undercooled melts that are Be-rich and Zr-rich.[81] However, a recent study,[82] using a three-dimensional atom probe (3DAP), small-angle x-ray scattering (SAXS), and transmission electron microscopy (TEM) has convincingly proved that there is no amorphous phase separation preceding the precipitation of a nanoscale icosahedral phase. Recently, a $La_{27.5}Zr_{27.5}Al_{25}Cu_{10}Ni_{10}$ alloy was found to show a strong separation into two glassy phases of different composition.[83] One of the phases is La–Cu-rich while the other one is Zr–Ni-rich. Microstructure studies revealed two clearly distinct amorphous phases. They are spherical in shape and their length scales vary from the nanometer to the micrometer range. In this case, it was proposed that the amorphous phase separation is due to the positive heat of mixing of the main alloying elements La and Zr.

Amorphous short-range order. To explain the observation that metallic liquids can be considerably undercooled below their melting temperatures without crystallization,[84] Frank proposed in 1952 that in liquids most of the groups of 12 atoms around one central atom may be arranged as an icosahedron.[85] As the icosahedron has fivefold rotational symmetry, such an icosahedral short-range order (ISRO) is incompatible with the translational symmetry of a crystal. Due to this structural difference, a substantial atomic rearrangement during cooling is needed to break the ISRO before the formation of the crystalline phases: a rearrangement that is costly in energy and thus may create a barrier to the nucleation of the crystalline phases, explaining the considerable undercooling. Frank's assumption was later supported by computer simulations for an undercooled Lennard-Jones liquid[86] and by mass spectra of free xenon clusters with numbers $n = 13$, 55, and 147, the numbers of spheres required for a complete icosahedral shell.[87]

The degree of ISRO is predicted to increase with increasing undercooling of the liquid[86] and as metallic glasses may be considered as deeply undercooled liquids, it has been proposed that in glassy alloys ISRO may be quenched-in during vitrification,[88] which subsequently may promote the formation of quasicrystals (QC) upon crystallization of the glass/SCL. The presence of the ISRO in metallic glasses has been recently corroborated by Saida et al.[89] By means of high-resolution transmission electron microscopy (HRTEM) they found ordered regions of about 2 nm in size in the as-quenched $Zr_{70}Pd_{30}$ glassy alloy, which forms quasicrystals as primary phase upon devitrification. After annealing far below the crystallization temperature the ordered regions grow and the nanobeam electron diffraction pattern shows clearly their icosahedral structure, strongly indicating that ISRO occurs also in the glassy state.

The idea that a particular quenched-in short-range order (i.e., icosahedral) in the glass is necessary for QC formation upon devitrification has been also indirectly supported by investigations of the crystallization behavior of Zr-based glassy powders produced by solid-state reaction (SSR).[90,91] As against quenching from the melt, glassy alloys can be produced by SSR processing, such as by MA of elemental powder mixtures.[59–61,92] It has been shown for Zr-based glassy powders produced by MA[91] that, in contrast to the corresponding alloys produced by quenching from the melt, the milled powders form the more stable crystalline phases upon crystallization instead of the metastable QC phase. This may suggest that MA, as a solid-state processing route that does not involve quenching from the melt,[92] precludes the formation of the quenched-in ISRO, thus leading to a different SRO than that obtained by liquid quenching and, consequently, to a different crystallization behavior. This hypothesis would imply that the production of QC-forming Zr-based glassy powders may not be possible via SSR. However, this is in contrast with what has been reported for amorphous Ni–Zr alloys prepared by MA and melt spinning (MS).[93] By evaluating the hydrogen storage properties it was found that the SRO of the differently prepared samples is essentially the same. In spite of that, so far no theoretical constraints or direct experimental evidence have demonstrated that the theorized ISRO in metallic glasses can be exclusively achieved by quenching from the melt. Instead, QC formation directly during MA has been reported for Al-based alloys[94] and recently in solid-state synthesized Zr-based glassy powders, which undergo QC formation upon crystallization.[95,96] This indicates that if a particular short-range order is necessary for the formation of quasicrystals, even if this short-range order is icosahedral, it can be achieved also by solid-state processing with no need of quenching from the melt. In addition, Mattern et al.[97] did not find among different ZrTiCuNiAl glassy alloys any special atomic arrangement for the special alloy forming quasicrystals upon heating, indicating that the occurrence of a particular quenched-in SRO is not the unique prerequisite for quasicrystal formation.

6.3 Variables Influencing the Crystallization of Metallic Glasses

6.3.1 Effect of Composition

The high GFA of multicomponent Zr-based alloys can be used for the production of fully glassy bulk samples with dimensions in the millimeter to centimeter range, or for the formation of bulk nanostructured materials. However, the phase selection upon crystallization is strongly affected by the chemical composition of the glassy phase. To obtain nanostructured materials, the glassy specimens are typically annealed at temperatures within the SCL region or close to the onset of crystallization. This procedure is based on the results first obtained for rapidly quenched thin ribbons, where sequential crystallization was observed for a variety of Zr-based bulk glass-forming alloys, such as Zr–Al–Ni–Cu–M (M = Ag, Au, Pt, Pd),[45,46,98] Zr–Cu–Al–Ni–Ti.[99–101] The stepwise crystallization behavior leads to primary precipitation of intermetallic or QC phases from the SCL, which are embedded in a residual glassy phase with changed composition.

Figure 6.4a displays the differential scanning calorimetry (DSC) scans for as-cast glassy $Zr_{62-x}Ti_xCu_{20}Al_{10}Ni_8$ bulk samples ($x = 0$, 3, 5, and 7.5) as typical examples for bulk glass-forming Zr-based alloys.[33,102] $Zr_{62}Cu_{20}Al_{10}Ni_8$ crystallizes via one sharp exothermic peak pointing to simultaneous formation of intermetallic compounds.[33,102] Upon Ti addition, the crystallization mode changes toward a double-step process indicating a successive stepwise transformation into the equilibrium compounds while maintaining an extended SCL region between the glass transition temperature T_g^{on} and the crystallization temperature T_x^{on} (T_g^{on} and T_x^{on} are defined as the onset temperatures of the endothermic glass transition and the exothermic crystallization events, respectively). With increasing Ti content, the first DSC peak shifts to lower temperatures and the enthalpy related to the second exothermic peak decreases.[33,102] To further investigate the crystallization process, the samples were isothermally annealed for different times below T_x^{on}. For $x = 0$, partial crystallization maintains almost the same T_g^{on} value while the crystallization peak shrinks corresponding to a smaller crystallization enthalpy and T_x^{on} is shifted to a lower temperature. For the Ti-containing samples, the first exothermic peak is eliminated upon annealing but the second peak remains unchanged, as checked by a subsequent DSC scan. This indicates a primary crystallization mode leading to precipitates embedded in a residual glassy matrix.

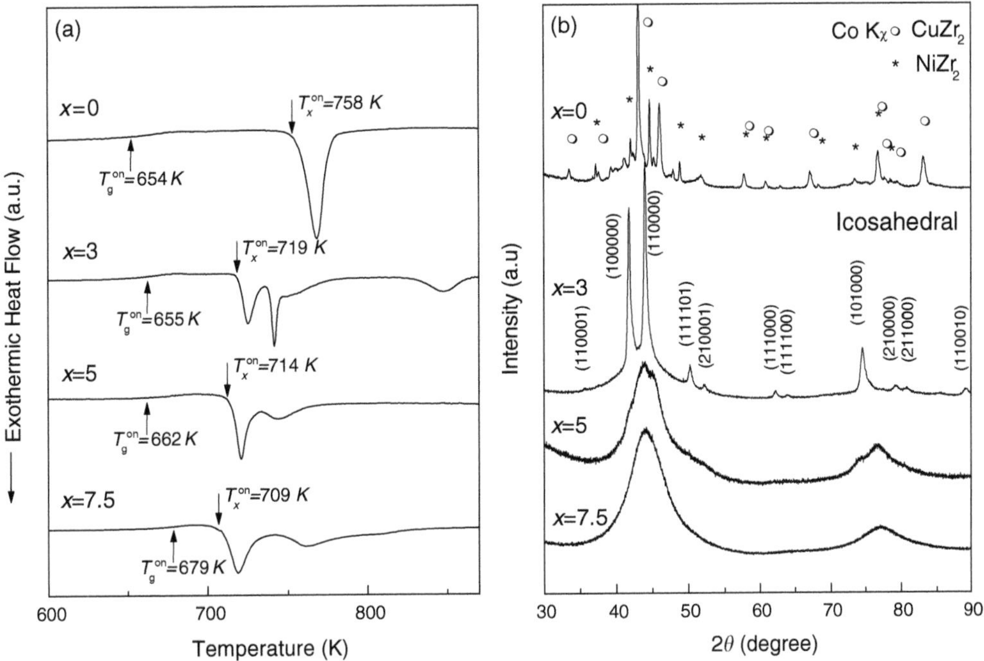

FIGURE 6.4 (a) DSC scans for $Zr_{62-x}Ti_xCu_{20}Al_{10}Ni_8$ glassy alloys. (b) corresponding XRD patterns after isothermal annealing: $x = 0$, annealed at 723 K for 30 min; $x = 3$, annealed at 703 for 5 min; $x = 5$, annealed at 683 K for 30 min and $x = 7.5$ annealed at 688 K for 40 min. (From Eckert, J., Kuhn, U., Mattern, N., Reger-Leonhard, A., and Heilmaier, M., *Scripta Mater.*, 44, 1587–1590, 2001. With permission from Elsevier Science Ltd.)

The nature of the crystallization products and the resulting microstructure after annealing were investigated by x-ray diffraction (XRD) (Figure 6.4b).[102] $Zr_{62}Cu_{20}Al_{10}Ni_8$ transforms into cubic $NiZr_2$- and tetragonal $CuZr_2$-type compounds. Annealing the alloy with $x = 3$ leads to primary precipitation of an icosahedral QC phase with spherical morphology and a size of about 50 to 100 nm.[33,102] For $x = 5$, the diffraction peaks are weaker in intensity and broader because the precipitates are as small as about 5 nm. For $x = 7.5$, the precipitates are about 3 nm in size. At first glance the XRD pattern after annealing displays no obvious reflections but only broad amorphous-like maxima. However, careful examination of the annealed state by high intensity synchrotron radiation clearly shows differences in scattering intensity compared to the as-cast state, pointing to precipitation of a metastable cubic phase with a grain size of about 2 nm coexisting with a residual glassy phase. This is consistent with the TEM results and corroborates the finding that extremely fine grains/clusters are embedded in the glass. Similar results were reported for other Zr–Cu–Al–Ni multicomponent alloys containing Ti, Ag, Pd, or Fe.[46,103–105] This indicates that Zr-based multicomponent alloys are promising candidates for the production of bulk nanostructured quasicrystal-based two-phase materials.

Figure 6.5 shows the isothermal DSC curves for the $Zr_{57}Ti_5Cu_{20}Al_{10}Ni_5$ bulk glassy alloy annealed at 673 and 703 K.[22] Three significantly different regions can be distinguished. Initially, there is a short period of incubation without exothermic heat flow, followed by a sharp drop of the curve indicating pronounced exothermic heat flow and rapid crystallization. Finally, a third part with small exothermic heat flow is observed, which reveals the continuing crystallization of the residual glassy phase with lower crystallization rate. Annealing at 673 K yields an incubation time of about 4 min and a duration of rapid crystallization of about 40 min (Figure 6.5a). Increasing the annealing temperature to 703 K enhances diffusion and chemical redistribution, leading to reduced times for incubation (1 min) and rapid crystallization (7 min) (Figure 6.5b). This demonstrates that a desired volume fraction of crystallites can be obtained by adjusting the annealing temperature and time for a glass of given composition.[22,33,100]

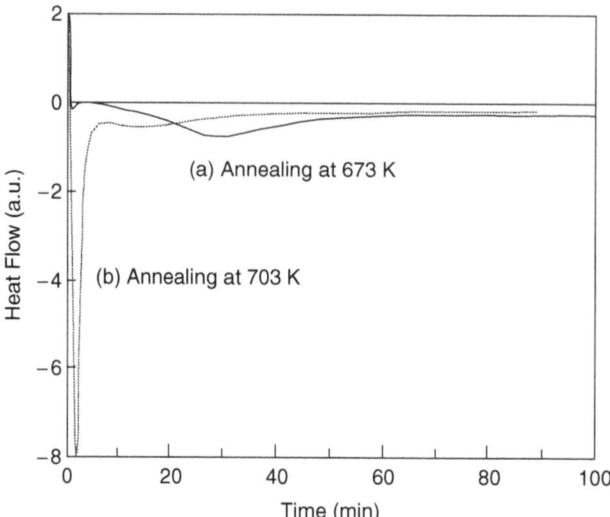

FIGURE 6.5 Isothermal DSC curves of a $Zr_{57}Ti_5Cu_{20}Al_{10}Ni_5$ amorphous alloy annealed (a) at 673 K and (b) at 703 K. (After Leonhard, A., Xing, L.Q., Heilmaier, M., Gebert, A., Eckert, J., and Schultz, L., *Nanostruct. Mater.*, 10, 805–817, 1998. With permission from Elsevier Science Ltd.)

Besides adjusting the annealing conditions, the desired microstructure can be achieved by choosing the appropriate chemical composition of the amorphous phase. When glass-composite materials are produced by partial crystallization of a glassy precursor, to predict and control the microstructure, knowledge of the crystallization mechanism is a necessary prerequisite. This is a crucial aspect for designing a material with the desired microstructure. In fact, with detailed knowledge of the crystallization mechanism, by choosing the appropriate composition and annealing conditions, it should be possible to tune and control the microstructure.

The effect of composition on the devitrification of Zr-based metallic glasses has been investigated in detail for melt-spun $Zr_{57}Ti_8Nb_{2.5}Cu_{13.9}Ni_{11.1}Al_{7.5}$ glassy ribbons. The $Zr_{57}Ti_8Nb_{2.5}Cu_{13.9}Ni_{11.1}Al_{7.5}$ alloy is of particular interest as it is characterized by different microstructures depending on the processing route used[95,96,106,107] and, therefore, it may offer interesting opportunities for a possible commercial application. For example, when the alloy is produced by copper mold casting, the as-cast microstructure consists of icosahedral QC particles of micrometer dimensions separated by a residual glassy phase.[106] On the other hand, when a higher cooling rate (i.e., MS) or SSR (i.e., ball milling) are used, the as-prepared samples are amorphous and devitrify by the precipitation of QC particles of nanometer dimensions embedded in the glass.[95,107]

Figure 6.6 shows the DSC scan of a $Zr_{57}Ti_8Nb_{2.5}Cu_{13.9}Ni_{11.1}Al_{7.5}$ as-spun ribbon.[107] In the range of temperature considered, the DSC curve is characterized by an endothermic event associated with the glass transition, T_g, indicating the transformation from the solid-state glass into the SCL, followed by two exothermic heat flow events due to the crystallization of the SCL at higher temperature, with onsets T_{x1} and T_{x2}, which reflect the thermal stability against crystallization of the SCL and of the primary phase, respectively.

The XRD patterns of the melt-spun $Zr_{57}Ti_8Nb_{2.5}Cu_{13.9}Ni_{11.1}Al_{7.5}$ ribbon heated to different temperatures are shown in Figure 6.7a. The as-spun ribbon shows only the typical broad maxima characteristic for amorphous materials with no crystalline peaks detected within the sensitivity limits of XRD. This is corroborated by the corresponding bright-field TEM image in Figure 6.7b. The image is featureless and characteristic of glassy material. In addition, the corresponding selected area electron diffraction pattern (inset in Figure 6.7b) shows the typical diffuse diffraction rings characteristic for this type of material. When the sample is heated up to 723 K, that is, above the first crystallization peak, the XRD pattern

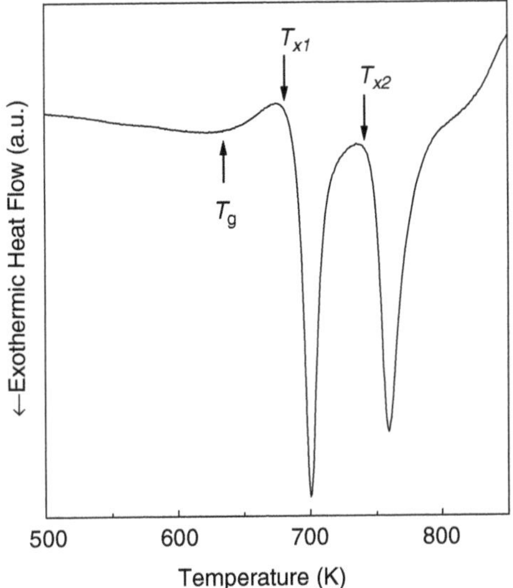

FIGURE 6.6 Constant heating rate DSC trace (heating rate 40 K min^{-1}) for the melt-spun $Zr_{57}Ti_8Nb_{2.5}Cu_{13.9}Ni_{11.1}Al_{7.5}$ glassy ribbon. (After Scudino, S., Kuehn, U., Schultz, L., Nagahama, D., Hono, K., and Eckert, J., *J. Appl. Phys.*, 95, 3397–3403, 2004. With permission from American Institute of Physics.)

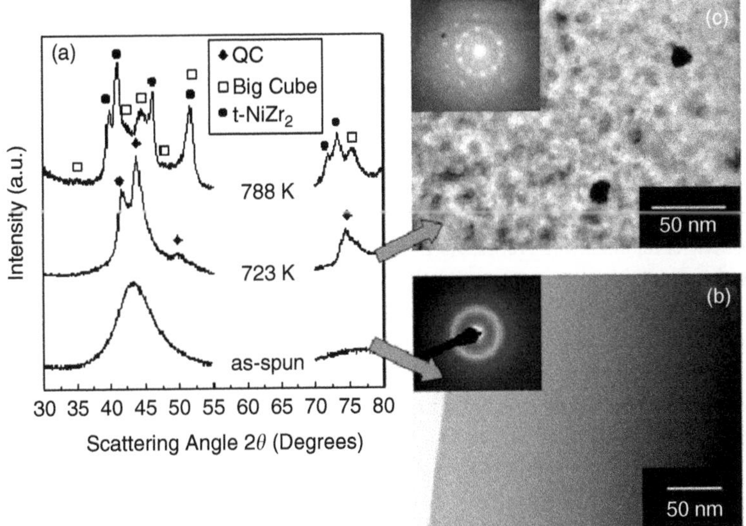

FIGURE 6.7 (a) XRD patterns (CoK_α radiation) for the $Zr_{57}Ti_8Nb_{2.5}Cu_{13.9}Ni_{11.1}Al_{7.5}$ as-spun ribbon and ribbon heated at 40 K min^{-1} up to different temperatures; Bright-field TEM images and corresponding nanobeam electron diffraction pattern (inset) of (b) as-spun ribbon and (c) ribbon heated up to 723 K.

displays some diffraction peaks, with positions and intensity ratios typical for powder diffraction patterns of icosahedral quasicrystals.[45,46,94,98–101,108,109] Besides the QC phase, broad halos due to the residual amorphous phase overlapping the diffraction peaks of the quasicrystals can be observed, indicating that the crystallization of the amorphous phase is not complete. The TEM micrograph of the ribbon after heating up to the completion of the first exothermic DSC event reveals a homogeneous distribution of particles with a size on the order of 5 to 10 nm. The nanobeam electron diffraction pattern presented in the inset in Figure 6.7c confirms the QC nature of the precipitated particles. It displays the features characterizing the fivefold rotational symmetry,[110] with no periodic distribution of spots, due to the icosahedral symmetry. Instead, groups of diffraction spots equidistant from the center are observed. The angles between the spots are $2\pi/10 = 36°$ and the distances along the radial directions are related to each other by $\tau = (1+\sqrt{5})/2$, the golden mean. In addition to the diffraction spots of the QC phase, a weak diffraction halo is visible, indicating the presence of a residual amorphous phase. Finally, heating up to the completion of the second exothermic DSC peak (788 K) leads to the formation of a tetragonal $NiZr_2$-type phase and a minor amount of fcc $NiTi_2$-type (*big cube*) phase.[111] The *big cube* peaks are remarkably broad, which suggests a crystallite size in the nanometer regime. No quasicrystal is detected in this stage of crystallization, indicating that the QC phase is metastable and transforms into more stable crystalline phase(s).

To completely clarify the crystallization process (i.e., polymorphous, eutectic or primary transformation), knowledge of the composition of the nucleating phase is of primary importance. The 3DAP investigations of the $Zr_{57}Ti_8Nb_{2.5}Cu_{13.9}Ni_{11.1}Al_{7.5}$ glassy ribbon revealed that the Al concentration is only about half of that in the matrix (~4.7 at.%). In this area, Zr is slightly enriched (from 57.5 at.% in the melt-spun glassy state to 61.7 at.% after annealing).[107] This proves that Al is depleted and Zr is enriched in the nanosized QC precipitates, revealing that partitioning of Al and Zr between the icosahedral and the amorphous phase occurs during crystallization. This indicates that QC formation in these alloys is a primary transformation (i.e., crystallization of a phase with a different composition than the amorphous matrix). This tendency is similar to that found for other Zr-based alloys.[112–114]

The redistribution of Al and Zr during QC formation indicates that the diffusion of these elements is an essential aspect of the mechanism of QC phase formation in the present alloy, suggesting the possibility to tune the thermal stability and the microstructure evolution of the glass by changing the amount of Al or Zr. For example, by increasing or decreasing the amount of Zr the formation of quasicrystals may be assisted or depressed, respectively. This behavior is important not only from a scientific point of view, to understand the effect of composition on the formation of quasicrystals, but it has also significant consequences on the possible engineering applications of such glassy–matrix composites due to the influence of the microstructure on the mechanical properties of the material.[22,33–35]

Another important aspect is the temperature range of stability of the quasicrystals. The utilization of such a partially devitrified material requires a wide temperature range between the formation of QC and the subsequent crystallization events. As thermal stability and microstructure evolution are strictly correlated with the chemical composition, detailed knowledge of the composition influence on both these features is extremely important.

The effect of Al[115] and Zr contents on the crystallization behavior of the $Zr_{57}Ti_8Nb_{2.5}Cu_{13.9}Ni_{11.1}Al_{7.5}$ alloy was analyzed by investigating two sets of melt-spun glassy ribbons with varying Zr and Al contents (Figure 6.8). All the DSC curves shown in Figure 6.8a and Figure 6.8b exhibit the same general behavior, namely, a distinct glass transition followed by two crystallization events at higher temperature. With decreasing Al or increasing Zr content the DSC peaks sharpen and shift toward lower temperatures. By decreasing the amount of Al the temperature of the glass transition (T_g) and the crystallization temperature related to the first crystallization event (T_{x1}) shifts by about 60 K to lower values, whereas by increasing Zr content T_g and T_{x1} are decreased by about 30 K. Therefore, varying the composition strongly affects the devitrification by lowering the temperature of the transformation from the solid-state glass to the SCL and by decreasing the stability of the SCL against crystallization.

When partially crystallized materials are considered, a fundamental parameter that has to be taken into account is the temperature range of stability of the phase formed, which in the present case is $\Delta T_{x2} = T_{x2} - T_{x1}$, the temperature interval between the first crystallization temperature, where the QCs form, and the subsequent crystallization event. For the alloy with $x = 7.5$, this interval was found to be about 51 K while for the alloy with $x = 0$ it increases to 84 K, which is 20 to 40 K wider compared with other multicomponent QC-forming alloys, such as $Zr_{65}Al_{7.5}Ni_{10}Cu_{12.5}X_5$ metallic glasses (X = Ag, Pd, Au, Pt, V, Nb, and Ta).[116]

The decrease of Al or increase of Zr contents strongly affects the devitrification process by decreasing the stability of the alloy against crystallization and giving rise to a strong increase of the temperature range

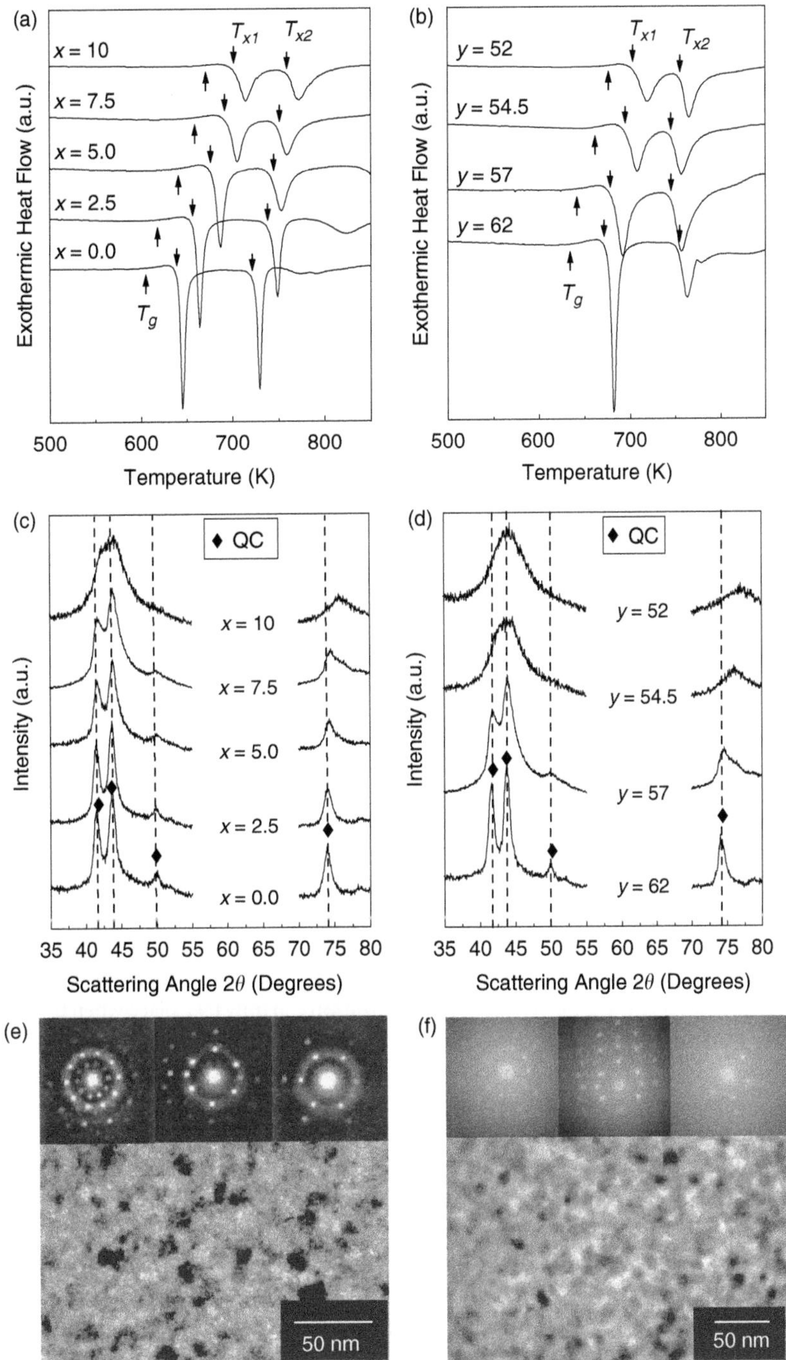

FIGURE 6.8 Constant-rate heating DSC scans (top) (heating rate 40 K min^{-1}), XRD patterns (CoK_α radiation) after heating up to the completion of the first crystallization event (middle) and bright-field TEM images and corresponding nanobeam electron diffraction patterns (bottom) of the melt-spun ribbons: (a, c, e) $(Zr_{0.616}Ti_{0.087}Nb_{0.027}Cu_{0.15}Ni_{0.12})_{100-x}Al_x$ with $x = 0$, 2.5, 5, 7.5, and 10; and (b, d, f) $Zr_y(Ti_{0.186}Nb_{0.058}Cu_{0.324}Ni_{0.258}Al_{0.174})_{100-y}$ with $y = 52$, 54.5, 57, and 62. The samples with $x = 7.5$ and $y = 57$ correspond to ribbons with composition $Zr_{57}Ti_8Nb_{2.5}Cu_{13.9}Ni_{11.1}Al_{7.5}$.

of stability of the QC phase. Therefore, if one is interested in glassy material it will be best to choose the alloys with $x = 10$ or $y = 52$ in which the glassy phase is more stable. On the other hand, if a partially crystallized material is required, ribbons with $x = 0$ or $y = 62$ are a better choice, as they are characterized by a wider temperature range of stability of the primary crystallized phase.

The phase formed after annealing the glassy ribbons up to the completion of the first exothermic peak was identified by XRD and the patterns are shown in Figure 6.8c and Figure 6.8d. With the exception of the samples with $x = 10$ and $y = 52$ and 54.5, which do not permit a univocal identification of the phase formed, the glassy ribbons devitrify by precipitation of a nanoscale QC phase. The formation of the QC phase during the first stage of the crystallization process even for the alloy with $x = 0$ implies that Al is not necessary for QC formation in the present alloy. With a decrease of the Al content the QC diffraction peaks sharpen, suggesting an increase of the QC grain size, and shift toward lower scattering angles, indicating an increase of the quasilattice constant, a_q. The same effect can be observed in the ribbons with high Zr content ($y = 57$ and 62). An overlapping broad diffraction background stemming from the residual amorphous phase can be observed in both groups of ribbons. The amorphous background is progressively less visible for decreasing Al or increasing Zr content, suggesting a decreasing amount of the residual glassy phase.

Figure 6.8e and Figure 6.8f display the bright-field TEM images of the ribbons with $x = 0$ and $y = 62$, respectively, after heating up to the completion of the first crystallization event. The images reveal a homogeneous distribution of particles with a size on the order of 10 to 20 nm. Obviously, these particles are larger than those observed after partial devitrification of the $Zr_{57}Ti_8Nb_{2.5}Cu_{13.9}Ni_{11.1}Al_{7.5}$ alloy (5 to 10 nm) (Figure 6.7c). The electron nanodiffraction patterns taken from the precipitated particles are presented in the insets in Figure 6.8e and Figure 6.8f. From left to right, the patterns exhibit five-, two-, and threefold rotational symmetries, which are the symmetries required for the icosahedral point group.[110,117] This confirms the results from XRD, corroborating that the particles precipitated in the first crystallization event have an icosahedral structure and that decreasing Al or increasing Zr contents yield larger QC grains. The crystallized volume fraction was found to be about 40 to 50 vol.% for both alloys with $x = 0$ and with $y = 62$, and is thus larger than that of the $Zr_{57}Ti_8Nb_{2.5}Cu_{13.9}Ni_{11.1}Al_{7.5}$ sample (20 to 30 vol.%).

The reduced thermal stability against crystallization and the increased size of the QC precipitates observed when the amounts of Zr and Al are appropriately varied can be attributed to the enhanced ease of precipitation of the QC phase from the SCL. If the nucleating crystalline phase(s) has a different composition with respect to the glass/SCL, then crystallization can only take place when the composition of a local region of the size of a critical nucleus satisfies the composition requirements for precipitation of the crystalline phase(s).[118] For multicomponent alloys, the probability of achieving a critical nucleus of the required composition is drastically reduced due to "confusion" of the system.[119] The nucleation of crystalline phases from the undercooled liquid requires significant atomic diffusion and redistribution.[118] As a result, the undercooled liquid is relatively stable against crystallization. The formation of quasicrystals in the ribbon with $y = 57$ requires a Zr enrichment of the nucleating phase and the simultaneous rejection of Al.[107] The increase of the Zr and the decrease of the Al content, therefore, favors the formation of the QC phase because it reduces the amount of elements that have to diffuse from/to the glassy phase/SCL to/from the arising quasicrystals. As a result, the formation of quasicrystals for the Zr-rich as well as for the Al-poor alloys may be shifted to a lower temperature. In the alloy with $x = 0$ and $y = 62$ the dimensions of the QC grains are larger than in the $Zr_{57}Ti_8Nb_{2.5}Cu_{13.9}Ni_{11.1}Al_{7.5}$ ribbon. This suggests a higher growth rate for the $x = 0$ and $y = 62$ alloys. Growth rates depend on the type of crystallization mechanism.[67] For example, primary crystallization reactions, which require long-range diffusion, are inherently more complex than a polymorphous transformation that involves only local topological atomic rearrangements, and thus are expected to give the finest microstructure.[69] Although for the alloys with the $x = 0$ and $y = 62$ the amorphous-to-QC transformation may still be primary, the increase of Zr and the reduction of Al contents reduces the required amount of atomic redistribution between the growing QC grains and the parent glassy phase/SCL. Most likely, this has a positive effect on the growth rate and, consequently, increases the dimensions of the QC precipitates.

6.3.2 Effect of Oxygen

Among the possible variables that can influence both the solidification and the crystallization behavior during heating of Zr-based alloys, oxygen contamination is of primary importance. For example, it has been reported for undercooled Zr–Ti–Cu–Ni–Al melts[120] that oxygen addition strongly affects crystal nucleation and can dramatically increase the necessary critical cooling rate for glass formation, thus limiting bulk glass formation and reducing the maximum attainable sample thickness. Over the range of oxygen content studied (300 to 5000 at. ppm), the time–temperature-transformation curves vary roughly by two orders of magnitude along the time axis. In other words, oxygen contamination ranging up to 0.5 at.% can increase the necessary cooling rate for glass formation by two orders of magnitude.[120]

The effect of oxygen on the devitrification of glassy materials has been studied in detail for melt-spun $(Zr_{0.65}Cu_{0.175}Ni_{0.10}Al_{0.075})_{100-x}O_x$ ribbons with $x = 0.2$, 0.4, and 0.8.[118] The $Zr_{65}Cu_{17.5}Ni_{10}\ Al_{7.5}$ is a well-known glass-forming system, which displays a SCL region of 127 K and can be cast into rods with diameter up to 16 mm, indicating a particularly low critical cooling rate for glass formation.[12] The addition of oxygen does not induce the formation of crystals during rapid solidification, as shown by XRD and TEM that proved that all the as-prepared samples are fully amorphous.[118] The DSC curves of the ribbons are characterized by an endothermic event associated with the glass transition, indicating the transformation from the solid-state glass into the SCL, followed by exothermic heat flow events due to the crystallization of the SCL at higher temperatures (Figure 6.9).[118] The ribbon with $x = 0.2$ exhibits a wide SCL region ($\Delta T_x = T_{x1} - T_g$) of more than 100 K above T_g and crystallizes by one sharp exothermic peak. With increasing oxygen content the DSC signal changes from a single sharp exothermic peak for $x = 0.2$ to two or three more or less well-resolved crystallization events, indicating a multistep transformation from the metastable SCL to the equilibrium crystalline intermetallic phases at different temperatures. Due to the splitting of the single crystallization peak observed for $x > 0.2$, the enthalpy release during the first step of crystallization decreases, whereas the heat release associated with

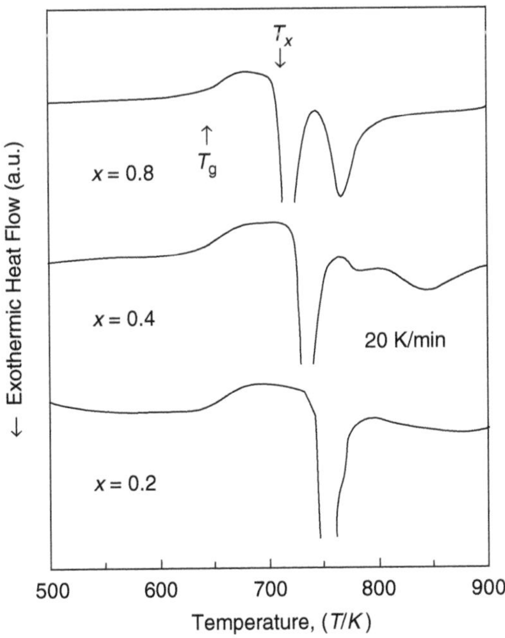

FIGURE 6.9 DSC traces of glassy $(Zr_{0.65}Al_{0.075}Cu_{0.175}Ni_{0.10})_{100-x}O_x$ melt-spun ribbons ($x = 0.2$, 0.4, and 0.8). (From Eckert, J., Mattern, N., Zinkevitch, M., and Seidel, M., *Mater. Trans. JIM*, 39, 623–632, 1998. With permission from Materials Transactions.)

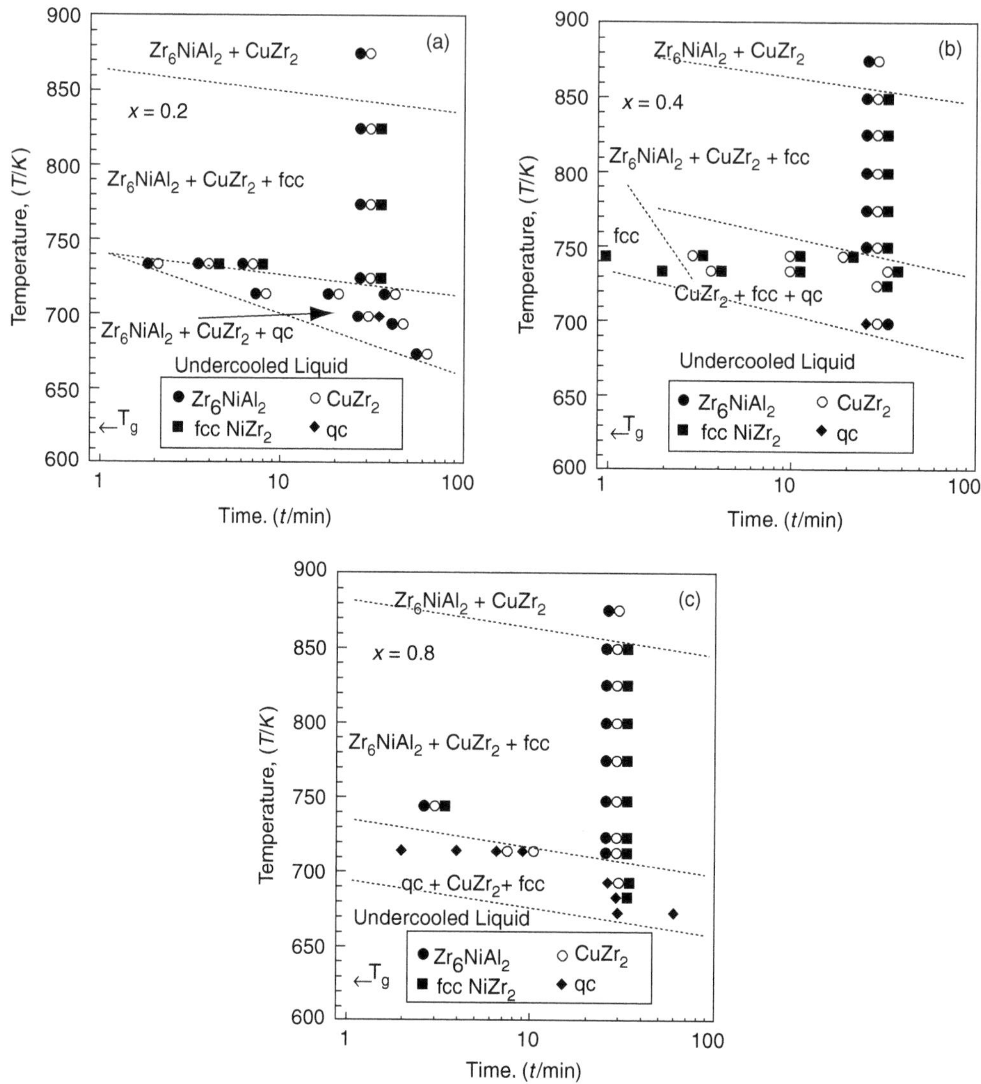

FIGURE 6.10 Schematic phase formation diagrams for $(Zr_{0.65}Al_{0.075}Cu_{0.175}Ni_{0.10})_{100-x}O_x$ ribbons with different oxygen content: (a) $x = 0.2$, (b) $x = 0.4$, and (c) $x = 0.8$. The dashed lines are only guidelines for the eye illustrating the temperature–time regimes where the different phases were detected. (From Eckert, J., Mattern, N., Zinkevitch, M., and Seidel, M., *Mater. Trans. JIM*, 39, 623–632, 1998. With permission from Materials Transaction.)

the exothermic events at higher temperatures increases with increasing oxygen content. Investigations on slowly cooled bulk samples of the same composition and comparable oxygen contents showed the same splitting of the DSC peaks for high oxygen contents.[121] The onset temperature of the glass transition, T_g, slightly increases whereas the onset temperature of the main exothermic crystallization event, T_x, decreases with increasing oxygen content. As a result, the extension of the SCL region, ΔT_x, is reduced for oxygen-rich samples.

The phases formed upon devitrification of the $(Zr_{0.65}Cu_{0.175}Ni_{0.10}Al_{0.075})_{100-x}O_x$ glassy ribbons were studied by temperature and time resolved *in situ* XRD.[118] Figure 6.10a to Figure 6.10c show schematic phase formation diagrams for the ribbons with different oxygen contents summarizing the *in situ* x-ray investigations.[118] The crystallization of the sample with $x = 0.2$ is governed by the simultaneous precipitation of QC, tetragonal Zr_2Cu and hexagonal Zr_6NiAl_2 phases, which are embedded in the residual amorphous matrix. A metastable fcc Zr_2Ni-type phase forms as an intermediate crystallization product. At elevated temperatures, the QC transforms into Zr_2Cu, and the fcc phase into hexagonal Zr_6NiAl_2. Sequential transformations from the SCL to the QC phase and intermetallic compounds are also observed for the samples with higher oxygen content. For the ribbon with $x = 0.4$ the investigations reveal that the first strong DSC peak (Figure 6.9) is related to the formation of quasicrystals, fcc Zr_2Ni-type and Zr_2Cu phases whereas the second small DSC signal is associated with the formation of the Zr_6NiAl_2 phase and with the transformation of the quasicrystals into Zr_2Cu. Finally, the third DSC peak is due to the crystallization of the residual amorphous phase and to the transformation of the fcc Zr_2Ni-type phase into Zr_6NiAl_2. For $x = 0.8$ the first crystallization event is due to the precipitation of the QC phase and the second DSC peak at higher temperatures corresponds to the transformation of the QC into Zr_2Cu together with the formation of Zr_6NiAl_2. The grain sizes of the different phases vary between 10 and 70 nm, depending on the annealing conditions.[118]

With increasing oxygen content the existence regions of the metastable fcc and QC phases become more extended. This supports the view that the presence of oxygen promotes the formation of these intermediate phases and stabilizes metastable structures against the transformation into the equilibrium compounds. However, the final crystallization products are the same regardless of the oxygen content of the material.

Similar results were obtained by different groups for multicomponent Zr-based alloys of different compositions.[112,120,122,123] In particular, Murty et al.[112] demonstrated that some amount of oxygen is needed for QC formation in $Zr_{65}Cu_{27.5}Al_{7.5}$ glass. By means of 3DAP, they found that the QC phase is enriched in oxygen, thereby demonstrating that the quasicrystals are stabilized by oxygen. The addition of oxygen lowers the thermal stability of the SCL against crystallization and is, therefore, detrimental. However, this leads to the precipitation of phases that cannot be obtained in highly pure metallic glasses and, by proper annealing, allows the synthesis of new composite materials with nanoscale microstructures from amorphous Zr-based alloys.

The oxygen-induced precipitation of the metastable fcc and QC phases can be rationalized by considering the effect of oxygen on the nucleation process. The high thermal stability of multicomponent Zr-based metallic glasses is generally attributed to the difficulty of precipitation of crystalline compounds from the undercooled liquid. The combination of elements with significantly different atomic sizes and negative enthalpies of mixing leads to a homogeneously mixed dense random packed structure of the liquid resulting in a large liquid–solid interface energy.[21] As already mentioned, if the nucleating phase has a different composition with respect to the homogeneous undercooled liquid, then the nucleation of the phase requires substantial atomic rearrangement.[118] Consequently, the undercooled liquid is relatively stable against crystallization. However, even small additions of oxygen drastically affect the thermal stability as well as the microstructure evolution upon heating. Increasing oxygen content shifts T_g to higher temperature (Figure 6.9). Most likely the small atomic size of oxygen is appropriate to fill up vacant sites in the disordered structure consisting of Zr, Cu, Ni, and Al atoms with large differences in atomic size, leading to an increased packing density of the amorphous solid.[118] This might enhance the thermal stability of the amorphous solid and, therefore, lead to an increase in T_g with increasing oxygen content. On the other hand, increasing amounts of oxygen lower the stability of the undercooled liquid with respect to crystallization, resulting in a reduced crystallization temperature and a smaller SCL region ΔT_x.[118]

Both the fcc and the QC phases are metastable and transform into equilibrium compounds at higher temperatures. As the formation of a metastable phase requires that kinetics allow the phase to nucleate and grow more rapidly than the stable crystalline phase, this suggests a higher nucleation rate for metastable phases than for equilibrium compounds. The nucleation is characterized by the work required to form a cluster of the critical size, W.[124] The work of critical cluster formation represents the barrier to nucleation, which arises from the interfacial energy σ between the nucleus and the parent glassy phase, and can be written as $W = 16\pi\sigma^3/3\Delta G^2$, where ΔG denotes the difference in Gibbs free energy between the glass and the nucleating phase, representing the driving force of the transformation.[69] In general, the nucleation rate can be influenced by the interfacial energy, the driving force, and the atomic mobility (diffusivity or

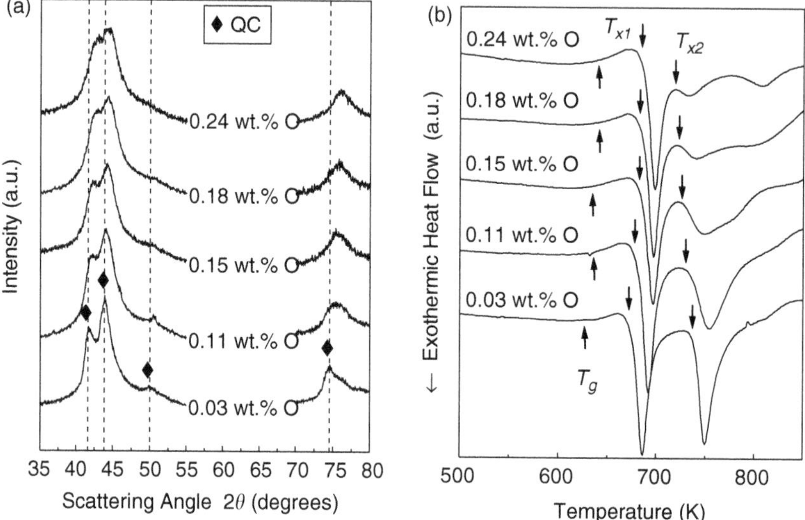

FIGURE 6.11 (a) XRD patterns (CoK_α radiation) for the melt-spun $Zr_{57}Ti_8Nb_{2.5}Cu_{13.9}Ni_{11.1}Al_{7.5}$ ribbons with different amounts of oxygen after heating up to completion of the first crystallization event. (b) DSC traces (heating rate 40 K min^{-1}) for the same melt-spun ribbon.

viscosity) of the system.[69] Most efficient for increasing the nucleation rate is a reduction of the interfacial energy, as the nucleation rate is extremely sensitive to the interfacial energy, σ, depending exponentially on σ^3. It is known that σ is smaller for metastable phases and can be reduced by small additions of other elements (e.g., oxygen).[69] For example, in $Co_{50}Zr_{50}$ metallic glasses, the shape of the stable CoZr phase changes significantly when the oxygen content of the glass increases to about 4 at.%.[69,74] This observation has been explained by a reduction of the interfacial energy by oxygen;[69] a significant increase of the driving force, ΔG, due to the small addition of oxygen is less likely. In addition, oxygen contamination in the range of about 1 at.% reduces the crystal growth rate by about one order of magnitude.[69] Hence, the metastable fcc and QC phases are expected to nucleate more easily in the presence of oxygen at lower temperatures than the equilibrium phases, thus reducing the thermal stability of the of the SCL against crystallization. As a result of the increased nucleation rate, the crystal growth is limited,[125] explaining the nanosized dimensions of the precipitates.

The addition of oxygen may eventually hinder the formation of quasicrystals. For example, oxygen contamination progressively suppresses QC formation in melt-spun $Zr_{57}Ti_8Nb_{2.5}Cu_{13.9}Ni_{11.1}Al_{7.5}$ glassy ribbons (Figure 6.11a). While the XRD pattern of the highly pure material (0.03 wt.%) heated up to the completion of the first crystallization event clearly shows the presence of an icosahedral QC phase, with increasing oxygen content the diffraction peaks belonging to the QC phase broaden, suggesting a decrease of the QC grain size, shift toward higher diffraction angles and finally are no longer clearly distinguishable for the sample with the highest oxygen content (0.24 wt.%). In addition, oxygen strongly influences the DSC scans (Figure 6.11b). With increasing oxygen content the first crystallization event is shifted to higher temperatures while the second peak is characterized by the appearance of a shoulder in the sample with 0.15 wt.% oxygen that further develops thus splitting the peak for the ribbons with higher oxygen content. These simultaneous effects give rise to a partial overlapping of the first two exothermic peaks with increasing oxygen content. While both the values of T_g and T_{x1} increase with increasing oxygen content, indicating that the addition of oxygen shifts the transformation from the solid-state glass into the SCL to higher temperatures and enhances the stability of the SCL against crystallization, T_{x2} noticeably decreases. This points out to a reduced stability of the primary phase formed upon crystallization.

As shown previously, the amorphous-to-QC phase transformation in the $Zr_{57}Ti_8Nb_{2.5}Cu_{13.9}Ni_{11.1}Al_{7.5}$ ribbon produced from highly pure materials is a primary transformation (i.e., crystallization of a phase with a different composition than the glassy matrix), characterized by the redistribution of Al and Zr between the amorphous and the QC phase. Al is depleted and Zr is enriched in the QC precipitates with respect to the parent glassy phase.[107] This implies that a variation of the composition of the amorphous phase involving an increase of Al or a decrease of Zr might affect the crystallization behavior, eventually inhibiting the formation of quasicrystals.

Such a variation of composition is most likely due to the preferential oxidation of Zr. During MS, in the liquid phase, oxygen might react with Zr giving rise to the formation of clusters rich in Zr–O bonds, which subsequently might be quenched-in during rapid solidification. As a result, the chemical composition of the glass will be depleted in Zr and, consequently, enriched in the other components (e.g., Al). This behavior of course opposes the mechanism for QC formation characterizing the present alloy, which in contrast requires the enrichment of Zr and the rejection of Al from the QC particles.[107]

This hypothesis is corroborated by analyzing the effect of Zr content on the crystallization behavior of the Zr–Ti–Nb–Cu–Ni–Al glassy system (Figure 6.8b and Figure 6.8d). The DSC scans shown in Figure 6.8b reveal that the decrease of the amount of Zr has a similar effect on the thermal stability data as increasing the oxygen content, that is, the values of T_g and T_{x1} both increase with decreasing Zr content. Furthermore, similar to the XRD patterns of the oxygen-contaminated ribbons (Figure 6.11a), the patterns of the samples with decreasing Zr content (Figure 6.8d) are characterized by a strong broadening of the diffraction peaks, suggesting a decrease of the dimensions of the QC precipitates below 5 to 10 nm. Due to such a line broadening the XRD patterns of the samples with $y = 54.5$ and 52 are characterized only by the presence of some modulations on the main amorphous maxima, indicating the formation of an ordered phase of nanoscale dimensions that cannot be clearly identified as QC, as it can be observed for the ribbon with the highest oxygen content (0.24 wt.%) in Figure 6.11a. Although some differences between the two sets of ribbons can be observed, for example, the absence of the splitting of the second DSC exothermic event in the ribbons with different Zr contents in contrast to the oxygen-contaminated samples, these findings strongly corroborate the earlier assumption that the variation of the crystallization behavior in the ribbons intentionally contaminated with oxygen originates from a compositional change of the glassy phase due to preferential oxidation of Zr.

6.3.3 Effect of Sample Preparation

Although several investigations have been performed since the discovery of a QC phase upon partial devitrification of Zr-based metallic glasses,[108] the question whether the lack of QC formation in Zr-based glassy alloys produced by solid-state processing routes, such as ball milling of intermetallic compounds or MA of elemental powder mixtures, is due to the absence of a special quenched-in short-range order in the glass or to the effect of impurities introduced during milling (e.g., oxygen) is still under discussion.[90,95] Generally, Zr-based alloys produced by solid-state techniques do not undergo QC formation upon crystallization.[90,91] This is in contrast to the devitrification behavior of the same alloy prepared by MS. A possible explanation for this difference might be the processing route dependence of the SRO in the glass. It has been proposed that in glassy alloys clusters of icosahedral SRO may be quenched-in during cooling from the melt,[88] which subsequently may promote the formation of quasicrystals. In this view, solid-state techniques, not involving quenching from the melt, should prevent the formation of these regions of icosahedral SRO, thus leading to a different crystallization behavior.

The effect of the processing route on the crystallization behavior, and in particular on the formation of quasicrystal upon partial devitrification of glassy alloys, has been investigated for multicomponent $Zr_{62}Ti_{7.08}Nb_{2.21}Cu_{12.28}Ni_{9.81}Al_{6.62}$ glassy alloys produced by different techniques.[96] The DSC scans of the $Zr_{62}Ti_{7.08}Nb_{2.21}Cu_{12.28}Ni_{9.81}Al_{6.62}$ glasses produced by MS as well as by MA are displayed in Figure 6.12 (left).[96] The onsets of the glass transition (T_g) and of the first crystallization event (T_{x1}) for the MA powder are shifted to higher temperatures with respect to the MS ribbon, indicating that the glassy phase formed for the powder is more stable against crystallization than that produced by MS.

When the melt-spun ribbon is heated up to the first crystallization peak, the XRD pattern ([a] in Figure 6.12 right)[96] reveals the formation of an icosahedral QC phase. Differently, the XRD pattern of the

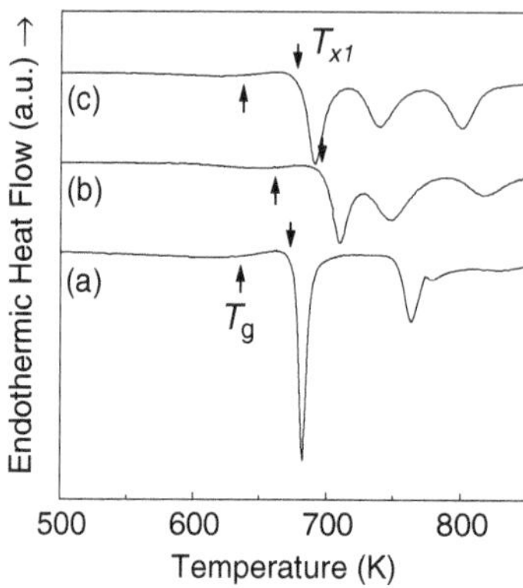

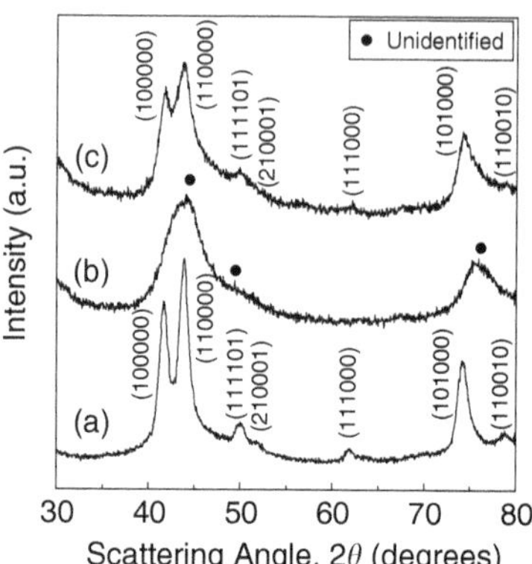

FIGURE 6.12 DSC traces (left) (heating rate 40 K min^{-1}) and XRD patterns (right) (CoK_α radiation) after heating up to the first exothermic DSC peak for: (a) $Zr_{62}Ti_{7.07}Nb_{2.21}Cu_{12.28}Ni_{9.81}Al_{6.62}$ melt-spun ribbon, (b) $Zr_{62}Ti_{7.07}Nb_{2.21}Cu_{12.28}Ni_{9.81}Al_{6.62}$ mechanically alloyed powder, and (c) $Zr_{67}Ti_{6.14}Nb_{1.92}Cu_{10.67}Ni_{8.52}Al_{5.75}$ mechanically alloyed powder. (After Scudino, S., Mickel, C., Schultz, L., Eckert, J., Yang, X.Y., and Sordelet, D.J., *Appl. Phys. Lett.*, 85, 4349–4351, 2004. With permission from American Institute of Physics.)

MA powder heated up to the completion of the first exothermic DSC peak ([b] in Figure 6.12 right)[96] shows some modulations on the main amorphous maxima that indicate the formation of an ordered phase of nanoscale dimensions that cannot be clearly identified.

The increase of T_g and T_{x1} with decreasing Zr content is a frequently reported feature of Zr-based amorphous alloys produced by MA as well as by MS. [14,126,127] It has also been reported that oxygen can have a similar effect on T_g and T_{x1} as decreasing the Zr content.[126,128,129] This is corroborated by the effect of Zr and oxygen on the thermal stability data presented in Figure 6.8b and Figure 6.11b, respectively, and indicates that the values of T_g and T_{x1} are strictly linked to the composition, suggesting a strong correlation between oxygen contamination and chemical composition or, more specifically between oxygen and Zr content. This correlation can explain the different crystallization behavior observed in the MA powder compared with the MS ribbon. It has been reported that QCs form in a narrow range of compositions [45,47,98–102,130]. Therefore, it is plausible that for the MA material oxygen selectively reacts with elemental Zr during milling, thus shifting the composition of the amorphous phase out of the range suitable for quasicrystal formation. Consequently, this leads to a different crystallization behavior compared to the ribbon with the same nominal composition. If this hypothesis is true, it should be possible to overcome the oxidation drawback and to produce MA powders, which undergo QC formation by adding an extra amount of Zr to the MA material. To verify this hypothesis, 5 at.% elemental Zr was added to the already amorphous MA powder to obtain a composition of $Zr_{67}Ti_{6.14}Nb_{1.92}Cu_{10.67}Ni_{8.52}Al_{5.75}$. This powder was then further milled for 100 h.[96]

The addition of elemental Zr has remarkable effects on both thermal stability and microstructure evolution.[96] The values of T_g and T_{x1} of the powder with extra Zr are shifted toward lower temperatures with respect to the parent MA material ([c] in Figure 6.12 left), corroborating the correlation between the thermal stability data and the Zr content of the alloy, as previously mentioned. The XRD pattern of the first crystallization product ([c] in Figure 6.12 right) reveals that the powder with extra Zr crystallizes by precipitation of a phase that, similar to the pattern of the MS ribbon (Figure 6.12a right), can be indexed as an icosahedral QC phase.[96] Figure 6.13[96] displays a bright-field TEM image of the powder with extra Zr after heating to the completion of the first crystallization event, revealing a homogeneous distribution of particles with a size on the order of 10 nm. The electron nanodiffraction pattern taken from the precipitated particles, presented in the inset in Figure 6.13, clearly exhibits a fivefold rotational symmetry.[96]

These results permit to depict the following scenario. During MA of the $Zr_{62}Ti_{7.08}Nb_{2.21}Cu_{12.28}$-$Ni_{9.81}Al_{6.62}$ MA powder, oxygen selectively reacts with elemental Zr. This preferential oxidation leads

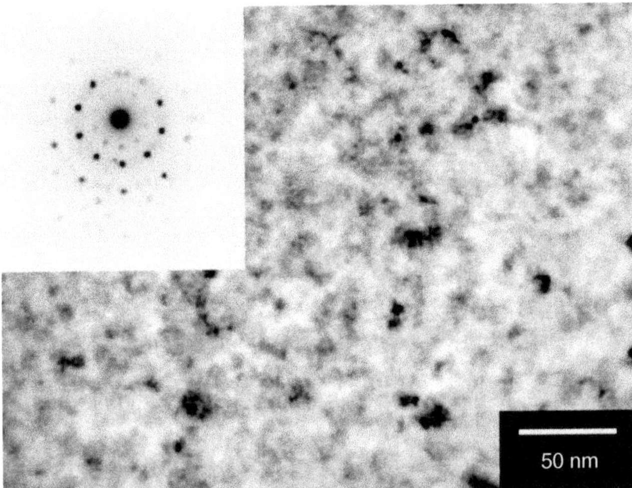

FIGURE 6.13 Bright-field TEM image and corresponding electron nanodiffraction pattern (inset) of the $Zr_{67}Ti_{6.14}Nb_{1.92}Cu_{10.67}Ni_{8.52}Al_{5.75}$ mechanically alloyed powder heated up to the first exothermic DSC peak, revealing icosahedral QC phase formation. (After Scudino, S., Mickel, C., Schultz, L., Eckert, J., Yang, X.Y., and Sordelet, D.J., *Appl. Phys. Lett.*, 85, 4349–4351, 2004. With permission from American Institute of Physics.)

to an amorphous phase depleted in Zr, which consequently has a different chemical composition and a different crystallization behavior with respect to the MS ribbon with the same overall composition. However, the oxidation problem can be bypassed and the quasicrystals can be formed by the addition of an appropriate amount of Zr. This demonstrates that in this multicomponent metallic glass the composition is the decisive factor for affecting QC formation and, in particular, it indicates that QC formation in the mechanically alloyed powder is crucially linked to the contamination of the starting material and to the real composition of the amorphous phase rather than to the question whether there is a special quenched-in SRO.

6.4 Summary

Besides direct synthesis during solidification of the melt, nanostructured glassy-matrix composites can be prepared by controlled crystallization of amorphous precursors. Although less simple than direct processing routes, due to the good flexibility in customizing the material properties, this two-stage technique (i.e., glass formation and subsequent crystallization) can be usefully employed when a material with specific properties, such as microstructure or thermal stability, is required. This is a very crucial aspect as the mechanical properties of composite materials are strongly dependent on their microstructure. For a given alloy composition the microstructure of the composite can be controlled by choosing the proper annealing conditions (i.e., temperature, time, and heating rate). On the other hand, if the crystallization mechanism is known, to obtain a material with the desired properties several features of the alloy can be tuned by appropriately varying the chemical composition. For example, by taking advantage of the primary character of the crystallization of $Zr_{57}Ti_8Nb_{2.5}Cu_{13.9}Ni_{11.1}Al_{7.5}$ glass, the thermal stability data can be varied within a broad range of values. The thermal stability against crystallization of the glassy phase can be tuned within a range of about 60 K and the temperature range of stability of the primary phase can be increased by about 30 K. As well, the range of accessible microstructures can be expanded from glassy-matrix composites characterized by a volume fraction of about 20 to 30 vol.% of precipitate particles with dimensions of less than 10 nm up to a volume fraction of about 40 to 50 vol.% and particle size of about 20 nm. Also, by knowing the crystallization mechanism, even a drawback like oxygen contamination can be turned into a positive factor for controlling the crystallization. In fact, oxygen not only induces the precipitation of phases that cannot be obtained in highly pure metallic glasses but it can be used to obtain extremely fine-grained precipitates. Finally, by investigating the devitrification of materials prepared by different processing routes, it is possible to analyze fundamental aspects of the crystallization process, such as the effect of postulated quenched-in nuclei on the phase selection during crystallization (i.e., on the formation of quasicrystals). The results indicate that if a particular short-range order is necessary for the formation of quasicrystals, even if this short-range order is icosahedral, it can be also achieved by solid-state processing without any need of quenched-in nuclei.

Acknowledgments

The authors gratefully acknowledge the support of Deutsche Forschungsgemeinschaft (DFG) (Grants No. EC 111/7, EC 111/8, EC 111/9, EC 111/10, EC 111/12, and Lu 217/17), EU within the framework of the research and training networks on Bulk Metallic Glasses (HPRN-CT-2000-00033) and Ductile BMG Composites (MRTN-CT-2003-504692) as well as Deutscher Akademischer Austauschdienst (DAAD). Special thanks are given to many coworkers, in particular, F. Baier, B. Bartusch, M. Calin, J. Das, S. Deledda, M. Frey, A. Gebert, R. Günther, G. He, W. Löser, N. Mattern, C. Mickel, H. Kempe, A. Kübler, U. Kühn, G. Kumar, A. Reger-Leonhard, H. Schulze, M. Seidel, M. Stoica, S. Venkataraman, T.G. Woodcock, and L.Q. Xing who have all contributed in one way or another to the results presented in this overview. Finally, the authors would like to thank R. Busch, A. Inoue, A.L. Greer, W.L. Johnson, D.H. Kim, C.C. Koch, U. Köster, J.H. Perepezko, K. Samwer, L. Schultz, D.J. Sordelet, A.R. Yavari, for many valuable discussions.

References

[1] Duwez, P., Willens, R.H., and Klement, W., Continuous series of metastable solid solutions in silver-copper alloys, *J. Appl. Phys.*, 31, 1136, 1960.

[2] Cohen, M.H. and Turnbull, D., Composition requirements for glass formation in metallic and ionic systems, *Nature*, 189, 131, 1961.

[3] Duwez, P., Willens, R.H., and Crewdson, R.C., Amorphous phase in palladium-silicon alloys, *J. Appl. Phys.*, 36, 2267, 1965.

[4] Ruhl, R.C. et al., New microcrystalline phases in the Nb–Ni and Ta–Ni systems, *Acta Metall.*, 15, 1693, 1967.

[5] Duwez, P. and Lin, S.C.H., Amorphous ferromagnetic phase in iron–carbon–phosphorous alloys, *J. Appl. Phys.*, 38, 4096, 1967.

[6] Ray, R., Giessen, B.C., and Grant, N.J., New non-crystalline phases in splat cooled transition metal alloys, *Scripta Metall.*, 2, 357, 1968.

[7] Chen, H.S., Thermodynamic considerations on the formation and stability of metallic glasses, *Acta Metall.*, 22, 1505, 1974.

[8] Kui, H.W., Greer, A.L., and Turnbull, D., Formation of bulk metallic glass by fluxing, *Appl. Phys. Lett.*, 45, 615, 1984.

[9] Inoue, A., Zhang, T., and Masumoto, T., Al–La–Ni amorphous alloys with a wide supercooled liquid region, *Mater. Trans. JIM*, 30, 965, 1989.

[10] Zhang, T., Inoue, A., and Masumoto, T., Amorphous Zr–Al–TM (TM = Co, Ni, Cu) alloys with significant supercooled liquid region of over 100 K, *Mater. Trans. JIM*, 32, 10005, 1991.

[11] Inoue, A. et al., Mg–Ni–La amorphous alloys with a wide supercooled liquid region, *Mater. Trans. JIM*, 30, 378, 1989.

[12] Inoue, A. et al., Preparation of 16 mm diameter rod of amorphous $Zr_{65}Al_{7.5}Ni_{10}Cu_{17.5}$ alloy, *Mater. Trans. JIM*, 34, 1234, 1993.

[13] Buschow, K.H., Thermal stability of amorphous Zr–Cu alloys, *J. Appl. Phys.*, 52, 3319, 1980.

[14] Altounian, Z., Tu, G.-H., and Strøm-olsen, J.O., Crystallization characteristics of Ni–Zr metallic glasses from $Ni_{20}Zr_{80}$ to $Ni_{70}Zr_{30}$, *J. Appl. Phys.*, 54, 3111, 1983.

[15] Peker, A. and Johnson, W.L., A highly processable metallic glass: $Zr_{41.2}Ti_{13.8}Cu_{12.5}Ni_{10.0}Be_{22.5}$, *Appl. Phys. Lett.*, 63, 2342, 1993.

[16] Kim, Y.J. et al., Metallic glass formation in highly undercooled $Zr_{41.2}Ti_{13.8}Cu_{12.5}Ni_{10.0}Be_{22.5}$ during containerless electrostatic levitation processing, *Appl. Phys. Lett.*, 65, 2136, 1994.

[17] Johnson, W.L., Bulk glass-forming metallic alloys: Science and technology, *MRS Bull.*, 24, 42, 1999.

[18] Busch, R., Bakke, E., and Johnson, W.L., Viscosity of the supercooled liquid and relaxation at the glass transition of the $Zr_{46.75}Ti_{8.25}Cu_{7.5}Ni_{10}Be_{27.5}$ bulk metallic glass forming alloy, *Acta Mater.*, 46, 4725, 1998.

[19] Busch, R., Liu, W., and Johnson, W.L., Thermodynamics and kinetics of the $Mg_{65}Cu_{25}Y_{10}$ bulk metallic glass forming liquid, *J. Appl. Phys.*, 83, 4134, 1998.

[20] Glade, S.C. et al., Thermodynamics of $Cu_{47}Ti_{34}Zr_{11}Ni_8$, $Zr_{52.5}Cu_{17.9}Ni_{14.6}Al_{10}Ti_5$ and $Zr_{57}Cu_{15.4}Ni_{12.6}Al_{10}Nb_5$ bulk metallic glass forming alloys, *J. Appl. Phys.*, 87, 7242, 2000.

[21] Inoue, A., Stabilization of metallic supercooled liquid and bulk amorphous alloys, *Acta Mater.*, 48, 279, 2000.

[22] Leonhard, A., Xing, L.Q., Heilmaier, M., Gebert, A., Eckert, J., and Schultz, L., Effect of crystalline precipitations on the mechanical behavior of bulk glass forming Zr-based alloys, *Nanostruct. Mater.*, 10, 805, 1998.

[23] Gilbert, C., Ritchie, R.O., and Johnson, W.L., Fracture toughness and fatigue-crack propagation in a Zr–Ti–Ni–Cu–Be bulk metallic glass, *Appl. Phys. Lett.*, 71, 476, 1997.

[24] Gebert, A. et al., Electrochemical investigations on the bulk glass forming $Zr_{55}Cu_{30}Al_{10}Ni_5$ alloy, *Mater. Corros.*, 48, 293, 1997.

[25] Kawamura, Y. et al., Superplastic deformation of $Zr_{65}Al_{10}Ni_{10}Cu_{15}$ metallic glass, *Scripta Mater.*, 37, 431, 1997.

[26] Masumoto, T. and Maddin, R., The mechanical properties of palladium 20 at.% silicon alloy quenched from the liquid state, *Acta Metall.*, 19, 725, 1971.

[27] Pampillo, C.A., Localized shear deformation in a glassy metal, *Scripta Metall.*, 6, 915, 1972.

[28] Wei, Q. et al., Evolution and microstructure of shear bands in nanostructured Fe, *Appl. Phys. Lett.*, 81, 1240, 2002.

[29] Inoue, A., Fan, C., and Takeuchi, A., Synthesis of high strength bulk nanocrystalline alloys containing remaining amorphous phase, *Mater. Sci. Forum*, 307, 1, 1999.

[30] Inoue, A. et al., Ductile quasicrystalline alloys, *Appl. Phys. Lett.*, 76, 967, 2000.

[31] Eckert, J. et al., Nanostructured materials in multicomponent alloy systems, *Mater. Sci. Eng. A*, 301, 1, 2001.

[32] Conner, R.D., Choi-Yim, H., and Johnson, W.L., Mechanical properties of $Zr_{57}Nb_5Al_{10}Cu_{15.4}Ni_{12.6}$ metallic glass matrix particulate composites, *J. Mater. Res.*, 14, 3292, 1999.

[33] Xing, L.Q. et al., High-strength materials produced by precipitation of icosahedral quasicrystals in bulk Zr–Ti–Cu–Ni–Al amorphous alloys, *Appl. Phys. Lett.*, 74, 664, 1999.

[34] Bian, Z., He, G., and Chen, G.L., Investigation of shear bands under compressive testing for Zr-base bulk metallic glasses containing nanocrystals, *Scripta Mater.*, 46, 407, 2002.

[35] Inoue, A. et al., High strength and good ductility of bulk quasicrystalline base alloys in $Zr_{65}Al_{7.5}Ni_{10}Cu_{17.5-x}Pd_x$ system, *Mater. Trans. JIM*, 40, 1137, 1999.

[36] Xing, L.Q., Eckert, J., and Schultz, L., Deformation mechanism of amorphous and partially crystallized alloys, *Nanostruct. Mater.*, 12, 503, 1999.

[37] Choi-Yim, H. and Johnson, W.L., Bulk metallic glass matrix composites, *Appl. Phys. Lett.*, 71, 3808, 1997.

[38] Kato, H. and Inoue, A., Synthesis and mechanical properties of bulk amorphous Zr–Al–Ni–Cu alloys containing ZrC particles, *Mater. Trans. JIM*, 38, 793, 1997.

[39] Choi-Yim, H., Busch, R., and Johnson, W.L., The effect of silicon on the glass forming ability of $Cu_{47}Ti_{34}Zr_{11}Ni_8$ bulk metallic glass forming alloy during processing of composites, *J. Appl. Phys.*, 83, 7993, 1998.

[40] He, G. et al., Novel Ti-base nanostructure–dendrite composite with enhanced plasticity, *Nat. Mater.*, 2, 33, 2003.

[41] Hays, C.C., Kim, C.P., and Johnson, W.L., Microstructure controlled shear band pattern formation and enhanced plasticity of bulk metallic glasses containing in situ formed ductile phase dendrite dispersions, *Phys. Rev. Lett.*, 84, 2901, 2000.

[42] Kühn, U. et al., ZrNbCuNiAl bulk metallic glass matrix composites containing dendritic bcc phase precipitates, *Appl. Phys. Lett.*, 80, 2478, 2002.

[43] Eckert, J. et al., Structural bulk metallic glasses with different length-scale of constituent phases, *Intermetallics*, 10, 1183, 2002.

[44] Szuecs, F., Kim, C.P., and Johnson, W.L., Mechanical properties of $Zr_{56.2}Ti_{13.8}Nb_{5.0}Cu_{6.9}$-$Ni_{5.6}Be_{12.5}$ ductile phase reinforced bulk metallic glass composite, *Acta. Mater.*, 49, 1507, 2001.

[45] Inoue, A., Zhang, T., and Kim, Y.H., Synthesis of high strength bulk amorphous Zr–Al–Ni–Cu–Ag alloys with a nanoscale secondary phase, *Mater. Trans. JIM*, 38, 749, 1997.

[46] Inoue, A. et al., Formation of icosahedral quasicrystalline phase in Zr–Al–Ni–Cu–M (M = Ag, Pd, Au or Pt) systems, *Mater. Trans. JIM*, 40, 1181, 1999.

[47] Chen, M.W. et al., Quasicrystals in a partially devitrified Zr65Al7.5Ni10Cu12.5Ag5 bulk metallic glass, *Appl. Phys. Lett.*, 75, 1697, 1999.

[48] Eckert, J. et al., Nanophase composites in easy glass forming systems, *Nanostruct. Mater.* 12, 439, 1999.

[49] Höland, W., Rheinberger, V., and Schweiger, M., Nucleation and crystallization phenomena in glass-ceramics, *Adv. Eng. Mater.*, 3, 768, 2001.

[50] Cahn, R.W. and Greer, A.L., Metastable states of alloys, in *Physical Metallurgy*, Cahn, R.W. and Haasen, P., Eds., Elsevier Science BV, Amsterdam, 1996, Chap. 19.

[51] Köster, U. and Janlewig, R., Microstructural design by controlled crystallization of metallic glasses, in *Mater. Res. Soc. Symp. Proc.*, Busch, R., Hufnagel, T., Eckert, J., Inoue, A., Johnson, W.L., and Yavari, A.R., Eds., 2004, 806.

[52] Luborsky, F.E., Eds., *Amorphous Metallic Alloys*, Butterworths, London, 1983.

[53] Güntherodt, H.J. and Beck, H., Eds., *Glassy Metals I*, Topics in Applied Physics Vol. 46, Springer-Verlag, Berlin Heidelberg, 1980.

[54] Güntherodt, H.J. and Beck, H., Eds., *Glassy Metals II*, Topics in Applied Physics Vol. 53, Springer-Verlag, Berlin Heidelberg, 1983.

[55] Güntherodt, H.J. and Beck, H., Eds., *Glassy Metals III*, Topics in Applied Physics Vol. 72, Springer-Verlag, Berlin Heidelberg, 1994.

[56] Weeber, A.W. and Bakker, H., Amorphization by ball milling. A review, *Phys. B*, 153, 93, 1988.

[57] Schwarz, R.B. and Johnson, W.L., Formation of an amorphous alloy by solid-state reaction of the pure polycrystalline metals, *Phys. Rev. Lett.*, 51, 415, 1983.

[58] Clemens, B.M., Solid-state reaction and structure in compositionally modulated zirconium–nickel and titanium–nickel films, *Phys. Rev. B*, 33, 7615, 1986.

[59] Koch, C.C. et al., Preparation of "amorphous" $Ni_{60}Nb_{40}$ by mechanical alloying, *Appl. Phys. Lett.*, 43, 1017, 1983.

[60] Hellstern, E. and Schultz, L., Amorphization of transition metal Zr alloys by mechanical alloying, *Appl. Phys. Lett.*, 48, 124, 1986.

[61] Schwarz, R.B., Petrich, R.R., and Saw, C.K., The synthesis of amorphous Ni–Ti alloy powders by mechanical alloying, *J. Non-Cryst. Solids*, 76, 281, 1985.

[62] Schultz, L. and Eckert, J., Mechanically alloyed glassy metals, in *Glassy Metals III*, Topics in Applied Physics Vol. 72, Güntherodt, H.J. and Beck, H., Eds., Springer-Verlag, Berlin Heidelberg, 1994, 69.

[63] Greer, A.L., Atomic transport and structural relaxation in metallic glasses, *J. Non-Cryst. Solids*, 62, 737, 1984.

[64] Chen, H.S., Structural relaxation in metallic glasses, in *Amorphous Metallic Alloys*, Luborsky, F.E., Ed., Butterworths, London, 1983, 169.

[65] Egami, T., Structure relaxation in metallic glasses, in *Structure and Mobility in Molecular and Atomic Glasses, Ann. New York Acad. Sci.* Vol. 371, New York, 1981, 238.

[66] Scott, M.G., Crystallization, in *Amorphous Metallic Alloys*, Luborsky, F.E., Ed., Butterworths, London, 1983, 144.

[67] Köster, U. and Herold, U., Crystallization of metallic glasses, in *Glassy Metals I*, Topics in Applied Physics Vol. 46, Güntherodt, H.J., and Beck, H., Eds., Springer-Verlag, Berlin Heidelberg, 1980, 225.

[68] Lu, K., Nanocrystalline metals crystallized from amorphous solids: Nanocrystallization, structure, and properties, *Mater. Sci. Eng. R*, 16, 161, 1996.

[69] Köster, U., Phase separation, crystallization and grain growth: Micromechanism in the design and stability of nanocrystalline alloys, *Mater. Sci. Forum*, 235–238, 377, 1997.

[70] Johnson, W.L., Thermodynamic and kinetic aspects of the crystal to glass transformation in metallic materials, *Prog. Mater. Sci.*, 30, 81, 1986.

[71] Clavaguera-Mora, M.T., The use of metastable phase diagrams in primary crystallization kinetics study, *Thermochim. Acta*, 314, 281, 1998.

[72] Ohlberg, S.M., Golob, H.R., and Stickler, D.W., Crystal nucleation by glass in glass separation, in *Symposium on Nucleation and Crystallization in Glasses and Melts*, Reser, M.K., Smith, G., and Insley, H., Eds., The American Ceramic Society, Columbus, OH, 1962, 55.

[73] Ramsden, A.H. and James, P.F., The effects of amorphous phase separation on crystal nucleation kinetics in BaO–SiO2 glasses. 1. General survey, *J. Mater. Sci.*, 19, 1406, 1984.

[74] Köster, U., Rüdiger, A., and Meinhardt, J., Influence of oxygen on nanocrystallization of Zr-based metallic glasses, *Mater. Sci. Forum*, 307, 9, 1999.

[75] Chou, C.P. and Turnbull, D., Transformation behaviour of Pd–Au–Si metallic glasses, *J. Non-Cryst. Solids*, 17, 169, 1975.

[76] Chen, H.S., Glass temperature, formation and stability of Fe,Co, Ni, Pd and Pt based glasses, *Mater. Sci. Eng.*, 23, 151, 1976.

[77] Tanner, L.E. and Ray, R., Phase separation in Zr–Ti–Be metallic glasses, *Scripta Metall.*, 14, 657, 1980.

[78] Nagahama, D., Ohkubo, T., and Hono, K., Crystallization of $Zr_{36}Ti_{24}Be_{40}$ metallic glass, *Scripta Mater.*, 49, 729, 2003.

[79] Löffler, J.F. and Johnson, W.L., Model for decomposition and nanocrystallization of deeply undercooled $Zr_{41.2}Ti_{13.8}Cu_{12.5}Ni_{10}Be_{22.5}$, *Appl. Phys. Lett.*, 76, 3394, 2000.

[80] Busch, R. et al., Decomposition and primary crystallization in $Zr_{41.2}Ti_{13.8}Cu_{12.5}Ni_{10}Be_{22.5}$ melts, *Appl. Phys. Lett.*, 67, 1544, 1995.

[81] Schneider, S. et al., SANS of bulk metallic ZrTiCuNiBe glasses, *Phys. B*, 241–243, 918, 1997.

[82] Martin, I. et al., Nanocrystallization of $Zr_{41.2}Ti_{13.8}Cu_{12.5}Ni_{10}Be_{22.5}$ metallic glass, *Acta Mater.*, 52, 4427, 2004.

[83] Kündig, A.A. et al., In situ formed two-phase metallic glass with surface fractal microstructure, *Acta Mater.*, 52, 2441, 2004.

[84] Turnbull, D., The subcooling of liquid metals, *J. Appl. Phys.*, 20, 817, 1949.

[85] Frank, F.C., Supercooling of liquids, *Proc. R. Soc. London A*, 215, 43, 1952.

[86] Steinhardt, P.J., Nelson, D.R., and Ronchetti, M., Bond-orientational order in liquids and glasses, *Phys. Rev. B*, 28, 784, 1983.

[87] Echt, O., Sattler, K., and Recknagel, E., Magic numbers for sphere packings: Experimental verification in free xenon clusters, *Phys. Rev. Lett.*, 47, 1121, 1981.

[88] Shen, Y., Poon, S.J., and Shiflet, G., Crystallization of icosahedral phase from glassy Pd–U–Si alloys, *Phys. Rev. B*, 34, 3516, 1986.

[89] Saida, J., Matsishita, M., and Inoue, A., Direct observation of icosahedral cluster in $Zr_{70}Pd_{30}$ binary glassy alloy, *Appl. Phys. Lett.*, 79, 412, 2001.
[90] Sordelet, D.J. et al., Synthesis route-dependent formation of quasicrystals in $Zr_{70}Pd_{30}$ and $Zr_{70}Pd_{20}Cu_{10}$ amorphous alloys, *Appl. Phys. Lett.*, 80, 4735, 2002.
[91] El-Eskandarany, M.S., Saida, J., and Inoue, A., Amorphization and crystallization behaviors of glassy $Zr_{70}Pd_{30}$ alloys prepared by different techniques, *Acta Mater.*, 50, 2725, 2002.
[92] Seidel, M. et al., Progress of solid-state reaction and glass formation in mechanically alloyed $Zr_{65}Al_{7.5}Cu_{17.5}Ni_{10}$, *Acta Mater.*, 48, 3657, 2000.
[93] Harris, J.H. Curtin, W.A., and Schultz, L., Hydrogen storage characteristics of mechanically alloyed amorphous metals, *J. Mater. Res.*, 3, 872, 1988.
[94] Eckert, J., Schultz, L., and Urban, K., Formation of quasicrystals by mechanical alloying, *Appl. Phys. Lett.*, 55, 117, 1989.
[95] Scudino, S. et al., Formation of quasicrystals by partial devitrification of ball-milled amorphous $Zr_{57}Ti_8Nb_{2.5}Cu_{13.9}Ni_{11.1}Al_{7.5}$, *Appl. Phys. Lett.*, 83, 2345, 2003.
[96] Scudino, S., Scudino, S., Mickel, C., Schultz, L., Eckert, J., Yang, X.Y., and Sordelet, D.J., Quasicrystal formation in mechanically alloyed Zr–Ti–Nb–Cu–Ni–Al glassy powders, *Appl. Phys. Lett.*, 85, 4349, 2004.
[97] Mattern, N. et al., Short-range order of $Zr_{62-x}Ti_xAl_{10}Cu_{20}Ni_8$ bulk metallic glasses, *Acta Mater.*, 50, 305, 2002.
[98] Inoue, A. et al., High strength and good ductility of bulk quasicrystalline base alloys in $Zr_{65}Al_{7.5}Ni_{10}Cu_{17.5-x}Pd_x$ system, *Mater. Trans. JIM*, 40, 1137, 1999.
[99] Xing, L.Q. et al., Effect of cooling rate on the precipitation of quasicrystals from the Zr–Cu–Al–Ni–Ti amorphous alloy, *Appl. Phys. Lett.*, 73, 2110, 1999.
[100] Xing, L.Q. et al., Crystallization behavior and nanocrystalline microstructure evolution of a $Zr_{57}Cu_{20}Al_{10}Ni_8Ti_5$ bulk amorphous alloy, *Phil. Mag. A*, 79, 1095, 1999.
[101] Fan, C. et al., Nanocrystalline composites with high strength obtained in Zr–Ti–Ni–Cu–Al bulk amorphous alloys, *Appl. Phys. Lett.*, 75, 340, 1999.
[102] Eckert, J., Kuhn, U., Mattern, N., Reger-Leonhard, A., and Heilmaier, M., Bulk nanostructured Zr-based multiphase alloys with high strength and good ductility, *Scripta Mater.*, 44, 1587, 2001.
[103] Fan, C. and Inoue, A., Improvement of mechanical properties by precipitation of nanoscale compound particles in Zr–Cu–Pd–Al amorphous alloys, *Mater. Trans. JIM*, 38, 1040, 1997.
[104] Fan, C. and Inoue, A., Preparation and mechanical properties of Zr-based bulk nanocrystalline alloys containing compound and amorphous phases, *Mater. Trans. JIM*, 40, 42, 1999.
[105] Mattern, N. et al., Formation of nanocrystals in Zr–Al–Cu–Ni alloys, *J. Metastable Nanocrystall. Mater.*, 8, 185, 2000.
[106] Kühn, U., Eckert, J., and Schultz, L., As-cast quasicrystalline phase in a Zr-based multicomponent bulk alloy, *Appl. Phys. Lett.*, 77, 3176, 2000.
[107] Scudino, S. et al., Microstructure evolution upon devitrification and crystallization kinetics of $Zr_{57}Ti_8Nb_{2.5}Cu_{13.9}Ni_{11.1}Al_{7.5}$ melt-spun glassy ribbon, *J. Appl. Phys.*, 95, 3397, 2004.
[108] Köster, U. et al., Formation of quasicrystals in bulk glass forming Zr–Cu–Ni–Al alloys, *Appl. Phys. Lett.*, 69, 179, 1996.
[109] Saida, J. and Inoue, A., Quasicrystals from glass devitrification, *J. Non-Cryst. Solids*, 317, 97, 2003.
[110] Janot, C., *Quasicrystals, A Primer*, 2nd ed., Oxford University Press, New York, 1994.
[111] Villars, P. and Calvert, L.D., Eds., *Pearson's Handbook of Crystallographic Data for Intermetallic Phases*, American Society for Metals, Metals Park, OH, 1985.
[112] Murty, B.S. et al., Influence of oxygen on the crystallization behavior of $Zr_{65}Cu_{27.5}Al_{7.5}$ and $Zr_{66.7}Cu_{33.3}$ metallic glasses, *Acta Mater.*, 48, 3985, 2000.
[113] Murty, B.S. et al., Direct evidence for oxygen stabilization of icosahedral phase during crystallization of $Zr_{65}Cu_{27.5}Al_{7.5}$ metallic glass, *Appl. Phys. Lett.*, 76, 55, 2000.

[114] Chen, M.W. et al., Redistribution of alloying elements in quasicrystallized $Zr_{65}Al_{7.5}Ni_{10}Cu_{7.5}Ag_{10}$ bulk metallic glass, *Phys. Rev. B*, 71, 092202, 2005.

[115] Scudino, S. et al., Influence of Al on quasicrystal formation in Zr–Ti–Nb–Cu–Ni–Al metallic glasses, Amorphous and Nanocrystalline Metals, in *Materials Research Society Symposium Proceedings*, Busch, R., Hufnagel, T., Eckert, J., Inoue, A., Johnson, W.L., and Yavari, A.R., Eds., 806, 83, 2004.

[116] Saida, J. and Inoue, A., Effect of Mo addition on the formation of metastable fcc Zr_2Ni and icosahedral phases in Zr–Al–Ni–Cu glassy alloys, *Jpn. J. Appl. Phys.*, 40, 769, 2001.

[117] Shechtman, D. et al., Metallic phase with long-range orientational order and no translational symmetry, *Phys. Rev. Lett.*, 53, 1951, 1984.

[118] Eckert, J., Mattern, N., Zinkevitch, M., and Seidel, M., Crystallization behavior and phase formation in Zr–Al–Cu–Ni metallic glass containing oxygen, *Mater. Trans. JIM*, 39, 623, 1998.

[119] Greer, A.L., Confusion by design, *Nature*, 366, 303, 1993.

[120] Lin, X.H., Johnson, W.L., and Rhim, W.K., Effect of oxygen impurity on crystallization of an undercooled bulk glass forming Zr–Ti–Cu–Ni–Al alloy, *Mater. Trans. JIM*, 38, 473, 1997.

[121] Gebert, A., Eckert, J., and Schultz, L., Effect of oxygen on phase formation and thermal stability of slowly cooled $Zr_{65}Al_{7.5}Cu_{17.5}Ni_{10}$ metallic glass, *Acta Mater.*, 46, 5475, 1998.

[122] Frankwicz, P.S., Ram, S., and Fecht, H.J., Observation of a metastable intermediate phase in water quenched $Zr_{65.0}Al_{7.5}Ni_{10.0}Cu_{17.5}$ cylinders, *Mater. Lett.*, 28, 77, 1996.

[123] Chen, M.W. et al., Impurity oxygen redistribution in a nanocrystallized $Zr_{65}Cr_{15}Al_{10}Pd_{10}$ metallic glass, *Appl. Phys. Lett.*, 74, 812, 1999.

[124] Christian, J.W., *The Theory of Transformations in Metals and Alloys: An Advanced Textbook in Physical Metallurgy*, Pergamon Press, Oxford, 1965.

[125] Altounian, Z., Volkert, C.A., and Strøm-Olsen, J.O., Crystallization characteristics of Fe–Zr metallic glasses from $Fe_{43}Zr_{57}$ to $Fe_{20}Zr_{80}$, *J. Appl. Phys.*, 57, 1777, 1985.

[126] Jang, J.S.C. and Koch, C.C., The glass transition temperature in amorphous $Cu_{1-x}Zr_x$ alloys synthesized by mechanical alloying/milling, *Scripta Metall.*, 23, 1805, 1989.

[127] Hellstern, E., Schultz, L., and Eckert, J., Glass-forming ranges of mechanically alloyed powders, *J. Less-Common Met.*, 140, 93, 1988.

[128] Altounian, Z. et al., The influence of oxygen and other impurities on the crystallization of $NiZr_2$ and related metallic glasses, *J. Appl. Phys.*, 61, 149, 1987.

[129] Seidel, M., Eckert, J., and Schultz, L., The effect of milling conditions and impurities on the properties of mechanically alloyed Zr-based metallic glasses with wide supercooled liquid region, *Mater. Sci. Forum*, 235, 29, 1997.

[130] Saida, J., Matsushita, M., and Inoue, A., Transformation in the initial crystallization stage of Zr–Al–Ni–Cu glassy alloys made with low oxygen concentrations, *J. Non-Cryst. Solids*, 312, 617, 2002.

7

Far-from-Equilibrium Processing of Nanostructured Ceramics

Bernard H. Kear
Rutgers University

Amiya K. Mukherjee
University of California, Davis

Abstract

Methods are described for the production and consolidation of metastable ceramic powders to yield fully dense nanostructured ceramics, including both single- and multiphase systems. Metastable powders produced by vapor condensation and rapid solidification methods are consolidated to full density by pressure-assisted and spark-plasma sintering methods. In all cases, the key to successful consolidation is control of a metastable-to-stable phase transformation during sintering, which promotes densification, enhances sintering kinetics, and minimizes grain growth. For single-phase or nanocrystalline ceramics, high pressures in the 1 to 8 GPa range are needed to achieve densification, without causing significant grain growth. For multiphase or nanocomposite ceramics, the pressure requirements are relaxed to the 0.1 to 0.5 GPa range. This is because the presence of one nanophase inhibits the growth of an adjacent nanophase(s), particularly when their volume fractions are comparable. Hence, the processing of nanocomposite ceramics is a more practical proposition. Nanocomposite ceramics also display high strain-rate superplasticity, which presents an unprecedented opportunity for near-net shape superplastic forming. In what follows, these and related aspects of this new far-from-equilibrium processing technology are discussed.

7.1 Introduction

For more than a decade, researchers have been attempting to harness the promise of nanostructured ceramics. The motivation for this work was the realization that reducing the grain size of single- or multiphase ceramics to nanoscale dimensions offered the potential of dramatic improvements in properties. Data are now emerging that support these expectations, especially in the area of mechanical properties. Although the property improvements in nanograined ceramics are significant, the ability to realize these increases faced a formidable barrier. When ceramic nanopowders are processed by conventional sintering methods, rapid grain growth occurs due to the high driving force caused by the large surface area. Thus, the high surface to volume ratio that gives rise to mechanical property improvements is also responsible for coarsening the nanoscale grain size during sintering. Thus, it was quickly recognized that unless the grain growth encountered during sintering could be mitigated, the promise of nanoceramic materials would not be realized.

Two methods have now been developed to mitigate grain growth during sintering: one for single-phase or nanocrystalline ceramics (NCs)[1–6] and the other for multiphase or nanocomposite ceramics (NCCs).[7–11] The first method makes use of a metastable *nano-sized* powder as the starting material and high-pressure sintering to develop a NC. As the pressure requirements are in the 1 to 8 GPa range, only small pieces can be fabricated. The second method utilizes a metastable *micron-sized* powder as the starting material and pressure-assisted sintering to develop a NCC. In this case, the pressure requirements are much lower, typically in the range of 0.1 to 0.5 GPa. As such pressures are well within the capabilities of today's hot pressing technologies, there is no reasonable practical limit to the size of sintered pieces that can now be fabricated. Similar work using a spark-plasma sintering (SPS) method has also demonstrated retention of nanoscale grain sizes in sintered composites.[12–17] Here, the grain growth problem is mitigated by exposing the nanopowder compact to a brief thermal transient at a high temperature, sufficient to achieve densification but not to permit significant grain coarsening.

In the original work on the processing of NCs, the starting materials were high surface area nanoceramic powders, prepared by inert gas condensation (IGC), chemical vapor condensation (CVC), and solution–precipitation (SP) methods (see also Chapter 1). Briefly, in IGC processing,[18–20] a metal source is evaporated in a resistively heated crucible or using a high enthalpy plasma. Metal nanoparticles are formed near the evaporative source by quenching the vapor stream with an inert cooling gas, or oxide nanoparticles are formed downstream by an additional oxidation treatment. In CVC processing,[21–25] oxide nanoparticles are formed directly by decomposing one or more metalorganic precursors in a flat-flame combustor, operating in a reduced pressure environment. In SP processing,[26–28] many procedures have been devised to obtain oxide nanoparticles by direct precipitation from an aqueous or organic solution. In all three cases, high-purity nanopowders are produced, with varying particle-size distributions and degrees of nanoparticle aggregation. For consolidation purposes, a *nonaggregated* nanopowder is required, as otherwise a powder compact contains large pores between the nanoparticle aggregates, which are difficult to remove during sintering. A simple test to ensure that a compacted nanoceramic powder has the desired characteristics is to examine its transparency. When there is little or no aggregation, the as-compacted nanopowder is transparent, as the nanopores are too small to scatter visible light.

In this review, methods for the production and consolidation of metastable single- and multicomponent powders are described, with the emphasis on processing multicomponent powders, as the resulting NCCs are of growing technical importance. In powder production, using an arc-plasma torch as heat source and an aggregated powder as feed material, a requirement is that all the feed particles experience complete melting and homogenization, prior to rapid quenching (solidification) to transform the molten particles into a metastable state. In powder consolidation, a requirement is control of a metastable-to-stable phase transformation during sintering, as this is the key to forming a uniform and fully dense NCC structure. Another option is to use an aerosol of a solution precursor as feed material, in which case the resulting metastable particles are of submicron- or nanoscale dimensions. Typically, such powders need additional processing before they can be used for consolidation purposes. However, an advantage of the aerosol method is the availability of inexpensive water-soluble salts of most elemental materials.

When fully developed, this new technology should be applicable to a broad range of materials, including oxides, nonoxides, and cermets. This will become clear later when the benefits of using a shrouded-plasma process to produce nanostructured powders or deposits are discussed. The important point is that the shroud excludes ambient air during plasma processing, so that carbides, nitrides, or borides can be produced in a controlled activity environment. For example, shrouded-plasma processing in a carburizing environment can be used to produce nano-WC/Co powder directly from a solution precursor, which contrasts with today's multistep spray conversion processing (SCP) technology.[29]

Over the past decade, a host of specialty applications for nanopowders of metals and ceramics have been developed, including additives to heat-transfer fluids and dispersants in polymer hosts. For the more general structural applications envisioned for bulk NCCs, where tonnage quantities of metastable starting powders are needed, we believe that shrouded-plasma spraying is the most attractive option (Section 7.2.2). Beyond that, we note that hot isostatic pressing (HIP) and SPS are both potential powder consolidation candidates (Section 7.3). At this time, however, powder consolidation by HIP appears to be the most promising, as large-capacity presses are available commercially. So far, only a few laboratory-scale NCC test pieces have been fabricated and evaluated. Nevertheless, a convincing demonstration has been made of the superplastic formability of NCCs (Section 7.4). This effect is a consequence of the resistance to grain coarsening displayed by NCCs, particularly when the volume fractions of their constituent phases are comparable. As there is a clear economic benefit to be derived from near-net shape forming of NCCs, the challenge of scaling this technology is now a high-priority objective.

7.2 Production of Metastable Powders

The production of metastable phases by rapid solidification processing methods has been widely studied and the fundamentals governing their formation well-established[30,31] (see Chapter 17). Many methods have been devised for processing metals, but few methods exist for processing ceramics, due to the difficulty of melting and containing such refractory materials. Here, we describe a modified plasma-spray process, which produces rapidly solidified ceramic powders or deposits, which, depending on composition and cooling rate, display a variety of metastable structures. As will be shown, using a conventional aggregated ceramic powder as feed material, the powder product consists of metastable micron-sized particles. On the other hand, using an aerosol of a solution precursor as feed material, the powder product consists of metastable submicron- or nano-sized particles. This size difference exerts an important influence on the choice of sintering parameters for the final consolidation step in the powder processing. In general, it is easier to handle micron-sized powders, as there is less surface area for contamination with water vapor and other impurities. Micron-sized particles also flow freely, so that a high compaction density can be achieved prior to sintering.

7.2.1 Conventional Plasma Spraying

In a typical plasma spray coating operation, Figure 7.1a, an aggregated ceramic powder (about 20 to 50 μm particle size) is fed into a high enthalpy plasma to achieve significant particle melting, prior to deposition on a chilled substrate.[32–34] The residual porosity in a well-controlled operation can be as low as 2%, despite the fact that the deposit is formed by the superposition of splat-quenched particles. Because of the rapid solidification experienced during splat quenching, a high fraction of the spray-deposited material has a far-from-equilibrium or metastable structure. In general, the feed powder comprises an aggregate of two or more ceramic phases. Hence, a uniform metastable deposit is developed only if the feed-particle residence time in the plasma is sufficient to cause complete melting *and* homogenization of all the particles prior to quenching. For reasons discussed below, this is seldom the case, so that a typical plasma-sprayed coating has a heterogeneous structure, which includes nonmelted or partially melted particles in an otherwise fully melt-quenched matrix. In another operational mode, known as "plasma densification," the plasma-melted particles are quenched in a bath of cold water, Figure 7.1b. Such powder

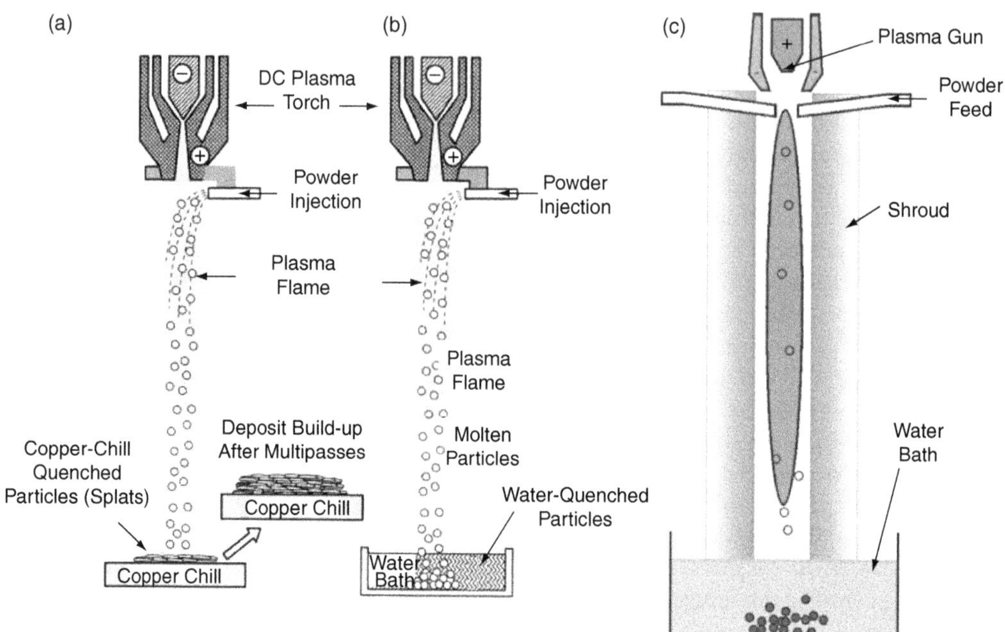

FIGURE 7.1 Schematics of conventional plasma-spray technology: (a) splat-quenching process and (b) water-quenching process. Note different paths of the feed particles when injected into the high enthalpy plasma flame. (c) Schematic of the shrouded-plasma-spray process, showing feed-particle confinement within an extended plasma hot zone, which increases the efficiency of particle melting and homogenization, prior to water quenching.

is also used as feed material in plasma spraying to obtain a more homogeneous and denser coating, which translates into improved properties and performance.

Quenching molten particles in cold water generates cooling rates $\sim 10^4$ K/s — the smaller the particle size the higher the cooling rate. On the other hand, quenching molten particles on a copper chill plate generates cooling rates $\sim 10^6$ K/s, as the impacting droplets spread out into thin splats in intimate contact with the heat-conducting substrate. Whatever the quenching mode, the effect is to generate a metastable structure, which may be an extended solid solution phase, a metastable intermediate phase, or an amorphous (glassy) phase. As indicated in Figure 7.2, a high cooling rate ($T = GR$, where G is the temperature gradient and R is the solidification rate) is needed to generate novel metastable phases. On the other hand, a high G/R ratio, sufficient to induce plane-front solidification, yields a segregation-free extended solid solution phase, in contrast to the segregated cellular or dendritic structures formed at lower G/R ratios. In many ceramic systems, water quenching of fine molten particles is sufficient to develop a homogeneous metastable state. When this is not the case, splat quenching on a water-cooled copper chill plate is used.

The susceptibility to metastable phase formation by rapid solidification depends not only on the cooling rate, but also on the composition. In general, ceramic systems are more readily transformed into metastable states by rapid solidification than metallic systems. In glass–ceramic processing technology, for example, silica-base compositions retain glassy structures even when cooled slowly from the melt.[35] However, for most of the ceramic systems of interest herein, amorphization or glass formation is not so readily achieved. Instead, the effect of rapid solidification is to develop an extended solid solution or metastable intermediate phase, usually with some *segregation*. For the present purpose, this is not ideal, as the essential requirement to obtain a *completely* uniform NCC structure by pressure-assisted sintering is that the starting material must be chemically homogeneous. Hence, in general, an amorphous or glassy powder is preferred, as there is no uncertainty regarding its chemical homogeneity. In this case, another positive effect is the large negative free volume change

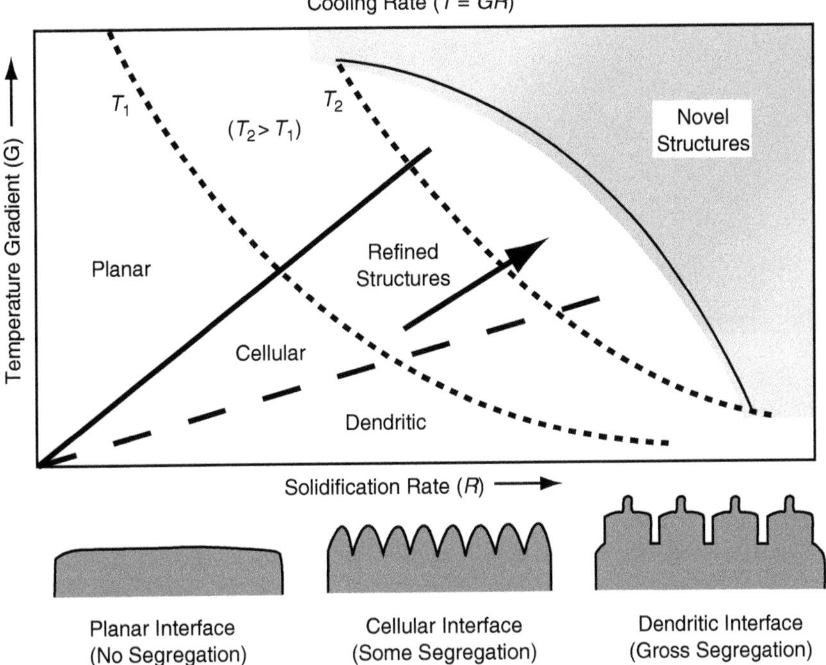

FIGURE 7.2 Dependence of solidification microstructure on cooling rate ($T = GR$), showing appearance of novel metastable structures at high cooling rates.

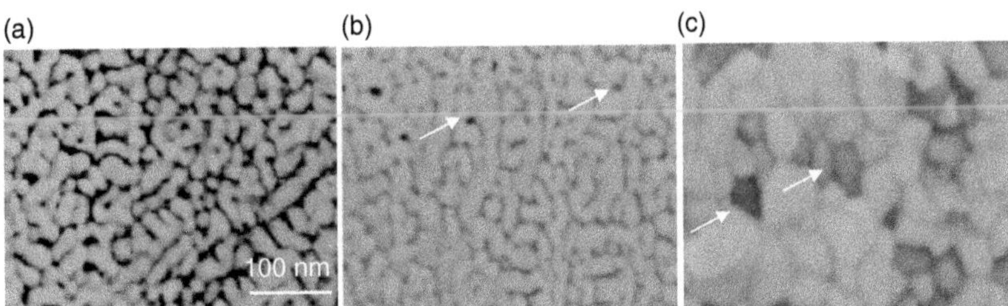

FIGURE 7.3 SEM micrographs of $ZrO_2(3Y_2O_3)/20Al_2O_3$ powder: (a) after water quenching, showing a metastable cellular-segregated structure, (b) after annealing at 1200°C for 1 h, showing nucleation of α-Al_2O_3 particles (arrows) in the cellular interstices, and (c) after annealing at 1400°C for 1 h, showing coarsening of the equilibrium α-Al_2O_3 + t-ZrO_2 structure; dark-contrasting phase (arrows) is α-Al_2O_3.

accompanying phase decomposition during hot pressing, which facilitates the formation of an NCC structure.

An extended solid solution phase,[36] with a fine-scale segregated cellular structure, has been observed in water-quenched $ZrO_2(3Y_2O_3)$-$20Al_2O_3$ powder, Figure 7.3a. Due to the wide melting range at this composition, the undercooling experienced during water quenching is not sufficient to prevent cellular segregation, but adequate to generate a metastable state. In this case, therefore, the as-quenched state comprises a metastable tetragonal-ZrO_2 phase, with a compositionally graded cellular structure. When the cooling rate is increased by splat quenching, the effect is to form a metastable nanocomposite, comprising a nanodispersion of ZrO_2-rich particles in an Al_2O_3-rich matrix, Figure 7.4. This is attributed to prolific nucleation of ZrO_2-rich nanoparticles in a drastically undercooled melt, that is, deep within the semisolid region of the phase diagram.

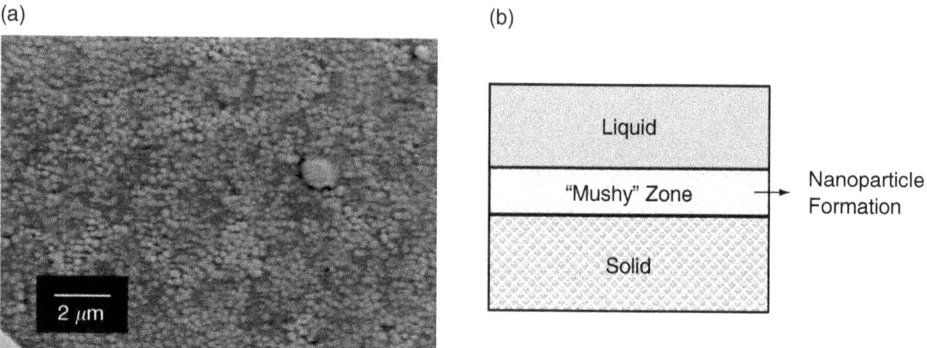

FIGURE 7.4 (a) Microstructure of splat-quenched deposit, showing fine-scale ZrO_2-rich particles (white-contrasting phase) uniformly dispersed in an Al_2O_3-rich matrix phase, and (b) illustrating mechanism of particle formation in the semisolid or "mushy" zone, due to drastic undercooling of the melt.

A powder with a metastable cellular-segregated structure is not ideal for consolidation purposes, as upon subsequent phase decomposition, the distribution of equilibrium phases (t-ZrO_2 + α-Al_2O_3) necessarily reflects the scale of the segregated structure. The effect is illustrated in Figure 7.3b, which shows that nucleation of the equilibrium α-Al_2O_3 phase occurs in the Al_2O_3-rich interstices of the cellular structure, thus forming a nanodispersion of α-Al_2O_3 particles in an otherwise nontransformed ZrO_2-rich matrix phase. Further decomposition at a higher temperature generates a uniform biphasic microcomposite structure, which consists of about 30 vol.% α-Al_2O_3 dispersed in a partially stabilized t-ZrO_2 matrix phase, Figure 7.3c. In this case, to access a true nanocomposite structure, it is necessary to start with a nonsegregated metastable powder. The best that can be accomplished for this particular composition is to use splat-quenched powder as the starting material, as this already has a metastable nanoscale structure, Figure 7.4. Other compositions in this system, notably those close to the eutectic are susceptible to amorphization by rapid quenching, and therefore, more suitable starting materials for processing into ideal NCC structures.

A similar behavior has been observed in water-quenched $ZrO_2(3Y_2O_3)$-27Al_2O_3-22$MgAl_2O_4$ powder. Upon phase decomposition, a triphasic nanocomposite is developed, in which the constituent phases (t-ZrO_2 + α-Al_2O_3 + c-$MgAl_2O_4$) have comparable volume fractions (40:30:30), Figure 7.5. Such a structure resists coarsening at high temperatures, as the presence of one phase quite naturally impedes the growth of an adjacent phase(s). As will be shown in Section 7.4, such thermally stable NCCs are of particular interest, as they display superplasticity at high temperatures, which opens a new opportunity for shape-forming NCCs.

7.2.2 Shrouded-Plasma Spraying

When conventional plasma spraying is used to generate a metastable powder or deposit, a particular challenge is to ensure that all the feed particles experience complete melting and homogenization prior to quenching, so that a uniform melt-quenched metastable product is produced. The nature of the problem is illustrated in Figure 7.1a and Figure 7.1b, which shows that different feed particles take different paths through the plasma, thus experiencing different thermal histories. One way to resolve this problem is to respray the water-quenched powder two or even three times, but this is both tedious and expensive.

To deal with this problem, we have developed a shrouded-plasma system,[37] Figure 7.1c, which enables efficient melting of the feed particles by increasing their residence times in the plasma flame. In this design, a water-cooled copper shroud incorporates a heat-resistant ceramic liner, which serves to maintain a very high temperature over the entire length of the refractory liner. Such a system enables an aggregated feed powder, irrespective of the melting points of its constituent ceramic phases, to be fully melted and homogenized in a single pass through the plasma. In one configuration, three radially symmetric feed

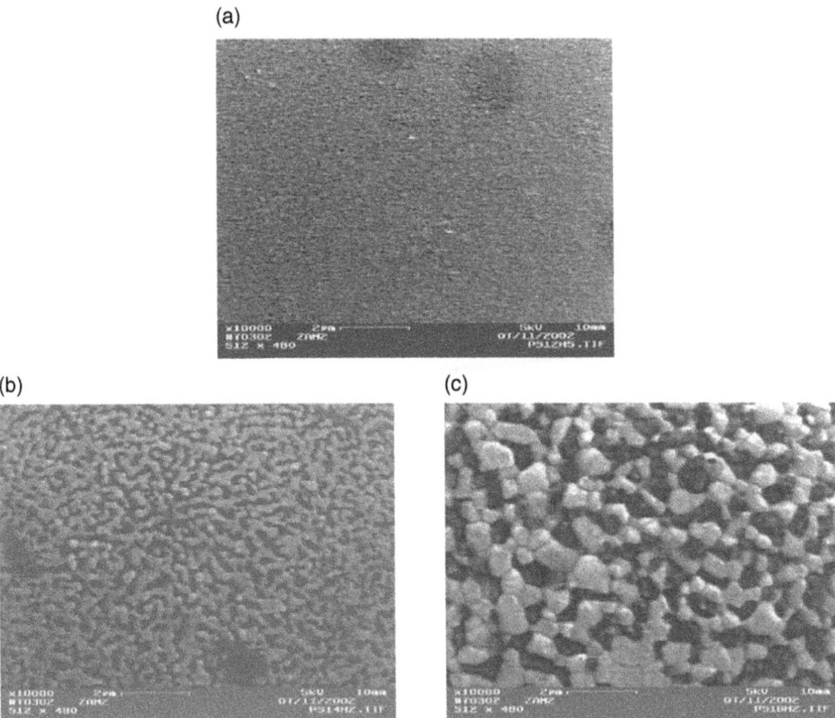

FIGURE 7.5 SEM micrographs of water-quenched $ZrO_2(3Y_2O_3)/27Al_2O_3/22MgAl_2O_4$ powder, after heat treatment for 1 h at (a) 1200°C, (b) 1400°C, and (c) 1600°C, showing significant coarsening of the triphasic composite structure at temperatures >1400°C.

units deliver the precursor powder to a steady-state reaction zone within the shrouded-plasma flame, where rapid particle melting and homogenization occurs. To obtain an optimal effect, the flow rates of the feed streams are adjusted to avoid deflecting or distorting the plasma flame, such that a steady-state reaction zone is created. In another configuration, one axially symmetric feed unit delivers the precursor powder to a steady-state reaction zone formed by the convergence of three plasma flames. Both systems are used to produce metastable ceramic powders. However, at this time, it appears that an axially fed triple arc-plasma torch (Mettech, Inc.), incorporating a shroud similar to the present design, is best suited for the high rate production of water-quenched metastable powders or splat-quenched deposits.

By shrouding the plasma in a heat-resistant ceramic tube, the thermal energy normally released to the surroundings is captured by that tube, which is rapidly heated to a high temperature. When the exterior of the tube is continuously cooled, a uniform temperature gradient is established through the tube wall. In effect, therefore, the processing unit becomes a "hot-wall reactor," where a very high inner-wall temperature is sustained by intense radiation from the plasma flame. Utilizing the high enthalpy within the plasma flame itself and the radiant energy from the hot reactor wall, rapid and efficient metastable processing of almost any feed material can be achieved, including the most refractory of oxide ceramics.

A feature of this so-called "radiantly coupled plasma" (RCP) process is the rapid heating of the tubular ceramic liner by the plasma itself, such that a very high inner-wall temperature is quickly achieved and then sustained. Using a conventional aggregated feed powder, comprising two or more ceramic phases, RCP processing parameters are adjusted to ensure complete melt-homogenization of all the feed particles prior to water quenching. As noted above, water quenching is often sufficient to generate a homogeneous metastable powder. When this is not the case, then splat quenching is used. This is best accomplished by spraying the molten particles onto a rapidly rotating copper chill plate, which removes the splats as fast as they are formed. As another option, the spray parameters are adjusted to build up a thick deposit or preform by superposition of splat-quenched particles. Rotation or translation of a substrate with respect to a fixed shrouded-plasma torch ensures the fabrication of a uniform metastable deposit.

When better control of the gaseous environment in RCP processing is required, then the shrouded-plasma reactor is enclosed within a water-cooled stainless-steel chamber. In the simplest case, the chamber is partially immersed in a bath of cold water, which serves to exclude ambient air. Thus, an inert environment is quickly established within the chamber when the system is operated using an Ar or N_2 plasma. Alternatively, the powder processing is carried out in a closed chamber, such that the powders are collected on the cold walls of the chamber or vented to an external particle collector. Clearly, an inert environment is essential when processing nonoxide ceramics, such as carbides, borides, or nitrides, which readily oxidize in air.

7.2.3 Solution Plasma Spraying

In another approach,[38–42] a solution precursor replaces an aggregated powder as feed material. The incentive is the relatively low cost of the precursor material, the flexibility afforded in adjusting its composition, and the ease with which it can be processed into a metastable powder or deposit. Typically, the solution precursor comprises an aqueous or organic solution of mixed salts: nitrates, acetates, chlorides, phosphates, and others. However, if metastable powders of exceptionally high purity are required, then semiconductor-grade metalorganic or organometallic precursors are used.

Using an aerosol-solution precursor as feed material in plasma spraying yields a metastable nano-sized powder, in contrast to the metastable micron-sized powder formed using an aggregated feed powder. Both types of metastable powders have their uses, with the choice for a particular application being determined largely by the requirements with respect to particle size, purity, and cost. In one application, a nanostructured ceramic coating is formed by spraying an aerosol-solution precursor directly onto a plasma-heated substrate. A relatively dense coating or deposit is formed by *in situ* sintering of the precursor decomposition products on the heated substrate surface. Typically, the spray-deposited material has a grossly heterogeneous structure, comprising a relatively small fraction of splat-quenched material dispersed in an otherwise pyrolyzed and sintered matrix phase. Even so, the spray-deposited material displays favorable properties, such as improved cracking resistance. This has led to a re-examination of the prospects for developing plasma-sprayed coatings with improved wear resistance and other properties.

A critical variable in solution plasma spraying is the aerosol feed rate. This is because a low feed rate barely affects the high enthalpy of the plasma, so that vaporization of all the precursor constituents occurs. Metastable nanoparticles are generated when the vapor stream is rapidly quenched in cold water or on a chill plate. Typically, the resulting nanoparticles have metastable crystalline structures, but favorable compositions can also have noncrystalline or amorphous structures. However, production rates are not particularly high. This is also the case when the feed rate is adjusted to give particle melting but not vaporization, in which case the metastable powder is generated by rapid solidification. In contrast, when the feed rate is high, the effect is to "cool" the plasma, so that varying degrees of precursor pyrolysis can be achieved. The resulting pyrolyzed powder product usually has an amorphous or partially crystallized structure, due to the retention of some of the precursor components. As the available plasma energy is used most efficiently in pyrolyzing the aerosol-solution precursor, and little or no energy is expended in its melting or vaporization, this particular operational mode gives a high powder production rate.

With the system operating in a vaporization mode, another option is to attach a supersonic nozzle to the bottom of the shroud. Metastable nanoparticles are then formed by rapid condensation of the vaporized species in the adiabatic cooling zone near the exit of the nozzle. The nanopowders are best collected on the chilled walls of a container to avoid surface hydrolysis by quenching in water. For most ceramic systems, it is necessary to minimize exposure of the nanopowders to water vapor and other contaminants, as the presence of these impurities adversely affects the final consolidation step in the processing of these high surface area materials.

7.3 Consolidation of Metastable Powders

A major challenge encountered in the consolidation of ceramic nanopowders is mitigation of grain growth during sintering. Typically, a nanopowder compact experiences rapid densification during the early stages of sintering, driven by the large surface to volume ratio. At this stage, the grain size of the partially sintered material remains small due to the presence of a uniform distribution of nanopores, which act as barriers to grain-boundary migration. However, in the final stages of sintering (>90% theoretical density or TD), when the nanopores are beginning to disappear, exaggerated and uncontrollable grain growth occurs, often leading to a micrograined sintered product. To some extent, this problem can be controlled by making an addition of a second phase that segregates to the grain boundaries, thus serving as an effective obstacle to grain growth. An additive can also form a low melting eutectic along the grain boundaries, which, since it lowers the sintering temperature, also reduces grain growth. All these different strategies have been tried, including programmed sintering cycles,[43] with varying degrees of success.

Here, we describe how metastable ceramic powders can be consolidated to full density by pressure-assisted and SPS methods. As will be shown, to mitigate grain growth during pressure-assisted sintering, much higher pressures are needed for single-phase systems than for multiphase systems. Hence the processing of NCCs is a more attractive commercial proposition.

7.3.1 High-Pressure Sintering

Figure 7.6a shows a schematic of the high-pressure sintering apparatus used to consolidate nanopowders of single component ceramics.[44] Briefly, it consists of matching pairs of WC/6% Co inserts and anvils, which are reinforced with prestressed 4340 steel rings. The arrangement enables pressures to be applied that are higher than the compressive strength of the anvil material. The reaction cell, Figure 7.6b, consists of a resistively heated graphite crucible and insulating ceramic layers to prevent overheating of the anvils. A critical component is the pressure-transmitting medium, which consists of deformable limestone or lava stone. A thick disk of this material, machined to match the profiles of the anvils, encases the reaction cell. Under high pressure, the plastically deformed ceramic exerts a strong restraining force on the reaction cell, thus generating a near-isostatic pressure in its working volume. Pressure is calibrated via known data for a number of phase transitions in solid substances, such as cerium, bismuth, and lead selenide. Temperature is calibrated via known values of the melting points of different materials under high pressure.

Metastable anatase-TiO_2 and γ-Al_2O_3 nanopowders, produced by IGC and CVC methods in sizes <50 nm, were consolidated by high-pressure sintering. The key steps in producing a fully dense nanograined sintered product are as follows: (1) the starting powder is dried to remove any moisture; (2) the powder is compacted at a low pressure (~500 MPa) to provide sufficient green strength; and (3) the green compact is placed in the reaction cell and subjected to 1 to 8 GPa at 400 to 800°C for 30 min. During the

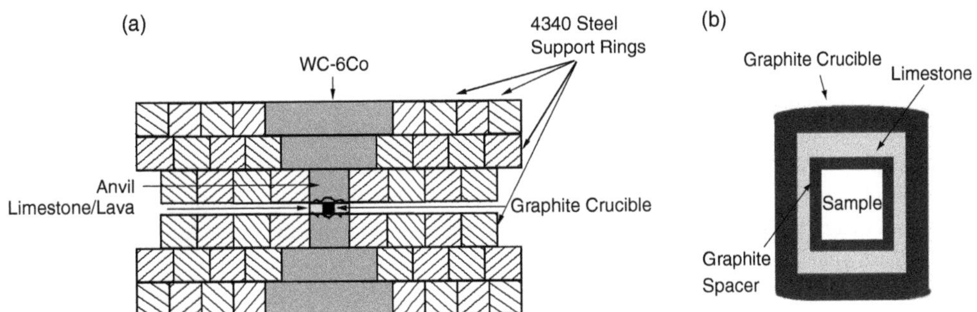

FIGURE 7.6 Schematic of toroidal-type high-pressure apparatus: (a) high-pressure unit showing arrangement of WC/Co anvils, and (b) reaction cell showing resistively heated graphite crucible and insulating ceramic layers.

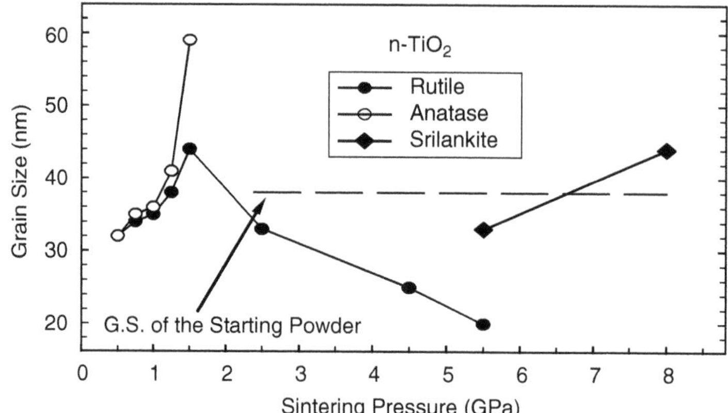

FIGURE 7.7 Effect of pressure on the grain size (GS) of sintered TiO_2; anatase-to-rutile phase transformation occurs at about 1.5 GPa. Sintering parameters are 400°C for 1 h.

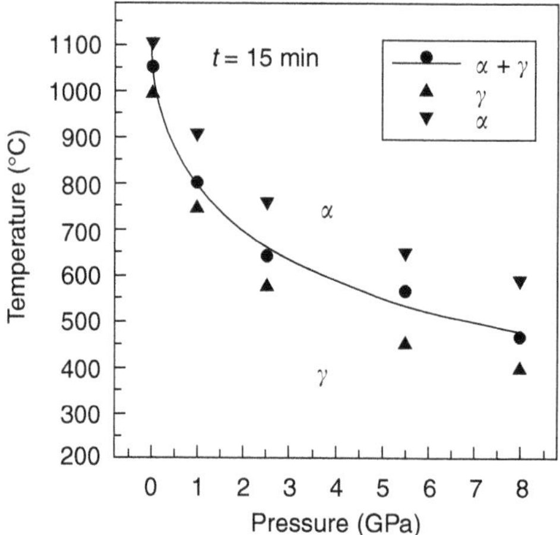

FIGURE 7.8 Effect of pressure on the γ–α phase transformation in nano-Al_2O_3.

compaction stage, neither grain growth nor phase transformations occur. Only during the final high pressure stage does the metastable powder begin the desired phase transformation via a nucleation process. Because of the high pressure, there is a marked increase in the number of nucleation events, and, at the same time, the high pressure reduces the diffusion rate. The net result is a slow, controlled transformation with minimal grain growth.

One of the most important effects observed in high-pressure sintering of nano-TiO_2 powder is the continued refinement in sintered grain size with increasing pressure. As shown in Figure 7.7, the starting phase of the metastable powder is anatase, with a grain size of 38 nm. As sintering progresses, the anatase phase initially increases in grain size during sintering. However, when the pressure is above about 1.5 GPa, the anatase phase transforms to the stable rutile phase, with a grain size that is increasingly smaller than the anatase phase. The most startling aspect of this curve is that above 2.0 GPa, the grain size is actually *smaller* than that of the starting powder. This is due to the increased number of nucleation events brought about by the higher pressures.

A similar result has been obtained for high-pressure sintering of metastable γ-Al_2O_3 nanopowder, which transforms under pressure to the stable α-Al_2O_3 phase. As shown in Figure 7.8, pressure lowers the transformation temperature to 450°C at 8 GPa. At the same time, a nanograin size is retained in the sintered product. As shown in Figure 7.9, the grain size is larger than the starting material (73 vs. 50 nm)

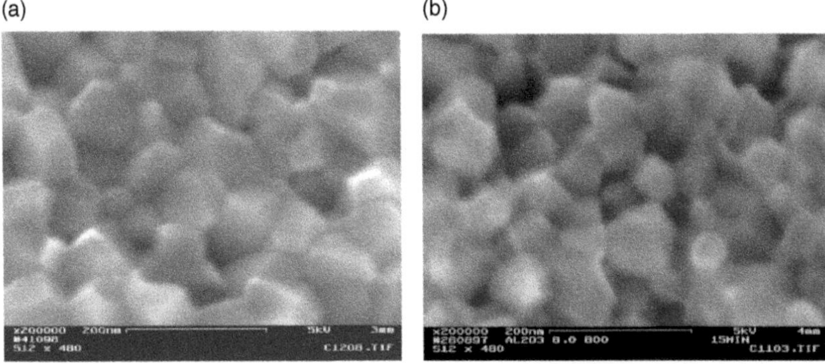

FIGURE 7.9 Effect of pressure on the grain size of α-Al_2O_3, after sintering at 800°C for 15 min at (a) 5.5 GPa and (b) 8.0 GPa.

at 5.5 GPa, whereas at 8 GPa the grain size is 49 nm, while achieving 98.2% density. Here again, pressure encourages nucleation of the stable α-Al_2O_3 phase while decreasing grain growth. This is analogous to the lowering of the anatase-to-rutile transformation temperature in TiO_2.

The theoretical basis for the influence of a phase transformation on sinterability has been described.[5] Briefly, when a volume reduction accompanies a phase transformation (as in the case of the anatase-to-rutile transformation), increasing the pressure reduces the nucleation barrier for the formation of the stable phase, and consequently increases its nucleation rate. In addition, the applied hydrostatic pressure also reduces the diffusion rate, and hence the grain growth rate. The overall effect is that the ratio of growth rate to nucleation rate decreases as the pressure increases. When the srilankite phase begins to form at 5.5 GPa, it exhibits a pressure dependence that is contrary to that observed during the anatase-to-rutile phase transformation. This may be, at least in part, due to the relatively small difference (~2.5%) in molar volume between rutile and srilankite, compared with 9% between anatase and rutile.

7.3.2 Hot Isostatic Pressing

As emphasized in Section 7.2, plasma spraying can be used to transform an aggregated powder or an aerosol-solution precursor into a homogeneous metastable powder. This is an important finding, since upon subsequent exposure to a high temperature, either by heat treatment or during hot pressing, a completely uniform fine-scale composite structure is realized. In general, the grain size of the resulting equilibrium composite structure is determined by the selected annealing temperature and time — the lower the decomposition temperature the smaller the composite grain size. On the other hand, the volume fractions of the constituent phases of the nanocomposite material are fixed by the selected starting composition, based on known phase diagram data. An example is shown in Figure 7.5, where 40:30:30 ratio of $ZrO_2(3Y_2O_3)$:Al_2O_3:$MgAl_2O_4$ phases is formed by postannealing a metastable ZrO_2-base ceramic powder. Information gained from such heat treatment experiments provides important guidance to the selection of suitable parameters for powder consolidation.

When a metastable-to-stable phase transformation occurs during pressure-assisted sintering, the effect is to promote densification at a relatively low temperature. The effect is particularly striking during HIP at a temperature where phase decomposition commences, apparently because the material experiences densification primarily by a superplastic-like flow mechanism, with grain-boundary and bulk diffusion playing a lesser role. Pressure is another variable that strongly influences the final grain size of a sintered product. Due to the negative free volume change that invariably accompanies a metastable-to-stable phase transformation, the application of pressure enhances the rates of nucleation of the equilibrium phases, thus enabling an NCC structure to be developed. Hence, pressure-assisted sintering is generally preferred to pressure-less sintering to produce NCCs. If a microcomposite structure is desired, an additional heat treatment at a high temperature is used to deliberately coarsen the original nanocomposite structure.

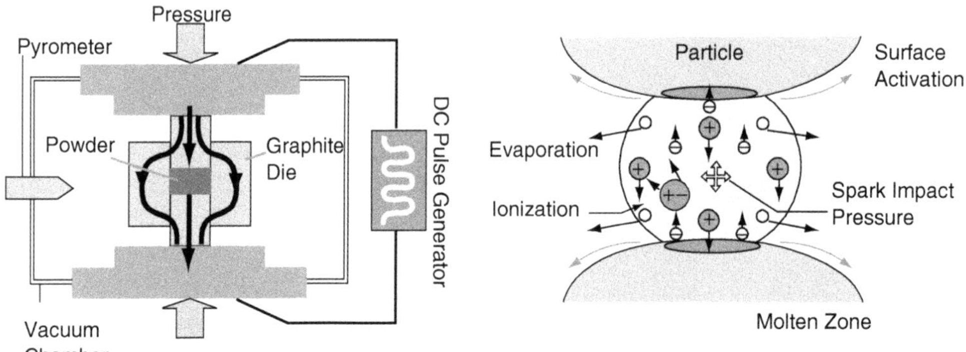

FIGURE 7.10 Schematic of SPS sintering apparatus and the accompanying activation effects.

Extensive work on ZrO_2-base systems has led to the following scaleable methodology for the production of NCCs: (1) selection of the smaller size fraction (<30 μm diameter) of melt-quenched metastable powder; (2) encapsulating the powder compact in a low-carbon steel container, with an inert material as parting compound; and (3) after degassing, HIP at 1250 to 1350°C for 4 h to develop a fully dense NCC structure. Although both nano- and micron-sized metastable powders have been used as starting materials in powder processing, micron-sized powder is preferred, as there is less surface area for impurity absorption. Another effect is the ease with which the plasma-densified powder can be poured into a steel container to obtain a high packing density, which facilitates consolidation by HIP.

7.3.3 Spark-Plasma Sintering

Spark-plasma sintering is a low-pressure sintering method that makes use of a plasma discharge through a powder compact to achieve rapid densification. The discharge is most effective when a DC current is applied in an on-off pulsing mode. It has been suggested that DC pulsing generates: (1) spark plasma, (2) spark impact pressure, (3) Joule heating, and (4) an electrical field diffusion effect.[12]

In a typical operation, powders are loaded into a graphite die and heated by passing an electric current through the assembly. The experimental setup and electric field effects are illustrated in Figure 7.10. The low heat capacity of the graphite die allows rapid heating, thus promoting heat and mass transfer. Hence, SPS rapidly consolidates powders to near TD through the combined actions of rapid heating rate, pressure application, and possibly powder surface cleaning. In most investigations, SPS is carried out under vacuum. Starting with a cold-pressed (~200 MPa) powder compact, typical processing parameters are: (1) an applied pressure of <100 MPa, (2) pulse duration of 12 ms and pulse interval 2 ms, and (3) pulse current of ~2000 A at a maximum voltage of 10 V. Typical heating rates range from 150 to 500°C/min.

SPS has been used to consolidate a wide variety of materials, including metals, intermetallics, ceramics, composites, and polymers. As for functionally graded materials and nanocrystalline materials, which are difficult to sinter by conventional methods, the advantage of SPS is more directly evident. For example, Zhan et al.[13] successfully sintered nanocrystalline α-Al_2O_3 with 5 vol.% carbon nanotubes at 1150°C in 3 min. Conventional hot pressing of α-Al_2O_3 requires 1500 to 1600°C for 3 to 4 h. Hence, with a lower sintering temperature and shorter sintering time, SPS enables better control of structure and properties of the consolidated material.

7.4 Shape-Forming of Ceramic Nanocomposites

Superplastic forming of metallic materials has become established industrial practice in the aerospace industry and is being applied increasingly in the automotive industry. The idea of using superplastic forming of ceramic materials is even more attractive, due to the increased complexity and cost of machining components. Unfortunately, in the one ceramic system in which superplasticity has been demonstrated, the processing temperature is much higher than that used for traditional shape-forming operations. Recent work, however, has shown that this barrier can be overcome when the ceramic to be processed has a thermally stable nanocomposite structure, where grain-boundary sliding (GBS) plays the dominant role in the superplastic deformation process.

7.4.1 Superplastic Deformation

Conventional fine-grain superplasticity, which occurs at strain rates of 10^{-4} to 10^{-3} s^{-1}, is due to GBS, accommodated by either dislocation activity or diffusion. High strain-rate superplasticity (HSRS), which occurs at strain rates $\sim 10^{-2}$ s^{-1} or greater, is less well understood. It appears to be primarily a grain size effect.[14] The most significant aspect of HSRS is the reduction in forming time due to the high deformation rate. More details on superplasticity may be found in Chapter 14.

The constitutive relationship for superplastic deformation usually takes the form of the following equation:[15]

$$\varepsilon = A\frac{D_0 G\mathbf{b}}{kT}\left(\frac{\mathbf{b}}{d}\right)^p\left(\frac{\sigma}{G}\right)^n \quad (7.1)$$

in which G is the elastic shear modulus, $\mathbf{b}$ is the Burgers vector, k is Boltzmann's constant, T is the absolute temperature, d is the grain size, p is the grain-size dependence coefficient, σ is the stress, n is the stress exponent, D_0 is the diffusion coefficient, and A is a constant. GBS is generally the predominant mode of deformation during superplastic flow. Plastic deformation by GBS is generally characterized by $n = 2$ and an apparent activation energy that is either equal to that for lattice diffusion or for grain-boundary diffusion. From this equation, it is clear that at constant temperature and stress, high strain-rate is more easily realized in materials with small grain sizes. Now that methods are being developed to make NCCs that have thermally stable grain structures, HSRS is becoming an attractive option.

Although HSRS in fine-grained metallic systems has been known since the 1970s, a similar behavior in ceramics was not reported until 2001, when Kim et al. demonstrated HSRS in a triphasic composite, comprising 40 vol.% t-ZrO_2, 30 vol.% α-Al_2O_3, and 30 vol.% $MgAl_2O_4$ (ZAM), see Figure 7.5. The observed strain-rate was an impressive 1 s^{-1}.[16] Unfortunately, the forming temperature was 1650°C, which may be too high for most commercial forging operations. Although the study did not provide any phenomenological data to account for a possible deformation mechanism, it seems likely that GBS of the fine-grained composite was the rate limiting step in the processing.

Because GBS involves the relative motion of adjacent grains, cavities must form at grain-boundary ledges and triple points, unless the stresses that nucleate and grow these cavities are accommodated. This accommodation can take the form of atomic transport along or through the grain boundaries. Without an accommodation mechanism, large and uniform macroscopic strain cannot be realized in the material. By reducing the grain size, hence the distance needed to accommodate GBS via atomic transport, the strain rate may be increased. Also, the deformation temperature may be reduced, as the shorter length scales for diffusional transport allow for easier accommodation.

In recent work, an attempt was made to lower the HSRS deformation temperature of ZAM to levels that would be commercially attractive. Conventional tests were conducted to establish the strain-rate sensitivity and activation energy of the deformation process. These two parameters are paramount in modeling the deformation behavior. SPS was used to enhance the deformation process, taking advantage of the faster atomic transport kinetics due to the pulsing electric field in the chamber.[12]

A scanning emission micrograph (SEM) of the as-sintered microstructure and the corresponding x-ray diffraction (XRD) pattern are shown in Figure 7.11. Final density was 98.5% of the TD of the composite (4.696 g/cm^3). Grain sizes of the three phases of the ZAM composite, each with an equiaxed grain morphology, were about 100 nm. The results of strain-rate jump tests performed at 1350 and 1400°C

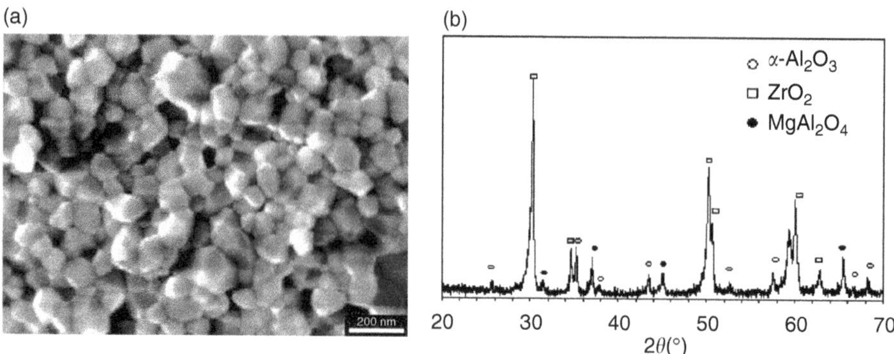

FIGURE 7.11 (a) SEM micrograph of fractured ZAM composite after SPS at 1150°C for 3 min. The grains are equiaxed and approximately 100 nm in size; (b) XRD pattern of the as-sintered ZAM compact. The MgO and Al_2O_3 react to form the intended $MgAl_2O_4$ phase; in addition the γ-Al_2O_3 transformed to α-Al_2O_3.

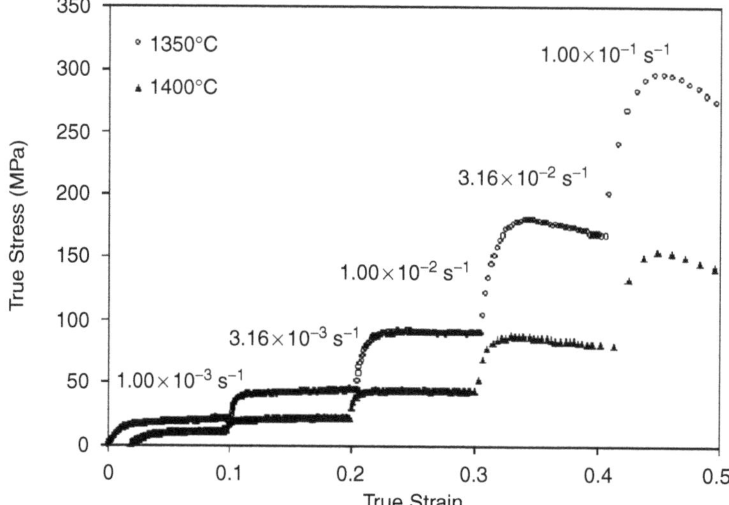

FIGURE 7.12 Strain-rate jump tests performed at 1350°C and 1400°C; identical tests were performed at 1300°C and 1450°C. The strain rates ranged from 10^{-3} s^{-1} to 10^{-1} s^{-1}. From these data the strain-rate sensitivity ($m = 0.55$) and activation energy of the deformation process ($Q = 620$ kJ/mol) were calculated.

are shown in Figure 7.12. By determining the linear slope of the log of the flow stress vs. the log of the strain-rate, the strain-rate sensitivity m was found to be 0.55. The inverse of m is the stress exponent n, and when $n \sim 2$ GBS tends to be the rate limiting deformation mechanism. This provides some solid data to support the hypothesis that GBS-mediated superplasticity is occurring in this system. From the data, the activation energy for the deformation process was calculated to be 620 kJ/mol. Unfortunately, there is no reliable data for lattice diffusion of the constituent phases of the composite, so that, at present, it is not possible to correlate superplastic deformation with a specific diffusional process in the material.

7.4.2 Consolidation and Forming

For future applications, the potential of combining consolidation and forming in a single operation is of particular interest. One of the benefits is the ease with which porous nanostructured preforms can be made by hot pressing nanopowder compacts. Plasma spraying can also be adapted to generate

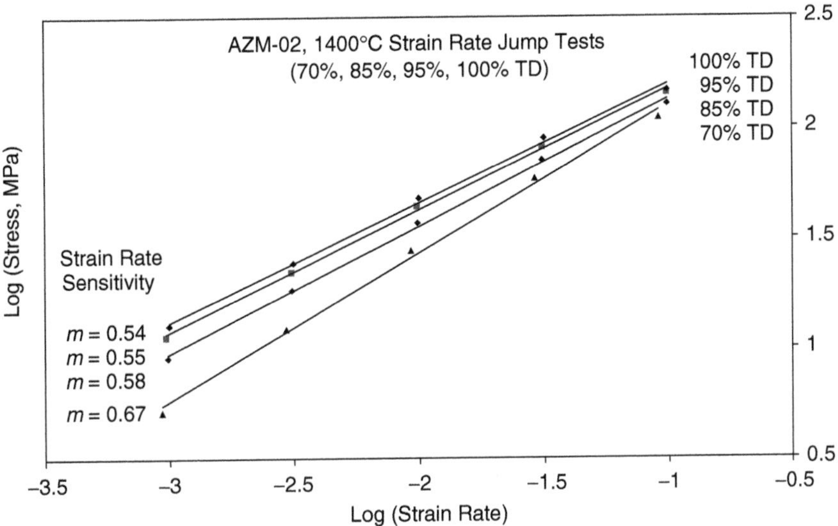

FIGURE 7.13 Plot of stress-rate relationship on a log–log scale to determine strain-rate sensitivity of porous specimens.

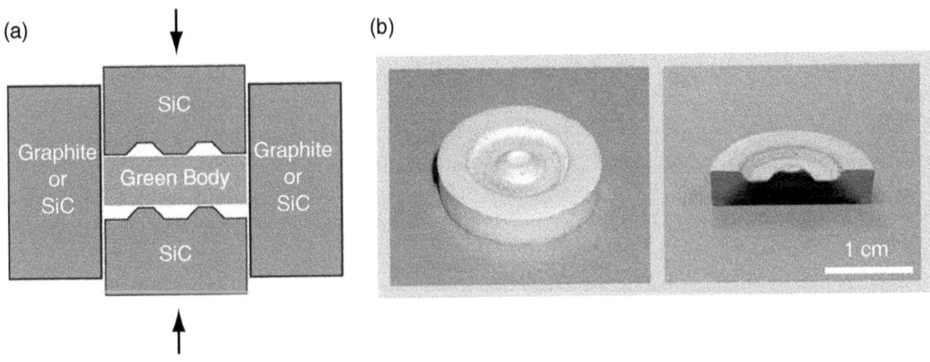

FIGURE 7.14 (a) Schematic of an axi-symmetric die design to induce radial constraint during superplastic forming; (b) Nanocomposite sample formed using this die design. Note that the cracks in the sectioned specimen were caused during sectioning.

nanostructured deposits or preforms of high density. In addition to being easily produced, porous preforms are appealing due to the fact that they are less susceptible to grain growth during heating than their fully dense counterparts, due to grain-boundary pinning by the pores. This limited grain growth leads to finer grain sizes during deformation and lower deformation temperatures and higher strain rates.

In preliminary testing, SPS was used to consolidate ZAM samples with a range of densities, 70 to 100% TD. The samples were subjected to strain rates of 10^{-3} to 10^{-1} s^{-1} to 40% engineering strain at 1400°C, Figure 7.13. As expected, the porous samples exhibited lower flow stresses due to lower density and finer grain size, and higher strain-rate sensitivity due to densification during deformation. Unfortunately, the samples did not densify as much as desired. For example, the sample with an initial density of 70% TD resulted in a final density of only 90% TD. This led to the concept that to achieve full densification during consolidation, an isostatic compressive stress is necessary. Ongoing work includes die set development and testing to add a radial constraint to the uniaxial deformation, thus introducing a near-isostatic compressive stress, Figure 7.14a. Using this die design, a porous nanoceramic composite sample was formed at 1300°C and an equivalent strain-rate of $\sim 10^{-3}$ s^{-1}. The initial density was 89% TD and the final density was 97% TD. The deformed sample and its cross-section are shown in Figure 7.14b. Current research suggests that superplastic deformation of this triphasic NCC is possible at temperatures as low as 1150°C and at a strain-rate of 10^{-2} s^{-1}.

7.5 Concluding Remarks

An important aspect of our current research on NCCs is the extension of the underlying processing methods to a wider range of materials. To date, the various processing methods described above have been applied to a limited range of oxide ceramic compositions. However, because of their versatility, there is no reason to believe that they cannot be extended to a very broad range of oxide and nonoxide ceramics. In particular, because it excludes ambient air, we note the potential of the shrouded-plasma process for producing Si_3N_4, SiC, B_4C, TiC, and WC, as well as their composites with metals, such as WC/Co or TiC/Fe. These opportunities are now being addressed. It is anticipated that this will lead to a host of structural and functional applications for a new generation of NCCs.

Although the emphasis has been placed herein on plasma processing of metastable powders, this same technology can also be used to generate metastable preforms. Using the shrouded-plasma process, for example, a metastable preform with little or no open porosity is readily obtained by directing the molten particle stream onto a rotating or translating substrate or mandrel. Thus, a final container-less HIP treatment at a temperature where phase decomposition commences is all that is needed to form a fully dense NCC. Clearly, this simplified procedure, as it involves no powder handling, has potential as a cost-effective production method. SPS processing of oxide and nonoxide powders also opens new opportunities for consolidation of ceramic powders at lower temperatures and in much shorter times. This enables the retention of nanoscale grain sizes in the sintered products. Such consolidated nanoceramics can then be superplastically deformed into near-net shape by a closed-die forming technique at significantly lower temperatures and at faster strain rates than hitherto attainable.

To summarize, alternative far-from-equilibrium processing technologies are described for the production and consolidation of metastable powders to form fully dense, bulk NCCs. These new processing technologies can be applied to a host of oxide and nonoxide ceramic composites, as well as metal-ceramic composites. Potential applications include machine tools and drill bits, blades and cutters, turbochargers and engine valves, IR-transparent windows and domes, high power lasers and optical amplifiers, and lightweight personnel and vehicular armor.

Acknowledgments

The authors would like to thank Dr. Lawrence Kabacoff of ONR for supporting this research. We would also like to acknowledge the exacting laboratory work conducted by our collaborators: at Rutgers University — Lin Gao, Shih-Chieh Liao, Rajendra K. Sadangi, Vijay Shukla, and Xinzhang Zhou — and at UC-Davis — Dustin Hulbert, Dong Tao Jiang, Joshua Kuntz, and Guodong Zhan.

References

[1] Liao, S.-C., Pae, K.D., and Mayo, W.E., High pressure and low temperature sintering of bulk nanocrystalline TiO_2, *Mater. Sci. Eng. A*, A204, 152, 1995.

[2] Liao, S.-C., Pae, K.D., and Mayo, W.E., The effect of high pressure on phase transformation of nanocrystalline TiO_2 during hot-pressing, *Nanostruct. Mater.*, 5, 319–323, 1995.

[3] Mishra, R.S., Lesher, C.E., and Mukherjee, A.K., High pressure sintering of nanocrystalline γ-Al_2O_3, *J. Am. Ceram. Soc.*, 79, 2989–2992, 1996.

[4] Liao, S.-C., Pae, K.D., and Mayo, W.E., Retention of nanoscale grain size in bulk sintered materials via a pressure-induced phase transformation, *Nanostruct. Mater.*, 8, 645–656, 1997.

[5] Liao, S.-C., Mayo, W.E., and Pae, K.D., Theory of high pressure/low temperature sintering of bulk nanocrystalline TiO_2, *Acta Mater.*, 45, 1063–1079, 1998.

[6] Liao, S.-C., Chen, Y.-J., Kear, B.H., and Mayo, W.E., High pressure/low temperature sintering of nanocrystalline Al_2O_3, *Nanostruct. Mater.*, 10, 1063–1079, 1998.

[7] Colaizzi, J., Kear, B.H., and Mayo, W.E., Breaking the nanograin barrier in sintered ceramics, in *Proceedings of the First International Conference on Advanced Materials*, Zhang, D.L., Pickering, K.L. and Xiong, X.Y. (Eds.), Institute of Materials Engineering, Australasia, 2000.

[8] Kear, B.H., Colaizzi, J., Mayo, W.E., and Liao, S.-C., On the processing of nanocrystalline and nanocomposite ceramics, *Scripta Mater.*, 44, 2065, 2001.

[9] Colaizzi, J., Mayo, W.E., Kear, B.H., and Liao, S.-C., Dense nanoscale single- and multi-phase ceramics sintered by transformation-assisted consolidation, *Int. J. Powder Metal.*, 37, 45–54, 2001.

[10] Sadangi, R.K., Shukla, V., and Kear, B.H., Processing and properties of $ZrO_2(3Y_2O_3)$-Al_2O_3 nanocomposites, *Int. J. Refract. Metals Hard Mater.*, 23, 363, 2005.

[11] Kear, B.H., Sadangi, R., Shukla, V., Stefanik, T., and Gentilman, R., Submicron-grained transparent yttria composites, in *Proceedings of SPIE Conference on Window and Dome Technologies and Materials IX*, Orlando, FL, 227, March 2005.

[12] Omori, M., Sintering, consolidation, reaction and crystal growth by the spark plasma system, *Mater. Sci. Eng.*, A287, 181–188, 2000.

[13] Zhan, G., Kuntz, J., Wan, J., and Mukherjee, A.K., Single-wall carbon nanotubes as attractive toughening agents in alumina-based nanocomposites, *Nat. Mater.*, 2, 38–42, 2003.

[14] Mishra, R.S., Bieler, T.R., and Mukherjee, A.K., Mechanism of high strain rate superplasticity in aluminum alloy composites, *Acta Metal. Mater.*, 45, 561–568, 1997.

[15] Mukherjee, A.K., High temperature creep, in *Treatise on Materials Science and Technology*, Arsenault, R.J. (Ed.), Academic Press, New York, Vol. 6, 163–224, 1975.

[16] Kim, B.N., Hirago, K., Morita, K., and Sakka, Y., A high strain rate superplastic ceramic, *Nature*, 413, 288–290, 2001.

[17] Morita, K., Hiraga, K., and Sakka, Y., High strain rate superplasticity in Y_2O_3 stabilized tetragonal ZrO_2 dispersed with 30% $MgAl_2O_4$ spinel, *J. Am. Ceram. Soc.*, 85, 1900–1902, 2002.

[18] Granquist, C.G. and Buhrman, R.A., *J. Appl. Phys.*, 47, 2200, 1976.

[19] Birringer, R., Gleiter, H., Klein, H.P., and Marquardt, P., Nanocrystalline materials an approach to a novel solid state structure, *Phys. Lett.*, 102A, 8, 365, 1984.

[20] Gleiter, H., Nanocrystalline materials, *Prog. Mat. Sci.*, 33, 4, 223–315, 1990.

[21] Glumac, N., Chen, Y.-J., Skandan, G., and Kear, B.H., Combustion Flame Synthesis of Nanophase Materials, U.S. Patent No. 5,876,683, March 2, 1999.

[22] Kear, B.H., Chang, W., and Skandan, G., Apparatus for making nanostructured powders and whiskers, U.S. Patent No. 5,514,350, May 7, 1996.

[23] Glumac, N.G., Chen, Y.-J., and Skandan, G., Diagnostics and modeling of nanopowder synthesis in low pressure flames, *J. Mater. Res.*, 13, 2572–2579, 1998.

[24] Chang, W., Skandan, G., Hahn, H., Danforth, S.C., and Kear, B.H., Chemical vapor condensation of nanostructured ceramic powders, *Nanostruct. Mater.*, 4, 345, 1994.

[25] Skandan, G., Hahn, H., and Parker, J.C., Nanostructured Y_2O_3: Synthesis and relation to microstructure and properties, *Scripta Met. Mater.*, 25, 2389, 1991.

[26] Uyeda, R., Studies of ultrafine particles in Japan: Methods of preparation and technological applications, *Prog. Mater. Sci.*, 35, 1–96, 1991.

[27] Harutyunyan, A., Gregorian, L., and Tokune, T., Method for synthesis of metal nanoparticles, U.S. Patent Publication, US 2004/0099092 Al, May 27, 2004.

[28] Miyao, T., Sawaura, T., and Naito, S., Preparation of silica-alumina ultrafine particles possessing nano-pore structure by means of reversed micelle technique, *J. Mater. Sci. Lett.*, 21, 867–70, 2002.

[29] McCandlish, L.E., Kear, B.H., and Kim, B.K., Chemical processing of nanophase WC-Co composite powders, *Mater. Sci. Technol.*, 6, 953, 1990.

[30] Cohen, M., Kear, B.H., and Mehrabian, R., Rapid solidification processing — an outlook, in *Proceedings of the Second International Conference on Rapid Solidification Processing*, Kear, B.H. and Mehrabian, R. (Eds.), Reston, VA, 1–23, 1980.

[31] Perepezko, J.H. and Wilde, G., Amorphization and alloy metastability in undercooled systems, *J. Non-Cryst. Solids*, 274, 271–281, 2000.

[32] Smith, R.W. and Knight, R., Thermal spraying I: Powder consolidation: From coating to forming, *JOM*, 47, 32–39, 1995; and Thermal spraying II: Recent advances in thermal spray forming, *JOM*, 48, 16–19, 1996.

[33] Thermal Spray 2003: Advancing the science and applying the technology, in *Proceedings of the International Thermal Spray Conference*, Marple, B.R. and Moreau, C. (Eds.), Orlando, May 2003. ASM, Materials Park, OH, 1708 pages in two volumes, 244 papers.

[34] Kear, B.H. and Strutt, P.R., Nanostructures: The next generation of high performance bulk materials and coatings, *Naval Res. Rev.*, XLVI, 4, 1994; and *KONA, Powder and Particle*, 45, 1995.

[35] Beall, G.H. and Pinckney, L.R., Nanophase glass–ceramics, *J. Am. Ceram. Soc.*, 82, 5–16, 1999.

[36] Zhou, X., Cannon, W.R., and Kear, B.H., Metastable phase formation in plasma sprayed $ZrO_2(Y_2O_3)$–Al_2O_3 powders, *J. Am. Ceram. Soc.*, 86, 1415–1420, 2003.

[37] Kear, B.H., Shukla, V., and Sadangi, R., Metastable materials and nanostructures derived therefrom, Rutgers docket #05-085, January 31, 2005.

[38] Karthikeyan, J., Berndt, C.C., Tikkanen, J., Reddy, S., and Herman, H., Plasma spray synthesis of nanomaterial powders and deposits, *Mater. Sci. Eng.*, A238, 275–286, 1997.

[39] Strutt, P.R., Kear, B.H., and Boland, R.F., Method of manufacture of nanostructured feeds, U.S. Patent No. 6,025,034, February, 15, 2000.

[40] Chow, G.-M., Kurihara, L.K., Xiao, T.D., Strutt, P.R., Strock, C.W., and Zatorski, R.A., Nanosize particle coatings made by thermally spraying solution precursor feedstocks, U.S. Patent No. 6,447,848, September 10, 2002.

[41] Ma, X., Murphy, S.P., and Roth, J.D., Apparatus and method for solution plasma spraying, PCT #WO 2004/063416 A2, July 29, 2004.

[42] Padture, N.P., Schlichting, K.W., Bhatia, T., Ozturk, A., Cetegen, B., Jordan, E.H., Gell, M., Xiao, T.D., and Strutt, P.R., Towards durable thermal spray coatings with novel microstructures deposited by solution-precursor plasma spray, *Acta Mater.*, 49, 2251–2257, 2001.

[43] Chen, I.-W. and Wang, X.-H., Sintering dense nanocrystalline ceramics without final-stage grain growth, *Nature*, 404, 168–171, 2000.

[44] Voronov, O.A., Tompa, G.S., and Kear, B.H., High pressure-high temperature device for making diamond materials, in *Proceedings of the Seventh International Symposium on Diamond Materials*, 25, 264–271, The Electrochemical Society, 2001.

[45] Strutt, P.R. and Kear, B.H., Thick film ceramic fuel cell and method of making same, U.S. Patent Application, April 29, 2004.

II

Deposition Processes

8

Physical and Chemical Vapor Deposition Processes

Chad Johns,
M. Saif Islam, and
Joanna R. Groza
University of California, Davis

Abstract

One of the most important processing steps for thin films that accelerated the semiconductor industry is that of physical and chemical vapor deposition. Over the years many different variants of both physical and chemical vapor deposition techniques have been created. The intention of each variant was to overcome inadequacies and solve specific problems, though the basic principles remain the same. Physical deposition techniques are grouped as those in which no chemical reactions occur. Deposition by chemical means, as the name implies, are those that require some sort of chemical reaction during the process.

Since the majority of the vapor deposition techniques are performed in vacuum, the chapter begins with the kinetic theory of gases. This is followed by the main physical vapor deposition techniques and applications, e-beam/vacuum deposition, pulsed laser deposition, DC/RF/magnetron sputtering, and molecular beam epitaxy. The common attribute of all these techniques is that the process is achieved through a physical mechanism where no chemical reaction occurs. The main difference between each technique is the energy source used to create the material vapor that will be deposited.

Chemical vapor deposition involves a chemical reaction; therefore, this section begins with some of the more common types of reactions followed by thermodynamics and kinetics. The main chemical deposition techniques and applications covered in this section include typical thermally activated chemical vapor deposition (CVD), plasma-enhanced CVD, photo and laser CVD, metal-organic CVD, atomic layer deposition, and metal-catalyzed

CVD for nanowire growth. Similar to physical deposition techniques, the main difference between the techniques is the energy source used to activate the reaction, though metal-organic CVD is classified owing to the use of metal-organic gasses.

The chapter finally concludes with a discussion on the nucleation and growth of the developing thin film microstructures. The three main initial growth modes, Volmer–Weber, Frank–Van der Merwe, and Stranski–Krastanov are discussed along with the structure zone model.

8.1 Introduction

Thin films have changed the way we live. Today, almost all electronic devices, whether it is a cell phone, or even a toy, employ some sort of thin film technology. Over the past century, numerous deposition techniques have been used and developed for the fabrication of thin films. Almost all these techniques can be classified into either deposition by physical or chemical means or a combination of both. Physical deposition techniques are those in which no chemical reactions occur. This includes and can be further distinguished into the various evaporation techniques and sputtering. Deposition by chemical means, as the name implies, are those that require some sort of chemical reaction during the process.

Just about all methods or techniques of thin film deposition and characterization are conducted in some type of reduced pressure or vacuum. So, for a better understanding of gas behavior in these conditions we start with the kinetic theory of gases that describes gases in a closed system. Building on the kinetic theory of gases, physical vapor deposition techniques such as e-beam evaporation, sputtering, and molecular beam epitaxy, are presented. This is followed by chemical vapor deposition and the more typical variations. Finally, microstructure formation upon thin film deposition is addressed.

8.2 Kinetic Theory of Gases

The kinetic theory of gases allows one to understand and predict the behavior of vapor and gases at an atomic level. The model allows one to get a good understanding of how changing certain process variables such as pressure, temperature, gas concentrations, and flow rates, will affect the overall deposition.

The model is a classical one that assumes a rather well-behaved billiard ball (or solid sphere) model in which the gas particles themselves are very small hard spheres, all collisions are elastic, and there are no other forces on the particles except those from collisions with other particles and the walls. The molecules in this model are in a continuous state of random motion that is strongly dependent on the temperature and pressure.

The statistical distribution of the velocities of simple monoatomic gas molecules is given by the Maxwell–Boltzmann formula:

$$f(v) = \frac{1}{n}\frac{\mathrm{d}n}{\mathrm{d}v} = 4\pi v^2 \left(\frac{m}{2\pi kT}\right)^{3/2} \exp\left(\frac{-mv^2}{2kT}\right) \tag{8.1}$$

where v is the speed or magnitude of the velocity vector, m is the molecular weight of the molecule, k is Boltzmann's constant, n is the number of molecules per unit volume otherwise known as the molecular density, and T is the temperature in Kelvin. This central equation shows that the fractional number of molecules $f(v)$ in the velocity range v to $v + \mathrm{d}v$ is related to the molecular weight of the molecule and the temperature. The higher the temperature the faster the molecule moves. Moreover, the heavier the billiard ball (molecular mass), the harder or slower it is to move. Using the ideal gas law, the molecular weight can further be linked to the pressure.

A few representative curves of the velocity distribution are shown in Figure 8.1. Noticeably, higher temperatures shift the curve to the right (molecules move faster) while broadening and lowering it. In addition, a lighter mass allows a molecule to move faster thus it also shifts the curve to the right, again broadening and lowering it.

From Equation 8.1, one can solve for a number of important parameters such as the average, mean square, root mean square, and the most probable velocities. For instance, the average and root mean

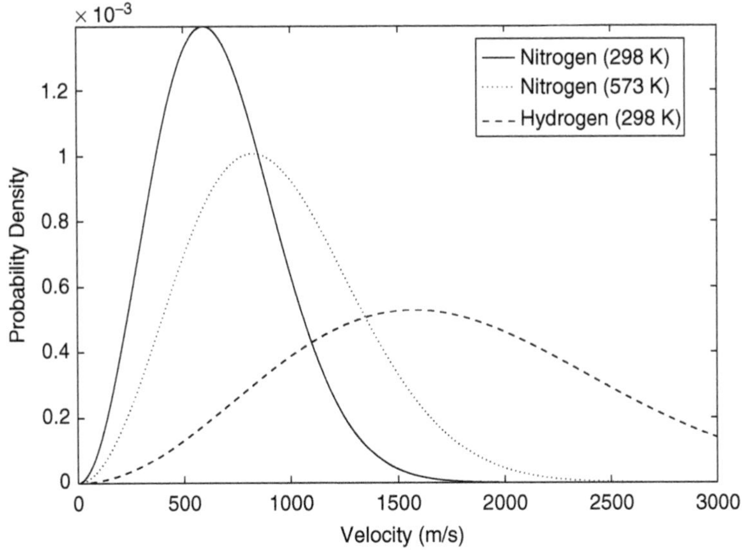

FIGURE 8.1 Probability density of velocity for N_2 at 298 and 573 K, and H_2 at 298 K. Higher temperature increases the velocity as seen for the N_2 at 298 and 573 K. Also heavier molecules move slower as seen between the heavier N_2 molecules compared to the lighter H_2molecule.

square velocities are given as:

$$v_{\text{avg}} = \frac{\int_0^\infty v f(v) \mathrm{d}v}{\int_0^\infty f(v) \mathrm{d}v} = \sqrt{\frac{8kT}{\pi m}} \tag{8.2}$$

$$v_{\text{rms}} = \sqrt{\frac{\int_0^\infty v^2 f(v) \mathrm{d}v}{\int_0^\infty f(v) \mathrm{d}v}} = \sqrt{\frac{3kT}{m}} \tag{8.3}$$

Again, this is not surprising: the hotter $(T\uparrow)$ it is or the lighter $(m\downarrow)$ they are, the faster the molecules move since $v \propto T/m$.

Momentum transfer from the gas molecules to the container walls gives rise to the forces that sustain the pressure in the system. Pressure is the most widely used unit in vacuum technology and most deposition processes; thus many parameters are dependent on pressure including the mean free path (the mean distance a molecule will travel, on average, without colliding with other molecules or λ). It is a very important parameter determining how fast transport of mass, energy, and momentum can take place in the gas. The mean free path can be simply calculated as the inverse of the collision rate giving the distance between each collision as:

$$\lambda = \frac{1}{\sqrt{2}\pi a^2 n} \tag{8.4}$$

where a is the diameter of the gas particle. As n is not commonly known, the above equation can also be expressed in a more useful way using the ideal gas law thus showing the pressure and temperature effects on the mean free path:

$$n = \frac{N}{V} = \frac{P}{kT} \quad \text{(molecular density)}$$

$$\lambda = \frac{kT}{\sqrt{2}\pi a^2 P} \xrightarrow{300K} \approx \frac{5}{P(\text{mtorr})}\text{cm} \tag{8.5}$$

The above equation is a good rule of thumb for quickly estimating the mean free path relative to the pressure. Typically, evaporation is carried out in a vacuum of 10^{-5} to 10^{-8} torr. In this pressure range the mean free path is large (10^2 to 10^5 cm) compared to the target to substrate distance. This allows the evaporated atoms to reach the substrate with little to no collisions. As will be explained later, this is what makes evaporation very directional. In comparison, sputtering uses argon at a pressure of about 10^{-2} torr that gives a mean free path of only 0.5 cm. Thus the material being deposited is scattered often and arrives from all directions onto the substrate.

Another relevant parameter for deposition processes is the gas impingement rate. This tells us the frequency with which molecules collide with the substrate. As only a singular surface in one direction is considered, the other directions are ignored. The number of molecules that strike a surface per unit time and area, also called the flux, is given by:

$$J = \int_{0}^{\infty} v_x \mathrm{d}n_x = \frac{n v_{\text{avg}}}{4} = \sqrt{\frac{P^2}{2\pi kTm}} = 3.5 \times 10^{22} \frac{P}{\sqrt{mT}} \quad \text{molecules/cm}^2\ \text{s} \tag{8.6}$$

where P is expressed in torr. The last equation, which relates the flux with the useful parameters P and T, is called the Knudsen equation. It tells us how many molecules strike the surface in a given time while setting an upper limit on the rate at which deposition will occur.

Another relevant issue is the time it takes for a surface to be coated with a monolayer of the gas molecules. This is also the time required to maintain a clean surface. It is basically the inverse of the flux equation (Equation 8.6) multiplied by the number of available surface atoms per unit area, N_0:

$$\tau = \frac{N_0}{J} \tag{8.7}$$

This becomes of great importance when trying to deposit pure thin films with the least amount of contamination. To minimize contamination during deposition, the time required should be less than the time it takes for deposition of a monolayer of the contaminant.

8.3 Physical Vapor Deposition

In physical vapor deposition (PVD) processes the material to be deposited is transformed into a gaseous state by a physical process such as thermal evaporation or an impact process. Typical PVD processes can be separated into two main categories: evaporation and sputtering. There is a third that can be considered as a mix between PVD and chemical vapor deposition (CVD), which is the reactive PVD process. The central focus of this chapter will be on the former two main categories.

One of the main advantages of PVD methods over CVD is that films can be deposited at relatively low temperatures. The vapor easily condenses onto most substrates with little to no heating, while high temperatures are typically required to activate the chemical reaction used in CVD. Also, just about any material can be deposited using PVD whereas CVD requires a chemical reaction.

Several techniques exist for evaporating the material to be deposited. The oldest method is simple thermal evaporation from a resistively heated crucible or current carrying wire. Quite simply, thermal energy is transferred to atoms in a liquid or solid source such that the material either evaporates or sublimes.

Disadvantages of thermal evaporation from a heated crucible include contamination by diffusion from crucibles, heaters, and support materials as they are all heated. There are also limitations with low input power levels that make it difficult to evaporate high melting point materials. To reach the higher temperatures required to melt these materials would require enormous amount of energy, which can become impractical.

Another alternative that has eliminated the disadvantages of thermal evaporation is the use of an electron beam. The electrons are typically accelerated by a voltage of 5 to 15 kV and focused onto the target or material to be evaporated. The kinetic energy of the electrons is then transformed into thermal energy that causes the material to sublime or melt and evaporate. A further more recent method called pulsed-laser deposition utilizes a high power laser to ablate or evaporate the material to be deposited.

Sputtering is different from evaporation in that sputtering uses ions to eject atoms from the source material at room temperature. This is a true physical process where atoms are physically knocked loose, similar to how pieces of a concrete wall are knocked loose by throwing rocks at it. The variations of sputtering include typical DC diode, AC or the more common radio frequency (RF), and magnetron.

8.3.1 PVD Processes and Applications

8.3.1.1 E-Beam/Vacuum Deposition

One of the most widely used forms of PVD is that of electron beam physical vapor deposition (EB-PVD). As seen in Figure 8.2, a high intensity electron beam is focused onto the target material to be evaporated. The schematic shows a typical setup in which electrons are bent from a source, 270°, using a bending magnet to focus onto the target. The energy from the electron beam, up to 15 keV, locally melts and evaporates a region of the target. As the target is locally heated from the electron beam, it can be placed in a water cooled crucible thus reducing contamination from diffusion. As a result, there is no alloying or reaction with the crucible material and consequently the purity of the deposited film is increased. Evaporated material then traverses the chamber condensing onto the substrate located some distance away. All of this is typically done under a vacuum of 10^{-5} to 10^{-8} torr to insure a large mean free path.

An important issue for evaporation is the conformal step coverage of nonplanar surfaces. The one exception is when liftoff is to be performed in which directionality is preferred. As already mentioned evaporation is the line-of-sight and very directional due to the large mean free path, so for very complicated structures shadowing from features and edges occurs and becomes a main limitation or concern.

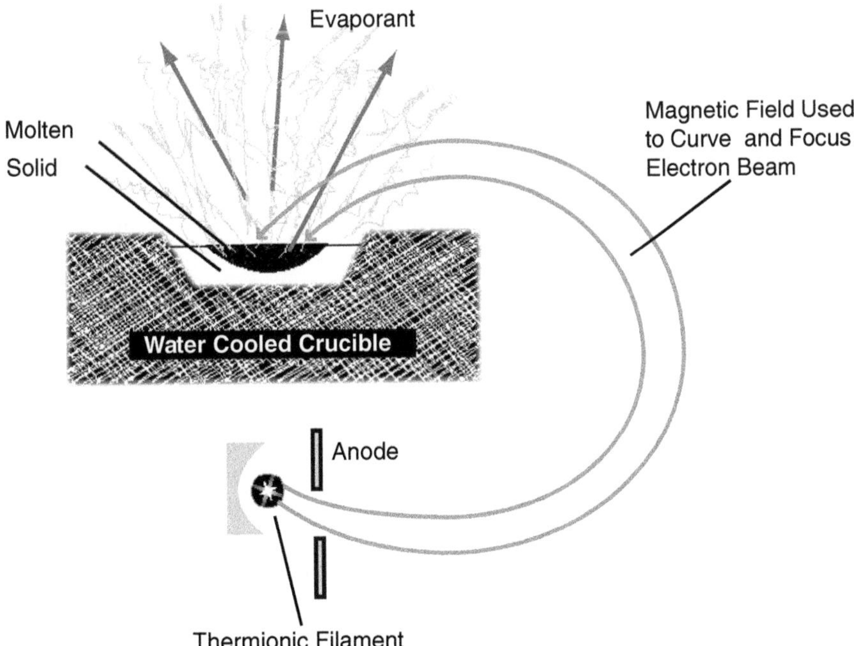

FIGURE 8.2 Schematic of an electron beam setup where electrons are boiled off a filament and bent using a magnetic field, then used to evaporate a solid target placed inside a water cooled crucible.

Two frequent methods used to improve the step coverage are rotation and heating of the sample. To accomplish the rotation, hemispherical chambers are used in which the wafers are placed on a planetary that rolls around during deposition. The second method involves heating the wafer. The heating is typically accomplished using banks of infrared lamps or refractory metal coils. By heating the wafers, atoms can diffuse across the surface before they become a permanent component of the film. Thus shadowing effects that create concentration gradients will result in a net movement of material toward the low concentration areas thus filling the shadowed areas. Surface diffusion coefficient, D_s, is defined by an Arrhenius type relation:

$$D_s = D_0 \exp\left(\frac{E_a}{kT}\right) \tag{8.8}$$

where D_0 is a constant and E_a is the activation energy. As the activation energy is much smaller than that of bulk diffusion, surface diffusion becomes very sensitive to changes in the temperature (T). Therefore, significant surface diffusion can occur at relatively low temperatures. The surface diffusion length is defined as:

$$L_s = \sqrt{D_s \tau} \tag{8.9}$$

where τ is the mean time before incorporation. Because of the exponential dependence of D_s on temperature, the surface diffusion length can be significantly increased by simply heating. It is a typical practice to have surface diffusion lengths greater than the feature sizes to allow conformal coatings over these features.

A further concern regarding evaporation systems is that of contamination. Materials with high vapor pressures at low temperatures are of major concern. These materials will sublime from the solid before melting. Generally, a material must have a vapor pressure of 10 mtorr or less at its melting temperature to insure a melt during evaporation. Most metals fall into this category and therefore melt during evaporation. In contrast, elements such as Cr, Fe, Mo, Si, and Ti reach suitably high vapor pressures below their melting point and could possibly sublime. As far as contamination, the above mentioned materials are not as much of a concern because they require temperatures well above room temperature. The main concern is with those elements with high vapor pressures near room temperature (e.g., P, Hg, S, K, Se, Cd, As, Na, Zn, Te, and Mg). All have sufficiently high vapor pressures at or slightly above room temperature. So if these elements somehow get into the chamber, they will exist as vapor once a vacuum is achieved and can therefore contaminate the depositing film.

Evaporation works well when only evaporating a single element, but problems tend to occur when an alloyed material is evaporated. As each element in the alloy has a different vapor pressure and melting temperature, they evaporate at very different rates, giving a final film composition, which is typically not what is expected. This is one of the main reasons alloyed films or doped films are predominantly deposited using CVD. To avoid the problem, multiple sources are used to better control the power levels individually and therefore the deposition rates of each element.

8.3.1.2 Pulsed-Laser Deposition

Pulsed-laser deposition (PLD) also called laser ablation is a flash evaporation method where lasers supply the required energy to remove material from the target. An intense laser pulse, typically with an energy density of a few J/cm^2, is focused onto the target where it is partially absorbed. Above a certain threshold power density (energy density divided by pulse time), material is removed from a thin surface region creating a conical vapor plume extending along the direction normal to the target surface. The threshold power density mainly depends on the target material and laser pulse duration and wavelength. Material released into this plume then travels across the chamber where it finally condenses onto the substrate. As high power densities can be obtained by focusing the laser beam, high melting temperature materials can easily be vaporized. This has led to the popularity of this technique for the deposition of superconductors[1–4] and diamond-like carbon.[5–12]

Growth of thin films by laser ablation has several key advantages. Small beam divergence allows the laser and associated equipment to be located far away. Typically the laser source is located outside of the vacuum chamber and focused through a quartz window. Also, lasers are clean thermal sources that introduce minimal contamination. Due to the high obtainable power densities, almost any condensed matter can be ablated. Moreover, the pulsed nature of the process permits precise control of the amount of deposited material. Due to this improved control, films can be deposited that better reproduce the stoichiometry and properties of the target than conventional evaporation or in some cases produce novel films that have quite different properties.

The greatest technological drawback of PLD, especially for commercial applications, is the production of unwanted micron-sized particles ejected during the ablation process. These can cause crystallographic defects from high kinetic energy particles bombarding the film during growth.

8.3.1.3 Sputtering

In sputtering, a gas discharge, typically argon, creates a plasma that is used to eject atoms from the target. In a truly physical process, accelerated ions collide with the material that is to be deposited thus releasing the atoms from the bulk. These atoms then make the journey across the chamber where they finally condense onto a substrate.

The first gas discharge studied for purposes of sputter coatings, and therefore the most understood, was the DC glow discharge. It is created when a high voltage (2 to 5 kV) is applied to a gas at low pressures, typically in the mtorr range for sputtering. A typical schematic of a sputter system is shown in Figure 8.3. The source material to be deposited (also called the target, as it is the target of the bombarding ions) is typically connected to the cathode, a negative voltage supply (DC and RF), while the substrate (material to be coated) is the positive anode. Positive argon ions are accelerated toward the negatively charged cathode where energy from the positive argon ions is transferred to the deposited material, which dislodges and ejects neutral target atoms among other things (ions, electrons, etc.). These atoms then make the journey across the chamber, colliding many times in the plasma, and finally land on the anode where they condense onto the substrate and become part of the film.

The DC diode plasma is a self-sustaining discharge because the charge carriers are created through a collision process between a relatively fast electron and a neutral argon atom. The collision results in the ionization of the argon atom leaving behind a positive argon ion and an electron plus the initial electron now slightly slowed:

$$e^- + Ar^0 \rightarrow 2e^- + Ar^+$$

Both these electrons are then accelerated by the applied electric field and can gain enough energy to further ionize more neutral atoms therefore resulting in an avalanche effect. This occurs spontaneously once the

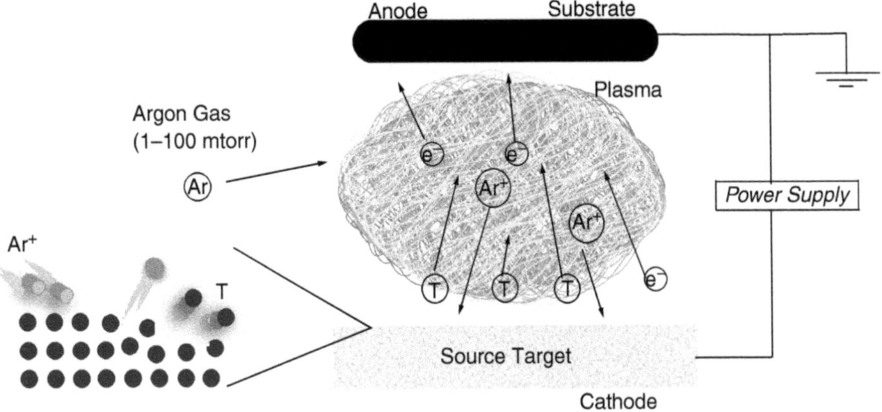

FIGURE 8.3 Typical diode plasma discharge showing ionization of argon from accelerated electrons. Incident Ar^+ ions physically knock loose target atoms, T, which then travel across the chamber where they condense on the substrate.

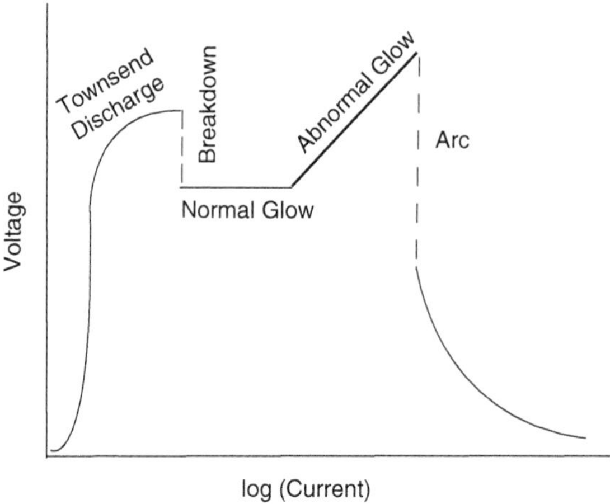

FIGURE 8.4 Current–Voltage relation of the various types of discharges between the electrodes in sputtering.

bias is applied to the low pressure gas as the initial electrons are present, which were created from natural radiation. Once sputtering has begun, the collisions of the target also release a small fraction of electrons, called secondary electrons. It is these secondary electrons that are accelerated back toward the plasma where collisions with neutral argon atoms produce more ions that further maintain the discharge.

Figure 8.4 shows the various types of discharges that can occur between the powered electrodes. In sputtering, the current is typically the independent variable and the voltage being dependent. After a threshold is reached, the electric field quickly sweeps all the available electrons from the space between the two electrodes where they collect at the anode. At this point the plasma is random and not self-sustaining. At the onset of ionization from these accelerated electrons, the voltage levels out due to the creation of positively charged ions. This region called the Townsend discharge is named after an early investigator of this phenomenon. An abrupt drop in the voltage signifies conduction through the gas and is known as gas breakdown. At this point the ions have enough kinetic energy to knock loose secondary electrons from the target material and the plasma becomes self-sustaining. It is called a glow discharge due to the electron excitation of the gas, which produces a glow of which the color depends on the gas. As current is increased further, the entire cathode area becomes covered by the plasma. Any further increase in current will be accompanied by an increase in voltage up until arcing occurs. Most sputtering systems operate in this abnormal glow region as it allows the highest sputtering rates.

Coming back to breakdown, there is a sudden drop in voltage at a fixed current. As mentioned above, this signifies the onset of secondary electron emission, but this is also accompanied by a spatial rearrangement of the discharge. The plasma re-arranges into densely charged regions with all the voltage dropping in a small region near the cathode. The neutral body of the plasma has a fairly constant voltage and thus current density as it becomes a good conductor once the plasma is formed. Of most importance is the small region near the cathode in which there is a large potential drop. It is at this area that positively charged ions are rapidly accelerated due to the large potential drop toward the cathode. Eventually they collide with the target that is placed on the cathode, releasing material and secondary electrons.

An important parameter of the sputtering process is the sputtering yield. It is defined as: $S =$ number of atoms ejected/number of incident ions. It determines the efficiency of the target material to be sputtered. Information is also given about the erosion rate of the targets and therefore mostly (there are other factors such as pressure and temperature) determines the deposition rate. Typically this can be measured by weight loss experiments. The sputter yield depends mostly on the target material and sputtering gas used. For the target material, the sputter yield depends on the binding energy and mass of the atoms. For the gas, sputtering yield increases with the mass and incident energy. Lighter atoms, like hydrogen, tend to bounce off the target without transferring momentum.

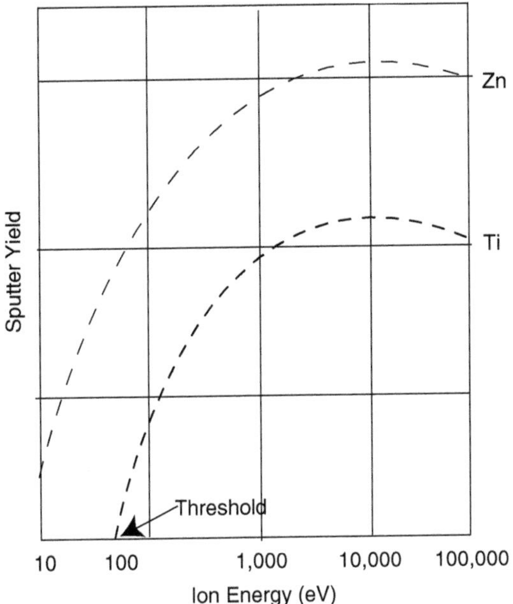

FIGURE 8.5 An illustration of the sputter yield as a function of Ar ion energy (not to scale). Sputtering only occurs once the threshold voltage has been passed.

There is a threshold for sputtering that is approximately equal to the heat of sublimation or binding energy of the most weakly bound surface atom. Below this threshold no sputtering occurs. After the threshold in the practical energy range for sputtering, the yield increases with incident ion energy, as shown in Figure 8.5.

8.3.1.4 RF Sputter Deposition

RF sputtering is used primarily for nonconducting targets (insulators). The only difference between RF sputtering and DC diode sputtering is that an alternating RF signal is used instead of a constant DC bias. A sputtering system is essentially a very large capacitor. So to overcome the problems caused by the nonconducting target, one must remember that the impedance of a capacitor drops as the frequency of the signal increases. Therefore, using a high frequency plasma, current can be passed through dielectrics in the same way current is passed through metals using DC plasmas.

When sputtering dielectrics, as the frequency increases above 50 kHz, it is observed [13] that the discharge can operate at increasingly lower pressures, finally leveling off for frequencies of a few MHz. This indicates that something other than secondary electrons is enhancing the ionization. This enhancement comes from the fact that electrons are very light compared to ions and Ar atoms, thus the electron can move quickly with the oscillating field. The end result of this extra electron motion is an increased probability of ionizing collisions with argon atoms. Therefore, at a given pressure, the density of the plasma is greater than that of a DC diode system.

Nowadays, most commercial RF sputtering systems operate at a frequency of 13.56 MHz. This frequency was established due to government communication regulations to minimize interference with other communication equipment.

Another important aspect of RF discharges that allows sputtering of insulators is the occurrence of a self-bias. In the plasma, electron mass is much less than that of the ions therefore they are relatively free to move with the applied voltage. Conversely, the massive ions cannot respond to the rapidly changing field and are therefore relatively static. Initially, during the first positive half-cycle as shown in Figure 8.6, the voltage over the plasma will be the same as the applied voltage and electrons will rapidly begin to

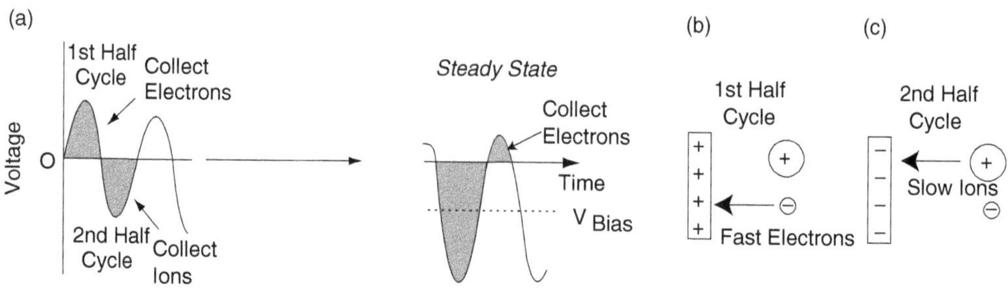

FIGURE 8.6 Establishment of a self-bias for an RF powered plasma. (a) During the first half-cycle electrons are collected at the electrode whereas on the second half-cycle ions are collected. A self-bias is formed due to the different mobilities of the charges during each time cycle. (b) The self-bias of an asymmetric RF plasma where sufficient electrons are collected when the voltage exceeds 0 V to counter the collection of the slower ions when the voltage is below 0 V. (c) An illustration showing what happens during each half-cycle. Electrons are much smaller than the ions and can therefore move faster.

move and collect on the electrode. With an insulator placed on the electrode, a capacitor is formed, which becomes charged by the accumulating electrons causing the voltage over the plasma to drop.

On the second half-cycle, the electrode becomes negative causing ions to move toward it. As already mentioned, the ions are too slow, therefore the ion current does not balance out the collected electron charge. On the next half-cycle, the electrode again collects more electrons, though slightly less than the first, causing the voltage over the plasma to drop even further. This process is continued with each cycle until the capacitive electrode is negatively charged enough that when the voltage exceeds 0 V, sufficient fast moving electrons are collected to counterbalance the number of slow moving ions collected during the remainder of the cycle. As seen in Figure 8.6b, the end result is a time-averaged negative DC bias at the RF powered electrode.

The negative self-bias allows the sputtering of insulating material. The slow massive ions cannot respond to the rapidly changing voltage, yet they do respond to the negative DC self-biased electrode. Therefore the ions are accelerated toward and collide with the RF powered electrode thus sputtering whatever material was placed onto the electrode.

8.3.1.5 Magnetron Sputter Deposition

Magnetron sputtering was developed to increase the deposition rates of both DC and RF sputtering. In normal sputtering systems, not all of the electrons contribute toward ionization. The wasted electrons fly around the chamber causing excess radiation and colliding with the chamber walls and target causing unwanted heating. Through the use of a magnetic field these wasted electrons can be captured. As shown in Figure 8.7, the magnetron is a magnetically assisted discharge that traps electrons near the target surface. This locally increases the ionization of the argon gas thereby enhancing the ion bombardment and sputtering rates. Thus the plasma is concentrated and intensified directly above the target. A direct consequence of this is higher sputter rates at lower argon pressures.

With the lower pressure there are fewer collisions, contamination decreases, and the deposition becomes more line-of-sight. In general the magnetron discharge is much more efficient than normal DC or RF sputtering.

8.3.1.6 Molecular Beam Epitaxy

Molecular beam epitaxy (MBE) is the coevaporation of elemental components through direct controlled beams. An important feature is that the deposition rate is very low which allows precise control of the thickness of each layer. MBE has been and is still used to epitaxially deposit very thin alternating layers of III–IV semiconductors used in optoelectronics. Now very thin layers can easily be placed between two very different bandgap materials thus allowing a very gradual transition of bandgaps, a mainstay in bandgap engineering. For more details on epitaxy, the reader is referred to Chapter 9.

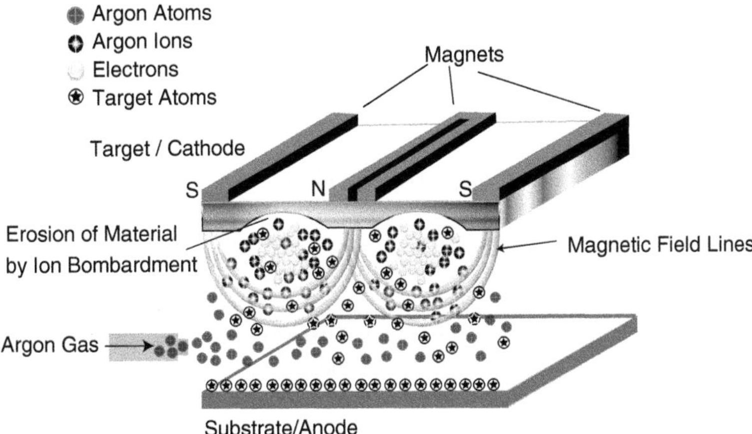

FIGURE 8.7 Side view schematic of a magnetron-assisted sputterer. The magnetron locally traps the electrons close to the substrate causing increased ionization.

MBE provides several advantages over other methods of epitaxy. First and foremost is the ability to perform *in situ* analysis during the growth of epitaxy layers. A typical MBE system includes an ultra high vacuum chamber, typically 10^{-10} torr. This ultra high vacuum makes it possible for *in situ* analysis tools like reflection electron diffraction, mass spectroscopy, and auger spectroscopy. As MBE uses several molecular beam sources to alloy films, it also avoids typical problems seen by evaporation of single multicomponent targets with each component having different vapor pressures. Also the deposition temperature can be lowered due to the slow deposition rates. Less heat is needed as plenty of time is allowed for the molecules to attach to the substrate.

The drawbacks can be attributed to the same component that gives it its advantages, the ultra high vacuum. Not only does it require time to pump down a system to 10^{-10} torr, the vacuum equipment and its maintenance is also very expensive. Another limitation can be attributed to the very slow deposition rate that has prevented widespread commercial use of this technique.

8.4 Chemical Vapor Deposition

Chemical Vapor Deposition is a method in which a thermally activated chemical reaction occurs from the vapor or gas phase depositing a solid thin film. The properties of the coating are controlled by process parameters such as temperature, pressure, flow rates, and input concentrations. With the proper selection of such parameters, the coating structure and properties (hardness, toughness, modulus, adhesion, thermal shock resistance, corrosion, wear and oxidation resistance, etc.) can be tailored to the desired application. The process can be understood and controlled through knowledge of mass transport, equilibrium thermodynamics, and the possible kinetic rate-limiting mechanisms of the system.

The deposition of thin films by CVD is much more complex than other techniques. The complexity arises from many factors that separate it from other methods. First and foremost, the fact that it requires a chemical reaction that generally includes multicomponent species with the possibility of forming several intermediate products complicates the modeling and understanding. Second, in comparison to PVD methods the growth process has numerous more independent variables. Precursor gas selection, concentration, flow rates, temperature, and pressure all affect the final quality of the thin film or nanostructure.

CVD has become a very important and popular technique for thin films (2-D), nanowires and nanotubes (1-D), and nanoparticles (0-D). As the deposition is controlled by chemical reaction rather than physical means, compounds can be deposited easier compared to PVD methods. Because of this control by chemical means, for III–V semiconductors CVD has become the preferred deposition method. Moreover, CVD provides films, which are of very high purity and dense, with accurate control of the stoichiometric composition and doping levels. As most reactions can occur at ambient pressures, the need for expensive high vacuum equipment — such as those used for PVD — can also be avoided. In contrast to PVD which is line-of-sight, CVD provides conformal coatings in which very complicated features can be coated as the gas reacts with the exposed surfaces. Furthermore, PVD techniques are not well suited for depositing insulating or semiconducting materials.

Methods to overcome the disadvantages of CVD have become the main topic of research for many groups over the past years. One of the main concerns associated with CVD is the high temperature needed for many of the reactions. This has led many to try alternate low temperature energy sources such as the use of plasmas or lasers. Another problem arises from the differences in thermal expansions of the substrate and the depositing film, which can lead to stress and possible delamination, though it is also a lesser concern with PVD methods. Yet another is that the precursors must be volatile near room temperature though metal-organic chemical vapor deposition (MOCVD) has eased this requirement. The gases used are usually highly toxic, hazardous, explosive, corrosive, and rather costly. Furthermore, CVD is not as simple as PVD methods in which simple analytical expressions can be used. Both the gas flow and the chemical reactions require numerical analysis that is both reactor and process dependent. To complicate things further, most reactions are not as simple as one would like. Several steps and various compounds and reactions can occur at the same time.

Generally, precursor molecules consist of the desired element or molecule chemically bonded to some other atoms, functional groups, or ligands. During the reaction, these other atoms become a by-product and react away leaving behind the desired material on the substrate. Small and simple molecules in which the central atom is bonded to another species such as hydrides, halides, or small organic radicals are generally preferred. The bonds of these small molecules are relatively easy to break, therefore more volatile and reactive than the larger molecules. This allows lower deposition temperatures that is very beneficial for most processes.

Pyrolysis (Thermal decomposition). The most common type of reaction is a thermal decomposition of gaseous species onto a hot susceptor/substrate. Films deposited through this type of reaction typically involve simple hydrides such as silane, carbonyls such as nickel carbonyl and organometallic compounds. One of the most common is the simple thermal decomposition of silane:

$$SiH_{4(g)} \rightarrow Si_{(s)} + 2H_{2(g)}$$

Hydrogen gas is the by-product. The decomposition of methane to produce diamond or diamond-like films has gained popularity over the past decade:

$$CH_{4(g)} \rightarrow C_{(s)} + 2H_2$$

Reduction. Reduction reactions typically involve halides with excess hydrogen as the reduction agent. Two commonly cited examples are the reduction of silicon tetrachloride and tungsten hexafluoride:

$$SiCl_{4(g)} + 2H_{2(g)} \rightarrow Si_{(s)} + 4HCl_{(g)}$$

$$WF6_{(g)} + 3H_{2(g)} \rightarrow W_{(s)} + 6HF_{(g)}$$

Oxidation. Oxidation reactions, as the name implies, involve the formation of oxides such as silicon dioxide and hard alumina coatings used for tooling. Typically oxygen is added to the system to oxidize the main precursor, removing hydrogen or halide atoms. A good example of this is the oxidation of silane to produce silicon dioxide:

$$SiH_{4(g)} + O_{2(g)} \rightarrow SiO_{2(s)} + 2H_{2(g)}$$

Or

$$SiH_{4(g)} + 2O_{2(g)} \rightarrow SiO_{2(s)} + H_2O_{(g)}$$

The less reactive tetraethoxysilane (TEOS) is also commonly used with ozone (O_3) to accelerate the oxidation reaction.

Hydrolysis. Hydrolysis reactions are often used with metal chloride precursors in which water vapor is used to remove the chloride atoms with HCl gas becoming the by-product of the reaction:

$$2AlCl_{3(g)} + 3H_2O_{(g)} \rightarrow Al_2O_{3(s)} + 6HCl_{(g)}$$

$$TiCl_{4(g)} + 2H_2O_{(g)} \rightarrow TiO_{2(s)} + 4HCl_{(g)}$$

Disproportionation. This reaction occurs when a nonvolatile material can form volatile compounds having different degrees of stability at various temperatures. It typically involves the dissociation of divalent halides into a solid and a four valence halide. An example is the reversible reaction:

$$2GeI_{2(g)} \leftrightarrow Ge_{(s)} + GeI_{4(g)}$$

in which, at 575 K the reaction proceeds to the right, thus depositing germanium, and at 875 K to the left allowing germanium to be transported.

This type of reaction allows solid compounds to be transported into the vapor phase by reacting it with its volatile, higher valence halide to form the more stable lower valence halide. This lower valence halide can be transported into the reactor where at lower temperatures it disproportionates depositing a solid film while converting back to its higher valence halide.

8.4.1 Thermodynamics — Driving Force

Using thermodynamics, a good understanding of the influence of process variables such as temperature, pressure, and inlet gas concentration on the deposition process can be understood. However, thermodynamics assumes the establishment of equilibrium which for most, especially flow reactors, is not true. Also, thermodynamics does not give any information about the speed of the reaction or the film growth rates. However, it provides a model that can be used to understand what each process variable does to the deposition.

Assuming a given CVD process involves a long time and many collisions between molecules, the system can be assumed as approaching equilibrium. For example, take the following reaction:

$$\underbrace{aA + bB}_{\text{Reactants}} \leftrightarrow \underbrace{cC + dD}_{\text{Products}}$$

where lower case letters represent the number of moles, for example aA means a moles of element A.

The feasibility of a certain reaction is determined by calculating the Gibbs free energy of the reaction for a given temperature and pressure. The standard Gibbs free energy change at a temperature T, ΔG_T, is the sum of the Gibbs free energies of formation of the products, less the sum of the Gibbs free energies of formation of the reactants at the same temperature:

$$\Delta G_T = \sum_{\text{Products}} n_p \Delta \underline{G}_{f,T} - \sum_{\text{Reactants}} n_r \Delta \underline{G}_{f,T} \tag{8.10}$$

where n is the number of moles of each species involved in the reaction, n_p for products and n_r for reactants.

A reaction will occur spontaneously if ΔG_T is negative, or the products have a lower energy state than the reactants. Some reactions have the possibility to form several possible products that are all thermodynamically feasible. When this is possible, the reaction with the most negative ΔG_T will dominate because it produces the most stable product. The free energy of an individual molecular species can also be represented as

$$G_i = G_i^o + RT \ln a_i \tag{8.11}$$

where R is the gas constant and G_i^o is the free energy of the species in its reference standard state. The activity a_i can be viewed as an effective thermodynamic concentration that reflects the change in free energy when the species is not in its standard state. If assuming an ideal gas, a_i can be replaced with the partial pressures of each species or, plugging Equation 8.11 into Equation 8.10 gives:

$$\Delta G = \Delta G^o + RT \ln \frac{a_C^c a_D^d}{a_A^a a_B^b} \tag{8.12}$$

The equilibrium constant of the reaction is defined as:

$$K = \frac{[C]^c[D]^d}{[A]^a[B]^b} = \frac{a_C^c a_D^d}{a_A^a a_B^b} = \frac{P_C^c P_D^d}{P_A^a P_B^b} \tag{8.13}$$

where the brackets ([]) are concentrations used for liquids and solids, a_i used for nonideal gases and P is the partial pressures that can be used if ideal gases are assumed. The equilibrium constant (K) can be related to the chemical equilibrium by the following:

$$\Delta G = \Delta H - T\Delta S = -RT \ln K \tag{8.14}$$

where H is the enthalpy and S is the entropy of the reaction, R the gas constant, T the temperature and K the equilibrium constant. Rearranging gives the following:

$$K = \frac{[C]^c[D]^d}{[A]^a[B]^b} = \exp\left(-\frac{\Delta G}{RT}\right) \tag{8.15}$$

Once the Gibbs free energy of the reaction is known along with the input gas concentrations, the partial pressures or concentration of each gaseous species can be determined.

8.4.2 Kinetics

Kinetics determines the rates/speed of the system. A reaction that is thermodynamically possible can still be considered useless due to slow kinetics. As far as importance, kinetics provides very important information about the deposition process. In contrast to thermodynamics which mainly considers just the reaction, kinetics encompasses the whole system and is, therefore, more complex than simple equilibrium thermodynamics. Parameters such as the reactor geometry, temperature and concentration gradients, and gas flows now must also be considered. There are several steps involved during deposition but they can be classified into two main categories, mass transport and surface kinetics. The slower of the two determines the deposition rate and is the rate-limiting mechanism.

As illustrated in Figure 8.8, films grown by CVD involve:

Arrival

1. Transport of reactant gases into the chamber
2. Intermediate reactant gases form from reactant gases
3. Transport and diffusion of reactant gases through the gaseous boundary layer to the substrate

Surface Reaction

4. Adsorption of gases onto the substrate surface
5. Surface diffusion
6. Single- or multistep reactions at substrate surface, nucleation

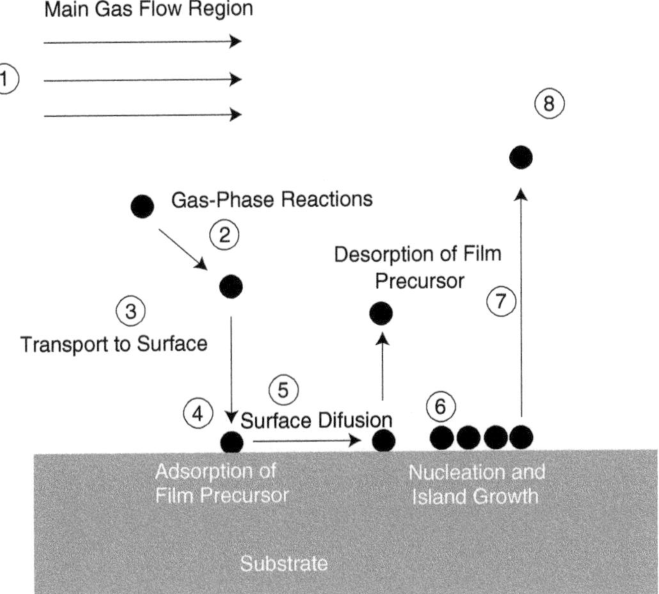

FIGURE 8.8 Various kinetic processes occurring during CVD.

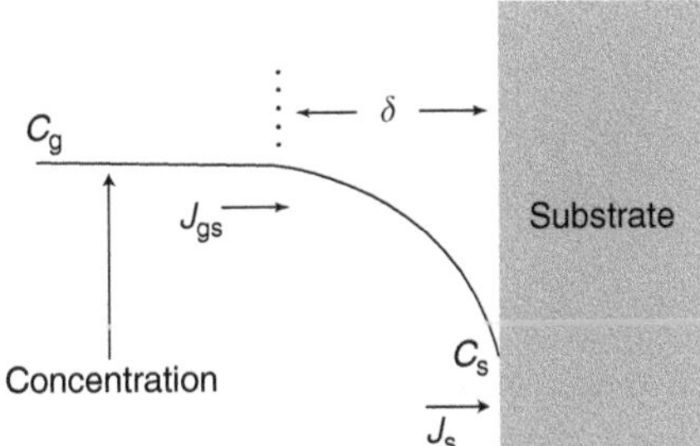

FIGURE 8.9 Gas concentration gradient at interface of film where the reactant concentration varies from the gas bulk, C_g, to the surface C_s, δ is the thickness of boundary layer and Js and Jgs are the fluxes at the surface and through the boundary layer respectively.

Removal of reactant by-products

7. Desorption of product gases from the substrate surface
8. Forced exit of product gases from the system

The slowest of the eight becomes the deposition rate-limiting step.

The model developed by Grove[14] best illustrates the temperature dependence of film growth. The essentials of the model, which is the environment near the gas–film interface, are shown in Figure 8.9.

Near the interface, the reaction consumes the reactants as the film grows. This causes a drop in the reactant concentration from the bulk C_g to the surface C_s. Thus, the flux of molecules diffusing to the surface from the gas stream is given by:

$$J_{gs} = h_g(C_g - C_s) = \frac{\overline{D}_g}{\delta}(C_g - C_s) \tag{8.16}$$

where h_g is called the gas-phase mass-transport coefficient. It is also described by the ratio D_g/δ where D_g is an effective diffusion constant for the gas-phase molecules and δ is the diffusion distance or boundary layer thickness. The flux at the surface is simply given by:

$$J_s = k_s C_s \tag{8.17}$$

where k_s is the surface-reaction rate constant. During steady state conditions, the fluxes must equal each other, $J_s = J_{gs}$ giving

$$C_s = \frac{C_g}{1+(k_s/h_g)} \tag{8.18}$$

This equation shows that the surface concentration drops to zero when $k_s \gg h_g$. When this condition is met, the system is mass-transport controlled. The slow gas transport through the boundary layer limits the reaction rate rather than the faster surface reaction. On the other hand, the system becomes surface-reaction controlled when $h_g \gg k_s$ and C_s approaches C_g. Here the surface-reaction rate is slow compared to the transfer of the reactants even with adequate reactant gas supplied. The film growth rate is equal to the flux at the surface J_s divided by the atomic density or number N of molecules per unit volume:

$$G = \frac{J_s}{N} = \frac{k_s h_g C_g}{(k_s + h_g)N} \tag{8.19}$$

With the case of $k_s \gg h_g$, the growth rate becomes mass-transport limited, and the deposition rate becomes:

$$G = h_g \frac{C_g}{N} \tag{8.20}$$

When $h_g \gg k_s$, the growth rate is surface reaction limited, and:

$$G = k_g \frac{C_g}{N} \tag{8.21}$$

As can be seen from the above equations, the temperature dependence of the growth rate comes from the properties of k_s and h_g. As seen in Equation 8.16, h_g is related to D_g/δ where D_g varies as T^2 at most and δ is weakly independent with T. On the other hand, k_s has an exponential temperature dependence, that is, $k_s \sim \exp(-E/RT)$ where E is the activation energy. Thus, at lower temperatures, film growth is dominated by the smaller exponential dependence of k_s and is therefore surface-reaction limited. As temperature increases, a point is reached where the exponential dependence of k_s passes that of h_g (Figure 8.10). At this point the growth rate changes to mass-transport limited and follows a much lesser dependence of $\sim T^{1.5\text{–}2}$. Most reactions are carried out in the mass-transport limited regime because of the relatively flat response to temperature fluctuations.

Also noteworthy in Figure 8.10 is the dependence of the deposition rate on pressure in the mass-transport limited area as well as the flow rate dependence. As this region is limited by the transfer (mass transport) of the gas molecules to the substrate, the system can be better controlled using the process variables. Higher flow rates allow more gas and therefore more molecules to react. So reaction rate increases as the flow rate is increased. Yet, as pressure is increased, the reaction becomes diffusion limited and the rate is slowed due to a decrease in diffusion ($D \propto (T/P)$) through the gas. The precursors must slowly diffuse through the higher pressure whereas for a low pressure system, there is little to collide or diffuse through.

8.4.3 CVD Processes and Applications

The traditional CVD method uses thermal energy in the form of heat to activate the chemical reactions. However, other sources or energy can be used such as laser energy or plasma. Furthermore, the CVD

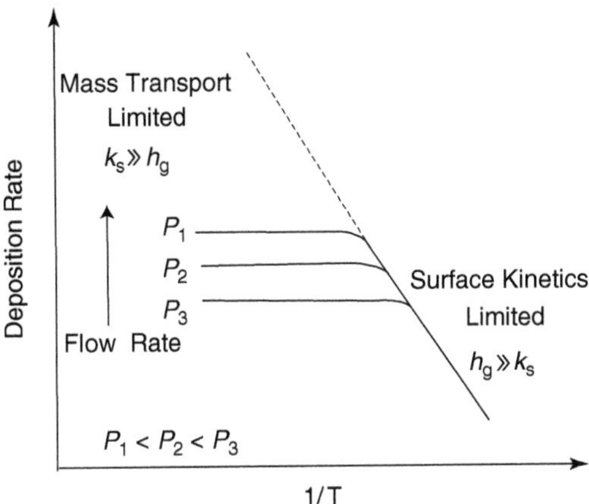

FIGURE 8.10 Deposition rates as a function of temperature. At low temperatures the deposition rate is controlled by reaction kinetics whereas at higher temperatures the rate is limited by the transport of the reactants. In this regime there is only a small dependence on temperature but a more significant dependence on the flow rates and pressure.

process can be separated even further depending on the pressure used. Two such CVD processes include atmospheric pressure chemical vapor deposition (APCVD) and low pressure chemical vapor deposition (LPCVD).

The main difference between these processes is that in APCVD due to the greater pressure, the rate of mass transfer or gas diffusion of the reactants and by-products through the boundary layer is slower than the rate of surface reaction. With the rate-limiting step being the slowest, the process becomes mass transport/diffusion limited. For LPCVD the reduced pressure alters the rate-limiting step toward that of surface-reaction limited. The lower pressure increases the precursor diffusion through the gas and the mass transfer rate of the gaseous reactants becomes higher than the surface-reaction rate. The advantages include better film uniformity, better film coverage over steps, and fewer defects.

Not only can CVD systems be separated depending on the pressure, they can also be divided into hot and cold wall reactors. In hot wall reactors the chamber walls contain the heating elements of the furnace. These reactors are frequently tubular in shape with heating elements surrounding the tube (similar to a tube furnace). Several temperature zones can be portioned to create various temperature gradients along the tube. As a consequence the whole chamber including the walls is heated to the desired temperature. The main advantage is that one can do large parts or many samples at once. Primary disadvantages are that coating occurs everywhere including the walls as everything is at an elevated temperature, requiring frequent cleaning and particulate problems. Also hot wall reactors involve higher thermal loads and energy usage. They are typically used in systems where the deposition reaction is exothermic and in many low pressure systems.

In cold wall reactors the susceptor, onto which the sample is placed, is the only heated area with the walls typically water cooled. Advantages include reduced deposition on the walls, less cleaning, lower thermal loads, faster heat up and cool down times, and lower energy consumption. Disadvantages are larger temperature nonuniformities that lead to variations in film thickness, smaller batch sizes, and possible thermal stresses on the substrates from nonuniformity and rapid heating and cooling. These reactors typically support endothermic reactions that proceed most readily on the hottest surfaces.

8.4.3.1 Plasma-Enhanced Chemical Vapor Deposition

Plasma-enhanced chemical vapor deposition (PECVD) arose from the demands of the semiconductor industry where an early application was the low temperature deposition of silicon nitride passivation capping layers. Today the applications have broadened to include dielectric films for optics along with amorphous and microcrystalline silicon.[15]

With PECVD a plasma rather than thermal energy is used to dissociate the precursor gas into smaller, more reactive, molecules. This allows deposition at higher rates and low temperatures than conventional thermal CVD. As dielectrics are typically deposited, an RF plasma is used. Disadvantages are poorer film quality in terms of rougher morphology, higher impurity incorporation, and ion damage to the film and substrate.

8.4.3.2 Photo and Laser Chemical Vapor Deposition

Photochemical vapor deposition (PCVD) utilizes electromagnetic radiation from lamps, typically short-wave ultraviolet, to activate the reactants. Instead of heat, selective absorption of the photonic energy by the reactant molecules initiates the desired reaction. Reactions can be selectively activated at lower temperatures than conventional CVD. There is also the advantage of driving reactions at surfaces of thin films.

Laser chemical vapor deposition (LCVD) is similar to photochemical vapor deposition except that instead of lamps it utilizes a laser beam. It also has the ability to locally heat a part of the substrate while passing the reactant gas, thereby inducing film deposition by locally driving the CVD reaction at the surface. In another variation, the lasers supply the required thermal energy for the desired reaction to proceed. One of the main advantages for use of lasers as a heat source is the capability of selective area deposition.

8.4.3.3 Metal-Organic Chemical Vapor Deposition

Metal-organic chemical vapor deposition is classified so because of the use of metalorganics as precursors, which are compounds containing metal atoms bonded to organic radicals. Compounds that have one or more direct metal–carbon covalent bond are called organometallics. For this reason some call this method organometallic CVD (OMCVD). The great advantage of MOCVD precursors are their high volatility at moderate to low temperatures, therefore reaction temperatures are lower (750 to 1100 K) than conventional CVD. Typically they undergo decomposition or pyrolysis reactions that occur at lower temperatures than the halides used in conventional CVD. The main disadvantages are the precursors tend to be very expensive and are very volatile. When reactive liquids are used they require accurate pressure control and are difficult to purify. Also carbon contamination of the films from the organics can be a problem.

8.4.3.4 Atomic Layer Deposition

Atomic layer deposition (ALD) is based on sequential, self-limiting surface reactions characterized by alternating the chemical reactants as illustrated in Figure 8.11. This unique CVD-based growth technique can provide atomic layer control and allow conformal films to be deposited on very high aspect ratio structures. It can produce unvarying, conformal thin films that have broad applications in contemporary technology, including semiconductor microelectronics, displays, optical filters, magnetic information storage, and catalysis. Because of its atomic layer control, it has become the preferred method for deposition of high-k gate oxides or diffusion barriers for back-end interconnects.

An ALD process deposits thin layers of solid materials by using two or more dissimilar vapor-phase reactants. It consists of an alternating series of self-limiting chemical reactions, called half-reactions, between gas phase precursors and the substrate. First, a dose of vapor from one precursor is brought to the surface of a substrate onto which a film is to be deposited. Using the example of Cu_2O with CuCl and H_2O as precursors, CuCl is first pulsed which bonds to the substrate (Figure 8.11a). Then any surplus un-reacted vapor of that reactant is pumped away. Subsequently, a vapor dose of the second reactant (H_2O) is brought to the surface and allowed to react, and the surplus is pumped away (Figure 8.11b). With the second vapor pulse, H has a stronger affinity for Cl thus forming HCl gas that is then purged. Similarly, O has a stronger affinity for Cu, thus it binds to form the desired Cu_2O film (Figure 8.11c). This cycle of steps normally deposits a monolayer or less of material though these cycles can be repeated to build up thicker films. Each of the reactants must set up the surface for its reaction with the other vapor by self-limiting further reaction with the present reactant vapor. This provides reproducible control of

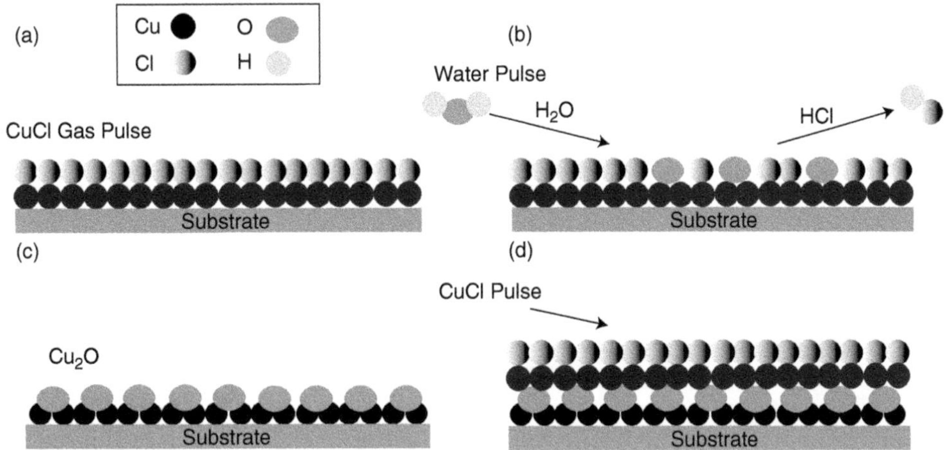

FIGURE 8.11 Schematic ALD of Cu_2O from CuCl and H_2O precursors. (a) Initial pulse of CuCl gas; (b) water vapor is pulsed when the O bonds to the Cu and the H bonds with Cl. (c) This leaves the desired atomic layer of Cu_2O. (d) Cycle is repeated until desired thickness.

the thickness, which can be found by simply counting the number of reaction cycles. ALD processes have been developed for several metal oxides, sulphides, nitrides, and fluorides.[16,17]

The main limitation of ALD, similar to MBE, is its speed or lack of it. Typically, only a few nanometers are deposited with this method so it is very impractical to deposit micrometer thick films, though the current trend is toward smaller dimensions and devices where ALD becomes very practical. As already mentioned, there are niche applications such as very thin diffusion barriers and high-k gate dielectrics that ALD provides advantages over other deposition techniques.

8.4.3.5 Metal-Catalyzed CVD for Nanowire Synthesis

Nanowires are often formed by a "metal-catalyzed" CVD process.[18] This technique forms nanowires with nano-dimensions without using fine-scale lithography. The synthesis method is most easily grasped pictorially (Figure 8.12). The key feature is the liquid catalyst that is able to absorb and decompose the vapor-phase reactants at a lower temperature than the substrate. As the catalyst absorbs more and more material it eventually becomes supersaturated, at which point material is deposited at the catalyst–substrate interface, thereby establishing nanowire formation. Extended formation of the nanowire can be maintained as long as a sufficient quantity of the reactant is present and the temperature of the catalyst is above a certain level. From the above description, it is clear that the growth method requires material in vapor phase and a catalyst in liquid phase to form a nanowire in solid phase and consequently the process is known as vapor–liquid–solid (VLS) growth mechanism. This mechanism was proposed by Wagner [19] during his studies of the growth of large single-crystalline whiskers in the 1960s.

Local growth on the nanometer scale area allows one to fabricate highly mismatched one-dimensional semiconductor heterostructures using the VLS method. Conventional planar growth of lattice mismatched layers leads to a deformation of the crystal structure and an accumulation of the strain in the growing layer, followed by a transition to the three-dimensional growth mode or by a generation of the misfit dislocations network that propagates through the layer along the growth direction. If the growth area is reduced to the scale where the accumulated lateral strain can be accommodated, the growth of a defect-free structure is feasible. In addition, along the length of the nanowire a gradual change of the lateral lattice constant may be possible starting with the strained lateral lattice constant of the substrate and relaxing within a few nanolayers to the unstrained lattice constant of the nanowire material. Successful realization of a few tens of nanometer length defect-free InAs nanowhiskers on GaAs substrate as well as InP/InAs/InP nanowires

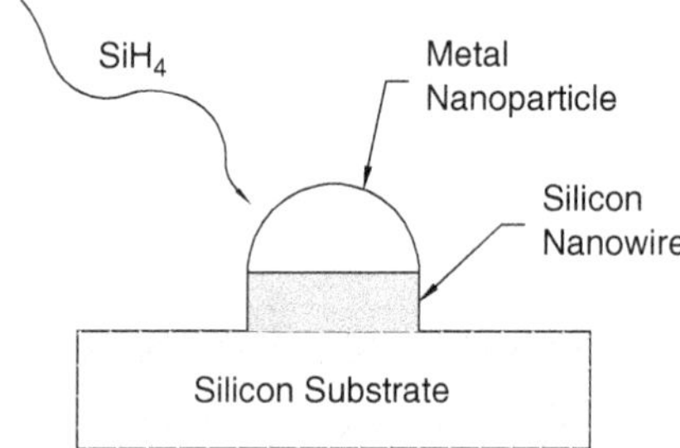

FIGURE 8.12 Vapor–Liquid–Solid growth of nanowires. Small metal nanoparticles that are liquid during the deposition process accelerate the decomposition of a silicon-containing gas; the silicon atoms precipitate between the nanoparticle and the substrate, forming a column (nanowire) of silicon.

with atomically sharp heterointerfaces was demonstrated.[20] Recently, InP nanowires were also grown on Si substrates.[21] The CVD application to carbon nanotube production is presented in Chapter 1.

8.5 Microstructure of Deposited Material

The final microstructure of a thin film depends on two stages: (1) the nucleation of small solid particles from the vapor phase and (2) the subsequent growth and coalescence of the nuclei. Compared to their bulk, the physical, electrical, and optical properties of thin films may strongly differ depending on their microstructure. Features like crystal orientation, grain size, voids, and others are mainly determined during the initial stages of nucleation and growth. Thus, knowledge of the initial nucleation and growth processes and how they can be controlled can lead to tailored properties for the desired application.

In the case of CVD or any other vapor-phase deposition, homogeneous nucleation is the self-nucleation of a pure solid or liquid from a supersaturated vapor phase in the absence of all foreign matter, wall surfaces, or substrate. True homogeneous nucleation is rare, as in most systems heterogeneous nucleation occurs in which the vapor phase will nucleate onto the solid substrate, walls, or even small particles in the vapor phase. But, in CVD systems the highly supersaturated vapor can cause homogeneous nucleation in which "snow" is formed that rains down onto the substrate. In most cases, this is undesirable as it can cause major defects or contamination in the film though there are a few applications such as production of ultra fine powders.

Typical homogeneous nucleation theory model assumes the solid or liquid phase nucleates from the supersaturated vapor as spherical clusters with radius r. More details on homogeneous nucleation may be found in Chapter 19. Assuming bulk thermodynamic quantities, the free energy minimization of the spherical nuclei from a supersaturated vapor involves an energy term for a volume transition and a surface formation. The volume transition from a gas phase to a solid or liquid phase results in a reduction of free energy (negative) of the system whereas the change in surface energy is always positive when forming surfaces. This leads to a net free energy change, ΔG:

$$\Delta G = -\frac{4}{3}\pi r^3 \Delta G_V + 4\pi r^2 \gamma \tag{8.22}$$

where γ is the surface free energy. Next, the change in volume free energy is given by:

$$\Delta G_V = \frac{kT}{\Omega} \ln \frac{P_V}{P_S} \quad \text{or} \quad \Delta G_V = \frac{kT}{\Omega} \ln S \tag{8.23}$$

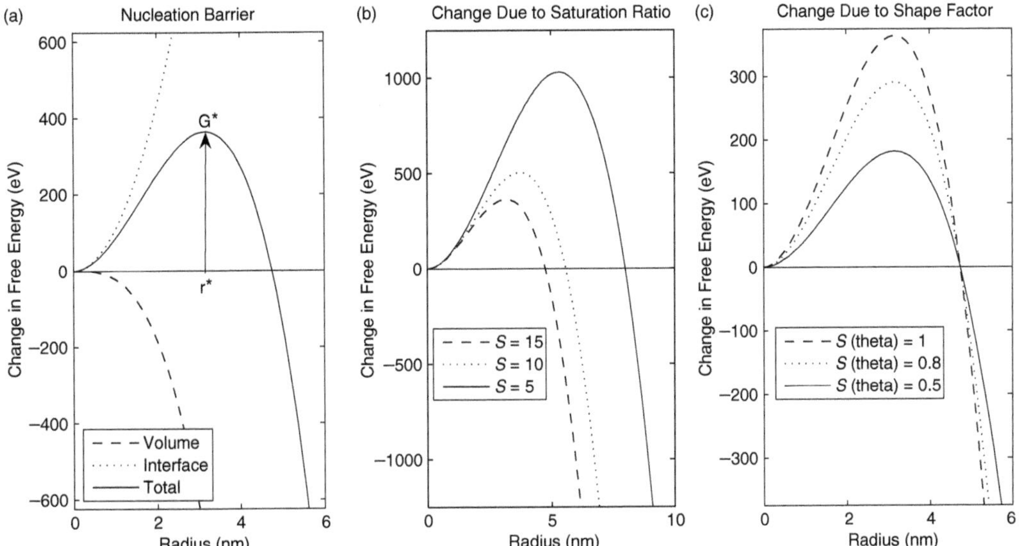

FIGURE 8.13 Nucleation barrier (a) shows the effect of the volume and interface term in homogeneous nucleation (Equation 8.22). (b) Change in free energy curve due to the saturation ratio. A higher saturation ratio reduces the critical radius making nucleation easier. (c) Change in free energy due to the shape factor for heterogeneous nucleation (Equation 8.28). The critical radius stays the same as homogeneous nucleation though the critical free energy is reduced.

Here P_V is the pressure of the supersaturated vapor, P_S is the vapor pressure above the solid or liquid, Ω is the atomic volume, and S is the vapor supersaturation ratio (P_V/P_S).

Homogeneous nucleation will not occur without supersaturation, if $S = 0$ then $\Delta G_V = 0$ and no change occurs. For homogeneous nucleation to occur the driving force becomes the supersaturation condition $P_V > P_S$, making ΔG_V negative and the process spontaneous.

Initial formation of a nucleus has an increase in free energy making it metastable. Initially small clusters are unstable and simply disappear until a critical radius is reached where the larger clusters grow. If the radius is less than the critical radius (r^*) then nuclei will shrink to lower the free energy (Figure 8.13a). Once the critical radius is passed the nuclei will continue to grow to lower its energy. Taking the derivative of Equation 8.22 with respect to r provides the critical nucleation radius at which the nuclei become stable, also called the Gibbs–Thompson equation:

$$r^* = -\frac{2\gamma}{\Delta G_V} \tag{8.24}$$

And plugging back into Equation 8.22 gives the critical energy barrier for homogeneous nucleation:

$$\Delta G^* = \frac{16\pi\gamma^3}{3\Delta G_V^2} \tag{8.25}$$

The more commonly occurring heterogeneous nucleation is the formation of nuclei on easier preferential sites such the substrate, the walls, impurities, catalyst, and the like. These are preferential sites because the nucleation energy barrier is reduced. As seen in Figure 8.14, the capillarity model assumes the formation of a spherical droplet on a solid plane surface.

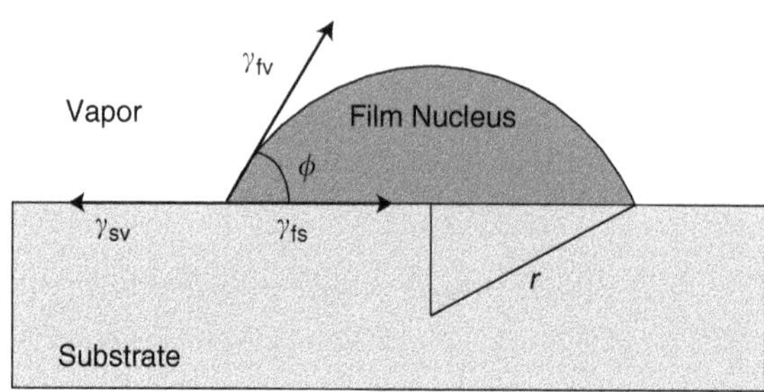

FIGURE 8.14 Capillarity model for heterogeneous nucleation where the film nucleus is modeled as a spherical cap. Each interface has an associated interfacial surface energy and therefore surface tension.

Static equilibrium can also be determined by summing the horizontal forces (surface tensions) and setting equal to zero. This yields Young's equation:

$$\gamma_{sv} = \gamma_{fs} + \gamma_{fv} \cos\phi \quad \text{and} \quad \cos\phi = \frac{\gamma_{sv} - \gamma_{fs}}{\gamma_{fv}} \tag{8.26}$$

The free energy of formation by heterogeneous nucleation of such a spherical droplet is given by

$$\Delta G = -V\Delta G_V + A_{vf}\gamma_{vf} + A_{fs}(\gamma_{fs} - \gamma_{sv}) \tag{8.27}$$

where V is the volume, A is the area, and γ is the interfacial surface energy between the vapor and film (vf), the film and substrate (fs) and the substrate and vapor (sv). Similar to homogeneous nucleation, there is a change in free energy caused by a volume transition and the creation of interfaces. For the spherical cap shown in Figure 8.14 the volume is given by: $V = (\pi/3)r^3(2 - 3\cos\phi + \cos^3\phi)$, the surface area between the vapor and film nucleus is $A_{vf} = 2\pi r^2(1 - \cos\phi)$, and the surface area between the substrate and film nucleus is $A_{fs} = 2\pi r^2 \sin^2\phi$. Plugging these back into Equation 8.27 with the static equilibrium conditions gives:

$$\Delta G = \left[-\frac{4}{3}\pi r^3 \Delta G_V + 4\pi r^2 \gamma_{fv}\right] S(\phi) \quad \text{where } S(\phi) = \frac{(2 + \cos\phi)(1 - \cos\phi)^2}{4} \tag{8.28}$$

Equation 8.28 shows heterogeneous nucleation is the same as homogeneous nucleation except for the wetting angle shape factor $S(\phi)$. The effect of this shape factor can be seen in Figure 8.13c. The critical radius is the same but the critical nucleation barrier has the added shape factor term:

$$\Delta G^* = \frac{16\pi\gamma_{fv}^3}{3\Delta G_V^2} S(\phi) \tag{8.29}$$

The shape factor has a maximum of one which would give spherical homogenous type nucleation. For heterogeneous nucleation the value is less than one, only dependent on the wetting angle. So it can be seen, for heterogeneous nucleation, the nucleation barrier is reduced.

It has been generally accepted that there are three main initial growth modes illustrated in Figure 8.15. More details on these growth modes are given in Chapter 9. In the island, or Volmer–Weber mode[22] (Figure 8.15a) small clusters nucleate with further growth occurring in three dimensions to form larger islands until they coalesce into a thin film. This typically occurs when the deposited atoms or molecules are more strongly bound to each other than the substrate and when surface diffusion is slow. This growth mode can be typically found with metal systems deposited onto amorphous substrates.

The layer, or Frank–Van der Merwe mode[23] is the ideal growth mode for epitaxial films. In contrast to the Volmer–Weber mode, the atoms are more strongly bound to the substrate than to each other and

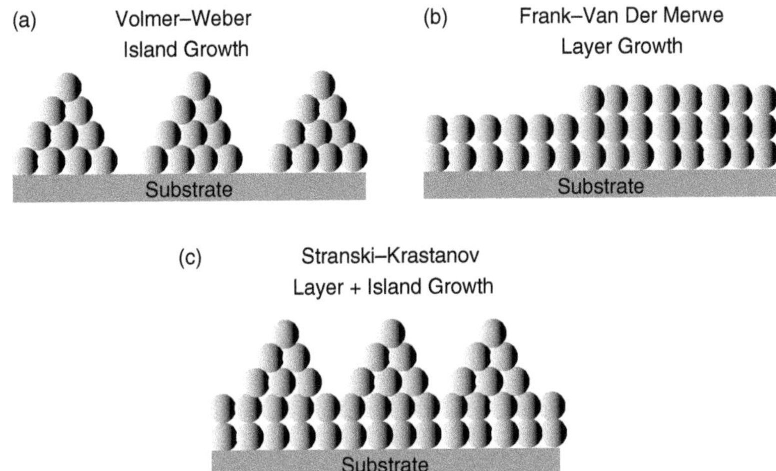

FIGURE 8.15 The three main growth modes (a) Volmer–Weber or island growth, (b) Frank–Van Der Merwe or layer growth, and (c) Stranski–Krastanov growth also called layer plus island.

there is fast surface diffusion. Here the first layer to condense and nucleate forms a complete monolayer on the substrate surface. Growth occurs from clusters in two dimensions on top of the substrate. This then continues with each subsequent layer, though somewhat less strongly bound than the previous. This is typically observed with epitaxial growth of semiconductor systems.

The third mode is a fairly common intermediate of the previous two with layers initially forming followed by islands. With the layer plus island, or Stranski–Krastanov mode,[24] a complete monolayer, or a few, is initially formed before some perturbation causes the growth to switch toward islands. Many things can cause the required perturbation including lattice strain and defects. It may not be possible for a certain molecular orientation of the first few monolayers to be continued into the bulk crystal. This causes a change toward a different lower energy orientation thus causing the growth mode to also change. This type of growth mode is common in all types of systems.

Similar to nucleation, several models have also been proposed for the subsequent growth of thin films. Some details on the microstructure development during growth of metallic films are given in Chapter 10. Among the most popular for evaporation and sputtered coatings is the structure zone model, first introduced by Movchan and Demchishin[25] during their study of thick evaporated metal oxide coatings. In the original model the film structures were categorized as belonging to one of three zones based on the ratio of substrate temperature (T_s) to the melting temperature of the evaporant T_m, (T_s/T_m). The temperature dependence was attributed to the interplay of three mechanisms: shadowing, adatom diffusion, and recrystallization and grain growth. In the low substrate temperature zone 1 ($T_s/T_m < 0.3$), adatom surface diffusion is insignificant; thus the film microstructure is controlled by shadowing effects and consists of open columnar grains with dome caps that taper toward the substrate (Figure 8.16). As the substrate temperature is increased surface diffusion begins to become dominant filling in some of the voids making a tighter film. Therefore, in zone 2 ($0.3 < T_s/T_m < 0.45$) the microstructure consists of tighter columnar grains with smooth mat surfaces. As temperature is further increased, bulk and grain boundary diffusion takes over as the dominant process thereby causing the grain sizes to increase. Therefore, in zone 3 ($T_s/T_m > 0.5$) the grains are equiaxed and the surface becomes smooth.

Several modifications have been made over the years since the introduction of the structure zone model. One of the most important was later made by Thornton[26] where he added the additional parameter of sputtering pressure that resulted in another zone called zone T between zone 1 and zone 2. Zone T was only noticeable when sputtering pressure was reduced.

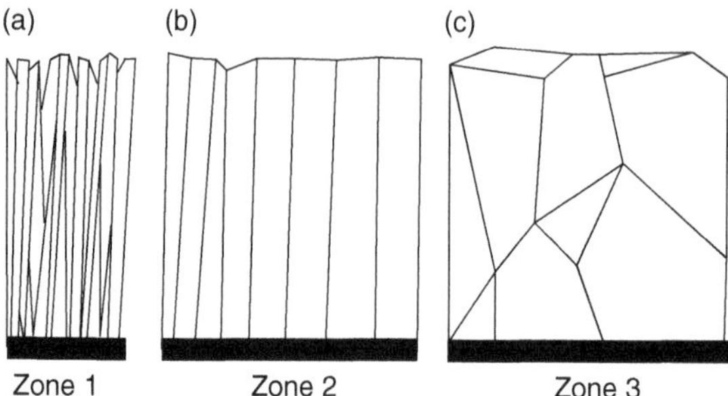

FIGURE 8.16 Structure Zone Model (a) Zone 1 ($T_s/T_m < 0.3$) characterized by columnar grains that taper toward the substrate. (b) Zone 2 ($0.3 < T_s/T_m < 0.45$) columnar grains grow (c) Zone 3 ($T_s/T_m > 0.5$) equiaxed grains.

8.6 Conclusions

This chapter was intended to give some of the basic underlying physics behind the deposition methods while serving to introduce some of the variants. It was in no way a comprehensive presentation as each deposition method could easily be a chapter or a book in itself. And this is not intended as a complete list of all possible PVD and CVD variants. In fact, with current research, most of these tools are being combined such that they no longer can easily be separated into a PVD or CVD method.

Commercially, both PVD and CVD methods are used. The main question of which method to be used will depend on the type of film, material, substrate, thickness, and quality. Certain applications or films tend to favor certain deposition processes. For example, MBE allows precise control of composition and thickness; therefore applications such as bandgap engineering tend to solely rely on this technique. Other examples include evaporation being the preferred method if liftoff will be performed as the deposition method is line-of-sight. Sputtering is preferred for its better adhesion and more conformal coatings over features. If dielectrics are to be deposited RF is needed instead of standard DC diode sputtering. CVD with its improved control, especially with epitaxy, is preferred when the film qualities can be controlled by chemical reactions, which can further be controlled by the process variables, temperature, pressure, flow rates, and so on. In most fabrication processes of integrated circuits one can expect to find both PVD and CVD techniques used somewhere in the process flow.

References

[1] Li, Q. et al., In-situ optical diagnosis during pulsed-laser deposition of high-Tc superconductor thin-films, *IEEE Trans. Appl. Superconduct.* 5, 1513–1516, 1995.

[2] Habermeier, H. U., Pulsed-laser deposition — a versatile technique only for high-temperature superconductor thin-film deposition, *Appl. Surface Sci.* 69, 204–211, 1993.

[3] Habermeier, H. U., Y-Ba-Cu-O high-temperature superconductor thin-film preparation by pulsed-laser deposition — recent developments, *Mater. Sci. Eng. B-Solid State Mater. Adv. Technol.* 13, 1–7, 1992.

[4] Habermeier, H. U. et al., Y-Ba-Cu-O high-temperature superconductor thin-film preparation by pulsed laser deposition and Rf-sputtering — a comparative-study, *Phys. C*, 180, 17–25, 1991.

[5] Minami, H. et al., Diamond-like-carbon films produced by magnetically guided pulsed laser deposition, *Appl. Phys. a-Mater. Sci. Proc.* 73, 531–534, 2001.

[6] Voevodin, A. A., Donley, M. S., and Zabinski, J. S., Pulsed laser deposition of diamond-like carbon wear protective coatings: A review, *Surf. Coat. Technol.* 92, 42–49, 1997.

[7] Hou, Q. R. and Gao, J., Pulsed laser deposition of diamond-like carbon films under a magnetic field, *J. Phys. Cond. Matter* 9, 10333–10337, 1997.

[8] Karuzskii, A. L. et al., Pulsed-laser deposition of "diamond-like" carbon coating on $YBa_2Cu_3O_7$ high-Tc superconductor films, *Appl. Surface Sci.* 92, 457–460, 1996.

[9] Voevodin, A. A. and Donley, M. S., Preparation of amorphous diamond-like carbon by pulsed laser deposition: A critical review, *Surf. Coat. Technol.* 82, 199–213, 1996.

[10] Voevodin, A. A. and Laube, S. J. P., In situ plasma monitoring of pulsed laser deposition of diamond-like carbon films, *Surf. Coat. Technol.* 77, 670–674, 1995.

[11] Voevodin, A. A. et al., Mechanical and tribological properties of diamond-like carbon coatings prepared by pulsed laser deposition, *Surf. Coat. Technol.* 77, 534–539, 1995.

[12] Pappas, D. L. et al., Pulsed laser deposition of diamond-like carbon-films, *J. Appl. Phys.* 71, 5675–5684, 1992.

[13] Koenig, H. R. and Maissel, L. I., Application of RF discharges to sputtering (Reprinted from IBM Journal of Research and Development, vol 14, 1970), *Ibm J. Res. Develop.* 44, 106–110, 2000.

[14] Grove, A. S., *Physics and Technology of Semiconductor Devices*, John Wiley & Sons, New York, 1967.

[15] Cabarrocas, P. R. I., Plasma enhanced chemical vapor deposition of amorphous, polymorphous and microcrystalline silicon films, *J. Non-Cryst. Solids* 266, 31–37, 2000.

[16] Suntola, T., Atomic Layer Epitaxy, *Thin Solid Films* 216, 84–89, 1992.

[17] George, S. M., Ott, A. W., and Klaus, J. W., Surface chemistry for atomic layer growth, *J. Phys. Chem.* 100, 13121–13131, 1996.

[18] Westwater, J. et al., Growth of silicon nanowires via gold/silane vapor-liquid-solid reaction, *J. Vacuum Sci. Technol. B* 15, 554–557, 1997.

[19] Wagner, R. S. and Ellis, W. C., Vapor-liquid-solid mechanism of single crystal growth, *Appl. Phys. Lett.* 4, 89–90, 1964.

[20] Bjork, M. T. et al., One-dimensional heterostructures in semiconductor nanowhiskers, *Appl. Phys. Lett.* 80, 1058–1060, 2002.

[21] Yi, S. S. et al., InP nanobridges epitaxially formed between two vertical Si surfaces, *Nanotechnology, 2005. 5th IEEE Conference on July 11–15, 2005*, 209–212, 2005.

[22] Weber, M. V. a. A., Nucleus formation in supersaturated systems, *Z. Phys. Chem.* 119, 227, 1926.

[23] Merwe, F. C. F. and van der Merwe, J. H., One-dimensional dislocations. I. Static theory, *Proc. R. Soc. London A* 198, 205, 1949.

[24] Stranski, I. N., Krastanov, L., Theory of orientation separation of ionic crystals, *Akad. Wiss. Lit. Mainz Math.-Natur. Kl. IIb* 146, 797, 1938.

[25] Demchishin, A. V. and Movchan, B. A., Investigation of the structure and properties of thick vacuum condensates of nickel, titanium, tungsten, aluminium oxide and zirconium dioxide, *Fiz. Met. Metalloved.* 28, 653, 1969.

[26] Thornton, J. A., Influence of substrate temperature and deposition rate on structure of thick sputtered Cu coatings, *J. Vacuum Sci. Technol.* 12, 830–835, 1975.

9

Epitaxial Processes

Peter Bjeletich
Intel Corporation

Abstract

Epitaxy is the process of extending a single crystal with new material. It can be carried out using the solid, liquid, or gaseous phase, and adds to the existing crystal by one of five modes of growth, depending on the conditions. Specifically, these modes are: (1) Frank–van der Merwe (FM), (2) Volmer–Weber (VW), (3) Stranski–Krastanov (SK), (4) columnar growth (CG), and (5) step flow (SF). Prevailing conditions during growth greatly influence the resulting epitaxial layer microstructure. Carefully grown epitaxial layers, free of defects, have many applications, especially in the semiconductor industry. By the addition of different materials, a process called heteroepitaxy can be achieved leading to more complex structures and opening up new possibilities for semiconductor devices and other applications.

9.1 Epitaxial Growth

The term epitaxy is derived from two Greek words, *epi*, meaning "upon," and *taxis*, meaning "ordered."[1–4] Epitaxy is the process of growing a thin, crystalline layer upon a crystalline substrate, such that the new layer's atoms are registered to the crystal structure of the substrate. The resulting structure is a larger single crystal.

Epitaxial growth is typically applied to the world of semiconductors and that will be the main focus of this section. However, it can also be applied to other materials. For example, the growth of individual grains in a metal during solidification (e.g., upon welding) is a form of epitaxial growth. The freezing atoms (as the weld fusion zone cools from liquid to solid) nucleate on the crystal structure of existing grains, elongating them.[5] Each of the individual grains in the polycrystalline metal increases its size through epitaxy.

When the epitaxial layer and substrate are the same material, the process is referred to as homoepitaxy, or simply epitaxy. In the extreme case, when the layer and substrate possess the same chemical potential (i.e., doping level in the case of semiconductors), in addition to being the same material, it is referred to as autoepitaxy; this is synonymous with simple crystal growth, and therefore this terminology is seldom used.[4] If, on the other hand, the new layer and substrate are different materials, the process is called heteroepitaxy.

Epitaxy can be performed using any of the states of matter as a source; solid, liquid, or gas. Deposition from the gaseous phase, however, is most common in semiconductors. This method allows for the greatest control of impurities, which is of paramount importance in the semiconductor world.

9.1.1 Solid Phase Epitaxy

Solid phase epitaxy (SPE) occurs when a metastable amorphous layer, in contact with a single crystal, crystallizes epitaxially at the interface between the two phases.[6] The amorphous layer acts as a source of atoms for epitaxy.

A transformation of this type occurs only in materials where the crystal phase is more energetically favorable than the amorphous phase, translating to an abrupt decrease in free energy of the system upon crystallization. [6] A typical example is silicon, where the amorphous layer is either deposited via chemical vapor deposition (CVD) or is formed by high dose ion implantation (see Chapter 8 and Chapter 3, respectively). Figure 9.1 demonstrates the latter process, where damage from a high dose implant results in an amorphous layer, which is subsequently crystallized by SPE.

This approach allows for the formation of high quality crystalline silicon at temperatures far below melting, thereby suppressing diffusion of existing doping profiles. Growth rates varying from 100 Å/s at 600°C to 1.4 μm/s at 745°C have been reported for silicon layers amorphized by ion implanting high doses of silicon.[7] Additionally, damage from the formation of heavily doped layers can be removed readily while preserving the crystal quality and dopant placement.

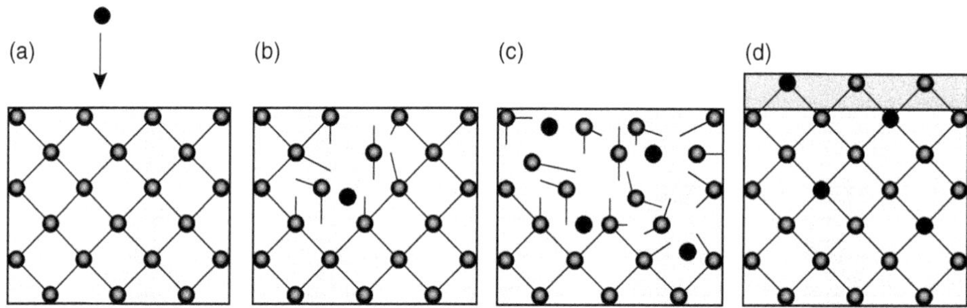

FIGURE 9.1 Solid phase epitaxy (SPE) through ion implantation. Ions are accelerated at the substrate at high energy (a). In part (b), they penetrate the substrate, dislodging atoms from their lattice positions and breaking bonds. More ions arrive in part (c), creating an amorphous layer. After annealing, in part (d), the amorphous material recrystallizes and the introduced atoms take substitutional sites on the lattice. The thickness of the crystal is increased, as shown by the darker region in part (d).

9.1.2 Liquid Phase Epitaxy

In its simplest form, Liquid Phase Epitaxy (LPE) is just growth of a melt (as in the Czochralski method of silicon ingot growth, detailed in Chapter 15).[8]

A more interesting liquid phase technique is the use of a solution. Epitaxy from solutions enables lower temperatures, high growth rates, better control of the growth, more abrupt junctions, fewer point defects (particularly for compound semiconductors), and more flexibility to grow complex or layered structures from successive solutions.[6]

Solution-based LPE involves the creation of a supersaturated solution of the semiconductor into a liquefied solvent, usually a metal. This is achieved by careful temperature control, as illustrated by the liquidus of the Ga–Si phase diagram in Figure 9.2. The substrate "seed" is introduced into the solution, wherein it facilitates epitaxial precipitation.

Unfortunately, the solvents tend to get incorporated into the growing layers, becoming dopants, thus severely limiting LPE's use for silicon layers. GaAs, however, is widely grown via LPE because gallium can serve as a solvent; consequently, incorporated solvent is not a dopant, but simply a component of GaAs.[7]

9.1.3 Vapor Phase Epitaxy

Vapor phase epitaxy (VPE) is the most widely used form of epitaxy. This makes use of either a vapor of the desired element(s) or of a compound containing the desired element(s). Because the use of these gaseous sources permits precise control over both growth and impurities, it is typically the best. Should the vapor consist of the elements themselves, then the process is termed physical vapor deposition (PVD). If, on the other hand, a compound is used, a chemical reaction is required to produce the epitaxial layer, hence the name CVD.[6] These processes are detailed in Chapter 8 and briefly discussed below.

In PVD systems, such as molecular beam epitaxy (MBE), the elements or compounds are either evaporated, sputtered, or laser ablated from a heated source. The vaporized material is transported to the substrate for deposition, without any chemical change. It must then adsorb to the substrate surface and migrate to an available site on the crystal face to produce epitaxial growth.[9,10]

CVD systems use volatile compounds of the desired elements. These vapors are transported to the substrate where they adsorb to the crystalline surface. Here, they react, freeing the atoms needed. These atoms

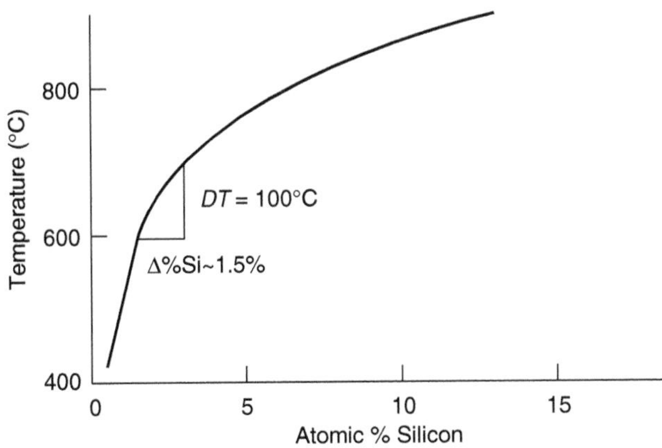

FIGURE 9.2 Supersaturation of silicon in gallium (liquidus line in the Ga–Si phase diagram). At 700°C, far less than silicon's melting point of 1414°C, a saturated solution that is approximately 3% Si can form. If the temperature is dropped by 100°C, the solution becomes supersaturated and roughly 1.5% of the dissolved silicon is available for precipitation in the form of LPE. (Data taken from Runyan, W.R. and Bean, K.E., *Semiconductor Integrated Circuitry Processing Technology*, Addison-Wesley, Reading, MA, 1990, chap. 7.)

migrate to the appropriate sites on the crystal face and begin to form the epitaxial layer. The by-products (remaining portions of the volatile compound) desorb and are carried away.[7,11]

For an epitaxial film to grow, three processes must occur; (1) transport of the material to the substrate surface, (2) its adsorbtion, and subsequent reaction, and (3) its nucleation and growth. As the first two steps for PVD (MBE) and CVD epitaxy differ, they will be discussed separately. However, the last step, (3) nucleation and growth, occurs identically for all forms of epitaxy, depending on the growth conditions and substrate, and not the technique (PVD or CVD) used; this will be addressed first.

9.2 Growth Modes

Epitaxy, of both forms (homoepitaxy and heteroepitaxy), occurs in much the same way as a bulk crystal except for the influence of the substrate. Because of the substrate's effects, epitaxy will proceed by one of following five growth modes: (1) Frank–van der Merwe (FM mode), (2) Volmer–Weber (VW mode), (3) Stranski–Krastanov (SK mode), (4) columnar growth (CG mode), or (5) step flow (SF mode). The mode by which a particular film grows is governed by its temperature, deposition speed, adhesion energy, and lattice mismatch between substrate and film. These modes are depicted in Figure 9.3, along with their evolution in time.[6]

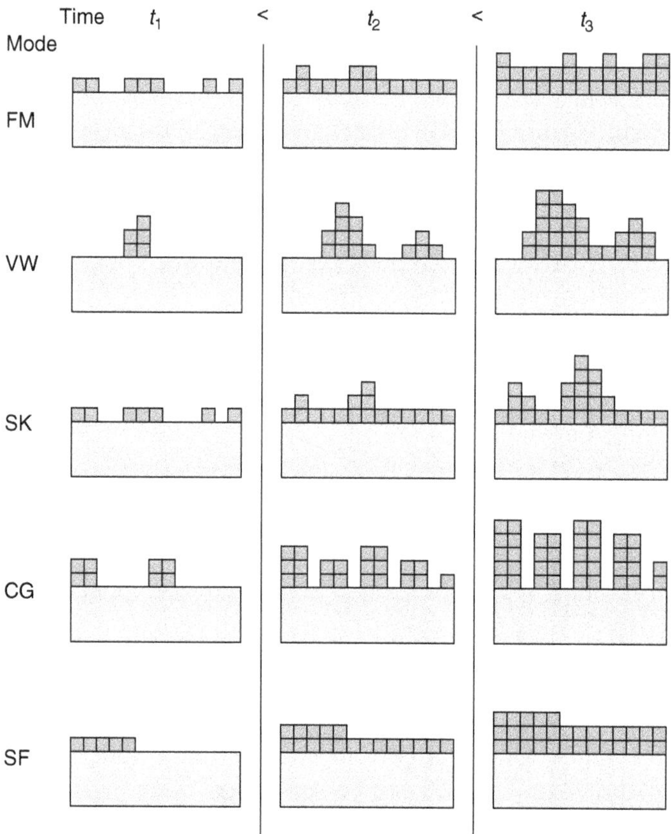

FIGURE 9.3 Growth modes of epitaxy. The five growth modes of epitaxy are displayed as they evolve in time. At time t_1, the initial stages of growth are shown. As the film grows (time t_2) more than a monolayer's worth of material has been incorporated. Finally, at time t_3, several monolayer's worth of material has grown. The growth modes are (from top to bottom); Frank–van der Merwe (FM) or layer-by-layer, Volmer–Weber (VW) or island, Stranski–Krastanov (SK) or layer-plus-island, columnar growth (CG), and step flow (SF).

In the FM mode, also known as the layer-by-layer mode, atoms of the epitaxial layer have a greater affinity for the substrate than they do for each other. Accordingly, the atoms strongly bond to the substrate, covering it completely with a monolayer before creating a second layer. Successively, one monolayer is completed before the next layer is started, and so on. Generally, the second layer is more weakly bonded than the first and the third layer is more weakly bonded than the second. In some cases, with large differences in the levels of attraction between epitaxial and substrate atoms, this can lead to a self-limiting thickness of a few monolayers.[6]

Provided the difference in the levels of attraction is strong enough, FM mode growth can provide a one monolayer epitaxy that cannot become any thicker. This is a desired trait for atomic layer epitaxy, described later in this chapter.

FM mode growth can be observed in the adhesion of some rare gases on metals, in some metal to metal epitaxy, and in some semiconductor to semiconductor epitaxy.

The VW mode is characterized by the opposite extreme in atomic attraction, that is, the epitaxial atoms are more strongly attracted to each other than to the substrate. Bonding, or nucleation, on the substrate is difficult; however, once it occurs, the epitaxial atoms easily adhere to the newly attached atoms, creating rapidly growing clusters, or islands. For this reason, this mode is often referred to as island mode or 3D growth.

Over time, the VW mode islands can grow together and merge over the entire substrate, creating a film over the surface that is not limited in thickness as in the FM mode case. However, as might be expected, the resulting film is nonuniform in thickness and defects are often present at the interface where islands grew together.

The VW growth mode is not typically seen in semiconductors. More commonly, it is found in certain metallic systems growing on insulators, such as mica.[6]

The SK, or layer plus island, mode is a hybrid between the VW and FM modes of growth. This growth mode typically starts as an FM mode type growth, with epitaxial atoms being more attracted to the substrate than each other. However, after the first monolayer (or few monolayers) is grown, the mechanisms change to favor island formation and growth (VW mode) atop of the newly formed intermediate layer.

SK mode growth often occurs where interfacial energy and strain energy are both high. The interfacial energy initially dominates, leading to layer-by-layer (FM) growth, but as the film grows, strain energy builds up. To reduce the strain, islanding, or VW mode growth takes over. VW mode growth results in lower strain energy, as the individual islands do not connect and therefore do not pass the strain to neighboring islands.[12,13]

Strained layer semiconductor on semiconductor layers often exhibit SK mode behavior under certain growth conditions. Though typically, the SK mode is avoided in strained semiconductor epitaxy, there are some cases where it is desired. In the case of InAs on GaAs, with the growth conditions carefully controlled, SK mode growth can lead to the formation of arrays of dots (islands).[6]

Upon first inspection, the CG mode looks similar to the VW and SK modes. It, too, has greater attraction between epitaxial atoms than to the substrate and forms island-like structures, called columns. The difference between islands and columns, however, is that the islands eventually grow together and merge into one film (though defects/dislocations may be present at the interfaces between islands), whereas columns do not. The columns of the CG mode remain separate, unattached entities throughout growth. After growth, they are easily fractured and separated.[6]

Typically, CG mode growth results from very low surface mobility during formation. The columns tend to be of poor crystalline quality and this mode is usually avoided in semiconductors. However, under certain growth conditions, an array of high quality nanocrystals can be grown using this mode, using a highly lattice mismatched substrate. This has been demonstrated using GaN on silicon to produce an array of 60 nm whiskers.[6]

The last growth mode, SF mode, is the most common type found in high quality epitaxy in the semiconductor industry. Semiconductor substrates are formed such that their surfaces are slightly misoriented to low Miller index planes (typically [100]). The misalignment between the surface and the crystal plane is

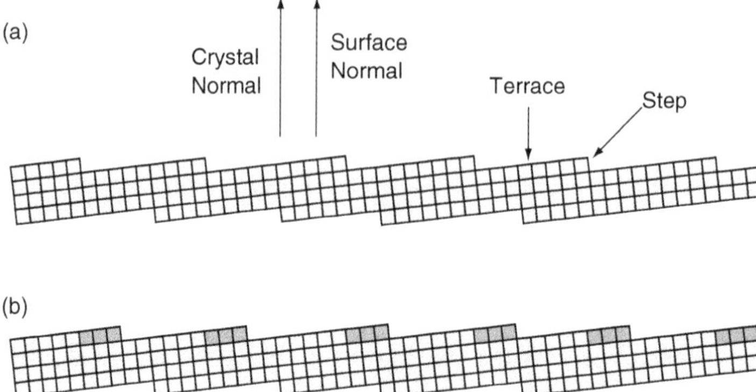

FIGURE 9.4 Misoriented surface plane and step flow (SF) growth. There is always some misalignment between the surface of the semiconductor wafer and the crystal plane. This misorientation is shown in (a), as the difference between the surface normal and crystal normal. A system of crystallographic steps and terraces results on the wafer surface. These steps are natural nucleation points for step flow (SF) mode growth, depicted in (b). New growth is shown as darker squares.

shown in Figure 9.4a. This provides a series of terraces, oriented to the desired plane and a series of steps between them.[7]

If the temperature of the epitaxial atom is high enough, or the growth rate is low enough, the adatom (epitaxial atom adsorbed to the surface) will have time to migrate to a step before incorporation into the substrate or meeting another adatom. At the step, the conditions are more favorable for bonding, that is, bonding in two directions rather than one. Growth occurs at these steps between terraces; this defines the SF mode.[6] The abundance of steps on the wafer surface enables rapid and uniform growth, as illustrated in Figure 9.4b.

If on the other hand, the adatom temperature is too low or growth rate too high, the adatom may meet another adatom before getting to the step; FM or SK mode growth will ensue and nucleation will start on the terrace rather than at a step.

9.3 Mass Transport and Surface Adsorption

9.3.1 Mass Transport and Surface Adsorption in MBE

Mass transport in an MBE system is provided by a beam of molecules evaporated from a source. This is carried out under ultra high vacuum pressures (10^{-10} torr). The pressure has to be kept low such that the mean free path of the molecules in the beam is much greater than the distance traveled from the source to the target (substrate). This assures that the molecules will arrive at the surface of the substrate without having collided or reacted with an impurity or chemically changed.[6,7]

Molecules that do collide with an impurity can react with or propel that impurity and potentially incorporate it into the growing epitaxial layer as a defect. Typically, for quality film growth, the rate of desired molecules arriving at the substrate should be more than 10^5 times greater than the rate of impurities arriving at the substrate.[14]

The molecules impinging on the substrate surface possess significant energy from the evaporation process during their formation. As a result, once the molecule adsorbs to the surface, less heating is required to assure adequate surface migration. In this way, the MBE process allows for nonequilibrium growth conditions; temperatures can be kept significantly lower than required by a CVD process.[14]

It is important to note that the adsorbed molecules are only physisorbed to the surface, meaning that only van der Waals type bonds hold the molecule to the surface. Migration under these weak bonds is relatively easy. After migrating to the proper site (step site in SF mode), the molecule disassociates and the atoms are incorporated, or chemisorbed, into the epitaxial film.

9.3.2 Mass Transport and Surface Adsorption/Reaction in CVD

Mass transport in a CVD epitaxial system requires the use of a transport gas, sometimes called a precursor. This transport gas is a volatile compound of the desired atom for epitaxy and typically falls into one of the following four categories: halides, oxides, hydrides, or metalorganic compounds. Halides are used for growing metals, elemental semiconductors ($SiCl_4$, SiH_2Cl_2, $SiHCl_3$), and compound semiconductors (GaCl); oxides are used for growing compound semiconductors (Ga_2O); hydrides are for elemental semiconductors (SiH_4); and metalorganics are for compound semiconductors ($Ga(CH_3)_3$, $Al(C_4H_9)_3$).[6]

The transport gas is combined with a carrier gas (usually hydrogen) that aids in the decomposition reaction and acts as a diluent (see more details in Chapter 8). The gases are introduced to the epitaxial chamber such that laminar flow is maintained. Flow conditions depend heavily on the Reynolds and Rayleigh numbers, which are strong functions of the chamber geometry, temperature gradient, and gas species. Chambers are typically designed to maximize laminar flow using these fluid dynamics principles.[1]

Under steady state conditions, the area immediately above the substrate surface becomes depleted of transport gas as material is consumed by the growing film. This area is referred to as the boundary layer.[15] Diffusion of reactant across the boundary layer is the rate limiting step in mass transport.

Once across the boundary layer, the transport gas adsorbs to the surface, migrates to an appropriate site on the crystal, and decomposes to release the atomic species. The atomic species is incorporated into the growing epitaxial layer and the reaction by-products desorb and are carried away.[10] As in the case of MBE, adequate time and energy to allow migration to the terrace step is required for epitaxial growth.

In silicon, the chlorine-based precursors are particularly interesting because of the ease of manipulation of the surface decomposition reaction and its bidirectionality:

$$\mathrm{Si}(s) + 2\mathrm{HCl}(g) \leftrightarrow \mathrm{SiCl_2}(g) + \mathrm{H_2}(g) \tag{9.1}$$

The surface can be etched with the same chemistry, allowing for *in situ* precleaning prior to epitaxial growth. Defects, native oxides, and the like, can all be removed before growth to ensure high quality epitaxy. The reaction direction is determined by the degree of supersaturation (σ), defined as:

$$\sigma = [p_{\mathrm{Si}}/p_{Cl}]_{\mathrm{Feed}} - [p_{\mathrm{Si}}/p_{Cl}]_{\mathrm{Equilibrium}} \tag{9.2}$$

where p_{Si} and p_{Cl} are the partial pressures of silicon and chlorine, respectively.[10] The first term is the ratio of the feed gases and the second is the equilibrium ratio, a function of temperature and pressure. If $\sigma < 0$, etching will result, while if $\sigma > 0$ deposition will result. It should be pointed out that if $\sigma \gg 0$, the reaction will proceed too quickly, that is, the adsorbed species will not have sufficient time to migrate to a step before incorporation. In this case, SF mode growth is replaced by VW or SK, leading to island formation. In the extreme case, with even greater σ, a polycrystalline or amorphous film is grown.[7,9,10]

The growth rate of a CVD grown epitaxial layer is shown schematically in the Arrhenius plot in Figure 9.5. The plot shows two distinct regimes: the surface reaction limited regime and the mass transport limited regime. The former is determined by the reaction rate at which the precursor decomposes on the surface of the substrate, while the latter is determined by the diffusion across the boundary layer (as mentioned previously). Because of the strong temperature dependence in the reaction limited regime and the poor control over temperature variation in real-world epitaxial chambers, processing is typically kept in the mass transport limited regime. As mass transport is a much weaker function of temperature, operation in this regime provides superior control and uniformity.[1,10,15]

Metalorganic-based epitaxy (MOVPE) works in much the same way as ordinary CVD-based epitaxy; however, the decomposition temperatures are much lower. This translates to a far lower temperature for

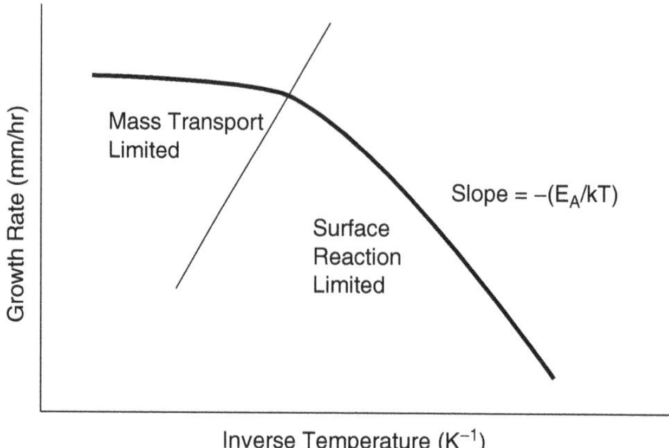

FIGURE 9.5 Chemical vapor deposited (CVD) epitaxy growth rate dependence on temperature. The growth rate of CVD epitaxy's dependence on temperature is shown symbolically. Two distinct regimes are indicated, separated by the thin line. The surface reaction limited regime is Arrhenius in nature (i.e., slope $= -(E_A/kT)$, where E_A is an activation energy, k is the Boltzmann's Constant, and T is temperature). The mass transport limited regime is less temperature dependent, and thus more often used for CVD epitaxial growth.[1,6,10,15,18]

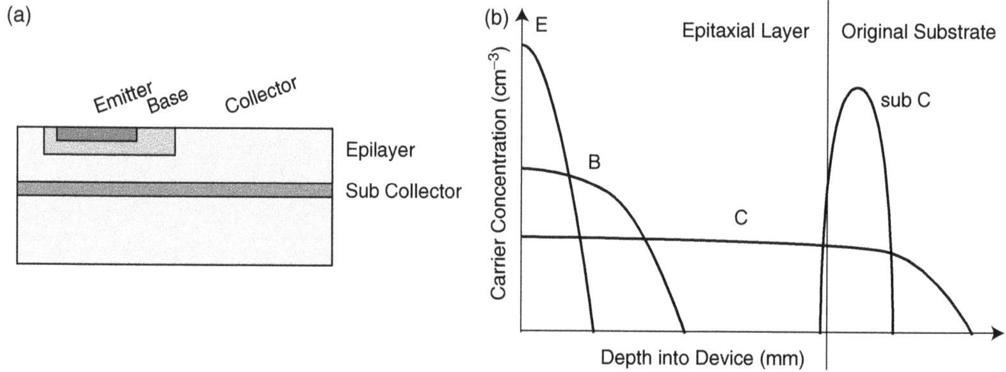

FIGURE 9.6 Doping profile in the bipolar junction transistor (BJT). The cross section in part (a) shows the elements of the BJT; the emitter, base, and collector. Additionally, epitaxy allows for the formation of a subcollector; a region of high doping located beneath low doping. This improves device performance. Part (b) shows the doping profile. The emitter, base, and collector are created in the epitaxial layer. The subcollecter is formed in the original substrate before epitaxy.

the mass transport limited regime, and thus lower temperature for the epitaxial growth. Growth at lower temperatures means less diffusion in existing structures on the substrate.[16,17]

9.4 Applications in Semiconductor Processing

Epitaxial layers have long been used in bipolar junction transistors (BJTs) in silicon. A very lightly doped collector (enabling a high breakdown voltage) is epitaxially grown atop a heavily doped subcollector (providing low resistance), as seen in Figure 9.6. Without epitaxy, the doping profile in Figure 9.6b could not be attained.[7,18]

Increasingly, epitaxy is being used in silicon metal-oxide-semiconductor (MOS) transistors, as well, because epitaxial layers are more latch-up resistant (interactive feedback between neighboring transistors),

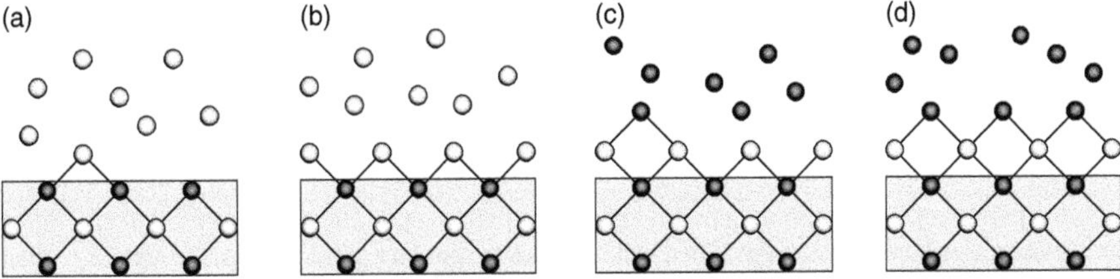

FIGURE 9.7 Atomic layer epitaxy (ALE). An extreme case of FM mode growth, ALE allows only one monolayer of growth, after which, no further growth is seen. This is especially useful in compound semiconductors (III–V and II–VI), where the atoms must arrange themselves in a particular order. In the representation above, part (a) shows an existing compound substrate with the darker colored atom group on the surface. Light colored atoms easily grow epitaxially by the FM growth mode. In part (b), however, a monolayer is formed and no further growth occurs. At this point, the supplied atoms change to the darker colors atoms, in part (c). This again, enables a monolayer of growth of dark colored atoms, shown in part (d). Repeating this sequence allows for controlled epitaxial growth of compound semiconductors.

have fewer of certain types of impurity-related defects, and have a smoother surface than ordinary bulk silicon wafers.

Bulk silicon wafers always contain impurity defects, mostly due to the oxygen and carbon incorporated during growth. Epitaxially grown silicon however, particularly from high purity vapor sources, contains significantly lower levels of these. Additionally, surface roughness, due to polish damage during wafer formation, is smoothed out by the SF growth mode of epitaxial silicon. The end result is a more defect-free, smooth surface enabling faster, smaller, less leaky MOS transistors.[18]

Epitaxy's applications extend beyond the silicon world; compound semiconductors rely on epitaxy for their formation. Most compound semiconductors that are grown from a melt are too defect filled to be of use. For example, in bulk III–V semiconductors (like GaAs) grown from melt, group V elements often fill group III locations on the crystal, forming a crystal of inferior quality and thus inferior devices. High quality epitaxial layers grown on top of the bulk create a medium in which to build high quality devices. Atomic layer epitaxy (ALE) assures that each species of element is placed properly, as depicted in Figure 9.7.

Unlike most high quality epitaxy, which makes use of the SF mode of growth, ALE uses an extreme of the FM mode. Here, each species of the epitaxial material can only form a monolayer due to repulsive forces, then the next species is introduced; this is repeated again and again to produce precise placement of species on a crystal.[6]

Once these high quality compound semiconductors are available, they can be combined in various combinations to form heterostructures, which will be discussed in detail in the last section.

9.4.1 Difficulties and Defects in Epitaxy

Despite the advantages and abilities that epitaxy provides, there are some complications and problems that it creates, the first of which is autodoping (for more details on doping, see Chapter 18).

Autodoping is the process of unintentionally redistributing dopant from adjacent areas into the epitaxial layer. As mentioned previously, the reactions associated with VPE etch under certain conditions, enable native oxide removal among other benefits. Additionally, however, during epitaxial growth, dopant can leach out of existing layers and incorporate into the growing region, resulting in undesired doping. The opposite effect has also been observed, where silicon is leached out of existing layers unintentionally increasing the doping of underlying layers. This phenomenon has also been found to occur across the wafer and from wafer to wafer within a chamber, removing dopant from an existing area and placing it in another existing area; this is referred to as lateral autodoping.[7,9,10]

Autodoping can be managed by careful control over the deposit or etch surface reactions, though complete control is not possible. Additionally, long purge steps following an etch cycle help to minimize

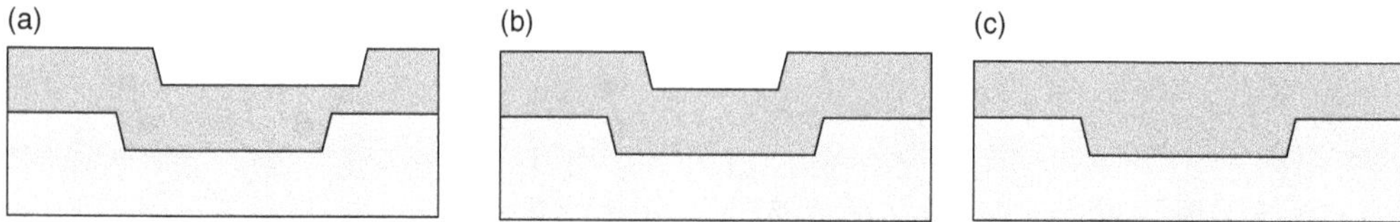

FIGURE 9.8 Geometric effects in epitaxy. When epitaxy is carried out on a pattern wafer, the pattern is often changed by the growth. The three typical geometric effects that occur in epitaxy are shown above. They are (a) pattern shift, (b) pattern distortion, and (c) washout. The original substrate with its etched pattern is the light colored area, while the epitaxy is darker.

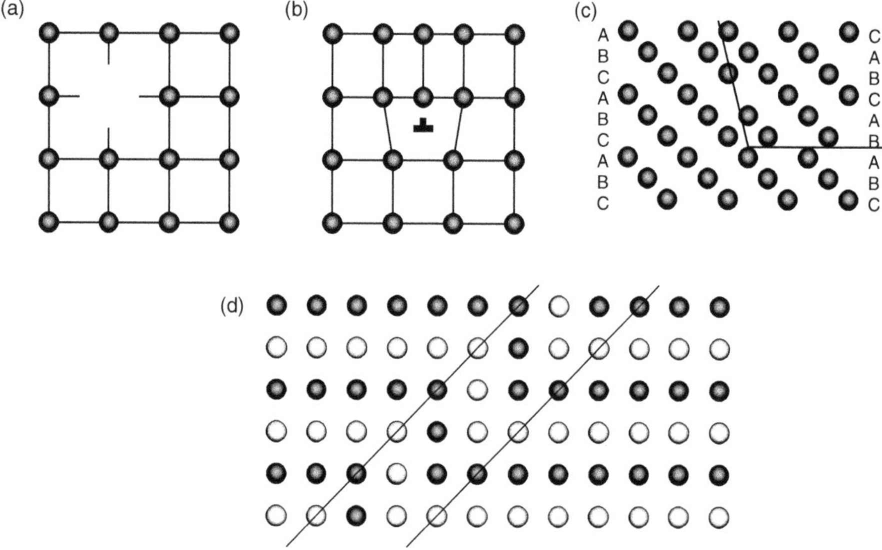

FIGURE 9.9 Common crystal defects suffered by epitaxy. (a) Point defects are voids in the crystal lattice. (b) Dislocations are linear defects, indicated by the inverted "T" in the figure. (c) Stacking faults are misordering of layers (note the missing C plane on the right side). (d) Twins are characterized by mirror planes (shown as thin lines), where unit cells grow unparallel to the substrate; these are called twinning planes.

the damage from autodoping.[9] Boron doped layers exhibit reduced autodoping at lower temperatures, and certain species of n-type dopant exhibit lower autodoping in general, such as antimony.[18]

Epitaxy can also result in a wide range of geometric effects when epitaxial growth is carried out on a patterned substrate. The general effects, pattern shift, pattern distortion, and washout, are shown in Figure 9.8.[7,18] These become particularly deleterious when careful alignment is required for subsequent layers during device fabrication.

Pattern shift is the spatial displacement of the feature in subsequent epitaxial layers; while pattern distortion is the uneven spatial displacement of a feature during epitaxy. Both these phenomena have their origins in the SF mode of epitaxial growth and the crystal orientation of the substrate surface. Depending on orientation and kinetics, any combination of these two effects can occur during growth. The extreme case of this results in washout, where the pattern is lost completely.

The epitaxial crystal, itself, can experience a number of problems, generally referred to as crystal defects. These come in three types; zero-dimensional, or point defects; one-dimensional, or dislocations; and two-dimensional, or twins and stacking faults. These are shown in Figure 9.9.[6,19]

When the point defect involves only the main component atoms of the crystal, it is referred to as an intrinsic point defect. These can be vacancies (missing atoms), exchanged atoms (atoms of one species on another species' site), or interstitials (atoms not on a crystal site, but squeezed between sites). If another species of atom is involved, it is called an extrinsic point defect. This can be desired, as in the case of dopant atoms, or undesired, as in the case of impurities.[6,19,20]

The severity of the point defect is generally determined by its size and abundance. Large clusters of atoms that agglomerate interstitially or high concentrations of small defects can degrade the performance of devices nearby. Impurity control and conditions, which favor pure SF mode growth (adequate migration before incorporation) generally decrease point defect formation, but cannot eliminate it.

One-dimensional defects, or dislocations,[19–21] are shown in Figure 9.9b (i.e., an edge dislocation denoted by the inverted "T" symbol). Dislocations occur in epitaxy for one of the following four reasons:

1. As an extension of an existing dislocation in the substrate
2. At the boundaries between merging islands in SK or VW mode growth when each crystal is only slightly misaligned
3. By the agglomeration of point defects into loops (dislocation loops)
4. By the plastic deformation of the epilayer, usually to accommodate strain (misfit dislocation). This case will be discussed in the last section on heteroepitaxy[6]

Stacking faults are planar defects where two adjacent planes of atoms are not in the sequence characteristic of a perfect lattice, as shown in Figure 9.9c. The normal stacking sequence (layer A, B, then C in this example) is interrupted, and a stacking fault occurs. If the stacking fault is located within a perfect crystal, it must be bounded completely by a dislocation loop.[6,20]

Stacking faults often occur in silicon epitaxy when growth conditions are nonideal. The defect is usually due to one of the following:

1. Misfit between merging islands in SK or VW mode growth, when the misfit is greater than in the dislocation case
2. When large numbers of point defects aggregate to form dislocation loops, creating a stacking fault[6]

A twin defect is characterized by a reflection of atomic positions across the twinning plane, as shown in Figure 9.9d. In this case, only one of the sides of the crystal unit cell is shared with the substrate. Evidence suggests that these too originate from misfit (either translational or rotational) between merging islands in SK or VW mode growth, though this is still the subject of current investigation.[6]

In addition to the crystalline defects, several gross defects are seen on the surface of epitaxially grown layers. Haze is a common effect found in epitaxy where there is a cloudy appearance to the film. This can be caused by oxygen leaking into the reaction chamber and forming oxide that gets trapped in the film. Trapped by-products or precursor gases in VPE can cause the same effect.[7,9]

"Orange peel," another gross defect found in epitaxy, is characterized by a surface covered in subtle bumps, like the surface of an orange. This is typically caused by uneven preferential etching during the etch or deposit cycle of VPE. Extremely high growth rates, where growth modes other than SF begin to appear, can also result in this effect; local islanding creates ripples on the surface of the film.[7,9]

9.4.2 Film Characterization

Epitaxial semiconductor films are most often grown with some difference in doping than the substrate on which they are grown (homoepitaxy, not autoepitaxy). Because of this doping difference, there are slightly differing properties between the substrate and epitaxial layer; these are exploited to measure the film's thickness.

The most common technique used to determine the thickness of an epitaxial film is Fourier transform infrared (FTIR) spectroscopy. Because of the differing dopant levels (usually a highly doped substrate with a lightly doped epilayer), the index of refraction (n) also is different for each layer. The interface between the layers, therefore, has some reflection associated with it. FTIR measures the interference pattern, over the infrared spectrum, of the reflections from the top surface (air-epi interface) and the epi-substrate interface, as seen in Figure 9.10. From this data, the path length through the epitaxial film is revealed.[10]

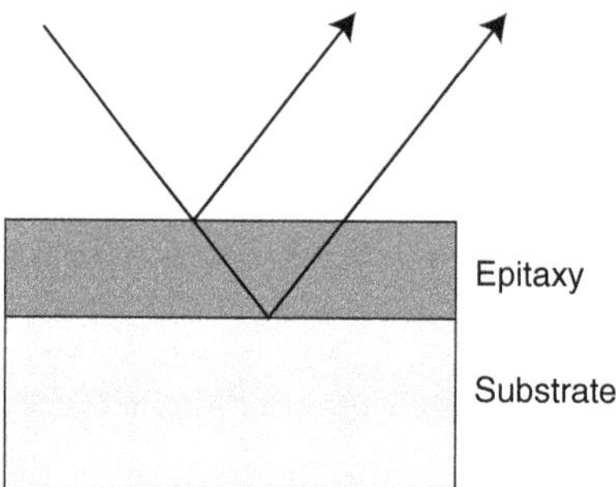

FIGURE 9.10 Reflections from surface and interface used in Fourier transform infrared (FTIR) spectroscopy. FTIR spectroscopy uses the infrared spectrum and projects it onto the epilayer. Constructive and destructive interference between the signals from the reflections from the two interfaces shown above enable the determination of the layer thickness.

Electrical methods of film characterization are also used, under the proper doping conditions. Four-point probe measurements determine the sheet resistance of the layer. Other, more detailed methods include capacitance–voltage (CV) profiling, spreading resistance probing (SRP), and groove-and-staining. Electrical methods are limited in accuracy due to autodoping.[10]

9.4.3 Selective Epitaxial Growth and Extended Lateral Overgrowth

Selective epitaxial growth (SEG) is a means by which an epitaxial layer is only grown in certain locations, as opposed to over the entire substrate. This is useful for certain semiconductor device applications, such as isolated devices, contact planarization, and elevated source or drain MOS transistors.

SEG in silicon takes advantage of the fact that silicon nucleates differently on different substrates. Nucleation on silicon dioxide (SiO_2) is more difficult than nucleation on silicon. Areas where growth is undesired are covered with oxide, while the desired areas are left bare. Under proper growth conditions (specifically, the increased partial pressure of the chlorine precursor), epitaxy occurs in the bare regions, leaving the oxide untouched.[9,10]

Unfortunately, SEG has a number of inherent challenges. The growth rate in SEG varies with the area of the window in the oxide; therefore thickness variation is inevitable in differing sized SEG areas. The vertical oxide–epitaxy interfaces are prone to defects that affect device performance. The corners of the SEG areas tend to be faceted due to growth rate variation in different crystal planes. Finally, some nucleation can occur on the oxide masking layers, leading to the growth of small silicon nodules. These effects are shown in Figure 9.11a.[10]

Extended lateral overgrowth (ELO) takes SEG to the extreme, growing the epitaxial region up and over the masking layer, as seen in Figure 9.11b. This technique is based on the same phenomena as SEG, and faces the same drawbacks. The ELO region, labeled L in the figure can extend for a few microns before defects corrupt the film.[6,10]

ELO has been successfully used to generate small areas of insulated silicon for silicon-on-insulator (SOI) technology, as well as in GaAs, GaP, and GaN. This technique has also been used to generate heteroepitaxial layers on substrates with extreme misfit, as the inherent dislocations generated by this misfit do not get translated through the oxide.[10]

9.5 Heteroepitaxy

Perhaps the most exciting and challenging field in semiconductor epitaxy today is that of heteroepitaxy, the growth of one material atop another, while maintaining crystal structure registry. This configuration

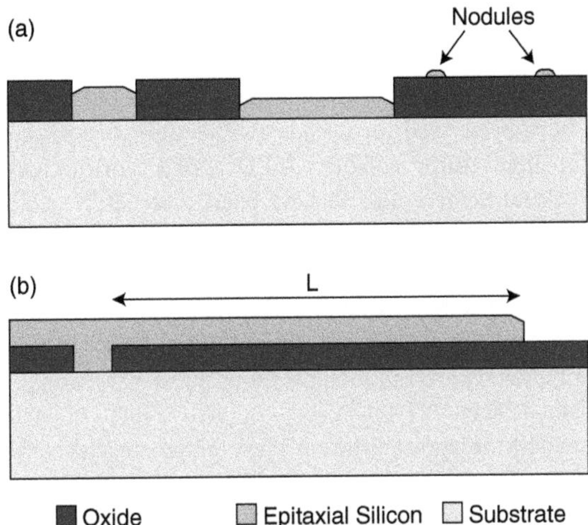

FIGURE 9.11 Selective epitaxial growth (SEG) and extended lateral overgrowth (ELO). SEG is shown in part (a). Note the unequal thicknesses of the large and small windows, along with the facet formations at the corners of the epitaxy. Silicon nodule formation on the oxide is also shown. Part (b) shows ELO, where SEG is used to grow a layer up and over the masking oxide. This is useful for silicon-on-insulator (SOI) applications, but is limited to small overlaps (labeled L above).

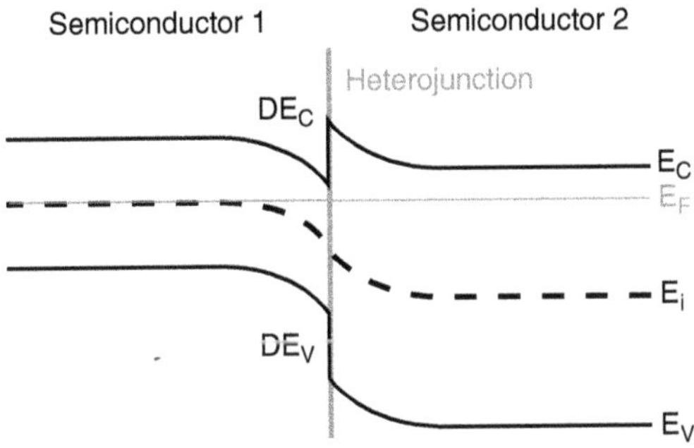

FIGURE 9.12 Heterojunctions and the energy band diagram. The energy band diagram of a heteroepitaxially formed heterojunction is shown above. The separate valence (E_V) and conduction (E_C) bands of each material form offsets at the heterojunction, denoted by ΔE_V at the valence band and ΔE_C at the conduction band. These offsets allow for carrier confinement. In the case above, electrons are trapped at ΔE_C, due to its proximity to the Fermi Level (thin grey line denotes E_F). This is beneficial because Semiconductor 1 is undoped (as indicated by the position of E_F) and thus has lowered impurity scattering; the electrons are trapped in a high mobility region. The intrinsic level (E_i) is shown for reference.

allows the formation of a heterostructure; a structure that is comprised of at least two layers of differing compositions, each with different properties (energy band structures, bandgaps, carrier transport properties).[8]

When the lattice parameters of the materials are relatively closely matched, defect-free interfaces can be formed allowing for the fabrication of high quality devices. Of particular interest is the energy band structure of the heterostructure, as shown in Figure 9.12. Each semiconductor has a characteristic energy gap (E_{G1}, E_{G2}), where E_C, E_V, and E_i are the conduction, valence, and intrinsic energy levels, respectively. The two semiconductors are placed together to form a heterojunction. The difference in band structure on either side of a heterostructure interface gives rise to energy band discontinuities, abrupt changes in the conduction (ΔE_C or conduction band offset) and valence band (ΔE_V or valence band offset) of structure.[22,23] These discontinuities, formed at a heterojunction, can be used as potential wells for carrier confinement, that is, the confinement of holes or electrons to a particular region within the heterostructures.

Column III–V (referring to column numbers on the periodic table of elements) heterostructure devices utilizing these properties have been used for decades with impressive results. Quantum well devices, arrayed waveguides (AWG), light emitting diodes (LEDs), and semiconductor lasers all make use of heterojunctions and carrier confinement and all have been successfully produced using heteroepitaxy. Additionally, high mobility transistors are an established technology in III–V material, utilizing band offsets to confine carriers to high mobility areas, away from interfacial effects.

Difficulties, however, arise in processing Column III–V alloys making them unattractive as a replacement for silicon, chief of which is the complete lack of a natural insulator (like silicon's SiO_2). Without a natural insulator, even a moderate level of integration becomes exceedingly complex, translating into high cost and more simplistic devices.[24,25]

Much effort has been made to integrate Column III–V heterostructures with silicon, to incorporate the strengths of each to enhance device performance. Although some successes have been demonstrated, this, too, has many processing complications arising from their incompatibility with silicon (autodoping, lattice mismatch). With such issues, it is unlikely that Column III–V alloys will be able to extend device performance for the next generation.[24,25]

9.5.1 Silicon Germanium

Silicon Germanium ($Si_{1-x}Ge_x$ or SiGe), a Column IV alloy, is also capable of forming heterostructures and high mobility devices. Unlike the Column III–V heterostructures, SiGe is fully compatible with silicon and silicon-based processing. The properties of $Si_{1-x}Ge_x$ depend on the composition (x) of the alloy; x can vary from zero to one. Only one phase is present, as silicon and germanium are completely miscible in the solid state.[26]

Heteroepitaxial growth of SiGe on silicon alters the crystal structure of the SiGe, as depicted in Figure 9.13. Bulk SiGe has a diamond cubic structure, shown in Figure 9.13a, where each side of the unit cell is equal ($a = b = c$) and is given by:

$$a_{\text{SiGe}} = a_{\text{Si}} \times (1 - x) + a_{\text{Ge}} \times (x) \tag{9.3}$$

where the lattice constant of silicon (a_{Si}) is 5.4305 Å, the lattice constant of germanium (a_{Ge}) is 5.6576 Å, and x is the germanium fraction.[27]

As SiGe is grown epitaxially on a silicon substrate, it is confined in the a and b directions (100 plane) by the substrate (a_{Si}). From Equation 9.3, it can be seen that, as long as there is some germanium present,

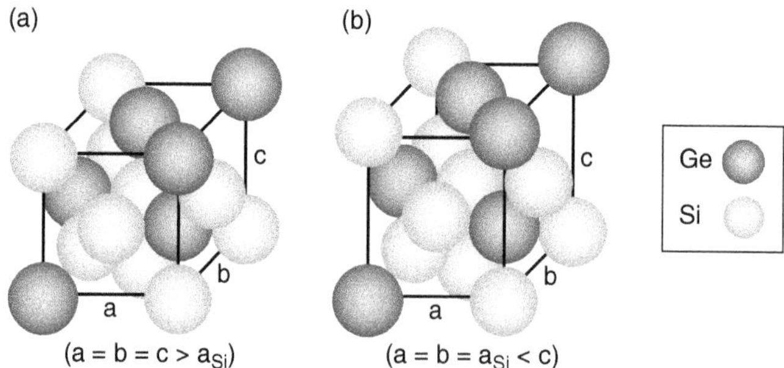

FIGURE 9.13 SiGe crystal structure. (a) Bulk SiGe is diamond cubic, with all unit cell sides equal in length. (b) Strained SiGe on silicon is body centered tetragonal, such that a and b are fixed to the unit cell size of silicon, but c expands to some greater value to accommodate the additional volume. Note that germanium has a 4% larger atomic radius than silicon.

the lattice constant of the SiGe will be larger than that of silicon. To accommodate this misfit, the growing film compresses in the a and b directions, while expanding in the c direction, as in Figure 9.13b.[28–31] The incorporation of strain to accommodate lattice misfit is known as pseudomorphic growth.[29] The resulting SiGe has body centered tetragonal (BCT) crystal structure.

9.5.2 Critical Thickness (h_{crit})

Unfortunately, the pseudomorphic layer can only tolerate a certain amount of strain before it becomes unstable. The point at which this occurs is a function of the lattice constant mismatch, between the layer and substrate, and the film thickness. As mismatch is solely compositionally dependent, instability is usually defined as a critical thickness (denoted h_{crit} or h_c). If the critical thickness is exceeded, the crystal will relax, relieving the strain by introducing misfit dislocations in the crystal structure. Misfit dislocations are defects extremely deleterious to device performance.[32–34] Figure 9.14 shows a schematic representation of a misfit dislocation illustrated in SiGe on silicon.

In reality, h_{crit} for most compositions of SiGe is too thin to be of use in a device. However, metastable layer growth is possible; layers that exceed h_{crit}, but are grown and processed at lower temperatures so as to avoid the kinetics of strain relaxation. Stringent restrictions must be made on the thermal budget of metastable films. As layers grow thicker, more strain accumulates and the activation energy of misfit dislocations drops precipitously. Thus, the concept of a critical thickness in pseudomorphic SiGe cannot be avoided, just extended to thicker films.[29,35]

9.5.3 Strain Compensation and Silicon Germanium Carbon

In the mid-1990s, it was first proposed to deal with the problem of critical thickness in pseudomorphic SiGe through strain compensation with substitutionally incorporated carbon. Carbon, as diamond, is another Column IV element sharing the same crystal structure and forming the same tetrahedral bond as silicon and germanium, though with a much smaller lattice parameter (3.5670 Å).[27]

The smaller lattice parameter of carbon compensates for the larger lattice parameter of germanium relative to the silicon substrate. At the ratio of 8.3 germanium atoms to 1 carbon atom, strain is fully compensated; the lattice parameter of the $Si_{1-x-y}Ge_xC_y$, or SiGeC becomes that of silicon. As all lattice constants return to equality, the material's crystal structure is diamond cubic and the critical thickness is infinite, as seen in Figure 9.15.

If more carbon (beyond the 8.3:1 ratio) is added, the strain in the film becomes tensile. Thus, SiGeC allows for strain tailoring, in much the same way as SiGe allowed for bandgap tailoring.[36–42]

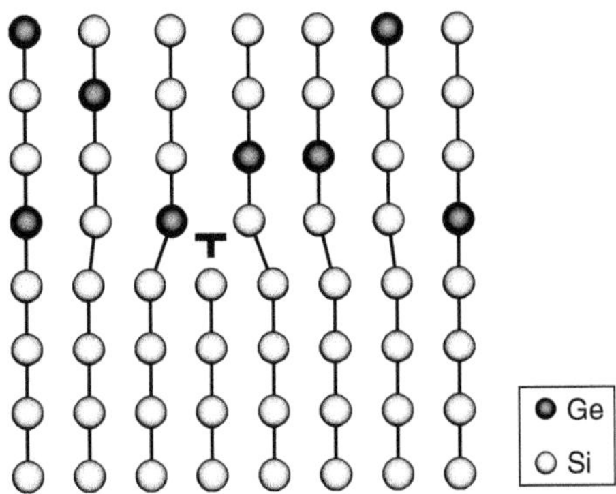

FIGURE 9.14 Misfit dislocations in SiGe on silicon. Schematic representation of a misfit dislocation in relaxed SiGe. The T-shaped mark denotes the misfit dislocation and marks the location of the missing plane of atoms in the crystal structure.

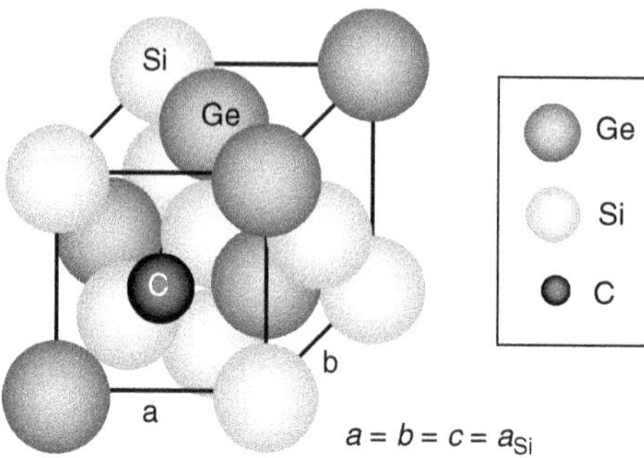

FIGURE 9.15 Fully strain compensated SiGeC. Substitutional carbon incorporated at a ratio of 1:8.3 to germanium, returns the lattice constant to that of silicon. This increases critical thickness to infinity and returns the crystal structure to Diamond Cubic.

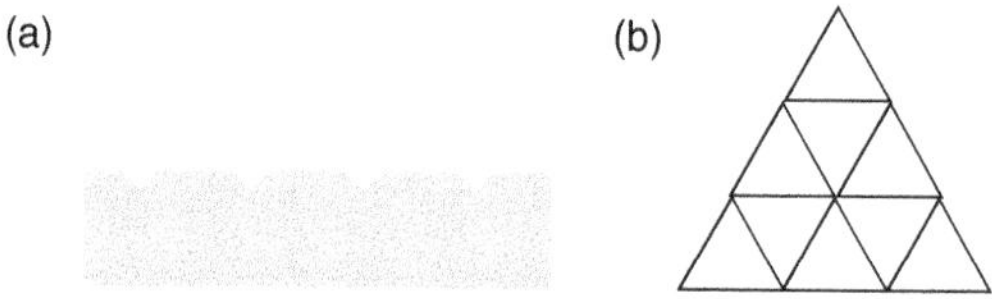

FIGURE 9.16 Graphoepitaxy. Graphoepitaxy allows for the growth of a single crystal layer atop an amorphous layer. Registration is achieved by patterning of the amorphous layer. The cross section of a patterned amorphous layer is shown in (a). A sixfold symmetric pattern, like that required for graphoepitaxially grown (111) silicon is shown in (b).

9.5.4 Hard Heteroepitaxy and Graphoepitaxy

When the misfit between the lattice constants is greater, or the crystal structure or chemical bonding is different, the growing of a solid film on a single crystal substrate is referred to as "hard heteroepitaxy." This is often accomplished by lattice matching in only one plane or the lattice constant of the (a b c) plane of the substrate matches the (x y z) plane of the epilayer. Careful control over surface orientation is required, as is precise information about the crystal planes.

The [110] directions of silicon have been matched to the [001] ([$\bar{1}$101] in hexagonal notation) direction of alumina (Al_2O_3) to create silicon-on-sapphire (SOS), a form of the earlier mentioned SOI. Sapphire's crystal structure and lattice constant are vastly different than that of silicon; however, by forming a super-lattice over several unit cells and rotating the orientation, a match is possible. The hard heteroepitaxy requires 21 sapphire unit cells to create 40 silicon unit cells, in this super-lattice formation.[12]

Graphoepitaxy, or artificial epitaxy, takes this one step further, enabling single crystal growth atop an amorphous material. In graphoepitaxy, the orienting influence on the epitaxial layer can stem from a variety of sources (geometrical, mechanical, electrical, etc.) that are macroscopic in nature, unlike ordinary epitaxy that relies on microscopic influences.[12]

The typical method of graphoepitaxy is achieved by patterning (geometric influence) an amorphous substrate such that the pattern is some super-lattice of the desired epilayer. The pattern acts to orient the film, as depicted in Figure 9.16. This has been demonstrated with the growth of single crystal silicon (111) and germanium (111) on sixfold symmetric patterned amorphous substrates.[12]

9.6 Conclusions

In its many forms (SPE, LPE, VPE, and MBE), epitaxy is widely used, especially in the semiconductor world. It endows transistors and devices, upon which the modern world relies so heavily, with the characteristics of high speed and low current leakage.

Epitaxy, though an established tool in semiconductor technology, has many new and promising frontiers left in the fields of heteroepitaxy and graphoepitaxy. These provide many new potential combinations of materials that a single material cannot accomplish. With multiple hetero/graphoepitaxial layers being used, the possibilities for new applications and structures are literally endless.

References

[1] Pearce, C.W., Epitaxy, in *VLSI Technology*, Second Edition, Sze, S.M., Ed., McGraw-Hill Publishing, New York, 1988, chap. 2.

[2] Brodie, I. and Murray, J.J., *The Physics of Micro/Nano-Fabrication*, Plenum Press, New York, 1992, p. 11.

[3] Plummer, J.D., Deal, M.D., and Griffin, P.B., *Silicon VLSI Technology; Fundamentals, Practice, and Modeling*, Prentice Hall, Upper Saddle River, NJ, 2000, p. 556.

[4] Markov, I.V., *Crystal Growth for Beginners; Fundamentals of Nucleation, Crystal Growth, and Epitaxy*, Second Edition, World Scientific, NJ, 2003, chap. 4.

[5] Askleland, D.R., *The Science and Engineering of Materials*, Second Edition, PWS-Kent Publishing, Boston, 1989, p. 237.

[6] Herman, M.A., Richter, W., and Sitter, H., *Epitaxy; Physical Principles and Technical Implementation*, Springer, Berlin, 2004, part I and II.

[7] Runyan, W.R. and Bean, K.E., *Semiconductor Integrated Circuitry Processing Technology*, Addison-Wesley, Reading, MA, 1990, chap. 7.

[8] Mayer, J.W. and Lau, S.S., *Electronic Materials Science: For Integrated Circuits in Si and GaAs*, Macmillan Publishing, New York, 1990, p. 16.

[9] Ghandi, S.K., *VLSI Fabrication Principles; Silicon and Gallium Arsenide*, John Wiley & Sons, New York, 1983, chap. 5.

[10] Campbell, S.A., *The Science and Engineering of Microelectronic Fabrication*, Second Edition, Oxford University Press, New York, 2001, chap. 14.

[11] Brodie, I. and Murray, J.J., *The Physics of Micro/Nano-Fabrication*, Plenum Press, New York, 1992, p. 319.

[12] Herman, M.A., Richter, W., and Sitter, H., *Epitaxy; Physical Principles and Technical Implementation*, Springer, Berlin, 2004, chap. 14.

[13] Jesson, D.E., Strain-induced morphological evolution of SiGe thin films, in *Properties of Silicon Germanium and SiGe:Carbon*, Kasper, E. and Lyutovich, K., Eds., INSPEC, London, 2000, p. 3.

[14] Herman, M.A., Richter, W., and Sitter, H., *Epitaxy; Physical Principles and Technical Implementation*, Springer, Berlin, 2004, chap. 7.

[15] Jaeger, R.C., Crystallization of amorphous material, *Volume V: Introduction to Microelectronic Fabrication*, Neudeck, G.W. and Pierret, R.F., Eds., Addison-Wesley Publishing, Reading, MA, 1988, p. 121.

[16] Stringfellow, G.B., *Organometallic Vapor-Phase Epitaxy: Theory and Practice*, Second Edition, Academic Press, San Diego, 1999, p. 469.

[17] Herman, M.A., Richter, W., and Sitter, H., *Epitaxy; Physical Principles and Technical Implementation*, Springer, Berlin, 2004, chap. 8.

[18] Wolf, S. and Tauber, R.N., *Silicon Processing for the VLSI Era; Volume 1: Process Technology*, Second Edition, Lattice Press, Sunset Beach, CA, 2000, chap. 7.

[19] Shackelford, J.F., *Introduction to Materials Science for Engineers*, Third Edition, Macmillan Publishing, New York, 1992, chap. 4.

[20] Guy, A.G., *Introduction to Materials Science*, McGraw-Hill, New York, 1972, chap. 4.

[21] Guy, A.G., *Essentials of Materials Science*, McGraw-Hill, New York, 1976, chap. 3.

[22] Braunstein, R., Valence band structure of germanium silicon alloys, *Phys. Rev.*, 130, 869, 1963.

[23] Van de Walle, C.G. and Martin, R.M., Theoretical calculations of heterojunction discontinuities in the Si/Ge system, *Phys. Rev. B*, 34, .5621 1986.

[24] Bean, J.C., Silicon based semiconductor heterostructures: Column IV bandgap engineering, *Proc. IEEE*, 80, 571, 1992.

[25] Scaffler, F., High-mobility Si and Ge structures, *Semiconductor Sci. Technol.*, 12, 1515, 1997.

[26] Herbots, N. et al., The onset of secondary phase precipitation during synthesis of heteroepitaxial SiGeC on Si(100), *Appl. Phys. Lett.*, 68, 782, 1996.

[27] Kittel, C., *Introduction to Solid State Physics*, Seventh Edition, John Wiley & Sons, Inc., New York, 1996, p. 23.

[28] Hinckley, J.M. and Singh, J., Hole transport theory in pseudomorphic SiGe alloys grown on Si(001) substrates, *Phys. Rev. B*, 41, 2912, 1990.

[29] Banerjee, S., Applications of silicon germanium carbon in MOS and bipolar transistors, in *Proceedings of SPIE: Microelectronic Device Technology* v.3212, Rodder, M. et al. Eds., SPIE — The International Society for Optical Engineering, 1997, p. 118.

[30] Whall, T.E., Fully pseudomorphic Si/SiGe/Si heterostructures for p-channel field effect devices, *Thin Solid Films*, 294, 160, 1997.

[31] Hinckley, J.M. and Singh, J., Influence of substrate composition and crystallographic orientation on the band structure of pseudomorphic SiGe alloy films, *Phys. Rev. B*, 42, 3546, 1990.

[32] Tersoff, J. and LeGoues, F.K., Competing relaxation mechanisms in strained layers, *Phys. Rev. Lett.*, 72, 3570, 1994.

[33] People, R. and Bean, J.C., Calculation of critical layer thickness versus lattice mismatch for GeSi/Si strained layer heterostructures, *Appl. Phys. Lett.*, 47, 322, 1985.

[34] Osten, H.J., Kim, M., Lippert, G., and Zaumseil, P., Ternary SiGeC alloys: Growth and properties of a new semiconducting material, *Thin Solid Films*, 294, 93, 1997.

[35] Hull, R., Metastable strained layer configurations in the SiGe/Si system, in *Properties of Silicon Germanium and SiGe:Carbon*, Kasper, E. and Lyutovich, K., Eds., INSPEC, London, 2000, 9.

[36] Osten, H.J. et al., Carbon containing group IV heterostructures on Si: Properties and device applications, *Thin Solid Films*, 321, 11, 1998.

[37] Powell, A.R., Eberl, K., Ek, B.A., and Iyer, S.S., SiGeC growth and properties of the ternary system, *J. Crystal Growth*, 127, 425, 1993.

[38] Sturm, J.C., Yang, M., Carroll, M.S., and Chang, C.L., SiGeC alloys: An enabling technology for scaled high performance silicon based heterojunction devices, in *Proceedings of the 1st Meeting on Silicon Monolithic Integrated Circuits in RF Systems*, Kayali, S., Ed., IEEE 1998, 3.

[39] Stein, B.L. et al., Electronic properties of Si/SiGeC heterojunctions, *J. Vacuum Sci. Technol. B*, 16, 1639, 1998.

[40] DeSalvador, D., Lattice parameter of SiGeC alloys, *Phys. Rev. B*, 61, 13005, 2000.

[41] Lee, Y.T., Miyamoto, N., and Nishizawa, J., The lattice misfit and its compensation in the Si-Epitaxial layer by doping with germanium and carbon, *J. Electrochem. Soc.*, 122, 530, 1975.

[42] Guedj, C. et al., Carbon and germanium distributions in SiGeC layers epitaxially grown on Si (100) by RTCVD, *Thin Solid Films*, 294, 129, 1997.

10

Ion Beam Assisted Deposition

Michael Nastasi and
Amit Misra
Los Alamos National Laboratory

James W. Mayer
Arizona State University

Abstract

The bombardment of a growing film with energetic particles has been observed to produce beneficial modifications in a number of microstructural features and physical and mechanical properties critical to the performance of thin films and coatings. Such modifications include: improved adhesion; densification of films grown at low substrate temperatures; magnitude and sign of growth-induced residual stresses; control of texture (orientation); grain size and morphology; optical and other physical properties; and mechanical properties such as hardness and ductility. In this chapter, we provide an overview of the role of ion–solid interactions on the microstructures and properties of vapor deposited thin films.

10.1 Introduction

The bombardment of a growing film with energetic particles has been observed to produce beneficial modifications in a number of characteristics and properties critical to the performance of thin films and coatings such as: improved adhesion; densification of films grown at low substrate temperatures; modification of residual stresses; control of texture (orientation); modification of grain size and morphology; modification of optical properties; and modification of hardness and ductility.

The process of simultaneous thin film deposition and directed ion bombardment from an ion source has been labeled by a variety of terms including: ion assisted coating (IAC); ion assisted deposition (IAD); ion vapor deposition (IVD); ion beam enhanced deposition (IBED); dynamic recoil mixing (DRM) at high energies; and ion beam assisted deposition (IBAD). This term *ion beam assisted deposition* or IBAD will be used here in favor of its growing acceptance by the energetic-particle/solid interaction research community.

The important role of ions in thin film deposition techniques has long been realized by the coating community. It is difficult, however, in many of the plasma based coating techniques, to separate out the degree to which ion and neutral particle fluxes as well as ion energies affect resultant coating properties. Mattox showed[1] as early as 1963 that energetic ions within plasmas have an important influence on coating properties in his early development of ion plating. In addition, other plasma based deposition processes, such as activated reactive evaporation (ARE), developed by Bunshah and coworkers,[2] employ ionization to promote film properties. Research on the role of ions in film deposition increased in the 1970s and early workers in the field include Weissmantel,[3] Pranevicius[4] and Harper, Cuomo, and associates at IBM.[5,6] The concept of using ionized energetic clusters of atoms to deposit thin films has been mainly researched by Takagi, Yamada, and associates at the University of Kyoto since 1972[7,8] and more recently in other laboratories.[9] The process of utilizing low-energy ion beam impingement alone to produce thin film deposition is sometimes termed Ion Beam Deposition and has been reviewed by Herbots et al.[10]

Over the past several years, the international ion beam community has applied increased interest to the use of simultaneous ion bombardment and physical vapor deposition. The motivation for this attention, in part, is the increased control of directionality and ion energies as compared to plasma based processes. Ion-assisted deposition has been studied in several laboratories worldwide since the early 1980s to study electronic materials, to understand the role of ions in plasma-assisted deposition processes, to prepare dense optical coatings, as well as to prepare tribological and corrosion resistant coatings. It should be noted that, with some exceptions, most of the work in this area has utilized much lower energy beams (typically 100 to 2000 eV) than used for ion implantation.

In this chapter, we will focus only on the general features of the IBAD process. For an extended review of the detailed experimental results and references associated with the IBAD process and other ion beam deposition phenomena the reader is referred to the literature.[11–15] In this chapter, we will first examine the details of film growth without the assistance of ion bombardment. We then examine the influence of ion bombardment on thin film growth, followed by a discussion on compound formation by IBAD processing. In general, our approach will be an empirical one, based on experimental results and observations.

10.2 Microstructure Development During the Growth of Metallic Films

The early stages of film nucleation and growth ultimately influence the microstructure of a thick film or coating. It is therefore important to understand how films grow and the mechanisms by which grain structures develop during film deposition (see also Chapter 8, Chapter 9, and Chapter 11). The morphologies of metallic films grown by thermal evaporation has been examined by Grovenor et al.[16] and they concluded that the microstructure developed in coatings is strongly influenced by the substrate temperature, T_s, expressed as a fraction of the absolute melting point (T_s/T_m). Analysis of the microstructures (Figure 10.1a) showed that it is possible to classify the film morphology into four characteristic zones.

10.2.1 Zone I, $T_s < 0.15\ T_m$

Film grown on substrates held at temperatures less than 0.15 T_m, where T_m is the melting point of the deposited material in K, are formed of equiaxed (i.e., equal extent in all three directions) grains 5 to 20 nm in diameter. At this temperature, the mobility of the deposited atoms is low, and the atoms stick where they land. Calculations indicate that the critical nucleus size at this temperature is a single atom

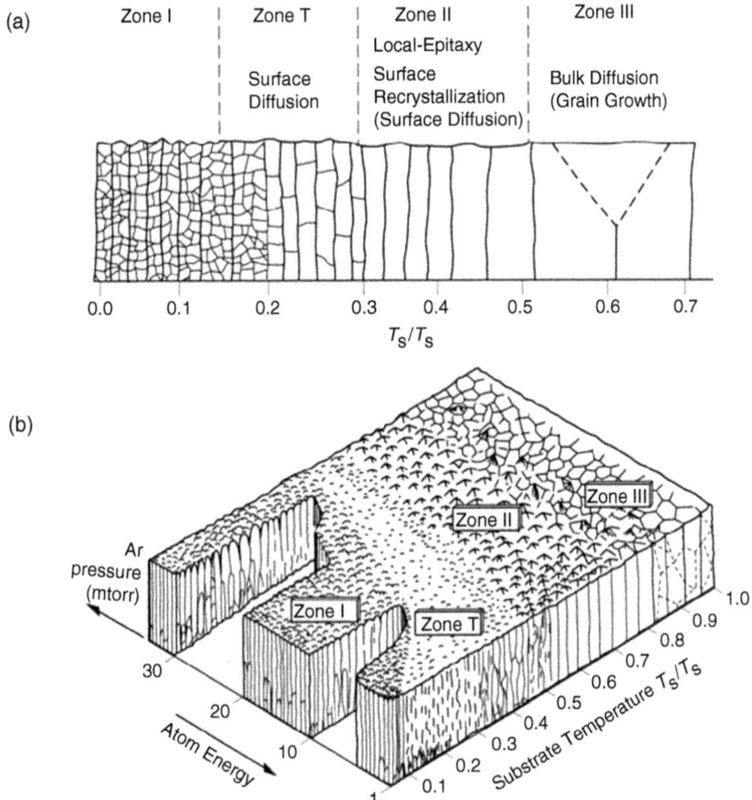

FIGURE 10.1 (a) Zone model for the grain structure of vapor deposited metal films (after Grovenor, C.R.M., Hentzell, H.T.G., and Smith, D.A., *Acta Metall.*, 32, 773, 1984. With permission) (b) Schematic representation of a film structure as a function of sputtering gas pressure or IBAD ion energy and the substrate temperature T_s normalized to the film's melting temperature T_m (After Thornton, J.A., in *Deposition Technologies for Film and Coatings*, Bunshah, R.F., Ed., Noyes Publications, Park Ridge, NJ, 1982, Chapter 5. With permission from Noyes).

and that the maximum distance a deposited atom can diffuse before being covered by further deposited atoms is about one atomic distance, suggesting that a noncrystalline or amorphous microstructure is favored. However, crystalline grains and not an amorphous microstructure are observed experimentally, indicating that some type of athermal recrystallization occurs from a precursor phase to produce the final crystallite size.

10.2.2 Zone T, 0.15 T_m < T_s < 0.3 T_m

In this temperature regime a microstructure is observed that can be expressed as a transition microstructure between the equiaxed microstructure of Zone I and the columnar structure found in Zone II. This transition is attributed to the onset of surface diffusion that allows the deposited atoms to migrate on the deposit surface before being covered by the arrival of further material. Under such conditions grain boundaries are becoming mobile, and the start of grain growth is anticipated.

10.2.3 Zone II, 0.3 T_m < T_s < 0.5 T_m

In this temperature regime, the deposited atoms have sufficient mobility to diffuse and increase the size of grains before being covered by the arrival of further material. In addition, grain boundaries become mobile. Films grown at substrate temperatures in this regime have a uniform columnar grain structure.

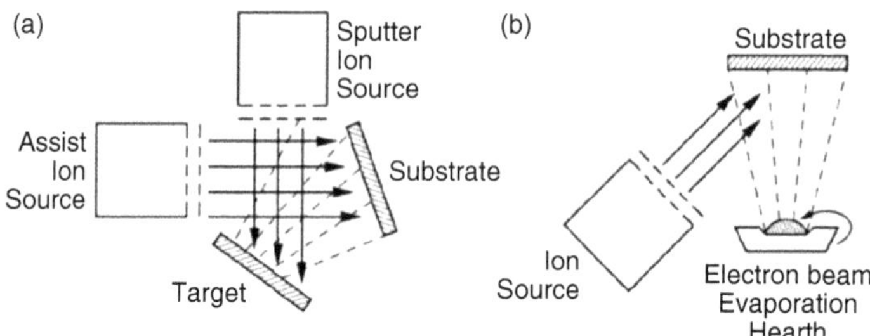

FIGURE 10.2 Schematic of geometry employed for energetic ion bombardment of thin films deposited by physical vapor deposition (PVD), commonly referred to as ion beam assisted deposition (IBAD). (a) employing sputter deposition, (b) electron beam evaporation (after Harper, J.M.E., Cuomo, J.J., Gambino, R.J., and Kaufman, H.E., in *Ion Bombardment Modification of Surfaces: Fundamentals and Applications*, Auciello, O., and Kelly, R., Eds., Elsevier Science Publ., Amsterdam, 1984, Chapter 4. With permission from Elsevier.)

In general, the grain diameters are less than the film thickness but increase with both temperature and film thickness.

10.2.4 Zone III, $T_s > 0.5\ T_m$

Films grown in this temperature regime have squat uniform columnar grains, with grain diameters larger than the film thickness. The grain structure in this zone is attributed to grain growth that results from bulk rather than surface diffusion.

10.3 Nonreactive IBAD Processing: Effect of Ions on Film Growth

There are several aspects of film growth that are beneficially influenced by ion bombardment during thin film deposition including: (1) adhesion, (2) nucleation or nucleation density, (3) control of internal stress, (4) morphology, (5) density, (6) composition, and (7) the possibility of low temperature deposition.

The IBAD process allows thicker alloyed regions to be attained than by either direct ion implantation or ion beam mixing but still incorporates advantages attributed to ion beams, such as superior adhesion due to precleaning and ion mixing during the initial stages of deposition.

A typical geometry of IBAD equipment is shown schematically in Figure 10.2. The neutral species (deposited atoms) is normally delivered via physical vapor deposition as shown by either sputter deposition (Figure 10.2a) or by evaporation (Figure 10.2b). The ion species is typically produced by a low-energy (0.2 to 2 keV) broad-beam gridded ion source producing beam currents up to 1 to 2 mA/cm^2 ($\approx 10^{16}$ions/s/cm^2).

Figure 10.3 is schematic of the IBAD process showing the interaction of the deposited coating atoms with energetic ions and the substrate. The figure depicts a physically mixed zone between the original substrate surface and coating produced by ion beam mixing (more prevalent at higher ion beam energies). Such a zone helps to avoid sharp interfaces and can improve adhesion if the coating and substrate elements have a mutual affinity in terms of solubility or compound formation.[17] The effect of an ion beam on film growth and its resultant physical properties will depend on the ion species, energy, and the relative flux ratio of the ions, J_I, and deposited atoms, J_A, customarily defined as R_i. Data showing the effect of ion bombardment on film properties is either expressed simply as ion flux (assuming constant deposited atom flux), as a relative ion/metal atom flux at the substrate (i.e., R_i), or as the average energy deposited per atom, E_{ave}, in eV/atom, which is simply the product of the relative ion/atom flux and the average ion energy, E_{ion}.

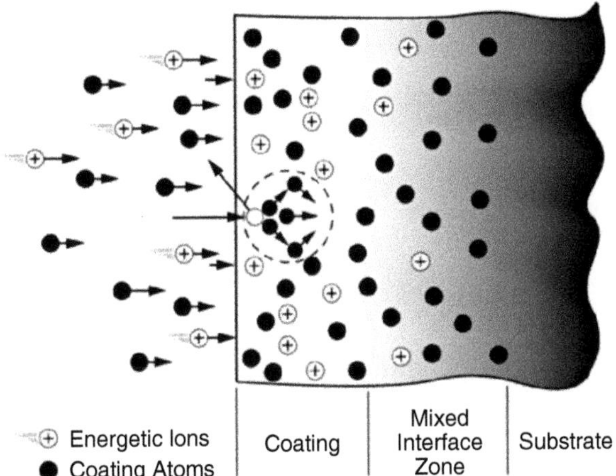

FIGURE 10.3 Schematic of the nonreactive IBAD process indicating the interaction of coating atoms in the collisional cascade of the ions and the possibility of a mixed interface zone.

The mathematical description of these two IBAD parameters is given by

$$R_i = \frac{\text{ion flux}}{\text{flux of deposited atom}} = \frac{J_I}{J_A} \tag{10.1}$$

and

$$E_{ave} = E_{ion} \cdot \frac{J_I}{J_A} = E_{ion} \cdot R_i \tag{10.2}$$

It will be seen that a R_i value as low as $R_i = 0.001$ can have a significant effect on thin film growth, especially at higher energies.

From Figure 10.3 we see that some ion implantation of the low-energy (typically <2 keV) assist ions takes place during the IBAD process. Reasonable estimates of the range can be obtained from the following range equation

$$R\ (\text{nm}) = 20.44\varepsilon^{0.36} \frac{(Z_1^{0.23} + Z_2^{0.23})^2 (M_1 + M_2)^2}{M_1 M_2 N\ (\text{atoms/nm}^3)} \tag{10.3}$$

where M_1 and Z_1 are the mass and atomic number of the incident ion, M_2 and Z_2 are the mass atomic number of the target atom, and ε is a reduced energy given by

$$\varepsilon = \frac{32.53 M_2 E_0}{Z_1 Z_2 (M_1 + M_2)(Z_1^{0.23} + Z_2^{0.23})} \tag{10.4}$$

where E_0 is the ion energy in keV.

As an example of Equation 10.3 consider a Ge IBAD experiment with 500 eV Ar ions. From Equation 10.4 we have $\varepsilon = 0.00436$. The atomic density for Ge is $N = 4.41 \times 10^{22}$ atoms/cm^3 = 44.1 atoms/nm^3, which gives an Ar range in Ge of 4.9 nm.

In an IBAD system, deposition and ion bombardment usually operate at pressures higher than typically employed for thermal deposition alone. As the operating pressure increases, the probability of incorporating unwanted contamination from residual gasses in the deposited film also increases. To minimize

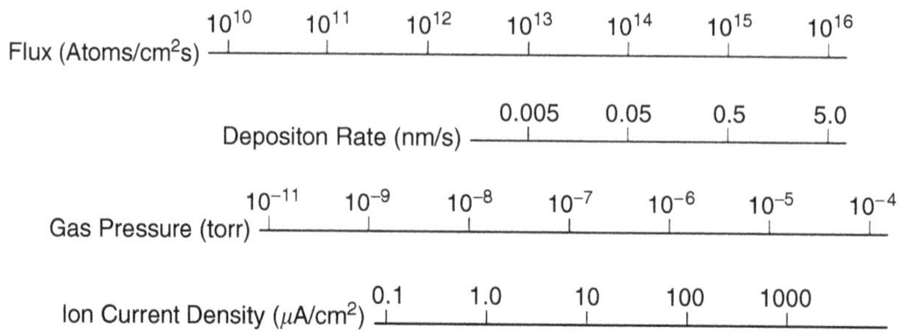

Flux (Atoms/cm²s) 10^{10} 10^{11} 10^{12} 10^{13} 10^{14} 10^{15} 10^{16}

Depositon Rate (nm/s) 0.005 0.05 0.5 5.0

Gas Pressure (torr) 10^{-11} 10^{-9} 10^{-8} 10^{-7} 10^{-6} 10^{-5} 10^{-4}

Ion Current Density (μA/cm²) 0.1 1.0 10 100 1000

FIGURE 10.4 Relationship of impinging particle (atoms, ions, and residual gas) fluxes to deposition rate, gas pressure, and ion current density. Here the sticking probabilities of the impinging particles is assumed to be unity. (After Miyake, K. and Tokuyama, T., in *Ion Beam Assisted Film Growth*, Itoh, T., Ed., Elsevier, Amsterdam, 1989, Chapter 8. With permission.)

contamination from residual gasses during deposition the following condition must be satisfied[18]:

$$S_A J_A \gg S_r J_r \tag{10.5}$$

where J_A and J_r are the deposited atom flux and the residual gas atom flux at the substrate, respectively, and S_A and S_r are the respective sticking probabilities. The residual gas flux can be represented as

$$J_r = 5.3 \times 10^{20} P\ (\mathrm{cm^{-2}\,s^{-1}}) \tag{10.6}$$

where P is the residual gas pressure in torr. In IBAD experiments one must also consider the ion flux at the surface and the ratio J_I/J_A. The ion flux J_I is related to the ion beam current density j_i through the expression

$$J_I = 6.25 \times 10^{12} j_i\ (\mathrm{cm^{-2}\,s^{-1}}) \tag{10.7}$$

where j_i is in units of μA/cm².

Figure 10.4 shows the relationship (i.e., the equivalency) of particle fluxes given in Equation 10.5 to Equation 10.7, assuming that the sticking probability of impinging ions, residual gas atoms, and deposited atoms is unity. As an example of the use of Figure 10.4 consider the IBAD processing condition of $R_i = 1$. If the thermal deposition rate of atoms is set to 0.5 nm/s, the atom arrival flux at the substrate will be 1×10^{15} atoms/cm²/s. For the condition $R_i = 1$, Equation 10.1 gives that the ion flux at the substrate is 1×10^{15} ions/cm²/s which, assuming singly charged ions, translates to an ion current density of 160 μA/cm². Correspondingly, a low base pressure is required to avoid incorporation of contaminants from unwanted background species (e.g., H_2O), as a background (partial) pressure of 10^{-6} torr corresponds to an arrival rate of about 5×10^{14} per cm²/s of a particular constituent within the vacuum. When performing IBAD processing of reactive elements (Ti, Cr, Nb) for example, the partial pressures of the vacuum species will influence the choice of deposition rates to minimize impurity incorporation.

10.3.1 Microstructure Development During IBAD

Several distinct modes of film growth have been observed in traditional (non-IBAD) thin film experiments. These include: (1) Frank–van der Merwe type in which the film grows layer-by-layer with complete coverage, (2) Volmer–Weber growth in which islands form and grow before coalescing, and (3) Stranski–Krastanov growth in which a wetting layer first forms followed by island growth. A discussion of these different growth modes is given by Greene et al.,[19] Smidt,[14] and Tu et al.[20] (see also

Chapter 8 and Chapter 9). Pranevicius showed in an early experiment the effect of ions on initial film growth and coalescence.[4] In that experiment, aluminum was evaporated onto an insulator at a constant rate and the electrical conductivity between two separated electrodes on the surface was measured as a function of time by observing the current flow between them. For no ion current, there is a considerable incubation time required before the growing Al islands overlap and provide electrical continuity. The presence of an ion beam serves to significantly shorten the time period for overlap to occur, attributed to increased adatom mobility and nucleation sites produced for island formation and growth. Studies performed with electron microscopy showed that there was a fourfold increase in nuclei density and a factor of 5 to 15 decrease in nuclei size in the presence of an ion beam.

In Section 10.2 and Figure 10.1a, we showed that the microstructure developed in non-IBAD evaporated coatings is strongly influenced by the substrate temperature expressed as a fraction of the absolute melting point (T_s/T_m). Thornton[21] further developed the T_s/T_m concept for sputtered films and included the effects of sputtering gas pressure. Messier et al.[22] have also examined the microstructural evolution during film growth and have explained Thornton's gas pressure effects during the sputter deposition of films as a change in the energy of particles that bombard the substrate during the deposition process. These bombarding particles are neutral atoms, created when an energetic sputtering ion does not penetrate the sputtering target surface, but instead experiences a collision with a surface atom that results in the ion being reflected (recoiled) away from the surface. During such reflection of collision events, the ion picks up the necessary electrons needed to turn it into a neutral atom. The kinetic energy of this reflected neutral atom will be determined by the kinematic details of the collision and is not influenced by any electric fields used to stimulate the sputtering process. The kinetic energy of such reflected neutrals can be reduced by collision events with other gas molecules or ions found in the sputtering plasma. The higher the sputtering gas pressure, the greater the density of possible collision centers, and the greater the probability of the reflected neutral atom losing energy through a collision event while traveling toward the substrate. Thus, there is an inverse relationship between the energy of the bombarding particle (reflected neutral) at the growing film surface and the sputtering gas pressure, making the sputtering process directly comparable to the average ion energy E_{ion} applied in the IBAD process. From the above discussion we see that the sputtering process can be directly compared to the IBAD process in terms of the average ion energy E_{ion} applied to the film growth process.

The relationship between microstructure and sputtering or IBAD processing conditions can be expressed in the structure zone diagram shown in Figure 10.1b.[21,22] The diagram is similar to the evaporation Zone diagram described in Section 10.2, but includes the Ar sputtering pressure or Ar bombarding energy as a processing parameter in addition to the substrate temperature (expressed as a fraction of the melting temperature in degrees Kelvin).

10.3.2 Densification

One of the most important microstructural modifications produced by IBAD is the densification of films deposited under conditions that would nominally fall within Zones T and II of the structure diagram, Figure 10.1a. The common occurrence of (columnar) microstructural features in coatings formed at low temperatures is well documented.[16,21,23] Films deposited in this regime have high porosity and low mechanical strength because the deposited atom diffusivity is low. Film density can be improved by increasing the substrate temperature during deposition but this is sometimes detrimental to the substrate on which the film is deposited. The use of energetic ion bombardment during low temperature deposition offers a means of essentially eliminating this mode of microstructural evolution. This densification effect has found many practical applications in producing compact coatings, such as optical coatings.

Yehoda et al.[24] have examined the influence of low-energy Ar ions on the void structure of evaporated Ge films. They varied the ion energy from 15 to 110 eV and the ion-to-atom arrival ratio between 0 and 25%. They found that the void fraction, determined by spectroscopic ellipsometry, decreased rapidly as a function of energy deposited per arriving ion, with a minimum occurring at about 5 eV for all ion energy and ion current combinations.

While the complex nature of the IBAD process makes detailed analytical modeling difficult, a great deal of physical insight into the evolution of IBAD films and the mechanisms responsible for densification can be obtained using computer based Monte Carlo and molecular dynamic calculations. In the remainder of this section, we will examine the calculated data obtained from the computer simulation of the IBAD process.

10.3.2.1 Densification: Monte Carlo Calculations

The effect of energetic ion bombardment on Zone 1 film microstructure was first modeled by Müller[25] using Monte Carlo calculations and a thermal spike model. Müller examined the role of ion-induced temperature spikes to induce atomic motion in the porous columnar network of a growing film. The initial spike-temperature distribution was chosen to have a maximum at a depth just below the surface corresponding to the ion range, and a width equal to the distribution of damage, both of which were determined from Monte Carlo collision cascade calculations. The heat in the spike was dissipated by thermal diffusivity. The motion of atoms in the vicinity of the spike center during the cascade life-time ($\cong 10^{-11}\,s$) was modeled using temperature-dependent atomic hopping rate ν, given by:

$$\nu = (8k_B T/\pi M d^2)^{1/2} \exp(-\Delta E/k_B T)$$

where, T is the spike temperature, M is the mass of the species being deposited to form the film, d the average atomic spacing, and ΔE is the activation barrier to atomic hopping. Müller's simulation showed that IBAD disrupts the porous columnar growth and stimulates bridging between neighboring grains, causing the closure of open voids. However, these simulations did not show any significant densification of the film. The primary reason for this is that the thermal spike model employed only local isotropic atomic rearrangements stimulated in the spike phase of the collision cascade, and that motion of atoms due to multiple collisions and momentum transfer during the collisional phase of the cascade were not accounted for.

When an energetic ion interacts with a material, the interacting ion produces a collision cascade where knock-on atoms, vacancies and interstitials, phonons, and electronic excitations result. These processes are predominant in the collisional phase of the cascade[26] and can be reasonably well modeled by Monte Carlo calculations. For ion bombardment in the low-energy regime of a few hundred eV, most of the collision cascade processes are confined to the material surface. For low ion energies the ions can be either backscattered or incorporated into the surface, while knock-on atoms may either leave the surface as sputtered atoms or be recoiled implanted below the surface and become trapped as interstitials. Under such ion bombardment conditions the recoil implanted target atoms produce an atomic density increase in the vicinity where they come to rest (i.e., at the end of their range) while sputtered atoms and the vacant sites left by the recoil implanted atoms result in a high concentration of vacancies at the surface and a reduction in surface density. When low-energy ion bombardment is combined with thermal evaporation, as in IBAD, the vacancies created near the surface by the bombarding ions are partially filled by the vapor deposited atoms. At large ion-to-vapor-atom ratios the IBAD process will result in an inward packing of the atoms in the film, eliminating the porous columnar network commonly observed in Zone I films (Figure 10.1b), and favor the growth of a densely packed structure.

This process of collision cascade induced densification during IBAD was studied by Müller[27] using the Monte Carlo program TRIM.[28] Monte Carlo calculations were used to determine mass transport of target atoms and implantation distributions and the results were applied to a two-dimensional IBAD model. Figure 10.5 shows the calculated mass density distribution $\rho(z)$ as a function of depth z for a ZrO_2 film grown under a 600 eV O^+ ion bombardment. For this IBAD simulation an oxygen ion of current $J_I = 200\ \mu A/cm^2 = 1.25 \times 10^{15}$ oxygen ions/cm^2/s was assumed. The ratio of ion-to-vapor-atom arrival rate was $R_i = J_I/J_A = 0.22$, which gives a vapor atom deposited rate of 2.75×10^{14} atoms/cm^2/s. The ion beam angle of incidence, relative to the surface normal, was $\alpha_I = 30°$. At times less than 0 s, the ZrO_2 film was grown without ion bombardment at the experimentally observed mass density in thin films of 4.41 g/cm^3. At times ≥ 0 s the growing film was assisted by 600 eV oxygen ions. As seen in the figure,

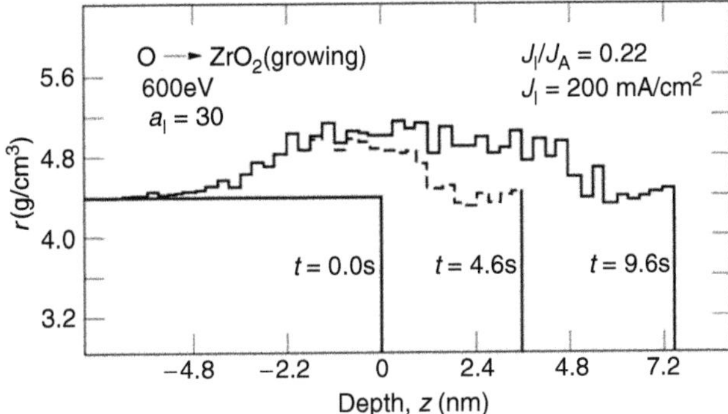

FIGURE 10.5 The calculated mass density distribution $\rho(z)$ as a function of depth z for a ZrO_2 film grown under a 600 eV O^+ ion bombardment. For this IBAD simulation an oxygen ion of current $J_I = 200\ \mu A/cm^2$and a ratio of ion-to-vapor-atom arrival rate of 0.22. The ion beam angle of incidence, relative to the surface normal, was $\alpha_I = 30°$. For $t < 0$ s the film was grown under non-IBAD conditions. (After Müller, K.-H., *J. Appl. Phys.*, 59, 2803, 1986. With permission.)

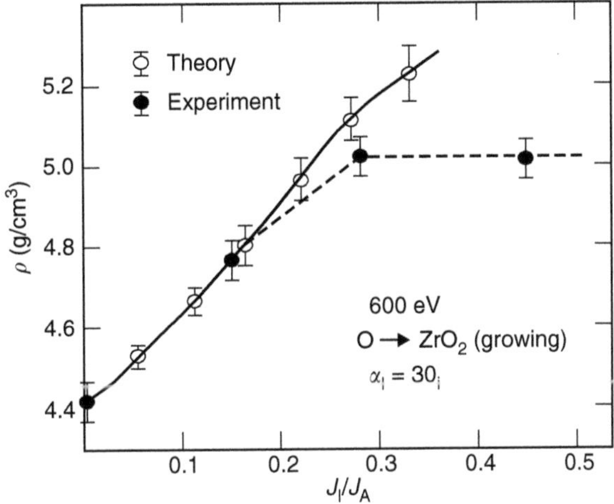

FIGURE 10.6 The calculated and experimentally observed average mass density of ZrO_2 films grown under a 600 eV O^+ ion bombardment as a function of J_I/J_A. ZrO_2 films have a typical mass density of 4.41 g/cm^3 while bulk ZrO_2 has a mass density of 5.7 g/cm^3. (After Müller, K.-H., *Phys. Rev. B*, 35, 7906, 1987. With permission.)

the calculations indicate that the region of densification starts at approximately 2.5 nm below the surface and the non-IBAD portion of the film ($t = 0$) becomes redensified up to a depth of about 3 nm.

The influence of ion-to-vapor atom flux ratio J_I/J_A on the ZrO_2 film average mass density is shown in Figure 10.6 and compared to experimental data. There is good agreement between calculation and experiment up to a ratio of $J_I/J_A \approx 0.3$ after which the experimental data levels off and the calculation predicts further density increases with increasing J_I/J_A. At $J_I/J_A = 0.22$ the mass density is close to the bulk density of ZrO_2, 5.7 g/cm^3. The leveling off of the experimental data presumably indicates that the atomic saturation density has been reached and the system resists further densification due to repulsive forces between atoms in the film, a mechanism that was not included in the model calculation. Such repulsive forces are expected to become important when atom and ion energies in the IBAD film approach a few eV. These are the atom energies found in the spike and cooling phases of the collision cascade, and cannot be modeled effectively with Monte Carlo simulations, and must be examined using molecular dynamic calculations.

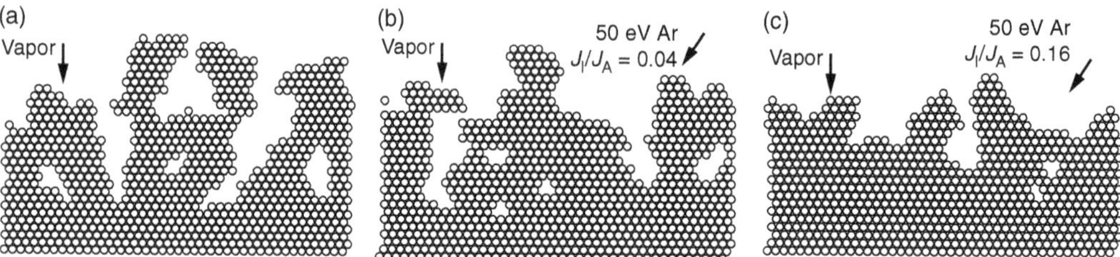

FIGURE 10.7 A typical microstructure obtained by two-dimensional molecular dynamic simulations for condensing Ni vapor atoms arriving under normal incidence (a) without ion bombardment, (b) with Ar ion bombardment of $E = 50$ eV, vapor impingement angle (α) of 30° and $J_I/J_A = 0.04$, (c) with Ar ion bombardment of $E = 50$ eV, $\alpha = 30°$ and $J_I/J_A = 0.16$ (After Müller, K.-H., *Phys. Rev. B*, 35, 7906, 1987. With permission.)

10.3.2.2 Densification: Molecular Dynamics Calculations

Müller[29] has used a two-dimensional molecular dynamics (MD) simulation to investigate the IBAD growth of Ni films on zero temperature Ni substrates under Ar ion bombardment. In these MD simulations the ions interacted with the film atoms through a screened Coulomb interaction potential with a Moliere screening function. The Ni atom–atom interactions in the growing IBAD films were calculated at high energies (i.e., small interaction distances, less than approximately 0.20 nm) with the Moilere potential, and at energies of a few eV (interaction distances greater than 0.2239 nm) with a Lennard–Jones potential. For intermediate energies a spline fit between the two interaction potentials was used.

The MD calculations of the Ni film growth using an IBAD process with Ar ions showed the following: Ar ion bombardment during Ni vapor deposition removes overhanging atoms, which cause shadowing effects, leaving void regions open until they are filled by new depositing atoms. Ni atoms sputtered by Ar ions are redeposited mainly in voids; and ion bombardment of films growing on substrates held at zero temperature (i.e., 0 K) induces: (1) surface diffusion with diffusion distances of a few interatomic spacings, (2) local heat spike heating, (3) collapse of voids, and (4) recrystallization.

An example of the MD calculated microstructure that evolves during the IBAD growth of about 500 Ni atoms with 50 eV Ar^+ ions is shown in Figure 10.7. In these calculations the Ni vapor impingement angle is 0°, the Ar ion bombardment angle is 30°, and the ion-to-atom flux ratio, J_I/J_A, was varied. Figure 10.7a shows the typical open, columnar, microstructure obtained without ion bombardment. The simulation of the deposited Ni assumes that the atoms in the vapor are monoenergetic with an energy of 0.1 eV/atom. Figure 10.7a and Figure 10.7b, which include concurrent ion bombardment, show that densification increases with increasing ion-to-atom ratio. At low J_I/J_A ratios, ion bombardment causes closure of the long open voids but leaves behind micropores, while large J_I/J_A values result in the total disappearance of microporosity.

The influence of ion-to-atom ratio on the density of Ni films at Ar bombardment energies of 10 and 50 eV is shown in Figure 10.8. The density of the Ni films is represented by the packing density, which is defined as the fraction of atoms occupying the total number of Ni-lattice sites in the first nine layers above the substrate. These data show that the film density increases linearly with J_I/J_A and also increases with bombarding energy.

10.3.3 Residual Stress

Residual stresses are ubiquitous in vapor deposited thin films on rigid substrates. These stresses may originate from lattice mismatch between films and substrate (coherency stress), coefficient of thermal expansion mismatch, phase transformation or film growth. The following discussion is focused on growth stresses that can be related to the film microstructure, as well as to incorporated impurities. Typically, thin films formed by thermal evaporation have a high void fraction (i.e., low atomic density) and are in a state of residual tensile stress. The mechanisms of stress generation in nonenergetic deposition (i.e., thermalized

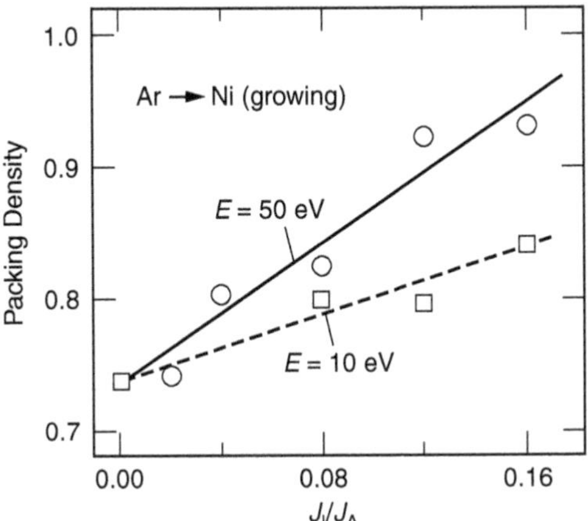

FIGURE 10.8 Packing density of Ni atoms with Ar ion assist, as a function of the ion-to-atom flux ratio, , for Ar ion energies of $E = 10$ and 50 eV. (After Müller, K.-H., *Phys. Rev. B*, 35, 7906, 1987. With permission.)

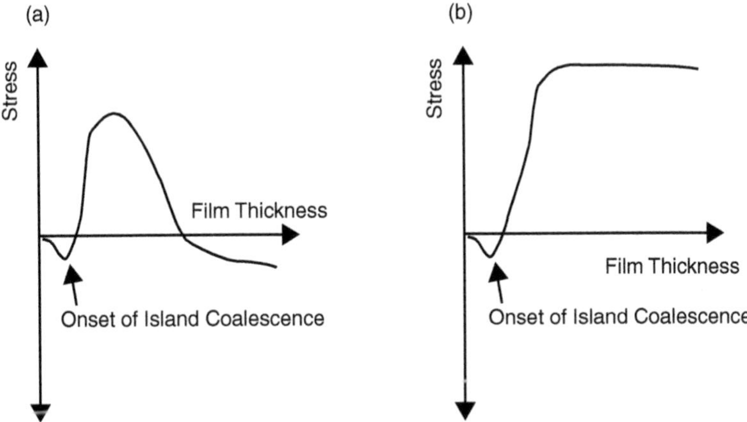

FIGURE 10.9 Typical trends of growth stress evolution as a function of film thickness for nonenergetic deposition: (a) materials with high atomic mobility, and (b) materials with low atomic mobility at the deposition temperature. (Adapted from Thompson, C.V. and Carel, R., *J. Mech. Phys. Solids*, 44, 657, 1996.)

flux) as a function of film thickness are briefly described below, followed by the mechanisms of stress generation in IBAD films.

10.3.3.1 Stress evolution in non-IBAD films

In the absence of bombardment from hyperthermal particles, the general trend of the evolution of intrinsic stress in island growth thin films as a function of film thickness is shown in Figure 10.9.[30,31] For materials with high atomic mobility at room temperature,[32–36] such as Ag, Al, Au, Cu, etc, stress evolution with increasing film thickness has three stages: (1) initial compressive stress, (2) rapid build up of tensile stress to a peak, and (3) relaxation of tensile stress and eventual build up of compressive stress (Figure 10.9a). Often these measurements are performed *in situ* and some relaxation of the compressive stress is observed when the deposition flux is stopped. For refractory metals with high melting points (e.g., Cr, Mo, W), diffusion at room temperature is limited and the transition from tensile to compressive stress with increasing film thickness is typically not observed during deposition, Figure 10.9b.[30]

The initial compressive stress that occurs when the film is discontinuous has been attributed to surface stress.[31] For low index surfaces of metals, the atoms would prefer to adopt a lower equilibrium spacing than bulk to increase the local electron density. In this case the atomic registry between the surface and underlying bulk atoms would result in stretching of the atomic bonds at the surface and shrinkage of the bonds in the underlying bulk (i.e., induced compressive stress in the bulk of the island). It should be emphasized that the surface stress represents the work done to elastically *strain* the surface atoms, and hence is different from surface energy that is the work done in *creating* a free surface. For fcc metals with high surface mobility (Figure 10.9a), island coalescence occurs early on in the deposition process and hence, large compressive stresses typically do not develop. However, in bcc refractory metals (Figure 10.9b) such as Mo, compressive stress on the order of 1 GPa has been observed in the early stages of deposition.[37]

A model for the evolution of tensile stress as the islands coalesce to form a continuous film was proposed by Nix and Clemens.[38] The driving force for coalescence is the removal of two free surfaces at the expense of a grain boundary (note that grain boundary energy is typically about one-half to one-third of surface energy in fcc metals) and elastic strain energy that results from the biaxial tensile straining of the film as the islands coalesce. This energy balance can be used to estimate an upper bound stress in the film from the island coalescence mechanism. The reason for the overestimation of stress was attributed to the assumption in the model that all crystallites are imagined to coalesce at the same time, with the consequence that no shear stress can be developed on the film–substrate interface. A more realistic picture would be to allow different crystallites to coalesce at different times and some sliding at the interface to occur. Subsequent investigations have modeled the gradual build up of tensile stress during film growth is the grain boundaries are formed after the initial island coalescence event.[34,39] Several other investigators have also attempted to refine the Nix–Clemens island coalescence model to yield quantitative predictions of tensile stress that compare reasonably with experimentally measured stresses. Following the idea that the island coalescence follows a crack zipping process, Seel et al.[36] used a finite element method to model it. By minimizing the sum of the positive strain energy and associated reduction in the boundary energy, the equilibrium configuration resulting from island coalescence was determined as a function of island radius. For a given island radius at impingement, this approach yielded average stress values that were an order of magnitude lower than the Nix–Clemens model. Freund and Chason[40] modeled the zipping process as a Hertzian contact mechanics problem. Their model is based on the theory of elastic contact of solids with rounded surfaces (Hertz contact theory). This analytical approach also predicted residual stresses on the order of a couple hundred MPa, as opposed to several GPa from the Nix–Clemens model. Floro et al.[35] and Seel et al.[36] also considered the effect of stress relaxation during growth, pointing out that the observed stress is due to a dynamic competition between stress generation due to island coalescence and stress relaxation due to surface diffusion.

Finally, the evolution of compressive stresses during film growth for conditions of nonenergetic deposition is described. Chason et al.[41] have developed a model for compressive stress generation based on the idea that an increase in the surface chemical potential caused by the deposition of atoms from the vapor drives excess atoms into the grain boundaries. Plating of extra atoms at grain boundaries would lead to an in-plane expansion of the film were it not rigidly bonded to the substrate, and the constraint from the substrate leads to compressive stress in the film. As the compressive stress raises the chemical potential of atoms in the grain boundary, the driving force for additional flow of atoms decreases with increasing stress and eventually a steady state is reached. The model also explains the relaxation of compressive stress, often observed during *in situ* measurements of stress of metals such as Ag with high mobility of surface atoms at room temperature, as the reverse flow of excess atoms from boundaries to surfaces.

10.3.3.2 Stress evolution in IBAD films

The stress evolution in films, of a constant thickness, bombarded with energetic particles (neutral atoms or ions) during deposition is described in this section. Typical trends are shown in Figure 10.10 where a positive stress value indicates a tensile stress and a negative stress value indicates a compressive stress. For e-beam evaporated W films (Figure 10.10a), bombardment with 400 eV Ar ions[42] resulted in a rapid build up of tensile stress and transition to compressive stress with increasing ion-to-atom flux ratio. For IBAD at higher temperatures (i.e., higher atomic mobilities), the peak tensile stress and neutral (zero) stress states occur for lower beam currents. Similar effects are seen as a function of ion energy,

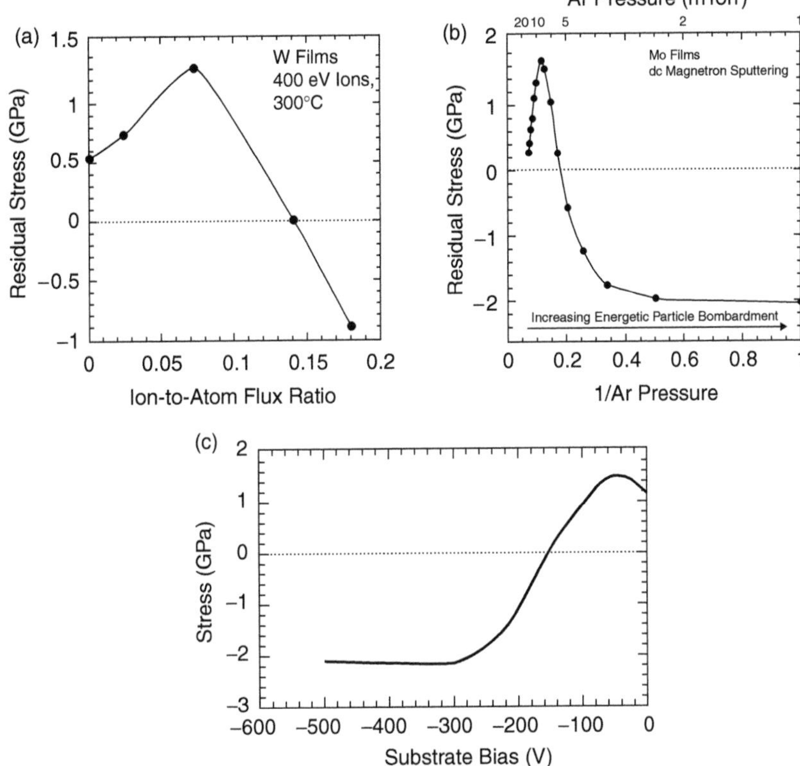

FIGURE 10.10 Typical trends in the evolution of thin film residual stresses during energetic particle bombardment; (a) IBAD W films (after Roy, R.R. and Yee, D.S., in *Handbook of Ion Beam Processing Technology*, J.J. Cuomo, S.M. Rossnagel, and H.H. Kaufman, Eds., Noyes Publications, Park Ridge, NJ, Chapter 11, 1989. With permission from Noyes), (b) sputtered Mo films with no substrate bias, showing the effect of argon pressure on growth stress (Hoffman, D.W., *Thin Solid Films*, 107, 353, 1983. With permission), and (c) effect of negative substrate bias (after Misra, A., and Nastasi, M., *Appl. Phys. Lett.*, 75, 3123, 1999. With permission). Positive values of stress indicate tensile stress and negative values compressive stress.

with lower critical current values needed to arrive at neutral stresses at higher energies. For magnetron sputtered films, decreasing Ar pressure or applying a negative substrate bias voltage results in an increase in bombarding particle energy. Hence, a transition in residual stress from tensile to compressive, similar to IBAD films (Figure 10.10a), is also seen in sputtered films with reducing Ar pressure during deposition (Figure 10.10b)[43] or increasing subtrate bias (Figure 10.10c).[44]

For the tensile to compressive stress transition, the ratio of the critical current value to the evaporant flux rate to produce a zero-stress state is termed the critical arrival rate ratio. Wolf[45] has determined that the conditions required for this zero-stress state depend on the composition of the film as well as the energy of the ions. For example, Cr requires 100 eV per atom for stress relief, whereas C or B require an order of magnitude less energy, attributed to their weaker and different type of bonding. In addition, their energy dependence is also different.[45] Preferential sputtering of background impurities (e.g., O_2) from thin Nb films has also been seen by Cuomo et al.,[46] as a factor in determining the resultant stress.

Brighton and Hubler[47] have used Monte Carlo collision cascade simulations to predict a critical arrival rate ratio for annealing stress in vapor deposited germanium. They postulated that stress annealing requires each atom in the growing film to be involved in at least one collision cascade. If the average volume affected by a cascade is V_{cas}, and the average atomic density in the film is N, then the average number of atoms affected per cascade is NV_{cas}. Ignoring cascade overlap, a lower limit for the critical ion-to-atom flux ratio, $(J_I/J_A)_c$, was obtained as $(NV_{cas})^{-1}$.

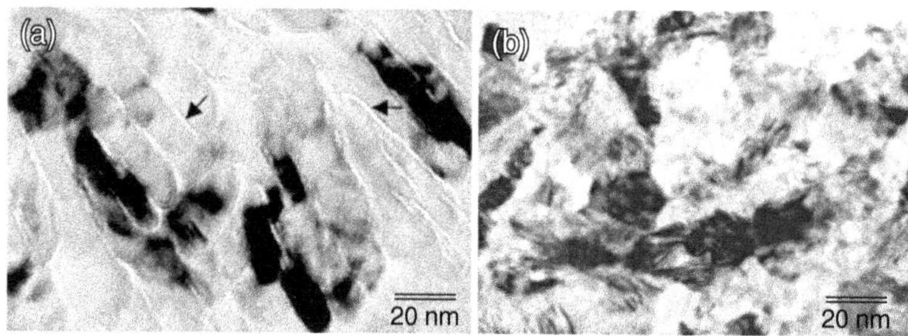

FIGURE 10.11 Plan view TEM images of sputtered 150 nm Cr films; (a) a film with tensile growth stresses, deposited without substrate bias, and (b) a film with compressive growth stress, deposited with negative substrate bias. Note the intergranular porosity, marked by arrows, revealed as bright fringes along the grain boundaries in the under-focused bright-field TEM image. (After Misra, A., and Nastasi, M., *Appl. Phys. Lett.*, 75, 3123, 1999. With permission.)

Based on the trends shown in Figure 10.10 the effect of increasing energetic particle bombardment on the stress evolution in thin films can be categorized into two stages. The first stage is the build up of the tensile stress to a maximum and then rapid relaxation. The second stage, at higher energies, is the build up of compressive stress to a maximum and then a gradual relaxation.

The increase of the tensile stress to a maximum followed by rapid relaxation with increasing particle bombardment can be interpreted in terms of the accompanying densification of the film. Starting with a partially coalesced film, as the gap between the adjoining islands (typically, <1 nm) reduces, the interatomic forces increase tending to close the intercolumnar gap. The driving force for coalescence is the reduction of potential energy of the solid by reducing the interatomic distances in the voided intercolumnar region, at the expense of the elastic strain energy from stretching the film in biaxial tension. The elastic strain continues to increase until the islands coalesce completely to form a grain boundary. In other words, significant tensile stress may be generated even before the two free surfaces are replaced by a grain boundary. This basic concept was first proposed by Hoffman,[43] referred to as the grain boundary relaxation model, and discussed in detail recently by Machlin[48] and Nastasi et al.[26] The idea that energetic particle bombardment leads to the generation of tensile stress via a "densification" process where intercolumnar voids are closed is supported by TEM observations, Figure 10.11[44,49] and MD simulations.[25,27,29,50]

The compressive stress generation during energetic particle deposition is due to the production of irradiation-induced point defects that add positive volume to the film causing it to swell.[44,51–54] However, the constraint that the film must remain rigidly bonded to the substrate prevents any in-plane expansion of the film leading to compressive stress generation in the plane of the film. Developing a model for compressive stress generation, based on the above hypothesis, requires knowledge of defect concentrations and the magnitude of local expansion in the lattice introduced by the defects. Interstitials add positive volume to the material and vacancies add a negative volume. However, an interstitial-vacancy pair (Frenkel defect) has a net positive relaxation volume. Thus, compressive stress can be generated if the number fraction of interstitials induced by irradiation either equals or exceeds the number fraction of vacancies generated. The maximum values of the elastic residual stresses, tensile or compressive, are set by the plastic yield strength of the film (diffusive relaxation is not considered).

10.3.4 Grain Size

Grain sizes show complicated dependencies on increasing ion flux, ion energy, and substrate temperature. Many metals show a significant decrease in grain size with increasing ion-to-atom ratio, J_I/J_A, and increasing average energy deposited per atom, E_{ave}. Roy and coworkers[42,55] have performed comprehensive studies on the grain size of IBAD deposited Cu films as a function of temperature, J_I/J_A, and E_{ave}. The crystallite size of Cu films formed under Ar bombardment at energies of 62 and 600 eV and at substrate temperatures ranging between 62 and 230°C are shown as a function of ion-to-atom ratio

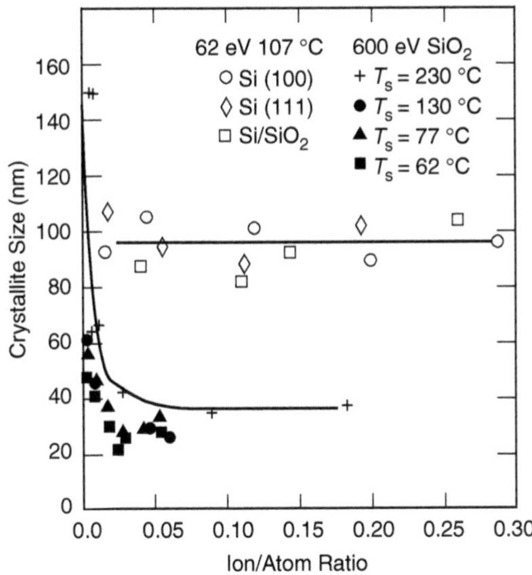

FIGURE 10.12 Crystallite size for evaporated copper films as a function of Ar-ion/Cu-atom ratio, Ar-ion energy, and substrate temperature. Film thicknesses range between 5 and 6 μm (After Roy, R.A., Cuomo, J.J. , and Yee, D.S., *J. Vac. Sci. Technol.*, A6, 1621, 1988. With permission.)

in Figure 10.12. Films deposited without an ion beam assist had grain sizes between 100 and 150 nm, depending on substrate temperature. Figure 10.12 shows that an ion-to-flux ratio of as little as 0.03 with 600 eV Ar ions reduced the grain size from 150 to 60 nm at 230°C. From Equation 10.2 we see that this value of J_I/J_A corresponds to an average energy deposited of $E_{ave} = 18$ eV/atom. For 600 eV bombardment at substrate temperatures of 62 to 103°C, the grain size saturates at about 30 nm for an ion-to-atom ratio of about 0.02 (i.e., $E_{ave} = 12$ eV/atom). At 230°C a slightly larger saturation size of 40 nm is observed at an ion-to-atom ratio of about 0.08 that corresponds to $E_{ave} = 48$ eV/atom. At 125 eV a trend similar to that observed at 600 eV was reported but with a saturated crystallite size of 70 to 80 nm. In contrast, at 62 eV the Cu crystallite size is slightly reduced but remains nearly constant ($\approx$100 nm) as a function of ion-to-atom ratio and substrate temperature.

At present no single mechanism has been confirmed to explain the grain size decrease with increasing ion bombardment. It is argued that Ar is incorporated at the grain boundaries in growing Cu films and that as J_I/J_A increases the Ar concentrations at the grain boundaries may reach levels sufficient to inhibit grain growth. In addition, as the beam energy increases the amount of irradiation-induced lattice disorder increases rapidly, which may also limit grain growth.

10.3.5 Texture

The orientation and the size of grains in films deposited under IBAD conditions can be a strong function of ion energy and the ion-to-atom flux ratio. An example from Nb films grown under 200 eV glancing angle (70° from normal incidence) Ar bombardment,[56,57] is shown in Figure 10.13. These films developed a (110) fiber texture with a restricted set of azimuthal orientations. The Nb films deposited without ion bombardment also showed a (110) fiber texture but without azimuthal ordering. As the data show, the degree of orientation (i.e., azimuthal ordering) increases with increasing ion-to-atom flux ratio. It was observed that the textured grains tended to be oriented so that the incident Ar ions are aligned along the (110) planar channeling direction. About half of the Nb grains are aligned to within 5° of the (110) planar channeling direction at a J_I/J_A ratio of 1.3. It was suggested that a low-energy channeling effect was responsible for the development of a preferred orientation.

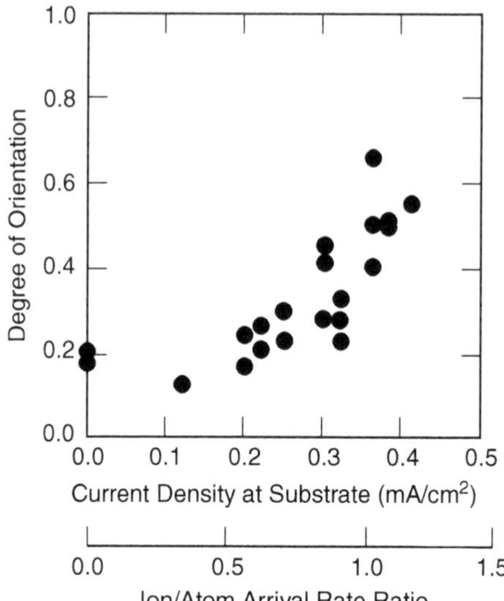

FIGURE 10.13 The degree of orientation of deposited Nb films as a function of glancing incidence ($\alpha = 70°$) ion bombardment current or ion-to-atom arrival ratio (After Yu, L.S., Harper, J.M.E., Cuomo, J.J., and Smith, D.A., Control of thin film orientation by glancing angle ion bombardment during growth, *J. Vac. Sci. Technol.*, A4, 443, 1986. With permission.)

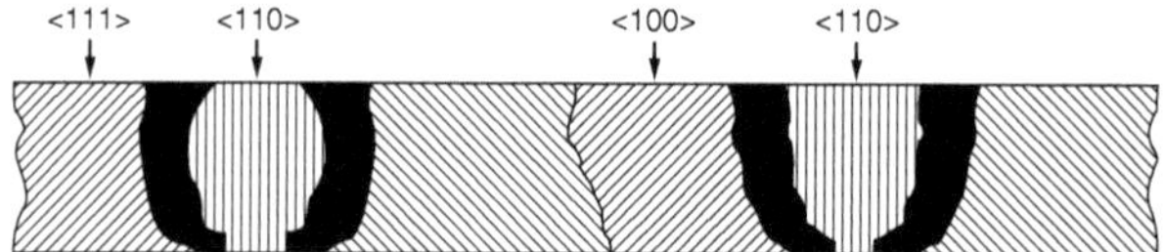

FIGURE 10.14 A schematic illustration of the way in which ⟨110⟩ texture develops in an fcc film grown under IBAD conditions. Grains with ⟨110⟩ planes parallel to the ion beam provide easy channeling and serve as seeds for epitaxial regrowth of the surrounding grains that are preferentially dissolved under ion bombardment (dark regions in the figure). (After Dobrev, D., *Thin Solid Films* 92, 41, 1982. With permission.)

A model of texturing in IBAD films due to preferential channeling of bombarding ions along the open crystalline channeling directions in the growing film has been developed by Dobrev.[58] Ions traveling in planar channels lose energy primarily through electronic energy loss processes, and as a result the ion's nuclear energy deposition at the film surface would be inversely proportional to the planar spacing. For fcc crystals, the ease of planar channeling is in the order of most open planes (i.e., large planar spacing): ⟨110⟩, ⟨200⟩, ⟨111⟩. Dobrev suggested that in a polycrystalline film, the crystallites with the most open channeling directions (easiest channeling direction) aligned with the ion beam would experience the least nuclear energy loss and thus remain the *coolest* and thereby experience the lowest lattice disorder. These grains would then serve as the seeds for the regrowth and of neighboring grains, which were initially not well aligned with the ion beam and therefore would experience high levels of nuclear energy deposition (thermal spike) and radiation damage. This spike mechanism of texture development is schematically displayed in Figure 10.14.

An alternate model based on the differences in sputtering yields at different crystallographic orientations has been proposed by Bradley et al.[59] to explain the development of texturing during IBAD. This model assumes that the film is deposited such that one crystal axis is normal to the film and that the azimuthal orientations are random. The orientation sputtering difference, which can be as high as a factor of 5 in some materials, leads to the preferential removal of grains oriented with a high sputtering yield and the newly deposited material grows epitaxially on the low sputtering yield orientations.

Calculation from this sputtering based model reproduced the general features of the Nb data from Yu et al.[56]

Both the IBAD texture models presented above, the spike regrowth model of Dobrev, and the preferential sputtering model of Bradley et al., are based on the same ion bombardment phenomenon; the deposited energy density due to nuclear stopping varies inversely with the openness of the channeling orientations. The sputtering yield is directly proportional to the nuclear stopping. Thus, we expect both reduced sputtering and the absence of dense cascades when the bombarding ion is aligned along planes with large interplanar spacings. Based on these arguments, Smidt[14] has summarized typical IBAD texture behavior. In general, thin films deposited by evaporation processes, without ion beam assistance, will normally be deposited with planes of highest atomic density parallel to the substrate surface. This crystallographic orientation corresponds to a fast growth geometry and will only be obtained when substrate temperatures are high enough for the depositing atoms to have surface mobility, that is, $T_s > 0.15\ T_m$. Under such conditions fcc films have a $\langle 111 \rangle$ texture, bcc films have a $\langle 110 \rangle$ texture, and hcp films with an ideal c/a ratio will have a $\langle 0002 \rangle$ texture. During IBAD processing, the crystallographic orientation will be shifted so that widely spaced planes (i.e., planes of easiest planar channeling) will be aligned along the ion beam axis.[26]

Based on the above discussion we expected that the $\langle 111 \rangle$ texture in an evaporated fcc film will be shifted toward a $\langle 110 \rangle$ for an IBAD process with the ion beam aligned with the film normal. For an IBAD process with an off-normal incident ion beam, different textures will be produced.

10.3.6 Epitaxy

Epitaxy is the phenomenon of growing a single crystalline film that is coherently oriented with the underlying single crystal substrate (for more details see Chapter 9). As the growth of an epitaxial film is kinetically limited, an optimum temperature, the epitaxial temperature T_e, will exist for a film/substart pair. The concurrent ion bombardment during the growth of thin films has been shown to have a strong influence on the film's epitaxial quality. In many systems a lowering of the epitaxial temperature T_e by an amount ΔT_e from its non-IBAD value is observed. This topic has been recently reviewed by Atwater[60] and Herbots et al.[10] A partial summary of experimental results is given in Table 10.1. In many systems a lowering of the epitaxial temperature T_e by an amount ΔT_e from its non-IBAD value is observed.[26]

Several mechanisms are proposed for the ion induced enhancement in film epitaxy[14] that include: (1) the removal of native surface oxide that makes the homoepitaxy possible in self-systems (i.e., Si vapor on Si substrates), (2) the addition of surface damage that provides heterogeneous nucleation sites, and (3) the addition of energy to stimulate the surface diffusion of the deposited atom. The effect of this latter mechanism of enhanced epitaxy has been examined for the IBAD growth of Ni films on zero temperature Ni substrates (homoepitaxy) under Ar ion bombardment by Müller[29] using molecular dynamic simulations. An example of the MD calculated microstructure that evolves for 50 eV Ar^+ ions at a bombardment angle of 30° is shown in Figure 10.7. The influence of IBAD parameters J_I/J_A and E_{ion} on the degree of homoepitaxy α_e is presented in Figure 10.15. The parameter α_e is defined such that a value of

TABLE 10.1 IBAD Enhanced Epitaxy

Vapor	Substrate	Ion/energy (eV)	T_e (K)	ΔT_e	Reference
Si	Si	Si/100	620	100	61
Si	Al_2O_3	Si/100	700	150	61
Ge	Si	Ge/50	503	270	62
Ge	Si	Ge/50	400	375	63
Ge	Ge,Si	Ge/100	573		64
Si	Si	Si/50	550	375	4
Sb	NaCl	Ne,Ar/400	423	80	65
Ag	Si	Ag, Si/25–100	300		62

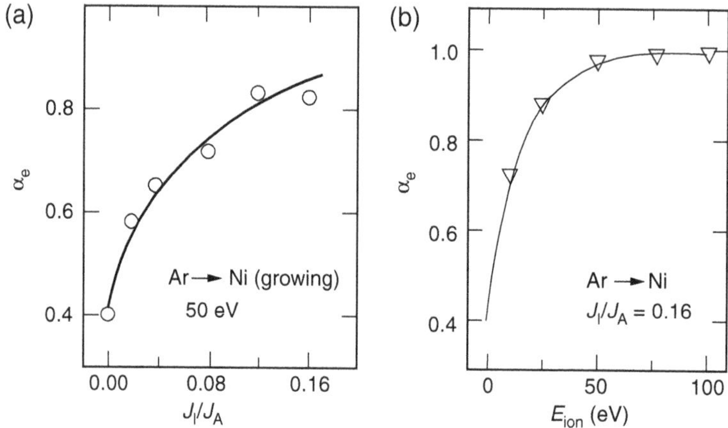

FIGURE 10.15 The degree of Ni homoepitaxy, α_e, determined from molecular dynamic simulations. (a) The dependence of α_e on Ar-ion-to-Ni-atom flux ratio at an Ar ion energy of 50 eV. (b) The dependence of α_e on Ar ion energy for an ion-to-atom ratio of 0.16. The parameter α_e gives a measure of homoepitaxy and is defined so that a value of $\alpha_e = 0.2$ implies no homoepitaxy while a value of $\alpha_e = 1$ indicates perfect epitaxy (After Müller, K.-H., *Phys. Rev. B*, 35, 7906, 1987. With permission.)

$\alpha_e = 0.2$ implies no homoepitaxy while a value of $\alpha_e = 1$ indicates perfect epitaxy. The data presented in Figure 10.15a indicate that the epitaxy improves for increasing J_I/J_A for E_{ion} fixed at 50 eV. Figure 10.15b shows that under a constant ion-to-atom flux ratio of 0.16 the degree of epitaxy rises sharply at low ion energies and approaches a plateau at $E_{ion} \approx 50$ eV.

In the epitaxial growth process, the migration and ordering of the atom deposited on the surface is critical to the epitaxial quality, and the displacement of atoms by concurrent low-energy ion irradiation is expected to play an important role. However, ion irradiation can induce a substantial amount of radiation damage and disorder in the bulk which in turn can promote the formation of metastable phases, such as the amorphous phase. Brice et al.[66] have used an analytical approach to model the partitioning of surface and bulk displacements for low-energy ions incident on a solid. They view the importance of ion beams in aiding epitaxy by being able to provide an appropriate density of mobile surface defects while avoiding damage in the bulk.

10.3.7 Adhesion

The adhesion of thin films on substrates depends on a variety of factors including: interface chemistry, film stress, differential thermal expansion, contaminant levels at the interface, and surface morphology (see more details in Chapter 19). This area has been studied extensively by Baglin.[67] *In situ* vacuum ion beam cleaning offers an excellent means of preparing substrates for coating by removing contaminant layers (e.g., adsorbed water, hydrocarbons, oxides, etc.) and sometimes by selective removal of surface material (texturing) to leave a favorable high bonding surface for either chemical or morphology (texture) influenced adhesion. The latter has often been observed for metal to polymer (e.g., Teflon) bonding improvements.

The use of high energy ions to mix or "stitch" a metallization onto a substrate has been realized for many years. This technique however requires an accelerator capable of producing ions of high enough energy to penetrate the film to be stitched with the substrate that limits practical film thicknesses to tens to hundreds of nanometers for medium mass ions of energies up to 200 keV.

Low-energy ions such as used in conventional IBAD processing (ion energies typically less than 2 keV) also appear to work effectively for enhancing adhesion. For example, Baglin and colleagues have previously conducted many adhesion studies of Cu on Al_2O_3 by using high energy ions of He and Ne to stitch

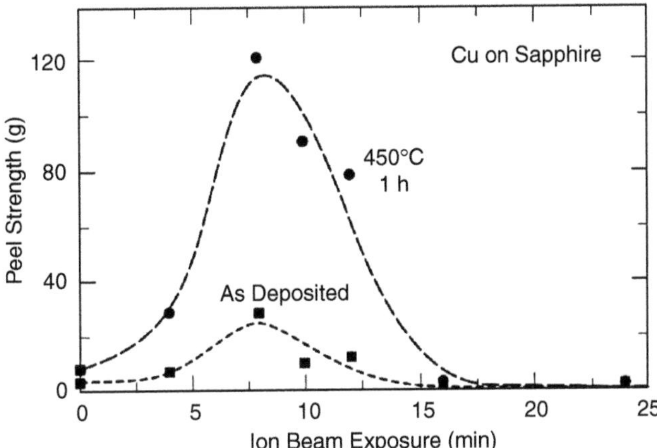

FIGURE 10.16 Adhesion for Cu film on Al_2O_3 substrate by means of presputtering the substrate prior to deposition of Cu. Sputtering was done with Ar^+ ions (500 eV, 50 μA/cm^2). (After Baglin, J., in *Handbook of Ion Beam Processing Technology*, J.J. Cuomo, S.M. Rossnagel, and H.R. Kaufman, Eds., Noyes Publications, Park Ridge, NJ, 1989, Chapter 14. With permission from Noyes.)

the films.[68] They found that the improved adhesion could not simply be attributed to a physical intermixing of Cu and Al_2O_3 . In later studies[62] similar results were achieved by preexposing the Al_2O_3 to a low energy argon beam (500 eV, 50 μA/cm^3) and then depositing the copper. It was observed, however, that there was an optimum sputtering dose requiring a time (7–8 min) for the given beam conditions, to obtain optimum adhesion as can be seen in Figure 10.16. This effect was attributed to preferential sputtering of oxygen from Al_2O_3, such that at a given composition, (Al_2O_{3-x}), copper will most strongly react to form stable ternary bonding configurations, or interface phases. X-ray photoelectron spectroscopy (XPS) analysis of a sample prepared under these same conditions yielding the best adhesion showed a new line, not identifiable with known binary compounds, and this line is thought to correspond to a ternary compound. Baglin[62] notes that Cu/Al_2O_3 samples prepared using presputtering showed no systematic increases in adhesion after higher energy "stitching" experiments using Ne or He ion at 250 keV, implying that bombardment reconfigures interface chemical bonds. Kellock et al.[69] have reported on the adhesion improvement of Au-GaAs interfaces by the use of low-energy (300 eV) ion beams at low temperatures (120 to 150°C). The adhesion depends critically on the substrate temperature and on the arrival ratio of Ar ions to Au atoms at the substrate. They postulate that the ion beam serves to enable a displacement reaction in the GaAs surface layer whereby As is liberated and a new equilibrium complex is formed at the surface.

Another example of a chemically induced adhesion enhancement involves the adhesion of gold onto glass substrates for optical reflectors. Martin et al.[70] found that IAD of Au onto glass with an Ar^+ beam made only an insignificant improvement, whereas the use of an oxygen beam by itself or an oxygen plus argon ion beam made adhesion improvements of 100 times to over 400 times that of nonassisted or argon-assisted deposition. The relative film adhesion was assessed with a diamond tipped scratch tester. The improvement was attributed to some (unidentified) form of chemical bonding.

There appears to be clear evidence for the need to produce a broad interfacial region between layers for good adhesion if there is no or little chemical affinity between them.[45] For IBAD processing this would require high energy beams. However, the substrate involved also has an important bearing on the appropriate ion energy to use for improving adhesion. For example, Ebe et al.[71] found that the adhesion of Cu films on polyimide substrates was better for 0.5 keV Ar ions than for 5 or 10 keV ions. This was attributed to carbonization of the polyimide at the higher energies. Another approach that has been successfully used is to use an intermediate layer between a chemically incompatible film and substrate, including metal on polymer systems.[69] Ideally, this *adhesor* layer would form strong bonds with both the outermost film and the substrate. This concept has been reviewed by Baglin.[67]

TABLE 10.2 Reactive IBAD

Material	Vapor	Ion/energy (eV)	J_I/J_A	Temp (K)	References
ZrO_2	ZrO_2	$O_2^-/1200$		550	72
TiO_2	TiO_2	$O_2^-/1200$	0.1–0.4	450	73
SiO_2	SiO	$O^+, O_2^-/300{,}500$	0.25–1.70	325–550	72
Al_2O_3	Al	$O_2^-/500$	0.16	300	80
Ta_2O_5	Ta_2O_5	$O_2^-/1200$	2.6	575	74
Si_3N_4	Si	$N_2^-/60, 100$	2.1	300	75
AlN	Al	$N_2^-/250\text{-}1000$	0.5–2.7	375	76
TiN	Ti	$N_2^-/1000$	0.01–0.03		77
BN	B	$N_2^-/25\text{–}40{,}000$	0.7	475	78

These measurements point to the complexity of adhesion and how effective ion beam techniques can be in enhancing adhesion. However, these examples also show how each system must be evaluated individually with respect to optimizing adhesion.

10.4 Reactive IBAD Processing: Compound Synthesis

Compound synthesis by the IBAD process is accomplished by adding one or more of the compound components in the vapor flux and adding one or more of the components in the ion beam flux. The use of reactive IBAD to form nitrides, carbides, and oxides offers a high degree of control of the film's composition while also maintaining the ability to induce IBAD modification of the film microstructure. Additional information about reactive IBAD can be found in Table 10.2 and the following References 26, and 79 to 82.

References

[1] Mattox, D.M., Ion plating, in *Handbook of Deposition Technologies for Film and Coatings*, Bunshah, R.F., Ed., Noyes Publications, Park Ridge, NJ, 1994. Chapter 6.

[2] Bunshah, R.F., Evaporation, in *Deposition Technologies for Film and Coatings*, by Bunshah, R.F., Ed., Noyes Publications, Park Ridge, NJ, 1982, Chapter 4.

[3] Weissmantel, C., Reisse, G., Erler, H-J., Henny, F., Bewilogua, K., Ebersbach, U., and Schurer, C., Preparation of hard coatings by ion beam methods, *Thin Solid Films*, 63, 315, 1979.

[4] Pranevicius, L., Structure and properties of deposits grown by ion-beam-activated vacuum deposition techniques, *Thin Solid Films*, 63, 77, 1979.

[5] Harper, J.M.E., Cuomo, J.J., Gambino, R.J., and Kaufman, H.E., Modification of thin film properties by ion bombardment during deposition, in *Ion Bombardment Modification of Surfaces: Fundamentals and Applications*, Auciello, O., and Kelly, R., Eds., Elsevier Science Publ., Amsterdam, 1984, Chapter 4.

[6] Harper, J., Particle bombardment effects in thin film deposition, in *Plasma-Surface Interactions and Processing of Materials*, Auciello, O., Gras-Marti, A., Valles-Abarca, J. A., and Flamm, D. L., Eds., Kluwer Academic Publishers, The Netherlands, 1990, p. 251.

[7] Takagi, T. and Yamada, I., Ionized cluster beam deposition, in *Ion Beam Assisted Film Growth*, Itoh, T., Ed., Elsevier, Amsterdam, 1989, Chapter 7B.

[8] Yamada, I. and Takaoka, G.H., Ionized cluster beams: Physics and technology, *Jpn. J. Appl. Phys.*, 32, 2121, 1993.

[9] Brown, W.L., Jarrold, M.F., McEachern, R.L., Sosnowski, M., Takaoka, G., Usui, H., and Yamada, I., Ion cluster beam deposition of thin films, *Nucl. Instrum. Methods Phys. Res.* B, 59/60, 182, 1991.

[10] Herbots, N., Hellman, O.C., Vancauwenberghe, O., Ye, P., and Wang, X., Chemical reactions and surface modifications stimulated by low energy ions, in *Low Energy Ion Surface Interaction*, Wayne Rabalais, J.W., Ed., John Wiley & Sons, New York, 1994.

[11] Mattox, D.M., Particle bombardment effects on thin-film deposition, *J. Vac. Sci. Technol.* A7, 1105, 1989.

[12] Cuomo, J.J., Rossnagel, S.M., and Kaufman, H.H., Eds., *Handbook of Ion Beam Processing Technology*, Noyes Publications, Park Ridge, NJ, 1989.

[13] Cuomo, J.J. and Rossnagel, S.M., Property modification and synthesis by low energy particle bombardment concurrent with film growth, *Nucl. Instrum. Method. Phys. Res.*, B19/20, 963, 1987.

[14] Smidt, F.A. Use of ion beam assisted deposition to modify the microstructure and properties of thin films, *Int. Mater. Rev.*, 35, 61, 1990.

[15] Hirvonen, J.K., Ion beam assisted thin film deposition, *Mater. Sci. Reports*, 6, 215, 1991.

[16] Grovenor, C.R.M., Hentzell, H.T.G., and Smith, D.A., The development of grain structure during growth of metallic films, *Acta Metall.*, 32, 773, 1984.

[17] Galuska, A.A., Ion-induced adhesion enhancement of Ni film on polyester: Silicon intermediate layer and Kr+ implantation, *Nucl Instrum. Meth. Phys. Res.*, 44, 428, 1990.

[18] Miyake, K. and Tokuyama, T., Direct ion beam deposition, in *Ion Beam Assisted Film Growth*, Itoh, T., Ed., Elsevier, Amsterdam, 1989, Chapter 8.

[19] Greene, J.E., Barnett, S.A., Sundgren, J.-E., and Rockett, A., Low energy ion/surface interaction during film growth from the vapor phase, in *Ion Beam Assisted Film Growth*, Itoh, T., Ed., Elsevier, Amsterdam, 1989, Chapter 5.

[20] Tu, K.-N., Mayer, J.W., and Feldman, L.C., *Electronic Thin Film Science for Electrical Engineers and Materials Scientists*, Macmillan Publishing Company, New York, 1992.

[21] Thornton, J.A., Coating deposition by sputtering, in *Deposition Technologies for Film and Coatings*, Bunshah, R.F., Ed., Noyes Publications, Park Ridge, NJ, 1982, Chapter 5.

[22] Messier, R., Giri, A.P., and Roy, R.A., Revised structure zone model for thin film physical structure, *J. Vac. Sci. Technol.*, A2, 500, 1984.

[23] Movchan, B.A. and Demchishin, A.V., Investigation of the structure and properties of thick vacuum deposited films of nickel, titanium, tungsten, alumina and zirconium dioxide, *Fiz. Metallov Metalloved.*, 28, 653, 1969.

[24] Yehoda, J.E., Vedam, B., and Messier, R., Investigation of the void structure in amorphous germanium thin films as a function of low energy ion bombardment, *J. Vac. Sci. Technol.*, A6, 1631, 1988.

[25] Müller, K.-H., Monte carlo calculations for structural modification in ion assisted thin film deposition due to thermal spikes, *J. Vac. Sci. Technol.*, 4, 184, 1986.

[26] Nastasi, M., Mayer, J.W., and Hirvonen, J.K., *Ion-Solid Interactions: Fundamentals and Applications*, Cambridge University Press, Cambridge, 1996.

[27] Müller, K.-H., Model for ion-assisted thin film densification, *J. Appl. Phys.*, 59, 2803, 1986.

[28] Ziegler, J.F., Biersack, J.P., and Littmark, U. *The Stopping and Range of Ions in Solids*, Pergamon Press, New York, 1985.

[29] Müller, K.-H., Ion-beam-induced epitaxial vapor-phase growth: A molecular dynamics study, *Phys. Rev. B*, 35, 7906, 1987.

[30] Thompson, C.V. and Carel, R., Grain growth and texture evolution in thin films, *J. Mech. Phys. Solids*, 44, 657, 1996.

[31] Cammarata, R.C., Trimble, T.M., and Srolovitz, D.J., Surface stress model for intrinsic stresses in thin films, *J. Mater. Res.*, 15, 2468, 2000.

[32] Shull, A.L. and Spaepen, F., Measurements of stress during vapor deposition of copper and silver thin films and multilayers, *J. Appl. Phys.*, 80, 6243, 1996.

[33] Ramaswamy, V., Clemens, B.M., and Nix, W.D., Stress evolution during growth of sputtered Ni/Cu multilayers, *Mater. Res. Soc. Symp. Proc.*, 528, 161, 1998.
[34] Phillips, M.A., Ramaswamy, V., Clemens, B.M., and Nix, W.D., Stress and microstructure evolution during initial growth of Pt on amorphous substrates, *J. Mater. Res.*, 15, 2540, 2000.
[35] Floro, JA., Hearne, S.J., Hunter, J.A., Kotula, P., Chason, E., Seel, S.C., and Thompson, C.V., The dynamic competition between stress generation and relaxation mechanisms during coalescence of Volmer-Weber thin films, *J. Appl. Phys.*, 89, 4886, 2001.
[36] Seel, S.C., Thompson, C.V., Hearne, S.J., and Floro, J.A., Tensile stress evolution during deposition of Volmer-Weber thin films, *J. Appl. Phys.*, 88, 7079, 2000.
[37] Adams, D.P., Parfitt, L.J., Bilello, J.C., Yalisove, S.M., and Rek, Z.U., Microstructure and residual stress of very thin Mo films, *Thin Solid Films*, 266, 52, 1995.
[38] Nix, W.D. and Clemens, B.M., Crystallite coalescence: A mechanism for intrinsic tensile stresses in thin films, *J. Mater. Res.*, 14, 3467, 1999.
[39] Sheldon, B.W., Lau, K.H.A., and Rajamani, A., Intrinsic stress, island coalescence, and surface roughness during the growth of polycrystalline films, *J. Appl. Phys.*, 90, 5097, 2001.
[40] Freund, L.B. and Chason, E., Model for stress generated upon contact of neighboring islands on the surface of a substrate, *J. Appl. Phys.*, 89, 4866, 2001.
[41] Chason, E., Sheldon, B.W., Freund, L.B., Floro, J.A., and Hearne, S.J., Origin of compressive residual stress in polycrystalline thin films, *Phys. Rev. Lett.*, 88, 156103, 2002.
[42] Roy, R.R. and Yee, D.S., Control of film properties by ion-assisted deposition using broad beam sources, in *Handbook of Ion Beam Processing Technology*, Cuomo, J.J., Rossnagel, S.M., and Kaufman, H.H., Eds., Noyes Publications, Park Ridge, NJ, 1989, Chapter 11.
[43] Hoffman, R.W., Stresses in thin films: The relevance of grain boundaries and impurities, *Thin Solid Films*, 34, 185, 1976.
[44] Misra, A. and Nastasi, M., Limits of residual stress in Cr films sputter deposited on biased substrates, *Appl. Phys. Lett.*, 75, 3123, 1999.
[45] Wolf, G.K., Modification of chemical properties by ion beam assisted deposition, *Nucl. Instrum. Methods Phys. Res.* B, 46, 369. 1990.
[46] Cuomo, J.J., Harper, J.M.E., Guarnieri, C.R., Yee, D.S., Attanasio, L.J., Angilello, J.C., Wu, R., and Hammond, R.H., Modification of niobium film stress by low energy ion bombardment during deposition, *J. Vac. Sci. Technol.*, 20, 349, 1982.
[47] Brighton, D.R. and Hubler, G.K., Binary collision cascade prediction of critical ion-to-atom arrival ratio in the production of thin films with reduced intrinsic stress, *Nucl. Instr. Meth.*, B28, 527, 1987.
[48] Machlin, E.S., *Materials Science in Microelectronics — The Relationships Between Thin Film Processing and Structure*, GIRO press, New York, 1, 157, 1995.
[49] Misra, A., Fayeulle, S., Kung, H., Mitchell, T.E., and Nastasi, M., Effects of ion irradiation on the residual stresses in Cr thin films, *Appl. Phys. Lett.*, 73, 891, 1998.
[50] Müller, K.-H., Stress and microstructure of sputter-deposited thin films: Molecular dynamics investigation, *J. Appl. Phys.*, 51, 1799. 1987.
[51] Misra, A. and Nastasi, M., Intrinsic residual stresses in metal films synthesized by energetic particle deposition, *Nucl. Instrum. Methods Phys. Res. B*, 175/177, 688, 2001.
[52] Windischmann, H., An intrinsic stress scaling law for polycrystalline thin films prepared by ion beam sputtering, *J. Appl. Phys.*, 62, 1800, 1987.
[53] Davis, C.A., Simple model for the formation of compressive stress in thin films by ion bombardment, *Thin Solid Films*, 226, 30, 1993.
[54] Knuyt, G., Lauwerens, W., and Stals, L.M., Unified theoretical model for tensile and compressive residual film stress, *Thin Solid Films*, 370, 232, 2000.
[55] Roy, R.A., Cuomo, J.J., and Yee, D.S., Control of microstructure and properties of copper films using ion-assisted deposition, *J. Vac. Sci. Technol.*, A6, 1621, 1988.

[56] Yu, L.S., Harper, J.M.E., Cuomo, J.J., and Smith, D.A., Alignment of thin films by glancing angle ion bombardment during deposition, *Appl. Phys. Lett.*, 47, 932, 1985.

[57] Yu, L.S., Harper, J.M.E., Cuomo, J.J., and Smith, D.A., Control of thin film orientation by glancing angle ion bombardment during growth, *J. Vac. Sci. Technol.*, A4, 443, 1986.

[58] Dobrev, D., Ion-beam-induced texture formation in vacuum-condensed thin metal films, *Thin Solid Films*, 92, 41, 1982.

[59] Bradley, R.M., Harper, J.M.E., and Smith, D.A., Theory of thin film orientation by ion bombardment during deposition, *J. Appl. Phys.*, 60, 4160, 1986.

[60] Atwater, H.A., Low energy ion-solid interactions during epitaxial growth, *Solid State Phenom.*, 27, 67, 1992.

[61] Narusawa, T. et al., Simultaneous Rheed-AES-QMS study on epitaxial Si film growth in Si (111) and Sapphire (1102) surfaces by partially ionized vapour deposition, *J. Vac. Sci. Technol.*, 16, 366, 1979.

[62] Thomas G.E. et al., Ion beam epiplantation, *J. Crystal Growth*, 56, 557, 1982.

[63] Zalm, P.C., Ion beam epitaxy of Si on Ge and Si at 400 K, *Appl. Phys. Lett.*, 41, 167, 1982.

[64] Yagi, K. et al., Ge and Si film growth by low energy ion beam deposition, *Jpn. J. Appl. Phys.*, 16, 245, 1977.

[65] Babaev V.O. et al., Effect of ion irradiation on the formation, structure and properties of thin films, *Thin Solid Films*, 38, 1, 1976.

[66] Brice, D.K., Tsao, J.Y., and Picraux, S.T., Partitioning of ion-induced surface and bulk displacements, *Nucl. Instrum. Methods Phys. Res.* B, 44, 68, 1989.

[67] Baglin, J., Interface structure and thin film adhesion, in *Handbook of Ion Beam Processing Technology*, J.J. Cuomo, S.M. Rossnagel, and H.R. Kaufman, Eds., Noyes Publications, Park Ridge, NJ, 1989, Chapter 14.

[68] Baglin, J.E.E. and Clark, G.J., Ion beam bonding of thin films, *Nucl. Instrum. Methods Phys. Res.*, B 7/8, 881, 1985.

[69] Kellock, A.J., Baglin, J.E.E., and Barlin, T.T., Adhesion improvement of Au on GaAs using ion beam assisted deposition, *Nucl. Instrum. Methods Phys. Res.* B, 59/60, 249, 1991.

[70] Martin, P.J., Ion-enhanced adhesion of thin gold films, *Gold Bull.* 19, 102, 1986.

[71] Ebe, A., Kuratani, N., Nishiyama, S., Imai, O., and Ogata, K., Metallization on polyimide film by ion and vapor deposition (IVD) method, *Trans. Mater. Res. Soc. Jpn.*, 17, 431, 1994.

[72] McNally, J. et al., Properties of optical thin films deposited using ion assisted deposition, *Proc. SPIE*, 678, 151, 1986.

[73] Williams, F.L. et al., Optical coatings deposited using ion assisted deposition, *J. Vac. Sci. Technol.*, A5, 2159, 1987.

[74] Williams, F.L. et al., Optical characteristics of thin films deposited at low temperature using ion assisted deposition, *J. Vac. Sci. Technol.*, A6, 2020, 1988.

[75] Netterfield, R.P. et al., Synthesis of silicon nitride and silicon oxide films by ion-assisted deposition, *Appl. Opt.*, 25, 3808, 1986.

[76] Targove, J.D. et al., Preparation of aluminium nitride and oxynitride thin films by ion-assisted deposition, *Mater. Res. Soc. Symp. Proc.*, 93, 311, 1987.

[77] Kant, R.A. et al., The structure and properties of Ni films grown by ion-beam assisted deposition, *Mater. Res. Soc. Symp. Proc.*, 128, 427, 1989.

[78] Satou, M. et al., Nitride film formation by ion and vapor deposition, *Nucl. Instrum. Methods* B, 7/8, 910, 1985.

[79] Erler, H.-J., Reisse, G., and Weissmantel, C., Nitride deposition by reactive ion beam sputtering, *Thin Solid Films*, 65, 233, 1980.

[80] Netterfield, R.P., Muller, K.H., McKenzie, D.R., Goonan, M.J., and Martin, P.J., Growth dynamics of aluminum nitride and aluminum oxide thin films synthesized by ion-assisted deposition, *J. Appl. Phys.*, 63, 760, 1988.

[81] Hubler, G.K., van Vechten, D., Donovan, E.P., and Carosella, C.A., Fundamentals of ion-beam-assisted deposition. II. absolute calibration of ion and evaporant fluxes, *J. Vac. Sci. Technol.*, A8, 831, 1990.
[82] Van Vechten, D., Hubler, G.K., Donovan, E.P., and Correll, F.D., Fundamentals of ion-beam-assisted deposition. I. Model of process and reproducibility of film composition, *J. Vac. Sci. Technol.*, A8, 821, 1990.

11

Spray Deposition and Coating Processes

Jean-Pierre Delplanque, Samuel Johnson, and Yizhang Zhou
University of California, Davis

Leon Shaw
University of Connecticut

Abstract

This chapter offers a concise overview of the aspects of spray deposition and coating processes that can shed some light on the processing–structure–property relationship. It is organized as follows. First, specific examples of spray deposition and coatings techniques are outlined in Section 11. 2 to illustrate the variety of the implementations that have been developed from the basic concept. Section 11.3 focuses on the process stages that characterize spray techniques: spray flow and spray deposition. Injection is not discussed here because, in the case of solid particles (powders), injection can be considered a part of the spray flow. Microstructure evolution and resulting properties of spray deposits and coatings are treated in Section 11.4. Finally, some comments regarding potential future developments in this area are also provided in Section 11.5. Multiple references to the relevant literature are included throughout the chapter to allow the reader to explore specific topics in more detail than this overview can provide.

11.1 Overview of Spray Processing

In general, spray processes aim at taking advantage of the increased surface area that results from breaking up a bulk material, either liquid (during processing) or solid (prior to processing), into a large number of smaller pieces (particles/droplets). As heat and mass transfer (including chemical reactions) between two different physical phases (e.g., liquid/gas, solid/gas) occur at the interface between the two phases, increasing the surface area of that interface enhances the rate at which these exchanges can occur. A case in point is that of a fuel injector; for given ambient conditions (temperature and pressure), the time required to evaporate a given mass of liquid fuel would be much larger if the liquid fuel was left in bulk than if it was broken up into small droplets (i.e., 50 μm and below). Spray-based materials processes are no exception; a dispersed liquid or solid phase enables heat transfer rates that would not be attainable with the bulk material.

Spray-based materials processes encompass a wide variety of techniques used to produce bulk materials or coatings by projection and consolidation of a dispersed phase onto a target. These processes form a subset of the family of aerosol processes (see Kodas and Hampden-Smith's exhaustive reference on that topic).[1] The scope of this chapter is limited to spray-deposition processes and thermal spray processes. Solid-state spray processes are also briefly discussed. These processes typically comprise three major stages: injection of the disperse phase (including jet breakup or atomization if that phase is liquid), transport of the resulting spray (both spray flow and deposition on the target), and consolidation. Techniques in which the dispersed phase is nucleated from the flowfield (e.g., flame synthesis) or precipitated from a possibly dispersed solution (e.g., spray pyrolisis or spray drying) are not discussed here. Over the past few decades, the importance of these nonequilibrium processing techniques has considerably grown because of their efficiency, the flexibility that they provide both in terms of microstructural control and *in situ* processing, and their relatively low environmental impact (scrap reduction and energy savings compared to more traditional processing techniques such as ingot metallurgy).[2] The practical success of these methods should not conceal the fact that these are complex processes (multidimensional multiphase turbulent reactive flow with heat transfer and phase change) and that despite the significant progress made in recent years there is still much to learn regarding the processing–structure–property relationship.

11.2 Spray Deposition and Coating Strategies

The processes discussed herein use the same building blocks: a disperse phase, a heat source (except for solid-state sprays), a carrier flow, and a target. They differ in the type of heat source used (e.g., Joules heating, induction, combustion, electric arc, or none), the magnitude of the carrier flow velocity, and whether the disperse phase is injected liquid or solid in the process.

11.2.1 Spray Deposition

In this section, the basic principles of spray-deposition processes are outlined. The comprehensive monograph by Lavernia and Wu[2] should be consulted for a detailed treatment.

11.2.1.1 Spray Atomization and Deposition

Spray-deposition processes start from a solid bulk material that is heated (inductively or conductively) in a crucible (Figure 11.1). Aluminum and iron alloys make up the bulk of spray-deposited materials but copper and magnesium alloys as well as superalloys have also been successfully processed using this technique. The melt is disintegrated into a fine dispersion of droplets to form a spray (dispersed liquid phase) using high-energy inert gas jets (e.g., Ar, He, N_2) with velocities ranging from subsonic (50 m/s) to supersonic (up to 500 m/s).[2,3] During their trajectory toward the target (stationary or mobile), the droplets undergo convective cooling, partial rapid solidification, and possibly chemical reaction with the surrounding atmosphere. They finally impinge on the target (ideally while still in a partially solidified state) where they consolidate into a deposit (or preform). High rates of solidification are achieved both in flight (fast heat extraction by convection) and at impact as a result of the small thickness of the splats.

The effective cooling rate that may be achieved in spray deposition is on the order of 10^3 to 10^4 K/s, which is several orders of magnitude higher than conventional ingot solidification. A number of variations on this base have been explored by modifying the type of substrate (flat translating for the spray forming

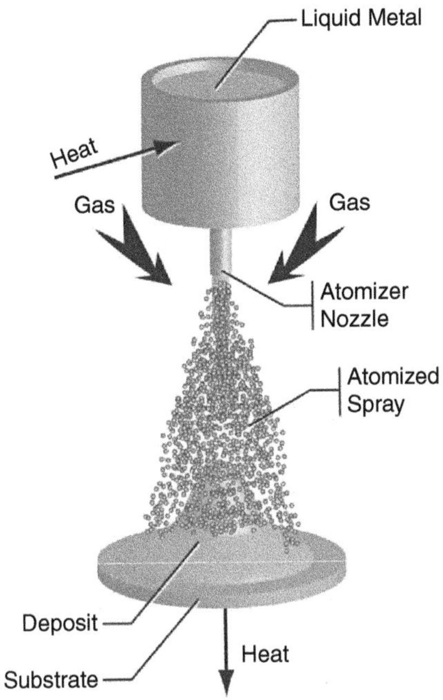

FIGURE 11.1 Sketch of the spray atomization and deposition process.

of sheets, flat rotating for the spray forming of billets, cylindrical rotating for the spray forming of tubes); co-injecting ceramic particles (spray atomization and codeposition of Metal Matrix Composites, MMCs, see Section 11.4.1.3) or shots (spray peening); or adding in-line postprocessing such as heating and forging (spray forging) or twin-roll casting (spray rolling, see Section 11.2.1.2). More details about these specific techniques are available in Lavernia and Wu's book[2] and in related references. Tubes, billets, and disks are the most common preform shapes of spray-deposited materials.[4] With a mobile substrate, continuous sheet or strip with predesigned thicknesses may also be produced.[5] It is worth noting that spray deposition has also been successfully applied in the manufacture of tool steel dies and molds with enhanced performance.[6]

Spray deposition allows microstructure refinement (grain size, uniformity, and intermetallic/secondary phases) and extended solid solubility limits characteristic of nonequilibrium solidification. This can result in improvement of the mechanical properties of spray-deposited materials. For example, room temperature yielding strength and ultimate tensile strength of the sprayed formed 2024 Al alloy can be increased by 27 and 25%, respectively, compared to its counterpart made by ingot metallurgy (IM).[7] Porosity had been a concern in the early developments of spray deposition processes but recently densities greater than 98% of the theoretical value have been achieved.[8–10] Overspray, droplets that do not hit the substrate or bounce off from deposition surface, is still a major concern in the commercial development of spray-deposition processes; research is ongoing for a goal of zero overspray.

11.2.1.2 Spray Rolling

Spray rolling was originally used to describe the process imagined by Singer.[5] In that process, a spray of liquid metal droplets was formed by breaking up ("atomizing") a liquid metal jet. The spray was projected onto a substrate to consolidate into a strip that was then rolled. Recently, a spray rolling process that combines spray deposition with twin-roll casting has been developed for the production of aluminum strips/sheets.[11] In this process, molten aluminum is atomized into a spray of droplets that is directed to the nip of a twin-roll caster (Figure 11.2). The deposit thus formed on the rotating rolls is consolidated and rolled into a strip. Results obtained at the laboratory scale have shown that spray rolled materials exhibit

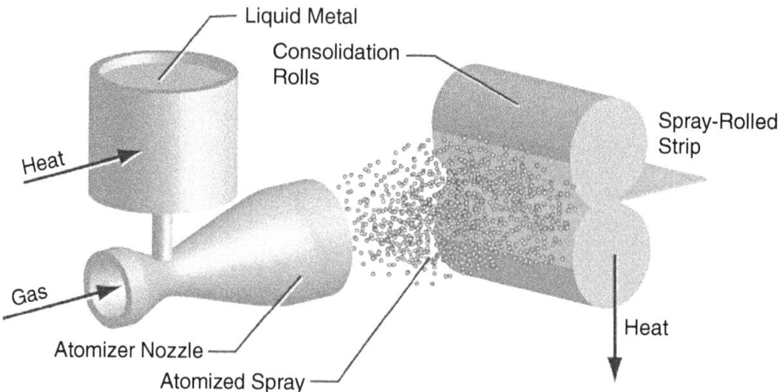

FIGURE 11.2 Sketch of the spray rolling process.

enhanced properties (e.g., tensile strength).[11] Furthermore, spray rolling has been successfully used to process alloys with wide freezing ranges such as 7050 or 2124 that are challenging to twin-roll cast.[12]

11.2.1.3 Reactive Spray Deposition

Reactive spray-deposition processes provide a means to allow the disperse phase to react chemically, in-flight, with the carrier flow. This modification of the basic spray-deposition process has been explored by Unigame et al.[13] and by Lavernia and coworkers.[14] Surface reaction can change the particle solidification pattern by providing additional nucleation sites. The dispersoids resulting from reaction contribute to a reduced grain size by pinning grain boundaries, as discussed in Section 4.1.3. In the case of aluminum alloys, reactive atomization and deposition (10% vol. O_2 added to the N_2 used for atomization) has been used to obtain equiaxed microstructures with a grain size under 18 μm.[15]

11.2.2 Thermal Sprays

The term "thermal spray" refers to an entire family of coating processes in which a material is heated and accelerated to form a coating on a target or substrate. Each process in the family is distinguished by the way and degree to which the material is heated and by the method used to impart velocity to the material. The family of thermal spray includes flame spray, plasma spray, electric-arc spray, detonation gun (D-Gun) spray, high velocity oxy-fuel (HVOF) spray, and their various modifications. Each of these thermal spray methods fulfills the need for a certain combination of temperature and velocity required to successfully process a wide range of coatings. Their distinct features are described below.

11.2.2.1 Flame Spray

It utilizes combustible gas as a heat source to melt the coating material. A combustion flame can be produced by burning any fuel such as hydrogen, acetylene, propylene, propane, and the like, with either air or oxygen. The flame characteristics depend on the oxygen to fuel gas ratios. Typically, oxygen to fuel ratio may vary from 1:1 to 1.1:1, resulting in reducing to oxidizing atmosphere respectively, and the flame temperature varies correspondingly from 3000 to 3350°C with the flame velocity changing from 80 to 100 m/s.[16,17] Due to its relatively low achievable temperatures, flame spray is well suited for producing polymer-based coatings,[18,19] and widely used by industry to deposit metallic coatings for restoration of worn or out-of-tolerance parts.[16,20] Flame-sprayed coatings typically exhibit lower bond strengths, higher porosity, and higher heat transmittal to the substrate than plasma and electric-arc spray. The bond strength reached is normally in the range of 15 MPa for ceramics coatings, while it is higher for metallic coatings (~30 MPa). Bond strengths as high as 60 MPa for NiAl coatings have been reported.[17] Porosity is in the range of 10 to 20% (could be no-porosity for self-fluxing alloys).[17]

To increase the wear resistance of flame-sprayed coatings, flame spray is often modified to incorporate a fusing treatment after flame spray.[16,21] Such a flame spray and fuse process can allow a coating with good workability to be deposited and machined, then followed by the fusing treatment to improve wear and corrosion resistance. Other modifications of flame spray include reactive flame spray[22,23] and liquid flame spray.[24] The former can produce metallic coatings with finely dispersed ceramic reinforcement generated *in situ* during spray, and the coatings so produced exhibit substantial improvements in wear and abrasion resistance.[22,23] Liquid flame spray (LFS) is also a reactive spray technique except that a liquid precursor is employed. Because of the formation of finely atomized liquid droplets, LFS has the potential to improve the homogeneity and properties of coatings.[24]

11.2.2.2 Plasma Spray

In this process the coating material is heated and propelled by a plasma, which is produced by ionizing an arc gas (usually argon or nitrogen) by the electrical discharge from a high frequency arc starter.[16] Plasma temperature can be increased by simply passing more current through it. To achieve even higher temperatures, secondary gases such as helium and hydrogen are added to the plasma. These gases raise the ionization potential of the arc gas mixture, thereby raising the plasma heat content to produce higher temperatures at lower power levels. The temperature of the plasma gas at the nozzle may reach as high as 6,000 to 12,000°C, and the gas velocity may reach 200 to 600 m/s.[25,26] Due to its high temperatures, plasma spray is widely used for producing ceramic coatings as well as metal coatings. In particular, air plasma spray (APS) is most suitable for producing oxide coatings such as thermal barrier coatings (TBC)[27,28] and Al_2O_3-13 wt.% TiO_2 coatings for wear and corrosion resistance.[29,30] The bond strength of plasma-sprayed ceramic coatings typically ranges from 15 to 25 MPa. The bond strength of plasma-sprayed metal coatings (sometimes in excess of 70 MPa) is also higher than the counterpart produced using flame sprays. Owing to the high liquid droplet temperature and impact velocity, the porosity of plasma-sprayed coatings (1 to 7%) is lower than that obtained from flame sprays.

The oxide content of plasma-sprayed metal coatings is inherently low due to the use of inert arc gases. Nevertheless, oxidation of nonoxide materials (e.g., WC/Co and NiAl coatings) could still occur during APS because the fluid dynamics of the plasma jet leads to the development of turbulence with strong entrainment of ambient gas into the plasma jet.[26] Therefore, for minimum oxide contents vacuum plasma spray (VPS), sometimes called low-pressure plasma spray (LPPS), is frequently used, such as the case of MCrAlY coatings deposited in the TBC system as a bond coat between the yttria-stabilized zirconia (YSZ) top coat and the Ni-superalloy substrate.[31,32] The bond strength of alloy coatings deposited using VPS is frequently greater than 80 MPa because of the minimum oxidation during plasma spray.[17] Another important variant of the traditional plasma spray is the use of liquid feedstock, as in the case of solution precursor plasma spray (SPPS)[33,34] and plasma spraying of liquid precursor (PSLP).[35] TBCs deposited via SPPS exhibit superior durability because of their nontraditional microstructure (Figure 11.3), and improved YSZ adhesion to bond coat.[36]

11.2.2.3 Electric-Arc Spray

It is also known as arc spray (AS) and only applicable to forming coatings from electrically conductive materials because arc heating is achieved by moving two electrically opposed charged wires, which are consumable electrodes, to a critical separation. The arc melts the tips of the wires, and a compressed gas (either inert or reactive) acts to atomize the molten wire tips and propel the fine atomized particles to the substrate. The conductive wires can be made of pure metals, alloys, or two phases with sheath and filler (known as cored wires).[16,17,37–39] Dissimilar wires can also be used to deposit pseudo-alloys.[16] The distinct features for AS are (1) high deposition rates with 55 kg/h being reported,[16,39] (2) low substrate heating due to the absence of flame touching on the substrate,[16] and (3) low costs because of requirements of low electrical power and no expensive gas in most instances, and its high deposition rate.[16,39] AS can compete with plasma spray for metallic coatings as cored wires can be used to spray alloys and composites.[39] Nevertheless, AS is limited to electrically conductive materials, and coatings of ceramics are not currently practical.

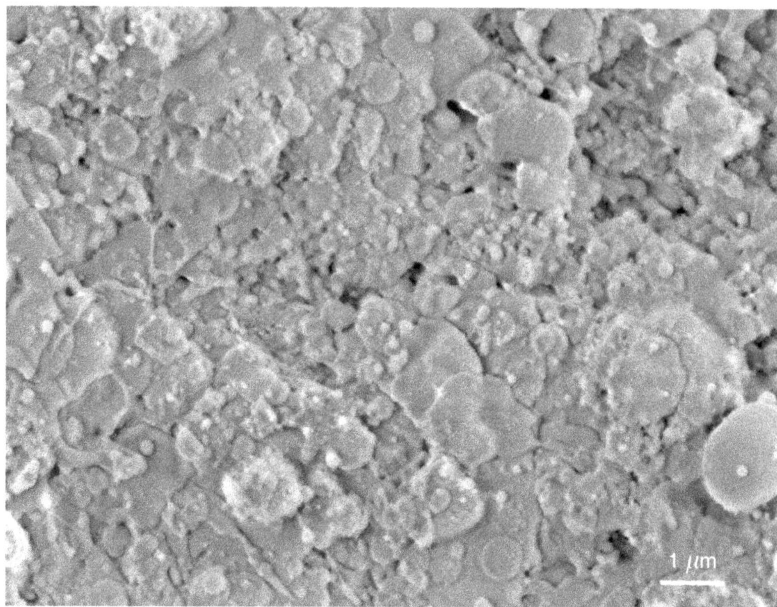

FIGURE 11.3 Ultra-fine splats of YSZ formed in the SPPS process with highly desirable microstructural features: splat diameter <2 μm, splat thickness <1 μm, and splat area is 1/2500 of splats in APS TBCs. (Courtesy of Professor Maurice Gell.)

11.2.2.4 HVOF Spray

HVOF spray is also a combustion-based technique.[40–46] However, it differs from flame spray because in the HVOF process, the flame stream is formed by combustion under high pressure (3 to 10 bars), resulting in flame velocities of more than 2000 m/s.[40] The particle velocities resulting from the extremely high flame speeds are in the range of 400 to 650 m/s, which are substantially higher than those in flame spray, arc spray, air plasma spray, and vacuum plasma spray. As a result, the porosity of HVOF sprayed coatings is very low (<1%).[17,40] The flame temperature of HVOF is similar to that of flame spray. Thus, in general HVOF is not well suited for depositing high temperature oxide coatings. However, in the HVOF top gun system, where the spray powder is injected axially and centrally into the combustion chamber, high particle temperatures (~2300°C) can be achieved, and ceramic coatings (e.g., Al_2O_3 and Al_2O_3–SiC composites) have been deposited using this method.[40]

The low flame temperature of HVOF can be advantageous. For example, WC–Co coatings are deposited using HVOF because low flame temperature can minimize oxidation and decarburization of WC.[41,42] In sharp contrast, substantial decarburization of WC occurs if APS is used.[41] By virtue of its low flame temperature, HVOF is also well suited for forming polymer and polymer composite coatings.[43,44] Finally, high particle velocities achieved in HVOF not only lead to high-density coatings, but also, in many cases, yield coatings with compressive stresses that prevent degradation of fatigue properties of coated components.[47]

11.2.2.5 Detonation Gun (D-Gun) Spray

The commercially available D-Gun™ system consists of a long, water-cooled barrel. The powder inside the barrel is propelled to a supersonic speed (760 to 1000 m/s) by a detonation wave generated from ignition of a gas mixture normally composed of oxygen and acetylene.[48] D-Gun spray is an intermittent process, but can repeat many times (up to 15) a second. The very high particle velocity and high impact of the powder particles against the work surface improve the bond between individual particles and to the surface being coated. As a result, the bond strength of D-Gun coatings is in excess of 140 MPa and Super D-Gun coatings in excess of 210 MPa.[48] High particle impacts on the substrate and the previous deposits also result in a coating with residual compressive stresses that are beneficial to the fatigue strength of the coated component. The hardness and wear resistance of WC–Co coatings and the fatigue strength of the components coated with the Super D-Gun process are all better than the counterparts formed by the HVOF process.[48]

11.2.3 Solid-State Spray Processes

Thermal sprays are primarily based on the thermal energy of impacting particles, that is, the feedstock should be partially or fully melted at the moment of impact to form the coating. These characteristics can impose some limitations on coating properties and performance, such as oxidation and phase transformation in the case of reactive materials and the loss of nanostructures in the case of nanograined feedstock. Solid-state spray processes can provide some solutions to these problems. Solid-state spray processes, including cold spray,[49,50] kinetic spray,[51,52] and kinetic metallization,[53,54] are relatively new processes in which coating formation does not depend on the thermal energy, but on the kinetic energy of impacting particles. The main features of solid-state sprays are described below.

11.2.3.1 Cold Spray

This technique employs a supersonic gas stream to accelerate particles to speeds of 500 to 1000 m/s.[49,50] The supersonic velocity is obtained via forcing a compressed gas to flow through a converging and then diverging nozzle.[55] When high-speed particles are impinged on the substrate or the previously deposited layer, the conversion of the kinetic energy makes it possible to plastically deform particles, causing very large strain (approximately 80% in the direction normal to impact). This deformation results in a huge increase in particle surface area (~400%), producing new surface that is oxide free. When these active surfaces come into contact under high interfacial pressures and temperatures, pure metallurgical bonds are formed. The bonding is achieved via the solid-state reaction as no impact-induced melting is observed.[56,57] As the deposition is achieved below the melting point of the powder particles, spray materials experience little change in microstructure and little oxidation and decomposition, while the coating formed has low porosity and thermal stresses.[50,55,56,58] However, cold spray is not suitable for brittle, nondeforming ceramics. The most important parameter for solid-state spray processes is the particle velocity prior to impact on substrate. Only particles with a velocity larger than the critical velocity can be deposited to produce coatings. Particles with a velocity lower than the critical velocity will lead to the erosion of the substrate.[59,60] The critical particle velocity changes with spray materials, and is also affected by the particle size, size distribution, substrate composition, and gun-substrate impact angle.[56,59,60] Many materials including Cu, Fe, Ni, Al, Ti, WC–Co, Al–SiC, and Al–Al_2O_3 have been deposited via cold spray.[49,50,55–58,61] All of these coatings are formed using the starting powder particles with sizes <50 μm because the deposition efficiency for particles larger than 50 μm is essentially zero in the cold spray process.[52,55,61]

11.2.3.2 Kinetic Spray

To form coatings from powder particles larger than 50 μm, the nozzle configuration of cold spray is modified.[51,52,55,62–64] The modified configuration imparts a reduction of the critical velocities above which the coatings start to form, and increased deposition efficiencies.[51,52,55]

11.2.3.3 Kinetic Metallization

Cold spray and kinetic spray use supersonic nozzles,[49,51] whereas kinetic metallization uses a "friction compensated sonic nozzle" which comprises an inner supersonic nozzle and an outer evacuator nozzle.[65] The inner nozzle accelerates and triboelectrically charges powder particles, while the outer nozzle is mechanically and fluidly dynamically coupled to the inner nozzle and the substrate to allow the inner nozzle to operate at exit pressures sufficiently below the ambient pressure. This way the inlet carrier gas pressure is reduced to maintain supersonic two-phase flow within the inner nozzle. Such nozzles allow particles to reach velocities of 500 to 1000 m/s at operating pressures of 0.35 to 0.7 MPa, while the cold spray supersonic nozzles require pressures of 2.1 to 3.45 MPa. As such, kinetic metallization is a more cost effective process than cold spray, kinetic spray, and HVOF.[65,66]

11.3 Heat, Mass, and Momentum Transfer Aspects of Spray Transport

Once the geometry of a particular spray deposition or coating process has been selected and once the working material is known, the processing parameters that can be used to control the properties of the sprayed material are: (1) the mass flow rate, chemical composition, velocity, temperature, and thermo-physical properties of the carrier gas or plasma at injection; (2) the mass flow rate, temperature, velocity and size distributions, and thermo-physical properties of the dispersed phase at injection; (3) the temperature, pressure, and chemical composition of the ambient; and (4) the temperature and displacement of the target. Consequently, heat, mass, and momentum transfer (transport phenomena) play a critical role and they are discussed briefly in this section.

11.3.1 Spray Flow

The analysis of spray flows in materials processing has historically benefited from the tools developed to investigate other applications such as-spray combustion.[67] The recent monograph by Fritsching[68] shows how these tools have been applied and adapted to the case of spray forming and provides a detailed account of the progress that has been made in that area over the past few years.

Spray flow is the process stage during which the particles interact with the carrier gas, it occurs after injection (and breakup if appropriate) and before deposition (see, for instance, Figure 11.1). The complexity of this stage stems from the nature of the flow: a multidimensional, multiphase turbulent jet with a gaseous continuous phase (carrier) and a solid or liquid disperse phase. A convenient perspective is provided by the Eulerian/Lagrangian simulation technique, where the flow of the continuous phase is evaluated in an Eulerian (fixed) frame of reference, while the individual particles are followed in a Lagrangian way. This type of perspective is better suited to the case of dilute sprays (i.e., away from the injection plane). The characteristics and behavior of both continuous and disperse phases as well as their coupling are discussed below.

11.3.1.1 Carrier Flow Characteristics

Typically, the function of the carrier flow is to condition the particles (i.e., bring them to a specific thermal state, solid fraction, and velocity) and to transport them to their target. In spray-deposition processes the flow is mostly induced by the gas flow used in the atomization process. The flow of a single component incompressible fluid is governed by (see, for instance, Reference 69) mass conservation:

$$\nabla \cdot \mathbf{V} = 0 \tag{11.1}$$

where $\mathbf{V}$ is the velocity field, given by the Navier-Stokes equation:

$$\rho \frac{\mathrm{D}\mathbf{V}}{\mathrm{D}t} = -\nabla p + \mu \nabla^2 \mathbf{V} + \mathbf{F} \tag{11.2}$$

where ρ is the fluid density, μ its viscosity, p is the pressure field, and $\mathbf{F}$ represents externally applied forces per unit volume (e.g., $\rho\mathbf{g}$ for gravity). The thermal energy transport equation is

$$\rho C_{\mathrm{v}} \frac{\mathrm{D}T}{\mathrm{D}t} = k\nabla^2 T + \mu \Phi_{\mathrm{v}} \tag{11.3}$$

where T is the temperature field, C_{v} is the fluid heat capacity at constant volume, k is the fluid thermal conductivity and Φ_{v} is the viscous dissipation function. Note that the left-hand side term in Equation 11.2

and Equation 11.3 (D/Dt), is a substantial derivative. In the case of Equation 11.2, that term is the sum of the local acceleration and the convective acceleration, which is nonlinear. This already complex equation system (coupled nonlinear second-order partial differential equations) is further complicated if compressibility effects must be taken into account (for Mach numbers greater than 0.3), which is the case in the near nozzle region of some spray-forming processes,[2] or if species transport is relevant (which requires the solution of $n - 1$ additional partial differential equations where n is the number of species). The effect of particles on the flow can be incorporated by adding source or sink terms of mass, momentum, and energy to Equation 11.1 to Equation 11.3 (see following section). Other source terms must be added if the flow is chemically reactive.

Even in the relatively simple form presented in Equation 11.1 to Equation 11.3, the flow equation system typically requires a numerical solution using a computational fluid dynamics (CFD) approach. Only a limited number of analytical solutions exist, mostly for academic cases. The computational cost of detailed CFD solutions can be prohibitive, especially in multidimensional, multicomponent, reactive cases. Consequently, the type of solution method used depends on the information sought. For example, to predict overspray, an important matter in the practical implementation of spray processes, an accurate and detailed prediction of the carrier flow field is necessary, which requires a multidimensional CFD simulation. Such simulations can help identify recirculation features that carry particles away from the target.[70] If, on the other hand, the focus is primarily on the dispersed phase (as is the case in the quest for a processing–structure–property relationship), an approximate solution that provides enough information to define the environment in which the dispersed phase is evolving might be sufficient. Such approximate solutions, based both on measurements and analysis, have been proposed by Grant et al.[71] to describe the exponential decay of the carrier gas axial velocity (u):

$$u = u_0 \exp\left(-\frac{z}{\kappa}\right) \tag{11.4}$$

where u_0 is the initial gas velocity and κ is an exponential decay coefficient.

The turbulent nature of the carrier flow has a direct influence on the dispersed phase behavior both in terms of particle dispersion and the enhancement of heat and mass transfer to the particles. As presented above, the flow equations (Equation 11.1 to Equation 11.3) apply to both laminar and turbulent flow. However, the direct numerical solution of these equations in the case of a turbulent flow (Direct Numerical Simulation or DNS) requires a resolution on the order of the smallest turbulence scale, the Kolmogorov scale.[72] For spray processes that scale is on the order of a 10^{-4} m, which would result in prohibitive computation time. Indeed, while DNS can provide otherwise inaccessible information regarding the physics of turbulent flow at small scales, it is still not viable for the simulation of industrial processes.[73] A practically successful approach has been to average the instantaneous flow equations over time to capture the effect that turbulence has on the mean flow, which yields the Reynolds-Averaged Navier–Stokes, or RANS, equations. A key issue with the RANS approach is that the averaging process results in a loss of information and an equation system that has more unknowns than equations. This is remedied by specifying additional equations, called "closure equations." Unfortunately, these closure equations are highly dependent on the specifics of the flow studied which limits their applicability outside of the flow system for which they were designed. An intermediate approach that solves the large-scales but models the smaller scales, large eddy simulation (LES), is making some headway into industrial applications.[73]

More details on the specific description of the carrier flow characteristics in spray processes, including plasma spray, flame spray and HVOF, and electric-arc spray, are given in References 17, 74, and 78.

11.3.1.2 Spray and Particle Behavior

The characterization of spray behavior in spray deposition and coating processes necessarily involves two main scales: that of the spray and that of the individual particles. Spray scale phenomena have a direct influence on the overall geometry of the deposit or coating (shape, thickness) while the effect of particle scale phenomena is felt more directly in the microstructure (porosity, grain size). Because of the resulting interest in the fate of individual particles, this section focuses mainly on their behavior. The example of one possible strategy to reconstruct spray behavior from the information obtained at the particle scale is also outlined.

The behavior of individual particles in-flight during spray processes has been described in detail by Grant et al.,[71,79] Lavernia and coworkers,[2,80,81] and Bergmann et al.[82] The history of a particle during the spray flow stage is primarily governed by the ambient conditions experienced by that particle. Therefore, the particle trajectory must be determined accurately. The particle dynamics is approximately described by the following equation:[81]

$$m_P \frac{d\mathbf{V}_P}{dt} = \frac{1}{2}\rho A \|\mathbf{V} - \mathbf{V}_P\|(\mathbf{V} - \mathbf{V}_P) C_D + m_P \left(1 - \frac{\rho}{\rho_P}\right) \mathbf{g} + \mathbf{F}_P \tag{11.5}$$

where $\mathbf{V}_P$ is the particle velocity vector, m_P its mass, ρ_p its density, C_D its drag coefficient (typically a function of the particle Reynolds number, $Re = \rho \|\mathbf{V} - \mathbf{V}_P\| D/\mu$ for a sphere of diameter D), A is the cross-sectional area used to define the drag coefficient and $\mathbf{F}_P$ represents the other forces that can influence the particle trajectory in specific configurations, such as thermophoretic or electromagnetic forces. This form of the particle dynamics equation neglects added mass and history terms.[83] Particle trajectories are affected by the turbulent character of the carrier flow; this is turbulent dispersion. The nature of that effect depends on the size of the particle relative to the Kolmogorov scale. The prevalent case in spray processing is that of particles smaller than the Kolmogorov scale such that the effect of the turbulence is akin to that of an unsteady flow. This effect can be quantified by adding an estimated velocity fluctuation to the mean gas velocity ($\mathbf{V}$ in Equation 11.5). Gosman and Ioannides[84] proposed that the gas velocity fluctuation experienced by the droplet be approximated stochastically using a normal distribution with a variance equal to two-thirds of the turbulent kinetic energy. Most numerical simulations of the spray flow stage neglect particle–particle interaction because the dilute spray assumption holds well in that region. Advanced diagnostics have been developed to experimentally determine the in-flight droplet or particle size and velocity, including laser Doppler velocimetery (LDV), and phase Doppler interferometry (PDI) or phase Doppler anemometry (PDA). PDI/PDA technique is a combination of LDV (velocity only) with particle sizing method.[85] The following important information can be generated by PDI/PDA: size and velocity histograms, mean values and statistical mean diameters, droplet number density and volume flux, size-velocity correlation plots, time-of-arrival analysis of droplet size and velocity, and temporal to spatial conversion of size distribution. This information can be used to understand the droplet behavior in fluid flow, characterize the spray behavior, and then provide feedback for optimization of processing parameters.[86,87] Recently, in-flight pyrometry has been adapted to the PDA system to provide *in situ* droplet temperature measurement, in addition to droplet size and velocity.[88] The detailed characterization of the spray behavior via the aforementioned experimentation can be critical for verification and improvement of spray models.

A key outcome of the analysis of droplet and spray behavior in spray processes is the determination of the state of the droplet at impact (temperature, solid fraction, velocity). If the heat transfer at the particle surface is slower than the heat transfer by conduction inside the particle, that is, if the particle Biot number is much less than unity (a common occurrence for metal particles), the heat transfer behavior in individual particles can described using a lumped parameter approach (see for instance the review by Armster et al.):[89]

$$\frac{dH}{dt} = \dot{Q}_{rad} + \dot{Q}_{conv} \tag{11.6}$$

H is the droplet total enthalpy and $\dot{Q}_{conv}$ and $\dot{Q}_{rad}$ are the convective and radiative heat transfer rate. For a particle that includes both solid and liquid phases, the total enthalpy is:

$$\frac{H}{m_P} = h_0 + f_s C_{p,s}(T_P - T_L) + (1 - f_s)(C_{p,l}(T_P - T_L) + \Delta h_f) \tag{11.7}$$

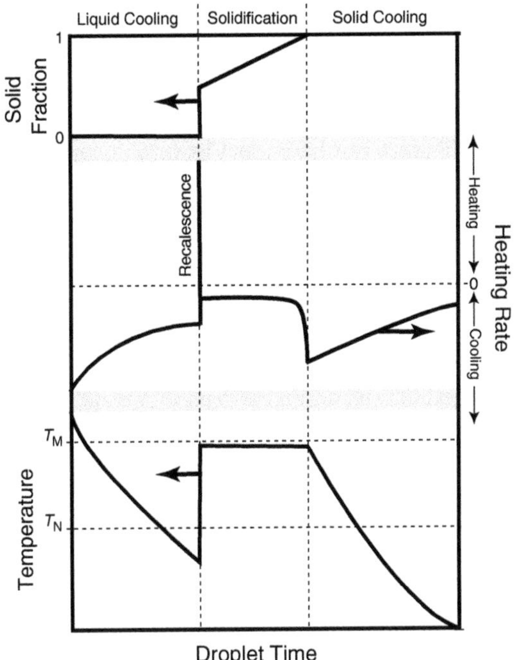

FIGURE 11.4 Solidification and thermal history of a droplet undergoing nonequilibrium solidification.

where h_0 is the specific enthalpy of the solid at the liquidus temperature (T_L), f_s the particle solid fraction, $C_{p,s}$ and $C_{p,l}$ are the specific heat of the solid and liquid phases, and T_P the temperature of the particle. The particle solid fraction history is evaluated differently for equilibrium and nonequilibrium phase change. For example, when a droplet undergoes rapid solidification during a spray deposition process, its temperature evolves as described in the example of Figure 11.4 (see Section 11.4.1 for more details on nonequilibrium solidification in droplets) and solidification is kinetically driven. A linearized version of the solidification is typically used.[80] This also requires calculation of the degree of undercooling and an adequate description of the recalescence stage. Because the characteristic time associated with recalescence is very small (see Figure 11.4), the assumption of small Biot number typically breaks down, the temperature distribution in the particle is no longer uniform, and using the lump parameter approach can lead to significant errors in cases where an accurate evaluation of the surface temperature is important. This is the case, for instance, when surface oxidation occurs.[81] Spatial resolution of the temperature in the particle can be obtained using a 1D approach in which the particle is approximated by an equivalent sphere.[81] Levi and Mehrabian[90] have obtained a 2D axisymmetric description of the thermal history of a droplet.

Sprays are polydispersed with distributed particle size and velocities (magnitude and direction). Various methods are available to reconstruct the spray from predicted droplet behavior, from the sectional methods commonly used in aerosol modeling[91] to stochastic methods, such as that recently presented by McHugh et al.[11] A caloric probe method has been developed to measure average specific enthalpy in molten metal spray,[92] which showed a good agreement with numerical calculations.

Another issue is that of coupling between phases. Spray flows in materials applications are undeniably two-way coupled, that is, the dispersed phase is affected by the thermo-physical conditions in the continuous phase and, in turn, constitutes sources or sinks of momentum, mass, and energy. However, spray density decreases significantly moving away from injection so that a one-way coupled approach, in which the effect of the dispersed phase on the continuous phase is neglected, can yield valuable insight.

Although computational tools and principles of heat, mass, and momentum transfer involved in spray deposition and thermal spray are the same, the thermal histories of disperse phases in spray deposition and thermal spray processes are quite different. In most cases, the disperse phase in thermal spray is solid particles that are heated in the plasma jet or flame and become liquid droplets before depositing on the substrate. However, due to the randomness of the turbulent flow and different sizes of particles, some particles are only partially molten and some particles never melt at all. Such diverse behavior of the disperse phase in thermal spray can be tracked by combining Equation 11.5 to Equation 11.7 into a computational scheme taking into account turbulent eddies with or without lifetimes of random values.[75,77] The thermal history and velocity profile of porous particles or particle agglomerates with pores have also been simulated with consideration of the effective thermal conductivity of porous particles and shrinkage during melting.[75–77,93]

11.3.2 Deposition

The dynamics of fluid flow and heat transfer during the deposition stage of spray-based materials processes significantly affect final material properties and quality. A large portion of the material's thermal and latent heat is extracted during this stage of processing, and hence, the heat extraction rate during deposition, coupled with the fluid flow of spreading droplets, plays a major role in determining the microstructure of the resulting material. Additionally, droplet–droplet and droplet–substrate interactions will determine the geometry and porosity of the deposited material.

11.3.2.1 Droplet-Scale Phenomena

The impingement of individual droplets is an important component of spray-based materials processes and has been studied experimentally,[94,95] numerically,[96–98] and analytically.[99–101] The experimental investigation of droplet impacts is typically accomplished with a droplet-on-demand generator and the droplet-spreading behavior is captured with either high-speed photography, or flash videography, where the dynamics of droplet spreading are reconstructed from individual photographs of many reproducible impacts. Analytical models for droplet impact generally take the approach introduced by Madejski.[101] In that model, droplet behavior is described by a mechanical energy balance:

$$\frac{\mathrm{d}}{\mathrm{d}t}(E_\mathrm{K} + E_\mathrm{P} + E_\mathrm{D}) = 0, \tag{11.8}$$

where E_K is the kinetic energy inside the droplet, E_P is the potential energy, associated with surface tension, and E_D is the mechanical energy dissipated through viscous effects.

There are a number of methods available for the numerical simulation of droplet impacts. These include volume of fluid methods, level set methods, front tracking methods, and Lagrangian finite element methods, each having its own strengths and weaknesses for simulating droplet impacts.

Both fluid flow and solidification influence the spreading behavior of a molten droplet impinging upon a substrate (Figure 11.5). If the fluid flow characteristic time is much shorter than that for solidification, fluid dynamics dominate the process, and the behavior of the spreading droplet is essentially decoupled from solidification. On the other hand, if the solidification timescale is comparable to or smaller than that for fluid flow the two processes are closely coupled.[89]

Parameters that influence the behavior of a droplet impinging on a substrate include the kinematic, transport, and thermodynamic properties of the droplet, as well as geometrical parameters, such as the size and shape of the droplet (Figure 11.6). Additionally, the conditions of the substrate (rough or smooth; liquid, solid, or mushy) must be considered. These parameters have been extensively explored by Rein[102] and by Armster et al.[89]

11.3.2.1.1 Fluid Mechanics

In cases where solidification plays only a minor role, the nondimensional fluid dynamical parameters of primary importance are the droplet Reynolds number, $\mathrm{Re} = \rho VD/\mu$, which provides a measure of the relative importance of inertia and viscous forces (V is the velocity of the droplet, D the diameter, ρ the density, μ the viscosity); and the droplet Weber number, $\mathrm{We} = \rho V^2 D/\sigma$, which quantifies the relative

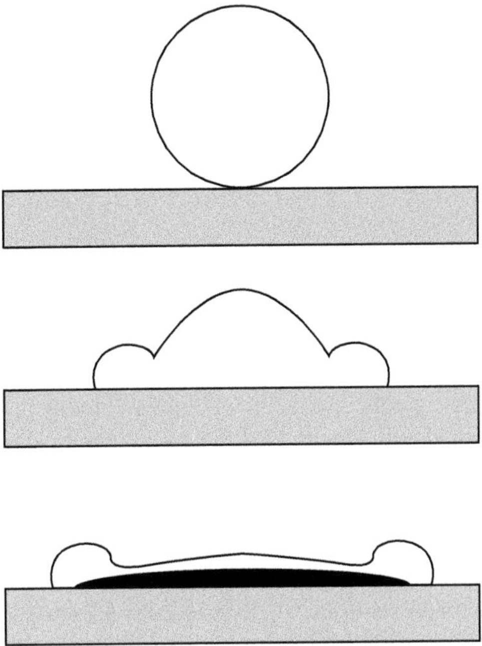

FIGURE 11.5 Schematic presentation of droplet spreading and solidification.

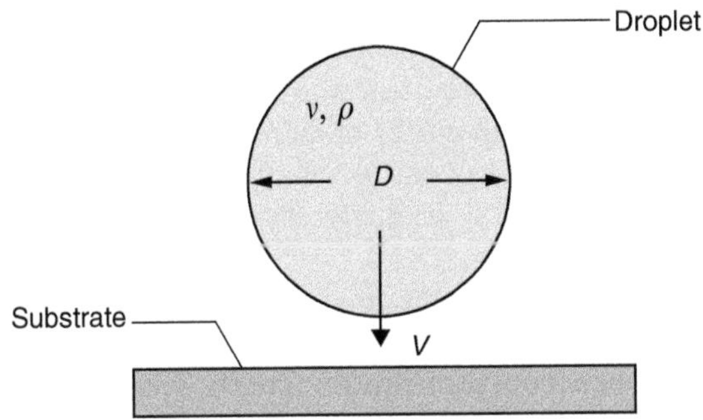

FIGURE 11.6 Parameters relevant to the normal impact of a droplet on a solid substrate.

magnitude of inertia and surface tension forces (σ is the surface tension coefficient between the droplet and the carrier gas). The wettability, expressed by the equilibrium contact angle θ_e, is also an important parameter for determining droplet-spreading dynamics. For the high-speed droplet impacts (on the order of 100 m/s) involved in some thermal spray applications, the compressibility of the droplet will influence spreading, and the Mach number ($Ma = V/c$) should be considered an important parameter as well,[102,103] where c is the speed of sound in the droplet.

As illustrated by Madejski[100] (see also Equation 11.8), the impingement of a droplet can be analyzed in terms of mechanical energy conservation. As the droplet deforms, kinetic energy is transferred to surface energy, and dissipated through viscous effects. At higher Reynolds and Weber numbers the influence of viscous and surface tension effects will be reduced and thinner splats will be produced, with a larger area of contact between the droplet and the substrate. Thinner splats will also be produced at lower contact angles, where the interfacial energy between the droplet and the substrate is relatively small. The droplet may also splash (Figure 11.7), forming secondary droplets,[104,105] or rebound from the substrate,[106,107] depending on the values of the Reynolds number, Weber number, and equilibrium contact angle. Typically, droplets with high velocities and heat contents splash and produce flower type splats having leaves and partial

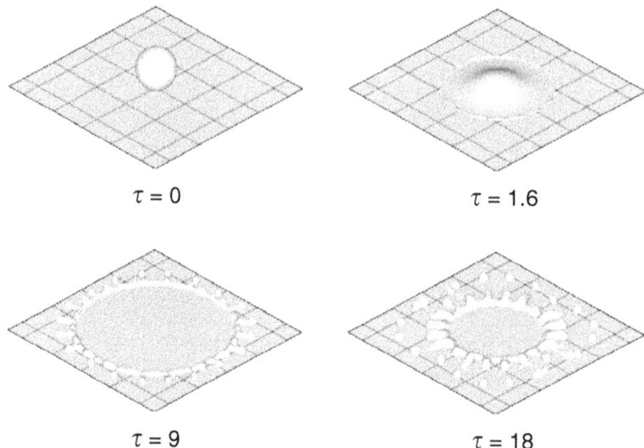

FIGURE 11.7 Numerical simulation of a 50 μm aluminum droplet impacting a hot solid substrate at 30 m/s, and breaking up. The numbers below each frame indicate the dimensionless time ($\tau = tV/D$).

contact with the substrate, whereas droplets with moderate velocities and heat contents yield pancake type splats in good contact with the substrate.[108] The propensity for a droplet to splash can be quantified with the splash parameter ($K = We^{0.5}Re^{0.25}$), and splashing will generally occur for values beyond a critical value of K, $K_{cr} \approx 57.7$.[109]

11.3.2.1.2 Solidification/Heat Transfer

The solidification rate of a droplet impinging on a substrate can be either heat transfer limited or kinetically limited. Kinetically limited solidification will occur when a droplet has experienced significant thermal undercooling. In such cases, the area and quality of thermal contact are important for maintaining a high solidification rate, but kinetic effects must be considered.[110] In the case of heat transfer limited solidification, the solidification rate will depend critically on the area and quality of thermal contact between the droplet and substrate.[111]

The quality of the thermal contact between the droplet and the substrate is typically quantified in terms of the heat transfer coefficient (h). This coefficient will depend on the ability of the droplet to wet the substrate, the impact velocity, the condition of the substrate, the degree of oxidation of droplets, and the inclusion of entrapped gas, and will vary through the solidification process, depending on the solidification state of the droplet at the substrate interface.[111] The change in the thermal contact resistance between the droplet and the substrate (or the previously deposited splats) can alter the splat microstructure substantially. In the most typical conditions, nucleation is heterogeneous and the solidification front moves from the interface between the droplet and the substrate toward the top of the splat, leading to the formation of a columnar microstructure.[112] However, when the thermal contact resistance is high, homogeneous nucleation with a large undercooling can occur, resulting in the formation of a fine-grained equiaxed microstructure in splats.[112,113]

11.3.2.2 Deposit/Coating-Scale Phenomena

Deposit scale phenomena are also an integral component of spray-based materials processes. In addition to the overall shape of the deposit or coating, residual stresses, porosity of the deposited material, and adhesion to the substrate and cohesion between splats have a significant influence on the material quality in spray deposition and spray coating technologies.

11.3.2.2.1 Deposit Shape

As the objective of spray-forming processes is typically the creation of net- or near-net-shape products such as tubes, billets, or sheets (Figure 11.8), the geometric development of the deposit is critical.

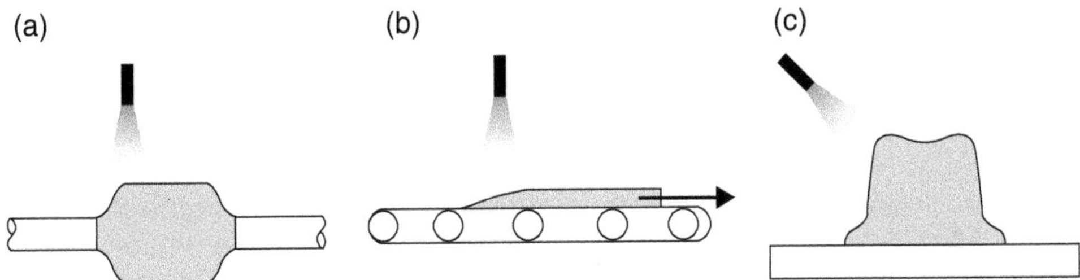

FIGURE 11.8 Schematic illustration of spray forming for (a) a tube, (b) a strip and (c) abillet.

This importance is compounded by the fact that the deposit shape, together with heat transfer, has a direct influence on the development of residual stresses during cooling, which may be detrimental to the final product.

A cylindrical billet is the most common shape produced with spray-forming processes. The shape of the billet will depend on a number of processing parameters, including: the melt flow rate, the atomization gas flow rate, the angle of the nozzle, and the withdrawal rate of the deposit.[114]

Several models have been presented in the literature to predict the development of the deposit in spray-forming processes. These models are purely geometric ones, which only investigate the shape of the developing deposit,[115] and models that combine the geometric evolution with heat transfer.[114] Those models that do not include heat transfer are not capable of evaluating the sticking efficiency of the incident spray, as it is related to the temperature. Additionally, the residual stress cannot be evaluated without accounting for heat transfer, though none of the models discussed here deal with residual stresses. A priori, three components are necessary to model the formation of the deposit geometry: a spray/atomization model, a deposition model, and a thermal model of the deposit. The spray model may be a simplified representation on the spray dynamics; Frigaard[115] used a specified field for the mass flux in the spray cone without including calculations of droplet energy or momentum conservation, and Hattel and Pryds[114] used a one-dimensional approximation to the droplet dynamics in the spray cone. The second component of the model is deposition, which involves evaluating the evolution of the surface of the billet from information regarding the mass flux from the incident spray. Droplet sticking efficiency and deposit shading significantly affect the deposit shape. The last component, thermal analysis, is used to calculate the temperature distribution in the deposit, which may be determined by numerically evaluating the thermal energy balance.

11.3.2.2.2 Residual Stresses

Residual stresses are usually present in most spray-deposited materials. They are caused, at least in part, by thermal gradients in the deposit and their distribution is typically nonuniform throughout the deposit.[116] The presence of compressive residual stresses in coating processes may be beneficial for enhancing corrosion resistance and fatigue life. In most cases, however, residual stresses are detrimental to material quality and performance. Excessive residual stress can lead to the formation of cracks as well as distortion of the deposited material. The presence of residual stresses is particularly detrimental with higher levels of porosity, as the pores provide crack initiation sites.[111] Additional discussion on the formation of residual stresses in coatings, from the microstructural perspective, is given in Section 11.4.2.2.

11.3.2.2.3 Deposit Porosity

The quality and performance of material produced by spray-based processing is significantly influenced by the porosity of the deposit. The presence of porosity is generally detrimental, as it reduces the strength of the material,[117] provides crack initiation sites,[111] and can result in the degradation of material properties, especially at high temperatures.[118] The three primary mechanisms influencing the formation of porosity in spray-deposited material are solidification shrinkage, gas entrapment, and interstitial porosity.[118,119] The formation of gas-related porosity is the result of entrapment of atomization gasses with limited solid solubility.[119] Interstitial porosity is the result of interstices formed from the deposition of partially solidified droplets. If the liquid fraction of the deposited spray is not sufficient to fill those interstices, pores will form. Solidification porosity is the result of a reduction in volume of the deposited material during solidification. Solidification porosity and interstitial porosity can be considered mutually exclusive, because solidification porosity occurs when the solid fraction of the incident spray is low, and interstitial porosity occurs when the solid fraction of the incident spray is too high.[118] This suggests that optimal processing parameters can be used to reduce porosity.

Sometimes pores are intentionally produced because they contribute to the low thermal conductivity of the coatings, as shown in Section 11.4.2.3.

11.3.2.2.4 Adhesion and Cohesion

Adhesion to the substrate and cohesion between splats are especially important for thermal-sprayed coatings many of which are used in the as-sprayed condition. The three major mechanisms that lead to adhesion between splats and the substrate are physical interaction, metallurgical interaction, and other interactions (e.g., mechanical interlocking and epitaxy). The physical interaction results from the action of *Van der Waals* forces and requires the gap between two surfaces to be less than 0.5 nm for *Van der Waals* forces to be effective.[120] The metallurgical interaction derives from diffusion or chemical reaction between two surfaces, whereas epitaxy occurs only when the sprayed material has the same or similar crystal structure as the substrate.[121,122]

The effectiveness of all these interactions is strongly influenced by the level of porosity and the area of contact, and thus affected by spray techniques and process parameters. Deposition in oxygen-free environments (e.g., VPS) reduces the content of oxides and improves metallurgical bonds in metallic coatings.[123] Addition of deoxidizing elements in the sprayed powder can also reduce oxidation, increase the contact area, change the flower type splat to that of pancake type, thus improving adhesion and cohesion.[124] Increasing the impact velocity of droplets via the use of HVOF or D-Gun™ spray can improve the pore-filling capacity of droplets, and thus substantially reduce the porosity and improve adhesion and cohesion.[47,48] Process optimization for a given spray technique can also improve bond strengths between splats and the interface between the coating and the substrate, as demonstrated in many studies.[29,125]

11.4 Microstructure and Properties of Spray-Processed Materials

11.4.1 Microstructure Evolution and Properties of Sprayed Deposits

It has been well documented[2,3] that spray-deposited materials show refined, equiaxed grain structure with less macrosegregation (Figure 11.9), uniform distribution of secondary phases and dispersoids, and increased solubility.[126] Significant property improvements have also been demonstrated due to the above microstructural features.[2,3] The evolution of microstructure during spray deposition can be attributed to two distinct but closely related stages. The first stage encompasses those phenomena that are primarily active in the atomized spray prior to impingement of droplets on the deposition surface. The second stage commences after the droplets have impinged the deposition surface, and alteration of the microstructure resulting from impingement must be considered.

11.4.1.1 Microstructural Evolution during Spray Transport

Following atomization, the spray is comprised of droplets with different sizes that are cooled by the atomization gas. The spherical or nearly spherical droplets continue to travel down the atomization chamber, rapidly losing thermal energy as a result of convection and radiation from the atomizing gas. In this stage, the important variables that affect heat transfer and solidification behavior of droplets

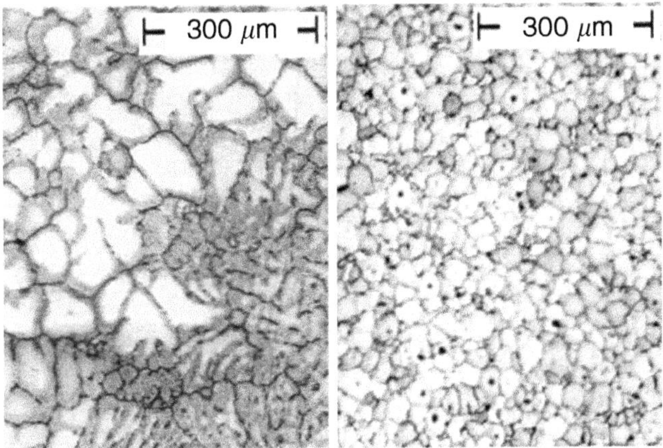

FIGURE 11.9 Optical micrographs of alloy Al 247 in: as-received ingot (left) and as-spray deposited conditions. (From Del Castillo, L. and Lavernia, E.J., *Metall. Mater. Trans. A*, 31A, 2287–2298, 2000. With permission.)

include: gas velocity, droplet size and spatial distribution, droplet temperature, solidification condition, and droplet flight distance.

The solidification structure of atomized powders is generally characterized by one or a mixture of the following: (1) featureless zones, (2) cellular, without secondary arms, (3) dendrites, and (4) equiaxed structures, depending on growth conditions and undercooling of the solidification front.[127]

The occurrence and extent of solidification in droplets depend on droplet size, superheat temperature, physical properties of the melt and the atomization gas, relative velocity between the droplets and the atomization gas, and the potency and distribution of nucleation agents. In general, nucleation of solid phases is either homogeneous or heterogeneous. In atomization it is very difficult to suppress heterogeneous nucleation as a result of the presence of droplet surfaces and other potential nucleation agents. During flight, solidification is catalyzed heterogeneously in all but the smallest droplets as a result of one or a combination of the following: (1) bulk heterogeneous nucleation within the droplet, (2) surface heterogeneous nucleation, (3) surface oxidation processes, or (4) interparticle collisions. Even for relatively small droplet sizes, attainment of undercooling levels sufficiently high to promote homogeneous nucleation is nearly impossible as a result of the presence of melt heterogeneities such as inclusions, undissolved phases, and crucible debris, which effectively catalyze solidification processes.

Nucleation is followed by the growth of solid phases at the expense of the liquid accompanying the release and dissipation of latent heat. A common phenomena, referred to as recalescence, accompanying nucleation is that the droplet temperature increases during nucleation when the rate of thermal energy released from the solid–liquid interface into the undercooled liquid is much faster than the rate of thermal energy dissipated from the droplet surface into the environment (Figure 11.4). Recalescence will terminate when the rate of heat release becomes comparable to the rate of heat extraction from the outer surface of the droplet to the surrounding environment. Therefore, at the end of recalescence, the rate of release of latent heat will be equal to the rate of heat extraction through the outer surface of the droplet.

The spray condition is characterized by droplets in liquid, semiliquid, and solid states. The overall fraction of liquid contained in the spray at the point of impingement critically influences the resultant microstructure during deposition. For example, if the liquid fraction at the deposit surface is too high, excessive liquid motion will promote the entrapment of gas ultimately yielding porosity. Moreover, an excessive amount of liquid during impingement will lead to microstructural coarsening. The overall liquid fraction in the spray arriving at the deposition surface depends on the dynamic and thermal behavior of the droplets which in turn are controlled by the processing variables, such as, the melt superheat, the gas–metal flow ratio, the flight distance, and the alloy composition.[3]

11.4.1.2 Microstructural Evolution during Deposition

Once the distribution of partially solidified droplets impinges a deposition surface, coalescence and solidification govern the microstructural evolution. The formation and refinement of equiaxed grains are critically important in terms of material uniformity (microstructure and properties across a bulk material), formability (high temperature deformation and superplasticity), and performance (strength and toughness). Equiaxed grains have been documented for numerous spray-deposited materials, regardless of composition.[128–132] The microstructural characteristics of spray-deposited materials depend to a great extent on the conditions of the droplets prior to impingement, that is, on the relative proportions of liquid and solid present, temperature, velocity, and scale of the microstructure in both the partially and fully solidified droplets.[129–132] Further studies are still needed to rationalize the precise influence of processing conditions (thermal, fluid, and solidification) on microstructural evolution observed microstructures. It is well documented[2,3] that when a dispersion of droplets in a mixed semisolid state impinge onto the deposition surface, dendritic arm fragmentation occurs due to a local strain and strain rate resulting from deformation in semisolid droplets, and this is attributed to the formation of fine, equiaxed grains by multiplication of crystalline nuclei for solidification during deposition. In addition to dendritic arm fragmentation, small solid droplets formed during the atomization stage can effectively enhance nucleation and further the formation of equiaxed grains during deposition, as well. As aforementioned, the overall fraction of liquid (or fraction of solid) contained in the spray at the point of impingement critically influences the resultant microstructure. Therefore, an optimum fraction of liquid fraction is required to maintain a desirable mushy layer on the deposition surface that is beneficial to the formation of equiaxed grains. It has been suggested[133] that during deposition, secondary nucleation and growth can occur and result in the formation of equiaxed grain morphology but smaller grain size, in comparison with the nucleation event during flight. The second nucleation depends on the volume fraction of remaining liquid and the degree of undercooling. It is worth noting that the grain coarsening in the mushy state is another important factor in controlling the formation of equiaxed grain morphology.[131,133]

An important characteristic of spray-deposited materials is the presence of a finite amount of noninterconnected pores.[9,10] The origin of porosity in spray-deposited materials may be attributed to one or a combination of the following mechanisms: (1) gas rejection, (2) solidification shrinkage, and (3) interparticle porosity. The existence of these pores is detrimental to the mechanical properties of the spray-formed materials and processing performance. Therefore, porosity should be reduced to the lowest possible value by optimizing the spray-forming conditions or by thermal mechanical processing.

11.4.1.3 Microstructural Evolution of MMCs by Spray Deposition

Spray deposition has been successfully applied in the manufacture of particle reinforced MMCs.[2] Conventional techniques of manufacturing MMCs such as squeeze casting, stir casting, and the blending of particles with metallic powders,[134] are usually associated with inhomogeneous distribution (segregation) of reinforcement particles as well as chemical reaction between melt and reinforcement phase, which may inhibit the full potential applications of particle reinforced MMCs.[135] For spray deposition of MMCs, there are two approaches available that can be implemented to introduce reinforcing particles or phases into metallic matrices, co-injection of reinforcements, and *in situ* formation of reinforcing phases when a reactive atomizing gas is used. In principle, the co-injection technique can reduce or eliminate the extreme thermal excursions, which may result in interfacial reaction and extensive macrosegregation of reinforcements, as well as porosity formed during slow solidification that normally exist in the conventional fabrication of MMCs. This will in turn improve the mechanical properties of MMCs . Typically, the mean diameter of particles is in the range of 5 to 20 μm.[2,136,137] The volume fraction of reinforcing particles incorporated into metal matrices using co-injection method ranges from 5 to 25 vol.%.[138] An interesting phenomenon in association with spray deposition of MMCs is the influence of the reinforcing particles on the refinement of grain size of the matrix materials. The solid reinforcing particles can reduce the grain size of the matrix when they act as heterogeneous nucleation catalyst for the matrix metal phase. A higher volume fraction of the reinforcing particles would result in a finer grain size due to more nucleation sites. However, the effects of particles on the grain size is generally evaluated on the basis of the

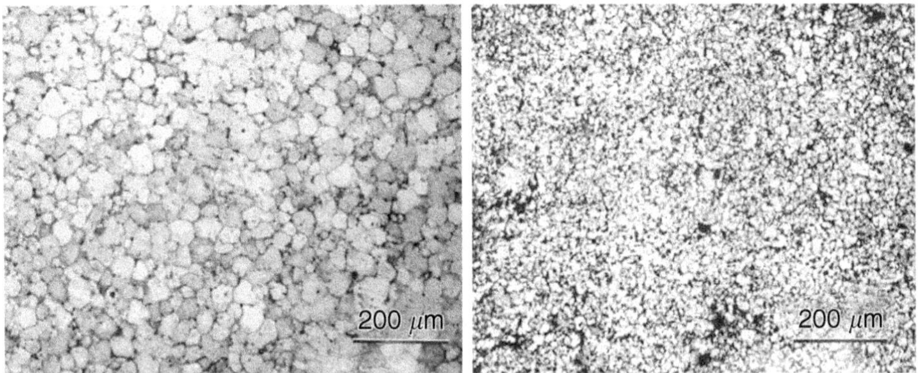

FIGURE 11.10 Optical micrographs showing grain size refinement resulting from the formation of oxide: (L) 5083 Al alloy processed using N_2 (R) 5083 Al alloy processed using N_2-10% O_2 (From Dai, S.L., Delplanque, J.P., and Lavernia, E.J., *Metall. Mater. Trans. A*, 29A, 2597–2611, 1998. With permission.)

Zener theory which can be used to elucidate the maximum achievable grain size that can be effectively stabilized by a dispersion of particles.[139] It was also found[140] that co-injection of SiC particles during spray-deposition process decreased the total enthalpy in a spray, and the presence of a distribution of SiC particles resulted in a reduction in grain size. This effect was attributed to a transfer of thermal energy from the atomized droplets to the co-injected SiC particles.

Reactive spray deposition, using a reactive gas mixture instead of an inert gas, has been used for the synthesis of materials reinforced with oxide (nitride or carbide) particles via *in situ* reactions.[15,141–143] Among them, the formation of oxide particles has attracted the most attention in understanding its influence on both microstructure evolution and mechanical performance of the deposited materials. For example, it has been reported[15,141–143] that grain refinement is achieved in the as-deposited materials resulting from the oxides that can enhance heterogeneous nucleation by serving as additional nucleation sites and inhibit grain growth by pinning of grain boundaries, the similar roles as described above for reinforcement particles introduced by co-injection during deposition (Figure 11.10). Further grain refinement resulting from the formation of fine oxide dispersoid enables the superplastic forming of these deposited materials.

11.4.1.4 Materials Performance Optimization by Spray Deposition

Spray deposition has been exploited to manufacture highly alloyed materials, which is otherwise difficult using conventional processing methods such as ingot metallurgy. One example[144,145] is that spray deposition can produce hypereutectic Al–Si alloys with Si phase uniformly distributed in fine-grained matrix, as compared to the formation of coarse Si blocks and eutectic phases usually associated with conventional casting methods. This dramatic change in microstructure can enhance the formability of the hypereutectic Al–Si alloys that demonstrate excellent wear resistance and strength with low thermal expansion coefficient. Another successful example is for the spray deposition of ultrahigh-carbon steels[146] and high-speed steels.[147] Due to the rapid solidification inherent in spray deposition, the carbide network, often observed in conventional steels, can be avoided, and instead, fine carbide is formed and uniformly distributed. This, accordingly, can lead to a significant improvement of materials properties such as mechanical strength, toughness, and hot workability. Spray deposition has also been explored for development of new materials, for example Cu–15Ni–8Sn alloy.[148] Due to the high content of Sn element, this alloy cannot be produced via conventional casting method. The Cu–15Ni–8Sn alloy by spray deposition is promising because it exhibits high strength with good conductivity and corrosion resistance, of particular interest for potential replacement of Cu–Be alloy.[148]

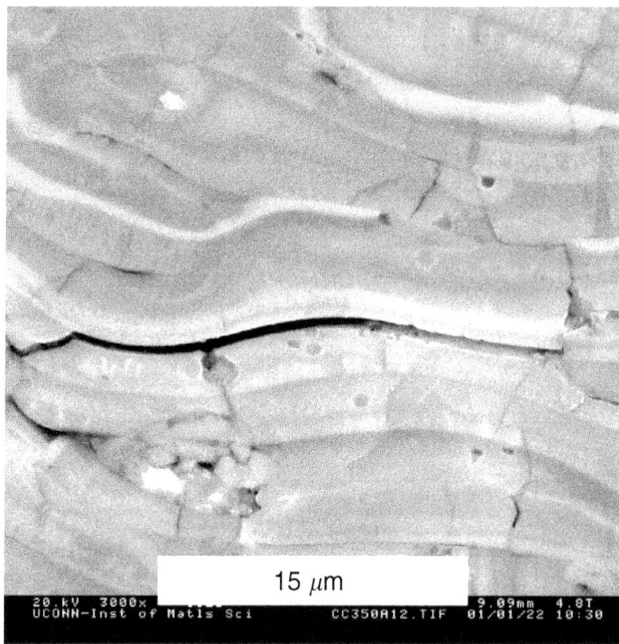

FIGURE 11.11 Typical microstructure of thermal-sprayed ceramic coatings. Shown here is an air plasma-sprayed Al_2O_3-13 wt% TiO_2 coating (Reprinted from Luo, H. et al., *Mater. Sci. Eng. A*, 346, 237–245. Copyright (2003) with permission from Elsevier).

11.4.2 Microstructure Evolution and Properties of Coatings

Shown in Figure 11.11 is the typical microstructure of thermal-sprayed ceramic coatings. It is very different from that of the corresponding bulk material. Similar to metals, the basic structural unit of the coating is pancake-shaped splat, ~1–5 μm thick and 10–50 μm in diameter, which is formed from a molten particle. Upon impact with the substrate, the molten droplet compresses vertically and spreads out horizontally due to the effect of momentum and reduced viscosity or stiffness in the droplet. The combined effect of many molten particles impacting the surface lead to the production of a coating made of many layers of splat. Within each splat, there are columnar grain structures, extending perpendicular to the splat boundaries, due to the rapid quenching of splats by the substrate and thus preferential solidification.[112,149]

Various types of defects are present in coatings, including pores, voids, and microcracks. Pores and voids are present due to the uneven spreading of splats, trapping of unmelted particles, or incomplete overlap of splats. The horizontal microcracks, lying along splat boundaries, form because of weak adhesion between splats and the presence of substantial amounts of micro- and nano-pores at the splat boundaries.[150] The vertical microcracks are present at low-adhesion splat boundaries or within the splats, resulting from the quenching stress and the subsequent stress relaxation during the cool-down process.[151]

The cooling rates during the solidification of splats are very high and range from 10^6 to 10^8 K/s, mainly depending on the substrate material and the interfacial condition between the coating and substrate.[113,152] Such high cooling rates frequently result in the formation of metastable phases and supersaturated solid solutions in the coating. The best-known examples are the formation of the γ-Al_2O_3 phase while spraying α-Al_2O_3 powder particles[153] and the formation of the γ-Al_2O_3 supersaturated solid solution in which 13 wt.% TiO_2 is dissolved.[154,155] Another technologically important example is the formation of the metastable t′-phase from APS of ZrO_2 with 15 wt.% Y_2O_3 powder.[156] The formation of these metastable phases alters the properties of coatings and may lead to instabilities and premature failure of the coating during service.[157,158]

11.4.2.1 Stress Development in Coatings

Residual stresses in the coating have been recognized as one of the most important characteristics and these affect the performance (e.g., themal cycling life, thermal shock resistance, adhesion strength, and erosion resistance) of the coating strongly.[159–161] Residual stresses in thermal-sprayed coatings are normally from three sources.[161–163] One is due to the rapid quenching of splats by the substrate and is referred to the "quenching stress," the other due to the thermal mismatch between the coating and substrate that takes place during the postdeposition cooling to room temperature and termed as the "thermal stress," and the last due to high-speed impacts of molten particles or semisolid particles which induce a "peening effect" on the underlying sprayed layer. The stress due to "peening effect" is normally observed at HVOF, D-Gun, and cold sprayed coatings, and not important in plasma spray.[163]

Quenching stresses at each layer are always tensile in nature and its maximum value, σ_{max}, can be described by:[160,161]

$$\sigma_{max} = \alpha_0 \Delta T E_0 \tag{11.9}$$

where α_0 and E_0 are the coefficient of thermal expansion (CTE) and elastic modulus of the coating respectively, and $\Delta T = T_m - T_s$ with T_m being the melting temperature of the coating and T_s the substrate temperature. Quenching stresses can be partially released by the formation of perpendicular microcracks, and partially compensated by the compressive stresses induced by the continual deposition of splats on top of the current layer.[160] Thermal stresses, σ_{TC}, can be estimated with the aid of:[163]

$$\sigma_{TC} = \Delta\alpha \Delta T E_0 \tag{11.10}$$

where $\Delta\alpha$ is the CTE mismatch between the coating and substrate and ΔT is the temperature difference upon postdeposition cooling. The residual stresses in plasma and flame-sprayed coatings would be the sum of quenching and thermal stresses as well as the stresses associated with solid-state phase transformations during postdeposition cooling.

11.4.2.2 Mechanical, Thermal, and Chemical Properties of Coatings

The properties of thermal-sprayed coatings depend strongly on their microstructure, phase contents, and residual stresses. Such coating structure/property relationships can be well illustrated with several prominent examples enumerated below. The formation of W and W_2C in WC–Co coatings due to the undesired decarburization results in lower hardness value in these wear and erosion resistance coatings.[41] Compressive residual stresses in coatings can prevent degradation of fatigue properties of coated components, while tensile residual stresses degrade fatigue properties.[47,48] Nanostructured coatings such as Cr_3C_2-25(Ni20Cr)[164] and Ni[165] exhibit higher hardness, strength, and corrosion resistance than the corresponding conventional coatings. Splat boundaries are weak links in thermal-sprayed coatings and often the path of macrocracks. However, microstructure engineering through the introduction of a bi-modal microstructure, a fully melted splat structure, and a partially melted particulate structure can improve the fracture toughness of ceramic coatings by 100%, as revealed in Al_2O_3-13 wt.%TiO_2 coatings (Figure 11.12 and Figure 11.13). The partially melted particulate region serves to trap and deflect the splat boundary cracks and thus the improved toughness.[166] This improved toughness translates into a 100% improvement in adhesion strength and 300% increase in abrasive wear resistance.[29] The life of TBCs can be extended by (1) increasing the contact between the splat and substrate and between the splats themselves through fine splat sizes, and (2) introducing a network of through-thickness vertical cracks, both of which can be achieved simultaneously via the solution precursor plasma spray process.[36] Pores in TBCs are intentionally produced because they contribute to the low thermal conductivity of the coatings.[167] The thermal conductivity of TBCs decreases with increasing the pore volume fraction, while it varies little with the sizes of micro-pores.[168] The corrosion resistance of properly coated components can be improved by 30 times, as demonstrated in various iron substrates with plasma-sprayed Al_2O_3 coatings.[169]

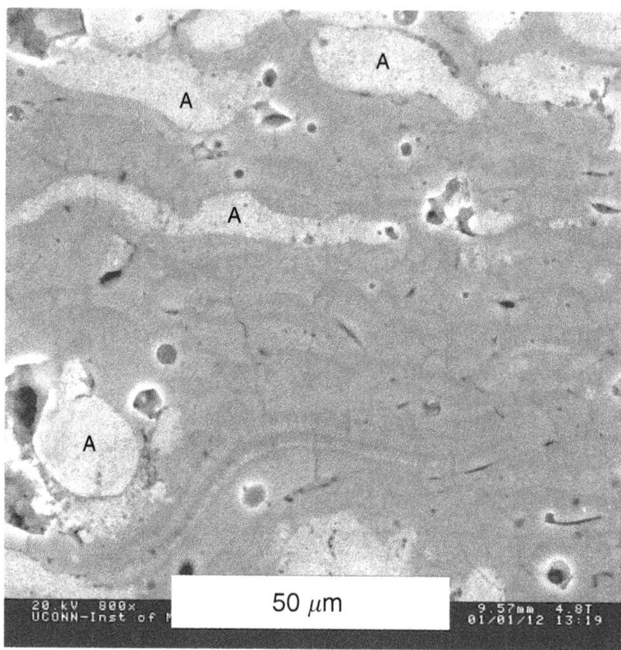

FIGURE 11.12 Air plasma-sprayed Al_2O_3-13 wt% TiO_2 coating with a bi-modal microstructure. Region A has a partially melted particulate structure, whereas the rest has a fully melted splat microstructure (Reprinted from Luo, H. et al., *Mater. Sci. Eng. A*, 346, 237–245. Copyright (2003) with permission from Elsevier).

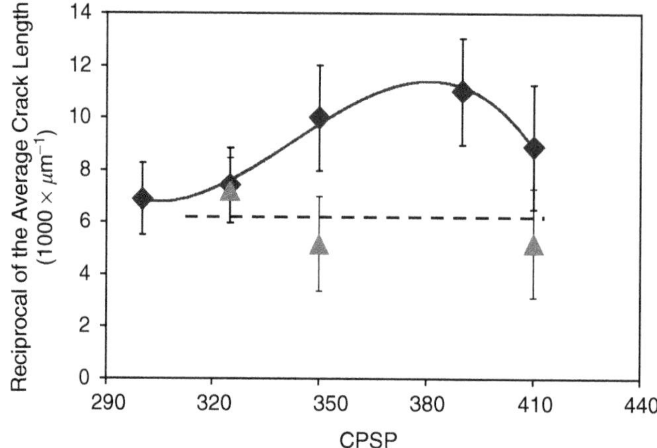

FIGURE 11.13 Fracture toughness (represented here by the reciprocal of the average indentation crack length) exhibits ~100% improvements for the coatings (diamond-shaped data) with a bi-modal microstructure shown in Fig. 3 over the coatings with all splat microstructure (triangle-shaped data) (Reprinted from Luo, H. et al., *Mater. Sci. Eng. A*, 346, 237–245. Copyright (2003) with permission from Elsevier).

11.4.2.3 Process Optimization for Desired Microstructure and Properties

The examples presented in Section 11.4.2.3 underscore the importance of obtaining the desired microstructure, which is in turn controlled by spray coating methods and processing conditions. The key characteristics of each thermal and cold spray process have been described in Section 11.2.2 and Section 11.2.3. Thus, the focus here will only be on the processing parameters of plasma sprays and their optimization for the desired microstructure and properties.

In general, the processing parameters of plasma spray can be categorized into three groups: (1) plasma condition, such as plasma power, plasma gas flow rate, plasma gas species, and geometry of the electrodes[17,25,26,170–174]; (2) powder feed parameters, including the size and shape of the powder, flow rates of the powder and the carrier gas, and powder injection site;[173,174] and (3) spray parameters, comprising standoff distance, velocity of the plasma torch and substrate movement, substrate roughness, substrate and coating surface temperature.[26,173,175–178] The effects of these processing parameters on the coating microstructure and properties have been discussed extensively in the references listed above. From a scientific viewpoint, particle velocity, particle temperature, and substrate temperature are the fundamental parameters, which determine the deposit build-up process and coating properties and will be highlighted here.

The most important parameter for the plasma condition is plasma power, which will increase the ability of the plasma jet to heat and accelerate the powder particles.[170,171] Recent studies[29,30,155] indicate that the ratio of the plasma power to the plasma gas flow rate, termed the critical plasma spray parameter (CPSP), is a good indicator of the plasma torch/particle temperature. By controlling CPSP with agglomerated nanoparticles, a bi-modal microstructure can be created that leads to substantial improvements in adhesion strength, fracture toughness, wear resistance, and spallation resistance to bend- and cup-tests.[29,166] The microstructure of powder particles can dramatically change the thermal conductivity of particles, and thus influence the particle temperature and coating microstructure.[155] Increasing plasma power and carrier gas flow rate at the same time can increase the particle temperature and velocity simultaneously, which leads to thinner splats, denser coatings, and higher hardness.[179] With an increase of substrate temperature, splat morphology changes from highly fragmented to contiguous disk-like shape, and coating integrity and property enhancement.[179,180] By increasing the substrate temperature, residual stresses in the molybdenum coating on steel substrates can be altered from tension to compression.[179] All of these examples illustrate the importance of process control and optimization to achieve the desired microstructure and properties.

11.5 Future Developments

Spray deposition has demonstrated its advantages in the development of materials with chemistries that makes them difficult to process by conventional manufacturing routes such as IM and powder metallurgy. It has also shown its potential for direct near net-shape manufacturing, even though mostly limited to relatively simple geometries such as sheet/plate, billet, ring, or tube shapes. However, fundamental studies on the spray-deposition process are still necessary to explore the full potential of spray-deposition technology. For instance, a better, more comprehensive understanding of grain refinement mechanisms can help guide the selection of suitable controlling processing parameters to achieve predefined microstructures. Efforts in spray processes[49–54] are expected to integrate with nanotechnology particularly in solid-state spray processes, which present unparalleled opportunities for the generation of nanostructured coatings with excellent properties.

Developments in thermal spray processes to achieve engineered microstructures such as bi-modal microstructures[29,166] for superior adhesion strength, fracture toughness, and spallation resistance are expected to continue in the near future. It is anticipated that efforts in reactive thermal sprays[22–24] that allow the formation of finely dispersed reinforcements or functional particles *in situ* during spray will increase in the next decade or two. Reactive thermal sprays offer new capabilities to deposit novel coatings that are not achievable currently by conventional thermal spray processes.

The ability to truly capture the processing–structure–property relationship in spray deposition and coatings processes hinges on the establishment of a strong link between the deposition stage and microstructure development and evolution in the deposits or coatings. This is a challenging endeavor in which modeling and simulation efforts are likely to play a major role. It will also require integral models to couple spray impacts models with models and simulation tools developed to predict microstructure evolution. As an illustration of the deficiencies in this area, optimized operation of a typical spray-deposition process requires that most droplets arrive on the target in a partially solidified state (see Section 11.2.1.1), while, to our knowledge, simulations of droplet impacts have been limited to completely liquid droplets.

Existing integral models can capture macroscopic features, such as macro-pores, in the deposit or coating, but micro-scale features are still elusive. This is an important issue as a complete integral model, one that incorporates the controlling mechanisms in all intermediate stages and thus provides a quantitative, physics-based connection between process parameters and resulting material properties, would provide experimentally inaccessible as well as invaluable insight for process optimization and material design. Furthermore, as demonstrated in other applications, integral process models can serve as a solid basis in the development of physics-based reduced-order models required to build efficient process control algorithms.[181]

Acknowledgments

JPD, SBJ, and YZ would like to thank Drs. E.J. Lavernia, K.M. McHugh, and Y. Lin for numerous and fruitful interactions in the field of Spray Atomization and Deposition and beyond. The support of the Office of Industrial Technologies, Energy Efficiency and Renewable Energy, U.S. Department of Energy under grant DE-FC07-00ID13816 is also gratefully acknowledged. LS would like to thank many of his past and current students who have contributed significantly to the thermal spray projects under ONR Contract N00014-98-C-0010. In particular, the contributions from Dr. D. Goberman, Dr. J. Villegas, and Mr. H. Luo are acknowledged. LS is also indebted to his colleagues, Drs. M. Gell, E. Jordan, N. Padture, T. Bergman, and B. Cetegen with whom he has the privilege to work together in the thermal spray projects sponsored by ONR.

References

[1] Kodas, T.T. and Hampden-Smith, M.J., *Aerosol Processing of Materials*, Wiley-VCH, New York, 1999.

[2] Lavernia, E.J. and Wu, Y., *Spray Atomization and Deposition*, John Wiley & Sons, Inc., New York, 1996.

[3] Grant, P.S., Spray forming, *Prog. Mater. Sci.*, 39, 497, 1995.

[4] Leatham, A., Commercial spray forming: Exploiting the metallurgical benefits, *Mater. World*, 4, 317, 1996.

[5] Singer, A.R.E., Sprayforming of sheet and strip metals, in *Casting of Near Net Shape Products*, Sahai, Y. et al., Eds. The Metallurgical Society, Inc., Warrendale, PA, 1988, pp. 245–256.

[6] McHugh, K.M., Microstructure transformation of spray-formed H13 tool steel during deposition and heat treatment, in *Solidification 1998*. The Minerals, Metals and Materials Society, Warrendale, PA, 1998.

[7] Lavernia, E.J., McKewan, G.W., and Grant, N.J., Structure and properties of rapidly solidified aluminum alloys 2024 and 2024 plus 1% lithium using liquid dynamic compaction, *Prog. Powder. Metall.*, 42, 457, 1986.

[8] Zhang, G.Q. et al., Spray forming and thermal processing for high performance superalloys, *Materials Science Forum*, 475–479, 2773, 2005.

[9] Cai, W. and Lavernia, E.J., Modeling of porosity during spray forming I. Effects of processing parameters, *Metall. Mater. Trans. B*, 29, 1085, 1998.

[10] Cai, W. and Lavernia, E.J., Modeling of porosity during spray forming II. Effects of atomization gas chemistry and alloy compositions, *Metall. Mater. Trans. B*, 29, 1097, 1998.

[11] McHugh, K.M. et al., Spray rolling aluminum alloy strip, *Mater. Sci. Eng. A*, A383, 96, 2004.

[12] Yuan, M., Lokyer, S., and Hunt, J.D., Twin roll casting of aluminium alloys, *Mater. Sci. Eng. A*, A280, 116, 2000.

[13] Unigame, Y., Lawley, A., and Apelian, D., In-situ spray casting of dispersion strengthened alloys — I: thermodynamics and reaction kinetics, in *Proceedings of the Powder Metallurgy World*

Congress, Bando, Y. and Kosuge, K., Eds. Japan Society of Powder and Powder Metallurgy, Kyoto, Japan, 14E-T6-6, 1993.
[14] Lavernia, E.J. and Delplanque, J.P. Spray casting: applications, in *Encyclopedia of Materials: Science and Technology*, Elsevier Science, 2001, pp. 8771–8776.
[15] Dai, S.L., Delplanque, J.P., and Lavernia, E.J., Microstructural characteristics of 5083 Al alloys processed by reactive spray deposition for net-shape manufacturing, *Metall. Mater. Trans. A*, 29A, 2597, 1998.
[16] Clare, J.H. and Crawmer, D.E., Thermal spray coatings, in *Metals Handbook, Vol. 5: Surface Cleaning, Finishing and Coating*, 9th ed., ASM International, Metals Park, OH, 1982, p. 361.
[17] Pawlowski, L., *The Science and Engineering of Thermal Spray Coatings*, John Wiley & Son Ltd., England, 1995.
[18] Simonin, L. and Liao, H., Characterization of flame-sprayed PEEK coatings by FT-IR-ATR, DSC and acoustic microscopy, *Macromol. Mater. Eng.*, 283, 153, 2000.
[19] Li, J., Liao, H., and Coddet, C., Friction and wear behavior of flame-sprayed PEEK coatings, *Wear*, 252, 824, 2002.
[20] Brantner, H.P., Pippan, R., and Prantl, W., Local and global fracture toughness of a flame sprayed molybdenum coating, *J. Therm. Spray Technol.*, 12, 560, 2003.
[21] Lin, L. and Han, K., Optimization of surface properties by flame spray coating and boriding, *Surf. Coat. Technol.*, 106, 100, 1998.
[22] Liu, C.S., Huang, J.H., and Yin, S., The influence of composition and process parameters on the microstructure of TiC-Fe coatings obtained by reactive flame spray process, *J. Mater. Sci.*, 37, 5241, 2002.
[23] Du, A.K., Wang, J.J., Ye, M.H., and Yan, J., Wear of the Al_2O_3-based multi-ceramic coatings produced by SHS flame spray, *Key Eng. Mater.*, 280–283, 1123, 2005.
[24] Tikkanen, J. et al., Characteristics of the liquid flame spray process, *Surf. Coat. Technol.*, 90, 210, 1997.
[25] Fauchais, P. and Verdelle, A., Thermal plasmas, *IEEE Trans. Plasma Sci.*, 25, 1258, 1997.
[26] Semenov, S. and Cetegen, B., Spectroscopic temperature measurement in direct current arc plasma jets used in thermal spray processing of materials, *J. Thermal Spray Technol.*, 10, 326, 2001.
[27] DeMasi-Marcin, J.T., Sheffler, K.D., and Bose, S., Mechanisms of degradation and failure in a plasma-deposited thermal barrier coating, *J. Eng. Gas Turbines Power*, 112, 521, 1990.
[28] Padture, N.P., Gell, M., and Jordan, E.H., Thermal barrier coatings for gas-turbine engine applications, *Science*, 296, 280, 2002.
[29] Jordan, E.H. et al., Fabrication and evaluation of plasma sprayed nanostructured alumina-titania coatings with superior properties, *Mater. Sci. Eng.*, A301, 80, 2001.
[30] Shaw, L. et al., The dependency of microstructure and properties of nanostructured coatings on plasma spray conditions, *Surf. Coat. Technol.*, 130, 1, 2000.
[31] Miller, R.A., Current status of thermal barrier coatings — An overview, *Surf. Coat. Technol.*, 30, 1, 1987.
[32] Brindley, W.J. and Miller, R.A., Thermal barrier coating life and isothermal oxidation of low-pressure plasma-sprayed bond coat alloys, *Surf. Coat. Technol.*, 43–44, 446, 1990.
[33] Xie, L. et al., Deposition mechanisms of thermal barrier coatings in the solution precursor plasma spray process, *Surf. Coat. Technol.*, 177–178, 103, 2004.
[34] Xie, L. et al., Processing parameter effects on solution precursor plasma spray process spray patterns, *Surf. Coat. Technol.*, 183, 51, 2004.
[35] Karthikeyan, J. et al., Nanomaterial deposits formed by DC plasma spraying of liquid feedstocks, *J. Am. Ceram. Soc.*, 81, 121, 1998.
[36] Gell, M. et al., Highly durable thermal barrier coatings made by the solution precursor plasma spray process, *Surf. Coat. Technol.*, 177–178, 97, 2004.

[37] Hedges, M.K., Newbery, A.P., and Grant, P.S., Characterisation of electric arc spray formed Ni superalloy IN718, *Mater. Sci. Eng.*, A326, 79, 2002.
[38] Newbery, A.P. et al., Electric arc spray forming of an Ni_3Al based alloy, *Scripta Mater.*, 35, 47, 1996.
[39] Sacriste, D. et al., An evaluation of the electric arc spray and HPPS processes for the manufacturing of high power plasma spraying McrAlY coatings, *J. Thermal Spray Technol.*, 10, 352, 2001.
[40] Uma Devi, M., Microstructure of Al_2O_3-SiC nanocomposite ceramic coatings prepared by high velocity oxy-fuel flame spray process, *Scripta Mater.*, 50, 1073, 2004.
[41] Kear, B.H. et al., Thermal sprayed nanostructured WC/Co hardcoatings, *J. Thermal Spray Technol.*, 9, 399, 2000.
[42] Factor, M. and Roman, I., Vickers microindentation of WC-12%Co thermal spray coating, Part 1: statistical analysis of microhardness data, *Surf. Coat. Technol.*, 132, 181, 2000.
[43] Schadler, L.S. et al., Microstructure and mechanical properties of thermally sprayed silica/nylon nanocomposites, *J. Thermal Spray Technol.*, 6, 475, 1997.
[44] Petrovicova, R. et al., Structure and properties of HVOF sprayed ceramic/polymer nanocomposite coatings, in *Thermal Spray: A United Forum for Scientific and Technological Advances*, Berndt, C.C., Ed. ASM International, Materials Park, Ohio, USA, 1997, 877.
[45] He, J., Ice, M., and Lavernia, E.J., Synthesis of nanostructured Cr Cr-25 ($Ni_{20}Cr$) coatings, *Metall. Mater. Trans. A.*, 31A, 555, 2000.
[46] He, J., Ice, M., and Lavernia, E.J., Synthesis of nanostructured WC-12% Co coating using mechanical milling and HVOF thermal spraying, *Metall. Mater. Trans. A.*, 31A, 541, 2000.
[47] Tucker, R.C., Considerations in the selection of coatings, *Adv. Mater. Proc.*, 162, 25, 2004.
[48] Quests, J. and Alford, J.R., Praxair thermal spray coatings for chrome plate replacement, *Document of Praxair Surface Technology, Inc.*, June 1997.
[49] Rocheville, C.F., Device for treating the surface of a workpiece, U.S. Patent 3,100,724, August 1963.
[50] Kim, H.-J., Lee, C.-H., and Hwang, S.-Y., Superhard nanoWC-12%Co coating by cold spray deposition, *Mater. Sci. Eng.*, A391, 243, 2005.
[51] Van Steenkiste, T.H. et al., Kinetic spray coating method and apparatus, U.S. Patent 6,139,913, October 2000.
[52] Van Steenkiste, T.H., Smith, J.R., and Teets, R.E., Aluminum coatings via kinetic spray with relatively large powder particles, *Surf. Coat. Technol.*, 154, 237, 2002.
[53] Tapphorn, R.M. and Gabel, H., The solid-state spray forming of low-oxide titanium components, *JOM*, 50, 45, 1998.
[54] Tapphorn, R.M. and Gabel, H., Kinetic energy metallization process feasibility study, Final report for the University of California Lawrence Livermore National Laboratory, 1996.
[55] Van Steenkiste, T. and Smith, J.R., Evaluation of coatings produced via kinetic and cold spray processes, *J. Thermal Spray Technol.*, 13, 274, 2003.
[56] Gilmore, D.L. et al., Particle velocity and deposition efficiency in the cold spray process, *J. Thermal Spray Technol.*, 8, 576, 1999.
[57] Lee, H.Y. et al., Cold spray of SiC and Al_2O_3 with soft metal incorporation: A technical contribution, *J. Thermal Spray Technol.*, 13, 184, 2004.
[58] Li, C.-J. and Li, W.-Y., Deposition characteristics of titanium coating in cold spraying, *Surf. Coat. Technol.*, 167, 278, 2003.
[59] Alkimov, A.P., Kosarev, V.F., and Papyrin, A.N., A method of cold gas-dynamic deposition, *Sov. Phys. Dokl.*, 35, 1047, 1990.
[60] Alkimov, A.P. et al., Gas dynamic spraying method for applying a coating, U.S. Patent 5,302,414, April 1994.
[61] Kim, H.J., Lee, C.-H., and Hwang, S.-Y., Fabrication of WC-Co coatings by cold spray deposition, *Surf. Coat. Technol.*, 191, 335, 2005.

[62] Choi, H. et al., Characterization of the spraying beads deposited by the kinetic spraying process, *Surf. Coat. Technol.*, 192, 374, 2005.

[63] Van Steenkiste, T. and Gorkiewicz, D.W., Analysis of tantalum coatings produced by the kinetic spray process, *J. Thermal Spray Technol.*, 13, 265, 2004.

[64] Van Steenkiste, T.H. et al., Kinetic spray coatings, *Surf. Coat. Technol.*, 111, 62, 1999.

[65] Tapphorn, R.M. and Gabel, H., Coating or ablation applicator with debris recovery attachment, U.S. Patent 6,074,135, June 2000.

[66] Gabel, H., Kinetic metallization compared with HVOF, *Adv. Mater. Proc.*, 162, 47, 2004.

[67] Sirignano, W.A. *Fluid Dynamics and Transport of Droplets and Sprays.* Cambridge University Press, 1999.

[68] Fritsching, U. *Spray Simulation: Modelling and Numerical Simulation of Sprayforming Metals.* Cambridge University Press, 2004.

[69] Kee, R.J., Coltrin, M.E., and Glarborg, P. *Chemically Reacting Flow: Theory and Practice.* Wiley-Interscience, 2003.

[70] Delplanque, J.-P. et al., Using CFD Simulations to Improve Spray-Based Materials Processes, 2002 International Conference on Process Modeling in Powder Metallurgy & Particulate Materials, October 28–29, 2002, Newport Beach, California.

[71] Grant, P.S., Cantor, B., and Katgerman, L., Modelling of droplet dynamics and thermal histories during spray forming-I. Individual droplet behavior. *Acta Metall. Mater.*, 41, 3097, 1993.

[72] Pope, S.B., *Turbulent Flows*, Cambridge University Press, 2000.

[73] Hanjalic, K., Will RANS survive LES? A view of perspectives, *J. Fluid Eng.-T ASME*, 127, 831, 2005.

[74] Mostaghimi, J. et al., Modeling thermal spray coating processes: A powerful tool in design and optimization, *Surf. Coat. Technol.*, 163–164, 1, 2003.

[75] Mingheng, L. and Christofides, P.D., Multi-scale modeling and analysis of an industrial HVOF thermal spray process, *Chem. Eng. Sci.*, 60, 3649, 2005.

[76] Zagorski, A.V. and Stadelmaier, F., Full-scale modeling of a thermal spray process, *Surf. Coat. Technol.*, 146–147, 162, 2001.

[77] Ahmed, I. and Bergman, T.L., Simulation of thermal plasma spraying of partially molten ceramics: Effect of carrier gas on particle deposition and phase change phenomena, *J. Heat Transfer*, 123, 188, 2001.

[78] Newbery, A.P., Grant, P.S., and Neiser, R.A., The velocity and temperature of steel droplets during electric arc spraying, *Surf. Coat. Technol.*, 195, 91, 2005.

[79] Grant, P.S., Cantor, B., and Katgerman, L., Modelling of droplet dynamics and thermal histories during spray forming-II. Effect of process parameters. *Acta Metall. Mater.*, 41, 3109, 1993.

[80] Liu, H., Rangel, R.H., and Lavernia, E.J., Modeling of reactive atomization and deposition processing of Ni3Al, *Acta Metall. Mater.*, 42, 3277, 1994.

[81] Delplanque, J.-P., Lavernia, E.J., and Rangel, R.H., Analysis of in-flight oxidation during reactive spray atomization and deposition processing of aluminum. *J. Heat Transfer*, 122, 126, 2000.

[82] Bergmann, D., Fritsching, U., and Bauckhage, K., A mathematical model for cooling and rapid solidification of molten metal droplets, *Int. J. Therm. Sci.*, 39, 53, 2000.

[83] Coimbra, C.F.M. and Rangel, R.H., General solution of the particle momentum equation in unsteady stokes flows. *J. Fluid Mech.*, 370, 53, 1998.

[84] Gosman, A.D. and Ioannides, E., Aspects of computer simulation of liquid-fueled combustors, *J. Energy*, 7, 482, 1983.

[85] Bachalo, W.D. and Houser, M.J., Phase/Doppler spray analyzer for simultaneous measurements of drop size and velocity distributions, *Opt. Eng.*, 23, 583, 1984.

[86] Bauckhage, K., The phase-Doppler-difference-method, a new laser-Doppler technique for simultaneous size and velocity measurements. Part 1: Description of the method, *Part. Part. Syst. Char.*, 5, 16, 1988.

[87] Zhou, Y. et al., Characterization of spray atomization of 3003 aluminum alloy during linear spray atomization and deposition, *Metall. Mater. Trans. B*, 29B, 793, 1998.

[88] Krauss, M. et al., In-situ particle temperature, velocity and size measurements in the spray forming process, *Mater. Sci. Eng. A-Struct.*, A326, 154, 2002.

[89] Armster, S.Q. et al., Thermo-fluid mechanisms controlling droplet based materials processes. *Int. Mater. Rev.*, 47, 265, 2002.

[90] Levi, C.G. and Mehrabian, R., Heat flow during rapid solidification of undercooled metal droplets. *Metall. Trans. A*, 13A, 221, 1982.

[91] Friedlander, S.K., *Smoke, Dust, and Haze: Fundamentals of Aerosol Dynamics*, Oxford University Press; 2nd ed., 2000.

[92] Buchholz, M. et al., Specific enthalpy measurement in molten metal spray, *Mater. Sci. Eng. A-Struct.*, A326, 165, 2002.

[93] Roychoudhary, S. and Bergman, T.L., Response of agglomerated, multi-ceramic particles to intense heating and cooling for thermal plasma spraying simulation, *Numer. Heat Transfer A-Appl.*, 45, 211, 2004.

[94] Attinger, D., Zhao, Z., and Poulikakos, D., An experimental study of molten microdroplet surface deposition and solidification: Transient behavior and wetting angle dynamics. *J. Heat Transfer*, 122, 544, 2000.

[95] Aziz, S.D. and Chandra, S., Impact, recoil and splashing of molten metal droplets. *Int. J. Heat Mass Transfer*, 43, 2841, 2000.

[96] Bussmann, M., Mostaghimi, J., and Chandra, S., On a three-dimensional volume tracking model of droplet impact. *Phys. Fluids*, 11, 1406, 1999.

[97] Delplanque, J.-P. and Rangel, R.H., Simulation of liquid-jet overflow in droplet deposition processes. *Acta Mater.*, 47, 2207, 1999.

[98] Waldvogel, J.M. and Poulikakos, D., Solidification phenomena in picoliter size solder droplet deposition on a composite substrate. *Inter. J. Heat Mass Transfer*, 40, 295, 1997.

[99] Bennett, T. and Poulikakos, D., Splat-quench solidification: Estimating the maximum spreading of a droplet impacting a solid surface. *J. Mater. Sci.*, 28, 963, 1993.

[100] Delplanque, J.-P. and Rangel, R.H., An improved model for droplet solidification on a flat surface. *J. Mater. Sci.*, 32, 1519, 1997.

[101] Madejski, J., Solidification of droplets on a cold surface. *Inter. J. Heat Mass Transfer*, 19, 1009, 1976.

[102] Rein, M., Phenomena of liquid drop impact on solid and liquid surfaces. *Fluid Dyn. Res.*, 12, 61, 1993.

[103] Haller, K.K., Ventikos, Y., Poulikakos, D., and Monkewitz, P., Computational study of high-speed liquid droplet impact. *J. Appl. Phys.*, 92, 2821, 2002.

[104] Bussmann, M., Chandra, S., and Mostaghimi, J., Modeling the splash of a droplet impacting a solid surface. *Phys. Fluids*, 12, 3121, 2000.

[105] Rieber, M. and Frohn, A., A numerical study on the mechanism of splashing. *Int. J. Heat Fluid Flow*, 20, 455, 1999.

[106] Kim, H.-Y. and Chun, J.-H., The recoiling of liquid droplets upon collision with solid surfaces. *Phys. Fluids*, 13, 643, 2001.

[107] Lee, H.J. and Kim, H.Y., Control of drop rebound with solid target motion. *Phys. Fluids*, 16, 3715, 2004.

[108] Houben, J.M., Relation of the adhesion of plasma sprayed coatings to the process parameters: Size, velocity and heat content of the spray particles, Ph.D. Thesis, Technical University of Eindhoven, Eindhoven, Holland, 1988.

[109] Mundo, C., Sommerfeld, M., and Tropea, C., Droplet-wall collisions: Experimental studies of the deformation and breakup process. *Int. J. Multiphase Flow*, 21, 151, 1995.
[110] Wang, G.X., Prasad, V., and Matthys, E.F., Solute distribution during rapid solidification into an undercooled melt. *J. Cryst. Growth*, 174, 35, 1997.
[111] Wang, G.X. and Matthys, E.F., Numerical modelling of phase change and heat transfer during rapid solidification processes: Use of control volume integrals with element subdivision. *Int. J. Heat Mass Transfer*, 35, 141, 1992.
[112] Wilms, V. and Herman, H., Plasma spraying of Al_2O_3 and Al_2O_3-Y_2O_3, *Thin Solid Films*, 39, 251, 1976.
[113] Moreau, C. et al., Temperature evolution of plasma sprayed niobium particles impacting on a substrate, *Surf. Coat. Technol.*, 46, 173, 1991.
[114] Hattel, J.H. and Pryds, N.H., A unified spray forming model for the prediction of billet shape geometry. *Acta Mater.*, 52, 5275, 2004.
[115] Frigaard, I.A., The dynamics of spray-formed billets. *SIAM J. Appl. Math.*, 55, 1161, 1995.
[116] Liang, X. and Lavernia, E.J., Residual stress distribution in spray atomized and deposited Ni3Al. *Scripta Metall. Mater.*, 29, 353, 1993.
[117] Hu, H. et al., Solidification — on the evolution of porosity in spray-deposited tool steels. *Metall. Mater. Trans. A*, 31, 723, 2000.
[118] Cai, W.D. and Lavernia, E.J., Modeling of porosity during spray forming. *Mater. Sci. Eng. A*, 226–228, 8, 1997.
[119] Payne, R.D., Moran, A.L., and Cammarata, R.C., Relating porosity and mechanical properties in spray formed tubulars. *Scripta Metall. Mater.*, 29, 907, 1993.
[120] Zaat, J.H., A quarter of a century of plasma spraying, *Annu. Rev. Mater. Sci.*, 13, 9, 1983.
[121] Kitahara, S. and Hasui, A., A study of bonding mechanism of sprayed coatings, *J. Vac. Sci. Technol.*, 11, 747, 1974.
[122] Ingham, H.S., Jr., Bonding of flame sprayed Ni-Al, *J. Vac. Sci. Technol.*, 12, 773, 1975.
[123] Braguier, M. and Tueta, R., Supersonic and vacuum arc plasma spraying apparatus characteristics and application on metallic coatings, in *Proceedings of the 9th International Thermal Spray Conference*, The Hague, Netherlands, pp. 167–172, 1980.
[124] McPherson, R. and Cheang, P., Microstructural analysis of Ni-Al plasma sprayed coatings, presented at the 12th International Thermal Spraying Conference, London, UK, June 4–9, 1989.
[125] Hasui, A. et al., Argon-helium as an operating gas for plasma spraying, *Surfacing J. Int.*, 1, 127, 1986.
[126] Del Castillo, L. and Lavernia, E.J., Microstructure and mechanical behavior of spray-deposited Al-Cu-Mg(-Ag-Mn) alloys, *Metall. Mater. Trans. A*, 31A, 2287, 2000.
[127] Lavernia, E.J., The evolution of microstructure during spray atomization and deposition, *Int. J. Rapid Solidif.*, 5, 47, 1989.
[128] Chu, M.G. et al., Evaluation of aluminum and nickel alloy materials produced by spray deposition, *Mater. Sci. Eng.*, 98, 227, 1988.
[129] Liang, X., Earthman J.C., and Lavernia, E.J., On the mechanism of grain formation during spray atomization and deposition, *Acta Metall. Mater.*, 40, 3003, 1992.
[130] Liang, X. and Lavernia, E.J., Evolution of interaction domain microstructure during spray deposition, *Metall. Mater. Trans. A*, 25, 2341, 1994.
[131] Manson-Whitton, E.D. et al., Isothermal grain coarsening of spray formed alloys in the semi-solid state, *Acta Mater.*, 50, 2517, 2002.
[132] Yu, F.X. et al., Fundamental differences between spray forming and other semisolid processes, *Mater. Sci. Eng. A-Struct.*, A304–306, 621, 2001.
[133] Xu, Q. and Lavernia, E.J., Influence of nucleation and growth phenomena on microstructural evolution during droplet-based deposition, *Acta Mater.*, 49, 3849, 2001.
[134] Clyne, T.W. and Withers, P.J., *An Introduction to Metal Matrix Composites*, University Press, Cambridge, U.K., 1993.

[135] Kennedy, A.R. and Clyne, T.W., in *Second International Conference on the Semi-Solid Processing of Alloys and Composites.* The Minerals, Metals & Materials Society, Cambridge, MA, 1993.
[136] Singer, A.R.E., Metal matrix composites made by spray forming, *Mater. Sci. Eng. A-Struct*, 135A, 13, 1991.
[137] Cantor, B., Optimizing microstructure in spray-formed and squeeze-cast metal-matrix composites, *J. Microsc.*, 169 (Pt 2), 97, 1993.
[138] Wu, Y., Interaction mechanisms between ceramic particulates and metallic matrices, and mechanical behavior of Al-Si-X (X=SiC, TiB_2) MMCs synthesized using spray atomization, co-injection and deposition, Ph.D. Dissertation, University of California, Irvine, 1994.
[139] Tszeng, T.C., The effects of particle clustering on the mechanical behavior of particle reinforced composites, *Composites B*, 29B, 299, 1998.
[140] Gupta, M., Mohamed, F.A., and Lavernia, E.J., The effect of ceramic reinforcements during spray atomization and codeposition of metal matrix composites: Part I. heat transfer, *Metall. Trans. A*, 23A, 831, 1992.
[141] Zeng, X., Nutt, S.R., and Lavernia, E.J., Microstructural characterization of Ni3Al processed by reactive atomization and deposition, *Metall. Mater. Trans. A*, 26A, 817, 1995.
[142] Lin, Y.J., Zhou, Y., and Lavernia, E.J., On the influence of in-situ reactions on grain size during reactive atomization and deposition, *Metall. Mater. Trans. A*, 35, 3251, 2004.
[143] Lin, Y.J., Zhou, Y., and Lavernia, E.J., Microstructural characteristics of oxides in 5083 Al synthesized by reactive atomization and deposition, *J. Mater. Res.*, 19, 3090, 2004.
[144] Zhou, J., Duszczyk, J., and Korevaar, B.M., As-spray-deposited structure of an Al-20Si-5Fe Osprey preform and its development during subsequent processing, *J. Mater. Sci.*, 26, 5275, 1991.
[145] Kim, W.J., Yeon, J.H., and Lee, J.C., Superplastic deformation behavior of spray-formed hyper-eutectic Al-25Si alloy, *J. Alloy. Compd.*, 308, 237, 2000.
[146] Zhang, J.G. et al., Microstructure and mechanical properties of spray formed ultrahigh-carbon steels, *Mater. Sci. Eng. A-Struct.*, A383, 45, 2004.
[147] Mesquita, R.A. and Barbosa, C.A., Spray forming high speed steel — properties and processing, *Mater. Sci. Eng. A-Struct.*, A383, 87, 2004.
[148] Zhang, J.G., Shi, H.S., and Sun, D.S., Research in spray forming technology and its applications in metallurgy, *J. Mater. Process. Technol.*, 138, 357, 2003.
[149] McPherson, R., The relationship between the mechanism of formation, microstructure and properties of plasma sprayed coatings, *Thin Solid Films*, 83, 297, 1981.
[150] McPherson, R. and Shafer, B.V., Interlamellar contact within plasma-sprayed coatings, *Thin Solid Films*, 97, 201, 1982.
[151] Nakamura, T., Qian, G., and Berndt, C.C., Effect of Pores on mechanical properties of plasma-sprayed ceramic coatings, *J. Am. Ceram. Soc.*, 83, 578, 2000.
[152] Fantassi, S. et al., Investigation of the splat formation versus different particulate temperatures and velocities prior to impact, in *Proceedings of 13th International Thermal Spray Conference*, Orlando, Florida, USA, 1992, pp. 755–760.
[153] McPherson, R., Formation of metastable phases in flame- and plasma-prepared alumina, *J. Mater. Sci.*, 8, 851, 1973.
[154] Kear, B.H. et al., Plasma-sprayed nanostructured Al_2O_3/TiO_2 powders and coatings, *J. Thermal Spray Technol.*, 9, 483, 2000.
[155] Goberman, D. et al., Microstructure development of plasma sprayed alumina — 13wt.% titania coatings derived from nanocrystalline powders, *Acta Mater.*, 50, 1141, 2002.
[156] McPherson, R., A review of microstructure and properties of plasma sprayed ceramic coatings, *Thin Solid Films*, 39–40, 173, 1989.
[157] Krishnan R., Dash S., Sole R.K., Tyagi A.K., and Raj B., Fabrication and characterisation of laser surface modified plasma sprayed alumina coatings *Surf. Eng.*, 18, 208, 2002.
[158] Pan M., Chen Y.B., and Gani M.S.J., Stress analysis of alumina coatings on silicon carbide based refractories during thermal cycling. *J. Wuhan Univ. Technol. Mater. Sci. Edition*, 14, 9, 1999.

[159] Callus, P.J. and Berndt, C.C., Relationships between the Mode II fracture toughness and microstructure of thermal spray coatings, *Surf. Coat. Technol.*, 114, 114, 1999.
[160] Qian, G., Nakamura, T., and Berndt, C.C., Effects of thermal gradient and residual stresses on thermal barrier coating fracture, *Mech. Mater.*, 27, 91, 1998.
[161] Kuroda, S. and Clyne, T.W., The quenching stress in thermally sprayed coatings, *Thin Solid Films*, 200, 49, 1991.
[162] Matejicek, J. et al., Quenching, thermal and residual stress in plasma sprayed deposits: NiCrAlY and YSZ coatings, *Acta Mater.*, 47, 607, 1999.
[163] Matejicek, J. and Sampath, S., Intrinsic residual stresses in single splats produced by thermal spray processes, *Acta Mater.*, 49, 1993, 2001.
[164] He, J., Ice, M., and Lavernia, E.J., Synthesis of nanostructured Cr_3C_2-25(Ni20Cr) coatings, *Metall. Mater. Trans.*, 31A, 555, 2000.
[165] Lau, M.L. et al., Thermal spraying of nanocrystalline Ni coatings, *Phys. Stat. Sol. (A)*, 166, 257, 1998.
[166] Luo, H. et al., Indentation fracture behavior of plasma-sprayed nanostructured Al_2O_3-13 wt.% TiO_2 coatings, *Mater. Sci. Eng.*, A346, 237, 2003.
[167] Zhu, D. and Miller, R.A., Thermal barrier coatings for advanced gas-turbine engines, *MRS Bull.*, 25, 43, 2000.
[168] Schlichting, K.W., Padture, N.P., and Klemens, P.G., Thermal conductivity of dense and porous yttria-stabilized zirconia, *J. Mater. Sci.*, 36, 3003, 2001.
[169] Dianran, Y. et al., The corrosion behavior of a plasma spraying Al_2O_3 ceramic coating in dilute HCl solution, *Surf. Coat. Technol.*, 89, 191, 1997.
[170] Pfender, E. Thermal plasma technology: Where do we stand and where are we going, *Plasma Chem. Plasma Process*, 19, 1, 1999.
[171] Malmberg, S.J., Analysis of the plasma jet structure, particle motion, and coating quality during DC plasma spraying, Ph.D. Dissertation, University of Minnesota, 1994.
[172] Planche, M.P., Coudert, J.F., and Fauchais, Pl. Velocity measurements for arc jets produced by a DC plasma spray touch, *Plasma Chem. Plasma Process*, 18, 263, 1998.
[173] Matejka, D. and Benko, B., *Plasma Spraying of Metallic and Ceramic Materials*, Chichester, West Sussex, UK; New York, John Wiley & Sons, 1989.
[174] Xia, L., The processing-structure relationships and failure mechanisms of thermal barrier coatings deposited using the solution precursor plasma spray process, Ph.D. Dissertation, University of Connecticut, 2003.
[175] Ahmed, I. and Bergman, T.L., Three dimensional simulation of thermal plasma spraying of partially molten ceramic agglomerates, *J. Thermal Spray Technol.*, 9, 215, 2000.
[176] Steffens, H.D., Babiak, Z., and Gramlich, M., Some aspects of thick thermal barrier coating lifetime prolongation, *J. Thermal Spray Technol.*, 8, 517, 1999.
[177] Bianchi, L. et al., Microstructural investigation of plasma-sprayed ceramic splats, *Thin Solid Films*, 299, 125, 1997.
[178] Vardelle, M. et al., Influence of particle parameters at impact on splat formation and solidification in plasma spraying processes, *J. Thermal Spray Technol.*, 4, 50, 1995.
[179] Jiang, X. et al., Process maps for plasma spray part II: Deposition and properties, in *Proceedings of the International Thermal Spray Conference 2000*, ASM International, Materials Park, OH, 2000, pp. 157–163.
[180] Jiang, X., Matejicek, J., and Sampath, S., Substrate temperature effects on the splat formation, microstructure development and properties of plasma sprayed coatings. Part II: Case study for molybdenum, *Mater. Sci. Eng.*, A272, 189, 1999.
[181] Raja, L.L. et al., Computational algorithm for dynamic optimization of chemical vapor deposition processes in stagnation flow reactors, *J. Electrochem. Soc.*, 147, 2718, 2000.

III

Dislocation-Based Processes

12

Metalworking

C.Y. Barlow
University of Cambridge, UK

N. Hansen
Risø National Laboratory, Denmark

Abstract

The shaping of metals and alloys by cold work is an ancient art that has largely been developed empirically. In recent years, understanding has grown of the parameters governing plastic deformation, and it has become possible to control mechanical properties with a very high degree of precision. Understanding the relationship between the working conditions and the microstructural evolution on one side, and the relationship between microstructure and mechanical properties on the other is the subject of this chapter.

We begin by summarizing the various metalworking processes that are available, comparing them on the basis of the nature and magnitudes of the strains that can be imparted. We then examine the nature and role of dislocations in plastic deformation. A section follows on the experimental techniques that are available for analyzing different aspects of the deformation microstructure. The development of deformation microstructures is then discussed in detail, together with models and hypotheses for their formation and evolution. In the following section, the way in which the microstructure leads to strengthening is examined, leading to a presentation of the different models and equations for the relationship between structural parameters and mechanical properties. Finally, some future trends are identified.

12.1 Metalworking Processes

The aim of metalworking is generally twofold, and may be summarized as achieving shape change by means of externally applied stress, together with an alteration in material mechanical properties. In this section, we will briefly review the main classes of metalworking processes, concentrating on their relationship with the development of microstructure.

12.1.1 Classification and Characterization

Plastic forming processes are traditionally divided broadly into hot and cold deformation, and into specific individual processes such as rolling, extrusion, wire-drawing, as summarized in Table 12.1. The characteristics of the principal traditional processes are given in Table 12.2. These metalworking processes

TABLE 12.1 Processes Used at Different Stages in Metalworking and Their Effects

Purpose and Nature of Processing	Macro-Scale and Micro-Scale Effects
First-stage shaping processes: Hot	
Forging, Rolling, Extrusion	Large-scale shape change Remove porosity Dynamic recrystallization Solutionize (chemically homogeneous; solid solution at elevated temperature); distribute second-phase particles parallel to working direction Generates uniform equiaxed crystallite structure of required grain size Generates texture
Second-stage shaping processes: Cold	
Forging, rolling, wire-drawing. Sheet forming processes, including deep drawing, stretch forming, bending, shearing	Finer-scale shape change Work-hardening; formation of deformation microstructure, including: single dislocations, dislocation walls, subgrain boundaries, twins Crystal structure generally has crystallites elongated parallel to straining directions Generates texture

TABLE 12.2 Characteristics of Principle Traditional Processes[1]

Process	Strain per pass	Strain rate	Stress state	Comments
Forging Hot or cold	0.1–0.5	$1–10^3$	Compression, may flow into shaped die	Open or closed die process
Rolling Hot or cold	0.1–0.5	$1–10^3$	Compress between rolls; elongate in rolling direction.	Rolls may be shaped to give round or channel sections
Wire drawing Cold	0.05–0.5	$1–10^4$	Tension to pull rod through die	Die may be round or more complex shape
Extrusion Hot	2–5	$10^{-1}–10^2$	Compression to force material through die	Die may be round or more complex shape. Extrusion may be direct, indirect, or impact
Sheet forming	0.1–0.5	$1–10^2$	Sheet stretched over former. May use uniaxial or biaxial tension alone, or more frequently in combination with compression, shear, bending	Includes such operations as deep drawing, ironing, sinking, bending, spinning Related processes can be used for tubes

TABLE 12.3 Principal Nontraditional Processes

Process	Process Outline
Accumulative roll bonding[2,3]	Sheet is rolled, stacked, rolled repeatedly. Wire-brushing assists bonding between stacked sheets
Cyclic extrusion compression[3]	Material extruded back and forth through die
Equal channel angular extrusion[3–5]	Material repeatedly extruded around bend (90° or other angle)
High-pressure torsion [3]	Sample rotated in die, introducing high shear strain without altering sample dimensions
Multidirectional deformation[3,6]	Typically compression, performed sequentially along three orthogonal axes
Repetitive corrugation and straightening[3,7,8]	Rolling process in which redundant deformation is introduced using shaped rolls and removed on subsequent pass on flat rolls.

may have different objectives; for example, breaking down a cast structure by hot-rolling, or shaping a component by stretch-forming. During hot-deformation or during the subsequent cooling the metal may recrystallize, leaving material that is almost dislocation-free on the microstructural scale. However, worked metal or alloy more typically contains crystallites and dislocation structures which are characteristic of the process parameters such as stress, strain, strain-rate, and temperature. The term crystallite is a structural parameter defined as a crystal volume with different orientation from its neighbors. Crystallites may be cells, subgrains or grains, and be surrounded by low, medium or high-angle boundaries, or a mixture. Besides these structures, a crystallographic texture characterizes the deformed metal. The microstructure and the texture may be homogeneous throughout the worked sample, but more typically gradients will build up giving rise to heterogeneity of both microstructure and texture. The characteristic length scales of the heterogeneous structures may vary, and can for example be determined by the sample size, the grain size, or a scale determined by the distribution of dislocations and second-phase particles. In all cases, the sample is then found to contain residual elastic stress.

In recent years, a number of nontraditional processes have been added. These have been developed specifically to produce very fine-grain structures in materials by applying very large plastic strains (5 to 10, or higher). These processes are characterized by the desire to introduce high internal strains while substantially retaining the original sample dimensions. The aim is therefore purely to introduce large amounts of redundant work into the materials. The characteristics of the main processes are outlined in Table 12.3. More details about such nontraditional processes are given in Chapter 13.

12.1.2 Process Parameters

The broad distinction between hot and cold working relates to the way in which the structure develops and the mechanical properties change. Hot working is favored when large amounts of deformation are required because the material does not work-harden, so the material reaches steady-state with a low or zero rate of damage accumulation. The absence of work-hardening means that hot-working processes predominantly involve compression and shear and not tension, as work-hardening is required to stabilize tensile deformation. With cold metalworking processes involving significant tension (such as wire-drawing) the amount of reduction is limited by work-hardening behavior. Cold-deformation processes involving high strain, such as those listed in Table 12.3, therefore avoid tensile stress and use only compression and shear.

Strain-rate effects are linked with temperature effects using the Zener–Holloman relationship

$$Z = \dot{\varepsilon} \exp\left(Q/RT\right) \tag{12.1}$$

where Z is the Zener–Holloman parameter, $\dot{\varepsilon}$ is the strain-rate, Q the activation energy, R the gas constant and T the absolute temperature.[9] In microstructural terms, it should be noted that altering the strain rate may change the nature of the deformation; this is discussed further in Section 12.2.2.

12.1.3 Process Efficiency

The minimum work to introduce a required shape change is given by the volume integral of stress over the strain range, $\int_V \sigma \, d\varepsilon$.[10]

In practice, the energy expended is always higher than this: *redundant work* is done inside the material. This relates to energy stored in the form of dislocations and walls,[11] and is accompanied by a temperature rise in the material. Extra energy is also expended in overcoming external friction, for example with dies. The amount of redundant work is dependent on the material, temperature, and also by the characteristics of the metalworking process. Very few metal-working processes produce completely homogeneous deformation of the material. When the levels of plastic deformation in cold-worked material do vary spatially, the material is always found to contain residual elastic stress. The residual stresses balance through the material: regions of tensile stress are matched by regions of compressive stress elsewhere. These stresses are frequently of the order of the material yield stress.

12.2 Role of Dislocations in Plastic Deformation[10]

12.2.1 Strain Using Dislocations

Plastic deformation of metals takes place predominantly by shearing: lattice planes in the material slide over each other, allowing macroscopic shape change without affecting the ordering and arrangements of atoms within the structure. The stress to cause plastic deformation can be reduced by a factor of 1000 if, rather than moving complete lattice planes simultaneously, deformation can be localized by the movement of line defects, which are dislocations.[12] For this reason, plastic deformation of metals depends on the movement of dislocations. Metals, even in the annealed state, contain a statistical density of dislocations (which can be determined using thermodynamic principles)[12] sufficient to allow plastic deformation to take place by this mechanism.

A characteristic intrinsic parameter of a dislocation is its Burgers Vector **b**, which is the amount of shear that it can produce by moving through the material. Dislocations can take a range of geometries,[12,13] and are classified into edge-type (moving and producing deformation normal to the line defect) and screw-type (moving and producing deformation parallel to the line defect). Mixed dislocations have intermediate character.

12.2.2 Crystallographic Effects

Dislocations have intrinsic elastic strain energy approximated by 0.5 $KG\mathbf{b}^2$ per unit length, where G is the shear modulus. K is a constant that takes the value 1 for screw dislocations, and $\{1/(1-\nu)\}$ for edge dislocations (ν is Poisson's ratio), a factor of about 1.5 greater than the energy of a screw dislocation.[12] A consequence of this is that dislocations tend to have the minimum available value of **b** in the structure, which is the shortest separation between atoms in a close-packed direction. To generate plastic strain, dislocations *glide* on close-packed planes. These planes are the most widely separated in a structure, so the stress required for shear deformation is lowest for sliding these planes over each other. Dislocations therefore generally move in close-packed directions, and on close-packed planes. *Conservative motion* of dislocations takes place when dislocations are able to glide in the required direction on a slip plane that contains both their line vector and the Burgers vector; the passage of the dislocation leaves perfect material behind.

Crystals contain a number of possible slip systems (crystallographically equivalent combinations of slip direction and slip plane containing that slip direction). The applied stress direction determines which systems are activated: they will normally be the direction and plane in which the resolved component of the shear stress reaches its highest value.[12] Crystals may deform by single slip (one slip system dominating), or multiple slip (two or more slip systems operating simultaneously). The two regimes are distinct: with single slip, dislocations pass through the material without significantly hindering each other; with multiple slip there is immediately conflict as dislocations run into each other. These give rise to different work-hardening rates. Dislocations cannot pass through each other without interacting and leaving segments of themselves on the other dislocation, *jogs*; these are in general incompatible with the slip plane of the host dislocation, and will leave trails of point defects if they are dragged through the lattice.[13] Dislocation movement of this type is referred to as *nonconservative*. Dislocations trapped within the material and intersecting a slip plane may be termed *forest dislocations*.

Slip systems in face-centered cubic (fcc) materials take the form $\langle 110 \rangle \{1\bar{1}1\}$; there are 12 distinguishable, equivalent systems. The large number of fully close-packed slip systems allows fcc materials to exhibit high ductility at all temperatures and under all loading conditions. Body-centered cubic (bcc) materials contain $\langle 111 \rangle$ close-packed directions, but there are no fully close-packed planes, so slip systems in these materials are more variable. The most commonly observed plane is $\{110\}$, but other planes that have been identified are $\{112\}$ (12 slip systems) and $\{123\}$(24 slip systems).[12,13]

The behavior of hexagonal close-packed (hcp) materials is heavily influenced by the geometry of the unit cell, in the form of the ratio between c and a, the principal crystallographic unit cell parameters. Slip is always in close-packed directions $\langle 11\bar{2}0 \rangle$, but the slip plane is not always the (0001) basal close-packed plane. A wide range of slip systems has been observed, using the pyramidal planes $\{10\bar{1}1\}, \{10\bar{1}2\}, \{11\bar{2}1\}$ and $\{11\bar{2}2\}$.[12,13]

The possibility of a range of crystallographically different slip systems results in materials demonstrating temperature dependence. A familiar example is that of pure or low-carbon iron, ferrite, which has a bcc structure, and which can switch between ductile and brittle behavior in the vicinity of room temperature. The effect occurs in bcc materials because operation of even the lowest-stress slip systems requires some thermal activation, and this can become quenched out at low temperatures.

12.2.3 Stress for Dislocation Movement

The stress to move a dislocation through an otherwise perfect lattice is given by the Peierls–Nabarro equation $\tau_P = 3G \exp - (2\pi w / b)$ where w is the dislocation width, typically up to about 10 atomic spacings in metals.[13]

The stress to move a dislocation is increased by any lattice imperfections, either through interaction of elastic strain fields of the dislocation and the imperfections, or through physical obstacles to dislocation movement, such as incoherent precipitates, high-angle grain boundaries and twin boundaries. These mechanisms are quantified in Section 12.5.1.

A feature of impedance to dislocation movement arising from elastic interactions is that given sufficient applied stress, the dislocations are still able to move, although other deformation or stress-release mechanisms may intervene, such as twinning or fracture. By contrast, dislocations cannot penetrate the physical obstacles, and must halt. They may form arrays in the matrix around the obstacles (e.g., pile-ups at grain boundaries; arrays around second-phase particles), or they may enter the interface and become incorporated into the structure (e.g., extrinsic dislocations in grain boundaries).

Inter-dislocation interaction to form jogs is a special case. As noted in Section 12.2.2 jogs can move, but if alternative mechanisms are available (such as dislocation multiplication by bowing) then they may be immobile.

12.2.4 Dislocation Multiplication and Interaction with Boundaries

12.2.4.1 Frank-Read Sources

As plastic strain increases, so also does the stress required to cause plastic deformation; this is the phenomenon of work-hardening. Dislocations do not disappear from the structure after passing though and causing plastic deformation; they become trapped inside it. Plastic deformation is therefore accompanied by an increase in dislocation density in the material. Annealed material typically contains a dislocation density of 10^{10}m/m^3; this will typically rise to about 10^{15} on straining to about 0.1, and continues to increase as the strain increases further.

Dislocations are nucleated at Frank-Read sources, which require a section of dislocation that is pinned at two points. Sources may be at boundaries (low-angle or high-angle boundaries; interphase boundaries), between incoherent particles, or at pinned dislocation sections within grains. The stress required to operate a Frank-Read source is approximately $\tau_{FR} = Gb/L$ where L is the separation between pinning points,[13] which may be the spacing of forest dislocations.

12.2.4.2 Formation of Boundaries

Dislocations within the structure can reduce their elastic strain energy by forming arrays or boundaries, by a recovery process referred to as polygonization.[12] The boundaries are associated with changes in the lattice orientation, and may form cell, subgrain, or grain boundaries within the material. For small angle boundaries, the misorientation across a dislocation boundary is approximated by $\theta = \mathbf{b}/h$ where h is the spacing of dislocations in the wall. Boundaries may be tilt or twist in character, depending on whether the dislocations are edge or screw respectively. Deformation-induced boundaries are discussed in more detail in Section 12.4.

12.2.4.3 Grain Boundaries as Sources and Sinks for Dislocations

Because of their importance to the mechanical behavior of polycrystalline samples, we will examine the interaction of dislocations with low-angle and high-angle boundaries in more detail. Experimentally, it can be observed that dislocations are able to pass through low-angle grain boundaries, with a stress that is inversely proportional to the spacing of the dislocations in the boundary.[14] Analytically, this is equivalent to dislocations passing through an array of forest dislocations. In both cases, dislocation traverse of the structure can take place by bowing between the pinning points, which are jogs. With high-angle grain boundaries, the spacing between intrinsic dislocations is so small (if indeed it still has physical reality) that dislocations cannot pass through by a bowing mechanism. However, high-angle boundaries can normally act as dislocation sources, though the stresses required for this are high enough that grain boundaries are regarded as a polycrystal strengthening mechanism.

In special cases, there is evidence that dislocations may enter a boundary from one grain and trigger release of dislocations from the boundary into the other grain. For particular orientation relationships, dislocations may effectively pass through the boundary. In both cases, dislocation fragments are left as extrinsic dislocations in the boundary.[15]

12.2.5 Influence of Stacking Fault Energy

The evolution of the dislocation structure depends on the nature of deformation, and in particular whether the material has a low or high stacking fault energy, and how many slip systems are operating. The stacking fault energy (SFE) determines how easy it is for dislocations to cross-slip: to move from one glide plane to another that also contains the dislocation's Burgers vector. High SFE materials are characterized by easy cross-slip; dislocations in low stacking-fault energy materials are *dissociated* (separated into two or more *partial* dislocations, which individually do not cause perfect lattice displacement) to a greater or lesser extent,[12,13] and cross-slip is much less easy. The number of slip systems is naturally relevant because there must be additional highly stressed slip systems available on to which dislocations can move.

12.2.6 Twinning

Shear strain can be achieved by the formation of mechanical twins, which is a very different mechanism from dislocation movement.[13,16] Twinning is a localized shear process, involving the cooperative shifting of lattice planes into crystallographically different configurations. It is a catastrophic process, occurring rapidly and abruptly, and often accompanied by audible clicks.

Mechanical twinning can take place in addition to or in preference to dislocation glide under certain conditions. It most commonly occurs when dislocation glide is difficult for whatever reason, and is a very process-sensitive deformation mechanism. Low dislocation mobility is typically found at low temperatures and high strain-rates. Crystal structure is a major factor: hcp materials frequently demonstrate twinning, and it may be the dominant mechanism. It is common also in bcc materials, particularly at low temperatures. It is less common in fcc metals, although twinning has been observed for deformation at very low temperatures (e.g., Cu at 4 K) or high strain-rates. Other systems in which dislocation mobility is impeded include solid solutions, and twinning is found for deformation of some fcc alloys, for example Ag-Au at room temperature.[16] A somewhat distinct type of twinning on a nano- or micro-scale can be found supplementing the dislocation structure in ultrahigh strained materials.[17]

12.3 Experimental Techniques for Characterization

Deformation of a metal causes changes in the microstructure and in the crystallographic texture. As illustrated in Figure 12.1, length scales associated with dislocation structures range from grain-size features (typically 0.1 to 10 mm) down to perhaps tens of atoms (1 to 10 nm). Because of this large range, characterization normally requires a combination of techniques.

Dislocation structures are composed of characteristic features that relate to the way dislocations are stored. Typically, the structure is subdivided by dislocation boundaries that are associated with different amounts of crystallographic misorientation. Low, medium, and high-angle boundaries are usually defined as providing misorientations of less than 5°, 5 to 15° and above 15° respectively. Loose dislocations are normally present between the dislocation boundaries, and their configurations may be affected by the presence of precipitates or other second-phase particles. The microstructural evolution may be very sensitive to the nature and distribution of such particles, which should therefore be included as part of the full microstructural characterization of the material. In parallel with the evolution of microstructure with increasing strain, a crystallographic texture develops that affects the microstructural evolution, and vice versa. Structural characterization of the deformed state should therefore include both the microstructure and texture.

12.3.1 Classification of Techniques

The complete analysis and characterization of deformation microstructures requires a range of techniques, spanning morphological aspects, crystallographic orientation, and chemical information. The key microstructural features and the principal techniques for their analysis are listed in Table 12.4. In addition to these, chemical microstructural analysis can be used to assist in phase identification, and in assessing the presence of chemical segregation. A range of techniques is available, often based on the emission of characteristic x-rays from atoms excited by electrons during imaging in Transmission Electron Microscopy, TEM, or Scanning Electron Microscopy, SEM. This aspect will not be further considered here.

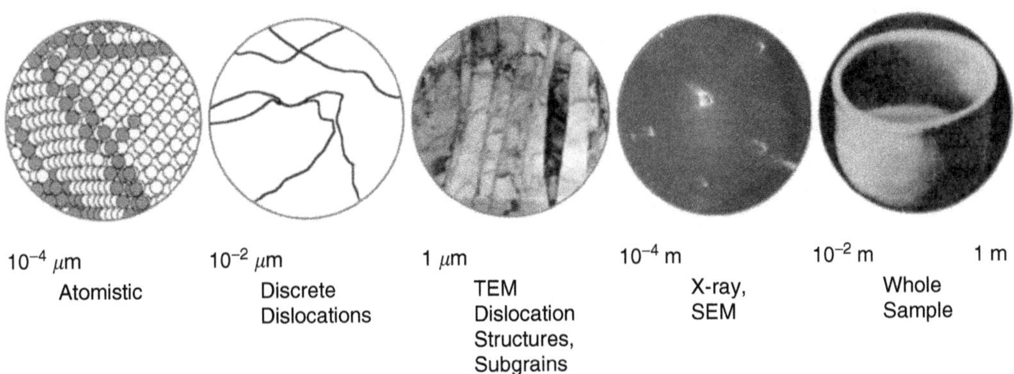

FIGURE 12.1 Length scales relevant to characterization of deformation structures. (Courtesy D.A. Hughes.)

TABLE 12.4 Key Microstructural Parameters and their Analytical Techniques

Structural or Microstructural Feature	Techniques Available	Size of Microstructural Feature
Grain, subgrain and cell sizes and aspect ratios	LM, SEM, TEM (Light, Scanning Electron, Transmission Electron Microscopy)	Grain sizes: Typically 10 μm up to mm Subgrain and cell sizes: 0.1–10 μm
Nature and distribution of second-phase particles	LM SEM TEM	Sizes 0.5–10 μm Sizes 0.1–10 μm Sizes 0.01–3 μm
Macro-texture	Neutron, x-ray,	Whole specimen
Micro-texture (local)	SEM, TEM	Subgrain size 0.1–5 μm
Macroscopic grain and subgrain orientation (with respect to external stress)	SEM, TEM	Grain sizes: Typically 10 μm up to mm Subgrain sizes: 0.1–10 μm
Orientation across grain and subgrain boundaries	SEM, TEM	Spatial resolution required to be at least that of grain or subgrain size.
Dislocations	TEM	5–20 nm
Structure of dislocation boundaries (width; tilt or twist misorientation)	TEM	2–100 nm
Macroscopic stress fields	Neutron, x-ray	Up to sample dimensions: 10 mm or more
Local stress fields	TEM, SEM, x-ray	2–20 nm

TEM and SEM form the basis of a large number of associated techniques.[18] This section is confined to an outline of experimental methods, so details of these will not be given here.

12.3.2 Examples of Use of the Principal Experimental Techniques for Key Parameters

12.3.2.1 Dislocation Density

Dislocation density in thin foils can be measured directly using TEM. Measurements are experimentally challenging (problems include sample preparation, and finding imaging conditions for all dislocations), and thin films may not be representative of bulk material (dislocations may move to foil surface and be lost). Indirect methods include the use of etching to produce pits where dislocations meet the specimen surface;[12,13] such methods can be much more straightforward than TEM but often of lower accuracy.

12.3.2.2 Subgrain Sizes

Subgrains are imaged in SEM and TEM by virtue of the orientation changes between adjacent subgrains.[19] Particularly in TEM, the sample may need to be observed at different tilt conditions to ensure that all subgrains have been revealed. Analysis is tedious but there are no particular problems to be overcome.

12.3.2.3 Second-Phase Particles

The mechanism for revealing particles and distinguishing them from the matrix varies between different techniques, and can include surface relief in LM and SEM (difference in hardness; chemical etching); crystallographic analysis (e.g., dark field imaging in TEM); atomic number contrast in SEM and TEM. In all cases, stereological analysis must be used to convert observed particle distributions to a true 3D distribution.[20]

12.3.2.4 Texture

Macroscopic texture measurements can be made using neutron scattering or x-ray diffraction, generating orientation data for the whole sample.[21] Microscopic or local-scale texture analysis is performed using electron diffraction: TEM for the highest resolution, SEM for larger areas but lower spatial and angular resolution.[22]

12.3.2.5 Local Stress Fields

Local stress fields cannot be revealed directly, but the elastic strain fields that they cause can be imaged directly in TEM as a result of effects on local orientation, and also by local changes to the lattice parameter.[23] High-resolution x-ray diffraction and electron back-scattering diffraction, EBSD are used in a similar way.[24]

12.3.2.6 Boundary Characterization

A complete description of a boundary requires:[19,25]

1. Knowledge of the orientation of the grains on either side of the boundary
2. Knowledge of the orientation of the boundary plane with respect to both grains
3. Description of the structure of the boundary

(1) and (2) are quantitative parameters; (3) is qualitative and is of secondary importance in its effect on mechanical properties.

1. Grain orientations can be measured with high accuracy (0.1°) in TEM using semi-automated analysis of Kikuchi diffraction lines,[26] or in SEM by EBSD with accuracy of up to about 1.5° [22] in a fully automated way. The misorientation between two crystals can be defined either by a rotation matrix, or by specifying an *axis–angle pair,* being a rotation axis and a clockwise rotation angle about that axis. It should be noted that because of the high degree of symmetry of metals, the description is not unique; in cubic crystals, for example, there are 24 equivalent descriptions of the same crystal pair relationship for example Reference 27.
2. Knowledge of the boundary plane can be combined with the misorientation data to allow the atomic-scale structure of the boundary to be identified. This can be helpful in understanding some aspects of dislocation interaction with boundaries.[15]
3. Low-angle boundaries evolve from loose tangles of dislocations; how far this has proceeded is indicated by the *width* of the boundary.[15]

12.4 Development of Deformation Microstructures and Texture

12.4.1 Classification and Terminology

During plastic deformation a small fraction of the mechanical energy (a few percent or less) is stored in the metal in the form of dislocations, which form characteristic 2D and 3D configurations, and which have a profound influence on mechanical properties.[11,31] The formation of 3D structures requires dislocation motion in directions other than those contained in the primary glide plane. This may be achieved by intersecting glide planes, or by dislocations *climbing* out of their primary glide plane. The key factors that affect dislocation mobility are:

- Stacking fault energy
- Presence of solute atoms
- Temperature and strain rate
- Deformation mode and grain orientation
- Strain path changes.

A classification of deformation microstructures in accordance with the mobility of dislocations is described in Reference 32. In this classification metals and alloys are separated into two groups: (1) Wavy glide materials in which 3D structures evolve, and (2) Planar glide materials in which dislocations are stored in 2D arrays. These two types of structures have been termed *Cell Block Structures* and *Taylor Lattice Structures,*

TABLE 12.5 Definitions of Dislocation Structures[32,33]

Term	Definition
Bamboo structure	Two roughly parallel *GNB*s linked by *IDB*s
Cell	Roughly equiaxed volume in which the dislocation density is well below the average and which is rigidly rotated from its neighboring volumes
Cell Block	Contiguous group of *cells* in which the same set of glide systems operates
Cell Boundary	Low-angle dislocation boundary surrounding a *cell;* classified as an *IDB*
Dense Dislocation Wall DDW	Single, nearly planar deformation-induced boundary, classified as geometrically necessary, enclosing a *cell block* at low or intermediate strains
Geometrically Necessary Boundary GNB	Boundary with angular misorientation controlled by the difference in glide-induced lattice rotations in the neighboring volumes
Incidental Dislocation Boundary IDB	Dislocation boundary formed by trapping of glide dislocations, and supplemented by forest dislocations
Lamellar Boundary LB	Single, nearly planar boundary, classified as a *GNB*, enclosing a narrow cell block at high strain
Microband MB	Plate-like region formed by two closely spaced *DDW*s and defining the edge of a *cell block*
S-band	Coarse slip band that intersects parallel groups of *DDW*s, *MB*s or *SB*s creating a string of S-shaped perturbations. Length generally less than grain diameter.
Shear band	Region of intense local shear that spans several grains
Subgrain	Nearly dislocation-free volume surrounded by higher angle boundaries
Taylor Lattice Structure	Relatively uniform distribution of mainly edge dislocations, comprising one or more sets of dislocations of alternating sign.

respectively. In parallel with this general classification, terminology for the individual microstructural features is proposed in Table 12.5.

12.4.2 Assemblies of Dislocations

It is a general observation that the dislocations generated during deformation assemble into characteristic configurations, which typically take the form of dislocation boundaries. These boundaries may form a cell structure, or may form characteristic extended planar boundaries. Observations indicate that there is an absence of significant stress/strain fields associated with these dislocation boundaries. It has therefore been suggested that they represent low-energy dislocation structures (LEDS). Their characteristics have been analyzed on the basis of the LEDS hypothesis,[14] which has proved to be a powerful tool. According to this hypothesis, the configurations of glide dislocations will alter to approach the lowest possible energy per unit length of dislocation line. The ability of dislocations to reach their lowest-energy configurations is constrained by a number of factors, mainly relating to dislocation mobility. The factors include the number of available slip systems, dislocation mobility (for both glide and climb), and the frictional stress.

12.4.3 Development of Dislocation Structures[34]

12.4.3.1 Dislocation Boundaries

During plastic deformation, dislocations multiply and are stored inside the material (Section 12.2.5). The dislocation strain energy is reduced by forming largely dislocation-free regions separated by dislocation-rich walls. The detailed structural evolution of a particle-containing material at low strain is shown schematically in Figure 12.2,[35] and Figure 12.3[33] shows a micrograph of a typical region of cells and dislocation walls. Schematics of the structures developing at higher strains are shown in Figure 12.4, and micrographs of typical regions in Figure 12.5.[32] The net effect is that the grain structure of deforming material becomes subdivided into regions that may deform in different ways. Macroscopic subdivision may be seen by bands traversing the structure; on a microscopic level, cells, cell blocks, and subgrains are formed. Figure 12.6 shows structures in a sample deformed to high strain by high-pressure torsion.

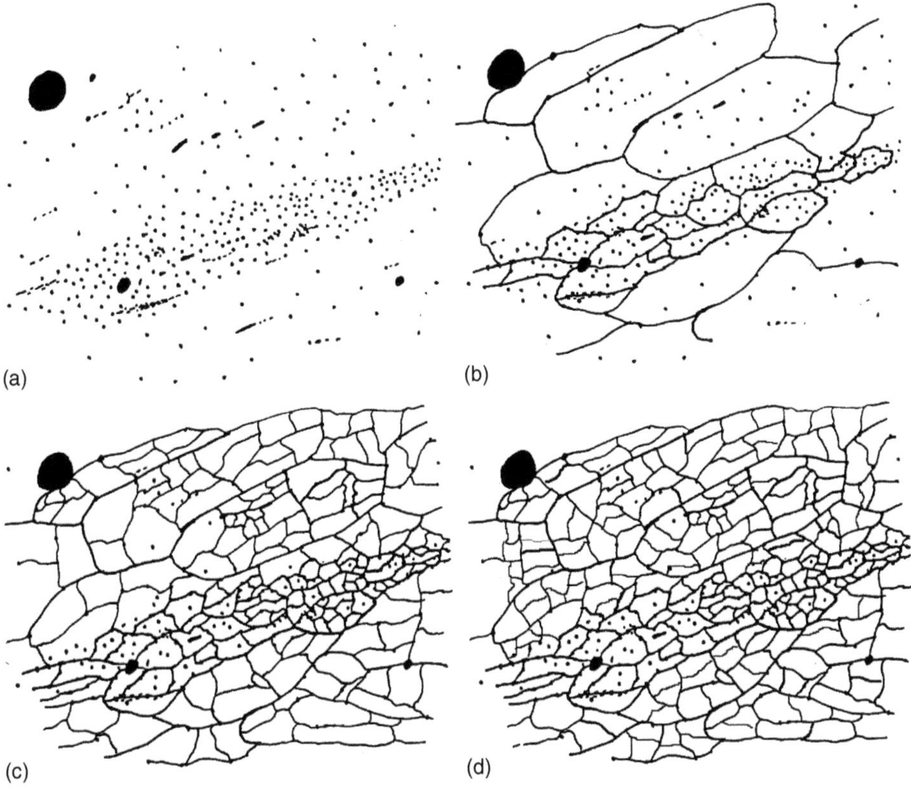

FIGURE 12.2 Schematic drawings of structural evolution in Al containing 4 vol% small Al_2O_3 particles; the particles accelerate the development of the deformation structure. (a) undeformed; strained to (b) ε_{VM} 0.04; (c) ε_{VM} 0.69; (d) ε_{VM} 2.30.

FIGURE 12.3 TEM of 99.98% Al cold-rolled to ε_{VM} 0.16. Well-defined GNBs labeled A and B are separated by a cell-block structure containing cell walls and loose dislocations. (After Barlow, C.Y.J., Bay, B. and Hansen, N., *Philos. Mag. A*, 51, 253, 1985, http://www.tandf.co.uk/journals/titles/14786435.asp. With permission.)

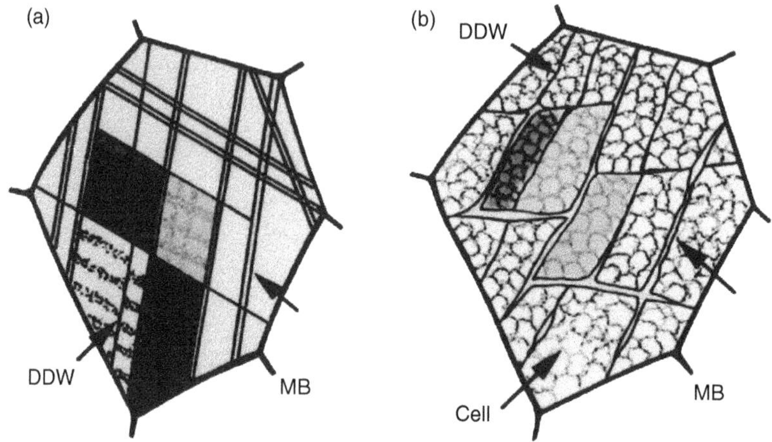

FIGURE 12.4 (a) Taylor lattice structure, typical of low SFE alloys such as Al-Mg, brass. (b) Cell-block structure, typical of Al, Ni, Interstitial Free steel. Courtesy D.A. Hughes.

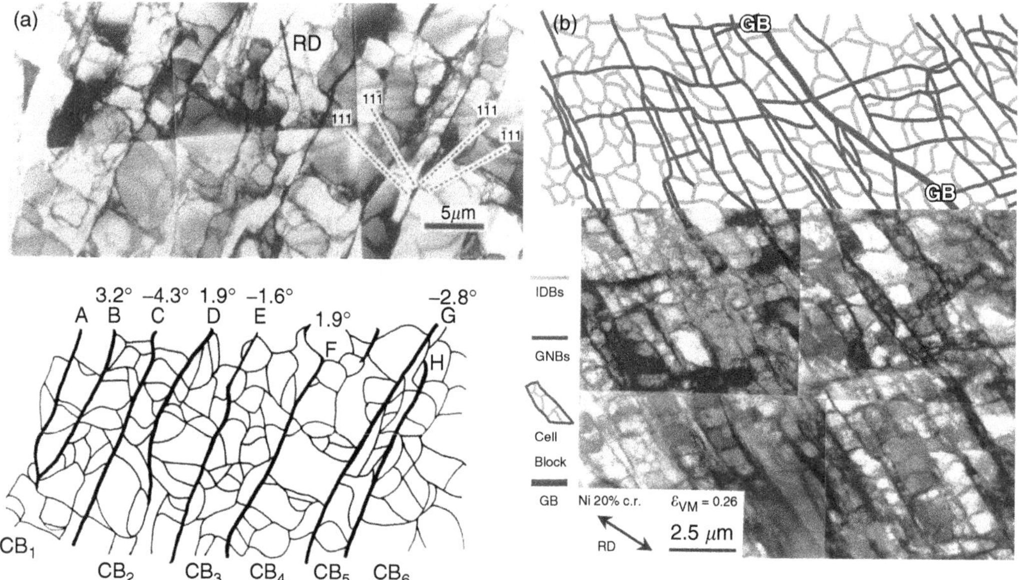

FIGURE 12.5 TEM Micrographs and schematics of cold-deformed structures showing cell blocks separated by GNBs (DDWs or MBs). (a) 99.96% Al ε_{VM} 0.12. (Courtesy Q. Liu.) (b) 99.99% Ni ε_{VM} 0.26. Two almost orthogonal sets of GNBs are seen. A grain boundary GB runs diagonally across the structure. (Courtesy D.A. Hughes.)

Figure 12.7[32] illustrates how the structure changes from a Taylor lattice structure to a cell-block structure and finally to a subgrain structure as the temperature is increased. Micrographs of boundary formation are shown in Figure 12.8.[36]

A distinction must be drawn between the two principal classes of dislocation boundary.[37]

Incidental Dislocation Boundaries, IDBs or cell boundaries, form by a fundamentally statistical or stochastic process of dislocation entrapment (see Figure 12.9a). As dislocations add to the boundary, the misorientation across it changes. Most of the dislocations originated from glide, and so will have Burgers vectors in the same sense. This means that the misorientation across a wall will steadily increase as dislocations add to it.

FIGURE 12.6 TEM of 99.99% Ni cold-deformed by high-pressure torsion to ε_{VM} 12. The shear direction is marked. (Courtesy X. Huang.)

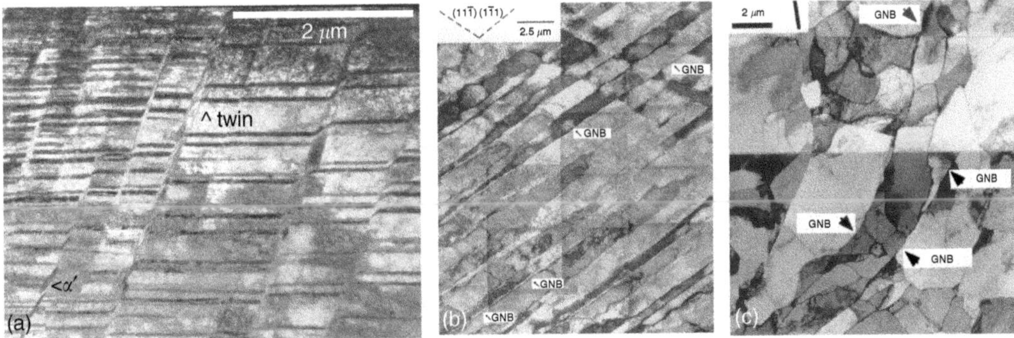

FIGURE 12.7 TEM of 304L stainless steel. (a) ε_{VM} 0.5 at 293 K, showing deformation twins and deformation-induced martensite. (b) ε_{VM} 0.2 at 1073 K. (c) ε_{VM} 0.26 at 1273 K. (After Hughes, D.A. and Hansen, N., *Phil. Mag. A*, 83, 3871, 2003, http://www.tandf.co.uk/journals/titles/14786435.asp. With permission.)

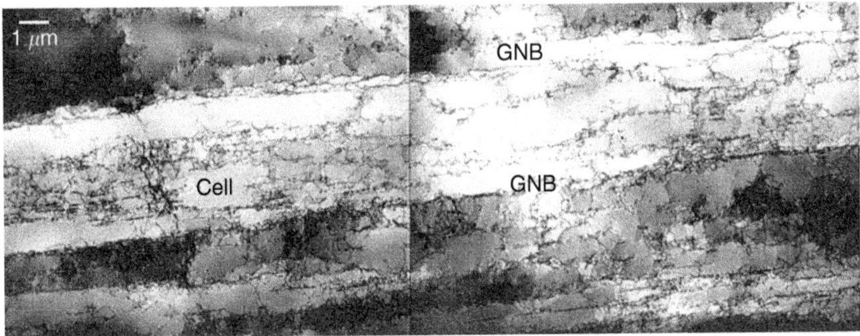

FIGURE 12.8 Extended planar GNBs in Cu single crystal oriented for single slip. Cells and loose dislocations are seen between the GNBs. (Courtesy D.A. Hughes.)

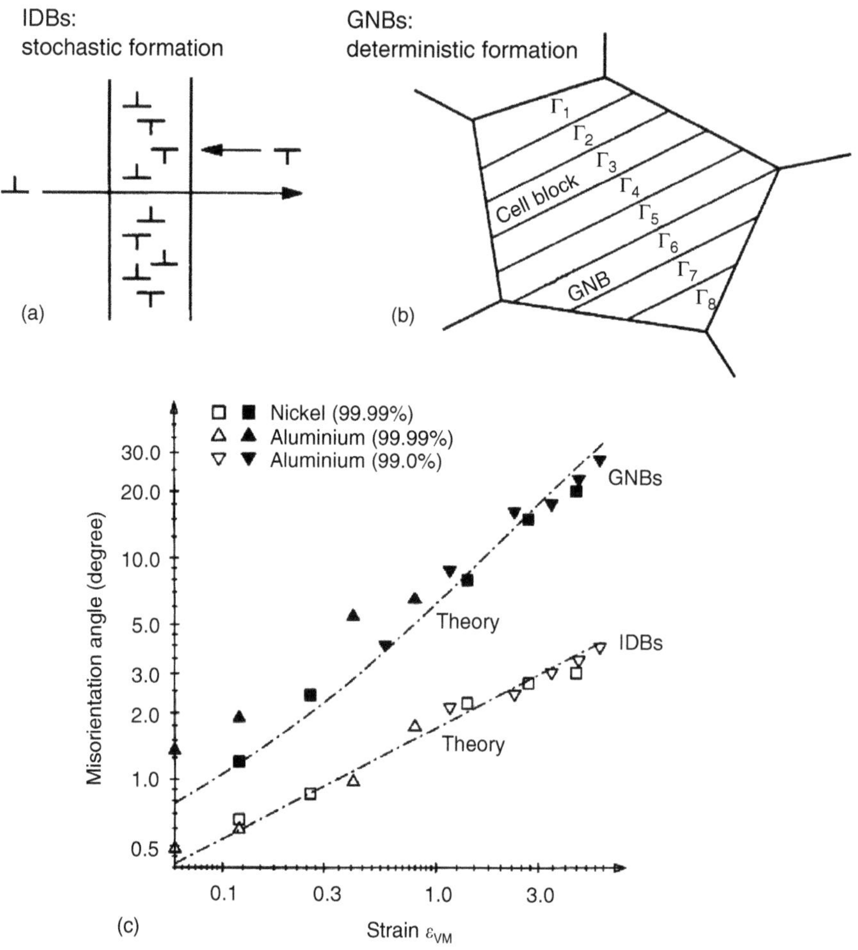

FIGURE 12.9 (a) IDB formation by random dislocation entrapment. (b) GNBs separating cell Blocks Γ. (c) Theoretical predictions of the increase in misorientation with strain match experimental observations for both IDBs and GNBs.

Geometrically Necessary Boundaries, GNBs usually form by a different mechanism from IDBs, and they fulfill specific functions within the structure, as shown in Figure 12.9b. They may separate cell blocks, so forming the boundary between regions deforming using different slip systems. They may also form in response to a variation in partition of shear between a common set of slip systems.

Based on the assumption that the deformation structures in both IDBs and GNBs are low-energy structures, the relationship between a wall's dislocation content and its misorientation axis/angle pair R, θ is given by Frank's formula:[15]

$$d = (r \times R)\, 2\, \sin \theta/2 \tag{12.2}$$

where vector **r** represents an arbitrary straight line lying in the plane of the boundary containing the dislocation network, and d is the sum of all the Burgers vectors of all the dislocations intersected by **r**. The energy associated with a dislocation wall rises with the misorientation until about 15°, and then forms a plateau, with local minima for special orientations and boundary planes.[15]

12.4.3.2 Microstructures

A classification of deformation microstructures is presented in Table 12.6.

TABLE 12.6 Classification of Deformation Microstructures

Category	Features	Examples
Easy 3D dislocation mobility	fcc metals, medium to high SFE, temperatures up to about $0.8T_M$	Al, Ni
	bcc metals at temperatures above about $0.2T_M$	Mild steel
	Low SFE fcc metals at temperatures above about $0.8T_M$	Austenitic stainless steel
Restricted 3D dislocation mobility	hcp materials at low T_M	Magnesium
	Alloys containing short-range order	γ' in Nickel-base superalloys

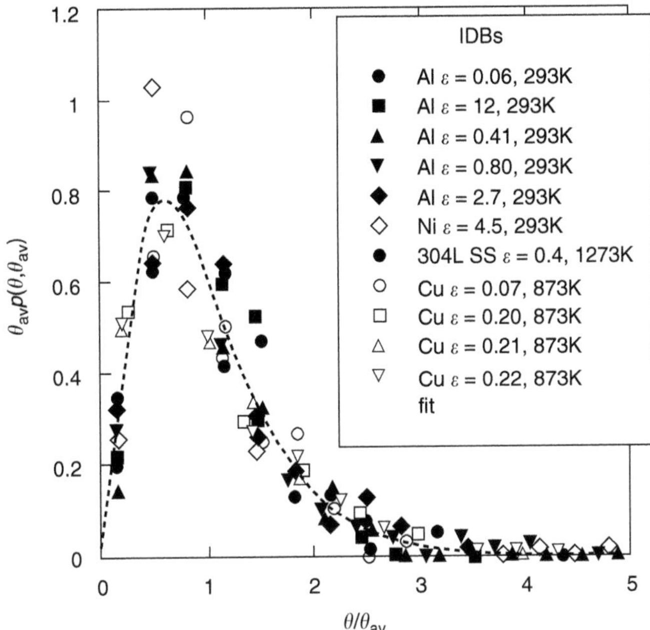

FIGURE 12.10 Probability distribution of misorientation angles θ for IDBs. Data from a wide range of materials and conditions collapse onto a single function when normalized by the average value of θ. (Courtesy D.A. Hughes.)

12.4.4 Dislocation Boundaries: Scaling and Similitude

The dislocation boundaries that subdivide deformed crystal have been defined in Table 12.5, using a classification based on bulk morphology and mechanisms for boundary formation and evolution. The distinction between IDBs and GNBs has been used[38] in a scaling analysis that shows that both misorientation angles and spacing can be determined by a single universal function, when each distribution is scaled by its mean angle or spacing. It is also found that the scaling only applies to IDBs and GNBs when they are analyzed separately, and not when they are combined and analyzed together. The distribution function, which has been derived empirically, is almost independent of material type, the temperature at which the plastic strain occurred, and the amount of deformation. An example is shown in Figure 12.10.[38] The observation of scaling has led to theoretical work that suggests the use of a Rayleigh distribution function for scaling analysis.[39] In general, scaling is very useful when analyzing large data sets and also in providing a rational basis for the comparison of similarities and differences between different deformation processes.

An alternative approach in the microstructural analysis has been to apply similitude,[14,40] expressing a fixed relationship between structural elements, for example, between boundary spacing D and dislocation spacing in a boundary, h. Taking h inversely proportional to the misorientation angle θ, similitude can be

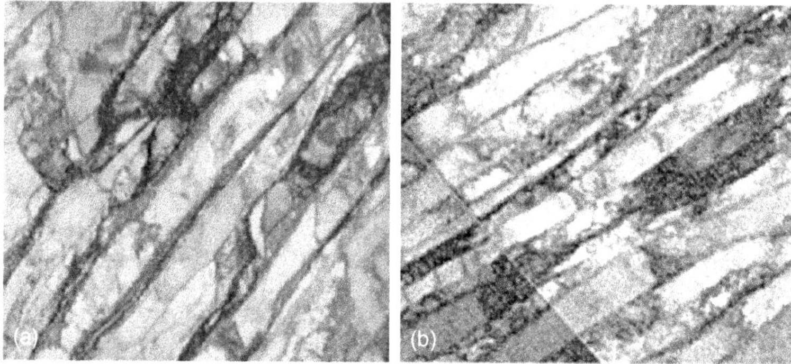

FIGURE 12.11 TEM showing structural refinement in cold-rolled Ni following a scaling relationship. (a) ε_{VM} 0.8; average spacing of GNBs 0.5 μm. (b) ε_{VM} 4.5; average spacing of GNBs 0.13 μm. The structure is refined by a factor of 4 but is otherwise very similar. (Courtesy D.A. Hughes.)

expressed by the function:

$$D\theta/b = C \tag{12.3}$$

where b is the magnitude of the Burgers vector and C is a constant. Different types of microstructures may be classified according to the value of C, which has been found to be almost constant for IDBs, independent of materials type and deformation mode (e.g., rolling, tension). Example structures are shown in Figure 12.11.

12.4.5 Experiments and Theory

The separation of deformation-induced boundaries into IDBs and GNBs has been the basis for the analysis of a number of experimental observations. Three examples are given here.

1. An analysis of the boundary structures has been based on the assumption that the microstructures are low-energy structures, that is, there is an absence of long-range stresses. This has allowed the use of Frank's formula (Equation 12.2) to analyze the character of the dislocations that form a GNB, based on its axis/angle pair, and relate this to the slip operating in the volumes on either side of the boundary.[41]
2. The experimental finding that the misorientation angle across GNBs and IDBs increases monotonically with the plastic strain has been used to calculate the angle-strain dependency using a deterministic formation model for GNBs and stochastic formation of IDBs (see Figure 12.9). For both types of boundary, good agreement between theory and experiment has been found.[42,43]
3. A correlation has been assumed between the deformation microstructure in grains/crystals and their crystallographic orientations. This correlation has been the basis of a polycrystal plasticity model relating the boundary plane of GNBs to the slip pattern that is calculated based on the external load conditions and the grain/crystal orientation.[44]

12.4.6 Local Crystallography and Microstructure

Studies of deformed single crystals show a strong relationship between crystal orientation and the characteristics of the evolving microstructure. This has, for example, been observed in copper, illustrated in Figure 12.12.[45] With the tensile axis in the middle of the stereographic triangle, extended boundaries (GNBs) develop almost parallel to active slip planes, forming what in single crystal studies has been termed

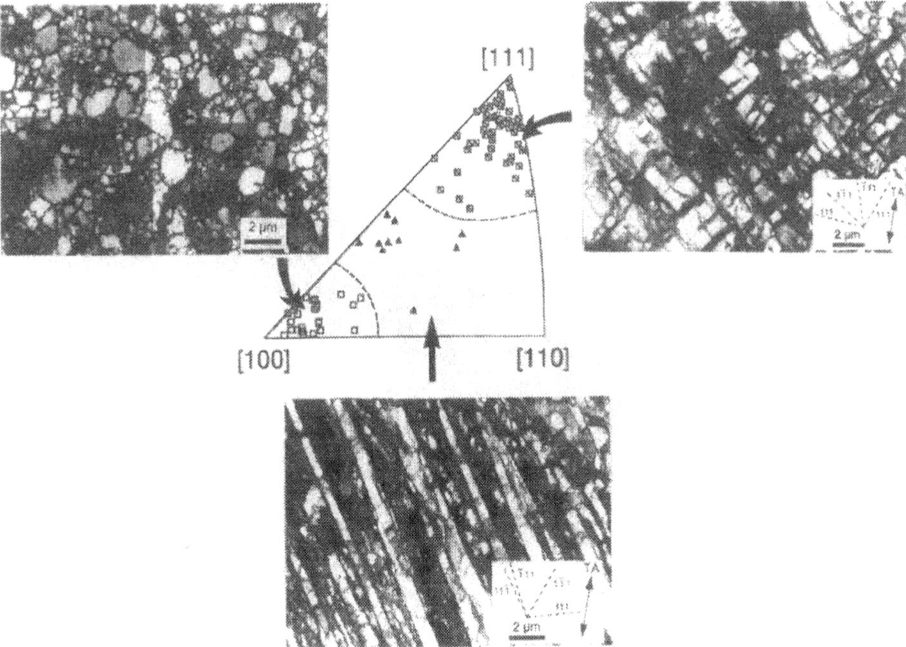

FIGURE 12.12 Microstructural types in grains of different orientations in polycrystalline 99% Cu deformed in tension to ε_{VM} between 0.05 and 0.34. (Courtesy X. Huang.)

a *carpet structure.* In crystals in which the tensile axis is near the [111] corner, extended boundaries are observed that are not parallel to active slip planes. Equiaxed structures develop in crystals with tensile axis near the [100] corner. A feature common to all three orientation groups is that the crystals deform by multiple glide in the strain range considered. This is the case for both single crystal and polycrystalline materials.[46] Studies of samples deformed in tension have shown that the grains can be classified according to their deformation structure into three different types, corresponding to those observed in single crystals. This is illustrated in Figure 12.13[47] which is based on characterization of a number of grains (grain size 300 μm) in polycrystal specimens strained in tension in the range 0.05 to 0.34. In this triangle, each structure type dominates fairly large areas, suggesting that different combinations of slip systems may lead to the evolution of structures with almost the same morphological appearance. These results support the existence of a general correlation between crystallographic orientation (which ultimately determines which slip systems will operate) and the microstructural characteristics (slip patterns).[44,46,48]

The slip pattern analysis is based on a polycrystal plasticity model using external local conditions as input parameters. It is found[49] that grains typically deform using fewer than the five systems that the Taylor model requires for strain compatibility between neighboring grains.

For the materials discussed here (medium to high stacking fault energy), the combined analysis of microstructure and slip pattern suggests that grains may maintain compatibility with their neighbors without a specific requirement for grain interaction. Interaction of this type is required in a number of polycrystal plasticity models, which would imply that grains of similar orientations in different locations within the polycrystal would show different microstructures and rotations. This is not in agreement with the observations in Figure 12.12 and Figure 12.13. Grain interaction effects may manifest themselves by changes in the microstructure and local crystallography near grain boundaries and triple junctions, as illustrated schematically in Figure 12.14.[33] However a much more typical observation (by EBSD and TEM) is that the interior structure extends all the way to the grain boundaries. These observations can be coupled with an analysis of grain interaction in high purity polycrystalline copper deformed in tension to a strain of 25%.[50] A relatively coarse grain size (about 100 μm) was chosen, allowing EBSD analysis

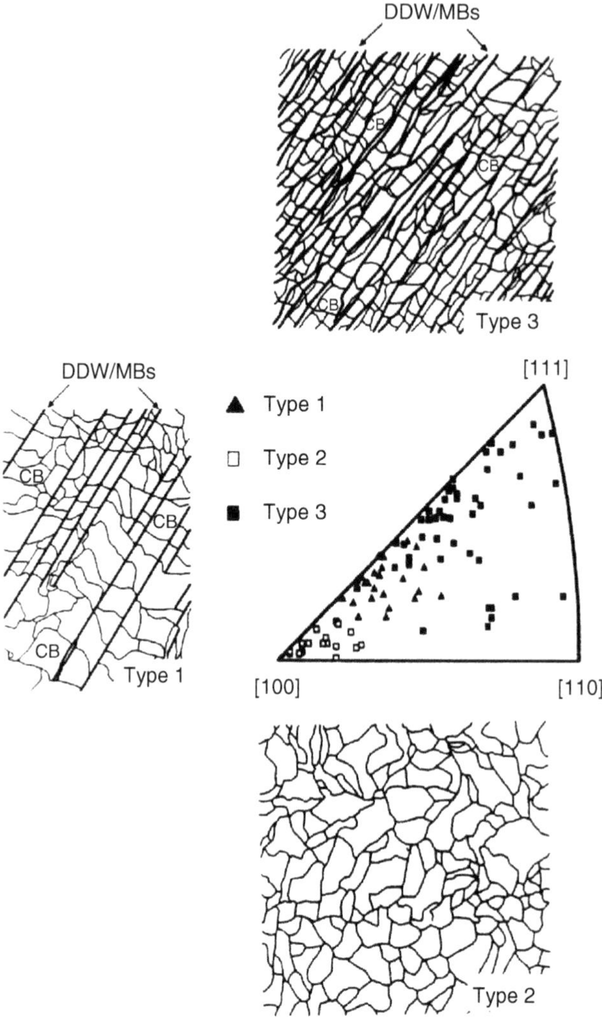

FIGURE 12.13 Relationship between deformation structure (shown as schematics) and orientation for 99.99% Al deformed in tension. (Courtesy X. Huang.)

to be used to follow grain break-up and rotation as a function of strain. A few small domains of special orientations do provide evidence of grain interaction. However, the crystal rotation of the large domains covering most of the grains is in some, but not all cases, consistent with the rotation direction predicted by the Taylor model. The observed correlation between crystallography and microstructure at small and medium strains is difficult to verify at high strains due to the fine scale of the deformation microstructure.

This problem has been investigated in cold-rolled samples in which the starting grain size is larger. With increasing deformation, the microstructure refines continuously and both the average misorientation angle and the fraction of high-angle boundaries increase. These high-angle boundaries include both the original boundaries and those that are deformation-induced. A typical structure assumed in aluminum cold-rolled to a large strain is illustrated in Figure 12.15, together with the misorientation angle across the lamellar boundaries and the rolling texture components in the lamellar structure.[43]

This figure illustrates the break-up of the structure into lamellae in the different crystallographic orientations: it is not an entire grain that rotates into a new position as the deformation texture evolves. Individual parts of the grain can rotate toward different orientations and the mutually misoriented cell block comprises the macroscopic texture components.[43]

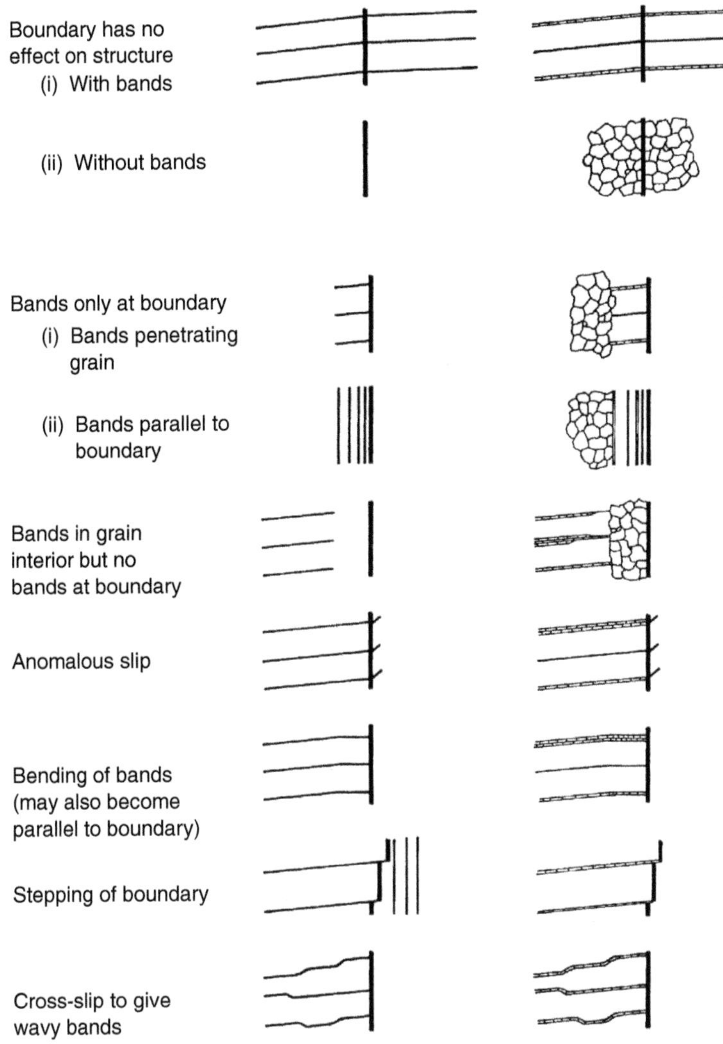

FIGURE 12.14 Classification of structures found at grain boundaries in cold-deformed Al. Surface structures seen by SEM or light microscopy in the left-hand column correspond to structures seen in TEM in the right-hand column. (After Barlow, C.Y.J., Bay, B. and Hansen, N., *Phil. Mag. A*, 51, 253, 1985, http://www.tandf.co.uk/journals/titles/14786435.asp. With permission.)

12.4.7 Large Strain Deformation

The dislocation structures formed during plastic deformation refine with increasing strain, and crystallites as small as 5 to 10 nm have been observed at ultrahigh strains (greater than about 5, but as high as 50 to 100 for processes such as high-pressure torsion, milling, and friction).[5,51–53] In parallel with the structural refinement more and more dislocations are stored in boundaries, leading to an increase in the average misorientation angle across boundaries and an increase in the number of high-angle boundaries (misorientation >15°). The structural morphology also changes, and this follows the shape change of the bulk material, for example, forming a lamellar structure in rolling and a fibrous structure in wire-drawing. The structural refinement is material dependent. For example in aluminum with a high SFE and easy cross-slip the smallest crystallite size is about 200 to 300 nm after rolling to ultrahigh strain, whereas in copper the minimum size may be about 100 nm. Addition of elements in solid solution can, however,

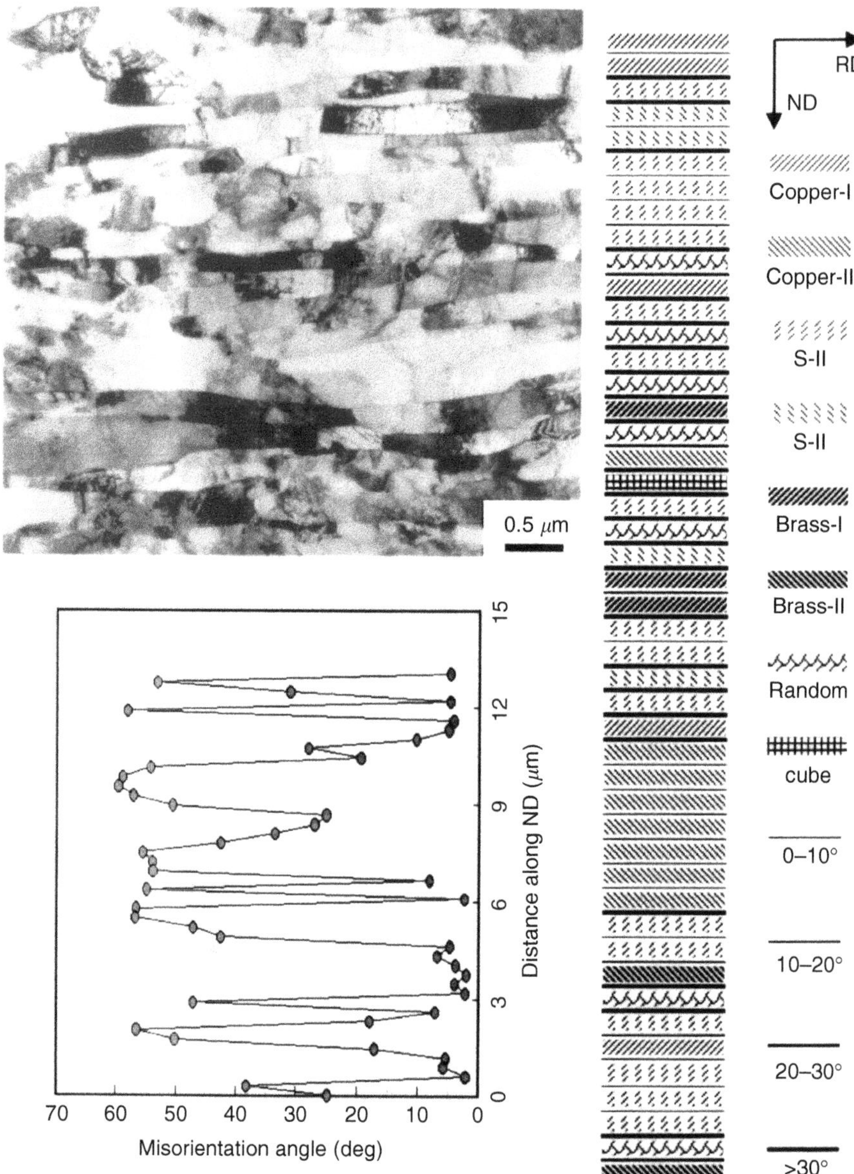

FIGURE 12.15 Commercial purity Al rolled to thickness reduction of 99%, longitudinal section. The lamellar structure is shown in TEM. Corresponding misorientation angles for successive boundaries in the normal direction are shown, with textures for each of the cell blocks. (Courtesy Q. Liu.)

increase dislocation pinning, and hence reduce the amount of annihilation during deformation. Using this mechanism it has been possible to reduce the crystallite size in aluminum to about 100 nm and in copper to about 5 to 10 nm, by cryomilling[54] and friction wear,[55] respectively. Figure 12.16[55] illustrates a graded nanostructure found in copper following friction deformation.

Analysis of the structural evolution by applying scaling of the boundary spacing at different strains has demonstrated that the same scaling law can apply over a structural scale from the micrometer dimension to the nanometer scale. This indicates that the basic mechanisms for grain subdivision by dislocation multiplication and storage apply independent of the structural scale. This means that the basic dislocation theory may also be applicable for nanostructured metals processed by plastic deformation.

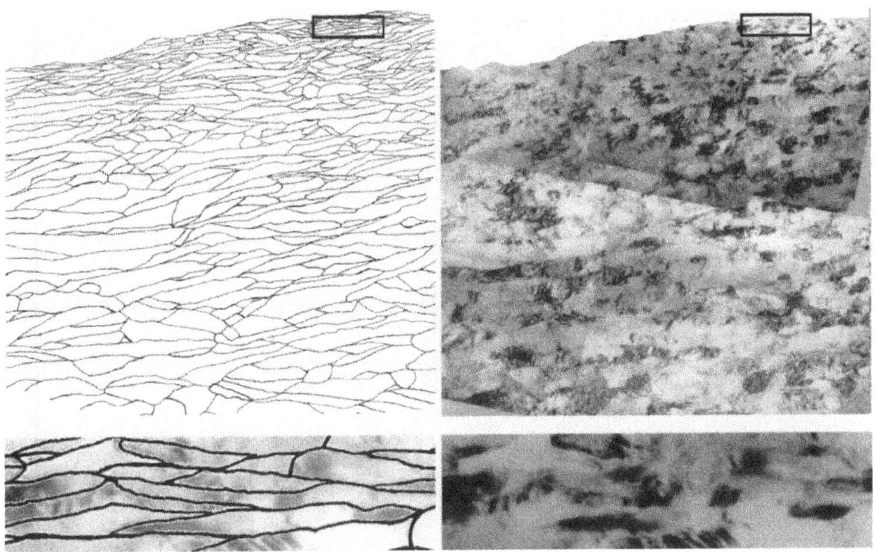

FIGURE 12.16 Graded nanostructure in Cu produced by 127mm friction deformation under 12 MPa viewed in cross-section by TEM, with schematic. (Reprinted figure with permission from Hughes, D.A. and Hansen, N., *Phys. Rev. Lett.*, 87, 135503, 2001. Copyright (2001) by the American Physical Society.)

12.5 Structure/Mechanical Property Relationships, and Modeling

12.5.1 Polycrystal Flow Stress

The stress to move a dislocation in perfect material has been introduced in Section 12.2.3, and the formation of dislocation arrays has been discussed in Section 12.4.[43,56–62] Table 12.7 summarizes and quantifies the main parameters that influence the stress required to move dislocations in real materials.

12.5.2 Microstructure and Mechanical Properties

As discussed in Section 12.4, metalworking is accompanied by changes in mechanical properties that show strong relationships with structural parameters. There is a general assumption that structural refinement enhances the strength of the metal (i.e., raises the yield stress), and this refinement can be expressed as a change in a number of structural parameters:

- The dislocation density within cells, ρ_0
- The spacing between IDBs D_{IDB} and between GNBs D_{GNB}
- The misorientation angle θ across IDBs and GNBs
- The fraction of high-angle boundaries, HABs

The dislocation density is an important parameter in determining mechanical properties. The density of loose dislocations can be measured, as mentioned in Section 12.3.2.1, but because dislocations are also present in walls and boundaries, the total dislocation density must be estimated by indirect methods. One such method is to estimate the dislocation density in a boundary, ρ, as the density required to form a boundary with misorientation angle θ. For a mixed tilt and twist equiaxed boundary ρ_b is equal to

TABLE 12.7 Microstructural Parameters Determining Polycrystal Flow Stress

Parameter	Microstructural Elements	Equation	Comments
Lattice flow stress: Peierls-Nabarro stress	Lattice parameters; Bond strength	$\tau_P = 3G\exp-(2\pi w/b)$	G = Shear modulus w = dislocation width b = Burgers vector
Point defects; solid solution	Elastic strain fields, determined by lattice parameter differences	$\tau_y = G\delta^{3/2}C^{1/2}$	$\delta = (\Delta r/r)$ where r is solvent atomic radius and Δr difference between solvent and solute radii. C = atomic concentration
Coherent precipitates	Elastic strain fields; size; spacing	$\tau_y = G\varepsilon C$	C = atomic concentration ε = elastic strain field, which is a function of precipitate diameter
Incoherent precipitates	Size, spacing	$\tau_y = KGb/L$	K = constant L = edge-to-edge spacing of precipitates
Forest dislocations and low-angle boundaries; cells	Dislocation density	$\Delta\sigma_\rho = M\alpha\mathbf{b}G\sqrt{\rho}$	M is Taylor factor; α = 0.2–0.5 ρ = dislocation density
Grain size	Boundary strength Grain size	$\Delta\sigma_{HP} = k_{HP}\sqrt{1/d}$	Hall-Petch hardening k_{HP} = Hall-Petch constant d = grain size
High-angle subgrain boundaries, deformation-induced	Boundary strength Boundary spacing (subgrain size)	$\Delta\sigma_{HAB} = k_{HP}\sqrt{1/D_{HAB}}$	Hall-Petch hardening D_{HAB} = HAB spacing k_{HP} = equivalent Hall-Petch constant
Extended wall	Spacing of GNBs	$\Delta\sigma_{GNB} = k_{HP}\sqrt{1/D_{GNB}}$	Hall-Petch hardening D_{GNB} = GNB spacing k_{HP} = equivalent Hall-Petch constant
Note: k_{HP} takes different values in these three Hall-Petch-type relationships.			
Texture	Crystallographic orientations averaged over polycrystal	M	Taylor factor

$1.5\theta/b$ and the total density contained in the boundaries is:

$$\rho_b = \frac{1.5\theta}{b} S_V \tag{12.4}$$

where S_V is the boundary area per unit volume. The total dislocation density within boundaries can therefore be defined as $\rho_b = \rho_{IDB} + \rho_{HAB}$ where the density stored in IDBs and cell structures is ρ_{IDB} and the equivalent dislocation density for HABs is ρ_{HAB}.

In addition to this density, the IDBs and GNBs also contain redundant dislocations that do not contribute to the misorientation angle, so the density estimated from this equation is a minimum value. As strain increases and the misorientation angle rises, the density of the redundant dislocations decreases.

The total dislocation density is now given by:

$$\rho_t = \rho_0 + \rho_b \tag{12.5}$$

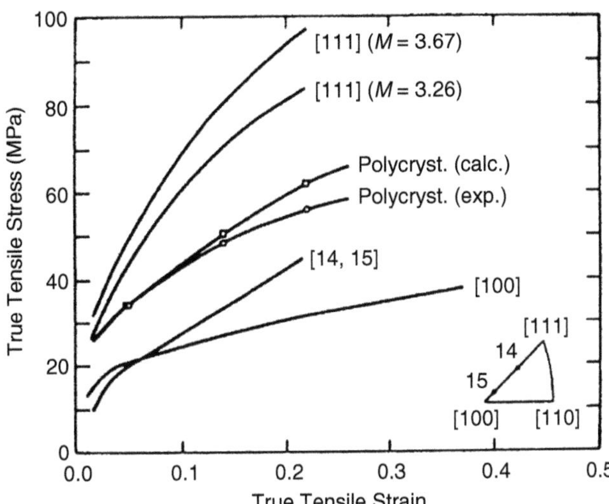

FIGURE 12.17 Comparison between experimental and calculated stress–strain curves for polycrystalline 99.99% Al.

The quantification of these different microstructural parameters is the basis for the formulation of strength–structural relationships. The flow stress can be related to the total dislocation density ρ_t using the relationship:

$$\tau - \tau_0 = \alpha G b \sqrt{\rho_t} \tag{12.6}$$

where α is a number of the order 0.3 to 0.5. Apart from a variation in α this equation is insensitive to changes in dislocation structure, grain size, strain rate, and temperature. As an example, the flow stress of polycrystalline aluminum analyzed in Figure 12.17 has been calculated using the relationship:

$$\sigma = M \alpha G b \sqrt{\rho_t} \tag{12.7}$$

where M is the Taylor factor that takes the crystallographic texture into account. Good agreement between calculation and experimental values has been found.[63] In this experiment, the microstructural evolution has been followed in grains of different crystallographic orientation, which therefore have different stress–strain relationships. For each structure type, a shear stress–strain relationship has been calculated and found to be in good agreement with data taken from single crystals with comparable orientations. This has allowed the stress–strain curve for the polycrystalline sample to be calculated with good accuracy from single crystal data that have been weighted according to quantitative analysis of the texture of the polycrystal,[63] see Figure 12.17. This example illustrates that, in agreement with the Taylor model, grain accommodation during straining may take place without a significant effect of grain interaction (see Section 12.4.6).

These equations based on the single parameter of the total dislocation density are suitable for use at low and medium strains, but cannot be directly extended to high strains where the microstructure is subdivided by boundaries with a range of misorientation angles. For such structures, the use of ρ_t will lead to a gross overestimate of the flow stress at high strains. The reason for this is that for boundaries above 3 to 5° an increase in boundary misorientation, which is translated into a large increase in dislocation density, results in only a small increase in flow stress. A more appropriate model for high strains looks at the strengthening effect of boundaries themselves, using a Hall-Petch relationship (see Table 12.7) in which the yield stress is inversely proportional to the square root of the boundary spacing. In addition, it is assumed that a critical angle exists for dislocation boundaries above which they effectively form a barrier to moving dislocations, acting in the same way as grain boundaries. This angle is material-dependent, and has been calculated to be 0.5 to 2°.[64] Boundaries of this type will predominantly be GNBs, as for IDBs the rate of increase of misorientation angle with applied strain is relatively low. The contribution of GNBs of this type can be expressed by:

$$\Delta\sigma_{GNB} = K_{HP}\sqrt{1/D_{GNB}} \tag{12.8}$$

where D_{GNB} is the spacing of boundaries with angle above this critical value. The total flow stress is thus:

$$\sigma = \sigma_0 + M\alpha Gb\sqrt{\rho_0 + \rho_{IDB}} + K_{HP}\sqrt{1/D_{GNB}} \tag{12.9}$$

where σ_0 is the flow stress of dislocation-free material in the interior of subgrains. This type of equation is favored for analysis of structures containing a mixture of low and high-angle boundaries, and is relatively straightforward to apply. Good agreement has been found with experimental results from aluminum and nickel cold-rolled over a large strain range, up to strains of 4 to 5.

The flow stress analysis of deformed materials is, however, often performed using only one parameter, D_B, which is the spacing between any of the boundaries that subdivide the material, making no distinction between different boundary types even though they interact very differently with dislocations. The parameter D_B is used in place of the grain size d in the Hall-Petch equation (Table 12.7), using a much higher value of the Hall-Petch constant k_{HP} than observed for grain boundary strengthening in polycrystalline material. Although a good empirical relationship is found between experiment and theory, the Hall-Petch constant used here has no physical meaning[63] as it represents an average resistance of low, medium, and high-angle boundaries.

In the flow stress calculation above, the effect of the crystallographic texture is introduced using the Taylor factor M. It is however possible to combine the effect of microstructure and texture by introducing a microstructural parameter into a polycrystal plasticity model. For example, flow stress anisotropy in deformed metals can be caused by the texture and by the strong directionality of the parallel GNBs. This can be modeled by introducing an anisotropic critical resolved shear stress τ_{CRSS} into the Taylor model. For the ith slip system:

$$\tau_{CRSS} = \tau_0 + XD_i^{-1/2} \tag{12.10}$$

where τ_0 is the isotropic part that is the same for all slip systems, and $XD_i^{-1/2}$ is the anisotropic part that differs between slip systems owing to the variation in the spacing D_{GNB}. The anisotropic hardening is calculated on the assumption that the strengthening parameter X depends on the boundary resistance to passage of dislocations, which is proportional to the boundary misorientation for small angles (<0.5 to 5°). Good agreement has been found between calculated and experimental values for the flow stress anisotropy of rolled sheet of aluminum, copper, and interstitial-free steel.[61,65] This agreement lends support for the estimated strengthening parameters, and the underlying assumptions of this structure–strength analysis.

12.6 Future Trends

An important trend in the materials area is the global interest in metals and alloys with ever-finer structures, down to nanometer dimensions (see Chapter 13, as well). Advanced and new processes are under development, and some materials with unexpected properties are emerging. Metalworking and the structural evolution during plastic deformation is an integral part of this global effort. However, characterization and quantification of structures on this ever-finer scale requires more techniques that can describe structures in 2D and 3D, and can do this quickly and accurately. Both hardware and software development for structural characterization are therefore areas of great activity. To combine processing conditions with material properties, structural characterization must be combined with modeling on different length scales, including multiscale modeling. This research and development covering a wide range of length scales offers unique opportunities for development of new materials and products. In parallel with this, it will advance our understanding of the basic physical and mechanical principles that govern the structural evolution during metal working.

Acknowledgments

This work (NH) was supported by the Danish National Research Foundation through the Centre for Fundamental Research: Metal Structures in Four Dimensions.

References

[1] Kalpakjian, S. and Schmid, S.R., *Manufacturing Processes for Engineering Materials*, 4th Edition, Prentice Hall, Upper Saddle River, NJ, 2003, Chapter 2.
[2] Saito, Y. et al., Novel ultra-high straining process for bulk materials — Development of the accumulative roll-bonding (ARB) process, *Acta Mater.*, 47, 579, 1999.
[3] Tsuji, N. et al., ARB (accumulative roll-bonding) and other new techniques to produce bulk ultrafine grained materials, *Adv. Eng. Mater.*, 5, 338, 2003.
[4] Segal, V.M., Materials processing by simple shear, *Mater. Sci. Eng. A*, 197, 157, 1995.
[5] Valiev, R.Z., Islamgaliev, R.K., and Alexandrov, I.V., Bulk nanostructured materials from severe plastic deformation, *Prog. Mater. Sci.*, 45, 103, 2000.
[6] Armstrong, P.E., Hockels, J.E., and Sherby, O.D., Large strain multidirectional deformation of 1100 Aluminum at 300K, *J. Mech. Phys. Solids*, 30, 37, 1982.
[7] Huang, J.Y. et al., Microstructures and dislocation configurations in nanostructured Cu processed by repetitive corrugation and straightening, *Acta Mater.*, 49, 1497, 2001.
[8] Zhu, Y.T.T. et al., A new route to bulk nanostructured metals, *Metall. Mater. Trans. A*, 32, 1559, 2001.
[9] Stephen Yue, *ASM Handbook: Bulk forming, Metalworking*, vol. 14A, ASM International, Materials Park, Ohio, 2006, 290.
[10] Dieter, G.E., *Mechanical Metallurgy*, McGraw-Hill book Co., Inc., New York, 1961.
[11] Bever, M.B., Holt, D.L., and Titchener, A.L., *Stored Energy in Cold Rolled Work; Progress in Materials Science vol. 17.* Pergamon Press, Oxford, 1973.
[12] Honeycombe, R.W.K., *The Plastic Deformation of Metals*, Edward Arnold, UK, 1975.
[13] Hull, D. and Bacon, D.J., *Introduction to Dislocations*, 3rd Edition, Pergamon Press, Oxford, 1984, 1.
[14] Kuhlmann-Wilsdorf, D., Theory of plastic-deformation — Properties of low-energy dislocation-structures, *Mater. Sci. Eng. A*, 113, 1, 1989.
[15] Sutton, A.P. and Balluffi, R.W., *Interfaces in Crystalline Materials*, Oxford University Press, Oxford, 1995. Sections 12.5, 12.6.
[16] Cottrell, A.H., *Dislocations and Plastic Flow in Crystals*, 2nd Edition. Oxford, 1963.
[17] Huang, X. et al., Microstructures of nickel deformed by high pressure torsion to high strains, *Materials Science Forum: Thermec'2003*, 426–432, 2819, 2003.
[18] Loretto, M.H., *Electron Beam Analysis of Materials*, Springer Verlag, Heidelberg, 1993.
[19] Forwood, C.T. and Clarebrough, L.M., *Electron Microscopy of Interfaces in Metals and Alloys*, IOP Publishing Ltd., Bristol, 1991.
[20] Kurzydlowski, K.J. and Ralph, B., *The Quantitative Description of the Microstructure of Materials*, CRC Press, Boca Raton, FL,1995.
[21] Randle, V. and Engler, O., *Introduction to Texture Analysis*, Taylor & Francis, Amsterdam, 2000.
[22] Randle, V., *Microtexture Determination and its Applications*, Institute of Materials, London, 1992.
[23] de Graef, M., *Introduction to Conventional Transmission Electron Microscopy*, Cambridge University Press, Cambridge, 2003.

[24] Poulsen, H.F., *Three-Dimensional X-Ray Diffraction Microscopy. Mapping Polycrystals and their Dynamics*, Springer, Berlin, 2004.
[25] Randle, V., *The Measurement of Grain Boundary Geometry*, Institute of Physics Publishing, Bristol, 1993.
[26] Liu, Q., A simple method for determining orientation and misorientation of the cubic-crystal specimen, *J. Appl. Crystall.*, 27, 755, 1994.
[27] Bhadeshia, H.K.D.H., *Worked Examples in the Geometry of Crystals*, Institute of Materials, London, 1987.
[28] Bay, B. et al., Overview No-96 — Evolution of Fcc deformation structures in polyslip, *Acta Metall. Mater.*, 40, 205, 1992.
[29] Hansen, N. and Juul Jensen, D., Development of microstructure in FCC metals during cold work, *Phil. Trans. R. Soc. London Series A-Math. Phys. Eng. Sci.*, 357, 1447, 1999.
[30] Ashby, M.F. et al., Deformation processing of metals, *Phil. Trans. R. Soc.* London Series A-Math. Phys. Eng. Sci., 357, 1443, 1999.
[31] Godfrey, A. et al., Stored energy, microstructure, and flow stress of deformed metals, *Metall. Trans. A*, 36A, 2371, 2005.
[32] Hughes, D.A. and Hansen, N., *ASM Handbook: Plastic Deformation Structures, vol. 9; Metallography and Microstructures*, ASM International, Materials Park, Ohio, 2004, 292.
[33] Barlow, C.Y.J., Bay, B., and Hansen, N., A comparative investigation of surface relief structures and dislocation microstructures in cold-rolled aluminum, *Phil. Mag. A*, 51, 253, 1985.
[34] Winther, G. et al., Critical comparison of dislocation boundary alignment studied by TEM and EBSD: Technical issues and theoretical consequences, *Acta Mater.*, 52, 4437, 2004.
[35] Barlow, C.Y. and Hansen, N., Deformation structures in aluminum containing small particles, *Acta Metall.*, 37, 1313, 1989.
[36] Hansen, N., Deformation microstructures with a structural scale from the micrometre to the nanometre dimension, *Evolution of Deformation Microstructures in 3D*, Gundlach, C. and others, Eds., Risø National Laboratory, Roskilde, Denmark, 2004, 13.
[37] Kuhlmann-Wilsdorf, D. and Hansen, N., Geometrically necessary, incidental and subgrain boundaries formed during cold deformation, *Scripta Metall. Mater.*, 25, 1557, 1991.
[38] Hughes, D.A. et al., Scaling of misorientation angle distributions, *Phys. Rev. Lett.*, 81, 4664, 1998.
[39] Pantleon, W. and Hansen, N., Dislocation boundaries — The distribution function of disorientation angles, *Acta Mater.*, 49, 1479, 2001.
[40] Hansen, N. and Hughes, D.A., Analysis of large dislocation populations in deformed metals, *Phys. Status Solidi (a)*, 149, 155, 1995.
[41] Wert, J.A., Liu, Q., and Hansen, N., Dislocation boundaries and active slip systems, *Acta Metall. Mater.*, 43, 4153, 1995.
[42] Pantleon, W., Disorientations and work-hardening in Stage IV., *Evolution of Deformation Microstructures in 3D,* Gundlach, C. and others, Eds., Risø National Laboratory, Roskilde, Denmark, 2004, 481
[43] Hansen, N., New discoveries in deformed metals, *Metall. Trans. A*, 32, 2917, 2001.
[44] Winther, G., Slip as a mechanism for structural evolution, *Evolution of Deformation Microstructures in 3D,* Gundlach, C. and others, Eds., Risø National Laboratory, Roskilde, Denmark, 2004, 211.
[45] Huang, X., Grain orientation effect on microstructure in tensile strained copper, *Scripta Mater.*, 38, 1697, 1998.
[46] Hansen, N. et al., Grain orientation and dislocation patterns, *Phil. Mag.*, 86, 3981, 2006.
[47] Huang, X. and Hansen, N., Grain orientation dependence of microstructure in aluminium deformed in tension, *Scripta Mater.*, 37, 1, 1997.
[48] Winther, G., Huang, X., and Hansen, N., Crystallographic and macroscopic orientation of planar dislocation boundaries — Correlation with grain orientation, *Acta Mater.*, 48, 2187, 2000.
[49] Wert, J.A. and Huang, X., Extended planar boundary inclinations in fcc single crystals and polycrystals subjected to plane strain deformation, *Phil. Mag.*, 83, 969, 2003.

[50] Thorning, C., Somers, M.A.J., and Wert, J.A., Grain interaction effects in polycrystalline Cu, *Mater. Sci. Eng. A*, 397, 215, 2005.

[51] Langford, G. and Cohen, M., Microstructural analysis by high-voltage electron-diffraction of severely drawn iron wires, *Metall. Trans., A* 6, 901, 1975.

[52] Hughes, D.A. and Hansen, N., Deformation structures developing at fine scales, *Phil. Mag.*, 83, 3871, 2003.

[53] Tsuji, N., Huang, X., and Nakashima, H., Microstructure of metals and alloys deformed to ultrahigh strains, *Evolution of Deformation Microstructures in 3D*, Gundlach, C. and others, Eds., Risø National Laboratory, Roskilde, Denmark, 2004, 147.

[54] Han, B.Q. and Lavernia, E.J., Deformation mechanisms of nanostructured alloys, *Adv. Eng. Mater.*, 7, 457, 2005.

[55] Hughes, D.A. and Hansen, N., Graded nanostructures produced by sliding and exhibiting universal behavior, *Phys. Rev. Lett.*, 87, article 135503, 2001.

[56] Hansen, N. and Kuhlmann-Wilsdorf, D., Low-energy dislocation-structures due to unidirectional deformation at low-temperatures, *Mater. Sci. Eng.*, 81, 141, 1986.

[57] Liu, Q. et al., Microstructure and strength of commercial purity aluminium (AA 1200) cold-rolled to large strains, *Acta Mater.*, 50, 3789, 2002.

[58] Thompson, A.W., Yielding in nickel as a function of grain or cell-size, *Acta Metall.*, 23, 1337, 1975.

[59] Kocks, U.F. and Mecking, H., Physics and phenomenology of strain hardening: The FCC case, *Prog. Mater. Sci.*, 48, 171, 2003.

[60] Hansen, N., Flow stress and microstructural parameters, *Numerical Predictions of Deformation Processes and the Behaviour of Real Materials,* Andersen, S.I. and others, Eds., Risø National Laboratory, Roskilde, Denmark, 1994, 325.

[61] Li, Z.J., Winther, G., and Hansen, N., Anisotropy in rolled metals induced by dislocation structure, *Acta Mater.*, 54, 401, 2006.

[62] Gil-Sevillano, J., van Houtte, P., and Aernoudt, E., Large strain work-hardening and textures, *Prog. Mater. Sci.*, 25, 69, 1980.

[63] Hansen, N. and Huang, X., Microstructure and flow stress of polycrystals and single crystals, *Acta Mater.*, 46, 1827, 1998.

[64] Hansen, N., Boundary strengthening in undeformed and deformed polycrystals, *Mater. Sci. Eng.*, A409, 39, 2005.

[65] Juul Jensen, D. and Hansen, N., Flow-stress anisotropy in aluminum, *Acta Metall. Mater.*, 38, 1369, 1990.

13

Mechanical Alloying and Severe Plastic Deformation

A. P. Newbery, B. Q. Han, and E. J. Lavernia
University of California at Davis

C. Suryanarayana
University of Central Florida

J. A. Christodoulou
Office of Naval Research

Abstract

Recently developed methods for materials processing, such as mechanical alloying, cryomilling, friction stir processing, equal-channel angular pressing, and high-pressure torsion, are reviewed in the present chapter. Over the last two decades, there has been increasing interest on the synthesis of advanced materials using mechanical alloying, and currently this represents a vital area of research in the area of materials processing. Typically, mechanical alloying introduces repeated plastic deformation of powders via collisions with a grinding medium. Because the heavy plastic deformation introduces a high density of crystalline defects, alloy formation and microstructural evolution, especially grain size reduction, are readily attained during milling. Hot or cold isostatic pressing followed by extrusion and forging are examples of routes that are commonly used to consolidate the milled powders. Mechanically alloyed materials usually exhibit higher strength and higher creep resistance, as compared to those of equilibrium materials. As a variant of mechanical alloying, cryomilling is carried out with a slurry formed by the powder, milling balls, and a cryogenic liquid. Cryomilling results in the formation of a highly thermally stable microstructure due to the presence of nanoscale dispersions.

In addition to mechanical alloying and cryomilling, this chapter covers three techniques that also belong to the family of severe plastic deformation methods: friction stir processing, equal-channel angular pressing, and high-pressure torsion. Friction stir processing is a rapidly maturing, solid-state thermo-mechanical technique for modifying the near-surface microstructure. In the section on friction stir processing, microstructural evolution and two major applications are discussed; these being the defect repair of cast alloys, and its use to make a precursor for superplastic forming. During equal-channel angular pressing, the material is pressed through a die that has two channels, with an identical cross-section. Equal-channel angular pressing refines the microstructure via deformation under high-strain conditions, resulting in enhanced strength, but usually low ductility. Finally, as one of most effective methods for processing nanostructured materials, the characteristics of high-pressure torsion are briefly outlined.

13.1 Introduction

When severe plastic deformation (SPD) is imparted to a material, it can have a very beneficial effect on microstructure, which in turn can lead to a significant enhancement of that material's properties. In general, SPD homogenizes and refines, reducing compositional variations and the size of microstructural features, such as second-phase particulates. Grains, often on the nanoscale, can be obtained that are much smaller than those obtained with conventional materials processing. SPD can also extend solid solubility limits, synthesize nonequilibrium crystalline and amorphous phases, and disordered intermetallics. Given that it is the most technologically advanced SPD process, most of the chapter is devoted to mechanical alloying (MA), and the low-temperature variant of MA, cryomilling. The remainder of the chapter concentrates on friction stir processing (FSP) and equal-channel angular pressing (ECAP). Apart from the MA of oxide-dispersion strengthened materials for high-temperature use, because SPD processes are so new, and the extra costs associated with the processing high, up until now they have largely been restricted to research and development, rather than commercial application. However, SPD shows great promise for the future in its ability to extend the bounds of material properties.

13.2 Mechanical Alloying

The process of ball milling has been traditionally used to reduce particle size, in the crushing of minerals for further processing and mineral dressing operations, for example. It was only in the later half of the 1960s that high-energy milling evolved into MA to produce materials with improved microstructure and properties for applications in the aerospace industry.[1,2] MA was developed as a result of an industrial necessity — to produce a Ni-based superalloy that combined the high-temperature strength of an oxide-dispersion and the intermediate-temperature strength imparted by the presence of the γ' phase. The required corrosion and oxidation resistance could also be obtained by suitable alloying additions. The great advantage of using MA is to obtain a highly uniform dispersion of the oxide. This is possible because MA is carried out in the solid state and such uniform dispersion has not been easily achieved by other techniques.

From the mid-1980s, it was realized that the technique of MA is capable of producing a wide variety of alloys starting from blended elemental powders. These include equilibrium and nonequilibrium alloys composed of solid solutions, intermetallics, quasi-crystalline phases, and amorphous phases. When the phase produced is crystalline, the grain size of the powder is often on the nanometer scale. Another advantage of this technique is that since the process is carried out at, near, or below room temperature, restrictions imposed on alloying by phase diagrams do not apply. Consequently, it is possible to produce true alloys between metals that are not even miscible under equilibrium conditions. Over the last two decades, there have been an increasing amount of investigations into the synthesis of advanced materials using MA, and currently this is one of the most active research fields in the area of materials processing.

13.2.1 Processing

Typically, the MA process starts with the blending of elemental and prealloyed powders in the correct proportion and loading of this blend into a mill along with the grinding medium, generally steel or tungsten carbide balls.[3] The powder and balls are then agitated for the time required to reach a steady state, at which point the composition of all of the powder particles is the same. The powder is then consolidated into a bulk shape and heat treated to obtain the desired microstructure and mechanical properties. Thus, the important components of MA are the raw materials, the ball mill, and the milling, consolidation, and heat treatment parameters.

A number of variants of MA have been developed, the major ones being: mechanical milling or grinding, in which an alloy is milled to induce phase transformations and reduce the size of microstructural features, such as the grain size;[4] reactive milling, during which a chemical reaction is induced;[5] and cryomilling, discussed in greater detail later in this chapter, carried out at low temperatures obtained by adding a cryogenic liquid to the powder and milling balls.[6]

It is usual that the raw materials used for MA are commercially available powders that have particle sizes in the range of 1 to 200 μm. These fall into the broad categories of pure metals, master alloys, prealloyed powders, and refractory compounds. In some cases, especially when milling ductile metal powders, a process control agent (PCA), usually stearic acid, up to 2 wt.% of the powder, is added to minimize the extent of particle agglomeration. Otherwise, large particles, several millimeters in size, can result. The PCA decomposes during milling and, for the milling of Al-containing alloys, can form compounds such as Al_2O_3 and Al_4C_3, which are incorporated into the Al matrix,[7] adding to the strength and thermal stability.

The types of ball mills available differ in their geometry, arrangement, capacity, and efficiency of milling. The three main types of ball mill used for MA are shown schematically in Figure 13.1. High-energy mills,

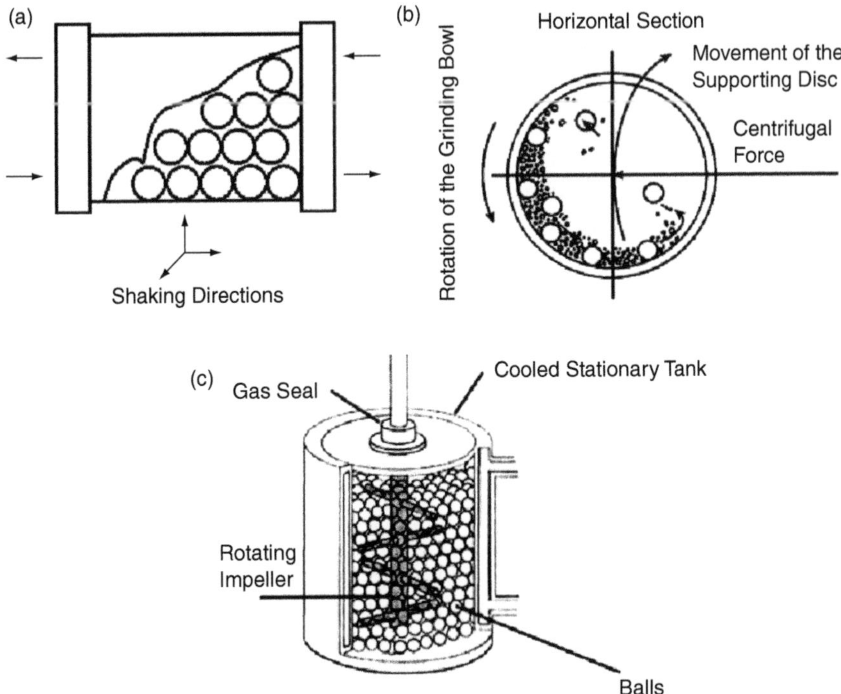

FIGURE 13.1 Schematic diagrams showing the main types of mills used for mechanical alloying: (a) SPEX, (b) planetary, and (c) attritor. (From Suryanarayana, C., *Prog. Mater. Sci.*, 46, 1, 2001. Reproduction by permission of Elsevier.)

such as SPEX shakers, capable of milling about 10 to 15 g of powder at a time, are the most common for alloy screening purposes. Due to the high amplitude of the clamp motion, the ball velocities are high, ~5 m/s, and consequently the force imparted by the balls when they impact is high. Planetary ball mills, such as the Fritsch Pulverisette, are less energetic than SPEX mills. However, this type of mill can accommodate two or four containers, each with a capacity of about 250 g. In the third type of mill, an attritor, the milling balls, contained within a stationary vessel, are energized by the rotation of a motor-driven impeller. The velocity of the attritor grinding media is much lower, ~0.5 m/s. However, the available range of powder capacity for attritors is wide; from less than 100 g, to several hundred kg for industrial mills. It is important to realize that the time taken to achieve steady-state conditions is longer as the milling energy is reduced. For example, a process that might take minutes to complete in a SPEX mill would take hours in a small attritor and days in an industrial attritor. In addition to the established types of mills described above, novel designs of mills used for MA have also been reported in the literature.[8–12]

MA is a complex process and hence involves optimization of a number of variables to achieve the desired product phase and microstructure.[12] For a given type of ball mill,[13] some of the important process variables are: the size, shape, and material of the container;[14] milling speed,[15,16] temperature,[17] and time;[18] type, size, and size distribution of the grinding balls;[19] ball-to-powder weight ratio (BPR);[20] extent of filling the container; milling atmosphere;[21] and the nature and amount of the PCA.[22] As examples of the effect of processing parameters, Figure 13.2a shows that the time it takes for the crystallite/grains of a Ni-base alloy to reach a minimum size decreases as the BPR is increased,[12] and Figure 13.2b shows that an increase in the amount of the PCA used tends to increase the grain size of a milled Al alloy.[22]

As diffusional processes are involved in the formation of alloy phases, irrespective of the nature of the phase formed, milling temperature plays an important role. However, only a few investigations have been carried out where the temperature of milling has been intentionally varied. Low-temperature milling has been carried out by dripping liquid N_2 on the milling container, while high-temperature milling has been achieved by the use of resistance tapes, for example. It has been found that the root mean square strain was lower and the grain size larger for materials milled at higher temperatures.[23] In addition, the extent of solid solubility was found to decrease on milling the powder at higher temperatures.[24,25] It has also been reported that the time required for amorphization increases linearly with the normalized milling temperature, defined as the ratio of the milling temperature and the melting temperature of the alloy, as shown in Figure 13.3 for three intermetallic compounds.[26]

Increasing the milling speed and BPR has generally been found to reduce the time for the formation of a particular alloy phase.[27] On the other hand, "soft" milling conditions, using a lower milling speed or BPR, are known to produce phases that are farther from equilibrium. For example, an amorphous phase has been reported to form in a Zr-33 at.% Co powder under "soft" conditions, while a mixture of crystalline phases was produced under more aggressive, "hard" conditions.[28] Similarly, a metastable phase has been synthesized under "soft" conditions with a Cu–In–Ga–Se powder blend, while an equilibrium phase formed using "hard" parameters.[29] A fuller description of the effect of process variables on the constitution and properties of MA alloys may be found in Reference 12.

13.2.2 Alloy Formation

A variety of crystalline, quasi-crystalline, or amorphous alloy phases have been reported to form by MA.[11,12,30] Even though the reasons for the formation of the different types of phases differ, a basic factor in the formation of alloy phases by MA is that the structure of the powder particles is refined, and that the constituent elements are very intimately mixed and uniformly distributed in each other. Additionally, the heavy plastic deformation introduces a high density of crystalline defects, such as dislocations, stacking faults, grain boundaries and vacancies, which enhance diffusion of the constituent elements.[11,12] If sufficient alloying has been found not to occur in the as-milled condition, a suitable postmilling heat treatment can increase the amount of diffusion further and greatly aid alloy formation.

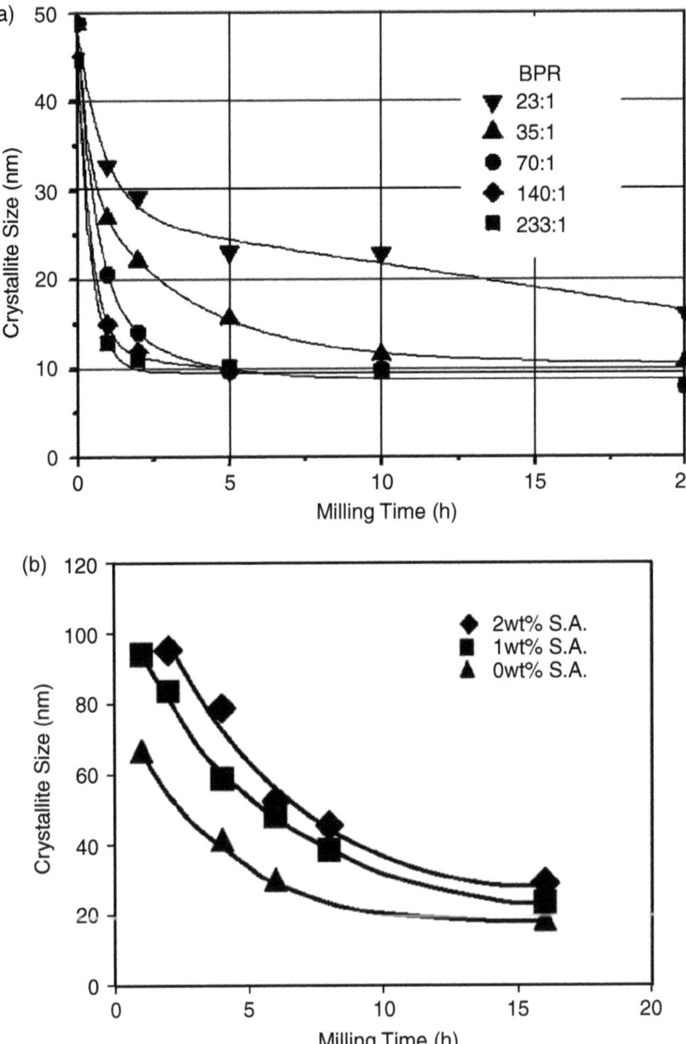

FIGURE 13.2 (a) Graph showing the variation of crystallite size with milling time as a function of ball-to-powder weight ratio (BPR). (From Suryanarayana, C., *Mechanical Alloying and Milling*, Marcel-Dekker, Inc., New York, NY, 2004. Reproduction by permission of Taylor & Francis.) (b) Graph showing the variation of crystallite/grain size with milling time for an Al alloy ($Al_{93}Fe_3Ti_2Cr_2$) as the amount of stearic acid (S.A.), used as a process control agent, is varied. (From Shaw, L. et al., *Metall. Mater. Trans. A*, 34, 159, 2003. Reproduction by permission of TMS.)

The nature of the phase formed is related to the thermodynamic stability of the different competing phases. For example, it has been reported that accumulation of lattice defects in a crystalline lattice raises its free energy to a level above that of the amorphous phase and consequently, the amorphous phase gets stabilized in preference to the crystalline lattice.[30] Formation of solid solutions and intermediate phases has been explained on the basis that the grain and crystallite sizes of the milled powder particles are usually on a nanometer scale and that such materials have a substantially large grain boundary area. Enhanced diffusion along the grain boundaries has been shown to increase the solid solubility levels.[31] It is generally expected that intermetallics synthesized directly by MA will be disordered in nature because of the heavy deformation involved, but both disordered and ordered intermetallics can form. If the difference in energy between the ordered and disordered states is small, then the MA alloy will exist in the disordered state; but, if it is large, the alloy will be in the ordered state.[32]

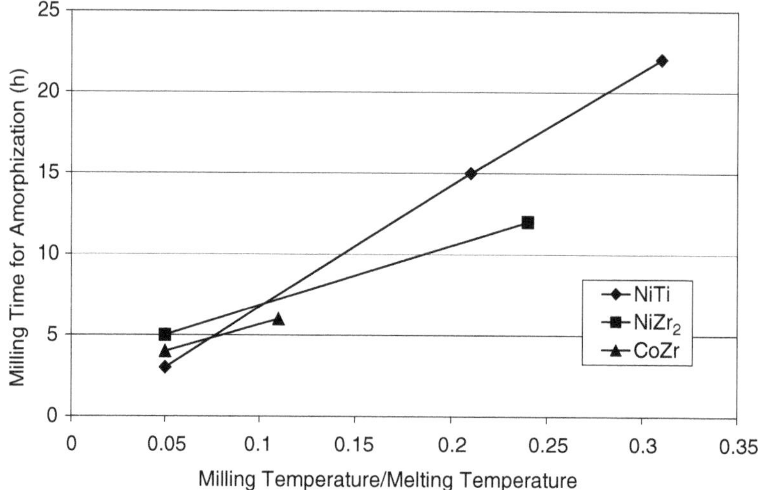

FIGURE 13.3 Graph showing the relationship between milling time for amorphization vs. normalized milling temperature for CoZr, NiTi, and $NiZr_2$ intermetallics. (From Koch, C.C., *Mater. Trans. Japan Inst. Metals*, 36, 85, 1995. Reproduction by permission of the Japan Institute of Metals.)

13.2.3 Microstructural Evolution

In the early stages of milling, the powder particles are relatively soft and therefore their tendency to cold weld together and form larger particles, under the impact forces exerted by the balls, is high. It has been suggested that typically around 1000 particles, with an aggregate weight of about 0.2 mg, are trapped between balls during each collision.[33] Upon continued milling and extreme deformation, the particles are work hardened, become increasingly brittle, and eventually fracture. With the generation of fresh, atomically clean particle surfaces, with increased activity, cold welding can again occur. When the cold welding and fracturing events balance each other, a steady-state exists, the particle size gets stabilized, and a minimum grain size, d_{min}, with typical dimensions of a few tens of nanometers, is obtained.

The value of d_{min} that is obtained during milling operations has been correlated with the crystal structure and the physical and mechanical properties of the material. Independent of other parameters, it has been shown that d_{min} decreases with an increase in the melting temperature, T_M,[34] unless the T_M is high, above ~1200°C, when d_{min} is virtually independent of T_M.[12] Consequently, one does not observe the inverse relationship between d_{min} and T_M in metals with a BCC structure, which tend to have very high melting points.[12] Inverse relationships have recently been established between the normalized minimum grain size (the ratio of d_{min} to the Burgers vector of the dislocations in the crystal structure) and the T_M, hardness, stacking fault energy, activation energy for self diffusion, bulk modulus, and the equilibrium distance between two edge dislocations.[35]

The final particle size, shape, and size distribution are dictated by the initial powder blend, which may be a combination of ductile–ductile, ductile–brittle, or brittle–brittle components. When a combination of ductile–ductile components is milled, a lamellar structure of the constituent elements is produced in the initial stages. When ductile–brittle components are milled, a typical example being oxide-dispersion strengthened materials, the brittle oxide particles become uniformly distributed in the soft metallic matrix, as shown in Figure 13.4. If brittle–brittle components are milled,[36] one would obtain a uniform dispersion of the harder, more brittle component in the softer, less brittle matrix, similar to the ductile–brittle combination.

Careful high-resolution TEM studies have revealed the following mechanism for the formation of nanoscale grains during milling:[37] In early stages, due to the high deformation rates experienced during MA, deformation is localized within shear bands. These shear bands, which contain a high density of

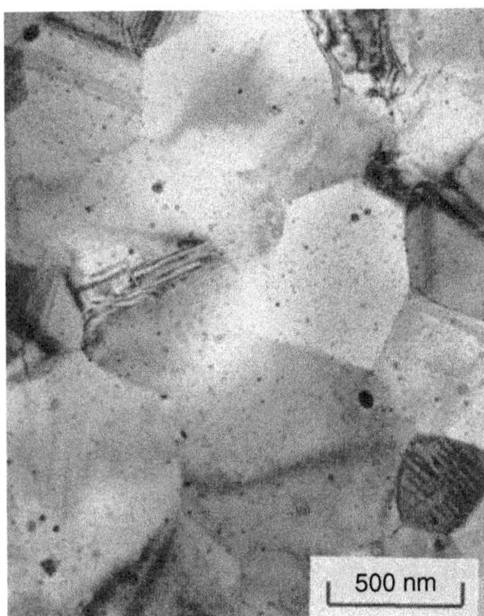

FIGURE 13.4 TEM micrograph showing a uniform dispersion of Er_2O_3 particles in an α_2-Ti_3Al alloy matrix after milling. (From Suryanarayana, C., *Prog. Mater. Sci.*, 46, 1, 2001. Reproduction by permission of Elsevier.)

dislocations, have a typical width of approximately 0.5 to 1.0 μm. Small grains, with a diameter of 8 to 12 nm, which electron diffraction patterns suggest have a significant preferred orientation, form within the shear bands. With continued milling, the average atomic level strain increases within the shear bands. At a certain dislocation density, sub-grains, usually between 20 and 30 nm, separated by low-angle grain boundaries, are formed, resulting in a subsequent decrease in the lattice strain. On further milling, deformation occurs in shear bands located in previously unstrained parts of the material. Sub-grains are again formed and the shear bands coalesce. Eventually, the small-angle boundaries are replaced by higher angle grain boundaries, accompanied by grain rotation, leading to the final structure composed of randomly orientated, dislocation-free, nanosized grains.

Another model for the refinement of grain size, d, after milling for a time, t, assumes that the interfaces of the shear bands directly lead to grain boundary formation and this results in the relationship $d = ct^{-2/3}$, where c is a constant.[38] This is in approximate agreement with the experimentally observed variation of grain size in the early stages of milling, termed Stage I. However, for the latter stages of milling, Stage II, the exponent is much reduced, being about -0.075, and there is a relatively small variation in grain size with respect to time.[38]

13.2.4 Consolidation of Milled Powders

Widespread application of MA materials requires efficient methods of consolidating the milled powders into useful shapes with close to 100% density. Prior to consolidation, the powder often requires canning and degassing. The consolidation pressures applied could be very low, a few tens of MPa, or very high, a few GPa. Conventional consolidation processes that have been used include: uniaxial cold pressing, cold isostatic pressing (CIP), vacuum hot pressing, and, most commonly, hot isostatic pressing (HIP).[39] A subsequent step, such as forging or extrusion, is generally needed for shear stresses to break up the prior particle structure and impart ductility.

Caution must be exercised in choosing the consolidation method and the parameters used, as exposure to high pressure and temperature for extended periods of time will result in extensive diffusion taking place, rapidly transforming and losing the beneficial structure achieved through MA. Precipitation of metastable or stable phases may occur in supersaturated solid solutions, metastable crystalline and quasi-crystalline phases may transform into equilibrium phases, the structure may coarsen, and an amorphous

TABLE 13.1 Room Temperature and Elevated Temperature Mechanical Properties of Commercial ODS Ni- and Fe-Based Superalloys Processed via MA

Alloy	Temperature (°C)	0.2% YS (MPa)	UTS (MPa)	Elong. (%)	RA (%)
MA 6000	RT	1220	1253	7.2	6.5
	871	675	701	2.2	4.6
	982	445	460	2.8	1.9
MA 754	RT	586	965	21	33
	871	214	248	31	58
	982	169	190	18	34
MA 956	RT	517	655	20	35
	1000	97	100	—	—
	1100	69	72	12	30

YS — yield strength; UTS — ultimate tensile strength; Elong. — elongation; RA — reduction in area.

Source: From Suryanarayana, C., *Prog. Mater. Sci.*, 46, 1, 2001. With permission.

phase could crystallize. Thus, the range of consolidation parameters are restricted and innovative methods of consolidation may be required.[40] Some newer methods to obtain consolidation of MA powders are quasi-isostatic forging, a process known by the trade name Ceracon,[41] and shock consolidation.[42]

However, not all MA alloys require a fine-grained structure, the prime example being the oxide-dispersion strengthened (ODS) alloys. In the extruded condition, they have a sub-micron, equiaxed grain structure. After extrusion, the alloy is worked to a final shape and then subjected to a secondary recrystallization anneal at, or above, 1300°C, to produce a coarse-grained structure that has better creep resistance.

13.2.5 Mechanical Properties

Mechanically alloyed ODS materials are strong at ambient and elevated temperatures, as shown in Table 13.1, their strength being derived from more than one mechanism. First, the uniform dispersion, with a spacing of the order of 100 nm, of very fine (5 to 50 nm) oxide particles, commonly Y_2O_3 (yttria), ThO_2 (thoria), or La_2O_3 (lanthana), which possess high thermal stability, inhibits dislocation motion and increases the resistance to creep deformation. Another function of the oxide dispersoids is to inhibit recovery and recrystallization processes, because of which a very stable structure is obtained, which resists grain growth and rotation during high-temperature deformation. Second, the homogeneous distribution of alloying elements conferred by MA gives both the solid-solution strengthened and precipitation-hardened alloys more stability at elevated temperatures and an overall improvement in properties. MA materials also have excellent oxidation and hot corrosion resistance, arising from the homogeneous distribution of the alloying elements and improved scale adherence due to the dispersoids.[43,44]

MA of rapidly solidified Al–Fe–Ce results in a uniform dispersion of sub-micron, thermodynamically stable carbides and oxides, which supplements the strengthening achieved by intermetallic particles. This leads to increased stiffness and strength at elevated temperatures, much higher than those obtained for the non-MA material,[45] the strength of the MA alloy being six times higher, even at 500°C. The MA material also has a significantly higher creep resistance, as shown in Table 13.2.[46] It is believed that the sub-micron dispersoids enhance the creep resistance by trapping dislocation clusters in cell walls, thereby effectively impeding recovery, and also by acting as a barrier to diffusion.[46]

13.2.6 Applications: Present and Future

The MA technique has now matured into an established materials processing method and several commercial Ni- and Fe-based ODS alloys are produced, the major customers being the thermal processing, glass

TABLE 13.2 Steady-State Creep Rates of MA and Non-MA Al-8Fe-4wt.%Ce at Two Different Temperatures and Stresses

Temperature (°C)	Stress (MPa)	Creep rate (s^{-1})	
		MA	Non-MA
350	103	8.3×10^{-10}	4.1×10^{-7}
380	83	1.6×10^{-9}	2.6×10^{-7}
380	103	1.4×10^{-9}	4.9×10^{-5}

Source: From Ezz, S.S., Lawley, A., and Koczak, M.J., in *Dispersion Strengthened Aluminum Alloys*, Y.-W. Kim and W.M. Griffith, Eds., TMS, Warrendale, PA, 1988, 243. With permission.

TABLE 13.3 Summary of Commercial Application for Some Selected MA ODS Ni- and Fe-Based Alloys

Alloy	Composition (wt.%)	Applications	Comments
MA 754	Ni-based 20Cr 0.3Al 0.5Ti 1.0Fe 0.6Y_2O_3	Nozzle guide vanes and band assemblies in military aero-engines — furnace fan blades	Strong ⟨100⟩ grain texture to enhance thermal fatigue resistance
MA 758	Ni-based 30Cr 0.3Al 0.5Ti 0.6Y_2O_3	Molten glass handling — fuel injection parts for internal combustion engines	High Cr gives extreme environment resistance
MA 6000	Ni-based 15Cr 4.5Al 2.5Ti 2.0Mo 4.0W 2.0Ta 1.1Y_2O_3	First and second stage turbine vanes and blades	Grains have aspect ratio greater than 10:1 giving flat rupture life curve
MA 956	Fe-based 20Cr 4.5Al 0.5Ti 0.5Y_2O_3	Furnace fixtures — molten glass handling — coke injection lances — nuclear waste disposal	Lower density, cost, and vapor pressure than Mo for furnace applications

manufacture, energy production, and aerospace industries, as shown for some selected alloys in Table 13.3. The two production facilities of Inco Alloys International in the United States have a combined annual milled powder production capacity approaching 300 tonnes, with a final wrought capacity greater than 200 tonnes. Applications in other areas are based on the chemical homogeneity, fine dispersion of one phase in the other, and enhanced chemical activity. These applications include: synthesis of targets for physical vapor deposition, magnets, superconductors, catalysts, hydrogen storage, food heaters, gas absorbers, solubility modification of organic compounds, waste management, and production of fertilizers.

The MA technique has been used for the production of super-corroding Mg-based alloys, containing Fe, Cu, C, Cr, or Ti, that operate as short-circuited galvanic cells, to react rapidly and predictably with an electrolyte, such as seawater, to produce heat and hydrogen gas.[47] Such a system is suitable as a heat source for warming deep-sea divers, as a gas generator to provide buoyancy, or as a fuel in hydrogen engines or fuel cells. Another application of MA Mg alloys, utilizing their high chemical activity, is in the provision of ready-to-eat meals (MRE) for U.S. soldiers. Finely ground, MA powders of Mg and Fe are pressed into a packet. In the field, water is added and heat is generated due to the galvanic couple existing between the two dissimilar metals, thus heating the food stored alongside.

One of the most promising applications appears to be for hydrogen storage using MA metal hydrides. Synthesis of nanocrystalline hydrides by MA has resulted in some remarkable developments, particularly in the kinetics of hydrogenation and dehydrogenation. Nanocrystalline Mg and Mg alloys absorb more hydrogen, more quickly, in comparison to their coarse-grained counterparts.[12,48,49] The fast kinetics, in combination with one of the highest reversible storage capacities, also qualifies these MA Mg-based materials for application in zero-emission vehicles.[48]

Potential applications for MA materials include: spray coatings, thermoelectric power generators, waste utilization, metal extraction, room temperature solders, and biomaterials. As MA is already an industrially viable process, it is reasonable to expect that the MA materials will continue to enjoy widespread application, in large based on the ability of the process to achieve fine powders, a much refined grain size, excellent chemical homogeneity, and a uniform distribution of the constituent phases. However, even though methods have been developed to minimize powder contamination, which can severely reduce mechanical properties, these methods are expensive and need to be developed further to make them suitable for industry. In addition, the milled powder is usually very hard and, as a consequence, traditional consolidation methods have not been effective in obtaining a fully consolidated and porosity-free product in a single step. Thus, several processes need to be carried out, greatly increasing the cost. Unless the powder can be used without consolidation, as a catalyst for example, and in low-end applications, where contamination is not a serious concern, extensive further commercial application of MA products will become feasible only when these issues are resolved.

13.3 Cryomilling

As stated earlier in this chapter, low temperatures can be beneficial to the process of MA and cryomilling is a variant in which milling is carried out with a slurry formed by the powder, milling balls and a cryogenic liquid, usually N_2.[50] The aim of the first investigations of cryomilling was that it could increase the envelope of what is possible with MA for the creation of dispersion strengthened alloys, such as an yttriated Fe alloy,[51] or Al_2O_3-reinforced Al.[6] This work indicated that shorter milling times were needed to obtain the smallest grain sizes. More generally, for other systems, the lower diffusion rates at cryogenic temperatures also means that the extent of solid solubility may be increased and the time for amorphization can be decreased. Recently, the main objective of development in this area is to produce the microstructural refinement that leads to consolidated structural alloys with greatly increased physical and mechanical properties. A range of cryomilled materials have been investigated, and Al-based alloys appear to offer the greatest commercial opportunities.[50]

13.3.1 Processing

The basic design of a cryomill consists of an attritor, schematically shown earlier in Figure 13.1c, that has been modified to allow the controlled flow of liquid N_2 into the container, thermally insulated to prevent excessive heat loss, so that the milling environment consists of a circulating slurry of milling balls, powder, and liquid N_2. Thermocouples are inserted through the lid of the container and positioned to monitor the temperature of the mill, ensuring that a constant level of liquid N_2 and a constant milling environment is maintained. Evaporated N_2 is exhausted through a blower equipped with a particle filter to retain any powder particles picked up by the gas flow. Upon the conclusion of cryomilling, the slurry is collected and the powder is transferred under the liquid N_2, to prevent contamination, to a glove box, where the N_2 is then allowed to evaporate.

13.3.2 Microstructure

Milling at cryogenic temperatures means that the annihilation of dislocations and the processes of recovery and recrystallization are suppressed, and the nanoscale grain structure is obtained in less time. Particle welding is also inhibited, so the amount of PCA, added to prevent excessive agglomeration and low yields, but a source of contamination,[22] is reduced. Still, ~0.2 wt.% stearic acid is generally required for softer metals. The presence of carbon from the PCA means that a fine dispersion of carbides can be formed during milling, in addition to oxides arising from the surface of the powder. A significant loss in ductility can be caused by the presence of hydrogen, most of which is removed by hot vacuum degassing.

The most significant positive aspect of milling in a liquid N_2 environment is the additional formation of fine, N-containing dispersoids, typically ~10 nm in diameter. These disk-shaped dispersoids maybe nitrides or oxy-nitrides, although their extremely fine nature makes their precise structure and composition difficult to determine. High-resolution transmission electron microscopy (HREM) of cryomilled Al-3%Al_2O_3 revealed platelets roughly 10 to 15 nm wide and only a few atoms thick.[52] Computed images

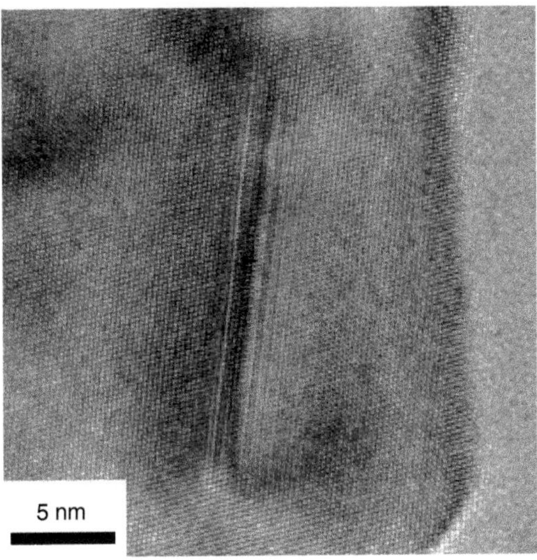

FIGURE 13.5 HREM image of an AlN dispersoid adjacent to a grain boundary in a cryomilled Al alloy. (Reproduction with permission of Patrick Berbon, DDG Cryogenics, Robert Hayes, Metals Technology Inc., and Rajiv Mishra, University of Missouri at Rolla.)

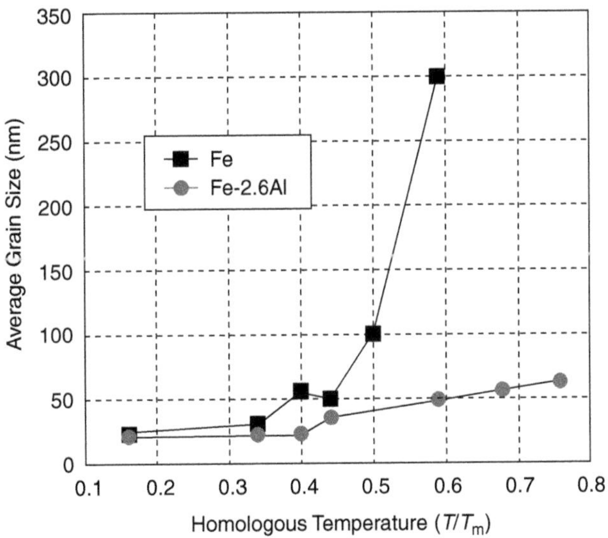

FIGURE 13.6 Graph showing the grain growth during isothermal annealing of cryomilled Fe, with and without Al. (From Perez, R.J. et al., *Metall. Mater. Trans. A*, 29, 2469, 1998.)

based on the HREM and convergent beam diffraction, indicate that the dispersoids are essentially layers of either N atoms in tetrahedral positions, or O atoms in the octahedral positions, both forming on the [111] planes in the Al FCC lattice. More recent HREM of cryomilled Al-10Ti-2Cu (wt.%) found nitride platelets, 3 to 5 atomic layers thick, an example of which is shown in Figure 13.5.[53]

As a consequence, cryomilled Al, or alloys containing some Al, exhibit substantial thermal stability. That is, grain growth is slow at high temperatures, especially when compared to Ni or Fe. Fe containing only 2.6%Al has much increased thermal stability compared to Fe without Al, as shown in Figure 13.6, where it can be seen that the Al-containing alloy retained a fine grain size up to almost 80% of its homologous melting temperature (T/T_M), whereas Fe exhibited substantial grain growth above $0.5T_M$.[54] The higher

TABLE 13.4 Tensile Properties of Cryomilled Al 5083 and Al-7.5wt.%Mg with Varying Amounts of Blended Unmilled Powder, Compared with Conventionally Processed Al 5083 in the Annealed (O) and Work Hardened (H34) Conditions

Unmilled (%)	Yield stress (MPa)	UTS (MPa)	Elongation (%)
Cryomilled Al 5083 (CIPped and extruded)[57]			
0	698	728	0.6
15	683	730	0.8
30	625	664	1.1
50	485	588	5.0
Cryomilled Al-7.5wt.%Mg (HIPped and extruded)[58]			
0	642	847	1.4
15	630	778	2.4
30	554	734	5.4
Conventional Al 5083[59]			
O	145	290	16
H34	285	345	9

resistance to grain growth for Al-containing alloys is probably due to AlN having a higher stability than Fe or Ni nitrides; Ni_3N, for example, decomposes above 500°C.[55] Evidence that the N-containing dispersoids is important has been emphasized by the cryomilling of NiAl in liquid Ar — annealing of the Ar-milled material led to grains that were an order of magnitude larger than NiAl milled in liquid N_2.[56]

Despite cyromilled Al alloy powder possessing excellent thermal stability, the time spent at temperature and pressure during degassing and consolidation does lead to some grain growth, and the typical grain size of conventionally consolidated material is in the range of 100 to 300 nm.

13.3.3 Mechanical Properties

Consolidated cryomilled materials have shown much greater strengths over their conventionally processed counterparts, as shown for Al 5083 in Table 13.4.[57–59] For example, the yield strength and ultimate tensile strength of cryomilled Al 5083 have been measured to be over twice that of relatively coarse-grained (~200 μm) material.[60] The main reason for the increased strength of cryomilled alloys is due to the small grain size, since, as described by the Hall–Petch relationship, the yield stress of metals is generally proportional to the inverse of the square root of the grain size. The presence of the AlN dispersoids also increases the yield stress, due to an Orowan strengthening mechanism. Two other microstructural effects, concentration differences, and solute atom/dislocation distributions also influence strength levels.[60] However, the increase in strength is usually accompanied by loss in ductility, as also shown in Table 13.4.

It has been realized that the presence of larger, micrometer-size grains within the nanoscale grains can lead to a considerable enhancement in ductility and fracture toughness.[61] This is due to the ability of the more ductile, larger grains to blunt cracks within the structure.[60] It is possible that these coarse grains, usually free from the nanoscale nitride dispersoids, can be introduced by a number of ways: some unmilled powder is blended in with the cryomilled powder prior to consolidation,[57,58] *in situ* formation during consolidation,[62,63] and the bulk material can be given a postconsolidation anneal.[64] Table 13.4 shows that the ductility of cryomilled Al 5083 and Al-7.5wt.%Mg is increased with increasing amounts of blended, unmilled powder, while still retaining high strength levels. In the case of HIP consolidation, diffusional processes can lead to coarse grains forming in the interstices between the prior powder particles, as shown in Figure 13.7.[62] Recent unpublished results suggest that the best performance in terms of ductility is

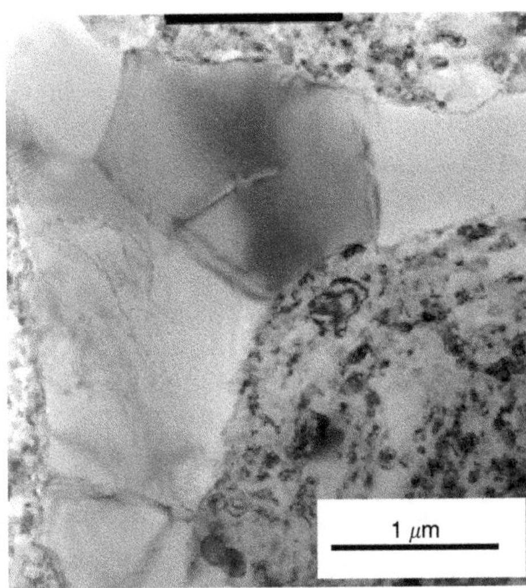

FIGURE 13.7 TEM image of as-HIPped Al 5083 showing the nanostructured regions originating from several prior particles and coarse-grained material, formed by diffusion, between them. (From Witkin, D. and Lavernia, E.J., in *Processing and Properties of Structural Nanomaterials*, Shaw, L.L., Suryanarayana, C., and Mishra, R.S., Eds., TMS, Warrendale, PA, 2003, 117. Reproduction by permission of TMS.)

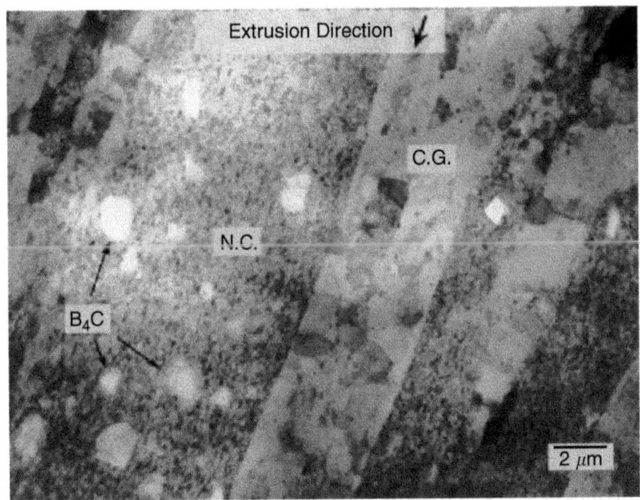

FIGURE 13.8 TEM image of CIPped and extruded Al 5083 + B_4C composite comprised of ultra fine-grained bands (N.C.), containing the B_4C particles, originating from cryomilled powder, and coarse-grained bands (C.G.) originating from blended unmilled powder. (From Ye, J. et al., *Scripta Mater.*, 53, 481, 2005. Reproduction by permission of Elsevier.)

obtained when H contamination is minimized, HIP is used to introduce coarse grains, and a secondary consolidation/deformation process, such as extrusion or quasi-isostatic forging,[41] is used to introduce shear stresses to break up the oxide on the prior powder particle boundaries.

The strength of cryomilled alloys can be increased considerably, at the further expense of ductility, by the addition of a ceramic reinforcing phase, such as SiC or B_4C, either before or after cryomilling. Cryomilled alloys, incorporating both coarse-grained material and ceramic particulate,[65] as shown in Figure 13.8, have the potential to be tuned to obtain the desired mechanical properties, depending on the specific application and need for strength or ductility.

13.3.4 Applications

As yet, there has been no commercial application of cryomilling technology, but there is considerable promise for the near future. Most of the development has been with Al alloys directed in two main areas — aerospace and military ground vehicles. Aerospace applications that have been targeted include the impeller in a rocket engine turbo pump, and tube for high-pressure propellant ducts, made from Al-7.5wt.%Mg.

For military ground vehicles, such as tanks and troop carriers, a substantial proportion of their considerable structural weight is typically comprised of an Al alloy. The increase in strength conferred by the cryomilled Al alloy will enable significant weight savings, leading to a greater overall vehicle performance, notably an improved fuel efficiency, better mobility, and an increase in operating range. Cryomilled Al 5083 has demonstrated markedly superior ballistic properties, meaning further weight savings by reducing the amount of steel armor used.

Naturally, a reduction in vehicle weight is also a major thrust for the automotive manufacturers, and there has been some interest in cryomilled materials. But the high cost associated with the extra processing required, along with a relatively low rate of material throughput, is likely to restrict large-scale commercial application in the near future.

13.4 Friction Stir Processing

Friction stir processing is a rapidly maturing solid-state thermo-mechanical technique for the modification of metallic structure to a depth of 20 to 25 mm. Akin to friction stir welding (FSW), the process exploits the microstructural refinement left in the wake of a friction stir tool and has been shown to enhance the fatigue resistance, strength, corrosion resistance, and formability of several alloys. Chief among the advantages of achieving the desired microstructures via FSP is simplicity. Depending on the preferred properties and application, single or multiple passes of the FSP tool can be made, producing a fine, equiaxed grain structure in specific regions of a component.

13.4.1 Processing and Microstructure

In contrast to FSW, where the workpiece is comprised of the two sections to be welded, the FSP workpiece is a single plate, sheet or, more interestingly, a complex-shaped cast component. As shown in Figure 13.9,[66] the FSP tool is rotated as it is lowered into contact with the workpiece. When sufficient frictional heat has been generated at the tool-pin–workpiece interface to soften the metal, the pin is plunged until the tool shoulder is in contact with the surface of the workpiece. The tool continues to rotate until enough heat has been generated to soften the area under the shoulder, at which time the rotating tool can be translated along the surface of the workpiece. As the tool moves forward, material is swept by the rotating pin, from the advancing side to the retreating side. The tool rotation and translation speeds used are largely determined by the workpiece alloy composition and initial microstructure, as well as the targeted microstructure.

FSP is an effective way to induce sufficient deformation to intimately mix material and form a relatively stable, localized region of fine recrystallized grains and a homogeneous distribution of microstructural constituents. The refined microstructure of the friction stir zone (SZ) is readily apparent in Figure 13.10,[67] which shows a typical workpiece in cross section after single-pass FSP. Generally, an interface region between the base metal and the SZ, known as the thermo-mechanically affected zone (TMAZ), also exists. Less desirable features of single-pass FSP, such as concentric rings, sometimes referred to as onion ring patterns, and an asymmetrical geometry that is attributed to the more intense deformation experienced on the advancing side, can also be seen in Figure 13.10. These negative features can be minimized by optimization of the processing parameters, tool material and design, and rastering protocol.

As in FSW, the optimized processing parameters (translational speed, rotational speed, z- and x-axis forces, processing depth, anvil temperature, etc.) are significantly influenced by the geometry and material

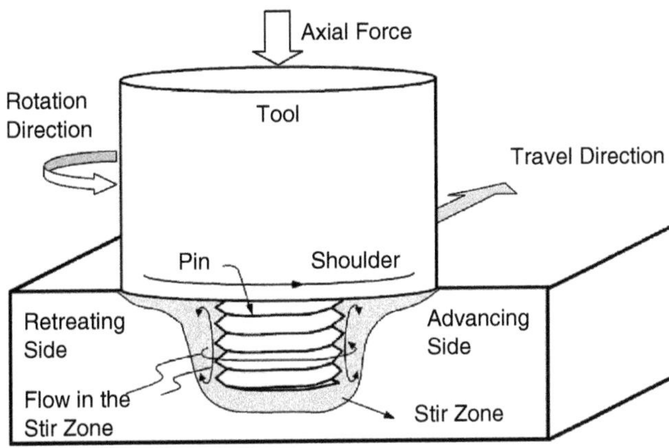

FIGURE 13.9 Schematic diagram of a typical FSP arrangement, showing the region of severe plastic deformation in the vicinity of the tool pin. (From Oh-Ishi, K. and McNelley, T.R., *Metall. Mater. Trans. A*, 35, 2951, 2004. Reproduction by permission of TMS.)

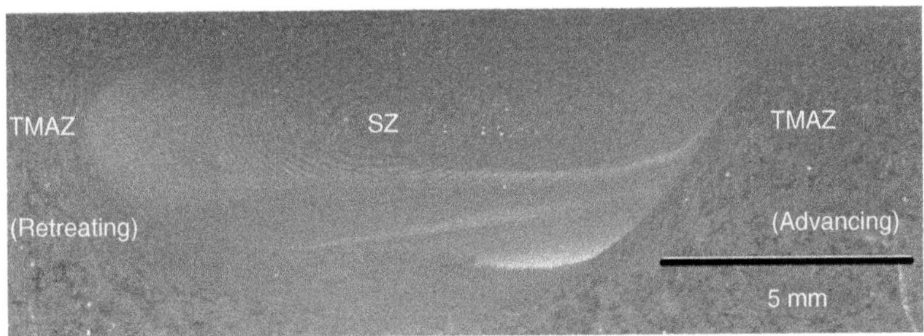

FIGURE 13.10 Optical micrograph of alloy A356 showing microstructural features in the friction stir zone (SZ) and thermo-mechanically affected zone (TMAZ) after single-pass FSP. (From Sharma, S.R., Ma, Y.Z., and Mishra, R.S., *Scripta Mater.*, 51, 237, 2004. Reproduction by permission of Elsevier.)

selected for the tool pin and shoulder. Ongoing research and modeling of tool design continues to improve the capabilities and affordability of the tools. A few FSP tool geometries that are currently used are exemplified in Figure 13.11.[68] Tool materials must be wear resistant and damage tolerant; they must remain chemically inert to relatively high temperatures, often within 70 to 90% of the workpiece melting temperature. A significant body of work concerning the FSP of Al alloys has been conducted with tools made from tool steels. For higher temperature alloys, including iron- and nickel-based alloys, tools made from tungsten–rhenium alloys and polycrystalline cubic boron nitride have been developed.[69]

An effective means of controlling the uniformity of the processed region is optimization of the raster pattern, which may be as simple as two passes made along the same path, but with opposite translation directions, and hence opposite rotation directions, to correct asymmetry. In other cases, a more sophisticated spiral raster pattern has proven to be efficient.[70]

13.4.2 Defect Repair of Cast Alloys

The earliest investigations illustrated the efficiency of microstructural refinement and homogenization by single-pass FSP for 7XXX series Al alloys,[71,72] and Al 1050.[73] Refined, recrystallized microstructures were obtained that were relatively stable at temperatures approaching 500°C, $\sim 0.9T_M$. This stability has been attributed to the distribution of fine oxides that were broken up and then redistributed during the severe

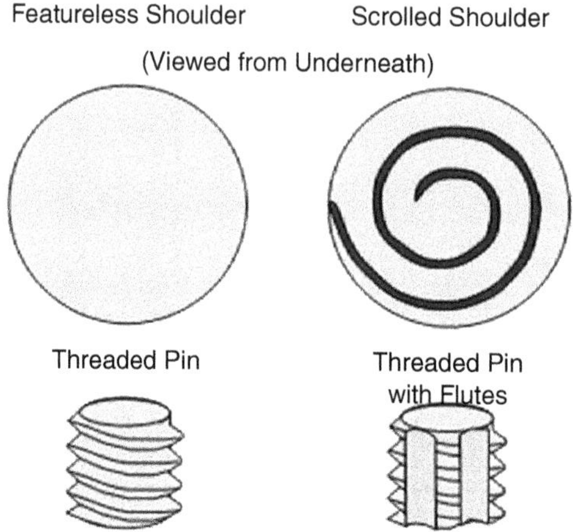

FIGURE 13.11 Schematic examples of FSP tool designs. (From Mishra, R.S., *Adv. Mater. Proc.*, October, 43, 2003. Reproduction by permission of ASM International.)

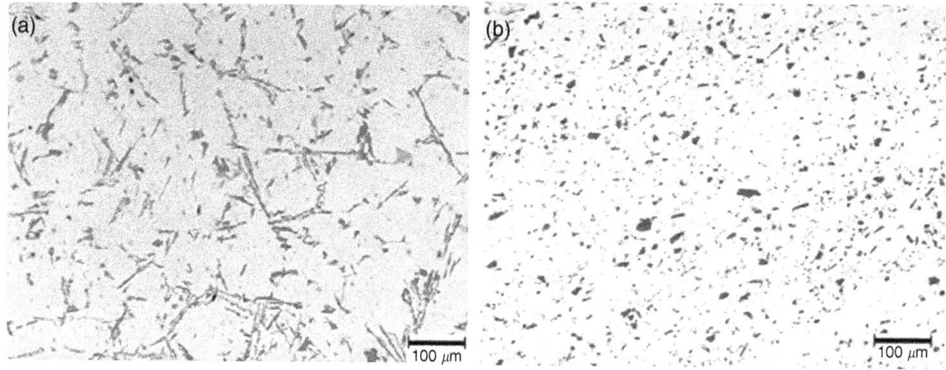

FIGURE 13.12 Optical micrographs of alloy A356: (a) as-cast and (b) after FSP. (From Sharma, S.R., Ma, Y.Z., and Mishra, R.S., *Scripta Mater.*, 51, 237, 2004. Reproduction by permission of Elsevier.)

FSP deformation and intermetallic grain boundary pinning agents that precipitated during cooling from the FSP working temperatures.[73]

As-sand-cast A356 contained coarse Al dendrites of size ~100 μm and large acicular Si with typical particle aspect ratios of 25, as shown in Figure 13.12a, and the improved microstructure after FSP is shown in Figure 13.12b.[67] The greatest reductions in porosity and the most extensive breakup of the microstructural constituents were achieved when a tri-flute pin tool was used at 700 rpm, translating at 3.4 mm/s. Compared with the as-cast material, optimally FSPed A356, subsequently aged at 155°C for 4 h to recover Mg_2Si precipitates dissolved during processing, exhibited a yield stress increase from 132 to 153 ± 20 MPa, an ultimate tensile strength increase from 169 to 212 ± 5 MPa, and an order of magnitude increase in elongation to failure from 3 to 26 ± 2%.[74] An increase in fatigue strength threshold of >80% for this alloy, attributed specifically to the reduced porosity and reduced Si particle size, has also been reported.[67]

A large body of FSP development has been conducted on nickel–aluminum–bronze (NAB) castings, UNS95800 (Cu-9Al-5Ni-4wt.%Fe) being an example, used for marine component applications; FSP removes porosity, reduces the size of large inclusions and intermetallics, and redistributes finer particles, thereby improving tensile and fatigue properties.[70] Microstructural refinement and homogenization also

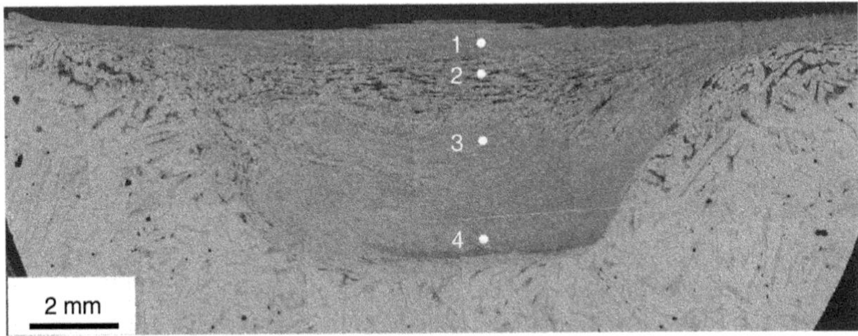

FIGURE 13.13 Composite optical micrographs of single-pass FSP NAB Cu-9Al-5Ni-4Fe—low magnification image of the entire stir zone. Lightly etched areas are α-Cu; darker, more heavily etched features are β-Cu transformation products, including Widmannstätten α, bainite, and martensite. (From Oh-Ishi, K. and McNelley, T.R., *Metall. Mater. Trans. A*, 35, 2951, 2004. Reproduction by permission of TMS.)

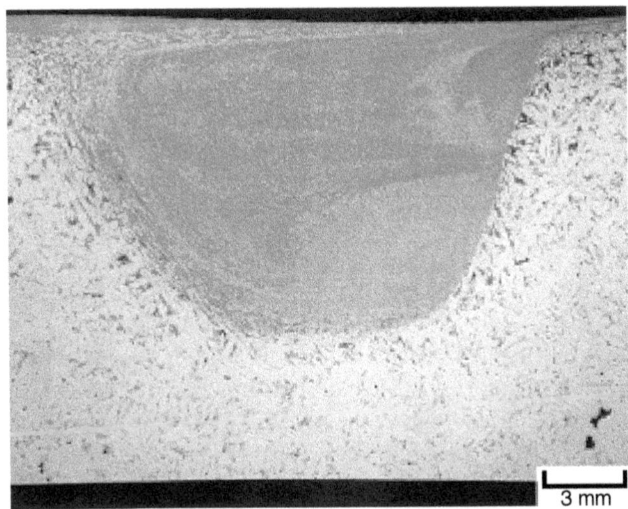

FIGURE 13.14 Optimized FSP NAB Cu-9Al-5Ni-4Fe with a stir zone comprised almost entirely of a homogeneous Widmannstätten microstructure. (From Mahoney, M.W. et al., *Mater. Sci. Forum*, 426–432, 2843, 2003. Reproduction by permission of Trans Tech Publications.)

appear to improve the corrosion resistance by reducing the size of grains susceptible to preferential attack, thus limiting pitting corrosion.[66]

Large NAB castings have near-equilibrium microstructures of Widmannstätten α-Cu and retained β-Cu, plus various β transformation products. Although the structure is significantly refined, the post-FSP microstructures in these multiphase, quaternary alloys can be highly complicated and be comprised of a range of microstructures, as shown in Figure 13.13,[66] where four distinct regions in the SZ can be identified. Parts of the SZ can experience markedly different thermo-mechanical processing histories with the highest temperatures, cooling rates, and strains occurring near the workpiece–tool-shoulder interface. Processed under conditions that allowed higher peak temperatures, nearly 1000°C, subsequent efforts to homogenize the stir zone and correct the asymmetry have been successful.[70] Two passes along the same FSP path, but in opposite directions, produced a more symmetric SZ with a completely Widmannstätten microstructure, as shown in Figure 13.14. This material had over double the strength ($\sigma_{UTS} = 823$ MPa) of the as-cast material ($\sigma_{UTS} = 399$ MPa), with similar ductility.

13.4.3 Precursor to Superplastic Forming

FSP generates a relatively stable, fine microstructure with high-angle grain boundaries (HAGBs) between grains that are typically in the size range 1 to 7 μm. For commercial aluminum alloys, this is often an ideal structure for high-strain-rate superplastic forming (SPF). After the first report of applying FSP to produce SPF-ready material,[71] numerous Al alloys have been investigated, including Al 7075,[75,76] an Al–Mg–Zr alloy developed particularly for SPF,[77] A356,[78] Al 5083,[79] Al 2095,[80] and an Al–Cu–Li alloy,[81] as well as some recent exploration of the Mg alloy AZ31.[82] These studies collectively emphasize that SPF is achieved at lower temperatures and flow stresses when the microstructure produced by FSP is both fine-scale and relatively stable, which is largely determined by alloy composition and FSP parameters. More information on superplasticity may be found in Chapter 14.

As an example, Al 7075 (Al-5.6Zn-3.5Mg-1.6Cu-0.23wt.%Cr), widely available as rolled plate, was FSPed to obtain material with two different grain sizes, 3.8 and 7.5 μm, through the implementation of different tool geometry, translation, and rotation speeds.[71,75,76] Subsequent heat treatment at 490°C for 1 h increased the mean grain size to 5.9 and 9.1 μm, respectively, demonstrating the reasonably slow grain growth necessary for SPF. This stability is attributed to the grain boundary pinning by fine Cr-rich dispersoids and $MgZn_2$-type precipitates. Between the strain rates of 3×10^{-3} to $10^{-1} s^{-1}$ and temperatures between 470 and 530°C, the flow stress was found to increase linearly with grain size. The stress exponent was ~2, indicating that grain boundary sliding was the dominant deformation mechanism. The optimal SPF parameters were determined to be 480°C for strain rates over the range 3×10^{-3} to 3×10^{-2} s^{-1} for the 3.8 μm grain material, achieving >1250% elongation, and 500°C and 3×10^{-3} s^{-1} for the 7.5 μm grain material, achieving 1042% elongation.[75] Higher strain rates, approaching 10^{-1} s^{-1}, were found to induce cavitation in both materials.[76] This represents an order of magnitude increase in strain rate capability over what has previously been achieved only with costly multiple-roll passes, typically employed to refine microstructure in sheet products, or equal-channel angular extrusion.[71]

13.4.4 Conclusions

Friction stir processing is proving to be an effective, flexible tool for modifying near-surface microstructure. Recent research and development have demonstrated the feasibility of the technique and inspired numerous ongoing activities to further explore the capabilities and limitations of FSP, which are only just beginning to be appreciated. Depending on the geometry of the tool and processing parameters, the depth of penetration can be controlled to that required for property enhancements, as dictated by the component design and application. There are numerous instances where the full thickness or surface area of a component need not be altered to gain substantial improvement in component performance as specific regions are likely to see increased loads or be vulnerable to environmental attack. In these cases, to reduce the cost of processing, FSP may be applied selectively to enhance mechanical and corrosion resistance. Similarly, some component designs may only require SPF deformation along particular patterns, and FSP could be limited to these areas. The flexibility of FSP to be used selectively, on thin sheet and thick sections, makes it very attractive for commercial applications where processing costs often dominate design decisions.

13.5 Equal-Channel Angular Pressing

Equal-channel angular pressing (ECAP), also known as equal-channel angular extrusion (ECAE), is the most developed severe plastic deformation method for producing ultrafine grains (UFG) directly in bulk metals.[83,84] Originally developed in the 1980s, ECAP has grown in the last decade to attract worldwide interest.

13.5.1 Processing

During ECAP, material is pressed through a die that has two channels, with an identical cross section, that intersect at an angle Φ (the channel-interaction angle), usually in the range 90 to 157.5°, as shown in

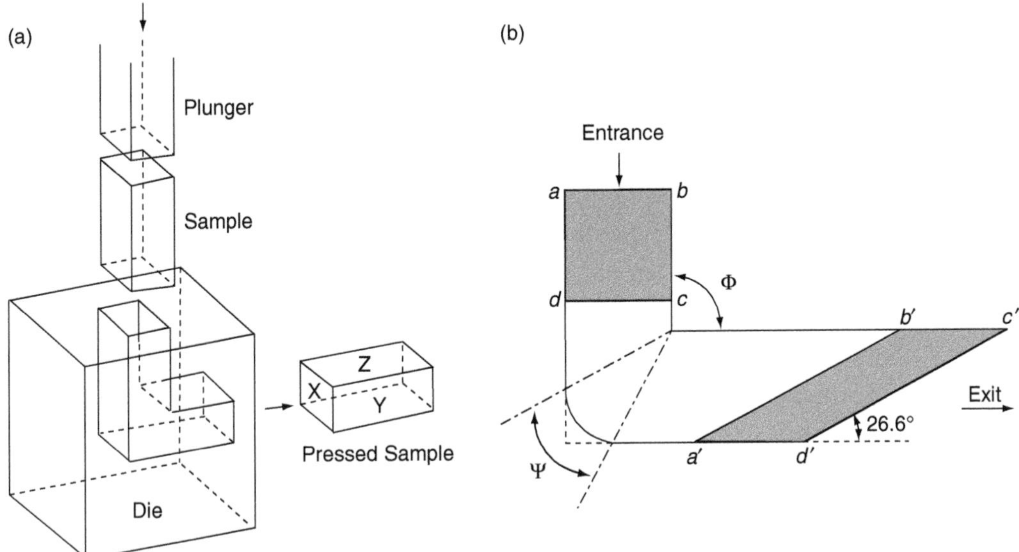

FIGURE 13.15 Schematic diagrams showing (a) typical ECAP operation, and (b) the shear deformation of a cube (*abcd*) into a parallelepiped (*a′b′c′d′*) as a result of ECAP. (From Komura, S. et al., *J. Mater. Res.*, 14, 4044, 1999. Reproduction by permission of MRS.)

Figure 13.15.[85] The principal whereby intense strain is imparted to the material is shown in Figure 13.15b, for $\Phi = 90°$, where a cube (*abcd*) is sheared into a parallelepiped (*a′b′c′d′*) at an angle of ~26.6° with the axis of the exit channel.[83,86,87] This shape change after one pass is in an excellent agreement with experiments, both on the macroscopic level, where a cubic shape is embedded in a matrix,[88] and on the microscopic level, where equiaxed grains of pure Fe have been observed to change into approximate parallelograms.[89]

The shear strain (γ) accumulated after each pass, usually in the range 1.15 to 0.23, decreases with increasing Φ. In view of the fact that material can be pressed numerous times, the total strain (γ_N) accumulated in the material after N passes through the die can be expressed by: $\gamma_N = N \cdot \gamma$. A general relationship considering the effect of both Φ and the angle of the outer arc, Ψ, is expressed as:[84,86]

$$\gamma_N = \frac{N}{\sqrt{3}}\left[2\cot\left(\frac{\Phi}{2}+\frac{\Psi}{2}\right) + \Psi\,\text{cosec}\left(\frac{\Phi}{2}+\frac{\Psi}{2}\right)\right]$$

From this equation, it follows that, at the most frequently used angles, $\Phi = 90°$ and $\Psi = 20°$, each pass corresponds to a strain of approximately 1.05. The grain size generally decreases with the amount of accumulated strain, and to obtain a UFG structure, repeated passes are required. Although a theoretical relationship between grain size and the accumulated strain during ECAP has not been established, experimental observation has revealed that a UFG structure can normally be achieved after eight passes.[85,90,91]

The structure of the material after multi-pass ECAP is significantly influenced by the orientation between passes. Usually, one of three routes is employed; A, B, and C, as shown in Figure 13.16.[86] For route A, the material is not rotated between passes. For route B, the material is rotated 90° between passes. If the material is rotated in the same direction around its axis on consecutive passes, the route is termed B_C. If the material is rotated in alternative directions between each consecutive pass, the route is termed B_A. For route C, the material is rotated 180° between consecutive passes. After even-numbered passes using route C, the shape of the material is restored to its original form. Investigations indicate that B_C is the most efficient route for $\Phi = 90°$.[86,87,92]

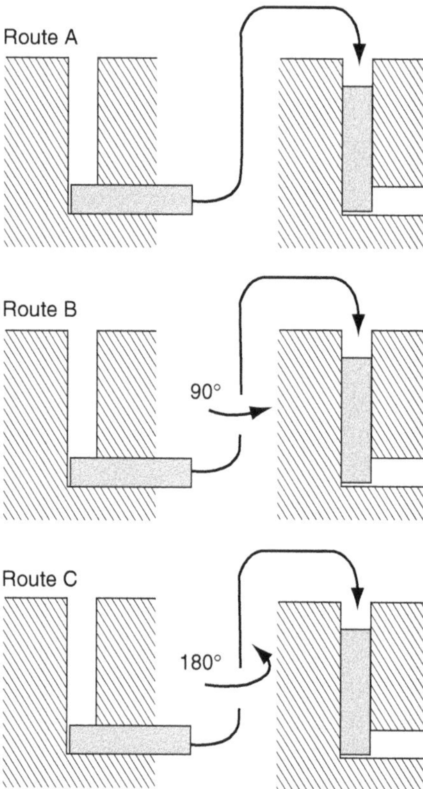

FIGURE 13.16 Schematic diagram showing the three different processing routes used in ECAP. For route A, there is no rotation between consecutive pressings, and for routes B and C the material is rotated by 90 and 180°, respectively. (From Iwahashi, Y. et al., *Metall. Mater. Trans. A*, 29, 2245, 1998. Reproduction by permission of TMS.)

13.5.2 Microstructure

Electron microscopy has revealed that ECAP materials have a typical grain size of 100 to 500 nm and ECAP has been used, for example, in the grain refinement of Al alloys.[93–97] Initially, after one or two passes, the original grains in the material become elongated, forming a banded structure, and there is a simultaneous increase in dislocation density. The high dislocation density leads to dynamic recrystallization, so that on subsequent passes, typically after a total of four, an equiaxed, UFG structure is formed, with many small-angle grain boundaries. Further passes lead to the formation of HAGBs, without any significant decrease in grain size. Despite having experienced dynamic recrystallization, ECAP material often has a high dislocation density, as shown in Figure 13.17a for Al 7075,[96] for which a dislocation density of 10^{15} m^{-2} has been estimated.

In alloys that are particularly susceptible to dynamic recrystallization, Al alloys such as Al-3%Mg, for example, the development of HAGBs may require less passes. The starting grain size of ~500 μm in Al-3%Mg was refined to ~230 nm after four passes at ambient temperature, as shown in Figure 13.17b.[97] The individual grains were generally separated by HAGBs, as evidenced by the continuous rings in the accompanying selected-area electron diffraction pattern (SADP), also shown in Figure 13.17b. However, in common with other materials processed by ECAP, some of the grain boundaries were not well-defined. HREM of the Al-3%Mg alloy, and an Al–Mg–Sc alloy processed via ECAP, showed that the grain boundaries were often wavy and faceted, that is, they were in high-energy, nonequilibrium configurations.[98]

For BCC or HCP metals, it is difficult to perform ECAP at room temperature due to the insufficient number of slip systems. Although eight passes at ambient temperature have been successfully performed

(a)

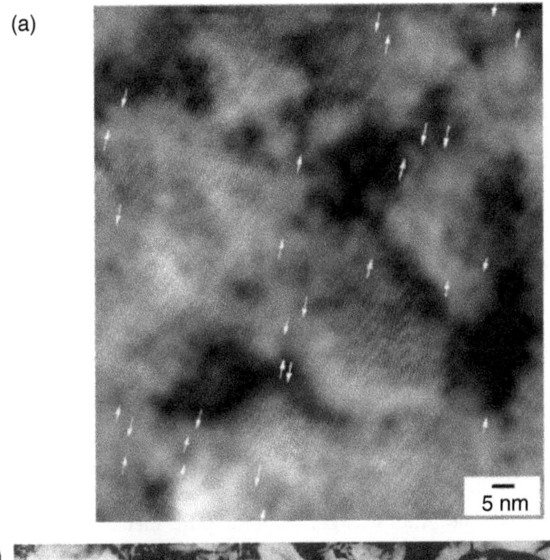

(b)

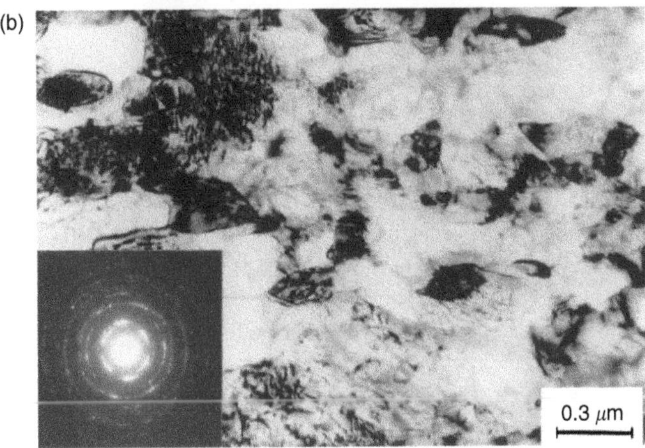

FIGURE 13.17 (a) One-dimensional HREM image, obtained by Fourier and inverse Fourier transformations of the original HREM image, of Al 7075 after 8-pass ECAP. The end of half atomic planes, the dislocation cores, are marked with white arrows (From Zhao, Y.H. et al., *Acta Mater.*, 52, 4589, 2004. Reproduction by permission of Elsevier); (b) TEM image, with SADP, of Al-3%Mg after 4-pass ECAP. (From Berbon, P.B. et al., *Metall. Mater. Trans. A*, 29, 2237, 1998. Reproduction by permission of TMS.)

on pure Fe, and an UFG microstructure obtained,[99] most BCC and HCP, plus some hard FCC metals have to be pressed at elevated temperatures to increase the number of slip systems or to soften the material.[94,96,100–102] For ECAP operated at elevated temperatures, the grain refinement still occurs, but there is increased dynamic recovery or recrystallization, leading to a lower number of nonequilibrium grain boundaries.

13.5.3 Mechanical Properties

Owing to the grain size refinement after ECAP, a significant increase strength is achieved, as shown in Figure 13.18, which compares the tensile yield strength of four pure metals before and after ECAP.[89,101,103,104] The FCC metals, Al and Cu, are about 2.5 times stronger and the HCP Ti is about 2 times stronger than the original coarse-grained material. However, the UFG Fe is over 10.5 times stronger.

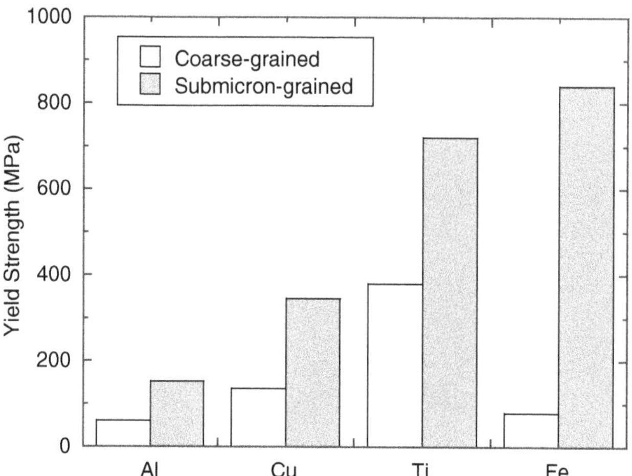

FIGURE 13.18 Comparison of tensile yield strength for four pure metals before (coarse-grained) and after (submicron-grained) ECAP. The ECAP was carried out at room temperature (unless stated): Al (8-pass),[103] Cu (12-pass)[105], Ti (8-pass at 450°C),[101] and Fe (8-pass).[89]

The plastic behavior of UFG ECAP materials is different from that of their coarse-grained counterparts. In coarse-grained materials, strain-hardening after yielding is generally observed. For UFG ECAP materials, necking with the occurrence of shear bands,[89,103] or nearly perfect plastic deformation,[98] is observed after only a brief strain-hardening region. Compared to coarse-grained materials, the tensile ductility of ECAP materials is low.

There is an obvious effect of strain rate, $\dot{\varepsilon}$, on yield strength in ECAP materials at ambient temperatures, in contrast to coarse-grained materials, where the strain rate has an insignificant effect on mechanical properties under quasi-static conditions, that is, a strain rate less than 1 s^{-1}. For instance, the strain rate sensitivity exponent, $m = \partial \ln \sigma / \partial \ln \dot{\varepsilon}$, determined by using tensile strain rate jump tests, at room temperature for ECAP Cu, is higher than that of coarse-grained Cu in the range of 6×10^{-7} to 10^{-1} s^{-1},[104] particularly at the lower end of this range. The high value of m for ECAP Cu was confirmed by another investigation,[105] which found that it was three-times higher than that of annealed coarse-grained Cu. The high strain rate dependency was attributed to additional thermally activated processes operating in the ECAP material due to the pinning role of the larger volume fraction of grain boundaries on dislocations.

Similar to that obtained by FSP described earlier in this chapter, the UFG microstructure created by ECAP can lead to superplasticity at higher strain rates and at lower temperatures than conventionally processed material. For example, for an Al–Mg–Sc alloy with an average grain size of ~200 nm after ECAP, has been reported to achieve elongations of >1000% at 400°C and at strain rates in the vicinity of 10^{-2} s^{-1}, with a maximum elongation of 2280% at 3.3×10^{-2} s^{-1}.[93] In another example, the superplasticity of ECAP Al 5083, with an average grain size of 300 nm, is reported at the low temperatures of 225, 250, and 275°C — a maximum elongation of 315 % was obtained at 275°C and a strain rate of 5×10^{-4} s^{-1}.[92]

13.5.4 Applications

Although ECAP is considered to be a candidate for many practical applications, there has been only one reported example of commercialization, for the fabrication of sputtering targets from Al and Cu alloys.[106] UFG sputtering targets provide excellent material uniformity and enhanced target life, compared to coarse-grained counterparts. ECAP can also reduce processing costs by forming large targets, up to 34 kg in weight, from one piece, removing the need for diffusion bonding. The successful commercialization of ECAP sputtering targets suggests that it is possible to apply ECAP technology to other industrial applications where the increased processing costs are offset by significant improvements in performance.

One potential application of ECAP is that the considerable shear deformation can be employed to consolidate powders to manufacture bulk material with benefits obtained from both the powder production route and ECAP.[107–109] ECAP consolidation at 300°C of CIPped Al 2024-3%Fe-5%Ni powder obtained a pore-free material with very high hardness.[107] ECAP at elevated temperatures has been used to consolidate a Zr-based metallic glass powder directly into a bulk form.[108] Experiments indicate that ECAP is also a viable method to consolidate Cu powder, achieving UFG material with a significant gain in strength and ductility.[109]

13.6 High-Pressure Torsion

In the technique commonly known as high-pressure torsion (HPT), a disk-shaped sample of material, typically 10 to 20 mm in diameter and 0.2 to 0.5 mm thickness, is held under an applied pressure of several GPa between two anvils.[84] By rotating the lower anvil, frictional forces are transferred to the material and it is subjected to very high strains due to the presence of the quasi-isostatic compressive stress state. Structural refinement can be observed after only half a rotation of the anvil, but generally several rotations are required to obtain a homogeneous grain size of approximately 100 nm. HPT can also be used for the consolidation of nanostructured powders, such as cryomilled Al-7.5wt.%Mg,[110] without excessive coarsening of the structure.

HPT is a useful process for the investigation of cumulative strain, applied pressure, and rotational speed on the structure and properties of the material concerned. However, the size of the product is small and, consequently, the range of potential commercial application of HPT is very limited.

Acknowledgments

The authors from UC Davis would like to thank the Office of Naval Research (under contract nos. N00014-03-C-0164 and N00014-D4-1-0370) and the Army Research Office (under contract W911NF-06-1-0230) for financial support.

References

[1] Benjamin, J.S., Mechanical alloying — A perspective, *Metal Powder Rep.*, 45, 122, 1990.

[2] Benjamin, J.S., Mechanical alloying — History and future potential, in *Advances in Powder Metallurgy and Particulate Materials*, vol. 7 (Novel Powder Processing), J.M. Capus and R.M. German, Eds., Metal Powder Industries Federation, Princeton, NJ, 1992, 155.

[3] Benjamin, J.S., Dispersion strengthened superalloys by mechanical alloying, *Metall. Trans.*, 1, 2943, 1970.

[4] Ermakov, A.E., Yurchikov, E.E., and Barinov, V.A., The magnetic properties of amorphous Y-Co alloy powders obtained by mechanical comminution, *Phys. Metall. Metallogr.*, 52, 50, 1981.

[5] Jangg, G., Kuttner, F., and Korb, G., Production and properties of dispersion hardened aluminum, *Aluminium*, 51, 641, 1975.

[6] Luton, M.J. et al., Cryomilling of nano-phase dispersion strengthened aluminum, in *Multicomponent Ultrafine Microstructures*, L.E. McCandlish, D.E. Polk, R.W. Siegel, and B.H. Kear, Eds., Mater. Res. Soc., Pittsburgh, PA, 1989, vol. 132, 79.

[7] Singer, R.F., Oliver, W.C., and Nix, W.D., Identification of dispersoid phases created in aluminum during mechanical alloying, *Metall. Trans. A*, 11, 1895, 1980.

[8] Calka, A. and Wexler, D., Electrical discharge assisted ball milling: a promising materials synthesis and processing method, *Mater. Sci. Forum*, 386–388, 125, 2002.

[9] Pochet, P. et al., Order-disorder transformation in Fe-Al under ball milling, *Phys. Rev.*, B52, 4006, 1995.

[10] Szymanski, K. et al., Friction-free mechanical grinder, *Mater. Sci. Forum*, 235–238, 223, 1997.

[11] Suryanarayana, C., Mechanical alloying and milling, *Prog. Mater. Sci.*, 46, 1, 2001.
[12] Suryanarayana, C., *Mechanical Alloying and Milling*, Marcel-Dekker, Inc., New York, 2004.
[13] Martinez-Sánchez, R. et al., Atmospheric and milling-device effects on the activation energy for crystallization of a partially amorphized Ni-Mo alloy, *Mater. Sci. Forum*, 386–388, 135, 2002.
[14] Harringa, J.L., Cook, B.A., and Beaudry, B.J., Effects of vial shape on the rate of mechanical alloying in $Si_{80}Ge_{20}$, *J. Mater. Sci.*, 27, 801, 1992.
[15] Mio, H. et al., Effects of rotational direction and rotation-to-revolution speed ratio in planetary ball milling, *Mater. Sci. Eng.*, A332, 75, 2002.
[16] Sherif El-Eskandarany et al., Cyclic phase transformations of mechanically alloyed $Co_{75}Ti_{25}$ powders, *Acta Mater.*, 50, 1113, 2002.
[17] Rochman, N.T. et al., Effect of milling speed on an Fe-C-Mn system alloy prepared by mechanical alloying, *J. Mater. Proc. Technol.*, 138, 41, 2003.
[18] Suryanarayana, C., Does a disordered γ-TiAl intermetallic exist in mechanically alloyed Ti-Al powders? *Intermetallics*, 3, 153, 1995.
[19] Gonzalez, G. et al., Effect of milling media on the phases obtained in mechanically alloyed equiatomic Fe-Co, *Mater. Sci. Forum*, 360–362, 355, 2001.
[20] Sá Lisboa, R.D. et al., Phase evolution and microstructural characterization of high-energy ball-milled Al-Si-Fe-Ni alloys, *Mater. Sci. Forum*, 386–388, 59, 2002.
[21] Goodwin, P.S. et al., Control of interstitial contamination during mechanical alloying of titanium-based materials, in *Titanium '95*, P. Blenkinsop et al., Eds., Institute of Materials, London, 1996, vol. 3, 2626.
[22] Shaw, L. et al., Effect of process control agents on mechanical alloying of nanostructured aluminum alloys, *Metall. Mater. Trans. A*, 34, 159, 2003.
[23] Hong, L.B., Bansal, C., and Fultz, B., Steady-state grain size and thermal stability of nanophase Ni_3Fe and Fe_3X (X = Si, Zn, Sn) synthesized by ball milling at elevated temperatures, *Nanostruct. Mater.*, 4, 949, 1994.
[24] Qin, Y., Chen, L., and Shen, H., In-situ X-ray diffraction examination of nanocrystalline $Ag_{37}Cu_{63}$ powders synthesized by mechanical alloying, *J. Alloys Compd.*, 256, 230, 1997.
[25] Klassen, T., Herr, U., and Averback, R.S., Ball milling of systems with positive heat of mixing: effect of temperature in Ag-Cu, *Acta Mater.*, 45, 2921, 1997.
[26] Koch, C. C., Research on metastable structures using high energy ball milling at North Carolina State University, *Mater. Trans. Jpn. Inst. Metals*, 36, 1995, 85.
[27] Suryanarayana, C., Chen, G.-H., and Froes, F.H., Milling maps for phase identification during mechanical alloying, *Scripta Metall. Mater.*, 26, 1727, 1992.
[28] Gerasimov, K.B. et al., Tribochemical equilibrium in mechanical alloying of metals, *J. Mater. Sci.*, 26, 2495, 1991.
[29] Suryanarayana, C. et al., Phase selection in a mechanically alloyed Cu–In–Ga–Se powder mixture, *J. Mater. Res.*, 14, 377, 1999.
[30] Weeber, A.W. and Bakker, H., Amorphization by ball milling. A review, *Physica*, B153, 93, 1988.
[31] Sui, H.X. et al., The enhancement of solid solubility limits of AlCo intermetallic compound by high energy ball milling, *J. Appl. Phys.*, 71, 2945, 1992.
[32] Schropf, H. et al., Ordering versus disordering tendencies in mechanically alloyed $(Ni_xFe_{1-x})Al$ alloys, *Scripta Metall. Mater.*, 30, 1569, 1994.
[33] Benjamin, J.S., Mechanical alloying, *Sci. Am.*, 234, 40, 1976.
[34] Eckert, J. et al., Structural and thermodynamic properties of nanocrystalline FCC metals prepared by mechanical attrition, *J. Mater. Res.*, 7, 1751, 1992.
[35] Mohamed, F.A., A dislocation model for the minimum grain size obtainable by milling, *Acta Mater.*, 51, 4107, 2003.
[36] Davis, R.M., McDermott, B., and Koch, C.C., Mechanical alloying of brittle materials, *Metall. Trans. A*, 19, 2867, 1988.

[37] Hellstern, E. et al., Mechanism of achieving nanocrystalline AlRu by ball milling, in *Multicomponent Ultrafine Microstructures*, L.E. McCandlish, D.E. Polk, R.W. Siegel, and B.H. Kear, Eds., Mater. Res. Soc., Pittsburgh, PA, 1989, vol. 132, 137.
[38] Li, S., Wang, K., Sun, L., and Wang, Z., A simple model for the refinement of nanocrystalline grain size during ball milling, *Scripta Metall. Mater.*, 27, 437, 1992.
[39] Atkinson, H.V. and Davies, S., Fundamental aspects of hot isostatic pressing: An overview, *Metall. Mater. Trans. A*, 31, 2981, 2000.
[40] Groza, J.R., Powder consolidation, in *Non-equilibrium Processing of Materials*, C. Suryanarayana, Ed., Pergamon, Oxford, UK, 1999, 347.
[41] Lynn-Ferguson, B. and Smith, O.D., Ceracon processing, in *ASM Metals Handbook*, Vol. 7 (Powder Metallurgy), ASM International, Materials Park, OH, 1984, 537.
[42] Korth, G.E. and Williamson, R.L., Dynamic consolidation of metastable nanocrystalline powders, *Metall. Mater. Trans. A*, 26, 2571, 1995.
[43] Hack, G.A.J., Dispersion strengthened alloys for aerospace, *Metals Mater.*, 3, 457, 1987.
[44] Fischer, J.J. and deBarbadillo, J.J., High temperature alloys minimize furnace downtime, *Heat Treating*, 23, 15, 1991.
[45] Őveçoglu, M.L. and Nix, W.D., Elevated temperature characterization and deformation behavior of mechanically alloyed rapidly solidified Al-8.4wt.%Fe-3.5wt.% Ce alloy, in *New Materials by Mechanical Alloying Techniques*, E. Arzt and L. Schultz, Eds., DGM, Oberursel, Germany, 1989, 287.
[46] Ezz, S.S., Lawley, A., and Koczak, M.J., Dispersion strengthened Al-Fe-Ce: A dual rapid solidification/mechanical alloying approach, in *Dispersion Strengthened Aluminum Alloys*, Y.-W. Kim and W.M. Griffith, Eds., TMS, Warrendale, PA, 1988, 243.
[47] Sergev, S.S., Black, S.A., and Jenkins, J.F., Supercorroding galvanic cell alloys for generation of heat and gas, U.S. Patent # 4 264 362, August 13, 1979.
[48] Klassen, T. et al., Tailoring nanocrystalline materials towards potential applications, *Z. Metallkde.*, 94, 610, 2003.
[49] Oelerich, W., Klassen, T., and Bormann, R., Metal oxides as catalysts for improved hydrogen sorption in nanocrystalline Mg-based materials, *J. Alloys Compd.*, 315, 237, 2001.
[50] Witkin, D.B. and Lavernia E.J., Synthesis and mechanical behavior of nanostructured materials via cryomilling, *Prog. Mater. Sci.*, 51, 1, 2006.
[51] Petkovic-Luton, R. and Vallone, J., Method for producing dispersion strengthened metal powders, U.S. Patent 4,647,304, Mar 3, 1987.
[52] Susegg, O. et al., HREM study of dispersoids in cryomilled oxide dispersion strengthened materials, *Phil. Mag. A*, 68, 367, 1993.
[53] Hayes, R.W., Berbon, P.B., and Mishra, R.S., Microstructure characterization and creep deformation of an Al-10wt%Ti-2wt%Cu nanocomposite, *Metall. Mater. Trans. A*, 35, 3855, 2004.
[54] Perez, R.J. et al., Grain growth of nanocrystalline cryomilled Fe-Al powders, *Metall. Mater. Trans. A*, 29, 2469, 1998.
[55] Lee, J. et al., Grain growth of nanocrystalline Ni powders prepared by cryomilling, *Metall. Mater. Trans. A*, 32, 3109, 2001.
[56] Huang, B., Vallone, J., and Luton, M.J., The effect of nitrogen and oxygen on the synthesis of B2 NiAl by cryomilling, *Nanostruct. Mater.*, 5, 631, 1995.
[57] Han, B.Q. et al., Deformation behavior of bimodal nanostructured 5083 Al alloys, *Metall. Mater. Trans. A*, 36, 957, 2005.
[58] Witkin, D. et al., Al-Mg alloy engineered with bimodal grain size for high strength and increased ductility, *Scripta Mater.*, 49, 297, 2003.
[59] Bauccio, M.L., Ed., *Metals Reference Book*, 3rd Edition, ASM International, Materials Park, OH, 1993.

[60] Han, B.Q. and Lavernia, E.J., Deformation mechanisms of nanostructured Al alloys, *Adv. Eng. Mater.*, 7, 457, 2005.

[61] Tellkamp, V.L., Melmed, A., and Lavernia, E.J., Mechanical behavior and microstructure of a thermally stable bulk, nanostructured Al alloy, *Metall. Mater. Trans. A*, 32, 2335, 2001.

[62] Witkin, D. and Lavernia, E.J., Processing-controlled mechanical properties and microstructures of bulk cryomilled aluminum-magnesium alloys, in *Processing and Properties of Structural Nanomaterials*, L.L. Shaw, C. Suryanarayana, and R.S. Mishra, Eds., TMS, Warrendale, PA, 2003, 117.

[63] Tang, F., Hagiwara, M., and Schoenung, J.M., Formation of coarse-grained inter-particle regions during hot isostatic pressing of nanocrystalline powder, *Scripta Mater.*, 53, 619, 2005.

[64] Han, B.Q. et al., Improvement of toughness and ductility of a cryomilled Al-Mg alloy via microstructural modification, *Metall. Mater. Trans. A*, 36, 2081, 2005.

[65] Ye, J. et al., A tri-modal aluminum based composite with super high strength, *Scripta Mater.*, 53, 481, 2005.

[66] Oh-Ishi, K. and McNelley, T.R., Microstructural modification of as-cast NiAl bronze by friction stir processing, *Metall. Mater. Trans. A*, 35, 2951, 2004.

[67] Sharma, S.R., Ma, Y.Z., and Mishra, R.S., Effect of friction stir processing on fatigue behavior of A356 alloy, *Scripta Mater.*, 51, 237, 2004.

[68] Mishra, R.S., Friction stir processing technologies, *Adv. Mater. Proc.*, October, 43, 2003.

[69] Collier, M. et al., Grain development of polycrystalline cubic boron nitride for friction stir processing of ferrous alloys, *Mater. Sci. Forum*, 426–432, 3011, 2003.

[70] Mahoney, M.W. et al., Microstructural modification and resultant properties of friction stir processed cast NiAl bronze, *Mater. Sci. Forum*, 426–432, 2843, 2003.

[71] Mishra, R.S. and Mahoney, M.W., Friction stir processing: a new grain refinement technique to achieve high strain rate superplasticity in commercial alloys, *Mater. Sci. Forum*, 357–359, 507, 2001.

[72] Mishra, R.S. et al., High strain rate superplasticity in a friction stir processed 7075 Al alloy, *Scripta Mater.*, 42, 163, 2000.

[73] Saito, N. et al., Grain refinement of 1050 aluminum alloy by friction stir processing, *J. Mater. Sci. Lett.*, 20, 1913, 2001.

[74] Ma, Z.Y. et al., Microstructural modification of cast aluminum alloys via friction stir processing, *Mater. Sci. Forum*, 426–432, 2891, 2003.

[75] Ma, Z.Y., Mishra, R.S., and Mahoney, M.W., Superplastic deformation of friction stir processed 7075Al alloy, *Acta Mater.*, 50, 4419, 2002.

[76] Ma, Z.Y. and Mishra, R.S., Cavitation in superplastic 7075Al alloys prepared via friction stir processing, *Acta Mater.*, 51, 3551, 2003.

[77] Ma, Z.Y. et al., High strain rate superplasticity in friction stir processed Al-Mg-Zr alloy, *Mater. Sci. Eng. A*, 351, 148, 2003.

[78] Ma, Z.Y., Mishra, R.S., and Mahoney, M.W., Superplasticity in cast A356 induced via friction stir processing, *Scripta Mater.*, 50, 931, 2004.

[79] Charit, I. and Mishra, R.S., Evaluation of microstructure and superplasticity in friction stir processed 5083 Al alloy, *J. Mater. Res.*, 19, 3329, 2004.

[80] Salem, H.G., Reynolds, A.P., and Lyons, J.S., Structural evolution and superplastic formability of friction stir welded AA 2095 sheets, *J. Mater. Eng. Perform.*, 13, 24, 2005.

[81] Attalah, M.M. and Salem, H.G., Influence of process parameters on superplasticity of friction stir processed nugget in high strength Al-Cu-Li alloy, *Mater. Sci. Technol.*, 20, 1370, 2004.

[82] Chang, C.I., Lee, C.J., and Huang, J.C., Relationship between grain size and Zener-Holloman parameter during friction stir processing in AZ31 Mg alloys, *Scripta Mater.*, 51, 509, 2004.

[83] Segal, V.M., Materials processing by simple shear, *Mater. Sci. Eng. A*, 197, 157, 1995.

[84] Valiev, R.Z., Islamgaliev, R.K., and Alexandrov, I.V., Bulk nanostructured materials from severe plastic deformation, *Prog. Mater. Sci.*, 45, 103, 2000.

[85] Komura, S. et al., Influence of stacking fault energy on microstructural development in equal-channel angular pressing, *J. Mater. Res.*, 14, 4044, 1999.

[86] Iwahashi, Y. et al., Microstructural characteristics of ultrafine-grained aluminum produced using equal-channel angular pressing, *Metall. Mater. Trans. A*, 29, 2245, 1998.

[87] Zhu, Y.T. and Lowe, T.C., Observations and issues on mechanisms of grain refinement during ECAP process, *Mater. Sci. Eng. A*, 291, 46, 2000.

[88] Kamachi, M. et al., A model investigation of shearing characteristics in equal-channel angular pressing, *Mater. Sci. Eng. A*, 347, 223, 2003.

[89] Han, B.Q., Lavernia, E.J., and Mohamed, F.A., Mechanical properties of iron processed by severe plastic deformation, *Metall. Mater. Trans. A*, 34, 71, 2003.

[90] Murayama, M., Horita, Z., and Hono, K., Microstructure of two-phase Al-1.7at.% Cu alloy deformed by equal-channel angular pressing, *Acta Mater.*, 49, 21, 2001.

[91] Vinogradov, A. et al., Structure and properties of ultrafine grain Cu-Cr-Zr alloy produced by equal-channel angular pressing, *Acta Mater.*, 50, 1639, 2002.

[92] Iwahashi, Y. et al., The process of grain refinement in equal-channel angular pressing, *Acta Mater.*, 46, 3317, 1998.

[93] Horita, Z. et al., Superplastic forming at high strain rates after severe plastic deformation, *Acta Mater.*, 48, 3633, 2000.

[94] Park, K.T. et al., Low-temperature superplasticity behavior of a submicrometer-grained 5083 Al alloy fabricated by severe plastic deformation, *Metall. Mater. Trans. A*, 33, 2859, 2002.

[95] Shin, D.H. et al., High-strain-rate superplastic behavior of equal-channel angular-pressed 5083 Al-0.2 wt pct Sc, *Metall. Mater. Trans. A*, 35, 825, 2004.

[96] Zhao, Y.H. et al., Microstructures and mechanical properties of ultrafine grained 7075 Al alloy processed by ECAP and their evolutions during annealing, *Acta Mater.*, 52, 4589, 2004.

[97] Berbon, P.B. et al., Fabrication of bulk ultrafine-grained materials through intense plastic straining, *Metall. Mater. Trans. A*, 29, 2237, 1998.

[98] Oh-ishi, K. et al., Grain boundary structure in Al-Mg and Al-Mg-Sc alloys after equal-channel angular pressing, *J. Mater. Res.*, 16, 583, 2001.

[99] Han, B.Q., Lavernia, E.J., and Mohamed, F.A., Dislocation structure and deformation in iron processed by equal-channel-angular pressing, *Metall. Mater. Trans. A*, 35, 1343, 2004.

[100] Shin, D.H. et al., Microstructural evolution in a commercial low carbon steel by equal channel angular pressing, *Acta Mater.*, 48, 2247, 2000.

[101] Stolyarov, V.V. et al., Influence of ECAP routes on the microstructure and properties of pure Ti, *Mater. Sci. Eng. A*, 299, 59, 2001.

[102] Zhu, Y.T. et al., Nanostructures in Ti processed by severe plastic deformation, *J. Mater. Res.*, 18, 1908, 2003.

[103] Yu, C.Y. et al., Mechanical properties of submicron-grained aluminum, *Scripta Mater.*, 52, 359, 2005.

[104] Wang, Y.M. and Ma, E., Temperature and strain rate effects on the strength and ductility of nanostructured copper, *App. Phys. Lett.*, 83, 3165, 2003.

[105] Torre, F.H.D., Pereloma, E.V., and Davies, C.H.J., Strain rate sensitivity and apparent activation volume measurements on equal channel angular extruded Cu processed by one to twelve passes, *Scripta Mater.*, 51, 367, 2004.

[106] Segal, V.M., Engineering and commercialization of equal channel angular extrusion (ECAE), *Mater. Sci. Eng. A*, 386, 269, 2004.

[107] Matsuki, K. et al., Microstructural characteristics and superplastic-like behavior in aluminum powder alloy consolidated by equal-channel angular pressing, *Acta Mater.*, 48, 2625, 2000.

[108] Karaman, I. et al., The effect of temperature and extrusion speed on the consolidation of zirconium-based metallic glass powder using equal-channel angular extrusion, *Metall. Mater. Trans. A*, 34, 247, 2003.

[109] Haouaoui, M. et al., Microstructure evolution and mechanical behavior of bulk copper obtained by consolidation of micro- and nanopowders using equal-channel angular extrusion, *Metall. Mater. Trans. A*, 35, 2935, 2004.
[110] Lee, Z. et al., Microstructure and microhardness of cryomilled bulk nanocrystalline Al-7.5%Mg alloy consolidated by high pressure torsion, *Scripta Mater.*, 51, 209, 2004.

14

Superplasticity and Superplastic Forming

Indrajit Charit and
Rajiv S. Mishra
University of Missouri-Rolla

Abstract

Superplasticity is the ability of a material to exhibit extensive tensile elongation. Its utilization can be realized through superplastic forming which is regarded as an enabling near-net shape manufacturing technique. The microstructural requirements for superplasticity are varied — most important being finer grain size. A number of processing techniques have been developed for producing superplastic materials. Grain boundary sliding, which is the predominant mechanism for superplasticity, is discussed in the framework of various theoretical models. An extensive review of superplastic properties is presented for a wide range of materials, including light alloys, ferrous alloys, intermetallics, ceramics, and metallic glasses. Increased use of superplastic forming would be possible through the utilization of high strain rate and low temperature superplastic properties and the introduction of innovative forming concepts.

14.1 Introduction

Superplasticity is described as the ability of a polycrystalline material to exhibit higher than normal tensile elongation (>200%) prior to failure. Superplasticity in most cases is observed to be a microstructure-dependent phenomenon and needs high homologous temperatures. Superplasticity has matured from an area of academic interest in the initial period to commercial superplastic forming (SPF) technique with several advantages over conventional forming operations. Currently it is an enabling technology for unitized structures.

In 1934, superplasticity was first observed by Pearson[1] in a eutectic Bi–Sn alloy that exhibited ~1950% elongation. These results did not generate much interest in the western countries. In the former Soviet Union, the phenomenon received much attention. In fact, the English word superplasticity comes from the Russian word "sverhplastichnost" meaning very high plasticity. In 1962, Underwood[2] compiled the Soviet work on superplasticity and it was a great driving force for renewed research effort on superplasticity in the West. In 1964, Backofen and coworkers[3] at the Massachusetts Institute of Technology demonstrated superplasticity in Zn–Al and Pb–Sn alloys, and practical application of superplasticity through simple forming operations. From that point on, superplasticity and SPF became an active field of study.

Most of the superplastic observations come from fine-grained materials. Superplasticity originating from fine microstructure is referred to as "structural superplasticity." Superplasticity can also be induced by other ways, such as transformation superplasticity and temperature cycling superplasticity.[4] However, there is very little technological relevance of these types of superplasticity, and this chapter will only deal with structural superplasticity, in accordance with the scope of this book. For a much more in-depth review of the topic, readers are encouraged to refer to various books and review articles available in the literature.[4–11] Here, we aim to succinctly review the science and application of superplasticity.

SPF is regarded as an important near-net shape technique capable of producing complex shapes in a cost-effective manner. At present, SPF is used in a number of manufacturing operations, most notably in aircraft industries. Diffusion bonding/SPF is an important technology that is used to make integrally stiffened structures. This enhances the design flexibility, and the ability to fabricate complex structures in a single operation that helps in reducing the number of parts in an assembly. The materials that are routinely used for SPF are titanium and aluminum alloys. Generally, the flow stresses associated with SPF are less (<10 MPa). Pilling and Ridley[11] summarized various SPF techniques, such as, simple female forming, reverse bulging and snap-back forming, to name a few.

Widespread use of SPF in high volume production sectors (such as automotive) is generally limited by two main factors, slow forming rates and higher cost of superplastic materials. It has been known that grain refining helps in increasing the optimum strain rate during SPF and reduces the superplastic temperature. For example, Figure 14.1 shows that with a decrease in grain size from 15 to 0.5 μm in aluminum alloys, the optimum strain rate increases by several orders of magnitude.[4] It has led to useful concepts of high strain rate superplasticity (HSRS; forming rates $\geq 10^{-2}\ s^{-1}$)[12] and low-temperature superplasticity (LTSP). Hence, in recent years research efforts have intensified in developing processing techniques that would give rise to fine-grained microstructure with enhanced superplastic properties.

14.2 Requirements for Structural Superplasticity

The prerequisites for superplastic behavior are well established for metallic alloys, which are as following:[4]

1. *Fine grain size*: One of the major requirements for superplasticity is that the grain size should be small, typically <15 μm. The optimum strain rate for superplasticity increases with decreasing grain size when grain boundary sliding (GBS) is the dominant process. For a given strain rate, finer grain sizes lead to lower flow stresses, which is beneficial for practical forming operations.
2. *Grain shape*: Grain boundaries can experience shear stresses easily promoting GBS if the grains are equiaxed.

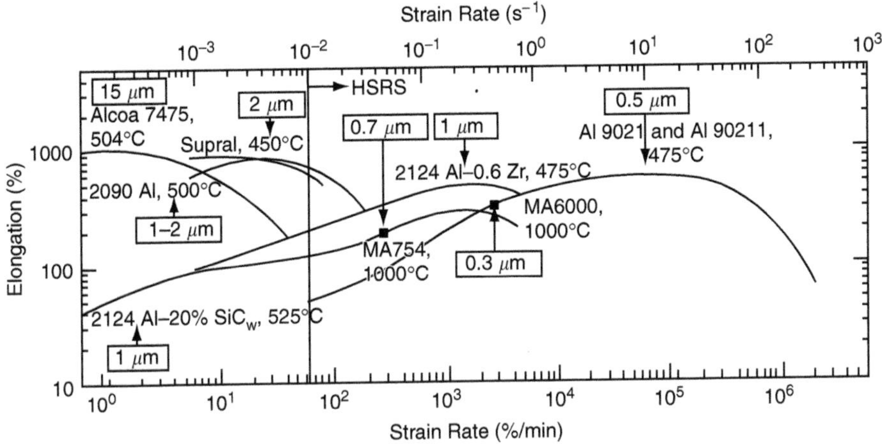

FIGURE 14.1 The effect of grain refinement on the optimum superplastic strain rate. (Taken from Sherby, O.D. and Wadsworth, J., *Prog. Mater. Sci.*, 33, 169, 1989. With permission from Elsevier Science.)

3. *Presence of second phases*: Single phase materials generally do not show superplasticity because grain growth occurs rapidly at elevated temperatures where GBS sliding is likely to occur. Hence, presence of second phases at grain boundaries is required to resist grain growth. Grain growth is inhibited if the volume fraction of second phases is increased while maintaining their fine size and uniform distribution. It is important to remember that pure metals seldom show any superplastic properties due to rapid grain growth. Large particles at grain boundaries can hinder GBS, and may nucleate cavities. Hence, it is important that the stress concentration is relaxed through some relaxation mechanisms (like diffusional). The stress concentration produced at small particles that are often used for retaining fine grain size relaxes easily under superplastic conditions as the relaxation distance remains fairly small.
4. *Nature of grain boundary*: Superplastic properties also depend on the type of grain boundaries present. It has been shown that high angle grain boundaries (>15°) exhibit enhanced GBS. The nature of grain boundaries may also influence the failure mechanisms that in part determine the extent of superplastic elongation.
5. *Mobility of grain boundaries*: Grain boundaries in superplastic materials need to be mobile. During GBS, stress concentration could be produced at various grain boundary discontinuities such as triple points. If the grain boundary can migrate during GBS, it could lead to reduction in stress concentrations. In this way, GBS can continue as a major deformation mechanism. Also, grain boundary migration during superplastic deformation can be indirectly evidenced in that the grain shape remains more or less equiaxed even after extensive deformation.

14.3 Mechanical Characteristics of Superplasticity and Relevant Models

Over the years, a number of superplasticity models have been proposed. A majority of experimental evidence suggests that GBS is the dominant deformation mechanism. Generally, superplasticity is obtained at homologous temperatures above ~0.5 to 0.6 because diffusion processes become rapid enough to impart superplastic property. The rate equation for diffusion-assisted deformation processes is given by the following form of equation:

$$\dot{\varepsilon} = \frac{ADGb}{kT}\left(\frac{\sigma}{G}\right)^n\left(\frac{b}{d}\right)^p \tag{14.1}$$

where $\dot{\varepsilon}$ the strain rate, G the shear modulus, b Burgers vector, σ is the applied stress, d the grain diameter (size), D the appropriate diffusivity, n stress exponent, p inverse grain size exponent, and A is a microstructure and mechanism dependent dimensionless constant. Often in superplasticity literature, strain rate sensitivity exponent ($m = \partial \log \sigma / \partial \log \dot{\varepsilon}$) is used instead of stress exponent (n), shown in Equation 14.1. However, "m" is just the reciprocal of "n." Higher m values mean larger resistance to external neck formation, and hence larger ductility. Generally, an m value of ~0.5 and a p value of 2 to 3 imply operation of GBS.

Any two contiguous grains in polycrystalline materials cannot slide indefinitely because they impinge on other grains. Then various processes like accommodation, grain rotation, grain boundary migration, grain switching events and so forth may take place to relax the stress concentrations generated, and thus continue GBS. The common school of thought assumes that the process that accommodates GBS would be the rate-controlling mechanism for superplasticity.[13] The accommodation mechanisms could be divided into two broad groups, diffusional accommodation and slip (involving dislocations) accommodation. The second group is further classified into accommodation processes due to dislocation pile-ups (either inside the grains or interfaces) and due to the motion of individual dislocations. However, it is worth noting that several superplasticity models have been developed over many decades, all of which could not be discussed within the scope of this chapter.

14.3.1 Diffusional Accommodation Model

Ashby and Verrall[14] proposed a model, which explains superplasticity as a transition region between a diffusion-accommodated flow at low strain rates and a diffusion-controlled dislocation climb at higher strain rates. At low strain rates, where diffusion-accommodated flow accounts for most of the total strain rate, the specimen elongation is obtained by grain rearrangement through GBS (and grain switching). To retain compatibility across the grain boundaries, the grains need to have transient yet complex shape changes that are achieved via diffusional transport. The final configuration retains the equiaxed grain shape. Due to the transient increase in grain boundary surface area during the grain rearrangement process, a threshold stress arises at very low strain rates. This model is depicted in Figure 14.2. However, at higher strain rates, where dislocation creep accounts for most of the total strain rate, specimen elongation is accompanied by change of shapes of individual grains. At the intermediate strain rates, both the processes (diffusion-accommodated flow and dislocation creep) would contribute. As these two mechanisms are parallel processes, the total strain rate ($\dot{\varepsilon}$) is given by the sum of the strain rates contributed by each process. The strain rate due to the diffusional processes is given by:

$$\dot{\varepsilon}_{\text{diffuse}} = \frac{100\Omega}{kTd^2}\left\{\sigma - \frac{0.72\Gamma}{d}\right\} D_{\text{L}} \left\{1 + \frac{3.3\delta}{d}\frac{D_{\text{b}}}{D_{\text{L}}}\right\} \tag{14.2}$$

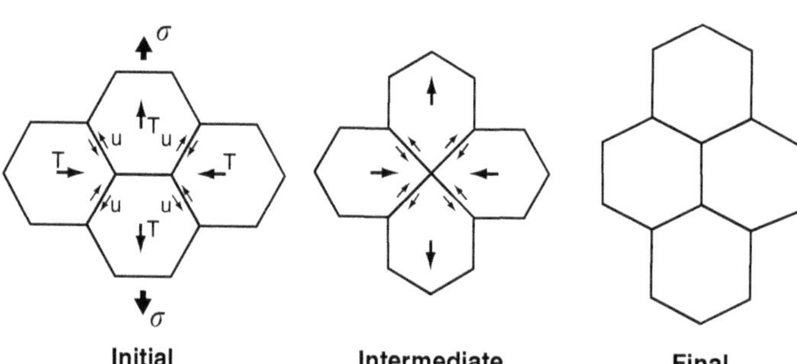

FIGURE 14.2 Schematic illustration of grain switching event, as described in Ashby–Verrall mechanism. (Taken from Ashby, M.F. and Verrall, R.A., *Acta Metall.*, 21, 149, 1973. With permission from Elsevier Science.)

where Ω is atomic volume, D_L lattice diffusivity, Γ grain boundary energy, δ grain boundary width, and D_b grain boundary diffusivity. The strain rate given by dislocation creep is:

$$\dot{\varepsilon}_{\text{disloc}} = \frac{AD_L Gb}{kT}\left\{\frac{\sigma}{G}\right\}^n. \tag{14.3}$$

When combined, the model predicts strain rate sensitivity (m) as a strong function of strain rate with a maximum approaching unity. The grain size dependence might vary from 0 to 3 depending on the strain rate and temperature. The activation energy predicted by this model will be intermediate between the activation energies of grain boundary diffusion and lattice diffusion.

The Ashby–Verrall model has attractive features and explains topological characteristics well. However, there are a few shortcomings in this model:

1. Differences between the predicted and experimentally observed strain rates are quite large.
2. Although the model predicts grain size exponent (p) values of 0 to 3, experimental results reveal that p lies between 2 and 3.
3. The prediction of the model that activation energy increases with temperature has no sound experimental support.

14.3.2 Dislocation Pile-Up Accommodation

14.3.2.1 Pile-Ups within Grains

Ball and Hutchison[15] proposed that groups of grains slide as a unit until they impinge on unfavorably oriented grains that obstruct the process. The resulting stress concentration is relieved by the generation of dislocations in the blocking grains. These dislocations tend to pile-up against the opposite grain boundaries, and deactivate the source of dislocation generation. No GBS occurs until dislocations from the pile-up climb into and along grain boundaries. The model leads to the following rate equation:

$$\dot{\varepsilon} = \frac{200 D_b Gb}{kT}\left(\frac{\sigma}{G}\right)^2\left(\frac{b}{d}\right)^2. \tag{14.4}$$

Mukherjee[16] proposed a modification to the Ball–Hutchison model in which it was assumed that grains slide individually instead of sliding as groups. Dislocations are generated at the grain boundary ledges, traverse through the grain, and pile-up at the grain boundaries. The rate of sliding is then governed by the climb rate of the leading dislocations and subsequent annihilation. The model leads to an equation, which is essentially similar to the Ball–Hutchison equation except for the value of the dimensionless constant. It is worth noting that the value of the constant, 2, in Mukherjee's model was calculated from theoretical considerations, whereas the value of 200 in Ball–Hutchison's model was from fitting the experimental data.

14.3.2.2 Pile-ups in Interfaces

Mukherjee[17] proposed a modification to his earlier model.[16] Here GBS is considered rate-limited by the motion of dislocations in the grain boundary by a combined glide–climb process, but not across the grains. The compatibility between the grains is maintained by the diffusion-controlled climb of lattice dislocations along the grain boundaries, and thus repeated accommodation is achieved. The rate equation is similar to that presented in Equation 14.4 but differs only in the value of the constant ($\sim$100).

Gifkins[18] treated the process in terms of the motion of grain boundary dislocations, with an approach known as the core-mantle model, which is schematically shown in Figure 14.3. In this model, grain boundary dislocations pile-up at the triple junctions. The resulting stress concentration is relieved by the dissociation of lead grain boundary dislocations into the other two adjoining boundaries of the triple junction or into lattice dislocations that accommodate sliding. These newly formed dislocations will climb into the two grain boundaries and annihilate or might form new grain boundary dislocations. However,

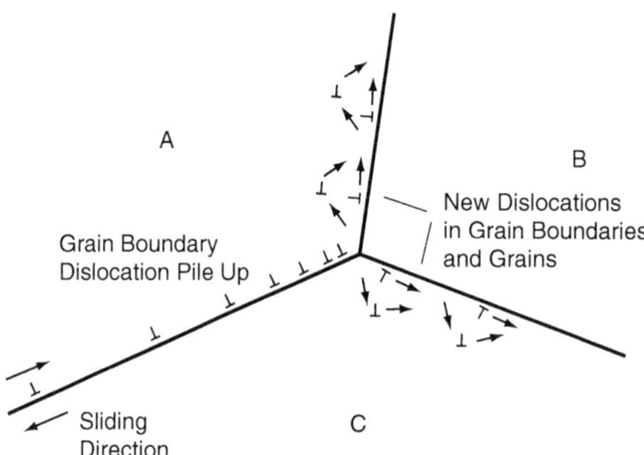

FIGURE 14.3 Core-mantle theory with accommodation by dislocation motion in and near grain boundaries. (Taken from Gifkins, R.C., *Metall. Trans. A*, 7, 1225, 1976, with permission from TMS, Warrendale, PA, USA.)

all these processes take place near the grain boundary or mantle. However, the core of the grain remains almost free of dislocations. This model would implicitly lead to grain rotation and rearrangement. The constitutive equation appropriate for this model is similar to Equation 14.4 but the theoretical constant parameter changes to 64.

14.3.2.3 Accommodation by Individual Dislocation Motion

Hayden et al.[19] put forward a model where GBS is rate controlled by intragranular slip. Dislocations are generated at grain boundary ledges and triple junctions, move inside the grain through glide and climb, and finally climb individually into the opposite grain boundaries (without forming pile-ups) before being annihilated. The model postulates that at a critical temperature, T_c, there will be a transition in the diffusion mechanism from pipe (at $T < T_c$) to lattice diffusion (at $T > T_c$). Moreover, they suggested that the rate of sliding is related to the rate of intragranular dislocation creep by a geometric constant that is independent of material and temperature, and the ratio of the two rates would vary inversely with grain size. For the superplastic regime, the model leads to the following equations:

$$\dot{\varepsilon} \propto \frac{D_b G b}{kT} \left(\frac{\sigma}{G}\right)^2 \left(\frac{b}{d}\right)^3 \tag{14.5}$$

for $T < T_c$, and

$$\dot{\varepsilon} \propto \frac{D_L G b}{kT} \left(\frac{\sigma}{G}\right)^2 \left(\frac{b}{d}\right)^2 \tag{14.6}$$

for $T > T_c$.

Spingarn and Nix[20] proposed that deformation would occur by intragranular slip along the slip bands, which are blocked by grain boundaries. The stress concentration at the boundaries is relaxed by diffusional flow in the boundaries. The slip band spacing varies with the strain rate, that is, decreases with increasing strain rate. When the stress is less, the slip band spacing equals the grain size, and the rate equation is given by:

$$\dot{\varepsilon} \propto \frac{D_b G b}{kT} \left(\frac{\sigma}{G}\right) \left(\frac{b}{d}\right)^3 \tag{14.7}$$

which is similar to the equation for Coble diffusional creep.[22] However, at very large stresses, the slip band spacing is taken to be equal to the subgrain size, and the rate equation is given by:

$$\dot{\varepsilon} \propto \frac{D_b Gb}{kT} \left(\frac{\sigma}{G}\right)^5 \left(\frac{d}{b}\right) \tag{14.8}$$

They suggested that the transition zone from $n = 1$ to $n = 5$ coincides with the superplastic regime. Also, the grain size exponent varies from -3 to $+1$. The activation energy is that for grain boundary diffusion.

Arieli and Mukherjee[21] proposed that individual lattice dislocations climb into and along the interfaces in a narrow region near the interfaces. During the climb process, dislocations multiply via Bardeen–Herring[22] mechanism. Due to the nearness of interfaces, this climb process is controlled by the grain boundary diffusion. The model predicts the rate equation as:

$$\dot{\varepsilon} = \left(\frac{4\pi\omega}{hZ^2 \tan(\theta/2)}\right) \frac{D_b Gb}{kT} \left(\frac{\sigma}{G}\right)^2 \left(\frac{b}{d}\right)^2. \tag{14.9}$$

The significance of this model is that the constant term is not considered as a geometric constant, rather it would vary with interface structure (through h and θ, where h is the proportionality constant between sliding distance and Burgers vector, θ the mismatch angle between adjacent grains), and the structure of the narrow zone near interface (through ω and Z which are substructure-related constants).

14.3.3 Other Superplasticity Models

Fukuyo et al.[23] have proposed a model that is similar to the Ball–Hutchison's model of GBS accommodated by slip. The slip accommodation process involves two sequential steps of glide and climb. When climb is the rate-controlling step, the strain rate sensitivity value is 0.5 ($\dot{\varepsilon} \propto \sigma^2$). On the other hand, when glide is the rate-controlling step, the strain rate sensitivity is equal to 1 ($\dot{\varepsilon} \propto \sigma$). As glide and climb processes are sequential, the slowest of the two processes is rate-controlling. So, this model predicts that at lower strain rates, the GBS process is accommodated by climb, whereas it is accommodated by glide at higher strain rates.

Langdon[24] proposed a unified model for Rachinger GBS[25] both in creep and superplastic conditions where the rate of sliding is controlled by the rate of accommodation through intragranular slip. The basic microstructural mechanism behind this model is shown in Figure 14.4. Some of the premises of the Ball–Hutchison model[15] were used in this analysis. Langdon[24] obtained two conditions. Rachinger sliding is given by the following relationship when d (grain size) $<\lambda$ (dislocation self-trapping distance):

$$\dot{\varepsilon}_{gbs(d<\lambda)} = \frac{A_{gbs(d<\lambda)} D_b Gb}{kT} \left(\frac{\sigma}{G}\right)^2 \left(\frac{b}{d}\right)^2 \tag{14.10}$$

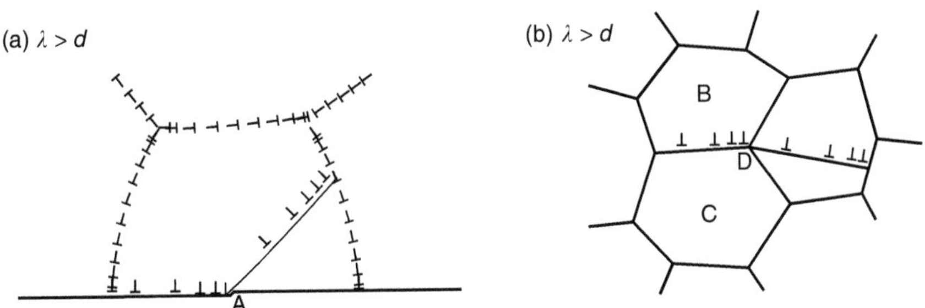

FIGURE 14.4 Schematic illustration of GBS mechanism at two different microstructural conditions. (Taken from Langdon, T.G., *Acta Metall. Mater.*, 42, 2437, 1994. With permission from Elsevier Science.)

where $A_{gbs(d<\lambda)}$ is a dimensionless constant having a theoretical value of ~10. When $d > \lambda$, the equation for Rachinger sliding is:

$$\dot{\varepsilon}_{gbs(d>\lambda)} = \frac{A_{gbs(d>\lambda)} D_L G b}{kT} \left(\frac{\sigma}{G}\right)^3 \left(\frac{b}{d}\right)^1 \tag{14.11}$$

where $A_{gbs(d>\lambda)}$ is a dimensionless constant having a value of ~1000, which was determined fitting experimental data.

Another interesting mechanism for superplasticity could be the solute-drag dislocation glide mechanism, similar to Class-I solid solution behavior in creep. In this mechanism, glide is the rate-controlling process of the glide–climb dislocation creep. Solute atoms tend to form solute atmospheres around the dislocations in certain temperature-strain rate regimes, thus imposing a dragging force on the gliding dislocations. This model predicts a strain rate sensitivity exponent value of 0.33. It does not predict any grain size dependence. The rate-equation[26] can be written as:

$$\dot{\varepsilon} = A D_{chem} \left(\frac{\sigma}{G}\right)^3 \tag{14.12}$$

where D_{chem} is the chemical diffusivity of solute atoms in the solvent matrix and A an approximate constant that depends on the misfit size parameter of solute with respect to solvent atoms and solute concentration. Although the strain rate sensitivity value is <0.5, it still has a high value (~0.33). That is why many coarse-grained Class-I solid solution alloys may also exhibit high ductility (>200%).

14.4 Processing Techniques

The best processing techniques for superplastic materials are those that produce the microstructural requirements detailed in Section 14.2, the most important being stable fine grains. Here, only the main features of a few processing techniques in relation to superplasticity are discussed.

14.4.1 Thermomechanical Processing

Although the processing techniques are discussed with particular reference to aluminum alloys, the principles herein are applicable to other alloys, too. The earliest attempts to impart superplasticity in aluminum alloys involved rolling-based thermomechanical processing (TMP) techniques (for more details, see Chapter 29). For example, Waldman[27] and Wert[28] developed multistep TMP techniques to produce fine-grained, superplastic aluminum alloys. A patent by Brown[29] to produce superplastic aluminum alloys states a method consisting of several processing steps, (1) homogenization, (2) hot rolling, (3) overaging, (4) hot rolling, (5) controlled cold rolling, and (6) recrystallization anneal (Figure 14.5). Due to this multistep complex procedure, the starting material becomes expensive. Also, to achieve grain size <15 μm, the true strain needs to exceed 2.3, and thus, the final sheet becomes thin (<3 mm), limiting its applications. Power metallurgy techniques are also viable routes to produce superplastic materials because of their ability to produce fine-grained materials stabilized by dispersoids. Various bulk deformation processes (rolling, extrusion, forging) are used to produce the desired form of these materials for SPF. It is important to note that superplastic composite and ceramic materials can only be processed through powder metallurgy (PM) techniques. For more details on powder processing, see Chapter 7 and Chapter 25.

14.4.2 Severe Plastic Deformation Techniques

The severe plastic deformation (SePD) techniques have been described in detail in Chapter 13 of this book. To overcome the drawbacks of TMP processing and to obtain enhanced superplasticity, different SePD techniques have been explored in the last two decades. They are (1) equal channel angular

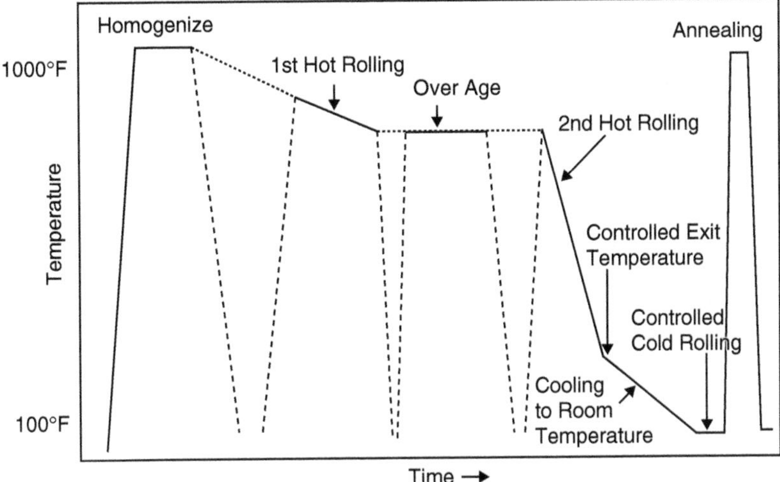

FIGURE 14.5 A graphic illustration of rolling-based TMP for developing superplastic properties in heat treatable aluminum alloys. (Taken from Brown, K.R., U.S. Patent No. 5772804, June 30, 1998.)

pressing (ECAP),[30] (2) torsional straining (TS) under pressure,[31] (3) multiaxial forging,[32] (4) accumulative roll bonding (ARB),[33] (5) multipass coin-forge,[34] (6) friction stir processing (FSP),[35–37] and (7) mechanical alloying.[38]

The principle of grain refinement in SePD techniques is grain fragmentation, that is, subdivision of grains by build-up of low angle boundaries, and conversion into high angle ones with increase in deformation. Although these processes have been successful in achieving a remarkable degree of grain refinement, they suffer from practical constraints. For instance, TS requires very high pressure (2 to 6 GPa) to obtain the desired grain refinement, and to date its application is limited to 20 to 25 mm diameter disks of ~0.5 mm thickness. In ECAP, several passes (6 to 8 passes) are required to obtain fine grain sizes with enough high grain boundary misorientations, which are essential for enhanced superplasticity. FSP is a relatively new technique and its practicality is currently being evaluated.

14.5 Superplasticity in Metallic Materials

To date, a majority of superplasticity studies have been carried out on metallic materials, primarily because of the industrial applications. The fundamental aspects of superplasticity were first investigated in eutectic and eutectoid alloys. In this chapter, we do not discuss those alloys because their superplastic properties are of little interest in modern structural applications.

14.5.1 Light Metals

Light metals are those metals that have low density. The densities of aluminum, magnesium, and titanium are 2.7, 1.76, and 4.2 g/cc, respectively. They have the highest potential for structural applications in transportation vehicles because of their high strength/weight ratio. For example, the use of light metals ensures improved fuel-economy and significant emission reduction. Although superplasticity studies of light alloys were initiated out of mere academic curiosity, currently these light alloys are perhaps the only group of metallic materials that are in commercial applications and utilize SPF as one of the fabrication steps.

14.5.1.1 Aluminum Alloys

Aluminum alloys have long been used as aerospace structural materials. Perhaps, the highest number of superplasticity papers is of aluminum alloys. Also, the automotive industry has interests in aluminum alloys, and SPF is regarded as a useful near-net shape technique that could be integrated into the manufacturing processes of various auto-components. Here, superplastic characteristics of a few commercial aluminum alloys are summarized. Most quasi-single phase alloys depend on dispersoids to retain microstructural stability at higher temperatures.

14.5.1.1.1 Al–Cu and Al–Cu–Mg Alloys

Al–Cu alloys (2XXX series) are used in aerospace and other structural applications. The nominal composition of AA 2024 Al is Al–4.6Cu–1.5Mg–0.6Mn (wt.%). Previous processing efforts to obtain superplasticity in 2024Al comprised of PM,[39] conventional TMP,[40] SePD,[41] and compositional modification.[42] Zheng and Baoliang[40] achieved very high elongations (1830%) in a TMP 2024 Al alloy at 500°C and at a strain rate of 1.3×10^{-3} s^{-1}. They used hot stamping followed by a high-temperature and high-ratio extrusion, which led to dynamic recrystallization in the alloy resulting in an average grain diameter of 2.5 μm. However, it is worth noting that the elongation at the highest strain rate (5×10^{-3} s^{-1}) dropped below 400%, and no HSRS was obtained. They observed high rate of concurrent grain growth (~15 μm) during superplastic deformation but the grain size remained equiaxed. An m value of ~0.4 was obtained. Recently, Langdon and coworkers[41] investigated the superplastic behavior of ECAP 2024 Al, where they obtained optimum elongations of 496% at 400°C and a strain rate of 10^{-2} s^{-1} with an m value of 0.3, implying the operation of dislocation glide process. Charit and Mishra[43] studied an FSP 2024 Al alloy, and achieved an elongation of 530% at a strain rate of 10^{-2} s^{-1} and 430°C. They attributed this to a GBS-related process ($m \sim 0.48$) that operates because of the presence of a fine grain size (~3.9 μm) and greater percentage of high angle grain boundaries (~95%). Figure 14.6a shows a temperature-strain rate space where results from various studies have been mapped. It shows that newer processing techniques (FSP and ECAP) have the potential for obtaining superplasticity at higher strain rates and lower temperatures. Flow stress data from different studies are also presented in a single plot to compare the deformation mechanisms operating therein (Figure 14.6b).

Kaibyshev et al.[47] studied a superplastic AA2219 Al. The composition of AA2219 Al is basically Al–Cu–Mn–Zr (without significant amount of Mg). The alloy was processed through a two-stage TMP as illustrated in Figure 14.7. A final grain size of 12 μm was obtained utilizing particle-stimulated recrystallization (PSN) mechanism. This fine-grained alloy exhibited a maximum elongation of 675% at 500°C and a strain rate of 2.2×10^{-4} s^{-1}. No HSRS or LTSP properties could be achieved. At 500°C, the strain rate sensitivity value varied from 0.48 at initial strain levels to 0.37 near terminal elongations. From the surface topography of superplastically deformed specimens, they noted cooperative GBS to be operating instead of only GBS during superplastic deformation. However, the cooperative grain boundary sliding (CGBS) was identified as coarse or quite nonhomogeneous. They also briefly studied the cavitation and grain growth behavior, and found them to influence the overall elongation.

Another study by Han and Langdon[48] was carried out in an ECAP 2219 Al alloy with 0.2 μm grain size. However, no evidence of superplasticity was revealed in this material due to the lower temperature range of 300 to 400°C. It was observed that the grain structure underwent excessive grain growth in this temperature range where enhanced superplasticity is expected for submicron materials. They noted the inability of θ and θ'' precipitates in hindering significant grain growth. This study highlighted that the alloy composition may play an important role in determining the extent of superplasticity from the fact that the ECAP 2024 showed enhanced superplasticity whereas ECAP 2219 did not.

AA2004 Al is an aluminum alloy with nominal composition of Al–6Cu–0.4Zr (wt.%). This is popularly known as Supral 100. It is one of the most well-known superplastic alloys. Grimes et al.[49] showed that elongations of >1000% can be attained in Supral at a strain rate of 10^{-4} to 10^{-3} s^{-1} in the vicinity of 450°C. The Supral alloys transform into a fine-grained microstructure during TMP through continuous dynamic recrystallization where fine grains evolve during the first stage of high-temperature deformation. Zirconium forms fine dispersoids that help in resisting grain boundary motion. Valiev et al.[50] investigated superplasticity in an ECAP 2004 Al (pressed for 12 passes) with a grain size of ~0.5 μm. They obtained an optimum elongation of 970% at a strain rate of 10^{-2} s^{-1} and 300°C. Further, they obtained 740%

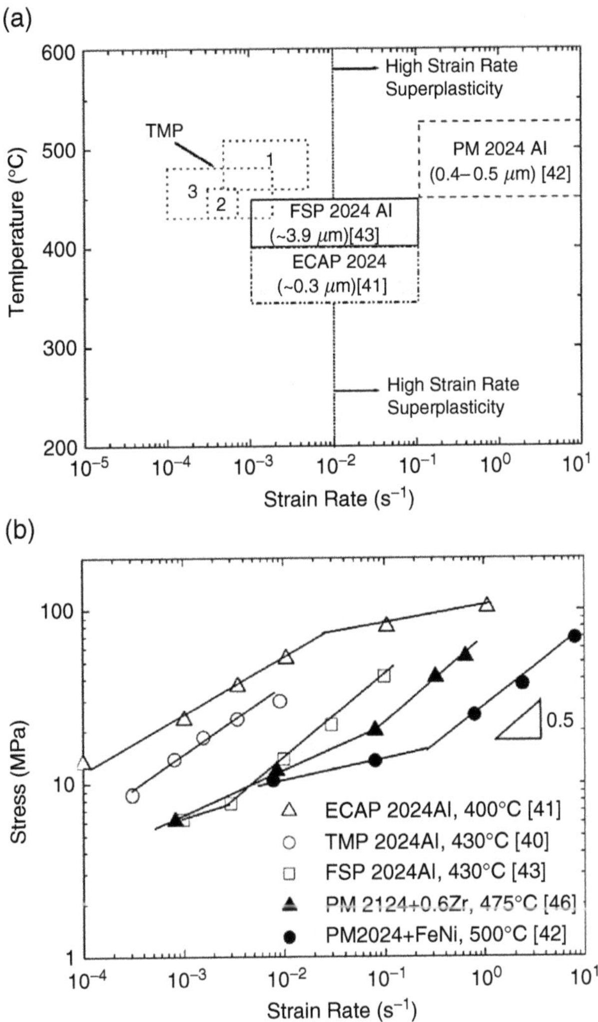

FIGURE 14.6 (a) A temperature-strain rate map showing the superplastic regimes for various 2024 Al alloys (region-1 from Reference 40, region-2 Reference 44, and region-3 Reference 45). (From Charit, I. and Mishra, R.S., *Mater. Sci. Eng.*, A359, 290, 2003. With permission from Elsevier Science.) (b) The variation of flow stress as a function of strain rate for various 2XXX alloys.

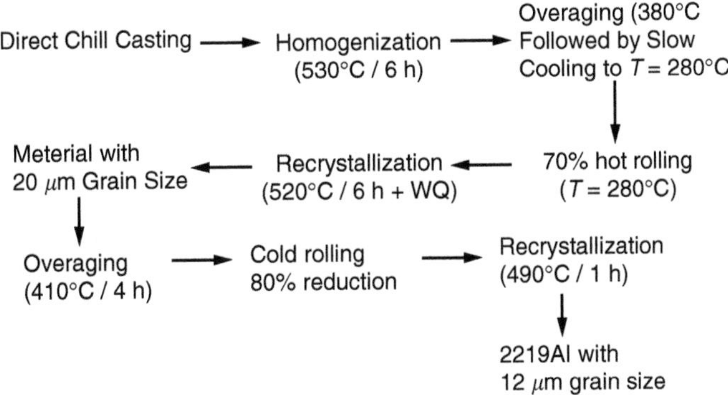

FIGURE 14.7 A two-stage TMP scheme for producing superplastic 2219 Al alloy.

elongation at a 10^{-1} s^{-1}. Although, no deformation mechanism was identified, the opportunity of developing HSRS and LTSP properties in submicron-grained alloys is promising.

14.5.1.1.2 Al–Mg Alloys

Al–Mg alloys (typically <7 wt.% Mg) are nonheat treatable. These alloys are quite popular for automotive structural applications. Some well-known alloys in this group are 5052 and 5083 Al which are basically of Al–Mg–Cr–Mn composition. These alloys generally have good corrosion resistance, good weldability, low density, and moderately high strength. Historically, two broad approaches have been followed to produce better superplastic 5083 Al. First, various conventional and novel processing techniques have been applied to the base composition of 5083 Al, including conventional rolling-based TMP,[51,52] ECAP,[53] ARB,[54] and FSP.[55] The drawbacks encountered in these approaches have been slower forming rates or instability of grain structures at superplastic temperatures. Second, alloy design concepts have been utilized for achieving enhanced superplasticity. Alloying additions such as Cu,[56] Mn + Sc,[57] Sc + Sn,[58] and Sc[59] in Al–Mg-alloy compositions have been reported. Significant improvements in ductility have been observed in the modified alloys.

Table 14.1 summarizes some significant results of superplasticity in various Al–Mg alloys. It includes a variety of processing techniques and compositions used for attaining superplasticity in Al–Mg alloys. Finer grain size always gave the best elongations. It can also be noted from Table 14.1 that the optimum superplastic strain rates for almost all Al–Mg alloys listed are less than 10^{-2} s^{-1}. Compositional modifications seem to raise the optimum strain rate to higher values mainly because the dispersoid formation helps in producing a fine-grained material and maintain the stability of the grain structure.

Several Al–Mg alloys (as evident from Table 14.1) show strain rate sensitivity values of ~0.5, which implies GBS mechanism. However, there are only a few p values reported. Generally, it has been found to be ~2. Another group of alloys show a strain rate sensitivity value of 0.3, which generally represents a solute-drag related dislocation glide mechanism. In Al–Mg alloys, Mg solute atoms in solid solution tend to form solute atmospheres around mobile dislocations, thus generating a drag effect on their movement. This is similar to the Class-I solid solution behavior. Taleff et al.[26] have observed this kind of mechanism to occur in several Al–Mg alloys. As this mechanism is not grain size-dependent (i.e., $p = 0$), even coarse-grained alloys may exhibit good elongations (300 to 400%).

14.5.1.1.3 Al–Mg–Si Alloys

The majority of commercial 6XXX series Al alloys are Al–Mg–Si alloys with Mg and Si content ranging between 0.2 and 1.4 wt.%. The ratio of Mg and Si contents are important because the strength of the alloy comes mainly from the precipitation of Mg_2Si particles. A few well-known alloys are 6061 Al, 6063 Al, 6013 Al, and the like. These alloys are becoming potential candidates for use in automotive structural applications.

In 1974, Otsuka et al.[69] first used TMP to produce superplastic Al–Mg–Si alloys. Three alloys were chosen; alloy-E (Al–8.2Mg–4.7Si), alloy-S (Al–6.5Mg–7.2Si), and alloy-M (Al–9.5Mg–3.4Si). These alloys were of monovariant eutectic compositions, and were much different from their wrought counterparts. Troeger and Starke[70] commented on the application of TMP to wrought variants of 6XXX Al alloy for obtaining suitable microstructure for superplasticity. Application of rolling-based TMP as proposed by Wert et al.[26] for 7XXX series alloys does not work as it produces unacceptable nonuniform distribution of overaged particles that are needed for particle-stimulated nucleation (PSN) of recrystallization. Earlier attempts by Washfold et al.[71] and Kovacs-Csetenyi et al.[72] could not produce fine-grained materials that are amenable for enhanced superplasticity. Later, Chung et al.[73] used the conventional TMP approach, achieving only marginal superplasticity (230%) at a strain rate of 3×10^{-4} s^{-1} with associated "m" value of 0.38. Troeger and Starke[70] were able to devise a two-stage TMP scheme utilizing PSN mechanism occurring at deformation bands to obtain a weakly textured, equiaxed microstructure with an average grain size of ~10 μm in a modified variant of 6013Al (only the Fe content was significantly less than the baseline 6013Al). An elongation of 375% was obtained at a strain rate of 3×10^{-4} s^{-1} and 540°C, along with an m value of ~0.5. Superplasticity in 6XXX series alloys are limited by their

TABLE 14.1 A Summary of Superplastic Properties of Various Al–Mg Alloys

Alloy (wt.%)	Processing	Grain size (μm)	Strain rate (s^{-1})	Temperature (°C)	Elongation (%)	m	p	Q (kJ mol^{-1})	Ref.
Al–5Mg 0.79Mn	TMP (1.6 mm)	~10	2×10^{-4}	525	>700	~0.5	—	—	51
Al–4.2Mg–0.71Mn–0.14Cr	TMP (1.6 mm)	6.5	2×10^{-4}	550	420	~0.6	—	—	52
Al–2.8Mg Al–5.5Mg	TMP (~5 mm)	30	1×10^{-4}	400	325	~0.3	0	~136	60
		250			254				
Al–4.7Mg–0.7Mn	TMP (0.5–3 mm)	0.5	2×10^{-3}	230	511	0.30	0–1.1	70–90	61
			1×10^{-3}	550	580	0.55	~2	82	
Al–4.7Mg–0.65Mn–0.13Cr*	TMP (2 mm)	11	5×10^{-5}	563	~460	0.50	2	140	62
Al–4.6Mg–0.63Mn–0.08Cr	ECAP (12 passes)[a]	~0.47/10.3[b]	1×10^{-3}	510	400	0.35	—	—	63
AA5083	ECAP (4 passes)	<2	5×10^{-5}	550	>350	—	—	—	64
Al–4.4Mg–0.7Mn–0.15Cr	ECAP (8 passes)	0.3	5×10^{-4}	275	315	0.4	—	63	53
Al–4.45Mg–0.57Mn–0.06Cr	ARB (5 cycles)	0.28/10	1.7×10^{-3}	400	430	0.43	—	—	54
Al–4.5Mg–0.6Mn–0.2Cr	TMP (1 mm)	~10	1×10^{-3}	570	~1000	0.75	—	—	65
Al–4.7Mg–0.7Mn–0.4Cu	TMP (1.8 mm)	~6.3	1×10^{-3}	550	>600	0.65	—	—	66
Al–4.15Mg–1Mn–0.52Sc	TMP (1.2–2 mm)	~5.4	1×10^{-4}	550	600	—	—	—	57
Al–4.8Mg–0.7Mn–0.13Cr–0.6Cu	TMP (1.6 mm)	<10	2.8×10^{-3}	545	~700	~0.7	—	—	56
Al–4.7Mg–1.6Mn–0.2Zr	TMP	6.2	2.8×10^{-3}	570	1150	~0.6	—	151	67
Al–4Mg–1Zr	FSP	1.5	1.0×10^{-1}	525	1280	~0.5	—	—	68
Al–4.1Mg–0.09Cr–0.62Mn	FSP	6.5	3.0×10^{-3}	530	590	~0.5	2	140–150	55

[a] A complex multi-step ECAP schedule with intermediate solution heat treatment.
[b] Preheating to 510°C grows the submicron grain structure to one with an average grain size of 10.3 micron.

not-so fine grain size and the inability of second phase particles to resist grain growth at high enough temperatures.

14.5.1.1.4 Al–Zn–Mg Alloys

7XXX series alloys are actively employed as aerospace structural materials, older versions of 7075 Al, 7475 Al and newer ones like 7055 Al. Wert et al.[26] first developed a four-step process that involved preferential recrystallization of grains at larger particles (PSN). In 7075 Al and 7475 Al, overaged $MgZn_2$ particles (diameters >0.75 μm) acted as the grain nucleation sites and Zener drag was provided by fine Al_3Zr particles. These techniques can produce microstructures with average grain sizes in the range of 8–14 μm. The 7075 Al exhibited superplastic behavior at a strain rate of 2×10^{-4} s^{-1} and 516°C. Table 14.2 summarizes some significant results in various Al–Zn–Mg alloys obtained by various workers. It is interesting to note there has been hardly any study on SePD 7075 Al alloys which exhibited enhanced superplasticity. Mishra and coworkers[37,74] have applied FSP to 7075 Al, and obtained significant superplasticity.

7475 Al alloys was developed as a high purity (low Fe and Si) version of 7075 Al. Mostly TMP has been employed to obtain fine grain sizes. A few significant results have been included in Table 14.2. AA 7475 Al is one of the alloys that are used in commercial SPF. But interestingly, all TMP (based on ingot metallurgy route) processing leads to microstructures that are not fine enough to generate optimum superplasticity at higher strain rates.

Although Matsuki and Yamada[84] attained enhanced superplasticity in an Al–11Zn–1Mg–0.4Zr alloy back in 1973, recently there has been a trend of developing high strength Al alloys with higher Zn content. Such an alloy is 7055 Al (nominal composition: Al–8.2Zn–2.1Mg–2.2Cu–0.2Zr). Kaibyshev et al.[82,83] have used a two-step TMP method to produce a fine-grained microstructure with good superplasticity. However, the grain structure was bimodal (11 and 2 μm), and continuous recrystallization occurred in the apparently recovered regions. ECAP was applied to the same alloy and a fine-grained microstructure (1 μm) was produced. However, there was a slight change in the optimum strain rate, temperature, and elongation as evident in Table 14.2.

14.5.1.1.5 Lithium-Containing Aluminum Alloys

Aluminum alloys containing lithium are considered potential structural materials for aerospace applications. The addition of Li reduces density, increases elastic modulus and results in a higher specific strength. In the 1950s, ALCOA developed the high strength alloy 2020 Al (Al–1.1Li–4.5Cu–0.2Cd–0.5 Mn).[85] Thus far, many Al–Li alloys have been developed. The alloys 2090 (Al–2.25Li–2.75Cu–0.12Zr), 8090 (Al–2.4Li–1.35Cu–1.1Mg–0.12Zr), and 8091 (Al–2.6Li–1.9Cu–0.85Mg–0.12Zr) are well-known commercial alloys, which have also been investigated for superplasticity. Wadsworth et al.[86] investigated superplasticity in an ingot metallurgy source Al–3Li–0.5Zr and Al–4Cu–3Li–0.5Zr alloys that were processed through a multistep TMP schedule. The grain sizes of the alloys were about 2 to 3 μm. In Al–3Li–0.5Zr alloy, a maximum elongation of 1035% was obtained at a strain rate of 3×10^{-3} s^{-1} and 450°C. The activation energy was measured to be 92.7 kJ mol^{-1}. In Al–3Li–4Cu–0.5Zr alloy, a maximum elongation of 850% was obtained at a strain rate of 7×10^{-3} s^{-1} and 450°C. Both alloys showed "m" values close to 0.5 at 450°C. Although several other investigations have been reported, some illustrative results are summarized in Table 14.3. Pu et al.[87] followed a TMP technique that resulted in very fine (sub)grain size of 0.7 μm. They obtained a maximum superplastic elongation of 710% at a strain rate of 8×10^{-4} s^{-1} and 350°C.

With the advent of various SePD techniques capable of producing ultrafine-grained materials, many Li-bearing aluminum alloys have been made highly superplastic. Russian 1420 Al (Al–5Mg–1Li–0.1Zr) and 1421 Al (Al–4.1Mg–2Li–0.16Sc–0.08Zr) alloys have exhibited exceptional superplasticity after TS and ECAP. Berbon et al.[91] processed 1420 Al alloy via 12 passes of ECAP, resulting in an average grain size of 1.2 μm. They obtained a maximum elongation >1180% at a strain rate of 10^{-2} s^{-1} and 350°C. It is worth noting that an elongation of ~630% was also noted at a low temperature of 250°C and a strain rate of 10^{-3} s^{-1}. Mishra and coworkers[92] reported superplasticity in a nanocrystalline 1420 Al alloy processed through TS (grain size 83 ± 33 nm). They obtained an elongation of 775% at 300°C and 10^{-1} s^{-1} with "m" values close to 0.37. They found that the kinetics in this alloy remained slow compared to

TABLE 14.2 A Summary of Superplasticity Data in 7XXX Al Alloys

Alloy (wt.%)	Processing	Grain size (μm)	Strain rate (s^{-1})	Temperature (°C)	Elongation (%)	m	p	Q (kJ mol^{-1})	Ref.
Al–5.77Zn–2.61Mg–1.33Cu–0.22Cr–0.15Fe–0.08Si	TMP	8	6×10^{-4}	516	>1200	—	—	—	75
Al–5.83Zn–1.39Cu–0.14Cr–0.25Mn–0.38Fe–0.17Si	TMP	10	8.3×10^{-4}	510	2100	—	—	—	76
Al–6.05Zn–1.91Mg–1.46Cu–0.15Cu–0.15Cr–0.09Mn–0.12Fe	TMP	12	1.0×10^{-3}	516	>730	0.7	—	163	77
7075-T651	FSP	3.3	1.0×10^{-2}	490	~1000	0.45	—	—	37
7075-T651	FSP	3.8	3.0×10^{-3}	480	~1450	0.5	—	—	74
7475	TMP	14	1.0×10^{-4}	517	1250	—	—	—	78
7475Al (Al–5.63Zn–2.25Mg–1.53Cu–0.2Cr)	TMP	8	5.0×10^{-4}	515	~1000	0.6	—	—	79
7475 + 0.7Zr	PM	2	1.0×10^{-1}	515	1000	0.5	—	160	80
7475 Al	TMP	6	2.8×10^{-3}	530	2000	0.67	—	95	81
7055 Al	TMP	11 (64%)/2 (36%)	3.3×10^{-4}	450	960	0.6	—	—	82
7055 Al	ECAP (Route A 4 passes)	~1	5.6×10^{4}	425	~750	0.46	—	—	83

TABLE 14.3 Data of Optimum Superplastic Elongations in Various TMP Al–Li Alloys

Alloy composition	Temperature (°C)	Strain rate (s^{-1})	Elongation (%)	Ref.
Al–3Li–2Cu–1Mg–0.15Zr (IM)	500	3.3×10^{-3}	798	88
Al–3Li–2Cu–1Mg–0.2Zr (PM)	500	3.3×10^{-3}	654	88
Al–2.5Li–2.0Cu–1Mg–0.2Zr	500	8.7×10^{-3}	875	89
Al–2.2Li–2.7Cu–1.2Mg–0.15Zr	500	8.3×10^{-4}	850	90

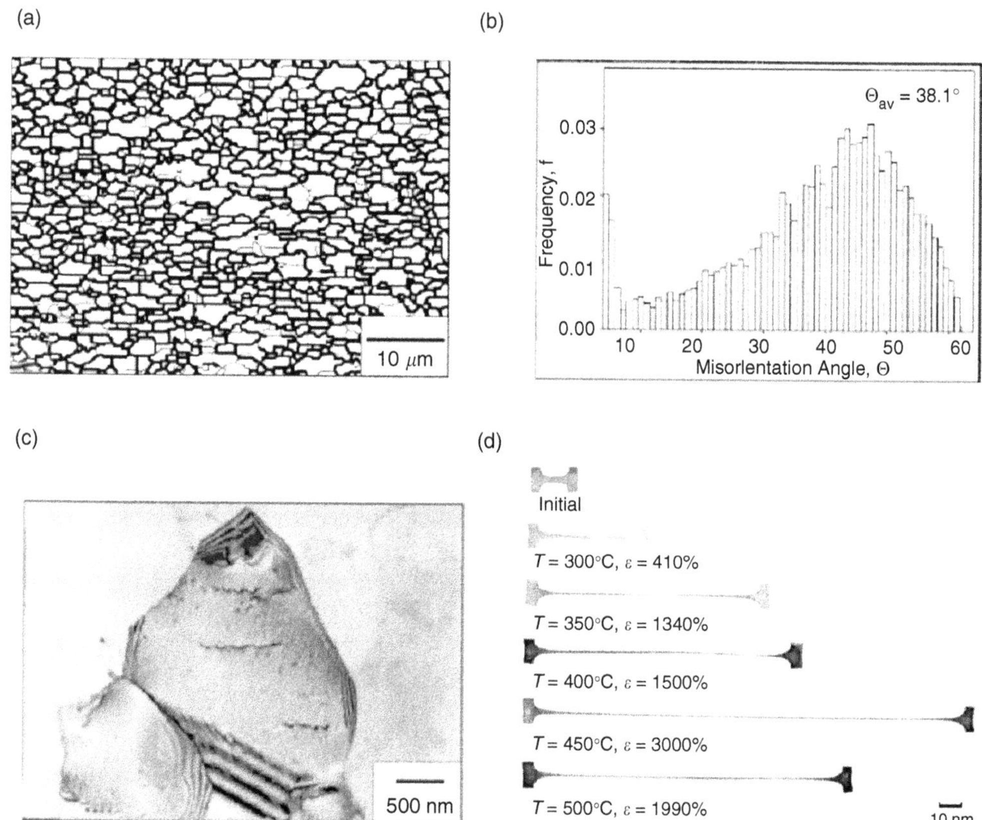

FIGURE 14.8 (a) A SEM-OIM misorientation micrograph, (b) misorientation distribution histogram, and (c) bright field TEM micrograph of 1421 Al ECAPed at 400°C. (d) appearance of superplastically deformed specimens. (From Kaibyshev, R. et al., *Mater. Sci. Technol.*, 21, 408, 2005. With permission from Maney Publishing on behalf of IOM, London, UK.)

the conventional constitutive relationship for superplasticity. They concluded that difficulty in producing enough dislocations in nanograins is likely to make the slip accommodation difficult, leading to slower kinetics. Kaibyshev et al.[93] ECAPed 1421 Al at three temperatures (240, 325, and 400°C) resulting in average grain sizes of 0.6, 0.8, and 2.6 μm. The alloy with 2.6 μm grain size has the highest fraction of high angle grain boundaries (~91%), and was found to be stable at elevated temperatures (Figure 14.8a to Figure 14.8c). This material showed the maximum elongation of ~3000% at a strain rate of 1.4×10^{-2} s^{-1} and 450°C (Figure 14.8d). The strain rate sensitivity value under optimum superplastic condition was ~0.68. Again, it is evident that the application of SePD techniques resulted in superplasticity at higher strain rates and lower temperatures.

14.5.1.2 Magnesium Alloys

In general, magnesium alloys possess good mechanical properties to be considered for various applications especially with great potential in automotive[94] and aerospace sectors.[95] Magnesium has a hexagonal close-packed (hcc) crystal structure with very limited slip systems operative at room temperature or moderate temperatures. Thus, cold forming is quite restricted in these alloys. Hot deformation (extrusion, rolling, forging) is one of the primary routes that may be used to make semifinished components as new slip systems become operative at higher temperatures. Further, if superplastic properties can be induced in these alloys, complex-shaped components can be made with ease through SPF as a secondary fabrication step. Of late, superplastic characteristics of many magnesium alloys have been investigated, especially by Higashi and coworkers.[96] Grain refinement in magnesium alloys can be achieved via various routes,[96] such as (1) combination of cold rolling plus static/dynamic recrystallization, (2) extrusion of cast alloys with high reduction ratios, (3) PM route utilizing rapidly solidified powders or machined chips, and (4) SePD techniques, such as ECAP.

Addition of aluminum to magnesium helps in improving the strength of the alloy. Mukai et al.[97] investigated an AZ31 (Mg–3Al–1Zn) alloy in two forms: rolled and extruded with average grain sizes of 130 and 5 μm. Tensile tests were carried out in the strain rate regime of 10^{-5} to 10^{-3} s^{-1}. The elongations were hardly close to 200% at a strain rate of 3×10^{-5} s^{-1} and 400°C, with the associated strain rate sensitivity value of ~0.33, and an activation energy for deformation of 127 kJ mol^{-1}. This is consistent with solute-drag creep process and basically it may be called extended ductility rather than superplastic. However, for the fine-grained AZ31, the maximum elongation was >600% at a strain rate of 10^{-4} s^{-1} and 325°C with a strain rate sensitivity value of ~0.5, an activation energy of 96 kJ mol^{-1} and a "p" value of 3. These parameters are consistent with a GBS-dominated mechanism. In the same study, an AZ61 alloy with a grain size of 171 μm exhibited a maximum elongation of ~400% at a strain rate of 10^{-5} s^{-1} and 325°C. The corresponding m, p, and Q values were estimated to be 0.5, 2, and 143 kJ mol^{-1}. Kim et al.[98] developed deformation mechanism maps for Mg–Al–Zn alloys. Tan and Tan[99] have proposed a two-stage deformation process to enhance superplasticity in a rolled AZ31 alloy with an initial grain size of 12 μm. First, the specimen was deformed at a lower temperature (250°C); up to a strain of 0.05, the grain size coarsens to ~25 μm. At a strain of 0.18, the grain boundaries started getting serrated and at a strain of 0.37, dynamic recrystallization begins, and by the strain of 0.47, almost 85 vol.% of grains became finer (~6 μm). Then the specimen was heated to 400/450°C and deformed to get a much better total elongation. The maximum elongations at 400 and 450°C were enhanced to 320% and 360%, respectively, at a strain rate of 10^{-4} s^{-1}.

Watanabe and coworkers[100] studied a coarse-grained rolled sheet of AZ31 alloy. The material showed high elongation (196%) at a temperature of 375°C and a strain rate of 3×10^{-5} s^{-1}. The "m," "p," and "Q" values were 0.33, 0, and 127 kJ mol^{-1}. The deformation mechanism was suggested to be viscous dislocation glide. Figure 14.9 shows diffusivity compensated strain rate vs. normalized stress taking into account the threshold stress. Also, data from other studies were included on the plot, and they superimpose on one another. The following equation expresses the relevant constitutive behavior:

$$\dot{\varepsilon} = 2 \times 10^{19} \left(\frac{\sigma - \sigma_0}{G} \right)^3 D_{\text{chem}} \tag{14.13}$$

where σ, σ_0, and G are in MPa and D_{chem} (solute diffusivity) in m^2 s^{-1}. Wu and Liu[101] also investigated superplasticity in coarse-grained AZ31 alloy with an initial grain size of 300 μm. The structure evolved during deformation resulting in a finer microstructure with 25 μm through continuous dynamic recrystallization process. A maximum elongation of 320% was obtained at a strain rate of 10^{-3} s^{-1} and 500°C. The mean "m" value and activation energy were 0.385 and 145 kJ mol^{-1}.

There are some reports on the superplasticity of Zr-bearing magnesium alloys. Mabuchi et al.[103] have investigated ZK series alloys (Mg–Zn–Zr). A PM source extruded ZK61 (grain size 0.6 μm) and IM source extruded ZK60 (grain size 1.2 μm) were investigated for superplasticity. A maximum elongation of 432%

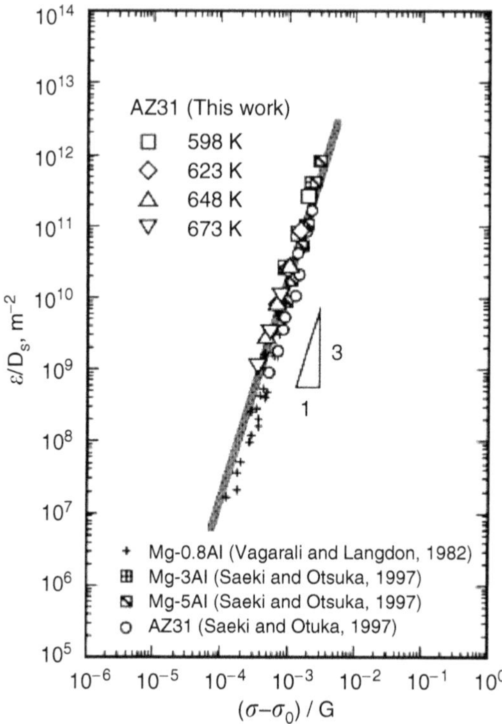

FIGURE 14.9 Solute-diffusivity compensated strain rate as a function of threshold stress compensated normalized stress in an AZ31 alloy. (From Watanabe et al., *Int. J. Plasticity*, 17, 387, 2001. With permission from Elsevier Science.)

at a strain rate of 10^{-1} s^{-1} and 300°C was obtained in the PM ZK61 whereas a maximum elongation of 730% was obtained at 4×10^{-3} s^{-1} and 300°C. It is clear from this study that PM processing may increase the superplastic strain rate significantly, however, with concomitant decrease in the extent of superplasticity.

Recently, SePD processes have been used to process Mg alloys to enhance their superplastic properties. As shown in Table 14.4, an ECAP processed AZ91 alloy (both as-ECAP and ECAP plus annealed condition) showed enhanced elongations. Unlike ECAP aluminum alloys, they did not show superplasticity at higher strain rates. However, the temperature was low enough (200°C) to term it as LTSP behavior. The annealed alloy showed better superplasticity than the as-ECAP one. Watanabe et al.[96] have reviewed and mapped superplastic elongations as a function of grain size for Mg-based alloys processed through various routes (Figure 14.10).

14.5.1.3 Titanium Alloys

Titanium alloys have the highest strength/weight ratio among all superplastic alloys. They also possess a good combination of creep and corrosion resistance, and SPF is important as a fabrication technique as it is difficult to impart complex shapes to titanium alloys via conventional methods. It is noteworthy that the conventional processing of titanium alloys leads to reasonable superplasticity.[108] Here, we briefly discuss the main features with a few illustrative examples.

Titanium has a hcp structure (α-phase) at lower temperatures and transforms to β-phase with a body-centered cubic (bcc) structure at temperatures >883°C. Alloying additions stabilize either α or β or a combination of both, depending on the alloy compositions and specific processing treatments.

TABLE 14.4 A Summary of Superplastic Characteristics in Various Mg Alloys

Alloy (wt.%)	Processing	Grain size (μm)	Strain rate (s^{-1})	Temperature (°C)	Elongation (%)	*m*	*p*	Q (kJ mol^{-1})	Ref.
AZ61	TMP hot rolling	8.7	2×10^{-4}	400	580	0.5		91	98
Mg–3Al–0.95Zn–0.28Mn (AZ31)	I/M extrusion	37.5/11 (DRX)	3×10^{-5}	400	314	0.4	—	—	102
Mg–9Al–1Zn–0.2Mn (AZ91)	I/M extrusion	1.4	$\sim 2 \times 10^{-4}$	300	~400	>0.3	—	—	103
AZ91	P/M extrusion	0.5	3×10^{-2}	300	~250	>0.3	—	—	103
AZ91	ECAP (strain of 6.34)	0.7	6.5×10^{-5}	200	661	0.5	3	95.6	104
AZ91	ECAP + annealing	3.1	7×10^{-5}	200	956	0.5	3	89.0	104
AZ91	RSP	1.2	3.3×10^{-3}	300	1480	0.5	—	49	105
Mg–8Li–0.2Zr	TMP	8	1.7×10^{-4}	300	639	—	—	—	106
Mg–8Li	Cast + extrusion + ECAP (2 passes)	1–3	1.0×10^{-4}	200	970	0.6	—	—	107

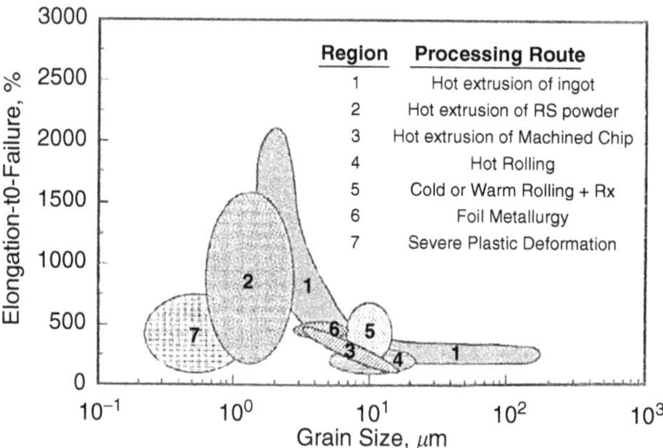

FIGURE 14.10 A schematic illustration of the variation of superplastic elongations with grain sizes in magnesium alloys produced by a variety of processing techniques. (From Watanabe et al., *Ultrafine Grained Materials II*, Zhu, Y.T. et al. (Eds.), TMS, Warrendale, PA, 2002, 469. With permission from TMS, Warrendale, PA, USA.)

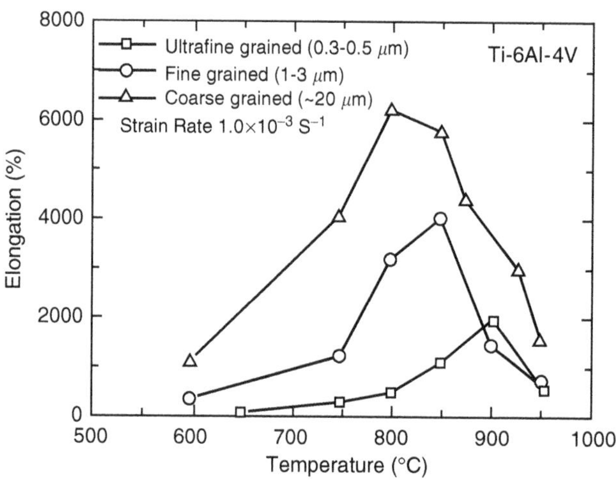

FIGURE 14.11 The variation of elongation as a function of temperature in Ti–6Al–4V with different grain sizes. (Taken from Yoshimura, H. and Nakahigashi, J., *Mater. Sci. Forum*, 426–432, 673–680, 2003. With permission from TransTech Publications, Switzerland.)

The diffusivity in the β-phase is almost two orders of magnitude faster than that in the α-phase. Most superplastic titanium alloys are dual phase (α/β) alloys. For superplastic alloys, grain size needs to be fine and equiaxed. Also, the ratio of phase volume fractions and grain size distributions become important for these dual phase alloys. In general, an alloy with 40 to 60 vol.% of α-phase is expected to have good superplasticity. Issues of dynamic grain growth and phase ratio change with temperature are also important. We take the example of superplastic Ti–6Al–4V alloy (which is an α/β alloy). In TMP Ti–6Al–4V alloy, the optimum superplastic temperatures are in the range of 800 to 1000°C and the corresponding strain rates vary from 10^{-4} to 10^{-2} s^{-1}. The strain rate sensitivity values have been observed to vary from as high as ~0.85 to ~0.3. Elongations ranging from 450 to 1700% are achieved easily. An interesting study has been carried out by Yoshimura and Nakahigashi.[109] They used protium (hydrogen) treatment to obtain an ultrafine grained microstructure with high angle grain boundaries in a Ti–6Al–4V alloy. Figure 14.11 shows the variation of elongation for protium-treated UFG Ti–6Al–4V as a function of temperatures along with data from a fine- and a coarse-grained alloy. The UFG alloy showed

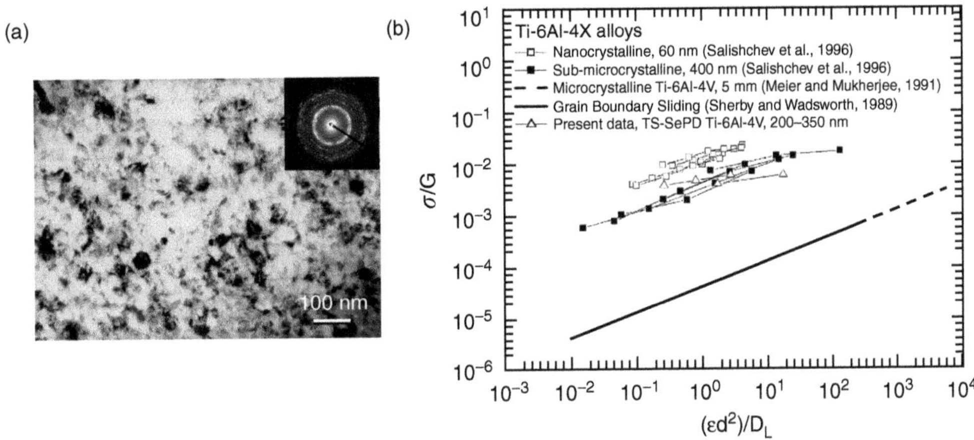

FIGURE 14.12 (a) A TEM image of as-TS Ti–6Al–4V alloy. (b) The variation of normalized stress with grain size and diffusivity compensated strain rate for alloys across microcrystalline to nanocrystalline Ti–6Al–4X alloys (X stands for Mo or V depending on the particular material). (From Mishra, R.S. et al., *Mater. Sci. Eng. A*, 298, 44–50, 2001. With permission from Elsevier Science.)

very high elongations >6000% at a strain rate of 10^{-3} s^{-1} and at temperatures in the vicinity of 800°C. Interestingly, at 850°C, they achieved an elongation of 9000% at slower strain rates (it may be the highest for any conventional superplastic material!).

The results for SePD α/β titanium alloys are also quite interesting. Salishchev et al.[110] investigated a Ti–6.7Al–4.7Mo processed by multiple forging with an average initial grain size of ~60 nm. The material contained 25% β phase. Maximum elongation of 1200% was achieved at a temperature of 575°C and a strain rate of 2×10^{-4} s^{-1}. The "m" value and the activation energy were found to be 0.45 and 315 kJ mol^{-1}, respectively. Sergueeva et al.[111] processed a hot rolled Ti–6Al–4V (10 μm grain size) via TS (total strain of 7), leading to a grain size ranging from 0.1 to 0.2 μm. They obtained a maximum elongation of 530% at 650°C and a strain rate of 10^{-4} s^{-1}, and the associated "m" value was 0.38. Ko et al.[112] employed ECAP for 4 and 8 passes for processing a Ti–6Al–4V alloy. The grain size obtained was ~0.3 μm in both cases. At 700°C and a strain rate of ~5×10^{-4} s^{-1}, they obtained higher elongations in the 8 pass material (>700%) than the 4 pass one (~400%). It was concluded that a higher fraction of high angle grain boundaries associated with lower activation energy was responsible for higher elongations although their initial grain size was same, as was the strain rate sensitivity value of ~0.45. Superplasticity in a torsionally strained Ti–6Al–4V alloy was reported by Mishra et al.[113] (30 to 50 nm grain size) (Figure 14.12a). A maximum elongation of >200% was obtained at 675°C and a strain rate of ~10^{-3} s^{-1}. It appeared that the onset of superplastic ductility coincided with the temperature for microstructural instability. Figure 14.12b shows a normalized plot showing the kinetics of deformation in nanocrystalline and UFG Ti–6Al–4V alloys. It was concluded that the slip accommodation process becomes difficult at very small grain sizes due to the difficulty in dislocation generation. The SePD processing techniques show high potentials for obtaining LTSP properties in α/β Ti alloys.

Hamilton also reviewed superplastic properties of beta-Ti alloys.[108] The beta-Ti alloys contain significant amounts of β stabilizer elements, and at suitable elevated temperatures, they become fully single phase alloys (β phase). However, due to the anomalously high diffusivity in the β phase and absence of any second phase, grain growth is very rapid. Consequently, the grain sizes are typically large (40 to 200 μm) and GBS is not expected to be effective. Griffiths and Hammond[114] investigated a Ti–15V–3Cr–3Sn–3Al alloy in which 200 to 400% elongations were achieved at 815°C and in the strain rate range of 10^{-3} to 2×10^{-4} s^{-1}. The superplastic flow was characterized by an initial regime with "m" value of ~0.3, which increased to ~0.4 to 0.5 in later stages. At lower temperatures, with the appearance of some α-phases, there appeared to be a slight increase in ductility. Furushiro et al.[115] achieved a maximum elongation

TABLE 14.5 A Summary of Superplastic Characteristics of Different δ–γ Stainless Steels

Alloy (wt.%)	Processing	Strain rate (s^{-1})	Temperature (°C)	Elongation (%)	Ref.
Fe–41Cr–22.5Ni–0.6Ti	Hot rolling	2.6×10^{-3}	982	>1000	126
Fe–26Cr–6.5Ni–0.4Ti	Cold rolling	3.2×10^{-4}	960	1050	127
Fe–23.4Cr–5Ni–1.5Mo–1Cu–0.15N	RSP + hot extrusion	2.0×10^{-4}	1070	500	128
Fe–25Cr–7Ni–3Mo–0.14N	Cold rolling	2.0×10^{-3}	950	2500	129
Fe–26Cr–6.5Ni–3Mo–0.12N	Cold rolling	4.0×10^{-3}	1000	>3000	130

of >350% at ~627°C and a strain rate of 4.2×10^{-4} s^{-1}. It was shown that the maximum superplasticity was achieved just below the beta-transus temperature. They noted that subgrain boundaries play a main role in obtaining higher elongations in place of GBS.

Evidence of superplasticity in near-α alloys is limited. Ti–6Al–2.5Sn alloy can show marginal superplasticity if deformed slightly below beta-transus temperature.[109] Lower diffusivity in the α-phase is a major reason for higher temperatures (>1000°C) for superplasticity in α alloys.

14.5.2 Ferrous Alloys

There have been several investigations of superplasticity in ferrous alloys.[116] Superplasticity in ferrous alloys can be divided into two main categories: carbon steels and alloy steels. It is interesting to note that although grain refinement studies involving various SePD techniques have been carried out in various ferrous alloys, no superplastic properties have yet been evaluated.

14.5.2.1 Carbon Steels

Schadler and coworkers,[117] for the first time, demonstrated an elongation of ~460% at 717°C and a strain rate of 3.3×10^{-4} s^{-1} in a 0.42C–1.9 Mn steel (average grain size of 1.5 μm) processed through a special TMP schedule. Many investigations have followed, some of which are summarized in Table 14.5. Various treatments ranging from rapid solidification processing to a variety of TMP techniques have been employed to render carbon steels fine grained, and thus superplastic. For an example, Torisaka et al.[118] developed a multistep TMP technique that can achieve grain refinement in high carbon steels amenable for superplasticity. Kim et al.[119] studied superplasticity in both I/M and P/M source Fe-C alloys with grain sizes of ~2 μm. Superplastic ductility in I/M alloys increases up to 2.1%C. At carbon levels <2.1 wt.%, fine cementite (Fe_3C) particles help in resisting dynamic grain growth, thus elongations of ~1000% can be achieved. However, above 2.1%C, the number of coarse cementite particles increase along with increasing cementite–cementite interfaces, leading to lower elongations. Interestingly, P/M alloys with higher carbon levels exhibit better superplastic ductility than similar I/M alloys. Superplasticity of these alloys was evaluated in the range of 10^{-4} to 10^{-3} s^{-1} and in the temperature range of 650 to 750°C. The superplastic elongation in these alloys was linked to the volume fraction of the cementite phase. A shaded regime where superplastic elongations over 400% can be obtained is shown on a Fe–C phase diagram in Figure 14.13.

14.5.2.2 High Alloy Steels

Stainless steels are the main high alloyed steels for which superplasticity data exist. In general, austenitic stainless steels consist of a single phase of austenite. They are not necessarily single phase at the superplastic temperatures. An investigation of a cold rolled Fe–1C–32Mn–11Al–1.5Si, an austenitic stainless steel, revealed elongations in excess of 500% at a strain rate of 7×10^{-5} s^{-1} and 800°C.[120] In this alloy, the Mn_3AlC precipitates were effective in resisting grain growth. Mineura and Tanaka[121] used repeated cold working and recrystallization, and obtained an average austenite grain size of 1 μm with a high number density of Cr_2N phases. They achieved an elongation of 527% at 800°C and a strain rate of 8×10^{-4} s^{-1}.

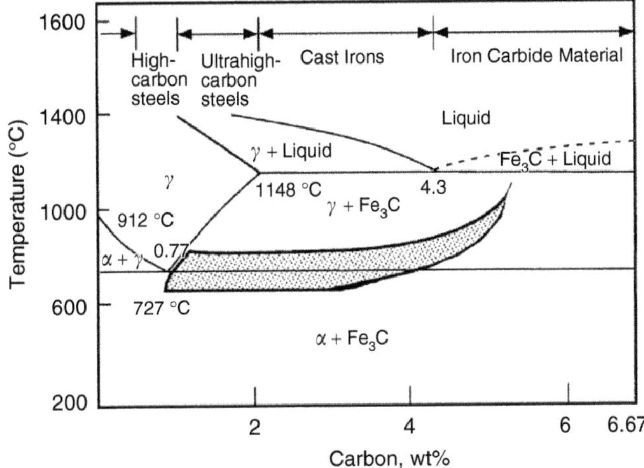

FIGURE 14.13 A Fe–C phase diagram showing a shaded regime where elongations >400% can be obtained at a strain rate of $\sim 1.6 \times 10^{-4}$ s^{-1}. (Taken from Kim, W.J., Taleff, E.M., and Sherby, O.D., *J. Mater. Sci.*, 33, 4977, 1998. With permission from Springer Publishers.)

An effort by Ray et al.[122] using PM processing resulted in an elongation of ~280% in a Fe–21Al–4B. In most studies, optimum elongations were obtained at lower strain rates. Taleff et al.[123] obtained optimum elongation of ~350% in a mechanically milled Fe–1.26C–1.6Cr–10.16Al–0.58 Mn alloy (with most grains in the size range of 0.5 μm) at 920°C and a strain rate of 10^{-1} s^{-1}.

Duplex (δ–γ) stainless steels have been investigated on several occasions, and impressive superplastic properties have been observed. Some of these results are summarized in Table 14.5. In duplex alloys, the grain growth is limited at elevated temperatures, and that is one of the reasons why these alloys exhibit good superplasticity. Maehara and Ohmori[124] reported an interesting superplastic feature in a Fe–25Cr–7Ni–3Mo–0.14N alloy. They observed two peak elongations (each ~2000%) at ~900 and ~1050°C. During deformation at 900°C, σ phase precipitation occurred dynamically from δ-ferrite phases through a eutectoid reaction leading to a γ/σ microstructure. The σ phase stabilized the microstructure leading to higher elongations. At a temperature of 1050°C, the microstructure was composed of δ and γ phases, and again higher elongation was obtained. Recently, Furuhara and Maki[125] have discussed the importance of grain boundary engineering for superplasticity in steels.

14.5.3 Metallic Glasses

Bulk metallic glasses (BMGs) have become an important class of materials with an attractive combination of novel properties (see Chapter 6 and Chapter 17 for more details on BMGs). Two temperatures are critical for BMGs: the glass transition temperature, T_g, above which the glass behaves as a supercooled liquid and the crystallization temperature, T_x. It has been demonstrated that Newtonian viscous flow ($m \sim 1$) could be achieved between the glass transition and crystallization temperatures. Therefore, the temperature range ($\Delta T = T_x - T_g$) in which the glass remains as the supercooled liquid with extreme flow ability could be utilized for forming of complex components out of BMGs.[131] Although no universal relationship has been observed between the glass-formability and ΔT, generally good glass formability corresponds to a large ΔT. Normalization of ΔT by the width of the undercooled liquid regime ($T_l - T_g$), where T_l is the liquidus temperature, gives the parameter $S = \Delta T/(T_l - T_g)$ that defines the relative formability of different BMGs.[132] Schroers[132] noted that a lower T_g is practically convenient for utilizing the supercooled regime to form components. T_g values are close to ~550°C for iron-based BMGs, ~350°C for zirconium-based, ~300°C for palladium-based, ~230°C for platinum-based and ~130°C for gold-based alloys.[132] The flow behavior in the supercooled regime also depends on strain rate. Typically,

TABLE 14.6 A Summary of Superplastic Properties of BMG Alloys

Alloy (at.%)	T_g (°C)	T_x (°C)	Test temperature (°C)	Strain rate (s^{-1})	m	Elongation (%)	Ref.
$Zr_{65}Al_{10}Ni_{10}Cu_{15}$	379	484	400	5×10^{-2}	0.8–1.0	340	134
$Pd_{40}Ni_{40}P_{20}$	305–324	378	347	1.7×10^{-1}	0.4–1.0	1260	134
$Zr_{52.5}Al_{10}Ni_{14.6}Cu_{17.9}Ni_{14.6}$	358–432	456	425	1.0×10^{-2}	0.45–0.55	650	135
$Zr_{65}Al_{10}Ni_{10}Cu_{15}$	379	484	423	6.3×10^{-3}	0.83	750	136
$Zr_{55}Al_{10}Ni_{5}Cu_{30}$	397	495	427	1.0×10^{-2}	0.5–0.9	800	137
$Zr_{41.25}Ti_{13.75}Ni_{10}Cu_{12.5}Be_{22.5}$	341	425	383	7.6×10^{-3}	0.4–1.0	~1624	138

the strain rate sensitivity approaches 1 at lower strain rates. Kawamura et al.[133] first demonstrated very large elongations (1800%) in a $La_{55}Al_{25}Ni_{20}$ alloy. The test temperature varied from 207 to 247°C. Some significant results are summarized in Table 14.6.

The classical definition of superplasticity does not apply to these materials because of the amorphous structure (no grain boundary- or dislocation-based mechanisms). However, superplasticity obtained in the supercooled regime can be exploited to make near-net shape components as it is done in conventional SPF.

14.6 Superplasticity in Ceramics

Ceramics generally have high grain boundary energy, and are prone to intergranular cracking at very low strains. Brittleness of ceramics renders them difficult-to-shape. Hence, superplasticity can greatly facilitate complex shape fabrication of ceramics. There are some characteristics that are unique to superplasticity observed in ceramics:[8]

1. A ceramic must have grain sizes smaller than 1 μm to be superplastic.
2. Strain rate sensitivity in ceramics can change with minor alterations in composition. This is related to the large effect of impurities on the diffusion in ceramics, or presence of amorphous phases at the grain boundaries.
3. The failure mechanism is controlled by cavitation rather than necking.
4. In ceramics, tensile ductility can be directly related to the Zener-Hollomon parameter $[(\dot{\varepsilon} \exp(Q/RT))]$.

Numerous superplasticity studies have been performed in yttria-stabilized zirconia (YTZ) and reviewed.[139] Yttria helps to stabilize the tetragonal phase, and yttria doping has been found particularly instrumental in retaining a fine grain size at elevated temperatures by way of solute partitioning and consequent solute-drag to the grain boundaries. The first discovery of ceramic superplasticity was made by Wakai et al. in an YTZ ceramic with a grain size of 0.3 μm.[140] They obtained ~170% elongation at 1450°C. Nieh and Wadsworth[141] demonstrated a maximum ductility of 800% at 1550°C and a strain rate of 2.7×10^{-4} s^{-1} with a true strain rate sensitivity of ~0.5. There has been much confusion about the deformation mechanisms involved, and it is beyond the scope of this chapter. The addition of additives has been found to influence superplastic properties to a great extent. Sakuma et al.[142] added 10 mol% silica to YTZ. They achieved an elongation of 1038% at a strain rate of 1.3×10^{-4} s^{-1} and a temperature of 1400°C. It has been proposed that silica creates an amorphous grain boundary phase that becomes viscous at elevated temperatures, and helps in GBS. Addition of 1 mol% GeO_2 has also been shown to improve superplasticity in zirconia. Nakano et al.[143] have investigated superplasticity in a wide array of zirconia ceramics with various additives. The grain sizes in their study varied from 0.23 to 0.47 μm. The elongations as a function of temperature are plotted in Figure 14.14. From this study, it seems that Nb and Ge

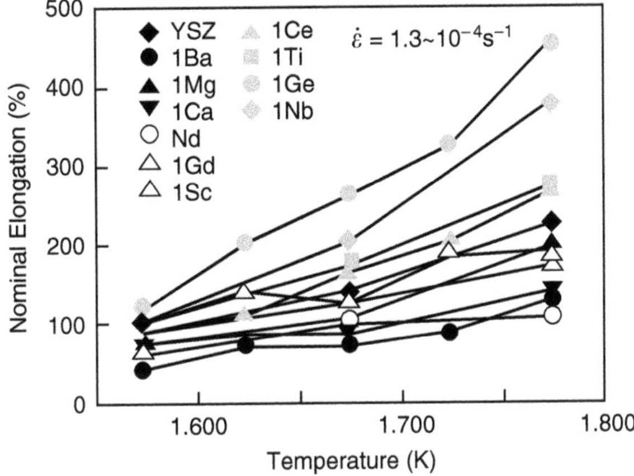

FIGURE 14.14 The variation of nominal elongations as a function of temperature for various zirconia ceramics. (Taken from Nakano, M., Nagayama, H., and Sakuma, T., *JOM*, 27, 2001. With permission from TMS, Warrendale, PA, USA.)

TABLE 14.7 A Summary of HSRS Properties in Oxide Ceramics

Alloy	Strain rate (s^{-1})	Elongation (%)	Temperature (°C)	Ref.
$20Al_2O_3$–ZrO_2 (3Y)	4×10^{-2}	300	1650	144
$5SiO_2$–ZrO_2 (2.5Y)	1×10^{-2}	360	1400	145
$2CaO$–$2TiO_2$–ZrO_2 (3Y)	1×10^{-2}	400	1400	146
$0.2Al_2O_3$–ZrO_2 (3Y)	3×10^{-2}	370	1450	147
$0.2Mn_3O_4$–$0.3Al_2O_3$ ZrO_2 (3Y)	1×10^{-2}	600	1450	148
$40ZrO_2$(3Y)–$30MgAl_2O_4$–Al_2O_3	8×10^{-2}	2500	1650	149
$30ZrO_2$(3Y)–$30MgAl_2O_4$–Al_2O_3	2×10^{-2}	660	1450	150

are the most effective in enhancing superplasticity among the elements examined. The activation energies obtained for these alloys were close to 450 kJ mol^{-1}.

However, superplasticity has mostly been observed at lower strain rates. So, a push has been made to examine the possibility of HSRS in ceramics. Hiraga et al.[144] have described the necessary attributes for obtaining HSRS in oxide ceramics. They also summarized HSRS results from various oxide ceramics as shown in Table 14.7.

There are also evidences of superplasticity in other ceramics, such as alumina-yttria and silicon nitride composites,[151] hydroxyapatites,[152] and so forth. However, they have not been as widely investigated as zirconia ceramics.

14.7 Superplasticity in Intermetallics

Structural intermetallics have high strength at elevated temperatures and good corrosion resistance. However, they have low ductility at ambient temperatures, thus making forming or machining difficult. Hence, SPF of intermetallics is considered important for near-net shape forming. Here, superplastic characteristics of titanium-, nickel-, and iron-based intermetallics are briefly presented.

14.7.1 Titanium-Based Intermetallics

Titanium-based aluminides have low density and high specific strength and elastic modulus, and thus have the potential to be used as structural components in aerospace applications. Ti_3Al is mainly composed of

TABLE 14.8 A Summary of Superplastic Properties of Ti_3Al- and TiAl-Based Intermetallics

Alloy (at.%)	Grain size (μm)	Strain rate (s^{-1})	Temperature (°C)	Elongation (%)	m	Q (kJ mol^{-1})	Ref.
(a) Ti_3Al-based intermetallics							
Ti_3Al–Nb	4	3.3×10^{-4}	980	520	0.6	—	153
Ti_3Al–Nb	5	8×10^{-5}	950	>800	0.6	—	154
Ti_3Al–Nb	—	1.5×10^{-4}	980	570	0.54	314	155
Ti–25Al–10Nb–3V–1Mo	3	$<1.0 \times 10^{-4}$	960–1020	1350	~0.5	—	156
(b) TiAl-based intermetallics							
Ti–47Al–Cr	18	5.4×10^{-4}	1200	383	0.57	—	157
Ti–50Al	5	8.3×10^{-4}	1025	250	0.43	—	158
Ti–(39–43)Al–V	0.8	3.0×10^{-4}	1147	600	0.8	—	159
Ti–48Al	0.5	5.6×10^{-3}	1050	555	0.5	350	160
Ti–46Al	0.20	1.3×10^{-3}	900	720	0.48	—	161
	1.5	6.4×10^{-4}		680	0.52		

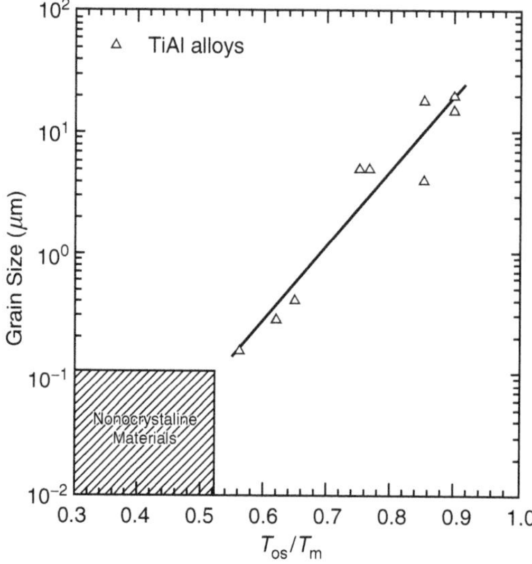

FIGURE 14.15 The variation of normalized optimum superplastic temperature as a function of grain size for various TiAl-based intermetallics. (From Mishra, R.S. and Mukherjee, A.K., *J. Mater. Sci.*, 35, 147, 2000. With permission from Springer Publishers.)

an intermetallic phase, called α_2 (with hexagonal DO_{19} structure). Additions of Nb, W, V, and so on are made to Ti_3Al to stabilize bcc β-phase and thus, improve room temperature ductility. For structural superplasticity, TMP of these alloys is carried out to obtain fine-grained, duplex microstructures with a phase fraction ratio (α_2:β) close to 1:1. Some illustrative superplastic properties of Ti_3Al-based intermetallics are summarized in Table 14.8a.

TiAl, often known as γ-phase, has a tetragonal $L1_0$ structure. However, superplastic properties are achieved when both α_2 and γ phases are present. Therefore, processing of TiAl alloys with suitable compositions is carried out to obtain microstructures with suitable phase proportions. Significant superplastic properties of TiAl-based intermetallics can be noted from Table 14.8b.

LTSP properties may have important implications in SPF of these intermetallics as it would help in reducing oxidation and tooling costs. Imayev et al.[161] applied multiple forging on various TiAl intermetallics, and developed ultrafine-grained microstructures that are amenable for LTSP. For instance, they obtained an elongation of 335% at a strain rate of 8.3×10^{-4} s^{-1}, and 800°C in a Ti–48Al–2Nb–2Cr (at.%) with an average grain size of 0.3 μm. Again, the importance of finer grain size has been demonstrated by Mishra and Mukherjee,[162] as shown in Figure 14.15.

TABLE 14.9 A Summary of Superplastic Properties of Ni-Based Alloys

Alloy (at.%)	Grain size (μm)	Strain rate (s^{-1})	Temperature (°C)	Elongation (%)	m	Q (kJ mol^{-1})	Ref.
Ni_3Si	15	1.3×10^{-3}	1050	500	0.5	555	163
Ni_3Si	14	6.0×10^{-4}	1090	665	0.5	526	164
Ni_3Al	6	9.0×10^{-4}	1100	641	0.8	256	165
Ni–45Al	270	2.5×10^{-4}	1075	245	0.3	—	166

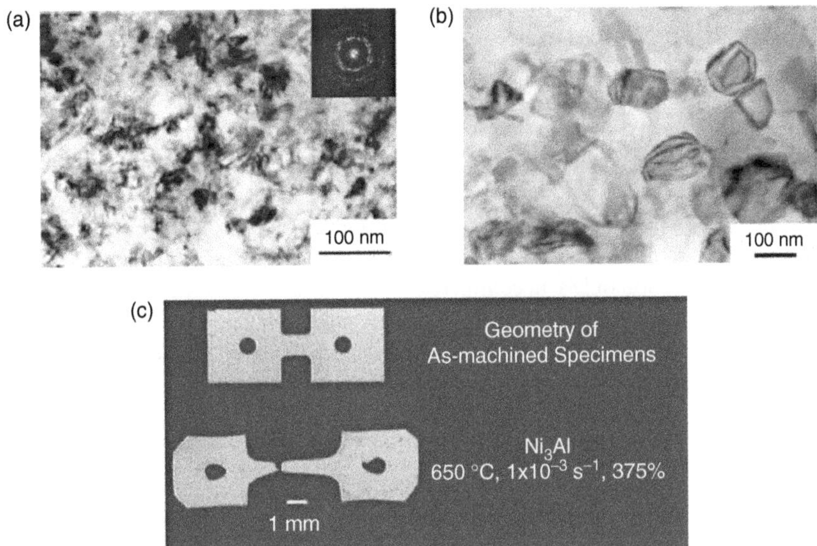

FIGURE 14.16 (a) The as-processed microstructure of Ni_3Al, and (b) the microstructure of the as-deformed specimen. (c) The appearance of tensile specimens before and after superplastic deformation. (Taken from McFadden, S.X., Mishra, R.S., Valiev, R.Z., and Zhilayev, A.P., *Nature*, 398, 684, 1999.)

14.7.2 Nickel-Based Intermetallics

Various nickel-based intermetallics (nickel silicides and nickel aluminides) have been developed. Ni_3Al and similar intermetallics have cubic $L1_2$ structure, whereas NiAl has a bcc B2 structure. Some illustrative results are summarized in Table 14.9.

Again, it is clear that high temperature in excess of 1000°C is required for reasonable superplasticity. Hence, to obtain superplasticity at lower temperatures, the grain size should be finer. McFadden et al.[167] examined LTSP properties in a torsionally strained Ni_3Al with Zr, B, and Cr additives (average grain size of 50 nm). Figure 14.16a shows the microstructure of the as-processed nanocrystalline Ni_3Al, whereas Figure 14.16b shows a microstructure of the superplastically deformed Ni_3Al. The grain size of the deformed specimen remained in the nanocrystalline range. Superplasticity was obtained at very low temperatures, that is, less than half of the melting point. At 650°C, an elongation of 375% was obtained at a strain rate of 10^{-3} s^{-1}. The deformed specimen is shown in Figure 14.16c.

14.7.3 Iron-Based Intermetallics

Iron-based intermetallics may be considered important substitutes to costly alloys. There have been several investigations to establish the superplastic behavior of Fe–Al-based intermetallics. Fe_3Al has a DO_3 structure and FeAl has a B2 structure. In a Fe–28Al–2Ti (at.%) intermetallic, an elongation of ~650% was achieved at 850°C and a strain rate of 10^{-3} s^{-1}.[168] Li et al.[169] evaluated superplasticity studies of

Fe–36.5Al–2Ti alloy with different grain sizes. It has been observed that the elongation can reach up to 500% with a grain size reduction from 60 to 20 μm. As expected, superplastic ductility increases with decreasing grain size. In general, Fe-based intermetallics contain coarser grain sizes (100 to 350 μm), yet they show good superplasticity. Superplastic behavior can be described by a continuous recovery and recrystallization process.[168] The strain rate sensitivity value is generally close to 0.33 to 0.42. Yang et al.[170] argued that the deformation process in these intermetallics is controlled by a viscous glide of dislocations. HSRS has been obtained in a FeAl–Ti intermetallic (300 μm grain size) where an elongation of 208% was obtained at 1000°C and a strain rate of 1.4×10^{-2} s^{-1}.

14.8 Concluding Remarks

Several aspects of superplasticity could not be covered due to the limited space and scope of this book. Superplastic properties in composites were not covered, although that is an important material subset. Good reviews are available in a few references.[4,8–10] Dynamic grain growth and cavitation tend to restrict achievable superplasticity, which must be addressed to improve superplastic properties. It is accepted that SPF can only cater to a niche market. However, it is expected that achieving LTSP and HSRS properties at lower flow stress values could open up new opportunities in the fabrication of a wide array of materials, leading to enhanced utilization of SPF in commercial applications. New concepts, such as selective superplasticity, thick plate superplasticity associated with FSP superplasticity may shape the future of SPF.[37] Barnes[171] has adequately described the techno-commercial challenges SPF faces, and it is expected that SPF will continue to be an important near-net shape technique.

References

[1] Pearson, C.E., Viscous properties of extruded eutectic alloys of lead-tin and bismuth-tin, *J. Inst. Metals*, 54, 111, 1934.

[2] Underwood, E.E., Superplasticity and related phenomena, *J. Metals*, 14, 914, 1962.

[3] Backofen, W.A., Turner, I.R., and Avery, D.H., Superplasticity in Al-Zn alloy, *Trans. ASM*, 57, 980, 1964.

[4] Sherby, O.D. and Wadsworth, J., Superplasticity — Recent advances and future directions, *Prog. Mater. Sci.*, 33, 169, 1989.

[5] Chokshi, A.H., Mukherjee, A.K., and Langdon, T.G., Superplasticity in advanced materials, *Mater. Sci. Eng.*, R10, 237, 1993.

[6] Edington, J.W., Melton, K.N., and Cutler, C.P., Superplasticity, *Prog. Mater. Sci.*, 21, 1, 1976.

[7] Padmanabhan, K.A., Vasin, R.A., and Enikeev, F.U., *Superplastic Flow: Phenomenology and Mechanics*, Springer, Berlin, NY, 2001.

[8] Nieh, T.G., Wadsworth, J., and Sherby, O.D., *Superplasticity in Metals and Ceramics*, Cambridge University Press, Cambridge, UK, 1997.

[9] Kaibyshev, O.A., *Superplasticity of Alloys, Intermetallides, and Ceramics*, Springer-Verlag, Berlin, NY, 1992.

[10] Presnyakov, A.A., *Superplasticity of Metals and Alloys*, British Library Lending Division, Boston Spa, 1976.

[11] Pilling, J. and Ridley, N., *Superplasticity in Crystalline Solids*, The Institute of Metals, London, 1989, 159.

[12] Glossary of terms used in metallic superplastic materials, JIS H 7007, Japanese Standards Association, Tokyo, 1995, 3.

[13] Arieli, A. and Mukherjee, A.K., The rate-controlling deformation mechanisms in superplasticity — A critical assessment, *Metall. Trans. A*, 13, 717, 1982.

[14] Ashby, M.F. and Verrall, R.A., Diffusion-accommodated flow and superplasticity, *Acta Metall.*, 21, 149, 1973.

[15] Ball, A. and Hutchison, M., Superplasticity in the aluminum–zinc eutectoid, *J. Met. Sci.*, 3, 1, 1969.

[16] Mukherjee, A.K., The rate-controlling mechanism in superplasticity, *Mater. Sci. Eng.*, 8, 83, 1971.

[17] Mukherjee, A.K., Role of grain boundaries in superplastic deformation, in *Grain Boundaries in Engineering Materials*, Walter, J.L. et al. (Eds.), Claitor Publishing, Baton Rouge, LA, 1975, 93–105.

[18] Gifkins, R.C., Grain boundary sliding and its accommodation during creep and superplasticity, *Metall. Trans. A*, 7, 1225, 1976.

[19] Hayden, H.W., Floreen, S., and Goodwell, P.D., Deformation mechanisms of superplasticity, *Metall. Trans. A*, 3, 833, 1972.

[20] Spingarn, J.R. and Nix, W.D., A model for creep based on the climb of dislocations at grain boundaries, *Acta Metall.*, 27, 171, 1979.

[21] Arieli, A. and Mukherjee, A.K., A model for the rate-controlling mechanism in superplasticity, *Mater. Sci. Eng. A*, 45, 61, 1980.

[22] Bardeen, J. and Herring, C., Diffusion in alloys and the Kirkendall effect, in *Imperfections in Nearly Perfect Crystals*, Shockley, W. (Ed.), John Wiley, New York, 1952, 261.

[23] Fukuyo, H., Tsai, H.C., Oyama, T., and Sherby, O.D., Superplasticity and Newtonian-viscous flow in fine grained class-I solid solution alloys, *ISIJ Int.*, 31, 76, 1991.

[24] Langdon, T.G., A unified approach to grain boundary sliding in creep and superplasticity, *Acta Metall. Mater.*, 42, 2437, 1994.

[25] Rachinger, W.A., Relative grain translations in plastic flow of aluminium, *J. Inst. Metals*, 81, 33, 1952–1953.

[26] Taleff, E.M., Nevland, P.J., and Yoon, S.J., The effect of ternary alloying additions on solute-drag creep in aluminum-magnesium alloys, in *Deformation, Processing and Properties of Structural Materials*, Taleff, E.M. et al. (Eds.), TMS, USA, 2000, 373.

[27] Waldman, J., Sulinski, H., and Markus, H., Effect of ingot processing treatments on the grain size and properties of Al alloy 7075, *Metall. Trans.*, 5, 573, 1974.

[28] Wert, J.A., Paton, N.E., Hamilton, C.H., and Mahoney, M.W., Grain refinement in 7075 aluminum by thermomechanical processing, *Metall. Trans.*, 12A, 1267, 1981.

[29] Brown, K.R., Method for Producing Aluminum Alloys having Superplastic Properties, U.S. Patent No. 5772804, June 30, 1998.

[30] Segal, V.M., Materials processing by simple shear, *Mater. Sci. Eng. A*, 197, 157, 1995.

[31] Valiev, R.Z., Structure and properties of ultrafine grained materials produced by severe plastic deformation, *Mater. Sci. Eng. A*, 168, 141, 1993.

[32] Ghosh, A.K., Method for Producing a Fine Grain Aluminum Alloy Using Three Axes Deformation, U.S. Patent No. 4721537, January 26, 1988.

[33] Tsuji, N., Saito, Y., Utsunomiya, H., Sakai, T., and Hong, R.G., Ultrafine grained bulk aluminum produced by accumulative roll bonding process, *Scripta Mater.*, 39, 1221, 1998.

[34] Ghosh, A.K. and Huang, W., Multipass deformation process to fabricate sheets and resulting microstructures, in *Ultrafine Grained Materials*, Mishra, R.S. et al. (Eds.), TMS, Warrendale, PA, 2000, 173.

[35] Thomas, W.M. et al., Friction Stir Butt Welding, U.S. Patent No. 5460317, October 1995.

[36] Mahoney, M.W., Rhodes, C.G., Flintoff, J.G., Spurling, R.A., and Bingel, W.H., Properties of friction-stir-welded 7075 T651 aluminum, *Metall. Mater. Trans. A*, 29, 1955, 1998.

[37] Mishra, R.S., Mahoney, M.W., McFadden, S.X., Mara, N.A., and Mukherjee, A.K., High strain rate superplasticity in a friction stir processed 7075 Al alloy, *Scripta Mater.*, 42, 163, 2000.

[38] Suryanarayana, C., *Mechanical Alloying and Milling*, Marcel Dekker, New York, 2004.

[39] Wilcox, B.A. and Clauer, A.H., Superplastic tendencies of two melt-spun Al alloys of commercial composition, in *The Microstructure and Design of Alloys*, Institute of Metals, London, UK, 1973, 227.

[40] Zheng, W. and Baoliang, Z., Superplasticity in the 2024Al alloy, *J. Mater. Sci. Lett.*, 13, 1806, 1994.

[41] Lee, S., Furukawa, M., Horita, Z., and Langdon, T.G., Developing a superplastic forming capability in a commercial aluminum alloy without scandium or zirconium additions, *Mater. Sci. Eng.*, A342, 294, 2003.

[42] Matsuki, K. et al., Effect of microstructure refinement on high strain rate superplasticity in a PM 2024Al-Fe-Ni alloy, *Mater. Trans. (JIM)*, 40, 737, 1999.

[43] Charit, I. and Mishra, R.S., High strain rate superplasticity in a commercial 2024 Al alloy via friction stir processing, *Mater. Sci. Eng.*, A359, 290, 2003.

[44] Huang, H., Wu, Q., and Hua, J., A study on superplasticity of commercial 2024Al Alloy, in *Superplasticity and Superplastic Forming*, Hamilton, C.H. and Paton, N.E. (Eds.), TMS, Warrendale, PA, 1988, 465.

[45] Liu, Y., Yang, G., Lian, J., and Zeng, X., Superplastic deformation and mechanisms of the commercial aluminium alloy Al-4Cu-1.6Mg-0.8Mn, in *Superplasticity in Metals, Ceramics, and Intermetallics*, Mayo, M.J., Kobayashi, M., and Wadsworth, J. (Eds.), *Material Research Society Symposium*, Pittsburgh, PA, 1990, 391.

[46] Nieh, T.G. and Wadsworth, J., High strain rate superplasticity of a powder metallurgy, Zr-modified 2124 Aluminum, in *The 4th International Conference on Aluminum Alloys*, Atlanta, Georgia, September 11–16, 1994, 48.

[47] Kaibyshev, R. et al., Superplasticity in a 2219 aluminum alloy, *Scripta Mater.*, 44, 2411, 2001.

[48] Han, B.Q. and Langdon, T.G., Significance of microstructural thermal stability in an Al-2219 alloy processed by severe plastic deformation, in *Ultrafine Grained Materials III*, Zhu, Y.T. et al. (Eds.), TMS, Warrendale, PA, 2002, 485.

[49] Grimes, R., Development of superplastic aluminum alloys, *Aluminum*, 51, 720, 1975.

[50] Valiev, R.Z. et al., Observation of high strain rate superplasticity in commercial aluminum alloys with ultrafine grain sizes, *Scripta Mater.*, 37, 1997, 1945.

[51] Imamura, H. and Ridley, N., Superplastic and recrystallisation behavior of a commercial Al-Mg 5083, in *Superplasticity in Advanced Materials*, Hori, S. et al. (Eds.), Tokyo, Japan, 1991, 453.

[52] Verma, R., Ghosh, A.K., Kim, S., and Kim, C., Grain refinement and superplasticity in 5083 Al, *Mater. Sci. Eng.*, A191, 143, 1995.

[53] Park, K.T., Hwang, D.Y., Chang, S.Y., and Shin, D.H., Low-temperature superplastic behavior of a submicron-grained 5083 Al alloy fabricated by severe plastic deformation, *Metall. Mater. Trans.*, 33A, 2859, 2002.

[54] Tsuji, N., Shiotsuki, K., and Saito, Y., Superplasticity of ultra-fine grained Al-Mg alloy produced by accumulative roll bonding, *Mater. Trans. (JIM)*, 40, 765, 1999.

[55] Charit, I. and Mishra, R.S., Evaluation of microstructure and superplasticity in friction stir processed 5083 Al alloy, *J. Mater. Res.*, 19, 3341, 2004.

[56] Watanabe, H., Ohori, K., and Takeuchi, Y., Superplastic behavior of Al-Mg-Cu alloys, *Trans. Iron Steel Inst. Jpn.*, 27, 730, 1987.

[57] Vetrano, J.S., Henager, C.H., Jr., and Bruemmer, S.M., Use of Sc, Zr and Mn for grain size control in Al-Mg alloys, in *Superplasticity and Superplastic Forming*, Ghosh, A.K. and Bieler, T.R. (Eds.), TMS, Warrendale, PA, 1998, 89.

[58] Henagar, C.H., Vetrano, J.S., Gertsman, V.Y., and Bruemmer, S.M., Effect of Sn additions on superplasticity in Al–Mg–Mn–Sc alloys, in *Superplasticity and Superplastic Forming — Current Status and Future Potential*, Berbon, P.B. et al. (Eds.), *Mater. Res. Soc. Symp.*, Warrendale, PA, 2000, 31.

[59] Park, K.T. et al., High strain rate superplasticity of submicron grained 5083 Al alloy containing scandium fabricated by severe plastic deformation, *Mater. Sci. Eng.*, 341, 273, 2003.

[60] Taleff, E.M., Nevland, P.J., and Krajewski, P.E., Solute-drag creep and tensile ductility in aluminum alloys, in *Creep Behavior of Advanced Materials for the 21st Century*, Mishra, R.S. et al. (Eds.), TMS, Warrendale, PA, 1999, 349.

[61] Hsiao, I.C. and Huang, J.C., Deformation mechanisms during low temperature and high temperature superplasticity in 5083 Al-Mg alloy, *Metall. Mater. Trans.*, 33A, 1373, 2002.
[62] Iwasaki, H. et al., Quantitative assessment of superplastic deformation behavior in a commercial 5083 alloy, *Mater. Sci. Eng. A*, 252, 199, 1999.
[63] Sinclair, J.W., Hartwig, K.T., Goforth, R.E., Kenik, E.A., and Voelkl, E., Microstructure and mechanical properties of aluminum 5083 processed by equal channel angular extrusion, in *Ultrafine Grained Materials*, Mishra, R.S. et al. (Eds.), TMS, Warrendale, PA, 2000, 393.
[64] Herling, D.R. and Smith, M.T., Superplastic microstructure of modified AA-5083 aluminum alloy processed by equal channel angular extrusion, in *Ultrafine Grained Materials*, Mishra, R.S. et al. (Eds.), TMS, Warrendale, PA, 2000, 411.
[65] Novikov, I.I., Portnoy, V.K., Levchenko, V.S., and Niki, A.O., Subsolidus superplasticity of aluminium alloys, *Mater. Sci. Forum*, 243–245, 463, 1997.
[66] Li, F., Bae, D.H., and Ghosh, A.K., Grain elongation and anisotropic grain growth during superplastic deformation in an Al–Mg–Mn–Cu alloy, *Acta Mater.*, 45, 3887, 1997.
[67] Kaibyshev, R.O. and Musin, F.F., Subsolidus superplasticity, *Doklady Phys.*, 45, 324, 2000.
[68] Ma, Z.Y. et al., High strain rate superplasticity in friction stir processed Al–Mg–Zr alloy, *Mater. Sci. Eng. A*, 351, 148, 2003.
[69] Otsuka, M., Miura, Y., and Horiuchi, R., Superplasticity in Al–Mg–Si monovariant eutectic alloys, *Scripta Metall.*, 8, 1405, 1974.
[70] Troeger, L.P. and Starke, E.A., Jr., Microstructural and mechanical characterization of a superplastic 6XXX aluminum alloy, *Mater. Sci. Eng.*, A277, 102, 2000.
[71] Washfold, J.S., Dover, I.R., and Polmear, I.J., Thermomechanical processing of an Al-Mg-Si alloy, *Met. Forum*, 8, 56, 1985.
[72] Kovacs-Csetenyi, E. et al., Superplasticity of AlMgSi alloys, *J. Mater. Sci.*, 27, 6141, 1992.
[73] Chung, Y.H., Troeger, L.P., and Starke, E.A., Jr., Grain refining and superplastic forming of aluminum alloy 6013, in *Proceeding of the 4th International Conference on Aluminum Alloys*, Sandars, T.H., Jr. and Starke, E.A., Jr. (Eds.), Atlanta, GA, I, 1994, 434.
[74] Ma, Z.Y., Mishra, R.S., and Mahoney, M.W., Superplastic deformation behavior of friction stir processed 7075Al alloy, *Acta Mater.*, 50, 4419, 2002.
[75] Shin, D.H. and Park, K.T., Directional cavity stringer formation in a superplastic 7075 Al alloy, *Mater. Sci. Eng.*, A268, 55, 1999.
[76] Xinggang, J. et al., A new grain-refinement process for superplasticity of high strength 7075 aluminium alloy, *J. Mater. Sci.*, 28, 6035, 1993.
[77] Malek, P., Superplasticity in an Al–Zn–Mg–Cu alloy, *Mater. Sci. Eng.*, A137, 21, 1991.
[78] Mahidhara, R.K., Superplastic flow and failure in a fine-grained 7475 Al alloy, *Mater. Lett.*, 25, 111, 1995.
[79] Smolej, A., Gnamus, M., and Slacek, E., The influence of the thermomechanical processing and forming parameters on superplastic behaviour of the 7475 aluminium alloy, *J. Mater. Proc. Technol.*, 118, 397, 2001.
[80] Kim, W.J., Higashi, K., and Kim, J.K., High strain rate superplastic behaviour of powder-metallurgy processed 7475Al+0.7Zr alloy, *Mater. Sci. Eng.*, A260, 170, 1999.
[81] Shin, D.H. et al., New aspects on the superplasticity of fine-grained 7475 aluminum alloys, *Metall. Trans.*, 21A, 2729, 1990.
[82] Kaibyshev, R., Superplastic behavior of a 7055 aluminum alloy, *Scripta Mater.*, 45, 1373, 2001.
[83] Kaibyshev, R., Superplasticity in a 7055 aluminum alloy subjected to intense plastic deformation, *Mater. Sci. Technol.*, 19, 1491, 2003.
[84] Matsuki, K. and Yamada, M., Superplastic behavior of Al–Zn–Mg alloys, *J. Jpn. Inst. Met.*, 37, 448, 1973.
[85] Murakami, Y., Aluminum-based alloys, in *Materials Science and Technology — A Comprehensive Treatment*, Vol. 8, VCH, Weinheim, Germany, 1996, 213.

[86] Wadsworth, J., Pelton, A.R., and Lewis, R.E., Superplastic Al–Cu–Li–Mg–Zr alloys, *Metall. Trans. A*, 16A, 2319, 1985.
[87] Pu, H.P., Liu, F.C., and Huang, J.C., Characterization and analysis of low-temperature superplasticity in 8090 Al–Li alloys, *Metall. Mater. Trans. A*, 26, 1153, 1995.
[88] Wadsworth, J. et al., Superplastic behavior of aluminum–lithium alloys, in *2nd International Symposium on Al–Li Alloys*, Starke, E.A., Jr. and Sanders, T.H., Jr. (Eds.), AIME-The Metallurgical Society, Warrendale, PA, 1983, 111.
[89] Grimes, R. and Miller, W.S., Superplasticity in Li-containing aluminum alloys, *2nd International Symposium on Al–Li Alloys*, Starke, E.A., Jr. and Sanders, T.H., Jr. (Eds.), AIME-The Metallurgical Society, Warrendale, PA, 1983, 153.
[90] Baojin, Z. et al., Grain boundary character distribution texture and superplastic deformation behavior of an Al–Li alloy, *Trans. Non-ferrous Met. Soc. China*, 9, 1, 1999.
[91] Berbon, P.B. et al., Fabrication of bulk ultrafine-grained materials through intense plastic straining, *Metall. Mater. Trans. A*, 29, 2237, 1998.
[92] Mishra, R.S. et al., High strain rate superplasticity from nanocrystalline Al alloy 1420 at low temperatures, *Phil. Mag. A*, 81, 37, 2001.
[93] Kaibyshev, R. et al., Achieving high strain rate superplasticity in an Al-Li-Mg alloy through equal channel angular extrusion, *Mater. Sci. Technol.*, 21, 408, 2005.
[94] Mukai, T., Application of superplasticity in commercial magnesium alloy for fabrication of structural components, *Mater. Sci. Technol.*, 16, 1314, 2000.
[95] Furuya, H. et al., Applications of magnesium alloys for aerospace structure systems, *Mater. Sci. Forum*, 350–351, 341, 2000.
[96] Watanabe, H., Mukai, T., and Higashi, K., Grain refinement and superplasticity in magnesium alloys, in *Ultrafine Grained Materials II*, Zhu, Y.T. et al. (Eds.), TMS, Warrendale, PA, 2002, 469.
[97] Mukai, T., Experimental study of the mechanical properties at elevated temperatures in commercial Mg–Al–Zn alloys for superplastic forming, *Key Eng. Mater.*, 171–174, 337, 2000.
[98] Kim, W.J. et al., Superplasticity in thin magnesium alloy sheets and deformation mechanism maps for magnesium alloys at elevated temperatures, *Acta Mater.*, 49, 3337, 2001.
[99] Tan, J.C. and Tan, M.J., Superplasticity in a rolled Mg-3Al-1Zn alloy by two-stage deformation method, *Scripta Mater.*, 47, 101, 2002.
[100] Watanabe, H. et al., Deformation mechanism in a coarse-grained Mg–Al–Zn alloy at elevated temperatures, *Int. J. Plast.*, 17, 387, 2001.
[101] Wu, X. and Liu, Y., Superplasticity of coarse-grained magnesium alloy, *Scripta Mater.*, 46, 269, 2002.
[102] Liu, M. et al., Superplastic behavior and microstructural evolution in a commercial Mg-3Al-1Zn magnesium alloy, *Mater. Trans. (JIM)*, 43, 2433, 2002.
[103] Mabuchi, M. et al., High strength and high strain rate superplasticity in magnesium alloys, *Mater. Sci. Forum*, 357–359, 327, 2001.
[104] Mabuchi, M. et al., Low temperature superplasticity of AZ91 magnesium alloy with non-equilibrium grain boundaries, *Acta Mater.*, 47, 2047, 1999.
[105] Solberg, J.K. et al., Superplasticity in AZ91 magnesium alloy, *Mater. Sci. Eng.*, A134, 1201, 1991.
[106] Fujitani, W. et al., Effect of zirconium addition on superplastic deformation of the Mg-8Li eutectic alloy, *J. Jpn. Inst. Met.*, 45, 333, 1995.
[107] Furui, M. et al., Improving the superplastic properties of a two-phase Mg-8%Li alloy through processing by ECAP, *Mater. Sci. Eng.*, 439, A410–A411, 2005.
[108] Hamilton, C.H., Superplasticity in titanium base alloys: a review, in *Mtg. on Adv. Mats.*, Kobayashi, M. and Wakai, F. (Eds.), Warrendale, PA, USA, Vol. 7, MRS, 1989, 59.
[109] Yoshimura, H. and Nakahigashi, J., Ultra-fine grain refinement, superplasticity, and its application of titanium alloys obtained through protium treatment, *Mater. Sci. Forum*, 426–432, 673, 2003.

[110] Salishchev, G.A. et al., Low temperature superplasticity of submicrocrystalline titanium alloys, *Mater. Sci. Forum*, 243–245, 585, 1997.
[111] Sergueeva, A.V. et al., Enhanced superplasticity in a Ti-6Al-4V alloy processed by severe plastic deformation, *Scripta Mater.*, 43, 819, 2000.
[112] Ko, Y.G. et al., Microstructural influence on low-temperature superplasticity of ultrafine-grained Ti-6Al-4V alloy, *Mater. Sci. Eng. A*, 410–411, 156, 2005.
[113] Mishra, R.S. et al., Mechanical behavior and superplasticity of a severe plastic deformation processed nanocrystalline Ti-6Al-4V alloy, *Mater. Sci. Eng. A*, 298, 44, 2001.
[114] Griffiths, P. and Hammond, C., Superplasticity in large grained materials, *Acta Metall.*, 20, 935, 1972.
[115] Furushiro, N. et al., Factors influencing ductility of superplastic Ti-6Al-4V alloy, in *Titanium '80, Science and Technology Proceedings, 4th International Conference on Titanium*, Kimura, H. and Izuma, O. (Eds.), AIME, New York, USA, 1980, 993.
[116] Maehara, Y. and Langdon, T.G., Superplasticity of steels and alloys, *Mater. Sci. Eng. A*, 128, 1, 1990.
[117] Schadler, H.W., Stress-strain rate behavior of manganese steel in temperature range of ferrite-austenite transformation, *Trans. Metall. Soc.*, 342, 1281, 1968.
[118] Torisaka, Y. et al., High-speed tool steel having recrystallized hyperfine grains and its application, *Tetsu-to-Hagane*, 71, 735, 1985.
[119] Kim, W.J., Taleff, E.M., and Sherby, O.D., Superplasticity of fine-grained alloys prepared by ingot and powder-processing routes, *J. Mater. Sci.*, 33, 4977, 1998.
[120] Toscano, E.H., Superplasticity in an austenitic stainless steel containing aluminum and manganese, *Scripta Metall.*, 17, 309, 1983.
[121] Mineura, K. and Tanaka, K., Superplasticity of 20Cr-10Ni-0.7N (wt.%) ultrahigh nitrogen austenitic stainless steel, *J. Mater. Sci.*, 24, 2967, 1989.
[122] Ray, R., Panchanathan, V., and Isserrow, S., Microcrystalline iron-base alloys made using a rapid solidification technology, *J. Met.*, 35, 541, 1983.
[123] Taleff, E.M. et al., High strain rate superplasticity in ultrahigh-carbon steel containing 10 wt.%Al (UHCS-10), *Scripta Mater.*, 34, 1919, 1996.
[124] Maehara, Y. and Ohmori, Y., Superplasticity of delta-ferrite austenite dulex stainless steels, in *Proceedings of MRS International Meeting on Advanced Material*, Vol. 7, MRS, Pittsburgh, PA, 1989, 93.
[125] Furuhara, T. and Maki, T., Grain boundary engineering for superplasticity in steels, *J. Mater. Sci.*, 40, 919, 2005.
[126] Hayden, H.W. and Brophy, J.W., Interrelation of grain size and superplastic deformation in Ni–Cr–Fe alloys, *Trans. ASM*, 61, 542, 1968.
[127] Smith, C.I., Norgate, B., and Ridley, N., Superplastic deformation and cavitation in a microduplex stainless steel, *Met. Sci.*, 10, 182, 1976.
[128] Zhang, Y., Dabkowski, F., and Grant, N.J., The superplastic response of three rapidly solidified micro-duplex stainless steels of varying ferrite-austenite content, *Mater. Sci. Eng.*, 65, 265, 1984.
[129] Maehara, Y., Superplasticity of delta-ferrite/austenite duplex stainless steels, *Trans. Iron Steel Inst. Jpn*, 25, 69, 1985.
[130] Osada, K., Uekoh, S., and Ebato, E., Superplasticity of as rolled duplex stainless steels, *Trans. Iron Steel Inst. Jpn*, 27, 713, 1987.
[131] Inoue, A., Viscous flow deformation in supercooled liquid state of bulk amorphous $Zr_{55}Al_{10}Ni_5$ alloy, *Mater. Trans. (JIM)*, 37, 1337, 1996.
[132] Schroers, J., The superplastic forming of bulk metallic glasses, *JOM*, May issue, 35, 2005.
[133] Kawamura, Y., High-strain-rate superplasticity due to Newtonian viscous flow in $La_{55}Al_{25}Ni_{20}$ metallic glass, *Mater. Trans. (JIM)*, 40, 794, 1999.
[134] Kawamura, Y., Nakamura, T., and Inoue, A., Superplasticity in $Pd_{40}Ni_{40}P_{20}$ metallic glass, *Scripta Mater.*, 39, 301, 1998.

[135] Nieh, T.G. et al., Superplastic behavior of a Zr-10Al-5Ti-17.4Cu-14.6Ni metallic glass in the supercooled liquid regime, *Scripta Mater.*, 40, 1021, 1999.

[136] Kim, W.J. et al., Superplastic flow in a $Zr_{65}Al_{10}Ni_{10}Cu_{15}$metallic glass crystallized during deformation in a supercooled regime, *Scripta Mater.*, 49, 1067, 2003.

[137] Chu, J.P. et al., Plastic flow and tensile ductility of a bulk amorphous $Zr_{65}Al_{10}Cu_{30}Ni_5$ at 700 K, *Scripta Mater.*, 49, 435, 2003.

[138] Wang, G. et al., Superplasticity and superplastic forming ability of a Zr–Ti–Ni–Cu–Be bulk metallic glass in the super-cooled liquid regime, *J. Non-Crystall. Solids*, 351, 209, 2005.

[139] Chokshi, A.H., Superplasticity in fine grained ceramics and ceramic composites, *Mater. Sci. Eng.*, A166, 119, 1993.

[140] Wakai, F., Sakaguchi, S., and Matsuno, Y., Superplasticity of yttria-stabilized zirconia polycrystals, *Adv. Ceram. Mater.*, 1, 259, 1986.

[141] Nieh, T.G. and Wadsworth, J., Superplastic behavior of a fine grained, ytrria-stabilized t-ZrO_2, *Acta Metall. Mater.*, 38, 1121, 1990.

[142] Sakuma, T., Kajihara, K., and Yoshizawa, Y., Superplasticity in SiO_2-containing TZP, *Scripta Metall. Mater.*, 28, 559, 1993.

[143] Nakano, M., Nagayama, H., and Sakuma, T., Superplasticity in cation-doped YSZ, *JOM*, March issue, 27, 2001.

[144] Chokshi, A.H., Nieh, T.G., and Wadsworth, J., Role of concurrent cavitation in the fracture of a superplastic zirconia alumina composite, *J. Am. Ceram. Soc.*, 74, 869, 1991.

[145] Kajihara, K., Yoshizawa, Y., and Sakuma, T., Enhancement of superplastic flow in tetragonal zirconia polycrystals with SiO_2 doping, *Acta Metall. Mater.*, 43, 1235, 1995.

[146] Oka, K., Tabuchi, N., and Takashi, T., High strain rate superplasticity in ceramics, *Mater. Sci. Forum*, 304–306, 451, 1999.

[147] Suzuki, T.S. et al., Enhanced superplasticity in an alumina-containing zirconia prepared by colloidal processing, *Scripta Mater.*, 43, 705, 2000.

[148] Sakka, Y. et al., Fabrication of high-strain rate superplastic yttria-doped zirconia polycrytsals by adding manganese and aluminum oxides, *J. Eur. Ceram. Soc.*, 24, 449, 2004.

[149] Kim, B.-N., Hiraga, K., and Morita, K., High strain rate superplastic ceramic, *Nature*, 413, 288, 2001.

[150] Morita, K., Hiraga, K., and Sakka, Y., High strain rate superplasticity in yttria-stabilized tetragonal ZrO_2 dispersed with 30 vol% $MgAl_2O_4$ spinel, *J. Am. Ceram. Soc.*, 85, 1900, 2002.

[151] Rouxel, T., Wakai, F., and Izaki, K., Tensile ductility of superplastic ceramics and metallic alloy, *Acta Metall. Mater.*, 39, 199, 1991.

[152] Wakai, F. and Kato, H., Superplasticity in HIP hydroxyapatite, *Adv. Ceram. Mater.*, 3, 71, 1988.

[153] Dutta, A. and Banerjee, D., Superplastic behavior of a Ti_3Al-Nb alloy, *Scripta Metall. Mater.*, 24, 1319, 1990.

[154] Ghosh, A.K. and Cheng, C.H., Superplastic deformation in titanium aluminides and modeling of transient deformation, in *Superplasticity in Advanced Materials*, Hori, S., Tokizane, M., and Frushiro, N. (Eds.), *The Jpn. Soc. Res. Superplast.*, 1991, 299.

[155] Yang, H.S., Superplasticity in a Ti_3Al-base alloy stabilized by Nb, V and Mo, *Scripta Metall. Mater.*, 25, 1223, 1991.

[156] Ridley, N., Islam, M.F., and Pilling, J., Superplasticity and diffusion bonding of microduplex super alpha 2, in *Structural Intermetallics*, Davolic et al. (Eds.), TMS, Warrendale, PA, 1993, 63.

[157] Masahashi, N. et al., High temperature deformation behavior of titanium aluminide based gamma plus beta, *ISIJ Int.*, 31, 728, 1991.

[158] Imayev, R.M., Imayev, V.M., and Salishchev, G.A., Formation of submicrocrystalline structure in TiAl intermetallic compound, *J. Mater. Sci.*, 27, 4465, 1992.

[159] Nobuki, M., Vanderschueren, D., and Nakamura, M., High temperature mechanical properties of vanadium alloyed gamma-base titanium aluminides, *Acta Metall. Mater.*, 42, 2623, 1994.

[160] Ameyama, K., Uno, H., and Tokizane, M., Superplastic behavior of Ti-48 at-percent-Al 2-phase titanium aluminide compacts made from mechanically alloyed powders, *Intermetallics*, 2, 315, 1994.
[161] Imayev, R. et al., Superplasticity and hot rolling of two-phase intermetallic alloy based on TiAl, *Scripta Metall. Mater.*, 34, 985, 1996.
[162] Mishra, R.S. and Mukherjee, A.K., An analysis of the role of grain size on superplasticity of γ titanium aluminides, *J. Mater. Sci.*, 35, 147, 2000.
[163] Altstetter, C. et al., Superplasticity in nanophase intermetallics, in *Mechanical Properties and Deformation Behavior of Materials having Ultrafine Microstructures*, Nastasi, M. (Ed.), Kluwer Academic Publishers, London, UK, 1993, 381.
[164] Cheng, S.C., Wolfenstine, J., and Sherby, O.D., Superplastic behavior of two-phase titanium aluminides, *Metall. Trans. A*, 23, 1509, 1992.
[165] Mishra, R.S. et al., High temperature deformation behavior of a nanocrystalline titanium aluminide, *Scripta Metall. Mater.*, 34, 1765, 1996.
[166] Jiang, D. and Lin, D., Superplasticity of single phase Ni-45Al intermetallics with large grains, *Mater. Lett.*, 57, 747, 2002.
[167] McFadden, S.X., Mishra, R.S., Valiev, R.Z., and Zhilayev, A.P., Low temperature superplasticity in nanostructured nickel and metal alloys, *Nature*, 398, 684, 1999.
[168] Lin, D., Shan, A., and Chen, M., Superplasticity in large-grained Fe_3Al alloys, *Intermetallics*, 4, 489, 1996.
[169] Li, D. et al., Superplasticity in Fe_3Al-Ti alloy with large grains, *Scripta Metall. Mater.*, 31, 1455, 1994.
[170] Yang, H.S., Jin, P., and Mukherjee, A.K., Superplastic behavior of regular alpha-2 and super alpha-2 titanium aluminides, *Mater. Sci. Eng.*, A153, 457, 1992.
[171] Barnes, A.J., Superplastic aluminum forming — Expanding its technical niche, *Mater. Sci. Forum*, 304–306, 785, 1999.

IV

Microstructure Change Processes

15

Single Crystal Growth

Roberto Fornari
Institute for Crystal Growth

Abstract

The development of many new devices is strictly connected with the availability of materials in single crystalline form, with low density of crystallographic defects and specific electrical, magnetic, optical, chemical, and mechanical properties. This implies the development of suitable growth processes by which the composition, the structure, as well as the physical properties of a given material are accurately controlled. Three major categories of crystal growth technologies have been applied to obtain large single crystals: melt growth, solution growth, and vapor transport. The basic characteristics of the material actually decide what approach is most convenient. This article reviews the theoretical basis of the crystal growth processes and the most important crystal growth techniques. Examples of growth techniques applied to specific materials are provided.

15.1 Introduction

Crystals are, with no exaggeration, the pillars of today's technology. Without crystalline materials, either bulk or thin films, there would be no electronic industry, no optical fiber communications, no solid-state lasers, no developments in the optical industry, no high-efficiency photovoltaic cells, just to mention some high-tech fields. But also more trivial things, like sugar and salt production, candy and drug preparation, the development of new pharmaceutical products, largely rely on sophisticated mass solution crystallization. The crystal growth science however remained a curiosity or, better, a tool to prepare materials for fundamental and academic studies till the second half of the twentieth century, when the discovery of the germanium, and later silicon, transistor opened up the field of modern micro-electronic and the related huge silicon industry.

Very often the advancement of a growth technology is driven by the needs of the crystal users. This is particularly true in the case of semiconductors where, for instance, the need for dislocation-free material has led to the development of conceptually new growth equipments with extremely low thermal gradients.

In this chapter, the theoretical basis of the crystal growth processes is briefly reviewed. The attention will then be focused on the three principal technologies applied to the growth of large single crystals: melt growth, solution growth, and vapor transport. The basic characteristics of the material actually decide the most appropriate technology. The growth from the melt is certainly the most attractive for industrial purposes, however, if the material of interest is a compound that does not melt congruently or if its melting point is too high (say above 2200°C) the melt growth techniques cannot be applied and other methods must be sought. The readers who wish to go in depth into theoretical and technological aspects of crystal growth may refer to the books in Reference 1 to Reference 5.

15.2 Fundamentals of Crystallization Processes

Crystal growth is a nonequilibrium process that implies a first-order (abrupt) phase transition between one nutrient (disordered) phase and one ordered solid phase. To achieve a stable growth, that is, an increase of the crystalline phase at the expense of the nutrient phase, a thermodynamic driving force is necessary. That means that the system has to be pushed slightly out of the equilibrium by acting on one or more thermodynamic entities (pressure, temperature, chemical potential). Practically, in the case of melt growth, temperature is lowered below the equilibrium temperature (melting point) by imposing a so-called "supercooling" whereas in the case of solution or vapor growth one can act on the mixture chemical potential by imposing a certain "supersaturation."

The growth of single crystals includes the following steps: (1) achievement of the necessary degree of supersaturation or supercooling, (2) formation of crystal nuclei possibly free of grain boundaries, (3) maintenance of the single crystal growth and production of a crystal with the desired shape and properties. Accordingly, for a full understanding of the growth processes one has to take up the basic concepts of supercooling and supersaturation, nucleation, interface kinetics, mass and heat transport, flow dynamic, and morphological stability. Let us examine separately such concepts.

15.2.1 Supercooling and Supersaturation

The thermodynamic driving force for crystallization is $\Delta\mu$, which is defined as the difference of chemical potential of the nutrient phase μ_{nutr} and the crystal μ_{cryst}. If

$$\Delta\mu = \mu_{\text{nutr}} - \mu_{\text{cryst}} > 0 \tag{15.1}$$

there is a gain of energy and the crystallization takes place, whereas when

$$\Delta\mu = \mu_{\text{nutr}} - \mu_{\text{cryst}} = 0 \tag{15.2}$$

the system is in equilibrium and $\mu_{\text{cryst}} = \mu^{\text{eq}}_{\text{nutr}}$

Finally, if

$$\Delta\mu = \mu_{\text{nutr}} - \mu_{\text{cryst}} < 0 \tag{15.3}$$

the solid phase will dissolve. Of course these three conditions are defined by external setting of macroscopic parameters such as temperature, partial pressure of reaction gases (vapor phase growth), or molar concentration (solution growth). In other words, $\mu = \mu(T, P, C)$, where C is the molar fraction of a generic species, T and P are the temperature and pressure, respectively. The condition 15.1 is anyway to be satisfied which, for the three bulk growth methods discussed in this article, leads to the following relations:

$$\Delta\mu = kT \ln(C/C^{eq}) > 0 \quad \text{(for solution growth in case of single component)} \tag{15.4}$$

where k is the Boltzmann constant, C the actual concentration of the solute and C^{eq} the respective concentration at the equilibrium. To have $\Delta\mu > 0$ the ratio C/C^{eq} must be >1 that means that the concentration of the solute must exceed the equilibrium value. Very often this is also expressed by a positive supersaturation $\sigma = (C - C^{eq})/C^{eq}$.

$$\Delta\mu = h\Delta T = h(T_{eq} - T) > 0 \quad \text{(for melt growth)} \tag{15.5}$$

where h is the latent heat of crystallization and ΔT is the supercooling. Note that the actual system temperature must be below the equilibrium one to have nucleation and stable growth.
Finally, for growth from the gas phase, in the case of a single component, one finds:

$$\Delta\mu = kT \ln(P/P^{eq}) > 0 \tag{15.6}$$

Here again it must be noted that the actual pressure of the component must exceed the equilibrium one, that is, that one again needs a supersaturation $\sigma = (P - P^{eq})/P^{eq} > 0$ to maintain stable growth conditions.

The equations reported above hold for single component systems. When the solution at the gas phase contains n components the ratios in Equation 15.4 and Equation 15.6 are modified to include the product of the partial concentrations or partial pressure, respectively:

$$\Delta\mu = kT \ln\left(\prod_{i=1}^{n} C_i \Big/ \prod_{i=1}^{n} C_i^{eq}\right) > 0 \tag{15.7}$$

$$\Delta\mu = kT \ln\left(\prod_{i=1}^{n} P_i \Big/ \prod_{i=1}^{n} P_i^{eq}\right) > 0 \tag{15.8}$$

where C_i and C_i^{eq}, P_i and P_i^{eq} are relative to the ith species.

15.2.2 Nucleation

The spontaneous nucleation of stable crystalline clusters within a metastable (supercooled or supersaturated) mother phase occurs when a critical level of supersaturation or supercooling is reached. The nucleation is called homogeneous if it occurs without the intervention of external bodies (e.g., container walls, impurities, dust particles or seeds). The homogeneous nucleation is assumed to be the result of two competing phenomena: the lowering of the free energy connected with the formation of a small regular solid and the corresponding increase of surface energy. This can be expressed as:

$$\Delta G = \Delta G^V + \Delta G^S = \frac{-4\pi r^3}{3\Omega}\Delta\mu + 4\pi r^2\sigma \tag{15.9}$$

where r is the radius of the crystal nucleus (assumed to be spherical), σ is the increase of surface energy per radius unit and Ω the corresponding volume increase. The volume contribution decreases monotonically with r while the surface contribution increases monotonically. As a result, the value of ΔG reaches a maximum corresponding to a certain value of radius r^*, as shown in Figure 15.1. In case of supersaturated

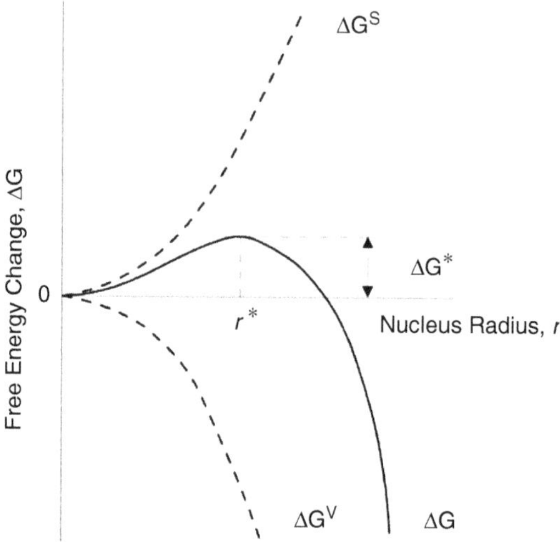

FIGURE 15.1 Variation of total free energy (ΔG) and free energies related to nucleus surface (ΔG^s) and nucleus volume (ΔG^v) as a function of the nucleus radius r.

solutions a certain number of clusters will statistically be generated due to microscopic temperature and concentration fluctuations but only those that will reach a dimension larger than r^* will have the possibility to grow. Those with size below r^* will probably dissolve again, therefore r^* is often referred to as critical radius. The maximum of ΔG can be obtained by taking the zero derivative $\partial(\Delta G)/\partial r = 0$, from which the critical radius results:

$$r^* = \frac{2\sigma\Omega}{\Delta\mu} \tag{15.10}$$

Combining the relations of Equation 15.9 and Equation 15.10 one obtains the critical nucleation energy for homogeneous spherical nuclei:

$$\Delta G^* = \frac{16\pi\sigma^3\Omega^2}{3(\Delta\mu)^2} \tag{15.11}$$

The heterogeneous nucleation is applied in most growth techniques, where seeds are employed to impose a preferred crystallographic orientation to the growing crystal. In this case the shape of the nucleus cannot be arbitrarily taken as spherical but depends on the seed morphology. Typically, when one has a flat substrate the nucleus can be approximated by a spherical cap (droplet) whose actual shape depends on the contact angle between the nucleus and the flat substrate. In any case, the free surface energy will be lower than for the homogeneous nucleation, which is reflected in a lower nucleation work:

$$f = \Delta G^*_{\text{Hetero}}/\Delta G^*_{\text{Homo}} < 1 \tag{15.12}$$

15.2.3 Interface Kinetics

Generally, crystals grow by the addition of "growth units," atoms, or molecules, from the nutrient phase. More details on growth processes may be found in Chapter 8 and Chapter 9. The path between the solid and mother phase is actually reversible and the rate at which the growth units join or leave the solid interface depends on external macroscopic factors such as supercooling or supersaturation as well as on the microscopic conditions of the crystal surface. In the growth from solution and from vapor it is simple to distinguish between atoms or molecules that are part of the solid or of the mother phase. This fact is not obvious in melt growth, particularly of metals, where the density difference between the solid and its melt is normally very small. In a few atomic layers at the interface, atoms can be either "solid" (vibrate about a lattice site) or "liquid" (have random trajectories). If the density of atoms is very different in the two phases, like in solution and vapor growth, the solid surface is likely to be smooth whereas in melt growth the occurrence of surface roughening is more probable. The growth rate, V, under the same macroscopic parameters, is very different in the two cases and is expressed as:

$$V = av^{+}Ju_k \tag{15.13}$$

where a is a distance proportional to the diameter of the growth units, v^{+} is the number of growth units that impinge the solid surface per time unit, J is a factor depending on the surface roughness and expresses the ratio between total surface sites and active sites, u_k is the normalized difference between growth units arriving and leaving the solid surface and is a function of thermodynamic conditions according to:

$$u_k = 1 - \exp \Delta\mu/kT \tag{15.14}$$

where $\Delta\mu$ is the difference of chemical potential between the two phases, that is, the thermodynamic driving force of the process.

Interface kinetics essentially addresses the construction of surface models containing steps, kinks, and singular (flat) areas and the way by which these surface elements evolve under different supercooling or supersaturation. The crystallographic structure and the growth conditions are thus important in determining the maximum growth rate on a given crystallographic plane. The shape of crystals grown freely from a supersaturated solution can be correctly interpreted by applying the correct kinetic models. This way, one can understand why crystals may assume completely different shapes (platelets, needles, cubes with or without edges, etc.) and can explain experimentally observed phenomena as step motion and step bunching on the surfaces of growing crystals.

15.2.4 Mass and Heat Transport

As a first-order phase transition, a certain amount of energy (latent heat of crystallization) is released during solidification when growth units are stably connected with the solid interface. If the heat is not driven away from the interface quickly enough the temperature will locally increase and the growth will not proceed regularly. Similarly, in the growth from solutions, a depletion of the solute concentration will occur in the vicinity of the interface, which can locally lead to insufficient supersaturation and stop of the growth. These examples show that in crystal growth processes energy and mass transport must be controlled for process continuity. The flows that occur are process and material specific and will not be treated in depth in this review. Some examples will be provided later for bulk crystal growth from the melt, solution, and vapor. Here we shall just mention that the heat transfer within the crystal (initially conduction from the interface to the colder volume of the crystal, then radiation and convection at the surface) is generally the rate-limiting factor in melt growth while in solution and vapor growth one can have either kinetically limited or mass transport-limited regimes.

15.2.5 Flow Dynamic and Morphological Stability

Practically, in all growth systems applied to bulk crystals there is a fluid (melt, solution or gas) in contact with the crystal. As the fluid is subjected to temperature fields convective flows normally take place. The fluid motion has important consequences for heat and mass transport especially in the melt growth of dielectric crystals where the convection heat transfer in the liquid phase is dominant. Vice versa, in metal melts the dominant heat transfer mechanism is conductivity, unless very strong melt flows are established due to excessive thermal gradients.

The most remarkable effect of the fluid motion is however given by the formation of a so-called "boundary layer," that is, a stagnant layer whose thickness is a function of the relative fluid-solid velocity and of the melt viscosity. In the case of solution or vapor growth, the boundary layer plays the role of physical barrier through which the growth species must diffuse up to reaching the solid surface, whereas in the melt growth it plays a fundamental role on the incorporation of dopants and impurities. The effective distribution coefficient K_{eff} of impurities is indeed governed by the well-known BPS[6] equation:

$$K_{eff} = \frac{K_0}{K_0 + (1 - K_0)\exp\left(-v_p\delta/D\right)} \tag{15.15}$$

where $K_0 = C_{sol}/C_{liq}$ is the equilibrium distribution coefficient, δ is the boundary layer thickness, D is the impurity diffusivity in the melt and v_p is the crystal growth speed. Equation 15.15 has two extremes: for ($v_p\delta \gg D$ then $K_{eff} \rightarrow 1$ and the solute transport is controlled by diffusion alone, while if $\delta = 0$ then $K_{eff} \rightarrow K_0$, that happens when the convective flows are very strong. Equation 15.15 also explains a typical feature of melt grown crystals: the so-called "growth striations," which are bands of impurities or composition bands (in multicomponent crystals) that are easily found when crystals are cut parallel to their growth axis. They are due to small variation of either v_p or δ produced by the thermal instabilities associated with convective flows. As the striations are undesired inhomogeneities, significant work has been devoted to damping the natural convective flows in melt growth. The experiments included improvement of the thermal stability and reduction of the gradients in the melt crucible, use of baffles within the melt, double-crucible arrangements, use of static and variable magnetic fields up to growth in microgravity conditions in space. For a detailed overview about this topic see Reference 7, Chapter 4.

It must be noted however, that melt convection has also some positive effects like, for example, the redistribution of dopants with $K_{eff} < 1$. These dopants are actually rejected at the growing front and give rise to a solute-rich layer before the solid–liquid interface. Solutes with $K_{eff} < 1$ lower the melting point of their solvent so that the equilibrium melting point away from the interface is higher than that in the proximity of the interface. Let us consider the solute concentration profile of Figure 15.2a to which corresponds the distribution of melting temperatures of Figure 15.2b. If the temperature gradient in the melt has a behavior like that depicted by the line A, a certain volume of liquid will be at temperatures below the thermodynamic equilibrium (liquidus line of the phase diagram) and therefore supercooled. This phenomenon, named "constitutional supercooling,"[8,9] is well known in the case of heavily-doped melts where an insufficient mixing leads to build up of a strong concentration profile.

The onset of constitutional supercooling brings about a type of morphological instability usually defined as cellular growth. Indeed, the statistical fluctuations of a planar solid–liquid interface can produce small protrusions that extend up to the supercooled region where they will grow faster. In doing so these protrusions will segregate laterally some solute, which in turn further lowers the melting point of the areas between the protrusions themselves, which finally will trap much more solute than the protrusions.

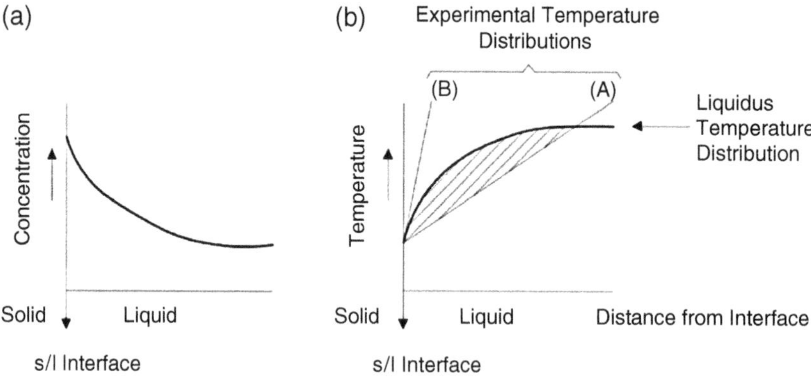

FIGURE 15.2 (a) Formation of a concentration gradient in front of the solid–liquid interface for impurities with $K_{eff} < 1$ and (b) corresponding liquidus temperature distribution.

FIGURE 15.3 Interface instability caused by constitutional supercooling during the Czochralski growth of a GeSi single crystal with 2% at Si. The crystal was pulled in the ⟨112⟩ direction (Courtesy of IKZ — SiGe group).

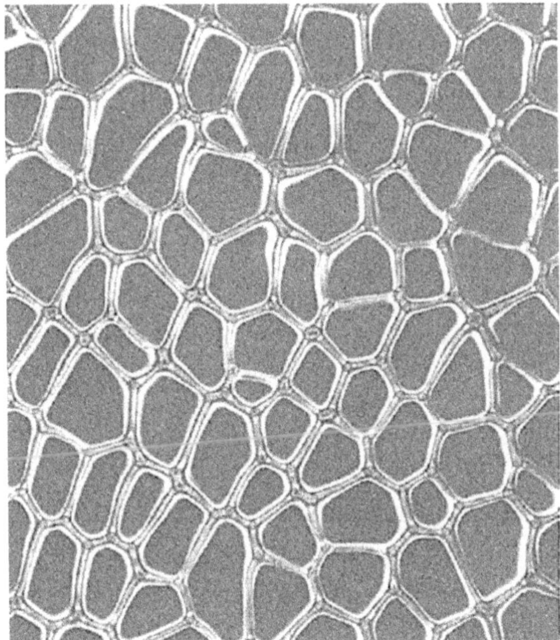

FIGURE 15.4 "Honeycomb" structure found on the cross-section perpendicular to the growth direction of a Sn crystal with 0.1% mol Sb doping.

The final result may be an ondulated interface like the one in Figure 15.3, which shows a portion of crystal cut parallel to the growth direction. A cut perpendicular to the growth direction instead shows a typical honeycomb structure like the one reported in Figure 15.4. There are several possibilities to prevent the onset of constitutional supercooling: increase the temperature gradient in the melt (line B in Figure 15.2b), reduce the growth rate so that the solute rejected at the interface will have time to redistribute into the bulk of the liquid and, finally, maintain a certain degree of beneficial convection in the melt.

This section has treated in a very simplified form the phenomenology of the crystal growth processes. In most cases the rigorous treatment is material-specific and does not lend itself to generalization. However, the few notes above should be sufficient to follow the subsequent description of the crystal growth methods and the discussion about the growth parameters–properties relationships of some chosen materials.

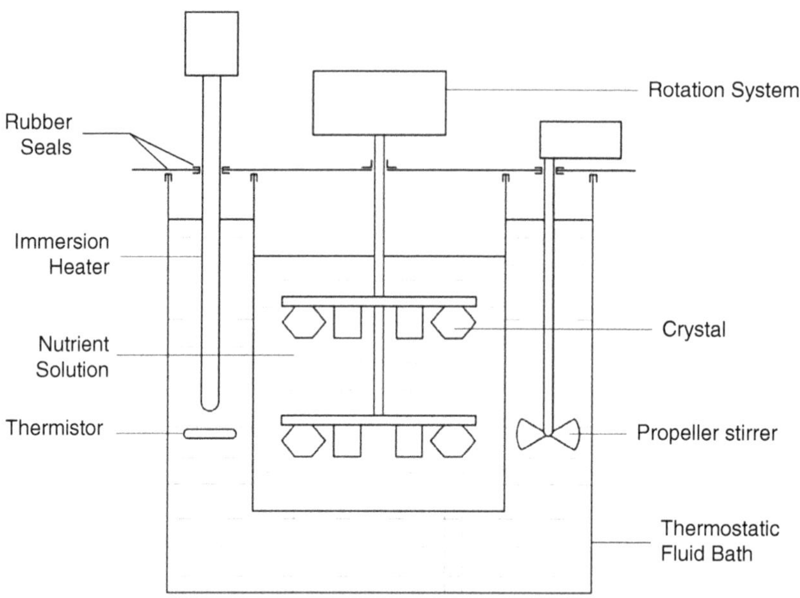

FIGURE 15.5 Schematic of low-temperature solution growth set up.

15.3 Solution Growth

Many types of crystals may be grown from supersaturated solutions. The solvents are normally water, organic liquids, metals, or mixtures of some chemical compounds. Aqueous and organic solutions are liquid under usual temperature and pressure conditions while for metals or chemical compound mixtures the temperature and pressure conditions may vary considerably. Consequently one can distinguish between low-temperature (<80°C) and high-temperature (up to 1500°C) solution growth. The upper limit is often set by the boiling point of the solvent used.

15.3.1 Low-Temperature Solution Growth

This popular method is often applied to substances that do not melt congruently, decompose below the melting point, or undergo phase transformation when moved from the crystallization temperature to room temperature. It is also worth mentioning that many crystals that form in the human body (kidneys stones, cholesterol, etc.) are the result of solution growth processes. The low-T solution growth has several interesting features and advantages: (1) the growth equipment is very simple and relatively inexpensive (see Figure 15.5), (2) the presence of defects induced by thermal stress is negligible, (3) the interface kinetic plays a fundamental role and the crystals have normally well-defined faces (typical growth habit). The supersaturation, that is metastability, is achieved in different ways: by solvent evaporation, by solution cooling, by imposing a temperature gradient across the solution. In all cases the growth rate is small. In solutions with high solute concentration (high solubility) the rate is essentially diffusion-controlled and the supersaturation at the solid–liquid interface is not uniform; it is higher close to the corners and edges and lower in the center of large facets. Consequently, the nucleation will occur more easily at the crystal edges and the growth steps will then run over the flat surfaces. Impurity segregation and defects will thus be maximized at the center of the faces, where the steps will coalesce. Moderate stirring of the solution, without inducing turbulence, can hinder this negative effect.

The choice of the proper solvent is always critical. At the growth temperature, the solvent must guarantee a sufficient solubility while having low volatility and low viscosity. A solubility in the 15 to 60% range is in most cases preferable as very concentrated or very diluted solutions lead to extremely low growth rates

FIGURE 15.6 Examples of large KDP crystals grown from solution. (After Li, G., Su, G., Zhuang, X., Li, Z., and He, Y., *J. Cryst. Growth* 269, 443–447, 2004, with permission from Elsevier).

owing, respectively, to diffusion-controlled or insufficient solute transport. Whenever possible water is chosen as it is cheap, very pure, inert to plastic or glass containers, nontoxic, or flammable and can be used over a wide temperature range.

Given the intrinsically low rate of crystallization, typically of the order of 1 to 2 mm/day, the low-temperature solution growth is mostly applied at the laboratory scale to produce new types of organic or inorganic crystals as well as for crystallization of complex biological molecules. However, today this technique is applied for the preparation of organic or inorganic crystals (e.g., large sizes of potassium dihydrogen phosphate (KDP- see Figure 15.6), potassium tytanil phosphate (KTP), and ammonium dihydrogen phosphate (ADP) for nonlinear optical applications[10–12] and to crystallization of biological macromolecules.[13] Although in principle simple, the solution growth requires some skills especially for the preparation of saturated solutions, their filtration and ageing, the addition of chemicals, which can enhance the metastable region width or inhibit the bacteria proliferation, the preparation and mounting of seeds.

15.3.2 High-Temperature Solution Growth

The high-T solution growth (or flux growth) allows the crystallization of many compounds at temperature much lower than their melting point. This is advantageous in terms of lower dislocation density and avoiding phase transitions upon cooling. This process has been reviewed in some books and articles[14,15] that can be useful references for the reader. Table 15.1 gives an overview of crystals grown from high-temperature solutions along with their solvent (flux). The latter can be a low melting point metal, an oxide or an ionic salt whose melting point is typically in the range 400 to 1000°C. A major disadvantage of this method is the high reactivity and corrosive nature of most fluxes that require the use of special furnaces with quartz or platinum liners. The flux must meet the following requirements: (1) high solubility of the crystal constituents with positive temperature coefficient for the solubility, (2) the required solid phase must be the only stable solid phase, (3) small vapor pressure at the growth temperature, (4) low viscosity, (5) low melting point, (6) availability in the needed purity and low cost, (7) low reactivity with the crucible and furnace materials, (8) easy to separate from the grown crystal after the final cool down, (9) low volatility and toxicity. It is almost impossible to find a solvent that possesses all these properties and therefore a compromise must be found. Successful results were so far obtained by employing lead or bismuth compounds, borates, vanadates, tungstates, molybdates, alkali halides, and carbonates (see Table 15.1).

High-T solution growth is in principle similar to low-T, in the sense that nucleation and subsequent crystal growth may be obtained by ramping the temperature down, by solvent evaporation or by establishing a temperature gradient across the solution. One problem that is however more difficult to solve is the solution seeding. Indeed if no seed is provided the spontaneous nucleation will produce a number of small

TABLE 15.1 List of Common High-Temperature Solvents, Their Melting Points and Examples of Oxides Crystallized Using These Solvents

Flux composition	Melting point/eutectic (°C)	Crystallized oxides
BaO-B_2O_3	870	$Y_3Fe_5O_{12}$
BaO-CuO	1015	$YBa_2Cu_3O_7$
BaO-Bi_2O_3-B_2O_3	600	$NiFe_2O_4$
$4Bi_2O_3$-V_2O_5	900	$CdCr_2O_4$, Cr_2O_3
KCl	776	$Bi_{2.2}Sr_2Ca_{0.8}Cu_2O_y$
Li_2O-MoO_3	532	$Be_3Al_2Si_6O_{18}$
NaCl	801	$Pb_2Sr_2(Y,Ca)Cu_3O_8$
$Na_2B_4O_7$	741	$NiFeO_4$
$Na_2W_4O_7$	620	RWO_6
PbF_2	855	$Al_2O_3:Cr^{3+}$
PbO-B_2O_3	500	$LaBO_3$
PbO-PbF_2	494	$RAlO_3$
PbO-PbF_2-B_2O_3	500	$R_3Al_5O_{12}$
$2PbO$-P_2O_5	500	RPO_4
$2PbO$-V_2O_5	824	RVO_4
PbO-PbF_2-V_2O_5	720	$MgCr_2O_4$
$Li_4B_2O_5$	726	$La_{2-x}Sr_xCuO_4$
CuO	1026	$Nd_{2-x}Ce_xCuO_4$, $(LaSr)_2CuO_4$

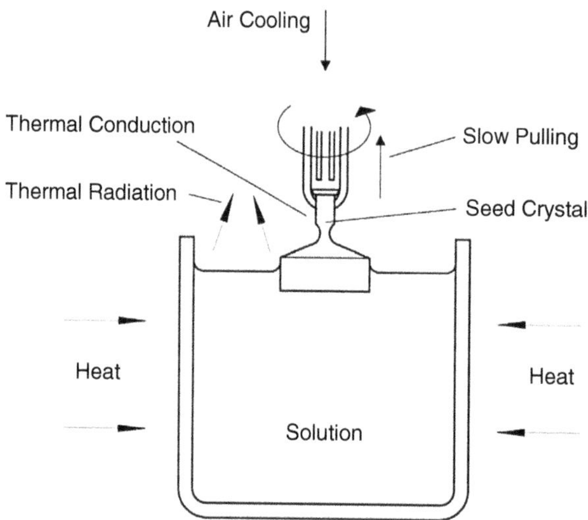

FIGURE 15.7 Schematic representation of a typical top-seeded solution growth equipment.

seeds that will limit the maximum size of the corresponding crystals. In low-T case dealing with aqueous solutions it is relatively simple to position one or more seeds in the bulk of the solution (Figure 15.5) but the problem becomes complicated when the solution must be heated to 1000°C. Top-seeded solution growth was successfully applied to the preparation of numerous garnets and perovskites. In the temperature gradient method with top seeding shown in Figure 15.7, a seed is put in contact with the upper surface of the solution maintained at the temperature T_1 and the bottom of the crucible kept at T_2. Solute transport to the seed is ensured by natural or forced convection (rotation of the crucible at constant or variable rates). A further advantage of the method is that the separation of the crystal from the solution is easy. Alternatively, there are closed crucibles that contain the solution in the lower part and have a seed fixed in the cover. These systems are turned 180° twice, once for melting, stabilizing the solution and crystal growth, then to separate the crystal from the mother bath.[16]

Flux growth has been applied to a variety of compounds such as: KTP, β-barium borate (BBO), lead zirconium titanate (PZT) and other ferroelectric compounds like $PbZ_{1/3}Nb_{2/3}O_3$, $PbSc_{1/2}Nb_{1/2}O_3$, $PbNi_{1/3}Nb_{2/3}O_3$, superconductors like YBCO, or different garnets like $Gd_3Ga_5O_{12}$ (GGG), and $Y_3Al_5O_{12}$ (YAG).

15.3.3 Hydrothermal Growth

In the high-T solution growth above, the flux is normally maintained at atmospheric pressure. There is however a variant that makes use of high-pressure high-temperature solutions. This is called hydrothermal growth, popular among geologists as it describes the action of water at high pressure and temperature on the formation of different crystals in the terrestrial crust. In crystal growth, the term hydrothermal refers to the use of autoclaves in which aqueous (sometimes nonaqueous) solutions containing a mineralizer are kept at high pressure and high temperature in order to dissolve, first, and then recrystallize substances that are insoluble under ordinary (room) conditions. This method was first applied to growth of quartz crystals. Many other oxides and compounds have been successfully prepared by this method: TiO_2, ZrO_2, Al_2O_3, Fe_2O_3, diamond, KTP and several different vanadates, tungstates, and phosphates. For a collection of articles regarding hydrothermal growth see Reference 17.

The autoclave is the core instrument for this growth process, capable of bearing pressures in the 5 Bar to 500 kBar range and reaching temperatures over 1500°C. An example of autoclave for industrial production of α-quartz is reported in Figure 15.8. Depending on the acidic or alkaline nature and thermodynamic

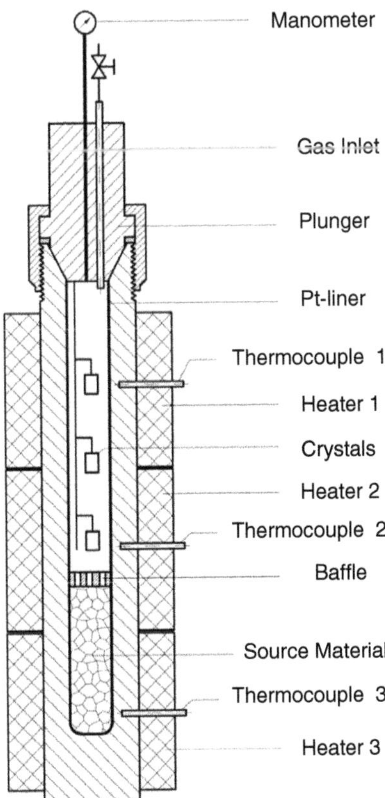

FIGURE 15.8 Autoclave assembly for the hydrothermal growth of α-quartz.

conditions of the solvent, a specific liner, made of Pyrex, quartz, silver-coated quartz, platinum, is interposed between the autoclaves walls and the crucible.

Because of its good piezoelectric properties, α-quartz is certainly the material most extensively studied among the many hydrothermally grown. The $\beta \rightarrow \alpha$ transition occurs at 573°C thus the direct growth of the α -phase must be carried out at temperatures below 570°C. On the other hand, the solubility of quartz at ordinary pressures and temperatures is negligible and only hydrothermal conditions provide a decent growth rate. The typical conditions are: dissolution temperature (bottom of the autoclave) of about 420°C, growth temperature (upper part of the autoclave) of about 375°C, pressure in the range 1000 to 1700 Bar, filling of the internal autoclave volume between 80 and 85%, and mineralizer NaOH with 0.5 to 1.0 M concentration. The baffle positioned between the saturation and the crystallization zones plays an important role, as it controls the convection and the mass transport toward the seeds, that is, the growth rate. Under these conditions the growth rate along the (0001) orientation is about 1 to 1.2 mm/day.

Other technologically important materials like gallium orthophosphate ($GaPO_4$), a piezoelectric material with $\beta \rightarrow \alpha$ transition at 976°C, were also grown by hydrothermal method. However, as $GaPO_4$ has a retrograde solubility, the arrangement is reversed with respect to quartz, that is, the seeds are placed at the bottom (higher T) and the nutrient in the upper part (lower T) of the autoclave, with a gradient of 20 to 40°C.

Recently, the hydrothermal method was successfully applied to the growth of large (up to 50 mm) crystals of ZnO,[18] a wide bandgap semiconductor with important electro-optical applications. Experiments are presently carried out to grow bulk nitrides (AlN, GaN), needed for the mass production of blue and green solid-state lasers. There is indeed a huge effort worldwide to grow GaN and AlN substrates with good electrical and crystallographic properties. The growth process from ammonia solutions[19] is however complicated and so far has not supplied crystals with the necessary size and purity. Therefore, other high-pressure approaches are attempted as reported below.

15.3.4 High-Temperature High-Pressure Solution Growth Processes

A main problem in the development of nitride-based lasers is the lack of suitable lattice-matched substrates. Devices fabricated on hetero-substrates contain a very high density of dislocations, of the order of 10^9 cm^{-2}, which limit the device yield and shorten the device lifetime. The preliminary attempts of homoepitaxy on tiny GaN substrates obtained by high-pressure solution growth demonstrated that the dislocation concentration is reduced by about three orders of magnitude, thus making more realistic the idea of long lifetime lasers. Such results triggered a lot of research in the field of bulk GaN, in spite of the thermodynamic properties of this semiconductor (melting point >2400°C, dissociation pressure at the melting point >40 kBar) that do not allow the use of traditional melt growth methods. An original high-temperature high-pressure solution growth approach has specifically been proposed for GaN, carried out in autoclaves with crucibles of about 30 mm diameter, under maximum pressure of 20 kBar and temperature of <1800°C.[20] The furnace can be accurately controlled to stabilize the temperature within 1°C and is designed to provide the temperature gradient necessary for the solution growth. A charge of gallium is contained in a boron nitride crucible while the growth chamber is filled with pure nitrogen. The nitrogen is reactant and pressure transmitting medium at the same time.

To understand the growth process, one must first analyze the phase diagram of GaN. Figure 15.9 shows the liquidus lines for nitride systems that is, the solubility of nitrogen into the different metals for increasing temperatures. The broken lines fix the experimental limits of the solution growth: the maximum fraction of nitrogen dissolved into liquid Al is set by the available experimental temperature of about 2000 K (i.e., below 0.1 at.%); on the other hand, in the case of InN, the maximum fraction is even smaller because the available pressure of 20 kBar is not sufficient to prevent the almost total decomposition of InN. The situation is more favorable for GaN as at 2000 K the decomposition pressure is about 20 kBar, therefore the experimental conditions, although critical, allow the achievement of a Ga:GaN solution saturated at about 1% at, from which GaN single crystals may be obtained, upon application of an appropriate temperature gradient.

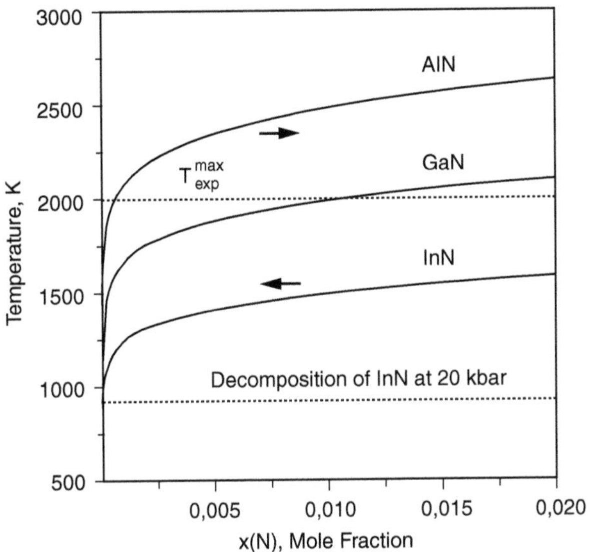

FIGURE 15.9 Liquidus lines for the Ga–GaN, Al–AlN, and In–InN systems. The horizontal dotted lines indicate that the main obstacle to solution growth of AlN is given by the maximum experimental temperature while for InN it is given by the maximum experimental pressure; in both cases it is impossible to dissolve an appreciable mole fraction of nitrogen. The situation is more favorable for the Ga–GaN system as the dissolved nitrogen can reach a molar fraction of about 1%.

In practice, the GaN crystals are grown under N_2 pressures of 10 to 17 kBar at temperatures in the 1500 to 1600°C range and gradients of 30 to 100°C/cm across the crucible. Low gradients correspond to low supersaturation conditions, whereas large gradients correspond to high supersaturation, that is, faster growth. The crystals grown under high supersaturation are bigger plates (with a surface of up to 200 mm^2) but of poorer crystallographic quality. At low supersaturation the growth rate is slow (around 0.1 mm/h). The high-pressure solution growth proceeds according to the well-known synthesis solute diffusion (SSD) mechanism: a thin polycrystalline GaN film forms on the external gallium surface (hottest part of the crucible) and is subsequently dissolved and transported as solute up to the colder part of the crucible where it may overcome the saturation threshold and initiate the nucleation of GaN seeds that are then enlarged by the incoming GaN solute. GaN crystals usually assume the shape of platelets with hexagonal boundaries and have a thickness of about 200 μm; the upper and lower mirror-like surfaces are (0001) planes with 20 to 40Å residual steps. The structural quality of these GaN crystals is very good as the width of the x-ray rocking curves for the (0004) Cu Kα reflection is below 30 arcsec, to be compared with values of $\geq$100 arcsec normally found in GaN epitaxially grown on foreign substrates, while the dislocation density is only 10^5 cm^{-2}, three to four orders of magnitude less than in conventional hetero-epitaxial layers. At high supersaturation the growth rate may reach 1 mm/h and, in particular when the size of the crystal is comparable with the crucible, the layer-by-layer growth mechanism is strongly disturbed. This leads to the formation of macrosteps on the (0001) surfaces or, in the worst cases, to dendritic growth.

15.4 Melt Growth Techniques

Melt growth is the most important industrial crystallization method. Most commonly, a given degree of supercooling must be maintained in the melt immediately before the solid–liquid interface, which implies a temperature gradient in the melt (between the interface kept approximately at the melting point and the bulk of the melt that is a few tenths of °C above the melting point). For a stable growth process, one must impose a temperature gradient across the growing solid capable of driving away the latent heat of solidification. Consequently, metals and semiconductors, which have high heat conductivity, may be grown at higher rates than dielectric crystals, for example oxides and halides. In case of multicomponent or heavily doped melts, however, the maximum growth rate may be decided by the occurrence of constitutional supercooling (see Section 15.2). The numerous melt growth techniques may be essentially subdivided in three groups: (1) pulling from the melt, (2) directional solidification in containers, (3) zone melting. They will be described in the following sections.

15.4.1 Pulling from the Melt

15.4.1.1 Czochralski Technique

In this technique, a melt contained in a heated crucible is contacted with an oriented seed attached to a cooled pulling shaft that is subsequently withdrawn leading to crystallization of an oriented boule. Depending on the material to be grown, the crucible can be made of quartz, alumina, graphite, or metals like platinum or iridium while the heating can be provided by a radio frequency coil or by a resistive heater surrounding the crucible. Most of the electronic silicon is produced by the Czochralski method, invented in 1916.[21] Silicon crystals with diameters up to 300 mm, ⟨100⟩ or ⟨111⟩-oriented, free of dislocations and weighing about 200 kg are routinely grown for the electronic industry. The schematic of the Czochralski (CZ) technique is reported in Figure 15.10. In the case of silicon, the heating of the quartz crucible is by electrical resistance and the seed is a small bar with a cross-section of about 1 cm^2. Crucible and seed are normally counter-rotated so that the crystal assumes a cylindrical shape. The first part of the crystal, 10 to 20 cm, is pulled at a high rate to form a thin round neck of 3 to 5 mm diameter by which the dislocations created by thermal shock at the time of seed dipping are readily eliminated. This step is crucial as it enables the further growth of an entirely dislocation-free boule.[22] The growth atmosphere is very important as it often determines the residual impurity concentration in the crystals. Argon is normally employed for Si crystal growth. Other inert gases are used in the growth of other semiconductors or metals, reducing atmospheres that are necessary for the growth of alkali halides; while oxygen, CO, or CO_2 are used for the pulling of oxides, some materials can be grown under vacuum.

A peculiarity of the CZ method is that the crystal grows freely, without contacts with solid walls, which avoids dangerous physical constrains due to different thermal expansion coefficients, that is, crystal

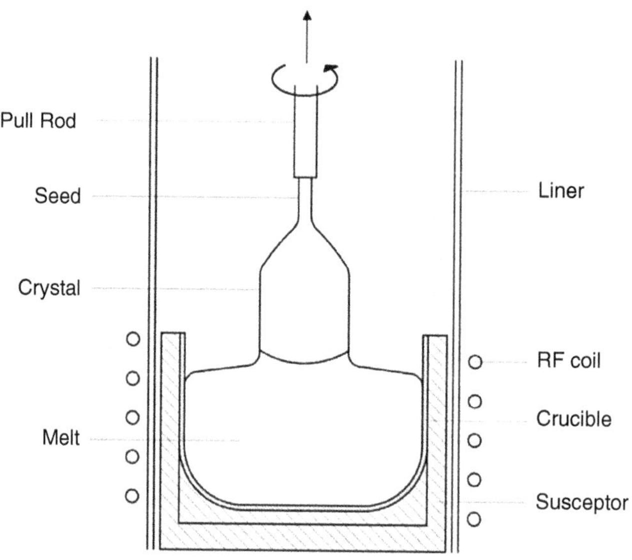

FIGURE 15.10 Schematic of Czochralski growth system with radio frequency heating.

deformation and generation of dislocations. At the same time, to save production cost and facilitate the crystal processing (cutting and polishing), it is highly desirable to have perfectly cylindrical ingots, which poses strict requirements for the crystal diameter control. As discussed in Section 15.2, the removal of the latent heat of solidification in most cases controls the growth rate. For a cylindrical crystal pulled at constant rate, this implies that the heat removal through the solid must exactly compensate for the latent heat released at the interface, according to:

$$k_s G_s - k_l G_l = Lv \tag{15.16}$$

where k_s and k_l are the thermal conductivities of crystal and melt, respectively, G_s and G_l are the thermal gradients, L is the latent heat, and v the pulling rate. However, owing to the different thermal boundary conditions in the course of the growth (strong radiation from the large free surface of the melt at the beginning, and radiation from crystal surface, which masks the liquid when the cone is completed and gas convection as the ingot length increases) the two gradients will vary considerably and the equilibrium condition (Equation 15.16) will be lost. To maintain L constant (cylindrical growth at constant pulling speed), one must therefore act on the power supplied to the crucible. Many efforts have thus been devoted to quickly detect even the smallest diameter variations. A central factor is the shape of the solid–liquid interface and the behavior of the liquid meniscus. The interface shape reflects the isotherm patterns in the liquid and is often slightly convex to the melt. This shape is, as experimentally observed, also favorable to reduce the occurrence of twins in many semiconductor and oxide crystals. The liquid meniscus instead is connected with the incomplete wetting of a solid by its own melt. Figure 15.11a schematically shows what happens at the solid–liquid transition region. The condition of incomplete wetting is expressed by the inequality:

$$\sigma_{SG} < \sigma_{SL} + \sigma_{LG} \tag{15.17}$$

where σ_{SG}, σ_{SL}, σ_{LG} are the crystal-gas, crystal-melt, melt-gas interfacial energies at the three phase boundary. The incomplete wetting gives rise to a contact angle Φ_0 characteristic of any given substance; for example, it is 11° for silicon, 13° for germanium and about 19° for sapphire. The complete wetting, on the other hand, is expressed by a contact angle of 0° and is observed in several metals. The value of Φ_0 ultimately depends on the surface tension of the melt and on the difference of densities between the solid and the melt.

The profile of the liquid meniscus for a cylindrical ingot is a function of the angle Φ_0 and of the presence of external fields (e.g., gravity) and is represented by the solution of a complex Laplace–Young differential equation. Useful approximate solutions to this problem are provided by Hurle.[23] For noncylindrical crystals the meniscus profile (i.e., height) depends, in addition to intrinsic material characteristics, on the growth angle Φ_G as graphically shown in Figure 15.11b. Crystal radius changes are immediately reflected in changes of the meniscus profile. Therefore, an effective diameter control may be achieved by taking the liquid meniscus as control variable. This may be practically done by directing a laser on the meniscus and

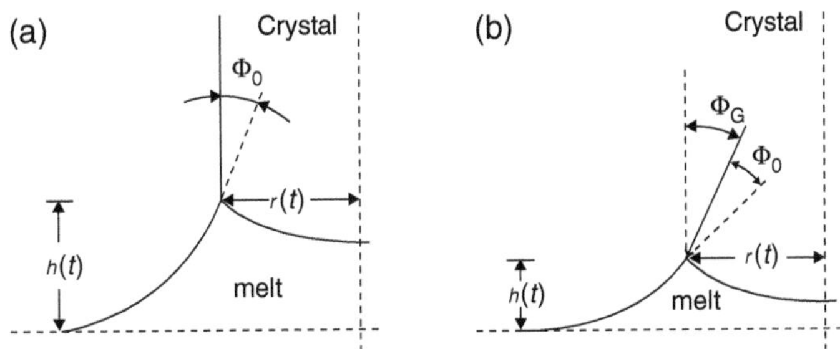

FIGURE 15.11 Meniscus profile in (a) a cylindrical crystal and (b) a crystal growing with increasing diameter.

detecting the corresponding reflected beam. Alternatively, one can monitor the meniscus shape by x-ray or, in a simpler way, by camera imaging the extension of the bright ring associated to the meniscus surface (the lifted surface reflects the incandescent crucible walls above the melt). The collected information is fed back to the power supplier to maintain a constant diameter.

When these optical techniques are much less sensitive (e.g., for oxides with low reflectivity) or even inapplicable (e.g., for compound semiconductors because of the presence of the liquid encapsulant, as discussed below), crystal weighing is used for an automatic diameter control.[24–26] Here the crystal weight, or its derivative, is taken as a control variable and used to drive the power supplier by applying suitable control algorithms.

15.4.1.2 Liquid Encapsulated Czochralski Growth

The CZ technique is applied to metals, silicon, germanium, and some oxides but never to compound semiconductors like GaAs, InAs, GaP, InP, GaSb, InSb (the so-called III–V compounds) as they dissociate at the melting point. The vapor pressure of the group V element ranges from below 1 Bar (as in the case of antimonides) to about 35 Bar (for GaP), and the volatile element would condense on the cold steel walls of the chamber. To avoid the melt decomposition, the growth system must be closed, as in the Bridgman configuration (see below Section 15.4.2), where the starting charge is contained in a heated sealed quartz tube, or virtually closed by using seals that effectively prevent the molten charge dissociation. The most popular version of the latter system is given by the Liquid Encapsulation Czochralski (LEC) pulling proposed by Mullin and coworkers.[27] A large fraction of the commercial GaAs wafers and most of the InP wafers are produced by the LEC technique.

The LEC technique is conceptually similar to the standard CZ except for the presence of a liquid encapsulant (boric oxide, B_2O_3) that completely seals the molten charge and prevents the escapes of the volatile elements (see Figure 15.12). The choice of boric oxide as the most popular encapsulant derives from the properties of this compound: lower melting point than that of III–V semiconductors and also lower density so that it can cover completely the charge before escape of the volatile element. It is chemically inert to the melt and transparent thus allowing the viewing and monitoring of seeding and pulling. Liquid encapsulation is often used in connection with high-pressure pullers, which makes

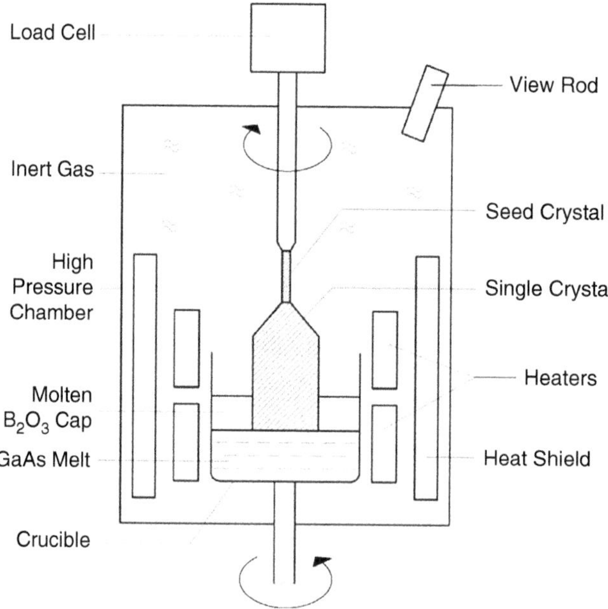

FIGURE 15.12 Liquid Encapsulated Czochralski growth of GaAs.

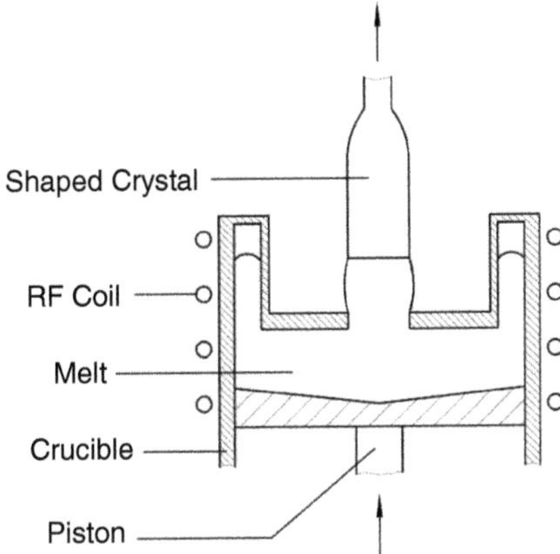

FIGURE 15.13 Shaping a crystal using a nonwetting die (Stepanov method).

the growth of phosphides (InP and GaP) possible and, in addition, allows the so-called synthesis *in situ*[28] of gallium and arsenic. This technology paved the way to the development of integrated circuits for high-speed telecommunications.

15.4.1.3 Stepanov and Edge-Defined Film-Fed Growth Techniques

As shown above, the pulling from a melt offers the attractive possibility of shaping the crystal during the growth (i.e., large semiconductor and oxide crystals may be grown with good cylindrical shape by using some close-loop control systems). Another approach consists in using a die floating on the melt or rigidly connected to the crucible. Two classes of dies have been employed: wetting and nonwetting. The pioneering work with nonwetting dies by Stepanov[29] in the former USSR was later applied to the growth of shaped semiconductor single crystals.[30] An example of a nonwetting system is given in Figure 15.13: a pressure is applied to the melt with a piston to force it to raise the die and achieve a control of the meniscus shape. The corresponding solid shape is relatively independent of small changes (piston movement, temperature fluctuations) in the height of the meniscus.

The use of wetted dies was introduced by La Belle and Mlansky[31] who named it edge-defined film-fed growth (EFG). The main applications of this technique are currently in the production of sapphire tubes with variable cross-section and silicon ribbons and polygons for solar applications (e.g., regular octagons with face width of about 10 cm, thickness of 200 to 400 μm and length over 400 cm). The principle of the technique is illustrated in Figure 15.14. The meniscus-shaping capillary forces control the shape and dimension of a crystal. The melt tends to rise along the capillary and to wet the upper surface of the die so that the shape of the meniscus, that is crystal, will ultimately be given by the upper surface of the shaper. Note that the shaper is normally floating on the melt surface and that, due to capillarity, the crystallization front is well above the surface of the remote melt in the crucible. The essential difference between wetting and nonwetting shapers is that in the former case the crystal shape is determined by the walls of the die while in the latter by the upper edges. A complete review about the pulling from the melt using different shapers was given by Tatarchenko.[32]

15.4.2 Directional Solidification in Containers

Slow directional solidification of melts in horizontal boats or vertical crucibles was pioneered by Bridgman,[33] and later perfectioned by Stockbarger.[34] The vertical method with a two-zone furnace having the upper zone at $T >$ melting point and lower at $T <$ melting point (Figure 15.15.a) is applied today for the preparation of different compound semiconductors and oxides. Horizontal crystallization in boats has also extensively been applied to the growth of compound semiconductors, in particular before the

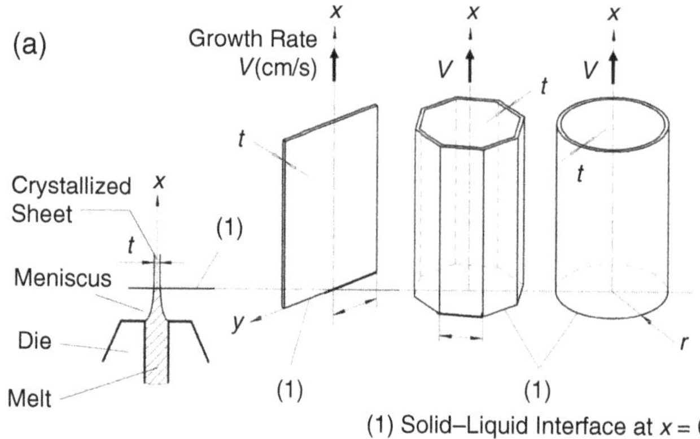

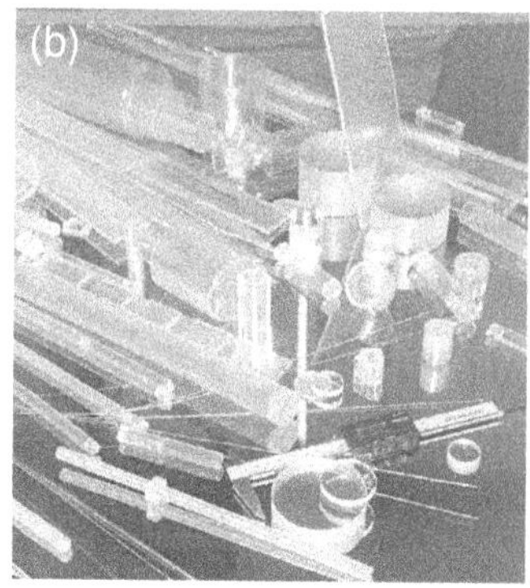

FIGURE 15.14 (a) Growth of sheets and tubes with thickness t by the EDGF method; (b) Examples of sapphire crystals with different shapes grown by using dies. (Reprinted from Kurlov, V.N., *Encyclopedia of Materials: Science and Technology*, Elsevier 2001, p. 8259, with permission from Elsevier).

development of the LEC technology (Section 15.4.1.2). The reason is that is was relatively easy to seal the boat and the reservoir of the volatile element in a quartz tube held in a two-zone furnace with the lower T zone at the temperature, which gave exactly the equilibrium vapor pressure of the volatile element at the melting point (see Figure 15.15b). All Bridgman–Stockbarger techniques have the common feature of requiring the translation of either furnace or crucible to obtain the phase transition by shifting the molten charge across the desired thermal gradient. Mechanical movement, even well controlled, may induce vibrations or small instabilities at the microscopic scale, therefore systems where the solidification is achieved by a change in the thermal profile of the furnace are preferred. This is achieved by building multizone furnaces (with 20 to 40 independently controlled zones) capable of providing a dynamic temperature profile. The vertical ampoule containing the charge (eventually encapsulated with boric oxide) is steadily positioned within the furnace while the profile changes with the time to initiate the solidification from the seed and continue until a full single crystal ingot is obtained. This system (dynamic gradient freeze, DGF) is today the most popular technique for the industrial growth of GaAs and used also for InP and GaSb.[35,36]

Nowadays, the horizontal Bridgman (HB) is mainly used for the synthesis of polycrystals as the D-shaped wafers cut from the single crystals are not readily usable in the modern electronic industry. On the contrary, the VB (Vertical Bridgman) is very popular at the moment, particularly for binary and multinary II–VI compounds like CdTe, CdZnTe, CdMnTe.[37] This is due to some specific peculiarities of the system: (1) it gives cylindrical crystals with no need of sophisticated diameter control devices, (2) the growth interface can be made very flat with evident advantages in terms of wafer uniformity, (3) the thermal gradients can be easily minimized with a consequent reduction of the dislocation density, (4) the melt composition can be controlled during growth by acting on the temperature of the volatile element reservoir.

Twin-free low dislocation density III–V single crystals have been prepared by VB and DGF. It must however be stressed that the yield of these processes is lower than for the classic LEC growth. In particular, several authors reported that twins occur very often in (100)-seeded crystals, especially in InP. To improve the yield of (100) VB-grown InP, Matsumoto et al. used liquid encapsulation and flat crucibles with a seed having the same diameter of the ingot to be grown.[38] This way the formation of the cone, which is a most critical zone for crystal twinning, may be avoided.

To conclude with directional solidification, the main advantages of this technology are in the possibility of very accurately controlling the thermal gradients in the liquid and in the solid, which implies a relatively good control of the interface shape, of the melt convection, and a low thermoelastic stress in the crystal. The main drawback is the relatively high twinning probability.

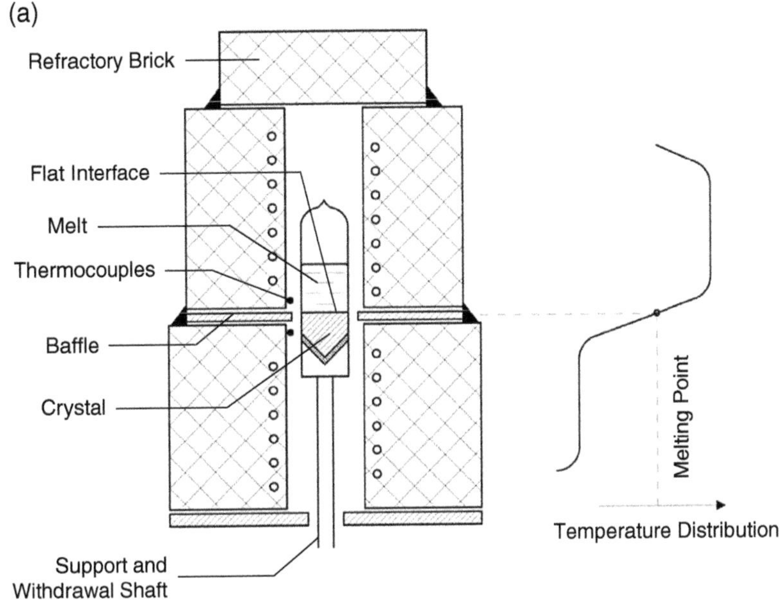

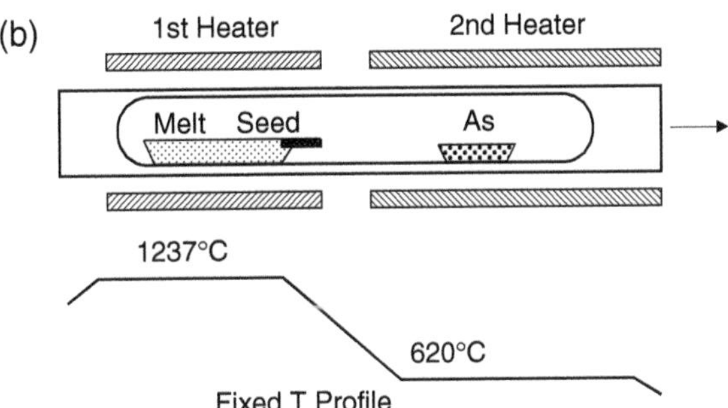

FIGURE 15.15 (a) Vertical Bridgman–Stockbarger assembly for the growth of single crystal within a crucible and corresponding temperature profile; (b) Horizontal Bridgman growth of GaAs.

15.4.3 Growth from the Melt without Use of Crucibles

15.4.3.1 Floating-Zone Technique

The floating-zone (FZ) technique is very popular in the bulk silicon industry to grow crystals without using a crucible. The absence of any contact between the melt and other materials results in Si single crystals of the highest purity. Schematically, the starting polycrystalline material as a free-standing rod is clamped at its upper end to a vertically mobile holder, while the other extreme of the feed rod is conically shaped and heated by a radio frequency coil (inductor) up to melting its tip that is then contacted with an oriented seed connected to the lower pulling shaft (Figure 15.16a). When a stable solid/melt contact is established the seed and the feed rod are simultaneously shifted downward to maintain a narrow molten zone corresponding to the fixed inductor and to crystallize a single crystal. The diameter of the inner hole of the inductor is normally smaller than the diameter of the feed rod and single crystal; therefore this method is known as needle-eye technique. For cylindrical ingots the seed and the feed rod are usually rotated. Generally, the diameter control in the FZ technique is much easier than in standard Czochralski

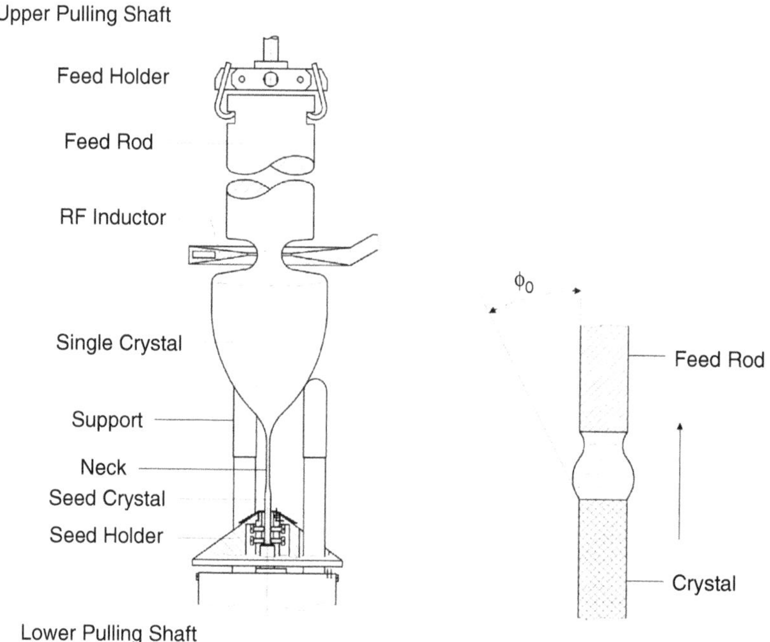

FIGURE 15.16 (a) Floating-zone (FZ) growth of silicon; (b) Shape of the molten zone under the effect of the gravitational force when both feed rod and crystal are translated downwards.

pulling due to an intrinsic dynamic stability that favors the self-compensation of growth perturbations, as discussed below.

It is important to mention that, by setting different speeds for feed rod and crystal, one can "squeeze" or "stretch" the crystal, that is, obtain a diameter respectively larger or smaller than that of the starting rod. A simple relation holds between the radius of feed rod r_f, the radius of the crystal r_c, the density of feed rod and crystal ρ_f, ρ_c and the speed of the upper and lower pulling shafts v_f, v_c:

$$v_f/v_c = (r_c/r_f)^2 \rho_c/\rho_f \tag{15.18}$$

The external shape and stability of the molten zone is crucial in the floating-zone technology and is the result of two contrasting factors: the gravity and the melt surface tension. In general, the molten zone will be more stable in materials with low density and high surface tension. In presence of a gravitational field a zone moving upward (equivalent to seed and feed rod being lowered) will assume a drop-like shape, as shown in Figure 15.16b. For a cylindrical crystal the angle between the liquid and the vertical side is the material-specific contact angle Φ_0, similar to that in the Czochralski growth (Section 15.4.1.1). The meniscus shape (upper and lower curvature of the drop) may be also found by taking a Laplace–Young equation with relevant boundary conditions. For a more detailed mathematical treatment, the interested readers can refer to Reference 39 and Reference 40. We shall simply consider the numerical solution of the Laplace–Young equation for a rotationally symmetric meniscus, which may be expressed as a function of dimensionless geometrical and material-specific entities shown below:

r_c/r_f ratio between crystal and feed rod radii
L/r_c ratio between length of the zone and crystal radius
$V_m/\pi r_c^2$ with V_m volume of the melt
$\varepsilon = r_c^2 \rho_m g/\gamma_m$ where ρ_m is the melt density, γ_m the melt surface tension, g the gravity acceleration.

To each value of crystal radius r_c corresponds just one value of ε for a given material and one maximum length of the molten zone L^*. Beyond L^* the zone becomes unstable and the contact between feed rod and crystal will be lost.

Another important question in FZ growth is the strong convection in the melt due to three effects: natural convection due to temperature gradients, Marangoni convection due to temperature-related surface tension gradients, and electromagnetic coupling from the inductor. These phenomena contribute to the formation of marked growth striations in the crystal.[41]

15.4.3.2 Zone Melting Technique

In the zone melting (ZM) techniques the compact polycrystalline charge is contained in quartz or glass tube that is translated across a heater giving a pronounced temperature spike. Resistance heaters or lamp-mirror assemblies are used to produce the sharp gradient necessary for the local melting. Like in the Si floating zone, only a reduced fraction of the charge is molten at any time of the growth. The principal advantage of this system is that the contamination from environment and crucible is strongly reduced. Furthermore, one can control much better the dopant segregation with respect to normal freezing or can use the zone melting for refining the single crystals and improving the purity. Although the longitudinal segregation parameters are better than in the Czochralski and Bridgman processes, there are very pronounced nonuniformities on a micro-scale. These are related to the strong liquid convection existing in the small molten zone, which of course leads to pronounced dopant striations on the longitudinal crystal cross-section. ZM can also be performed in horizontal configuration. Here, because of the thermal convection, the molten zone will be asymmetric, with the distance between the interfaces larger at the top of the zone with respect to the bottom. In this case, to have a symmetric radial distribution of the impurities, it is preferable to use a cylindrical container for the polycrystalline charge and rotate it around its axis.

The vertical ZM is obviously advantageous for doping uniformity and has received attention also from the III–V crystal growers.[42] They use round crucibles containing a seed in the bottom and B_2O_3 encapsulation, and pressurize the furnace with 1 to 2 Bar of inert gas. A molten zone of about 20 mm with nearly flat solid–liquid interfaces is created by adopting a temperature spike over an otherwise nearly flat temperature field (temperature gradients in the solid were in the range 4 to 9°C/cm). The single crystal growth is carried out with a first passage of the hot zone to melt and compact the polycrystalline charge (GaAs granules) and a second one to initiate the single crystal growth. Low dislocation densities GaAs (<1000 cm^{-2} for diameters of 35 mm) are reported.

For its unique peculiarities, the FZ technique has been applied for growing 20 mm diameter 90 mm long GaAs crystals in space experiments[43] while another version of the zone melting (called Traveling Heater Method, THM) has been used for the growth of ternary III–V bulk crystals.[44] The THM technique is in principle a continuous liquid phase epitaxial process that uses a mirror furnace capable of concentrating the heat in a very narrow zone of the solid charge. The charge is contained in a closed quartz ampoule and is normally formed by two rods of binary compounds (for instance GaP and InP) having cross-sections proportional to the desired composition of the ternary III–V crystals. The charge is thus shifted across the heated zone at a rate of approximately 0.5 to 1.5 cm/day.

15.4.3.3 Verneuil Growth Technique

The flame fusion technique is a method for growing crystals without crucibles developed by Verneuil to prepare single crystals of ruby ad sapphire at the beginning of the twentieth century.[45] Stable oxides and congruent melting oxides may be grown by flame fusion. In some cases, compounds with a volatile component (like for instance $SrTiO_3$) have been prepared, provided that an excess of the volatile element is added during growth. For a review of the Verneuil method and a list of the materials that can be prepared by this technique see Reference 46. The Verneuil technique is based on the use of a gas burner to melt a powder of the feed material that falls in controlled amount through the flame (Figure 15.17). The powder fuses and the corresponding liquid droplets fall on a crystalline seed, giving rise to a small ingot. Hydrogen combustion is employed to achieve the high temperatures needed for melting oxides like Al_2O_3 (about 2200°C). The experimental apparatus is simple and includes a conical powder box equipped with a

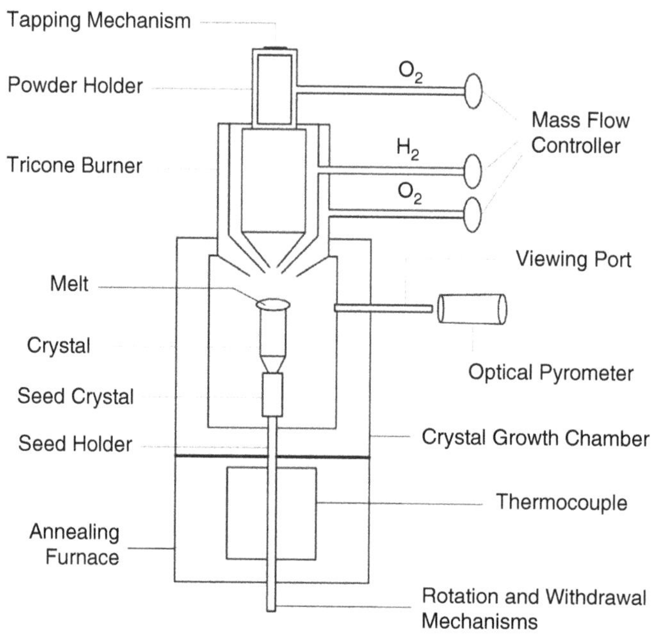

FIGURE 15.17 Schematic view of the traditional Verneuil growth system.

mechanical device that allows the dropping of a constant amount of powder, a seed holder connected to a pulling shaft that allows the crystal withdrawal and rotation, and a burner consisting of concentric tubes. The design of the burner is critical as it decides the thermal gradient within the combustion chamber, that is, the thermoelastic stress and the relative dislocation density of the grown crystals. Furthermore, the way the gases mix and react in the burner must be such that no turbulence occurs. In many cases, the burners may provide a reducing or oxidizing atmosphere. The diameter of the crystal may be controlled by acting on the powder feed rate, temperature of the flame, and speed of the seed down pulling. The Verneuil technique has several advantages, like low-cost and high growth rate and is crucible-free; however, the crystallographic quality of the grown crystals is normally lower than in more sophisticated techniques. The dislocation density is high and in extreme cases cracks were also observed. Some improvement in quality can be achieved by using thermal baffles or additional heaters, which slow down the cooling of the crystal. Postgrowth annealing is also often applied to relieve the thermal stress.

15.5 Vapor Transport and Sublimation for Growth of Bulk Crystals

The growth of bulk crystals from the vapor phase is a well-established technique often applied to materials that undergo phase transformations or melt incongruently or have too high melting point. The crystals grow very slowly therefore normally they contain a lower dislocation density and concentration of point defects in comparison with melt grown materials. The vapor phase growth techniques may be subdivided in two large groups: physical vapor transport (PVT) and chemical vapor transport (CVT). They are again to be divided in two groups that are: open tube or closed tube. Important is also the type of source used for the crystallization process. Solid sources (normally powders containing the crystal components) that are evaporated at a certain temperature and then re-crystallized at lower temperature are the basis of PVT. If the vapor pressure is too low a chemical agent is added to speed up the dissociation of the solid source and the material transport to the crystal surface. This is the case of CVT. The most popular

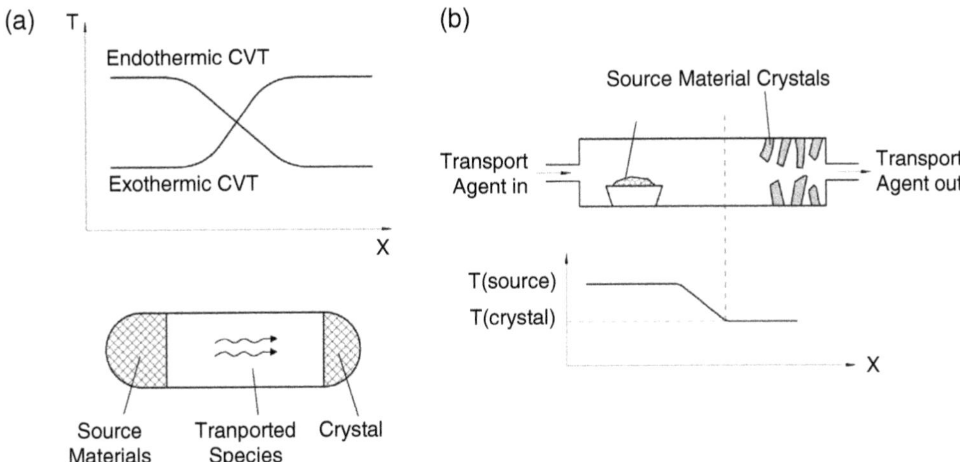

FIGURE 15.18 (a) Close-ampoule CVT and corresponding temperature profiles for exothermic and endothermic reactions between the carrier species and the compound species; (b) Open tube — CVT: a gaseous transport agent reacts with the source material at high temperature and dissociates at lower temperature giving rise to crystallization of the desired compound; the transport agent can be sometimes re-cycled.

transport agents used for these reversible reactions are halogens that tend to form halides with the source material. Halides are then transported via diffusion and convection to the deposition zone where, because of the temperature dependence of the reaction free energy, they dissociate making available the element or compound for crystal building and the halogen atom for another transport cycle. A schematic view of a closed-tube vapor transport apparatus is reported in Figure 18a, while detailed information about PVT and CVT techniques can be found in Reference 47 and Reference 48. A main advantage of the closed-tube is that there is no loss of the source material (i.e., yields close to 100%). However, because of the accumulation of reaction by-products and impurities, the growth of many materials is inhibited after a certain time. This deficiency can be overcome by applying the so-called open or semi-open system (Figure 15.18b). In this case, however, the loss of material can be significant. The use of open CVT systems, in conjunction with gas sources, is therefore extremely popular for the deposition of thin films (vapor phase epitaxy, see also Chapter 9) but less common for the production of bulk crystals.

Let us see some examples to illustrate the principles and functioning of the various vapor transport methods. The simplest PVT technique is the sublimation where the solid source is placed at the bottom of a vertical container or at one side of a horizontal ampoule and heated so that it sublimes and diffuses to the colder side of the container/ampoule where it crystallizes. The growth can be initiated using an oriented seed or by spontaneous nucleation on a substrate. The bulk silicon carbide crystals so important for high-temperature electronic devices are industrially produced by this technique (Figure 15.19). The source material is pure SiC powder kept at temperature >2200°C while the container is made of high-purity graphite. The seed is a thin (0001) 6H or 4H SiC wafer suspended from the container ceiling. The temperature difference between powder reservoir and seed may be of 20 to 40°C. To keep a vertical thermal gradient a hole is created in the thermal insulation around the graphite crucible so that the backside of the graphite/seed assembly is radiation-cooled. The low-pressure vapor (<100 mBar) is nonstoichiometric as it contains an excess of the highly volatile Si species; this silicon can however escape as the system is semi-open. The transport species are SiC_2 and Si_2C and the seeding is controlled by the seed surface polarity (Si or C face) and by the supersaturation within the container. Typical growth rates are about 200 μm/h. Nowadays 20 to 30 mm thick crystals with 50 to 100 mm diameter are routinely produced by this method.[49,50]

Cadmium chalcogenides (CdS, CdTe, and CdSe) provide a good example of the application of close-ampoule PVT. They are conveniently grown by this technique as their vapor pressure is relatively high at

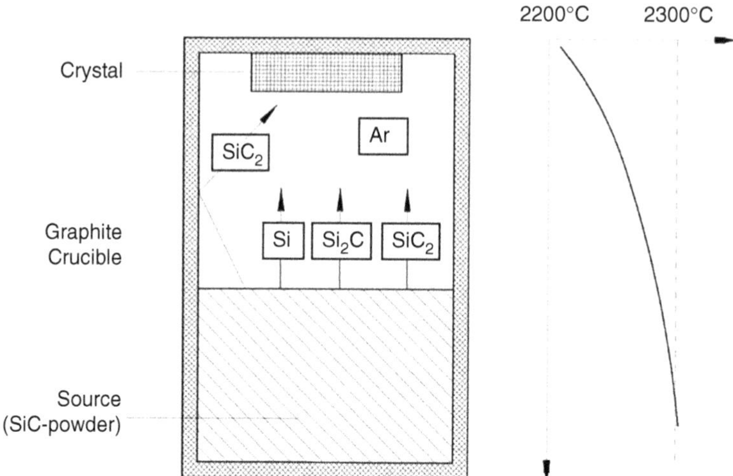

FIGURE 15.19 Schematic view of the typical assembly and temperature profile for sublimation of silicon carbide.

relatively low temperature. Two species are expected to evaporate at typical experimental temperatures of 800 to 900°C: pure cadmium and diatomic molecules of the chalcogen (X = S, Te, or Se):

$$CdX_{(solid)} \rightarrow Cd\ (vap) + 0.5X_2\ (vap) \tag{15.19}$$

The partial pressures of these gas species, $p(Cd)$ and $p(X_2)$, depend on the temperature of the source and solidification zones, as well as on the composition of the powder source. For the most effective transport conditions, one should have a stoichiometric ratio of the two gases, that is, a ratio $\alpha = p(Cd)/p(X_2) = 2$, but this is not possible for two reasons: (1) the starting charge is often nonstoichiometric, (2) the diffusion and advection of the two gas species are generally different. A practical result of this phenomenon is the formation of a "boundary layer" rich in one of the two components in proximity of the solid–gas interface of the growing crystal. As shown in Section 15.5.2., this boundary layer strongly affects the mass transport and ultimately limits the growth rate or even stops the transport. The growth rate is maximum when $\alpha = 2$, but rapidly decreases when $\alpha > 2$ (Cd-rich gas phase) or $\alpha < 2$ (chalcogen-rich gas phase), as experimentally observed in II–VI chalcogenides as well as in many other compounds. One begins now to understand that the closed-tube technique may have strong limitations that can be mitigated by using a semi-open system. Piper and Polich[51] pioneered this PVT version by applying a nearly sealed ampoule that allowed the out-diffusion of the excess gas until a stoichiometric ratio of the components was achieved; the ampoule was then fully sealed with a mechanical stopper and the growth of the single crystal was carried out. A refinement of this technique consisted in drilling a little effusion hole (diameter of 100 to 200 μm) in the growth ampoule and connecting it to the vacuum.[52,53] The hole permits the elimination of impurities and reaction by-products but also allows a small controlled amount of the charge to escape, thereby setting the composition of the gas phase and helping to maintain a nearly stoichiometric condition.

As shown above, in CVT solid substances can be transported through a vapor phase, in the presence of a temperature gradient and a suitable reactive transport agent in volatile form at the source temperature T_S. The solid source is then transformed into gaseous species via heterogeneous chemical reactions. By inverting the reaction at $T_C \neq T_S$, a gas movement takes place between the source and crystallization region. In the region of the ampoule at T_C the solid phase is again restored and crystallized.

In contrast to PVT, CVT does not always proceed with mass transfer from the hot to the cold region $T_S > T_C$, though this case seems to prevail over the reverse $T_S < T_C$ case. In fact, the mass transfer direction in CVT rather depends on the sign of the enthalpy change (ΔH) associated with the chemical transport reaction. When $\Delta H > 0$ (endothermic reaction) the transport takes place from the hot to the cold region. The opposite occurs when $\Delta H < 0$ (exothermic reaction). The choice of a transport agent for a given solid substance is in relation to the mass transfer yield (productivity) of the specific CVT process. Iodine is the most commonly employed transport agent, but many others substances, both elemental and compounds, have also been reported. When using a compound as a transport agent, it is often required that it contains an ion in common with the solid to be transported. This is the case of FeS with $FeCl_2$ as transport agent or the case of TeO_2 when transported with $TeCl_4$. However, cases in which the transport agent is a compound with no common ion with the substance to be transported have also been reported, for instance the transport of CdTe with NH_4Cl.

When dealing with CVT, one has to distinguish between "symmetric" and "asymmetric" (incongruent) processes. CVT is called "symmetric" when one and the same solid phase is simultaneously present in both the source and crystallization region, whereas "asymmetric" or "incongruent" systems are those in which more than one condensed phase is present in either region. The number of condensed phases may not be the same in the two regions. For example, in the CdTe/NH_4Cl system, at sufficiently low total pressure and temperature ≥450°C, two condensed phases coexist at T_s: CdTe(solid) and Te(liquid).

"Incongruent" is also a system, in which a solid compound is present in both regions with a different stoichiometric ratio. This is the case of transition metal oxides and monosulfides for which the possibility of controlling their homogeneity region by CVT has been extensively investigated (for a review, see Reference 54). Closed-tube CVT processes are especially employed for growing bulk crystals although, usually, more crystals are simultaneously grown per run and the reproducible preparation of "large crystals" remains difficult. Working in closed systems (usually ampoules loaded and sealed off under vacuum, see Figure 15.18a) allows one to obtain relatively pure crystals, though traces of the transport agent cannot be avoided. For instance, high concentrations of iodine (200 to 700 ppm) were found in CdS crystals obtained by CVT in a closed ampoule with iodine as transporter . Even larger amounts of bromine and chlorine are incorporated in the final crystals of II–VI binary compounds in the CVT with HBr and HCl as transport agents. It should be noticed that the closed-tube techniques are conceptually very simple: the experimental setup consists of a two-zone vertical/horizontal furnace where the only parameter to control is temperature. Additionally, one needs a vacuum system for evacuating the ampoules (Pyrex or quartz) and a flame welding system. As closed-tube CVT is self-regulating (once the temperature profile is established), they proved particularly appropriate for long-time growth experiments. The main drawback is that once the ampoule has been sealed off, the growth environment is no longer accessible and no way exists, besides changing the temperature settings, to modify the growth process from outside. Also, the temperature profile, that is, the difference (T_s–T_c), cannot be practically increased above a certain limit otherwise the supersaturation will become excessive and spurious nucleation and growth instability will occur. As a consequence, the growth rate is usually smaller than the predictable maximum.

The closed-tube configuration is by far the most popular CVT method for growing bulk crystals; however, when the reverse vapor–solid reaction is sufficiently fast, open-tube configurations are also considered. Rather simple flow techniques can be used when the transport agent that flows in the reaction tube is already in gaseous form at room temperature, as is the case, for example, with H_2 and O_2. RnO_2 and IrO_2 have been grown under O_2 flow and $T_s = 1000°C$ and $T_c = 900°C$.[55] Very often the transport agent is introduced into the reaction tube mixed with an inert gas, typically N_2 or Ar. For instance, HCl was diluted in Ar, for the growth of Al_2O_3 at about 1500°C[56] and YIG ($Y_3Fe_5O_{12}$) at about 1100°C.[57]

When the transport agent is not already in gaseous form at room temperature, it may be vaporized from a heated reservoir and introduced into the source region mixed with an inert carrier gas (Figure 15.18b). This accessibility of the open-tube configurations, where different reaction gases may be injected, after dilution with suitable carrier gases, into the system, is in principle useful for the growth of mixed crystals such as, for example, $Zn_xCd_{1-x}Te$, where x can be varied independently over a wide range of values.[58] Open-tube CVT was also reported to yield stoichiometric compound crystals. For instance, Mercier et al.[59] were able to grow strictly stoichiometric FeS crystals, while this compound has the tendency to crystallize as Fe_xS, with $x < 1$. As the gas flow rates are controlled from outside, independent of how large (T_s–T_c) is, the open-tube CVT techniques are generally characterized by elevated mass transport rates.

The drawbacks of the open-tube CVT techniques consist, first, in the large waste of carrier and transport gases, which it is sometimes obviated by means of recycling procedures. Next, one has to control many other working parameters such as flow rates, stability and composition of the vapor composition during growth (usually long periods), in addition to the temperature profile. All this makes the open-tube CVT unsuitable for long-run growth experiments. A further drawback is also the possible incorporation of a high concentration of impurities in the crystals, due to the large amounts of gas that flow around them during growth.

With regard to growth efficiency, two trends are observed. The first is toward investigating new CVT, but also PVT systems, able to yield preliminary samples of new crystals (e.g., mixed oxides, refractory materials, organics) to be employed in basic and applied research. The second trend is directed to develop more sophisticated facilities for growing, on a routine basis, large high-quality crystals, especially for electronic and optical use.

15.6 Conclusions

The main techniques for the growth of single crystal materials have been reviewed in this article. The growth of single crystals includes the following steps: (1) achievement of the necessary degree of supersaturation or supercooling, (2) formation of crystal nuclei possibly free of grain boundaries, (3) maintenance of the single crystal growth and production of a crystal with the desired shape and properties. Accordingly, for a full understanding and control of the growth processes, one has to get acquainted with the basic concepts of supercooling and supersaturation, nucleation, interface kinetics, mass and heat transport, flow dynamic, and morphological stability. These phenomena have been briefly presented and discussed in Section 15.2, while some typical setups for the growth from solutions, melts of vapor phase have been described, together with some selected example, in Section 15.3, Section 15.4, and Section 15.5, respectively.

It has been mentioned that melt growth, whenever possible, is to be preferred as it allows the fast growth of very large single crystals. However, when the materials to be grown do not melt congruently or have an extremely high melting point, one has to apply different methods such as solution growth or vapor transport. In these cases, the crystals grow at a much lower rate that may bring some advantages in terms of lower concentration of extended and point defects.

References

[1] Brice, J.C., *Crystal Growth Processes*, John Wiley & Sons, New York, 1986.

[2] Fornari, R. and Paorici, C., Eds., *Theoretical and Technological Aspects of Crystal Growth*, Trans Tech Publ., Zurich, 1998.

[3] Hurle, D.T.J., Ed., *Handbook of Crystal Growth*, Vol. 1 and 2, North-Holland, Amsterdam, 1993.

[4] Müller, G., Metois, J.J., and Rudolph, P., Eds., *Crystal Growth: From Fundamentals to Technology*, Elsevier, Amsterdam, 2004.

[5] Sato, K., Furukawa, Y., and Nakajima, K., Eds., *Advances in Crystal Growth Research*, Elsevier, Amsterdam, 2001.

[6] Burton, J.A., Prim, R.C., and Slichter, W.P., The distribution of solute in crystals grown from the melt.1. Theoretical, *J. Phys. Chem.*, 21, 1987, 1953.

[7] Müller, G., *Convection and Inhomogeneities in Crystal Growth from the Melt*, Series Crystals, Vol. 12, Freyhardt, H.C., Ed., Springer Verlag, Berlin, 1988.

[8] Tiller, W.A., Jackson, K.A., Rutter, J.W., and Chalmers, B., The redistribution of solute atoms during the solidification of metals, *Acta Metall.*, 1, 428, 1953.

[9] Hoglund, D.E., Thomson, M.O., and Aziz, M.J., Experimental test of morphological stability theory for a planar interface during rapid solidification, *Phys. Rev. B*, 58, 189, 1998.

[10] Sasaki, T., Miyamoto, A., Yokotani, A., and Nakai, S., Growth and optical characterization of large potassium titanyl phosphate crystals, *J. Cryst. Growth*, 128, 950, 1993.

[11] Ravi, G. et al., Growth and characterization of sulfate mixed L-arginine phosphate and ammonium dihydrogen phosphate potassium dihydrogen phosphate mixed crystals, *J. Cryst. Growth*, 137, 598, 1994.

[12] Penn, B.G. et al., Growth of bulk single-crystals of organic materials for nonlinear optical-devices — An overview, *Prog. Cryst. Growth Charact. Mater.*, 22, 19, 1991.

[13] Chernov, A.A., Protein crystals and their growth, *J. Struct. Biol.*, 142, 3, 2003.

[14] Elwell, D. and Scheel, H., *Crystal Growth from High-temperature Solutions*, Academic Press, London, 1975.

[15] Goernert, P., Kinetics and mechanism of flux crystal growth, *Prog. Cryst. Growth Charact. Mater.*, 20, 263, 1990.

[16] Tolksdorf, W., Growth of yttrium iron garnet single crystals, *J. Cryst. Growth*, 3/4, 463, 1968.

[17] Byrappa, K., Ed. *Hydrothermal Growth of Crystals*, Pergamon Press, Oxford, 1990.

[18] Ohshima, E., Ogino, H., Niikura, I., Maeda, K., Sato, M., Ito, M., and Fukuda, T., Growth of the 2-in-size bulk ZnO single crystals by the hydrothermal method, *J. Cryst. Growth*, 260, 166, 2004.

[19] Yoshikawa, A., Ohshima, E., Fukuda, T., Tsuji, H., and Oshima, K., Crystal growth of GaN by ammonothermal method, *J. Cryst. Growth*, 260, 67, 2004.

[20] Porowski, S., High pressure growth of GaN — New prospects for blue lasers, *J. Cryst. Growth*, 166, 583, 1996.

[21] Czochralski, J., Z. Ein neues Verfahren zur Messung der Kristallisationsgeschwindigkeit der Metalle, *Physik. Chem.*, 92, 219, 1917.

[22] Dash, W.C., Silicon crystals free of dislocations, *J. Appl. Phys.*, 29, 736, 1958.

[23] Hurle, D.T.J., Analytical representation of the shape of the meniscus in Czochralski growth, *J. Cryst. Growth*, 63, 13, 1983.

[24] Bardsley, W., Green, G.W., Holliday, C.H., and Hurle, D.T.J., Automatic-control of Czochralski crystal-growth, *J. Cryst. Growth*, 16, 277, 1972.

[25] Bardsley, W., Joyce, G.C., and Hurle, D.T.J., Weighing method of automatic Czochralski crystal-growth. 1. Basic theory, *J. Cryst. Growth*, 40, 13, 1977.

[26] Bardsley, W., Hurle, D.T.J., Joyce, G.C., and Wilson, G.C., Weighing method of automatic Czochralski crystal-growth. 2. Control equipment, *J. Cryst. Growth*, 40, 21, 1977.

[27] Mullin, J.B., Straughan, B.W., and Brickell, W.S., Liquid encapsulation techniques — Use of an inert liquid in suppressing dissociation during melt-growth of InAs and GaAs crystals, *J. Phys. Chem. Solids*, 26, 782, 1965.

[28] AuCoin, T.R., Wade, M.J., Ross, R.L., and Savage, R.O., Liquid encapsulated compounding and Czochralski growth of semi-insulating gallium arsenide, *Solid State Technol.*, 22, 59, 1979.

[29] Stepanov, A.V., New method of producing articles (Sheets, Tubes, Rods, Various Sections, etc.) directly from liquid metal. 1. *Sov. Phys. Techn. Phys.*, 4, 339, 1959.

[30] Antonov, P.I., Properties of profiled semiconductor single-crystals grown by Stepanovs method, *J. Cryst. Growth*, 23, 318, 1974.

[31] La Belle, H.E. and Mlavsky, A.I., Growth of controlled profile crystals from melt.1. — Sapphire filaments, *Mater. Res. Bull.*, 6, 571, 1971.

[32] Tatarchenko, V.A., Shaped crystal growth, in *Handbook of Crystal Growth*, Vol. 2b, Hurle, D.T.J., Ed., North-Holland, Amsterdam, Chapter 16, 1011, 1993.

[33] Bridgman, P.W., Certain physical properties of single crystals of tungsten, antimony, bismuth, tellurium, zinc. and tin, *Proc. Am. Acad. Arts. Sci.*, 60, 305, 1925.

[34] Stockbarger, D.C., The production of large single crystals of lithium fluoride, *Rev. Sci. Instrum.*, 7, 133, 1936.

[35] Birkmann, B., Rasp, M., Stenzenberger, J., and Müller, G., Growth of 3″ and 4″ gallium arsenide crystals by the vertical gradient freeze (VGF) method, *J. Cryst. Growth*, 211, 157, 2000.

[36] Gault, W.A., Monberg, E.M., and Clemans, J.E., A novel application of the vertical gradient freeze method to the growth of high-quality III-V crystals, *J. Cryst. Growth*, 74, 491, 1986.

[37] Rudolph, P. and Mühlberg, M., Basic problems of vertical Bridgman growth of CdTe, *Mater. Sci. Eng. B*, 16, 8, 1993.

[38] Matsumoto, F., Okano, Y., Yonenaga, I., Hoshikawa, K., and Fukuda, T., Growth of twin-free (100) InP single-crystals by the liquid encapsulated vertical Bridgman technique, *J. Cryst. Growth*, 132, 348, 1993.

[39] Surek, T. and Coriell, S.R., Shape stability in float zoning of silicon-crystals, *J. Cryst. Growth*, 37, 253, 1977.

[40] Coriell, S.R. and Cordes, M.R., Theory of molten zone shape and stability, *J. Cryst. Growth*, 42, 466, 1977.

[41] Riemann, H., Lüdge, A., Böttcher, K., Rost, J., Hallman, B., and Schröder, W., Silicon floating-zone process — Numerical modeling of RF field, heat-transfer, thermal-stress, and experimental proof for 4-inch crystals, *J. Electrochem. Soc.*, 142, 1007, 1995.

[42] Henry, R.L., Nordquist, P.E.R., Gorman, R.J., and Qadri, S.B., Growth of (100) GaAs by vertical zone melting, *J. Cryst. Growth*, 109, 228, 1991.

[43] Herrmann, F.M. and Müller, G., Growth of 20 mm diameter GaAs crystals by the floating-zone technique with controlled As-vapour pressure under microgravity, *J. Cryst. Growth*, 156, 350, 1995.

[44] Bischopink, G. and Benz, K.W., THM growth of AlxGa1-XSB bulk crystals, *J. Cryst. Growth*, 128, 470, 1993.

[45] Verneuil, A.V.L., Memoire sur la reproduction artificielle du rubis par fusion, *Ann. Chim. Phys.*, 3, 20, 1904.

[46] Ueltzen, M., The Verneuil flame fusion process — Substances, *J. Cryst. Growth*, 132, 315, 1993.

[47] Kaldis, E. and Piechokta, M., *Handbook of Crystal Growth*, Vol. 2a, Ed. Hurle, D.T.J., North Holland, Amsterdam, Chapter 11, 1994.

[48] Paorici, C. and Attolini, G., Vapour growth of bulk crystals by PVT and CVT, *Prog. Cryst. Growth Charact. Mater.*, 48, 2, 2004.

[49] Hofmann, D., Bickermann, M., Eckstein, R., Koelbl, M., Müller, St. G., Schmitt, E., Weber, A., and Winnacker, A., Sublimation growth of silicon carbide bulk crystals: Experimental and theoretical studies on defect formation and growth rate augmentation, *J. Cryst. Growth*, 198, 1005, 1999.

[50] Barrett, D.L., McHugh, J.P., Hobgood, H.M., Hopkins, R.H., Mc Mullin, P.G., and Clarke, R.C., Growth of large SiC single-crystals, *J. Cryst. Growth*, 128, 358, 1993.

[51] Piper, W.W. and Polich, S.J., Vapor-phase growth of single crystals of II-VI compounds, *J. Appl. Phys.*, 32, 1278, 1961.

[52] Faktor, M.M., Heckingbottom, R., and Garrett, I., Growth of crystals from gas phase. 1. diffusional limitations and interfacial stability in crystal growth by dissociative sublimation, *J. Chem. Soc.*, A 16, 2657, 1970.

[53] Faktor, M.M., Garrett, I., and Moss, R.M., Growth of crystals from gas-phase. 4. Growth of gallium-arsenide by chemical vapor transport, and influence of compositional convection, *J. Chem. Soc. Farad. Trans.*, 1, 1915, 1973.

[54] Mercier, J., Recent developments in chemical vapor transport in closed tubes, *J. Cryst. Growth*, 56, 235, 1982.

[55] Schäfer, H., *Chemical Transport Reactions*, Academic Press, New York, 1964.

[56] De Vriesm, R.C. and Sears, G.W., Growth of aluminum oxide whiskers by vapor deposition, *J. Chem. Phys.*, 31, 1256, 1959.

[57] Linares, R.C. and McGraw, R.B., Vapor growth of yttrium iron garnet, *J. Appl. Phys.*, 35, 3630, 1964.

[58] Hartman, H., Mach, R., and Selle, B., Wide gap II–VI compounds as electronic materials, in *Current Topics in Materials Science*, Vol. 9, Kaldis, E., Ed., North-Holland, Amsterdam, 1982.

[59] Mercier, J., Dalberto, G., Lacour, P., and Since, J.J., Vapor-phase reactions in an open tube for FeS crystallisation, *J. Cryst. Growth*, 29, 167, 1975.

16

Casting and Solidification

Rohit Trivedi
Iowa State University

Wilfried Kurz
Swiss Federal Institute of Technology Lausanne

Abstract

One of the key aspects of casting is the prediction of the complex microstructure and possible defect formation that is associated with the solidification process. The selection of microstructure and its characteristic length scales strongly depend on local conditions at the advancing solidification front. The selection of interface morphologies over a wide range of processing conditions encountered in different solidification technologies will be presented in terms of microstructure and phase selection maps. A brief discussion of the coupling of macroscopic heat flow with the microstructural selection will be given. During casting of alloys, dendritic and eutectic microstructures are most commonly observed. Both columnar and equiaxed dendrites are important in casting, and the basic modeling for these two cases will be presented with a discussion on the columnar to equiaxed transition.

16.1 Introduction

During casting, processing conditions play a key role in the formation of microstructure and defects. The prediction of microstructure requires macroscopic heat flow model that characterizes the local conditions at the advancing interface. However, the key aspect of the modeling is to correlate the conditions at the interface with the microstructure evolution. We shall therefore concentrate on developing basic ideas on how processing conditions influence microstructures that have a profound effect on the in-service behavior of cast, welded, and rapidly solidified materials. It is thus critical to know how to predict, modify, and control the internal structure of the material by designing appropriate composition and processing conditions. Once the correlation between the processing conditions and the microstructure is established, it is important to see how it can be applied to the complex conditions that exist in a real casting. During casting, most common microstructures consist of dendrites and eutectics, so that the effect of processing variables on these microstructures will be emphasized.

16.1.1 Columnar and Equiaxed Microstructures

The thermal field in a casting is very important in determining the microstructure of the cast alloy. Two distinctly different heat flow conditions may exist in a mold that would give rise to two different types of dendrite or eutectic morphologies: columnar and equiaxed microstructures.[1] When the temperature gradient in the liquid and solid are positive such that the latent heat generated at the interface is dissipated through the solid in the opposite direction to the interface displacement, columnar crystals form, as shown in Figure 16.1a for dendritic growth. In the second case, equiaxed dendrites or eutectic form when the liquid is supercooled, in which initially nuclei grow spherically and become unstable to form dendritic structures. In cubic metals, six dendrites originate from a nucleus so that a so-called equiaxed grain is produced upon solidification, as shown in Figure 16.1b. Directional and equiaxed eutectic microstructures are shown in Figure 16.1c and Figure 16.1d, and equiaxed eutectics have a smooth interface and grow as spherical polyphase nodules. For equiaxed crystals, the latent heat of fusion is dissipated through the surrounding liquid ahead of the dendrites or eutectics, and the heat flow direction is the same as the

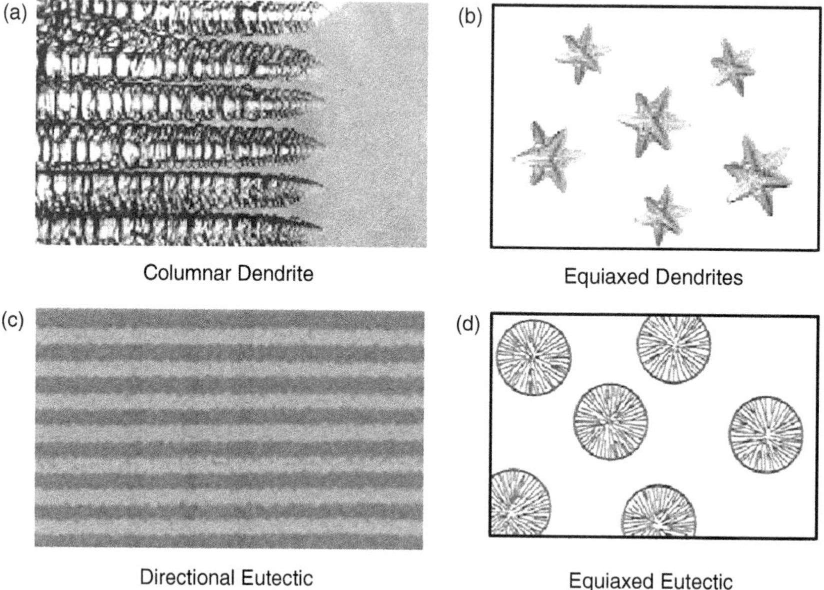

FIGURE 16.1 Columnar and equiaxed structures of dendrites and eutectic. Typical spacings for a, b, and d are 100 μm and for c 10 μm.

growth direction. Columnar structures form near the wall of the mold where heat flow from the liquid to the solid to the mold wall is important. Equiaxed structures generally form at the center of a casting where a substantial part of the liquid becomes constitutionally undercooled. The relative amounts of columnar and equiaxed structures are critical in governing the properties of a casting.

16.1.2 Length Scales of Microstructure

The properties of a solidified alloy are governed by characteristic lengths of the microstructure. For eutectics, Figure 16.1c, the important length is the lamellar or rod spacing, whereas the tip radius, R, and the secondary arm spacing, λ_2 are important for dendrites. For directional solidification, additional lengths that become important are the primary dendrite spacing, λ, and the length of the dendrite or the extent of the mushy zone. The formation of a complex dendritic microstructure causes solute segregation pattern as dendrites grow with a smaller composition than the alloy composition, and the solute accumulates in the interdendritic region for phase diagrams in which the solute segregation coefficient is less than one (decreasing liquidus temperature with concentration). From a practical viewpoint, the interarm spacings characterize the solute segregation pattern and often a second phase distribution in the interdendritic region results, which again has a strong influence on the properties of the material. In addition, void formation can also occur in the interdendritic region as the dissolved gas atoms are rejected in the interdendritic region. The dendrite tip radius is important as it determines the dendrite tip temperature and composition that strongly influence the selection of solidification microstructure, for example, dendrite or eutectic microstructure. The theoretical models of microstructure formation in a casting are thus directed toward quantitatively understanding how the processing conditions, alloy composition, and thermo-physical properties control the magnitudes of these spacings and of the tip temperature.

We shall first describe the basic model of dendrite and eutectic growth, and then discuss how these results are used to determine the microstructure in a casting through the coupling of macroscopic heat flow model with the microstructure evolution. We shall then discuss the columnar to equiaxed transition, and briefly consider the effect of convection on this transition. Finally, the solute segregation patterns in castings will be presented.

16.1.3 Microstructure and Processing Conditions

The characteristic lengths of a dendritic structure, which control the properties of the cast material, are governed by the processing conditions, G (temperature gradient) and V (velocity), or the cooling rate ($\dot{T} = GV$). Finer spacings are obtained when the solidification processing is carried out at increasing velocity. This is illustrated in Figure 16.2 for a single phase growth of an Al–Cu alloy directionally solidified under different G and V conditions. A correlation with the cooling rate (GV) is also shown in the figure. A planar front growth occurs at very low velocity, as in the single crystal casting, or at very high velocity, as in laser processing and atomization of very fine droplets. A cellular microstructure forms at low velocities when a planar front becomes unstable. It also forms at very high velocities before the planar interface becomes stable. Under normal casting conditions, dendritic structures are most commonly observed (gray area). The scale of microstructure becomes finer as one goes from single crystal growth to normal casting to continuous casting to strip casting to laser welding. We shall now examine how these microstructure scales depend on the processing conditions of V and G for dendritic growth.

16.2 Dendrite Growth

The shape of the crystals in the shrinkage cavity of a casting was reported for the first time by Grignon,[2] who described it as a tree-like structure with branches of fourfold symmetry. The term dendrite, which comes from the Greek word dendron, means tree and is now used to describe this structure. Like a tree such as a spruce, the dendrite consists of a branched structure with primary, secondary, tertiary, and

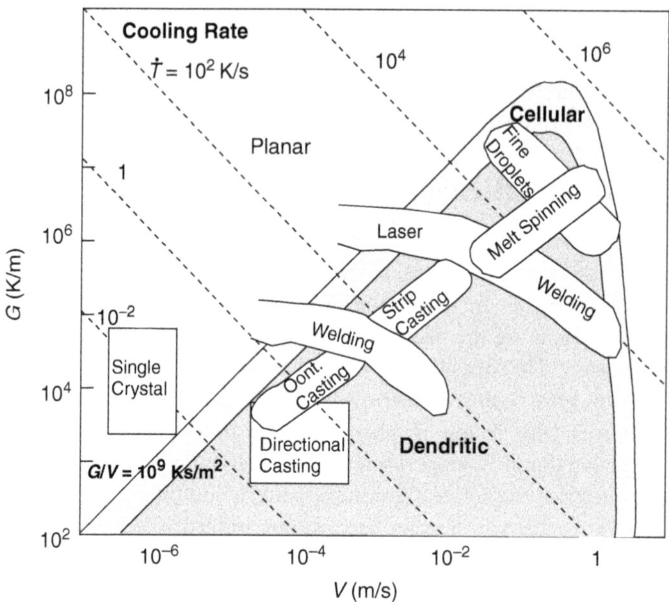

FIGURE 16.2 A microstructure — processing diagram showing typical processing conditions of G and V for different solidification processes (with $G > 0$). Different cooling rates conditions are shown as dotted lines while the continuous lines represent given G/V ratios.

eventually higher order branches. It has been well recognized that dendritic structures form because they can effectively allow rapid dissipation of the latent heat of fusion as well as any solute redistribution required for growth due to the difference in compositions in the solid and the liquid. Consequently, dendrite growth models have been based on the solutions of the thermal and solute diffusion problems. For directional solidification, the latent heat effect is negligible except under rapid solidification conditions, and the formation of dendrite is largely dictated by the solute transport process and the interfacial free energy.

To establish the correlation between the characteristics of dendrites and processing conditions, columnar dendrite growth is studied by directional solidification in which the thermal gradient and velocity are fixed, whereas equiaxed dendrites are examined by undercooling the melt to a known value. In both these cases, steady-state dendrite growth occurs, and the theoretical model gives the relationship between the dendrite tip composition (or temperature) with velocity, which is critical in the modeling of casting microstructure. There are two important aspects of the dendrite growth model based on the transport processes that one must address. First, the steady-state dendrite growth problem is a free-boundary problem in which the shape of the interface is not known a priori, but it must be determined in a self-consistent manner such that this shape is preserved during growth. Second, if only thermal and mass transport processes were important for the formation of equiaxed dendrites, the dendrite tip will keep splitting to form finer dendrites or nondendritic structures. Thus, the thermal and solute diffusion processes are not sufficient to describe the steady-state dendrite growth, and a stabilizing effect such as interfacial free energy needs to be included in the dendrite growth model along with the anisotropy in interface energy that controls the growth direction of dendrites.

16.2.1 Thermal and Solute Profiles in the Liquid

A steady-state growth model was first developed by Ivantsov[3] under the constraint of constant interface temperature and composition. He showed that a parabolic dendrite satisfies the shape-preserving constraint, and obtained the relationship between the interface composition and velocity as:

$$C_t = C_0/[1 - (1 - k)\text{Iv}(P)] \tag{16.1}$$

where C_t is the constant interface composition that is assumed to be the dendrite tip composition, C_0 is the initial alloy composition, k is the solute distribution coefficient, and P is the solute Peclet number equal to $VR/2D$ in which V is the constant steady-state velocity, R the tip radius and D the diffusion coefficient of solute in the liquid. The Ivantsov function is defined as: $\text{Iv}(P) = Pe^P E_1(P)$, where $E_1(P)$ is the first exponential integral.[4] The dendrite tip undercooling, ΔT_t, which is defined as the difference between the liquidus temperature of the alloy and the dendrite tip temperature, can now be obtained from Equation 16.1 and the phase diagram, as:

$$\Delta T_t = \left[\frac{k\Delta T_0 \text{Iv}(P)}{1 - (1 - k)\text{Iv}(P)}\right] + \left[\frac{2\Gamma}{R}\right] \tag{16.2}$$

The term $\Delta T_0 = mC_0(k - 1)/k$, in which m is the slope of the liquidus. ΔT_0 corresponds to the equilibrium freezing range of the alloy for linear liquidus and solidus lines. The second term, $2\Gamma/R$, is capillarity undercooling due to the curvature of the dendrite tip in which Γ is the Gibbs–Thomson coefficient. The above results for interface composition and interface temperature are valid for both the columnar and equiaxed dendrites, and they relate dendrite tip composition and temperature with the velocity and tip radius.

For undercooled melt, it is generally desired to relate velocity and tip radius with melt undercooling which is often controlled by the nucleation event. If T_∞ is the ar field melt temperature, then the temperature difference, $T_t - T_\infty$, causes heat flow into the liquid, and the Ivantsov solution of the thermal diffusion equation gives:

$$T_t - T_\infty = (c_l/\Delta H)[P_t \exp(P_t)E_1(P_t)] \tag{16.3}$$

in which c_l is the specific heat of the liquid, ΔH is the enthalpy change, and P_t is the thermal Peclet number of the dendrite tip, $P_t = VR/2a_l$, where a_l is the thermal diffusion coefficient in the liquid. The relation between dendrite tip velocity, radius, and bath undercooling for equiaxed growth, ΔT, is thus obtained as:

$$\Delta T = \left[\frac{c_l}{\Delta H}\text{Iv}(P_t)\right] + \left[\frac{k\Delta T_0 \text{Iv}(P)}{1 - (1 - k)\text{Iv}(P)}\right] + \left[\frac{2\Gamma}{R}\right] \tag{16.4}$$

The three terms in Equation 16.4 correspond to thermal, solutal, and capillarity undercoolings, respectively. Note that if the second term in the above equation is ignored, one obtains the relationship between the bath undercooling and thermal Peclet number for dendrite growing in a pure undercooled melt.[5,6]

The presence of capillarity effect will cause the composition to vary along the interface so that the Ivantsov model that assumes a constant composition needs to be modified. A more rigorous steady-state dendrite growth model has been developed by using the phase-field approach[7] in which the concentration and the temperature along the interface are allowed to vary due to the capillarity effect. The steady-state shape of the dendrite is found to be very close to the parabolic shape, and the capillary undercooling is observed to be negligible compared to the solutal undercooling for dendrite growth at low velocities that are generally present under casting conditions. The solute profile in the liquid can thus be adequately described by the Ivantsov solution.

16.2.2 Dendrite Tip Stability Criterion

The results of the dendrite growth model, given by Equation 16.2 or Equation 16.4, relate melt undercooling with the thermal and solutal Peclet numbers, that is, with the product of V and R. This indicates that for a given bath undercooling, infinite number of solutions are possible for V and R. For directional solidification, where V is known, Equation 16.2 predicts infinite solutions between ΔT_t and R. Experimental studies, however, show that a unique tip radius is selected under given processing conditions so that another criterion is required for the selection of a specific value of the dendrite tip radius.

Significant research in the dendrite tip radius selection has been carried out, and it has now been established that the interfacial energy effect plays a critical role in this selection process. This selection criterion can be viewed as a balance between the destabilizing effect of the solute and thermal gradients in the liquid for undercooled melt, and the stabilizing effect of the interfacial energy. For dendritic growth at low velocities, the results of different theoretical models can be summarized in the form of a general expression that balances the constitutional supercooling with the interface energy effect.[8]

$$mG_c - G = (\Gamma\sigma^*/R^2) \tag{16.5}$$

where G_c is the concentration gradient at the dendrite tip, and $G = (K_s G_s + K_l G_l)/(K_s + K_l)$, the mean temperature gradient in which K_s and K_l are the thermal conductivities of the solid and liquid, respectively. σ^* is a constant and its interpretation is the key to the dendrite tip selection process. We shall now examine this result first for an undercooled alloy melt, and then for directional solidification. Substituting the value of the concentration and thermal gradient from the Ivantsov solution, the dendrite tip selection criterion can be written as:

$$VR^2\left[\frac{k\Delta T_0}{\Gamma D}\right]\left[\frac{C_t}{C_o}\right] + \delta\left[\frac{(\Delta H/c_l)}{2\Gamma a_l \beta}\right] = \left[\frac{1}{\sigma^*}\right] \tag{16.6}$$

where $\delta = 1$ for undercooled melt and $\delta = 0$ for directional solidification, and $\beta = 0.5[1 + (K_s/K_l)]$. When a_l/D is large, the thermal term becomes small for concentrated alloys, and is often neglected, thereby giving identical results for the directional and equiaxed dendrites.

The key parameter in the above model is σ^* that was first introduced by Langer and Müller-Krumbhaar.[9] Its origin comes from the observation that no steady-state dendrite growth is obtained when isotropic interface energy effects are considered that give rise to nonisothermal or nonisoconcentrate dendrite. Rigorous phase field simulations have shown that a steady-state solution is obtained only when the anisotropy in interface energy is taken into account. A relationship between σ^* and the anisotropy has been numerically obtained by Karma and Rappel.[7] An experimental value of the anisotropy in interface energy in the Al–Cu system was measured by Liu et al.[10] who obtained the value of $\sigma^* = 0.024$ for dendrites growing along the [001] direction. Although this anisotropy value is very small, it is critical in determining the preferred dendrite growth direction as the dendrites grow in the direction of the largest interfacial energy value. In some aluminum alloys, such as Al–Zn and Al–Mg, the interfacial anisotropy changes with composition so that the dendrite orientation changes from [001] to [011] as the composition is increased.[11]

The above model, Equation 16.2 and Equation 16.6, can be numerically solved to obtain the relationship between the dendrite tip temperature and dendrite velocity. The variation in interface temperature with velocity is shown in Figure 16.3 for different single phase morphologies during the directional solidification of alloys. The interface temperature for planar front is the solidus temperature, whereas that for the low velocity cellular microstructure is given by the liquidus temperature reduced by the cell tip undercooling GD/V.[12]

For casting conditions where the velocity (or the Peclet number) is small, the dendrite growth model is often simplified by approximating the Ivantsov solution to give a much simpler relationship:[1]

$$\Delta T_t \cong K \cdot V^n \tag{16.7}$$

where K and n are constants whose values depend on the range of Peclet number that is relevant to the problem.[1] The value of n is of the order of 0.3 to 0.5 for small Peclet number conditions. Note that this form of relationship is valid for columnar dendrites, and also for equiaxed crystal if a_l/D is large. This simplified result is commonly used to obtain dendrite velocity from the interface temperature obtained from macroscopic heat flow models in numerical models.

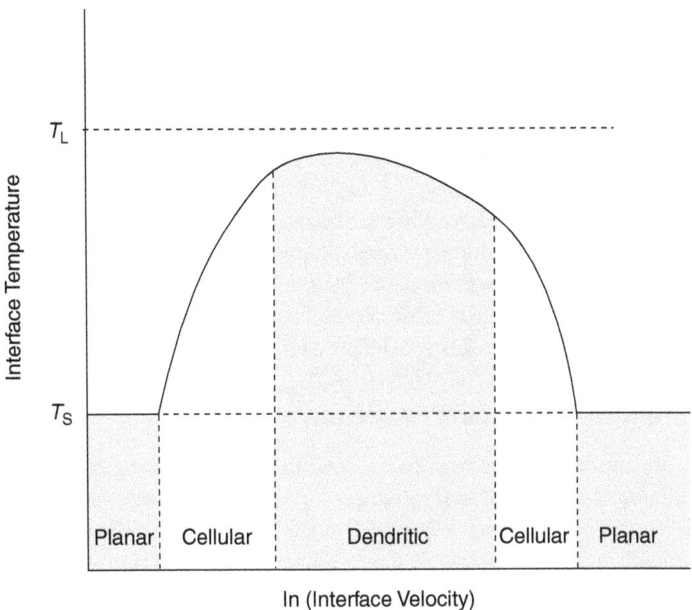

FIGURE 16.3 The variation in interface temperature with velocity, showing the regimes of different single phase microstructures under the assumption of local equilibrium condition at the interface.

16.2.3 Multicomponent Alloys

The dendrite growth model is described for binary alloys only, whereas industrial alloys are always multicomponent. Therefore, the equations developed above need to be generalized to the case of more than two components. This is generally done by linearizing the phase diagram in the composition range of interest and setting the cross diffusion coefficients equal to zero. Several authors have developed solutions to this problem; see Rappaz and Boettinger.[13] The columnar dendrite growth, given by Equation 16.6 (with $\delta = 0$) and Equation 16.1 is thus modified to:

$$R\sum_{j=1}^{N}\left[m_j(k_j-1)P_j\left(\frac{C_{t,j}}{C_{0,j}}\right)\right]=\left[\frac{\Gamma}{2\sigma^*}\right] \tag{16.8}$$

$$C_{t,j}/C_{0,j} = 1/[1-(1-k_j)Iv(P_j)] \tag{16.9}$$

where N is the number of solute components. The velocity and undercooling at the tip are then given by: $V = 2P_1D_1/R$ and $\Delta T_t = \sum_{j=1}^{n} m_j(C_{0,j} - C_{t,j})$.

16.2.4 Rapid Solidification

In solidification technologies such as laser processing and atomization, very high velocities or undercoolings are present that will introduce new effects that must be included in the model. The basic dendrite growth model for solute and heat transport is still given by the Ivantsov solution with capillary correction. However, the dendrite tip selection criterion is altered significantly at high growth rates, and Equation 16.5 is modified as[14,15]

$$mG_c\xi_c - \bar{G} = (\Gamma\sigma^*/R^2) \tag{16.10}$$

where $\bar{G}$ is the effective temperature gradient at the interface equal to $(K_sG_s\xi_s + K_lG_l\xi_l)/(K_s + K_l)$ in which the functions ξ_c, ξ_l, and ξ_s are given by Trivedi and Kurz,[16] which are unity at low growth rates but become zero at very high growth rates (or when the Peclet numbers become much larger than one).

This gives rise to a much faster increase in tip radius and tip composition at high velocities for columnar growth, and they sharply increase the velocity at high undercoolings for equiaxed dendrite growth.

The basic model of dendrite growth considered earlier assumed that local equilibrium is present at the interface. At very high growth rates, nonequilibrium effects at the interface become important. These lead to the velocity-dependence of the partition coefficient (k_V) and the effective slope of the liquidus line (m_V), which are given by:[17,18] $k_V = [k + (a_0 V/D)]/[1 + (a_0 V/D)]$ and $m_V/m = [1 - k_V\{1 - \ln(k_V/k)\}]/(1 - k)$. Under high growth rates, interface undercooling can become large so that the temperature of the interface becomes sufficiently low to alter the diffusion coefficient in the liquid. These rapid solidification effects are then included in the dendrite growth model by introducing ξ functions, k_V, m_V, and $D(T)$, and they have been examined in detail for both the columnar and equiaxed dendrite growth by Trivedi and Kurz.[8] These modifications are critical in correlating dendrite characteristics in processes such as laser welding.[19]

16.2.5 Columnar to Equiaxed Transition

This transition is an important phenomenon in solidification processing as it gives rise to different properties. In continuous casting of steel, for example, one prefers a mostly equiaxed microstructure that is produced by electromagnetic stirring, whereas in single crystal casting of turbine blades one wants a purely columnar growth, avoiding completely the columnar to equiaxed transition (CET).

In the columnar zone of a casting, dendrites with their preferred growth orientation, which is [001] in most cubic crystals, along the heat flow direction will grow at a higher temperature and thus will outgrow dendrites growing at an angle from the heat flow direction. These oriented columnar dendrites will keep growing as long as no nucleation ahead of the dendrite tip occurs. As the stable dendrite tip growth requires a balance between the constitutional supercooling and curvature effects, Equation 16.5, a constitutionally undercooled zone is always present ahead of the dendrite tip so that equiaxed growth can occur if the required nucleation undercooling is smaller than the undercooling ahead of the dendrite, as shown in Figure 16.4. If enough nuclei are present and the growth undercooling is sufficiently large, a transition from columnar to equiaxed structure can occur. In casting and welding, the nuclei often originate from the detachment of the secondary arms close to the columnar dendrite tip region due to the fluctuating thermal field caused by convection. These fragments grow at the slightest undercooling so that they are highly effective in producing CET.

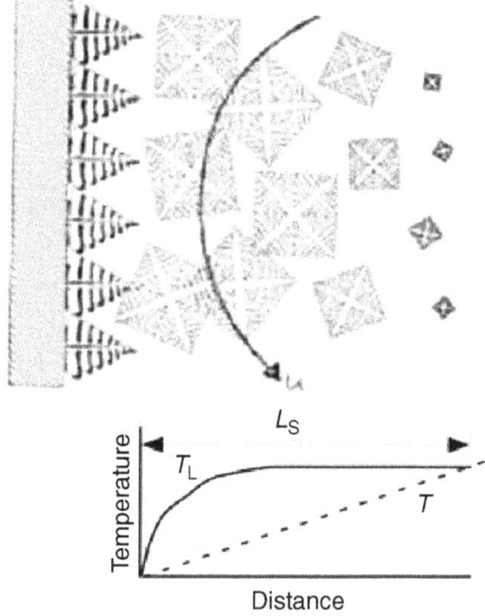

FIGURE 16.4 Equilibrium liquidus and local temperature profile ahead of columnar dendrites, showing the presence of constitutionally undercooled zone where equiaxed dendrites form.

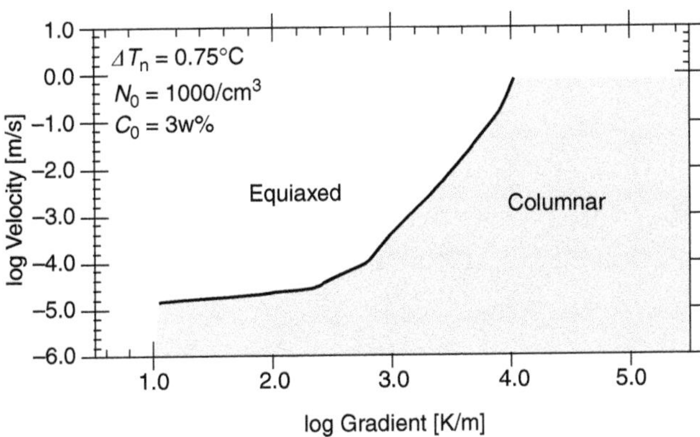

FIGURE 16.5 Columnar to equiaxed transition condition in Al-3.0 wt % Cu alloy predicted by the numerical model.

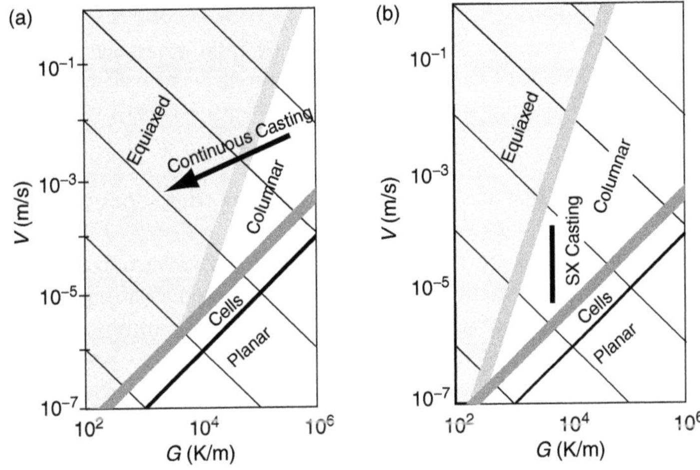

FIGURE 16.6 *G*–*V* diagram showing the regimes of planar, cellular, columnar, and equiaxed structures. The processing conditions are also shown in the figures for (a) continuous casting, and (b) single crystal casting.

To characterize the transition, the competition between columnar and equiaxed growth needs to be considered, as first proposed by Hunt.[20] The key aspect of this model (for a nucleation undercooling $\Delta T_n = 0$) can be understood by comparing two length scales: the length of the constitutionally undercooled zone ahead of the columnar dendrite, $L_T = \Delta T_t/G$, and the mean length between the nuclei forming the equiaxed zone, $L_N = N_0^{-1/3}$, where N_0 is the density of crystal nuclei or fragments. CET will occur when the distance between the nuclei is less than the length of the constitutionally undercooled liquid, that is, $L_N < L_T$ or $G < N_0^{1/3}\Delta T_t$ (Figure 16.5). Note that larger N_0 allows equiaxed zone to close up more rapidly and larger ΔT_t increases the columnar tip undercooling and thus the length over which the equiaxed grains can grow. The density of nuclei, N_0, can be estimated by considering the nucleation rate or by calculating the number of secondary arms in the open region close to the dendrite tip that can get detached, and redistributing them over the constitutionally supercooled zone. The results of this model, given in Figure 16.5, show the transition condition as a function of processing variables, V and G for a given composition and also for a given nucleation undercooling.[21] Equiaxed zone is favored at low G or high V values. In the CET model the transition is assumed to be continuous and it depends on the fraction of equiaxed grains.

The microstructure selection is shown in Figure 16.6 that maps out the regimes of planar, cellular, columnar, and equiaxed structures as a function of processing variables. Figure 16.6a and Figure 16.6b

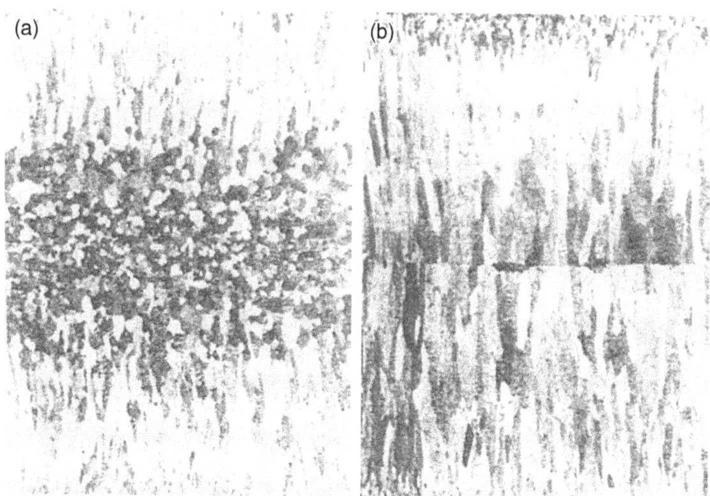

FIGURE 16.7 The effect of convection on the CET microstructure in (a) stirred, and (b) unstirred liquid during the continuous casting of steel. This figure shows the strong effect stirring can have on the fraction of equiaxed grain formation. (After Kobayashi, S., Tomono, H., and Toda, K., *Rev. Metall. – CIT* 80, 887, 1983. With permission.)

show the processing conditions that are generally present during continuous casting and single crystal casting, respectively.[22,23]

The transition from columnar to equiaxed dendrites is favored if convection effects are present during the solidification of a casting. The presence of stirring is found to produce an equiaxed zone, as shown in Figure 16.7a, whereas in the absence of stirring only a columnar region forms, as shown in Figure 16.7b for continuous casting of steel.[24]

16.3 Dendrite Spacing and Microsegregation

The growth of dendritic structures is associated with solute segregation as the solid rejects solute during growth (for $k < 1$) which accumulates in the surrounding liquid. In columnar growth, the periodicity of the solute profile depends on the primary and secondary arm spacings. In equiaxed crystals the solute accumulation will also result at the boundaries of neighboring grains so that the important parameter is the grain size.

16.3.1 Primary and Secondary Dendrite Arm Spacing

The analytical primary spacing model simplifies the complex structure in the mushy zone that contains secondary branches, and considers a dendrite envelope representing the mean cross-section of the trunk and branches. Detailed numerical simulations have been carried out to characterize the effects of composition and the processing variables on primary spacing, λ.[25] A range of primary spacing is found to be stable and the model predicts only the minimum stable spacing. Experimental results show that the maximum spacing is about 1.45 times the minimum spacing.[26,27] At low velocities that are generally present in a casting, simpler analytical models have been developed,[28–30] which show the following relationship between the minimum stable spacing and the processing variables.[30]

$$\lambda = AV^{-1/2}G^{-1/2}R^{-1/2} \tag{16.11}$$

where $A = [4\sqrt{2}\Gamma D/\sigma^*]^{1/2}$

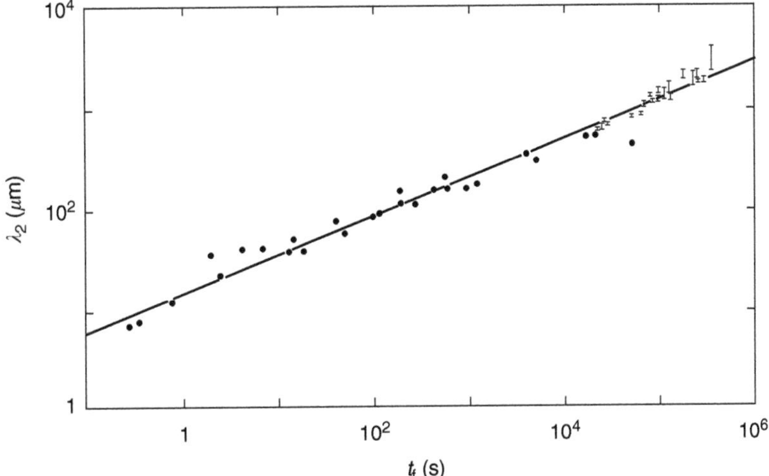

FIGURE 16.8 The variation in secondary arm spacing with local solidification time (After Bower, T.F., Brody, H.D., and Flemings, M.C., *Trans. TMS-AIME*, 236, 624, 1966. Reproduced with the permission of TMS.).

The secondary arms originate close to the dendrite tip, and their initial spacing scales with the dendrite tip radius. However, the final secondary spacing near the base of the dendrite, which is observed in the solidified material, is significantly larger due to the coarsening phenomenon, and follows the general coarsening law that predicts secondary spacing.[31,32]

$$\lambda_2 = \text{const.}t_f^{1/3} \tag{16.12}$$

where t_f is the local solidification time. In directional solidification, the local solidification time is inversely proportional to the cooling rate, which is a product of G and V, so that secondary arm spacing is generally related to the cooling rate.[12] Figure 16.8 shows the experimental results in the Al–Cu system on the variation in secondary arm spacing with the local solidification time. The secondary spacing is the smallest length scale over which solute segregation occurs, and it is correlated to the mechanical properties of single phase alloys. In reality these alloys are often not single phase but frequently contain small amounts of second phases such as eutectic, precipitates, pores, and the like, that form due to microsegregation and control the properties.

16.3.2 Microsegregation

In columnar growth, the temperature at the dendrite tip is higher than that at the base, so that the liquid (or solid) composition at the tip will be smaller than that at the base, which gives rise to a solute segregation profile in the solidified material. As the liquid region between the primary dendrites (or cells) is small compared to the diffusion distance in the liquid, the concentration is nearly uniform in the direction normal to the primary trunk growth. The solute profile in the solid is then given by the modified Scheil equation in which back diffusion in the solid is not negligible,[33,34] which is given by[34]:

$$C_s = kC_o[1 - f_s(1 - 2\alpha' k)]^{(k-1)/(1-2\alpha' k)} \tag{16.13}$$

in which the parameter α' is given in terms of the dimensionless parameter $\alpha = D_s t_f / L^2$ as:

$$\alpha' = \alpha[1 - \exp(-1/\alpha)] - 0.5\exp(-1/2\alpha)$$

D_s is the diffusion coefficient in the solid and L is the distance between the dendrite arms, λ_2.

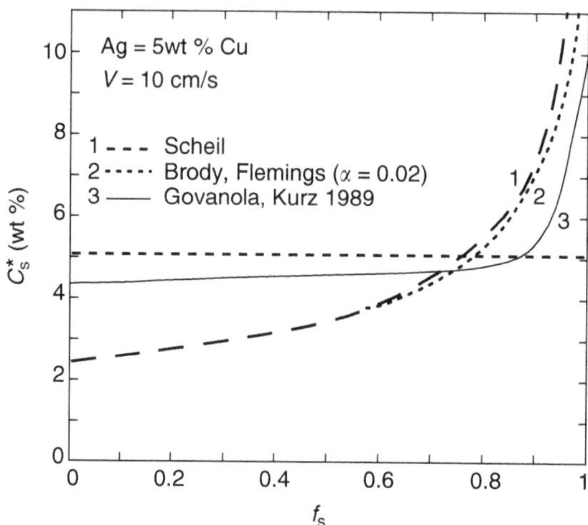

FIGURE 16.9 The rapid solidification effect significantly alters the interdendritic microsegregation profile compared to the results of the low velocity models.

The solute segregation in the interdendritic region can lead to the formation of a new phase when the local composition reaches the solubility limit. In this case, the remaining liquid will form a new phase, such as eutectic or peritectic phase or a compound depending on the phase diagram. The properties of the material are then governed by the fraction of the new phases, their spacing, and morphology, the latter two depending on the dendrite size. The fraction of the second phase can be obtained from Equation 16.13 by taking the composition in the solid as the solubility limit.

Under rapid solidification conditions, as in laser welding, the dendrite tip undercooling is large so that the tip composition is high leading to a smaller solute segregation. In addition, the assumption of uniform composition in the liquid in the normal direction occurs significantly behind the tip.[1,35] This is due to the fact that the diffusion length decreases with $1/V$ while the secondary spacing decreases with $1/(GV)^{1/3}$. The difference in the solute profiles between normal and rapid solidification conditions is examined by Giovanola and Kurz,[3] and their results are shown in Figure 16.9.

16.4 Eutectic Growth

Besides dendrites, eutectic is the most commonly observed microstructure in castings. Alloys of near eutectic composition have small freezing range, which gives them excellent flow properties that makes them very valuable for casting and soldering applications. In eutectics, two (or more) solid phases form simultaneously from the liquid with a nearly planar solid–liquid interface, as shown in Figure 16.1c. Two important parameters that govern the properties are eutectic spacing and the volume fraction of the two phases.[36] Compared to dendritic structures, the eutectic spacing is close to the dendrite tip radius, and thus it is much finer than the primary and secondary arm spacings, which gives eutectics improved mechanical properties.

During the growth of eutectic, solute diffusion effects are significantly larger than the thermal gradient effect so that eutectic growth is governed by solute diffusion and interface energy effects only. Thus, there is no difference in the model for columnar and equiaxed eutectics. The theoretical models, developed by Hillert,[37] and Jackson and Hunt,[38] predict infinite solutions with undercooling exhibiting a minimum (for directional growth), or the velocity exhibiting a maximum (for free growth) with eutectic spacing. The minimum observed spacing in both these cases corresponds to the extremum condition, that is, λ_m. The minimum spacing is found to scale with velocity as:

$$\lambda_m = K_e V^{-1/2} \tag{16.14}$$

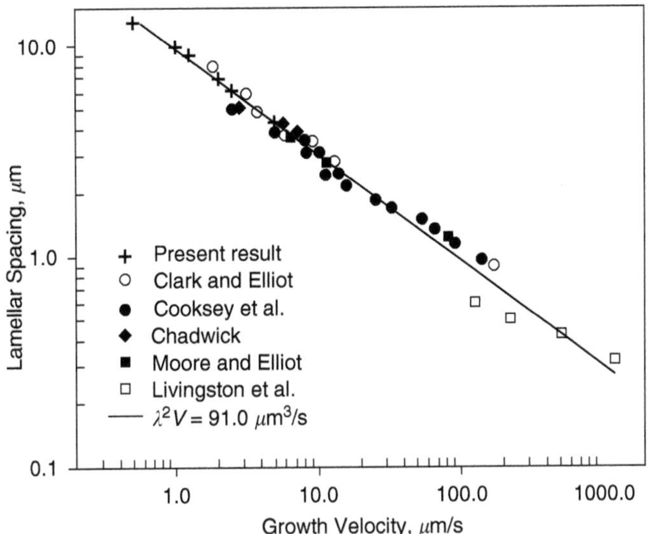

FIGURE 16.10 Experimental results and theoretical prediction of eutectic spacing in Al–Cu eutectic alloys.

where K_e is a constant that contains system parameters.[39] The above scaling law between the spacing and velocity has been confirmed experimentally in the Al–Cu eutectic alloy over several orders of magnitude in velocity, and a very good agreement is observed,[40] as shown in Figure 16.10. Unlike the dendrite tip radius, eutectic spacing selection is not unique, and a narrow range of spacing is generally observed. Experimental studies in the Al–Cu system have shown that the maximum spacing is 1.2 λ_m, so that the average spacing is about 10% larger than the minimum spacing.[41] A significantly larger value is observed in faceted–nonfaceted eutectics such as Fe–C and Al–Si, which are the two most important casting alloys. This large ratio of the average spacing and the minimum spacing has been characterized empirically.[42] A precise quantitative model still needs to be developed.

16.4.1 Microstructure Selection

Theoretical models discussed above relate microstructure characteristics with processing conditions. The key question that still remains is which microstructure is actually selected under the given processing conditions of G and V. An empirical approach is currently used for the selection of microstructures. For directional solidification, the formation of all microstructures of stable and metastable phases is first considered, and then it is assumed that the microstructure that has the highest interface temperature for a given G and V condition will be selected. In undercooled melt, the phase that has the highest nucleation temperature is assumed to be selected. If more than one phase nucleates, the phase with the highest growth rate becomes dominant in the final microstructure.[1] In both the cases, the interface temperature or the growth velocity, are considered for steady-state growth only. As most selection processes occur during varying growth conditions before the steady state is reached, a dynamical approach such as the phase-field method needs to be used to more precisely determine the selection of microstructures.

16.5 Macro-Micro Modeling

We have discussed the effects of processing conditions on microstructure formation and it is now important to see how they can be applied to the complex conditions that exist in a real casting. Detailed numerical codes have been developed for macroscopic heat and mass transfer in casting and welding processes,[43–46] and they are coupled with the microscopic aspects of interface structures.[47–52] We shall illustrate this approach for a simple case of columnar dendrite or eutectic growing under slowly varying thermal conditions (i.e., nearly steady state). In this case, the front of the dendrite tips or of the eutectic interface is moving at the same speed as that of the corresponding isotherm (liquidus or eutectic, respectively). The actual speed (and undercooling) of the dendrite tips depend on their relative orientation with respect to the thermal gradient: if ψ is the angle between the dendrite trunk direction (normally $\langle 100 \rangle$ direction for cubic metals) and the thermal gradient, the actual velocity of the dendrites is equal to $v_L / \cos \psi$. In other words, dendrites that are not well aligned with the thermal gradient have a higher speed to follow the liquidus isotherm and thus are slightly more undercooled. This mechanism is precisely at the basis of grain selection and texture evolution in directional solidification (e.g., of turbine blades).[53]

The velocity of columnar structures (whether dendrites or eutectic) is related to the speed of the corresponding isotherm, and the thermal gradient at the isotherm can also be obtained from the heat flow model when an appropriate f_s–T relationship is used. The knowledge of V and G at different time or different locations in the casting can now be used to evaluate the eutectic spacing or dendrite tip radius, primary and secondary spacing according to the models discussed earlier. The use of the microsegregation model also gives the length of the mushy zone. Note that theoretical models are based on steady-state growth conditions, but the velocity and gradient at the interface are not constant in castings. However, the adjustment of the tip radius and eutectic spacing is quite rapid so that local characteristics can be described by the local conditions of G and V.

For equiaxed microstructures, the problem is more delicate as the actual speed of the interface of the grains (dendritic or eutectic) cannot be simply related to that of the isotherms. In particular, it depends on the density of grains, and thus on nucleation. However, providing a nucleation law is given, the same growth kinetics laws described before can be used to predict the evolution of the grains from the local undercooling. This is achieved via a convolution of nucleation and growth, as for an Avrami description of phase transformations. A detailed discussion on the modeling of columnar and equiaxed crystals is given by Rappaz[51,52] and further extensions, including convection effects (see next section), have been provided by Beckermann et al.[54,55] Finally, it should be mentioned that such nucleation and growth mechanisms for both columnar and equiaxed microstructures have been implemented in the form of Cellular Automata (CA) coupled with Finite Elements (FE) thermal calculations.[56,57] This so-called CAFE model allows one to directly visualize the grain structure, the competition between columnar and equiaxed grains, and the selection and texture evolution of columnar grains, thus providing some type of computer-assisted solidification microstructure tool.

16.6 Convection Effects

Convection is always present in the mold, and it significantly influences microstructural characteristics due to the change in thermal and concentration profiles in the liquid.[58] Furthermore, fluid flow in the mushy zone can give rise to macrosegregation in a casting.[59,60] The flow is mainly induced due to thermal and concentration gradients in the liquid, and it depends on the direction of the thermal gradient and on the change in density with solute content. Flow also occurs to feed the solidification shrinkage and the shrinkage occurring during the cooling of the solid and liquid.

The flow causes detachment of the secondary arms near the dendrite tip that play a crucial role in the formation of the equiaxed zone. These detached fragments can float or settle in the liquid depending on its density relative to the liquid. The flow of liquid in or out of the mushy zone causes macrosegregation (over the dimensions of the casting), and this flow occurs through a complex network of dendrite branches.[61] A proper modeling of macrosegregation can only be done if microstructures and microsegregation are also considered. It thus requires the knowledge of thermal and solute profiles, fluid flow, and motion of solid particles, which must be coupled with the nucleation, microstructure evolution, and microsegregation.[62] This complex problem requires numerical methods. For example the phase-field method is now being extended to include fluid flow [63,64] so that the effect of fluid flow on detailed microstructural features can be properly established.

16.7 Conclusions

Processing conditions are shown to play a key role in characterizing the microstructure that ultimately determines the properties of a solidified material. Microstructure evolution has been shown to depend on the composition of the alloy as well as on the processing conditions that are characterized by the local thermal gradient and growth velocity. A general microstructure map that correlates processing conditions of G and V, or the cooling rate, in different solidification technologies has been given in Figure 16.2, How these changes in processing conditions influence critical microstructure scales such as secondary dendrite arm spacing and eutectic spacing is described. The processing conditions required to control the columnar and equiaxed microstructures are also emphasized. In castings, the values of G and V change with location and time, so that one needs to couple the macroscopic heat flow model with local conditions at the interface to determine the time and spatial evolution of microstructure. The microstructure, the solute segregation patterns, and defect formation are also strongly influenced by the presence of fluid flow in the casting, and appropriate references for these aspects are given.

Acknowledgments

This work was carried out in part at Ames Laboratory which is operated by Iowa State University for the Office of Basic Energy Science, Division of Materials Science, U.S. Department of Energy under Contract No. W7405-Eng-82. Appreciation is expressed to M. Rappaz for valuable comments on the manuscript.

References

[1] Kurz, W. and Fisher, D.J., *Fundamentals of Solidification*, Trans Tech. Publ. Ltd, Switzerland, 1989.

[2] Grignon, M., Mémoires de physique sur l'art de fabriquer le fer, d'en fondre et forger des canons d'artillerie; sur l'histoire naturelle, et sur divers sujets particuliers de physique et d'économie, Delalin, Paris, 1775, p. 71/ p. 480.

[3] Ivantsov, G.P., Temperature field around spherical, cylindrical and needleshaped crystals which grow in supercooled melt, Dokl. Akad. Nauk SSSR, 58, 567, 1947.

[4] Abramowitz, M. and Stegun, I.A., eds, *Handbook of Mathematical Functions*, Dover, New York, 1965.

[5] Langer, J.S., Instabilities and pattern formation in crystal growth, *Physicochem. Hydrodyn.*, 1, 41, 1980.

[6] Huang, S.C. and Glicksman, M.E., Fundamentals of dendritic solidification, *Acta Meall.*, 29, 701, 1981.

[7] Karma, A. and Rappel, W.J., Quantitative phase-field modeling of dendritic growth in two and three dimensions, *Phys. Rev.*, E57, 4323, 1998.

[8] Trivedi R. and Kurz, W., Dendritic growth, *Int. Mater. Rev.*, 34, 93, 1990.

[9] Langer, J.S. and Mueller-Krumbhaar, H., Stability effects in dendritic crystal growth, *J. Cryst. Growth*, 42, 11, 1977.

[10] Liu, S., Napolitano, R., and Trivedi, R., Measurement of anisotropy of crystal-melt interfacial energy for binary Al-Cu alloy, *Acta Mater.*, 49, 42, 2001. 71.

[11] Gonzales F. and Rappaz M., Dendrite growth direction in Al-Zn Alloys, *Metall. Mater. Trans.*, 37A, 2797, 2006.

[12] Bower, T.F., Brody, H.D., and Flemings, M.C., Measurements of solute redistribution in dendritic solidification, *Trans. TMS-AIME*, 236, 624, 1966.

[13] Rappaz, M. and Boettinger, W.J., Dendritic solidification of multicomponent alloys with unequal liquid diffusion coefficients, *Acta Mater.*, 47, 3205, 1999.

[14] Lipton, J., Kurz, W., and Trivedi, R., Rapid dendrite growth in undercooled melt, *Acta Metall.*, 35, 957, 1987.

[15] Kurz, W, Giovanola B., and Trivedi R., Theory of microstructural development during rapid solidification, *Acta Metall.*, 34, 823, 1986.

[16] Trivedi, R. and Kurz, W., Morphological stability of a planar interface under rapid solidification conditions, *Acta Metall.*, 34, 1663, 1986.

[17] Aziz, M.J., Model for solute redistribution during rapid solidification, *J. Appl. Phys.*, 53, 1158, 1982.

[18] Boettinger, W.J., Coriell, S.R., and Trivedi, R., Application of dendritic growth theory to the interpretation of rapid solidification microstructures, in: *Rapid Solidification Processing: Principles and Technologies*, ed. R. Mehrabian and P.A. Parrish, Claitor's, Baton Rouge, 1988, p. 13.

[19] Kurz, W., Solidification microstructure processing maps: Theory and application, *Adv. Eng. Mater.*, 3, 443, 2001.

[20] Hunt, J.D., Steady state columnar and equiaxed growth of dendrites and eutectic, *Mater. Sci. Eng.*, 65, 75, 1984.

[21] Gäumann, M., Trivedi, R., and Kurz, W., Nucleation ahead of the advancing interface in directional solidification, *Mater. Sci. Eng.*, A226–228, 763, 1997.

[22] Martorano, M.A., Beckermann, C., and Gandin, C.-A., A solutal interaction mechanism for the columnar-to-equiaxed transition in alloy solidification, *Metall. Mater. Trans.*, A, 34, 1657, 2003.

[23] Badillo A. and Beckermann C., Phase-field simulation of the columnar-to equiaxed transition in alloy solidification, *Acta Mater.*, 54, 2015, 2006.

[24] Kobayashi, S., Tomono, H., and Toda, K., Amélioration de la qualité des brames de coulée continue grâce au brassage électromagnétique par courant de conduction n, *Rev. Metall. – CIT* 80, 887, 1983.

[25] Hunt, J.D. and Lu, S.Z., Numerical modeling of cellular/dendritic array growth: Spacing and structure predictions, *Metall. Mater. Trans.*, 27A, 611, 1996.

[26] Han, S.H. and Trivedi, R., Primary spacing selection in directionally solidified alloys, *Acta Metall. Mater.*, 42, 25, 1994.

[27] Hui, J., Tiwaru, R., Wu, X., Tewari, S., and Trivedi, R., Primary dendrite distribution and disorder during directional solidification of Pb-Sb alloys, *Metall. Mater. Trans.*, 33A, 3499, 2002.

[28] Hunt, J.D., Primary dendrite spacing, *Solidification and Casting of Metals*, The Metals Society, Book 192, London, 1979, p. 3.

[29] Kurz, W. and Fisher, D.J., Dendrite growth at the limit of stability: Tip radius and spacing, *Acta Metall.*, 29, 11, 1981.

[30] Trivedi, R., Interdendritic spacing: II. A comparison of theory and experiments, *Metall. Trans.*, 15A, 977, 1984.

[31] Kattamis, T.Z. and Flemings, M.C., Dendrite morphology, microsegregation, and homogenization of low alloy steel, *Trans. TMS-AIME*, 233, 992, 1965.

[32] Feurer, U. and Wunderlin, R., Einfluss der zusammensetzung und der erstarrungs-bedingungen auf die dendritenmorphologie binarer Al-legierungen, Fachbericht, *Fachbericht DGM*, Oberursel, Germany, 1977.

[33] Brody, H. and Flemings, M.C., Solute redistribution in dendritic solidification, *Trans. TMS-AIME*, 236, 615, 1966.

[34] Clyne, T.W. and Kurz, W., Solute redistribution during solidification with rapid solid state diffusion, *Metall. Trans.*, 12A, 965, 1981.

[35] Giovanola, B. and Kurz, W., Modeling of microsegregation under rapid solidification conditions, *Metall. Mater. Trans.*, 21A, 260, 1990.

[36] Mollard, F. and Flemings, M.C., Growth of composites from the melt, *Trans. TMS-AIME*, 239, 1526, 1967.

[37] Hillert M., The role of interfacial energy during solid state phase transformations, *Jernkont. Annaler*, 141, 757, 1957.

[38] Jackson, K.A. and Hunt, J.D., Lamellar and rod eutectic growth, *Trans. TMS-AIME*, 236, 1129, 1966.
[39] Trivedi, R. and Kurz, W., Microstructure selection in eutectic alloy systems, in: *Solidification Processing of Eutectic Alloys*, ed. D.M. Stefanescu, G.J. Abbaschian, and R.J. Bayuzik, Metallurgical Society of AIME, Warrendale, PA, 1988, pp. 3–34.
[40] Lee, J.H., Liu, S., and Trivedi, R., The effect of fluid flow on eutectic growth, *Metall. Mater. Trans.*, 36A, 3111, 2005.
[41] Walker, H., Growth and stability of lamellar eutectic structures, M.S. thesis, Iowa State University, Ames, IA, 2005.
[42] Jones, H. and Kurz, W., Relationship if interphase spacing growth temperature to growth velocity in Fe-C and Fe-Fe_3C eutectic alloys, *Z. Metall.*, 72, 792, 1981.
[43] Viskanta, R. and Beckermann, C., Mathematical modeling of solidification, in: *Interdisciplinary Issues in Materials Processing and Manufacturing*, ed. S.K. Samanta, R. Komanduri, R. McMeeking, M.M. Chen, and A.Tseng, vol. 2, 1987, p. 501.
[44] Clyne, T.W., The use of heat flow modelling to explore solidification phenomena. *Metall. Trans.*, 13B, 471, 1982.
[45] Desbiolles, J.-L., Droux, J.J., Rappaz, J., and Rappaz, M., Simulation of solidification of alloys by the finite element method, *Comput. Phys. Rep.*, 6, 371, 1987.
[46] Rolph III, W.D. and Bathe, K.J., *Int. J. Num. Meth. Eng.*, 18, 119, 1982.
[47] Stefanescu, D.M. and Trufinescu, S., Crystallization kinetics of gray iron, *Z. Metallkunde*, 65, 610, 1974.
[48] Stefanescu, D.M. and Kanetkar, C., in: *State of the Art of the Computer Simulation of Casting and Solidification Processes*, ed. H. Fredriksson, Les Editions de Physique, Les Ulis, 1986, p. 255.
[49] Rappaz, M. and Thevoz, Ph., Solute diffusion model for equiaxed dendritic growth, *Acta Metall.* 35, 1487, 1987.
[50] Gandin, C.-A. and Rappaz, M., A coupled finite element cellular automata model for the prediction of grain refinement, *Acta Metall. Mater.*, 43, 2233, 1994.
[51] Rappaz, M., Modeling of microstructure formation in solidification processes, *Int. Mater. Rev.*, 34, 93, 1989.
[52] Rappaz, M., Microstructure modelling, in: *Encyclopedia of Materials: Science and Technology*, Elsevier Science Ltd., 2001, p. 5666.
[53] Gandin, Ch.-A., Rappaz, M., West, D., and Adams, B.L., Grain texture evolution during the columnar growth of dendritic alloys, *Metall. Mater. Trans.*, 26A, 1543, 1995.
[54] Wang, C.Y. and Beckermann, C., A unified solute diffusion model for columnar and equiaxed dendritic alloy solidification, *Mater. Sci. Eng. A*, 171, 199, 1993.
[55] Wang, C.Y. and Beckermann, C., Equiaxed dendritic solidification with convection, *Metall. Mater. Trans., 27A*, 2754,1996 and 27A, 2765, 1996.
[56] Gandin, Ch.-A, Desbiolles, J.-L., Rappaz, M., and Thévoz, Ph., A three-dimensional cellular automaton — Finite element model for the prediction of solidification grain structures, *Metall. Mater. Trans.* 30A, 3153, 1999.
[57] Rappaz, M., Jacot, A., and Gandin, C.-A, Modeling of dendritic grain formation during solidification at the level of macro- and microstructures, in *Continuum Scale Simulation of Engineering Materials*, ed. D. Raabe, F. Roters, F. Barlat, and L.Q. Chen, Wiley–VCH Weinheim, 2004. Chap. 10, p. 251.
[58] Flemings, M.C. and Nereo, G.E., Macrosegregation: Part I, *Trans. TMS-AIME*, 239, 1449, 1967.
[59] Mehrabian, R., Keane, M., and Flemings, M.C., Interdendritic fluid flow and macrosegregation: Influence of gravity, *Metall. Trans.*, 1, 1209, 1970.
[60] Voller, V.R., Brent, A.D., and Prakaxh, C., The modeling of heat, mass and solute transport in solidification systems, *J. Heat Mass Transfer*, 32, 1718, 1989.
[61] Ganeshan, A. and Poirier, D.R., Conservation of mass and momentum for the flow of interdendritic liquid during solidification, *Metall. Trans.* 21B, 173, 1990.

[62] Beckermann, C., Modeling of macrosegregation: Past, present and future, in *Proceedings of the Merton Flemings Symposium on Solidification and Materials Processing*, ed. R. Abbaschian, H. Brody, and A. Mortensen, TMS, Warrendale, PA, 2000, p. 297.
[63] Beckermann, C., Modeling of melt convection in phase-field simulations of solidification, *J. Comput. Phys.*, 154, 468, 1999.
[64] Diepers, H.J., Beckermann, C., and Steinbach, I., Simulation of convection and ripening in a binary alloy mush using the phase-field method, *Acta Mater.* 47, 3663, 1999.

17

Rapid Solidification and Bulk Metallic Glasses — Processing and Properties

Jörg F. Löffler,
Andreas A. Kündig, and
Florian H. Dalla Torre
Laboratory of Metal Physics and Technology, ETH Zurich

List of Symbols

Symbol	Description
a	Notch length
A_{v}	Constant
B	Bulk Modulus
$\Delta c_{\mathrm{p}}^{\mathrm{l-x}}$	Molar heat capacity difference between liquid and crystal at constant pressure
$c_{\mathrm{p}}\rho$	Heat capacity per unit volume at constant pressure
$\bar{c}$	Correction factor of $\sim$1.1
D	Atomic diffusivity
D^*	Fragility parameter
d	Dimension
d_{c}	Critical casting thickness
$D_{\mathrm{e}} = \dot{\gamma}\tau$	Deborah number
E	Young's modulus
f	Crystalline volume fraction
$f(Q)$	Atomic scattering factor
G	Shear modulus
G'	Storage modulus
G''	Loss modulus
G_{c}	Critical strain energy release rate, with c the critical point at fracture
ΔG^*	Energy barrier for forming of a nucleus
ΔG_{v}	Gibbs free energy difference (liquid to crystal) per unit volume
$\Delta H_{\mathrm{m}}^{\mathrm{f}}$	Molar enthalpy of fusion
$I(Q)$	Normalized elastic scattering intensity
I_{v}	Nucleation rate (per volume)
j_Q	Endothermic heat flow
k_{B}	Boltzmann constant
$K_{\mathrm{c}} = \sqrt{EG_{\mathrm{c}}}$	Fracture toughness, with c the crucial point at fracture
l	Atomic diameter
N_{A}	Avogadro's constant
$\mathbf{Q}$	Scattering vector
R	Gas constant
r	Distance (from a central atom)
r^*	Critical radius of a nucleus
R_{c}	Critical cooling rate
$S(Q)$	Atom–atom interference function
ΔS	Normalized thermal stability
$\Delta S_{\mathrm{m}}^{\mathrm{f}}$	Molar entropy of fusion
$\Delta T = T_{\mathrm{x}} - T_{\mathrm{g}}$	Undercooled liquid regime
t	Crystallization time
T_0	Vogel–Fulcher–Tammann temperature
T_{r0}	Reduced Vogel–Fulcher–Tammann temperature
T_{g}	Glass transition temperature
T_{liq}	Liquidus temperature
T_{r}	Reduced temperature
$\Delta T_{\mathrm{r}} = 1 - T/T_{\mathrm{liq}}$	Relative undercooling
T_{rg}	Reduced glass transition temperature
T_{x}	Crystallization temperature
v	Growth velocity
V_{m}	Molar volume
$\alpha = \kappa/(c_{\mathrm{p}}\rho)$	Thermal diffusivity
$\tilde{\alpha} \equiv \dfrac{(N_{\mathrm{A}} V_{\mathrm{m}}^2)^{1/3}\gamma}{\zeta(T)\Delta H_{\mathrm{m}}^{\mathrm{f}}}$	Constant

$\tilde{\beta} \equiv \frac{\zeta(T)\Delta S_m^f}{R}$	Constant
γ	Interfacial energy (per unit area)
$\dot{\gamma}$	Shear strain rate
ε	Strain
$\dot{\varepsilon}$	Strain rate
$\zeta(T)$	Temperature-dependent correction factor
η	Viscosity
$\tilde{\eta}$	Constant inversely proportional to the molar volume of the liquid
η^*	Complex viscosity (from dynamic measurements)
η_0	Viscosity extrapolated to shear strain rate $\dot{\gamma} = 0$
θ	Diffraction angle
κ	Thermal conductivity
λ	X-ray wavelength
ν	Poisson's ratio
ρ_0	Average atomic number density
$\rho(r)$	Pair distribution function
σ	Flow stress
σ_y	Yield strength
τ	Relaxation time
τ_c	Characteristic cooling time
τ_s	Shear stress
ω	Angular frequency

Abstract

This chapter describes various processing techniques for the preparation of metastable phases, with a focus on metallic glasses. Following a general introduction, Section 17.1 outlines various methods of amorphization and rapid solidification. Special attention is given to *bulk* metallic glasses, a new class of alloys that can be manufactured as massive amorphous pieces in dimensions of several centimeters at cooling rates less than 100 K/s. Section 17.2 describes the microstructure of metallic glasses and highlights the most relevant atomic models, while Section 17.3 relates the thermophysical properties of metallic glass-forming liquids with their glass-forming ability. A time–temperature–transformation diagram of undercooled liquids is derived theoretically and compared with experimental data to provide the basis for Section 17.4, in which the processing of metallic glass-forming melts is described. There, methods of injection casting and superplastic forming of undercooled liquids that can be applied to today's bulk metallic glass-formers are discussed; also presented are a concept and preliminary experimental results on the co-extrusion of metals and polymers to obtain a new form of metal/polymer composites. Section 17.5 outlines mechanical properties at elevated and ambient temperatures, and discusses homogeneous and inhomogeneous deformation in metallic glasses. This topic is extended in Section 17.6, where the synthesis and processing of bulk metallic glass composites and their potential use as structural material are presented. The chapter concludes with Section 17.7, which highlights the various applications of (bulk) metallic glasses.

17.1 Introduction

17.1.1 General Terms

Rapid quenching of metals in water has been a familiar process for about two millennia, ever since the first blacksmiths began making tools and weapons from steel and other alloys strengthened by solid-solution or work hardening. Nowadays rapid quenching is still a major technique in materials manufacturing, and cooling rates depend greatly on the size of the component and the techniques used.

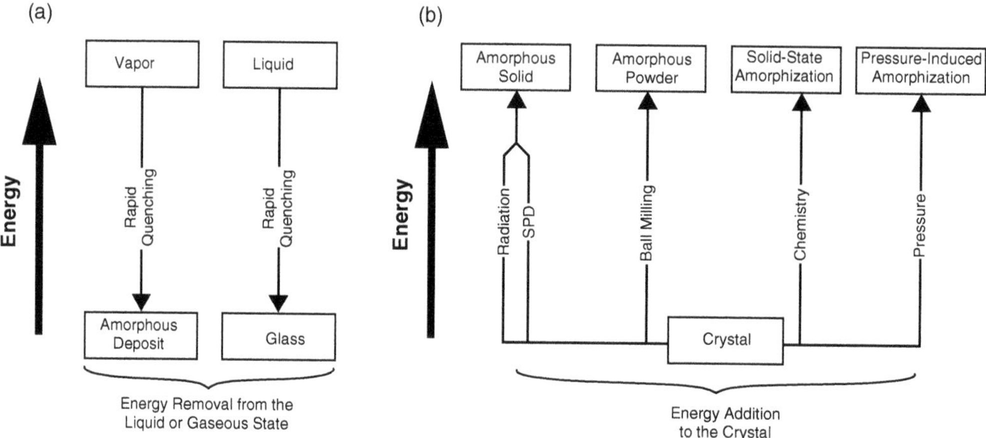

FIGURE 17.1 Schematic diagrams representing the various amorphization techniques, where (a) energy is extracted from a vapor or liquid phase, or (b) where energy has to be added to the system. Note that the term "glass" is restricted to an amorphous material created via cooling from the liquid state (SPD stands for severe plastic deformation).

As illustrated in Figure 17.1, amorphization techniques are basically divided into those methods where energy (heat) has to be rapidly extracted from the gaseous or liquid state[1–12] (Figure 17.1a), or where energy has to be added to the solid system (crystal)[5,7,13–17](Figure 17.1b). In the latter case, energy is provided to the crystal in the form of chemical reactions (solid-state amorphization[13]), radiation (amorphization by high-energy neutrons, protons, heavy ions or electrons[14] — see also Chapter 10), pressure (pressure-induced amorphization[15]), deformation and cold welding (mechanical milling[16]), or shear (severe plastic deformation[17]). More details about pressure and deformation effects are given in Chapter 13.

For historical reasons, only amorphous materials produced via rapid quenching from the liquid state are called *glass*. This does not apply to amorphous materials produced by techniques unrelated to quenching (Figure 17.1b). In this article, we have restricted ourselves to rapid solidification techniques from the liquid (and more rarely from the gaseous) state. For the amorphization techniques unrelated to quenching the reader is referred to References 13–17.

The first metallic glass (an eutectic Au–Si alloy[2]) was discovered in 1960 by P. Duwez and coworkers at CalTech, Pasadena. This discovery occurred by chance: the actual aim of the research project was to produce nonequilibrium solid solutions to verify whether the Hume–Rothery rule applies to solid solutions based on copper and noble metals.[18,19] Duwez successfully managed to quench Cu–Ag and Ag–Ge alloys[3] into a single phase, which showed that solid solubility can indeed be increased by rapid solidification from the melt, as shown earlier for vapor quenching.[1] Duwez's technique is still used today to produce metastable crystalline phases, as metastable phase formation can improve various properties as compared to the equilibrium crystalline state. Rapid solidification of crystalline alloys shows, for example, improved mechanical strength due to enhanced solid solubility and reduced grain size.[20] Examples for the formation of metastable phases upon rapid solidification are given in Reference 7.

The glassy Au–Si alloy discovered by Duwez was produced in dimensions of less than 50 μm, which is why a very high cooling rate of approximately 10^6 K/s resulted. Sample dimension and cooling rate are in fact closely related in rapid melt solidification. The lowest cooling rate required to produce a glassy material from the melt is termed "critical cooling rate," R_c. The maximum thickness at which a material can be produced glassy is termed "critical casting thickness," d_c. As heat must be extracted from the melt, the critical casting thickness can be calculated from the Fourier heat flow equation, $\partial T/\partial t = \alpha\nabla^2 T$, where $\alpha = \kappa/(c_p\rho)$ is the thermal diffusivity, κ is the thermal conductivity and $c_p\rho$ is the heat capacity per unit volume at constant pressure. The solution to this equation with the corresponding boundary conditions involves a characteristic cooling time, $\tau_c = d^2/\alpha$, required by a molten sample of typical dimension

d and initial temperature T_{liq} (liquidus temperature of the alloy) to achieve ambient temperature. The asymptotic cooling rate $\partial T/\partial t$ is then

$$\dot{T} = \frac{T_{liq}}{\tau_c} = \frac{\alpha T_{liq}}{d^2} = \frac{\kappa T_{liq}}{c_p \rho d^2}. \tag{17.1}$$

With typical values for a metallic melt [$\kappa \approx 4$ W/(m K), $c_p\rho \approx 2$ J/(cm^3 K), and $T_{liq} \approx 1000$ K], one obtains

$$\dot{T}(\text{K/s}) \approx \frac{20}{d^2(\text{cm}^2)}. \tag{17.2}$$

Thus, a sample of 50 μm thickness can be cooled at a rate of approximately 10^6 K/s, or, in other words, a liquid alloy needs to be cooled at a rate of 10^6 K/s to produce a glassy alloy with a critical casting thickness of $d_c = 50$ μm, while an alloy with critical cooling rate $R_c = 20$ K/s can be produced with a dimension of 1 cm.

Ever since the discovery of the first metallic glass, researchers have tried to improve glass-forming ability, the latter being synonymous with a reduction in the critical cooling rate or an increase in the critical casting thickness. In fact, a few years after Duwez's discovery, Chen and Turnbull were able to make glassy spheres of ternary Pd–Si–N with N = Ag, Cu, or Au,[21] and the alloy $Pd_{77.5}Cu_6Si_{16.5}$ could be quenched into a glass with a diameter of 0.5 mm. This alloy also showed a distinct glass transition, which motivated the Turnbull group to extend the general theories of the glass transition[22,23] to metallic glasses.[24,25] As outlined in more detail in Section 17.2, it was found that alloys with good glass-forming ability generally show a high "reduced glass transition temperature" T_{rg} ($\sim$2/3), which is defined as

$$T_{rg} = \frac{T_g}{T_{liq}}, \tag{17.3}$$

where T_g is the glass transition temperature and T_{liq} the liquidus temperature of the alloy.

Parallel to the discoveries of the first metallic glasses and the scientific interest they attracted, various production techniques were developed, such as gun quenching, splat quenching, or melt spinning, which facilitated the application of high cooling rates to metallic liquids (see details in Section 17.1.2). A strong interest in up-scaling emerged in the 1970s, when Fe-based metallic glasses with improved soft-magnetic properties were discovered (for an overview, see e.g. Reference 26). This led to the development of the planar flow casting technique (e.g. Reference 27), where amorphous ribbons of a few tens of microns in thickness, a few tens of centimeters wide, and several meters long could be produced and successfully implanted into transformer cores.

Further development and refinement of synthesis methods for ternary metallic glass systems (Pd–Cu–Si[28], Au–Pb–Sb[29], Pd–Ni–P[30,31]) showed the potential for metallic glasses to be produced in sample dimensions more closely representing a bulk material. However, such bulk metallic glasses were long regarded as laboratory curiosities and as nonreproducible on a larger scale. This situation changed in the 1990s when more complex quaternary and quinary metallic glasses, showing critical cooling rates of less than 100 K/s, were developed by the groups of Inoue (Tohoku University, Sendai, Japan)[32–34] and Johnson (CalTech, Pasadena).[35–37] These findings in turn motivated the development of liquid metal quenching methods, which enabled the casting of massive samples in a controlled manner and led to the first bulk metallic glasses.

Figure 17.2 shows various metallic glass-forming alloys plotted according to their critical cooling rate (and associated critical casting thickness) as a function of their reduced glass transition temperature (see also Reference 38). The rapid solidification techniques (discussed in more detail in Section 17.1.2) are also correlated with the quenching rate. The highest possible quenching rate needed for an elemental face-centered cubic (fcc) metal to amorphize cannot be reached experimentally, but has been achieved in computer simulations.[39,40] The highest experimental cooling rate has so far been attained using laser quenching (R_c estimated in Reference 38). However, this technique is restricted to small droplets corresponding to

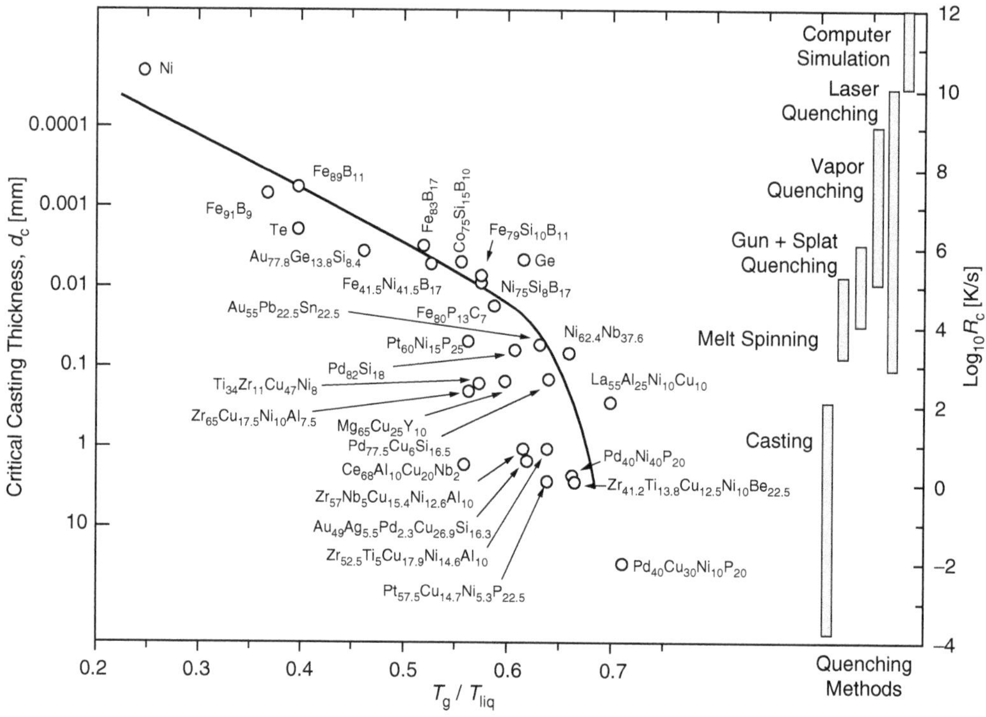

FIGURE 17.2 Critical casting thickness, d_c, and critical cooling rate, R_c, vs. reduced glass transition temperature T_g/T_{liq} for various metallic glass-forming alloys and correlated with the various quenching techniques.

the spot size of the high-intensity laser beam, and is thus not discussed further. In Section 17.1.2 the other techniques shown in Figure 17.2 are presented in the descending order of their achievable cooling rates.

17.1.2 Rapid Quenching Techniques

17.1.2.1 Vapor Quenching

Vapor quenching, or gas condensation, was initially used for research purposes and goes back to the work of Kramer,[41] and of Buckel and Hilsch.[1] They deposited metal films from an evaporation source onto a liquid-helium-cooled substrate, which led to increased solid solubility or, in a few cases, to amorphization of the metal film. Later work has concentrated on processes to increase the dimensions of possible products either by sputtering[42] or vapor quenching.[43]

With the investigation of nanocrystalline metals instigated by the work of Gleiter and coworkers,[44,45] vapor quenching via an inert-gas condensation technique (IGC) or via sputtering has now re-emerged in the research community. The goal of this ongoing research field is to study the size effects of mechanical, structural, and other physical properties in elemental metals and alloys. In contrast to the other methods mentioned above, in IGC the metal solidifies first as a powder and is then compacted to solid samples in a second step. Figure 17.3 shows a schematic drawing of an IGC facility. The system is divided into four parts: a loading unit (A), a main ultrahigh vacuum chamber (B), a compression unit (C), and a special charging device (D) to potentially press powders prepared by other techniques. The material is thermally evaporated from a resistive heating device (e.g., a tungsten boat [E]) into a 10^{-6} Pa base pressure vacuum system filled with 100 Pa of 99.999% pure He gas (B). During the IGC process (F) the atoms in the vapor phase condense to form small nanometer-sized crystallites that are collected on a rotating cylinder (G) cooled by liquid nitrogen. The powder is scraped off (H) into a funnel (I) and transferred to the compaction unit (C). After re-evacuation to the initial pressure of 10^{-6} Pa the powder is consolidated by pressing

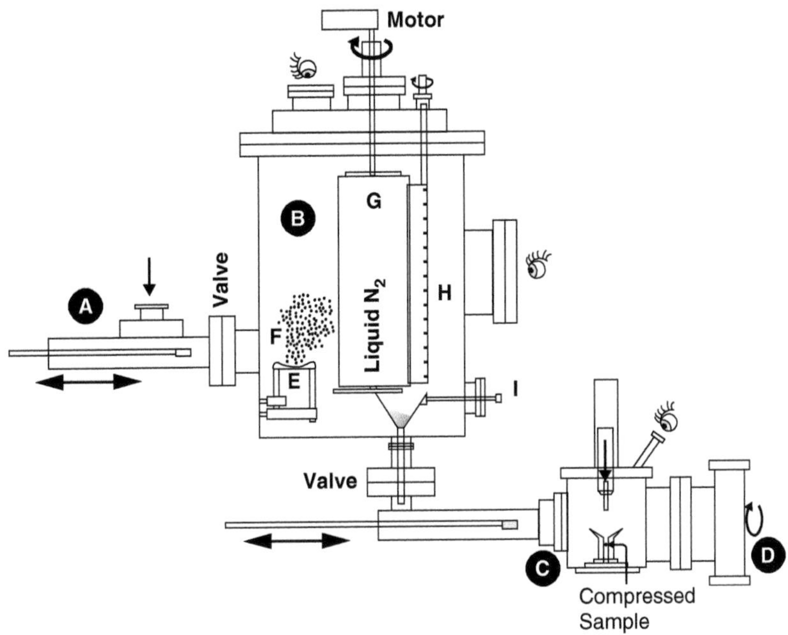

FIGURE 17.3 Schematic drawing of an inert-gas condensation facility to produce nanocrystalline metallic alloys by vapor quenching and consolidation.

(1 to 2 GPa) at room temperature or elevated temperatures. This technique allows bulk nanocrystalline materials to be produced with typical dimensions of 1 cm diameter and 1 mm height and crystal sizes of approximately 10 nm. Research into these nanocrystalline materials has contributed significantly to an understanding of the influence of small crystallites on physical properties, as compared to that of their coarse-grained counterparts.[46]

17.1.2.2 Gun and Splat Quenching, Melt Spinning

The first glassy alloys were prepared using a so-called "gun quenching" technique. With this technique, invented by Duwez and coworkers, it was possible to achieve cooling rates of $\sim 10^6$ Kelvin per second. A detailed description is given in Reference 47. The gun quenching technique led to the development of an apparatus capable of producing larger sample dimensions. The so-called "splat quencher" or "drop smasher," shown in Figure 17.4a, enables the production of discs of approximately 1 to 4 cm in diameter with thicknesses of up to 50 μm. A small mass is levitated in an induction coil using a high frequency current and melted. The current is then switched off; the molten droplet falls and triggers, via a laser signal, two pneumatically or electromagnetically driven flat pistons that collide at high speed and quench and flatten the molten drop to a "splat" (a thin foil) at a cooling rate of $\sim 10^6$ K/s. The configuration used today is adapted from that shown in Figure 17.4a.

Further studies of metallic glasses have increased the complexity of systems and the number of constituent elements, leading to a decrease in the critical cooling rate (Figure 17.2). Once the potential of metallic glasses for various applications (in particular in their promising soft-magnetic properties) was recognized, it was seen that a scaling-up method was needed to deploy metallic glasses commercially. In the 1970s several researchers developed continuous casting or melt spinning methods in response to this requirement.[27,48,49] These techniques facilitated the production of metallic ribbons cast onto a rotating copper wheel, as is shown schematically in Figure 17.4b. Further improvements and variations of the melt spinning apparatus resulted in glassy metal ribbons of several meters in length and a few centimeters in width, using either a single wheel or a twin-roller configuration where the melt is quenched in between (see also Reference 47). In principle there are two main differences in how the metal is fed onto the rotating

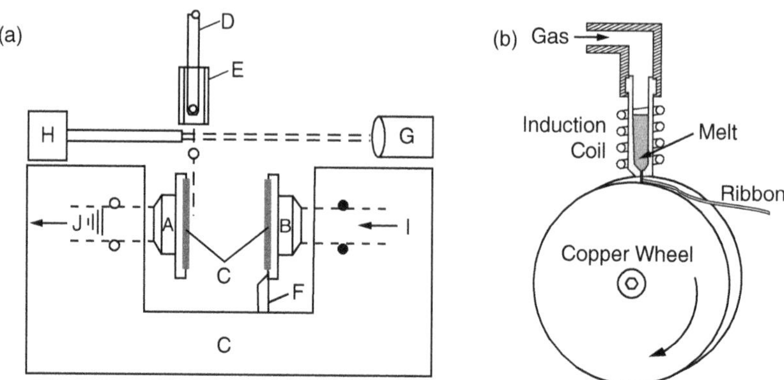

FIGURE 17.4 Schematic drawing of (a) splat quencher and (b) melt spinner. (17.4a from Johnson, W.L., *Metals Handbook* 2, 805, 1990. With permission.)

substrate. In chill-block melt spinning the liquid is pressed with an inert gas from the nozzle, from which a liquid stream of metal drops onto the spinning copper wheel (see Figure 17.4b).[50] In pendant-drop melt extraction all of the solid metal is fed from above and heated inductively just above the spinning wheel, where the molten phase is continuously dragged away by the rotor.[51] This method also makes possible the production of wires using a knife edge on the copper wheel.

In planar flow casting, an advanced version of the classical chill-block melt spinning, a narrow, precise distance between the nozzle and the copper wheel is maintained to constrain the melt between the wheel and the nozzle and generate a stable melt flow. Using this process it is possible to use rectangular nozzles of several tens of centimeters in width and manufacture large amorphous metal sheets.[52] This method has been applied commercially since the 1970s in the production of soft-magnetic ferrous foils for use in transformer cores (see also Section 17.7.1). A detailed description of this and other rapid quenching techniques can be found in Reference 53.

17.1.2.3 Copper Mold Casting and Injection Casting

In the early 1990s, the development of complex multicomponent metallic glasses[32–37] with very low critical cooling rates (even down to 1 K/s[35]) provided a new and so far uninterrupted scientific and commercial impetus to the search for new bulk metallic glasses and research into their physical properties and possible applications. It is now possible to quench bulk metallic glass samples in large copper molds with either inductive (radio frequency, RF) or inert-gas plasma arc heating. There are two basic techniques for forcing the liquid into a mold: either (1) the use of an inert-gas pressure difference (via inert-gas pressure injection or suction casting), or (2) piston injection, where a solid piston shoots the melt into a copper mold. These systems have so far mainly been used in configurations similar to that of Figure 17.5, where the material is kept in either a quartz or boron-nitride crucible during heating, before being cast into the mold via application of a pressure difference. Various versions of the setup shown have also been built to create the special conditions required for the successful casting of metals such as Ca-based or Mg-based bulk metallic glasses that are reactive and have a low vapor pressure. For more information on the varieties of copper mold casting techniques, the reader is referred to Reference 8. Apart from these synthesis methods, various processing methods for discovering new bulk metallic glass compositions have also been developed in recent years. The high-temperature centrifugation method,[54,55] for example, can in multicomponent alloys physically isolate deep eutectic compositions, which have a high potential for good glass-forming ability.

17.2 Microstructure of Metallic Glasses

It has been shown that different processing techniques generate different atomic structure and free volume in metallic glasses, and that these differences are very sensitive to the cooling rates with which they are

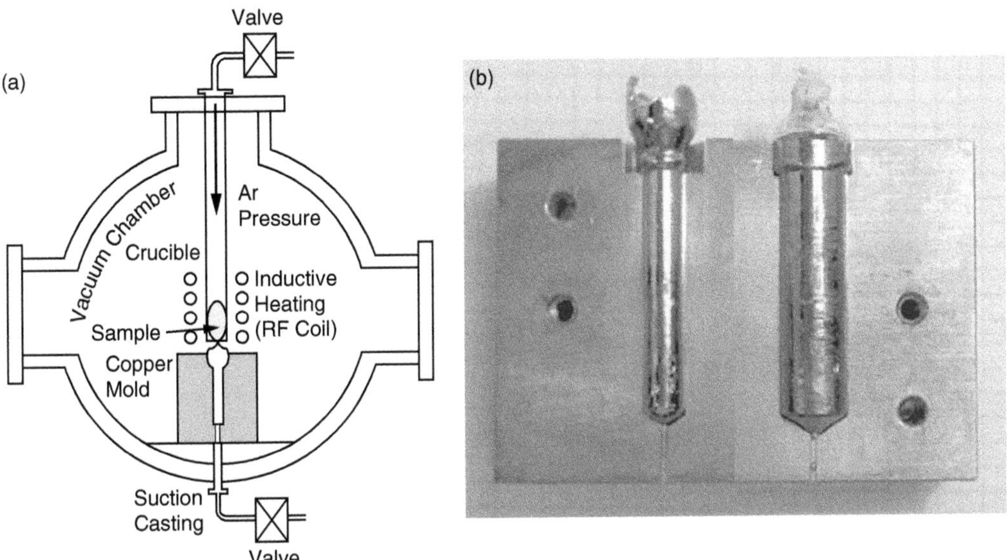

FIGURE 17.5 (a) Schematic drawing of a casting box, in which the samples are inductively heated and cast by either suction or pressure injection; (b) close-up photograph of a copper mold with cast Zr-based bulk metallic glass samples of diameters 5 and 10 mm.

produced. This in turn can have a drastic effect on mechanical and other physical properties: Mg-based bulk glasses, for example, are brittle if cast into larger samples, but bend in a ductile manner if produced at higher cooling rates.[56] To highlight this strong relationship between microstructure, processing, and properties we present some experimental methods for determining the microstructure of metallic glasses, point out the most relevant atomic models, and classify bulk metallic glasses according to their atomic structure.

17.2.1 Atomic Models for Metallic Glasses

17.2.1.1 Experimental Methods

Rapidly solidified amorphous alloys, as opposed to crystalline equilibrium alloys, are in a metastable state of nonperiodic atomic arrangement. This gives rise to typical diffuse rings in two-dimensional diffraction patterns, as shown in Figure 17.6a, or broad humps in a powder diffraction pattern (Figure 17.6b). In high-resolution transmission electron microscopy (HR-TEM) such metallic glasses show an irregular contrast resulting from the lack of any long-range periodicity in their atomic structure (Figure 17.6c). Metallic glasses show, however, some atomic short-range order on a length scale that is similar to the constituent atoms. The methods generally used to accurately detect such atomic short-range order are synchrotron radiation or neutron diffraction experiments.[57]

Using these techniques, the radial distribution function (RDF), describing the probability of interatomic distances, can be derived. Here the scattering vector **Q** arising from the x-ray diffraction of the amorphous solid is described by

$$Q = |\mathbf{Q}| = \frac{4\pi}{\lambda} \sin\theta, \qquad (17.4)$$

where θ is the diffraction angle and λ is the x-ray wavelength. The atom–atom interference function, $S(Q)$, is derived from the total scattering intensity after subtraction of inelastic x-ray (Compton scattering) and

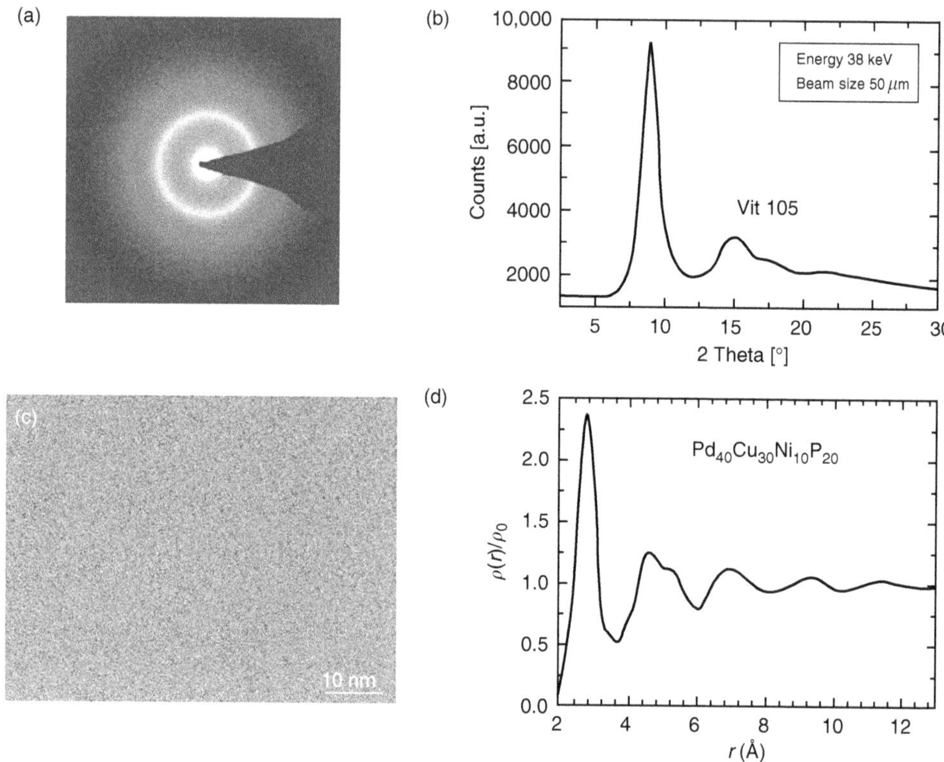

FIGURE 17.6 (a) Selected-area electron diffraction pattern of $Zr_{52.5}Ti_5Cu_{17.9}Ni_{14.6}Al_{10}$ (Vit105); (b) transmission synchrotron x-ray diffraction pattern of Vit105; (c) high-resolution transmission electron microscopy image of La–Cu–Al-based metallic glass; (d) normalized pair distribution function of amorphous $Pd_{40}Cu_{30}Ni_{10}P_{20}$. (17.6d from Yavari, A.R. et al., *Acta Mater.* 53, 1611, 2005. With permission.)

multiple scattering intensities,[58] that is

$$S(Q) = \frac{I(Q)}{\langle f(Q)\rangle^2} + 1 - \frac{\langle f(Q)^2\rangle}{\langle f(Q)\rangle^2}, \quad (17.5)$$

where $I(Q)$ is the normalized elastic scattering intensity after corrections for polarization and absorption, $f(Q)$ is the atomic scattering factor and $\langle \ldots \rangle$ refers to the composition average.[57] After normalizing $S(Q)$ to unity for large **Q** vectors, the atomic pair distribution function (PDF), $\rho(r)$, is obtained by Fourier transformation of $S(Q)$,

$$\rho(r) = \rho_0 + \frac{1}{2\pi^2 r}\int_{Q_{\min}}^{Q_{\max}} Q\,[S(Q) - 1]\sin(Qr)\,dQ, \quad (17.6)$$

where ρ_0 is the average atomic number density, $4\pi r^2\rho(r)$ is the RDF,[58–60] and the area under the first peak in the RDF provides the nearest-neighbor coordination number (see Figure 17.6d).[57] More recently, anomalous x-ray scattering techniques (measurement of the electron–photon resonance at the absorption edge), HR-TEM and TEM convergent-beam or nano-beam diffraction experiments have also been successfully applied to tackle the atomic structure of bulk metallic glasses experimentally.[10,61,62] Based on these and earlier results on the amorphous structure of solids and liquids, theoretical models were derived to assess the topology of the atomic structure.

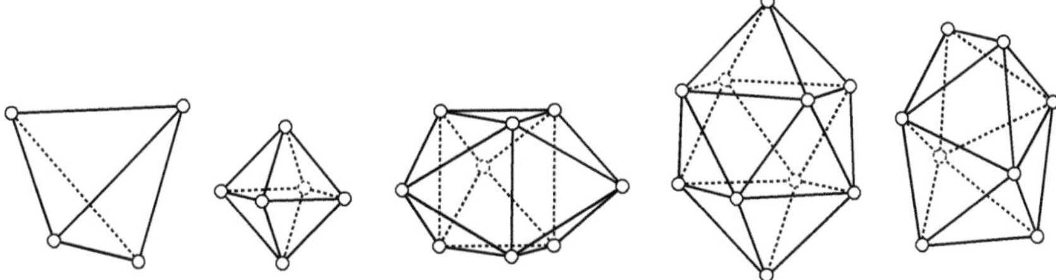

FIGURE 17.7 Bernal's "canonical holes," representing the different polyhedral void structures. These are tetrahedron, octahedron, trigonal prism, Archimedian anti-prism, and tetragonal dodecahedron (from left to right). (After Bernal, in Cahn, R.W. [Chap. 28], in *Physical metallurgy*, Cahn, R.W. and Haasen, P. (Eds.), Elsevier Science, New York, 1983, p. 1779. With permission.)

17.2.1.2 Theoretical Models

Due to the long-range order in crystalline solids all atoms form part of a unit cell, the latter being the smallest volume of atoms that can be translated along a lattice vector in the crystal without causing any overlap. Glasses, in contrast, can only be described in a statistical manner because of their lack of long-range order, and the RDF or PDF reflects the likelihood of finding the center of an atom in a distance r from a central atom. Therefore, models and simulations where the structure is precisely known and from which simulated PDFs may be derived and compared to experimental results are essential for an understanding of the structure of glassy solids and liquids.

Hard sphere models applied to crystalline solids have shown the closest packing orders for fcc and hexagonal closed-packed (hcp) structures, where in a 3-D space each sphere is coordinated by its 12 nearest neighbors, yielding a packing efficiency of 0.74. The void structure between the spheres forms tetrahedrons or octahedrons. In glassy materials, however, atoms are *randomly closed packed.* Bernal's[63–65] empirical model, the *dense random packing hard sphere* (DRP-HS) model, based on the structure of hard steel balls filling an irregular surface, yielded a packing efficiency much reduced compared to hcp and fcc structures of 0.60 to 0.63. The average likelihood of atomic positions resulted in a nearest-neighbor coordination number of 14.25 atoms and separated voids of different polyhedral shapes, whose vertices are defined by the (atomic) sphere centers, as shown in Figure 17.7. Polyhedral coordinated atomic clusters defined by Bernal's so-called "canonical holes" and the nearest-neighbor atoms can also be imagined.[66,67]

Regardless of the fairly good match of the PDF derived from the DRP-HS model with that of experimentally measured PDFs (e.g., on amorphous $Fe_{40}Ni_{40}P_{14}B_6$[68]), the DRP-HS model yields an unrealistically low density. This led to a modification of the DRP-HS model, which was consequently replaced by a *soft sphere model* increasing the packing efficiency to 0.69 to 0.70, in closer agreement with density measurements of amorphous metals and the observed PDFs.[69,70]

Despite this agreement with the experimental results for metallic glasses, the deficiency of these purely geometrical DRP models in being unable to cope with smaller metalloid atoms (whose presence generally improves glass-forming ability) and any variation in chemistry led to the so-called *local cluster models* (see, e.g., Reference 71). These models are particularly important for bulk metallic glasses, as they take into account the large variety of atomic species of different atomic diameter needed to frustrate crystallization. Following the proposal of Polk,[72] where small atoms such as those coming from metalloids (MO) fill the larger Bernal holes without distorting the DRP skeleton (formed by the larger transition metal [TM] atoms), Gaskell[70] showed that this was only possible if the smaller MO atoms had a diameter $\leq$0.48 that of the larger TM atoms. Such a structure is only possible for atoms like carbon and boron, but not for phosphorous or silicon. Gaskell therefore proposed a model where atoms are oriented in prisms analogous to the smallest stable crystalline phase.[70] Metallic glasses consisting of TM and MO atoms were thus suggested as forming clusters of cupped trigonal prisms among MO, which are loosely interconnected and possess orientations strongly dependent on the TM–MO attraction and their size. Today Gaskell's *network model*[70] is still applied to TM–MO glassy metals (see also next section) and has been proven to reflect satisfactorily the experimentally deduced PDFs for various bulk metallic glass systems.[73,74]

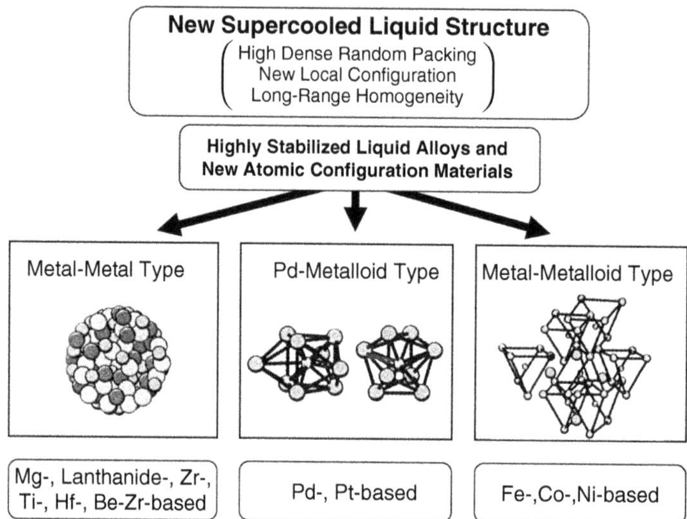

FIGURE 17.8 Classification of bulk metallic glasses according to their difference in local topological order. (From Inoue, A. and Takeuchi, A., *Mater. Trans. JIM* 43, 1892, 2002. With permission.)

17.2.2 Modern Classification

The findings concerning the atomic structure of amorphous alloys helped researchers to elaborate several rules with the goal of improving glass-forming ability,[9,11,75–78] leading to a decrease in the critical cooling rate and consequently to the generation of new bulk metallic glasses. The most-cited rules aiming to suppress crystallization are given in Reference 76, and can be summarized as follows: (1) the alloys must be composed of at least three elements; (2) the atomic size ratio of the constituent elements must be above 12%; and (3) a negative mixing enthalpy between the different elements must be present.

In an attempt to classify the wide range of recently-developed bulk metallic glasses, Inoue distinguished between three main groups. These are separated according to their chemical compositions and different atomic configurations into metal–metal-type alloys, metal–metalloid-type alloys (e.g., Fe–B or Ni–P), and Pd–metalloid-type alloys (see Figure 17.8).[79,80]

Experimental studies on the atomic structure of *metal–metal-type bulk metallic glasses* such as Zr–Al–Ni–Cu, Hf–Al–Ni–Cu and Zr–Ti–Cu–Ni–Be systems by means of HR-TEM, XRD, and neutron diffraction revealed an atomic structure consisting of icosahedral clusters,[61,62,81–85] as already suggested much earlier for liquids.[86] It was shown that an icosahedron is more stable than an fcc structure, because for a cluster of 13 atoms an icosahedron is connected through tetrahedrons making up 42 bonds, while an fcc structure is built by an octahedron and tetrahedron resulting in only 36 bonds.[81] On a local level where the nearest-neighbor atoms are bonded the stability is a function of the number of bonds, which can be increased by forming triangles in two dimensions and tetrahedrons in three dimensions. This leads to a geometric incompatibility in metallic glasses, because on a macroscopic level a 3-D structure consisting solely of tetrahedrons is, due to the strong mismatch and distortion energy required, less favorable than a mixture between tetrahedrons and octahedrons (i.e., closed-packed structure). This contributes to the stability of the glassy structure, as atoms do not know which configuration they should take.[87] Icosahedral structures have also been found in the supercooled liquid region of bulk metallic glass alloys prior to crystallization,[61,82,83,88,89] which may be explained by the lower activation energy needed for the nucleation of an icosahedral phase than for a closed-packed one.[90]

As mentioned earlier, *metal–metalloid-type systems* such as Fe–Lanthanide–B and Fe–M–B (M = Zr, Hf, Nb, Ta) consist of trigonal prisms where the smaller metalloids are located inside a skeleton made up by the TM atoms.[70] More recent studies on bulk metallic glasses consisting of TM–MO have (as for conventional TM–MO glasses) shown that the structure can be satisfactorily described by a variation of the DRP model. Consequently, such structures exhibit higher densities than metal–metal-type metallic glasses.[73,74]

Pd–metalloid-type bulk metallic glasses, among which the Pd–Cu–Ni–P system shows the highest glass-forming ability of all bulk metallic glasses, are composed mainly of two large structures: transformed tetragonal dodecahedrons, which comprise Pd, Cu, and P, and trigonal prisms capped with three half-octahedral polyhedrons occupied by Pd, Ni, and P.[10] The significantly greater stability of the supercooled liquid and therefore the improved glass-forming ability of the Pd–Cu–Ni–P system compared to that of Pd–Ni–P, the latter built by trigonal prism clusters only, has been associated with the difficulty of rearranging the two clusters in a crystallographically ordered structure, their stable coexistence, and the high bonding strength of metal–metalloid atomic pairs in the cluster units.[10]

17.3 Thermophysical Properties of Metallic Glasses

As mentioned above, metallic glasses are amorphous metals quenched from the liquid state at a rate high enough to suppress crystal nucleation and growth. In this section, the rate of nucleation and growth will be derived as a function of undercooling. It will become evident why the reduced glass transition temperature T_{rg} (see Figure 17.2) and other parameters such as viscosity and fragility are important in describing glass-forming ability. In the context of homogenous nucleation and growth, a time–temperature–transformation (TTT) diagram for undercooled liquids is described theoretically and then compared to experimental data. The TTT diagram provides the basis for understanding how metallic liquids need to be processed to form bulk metallic glass parts.

17.3.1 Nucleation Rate and the Ability to Form a Glass

The ability to form a glass by cooling from an equilibrium liquid is equivalent to suppressing crystallization within the undercooled liquid. If a liquid is cooled below the melting point, the free energy difference between liquid and crystal will provide a driving force for crystal nucleation, while the creation of the liquid–crystal interface will create a positive interfacial energy that disfavors nucleation. This results in an energy barrier that a local composition fluctuation must overcome to form a nucleus. Easily derived, for a spherical nucleus this energy barrier is

$$\Delta G^* = \left(\frac{16\,\pi\,\gamma^3}{3\,(\Delta G_v)^2}\right), \tag{17.7}$$

where ΔG_v is the difference in Gibbs free energy (per unit volume) between liquid and crystal, and γ is the interfacial energy (per unit area).[25]

ΔG_v can be calculated from the difference in molar heat capacity at constant pressure between liquid and solid, Δc_p^{l-x}, the molar enthalpy of fusion (at the melting temperature), ΔH_m^f, the molar entropy of fusion, $\Delta S_m^f (= \Delta H_m^f / T_{liq})$ and the liquidus temperature, T_{liq}, by (e.g., Reference 91)

$$\Delta G_v(T) = \frac{1}{V_m}\left\{\left(\Delta H_m^f - \int_T^{T_{liq}} \Delta c_p^{l-x}(T')\,\mathrm{d}T'\right) - T\left(\Delta S_m^f - \int_T^{T_{liq}} \frac{\Delta c_p^{l-x}(T')}{T'}\mathrm{d}T'\right)\right\}. \tag{17.8}$$

All the variables in the formula can be derived from quantitative heat capacity measurements using differential scanning calorimetry.

Experimental data concerning the molar Gibbs free energy difference, $\Delta G^{l-x}(T) = V_m \Delta G_v(T)$, with V_m the molar volume, are given in Figure 17.9 for selected metallic glasses.[92] The various alloys show different increases in ΔG^{l-x}. The best glass-formers, $Pd_{43}Cu_{27}Ni_{10}P_{20}$ (R_c = 0.2 K/s; Reference 93) and $Zr_{41.2}Ti_{13.8}Cu_{12.5}Ni_{10}Be_{22.5}$ (1 K/s; Reference 94), clearly show a flatter increase in ΔG_v with relative

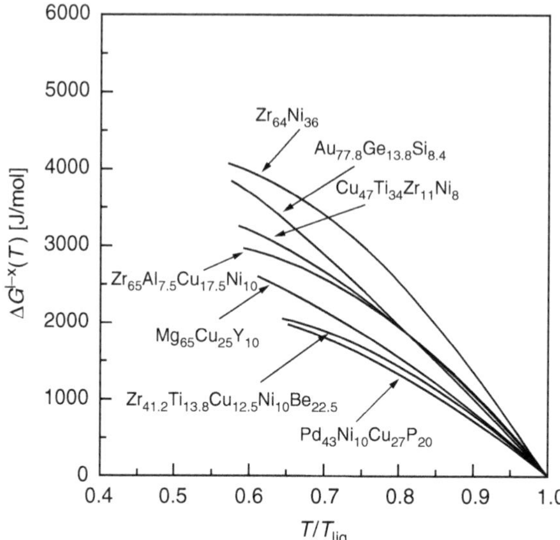

FIGURE 17.9 Molar Gibbs free energy differences between liquid and crystal $\Delta G^{l-x}(T) = V_m \Delta G_v(T)$ as a function of temperature for various metallic glass-forming melts. (From Fan, G.J., Löffler, J.F., Wunderlich, R.K., and Fecht, H.J., *Acta Mater.* 52, 667, 2004. With permission.)

undercooling, $\Delta T_r = 1 - (T/T_{liq}) = 1 - T_r$, than the marginal glass-former $Zr_{64}Ni_{36}$ ($R_c = 10^5$ K/s). The glass-formers $Mg_{65}Cu_{25}Y_{10}$, $Zr_{65}Al_{7.5}Cu_{17.5}Ni_{10}$, $Cu_{47}Ti_{34}Zr_{11}Ni_8$ and $Au_{77.8}Ge_{13.8}Si_{8.4}$ with critical cooling rates from 50 to 10^4 K/s lie in between.

In the following the temperature dependence of the Gibbs free energy difference is approximated as a function of undercooling, that is

$$\Delta G_v = \zeta(T)\frac{\Delta S_m^f}{V_m}(T_{liq} - T) = \zeta(T)\frac{\Delta H_m^f}{V_m}\Delta T_r, \qquad (17.9)$$

where $\zeta(T)$ is a temperature-dependent correction factor that decreases slowly from 1 at T_{liq} to $\sim$0.7 at T_g.

The nucleus must form statistically by random atomic movement into a close-packed structure within the liquid. If the nucleus is smaller than the critical radius $r^* = 2\gamma/\Delta G_v$, the nucleus will dissolve. If, however, the nucleus is (in a few rare cases) larger than r^* the total free energy decreases with the size of the nucleus, and the latter is able to grow. The resulting crystal nucleation rate per unit volume, I_v, is thus the product of a thermodynamic term that depends on the probability of a fluctuation to overcome the nucleation barrier, and a kinetic term that depends on the atomic transport (described by the atomic diffusivity, D, or viscosity, η), that is,

$$I_v = \frac{A_v}{\eta(T)}\exp\left(-\frac{\Delta G^*}{k_B T}\right), \qquad (17.10)$$

where A_v is a constant on the order of 10^{32} Pa s/(m^{3}s) and k_B is the Boltzmann constant.

Introducing the dimensionless parameters

$$\tilde{\alpha} \equiv \frac{(N_A V_m^2)^{1/3}\gamma}{\zeta(T)\Delta H_m^f} \quad \text{and} \quad \tilde{\beta} \equiv \frac{\zeta(T)\Delta S_m^f}{R}, \qquad (17.11)$$

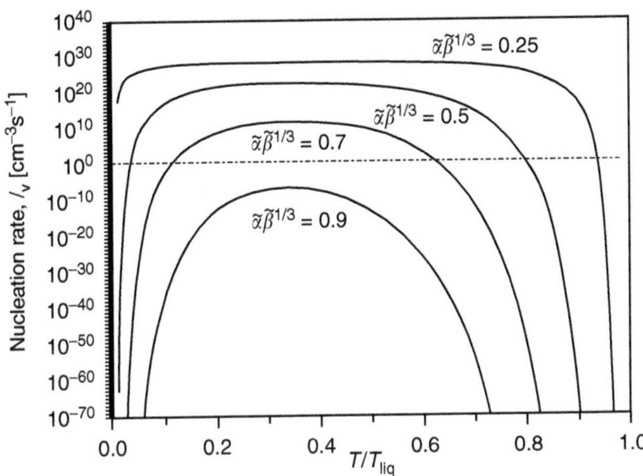

FIGURE 17.10 Homogeneous nucleation rate (Equation 17.14) as a function of reduced temperature with $\tilde{\alpha}\tilde{\beta}^{1/3}$ as parameter and $A_v/\eta = 10^{29}$ cm^{-3}s^{-1}. The dotted line indicates the range where one nucleus forms per second in a sample of volume $V = 1$ cm^3. This may be taken as the upper limit for obtaining a metallic glass by quenching.

where N_A is Avogadro's constant and R is the gas constant, the combination of Equation 17.7 and Equations 17.9–17.11 results in a nucleation rate of

$$I_v = \frac{A_v}{\eta} \exp\left(-\frac{16\pi}{3}\frac{\tilde{\alpha}^3\tilde{\beta}}{(\Delta T_r)^2 T_r}\right). \tag{17.12}$$

For metals, $\tilde{\alpha} \approx 0.4$–$0.5$, as $\gamma \approx 0.1$–0.5 J/m^2 and $\Delta H_m^f \approx 10$ kJ/mol (setting $\zeta(T) \approx 1$). Further, $\tilde{\beta} \approx 1$, as $\Delta S_m^f \approx R$. For nonmetals, $\tilde{\alpha} \approx 1/3$ and $\tilde{\beta}$ typically varies from 4 to 10.[95]

To obtain an upper bound for the nucleation rate, we first assume a constant viscosity of 10^{-3} Pa s, which is a typical value for the equilibrium melt viscosity of a metal, and plot the nucleation rate as a function of the reduced temperature $T_r = T/T_{liq}$ with $\tilde{\alpha}\tilde{\beta}^{1/3}$ as a parameter (Figure 17.10). For metals, $\tilde{\alpha}\tilde{\beta}^{1/3} \approx 0.4$ to 0.5; for nonmetals, $\tilde{\alpha}\tilde{\beta}^{1/3} \approx 0.5$ to 0.7. Apparently, the nucleation rate rises steeply with increasing undercooling, but nucleation is often negligible down to $\Delta T_r \approx 0.2$.

A more realistic approach is to assume that the viscosity follows a Vogel–Fulcher-type temperature dependence, $\eta(T) = \eta_0 \exp[B/(T - T_0)]$. According to Angell's fragility concept,[96,97] the viscosity of liquids is commonly described by a modification of the Vogel–Fulcher–Tammann (VFT) relation,

$$\eta(T) = \tilde{\eta} \exp\left(\frac{D^* T_0}{T - T_0}\right), \tag{17.13}$$

where D^* is the fragility parameter ($1 \leq D^* \leq 100$), T_0 the VFT temperature, and $\tilde{\eta}$ a constant inversely proportional to the molar volume of the liquid. Physically, T_0 is the temperature where the barrier would become infinite with respect to flow. The fragility describes the degree to which the viscosity of a supercooled liquid deviates from an Arrhenius behavior. Liquids are commonly referred to as "fragile" when $D^* < 10$ and "strong" when $D^* > 20$. Strong liquids have a high equilibrium melt viscosity and show a more Arrhenius-like temperature dependence in the viscosity than fragile liquids. For pure metals, D^* is close to unity, while for the network glass SiO_2, the classical example of a "strong" glass former, $D^* = 100$. The viscosity of SiO_2 shows an Arrhenius-like temperature dependence.

According to this concept, Equation 17.12 takes, in dimensionless parameters, the form

$$I_v = \frac{A_v}{\tilde{\eta}} \exp\left(-\frac{D^* T_{r0}}{T_r - T_{r0}}\right) \exp\left(-\frac{16\pi}{3} \frac{\tilde{\alpha}^3 \tilde{\beta}}{(\Delta T_r)^2 T_r}\right), \tag{17.14}$$

where $T_{r0} = T_0/T_{liq}$ is the reduced VFT temperature. As has recently been shown, the VFT temperature T_0 is significantly lower than the glass transition temperature T_g, and can be as low as 0.6 T_g.[98] Nevertheless, we assume in the following that $T_{r0} \approx T_{rg}$ (with T_{rg} the reduced glass transition temperature) to obtain an upper bound for the nucleation rate.

Apparently there are two important parameters that determine the nucleation rate: the reduced VFT temperature T_{r0} (or reduced glass transition temperature T_{rg}), and the fragility parameter D^*. While there are many experiments that determine the fragility parameter in metallic glasses,[98] its influence on the nucleation rate has not so far been parameterized. We therefore take into consideration the reduced glass transition temperature T_{rg}, as did Turnbull[25] (see Figure 17.11a), and the fragility parameter D^* (see Figure 17.11b). To visualize the influence of T_{rg} on the nucleation rate, Figure 17.11a

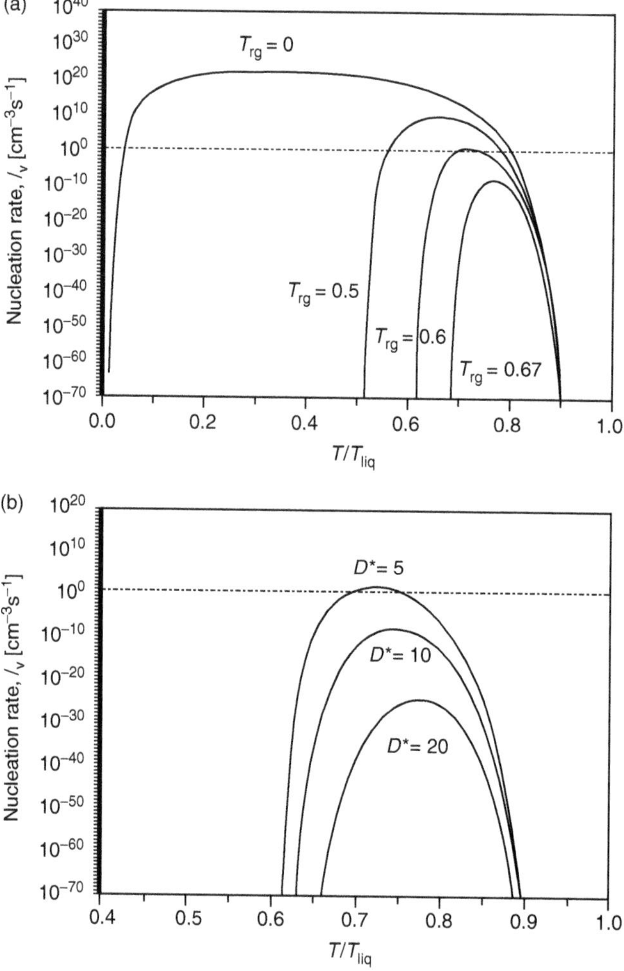

FIGURE 17.11 Homogeneous nucleation rate (Equation 17.14) as a function of the reduced temperature, with $\tilde{\alpha}\tilde{\beta}^{1/3} = 0.5$ and $A_v/\tilde{\eta} = 10^{-29}$ cm^{-3}s^{-1}. (a) $D^* = 5$ and T_{rg} ($\approx T_{r0}$) as parameter; (b) T_{rg} ($\approx T_{r0}$) $= 0.6$ and D^* as parameter.

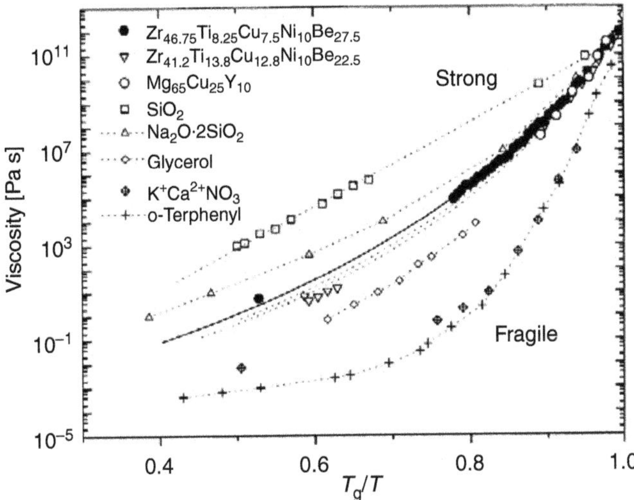

FIGURE 17.12 Fragility plot for various oxide glass-formers, organic glass-formers, and three metallic glass-formers. (Reproduced from Busch, R., Bakke, E., and Johnson, W.L., *Acta Mater.* 46, 4725, 1998. With permission.)

TABLE 17.1 Fragility Parameter, D^*, Critical Cooling Rate, R_c, and Reduced Glass Transition Temperature, T_{rg}, for Selected Metallic Glasses

Alloy	D^*	R_c [K s^{-1}]	T_{rg}	Additional Ref.
SiO_2	100	<0.1	0.67	
$Mg_{65}Cu_{25}Y_{10}$	23	20	0.68	103
$Zr_{41.2}Ti_{13.8}Cu_{12.5}Ni_{10}Be_{22.5}$ (Vit 1)	22.7	1	0.66	104
$Zr_{46.75}Ti_{8.25}Cu_{7.5}Ni_{10}Be_{27.5}$ (Vit 4)	20.4	18	0.62	98
$Pd_{48}Ni_{32}P_{20}$	15.1	10 (if fluxed)	0.66	105
$Zr_{65}Al_{7.5}Cu_{17.5}Ni_{10}$	13.8	50	0.60	
$Zr_{60}Ni_{25}Al_{15}$	11.6	10^2	0.60	32
$Pd_{77.5}Cu_6Si_{16.5}$	6.3	10^3	0.58	105
$Au_{77.8}Ge_{13.8}Si_{8.4}$	8.4	10^5	0.55	106
$Zr_{64}Ni_{36}$	7.9	10^6	0.50	

Source: Modified after Johnson, W.L., in *Science of Alloys for the 21st Century: A Hume-Rothery Symposium Celebration,* P.E.A. Turchi, R.D. Shull and A. Gonis (Eds.), The Minerals, Metals & Materials Society, 2000. With permission.

shows the nucleation rate as a function of the reduced temperature $T_r = T/T_{liq}$ with $\tilde{\alpha}\tilde{\beta}^{1/3} = 0.5$, $A_v/\tilde{\eta} = 10^{-29}$ cm^{-3}s^{-1}, and $D^* = 5$. To visualize the influence of D^*, Figure 17.11b shows the nucleation rate as a function of the reduced temperature T_r with $\tilde{\alpha}\tilde{\beta}^{1/3} = 0.5$, $A_v/\tilde{\eta} = 10^{-29}$ cm^{-3}s^{-1}, and $T_{rg} = 0.6$. Apparently, while fragile liquids with $D^* \leq 5$ crystallize when $T_{rg} = 0.6$, stronger liquids with $T_{rg} = 0.6$ can be quenched into a glass.

By fitting the VFT relation (Equation 17.13) to experimental data, values of $D^* = 22.7$ and $T_0 = 372$ K were obtained for the alloy $Zr_{46.75}Ti_{8.25}Cu_{7.5}Ni_{10}Be_{27.5}$.[98] The viscosity data as a function of T_g/T is visualized in the fragility plot[97] in Figure 17.12. The viscosity data were derived from beam-bending experiments up to a temperature of 100 K above T_g[99] and from additional melt viscosity data.[100] As the timescales for the experiments at temperatures between T_g and T_{liq} are very short, noncontact measurement techniques have also recently been used to study the viscosity of undercooled liquids.[101]

Table 17.1 shows experimental data for D^* and the critical cooling rate of several metallic glass-forming alloys. The data show that good glass-formers indeed have a high fragility index of $D^* > 20$.

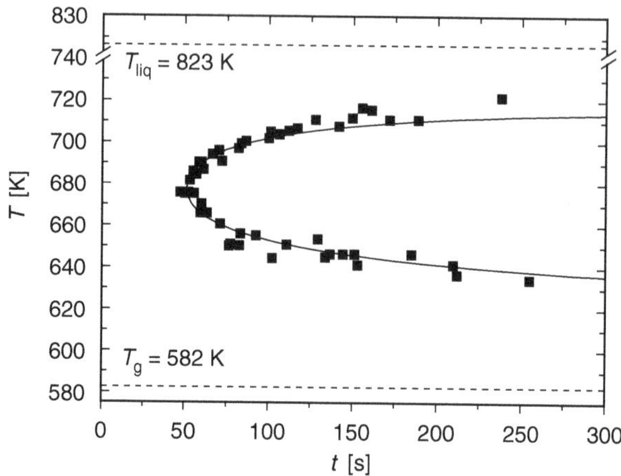

FIGURE 17.13 Time–temperature–transformation (TTT) diagram of the bulk metallic glass-former $Pd_{40}Cu_{30}Ni_{10}P_{20}$. (Reproduced from Löffler, J.F., Schroers, J., and Johnson, W.L., *Appl. Phys. Lett.* 77, 681, 2000. With permission.)

17.3.2 Time–Temperature–Transformation Diagrams

Crystallization in the undercooled melt preferably sets in if the temperature is low enough to result in sufficient undercooling (resulting in an increasing thermodynamic driving force for nucleation) and if it is high enough to allow sufficient atomic mobility. The onset times plotted for different temperatures form a typical nose-type curve with the so-called "nose temperature" at the shortest crystallization time, a lower branch representing decreasing kinetics, and a higher branch representing decreasing undercooling or decreasing driving force for nucleation when moving away from the nose temperature (see Figure 17.13). Stable metallic glasses, such as Vit 1 ($Zr_{41.2}Ti_{13.8}Cu_{12.5}Ni_{10.0}Be_{22.5}$) or $Pd_{40}Cu_{30}Ni_{10}P_{20}$, reach incubation times at the nose temperature of ~50 s.[94,107] The experiments are usually performed using containerless levitation techniques, for example, electrostatic levitation,[108] or in graphite crucibles using B_2O_3 flux.

The crystallization times can be fitted using the Johnson–Mehl–Avrami model.[109,110] For low values of crystallized volume fraction, f, the crystallization time, t, is

$$t = \left(\frac{3f}{\pi \; I_v v^3} \right)^{1/4}, \tag{17.15}$$

with the nucleation rate, I_v, given by Equation 17.12 and the growth velocity, v, given by

$$v = \frac{D}{l} \left\{ 1 - \exp\left(-\frac{(V_m/N_A)\,\Delta G_v(T)}{k_B T} \right) \right\}. \tag{17.16}$$

Here the diffusivity, D, may be related to the viscosity via the Stokes–Einstein relation $D = k_B T/(3\pi\eta(T)l)$, with l the (average) atomic diameter. Using measured values for the temperature dependence of the viscosity, $\tilde{\eta} = 9.34 \times 10^{-3}$ Pa s, $D^* = 9.25$ and $T_0 = 447$ K,[111] using the approximation $V_m \Delta G_v \approx \Delta S_m^f (T_{liq} - T)$ with $\Delta S_m^f = 8.55$ J/(mol K) and $T_{liq} = 823$ K, estimating the detectable volume fraction of crystals, f, to be ~0.5%, and using $l \approx 3.1$ Å, the curve in Figure 17.13 can be fitted to the data using A_v and γ as fitting parameters. The fit shown in Figure 17.13 gives $A_v = 4.4 \times 10^{31}$ Pa s/(m^3 s) and $\gamma = 0.067$ J/m^2. The latter is comparable to values for monatomic liquids with low melting temperature (e.g., $\gamma = 0.058$ J/m^2 for Cd and 0.108 J/m^2 for Al).

Quenching of a melt into a metallic glass can be seen as cooling it fast enough to pass from T_{liq} to T_g without crossing the nose-shaped crystallization curve. In this respect continuous-cooling–transformation diagrams (CCT) would be more suitable, but no experimental CCT diagrams have so far been reported. Note that for different primary crystallizing phases, different crystallization curves need to be drawn into the TTT diagram. This (together with a possibility of phase separation in the deeply undercooled liquid region[112–114]) is one of the reasons why the TTT diagram of Vit1 could not be successfully fitted with the procedure described above over the entire temperature range.[94]

17.3.3 Processing Possibilities

As outlined in Section 17.3.2, a metallic glass can be produced if the liquid is cooled quickly enough to bypass the nose of the TTT diagram. In principle there are two processing methods, as shown schematically in Figure 17.14.

In the "injection casting" procedure, the glass-forming melt is processed directly from the melt into its final shape, as shown in Figure 17.14a. The melt is first undercooled to a temperature below the nose at a cooling rate high enough to bypass the nose of the TTT diagram. Then, at the temperature below the nose (but above the glass transition temperature T_g), the material can be injected in a way similar to polymer melts, as the viscosity is higher than in conventional melts but still low enough ($<10^8$ Pa s) to allow for injection molding. It is interesting to note that with suitable processing equipment and process control, the injection casting method can deliver glassy parts with larger dimensions than the critical casting thickness, d_c, of their constituent material: for example, the glass-forming liquid can be cooled quickly to a temperature below the nose (e.g., by flowing through various thin, water-cooled channels with $d < d_c$) and then injection-cast into a part with $d > d_c$ at a temperature above T_g.

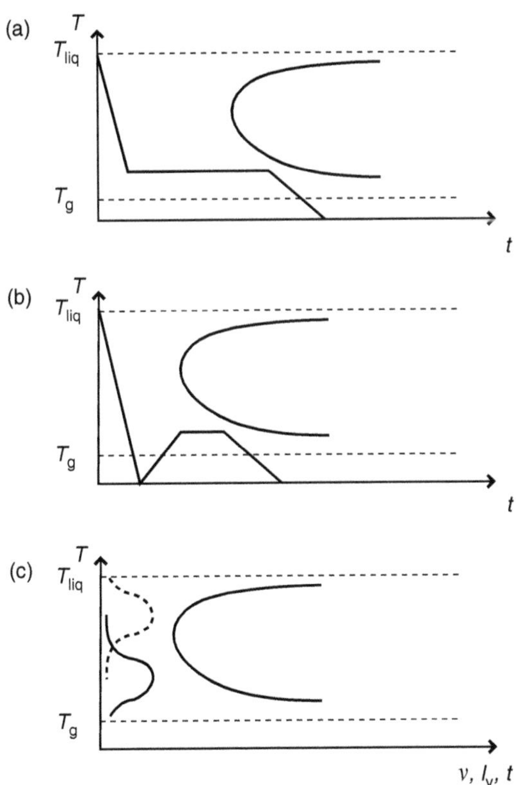

FIGURE 17.14 Schematic TTT diagrams and processing possibilities for metallic glasses in the undercooled liquid regime: (a) upon cooling from the melt ("injection casting") and (b) upon heating of a glass ("superplastic forming"). (c) Different temperatures for maxima of growth rate, v, (dashed) and nucleation rate, I_v, (solid line), explaining that the processing times in "injection casting" (a) are generally higher than in "superplastic forming" (b).

The "superplastic forming" process is shown schematically in Figure 17.14b. Here the glass-forming melt is quenched into a glass (i.e., to a temperature below T_g) at a cooling rate high enough to suppress crystallization. The glassy material can then be stored as feedstock material (in the form of small spheres, for example) at room temperature for extended periods of time without any effect on the microstructure. For processing in the supercooled liquid, the material is then heated to a temperature above T_g (but below the nose temperature) and formed.

As shown in Figure 17.14, the TTT diagrams differ for the processes of injection molding and superplastic forming, respectively. This is because crystallization in undercooled liquids is history-dependent, that is, different crystallization times exist at one temperature T depending on whether the temperature is reached by heating of the glass or undercooling of the melt.[115] This asymmetry results from the fact that the thermodynamic driving force for nucleation increases with increasing undercooling, while the kinetics simultaneously decreases. Consequently, there exist different temperatures for the maxima for growth rate and nucleation rate, as shown schematically in Figure 17.14c. As nucleation is the rate-dominating step for crystallization, the TTT diagram of "injection casting" (Figure 17.14a) generally shows longer crystallization times than that of "superplastic forming" (Figure 17.14b).

17.4 Processing in the Undercooled Liquid Region

17.4.1 Thermal Stability and Formability

For processing metallic glasses in the undercooled liquid, the onset of crystallization is the limiting factor. A measure for the stability of the undercooled liquid upon heating is the extension of the undercooled liquid regime, $\Delta T = T_x - T_g$, usually measured in differential scanning calorimetry (DSC) experiments performed at a certain heating rate, with T_x the first crystallization temperature (Figure 17.15). Despite the fact that thermal stability and glass-forming ability are not necessarily correlated,[116,117] conventional metallic glasses usually show only a small undercooled liquid region of $\Delta T \lesssim 20$ K (or show no glass transition at all), while the more recently developed bulk metallic glasses generally exhibit a large undercooled liquid regime of $50\text{ K} < \Delta T < 150\text{ K}$. It is this extension of the undercooled liquid range that enables bulk metallic glasses to be superplastically formed (from a feedstock material) in a way similar to polymers.

For current bulk metallic glasses, the processing temperature is, however, still a compromise between temperatures too high to suppress crystallization and too low to facilitate forming, as the viscosity increases

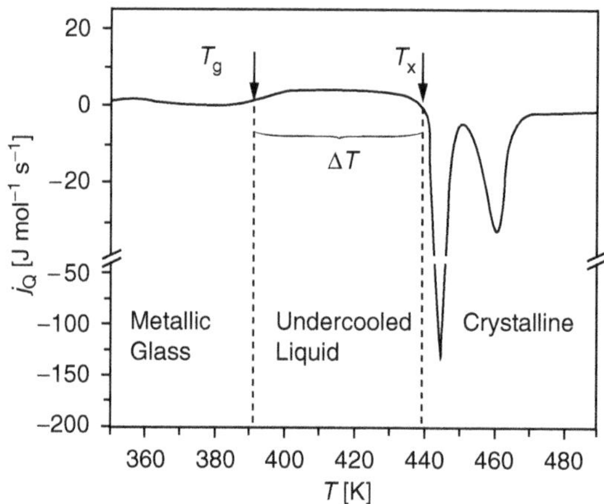

FIGURE 17.15 Differential scanning calorimetry scan of a metallic glass ($Au_{49}Ag_{5.5}Pd_{2.3}Cu_{26.9}Si_{16.3}$) with glass transition (T_g), crystallization temperature (T_x), and different regimes marked (positive j_Q values represent endothermic heat flow, measured at a heating rate of 10 K/min).

TABLE 17.2 Glass Transition Temperature, T_g, Thermal Stability, ΔT, Normalized Thermal Stability, ΔS, and Critical Casting Thickness, d_c, of Various Bulk Glass-Forming Alloys

Alloy	T_g(K)	ΔT(K)	ΔS	d_c(mm)	Ref.
$Ce_{68}Cu_{20}Al_{10}Nb_2$	341	81	0.26	8	119
$Au_{49}Ag_{5.5}Pd_{2.3}Cu_{26.9}Si_{16.3}$	401	58	0.24	5	120
$Mg_{58}Cu_{30.5}Y_{11.5}$	424	75	0.22	9	121
$La_{55}Ni_{20}Al_{25}$	472	73	0.16	3	122
$Pt_{57.5}Cu_{14.7}Ni_{5.3}P_{22.5}$	507	98	0.34	16	123
$Pd_{40}Cu_{30}Ni_{10}P_{20}$	577	95	0.37	72	124
$Zr_{44}Ti_{11}Cu_{10}Ni_{10}Be_{25}$	625	114	0.20	10	89
$Zr_{65}Cu_{15}Ni_{10}Al_{10}$	645	106	0.23	4	76

steeply with decreasing temperature and the maximum force to be applied during forming is often limited. This is particularly true for micro- and nanostructuring, where, for example, silicon molds with limited strength are used.

As the decrease in flow stress with temperature (above T_g) differs for the various alloys, it has been suggested that ΔT normalized to the temperature interval between T_g and the liquidus temperature T_{liq},

$$\Delta S = \frac{\Delta T}{T_{liq} - T_g}, \tag{17.17}$$

will estimate formability better than ΔT.[118] In Table 17.2 suitable alloys for superplastic forming representing different metallic systems are listed together with their values for T_g, ΔT, ΔS, and d_c.

The above-mentioned parameters for the stability of metallic glasses do not take into account the viscosity of the undercooled melt. It has indeed been shown that the viscosity at the crystallization temperature is lower for thermally less-stable glasses[125] (e.g., 10^5 Pa s for $Zr_{41.2}Ti_{13.8}Cu_{12.5}Ni_{10}Be_{22.5}$[3] and 7×10^5 Pa s for $Mg_{60}Cu_{30}Y_{10}$[126] with a lower ΔT of 55 and 60 K, respectively), but higher for thermally more stable glasses (e.g., 8×10^7 Pa s for $La_{55}Al_{25}Ni_{20}$[127] and 3×10^8 Pa s for $Zr_{65}Al_{10}Ni_{10}Cu_{15}$[128] with a larger ΔT of 70 and 105 K, respectively).

More exact forming properties that do take into account the various fragilities are obtained by measuring the flow stress of metallic glasses above the glass transition temperature. These measurements have been performed in detail on several alloys (see Section 17.5). The flow stress σ and the strain rate $\dot{\varepsilon}$ are related to the viscosity via

$$\eta = \frac{\sigma}{3\dot{\varepsilon}}. \tag{17.18}$$

The values described above for viscosity are good estimates of forming ability. For more detailed studies, however, dynamic measurements are required in which the viscosity is represented by a complex value,

$$\eta^* = \frac{G' + iG''}{\omega}, \tag{17.19}$$

where the real part of $\omega\eta^*$ is the storage modulus G', and the imaginary part of $\omega\eta^*$ is the loss modulus G''. Physically, G' represents the (time-independent) elastic contribution, while G'' represents the (time-dependent) viscous contribution for a given angular frequency, ω.

Figure 17.16a shows measurements of G' and G'' for the metallic glass $Ce_{68}Cu_{20}Al_{10}Nb_2$ as a function of temperature (measured at a heating rate of 10 K/min) using oscillating parallel plate geometry (not to be confused with earlier "parallel plate" indentation measurements).[129] Also shown is the modulus of the dynamic viscosity, $|\eta^*| = (1/\omega)\sqrt{(G')^2 + (G'')^2}$. Interestingly, the metallic glasses show a viscoelastic behavior in the superplastic regime with a dominant viscous component. Such dynamic viscosity measurements are particularly helpful for predicting the flow behavior in the forming of parts with complex

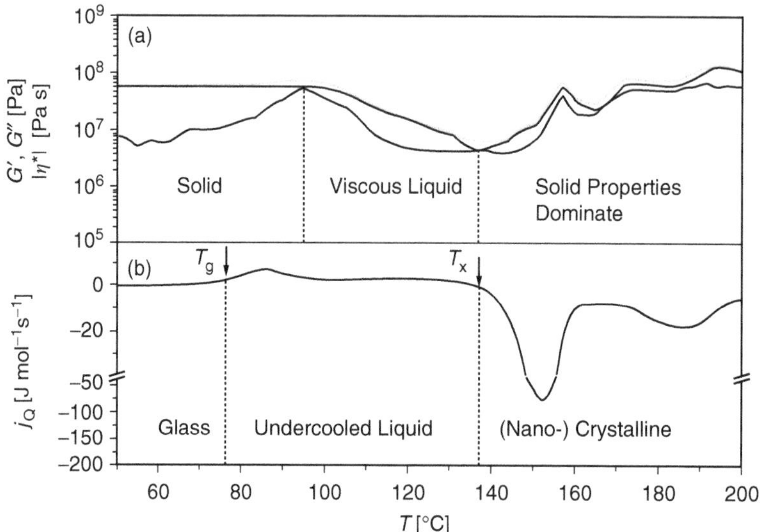

FIGURE 17.16 (a) Storage modulus (G'), loss modulus (G'') and absolute value of the dynamic viscosity ($|\eta^*|$) for the bulk metallic glass $Ce_{68}Cu_{20}Al_{10}Nb_2$; and (b) DSC trace for the same alloy, measured at a heating rate of 10 K/min.

geometries. To determine the structural changes during heating, a DSC curve has also been measured at the same heating rate as in the viscosity measurements, and is shown in Figure 17.16b.[129]

17.4.2 Injection Casting

Metallic glass-forming melts typically show a high melt viscosity and can be easily undercooled. Injection casting of such undercooled melt has several advantages: compared to crystalline alloys, the absence of crystallization shrinkage and of crystalline anisotropy in the final parts allows near-net-shape molding of parts with isotropic properties. In addition, the material exhibits homogenous properties on a length scale down to the nanometer range. As modern applications require parts in the micro- and nanometer range, the homogeneity of the material (e.g., the absence of grain boundaries) becomes crucial. Parts with micrometer-sized features or of complicated shape have been successfully prepared by injection casting of metallic glasses,[130] as shown in Figure 17.17. In addition, due to the available long processing times in the undercooled liquid regime, injection casting can deliver large parts in the range of several centimeters, as shown for example in Figure 17.18 (parts available commercially).

17.4.3 Superplastic Forming

The superplastic deformation behavior of metallic glasses above T_g is described by viscous flow (Newtonian or non-Newtonian; see Section 17.5.1) and should be distinguished from, for example, grain-boundary sliding or dynamic recrystallization in crystalline *superplastic* alloys.[131] More general details on superplasticity are given in Chapter 14.

One of the first successful superplastic forming experiments on metallic glasses was the extrusion of glassy $Zr_{65}Al_{10}Ni_{10}Cu_{15}$ metal powder at temperatures above T_g.[132] Here superplastic forming was possible because of the alloy's high thermal stability. This procedure generated the same mechanical properties as casting, which shows that superplastic forming in the undercooled liquid regime is possible without any degradation of the material's properties.

Superplastic forming of parts and surface structures in the micro- and nanometer range is also attracting great interest. Metallic glasses show good microformability, high strength, high elastic limit, and good corrosion properties combined with a homogeneity of the mechanical and chemical properties on the

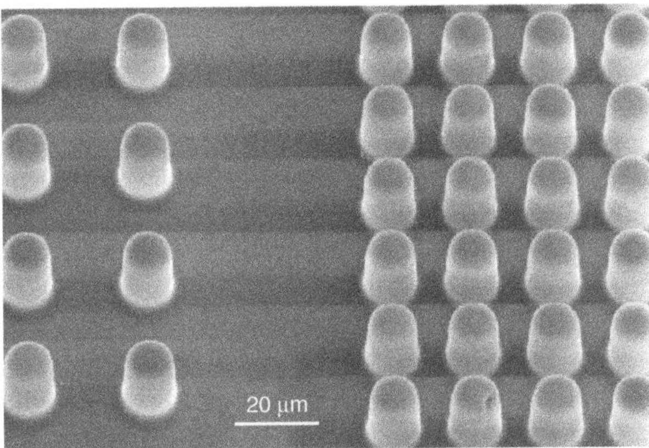

FIGURE 17.17 Pins of 20 μm in height and 8 μm in diameter of a bulk metallic glass, replicated from a surface-microstructured Si wafer by rapid cooling from the melt.

FIGURE 17.18 Photo of commercially available parts made of injection-molded metallic glass (SanDisk Cruzer® Titanium 512 MB USB memory stick and Liquidmetal® Technologies cell phone case).

micrometer and sub-micrometer scale due to the absence of grain boundaries.[133] The replication of surface topologies on the micrometer scale from, for example, silicon in metallic glasses by superplastic forming has been realized for $Zr_{46.8}Ti_{8.2}Cu_{7.5}Ni_{10}Be_{27.5}$ (Vit 4)[134] (see Figure 17.19), $Pt_{48.75}Pd_{9.75}Cu_{19.5}P_{22}$,[135] $La_{60}Al_{20}Ni_{10}Co_5Cu_5$[136] and $Pt_{57.5}Cu_{14.7}Ni_{5.3}P_{22.5}$.[118] With the Pt-based alloys, features in the range of 10 to 100 nm (with a low aspect ratio)[135] and a few micrometers (with a high aspect ratio)[118] were successfully replicated from silicon molds.

Superplastic forming can also be used to fabricate parts from metallic glass of a larger size than the critical cooling rate allows. If a feedstock material is first quenched in small droplets, these metallic glass billets can subsequently be forged into a large part. With the recent development of better glass-forming alloys, thermal stability has also increased, making superplastic forming of large parts similar to that of polymer materials feasible. Indeed, as can be seen in Figure 17.20, the forming of hand-size parts using pellets as feedstock material has already been realized with the alloy $Pt_{57.5}Cu_{14.7}Ni_{5.3}P_{22.5}$. Here the pellets were pressed into a cup for 100 s using a pressure of 20 MPa at a temperature of 260° C.[118]

17.4.4 Co-Extrusion with Polymers

With further development of metallic glasses and the resulting larger parameter ranges (wider range of T_g, higher thermal stability, lower critical cooling rates), additional processing methods will become available

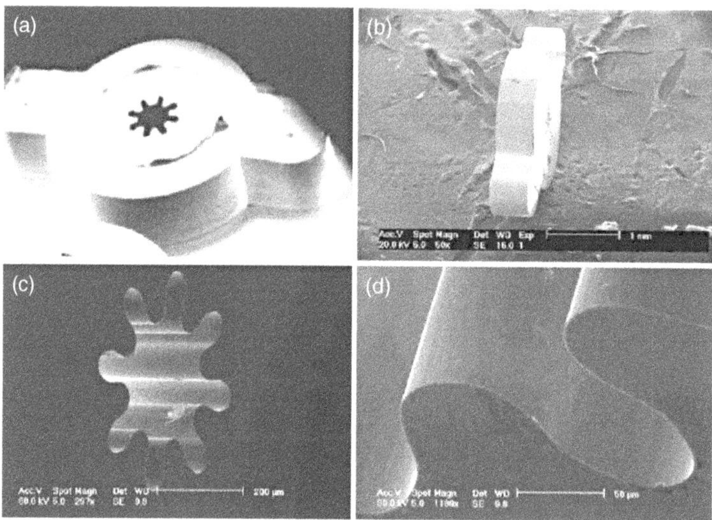

FIGURE 17.19 Microgears made of bulk metallic glass ($Zr_{46.8}Ti_{8.2}Cu_{7.5}Ni_{10}Be_{27.5}$, Vit 4) replicated from a $Ni_{85}Co_{15}$ mold by superplastic forging. (From Zumkley, T., Suzuki, S., Seidel, M., Mechler, S., and Macht, M.P., *Mater. Sci. Forum* 386–388, 541, 2002. With permission.)

FIGURE 17.20 Feedstock material and superplastically formed part made of the metallic glass $Pt_{57.5}Cu_{14.7}Ni_{5.3}P_{22.5}$. (Reproduced from Schroers, J., *J. Mater. JOM* 57, 35, 2005. With permission.)

for metallic glasses. One of these could be the co-extrusion of metallic glasses and polymers to form composites. With the recent development of metallic glasses with a T_g in the range of that of polymers, the two can be processed at temperatures where they show the same viscosity. In Figure 17.21 the dynamic viscosities for various metallic glasses are shown upon continuous heating and compared with the zero viscosities (i.e., viscosities extrapolated to $\dot{\gamma} = 0$) of polymers, η_0.[129] The first drop in viscosity in the metallic glasses is the softening of the metallic glass above T_g. An additional drop is observed in the Mg and Ce alloys after the onset of crystallization. This may be due to a change in composition or density of the remaining amorphous phase after the onset of crystallization. As Figure 17.21 shows, the viscosities of polymers and metallic glasses match closely, allowing co-processing of the two with similar flow behavior. Thus, even with today's metallic glasses, polymer–metallic glass pairs with similar viscosities can be picked out and treated together (e.g., the Ce-based metallic glass $Ce_{68}Cu_{20}Al_{10}Nb_2$ and high-density polyethylene HDPE6021 at 120°C; or the Au-based metallic glass $Au_{49}Ag_{5.5}Pd_{2.3}Cu_{26.9}Si_{16.3}$ and polystyrene, PS, at 160°C). With the development of these new metallic glasses the production of metal/polymer composites via co-processing may become a simple method for interesting composite materials.

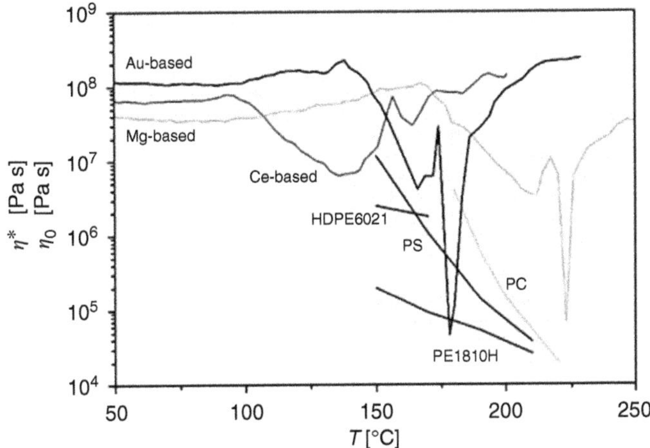

FIGURE 17.21 Dynamic viscosity data for three metallic glasses, Au-based ($Au_{49}Ag_{5.5}Pd_{2.3}Cu_{26.9}Si_{16.3}$), Ce-based ($Ce_{68}Cu_{20}Al_{10}Nb_2$) and Mg-based ($Mg_{58}Cu_{30.5}Y_{11.5}$) as a function of temperature (heating rate: 10 K/min). In addition, zero viscosities for four polymers are shown (HDPE6021: a high-density polyethylene; PE: low-density polyethylene; PS: polystyrene; PC: polycarbonate). Note that the zero viscosity for an amorphous material is slightly higher than the dynamic viscosity.

17.5 Mechanical Properties of Metallic Glasses

17.5.1 Introduction

In the earlier stages of research into metallic glasses, mechanical properties such as tensile strength, plastic deformation, fracture toughness and superplastic forming generated only minor interest due to the limited thickness of the glassy metal foils obtained.

Research into metallic glasses mostly concentrated on magnetic properties, due to their excellent soft-magnetic properties (explained by the so-called random-anisotropy model, e.g., References 137 and 138) and their potential commercial applications in transformer cores. However, with the advent of bulk metallic glasses in the 1990s, research into mechanical properties became increasingly important due to the potential of bulk metallic glasses as structural and functional materials. Here we describe the mechanical properties of bulk metallic glasses but exclude a description of their magnetic properties. (For overviews on the magnetism of metallic glasses we refer to References 26, 139, and 140. Some applications of magnetic glasses are, however, outlined in Section 17.7.1.)

The processing of metallic glass-forming liquids can strongly influence mechanical properties. It is thus important to understand the influence of deformation mode, temperature, and strain rate on the mechanical response of bulk metallic glasses. For instance, controlled processing can lead to a composite-type material, consisting of ductile dendritic crystals in a glassy matrix, which shows improved mechanical properties compared to the monolithic glass. Various production methods for bulk metallic glass composites are outlined in Section 17.6. In the current section, a general overview of the mechanical properties of bulk metallic glasses is presented, followed by a detailed discussion of their mechanical properties at ambient and elevated temperatures and a description of their inhomogeneous and homogeneous deformation behavior.

17.5.2 Temperature Dependence of Deformation in Metallic Glasses

At temperatures significantly lower than the glass transition temperature, T_g, plastic deformation is inhomogeneous and characterized by the propagation of a few highly localized shear bands. In contrast

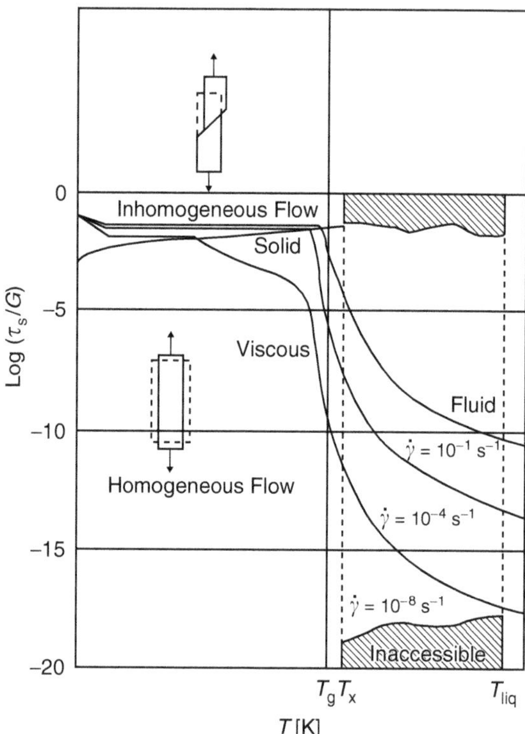

FIGURE 17.22 Deformation map (shear stress-temperature curve) for metallic glasses. Plotted is the normalized shear stress (shear stress divided by shear modulus) as a function of temperature, with the shear strain rate $\dot{\gamma}$ as parameter. (Reproduced from Spaepen, F., *Acta Metall.* 25, 407, 1977. With permission.)

to their crystalline counterparts, where resistance to plastic deformation is significantly lower due to the operation of 2-D lattice defects (dislocations), metallic glasses lack atomically even glide planes allowing for easy slip. Their disordered atomic arrangement represents a great obstacle to plastic deformation, such that slip is hindered unless very high stresses are applied which trigger the hopping of atoms and the diffusion of vacancy-like free volume, resulting in the formation of localized shear planes. In Spaepen's deformation map or stress-temperature diagram (Figure 17.22)[141] this deformation behavior generates a very low but negative slope, indicating that the shear or fracture stress decreases linearly with increasing temperature as the elastic modulus decreases.[142]

At temperatures close to T_g, plastic flow in metallic glasses becomes homogeneous and requires significantly lower stress than at lower temperatures. At low strain rates it is described by Newtonian viscous flow, where the viscosity η scales with $\tau_s/\dot{\gamma}$. In this case, the shear stress τ_s is directly proportional to the shear strain rate $\dot{\gamma}$ (cf. Equation 17.18 with $\sigma \approx 2\tau_s$ and $\dot{\varepsilon} \approx 2\dot{\gamma}$ for a fracture angle of 42°, according to the well-known Schmid's law) and η depends only on temperature. Thus, τ_s shows a similar temperature dependence as the viscosity (compare Figures 17.12 and 17.22).

In Figure 17.22, T_X represents the crystallization temperature, which would need to be shifted significantly closer to the liquidus temperature, if the good glass-forming ability of today's bulk metallic glasses were considered. With the onset of crystallization the rheology of the undercooled liquid changes drastically, leading first to a decrease and, after a critical volume fraction of crystals has nucleated and grown, to an increase in viscosity, η^* (see Figure 17.21). Measurements of viscosity in the Newtonian flow regime are used to evaluate the flow behavior in superplastic forming (see Section 17.4.3) and also to explain the good glass-forming ability of some alloys.

At higher strain rates or lower temperatures, non-Newtonian flow has been observed to depend strongly on temperature and strain rate.[143,144] From a microscopic point of view it is still under debate as to whether this regime can be classified as homogeneous or inhomogeneous flow.[145] Non-Newtonian flow in a compression test is characterized by an overshoot stress, which occurs upon yielding before steady-state flow at lower stress is reached.[146] This overshoot may be understood as a competition between a slower process, namely the structural breakdown governed by Newtonian flow, and a shear relaxation mechanism governing the steady-state viscosity.[146]

It has also been shown that the transition between Newtonian and non-Newtonian is rather abrupt and can be quantified by the Deborah number, $D_e = \dot{\gamma}\tau$, where $\dot{\gamma}$ is the shear strain rate and τ the relaxation time for the annihilation of the flow defects. A value of $D_e \sim 0.5$ has been measured for amorphous $Pd_{40}Ni_{40}P_{20}$, compared to the typical value of $D_e \sim 1$ known for the onset of flow instability.[145]

17.5.3 Mechanical Properties at Ambient Temperatures

At low temperatures where inhomogeneous deformation prevails, metallic glasses show consistently higher strength than crystalline materials of the same density.[12] This is shown in Figure 17.23, where the average strength of the crystalline and amorphous metals is plotted as a function of their density. Amorphous alloys generally show an average strength that is more than twice that of crystalline alloys. In addition, a linear fit to the data results in an average specific strength of 346 MPa/g/cm^3 for amorphous alloys and 155 MPa/g/cm^3 for crystalline alloys.

The high strength of metallic glasses results in an elastic strain limit of ~2%, which is approximately three to four times higher than that of their crystalline counterparts. Figure 17.24 summarizes data from various references on the yield strength, σ_y, and the Vickers hardness as a function of the Young's modulus, E, for bulk metallic glasses and crystalline metallic alloys.[10] For the same Young's modulus, a strength is reached in bulk metallic glasses that is about three times higher than that of their crystalline counterparts. It is also interesting to note that the metallic glasses show a linear dependence of the yield strength (basically following Hooke's law, $\sigma_y = E\varepsilon$, where ε is the strain) and hardness on the Young's modulus, while the crystalline metals deviate from this linear behavior. This reflects the strong dependency of the deformation behavior in crystalline metals on chemical and therefore microstructural change, which is less pronounced in amorphous materials.

17.5.4 Inhomogeneous Deformation

This section presents typical characteristics of the inhomogeneous flow of metallic glasses. These are important in understanding the general differences in deformation behavior (and their underlying

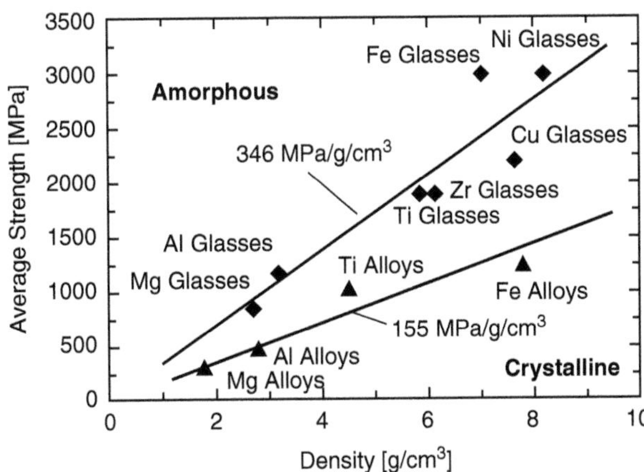

FIGURE 17.23 Average yield strength as a function of density for amorphous and crystalline alloys. (Reproduced from Löffler, J.F., *Z. Metallkd.* 97, 225, 2006. With permission.)

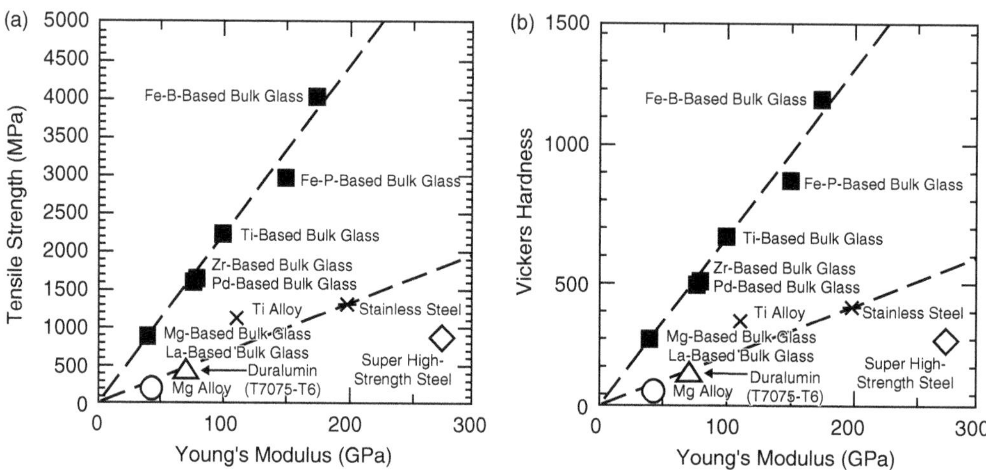

FIGURE 17.24 (a) Tensile strength and (b) Vickers hardness vs. Young's modulus for various amorphous and crystalline materials. (Reproduced from Inoue, A., *Acta Mater.* 48, 279, 2000. With permission.)

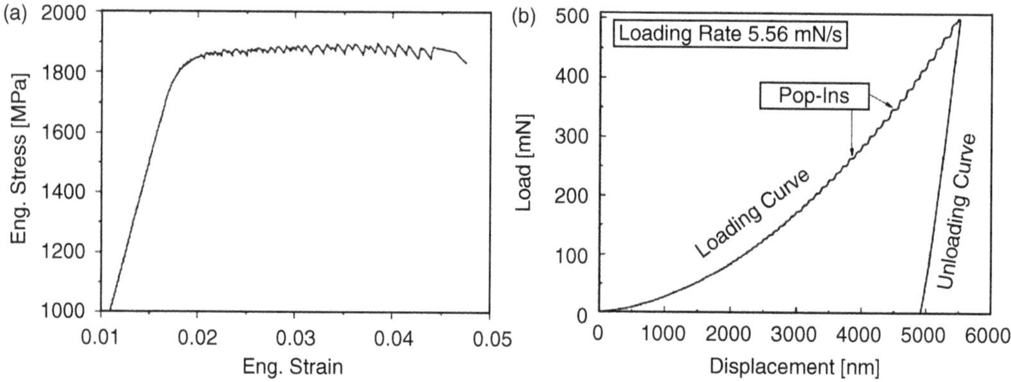

FIGURE 17.25 (a) Compression and (b) nanoindentation tests on glassy $Zr_{52.5}Ti_5Cu_{17.9}Ni_{14.6}Al_{10}$ (Vit105).

mechanisms) compared to those of their crystalline counterparts. The most obvious property observed during deformation is serrated flow (Section 17.5.4.1) and its manifestation in the form of "vein" patterns on fracture surfaces (Section 17.5.4.2). Other important parameters discussed are fracture toughness (Section 17.5.4.3) and plastic strain limits, measured in compression and tension (Section 17.5.4.4).

17.5.4.1 Serrated Flow

In Figure 17.25, a typical stress–strain curve tested in compression (a) and a nanoindentation curve (b) are shown for a Zr-based bulk metallic glass. Both types of curves exhibit nonuniform inhomogeneous deformation, represented by the serrated flow curves in the compression test (a) and the "pop-ins" in the indentation depth signal (b). These discrete steps correspond to single deformation bursts along preexisting or newly formed shear bands.

Within these highly localized shear bands, great shear displacements on the order of a few micrometers can take place and occur within short time intervals of tens to hundreds of milliseconds, depending on the deformation temperature and speed. Therefore, within such shear bands shear strains of greater than 1 have to be assumed. According to the deformation models of References 141 and 147, vacancy-like free volume is shuffled around and underlies a competing process of creation and annihilation of extra free volume. Due to the higher rate of free volume creation, this leads to a softening of the material within a shear band. Whether this deformation before failure is additionally triggered by a local change in the

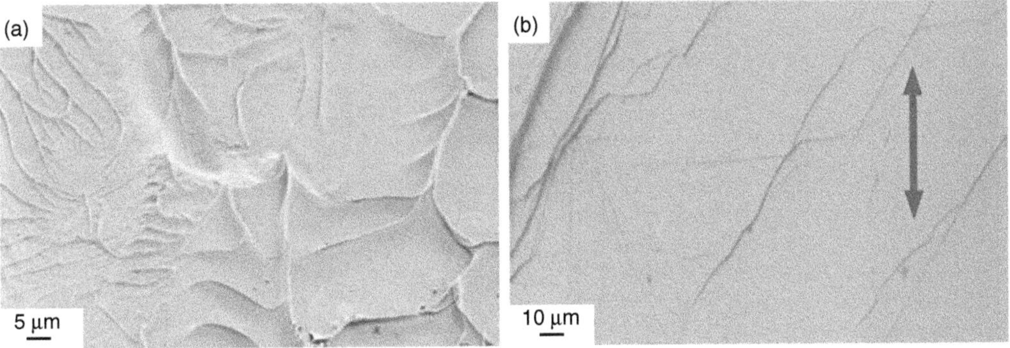

FIGURE 17.26 (a) Fracture surface and (b) shear-band offsets on the surface of a compression specimen of a Zr-based bulk metallic glass.

rheological behavior of the material, that is, a drastic drop in viscosity due to adiabatic heating, is a matter of current research.[148–150]

17.5.4.2 Fracture Surfaces and Shear Bands

Figure 17.26a shows a fracture surface of a Zr-based bulk metallic glass sample tested in compression. It shows a typical vein pattern arising from viscous flow, which occurs during the sudden relief of stress energy at failure and its conversion into heat. Re-solidified droplets are also found frequently, implying that temperatures are sufficiently high at fracture for local melting of the material to occur.

On the surrounding surfaces of compression specimens, shear bands appear as sharp, irregular sub-micrometer wide offsets from the surface, as shown in Figure 17.26b. They are generally oriented in a conjugated configuration, at an angle of ~40 to 45° to the compression direction. This angular dependency of shear bands reflects the stress states in the solid and its associated yield (Mohr–Coulomb) criterion.[151] Note that as shear bands are the main carriers of plastic deformation in metallic glasses, their density is a linear function of the amount of global plastic deformation the material will support before failure.

17.5.4.3 Fracture Toughness

Apart from the high strength and elastic limit observed in bulk metallic glasses, information on fracture toughness (K_c) is crucial for the successful deployment of bulk metallic glasses in structural and functional applications. Fracture toughness is the ability of a material with a preset flaw to withstand an applied load without immediate fracture. Materials exhibiting high fracture toughness generally show ductile fracture, whereas low fracture toughness values are indicative of brittle materials. The fracture toughness is defined as

$$K_c = \sqrt{EG_c} = \bar{c}\sigma_c\sqrt{\pi a} \tag{17.20}$$

where E is the Young's modulus and G is the critical strain energy release rate, with subscript c indicating the crucial point at fracture; σ is the stress on the workpiece, a is the notch length and $\bar{c}$ is a correction factor of ~1.1.[152]

As shown above in Figure 17.26b, bulk metallic glasses deform in shear bands and can exhibit very high plastic strains locally. This observation, combined with the fact that bulk metallic glasses exhibit an unusually high fraction of theoretical strength, generally leads to high fracture toughness ($K_c > 20$ MPa $\sqrt{\text{m}}$).[153] However, on a macroscopic level the material cannot be properly termed as ductile and should instead be defined as malleable, that is, with limited plastic strain only in the compressive or bending mode. The K_c values of various bulk metallic glasses are presented in Table 17.3.

Compared to silica glasses, metallic glasses generally show significantly higher fracture toughness. However, they span a large range in K_c because of their intrinsic and extrinsic embrittlement, that is, their

TABLE 17.3 Fracture Toughness Values of Various Metallic Glasses

Material	K_C-values [MPa $\sqrt{m}$]	Ref.
Fused silica	0.5	157
Window glass	0.2	157
$Mg_{65}Cu_{25}Tb_{10}$	2	158
$Cu_{60}Zr_{20}Hf_{10}Ti_{10}$	68	159
$Ti_{50}Ni_{24}Cu_{20}B_1Si_2Sn_3$	50	160
$Zr_{57}Ti_5Cu_{20}Ni_8Al_{10}$	80	161
$Zr_{55}Al_{10}Ni_5Cu_{30}$	67	8
$Zr_{41.2}Ti_{13.8}Cu_{12.5}Ni_{10}Be_{22.5}$	16–86	162–164
$Fe_{40}Ni_{40}P_{14}B_6$	38–48	165
$Ni_{43}Fe_{29}P_{14}Al_3B_6$	120	166
$Pt_{57.5}Cu_{14.7}Ni_{5.3}P_{22.5}$	79–84	167
$Ce_{60}Al_{20}Ni_{10}Cu_{10}$	10	158
$La_{55}Al_{25}Ni_5Cu_{10}Co_5$	5	168

high sensitivity to flaws.[154] In the best cases, where external flaws are suppressed, the fracture toughness reaches values close to those of conventional structural crystalline materials, such as crystalline Al alloys (~20 to 40 MPa$\sqrt{m}$), steels (~50 MPa $\sqrt{m}$),[153] Ti alloys (44 to 66 MPa $\sqrt{m}$)[155], and Ni alloys ~(5 to 30 MPa $\sqrt{m}$).[156] Upon aging at sufficiently high temperatures, however, the fracture toughness of metallic glasses decreases significantly,[153] limiting the application of these materials to environments where temperature fluctuations are low. At temperatures where crystallization occurs, fracture toughness may even drop to values as low as 1 MPa $\sqrt{m}$.[163]

17.5.4.4 Ductility and Malleability in Metallic Glasses

Several strategies have been employed to improve the intrinsic and extrinsic effects on the flow behavior of metallic glasses with respect to their fracture toughness and overall plastic strain in compression (malleability):

1. Similar to crystalline metals,[169–171] embrittlement in metallic glasses is strongly related to the elastic properties of the main component of the alloy and is favored by a high ratio of G/B, where G is the elastic shear modulus and B the bulk modulus, or low Poisson's ratio ν.[153,172] Fracture toughness tests and compression tests on various metallic glasses have shown an increase in fracture energy and plastic strain in compression with decreasing G/B and increasing ν. Since ν is generally only high for costly noble metals, applications of such systems may, however, be restricted.[153,167]
2. Another important aspect in choosing the main component is the relaxation behavior of the material as a function of its homologous glass transition temperature. Relaxation in an amorphous structure is generally linked to a topological rearrangement of atoms towards a denser atomic configuration,[173,174] which generates embrittlement due to a reduction of free volume. Metallic glasses with very low glass transition temperatures with respect to the ambient temperature (as with most Mg-based metallic glasses) show room temperature embrittlement after a short period of aging.[56] Therefore, unless relaxation processes can be suppressed by means other than temperature, applications will concentrate mainly on systems where the main components show high glass transition temperatures.
3. A third strategy for improving malleability involves mixing the brittle metallic glass with a more ductile second phase that either acts as an active carrier of plastic strain or passively enhances the multiplication of shear bands via shear-band splitting. This approach will be outlined in greater detail in Sections 17.6.1 and 17.6.2.

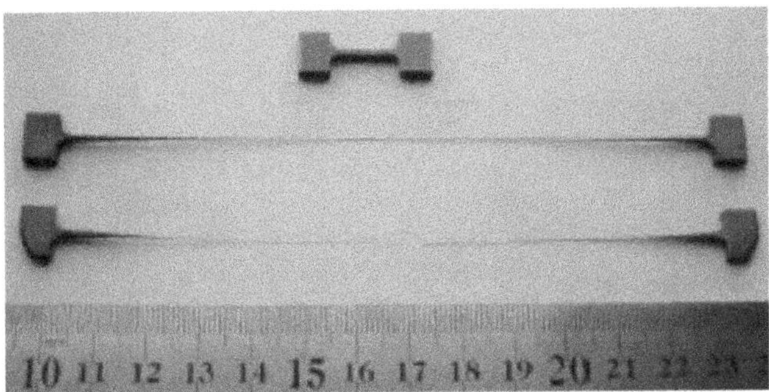

FIGURE 17.27 Tensile specimens of $Zr_{41.25}Ti_{13.75}Ni_{10}Cu_{12.5}Be_{22.5}$ deformed at 616 K. (Reproduced from Wang, G., Shen, J., Sun, J.F., Lu, Z.P., Stachurski, Z.H., and Zhou, B.D., *Intermetallics* 13, 642, 2005. With permission.)

17.5.5 Homogeneous Deformation at Elevated Temperatures

The mechanical behavior of bulk metallic glasses is very sensitive to a change in temperature close to the glass transition temperature. As discussed above, at temperatures far smaller than the glass transition, deformation is inhomogeneous and locally restricted to plastic deformation of a few shear bands, leading to only a few percent of plastic strain (in compression). At temperatures close to T_g, the deformation mechanism changes and becomes homogeneous, this being characterized by an evenly distributed strain over the whole volume. In this case, low flow stresses and large strains up to several hundred percentages can be achieved in tension and compression. At low strain rates the deformation can be described by Newtonian flow, where the viscosity η solely depends on temperature. At higher strain rates the behavior of metallic glasses becomes non-Newtonian and the viscosity depends on both temperature and strain rate[175] (see Section 17.5.2). As seen in Figure 17.27, superplastic flow with several hundred percent strain can be achieved in tension at strain rates of up to 10^{-1} per s at temperatures near T_g.[176,177] This unique behavior of metallic glasses close to the glass transition temperature, where a temperature change triggers a drastic change in strength and ductility, allows processing of the material in the undercooled liquid region to net-shaped parts that exhibit high strength and hardness (see Section 17.4).

17.6 Metallic Glass Composites

As outlined in Section 17.5.4.4, the ductility or malleability of metallic glasses may be greatly increased by introducing a second, more ductile phase into the metallic glass matrix. This section discusses several such processing routines and describes the resulting microstructure and mechanical properties.

17.6.1 *In Situ* Composites

In situ composite formation of a microstructure exhibiting a ductile crystalline phase and a hard metallic glass phase is based on the thermophysical properties of the undercooled melt, its composition, and the nucleation rate of the crystalline phase. The main requirement in such a process is to find a multicomponent composition that (1) leads upon cooling from the melt to the nucleation of a ductile crystalline phase (preferably not an intermetallic phase) of a certain volume percentage, and (2) causes the depletion of the remaining melt by the solidified elements in the direction of a composition that exhibits good glass-forming ability.[178,179] The processing methods (copper mold casting and injection molding) for the production of such composites do not differ from those producing monolithic metallic glasses, but may be altered slightly so that slower cooling rates allow crystallization. For the Zr–Ti–Cu–Ni–Be–Nb system a composition of this type has been found that upon cooling from the melt shows a microstructure of crystalline Ti–Zr–Nb (β) dendrites of bcc structure embedded in a Zr–Ti–Nb–Cu–Ni–Be glassy matrix (Figure 17.28).[178] This composite exhibits enhanced plastic strain under compression. Plastic strain of a

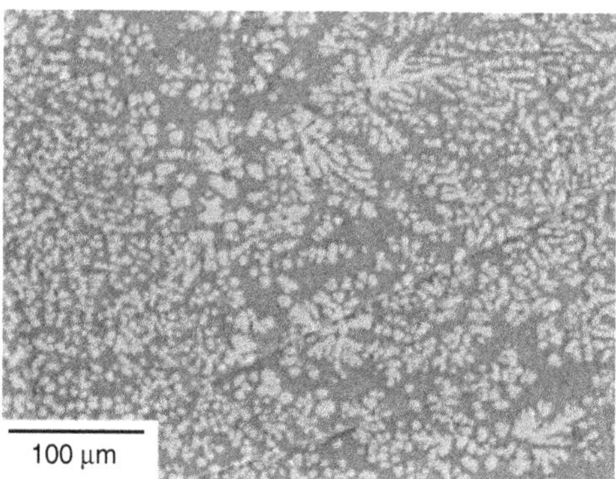

FIGURE 17.28 Dendritic structure of bcc Ti–Zr–Nb β phase (bright) embedded in a glassy Zr–Ti–Nb–Cu–Ni–Be matrix. (Reproduced from Hays, C.C., Kim, C.P., and Johnson, W.L., *Phys. Rev. Lett.* 84, 2901, 2000. With permission.)

low percentage was also seen in tension for this composite[178] when the percentage of the ductile phase was higher than a critical value of ~20 to 30 vol.%. This tensile ductility is important for structural applications.

Such microstructures, composed of dendritic ductile and amorphous tough phases, have meanwhile led to a drastic increase in compressive plastic strain without a significant loss in strength for several other metallic glass systems.[180–182] Compression tests performed on Ti-based bulk metallic glasses containing 20 vol.% of β-Ti(Ta,Sn) dendrites embedded in a nanocrystalline matrix revealed a plastic compressive strain of 15% and a maximum strength of 2.4 GPa. Strain-hardening behavior attributed to the crystalline phase has also been observed, raising expectations of achieving enhanced plastic strain in tension.[181] Similar enhanced mechanical strength and ductility have been observed in as-cast Cu–Zr samples with an equated, nanosized grain structure embedded in an amorphous matrix.[183]

17.6.2 *Ex Situ* Composites

A different approach to forming a composite material was guided by classical particle reinforcement methods, where — as opposed to crystalline solids — particles are meant to represent the phase which enhances plasticity and not increasing strength. Such composites are generally formed by adding solid particles to the liquid melt, forming a metallic glass matrix.[184] The addition of up to 20 vol.% of SiC, W, WC, or Ta particles with sizes ranging between 10 and 40 μm to a Zr-based glassy matrix has been shown to increase the plastic strain by up to 5% without reducing strength compared to the monolithic samples.[185] Much higher plastic strain of up to 18.5% has recently been achieved by adding soft graphite particles of 25 to 75 μm in size to Zr-based metallic glasses.[186] In Figure 17.29 the compression curves for monolithic Vit105 ($Zr_{52.5}Ti_5Cu_{17.9}Ni_{14.6}Al_{10}$) and the alloy reinforced with 3.5 vol.% graphite particles are shown, together with a scanning electron microscopy image of shear-band interaction with the graphite particles.

In contrast to the technique presented in Section 17.6.1, where a ductile phase is formed *in situ* during cooling from the melt, more control, and therefore reproducibility, over the type, size, and volume fraction of the ductile phase can be obtained in foreign-particle-reinforced bulk metallic glasses. Thus the approach presented in Reference 186 is of particular importance for the use of bulk metallic glasses as structural engineering material.

17.6.3 Amorphous Metallic Foams

A different approach to increasing plastic deformation has been explored by introducing a substantial volume fraction of micrometer-sized pores into the material.[187–189] Due to their reduced density such microstructures exhibit a substantial increase in strength vs. density combined with excellent malleability in compression. Metallic foam structures are used nowadays in several structural applications and are also attractive for bone replacements, due to their tunable elastic modulus (dependent on the pore volume[190]).

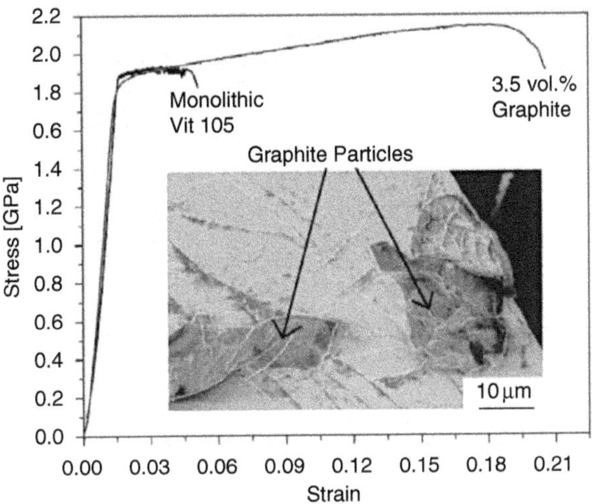

FIGURE 17.29 Compressive stress–strain curves for monolithic Vit105 and Vit105 reinforced with 3.5 vol.% graphite. Inset: Scanning electron microscopy image illustrating shear-band interactions with the graphite particles.

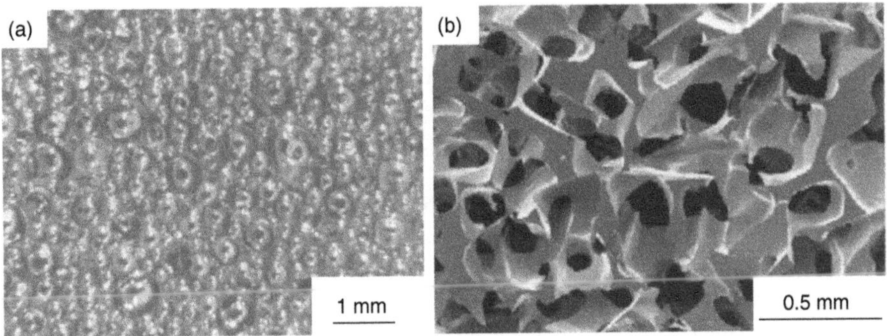

FIGURE 17.30 Metallic glass foams produced via (a) melting of a mixture of Pd-based alloy granulate with hydrated B_2O_3 and subsequent quenching during dehydration of B_2O_3 leading to a closed porosity, and (b) salt replication process (using Zr-based Vit106) leading to an open porosity structure. (Reproduced from Schroers, J., Veazey, C., and Johnson, W.L., *Appl. Phys. Lett.* 82, 370, 2003 and Brothers, A.H. and Dunand, D.C., *Acta Mater.* 53, 4427, 2005. With permission.)

To date several methods have been exploited and applied to metallic glasses, leading either to (1) a closed porosity[187,188] or to (2) an open-cell structure.[189]

Commercially available crystalline foams with a closed porosity are frequently produced by introducing a pressurized gas into the molten state of the metal.[190] A different approach to achieving a closed pore structure has been demonstrated for Pd-based metallic glass foams, where the alloy powder is mixed with hydrated B_2O_3. The latter releases gas at elevated temperatures and leads to an expansion of the molten metal volume via uniformly distributed gas bubbles[187] (Figure 17.30a). Yet another method takes advantage of the stability of the metallic glass in the undercooled liquid region, where within this temperature range gas bubbles preintroduced during the melting process expand to their final size.[188]

Open porous structures for metals are typically produced by a salt replication process, which has produced aluminum foams routinely since the 1960s.[190] This technique has now been successfully applied to Vit106 ($Zr_{57}Nb_5Cu_{15.4}Ni_{12.6}Al_{10}$), where compressive strains of 60% were reached in samples with a pore volume of 14 to 28 vol.%.[191]

FIGURE 17.31 HR-TEM image of the alloy $La_{27.5}Zr_{27.5}Cu_{10}Ni_{10}Al_{25}$ at the interface of a Zr-rich phase (on the left) and a La-rich phase (on the right). Fast Fourier-transformed images are shown for the left area (lower left-hand image, larger ring) and for the right area (lower right-hand image, smaller ring). The selected-area electron diffraction pattern (upper left-hand corner) shows two distinct rings. (Reproduced from Kündig, A.A., Ohnuma, M., Ping, D.H., Ohkubo, T., and Hono, K., *Acta Mater.* 52, 2441, 2004. With permission.)

17.6.4 Two-Phase Metallic Glasses

A recent approach to *in situ*-formed metallic glass composites are alloys that decompose in the liquid phase into two compositions both of which form a metallic glass, as in the systems La–Zr–Cu–Ni–Al,[192] Y–Ti–Al–Co,[193] and Ni–Nb–Y.[194] In the case of $La_{27.5}Zr_{27.5}Cu_{10}Ni_{10}Al_{25}$, the alloy separates into a La–Cu-rich and a Zr–Ni-rich phase, both compositions being close to bulk glass-formers. The interface is rather sharp and can be detected by HR-TEM due to a larger average atomic distance in the La-rich alloy, as seen in Figure 17.31.

As with crystalline/amorphous composites, the abrupt change in composition can halt, deflect, or multiply shear bands. In a bulk metallic glass of the Cu–Zr–Al–Ag system the separation into Cu-rich and Al-rich phases has been shown to exhibit enhanced malleability and strength compared to monolithic single-phase amorphous metals.[195] Similarly, a high plastic strain of $>16\%$ and a flow stress of 2 GPa has recently been reported for a $Cu_{47.5}Zr_{47.5}Al_5$ bulk metallic glass separating into two immiscible amorphous phases.[183]

17.7 Applications

Metallic glasses exhibit outstanding mechanical, magnetic, and chemical properties, and possess forming capabilities suitable for a large range of applications — of which only a small fraction have so far been commercially deployed. For successful commercialization a careful comparison of the advantages of a metallic glass product and its higher price is necessary. The cost of metallic glasses is determined by (1) the raw material costs and (2) the processing costs involved. For current bulk metallic glasses, very clean (and therefore expensive) raw materials are required that must be processed under clean (and likewise expensive) conditions. For applications where thin layers or ribbons of metallic glass suffice, a larger range of compositions may be applied and purity is less important. For this reason metallic glass ribbons for magnetic applications are currently their most important application industrially, followed by metallic glass coatings for corrosion protection and then bulk metallic glasses for structural applications. Metallic glass ribbons can be produced inexpensively as the raw materials may be similar in grade to the same elements (e.g. Fe, Ni, Co, Si and B) used in the high-volume steel industry. Thanks to the increase in glass-forming ability brought about by alloy development, not only has the maximum thickness of an amorphous part been increased, but also the process stability of parts with dimensions smaller than the maximum casting thickness. Thus the development of good metallic glass-formers is a key factor for the deployment of metallic glasses on an industrial scale.

17.7.1 Magnetic Metallic Glass Ribbons

Metallic glasses containing large amounts of ferromagnetic material often demonstrate superior soft-magnetic behavior and high permeability. For applications in transformer cores, metallic glasses such as Metglas® Alloy 2605SA1 can reduce transformation losses by 70% compared to (crystalline) iron cores, which clearly justifies their higher production cost.[196]

Also successful has been the deployment of a combination of metallic glass ribbons in article surveillance systems. The marker in this system consists of a ferromagnetic metallic glass ribbon (e.g. $Fe_{40}Ni_{38}Mo_4B_{18}$) which shows magnetostrictive behavior at a certain bias field. The ribbon is magnetically biased with another ribbon of a hard magnetic material, typically of an annealed metallic glass. If the marker is excited in an electromagnetic field at the correct frequency (e.g. 60 kHz), the magnetostrictive ribbon will vibrate mechanically and emit an electromagnetic RF signal of the same frequency as the activation, which decays exponentially with time. The marker is activated by a pulse from an electromagnetic field, placed for example at the exit of a retail shop, and the response is measured by a receiver coil. After demagnetizing the hard magnetic ribbon, the magnetostrictive behavior of the soft magnetic ribbon ceases to be activated by the RF coil.[197]

One potential future application of magnetic metallic glasses may lie in the area of shielding alternating magnetic fields from power lines. Efficient shielding is obtained, for example, by using two layers of the alloy $Fe_{67}Co_{18}B_{14}Si_1$ (Metglas 2605CO), which possesses good shielding ability of large fields.[196] Magnetic glasses lacking metalloids[198] may also be interesting for such applications.

17.7.2 Metallic Glass Coatings

A large variety of coating and deposition techniques, such as physical and chemical vapor deposition, are currently deployed commercially. Thermal spray coating in particular has become one of the most successful techniques for producing large-scale amorphous metal coatings for industrial applications.[199] Here the coating material is introduced as a wire, molted by a combustion or plasma torch far above the melting temperature, and sprayed via a noble gas jet as small droplets onto a substrate kept at room temperature. The droplets are a few hundred micrometers in size and arrive at the substrate at a velocity of several hundred meters per second. The hot droplets melt a thin layer of the surface, weld onto it, and are quenched at a high cooling rate (10^4 to 10^6 K/s). With the high cooling rates given, a large range of alloy compositions may be quenched into a glass and optimized for corrosion and mechanical properties, while the substrate properties remain unaffected and can be chosen independently.

Metallic glasses based on, for example, Fe, Ni, and Co show very good corrosion resistance and high strength compared to their crystalline counterparts. Their high corrosion resistance stems mainly from their composition (i.e., substantial Si and B contents) and chemical homogeneity. The absence of grain boundaries generates uniform protective oxide layers. Applications range from fireboxes for large boilers (1500 MW) to refinery cookers.[199] One project that may also lead to large-volume application of metallic glasses is the study of coating materials for radioactive waste containers, with special attention to long-term corrosion resistance.[200]

For wear resistance, again only a thin layer is needed, and the coating material can be optimized for mechanical strength (i.e., hardness), while the substrate material exhibits greater toughness. Due to similar base elements in the steel substrate and the amorphous coating, adhesion is usually very good.

Metallic glass ribbons and coatings can also be used as catalytic materials. In many cases the activity of the amorphous phase is greater than that of its crystalline counterpart. For example, the turnover frequency for semihydrogenation of a propargylic alcohol in "supercritical" carbon dioxide is 50 times higher on an amorphous Pd–Si catalyst than on conventional supported Pd on SiO_2.[201]

17.7.3 Bulk Metallic Glasses

Structural applications of fully bulk amorphous parts continue to gain momentum. The first applications of these materials were as golf club heads, tennis racket elements, and baseball bats, all benefiting from the excellent elastic properties of metallic glasses.[199] The development of processing equipment has accompanied new applications, and processing in the undercooled liquid region ("injection casting") has been studied and exploited. This procedure is more stable to crystallization (see Section 17.3.3) than heating the glass from room temperature ("superplastic forming"), but both processes present the advantage of very low solidification shrinkage and the potential for near-net-shape forming of parts (see Section 17.4.2 and Section 17.4.3). The resulting applications include electronic casings and hinges[199] (see, e.g., Figure 17.18). A low processing temperature in combination with high material strength is also a property unique to metallic glasses. High-strength parts can now be processed similarly to plastics (see, e.g., Figure 17.20).

Another advantage of metallic glasses is their combination of high yield strength and low Young's modulus; this has inspired applications such as a Coriolis force flow meter[202] and pressure sensor membranes.[203] As mentioned above, the improved processability of bulk metallic glasses also makes them attractive for applications where only thin parts are needed.

17.7.4 Micro- and Nano-Patterned Structures

Due to the absence of grain boundaries, metallic glasses are mechanically and chemically homogenous on all length scales down to a few nanometers. This makes glasses (and single crystals such as Si) preferred materials for micro- and nanostructured parts. Prototypes of microgears, in which high precision and a low abrasion rate are important, have been tested successfully.[204] Structures with dimensions down to the nanometer range such as test grooves and compact disc templates have been directly patterned[205] and replicated from Si wafers[130,135] with high precision (see Figure 17.17). These high-strength, high-precision structures can be used, for example, for the embossing of polymers, which can then be used in microfluidics or, if the structures are selectively filled with metals, in microelectronics.

17.7.5 Medical Applications

Deploying metallic glasses in surgical instruments (scissors, etc.) or as implant material takes advantage of their unique combination of low Young's modulus, high strength, high elastic limit, and good corrosion resistance. Zr-based metallic glasses form stable oxide layers and the release of alloy elements from a Zr-based implant is low. Excluding the toxic elements Ni and Cu will increase biocompatibility still further; a Ni-free, Zr-based metallic glass with a low critical cooling rate has recently been developed

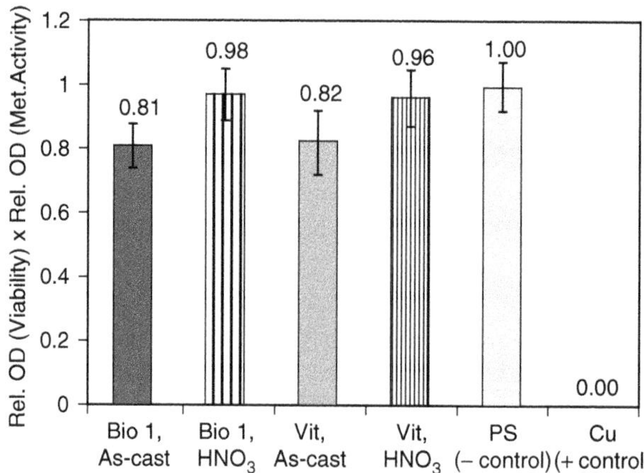

FIGURE 17.32 Direct cytotoxicity tests (performed by measuring the viability and metabolic activity of mouse fibroblasts via optical density measurements) on $Zr_{58}Cu_{22}Fe_8Al_{12}$ (Bio 1) and $Zr_{58.5}Cu_{15.6}Ni_{12.8}Al_{10.3}Nb_{2.8}$ (Vit106a) in the as-cast state and after surface treatment in nitric acid, in comparison to negative (PS, polystyrene) and positive (Cu) controls. (Reproduced from Buzzi, S., Jin, K., Uggowitzer, P.J., Tosatti, S., Gerber, I., and Löffler, J.F., *Intermetallics*, 14, 729, 2006. With permission.)

for this purpose.[206] Cytotoxicity measurements of the latter, $Zr_{58}Cu_{22}Fe_8Al_{12}$, and a Ni-bearing metallic glass, $Zr_{58.5}Cu_{15.6}Ni_{12.8}Al_{10.3}Nb_{2.8}$ (Vit 106a), show good biocompatibility for both alloys, which can be increased to the level of polystyrene (the negative control) after a passivation treatment in HNO_3[207] (see Figure 17.32). Bone plates and aneurism stents may be among the applications that result.

Acknowledgments

This work was supported by the Swiss National Science Foundation (Grant # 200021-105647 and Grant # 200021-108071), by an ETH Research Grant (Grant # TH-21/04-2), and by start-up funds from ETH Zurich (J.F.L.).

References

[1] Buckel, W. and Hilsch, R., Supraleitung und Widerstand von Zinn mit Gitterstörungen (Mitteilungen zur Supraleitung II). *Z. Phys.* 131, 420, 1952.

[2] Klement, W., Willens, R.H., and Duwez, P., Non-crystalline structure in solidified gold-silicon alloys. *Nature* 187, 869, 1960.

[3] Duwez, P., Structure and properties of alloys rapidly quenched from the liquid state. *ASM Trans. Quart.* 60, 607, 1967.

[4] Cahn, R.W., Metallic glasses. *Contemp. Phys.* 21, 43, 1980.

[5] Turnbull, D., Metastable structures in metallurgy. *Metall. Trans. A* 12, 695, 1981.

[6] Johnson, W.L., Metallic glasses. *Metals Handbook* 2, 805, 1990.

[7] Johnson, W.L., Metastable phases (Chap. 29), in *Intermetallic Compounds: Vol. 1, Principles*, J.H. Westbrook and R.L. Fleischer (Eds.), John Wiley & Sons Ltd., Chichester, 1994, p. 687.

[8] Inoue, A., *Bulk Amorphous Alloys, Preparation and Fundamental Characteristics.* Trans. Tech. Publ., Switzerland, 1998.

[9] Johnson, W.L., Bulk glass-forming metallic alloys: Science and technology. *MRS Bull.* 24, 42, 1999.

[10] Inoue, A., Stabilization of metallic supercooled liquid and bulk amorphous alloys. *Acta Mater.* 48, 279, 2000.

[11] Löffler, J.F., Bulk metallic glasses. *Intermetallics* 11, 529, 2003; reprinted from J. F. Löffler, in *The Encyclopedia of Materials: Science and Technology*, K.H.J. Buschow, R.W. Cahn, M.C. Flemings, B. Ilschner, E.J. Kramer, and S. Mahajan (Eds.), Pergamon/Elsevier Science, Amsterdam, 2003.

[12] Löffler, J.F., Recent progress in the area of bulk metallic glasses. *Z. Metallkd.* 97, 225, 2006.

[13] Schwarz, R.B. and Johnson, W.L., Formation of an amorphous alloy by solid-state reaction of the pure polycrystalline metals. *Phys. Rev. Lett.* 51, 415, 1983.

[14] Russell, K.C., Phase stability under irradiation. *Prog. Mater. Sci.* 28, 229, 1984.

[15] Ponyatovsky, E.G. and Barkalov, O.I., Pressure-induced amorphous phases. *Mater. Sci. Eng. Rep.* 8, 147, 1992.

[16] Schultz, L. and Eckert, J., Mechanically alloyed glassy metals. *Topics Appl. Phys.* 72, 69, 1994.

[17] Valiev, R.Z., Islamgaliev, R.K., and Alexandrov, I.V., Bulk nanostructured materials from severe plastic deformation. *Prog. Mater. Sci.* 45, 103, 2000.

[18] Duwez, P., Willens, R.H., and Klement Jr., W., Continuous series of metastable solid solutions in silver-copper alloys. *J. Appl. Phys.* 31, 1136, 1960.

[19] Duwez, P., Metallic glasses-historical background (Chap. 2), in *Glassy Metals I (Topics in Applied Physics, Vol. 46)*, H.J. Güntherodt and H. Beck (Eds.), Springer, Berlin, 1981, p. 19.

[20] Jones, H., Splat cooling and metastable phases. *Rep. Prog. Phys.* 36, 1425, 1973.

[21] Chen, H.S. and Turnbull, D., Formation, stability and structure of palladium-silicon based alloy glasses. *Acta Metall.* 17, 1021, 1969.

[22] Gibbs, J.H. and DiMarzio, E.A., Nature of the glass transition and the glassy state. *J. Chem. Phys.* 28, 373, 1958.

[23] Cohen, M.H. and Turnbull, D., Molecular transport in liquids and glasses. *J. Chem. Phys.* 31, 1164, 1959.
[24] Cohen, M.H. and Turnbull, D., Composition requirements for glass formation in metallic and ionic systems. *Nature* 189, 131, 1961.
[25] Turnbull, D., Under what conditions can a glass be formed? *Contemp. Phys.* 10, 473, 1969.
[26] Smith, C.H., Applications of rapidly solidified soft magnetic alloys (Chap. 19), in *Rapidly Solidified Alloys: Processes, Structure, Properties, Applications*, H.H. Liebermann (Ed.), Dekker Inc., New York, 1993, p. 617.
[27] Liebermann, H.H. and Graham Jr., C.D., Production of amorphous alloy ribbons and effects of apparatus parameters on ribbon dimensions. *IEEE Trans. Mag.* 12, 921, 1976.
[28] Chen, H.S., Thermodynamic considerations on the formation and stability of metallic glasses. *Acta Metall.* 22, 1505, 1974.
[29] Lee, M.C, Kendall, J.M., and Johnson, W.L., Spheres of the metallic glass $Au_{55}Pb_{22.5}Sb_{22.5}$ and their surface characteristics. *Appl. Phys. Lett.* 40, 383, 1982.
[30] Drehman, A.L., Greer, A.L., and Turnbull, D., Bulk formation of a metallic glass — $Pd_{40}Ni_{40}P_{20}$. *Appl. Phys. Lett.* 41, 716, 1982.
[31] Kui, H.W., Greer, A.L., and Turnbull, D., Formation of bulk metallic glass by fluxing. *Appl. Phys. Lett.* 45, 615, 1984.
[32] Inoue, A., Zhang, T., and Masumoto, T., Zr–Al–Ni amorphous alloys with high glass-transition temperature and significant supercooled liquid region. *Mater. Trans., JIM* 31, 177, 1990.
[33] Inoue, A. et al., Mg–Cu–Y amorphous alloys with high mechanical strengths produced by a metallic mold casting method. *Mater. Trans., JIM* 32, 609, 1991.
[34] Inoue, A. et al., Bulk La–Al–TM (TM = transition-metal) amorphous alloy with high tensile strength produced by a high-pressure die-casting method. *Mater. Trans., JIM* 34, 351, 1993.
[35] Peker, A. and Johnson, W.L., A highly processable metallic glass — $Zr_{41.2}Ti_{13.8}Cu_{12.5}Ni_{10}Be_{22.5}$. *Appl. Phys. Lett.* 63, 2342, 1993.
[36] Peker, A. and Johnson, W.L., Be bearing amorphous metallic alloys formed at low cooling rates, U.S. Patent No. 5,288,344, 1993.
[37] Lin, X.H. and Johnson, W.L., Formation of Ti–Zr–Cu–Ni bulk metallic glasses. *J. Appl. Phys.* 78, 6514, 1995.
[38] Davies, H.A., Metallic glass formation (Chap. 2), in *Amorphous Metallic Alloys*, F.E. Luborsky (Ed.), Butterworths Monographs in Materials, London, 1983, p. 8.
[39] Bennet, C.H., Serially deposited amorphous aggregates of hard spheres. *J. Appl. Phys.* 43, 2727, 1972.
[40] Rahman, A., Mandell, M.J., and McTague, J.P., Molecular-dynamics study of an amorphous Lennard-Jones system at low-temperature. *J. Chem. Phys.* 64, 1564, 1976.
[41] Kramer, J., Non conductive transformations in metal. *Ann. Phys.* 19, 37, 1934.
[42] Allen, R.P., Dahlgren, S.D., and Merz, M.D., in *Rapidly Quenched Metals II*, N.J. Grant and B.C. Giessen (Eds.), MIT Press, Cambridge, 1976, p. 37.
[43] Granqvist, C.G. and Buhrman, R.A., Ultrafine metal particles. *J. Appl. Phys.* 47, 2200, 1976.
[44] Birringer, R. et al., Nanocrystalline materials — An approach to a novel solid structure with gas-like disorder. *Phys. Lett. A* 102, 365, 1984.
[45] Gleiter, H., Nanocrystalline materials. *Prog. Mater. Sci.* 33, 223, 1989.
[46] Gleiter, H., Nanostructured materials: Basic concepts and microstructure. *Acta Mater.* 48, 1, 2000.
[47] Cahn, R.W., Background to rapidly solidification processes (Chap. 1), in *Rapid Solidified Alloys*, H.H. Liebermann (Ed.), Dekker Inc., New York, 1993, p. 5.
[48] Polk, D., Method of continuous casting metal filament on interior groove of chill roll, U.S. Patent 3,881,542 (1975).
[49] Kavesh, S., Method of forming metallic filament cast on interior surface of inclined annular quench roll, U.S. Patent 3,881,540 (1975).

[50] Pond, R.B. and Maddin, R., A method of producing rapidly solidified filamentary castings. *Trans, Met. Soc. AIME* 245, 2475, 1969.
[51] Ohnaka, I. and Fukusako, T., Production of metallic filament by ejecting molten lead into water or atmosphere and stability of molten lead jet. *J. Japan Inst. Met.* 42, 415, 1978.
[52] Narasimhan, M.C., Continuous casting method for metallic strips, U.S. Patent 4,142,571 (1979).
[53] Anantharaman, T.R. and Suryanarayana, C., *Rapidly Solidified Metals.* Trans. Tech. Publ., Switzerland, 1987.
[54] Löffler, J.F. and Johnson, W.L., Crystallization of Mg–Al and Al-based metallic liquids under ultra-high gravity. *Intermetallics* 10, 1167, 2002.
[55] Löffler, J.F. et al., Eutectic isolation in Mg–Al–Cu–Li(–Y) alloys by centrifugal processing. *Phil. Mag.* 83, 2797, 2003.
[56] Castellero, A. et al., Critical Poisson's ratio for room temperature embrittlement of amorphous $Mg_{85}Cu_5Y_{10}$. *Phil. Mag. Lett.* (in press).
[57] Egami, T., Atomic structure of rapidly solidified alloys (Chap. 9), in *Rapidly Solidified Alloys*, H.H. Liebermann (Ed.), Dekker Inc., New York, 1993, p. 231.
[58] Warren, B.E., *X-ray diffraction.* Dover, New York, 1969.
[59] Wagner, C.N.J., Direct methods for the determination of atomic-scale structure of amorphous solids (x-ray, electron, and neutron scattering). *J. Non-Cryst. Solids* 31, 1, 1978.
[60] Löffler, J.F. and Weissmüller, J., Grain-boundary atomic structure in nanocrystalline palladium from x-ray atomic distribution functions. *Phys. Rev. B* 52, 7076, 1995.
[61] Saida, J., Matsushita, M., and Inoue, A., Direct observation of icosahedral cluster in $Zr_{70}Pd_{30}$ binary glassy alloy. *Appl. Phys. Lett.* 79, 412, 2001.
[62] Saida, J., Matsushita, M., and Inoue, A., Nanoicosahedral quasicrystalline phase in Zr–Pd and Zr–Pt binary alloys. *J. Appl. Phys.* 90, 4717, 2001.
[63] Bernal, J.D., Geometrical approach to the structure of liquids. *Nature* 183, 141, 1959.
[64] Bernal, J.D., Co-ordination of randomly packed spheres. *Nature* 188, 910, 1960.
[65] Scott, G.D., Packing of equal spheres. *Nature* 188, 908, 1960.
[66] Finney, J.L., Random packings and structure of simple liquids. 1. Geometry of random close packing. *Proc. Roy. Soc. London Ser. A-Math. Phys. Sci.* 319, 479, 1970.
[67] Finney, J.L., Random packings and structure of simple liquids. 2. Molecular geometry of simple liquids. *Proc. Roy. Soc. London Ser. A-Math. Phys. Sci.* 319, 495, 1970.
[68] Egami, T., Structural relaxation in amorphous $Fe_{40}Ni_{40}P_{14}B_6$ studied by energy dispersive x-ray diffraction. *J. Mater. Sci.* 13, 2587, 1978.
[69] Srolovitz, D., Egami, T., and Vitek, V., Radial distribution function and structural relaxation in amorphous solids. *Phys. Rev. B* 24, 6936, 1981.
[70] Gaskell, P.H., A new structural model for amorphous transition metal silicides, borides, phosphides and carbides. *J. Non-Cryst. Solids* 32, 207, 1979.
[71] Zallen, R., *The Physics of Amorphous Solids.* Wiley-VCH, Weinheim, 2004.
[72] Polk, D.E., Structural model for amorphous metallic alloys. *Scripta Metall.* 4, 117, 1970.
[73] Inoue, A. et al., High packing density of Zr- and Pd-based bulk amorphous alloys. *Mater. Trans., JIM* 39, 318, 1998.
[74] Wang, W.H. et al., Microstructural transformation in a $Zr_{41}Ti_{14}Cu_{12.5}Ni_{10}Be_{22.5}$ bulk metallic glass under high pressure. *Phys. Rev. B* 62, 11292, 2000.
[75] Egami, T. and Waseda, Y., Atomic size effect on the formability of metallic glasses. *J. Non-Cryst. Solids* 64, 113, 1984.
[76] Zhang, T., Inoue, A., and Masumoto, T., Amorphous Zr–Al–Tm (Tm = Co, Ni, Cu) alloys with significant supercooled liquid region of over 100 K. *Mater. Trans., JIM* 32, 1005, 1991.
[77] Miracle, D.B., Sanders, W.S., and Senkov, O.N., The influence of efficient atomic packing on the constitution of metallic glasses. *Phil. Mag.* 83, 2409, 2003.
[78] Lu, Z.P. and Liu, C.T., Glass formation criterion for various glass-forming systems. *Phys. Rev. Lett.* 91, 115505, 2003.

[79] Inoue A. and Takeuchi, A., Recent progress in bulk glassy alloys. *Mater. Trans. JIM* 43, 1892, 2002.
[80] Inoue, A., Bulk glassy and nonequilibrium crystalline alloys by stabilization of supercooled liquid: fabrication, functional properties and applications (Part 1), *Proc. Jpn. Acad. Ser. B* 81, 156, 2005.
[81] Boudreaux, D.S., Structure of metallic glass alloys (Chap. 1), in *Glassy Metals: Magnetic, Chemical and Structural Properties*, R. Hasegawa (Ed.), CRC Press, Boca Raton, FL, 1983, p. 1.
[82] Eckert, J. et al., Crystallization behavior and phase formation in Zr–Al–Cu–Ni metallic glass containing oxygen. *Mater. Trans., JIM* 39, 623, 1998.
[83] Wanderka, N. et al, Formation of quasicrystals in $Zr_{46.8}Ti_{8.2}Cu_{7.5}Ni_{10}Be_{27.5}$ bulk glass. *Appl. Phys. Lett.* 77, 3935, 2000.
[84] Löffler, J.F. et al., Structural and electronic properties of Zr–Ti–Cu–Ni–Be alloys. *Mat. Res. Soc. Symp. Proc.* 644, L1.9.1, 2001.
[85] Pekarskaya, E., Löffler, J.F., and Johnson, W.L., Microstructural studies of crystallization of a Zr-based bulk metallic glass. *Acta Mater.* 51, 4045, 2003.
[86] Frank, F.C., Supercooling of liquids. *Proc. Roy. Soc. A*, 215, 43, 1952.
[87] Nelson, D.R., Order, frustration, and defects in liquids and glasses. *Phys. Rev. B* 28, 5515, 1983.
[88] Tang, X.P. et al., Devitrification of the $Zr_{41.2}Ti_{13.8}Cu_{12.5}Ni_{10.0}Be_{22.5}$ bulk metallic glass studied by XRD, SANS, and NMR. *J. Non-Cryst. Solids* 317, 118, 2003.
[89] Waniuk, T.A., Schroers, J., and Johnson, W.L., Timescales of crystallization and viscous flow of the bulk glass-forming Zr–Ti–Ni–Cu–Be alloys. *Phys. Rev. B* 67, 184203, 2003.
[90] Holland-Moritz, D., Herlach, D.M., and Urban, K., Observation of the undercoolability of quasicrystal-forming alloys by electromagnetic levitation. *Phys. Rev. Lett.* 71, 1196, 1993.
[91] Glade, S.C. et al., Thermodynamics of $Cu_{47}Ti_{34}Zr_{11}Ni_8$, $Zr_{52.5}Cu_{17.9}Ni_{14.6}Al_{10}Ti_5$ and $Zr_{57}Cu_{15.4}\ Ni_{12.6}Al_{10}Nb_5$ bulk metallic glass forming alloys. *J. Appl. Phys.* 87, 7242, 2000.
[92] Fan, G.J. et al., Thermodynamics, enthalpy relaxation and fragility of the bulk metallic glass-forming liquid $Pd_{43}Ni_{10}Cu_{27}P_{20}$. *Acta Mater.* 52, 667, 2004.
[93] Lu, I.R. et al., Thermodynamic properties of Pd-based glass-forming alloys. *J. Non-Cryst. Solids* 252, 577, 1999.
[94] Kim, Y.J. et al., Experimental determination of the time–temperature–transformation diagram of the undercooled $Zr_{41.2}Ti_{13.8}Cu_{12.5}Ni_{10.0}Be_{22.5}$ alloy using the containerless electrostatic levitation processing technique. *Appl. Phys. Lett.* 68, 1057, 1996.
[95] Uhlmann, D.R., Glass formation. *J. Non-Cryst. Solids* 25, 42, 1977.
[96] Angell, C.A. and Sichina, W., Thermodynamics of the glass transition: empirical aspects. *Ann. New York Acad. Sci.* 279, 53, 1976.
[97] Angell, C.A., Formation of glasses from liquids and biopolymers. *Science* 267, 1924, 1995.
[98] Busch, R., Bakke, E., and Johnson, W.L., Viscosity of the supercooled liquid and relaxation at the glass transition of the $Zr_{46.75}Ti_{8.25}Cu_{7.5}Ni_{10}Be_{27.5}$ bulk metallic glass forming alloy. *Acta Mater.* 46, 4725, 1998.
[99] Waniuk, T.A. et al., Equilibrium viscosity of the $Zr_{41.2}Ti_{13.8}Cu_{12.5}Ni_{10.0}Be_{22.5}$ bulk metallic glass-forming liquid and viscous flow during relaxation, phase separation, and primary crystallization. *Acta Mater.* 46, 5229, 1998.
[100] Masuhr, A. et al., Time scales for viscous flow, atomic transport, and crystallization in the liquid and supercooled liquid states of $Zr_{41.2}Ti_{13.8}Cu_{12.5}Ni_{10.0}Be_{22.5}$. *Phys. Rev. Lett.* 82, 2290, 1999.
[101] Mukherjee, S., Johnson, W.L., and Rhim, W.K., Noncontact measurement of high-temperature surface tension and viscosity of bulk metallic glass-forming alloys using the drop oscillation technique. *Appl. Phys. Lett.* 86, 014104, 2005.
[102] Johnson, W.L., Physics and metallurgy of bulk glass forming alloys, in *Science of Alloys for the 21st Century: A Hume-Rothery Symposium Celebration*, P.E.A. Turchi, R.D. Shull, and A. Gonis (Eds.), The Minerals, Metals & Materials Society, 2000, p. 183.
[103] Busch, R., Liu, W., and Johnson, W.L., Thermodynamics and kinetics of the $Mg_{65}Cu_{25}Y_{10}$ bulk metallic glass forming liquid. *J. Appl. Phys.* 83, 4134, 1998.

[104] Busch, R., Kim, Y.J., and Johnson, W.L., Thermodynamics and kinetics of the undercooled liquid and the glass transition of the $Zr_{41.2}Ti_{13.8}Cu_{12.5}Ni_{10.0}Be_{22.5}$ alloy. *J. Appl. Phys.* 77, 4039, 1995.
[105] Chen, H.S., A method for evaluating viscosities of metallic glasses from the rates of thermal transformations. *J. Non-Cryst. Solids* 27, 257, 1978.
[106] Marcus, M. and Turnbull, D., On the correlation between glass-forming tendency and liquidus temperature in metallic alloys. *Mater. Sci. Eng.* 23, 211, 1976.
[107] Löffler, J.F., Schroers, J., and Johnson, W.L., Time–temperature–transformation diagram and microstructures of bulk glass forming $Pd_{40}Cu_{30}Ni_{10}P_{20}$. *Appl. Phys. Lett.* 77, 681, 2000.
[108] Rhim, W.K. et al., An electrostatic levitator for high-temperature containerless materials processing in 1-g. *Rev. Sci. Instrum.* 64, 2961, 1993.
[109] Johnson, W.A. and Mehl, R.F., Reaction kinetics in processes of nucleation and growth. *Trans. Am. Inst. Min. Metall. Eng.* 135, 416, 1939.
[110] Avrami, M., Kinetics of phase change III: Granulation, phase change an microstructures. *J. Chem. Phys.* 9, 177, 1941.
[111] Nishiyama, N. and Inoue, A., Supercooling investigation and critical cooling rate for glass formation in Pd–Cu–Ni–P alloy. *Acta Mater.* 47, 1487, 1999.
[112] Schneider, S., Thiyagarajan, P., and Johnson, W.L., Formation of nanocrystals based on decomposition in the amorphous $Zr_{41.2}Ti_{13.8}Cu_{12.5}Ni_{10.0}Be_{22.5}$ alloy. *Appl. Phys. Lett.* 68, 493, 1996.
[113] Löffler, J.F. and Johnson, W.L., Model for decomposition and nanocrystallization of deeply undercooled $Zr_{41.2}Ti_{13.8}Cu_{12.5}Ni_{10}Be_{22.5}$. *Appl. Phys. Lett.* 76, 3394, 2000.
[114] Löffler, J.F. et al., Comparison of the decomposition and crystallization behavior of Zr and Pd based bulk amorphous alloys. *Mater. Sci. Forum* 343, 179, 2000.
[115] Schroers, J. et al., Pronounced asymmetry in the crystallization behavior during constant heating and cooling of a bulk metallic glass-forming liquid. *Phys. Rev. B* 60, 11858, 1999.
[116] Löffler, J.F. et al., Crystallization of bulk amorphous Zr–Ti(Nb)–Cu–Ni–Al. *Appl. Phys. Lett.* 77, 525, 2000.
[117] Kündig, A.A. et al., Influence of decomposition on the thermal stability of undercooled Zr–Ti–Cu–Ni–Al alloys. *Scripta Mater.* 44, 1269, 2001.
[118] Schroers, J., The Superplastic Forming of Bulk Metallic Glasses. *J. Mater. JOM* 57, 35, 2005.
[119] Zhang, B. et al., Amorphous metallic plastic. *Phys. Rev. Lett.* 94, 205502, 2005.
[120] Schroers, J. et al., Gold based bulk metallic glass. *Appl. Phys. Lett.* 87, 061912, 2005.
[121] Ma, H. et al., Doubling the critical size for bulk metallic glass formation in the Mg–Cu–Y ternary system. *J. Mater. Res.* 20, 2252, 2005.
[122] Inoue, A., Zhang, T., and Masumoto, T., Al–La–Ni amorphous alloys with a wide supercooled liquid region. *Mater. Trans. JIM* 30, 965, 1989.
[123] Schroers, J. and Johnson, W.L., Highly processable bulk metallic glass-forming alloys in the Pt–Co–Ni–Cu–P system. *Appl. Phys. Lett.* 84, 3666, 2004.
[124] Inoue, A., Nishiyama, N., and Kimura, H., Preparation and thermal stability of bulk amorphous $Pd_{40}Cu_{30}Ni_{10}P_{20}$ alloy cylinder of 72 mm in diameter. *Mater. Trans. JIM* 38, 179, 1997.
[125] Wert, J.A., Pryds, N., and Zhang, E., Rheological properties of a $Mg_{60}Cu_{30}Y_{10}$ alloy in the supercooled liquid state. *Proceedings of the 22nd Risø International Symposium on Materials Science*, 423, 2001.
[126] Pryds, N.H. et al., Preparation and properties of Mg–Cu–Y bulk amorphous alloys. *Mater. Trans., JIM* 41, 1435, 2000.
[127] Kawamura, Y. et al., High strain-rate superplasticity due to Newtonian viscous flow in $La_{55}Al_{25}Ni_{20}$ metallic glass. *Mater. Trans., JIM* 40, 794, 1999.
[128] Kawamura, Y. et al., Deformation behavior of $Zr_{65}Al_{10}Ni_{10}Cu_{15}$ glassy alloy with wide supercooled liquid region. *Appl. Phys. Lett.* 69, 1208, 1996.
[129] Kündig, A.A., Schweizer, T., Schafler, E., and Löffler, J.F., Metallic glass/polymer composites by co-processing at similar viscosities. *Scripta Mater.* 56, 289, 2007.

[130] Kündig, A.A. et al., Preparation of high aspect ratio surface microstructures out of a Zr-based bulk metallic glass. *Microeletr. Eng.* 67, 405, 2003.

[131] Chokshi, A.H., Mukherjee, A.K., and Langdon, T.G., Superplasticity in advanced materials. *Mater. Sci. Eng. R* 10, 237, 1993.

[132] Kawamura, Y. et al., Full strength compacts by extrusion of glassy metal powder at the supercooled liquid state. *Appl. Phys. Lett.* 67, 2008, 1995.

[133] Saotome, Y., Zhang, T., and Inoue, A. Microforming of MEMS parts with amorphous alloys. *Mat. Res. Soc. Symp. Proc.* 554, 385, 1999.

[134] Zumkley, T. et al., Superplastic forging of ZrTiCuNiBe-bulk glass for shaping of microparts. *Mater. Sci. Forum* 386–388, 541, 2002.

[135] Saotome, Y. et al., The microformability of Pt-based metallic glass and the nanoforming of three-dimensional structures. *Intermetallics* 10, 1241, 2002.

[136] Saotome, Y. et al., Superplastic micro/nano-formability of $La_{60}Al_{20}Ni_{10}Co_5Cu_5$ amorphous alloy in the supercooled liquid state. *Mater. Sci. Eng. A* 304–306, 716, 2001.

[137] Herzer, G., Nanocrystalline soft magnetic materials. *J. Magn. Magn. Mater.* 112, 258, 1992.

[138] Löffler, J.F., Braun, H.B., and Wagner, W., Magnetic correlations in nanostructured ferromagnets. *Phys. Rev. Lett.* 85, 1990, 2000.

[139] Mizoguchi, T., Intrinsic magnetic properties of rapidly solidified alloys (Chap. 16), in *Rapidly Solidified Alloys: Processes, Structure, Properties, Applications*, H.H. Liebermann (Ed.), Dekker Inc., New York, 1993, p. 461.

[140] Hernando, A. and Vázquez, M., Engineering magnetic properties (Chap. 17), in *Rapidly Solidified Alloys: Processes, Structure, Properties, Applications*, H.H. Liebermann (Ed.), Dekker Inc., New York, 1993, p. 553.

[141] Spaepen, F., A microscopic mechanism for steady state inhomogeneous flow in metallic glasses. *Acta Metall.* 25, 407, 1977.

[142] Masumoto, T. and Maddin, R., Mechanical properties of palladium 20 a/o silicon alloy quenched from liquid state. *Acta Metall.* 19, 725, 1971.

[143] Kato, H. et al., Newtonian to non-Newtonian master flow curves of a bulk glass alloy $Pd_{40}Ni_{10}Cu_{30}P_{20}$. *Appl. Phys. Lett.* 73, 3665, 1998.

[144] Lu, J., Ravichandran, G., and Johnson, W.L., Deformation behavior of the $Zr_{41.2}Ti_{13.8}Cu_{12.5}Ni_{10}Be_{22.5}$ bulk metallic glass over a wide range of strain-rates and temperatures. *Acta Mater.* 51, 3429, 2003.

[145] Safarik, D.J., Cady, C.M., and Schwarz, R.B., Shear processes in bulk metallic glasses. *Acta Mater.* 53, 2193, 2005.

[146] Chen, H. S., Kato, H., and Inoue, A., A fictive stress model and nonlinear viscoelastic behaviors in metallic glasses. *Mater. Trans.* 42, 59, 2001.

[147] Argon, A.S., Plastic deformation in metallic glass. *Acta Metall.* 27, 47, 1978.

[148] Lewandowski, J.J. and Greer, A.L., Temperature rise at shear bands in metallic glasses. *Nature Mater.* 5, 15, 2006.

[149] Yang, B. et al., Dynamic evolution of nanoscale shear bands in a bulk metallic glass. *Appl. Phys. Lett.* 86, 141904, 2005.

[150] Dalla Torre, F.H. et al., Negative strain rate sensitivity in bulk metallic glass and its similarities with the dynamic strain aging effect during deformation. *Appl. Phys. Lett.* 89, 091918, 2006.

[151] Wright, W.J., Saha, R., and Nix, W.D., Deformation mechanisms of the $Zr_{40}Ti_{14}Ni_{10}Cu_{12}Be_{24}$ bulk metallic glass. *Mater. Trans., JIM* 42, 642, 2001.

[152] Bowman, K., *Mechanical Behaviour of Materials*, John Wiley & Sons, Inc., Hoboken, USA, 2004.

[153] Lewandowski, J.J., Wang, W.H. and Greer, A.L., Intrinsic plasticity or brittleness of metallic glasses. *Phil. Mag. Lett.* 85, 77, 2005.

[154] Lewandowski, J.J., Shazly, M., and Nouri, A.S., Intrinsic and extrinsic toughening of metallic glasses. *Scripta Mater.* 54, 337, 2006.

[155] Callister, W.D., *Materials Science and Engineering: An Introduction.* John Wiley & Sons, Inc., New York, 1994, p. 194.

[156] Cockeram, B.V., The fracture toughness and toughening mechanisms of nickel-base wear materials. *Metall. Mater. Trans. A* 33, 33, 2002.

[157] Wang, R.J. et al., Responses of glassy structure and properties to pressure and devitrification. *Appl. Phys. Lett.* 83, 2814, 2003.

[158] Xi, X.K. et al., Fracture of brittle metallic glasses: Brittleness or plasticity. *Phys. Rev. Lett.* 94, 125510, 2005.

[159] Wesseling, P. et al., Preliminary assessment of flow, notch toughness, and high temperature behavior of $Cu_{60}Zr_{20}Hf_{10}Ti_{10}$ bulk metallic glass. *Scripta Mater.* 51, 151, 2004.

[160] Saida, J. et al., Crystallization and grain growth behavior of $Zr_{65}Cu_{27.5}Al_{7.5}$ metallic glass. *Mater. Sci. Eng.* A 304, 338, 2001.

[161] Wang, W.H., Dong, C., and Shek, C.H., Bulk metallic glasses. *Mater. Sci. Eng.* R 44, 45, 2004.

[162] Lowhaphandu, P. and Lewandowski, J.J., Fracture toughness and notched toughness of bulk amorphous alloy: Zr–Ti–Ni–Cu–Be. *Scripta Mater.* 38, 1881, 1998.

[163] Gilbert, C.J., Schroeder, V., and Ritchie, R.O., Mechanisms for fracture and fatigue-crack propagation in a bulk metallic glass. *Metal. Mater. Trans. A.* 30, 1739, 1999.

[164] Flores, K.M., Johnson, W.L., and Dauskardt, R.H., Fracture and fatigue behavior of a Zr–Ti–Nb ductile phase reinforced bulk metallic glass matrix composite. *Scripta Mater.* 49, 1181, 2003.

[165] Wetzig, K. et al., The relaxation behavior of strength and fracture-toughness of amorphous metal–metalloid alloys. *Cryst. Res. Technol.* 18, 1181, 1983.

[166] Davis, L.A., Fracture of Ni–Fe based metallic glasses. *J. Mater. Sci.* 10, 1557 , 1975.

[167] Schroers, J. and Johnson, W.L., Ductile bulk metallic glass. *Phys. Rev. Lett.* 93, 255506, 2004.

[168] Nagendra, N. et al., Effect of crystallinity on the impact toughness of a La-based bulk metallic glass. *Acta Mater.* 48, 2603, 2000.

[169] Pugh, S.F., Relations between the elastic moduli and the plastic properties of polycrystalline pure metals. *Phil. Mag.* 45, 823, 1954.

[170] Kelly, A., Tyson, W.R., and Cottrell, A.H., Ductile and brittle crystals. *Phil. Mag.* 15, 567, 1967.

[171] Rice, J.R. and Thomson, R., Ductile versus brittle behavior of crystals. *Phil. Mag.* 29, 73, 1974.

[172] Johnson, W.L. and Samwer, K., A universal criterion for plastic yielding of metallic glasses with a $(T/T_g)^{2/3}$ temperature dependence. *Phys. Rev. Lett.* 95, 195501, 2005.

[173] Van den Beukel, A. and Radelaar, S., On the kinetics of structural relaxation in metallic glasses. *Acta Metall.* 31, 419, 1983.

[174] Van den Beukel, A., Analysis of structural relaxation data in metallic glasses in terms of different models. *Acta Metal. Mater.* 39, 2709, 1991.

[175] Kato, H. et al., Newtonian to non-Newtonian master flow curves of a bulk glass alloy $Pd_{40}Ni_{10}Cu_{30}P_{20}$. *Appl. Phys. Lett.* 73, 3665, 1998.

[176] Kawamura, Y. and Inoue, A., Newtonian viscosity of supercooled liquid in a $Pd_{40}Ni_{40}P_{20}$ metallic glass. *Appl. Phys. Lett.* 77, 1114, 2000.

[177] Wang, G. et al., Tensile fracture characteristics and deformation of a Zr-based bulk metallic glass at high temperatures. *Intermetallics* 13, 642, 2005.

[178] Hays, C.C., Kim, C.P., and Johnson, W.L., Microstructure controlled shear band pattern formation and enhanced plasticity of bulk metallic glasses containing in situ formed ductile phase dendrite dispersions. *Phys. Rev. Lett.* 84, 2901, 2000.

[179] Lee, M.L., Li, Y., and Schuh, C.A., Effect of a controlled volume fraction of dendritic phases on tensile and compressive ductility in La-based metallic glass matrix composites. *Acta Mater.* 52, 4121, 2004.

[180] Kühn, U. et al., ZrNbCuNiAl bulk metallic glass matrix composites containing dendritic bcc phase precipitates. *Appl. Phys. Lett.* 80, 2478, 2002.

[181] He, G. et al., Novel Ti-base nanostructure–dendrite composite with enhanced plasticity. *Nature Mater.* 2, 33, 2003.

[182] Xu, Y. et al., Mg-based bulk metallic glass composites with plasticity and gigapascal strength. *Acta Mater.* 53, 1857, 2005.
[183] Das, J. et al., Work-hardenable ductile bulk metallic glass. *Phys. Rev. Lett.* 94, 205501, 2005.
[184] Choi-Yim, H. and Johnson, W.L., Bulk metallic glass matrix composites. *Appl. Phys. Lett.* 71, 3808, 1997.
[185] Choi-Yim, H. et al., Synthesis and characterization of particulate reinforced $Zr_{57}Nb_5Al_{10}Cu_{15.4}Ni_{12.6}$ bulk metallic glass composites. *Acta Mater.* 47, 2455, 1999.
[186] Siegrist, M.E. and Löffler, J.F., Bulk metallic glass-graphite composites. *Scripta Mater.* (submitted).
[187] Schroers, J., Veazey, C., and Johnson, W.L., Amorphous metallic foam. *Appl. Phys. Lett.* 82, 370, 2003.
[188] Wada, T., Greer, A.L., and Inoue, A., Enhancement of room-temperature plasticity in a bulk metallic glass by finely dispersed porosity. *Appl. Phys. Lett.* 86, 251907, 2005.
[189] Brothers, A.H. and Dunand, D.C., Plasticity and damage in cellular amorphous metals. *Acta Mater.* 53, 4427, 2005.
[190] Ashby, M.F. et al., *Metal Foams: A Design Guide.* Butterworth Heinemann, Woburn, 2000.
[191] Brothers, A.H. and Dunand, D.C., Processing and structure of open-celled amorphous metal foams. *Scripta Mater.* 52, 335, 2005.
[192] Kündig, A.A. et al., In situ formed two-phase metallic glass with surface fractal microstructure. *Acta Mater.* 52, 2441, 2004.
[193] Park, B.J. et al., In situ formation of two amorphous phases by liquid phase separation in Y–Ti–Al–Co alloy. *Appl. Phys. Lett.* 85, 6353, 2004.
[194] Mattern, N. et al., Microstructure and thermal behavior of two-phase amorphous Ni–Nb–Y alloy. *Scripta Mater.* 53, 271, 2005.
[195] Oh, J.C. et al., Phase separation in $Cu_{43}Zr_{43}Al_7Ag_7$ bulk metallic glass, *Scripta Mater.* 53, 165, 2005.
[196] www.metglas.com (Metglas® Magnetic Materials).
[197] Andersen III, P.M., Bretts, G.R., and Kearney, J.E., Surveillance system having magnetomechanical marker. U.S. Patent 4,510,489, 1985.
[198] Mastrogiacomo, G., Kradolfer, J., and Löffler, J.F., Development of magnetic Fe-based metallic glasses without metalloids. *J. Appl. Phys.*, 99, 023908, 2006.
[199] http://www.liquidmetal.com (Liquidmetal® Technologies).
[200] www.ocrwm.doe.gov/osti/pdf/43757.pdf (Advanced Technologies, FY2004 Project: High-performance corrosion-resistant coatings, U.S. Department of Energy, Office of civilian radioactive waste management).
[201] Tschan, R. et al., Semihydrogenation of a propargylic alcohol over highly active amorphous $Pd_{81}Si_{19}$ in "supercritical" carbon dioxide. *Appl. Catal. A* 223, 173, 2002.
[202] Ma, C., Nishiyama, N., and Inoue, A., Fabrication and characterization of Coriolis mass flowmeter made from Ti-based glassy tubes. *Mater. Sci. Eng. A* 407, 201, 2005.
[203] Inoue, A., Bulk glassy and nonequilibrium crystalline alloys by stabilization of supercooled liquid: fabrication, functional properties and applications (Part 2). *Proc. Jpn. Acad. Ser. B* 81, 172, 2005.
[204] Ishida, M. et al., Fillability and imprintability of high-strength Ni-based bulk metallic glass prepared by the precision die-casting technique. *Mater. Trans., JIM* 44, 1239, 2004.
[205] Sharma, P. et al., Nanoscale patterning of Zr–Al–Cu–Ni metallic glass thin films deposited by magnetron sputtering. *J. Nanosci. Nanotechnol.* 5, 416, 2005.
[206] Jin, K. and Löffler, J.F., Bulk metallic glass formation in Zr–Cu–Fe–Al alloys. *Appl. Phys. Lett.* 86, 241909, 2005.
[207] Buzzi, S. et al., Cytotoxicity of Zr-based bulk metallic glasses. *Intermetallics* 14, 729, 2006.

18

Diffusion-Based Processes

A. Lupulescu
Union College, Schenectady, NY

J. P. Colinge
University of California at Davis

Abstract

This chapter focuses on diffusion based processes related to surface heat treatment. It first discusses diffusion coefficients, diffusion types and diffusion laws. It continues with surface heat treatment diffusion controlled processes, with special emphasis on the error function solution and thin-film solution. Applications and modeling of the surface diffusion processes conclude the topic.

18.1 Introduction

Surface heat treatments or doping refer to high temperature processes that modify the chemistry and structure of surfaces of materials such as pure components, alloys, or minerals. The surface achieves properties that are different from those of the bulk. The purpose of surface treatments varies widely from promoting radiation electronic interactions to providing thermal, diffusion, and radiation barriers or reducing corrosion, frictional energy, and mass losses, to "improving aesthetic appearance or economics of production."[1] The surface heat treatments are controlled by: (1) the roughness or smoothness of the material's surface, (2) the way the diffuser is produced, (3) diffusivity characteristic of the diffusing species itself and (4) processing parameters. This chapter *focuses* first upon the diffusion aspects and solutions involved in changing the surface layers by ultimately altering the material's chemistry. Next, specific applications of the diffusion laws and kinetics are given for processes used in semiconductors and structural materials, with reference to phenomena encountered in geological systems.

18.2 Solid–State Diffusion

Diffusion is an out-of-equilibrium kinetic process.[2–7] Usually, diffusion is considered a transport process at the atomic or molecular scale that leads to homogenization, or mixing of the chemical components in a phase. Diffusion acts to bring the system toward chemical equilibrium. Diffusion — not involving an external field — is a natural spontaneous process, and, therefore, always occurs by reduction in the Gibbs free energy.[2–7] Convection, by contrast, requires energy to overcome viscous resistance. However, crystals and amorphous solids (e.g., glass) resist flow, and are therefore seldom subject to convection. Diffusion is the only *significant* transport process that is possible in solids. In solids, diffusion refers to the motion of the atoms and molecules dissolved in a phase while considering the lattice symmetry of the phase of interest.[6]

Diffusion is a thermally activated process. It becomes exponentially more rapid at higher temperatures.[2–7] The diffusion coefficient, D, obeys an Arrhenius relationship:[8]

$$D = D_0 \exp\left(\frac{-E_a}{R_g T}\right) \tag{18.1}$$

where D_0 is the prefactor or preexponential constant, E_a is the activation energy for diffusion, R_g is the gas constant, and T is the temperature. The prefactor, D_0, has the same units as the diffusivity, D, that is, distance (length) square per second. The activation energy has units of energy (e.g., Joules [J] per mole). For processes involving elemental particles (atoms or electrons), Boltzmann's constant k_B is used instead of the universal gas constant, R_g.

An atomistic interpretation[4,7] of the diffusion coefficient in terms of particles jumping at random in three dimensions gives the following expression for the diffusion coefficient in a cubic system:

$$D = \frac{1}{6}\Gamma \cdot \lambda^2 \tag{18.2}$$

where Γ is the jump frequency, and λ is the jump distance. The jump frequency is expressed in units of $[s^{-1}]$, and represents the jump of a dissolved atom from a given position or site. The general expression of jump frequency is:

$$\Gamma = \nu \cdot \exp(-E_a/k_B T) \tag{18.3}$$

where ν is the oscillation frequency ($\approx 10^{14}$ Hz). The atom of interest would make periodic jumps (Figure 18.1) in random directions (known as "random walk") and sometimes revisiting the sites.[4,6,7,9] The random walk theory connects the root-mean square displacement, $\langle R^2 \rangle$, to the number of steps, n, and the displacement, λ:[4,7]

$$\langle R^2 \rangle = n \cdot \lambda^2 \tag{18.4}$$

The continuum diffusion theory connects $\langle R^2 \rangle$ as second moment, D as the diffusivity, and t as time:

$$\langle R^2 \rangle = 6Dt \tag{18.5}$$

Einstein combined the random-walk theory (Equation 18.4) with the continuum diffusion theory (Equation 18.5) and obtained the purely atomistic expression of the diffusivity:[7]

$$D = \frac{1}{6}\frac{n}{t}\lambda^2 = \frac{1}{6}\Gamma\lambda^2 \tag{18.6}$$

For cubic crystals (Figure 18.2), the random-walk properties are independent[7,9] of crystallographic Bravais lattice type (i.e., simple cubic [sc], body- or face-centered cubic [bcc or fcc]).

In crystalline solids, there are several mechanisms to account for the random jumps. The most important is the *vacancy* mechanism where the diffusion atom jumps from its original position into a near-neighbor

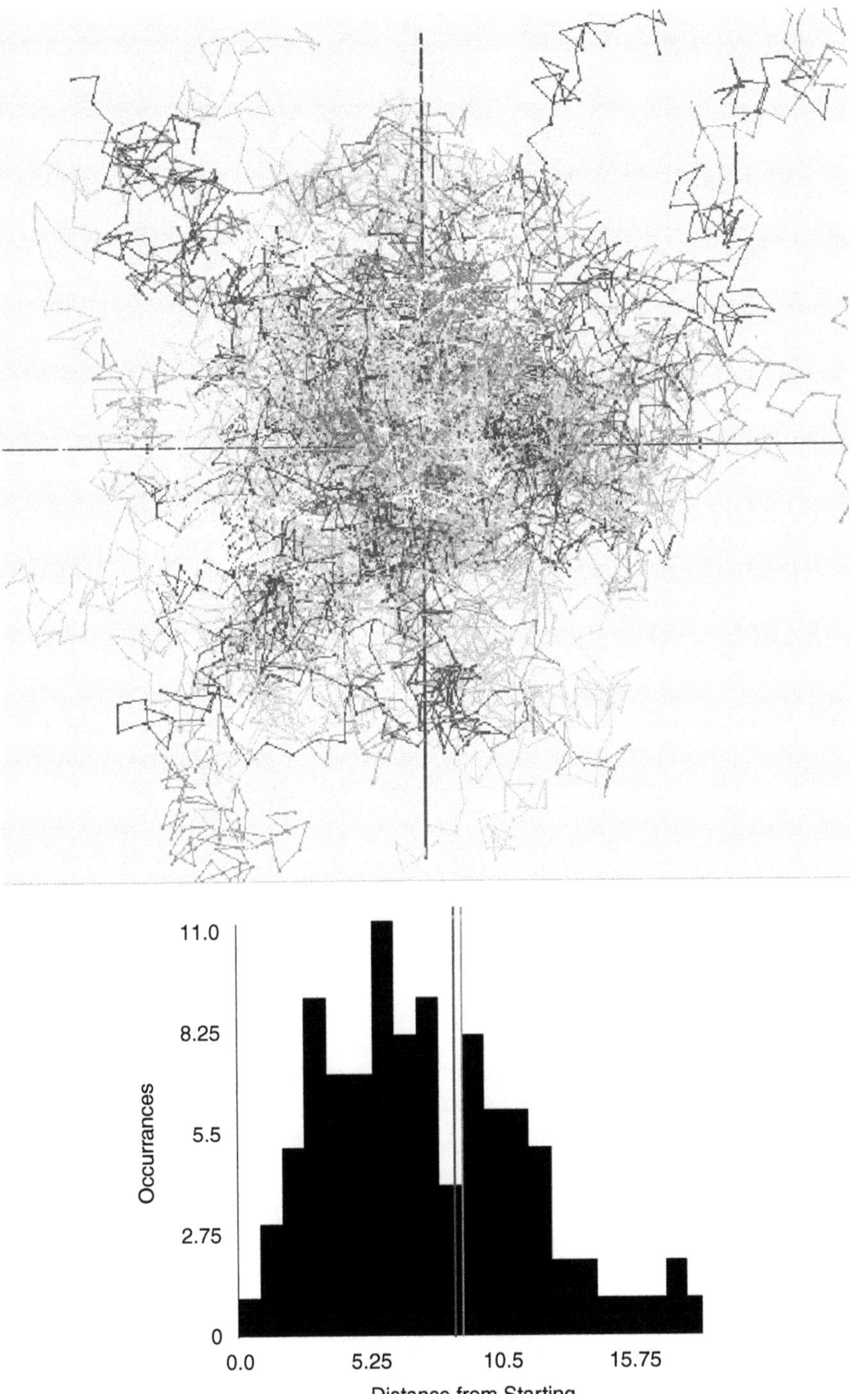

FIGURE 18.1 Example of a 2-d random walk in arbitrary directions consisting of 100 steps and 100 walks.[9]

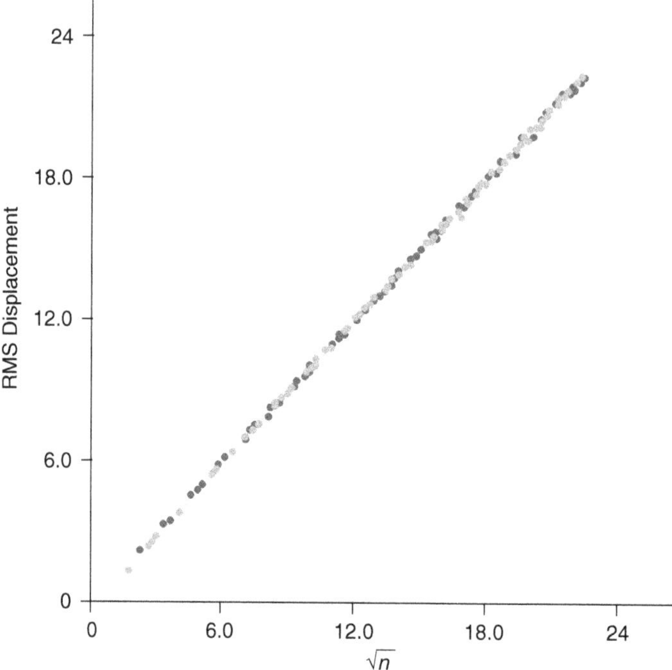

FIGURE 18.2 Three-dimensional random-walk data in simple cubic (SC), body-centered cubic (bcc), and face-centered cubic crystals (fcc). Number of realizations is 100. Number of random walks is 5000. These data show convincingly that the statistical random-walk formula is the same in all cubic crystal structures.[9]

vacant site.[4,6,7] The other possible mechanisms are: (1) *the interstitial* mechanism where the diffusant occupies a position between normal lattice sites, (2) *the ring* mechanism involves a simultaneous rotation of a cluster of atoms, (3) *interstitialcy* or *crowdion* mechanism implies a local defect of the lattice.[4,6,7]

18.2.1 Pressure Effect on Diffusion

For both materials and geological systems, the diffusivity dependence on the pressure is important. The effect of pressure, P, on diffusivity is characterized by the activation volume, ΔV. The activation volume is considered as a sum of the formation volume, ΔV_f, and the migration volume, ΔV_m:[6,10]

$$\Delta V = \Delta V_f + \Delta V_m = -kT\left(\frac{\partial \ln D}{\partial P}\right)_T \tag{18.7}$$

The atomistic significance of the activation volume depends on the diffusion mechanism. Shewmon[3] showed that for a *vacancy* mechanism, the activation volume represents the sum of partial volumes and activated complexes.[3] The activation volume could be positive or negative depending upon the diffusion mechanism.[6,10] Considering the pressure effect, a more complete Arrhenius equation could be now written:

$$D = D_0 \exp\left(\frac{-E_a + \Delta V * P}{R_g T}\right) \tag{18.8}$$

The activation volume has units of [cm^3/mole] or could be expressed as Ω, where Ω is the atomic volume at standard temperature and pressure.[10]

18.2.2 Diffusion Coefficient Types

There are numerous types of diffusion depending on the nature of the diffusing species and the matrix or host. Consequently, the distinction between different diffusion coefficients is important. As the *defect diffusion, grain boundary diffusion,* and *multicomponent diffusion* are not included in this chapter, we simply refer to the terms known and used mostly in binary diffusion in solid media.[3,6,7]

Self-diffusion and self-diffusion coefficient, D_{self}, describes the atomic transport of a pure element. Self-diffusion is best described as a random-walk process (Equation 18.2 and Equation 18.6). Tracer diffusion experiments are considered the most direct method of measuring self-diffusivities.

Tracer diffusion and tracer diffusivity, D^* or D^{tr}, refers to self-diffusivity. More exactly, it describes the movement of the radioactive atoms of a material (element) within a nonradioactive host of the same composition. Measurement of D^{tr} requires a flux of the tracer, but not a flux of the solvent. Although there is no universal agreement, it is generally considered that $D^{\text{tr}} \approx D_{\text{self}}$. Almost all self-diffusivity studies are based on tracer diffusion experiments. The quantity of the added tracer is very small, the concentration gradient is very small, and therefore the tracer diffusivities are almost indistinguishable from the self-diffusivity coefficients.

Intrinsic diffusivity and its correspondent coefficients, D_{A}^{in}, D_{B}^{in} describes species mobility in a binary solid. These quantities are in general not equal. In contrast to self-diffusivity or tracer diffusivity, these intrinsic diffusivities imply concentration gradients of both A and B species.

Chemical (bulk) diffusion and the interdiffusion coefficient, $\tilde{D}$, is not an intrinsic property of a substance. The diffusion occurs because of a chemical potential gradient, which in turn is caused by a composition gradient. However, we must consider as well the activity coefficient, $\gamma_{\text{A}}, \gamma_{\text{B}}$, of each discussed species (A and B), the diffusivity of each species, D_{A}, D_{B}, and the correspondent molar concentration, $X_{\text{A}}, X_{\text{B}}$. The intrinsic diffusivities are then related by the so-called Darken equation:[7,11,12]

$$\tilde{D} = X_{\text{A}} D_{\text{B}}^{\text{in}} + X_{\text{B}} D_{\text{A}}^{\text{in}} \tag{18.9}$$

The variation of the interdiffusion coefficient with the composition brings into picture the solution ideality. Recalling that $D_{\text{i}} = M_{\text{i}} R_{\text{g}} T$ (M_{i} — mobility, R_{g} — gas constant, T — temperature), for an ideal solution, the presence of a chemical gradient in a concentrated solid solution alters the diffusivity expression, now given by the Darken–Hartley–Crank equation:

$$\tilde{D} = (X_{\text{A}} D_{\text{B}}^{*} + X_{\text{B}} D_{\text{A}}^{*}) \left[1 + \frac{\partial \gamma_{\text{B}}}{\partial X_{\text{B}}} \right] \tag{18.10}$$

where $[1 + \partial \gamma_{\text{B}} / \partial X_{\text{B}}]$ is the thermodynamic factor for B species.[7,11–13] In surface diffusion, tracer and intrinsic diffusivities are most commonly used. The interdiffusion coefficients are mainly used for bulk diffusion.

18.2.3 Fick's Laws

Diffusion laws were discovered by Adolph Fick in 1855 by noting the similarities with heat conduction (Fourier's law) and electricity transport (Ohm's law).[2–7,14–19] *Fick's first law* connects the flux or mass transport, J, to material's diffusivity, D, and the concentration gradient, $\mathrm{d}C/\mathrm{d}x$.

$$J = -D \frac{\mathrm{d}C}{\mathrm{d}x} \tag{18.11}$$

The minus sign comes from the fact that matter flows, with a few exceptions, down the concentration gradient. The conventional unit of diffusion flux is [mass/area-time]. *Fick's second law of diffusion* is a more general conservation equation given as:[2–7,14–19]

$$\frac{\partial C}{\partial t} = D\left(\frac{\partial^2 C}{\partial x^2} + \frac{1}{A}\frac{\partial A}{\partial x}\frac{\partial C}{\partial x}\right) \tag{18.12}$$

where A represents the cross-sectional area, $\partial A/\partial x$ is the change in cross-sectional area, and $\partial C/\partial t$ represents the change in concentration with time. This equation is mainly known as a second order partial differential equation and describes the change in concentration with time:

$$\frac{\partial C}{\partial t} = D\left(\frac{\partial^2 C}{\partial x^2}\right) \tag{18.13}$$

Solutions to this equation have been obtained and published for numerous boundary conditions.[2–7]

18.2.4 Solutions to Fick's Second Law Applicable to Surface Treatment Processes

For surface heat treatment problems, two of Fick's second law solutions are most applicable: (1) *error function solution*, and (2) *thin-film solution*. Rather than discussing the derivations of these two different solutions, we concentrate on showing some details regarding their behavior. Some applications of these solutions may be also found in Chapter 21.

18.2.4.1 Error Function Solution

Error function solution is a solution to the linear diffusion equation (Equation 18.13) in a semi-infinite medium. The boundary conditions are specified as follows: (1) the initial state of the system is described as $C = 0$ for $x > 0$, at $t = 0$, and (2) at $x = 0$, a fixed concentration, C_0, is maintained, for all $t > 0$. The error functions are used when we discuss the thick diffusion couple solution also known as Grube–Jedele solution, the fictitious image sources, and error function approximation.[5,7,12,14–19] The error function solution is:

$$C(x,t) = C_0 \operatorname{erfc}\left(\frac{x}{2\sqrt{Dt}}\right) = C_0(1 - \operatorname{erf})\left(\frac{x}{2\sqrt{Dt}}\right) \tag{18.14}$$

where C is concentration at various depths, x and times, t. For an infinitely long bar, subject to a steady-state concentration, C_0, at $x = 0$, the concentration gradient, $\nabla C(x,t)$, to satisfy Fick's first law is the following:[7,12]

$$\nabla C(x,t) = \left(\frac{\partial C(x,t)}{\partial x}\right)_t = \frac{\partial}{\partial x}\left(C_0 - C_0 \operatorname{erf}\left(\frac{x}{2\sqrt{Dt}}\right)\right) \tag{18.15}$$

The error function, $\operatorname{erf}(z)$, equals:

$$\operatorname{erf}(z) \equiv \frac{2}{\sqrt{\pi}}\int_0^z e^{-u^2}\,du \tag{18.16}$$

Taking the gradient, $\partial \operatorname{erf}(z)/\partial x$ of Equation 18.16:

$$\frac{\partial \operatorname{erf}(z)}{\partial x} = \frac{2}{\sqrt{\pi}} e^{-z^2}\frac{\partial z}{\partial x} \tag{18.17}$$

We obtain a final expression of the concentration gradient:

$$\nabla C(x,t) = \left(\frac{\partial C(x,t)}{\partial x}\right)_t = -\frac{C_0}{\sqrt{\pi Dt}} e^{-x^2/4Dt} \tag{18.18}$$

Consequently at $x = 0$,

$$\nabla C(0, t) = -\frac{C_0}{\sqrt{\pi D t}} \tag{18.19}$$

The error function, erf(z), and its complement, erfc(z) = 1 − erf(z), appear frequently in solving linear diffusion problems involving infinite and semi-infinite experimental geometries.[5,7,12] These two functions are connected through the following relations:[7,12]

$$\mathrm{erf}(z) = 1 - \mathrm{erfc}(z) = 1 - \frac{2}{\sqrt{\pi}} \int_0^z \mathrm{e}^{-\eta^2} \mathrm{d}\eta \tag{18.20}$$

One interested in using the error function solution should be very careful of the subtle differences between the results obtained using different methods. There is a simple polynomial calculation provided by Abramovitz and Stegun[20] and discussed by Glicksman:[7]

$$\mathrm{erf}(z) \cong 1 - [1 + 0.278393z + 0.230389z^2 + 0.000972z^3 + 0.078108z^4]^{-4} + \varepsilon(z) \tag{18.21}$$

where $\varepsilon \leq (z) = 10^{-5}$.

Most math software packages (Maple, Mathematica, Mathcad, Matlab) include a simple call for the error function, and allow convenient manipulation and graphing of the error function and its complement. As they are built-in functions, the user eliminates most programming.[7,12] The error function, erf(z), is an antisymmetric function as erf(z) = erf($-z$)[7,12] (Figure 18.3). The complementary error function, erfc(z) is a nonantisymmetric function and consequently erfc(z) ≠ −erfc($-z$)[7,12] (Figure 18.4).

Attention is required when using the diffusivity values obtained from error function solution. The above mentioned polynomial solution when plotted against the Mathematica solution (Figure 18.5) shows an

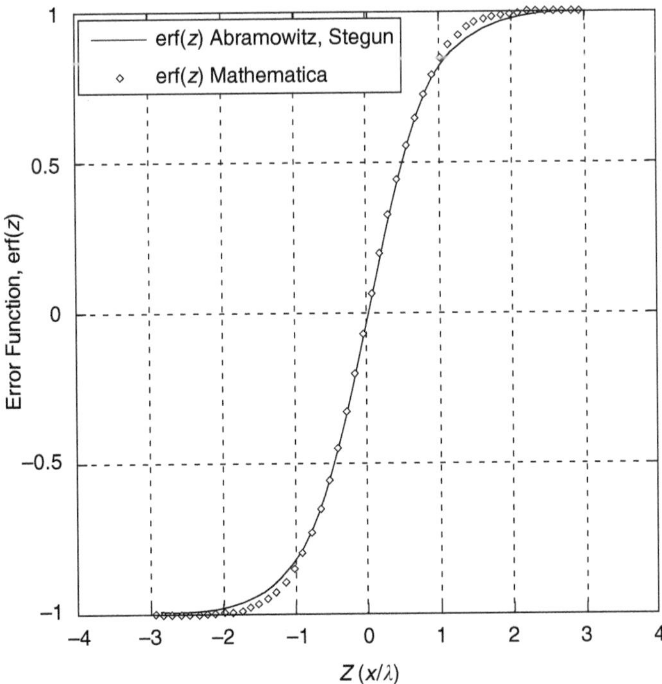

FIGURE 18.3 Error function profiles using: (a) Abramovitz, Stegun approach, (b) Mathematica approach (see Reference 12). Error Function is an antisymmetric function, erf(z) = −erf($-z$).[3]

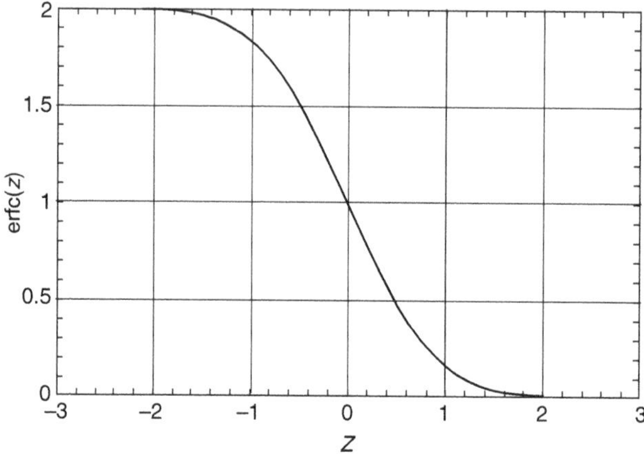

FIGURE 18.4 Complementary error function. This is a nonantisymmetric function, $\mathrm{erfc}(z) \neq -\mathrm{erfc}(-z)$.[12]

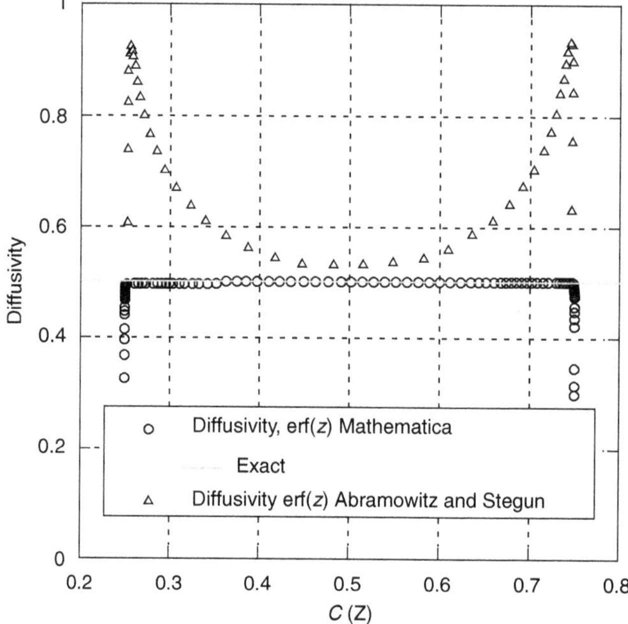

FIGURE 18.5 Diffusivity versus concentration:[12] (a) Empty circles show diffusivity produced by applying error function solution in Mathematica; (b) exact solution; (c) diffusivity obtained by applying error function expression produced by Abramowitz and Stegun (see Reference 20).

error greater than the one claimed by the polynomial solution offered by Abramovitz and Stegun.[20] The relative error in concentration between the two mentioned methods is shown in Figure 18.6.

The function's argument, z, is also subject for discussion. This argument, also called "similarity variable," $z = \xi = x/2(Dt)^{1/2}$, "is a special combination of the space-time field variable."[7] The nominator, $2(Dt)^{1/2}$, is called a "time tag." The "similarity variable" is dimensionless, while the "time tag" has in fact unit length:

$$z = \xi = \frac{x}{2\sqrt{Dt}} \rightarrow \left[\frac{m}{2\sqrt{m^2/s}}\right] \tag{18.22}$$

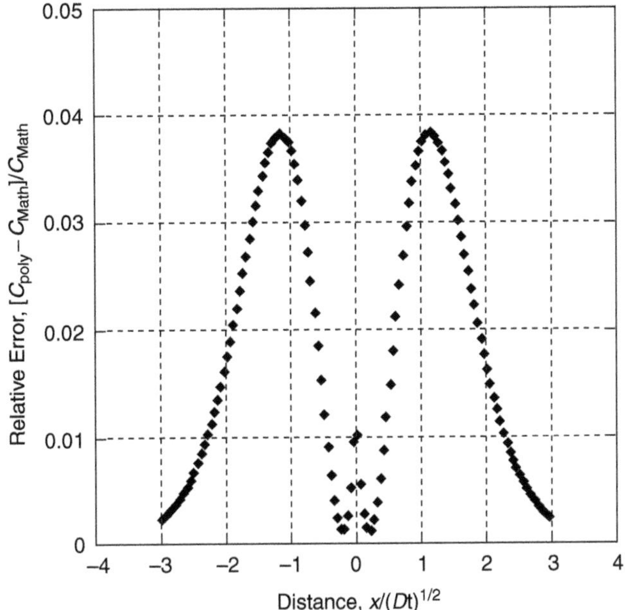

FIGURE 18.6 Relative error, $[(C_{\text{poly}} - C_{\text{Math}})/C_{\text{poly}}]$, versus distance, $x/Dt^{1/2}$ [known as "time tag"]: C_{poly} is obtained using Abramovitz and Stegun approach; C_{Math} is obtained using Mathematica approach.[12]

18.2.4.2 Instantaneous Planar Diffusion Source in an Infinite Medium and Thin-Film Configuration

This solution is also known as "Gaussian" solution given as an exponential term. The experimental geometry for this problem considers a source, M, instantaneously releasing diffusing material, at the plane $x = 0$, and time $t = 0$. The material will be transported in both directions, $-\infty < x < 0$, and $0 < x < \infty$. The solution derived elsewhere[5,7] becomes:

$$C(x,t) = \frac{M}{2\sqrt{\pi Dt}} e^{-x^2/4Dt} \tag{18.23}$$

and the curves represent the concentration field versus distance are bell-shaped curves.

As detailed below, this solution is typically applied in the impurity diffusion in semiconductors, termed "doping." The doping situations only consider half of the geometry discussed above, ending up with a semi-infinite system in which the matter flows only in the direction $x > 0$:[5–7]

$$C(x,t) = \frac{M_{\text{thin film}}}{\sqrt{\pi Dt}} e^{-x^2/4Dt} \tag{18.24}$$

This thin-film diffusion solution, except for the prefactor is identical to the "instantaneous" planar source solution. In other words, the planar source problem "splits" exactly into two halves (Figure 18.7).[7]

18.2.4.3 General Approach for Solving a "Thin-Film" Problem

After performing the experiment containing the tracer,[12] we are looking into slicing the experiment — each slice with a thickness of Δx — and measuring the radioactivity, A^* (counts/s). At different depths, x, A^* is proportional with the concentration. The equation connecting the measured radioactivity, A^*,

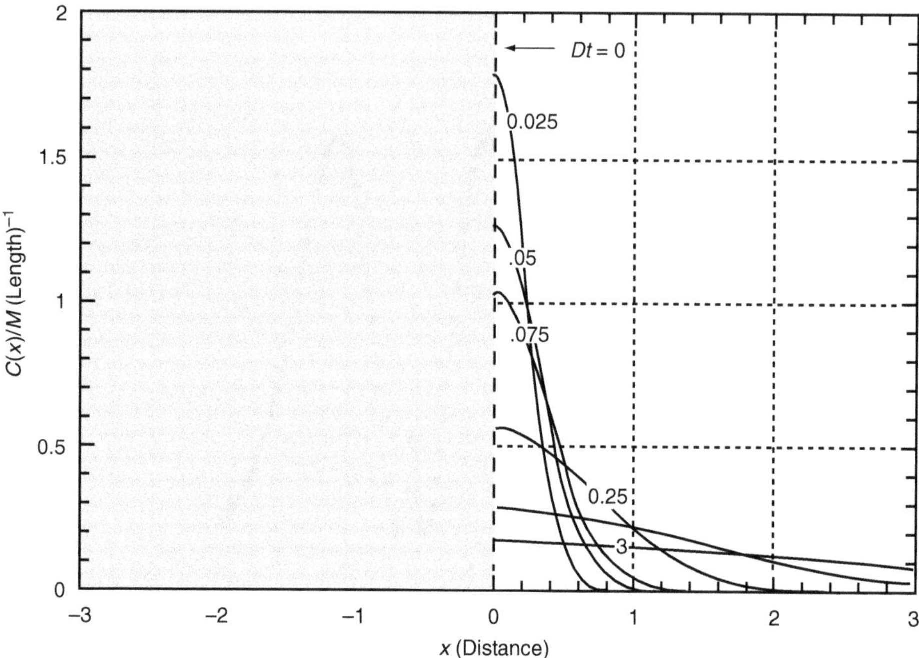

FIGURE 18.7 Thin-film solution is just half of the planar source solution [Adopted from Ref. 7]. Normalized plot of Equation 18.24.

the concentration, $C(x, t)$, and the diffusion solution is as follows:[12]

$$C(x,t) = \frac{A^*}{\beta} = \frac{M_{\text{thin film}}}{\sqrt{\pi Dt}} e^{-x^2/4Dt} \tag{18.25}$$

If we take natural logarithm of both sides we obtain:

$$\ln A^*(x) = \ln\left(\frac{\beta M_{\text{thin film}}}{\sqrt{\pi Dt}}\right) - \frac{x^2}{4Dt} \tag{18.26}$$

By plotting ln A^* versus distance squared (x^2), the result is a line the slope of which, $m = -(4Dt)^{-1}$. From this relation, we could extract the diffusivity, D:[12]

$$D = -\frac{1}{4t \cdot m}. \tag{18.27}$$

18.2.5 Applications of the Error Function Solution

18.2.5.1 Industrial Processes

An English metallurgist, W.C. Roberts-Austen measured in 1896, for the first time, diffusion coefficients in molten and solid metals.[17,21] Diffusion processes have been considered to change the properties of materials, especially alloys, for use in engineering applications. There are several diffusion processes applied to improve materials properties. One of the applications is to harden a material to resist wear and reduce mass losses and corrosion (e.g., in carburizing, nitriding, or boriding). Generally, while hardening a material, its ductility is compromised. Because of this, surface hardening is preferred over through hardening. *Through hardening* often causes distortion and cracking when it involves thick sections of

TABLE 18.1 Engineering Methods for Surface Hardening

Method	Surface treatment
Diffusion Methods	Carburizing Nitriding Carbonitriding Nitrocarburizing Boriding Titanium- or Tungsten-Carbide Formation (or Toyota Diffusion Process) Electrolytic deposition followed by diffusion
Selective Hardening Methods	Flame Hardening Induction Hardening Laser Hardening Electron Beam Hardening Ion Implantation Chemical or Physical Vapor Deposition

Source: Modified from http: www.Key-to-steel.com

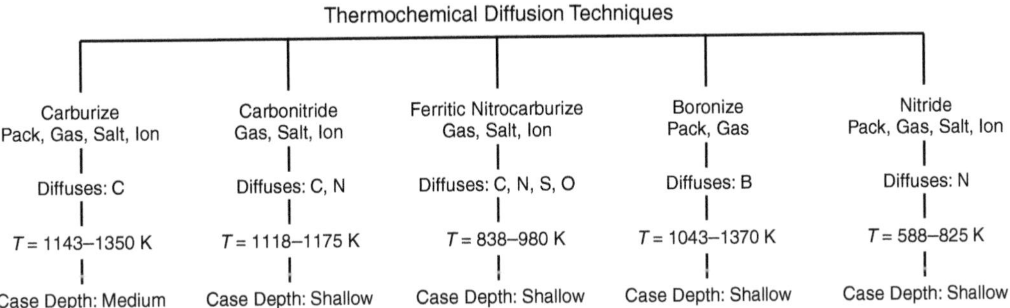

FIGURE 18.8 Diffusion methods and processes for heat treatments — thermochemical diffusion techniques. (From http://www.industrialheating.com)[23]

a body. In *surface hardening*, only a certain depth into the surface is hardened. In this case, the piece acquires a desirable combination of a hard surface to resist wear while having a tough but soft interior that resists shock and impact. Typical examples of parts that need the "*tough on the outside but soft in the inside*" properties are steel cams and gears.[22] Table 18.1 indicates common engineering methods for surface hardening steels.[22]

The selective hardening methods generally imply surface modification without altering the substrate chemistry (i.e., surface heat treatments — see also Chapter 19). As discussed below, ion implantation is used to introduce diffusing species and enhance their diffusion. Chemical and Physical Vapor Deposition (CVD, PVD) processes modify the surface by adding layers to the surface. These processes are discussed in Chapter 8. Some common engineering methods for surface hardening by diffusional processes are presented below.[1]

1. *Thermochemical diffusion treatments* allow introducing interstitially elements such as carbon (C), nitrogen (N), boron (B), or combinations of C and N, into a material (commonly metal but sometimes ceramic) surface at high temperatures.[1] Typical thermochemical processes for steels are *carburizing, nitriding, boriding, or boron-nitriding.* A short overview of these diffusive methods used in industrial processes is presented in Figure 18.8.[23] Plasma, vacuum, and gas carburizing are common methods applied to steels.

TABLE 18.2 Diffusion Data in Pure Fe

Diffuser	Radius (Å)	*D* for bcc (at 1183 K)	*D* for fcc (at 1183 K)
C	0.77	3.9×10^{-5}	1.1×10^{-7}
N	0.71	7.5×10^{-5}	—
B	0.97	—	6.1×10^{-7}
Cr	1.85	6.0×10^{-11}	1.4×10^{-12}
S	1.27	2.8×10^{-8}	2.7×10^{-10}

Influence of the diffuser atom radius. Diffusivity, D, is in cm^2/s.
Note: 1183 K is the bcc-to-fcc transition in pure Fe.[7]

Chromizing, aluminizing, and *siliconizing* involve substitutional diffusion of elements like chromium (Cr), aluminum (Al), silicon (Si).[1] Table 18.2 gives a short overview of the interstitial and substitutional diffusion data and the influence of the diffuser atom radius in bcc and fcc materials.

Some diffusion treatments involve surface carbide formation by diffusion of Ti or W to react with the C in steels (usually with >0.4%C). This is known as the Toyota diffusion process (Table 18.1) and is used to increase the wear resistance of tools and machine parts. Most commonly, the diffusers are supplied at the surface by a gaseous, liquid (e.g., a salt), or solid (termed as "pack") medium. In steels we consider either (1) the diffusion carried out at low temperatures within *ferritic* phase (i.e., no phase transformation of the initial ferrite-bcc steel structure), or (2) high temperatures diffusion in the *austenitic* phase (i.e., heating is associated with a phase transition into fcc austenite).

(a) The *ferritic* processes produce usually case depths of up to 250 microns in alloy steels.[1] Ferritic nitrocarburizing methods can be used for a much wider variety of ferrous alloys. For low-carbon mild steel, a thin layer of approximately 10 microns can improve both wear and corrosion resistance. The well-known processes include *gas nitriding* (at 700 K), *plasma nitriding* (at 675 to 875 K), and *nitrocarburizing* (starting at 775 K).

(b) The *austenitic* treatments are performed at temperatures around 1173 K and they produce a thicker "layer" (up to several mm) than the *ferritic* processes.[1] The *austenitic* treatments also create greater surface distortion and growth. For nickel based superalloys and nickel/chromium (Ni–Cr) alloys the most common surface layers are: C, C–N, B, Al, Cr.

2. *Ion implantation* is a process that enhances diffusion and, therefore, takes place at low temperatures, typically, 425 K.[1] It consists of atoms of gaseous or metallic elements that are further ionized and passed to a high vacuum chamber. In the vacuum chamber, the elements are accelerated using a mass separator. Selected ions are further implanted into the target component. The layer obtained using this method is very shallow, approximately 0.2 microns. However, the wear resistance, friction and oxidation/corrosion resistance are enhanced. More details on ion implantation are given in Chapter 3 and Chapter 10.
3. *Electroplating and thermal diffusion* processes[1] involve first the electrolytic deposition of the diffuser, then the diffusion or reactive diffusion step. They are used to create, for instance, a thin layer of tin (Sn) onto ferrous materials. Diffusion treatments are performed at 675 to 875 K to form Fe–Sn compounds. These compounds increase scuffing and corrosion resistance. Bronze coatings are also created through this procedure. Details about electrolytic deposition may be found in Chapter 22.
4. *Sulfur (S) treatments* are used to incorporate S onto ferrous substrates.[1] The process is usually held at temperatures below 475 K, because of sulfur's low melting point. The sulfides formed during this process confer good lubricating properties. Cylinder liners, gears, constant velocity joints, heavy duty rear axle spiders, textile machinery parts benefit from this kind of surface treatment.[1]
5. *Phosphating* is a process that creates phosphate coatings at 315 to 345 K on iron (Fe), zinc (Zn), and manganese (Mn).[1] The phosphate layers absorb oil and grease more readily, and they are "effective in reducing galling, pick up, and scuffing." Consequently, they prevent adhesive wear and fretting.[1]

6. *Oxide coatings* or films could be obtained on ferrous or other substrates by thermal exposure at 575 to 875 K.[1] For this purpose, steam tempering or autoclaving is applied to high-speed steel drills and zirconium alloy components. If oil is present, the oxide coatings prevent scuffing, wear, and metal transfer.
7. *Anodizing treatments* produce oxide layers for Al alloys.[1] The presence of the oxide layer reduces adhesive wear and significantly increases the hardness of the oxidized portion of the material. Through *hard anodizing* in an oxidizing acid at 273 K, up to a 500 microns oxide layer could be "grown." Thinner layers could also be produced for decorative and corrosion protection purposes.

18.2.5.2 Surface Treatments of Steels

The most widely applied thermochemical treatments are for steels. There are different classifications of steels that are based on the weight percent of the base carbon. Low-carbon steels are characterized as having less than 0.3 wt% carbon and are fairly ductile while high-carbon steels contain 0.5% to 0.8 wt% carbon and are characterized by high hardness and wear resistance with little ductility (Table 18.3). In industrial practice, the base carbon content is typically at 0.2 wt% C and the aim of carburization processes is for the surface case carbon content to be controlled up to 0.8 wt% C. Table 18.4 illustrates comprehensively the connection between the steel type and applicable diffusion processes as surface treatment processes.

TABLE 18.3 Steel Classification and Characteristics Depending on Carbon Content

Steel	C content (wt%)	Characteristics
Low-Carbon	<0.3	Ductility, low cost, used for stamping
Medium-Carbon	0.3–0.5	High strength, toughness in small components. Used in cases of high stress
High-Carbon	0.5–0.8	High hardness, wear resistance, little ductility. Used for precision such as cutting tools, so it would break before yielding

TABLE 18.4 Steel Types and Applicable Diffusion Processes

Diffusion substrate	Surface treatment process
Low-Carbon Steels	Carburizing Cyaniding Ferritic nitrocarburizing Carbonitriding
Alloy Steels	Nitriding Ion Nitriding
Tool Steels	Titanium Carbide Boriding Salt Nitriding Ion Nitriding Gas Nitriding
Stainless Steels	Gas Nitriding Titanium Carbide Ion Nitriding Ferritic Nitrocarburizing

Source: From http: www.Key-to-steel.com

TABLE 18.5 Depths of Carbon Diffusion for Different Carburizing Applications

Applications	Depth of diffusion (mm)
High wear resistance, low to moderate loading-small and delicate machine parts subject to wear	<0.51
High wear resistance, moderate to heavy loading-light industrial gearing	0.51–1.02
High wear resistance, heavy loading, crushing loads or high magnitude alternating bending stresses-heavy duty industrial gearing	1.02–1.52
High wear resistance, shock resistance, high crushing loads-bearing surfaces, mill gearing, rollers	1.52–6.4

Source: From http: www.Key-to-steel.com

For any surface treatment, the diffusion depth depends on the diffusion coefficient and the time while heated in an atmosphere rich in hardening elements (i.e., diffusers). The diffusion coefficient is dependent on temperature according to the Arrhenius relationship (Equation 18.1). Diffusion and the preexponential factor can be easily found by running several tests at different temperatures. The diffusion depth is important for hardening. The required diffusion depth increases as the expected load on the object increases. There are tables of *recommended case depths of diffusion* for certain materials and applications (e.g., Table 18.5 for the carburization case).

Diffusion case depths are affected by temperature, time of heating, initial diffusing species concentration, and the desired final surface concentration, as shown in Table 18.6. Final case hardness values are also shown in this Table. Heat treatments such as quenching and tempering are typically applied after steel carburizing. More details about heat treatments may be found in Chapter 19. Knowing the case depth needed, the desired concentration, and the diffusivity, D, time can be determined easily by solving the error function solution (Equation 18.14). In conclusion, the Arrhenius equation (Equation 18.1) and the error function solution (Equation 18.14) are combined to find the diffusivity of the hardening element, and the time to treat it with hardening elements at an elevated temperature.

18.2.6 Applications of the Thin-Film Solution

The fabrication of semiconductor devices such as transistors and integrated circuits employs doping steps during which impurity atoms are introduced in the semiconductor material, followed by thermal annealing during which these impurities diffuse across some distance to create controlled doping profiles. In the most commonly used semiconductor, silicon, impurity atoms such as boron, phosphorous, or arsenic are usually introduced by ion implantation. The technique allows for the introduction of doping impurities in a semiconductor with a high level of accuracy. An ion implanter is basically a particle accelerator composed of an ionization chamber called the source, an acceleration stage in which the ions are accelerated to a given energy, a mass separation stage, an electrostatic deflection system, and a target chamber where the silicon wafers are placed for implant (see, for instance, Figure 10.2 in Chapter 10). A substance containing the doping element is introduced in the source, where filament heating and microwave energy produce a plasma containing ions of the desired implant species. The ions are accelerated by an electric field and deflected by an electromagnet depending on their mass. The current in the electromagnet is chosen so that only the desired ions continue their flight toward the semiconductor target. Ions accelerated by the implanter penetrate into the silicon and stop at a given depth in the crystal depending on the chosen implant energy. The higher the ion energy, the deeper the ions are implanted (for more details, see Chapter 10).

TABLE 18.6 Examples of Diffusion Treatments and Results

Substrate	Method	Diffusing species	Temperature, K	Case depth (mm)	Case hardness, Rc
Low-carbon steel or alloy	Carburizing solid (pack)	C	1090–1365	0.125–1.5	50–63[a]
Low-carbon steel or alloy	Gas	C	1090–1250	0.075–1.5	50–63[a]
Low-carbon steel or alloy	Liquid	C, N	1090–1250	0.05–1.5	50–65[a]
Low-carbon steel or alloy	Vacuum	C	1090–1365	0.075–1.5	50–63[a]
Alloy steel, nitriding steel, stainless steel	Nitriding gas	N	755–865	0.012–0.075	50–70
Ferrous metals and cast iron	Liquid (salt)	N	785–840	0.002–0.75	50–70
Alloy steels, nitriding steels, stainless steels	Ion	N	615–840	0.075–0.75	50–70
Low-carbon steel or alloy, stainless steel	Carbon nitriding gas	C, N	1035–1145	0.075–0.75	50–65[a]
Low-carbon steel	Liquid (cyaniding)	C, N	1035–1145	0.002–0.125	50–65
Low-carbon steel	Ferritic nitrocar-burizing	C, N	840–950	0.002–0.025	40–60[a]
Low-carbon steel	Aluminizing pack	Al	1145–1250	0.025–1	<20
Low-carbon steel	Siliconizing by CVD	Si	1200–1315	0.025–1	30–50
High (H)–Low (L) Carbon steel	Chromizing by CVD	Cr	1255–1365	0.025–0.050	50–60 (H) <30 (L)
Alloy and tool steel	Titanium carbide	C, Ti, TiC compound formation	1175–1285	0.002–0.012	>70[a]
Alloy and tool steel, cobalt (Co) and nickel (Ni) alloys	Boriding	B	675–1425	0.012–50	40–>70

[a] Requires quench from austenitizing temperature.
Source: From http: www.Key-to-steel.com

The concentration profile of the implanted atoms can be described within reasonable accuracy by a Gaussian distribution. The peak of the Gaussian distribution is located at a depth beneath the silicon surface called the "projected range," noted R_p. The width of the distribution is characterized by a standard deviation called the "straggle" and noted ΔR_p. The concentration of implanted impurities is described by the following relationship:

$$C(x) = C_p \exp - \frac{(x - R_p)^2}{2\Delta R_p^2} \tag{18.28}$$

where C_p is the concentration at the peak of the Gaussian distribution. By integrating the doping concentration over the entire Gaussian distribution one obtains the total implanted dose, which yields a relationship between the peak concentration, C_p (cm^{-3}) and the implanted dose N' (cm^{-2}):

$$C_p = \frac{N'}{\sqrt{2\pi}\,\Delta R_p} \tag{18.29}$$

The projected range R_p and the standard deviation ΔR_p can be expressed empirically as a function of the implantation energy, E. These expressions are found in Table 18.7.

TABLE 18.7 Empirical Expressions for the Projected Range R_p and the Standard Deviation ΔR_p as a Function of the Implantation Energy, E (keV), in Silicon

Boron	$R_p = 5.2629\ E^{0.8909}$	$\Delta R_p = 5.34216\ E^{0.5610}$
Phosphorus	$R_p = 1.23612\ E^{1.0000}$	$\Delta R_p = 0.76046\ E^{0.8287}$
Arsenic	$R_p = 1.09590\ E^{0.8638}$	$\Delta R_p = 0.30303\ E^{0.8038}$

Source: From Colinge, C.A. and Colinge, J.P., *Physics of Semiconductor Devices*, Kluwer Academic Publishers, USA, May 2002, p. 369. With permission.

In silicon, impurity diffusion is generally carried out at high temperature (1075 to 1375 K) in a furnace. At those temperatures the impurity atoms can diffuse throughout the crystal lattice through interactions with point defects (interstitials and vacancies). The diffusion of implanted impurities is governed by Fick's second law of diffusion (Equation 18.13) with the solution given by Equation 18.23 (M in Equation 18.23 is equivalent to N' in Equation 18.29). If the impurity concentration distribution before diffusion is Gaussian, it remains Gaussian after diffusion. The depth of the peak concentration, R_p, remains unchanged, but the peak concentration, C_p, decreases, and the distribution spreads out, such that the standard deviation increases. It is convenient to characterize the spread of the profile brought about by the diffusion process by introducing a value called the "characteristic diffusion length," L, which is a function of temperature and time of diffusion: $L(T,t) = 2\sqrt{Dt}$. This characteristic diffusion length is equivalent to the "time tag" defined earlier (see Equation 18.22). The standard deviation of the Gaussian profile after implantation and diffusion, noted L', is a function of the prediffusion standard deviation, ΔR_p, and the diffusion length, $2\sqrt{Dt}$ and it is equal to:

$$L' = \sqrt{2\Delta R_p^2 + 4Dt} \tag{18.30}$$

The impurity concentration profile after diffusion is given by:

$$C(x,t) = \frac{N'}{L'\sqrt{\pi}} \exp -\frac{x - R_p^2}{L'} \tag{18.31}$$

It is worthwhile noting that the concentration profiles of implanted impurities before and after diffusion are described by the same equation, $C(x) = (N'/\sqrt{\pi}L)\exp -((x - R_p^2)/L)$, in which $L = \sqrt{2}\Delta R_p$ before diffusion and $L = \sqrt{2\Delta R_p^2 + 4Dt}$ after diffusion.

If several annealing steps are carried out (the number of annealing steps being n) the final characteristic diffusion length is equal to:

$$L' = \sqrt{2\Delta R_p^2 + \sum_{i=1}^{n} L_{i^2}} = \sqrt{2\Delta R_p^2 + \sum_{i=1}^{n} 4D_i t_i} \tag{18.32}$$

A typical example of a sequence of implantation and annealing steps can be found in the fabrication of NPN bipolar junction transistors. Such a device necessitates the formation of three regions with alternate doping polarities: a heavily doped N-type region formed by the implantation of arsenic atoms at the silicon surface, a P-type region formed by implantation and diffusion of boron atoms, and a second N-type region deeper in the silicon. The device can be formed using a silicon wafer containing an initial constant phosphorous doping concentration (Figure 18.9). Then, boron and arsenic atoms are implanted with energy values chosen to create impurity profiles centered around projected ranges $R_{p,B}$ and $R_{p,As}$, respectively. Thermal annealing is then used to place the implanted atoms in substitutional sites in the

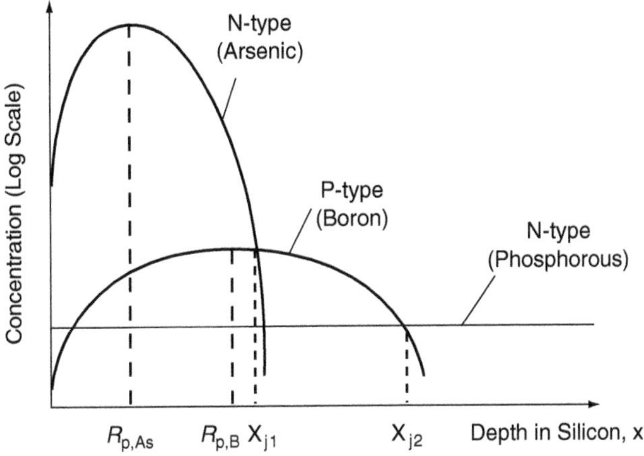

FIGURE 18.9 Typical doping impurity profile in a silicon bipolar junction transistor.

silicon matrix and activate them electrically. P–N junctions are formed when the material changes from being mainly P-type to mainly N-type, and reciprocally. In a bipolar junction transistor, careful control of the ion implantation and annealing parameters allow one to precisely position the depth of two junctions X_{j1} and X_{j2}. The distance between those two junctions is a parameter that influences greatly device performance.

18.3 Modeling of Surface Heat Treatment Diffusion Processes

Modeling of any heat treatment process is of tremendous importance for controlling the process itself and consequently obtaining the desired properties of the material. These efforts stretched from diffusivity rates of various elements in binary or multicomponent systems to comprehensive models of diffusional processes. Carburization most of all, was the subject for many modeling approaches. Most recently, Sisson and Arimoto developed comprehensive models of the evolution of the heat treatment processes.[24,25] Sisson underlined the database needs for diffusion coefficients as a function not only of temperature, but of gas composition and steel surface condition in carburizing.[24]

Early modeling efforts calculated the rate of diffusion of carbon in austenitic Ni and Mn steels and provided nonsteady-state diffusivities data obtained at temperatures between ≈1075 and 1475 K.[26,27] Later, Harris calculated case depths, arithmetic averages, and correspondent standard deviations claiming that 3/4 of the observed values are expected to fall within the reported range.[28,29] He developed an empirical formula for the effect of time and temperature on case depth for normal carburizing:

$$\text{Case depth} = \frac{31.6\sqrt{t}}{10^{6700/T}} \tag{18.33}$$

where case depth is in inches, t is the time in hours, and T is the absolute temperature, in degrees Rankine (°F + 460). For a specific carburizing temperature, the relationship becomes simply the square-root diffusive dependence, similar to the "time tag" definition:

$$\text{Case depth} = K'\sqrt{t} \tag{18.34}$$

where the constant K' was temperature dependent.

Carbon diffusivity in γ iron in steels or other alloys has been the subject of intense research. For instance, Smith reported on steady-state diffusivities of carbon in γ iron–cobalt alloys[30,31] and compared his results with those previously shown by Wells and Mehl, and Wells et al.:[26,27]

$$D_{0.1\%C}^{\gamma\,Fe} = 0.668 \cdot \exp\left(\frac{-37460}{RT}\right) \tag{18.35}$$

Kaufman et al. developed the following exponential expression while solving the thermodynamics of the bainite reaction:[32]

$$D = 0.5\exp(-30x_C) \cdot \exp\left(-\frac{38300 - 1.9 \cdot 10^5 x_C + 5.5 \cdot 10^5 x_C^2}{RT}\right) \tag{18.36}$$

where x_c is the carbon fraction, R is the gas constant expressed in (cal/mol*K), and T is temperature calculated in Kelvin degrees.[32]

Tibbets created an apparatus for steady-state measurements and derived the following diffusivity expression for carbon diffusion in iron and steel at high temperatures:[33]

$$D = 0.47 \cdot \exp(-1.6C) \cdot \exp\left(\frac{-37000 + 6600\%C}{RT}\right) \tag{18.37}$$

In 1986, Agren revised the expression for the diffusivity, D, of carbon in binary Fe–C austenite as a function of temperature, T:

$$D = 4.53 \cdot 10^{-3}\left\{1 + y_C(1 - y_C)\frac{8339.9}{T}\right\} \cdot \exp\left\{-\frac{1}{T} - 2.221 \cdot 10^{-4}\right\}(17767 - 26436y_C) \tag{18.38}$$

where $y_C = 1 - x_C$ and x_C is the carbon fraction.[34]

Mathematical models to simulate surface diffusion processes have covered carburization most extendedly. Goldstein and Moren[35] simulated the carburization process in low alloy steels. Two-step and vacuum carburization are the heat treatments considered. A second model simulates the behavior of major ternary alloying addition such as Mn, Cr, Ni, Si during carburization. More comprehensive models include diffusant changes (e.g., carburizing gas composition), microstructural evolution, and interactions with heat treatment furnaces. For instance, Stickles developed two models for steady-state and time-dependent carburizing in batch and continuous furnaces.[36] The models describe the changes in the furnace atmosphere and composition during carburizing. All changes are the result of interaction between the instantaneous carbon demand and the rate of supply of carburizing gases to the furnace.

Nitrocarburizing and nitriding have also received modeling attention. Agren and coworkers studied the microstructural and compositional evolution of compound layers during gas nitrocarburizing and nitriding.[34,37] The results indicate that fast C and slow N absorption at the surface govern the microstructure in the early stages of gaseous nitrocarburizing. To simulate the nitriding, Larsson and Agren[38] used the DICTRA software and emphasized the drawbacks related to a nonequilibrium process.

Recently, Bernal et al.[39] developed an algorithm to estimate the nitrogen diffusion coefficient during the growth of nitrided layers produced by microwave postdischarge nitriding. An inverse problem is set in order to calculate the diffusion coefficients in each phase, while a compact nitrided layer γ'- Fe_4N_{1-x} and an austenite layer are formed. The authors consider a moving boundary problem, like a Stefan type problem, where the diffusion coefficient of ferrite is assumed to be known.

Internal or external oxidation is also a profound problem that interested a lot of researchers.[40,41] The authors characterize the internal oxidation defining the moving boundary between internal oxidation region and unoxidized region for both oxygen saturated and unsaturated cases.

Commercial models are currently available to solve the heat treatment process problems. DANTE, DEFORM, DICTRA, SYSWELD, FACTSAGE are the most used commercial packages. They all function by being connected to a specific database. FEA is also required for heat and diffusion as well as deformation and strains problems.

The heat treatment processes modeling is generally controlled by the mass transfer coefficients at the surface and diffusion coefficients in the substrate (steel most commonly). Every model, if properly done, has to include the surface boundary condition and the flux balance equation.[24]

$$h_m\left(a_C^{gas} - a_C^{surface}\right) = -D\frac{da_C^{surface}}{dx} \tag{18.39}$$

where h_m is the mass transfer coefficient, and a_C represents the carbon activity.

Usually, it is desirable to enhance the surface reactions between the gas and the substrate by controlling surface condition and chemistry. The diffusion coefficient needs to be maximized by increasing temperature and controlling alloy conditions. The latter refers to transformation diagrams, elemental diffusion, surface chemistry, and mass transfer coefficient. The surface condition introduces a few more complications like surface chemistry, surface roughness, and oxide layer formation.[24]

The modern modeling challenges are as follows: (1) how well do we know the diffusivity coefficient dependence on temperature and composition; (2) databases; (3) how well do we know mass transfer coefficient and its temperature dependence, h_m (Equation 18.37); (4) effects of phase transformation kinetics.[24] It is interesting that diffusion laws applied to geological processes may provide similarities in overcoming some of these challenges.

The diffusion couple and tracer experiments are well known and used in experimental geochemistry and the results are applied to field geology.[6] Although the time frames are significantly longer, the geologist approach is similar to the metallic system experiments. In fact, for the Earth's core or meteorites, the geologists are faced with metallic systems (FeNi) and the numerous trace elements involved in Earth's core composition. The "*error function*" solution and "*thin-film*" solution have been widely used to obtain tracer diffusivities, self-diffusivities, and interdiffusivities in geological systems (e.g., silicates). Problems like melt connectivity, mineral dissolution, diffusion of dissolved volatiles (CO_2, S, F, Cl) in magmatic liquids, noble gases diffusion in amorphous silicates are very well experimented and the diffusivity database is developing continuously. The interrelationships between phase transformation, transport properties, volatile abundance, in both crustal and mantle silicate systems are useful approaches to solve diffusion problems in materials processing such as phase changes and diffusion dependence on composition, surface conditions, or environments.

References

[1] The Surfaceweb Guide to Surface engineering. Author Keith Stevens. Available at http://www.it-innovation.soton.ac.uk/surfaceweb

[2] Cussler, E.L., *Diffusion: Mass Transfer in Fluid Systems*, Cambridge University Press, NY, NY, USA, 1994.

[3] Shewmon, P.G., *Diffusion in Solids*, 2nd edition, The Minerals, Metals and Materials Society, Warrendale, Pennsylvania, 1989.

[4] Philibert J., *Atom Movements: Diffusion and Mass Transport in Solids*, trans. by S.J. Rothman, Les Editions de Physique, 1991.

[5] Crank, J., *The Mathematics of Diffusion*, 2nd edition, Clarendon Press, Oxford, 1994.

[6] Watson, E.B., Diffusion in volatile-bearing magmas, *Rev. Mineral.*, 30, 379–401, 1994.

[7] Glicksman, M.E., *Diffusion in Solids: Field Theory, Solid-State Principles, and Applications*, John Wiley & Sons, Inc., New York, 2000.

[8] Arrhenius, S., Über die Reaktionngeschwindigkeit bei des Inversion von Rohrzucker durch Säuren, *Zeit. Phys. Chem.*, 4, 224–248, 1889.

[9] Lupulescu, A., Glicksman, M.E., and Yang, W., *Java Applets*, http://www.rpi.edu//locker/72/001272.

[10] Salman, M., et al., Activation volume for arsenic diffusion in germanium, *Appl. Phys. Lett.*, 69, 922–924, 1996.
[11] Darken, L.S., Diffusion, mobility and their interrelation through free energy in binary metallic systems, *Trans. AIME*, 175, 184–201, 1948.
[12] Glicksman, M.E., Lupulescu, A., *Kinetics Lecture Modules*, Spring 2005. Available at http://www.rpi.edu/dept/materials/in_pe.htm, and http://www.rpi.edu//locker/72/001272.
[13] Hartley, G.S. and Crank, J., Some fundamental definitions and concepts in diffusion processes, *Trans. Faraday Soc.*, 45, 801, 1949.
[14] Gottstein, G., *Physical Foundation of Materials Science*, Springer-Verlag, Berlin, 2004.
[15] Fourier, J.B.J., *The Analytical Theory of Heat*, trans. by A. Freeman, University Press, Cambridge, 1878.
[16] Carslaw, H.S. and Jaeger, J.C., *Conduction of Heat in Solids*, Clarendon Press, Oxford, 1959.
[17] Phillibert, J., *Adolph Fick and Diffusion Equations*, Proceedings of DiSo Conference, Moscow, Russia, 2005. In print.
[18] Fick, A., Über Diffusion, *Poggendorff's Annalen*, 94, 59–86, 1855.
[19] Fick, A., On Liquid Diffusion, *Phil. Mag.*, 10, 30–39, 1855.
[20] Abramovitz, M. and Stegun, I.A., eds., *Handbook of Mathematical Functions*, Applied Mathematics Series 55, National Bureau of Standards, U.S. Government Printing Office, Washington, DC, 1964.
[21] Roberts-Austen, W.C., Bakerian lecture on the diffusion in metals, *Phil. Trans. R. Soc. London*, A187, 1896.
[22] http: www.Key-to-steel.com
[23] Pye, D., *Diffusion Surface Treatment Techniques: A Review*, Pye Metallurgical Consulting Inc., Meadville, PA, 2005. http://www.industrialheating.com
[24] Sisson, D.R., Jr., *Carburization Process Modeling*, Multicomponent Multiphase Diffusion Symposium in Honor of John E. Morral: Applications of Multicompnent Multiphase Diffusion, TMS Annual Meeting, San Francisco, CA, 2005, 313.
[25] Arimoto, K., *Personal Communication*, 2005.
[26] Wells, C. and Mehl, R.F., Rate of diffusion of carbon in austenite in plain carbon, in nickel and in manganese steels, *Trans. AIME*, 140, 279–306, 1940.
[27] Wells, C., Batz, W., and Mehl, R.F., Diffusion coefficient of carbon in austenite, *Trans. AIME*, 188, 553–560, 1950.
[28] Harris, F.E., Case Depth — An attempt at a practical definition, *Metal Progress*, 265–272, 1943.
[29] Harris, F.E., Carburizing and diffusion data, *Metal Progress*, 910B, 1944.
[30] Smith, R.P., The diffusivity of carbon in iron by the steady-state method, *Acta Metall.*, 1, 578, 1953.
[31] Smith., R.P., The diffusivity of carbon in gamma iron-cobalt Alloys, *Trans. AIME*, 230, 476–480, 1964.
[32] Kaufman., L.R., Radcliffe, S.V., and Cohen., M., Thermodynamics of the bainite reaction, in *Decomposition of Austenite by Diffusional Processes*, ed. V.F. Zackay and H.I. Aaronson, AIME Interscience Publishers, pp. 313–352, 1962.
[33] Tibbetts, G., Diffusivity of carbon in iron and steels at high temperature, *J. Appl. Phys.*, 51, 4813–4816, 1980.
[34] Agren, J., A revised expression for the diffusivity of carbon in binary Fe-C austenite, *Scripta Metall.*, 20, 1507–1510, 1986.
[35] Goldstein, J.I. and Moren, A.E., Diffusion modeling of the carburization process, *Metall. Trans.*, 9A, 1515–1525, 1978.
[36] Stickles, C.A., Analytical models for the gas carburizing process, *Metall. Trans.*, 20B, 535–546, 1988.
[37] Du, H., Sommers, M.A.J., and Agren, J., Microstructural and compositional evolution of compound layers during gaseous nitrocarburizing, *Metall. Trans.*, 31A, 195–211, 2000.

[38] Larsson, H. and Agren, J., Gas nitriding of high vanadium steels-experiments and simulations, *Metal. Mater. Trans.*, 35A, 2799, 2004.

[39] Bernal, J.L., et al., Growth kinetics of nitride layers during microwave post-discharge nitriding, *Surface Coat. Technol.*, 200, 1458, 2005.

[40] Li, Y. and Morral, J.E., A local equilibrium model for internal oxidation, *Acta Mater.*, 50, 3683, 2002.

[41] Morral, J.E. and Li, Y., Internal oxidation and local equilibrium, *Mater. High Temp.*, 20, 273, 2003.

19

Basic Phase Transformations in Heat Treatment

John Ågren
Department of Materials Science and Engineering, Royal Institute of Technology

Abstract

The properties of metallic materials may be varied over a wide range by alloying and heat treatment. The reason for this variability, which is of tremendous practical importance, is that the behavior of a material to a large extent depends on its structure that evolves as a result of the phase transformations during heat treatment. The study of microstructure evolution and its relation to the mechanical properties is the essence of physical metallurgy.

In this text, we shall discuss the phase transformations that occur during heat treatment and may be controlled to yield a suitable microstructure. From a more fundamental point of view the following questions will be addressed: (i) Why does a structural change occur? (ii) How does the morphology evolve? (iii) What is the rate of the change?

The first question may be tackled by application of thermodynamics and will be dealt with in the first section. The remaining two questions will be dealt with in the following sections. The emphasis will be on the quantitative theory and relations that allow prediction of important quantities.

19.1 Thermodynamics of Phase Transformations

19.1.1 Equilibrium — Internal and External Variables

If an alloy with a fixed composition is kept at constant temperature and pressure for a sufficiently long time all its internal reactions will come to an end and it approaches a state of thermodynamic equilibrium. At equilibrium nothing more happens regardless of the time as long as the external conditions (i.e., temperature and pressure) are kept constant. On the other hand, a change in the conditions will lead to internal changes but if the conditions are kept constant at their new values a new state of equilibrium will be eventually approached.

It is convenient to distinguish between external and internal variables. The external variables are the ones that are controlled by the experimentalist, for example, temperature, pressure, and alloy composition. The internal variables are the ones that are changed by spontaneous reactions until equilibrium is reached. Examples of internal variables are the amount and the composition of the different phases that may appear. For example, at room temperature a simple steel with 0.1 mass % carbon consists of almost pure iron and particles of cementite; Fe_3C. As the temperature is increased some of the cementite dissolves and the carbon content of the iron matrix increases accordingly. The fraction of cementite and the carbon content in the iron matrix may then be regarded as internal variables whereas the temperature and the pressure (kept at 1 at.) are external variables. In some cases, the alloy composition is not kept constant and is not an external variable. An important example is a steel component that is kept in a carburizing atmosphere and carbon is added to the steel surface by reactions with the gas. In such cases the steel composition is an internal variable whereas the carburizing potential is an external variable.

19.1.2 Conditions to Calculate the State of Equilibrium

In practice the most common situation is that a given material (i.e., with a given chemical composition) is kept at a given temperature and pressure. The external variables are then the composition of the alloy, which may be given in several different ways, the pressure P and the temperature T. In that case the internal variables will take the values that give the lowest Gibbs energy G. In industrial practice the composition is usually given as mass percent of different elements k but in a scientific analysis it is often more convenient to use the mole fraction x_k. It may be calculated from the mass percentage m_k and the atomic masses M_k from:

$$x_k = \frac{m_k/M_k}{\sum_{i=1}^{n} m_i/M_i} \tag{19.1}$$

where the summation in the denominator is taken over all elements n, (i.e., the solvent must be included). For low alloy contents one may use the approximative formula:

$$x_k = \frac{m_k M_i}{100 M_k} \tag{19.2}$$

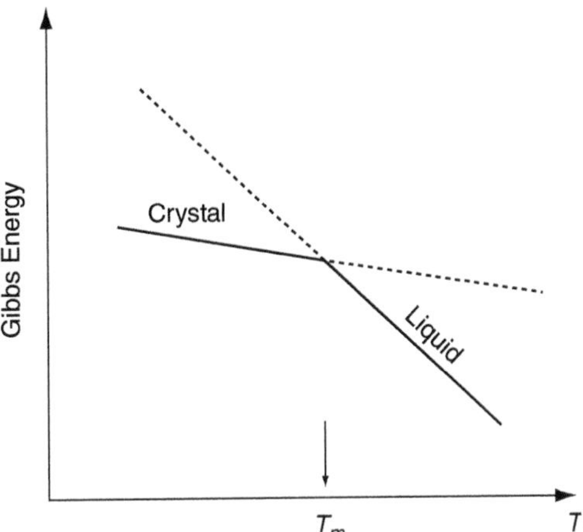

FIGURE 19.1 Gibbs energy as a function of temperature.

where i now stands for the solvent element, for example, iron in the case of low-alloy steels. In thermodynamic formulas temperature is usually given in Kelvin K.

Here we will not discuss the meaning of Gibbs energy in any depth. For such a discussion we refer to some text book on thermodynamics, for example, Hillert 1998 [1]. We simply remind the reader that the formal definition of Gibbs energy is:

$$G = H - TS \tag{19.3}$$

where H is the enthalpy and S the entropy. As a simple illustration one may consider the melting and crystallization of a pure substance, see Figure 19.1. In that case we may regard the fraction of liquid or solid as an internal variable that takes the value that minimizes the Gibbs energy. Below the melting point, where the Gibbs-energy curves of the liquid and the crystalline phase intersect, the lowest Gibbs energy is given by the fully crystalline state and above the melting point by the fully liquid state. For a binary alloy the composition of both the liquid and the crystalline phase may vary in addition to the fraction. This leads to the well-known common tangent construction and the existence of a two-phase field in the binary phase diagram.

If Gibbs energy for the individual phases in a system is known as functions of pressure, temperature, and their composition, one may calculate the equilibrium state for various conditions by a minimization procedure. The Gibbs energy may be evaluated from various thermochemical measurements, for example, calorimetry. Experimental information on phase equilibria also give important information on the Gibbs energy but usually it is not sufficient to evaluate it fully. Consider for example the transformation from austenite to ferrite in pure iron. At atmospheric pressure the transformation occurs at 910°C. At this temperature the Gibbs energy of austenite and ferrite must have the same value and from Equation (19.3) we conclude that at $T_0 = 910 + 273 = 1283$ K:

$$\Delta H = T_0 \Delta S \tag{19.4}$$

The information about the transformation temperature thus only gives a relation between ΔH and ΔS rather than individual values. However, if ΔH is measured by calorimetry this relation may be used to evaluate ΔS. Obviously a combination of thermochemical measurements and phase-equilibrium information is needed to obtain a full description. The example with the melting point of a pure substance is very simple and for more general cases one more essential part is needed and that is the mathematical representation of the experimental data, that is, the types of mathematical expressions that are used. This will be discussed in the next section.

19.1.3 Thermodynamic Modeling and CALPHAD

The procedure to choose an expression $G_m(P, T, x_1, x_2...)$ for each phase in the system, and to adjust the involved parameters to fit experimental information on both thermochemistry and phase equilibria is called CALPHAD, an acronym for CALculation of PHASe Diagrams [2]. The CALPHAD technique was introduced in academia in the 1970s [3] but already within a decade the first industrial applications were reported. Now it is rather widely used in materials industry. Its advantage, compared to the conventional use of experimental data found in tables or phase diagram compilations, is that once the Gibbs energy expressions are available they can be combined and used far beyond the often narrow areas where direct experimental data are available. Thus CALPHAD may be regarded as the most efficient use of expensive experimental data.

The simplest thermodynamic model is based on a random mixture between the atoms and yields the regular solution type of expression. For a given phase with n components it yields the following expression:

$$G_m = \sum_{j=1}^{n} x_j({}^{\circ}G_j + RT \ln x_j) + \sum_{j=1}^{n} \sum_{i>j}^{n} x_j x_i L_{ij} \tag{19.5}$$

The parameters ${}^{\circ}G_j$ and L_{ij} are properties of the phase under consideration. The first type of quantity represents the Gibbs energy of pure j in the phase under consideration and is often called the lattice stability. If this is not a stable phase for pure j it is difficult to measure directly ${}^{\circ}G_j$ and it can only be extracted indirectly from measurements on alloys containing j. However, it should be emphasized that this should be regarded as a true physical property and thus, at given P and T it must have a unique value regardless of what alloy is under consideration. The second type of quantity L_{ij} represents the interaction between i and j and may be temperature dependent. It may also vary with composition and it is common to talk about subregular solution if the variation is linear and subsubregular solution if it is a second order polynomial. Higher-order polynomials are possible in principle and may be required to fit all the experimental data. However, as their behavior is much more complex when extrapolating the experimental data they should be avoided. Another alternative that is usually much better is to apply a different physical model. For example, in the sublattice model similar atoms are assumed to mix randomly on one type of lattice sites. The simplest example is a solution with both interstitial and substitutional elements. For a detailed discussion on modeling the reader is referred to the book by Hillert.

19.1.4 The Chemical Potential and Driving Force

Temperature and pressure are thermodynamic potentials. A potential has the important property that its gradient (e.g., $\partial T/\partial z$), yields a force to move the quantity corresponding to the potential. Thus a temperature gradient $\partial T/\partial z$ yields a driving force for heat transport. The chemical potential μ_k of an element k is defined in a similar way and consequently its gradient $\partial \mu_k/\partial z$ yields a force to move that element. Formally it is defined from the Gibbs energy G as:

$$\mu_k \equiv \left(\frac{\partial G}{\partial n_k}\right)_{n_j,P,T} \tag{19.6}$$

that is, the change in Gibbs energy upon adding an infinitisimal amount dn_k of k at fixed P and T is given by $dG = \mu_k dn_k$. In a system with n components it may be calculated from the molar Gibbs energy expression $G_m(P, T, x_1, x_2, \ldots)$ by means of:

$$\mu_k = G_m + \frac{\partial G_m}{\partial x_k} - \sum_{j=1}^{n} x_j \frac{\partial G_m}{\partial x_j} \tag{19.7}$$

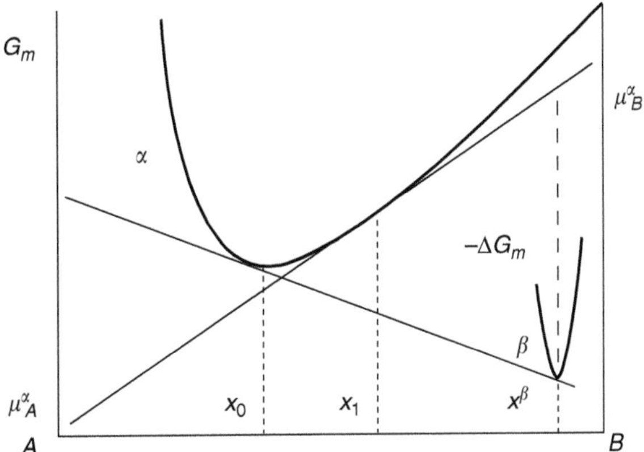

FIGURE 19.2 Gibbs energy as a function of mole fraction x_B at fixed P and T for phases α and β.

Equation (19.7) has a simple geometrical interpretation in a binary system A–B. It gives the intersection between the tangent to the G_m curve and the G_m axes at pure A and at pure B, see Figure 19.2.

Phase transformations are usually dissipative processes, which means that a driving force is needed for them to proceed. For example, consider the parent phase α with a composition $x_B = x_1$, see Figure 19.2. As can be seen that composition falls in the two-phase field $\alpha + \beta$ and the equilibrium composition of α is given by the common tangent as x_0. During the growth of β the atoms are transferred from one level of chemical potentials in the parent phase to another level in the β phase. This gives the total driving force available for the transformation α to β counted per mole of β formed. If β grows with a fixed composition x^β the driving force is:

$$\Delta G_m = \sum_{j=1}^{n} x_j^\beta \Delta\mu_j \tag{19.8}$$

where $\Delta\mu_j = \mu_j^\alpha - \mu_j^\beta$, see Figure 19.2. ΔG_m is thus defined as positive for a spontaneous reaction. This driving force is dissipated by various processes, for example, diffusion in the parent phase ahead of the migrating interface and sometimes also by interfacial reactions.

For a particular case one needs the detailed thermodynamic properties to calculate the driving force. Such calculations are easily performed using commercial softwares provided that a suitable database is available. For approximate calculations one may use equations that only involve what can be read directly from a phase diagram. For a binary alloy and low B content in α we may derive:

$$\Delta G_m = (1 - x^\beta)RT\ln\left(\frac{1 - x_1}{1 - x_0}\right) + x^\beta RT\ln\left(\frac{x_1}{x_0}\right) \tag{19.9}$$

19.1.5 Effect of Interfacial Energy

If the growing β phase in the previous example has a curved interface a part of the driving force must balance that effect. For a spherical β particle with radius r and the isotropic interfacial energy σ one needs to subtract $2\sigma V_m^\beta/r$ from the driving force. Thus there is a critical radius where the interfacial energy contribution exactly balances the driving force. If the β particle were smaller the difference between driving force and interfacial energy contribution would turn negative and the particle would dissolve rather than grow and if it were bigger the difference would become positive and the particle would grow. In an accurate thermodynamic calculation we should simply add the interfacial energy contribution to the Gibbs energy of the β phase but again approximate formulas may be useful. For the case of low supersaturation, that is, when the quantity $x_1 - x_0$ is much smaller than unity the expression for the driving force may be written:

$$\Delta G_m \cong RT(x_1 - x_0)(x^\beta - x_0)/x_0 \tag{19.10}$$

By subtracting the interfacial energy contribution and rearranging we obtain the well-known Gibbs–Thomson equation:

$$x_1 - x_0 \cong \frac{2\sigma V_m^\beta x_0}{RTr(x^\beta - x_0)} \tag{19.11}$$

The Gibbs–Thomson equation gives an approximate value for the solubility limit for β in the parent phase that is displaced due to the extra energy of a curved interface. For large β particles the parent phase composition would fall in the $\alpha + \beta$ two-phase field and β grows whereas for small particles it would fall inside the α one-phase region and the small particles would thus dissolve.

Two important consequences of the critical size should be mentioned. First, even in a supersaturated solution (i.e., where there is a positive driving force) the precipitation reaction would not start until the critical size is bypassed by some fluctuation (i.e., nucleation is needed). The second consequence is that particles of different size could never be in equilibrium with each other. The larger one would always grow at the expense of the smaller one. Even if they had the same size and were in equilibrium that would be an unstable equilibrium because the slightest fluctuation in size would make one of them bigger and it would start to grow at the expense of the other particle. This phenomenon is called coarsening or Ostwald ripening and is discussed in Chapter 5 and in Section 19.3.6.

19.1.6 Fluctuations

The driving force for phase transformations at given P and T and composition involves a decrease in Gibbs energy. This is a consequence of the second law of thermodynamics. However, the second law does not completely rule out that the Gibbs energy increases for a short time in a small part of the system. In an equilibrium system there are always fluctuations due to the thermal disorder. The probability that a fluctuation would lead to an increase ΔG in Gibbs energy is given by:

$$p = \exp\left(-\frac{\Delta G}{k_B T}\right) \tag{19.12}$$

where k_B is Boltzmann's constant. As k_B is very small ($\approx 1.38 \times 10^{-23}$) only fluctuations that yield an exceedingly small increase ΔG have a finite probability.

19.2 Diffusion

19.2.1 Fick's First and Second Law

In many phase transformations the phases that form have different composition than the parent phase. Such changes in composition occur by diffusion. Diffusion is extensively covered in Chapter 18. The diffusional flux J_k of a species k is defined as the amount of k (mole, gram, etc.) that is transported through a cross section with unity area during a time unit. According to Fick there is proportionality between flux and concentration gradient. For the one-dimensional case with diffusion along the z axis Fick's first law takes the form:

$$J_k = -D_k \frac{\partial c_k}{\partial z} \tag{19.13}$$

where D_k is the diffusion coefficient of k. The concentration c_k is the number of moles of element k per unit volume. It is related to the mole fraction by means of:

$$c_k = \frac{x_k}{V_m} \tag{19.14}$$

where V_m is the molar volume. If it is approximated as constant Fick's law takes the form:

$$J_k \cong -\frac{D_k}{V_m}\frac{\partial x_k}{\partial z} \tag{19.15}$$

By combining Fick's first law with the conservation condition one obtains Fick's second law:

$$\frac{\partial c_k}{\partial t} = \frac{\partial}{\partial z} D_k \frac{\partial c_k}{\partial z} \tag{19.16}$$

It should be emphasized that Fick's second law is not a "new" law but just a mathematical consequence of the conservation law.

19.2.2 Analytical Solutions to Fick's First and Second Law

When solving Equation (19.16) one seeks for solutions that obey some special initial and boundary conditions. In general the equation must be solved by numerical methods, for example, finite-difference or finite-element techniques, to allow realistic conditions and take into account the fact that the diffusion coefficient is usually not constant. For the case when the diffusion coefficient may be approximated as constant one may try the error-function type of solution;

$$c_k = A + B\,\mathrm{erf}(z/\sqrt{4D_k t}) \tag{19.17}$$

where $\mathrm{erf}(\beta)$ is the error function defined as:

$$\mathrm{erf}(\beta) = \frac{2}{\sqrt{\pi}}\int_0^{\beta} \mathrm{e}^{-y^2}\,dy \tag{19.18}$$

The constants A and B are determined from the boundary conditions. When the diffusion coefficient is constant Equation (19.16) becomes linear and the so-called superposition principle holds. That means that any sum of error-function solutions is also a solution. For short times the error-function solution gives a step in the concentration profile, one step for each term containing an error function. A typical example of application of the error-function solution is diffusion in a joint between two different materials, a so-called diffusion couple. Another example is when adding material from the outside(e.g., during carburizing).

The error-function solution is usually not convenient at very long times or if the initial concentration profile has a complex shape becuase that would require many terms. In such cases the sine-solution may be used instead:

$$c_k = A + B\exp\left(-\frac{4\pi^2}{\lambda^2}D_k t\right)\sin\left(\frac{2\pi}{\lambda}z\right) \tag{19.19}$$

where λ is the wavelength of the composition variation. Again, we may conclude from the superposition principle that a sum of such sine functions is a solution if the given initial and boundary conditions are obeyed. As rather arbitrary functions may be expanded in infinite series of such terms having different λ by means of Fourier expansion this method is quite general. On the other hand, it is inconvenient to handle many terms and usually one thus restricts the practical use to such long times that only one or two terms are needed. One example, is diffusion during homogenization of microsegregations. As can be seen

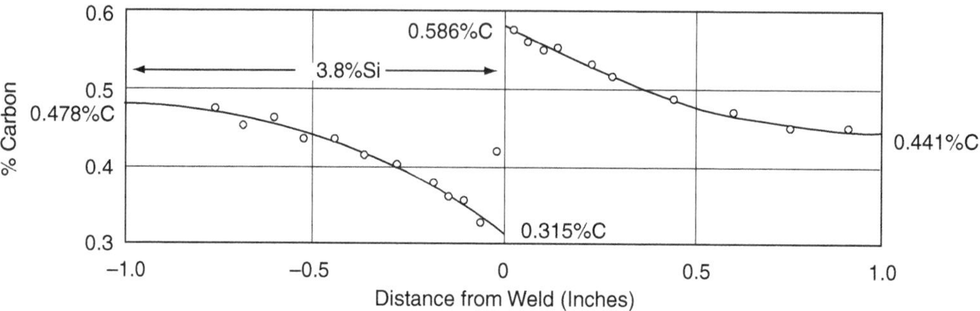

FIGURE 19.3 Measured carbon concentration profile in welded joint (after Darken, L. S., *Trans. AIME*, 180, 1949, 430–438).

from Equation (19.19) concentration variations represented by short wavelength will decay rapidly and in order to study homogenization it is often enough to only consider the longest wavelength.

19.2.3 Multicomponent Diffusion

Real alloys often contain high contents of many alloy elements. In such cases Equation (19.13) does not account for the experimental observations satisfactorily. It is often found that there is a coupling between the different elements and Equation (19.13) is replaced by the so-called Fick–Onsager law:

$$J_k = -\sum_{i=1}^{n-1} D_{ki}^n \frac{\partial c_i}{\partial z} \tag{19.20}$$

There are now several diffusion coefficients that may be arranged in a matrix. The diagonal elements of that matrix describe the "normal" diffusion whereas the off-diagonal elements represent the different couplings. A classical example of such coupling is Darken's experiment [4] in which two steels with roughly the same carbon content but different silicon contents were welded together and heat treated for some time at a temperature where both alloys were austenitic. During the heat treatment the mobile carbon atoms diffused from the Si rich to the Si poor alloy and in the joint carbon even diffused against its own concentration gradient, so-called up-hill diffusion, see Figure 19.3. These coupling effects are very important in technological applications. In order of decreasing strength the physical basis for the coupling effects are:

- Thermodynamic interactions
- Frame of reference
- Correlation effects

The first effect can be understood from the fact that the thermodynamic driving force for diffusion of an element is the gradient of its chemical potential. One usually introduces the mobility M_k of element k and writes:

$$J_k = -c_k M_k \frac{\partial \mu_k}{\partial z} \tag{19.21}$$

where μ_k is the chemical potential of element k, see Section 19.1.4. As the chemical potential is a function of composition its gradient may be expanded in terms of all the concentration gradients:

$$J_k = -c_k M_k V_m \sum_{i=1}^{n-1} \frac{\partial \mu_k}{\partial x_i} \frac{\partial c_i}{\partial z}. \tag{19.22}$$

By comparing with Equation (19.20) we find that:

$$D_{ki}^{n} = c_k M_k V_m \frac{\partial \mu_k}{\partial x_i} \tag{19.23}$$

that is, nonzero values of the off-diagonal ($k \neq i$) derivatives $\partial \mu_k / \partial x_i$ will cause coupling.

The second effect may occur even in an ideal system where derivatives $\partial \mu_k / \partial x_i$ vanishes unless $i = j$. If the mobilities are unequal, which is usually the case, there will be a net flow of atoms relative to the lattice. In a so-called number-fixed frame of reference, where only interdiffusion is considered, it will appear as there is a coupling between the different components. The interdiffusion fluxes are obtained as $J'_k = J_k - x_k \sum_{i=1}^{n} J_i$ and consequently obey $\sum_{k=1}^{n} J'_k = 0$. It is found that:

$$J'_k = -\sum_{i=1}^{n-1} L_{ki} \frac{\partial (\mu_i - \mu_n)}{\partial z} \tag{19.24}$$

where L_{kj} is a combination of the different mobilities and $\mu_i - \mu_n$ is often referred to as the interdiffusion potential.

This effect is usually much weaker than the thermodynamic coupling. The third effect is caused by the fact that the consecutive diffusive jumps are not independent but there is a correlation. This effect is difficult to estimate except in very simple cases like self-diffusion in a pure metal but is believed to be negligible compared to the other effects and the uncertainty of the experimental data.

The multicomponent version of Fick's second law is obtained by inserting the expression for J_k in the conservation equation. This gives a system of coupled partial differential equations. It is now more difficult to find analytical solutions, but if the elements of the diffusivity matrix may be approximated as constants it is possible to diagonalize the matrix by transforming into a new set of concentration variables. One then obtains a system of independent partial differential equations that may be solved in the same way as in the dilute case. Unfortunately, we would not in general expect the D_{ki}^{n}:s to be constant. Even the simplest atomistic models lead to the conclusion that the off-diagonal elements (i.e., $k \neq i$), would be proportional to the product $x_k x_i$. Analytical solutions can still be applied as reasonable approximations provided that only a narrow concentration range is considered and $x_k x_i$ is sufficiently constant. In general one would then have to apply numerical calculations based on finite elements or finite differences.

19.3 Precipitation and Dissolution Processes

19.3.1 The Concept of Nucleation and Growth

A precipitation process starts by nucleation where the new phase starts to grow from one or several nuclei and form new particles imbedded in the parent matrix. The resulting microstructure will thus depend on both the density and the location of the nuclei as well as their growth characteristics. The overall rate of the transformation depends on both nucleation and the growth of individual particles. It is convenient to introduce an overall reaction-path coordinate ϕ defined in such a way that it is zero before the transformation has taken place and unity after it has come to an end. We may thus denote ϕ as the extent of the transformation. The nucleation and growth concept has been applied to different materials processes described in Chapters 8, 10, 11, 15–17, and 28.

Consider the precipitation of β phase particles from a supersaturated α phase and define $\phi = f^{\beta}/f_{\text{eq}}^{\beta}$, where f^{β} and f_{eq}^{β} are the volume fraction of β at some instant t and at equilibrium, respectively. If ϕ is plotted against time a typical S curve is obtained, see Figure 19.4. This behavior is typical for precipitation reactions.

The overall rate will increase as the particles grow bigger, and we will enter the expansive growth regime of the S curve. As the growth proceeds the growth rate steadily increases, typically as $\mathrm{d}\phi/\mathrm{d}t = bt^{n}$,

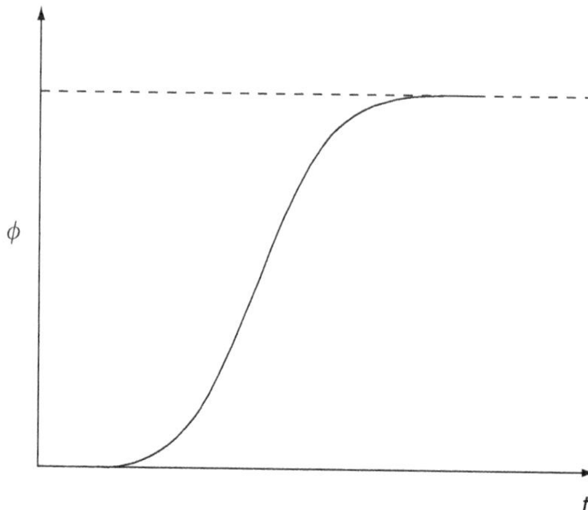

FIGURE 19.4 Extent of transformation as function of time for a heterogeneous phase transformation.

with n in the range 1.5 to 5 and b being a constant. After some time the effect of the finite driving force will gradually slow down the reaction. The growing particles will impinge on each other, either by hard impingement, where they grow until they collide and the growth abruptly stops at the collision points, or by a more gradual process called soft impingement. As a first approximation we may modify the above expression by multiplying it with the phenomenological factor $1 - \phi$, (i.e., $d\phi/dt = bt^n(1 - \phi)$), that would make the rate vanish as ϕ approaches unity. We may then easily integrate the equation and obtain:

$$\phi = 1 - \exp\{-bt^{n+1}/(n + 1)\} \tag{19.25}$$

Several authors have derived this equation or similar ones by various methods. The most elegant derivation is due to the Russian mathematician Kolmogorov but the equation was also derived by Johnson and Mehl and by Avrami. In western literature it is usually called the Johnson–Mehl or the Avrami equation although one nowadays also finds the notation KJMA equation in honor of all the authors.

Kolmogorov considered the case of hard impingement in a situation where the equilibrium volume fraction is unity (i.e., $\phi = f^\beta$). This may serve as a reasonable approximation for transformations that do not involve long-range diffusion, like eutectoid growth and recrystallization. If there is no soft impingement but only hard impingement the particles will grow without feeling the presence of each other unless they collide. Kolmogorov obtained by a simple statistical analysis:

$$f^\beta = 1 - \exp\left(-\sum_k V_k/V\right) \tag{19.26}$$

where V_k is the volume of the kth particle if it had grown without impingement and V is the total volume. Equation (19.26) is quite a general expression that holds for a random distribution of particles and independent growth until hard impingement sets in. If all particles are spherical and nucleated at $t = 0$ and grow with a constant growth rate g we have:

$$f^\beta = 1 - \exp\left(-N_V \frac{4}{3}\pi g^3 t^3\right) \tag{19.27}$$

where N_V is the number of nuclei per unit volume. In general we may expect that new nuclei are formed continuously (i.e., N_V increases in time). If $\dot{N}_V$ denotes the nucleation rate $\dot{N}_V\, d\tau$ new nuclei are formed during the time interval between τ and $\tau + d\tau$. At the time t those have grown to the volume $\frac{4}{3}\pi g^3(t-\tau)^3$

TABLE 19.1 Exponents in KJMA Equation

	Exponent n	
Geometry	Constant N	Constant $\dot{N}_V$
Spheres, constant growth rate	3	4
Spheres, parabolic rate	3/2	5/2
Plates, parabolic thickening	5/2	7/2
Needles, parabolic thickening	2	3

and one obtains:

$$\sum V_k/V = \frac{4}{3}\pi g^3 \int_0^t \dot{N}_V\,(t-\tau)^3\,\mathrm{d}\tau \tag{19.28}$$

If $\dot{N}_V$ is constant it may be moved outside the integral and we obtain by inserting in Equation (19.26):

$$f^\beta = 1 - \exp\left(-\frac{1}{3}\pi g^3 \dot{N}_V t^4\right) \tag{19.29}$$

Thus simple assumptions concerning particle shape and nucleation behavior yield the time exponents $n = 3$ and $n = 4$ as exact solutions to the problem. Kolmogorov's solution is strictly valid only for hard impingement but we may apply it as an approximation also for cases with soft impingement. Consider for example the precipitation of spherical particles obeying a parabolic growth law, that is, the radius varies as $\sqrt{t}$. In that case we obtain, by applying the above method $n = 5/2$ if $\dot{N}_V$ is constant. The edgevise growth of platelets usually proceeds with a constant rate whereas the thickening is parabolic. The volume of an individual particle would thus vary as $t^{5/2}$. A constant $\dot{N}_V$ would then yield $n = 7/2$. The situation for simple shapes and nucleation conditions is summarized in Table 19.1.

The temperature dependence of the rate of heterogeneous transformations comes from the temperature dependence of the growth rate as well as the nucleation rate. Both processes have a complex temperature dependence that will be discussed in detail in the chapters that follow. We end this section by noting that the temperature dependence of each rate is composed of the temperature dependence of the driving force and the kinetic behavior. The kinetic behavior is usually controlled by thermally activated processes and obeys the Arrhenius relation. As a first approximation the driving force is related to the supercooling or superheating. For transformations occurring during cooling we will then have the typical C curve, Figure 19.5a, whereas transformations that occur on heating would yield a curve without a nose (i.e., an L curve, Figure 19.5b).

19.3.2 Classical Theory of Homogeneous Nucleation

In homogeneous nucleation the formation of nuclei occurs with the same probability everywhere in the system. Heterogeneous nucleation, which occurs mostly in practice, will be discussed in the next section. As already mentioned, the effect of interfacial energy yields a Gibbs energy barrier that may be bypassed by thermal fluctuations if it is low enough. This is the physical picture underlying the classical nucleation theories. Moreover, to make the problem mathematically tractable, it is assumed that both phases have their normal properties determined in bulk samples and that the extra Gibbs energy due to the small particle size is represented by the isotropic interfacial energy σ, that is, Gibbs energy per unit area of phase interface. Of course, these assumptions are highly questionable in the case of nucleation where we usually consider a very small nucleus with a large fraction of its atoms in the interfacial region. Nevertheless, these difficulties are ignored and the change in Gibbs energy of the system due to the formation of a spherical

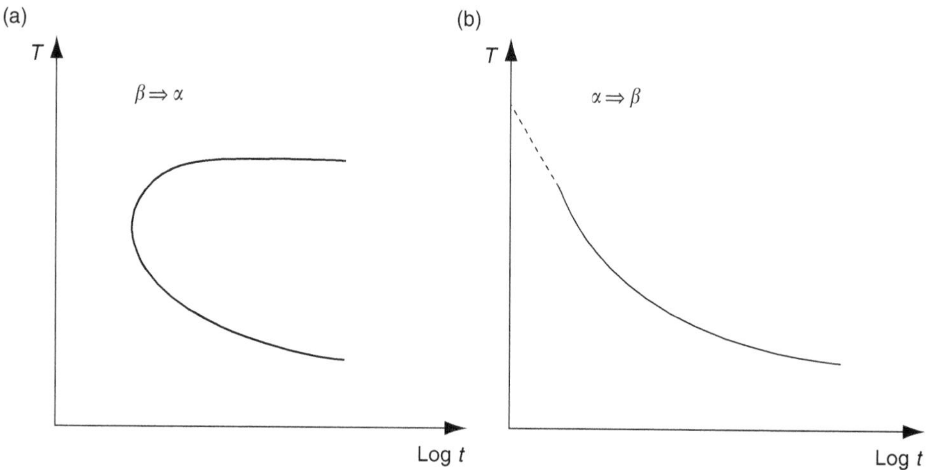

FIGURE 19.5 (a) The curve denoting 1% formed from β during cooling. (b) Curve for 1% β formed from α during heating.

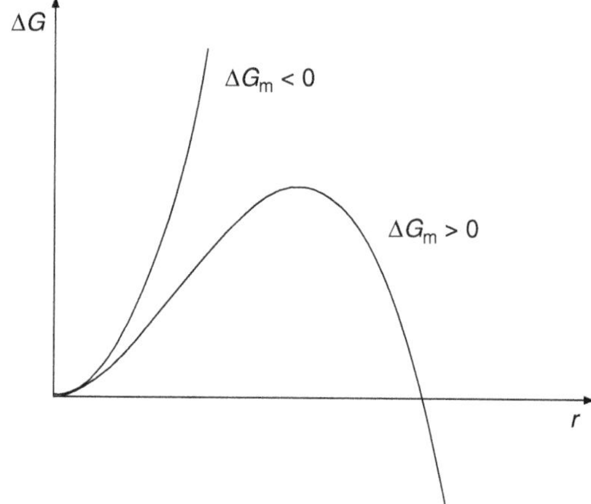

FIGURE 19.6 Variation in Gibbs energy for formation of a spherical liquid droplet from vapor.

droplet of radius r is written as:

$$\Delta G = -\frac{4\pi}{3} r^3 \frac{\Delta G_m}{V_m} + 4\pi r^2 \sigma \tag{19.30}$$

where ΔG_m is the driving force and defined as positive in a supersaturated solution. It should be mentioned that Equation (19.30) only holds if the new phase is incompressible. For example, the equation holds approximately for formation of liquid droplets from a vapor but not for formation of gas bubbles in a liquid. In the latter case a different approach must be taken. For a pure substance ΔG_m is simply the difference in molar Gibbs energy between the parent phase and the new phase. V_m is the molar volume of the new phase. Consider formation of liquid droplets from a supersaturated vapor. A schematic plot of ΔG vs. r is shown in Figure 19.6 for the two cases of positive and negative ΔG_m.

The position of the maximum is given by $\mathrm{d}\Delta G/\mathrm{d}r = 0$, that is:

$$r_{\mathrm{cr}} = \frac{2\sigma}{\Delta G_m / V_m} \tag{19.31}$$

and the corresponding height of the barrier is:

$$\Delta G^* = \frac{16\pi}{3} \frac{\sigma^3}{(\Delta G_m / V_m)^2}. \qquad (19.32)$$

It is obvious that $r = 0$ corresponds to a metastable state because small fluctuations would lead to an increase in Gibbs energy. However, $r = r_{cr}$ represent an unstable equilibrium, that is, all possible changes of the system lead to a decrease in Gibbs energy. States with $r > r_{cr}$ represent the supercritical nucleus and only further growth of the droplet leads to a decrease in Gibbs energy. Thus it is necessary to form a droplet larger than r_{cr} to have a spontaneous growth of liquid from vapor.

How can a supercritical nucleus form? It should be noticed that the process of growth from $r = 0$ to $r = r_{cr}$ actually violates the second law of thermodynamics because it yields an increase in Gibbs energy and the only possibility thus is that the growth occurs as a result of thermal fluctuations. In a stable equilibrium system under given P and T the probability p of a fluctuation that would give an increase ΔG^* in Gibbs energy is given by:

$$p = \exp(-\Delta G^* / k_B T). \qquad (19.33)$$

It should be emphasized that the equation is strictly valid only for fluctuations in stable systems but should be approximately applicable to metastable systems provided that the fluctuations are small enough, and certainly not larger than ΔG^*. Volmer and Weber [5] and Becker and Döring [6] presented the first satisfactory theory for homogeneous nucleation and gave the following equation for the nucleation rate:

$$I = A\sqrt{\frac{\Delta G^*}{3\pi kT}} \exp\left(-\frac{\Delta G^*}{k_B T}\right) \qquad (19.34)$$

where A is a frequency factor. Often a number of parameters are lumped together as $1/n^* \sqrt{\Delta G^* / 3\pi kT}$, the so-called Zeldovich factor denoted by Z, n^* is here the number of atoms in the critical nucleus. Equation (19.34) has a number of limitations. It was derived for homogeneous nucleation of a pure substance from its supersaturated vapor. Thus the analysis did not take into account nucleation on surfaces, grain boundaries, phase interfaces or dislocations, and the like, nor did it account for the difference in composition between parent phase and nucleus. Consequently the analysis has been modified in various ways. For solid-state nucleation the frequency factor is usually written $A = N\nu$ where N is the number of possible nucleation sites per volume, typically around 10^{28} m^{-3}, and ν is the phonon frequency around 10^{13} s^{-1}. Another straightforward modification is to express the driving force according to the discussion in Section 19.1.4 which would yield $\Sigma x_j^\beta \Delta\mu_j$, where β stands for the nucleus. This allows multicomponent alloys to be considered. Equation (19.34) involves no explicit time dependence in contrast with the belief that during solid-state transformations there is a pronounced incubation period, that is a period before nucleation starts. Turnbull analyzed the derivation of Equation (19.34) and rather than taking the steady-state solution for granted he solved the differential equations numerically and obtained a transient state during which the nucleation rate increased to approach Equation (19.34). Later Turnbull [7] suggested that this behavior could be represented by a correction factor $\exp(-\tau/t)$, where τ is the incubation period and t the time. Further modifications of the nucleation have been suggested over the years, for example by Kelton [8] who presented a numerical approach allowing for more general time dependent nucleation.

The classical theory is often criticized because it is based on the assumption that the Gibbs energy of a nucleus is built up of a bulk and surface contribution. The interior of the nucleus is then assumed to have the same thermodynamic properties as a bulk phase of the same structure and the surface energy is assumed the same as that of a much larger particle. If the Gibbs energy per atom in the nucleus is plotted as a function of $n^{-1/3}$, n being the number of atoms, one would obtain a straight line. Spaepen [9] analyzed data for small clusters that had been calculated theoretically by considering the interatomic potentials.

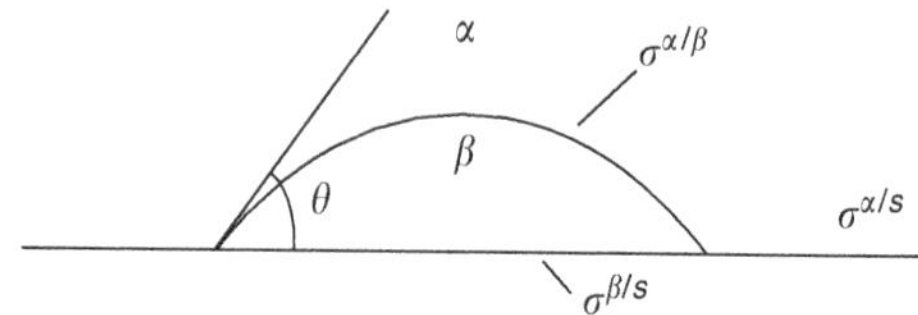

FIGURE 19.7 Formation of β from α on a surface.

Spaepen concluded a linear behavior down to cluster sizes involving five atoms but the intersection at $n = 0$ fell above the corresponding bulk energy for the stable face-centered cubic (FCC) phase indicating that the structure of the clusters are more liquid like.

19.3.3 Heterogeneous Nucleation

In practice homogeneous nucleation rarely occurs. Various surfaces, for example, impurity particles or the walls of the container in a gas or liquid, and grain boundaries or phase interfaces in a solid, will promote nucleation before the supersaturation needed for homogeneous nucleation is reached. A simple picture of heterogeneous nucleation is shown in Figure 19.7.

This model was first considered by Volmer [10]. The segment of β is a part of a sphere in this model because the surface tension is constant. The equilibrium condition between the different surface tensions then is:

$$\sigma^{\alpha/s} = \sigma^{\beta/s} + \sigma^{\alpha/\beta} \cos\theta \tag{19.35}$$

where $0 < \theta < \pi$. If the calculation yields a θ outside the range either β will wet the surface and spread as a thin film over it and the nucleation becomes barrierless or β will form as a sphere and touch the surface at a point only and the surface does not promote the nucleation. In principle the Gibbs energy as a function of particle size will show a maximum as for the homogeneous nucleation. The following Gibbs energy barrier is obtained:

$$\Delta G_c = \frac{4\pi}{3} \frac{\sigma^3}{(\Delta G_m / V_m)^2} \left[2 - 3\cos\theta + \cos^3\theta\right] \tag{19.36}$$

Compared to homogeneous nucleation the barrier is reduced with the factor:

$$[2 - 3\cos\theta + \cos^3\theta] \tag{19.37}$$

For $\theta = \pi$ it is unity and the barrier is not reduced at all. As θ approaches zero the barrier disappears and the nucleation becomes barrierless.

Similar analyses may be performed for nucleation at a grain boundary between two grains, at a three-grain junction and at a corner common to four grains. Such analyses have been presented by Cahn [11] and the result is summarized in Figure 19.8 showing the correction factor as a function of $\cos\theta$.

A lower nucleation barrier does not necessarily mean a higher nucleation rate because the number of possible nucleation sites is decreased when going from grain boundaries to grain edges and grain corners.

Also Clemm and Fisher [12] derived expressions for ΔG^* for nucleation at two-, three- and four-grain junctions and Cahn considered nucleation on dislocations. In all these analyses the results are presented as a reduction of ΔG^*. An interesting feature is that ΔG^* is predicted to vanish under some conditions, that is, the secondary phase can start to grow without the need of thermal activation.

If the phase interface between substrate *s* and the new particle β is coherent the corresponding interfacial energy may be particularly low and nucleation should be much favored. As discussed in a previous section the fit between nucleus and substrate is usually not perfect but there is some misfit that is accommodated

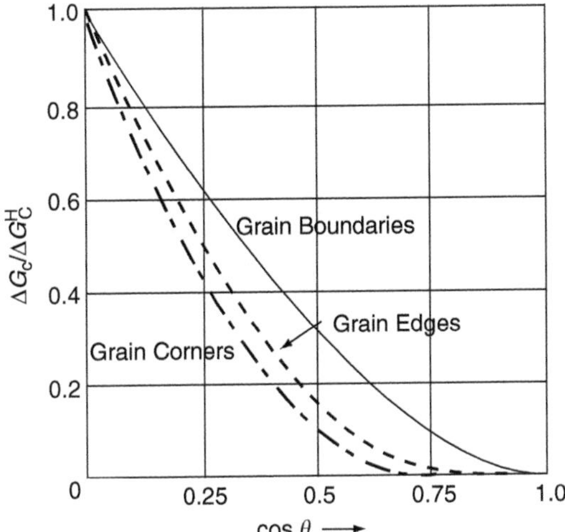

FIGURE 19.8 Correction factor for homogeneous nucleation barrier due to heterogeneous nucleation.

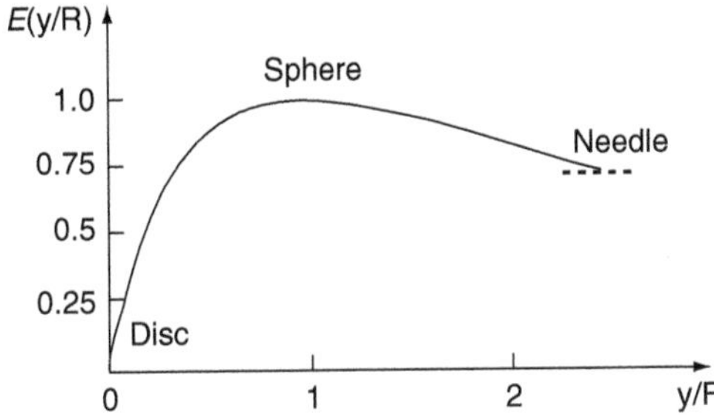

FIGURE 19.9 Shape correction factor for elastic energy of nonspherical particles.

by elastic stresses or by dislocations or a combination of both. We will now modify the nucleation theory to take into account also an elastic contribution. The major difference now is that we have to subtract the elastic energy ΔG_m^{el}, always positive, from the driving force ΔG_m, defined as positive for a reaction $\alpha \rightarrow \beta$ to occur. The subtraction of the elastic energy will thus decrease the driving force available. Nabarro [13] considered the case with no coherency stresses but just a dilatational strain effect due to a difference in molar volume between the two phases. For a β particle in an α matrix with a shape given by ellipsoids of revolution having semiaxes R and y Nabarro derived for the strain energy:

$$\Delta G_m^{el} = 2\mu^{\alpha} V_m^{\alpha} \left(\frac{V_m^{\beta} - V_m^{\alpha}}{V_m^{\beta}} \right)^2 E(y/R) \tag{19.38}$$

The equation holds for the case when the particle is much stiffer than the matrix and all the elastic strain is taken up by the matrix. V_m^{β} and V_m^{α} are the molar volumes of the β and α phases, respectively. For the sphere the $E(y/R) = 1$. The strain energy according to Nabarro is plotted for some shapes in Figure 19.9. As can be seen minimum strain energy is obtained by making very thin plates. However, thin plates would have a large surface energy and the state should be found that minimizes the total Gibbs energy. The Gibbs energy of the nucleus would now be a function of both r and the shape given by the ratio y/R (i.e., $\Delta G(r, y/R)$).

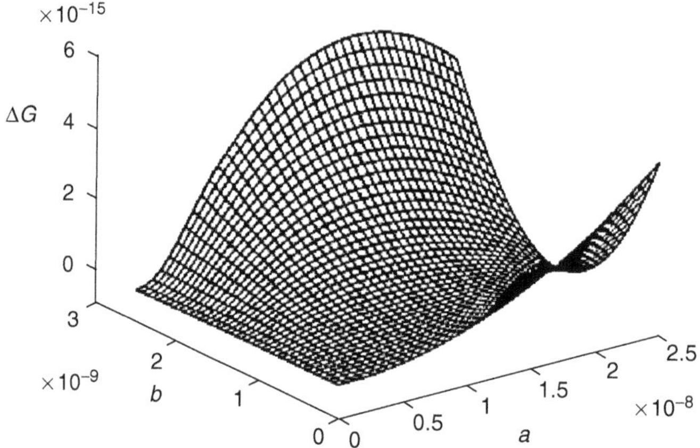

FIGURE 19.10 Gibbs energy surface for general ellipsoids with elastic energy and polar radii a and b.

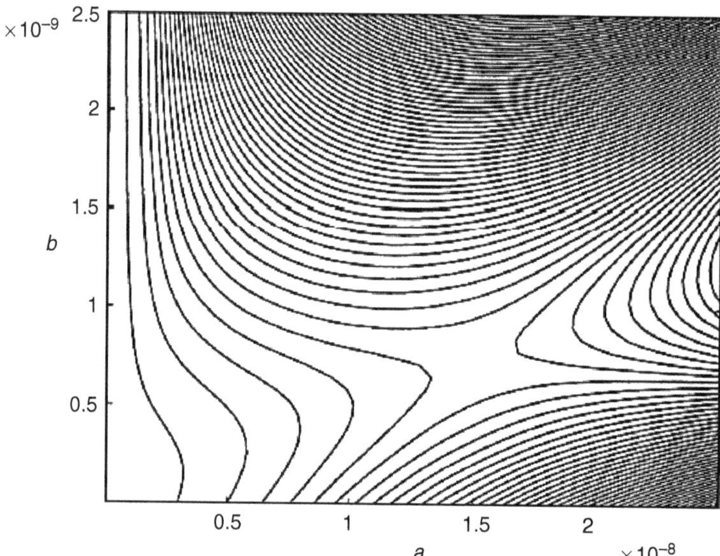

FIGURE 19.11 Gibbs energy iso-energy lines for general ellipsoids with elastic energy and polar radii a and b.

The Gibbs energy surface is plotted as a function of the two radii a and b in Figure 19.10 and Figure 19.11. As can be seen the lowest barrier is now a saddle point obtained by solving:

$$\frac{\partial \Delta G}{\partial r} = 0 \quad \frac{\partial \Delta G}{\partial (y/r)} = 0 \tag{19.39}$$

Let us now consider a coherent nucleus with a crystallographic plane parallel with a crystallographic plane of similar atomic arrangement in the parent phase but with a somewhat different atomic spacing. For the simple case when there is a mismatch in one-dimension only we may define:

$$\delta = \frac{a^{\alpha 0} - a^{\beta 0}}{a^{\beta 0}} \tag{19.40}$$

where $a^{\alpha 0}$ and $a^{\beta 0}$ are the atomic spacings in the direction of mismatch of the α and β phases in the strain-free state. The strain energy for the fully coherent interface is then given by:

$$\Delta G_m^{coh} = \frac{\mu V_m^\beta}{1-\nu^2}\varepsilon^2 \tag{19.41}$$

where ν is Poisson's ratio and the strain $\varepsilon = \delta$ for a fully coherent interface. For a coherent interface the driving force available is decreased by the amount ΔG_m^{coh} and the effective driving force thus is generally lower in coherent nucleation than in incoherent nucleation. However, it should be kept in mind that the surface energy σ is much lower for a coherent interface and the resulting nucleation barrier may thus be much lower than in incoherent nucleation. The surface energy of a coherent interface may typically be one order of magnitude lower than the surface energy of an incoherent interface. We may further note that there is a critical supersaturation at which the effective driving force vanishes and at lower supersaturations coherent nucleation is not possible.

If the elastic energy is large, there is a possibility to relax some of the energy by introducing dislocations at the interface. In that case the coherency will not be perfect and the interface is said to be semicoherent. The elastic strain will be lowered, that is, $\varepsilon = \delta - d$ where d is the part of the strain taken up by dislocations. There is some extra energy connected with the dislocations and the decrease in elastic energy is to some extent opposed by the dislocation energy. It is thus possible to find a dislocation content that would minimize the total Gibbs energy. To find the lowest nucleation barrier we would have to write the total Gibbs energy of the system containing a nucleus as a function of its size r, its shape y/R and the coherency strain ε and find the saddle point:

$$\frac{\partial \Delta G}{\partial r} = 0 \quad \frac{\partial \Delta G}{\partial \left(y/r\right)} = 0 \quad \frac{\partial \Delta G}{\partial \varepsilon} = 0 \tag{19.42}$$

In general there will be a misfit not only in one direction and the principal strains $\varepsilon_1, \varepsilon_2$ and ε_3 are needed to define the degree of coherency.

19.3.4 Sharp-Interface Approach of Growth and Dissolution

Precipitation and dissolution reactions proceed by migration of a phase interface. The traditional approach is to consider the interface as a mathematically sharp boundary between the two phases. In this approach the interface has no particular properties at all except for an interfacial energy. From a mass balance around the phase interface the following general condition can be derived:

$$\frac{v}{V_m}\left(x_k^\alpha - x_k^\beta\right) = J_k^\alpha - J_k^\beta \tag{19.43}$$

where v is the migration rate of the phase interface, V_m the molar volume and x_k^α and x_k^β are the contents of k on each side adjacent to the phase interface and J_k^α and J_k^β are the corresponding fluxes given by Equation (19.13) or Equation (19.20).

Once the contents x_k^α and x_k^β are known the diffusion problem in each phase may be solved, the fluxes evaluated and the migration rate calculated from Equation (19.43). As boundary conditions for the diffusion problem have to be imposed at the interface but its position is unknown the problem is by no means trivial. It is a so-called free-boundary problem and is often referred to as the Stefan problem. The traditional way of evaluating x_k^α and x_k^β is to assume that the interface is locally in thermodynamic equilibrium. This is not always a good approximation but in the cases when it is valid it is very powerful because it allows the prediction of the migration rate from the bulk properties of the involved phase and does not require any insight about the details of the interface.

It is quite straightforward though, to modify the approach to account for so-called interfacial friction. In that case part of the available driving force would be used to "push" the phase interface itself. It is common to write that part $\Delta G_m^i = (v/M)V_m$, M being the interfacial mobility. For approximate calculations one may proceed as when deriving Equation (19.11) and obtain:

$$x_1 - x_0 \cong \frac{vV_m^\beta x_0}{RTM(x^\beta - x_0)} \tag{19.44}$$

If considerable parts of the driving force is used for both diffusion and to overcome the interface friction one sometimes talks about mixed-mode transformations.

In general the Stefan problem may be solved exactly by analytical methods only under a number of simplifying assumptions. The most important ones are that the diffusivity is constant and that the conditions at the phase interface do not change in time. The first assumption is usually not strictly valid. Often the experimental measurements reveal strong concentration dependence in the diffusivity. One may then hope to solve the problems analytically with a constant diffusivity chosen according to some kind of averaging procedure. It may be shown that such an averaging predicts the correct flux under steady-state conditions but not in the general time-dependent case. The assumption of constant conditions at the phase interface excludes, for example, the treatment of nonisothermal processes. During isothermal processes the conditions will be constant at the phase interface in binary systems and, at short times, in multicomponent systems.

We shall now first consider some cases where exact solutions can be given analytically. Thereafter, we will turn to some approximate methods yielding rather simple analytical solutions and finally we will demonstrate some results of numerical calculations. We will only treat the case with one diffusing species. Suppose that we want to calculate the growth rate of one single planar β particle in an infinite α matrix. Let us consider the case when the particle grows both in the positive and negative directions. The concentration field is thus symmetric with a plane of symmetry at $z = 0$. We may then treat only the half side for which $z > 0$. The concentration in the matrix is $x^{\alpha\infty}$ far away from the particle, $x^{\alpha/\beta}$ close to the interface and x^β inside the particle. Assuming a parabolic time dependence, that is, that the concentration is a function of $z/\sqrt{t}$, it seems reasonable to assume also that the half thickness s should vary as $s = k\sqrt{Dt}$, where k is a constant to be determined. The error function is a solution of Fick's second law and the expression that obeys the boundary conditions is:

$$x = x^{\alpha\infty} + (x^{\alpha/\beta} - x^{\alpha\infty})\frac{1 - \text{erf}\left(z/\sqrt{4D^\alpha t}\right)}{1 - \text{erf}\,(k/2)} \qquad z > s \tag{19.45}$$

From this expression the flux at the α/β phase interface is evaluated and inserted into the flux-balance Equation (19.43). It is then found that t does not enter into the resulting equation from which k can be obtained:

$$\Omega = \frac{1}{2}k\sqrt{\pi}[1 - \text{erf}(k/2)]\exp(k^2/4) \tag{19.46}$$

where Ω is a measure of the supersaturation and is defined as:

$$\Omega = \frac{x^{\alpha/\beta} - x^{\alpha\infty}}{x^{\alpha/\beta} - x^\beta} \tag{19.47}$$

For the growth of a spherical particle with radius $r = k\sqrt{Dt}$ a similar analysis yields k as the solution of the equation:

$$\Omega = \frac{1}{2}k^3F(k)\exp(k^2/4) \tag{19.48}$$

where:

$$F(k) = \frac{1}{k}\exp(-k^2/4) - \frac{\sqrt{\pi}}{2}[1 - \mathrm{erf}(k/2)] \tag{19.49}$$

It is possible, by a similar method, to derive exact solutions for a number of different shapes (e.g., ellipsoids). However, it should be emphasized that the shape of the growing particles does not come out as a result of the calculation but is rather imposed as a constraint.

The problem of a shrinking particle is more difficult except in the planar case, where there is no difference if the interface moves to the right or to the left. For dissolution of a spherical particle no exact solution exists and approximate methods have to be used.

In those methods the concentration profile is not obtained by solving Fick's second law. One rather suggests ad hoc some profile that obeys the boundary conditions and then evaluates fluxes and solves the flux balance equation to obtain the growth rate. Several methods of defining the profile have been suggested. In one method one uses the profile that is a solution to the Laplace equation, which defines the steady-state diffusion field, rather than Fick's second law. In another method the profile obtained by solving Fick's second law but neglecting the interface motion is applied (i.e., the interface is approximated as stationary). For the dissolution of a spherical particle the steady-state field that obeys the boundary conditions is $x = x^{\alpha\infty} + R(x^{\alpha/\beta} - x^{\alpha\infty})/r$ for $r \geq R$. The radius of the particle is R and r is the radial distance from its center. The final result is:

$$R = \sqrt{R_0^2 - kDt} \tag{19.50}$$

where $k = -2\Omega$ (i.e., $k > 0$ for dissolution and $k < 0$ for growth). This approximation can thus handle both growth and dissolution.

When applying the second method the result is:

$$\frac{\mathrm{d}R}{\mathrm{d}t} = -\frac{k}{2}\left(\frac{D}{R} + \sqrt{\frac{D}{\pi t}}\right) \tag{19.51}$$

Also here $k = -2\Omega$. This equation cannot be integrated analytically but has to be solved by numerical methods. In Figure 19.12 the particle size has been calculated according to the steady-state method and the stationary interface method and compared with a finite-difference simulation of the dissolution.

19.3.5 Approximate Treatment of Plate-Like Growth

At sufficiently large supersaturations precipitates grow with a plate or needle-like shape. An approximate analysis of such growth will now be given. Experimentally it is found that such growth occurs with constant lengthening rate. One defines the Peclet number $P_c = (\rho v)/(2D)$, ρ being the radius of curvature at the tip. The solution of the diffusion problem gives a relation between the supersaturation Ω and the Peclet number. Thus the individual values of ρ and v are not obtained. How could we know the operating tip radius? As a first approximation one may assume that the edgewise growth of a plate would be only half that of a sphere having the same radius, that is:

$$P_c = \frac{1}{2}\Omega \tag{19.52}$$

From the definition of the Peclet number it is evident that this solution predicts that, for a given Ω, the growth rate increases the sharper the tip is. In particular it diverges as $\rho \to 0$. The sharper the tip is the more important the effect of interfacial energy is and actually the supersaturation will decrease due to the Gibbs–Thomson Equation (19.11) (i.e., Ω will decrease). Zener [14] suggested that this effect could be approximately taken into account by multiplying Ω with the factor $(1 - \rho_c/\rho)$. It should be noticed that

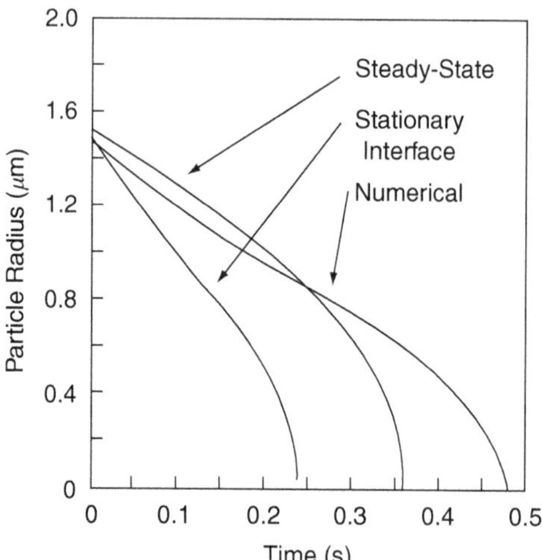

FIGURE 19.12 Particle size as function of time for dissolution of spherical particle according to various methods.

Ω is here constant and has the value corresponding to $\rho \to \infty$. It is then clear that v has a maximum at $\rho = 2\rho_c$ and Zener suggested that this would be the tip radius observed in nature (the maximum growth rate hypothesis). It is readily shown that the correction factor holds exactly provided that the effect of surface tension may be represented by the Gibbs–Thomson relation and we obtain for the case of the tip of a plate:

$$\rho_c = \frac{\sigma x^{\alpha/\beta}}{RT(x^{\alpha} - c^{\beta})(x^{\alpha/\beta} - x^{\alpha\infty})} \tag{19.53}$$

Equation(19.52) is a good approximation at low Ω but does not give a reasonable result as $\rho \to 1$ where v should diverge. The equation was thus modified by Hillert [15] to cover also high supersaturations. With the Zener correction factor the final result is:

$$P_c = \frac{1}{2}\frac{\Omega}{1-\Omega}\left(1 - \frac{\rho_c}{\rho}\right) \tag{19.54}$$

However, if the effect of interfacial energy is neglected the diffusion problem may be solved exactly by the Ivantsov [16] method. The Peclet number is obtained by solving the equation:

$$\Omega = \sqrt{\pi P_c}\exp(P_c)(1 - \operatorname{erf}\sqrt{P_c}) \tag{19.55}$$

The interfacial energy may be taken into account approximately by multiplying Ω with the Zener correction factor, that is, instead of Equation (19.55) we solve:

$$\Omega\left(1 - \frac{\rho_c}{\rho}\right) = \sqrt{\pi P_c}\exp(P_c)(1 - \operatorname{erf}\sqrt{P_c}) \tag{19.56}$$

If we use Equation (19.55) the maximum growth rate at $\rho = 2\rho_c$ is:

$$v_{\max} = \frac{D}{4\rho_c}\frac{\Omega}{1-\Omega} \tag{19.57}$$

This is thus the growth rate that one should expect experimentally.

19.3.6 Coarsening — LSW Theory

As a precipitation process proceeds the supersaturation will eventually start to decrease when the diffusion fields of nearby particles start to impinge. As the supersaturation, which we may represent by the driving force ΔG_m, decreases the critical size increases and further nucleation will be extremely unlikely and in reality impossible. Suppose that all the particles were nucleated at the same time and have exactly the same size. At some instant the supersaturation has decreased to such a level that the critical size is the same as the size of the particles and equilibrium is established. However, as has been mentioned, this is an unstable equilibrium and if some particles were just a little bit smaller they would start to dissolve and particles just a little bit bigger would grow. In practice we would thus have a situation where large particles grow and small particles dissolve, leading to fewer particles and a gradual increase in their average size. This process was first understood by Ostwald who studied solid precipitates in aqueous solutions. See Chapter 5 on Ostwald ripening. It is now realized that it occurs during all type of precipitation processes (i.e., also during solid-state reactions). For solid-state reactions at low temperatures there may be an extra difficulty due to the effect of stresses which will not be discussed further in this text. At high temperatures the stresses are expected to relax quickly enough to be less important.

Consider now β particles in an α matrix. At equilibrium the effect of the interfacial energy would cause a concentration gradient between two neighboring particles having different sizes ρ_1 and ρ_2. The local equilibrium concentration at the phase interface would be displaced in accordance with the Gibbs–Thomson equation. Let us denote with $\bar{\rho}$ the average size of the particles. Adopting the steady-state approximation for growth and dissolution of particles, see Section 19.3.4, we obtain the growth rate of a particle having radius ρ_1:

$$\frac{\mathrm{d}\rho_1}{\mathrm{d}t} = \frac{2DV_m^\beta x_\circ^{\alpha/\beta}\sigma}{RT(x^\beta - x^\alpha)^2}\frac{1}{\rho_1}\left(\frac{1}{\bar{\rho}} - \frac{1}{\rho_1}\right) \tag{19.58}$$

where $x_\circ^{\alpha/\beta}$ is the normal solubility of β in α. Particles larger than $\bar{\rho}$ will thus grow and smaller ones will shrink. The problem now is to derive how the distribution of particle sizes will develop in time and in particular how the average size $\bar{\rho}$ increases. The problem was solved rigorously by Lifshitz and Slyozov [17] and independently by Wagner [18] who introduced a size distribution function $f(\rho, t)$ and $f(\rho, t)\mathrm{d}\rho$ thus is the number of particles having a size in the range ρ and $\rho + \mathrm{d}\rho$. The total number of particles is:

$$N = \int_0^\infty f\,\mathrm{d}\rho \tag{19.59}$$

and their mean radius:

$$\bar{\rho} = \frac{1}{N}\int_0^\infty \rho f\,\mathrm{d}\rho \tag{19.60}$$

If we assume that no new particles form by nucleation the time evolution of the distribution must obey a continuity equation of the form:

$$\frac{\partial f}{\partial t} + \frac{\partial}{\partial \rho}\left(f\frac{\mathrm{d}\rho}{\mathrm{d}t}\right) = 0 \tag{19.61}$$

where $\mathrm{d}\rho/\mathrm{d}t$ is given by Equation (19.58). The total number of particles decreases in time because as $\rho \to 0$ the particles dissolve completely. We have:

$$\frac{\mathrm{d}N}{\mathrm{d}t} = \frac{\mathrm{d}}{\mathrm{d}t}\left(\int_0^\infty f\,\mathrm{d}\rho\right) = \lim_{\rho\to 0}\left(f\frac{\mathrm{d}\rho}{\mathrm{d}t}\right) \tag{19.62}$$

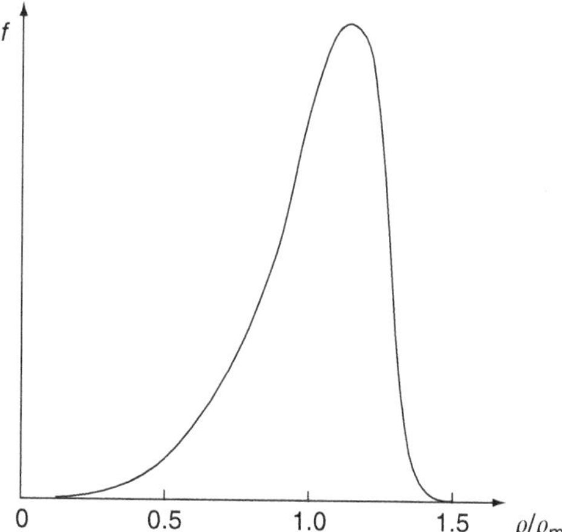

FIGURE 19.13 The distribution function obtained by Lifshitz, Slyozov, and Wagner.

It was then possible to show that at long times any distribution would approach a constant shape if plotted against $\rho/\bar{\rho}$. The asymptotic distribution is given by:

$$f(\rho, t) = N(t)\psi(u)/\bar{\rho} \tag{19.63}$$

where $u = \rho/\bar{\rho}$ and the number of particles per unit volume $N(t) \approx ((4DV_m^\beta x_\circ^{\alpha/\beta}\sigma/RT(x^\beta - x^\alpha)^2)t)^{-1}$ and the average size is:

$$\bar{\rho}^3 - \rho^3 = \frac{8}{9}\frac{DV_m^\beta x_\circ^{\alpha/\beta}\sigma}{RT(x^\beta - x^\alpha)^2}t \tag{19.64}$$

and

$$\psi(u) = \frac{4}{9}u^2\left(\frac{3}{3+u}\right)^{7/3}\left(\frac{3/2}{3/2-u}\right)^{11/3}\exp\left(\frac{-u}{3/2-u}\right) \tag{19.65}$$

The distribution function $f(\rho, t)$ is depicted in Figure 19.13. Later on the theory has been modified and refined in various aspects.

19.3.7 Multicomponent Treatment

In a multicomponent system with $n > 2$ elements there is an infinite number of possible tie lines that describe the local equilibrium at the phase interface. The question is how to determine which one is actually the operating one. Can it be determined from the $n - 1$ independent flux balance equations of type (19.43)? To answer this question it is instructive to consider Gibbs' phase rule that gives the number of potentials that must be fixed to have a unique equilibrium state. For given P and T and a two-phase equilibrium the phase rule states $f = n - 2$. In a binary system $n = 2$ and $f = 0$. The equilibrium state can be directly calculated (or read from a binary phase diagram) and the single flux balance equation may be used to determine the migration rate of the phase interface.

In a ternary system $n = 3$ and $f = 1$. To have a unique equilibrium one must thus fix one chemical potential. The value of this potential as well as the phase interface migration rate may be determined from the two flux balance equations.

One thus finds that in the general case the $n-2$ chemical potentials and the phase interface migration rate may all be determined from the $n-1$ flux balance equations. In the analytical solutions one should thus introduce one Ω_k and an equation, e.g., Equation (19.46), Equation (19.55), or Equation (19.58), for each independent component k.

19.4 Partitionless Reactions

19.4.1 Finite Interface Mobility

In a partitionless transformation there is no difference in composition between the new phase and the parent phase. Such transformations are often called diffusionless transformations because there is no need for long-range diffusion. It is obvious that local equilibrium cannot always prevail during a partitionless transformation. For example, consider the schematic binary phase diagram in Figure 19.14. A series of alloys 1 to 4 are first homogenized in the β one-phase field at the temperature T_1 and then quenched down to the temperature T_2 where α phase may form. According to the local equilibrium assumption the lines of the phase diagram give the compositions in α and β close to the interface and we obtain $x^{\alpha/\beta}$ and $x^{\beta/\alpha}$, respectively. As soon as these compositions are known the growth rate can be calculated by solving the diffusion problem as was discussed previously. It is then found that the growth rate increases as the alloy composition moves to the left and the rate tends to very high values as the $\alpha/(\alpha+\beta)$ phase boundary is approached and in particular at the phase boundary itself it becomes infinite. At this point the growing phase and the initial supersaturated phase have the same composition and consequently no diffusion is required for the transformation to proceed. If, for some reason, an alloy is quenched directly from the β phase into the α one-phase region, see alloy 4, the α phase may form with the same composition as the β phase. However, if the same method as previously is tried we will find that there is no solution of the diffusion problem. It is thus evident that this situation cannot be handled using the local equilibrium assumption.

For partitionless growth of α into β at given P and T the driving force, see Section 19.1.4, is simply the difference in Gibbs energy between the two phases having the same composition. For low ΔG_m values it is common to approximate the dependence between growth rate v for a partitionless reaction and driving force as linear and write:

$$v = M\frac{\Delta G_m}{V_m} \tag{19.66}$$

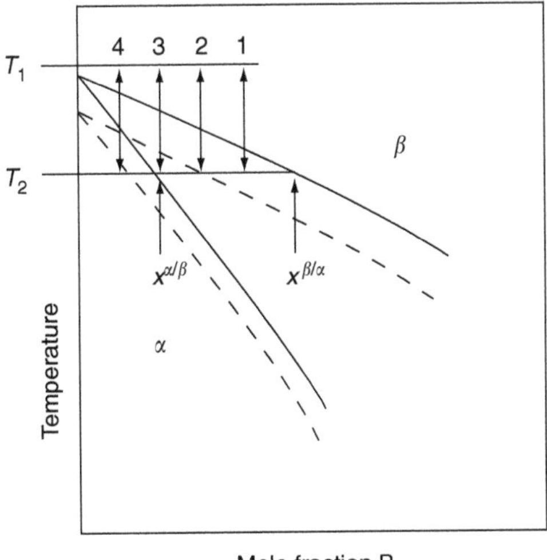

FIGURE 19.14 Schematic phase diagram A–B. Dashed lines show the effect of low interface mobility.

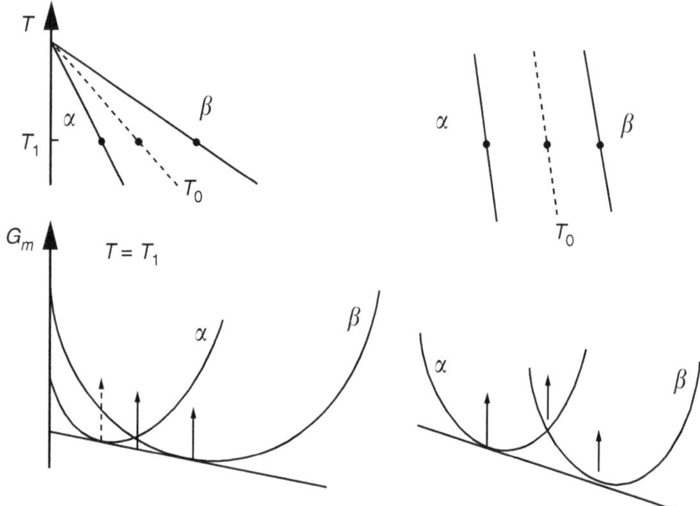

FIGURE 19.15 Details of phase diagrams with the corresponding molar Gibbs energy diagrams.

where M is the interface mobility. This relation was used in Section 19.3.4. However, in that case, a part of the total driving force was used to drive diffusion ahead of the interface and the part used to overcome interface friction was denoted ΔG_m^i.

Suppose an alloy is cooled from the homogeneous high-temperature state β. When the temperature has fallen inside the $\alpha + \beta$ two-phase field the α precipitation starts at favorable nucleation sites, for example, in grain corners, and will proceed with a rate controlled by diffusion. An important question is: How large undercooling is needed for the partitionless reaction to start? Thermodynamically this is possible as soon as Gibbs energy for the α phase is lower than for β at the same composition. There is a line in the two-phase regions of a binary phase diagram on which the two phases have the same value for the Gibbs energy if they are compared at the same composition (this is of course true only if the phases have variable compositions). It is usually called the T_0 line and sometimes the allotropic phase boundary if the two-phase region starts from an allotropic transformation in one of the components. Approximately it falls in the middle of the two-phase field. The two cases are demonstrated in Figure 19.15 showing details of phase diagrams together with the corresponding molar Gibbs energy diagrams at some temperature.

19.4.2 Massive Transformation

In the solid state a partitionless transformation can proceed in two different ways. If the transformation proceeds by the lateral motion of a coherent interface by means of ledges or dislocations it is called a martensitic transformation. Such transformations will be discussed in a separate section. If the interface moves by the random jumps of individual atoms the transformation is called massive.

According to one suggestion, the massive transformation cannot start until the alloy has been cooled down into the one-phase field of the new phase. Recent experimental information on $\gamma \rightarrow \alpha$ in Fe–Ni alloys [19] shows that the massive transformation starts close to the $\alpha/(\alpha + \gamma)$ phase boundary at high temperatures and well inside the two-phase field at low temperatures.

On the other hand, the martensitic transformation starts at a temperature M_S which is approximately given by:

$$M_S = T_0 - \Delta T \tag{19.67}$$

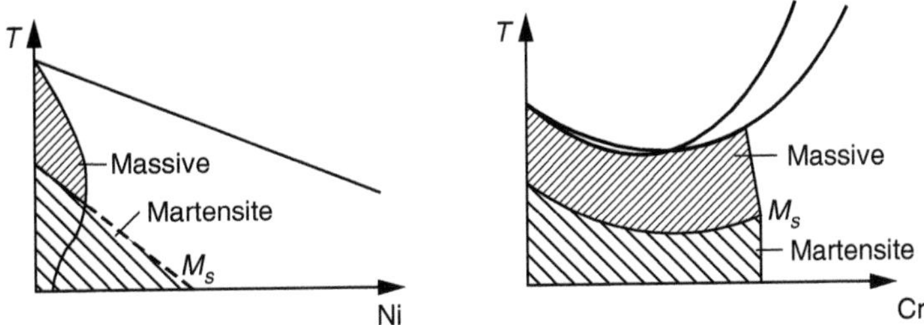

FIGURE 19.16 Regimes in phase diagram for massive and martensitic transformation.

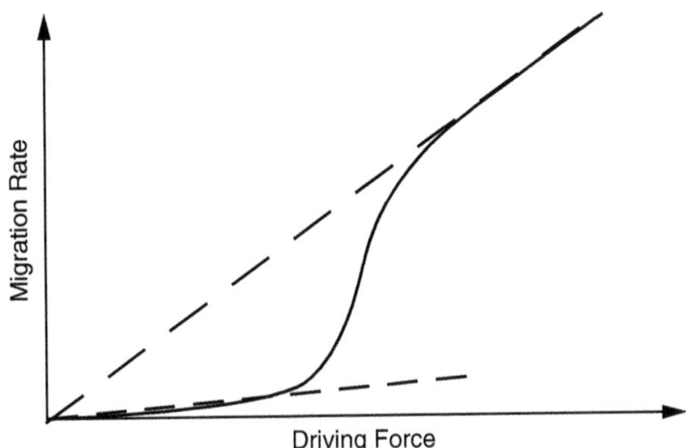

FIGURE 19.17 Schematic variation of grain boundary migration rate with driving force when there is a solute drag effect.

where ΔT is related to elastic stresses and mobility of interfacial dislocations. For alloys of iron with nickel and chromium, respectively, one would thus get the regimes depicted in Figure 19.16 for the two types of partitionless transformation.

19.4.3 Solute Drag in Grain Growth

To understand the transition to partitionless reaction it is instructive to consider grain-boundary migration. It is well-known that already small alloy additions, well below the solubility limit, may decrease the rate of recrystallization with several orders of magnitude. Lücke and coworkers [20] suggested that this strong effect is due to the interaction between the moving grain boundary and the solute atoms. In general, one may expect that solute atoms would segregate to a stationary grain boundary. If the grain boundary starts to move the atoms would be left behind but the grain boundary would exert an attractive force on them and cause them to diffuse toward the grain boundary.

The segregation would decrease with increasing migration rate v of a grain boundary and the solute drag would then vanish if v is large enough because diffusion cannot keep up with the grain boundary migration. This indicates that the drag effect would be most severe for intermediate driving forces. Above a critical driving force the rate would be very high and limited by the finite mobility of the grain boundary. The situation is depicted schematically in Figure 19.17. The driving force for recrystallization is directly related to the degree of plastic deformation.

Cahn [21] and Hillert and Sundman [22, 23] considered the whole concentration profile across the grain boundary. The solute drag under constant pressure and temperature is then simply the Gibbs energy dissipated by solute interdiffusion:

$$\Delta G_m^{sd} = -\frac{V_m}{v}\int J\frac{\partial(\mu_B - \mu_A)}{\partial z}\mathrm{d}z \qquad (19.68)$$

where J is the interdiffusive flux. Hillert and Sundman considered a binary system with the substitutional elements A and B. They wrote the interdiffusion of A and B as:

$$J = -J_A = J_B = -\frac{D}{RTV_m}(1 - x_B)x_B\frac{\partial(\mu_B - \mu_A)}{\partial z} \tag{19.69}$$

As already mentioned the quantity $(\mu_B - \mu_A)$ is called the interdiffusion potential. It is a function of the alloy content x_B. For example, if the phase under consideration is a regular solution with a molar Gibbs energy given by Equation (19.5). Then the interdiffusion potential is given by:

$$\mu_B - \mu_A = ^\circ G_B - ^\circ G_A + RT\ln\left(\frac{x_B}{1 - x_B}\right) + (1 - 2x_B)L_{AB} \tag{19.70}$$

and its gradient becomes:

$$\frac{\partial(\mu_B - \mu_A)}{\partial z} = \frac{RT}{(1 - x_B)x_B}\left[1 - 2(1 - x_B)x_B\frac{L_{AB}}{RT}\right]\frac{\partial x_B}{\partial z} + \frac{\partial(^\circ G_B - ^\circ G_A)}{\partial z} + (1 - 2x_B)\frac{\partial L_{AB}}{\partial z} \tag{19.71}$$

The first term, present also in a system without a grain boundary, comes from the concentration gradient whereas the remaining terms come from the variation in the thermodynamic properties across the grain boundary.

Assuming that the concentration profile is stationary in a frame of reference fixed to the grain boundary moving with the constant velocity v we have the following diffusion equation:

$$\frac{v}{V_m}\frac{\partial x_B}{\partial z} = \frac{\partial J_B}{\partial z} \tag{19.72}$$

The concentration profile for a certain v may now be calculated from the diffusion equation by inserting the expressions for the flux and the gradient in diffusion potential provided that the variation in thermodynamic properties and diffusivity across the grain boundary are given. The solute drag ΔG_m^{sd} may subsequently be calculated by integration of the concentration profile and compared with the driving force available. If they do not agree we must try another v, and so on, until the two values agree. In this way it is possible to calculate v as a function of driving force. In general the grain boundary will have a finite interface mobility M and the unique solution is obtained by comparing the available driving force with the total driving force dissipated by both solute drag and the interface motion that is,

$$\Delta G_m^{sd} + V_m\frac{v}{M} \tag{19.73}$$

It must be emphasized that ΔG_m^{sd} may be given as an explicit function of v only for some special cases. In general, one has to solve the diffusion equation numerically and perform the integration of the concentration profile numerically.

19.4.4 Solute Trapping

We shall now return to the transformation of a supercooled phase β into a more stable phase α, see Figure 19.14. If the supersaturation is low the rate v will be low and local equilibrium will more or less prevail at the phase interface. As the supersaturation increases a larger portion of the driving force is dissipated by motion of the phase interface and there is a deviation from equilibrium at the phase interface. However, the effect of a finite interface mobility will only suppress the stability of the α phase and could never explain a transition to a partitionless transformation.

During rapid solidification it has been observed that there is a transition to partitionless solidification somewhere in the two-phase field between the T_0 and solidus lines. This effect is of practical importance because it makes it possible to extend the solubility by rapid solidification processing.

The observation of a partitionless solidification has caused confusion and attracted some attention. The reason for the confusion is the following. When a pure substance A solidifies the A atoms exhibit a drop in their chemical potential as they cross the interface between liquid and solid. A naive guess would now be that if a solute B is added the B atoms would exhibit a similar drop in their chemical potential by jumping from liquid to solid. Several theories have been developed from this idea, for (e.g., the theory by Borisov [24]). However, as is easily concluded from the experimental information and some basic solution thermodynamics, this is not the case in reality and the naive idea must thus be incorrect. During the partitionless solidification the solvent atoms decrease their chemical potential by jumping from liquid to solid but the solutes actually increase their chemical potential. One may thus regard the solute atoms as being reluctantly trapped into the growing solid although they would have preferred to stay in the liquid. It is thus obvious that all theories that presume that the flux of atoms across the phase interface is proportional to the chemical potential difference will always fail to predict a partitionless transformation. In the next section, we will show that the transition to partitionless transformation may be understood by taking into account the solute-drag effect on the moving phase interface.

19.4.5 Transition to Partitionless Transformation

If two phases have different compositions and the moving interface between them has a finite thickness then there must be a concentration gradient inside the interface. The situation thus is similar to the one with a moving grain boundary, the main difference being that the concentration on the two sides of the interface will in general not be the same during a phase transformation, see Figure 19.18.

As the interface moves the gradient must move with the interface if steady-state conditions are to be maintained. The motion of the gradient requires diffusion and consequently there must be a driving force for this diffusion. This driving force must be taken from the total driving force available and consequently there will be less driving force left for the diffusion in front of the growing phase and for the motion of the interface itself. This means that diffusional equilibrium can no longer prevail at the interface.

The effect of the solute drag is shown schematically in Figure 19.19. The B component diffuses from the α side of the interface to the β side and further in front of it. It is thus necessary to have a monotonically decreasing diffusion potential for B in this direction. The chemical potential of B must have its highest value $\mu_B^{\alpha/\beta}$ at the α side of the interface and a somewhat lower value $\mu_B^{\beta/\alpha}$ at the β side and the lowest value $\mu_B^{\beta 0}$ far ahead of the interface. As can be seen in Figure 19.19 this results in the composition of α being displaced into the two-phase field.

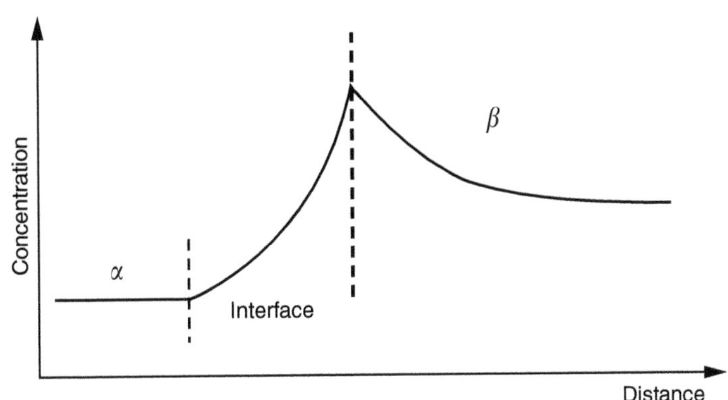

FIGURE 19.18 Schematic concentration profile across a phase interface.

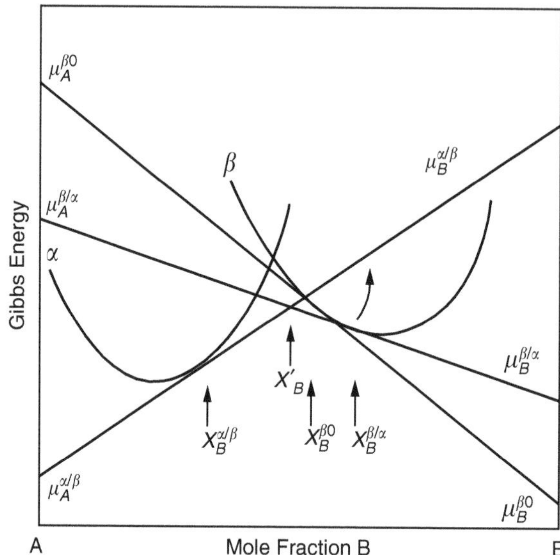

FIGURE 19.19 The Gibbs energy diagram for alloy system *A*–*B*. The effect of a driving force for diffusion acting over an interface between α and β. The tangents describing the chemical potentials on both sides of the interface are rotated relative to each other.

The ambitious treatment by Hillert and Sundman [22, 23], based on the numerical solution of a diffusion equation inside the phase interface, showed that the growing phase could get a composition inside its one-phase field as well as inside the two-phase field and the experimental observations could thus be satisfactorily explained.

A small composition change in a one-phase system corresponds to a rotation of the tangent to the Gibbs energy curve. In Figure 19.19 the tangents have been plotted for the initial composition of the β phase $x_B^{\beta 0}$, for the composition of β close to the phase interface $x_B^{\beta/\alpha}$, and for the growing α phase $x_B^{\alpha/\beta}$. As can be seen, the diffusion in front of the phase interface corresponds to a rotation of the tangents around a point having a composition between $x_B^{\beta 0}$ and $x_B^{\beta/\alpha}$. The diffusion across the phase interface corresponds to a rotation around a point having its composition somewhere between $x_B^{\beta/\alpha}$ and $x_B^{\alpha/\beta}$ and denoted as x'_B.

A complete treatment of the problem should yield the compositions on both sides of the interface as functions of the phase interface migration rate. Moreover, the solution must degenerate to the equilibrium solution as the rate vanishes.

We may take the effect of the extra energy of a curved interface and finite interface mobility into account by:

$$\Delta G_m = \Delta G_m^{sd} + \sigma V_m^{\alpha}\left(\frac{1}{\rho_1} + \frac{1}{\rho_2}\right) + \Delta G_m^i \tag{19.74}$$

where ρ_1 and ρ_2 are the principal radii of curvature for a curved interface and ΔG_m^i is the driving force needed to displace the phase interface, for example, $(v/M)V_m$ if the simple linear expression is applied.

Let us consider the growth of a plate-like particle and take the effect of solute drag, interfacial energy, and finite interface mobility into account. We may proceed as in Section 19.3.4 but the local equilibrium conditions have to be modified. The results of such a calculation are summarized in Figure 19.20. The sharpness of the edge, defined as $1/\rho$, is plotted on the horizontal axis and the growth rate on the vertical axis using a logarithmic scale. Each curve is calculated using a specific overall composition.

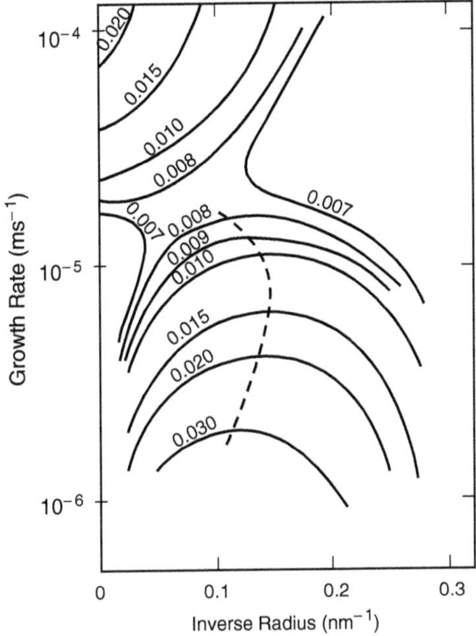

FIGURE 19.20 Calculated edgewise growth rate of α as a function of sharpness $1/\rho$ of growing edge. Each curve holds for the initial composition of β indicated. Dashed curve shows the locus of maximum growth rate.

For low supersaturation (i.e., concentration larger than 0.025), there is a family of solutions shown in the lower part of the diagram, corresponding to low growth rates. These curves yield the growth rate as a unique function of the edge radius and each curve has a maximum for some particular radius of curvature. The growth is basically controlled by diffusion and the deviation from local equilibrium is small. The occurrence of a maximum is caused by the effect of interfacial energy. As the growing edge becomes sharper the concentration gradient controlling the diffusion ahead of the edge becomes steeper and the growth rate increases. The energy of the curved interface has the opposite effect because it destabilizes the growing α phase, and causes the compositions at the phase interface to be displaced toward lower values. The dashed curve has been drawn through the locus of the maxima.

As the supersaturation is increased and the concentration becomes smaller than 0.025 a second family of solutions appear in the upper part of the diagram corresponding to high growth rates. These solutions correspond to a large deviation from local equilibrium and require a lower amount of diffusion. It seems natural to associate the occurrence of the new family of solutions with the crossing of the T_0 line because it is situated at 0.025 in this particular case.

The solutions in the lower part of the diagram did not show any drastic change when the composition 0.025 was crossed. Actually, it was possible to obtain curves of the same shape inside the allotropic phase boundary. These curves represent mainly diffusion-controlled growth although there is an increasing deviation from local equilibrium as the supersaturation is increased.

At the composition 0.0075 a critical point is reached and there is a drastic change in the behavior of the solutions. The maximum disappears and instead the solutions diverge into two branches. The part that can be seen in the left hand side of the diagram, see the curve for $x_C^{\gamma 0} = 0.007$, bows toward the left and the maximum growth rate is achieved for a flat interface. A close examination reveals that the growth occurs close to local equilibrium conditions and that the growth rate is limited by the rate of diffusion inside the interface (i.e., the solute drag).

In the other branch the growth rate increases to very high values. A closer examination reveals that the concentration gradient inside the interface is gradually diminished as the growth rate is increased. The solute-drag effect will thus disappear and there is a gradual transition to the partitionless mode of transformation.

From various assumptions Aziz [25] has derived the following relation between the partition coefficient $k_B^{S/L}$ and the growth rate v:

$$k_B^{S/L} = \frac{x_B^{S/L}}{x_B^{L/S}} = \frac{{}^0k_B^{S/L} + v/v_D}{1 + v/v_D} \tag{19.75}$$

where ${}^0k_B^{S/L}$ is the equilibrium partition coefficient and v_D is a parameter that Aziz called "maximum diffusive speed."

The main part of the discussion has been concerned with binary systems but the principles may be straightforwardly extended to multicomponent alloys.

19.4.6 Martensite in Steel

The hard and brittle structure that forms during hardening of steels (i.e., as a result of quenching a steel from a sufficiently high temperature) is called martensite. The martensitic structure is a result of a particular class of transformations that is now denoted as martensitic transformations and has been observed in a variety of materials including not only metals but also ceramics and protein crystals as well. Classically, the martensitic transformation exhibits a number of characteristic features. The most important ones are:

- No solute partitioning
- Very high growth rates
- Special crystallographic relations between matrix and particles
- Very complex microstructures are formed

Like the massive transformation that was discussed in the previous section, the martensitic transformation does not include any redistribution of solute atoms. The transformation may thus proceed without the need of long-range diffusion and the rate will be controlled by the mobility of the migrating phase interface. Thus the massive transformation can be very rapid, even though not as rapid as the martensitic one. Whereas the massive transformation occurs by the uncorrelated jumps of individual atoms across an essentially incoherent phase interface and thereby causing it to migrate, the martensitic transformation involves the motion of a coherent or semicoherent interface. The coherency leads to special orientational relationships between matrix and particle that may lead to big shape changes. To minimize the strain caused by such shape changes complex structures may form. For example, in steels with low carbon content one observes so-called lath martensite, see Figure 19.21 and Figure 19.22, which consists of packets of fine plates having very high internal dislocation density (10^{16} m^{-2}). The individual plates in a packet are separated by high- or low-angle grain boundaries or twin boundaries.

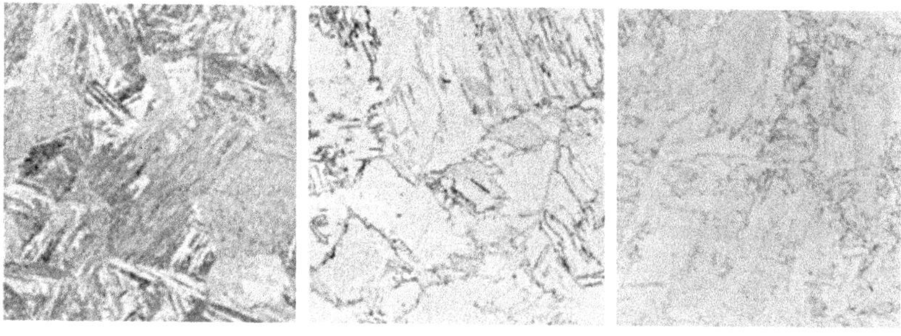

FIGURE 19.21 Lath martensite in Fe-0.2%C, Fe-1.94%Mo and Fe-0.49%V (from Krauss, G. and Marder, A. R. *Metall. Trans.*, 2, 1971, 2342–2357).

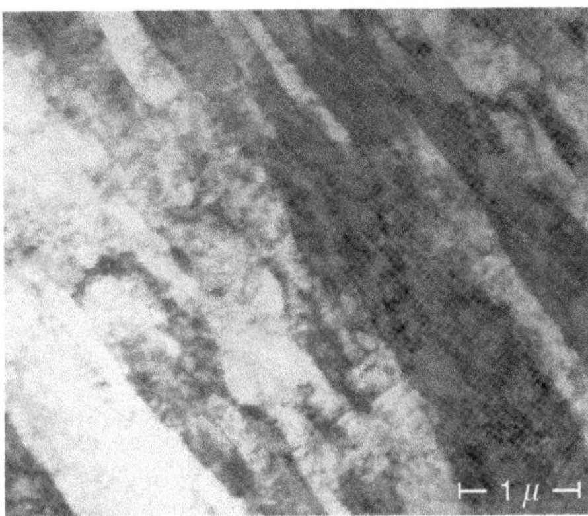

FIGURE 19.22 Dislocation structure in lath martensite in Fe-0.49%V (from Krauss, G. and Marder, A. R. *Metall. Trans.*, 2, 1971, 2342–2357).

FIGURE 19.23 Lenticular martensite in Fe-1.86%C (from Krauss, G. and Marder, A. R. *Metall. Trans.*, 2, 1971, 2342–2357).

At higher carbon contents one observes instead the plate martensite (sometimes called lenticular or acicular martensite), see Figure 19.23 and Figure 19.24. In this case individual plates form and grow very rapidly until they reach an obstacle (e.g., a grain boundary or another martensitic plate). When precipitates form as plate-like particles they are usually oriented parallel to well-defined crystallographic planes of the parent phase. These planes are called habit planes and denoted by the corresponding crystallographic indices of the parent phase. The habit plane also represents the orientation of the coherent or semicoherent phase interface. On closer investigation of the plate martensite it is found that each plate consists of a series of very fine parallel twins.

It seems likely that the occurrence of a high density of dislocations in the lath martensite in contrast to the predominance of twins in the plate martensite is connected with the shape changes during the martensitic transformation and the methods nature uses to accommodate those shape changes. It is well-known that deformation by dislocation glide becomes more difficult at lower temperatures and in BCC iron one may then observe deformation twins, sometimes called Neuman bands. The transition to plate martensite would then stem from the lower M_S temperature at higher carbon contents. All austenite stabilizing elements that lower M_S, would thus favor formation of plate martensite.

In low alloy steels, the martensitic transformation is connected with the formation of body-centered cubic (BCC) (or rather body-centered tetragonal [BCT]) from the high temperature FCC phase (called Austenite). In most other systems with allotropic transformations, BCC rather than FCC is the high temperature phase. Also in iron BCC is stable at high temperatures and FCC at lower temperatures but BCC becomes stable again at still lower temperatures due to the magnetic ordering. In other systems

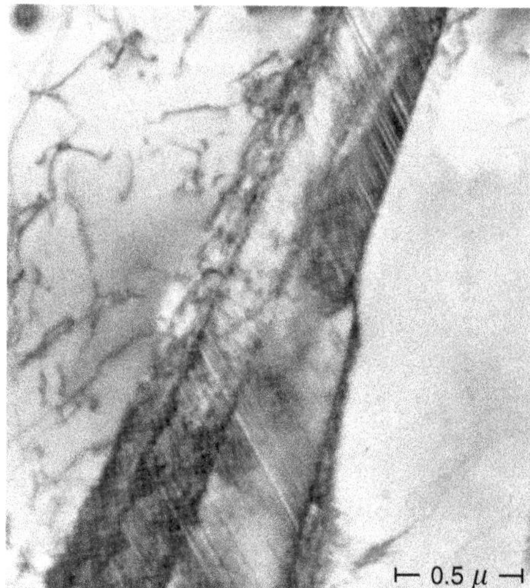

FIGURE 19.24 Twin structure in lenticular martensite in Fe-32%Ni (from Krauss, G. and Marder, A. R. *Metall. Trans.*, 2, 1971, 2342–2357).

the martensitic transformation yields a change from BCC to a typical low temperature phase like HCP or FCC.

From a practical point of view the martensitic transformation is important not only due to its hardening effect in steel but due to its shape change. This shape change is the basis for the behavior of shape memory alloys and transformation toughening of ceramics.

As we shall see later, the character of the edgewise growth of a martensite unit results in restrictions for the sidewise growth due to the shape changes. When the edge has stopped, for instance at the grain boundaries in the γ-Fe, the sidewise growth can only continue until the sides have developed some curvature. The continued reaction will then depend upon the formation of new nuclei. In most cases, the martensitic transformation will come to a stop before completion and will continue only if new nucleation is stimulated by a further decrease of the temperature. The amount of martensite is thus a function of temperature rather than time. However, there are some cases where a slow continued nucleation at constant temperatures has been observed. Such "isothermal" martensite can go to completion without further lowering of the temperature. The growth of each unit during a martensitic transformation occurs with a high rate. It is thus impossible to inhibit the martensitic transformation by rapid quenching and thus to preserve the high-temperature state to low temperatures. In general, the motion of a coherent interface occurs by the movement of ledges. It is evident that the ledges in the rapid martensitic growth must be composed of glissile dislocations.

The complicated mechanism of the ordinary martensitic transformation and, in particular, the large stresses around the edge, leads to the requirement of an extra driving force. A certain undercooling is thus required before the martensitic transformation can start. If the equilibrium temperature between the parent phase and the low-temperature phase is so high that the atoms are very mobile, then the transformation may take place in a massive way before the material has cooled to a sufficient undercooling to allow the martensitic transformation to start. The cooling rate will then be the decisive factor that determines how the transformation takes place. On the other hand, the massive transformation can never occur if already the equilibrium temperature is low.

Thermodynamically a partitionless transformation $\gamma \rightarrow \alpha$ can occur if the initial γ phase is cooled below the so-called T_0 temperature. This temperature is defined as the temperature where the Gibbs energy of the two phases are the same for the same alloy content, see Section 19.4.2. If the reaction requires some driving force further undercooling is needed. The starting point of the martensitic transformation is often called M_S and may approximately be regarded as a parallel displacement of the T_0 line. In iron base alloys the parallel displacement for the $\gamma \rightarrow \alpha$ transformation corresponds to approximately 1000 J/mol.

As the alloy content usually has an effect on the lattice parameter, it may also have an effect on the possibilities of forming an interface where the martensite and the parent phase can fit together. This may be why the habit plane of martensite changes with the steel composition. As mentioned, in the Fe–C system there is a change from the type that occurs by plastic deformation of the martensite and to the twinning type as the carbon content is increased. The first type (i.e., the lath martensite) is thus found at low carbon contents and it has the habit planes (111). The second type (i.e., the plate martensite) starts to predominate at 0.6%C. It usually has the habit plane (225) but changes to (259) at very high carbon contents.

At very low temperature, pure FCC Fe would transform to HCP if the transformation to BCC could be prevented. There is thus a metastable equilibrium temperature between FCC and HCP Fe that is believed to be at about 400°C. The Gibbs energy conditions can be changed by alloying additions in such a way that the BCC phase is suppressed and it is thus possible to observe the martensitic transformation from FCC to HCP. That transformation is identical to the corresponding transformation in Co from FCC to HCP. This type of martensite is called ε martensite to distinguish it from the ordinary α martensite. The HCP phase itself is often denoted by ε. The ε martensite can form as thin, straight plates in high-manganese steels and in austenitic stainless steels. It can form at a lower undercooling than α martensite due to the good matching of the close-packed planes that permits a transformation mechanism with a very slight shape change. It is thus possible to form ε martensite even in cases where the α phase is actually the stable phase. In fact the formation of α martensite is sometimes initiated by the formation of ε martensite.

19.4.7 Crystallographic Aspects of the Martensitic Transformation — The Invariant Plane Strain

The martensitic transformation proceeds by the motion of coherent or semi-coherent phase interfaces. A characteristic feature of such a coherent interface is that the atoms belonging to the plane of the interface have the right position with respect to both the lattice of the martensite unit and the lattice of the parent phase. If we regard the martensitic transformation as a deformation of the lattice of the parent phase, the deformation must occur in such a way that the phase interface is a plane of no deformation. Such a deformation is regarded as an invariant plane strain (IPS). The simplest example of an IPS deformation is twinning, the invariant plane being the twin boundary. During twinning the deformation yields a new crystal of the same phase but of a new orientation. The atoms are displaced in such a way that the lattices on each side of the boundary become mirror images of each other.

A characteristic feature of an IPS is that the displacement of the atoms increases linearly with the distance from the invariant plane. If the solid line in Figure 19.25 shows the intersection with the external surface before the IPS, it is evident that the strain will cause a relief at the surface. In the case of twinning the displacement is pure shear, that is, the atoms are displaced parallel to the invariant plane with the

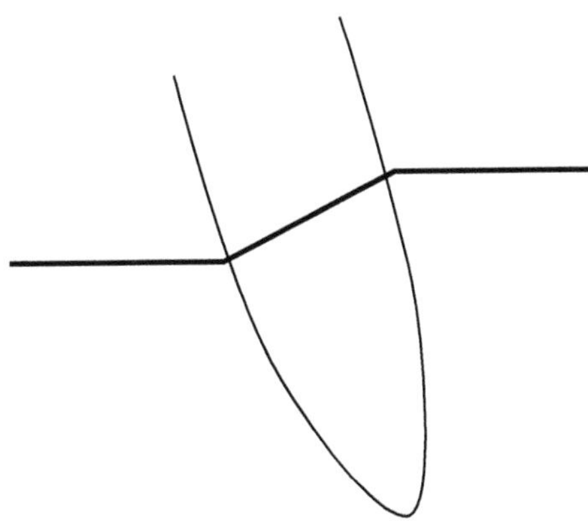

FIGURE 19.25 Invariant plane strain. The solid line shows the direction of the same lattice plane.

displacement proportional to the distance from the plane. As discussed below, the FCC→HCP martensitic transformation may be achieved by pure shear but in general a change in crystal structure is represented by an IPS that also contains a component normal to the plane. In general the martensitic transformation will thus be composed of both a shear and a normal strain component.

19.4.8 Crystallographic Aspects of the Martensitic Transformation — The Bain Distortion

There has been much speculation regarding the detailed mechanism for the change from the atomic arrangement of the parent phase to that of the martensitic phase and the hope has been to be able to predict the orientation relationship. To demonstrate how complex this problem may be, it is convenient to consider the classical case of the martensitic transformation that is the transformation from γ-Fe to α-Fe. At first glance this may appear to be a very simple case because the two structures are closely related. As demonstrated by Figure 19.26, it is possible to choose a rectangular cell in the FCC structure which has a BCC atomic arrangement, the so-called Bain correspondence. One axis is longer than the other two and the FCC structure may thus be described as a BCT structure just as well as an FCC structure. It could be transformed directly into the BCC structure of α-Fe by compression in one direction and expansion in the other two. This mechanism for the martensitic transformation is called the Bain distortion. If it would really operate it would result in an orientation relationship where one of the cube axes in α-Fe is parallel with one in γ-Fe and the others fall at 45° between those of the γ-Fe. This is not what has been observed. The reason why the Bain distortion does not operate in practice may be that it gives such a large shape change that it would yield very strong stresses in the material. Otherwise, the two crystals would not fit together as demonstrated by the very simple sketch shown in Figure 19.27 where the martensitic α-Fe crystal has been drawn with solid lines and the hole in the parent γ-Fe crystal with dashed lines.

The situation would improve considerably if one instead considers the transformation of a thin plate-like volume element at 45° from the direction of compression. The Bain distortion would deform such an element by shear. It would thus be able to fit into the hole along the long sides, see Figure 19.28. The martensitic transformation makes use of this advantage and it may thus properly be described as a shear

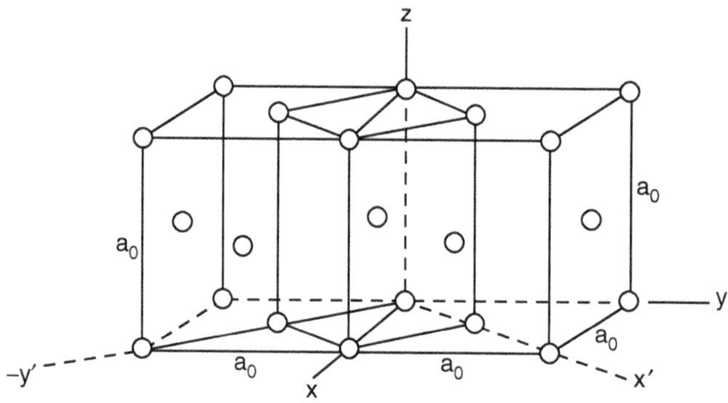

FIGURE 19.26 Bain correspondence between FCC and BCC.

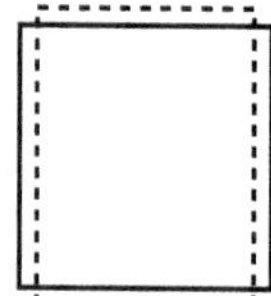

FIGURE 19.27 Shape change due to Bain distortion.

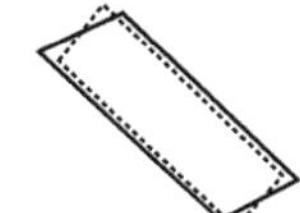

FIGURE 19.28 Bain distortion by shear.

FIGURE 19.29 Lenticular morphology of martensite.

transformation. However, two problems remain, one of which is evident from Figure 19.28. The short sides have been rotated relative to the long sides and can no longer fit into the hole unless there is a strong additional deformation. This difficulty is further exaggerated by the fact that the martensitic transformation from γ-Fe to α-Fe results in a net change of volume. If one would like to keep the original length along the long sides, one should have to accept that the whole volume change results in an expansion at right angle to the long sides. The requirement for an additional deformation will be less if the volume element is not quite plate like but rather shaped like a lens with a sharp edge. This is probably why the lenticular shape is natural for martensitic transformations, see Figure 19.29. On the other hand, the martensite crystal may widen as soon as it reaches a free surface. It should be noticed that widening does not occur at a grain boundary due to the resistance from the adjoining grain.

The second problem depends on the fact that there may also be a dimensional change at right angle to the picture we have examined so far. In reality, there is no plane that does not change in at least some direction. As a consequence, there can be no plane on which the martensitic and the parent crystal can fit together without an additional deformation. One has found that nature has at least two methods to solve this problem. One method makes use of a plastic deformation of the martensite crystal by gliding along closely spaced, parallel planes, see Figure 19.22. According to the other method, the transformation volume is divided into thin, parallel regions and they are subjected to Bain distortions in alternating directions. The resulting unit of martensite will thus consist of a series of parallel twins, see Figure 19.24. Both these methods may result in a plane that does not change its length in any direction. That plane will become the interface between the parent crystal and the martensitic unit. This plane thus is the habit plane of the martensite and it is common to give its crystallographic indices with reference to the parent crystal.

From a crystallographic point of view the martensitic transformation may thus be regarded as composed of three strain tensors: the Bain strain **B**, the rotation **R** to obtain the invariant plane, and the internal deformation (lattice invariant shear) **S** to accommodate the shape changes. If we write these deformations as a product of deformation tensors we have:

$$\mathbf{E} = \mathbf{BRS} \tag{19.76}$$

For a deformation to be IPS special requirements must be fulfilled. In terms of the principal strains of the deformation ε_1, ε_2 and ε_3 it is required that $\varepsilon_1 < 0$, $\varepsilon_2 = 0$ and $\varepsilon_3 > 0$. It is easy to check that the Bain strain itself does not obey these relations because it is characterized by $\varepsilon_1 = 0.018$ and $\varepsilon_2 = \varepsilon_3 = 0.012$. However, the combined deformation **E** fulfils the requirements.

19.4.9 Migration of the Martensitic Phase Interface

The migration of a coherent interface poses special difficulties. If it occurred by uncorrelated displacements of individual atoms the coherency would soon be lost. On the other hand, it is extremely unlikely that a whole lattice plane would move by the shear that is required, for example, for the growth of a twin. This difficulty is in fact very similar to the problem of understanding the plastic deformation in metals. On the atomic scale plastic deformation occurs by the gliding of atomic planes. But rather than sliding the entire planes, which would require too high stress, the deformation is achieved by migration of a dislocation on the plane. The dislocation is the boundary between the part of the plane that has glided and the part that has not yet glided. In fact the solution is the same in the motion of a coherent interface. This is illustrated by considering the motion of a twin boundary in an FCC lattice. The FCC structure has the stacking ABCABC… where A, B and C denote the densely packed plane (i.e., (111) in FCC). Using the same notation the stacking of HCP may be written ABAB… If we have a twin boundary in FCC the stacking thus is as shown in Figure 19.30. The arrow denotes the position of the twin boundary and the plane denoted by A is an invariant plane and its atoms belong to both twins. We may also say that the habit plane of the twin is (111). This stacking can be compared with the stacking fault in FCC shown in Figure 19.31. The arrow shows the position of the stacking fault that is regarded as being between the A and C plane. On the right side of the stacking fault there is a C plane instead of a B plane. By introducing a second stacking fault we obtain the situation depicted in Figure 19.32.

A stacking fault between each dense-packed plane would yield Figure 19.33. Stacking faults, between all the dense-packed planes inside an FCC crystal, would thus yield a twin crystal.

A stacking fault in an FCC lattice is bounded by a loop of a Shockley partial dislocation with Burgers vector of the type:

$$\mathbf{b} = \frac{a}{6}\left[11\bar{2}\right] \tag{19.77}$$

FIGURE 19.30 Twin boundary in FCC depicted by stacking sequence.

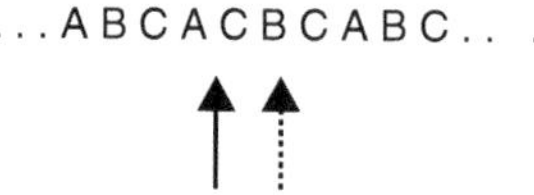

FIGURE 19.31 Stacking fault in FCC depicted by stacking sequence.

FIGURE 19.32 Double stacking fault in FCC depicted by stacking sequence.

. . . A B C A C B A C B A . . .

FIGURE 19.33 Series of stacking faults in FCC depicted by stacking sequence.

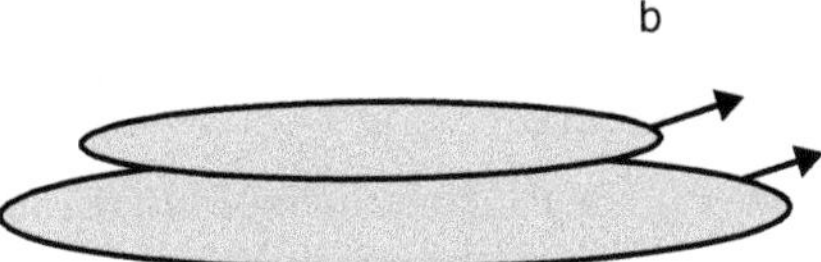

FIGURE 19.34 Loops of partial dislocations bounding a twin crystal. The gray areas denote the stacking faults.

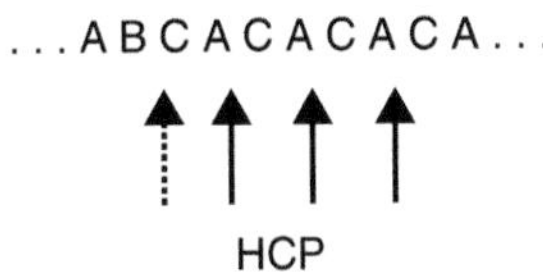

FIGURE 19.35 Formation of HCP from FCC by introducing stacking faults.

where a is the lattice parameter of FCC. The stacking fault grows or shrinks by the motion of the partial dislocation. Evidently, we may thus consider a twin as a pile of stacking faults between each (111) plane, and the twin boundary on the edge as a series of dislocation loops that bound each stacking fault, see Figure 19.34. The twin may grow by the expansion of the individual dislocation loops. Growth in the perpendicular direction requires nucleation of new dislocation loops and is more difficult.

We can repeat the same exercise but introduce a stacking fault between each second (111) plane. This would yield the situation shown in Figure 19.35. It is evident that this procedure would yield a plate-like HCP crystal. As in the case of the twin in FCC, the HCP crystal will be bounded by the Shockley partial dislocations. It is also evident that the interfaces on the sides are coherent and that the plane denoted by the dashed arrow is the invariant plane.

It is thus clear that one may have a transformation FCC→HCP that obeys all the characteristics of a martensitic transformation and is governed by the motion of Shockley partial dislocations. As already mentioned such a martensitic transformation does occur, for example in Fe–Mn alloys and in austenitic stainless steels, and the product phase is called ε martensite.

We can calculate the shear introduced by this IPS as:

$$\gamma = \frac{1}{2}|b|/h \tag{19.78}$$

where b is the magnitude of the Burgers vector of the partial dislocation and h the distance between the dense-packed planes in FCC. We thus find a shear of 0.353 corresponding to the angle 19.47°. This is a very large shear and the elastic energies would be very large when the shear is accommodated. In FCC there are four different (111) planes and three Schockley partials on each. There are thus 12 possible variants of the martensitic transformation and the strain energy may be minimized by combining the different variants. On the other hand, we may conclude that the FCC→HCP martensitic transformation can be an efficient deformation mechanism if a particular variant is chosen.

One may generalize the above reasoning to apply also for other transformations, for example, the classical martensitic transformation FCC→BCC by introducing more complex arrangements of dislocations at the interface. The partial dislocations that govern the transformation are often called *transformation dislocations* and *twinning dislocations* in the special case of twinning. However, they also have the important function of maintaining the coherency at the interface during the transformation. Consequently the name *coherency dislocations* has been suggested to emphasize that they have a different role than the dislocations that occur to relax coherency stresses, that is, to destroy the coherency, and should be called *anticoherency dislocations* although the term *misfit dislocations* is more common.

19.4.10 Thermodynamic Aspects of the Martensitic Transformation

19.4.10.1 The Gibbs Energy Difference

Consider a martensitic transformation $\gamma \rightarrow \beta$. Upon cooling from the austenitic state the transformation starts at the M_S temperature and upon heating the reverse transformation $\beta \rightarrow \gamma$ starts at the A_S temperature. Suppose for the sake of simplicity that the conditions are such that we do not have any influence of diffusional reactions. In steels this implies sufficiently rapid cooling and heating. Moreover, to avoid diffusional reactions upon heating the content of fast diffusing interstitials must be very low. Otherwise there will be tempering of the martensite during heating. As already mentioned, the T_0 temperature is defined as the temperature where the normal Gibbs energy functions of γ and β have the same value (if evaluated for the same composition). The Gibbs energy of a system with martensitic β will generally be higher than predicted from the Gibbs energy of the normal β. The reason is that there is an elastic contribution to the internal energy due to elastic accommodation of the transformation strain (i.e., the shape and volume change) as well as a contribution caused by the large number of defects. These defects, dislocations and interfaces, form because part of the transformation strain is plastically accommodated by dislocation glide in γ and twinning or dislocation glide in β. We may thus divide the change in Gibbs energy accompanying the reaction into a chemical part, which changes sign at the T_0 temperature, and an elastic and a plastic part. As a first approximation we shall assume that the chemical part of the change in Gibbs energy varies linearly with temperature, that is:

$$G_m^{\gamma} - G_m^{\beta} = b\,(T_0 - T) \tag{19.79}$$

where $b > 0$. Moreover, we shall assume that the elastic and defect contributions do not vary with temperature. For a system containing both austenite and martensite we may thus write its total molar Gibbs energy as:

$$G_m = fG_m^{\beta} + (1-f)\,G_m^{\gamma} + \phi\,(f) \tag{19.80}$$

where f is the fraction of martensite and $\phi\,(f)$ is a function representing the extra contributions of the elastic energy and the internal defects. The change in Gibbs energy per mole β martensite formed is:

$$\frac{\mathrm{d}G}{\mathrm{d}n^{\beta}} = \frac{\mathrm{d}G_m}{\mathrm{d}f} = G_m^{\beta} - G_m^{\gamma} + \frac{\mathrm{d}\phi}{\mathrm{d}f} \tag{19.81}$$

This is the driving force for the martensitic transformation when it, or a part of it, occurs at a given temperature and external pressure. We may write $\mathrm{d}\phi/\mathrm{d}f = U^{\mathrm{elastic}} + U^{\mathrm{defects}}$, where the U:s are the energies stored in the material after the martensitic transformation. In general they depend on f, and we thus have:

$$\frac{\mathrm{d}G}{\mathrm{d}n^{\beta}} = G_m^{\beta} - G_m^{\gamma} + U^{\mathrm{elastic}} + U^{\mathrm{defects}} \tag{19.82}$$

In addition, the martensitic reaction does require some driving force $D = \mathrm{d}(T\Delta S_i)/\mathrm{d}n^{\beta} > 0$ due to finite interface mobility and various other irreversible effects (e.g., plastic work when accommodating the transformation strain). For the sake of simplicity we shall assume that both the forward and reverse reaction require the same driving force D and for the forward reaction $\gamma \rightarrow \beta$ we thus add a term D but for the reverse reaction $\beta \rightarrow \gamma$ we subtract the same term.

19.4.10.2 The Heat Effect

At constant pressure P we have $\Delta Q = \Delta H = \Delta G + T\Delta S = \Delta G - T\,\mathrm{d}\Delta G/\mathrm{d}T$. From the previous equations we thus obtain (per mole of martensite formed):

$$\Delta Q = \Delta H = -bT_0 + U^{\text{elastic}} + U^{\text{defects}} \tag{19.83}$$

where we have approximated the elastic and plastic contributions as temperature independent. Since b is positive ΔQ must be negative for the special case when both U:s vanish (i.e., for $\gamma \to \beta$ heat is evolved and the reaction is exothermic). Normally both U:s are positive and there would be a decrease in the amount of heat evolved.

19.4.10.3 Relations between T_0, M_S and A_S

At M_S we have for the $\gamma \to \beta$ transformation:

$$\frac{\mathrm{d}G}{\mathrm{d}n^{\beta}} = -b(T_0 - M_S) + U^{\text{elastic}} + U^{\text{defects}} + D = 0 \tag{19.84}$$

and at A_S we have for the reverse transformation $\beta \to \gamma$, assuming that the same driving force D is needed:

$$\frac{\mathrm{d}G}{\mathrm{d}n^{\beta}} = -b(T_0 - A_S) + U^{\text{elastic}} + U^{\text{defects}} - D = 0 \tag{19.85}$$

Subtraction of the former equation from the latter and rearrangement yield:

$$\frac{M_S + A_S}{2} = T_0 - \frac{U^{\text{elastic}} + U^{\text{defects}}}{b} \tag{19.86}$$

whereas summation of the equations and rearrangement yield:

$$A_S - M_S = 2\frac{D}{b} \tag{19.87}$$

Often one evaluates the T_0 temperature from experimental information as the average of M_S and A_S. As can be seen one would expect T_0 to be higher unless the U:s are small enough. It is also evident that the hysteresis $A_S - M_S$ depends on the irreversible processes accompanying the transformation. We may rearrange the expression for the forward reaction to obtain:

$$M_S = T_0 - \frac{1}{b}(D + U^{\text{elastic}} + U^{\text{defects}}) \tag{19.88}$$

and the backward reaction to yield:

$$A_S = T_0 + \frac{1}{b}(D - U^{\text{elastic}} - U^{\text{defects}}) \tag{19.89}$$

Obviously M_S is suppressed by the irreversible processes as well as the internal elastic stresses whereas A_S is displaced to lower temperatures by the elastic energy (i.e., the reaction goes easier) and displaced to higher temperatures by D (the reaction becomes more difficult).

19.4.10.4 Thermoelastic Martensite

If the martensitic transformation has started upon cooling but is reversed upon already a small temperature increase the martensite is said to be thermoelastic. The hysteresis of thermoelastic martensite is thus very small, only a few K, whereas in nonthermoelastic martensite it may be more than 100 K. In the thermoelastic case both M_S and A_S may be much lower than T_0. This behavior is explained from the observation that the hysteresis is proportional to D (i.e., thermoelastic martensite has a low D). We thus find in the limit when D vanishes:

$$M_S = T_0 - \frac{1}{b}(U^{\text{elastic}} + U^{\text{defects}}) = A_S \tag{19.90}$$

It should be emphasized that if cooling continues to some temperature below M_S and some fraction of martensite has formed the reverse reaction would start almost as soon as the temperature is increased (i.e., strictly speaking A_S falls below M_S). This is the case when $D = 0$ and the two terms $(G_m^\beta - G_m^\gamma)$ and $(U^{\text{elastic}} + U^{\text{defects}})$ balance (i.e., there is a thermoelastic equilibrium between martensite and parent phase).

19.4.10.5 Stress-Induced Martensite

The martensitic transformation in ferrous alloys is accompanied by a large shape change. This implies that the martensitic transformation could serve as a deformation mechanism. Under an external load martensitic variants having a favorable orientation may form. In general martensite will form above M_S when an external load is applied, so called stress-induced martensite. It is even possible in some cases to form martensite in alloys where martensite would never form at any temperature if there was no external load.

The thermodynamic effect of an external load may be derived by a combination of the first and second laws:

$$\mathrm{d}U = T\,\mathrm{d}S - P\,\mathrm{d}V + \sigma'_{ij}\,\mathrm{d}\Omega'_{ij} - T\,\mathrm{d}S_i \tag{19.91}$$

where U and S are the internal energy and entropy, and T and P the temperature and pressure respectively. σ'_{ij} is the deviator stress and defined as $\sigma'_{ij} = \sigma_{ij} + P\delta_{ij}$, where δ_{ij} is the Kronecker delta (i.e., $\delta_{ij} = 1$ when $i = j$ and 0 otherwise). σ_{ij} is the normal stress tensor. The mechanical work has thus been divided into its *dilatoric* and *deviatoric* components. $\mathrm{d}\Omega'_{ij}$ is defined as $\mathrm{d}\Omega'_{ij} = \mathrm{d}\varepsilon'_{ij}V$ where ε'_{ij} is the deviator strain defined as $\varepsilon'_{ij} = \varepsilon_{ij} - \delta_{ij}\,\mathrm{d}V/V$ and ε_{ij} is the normal strain tensor. $T\,\mathrm{d}S_i$ is the entropy production and the second law requires $T\,\mathrm{d}S_i \geq 0$.

In the usual way we may change variables and introduce the Gibbs energy G:

$$\mathrm{d}(G - \sigma'_{ij}\Omega'_{ij}) = -S\,\mathrm{d}T + V\,\mathrm{d}P - \Omega'_{ij}\,\mathrm{d}\sigma'_{ij} - T\,\mathrm{d}S_i \tag{19.92}$$

Under constant T, P and σ'_{ij} we thus have:

$$\mathrm{d}(G - \sigma'_{ij}\Omega'_{ij}) = -T\,\mathrm{d}S_i \tag{19.93}$$

It must be emphasized that T, P and σ'_{ij} are external variables, that is, the variables that are fixed by the experimental conditions. Under external load we should thus replace the G with the function $G - \sigma'_{ij}\Omega'_{ij}$ in our previous equations. For the forward reaction we then have:

$$\frac{\mathrm{d}(G - \sigma'_{ij}\Omega'_{ij})}{\mathrm{d}n^\beta} = -b(T_0 - M_S) + U^{\text{elastic}} + U^{\text{defects}} - \sigma'_{ij}(\Omega'^{\beta}_{ij} - \Omega'^{\gamma}_{ij}) + D = 0 \tag{19.94}$$

For a single crystal of austenite we have:

$$\sigma'_{ij}(\Omega'^{\beta}_{ij} - \Omega'^{\gamma}_{ij}) = \tau\gamma V_m \tag{19.95}$$

where τ is the resolved shear stress for the martensite variant under consideration and γ is the shear due to the martensitic transformation. For a single crystal we may thus write:

$$b(T_0 - M_S) - U^{\text{elastic}} - U^{\text{defects}} + \tau\gamma V_m - D = 0 \tag{19.96}$$

and we obtain:

$$M_S = T_0 - \frac{1}{b}(D + U^{\text{elastic}} + U^{\text{defects}}) + \frac{\tau\gamma V_m}{b} \tag{19.97}$$

A shear stress would thus displace M_S to higher temperatures. However, the resolved shear stress cannot exceed τ_c the critical shear stress for dislocation glide and thus there is a maximum value of M_S, the so-called M_d, above which martensite cannot form by application of an external stress. However, if the material is deformed by dislocation glide there will be a deformation hardening and τ_c may increase allowing a higher load and a further increase in M_S.

19.4.10.6 The Superelastic Effect

If the conditions are such that M_S has increased by external loading to a temperature above the ordinary A_S then the reverse transformation would start when the sample is unloaded. The macroscopic deformation accomplished by the martensite formation will then be elastic. This effect is very pronounced and can yield rubber-like behavior in some metallic materials.

19.5 Applications — Examples

19.5.1 Homogenization

Conventional casting results in an inhomogeneous distribution of alloy content (i.e., segregation). As a consequence the properties will be different in different parts of the ingot. One distinguishes between long-range variations extending over distances comparable to the size of the ingot, macrosegregation and short-range variations extending over a single dendrite arm, microsegregation, typically in the range 10 to 100 μm. The distances involved in microsegregations are short enough to allow homogenization by diffusional processes whereas the macrosegregations extend over too long distances. Such a homogenization process may approximately be represented by Equation (19.19). Even though the concentration profile may have a very complicated and irregular shape at the start higher order terms decay rapidly and for practical purposes it may be sufficient to use only the first term and the wavelength λ would then be identified with dendrite arm spacing. The coefficient A is the average alloy content and in principle the amplitude B could be evaluated by a Fourier expansion of the initial concentration profile. However, in practice it is often sufficient to estimate B from much simpler methods. For example, if the concentration profile were given as a square wave, with the variation $c^{\text{max}} - c^{\text{min}}$, the detailed Fourier analysis would give $B = 2(c^{\text{max}} - c^{\text{min}})/\pi$. As an approximation, one could also use that value for cases where only c^{max} and c^{min} are known.

From Equation (19.19) one would then predict that an initial concentration difference Δc_k° will decay according to:

$$\Delta c_k \simeq 1.27\Delta c_k^\circ \exp\left(-\frac{4\pi^2}{\lambda^2}D_k t\right) \tag{19.98}$$

If the distance λ is reduced by plastic deformation, it has a very strong effect on the time needed for homogenization. A reduction with a factor 4 (e.g., by rolling), would reduce the time needed to achieve a certain degree of homogenization with a factor 16.

In steels with slow diffusing substitutional elements and rapid interstitials like carbon the latter homogenize quite quickly. However, this does not necessarily mean that carbon concentration difference levels out quickly. It rather means that carbon chemical-potential differences level out quickly. As many substitutional alloy elements have a strong effect on the carbon chemical potential carbon will be inhomogeneously distributed as long as the substitutional element are inhomogeneously distributed.

19.5.2 Austenitizing of Steel

Heat treatment of steel often involves austentizing in which the BCC structure (ferrite), stable at low temperatures, transforms to the FCC structure (austenite), stable at high temperatures. In low alloy steels austenitizing usually involves a dissolution of the cementite (M_3C) and leads to a one-phase structure of FCC. As both the driving force and the transformation kinetics is favored by a higher temperature the transformation becomes quicker the higher the temperature is, see Figure 19.5b.

If the temperature is not so high the transformation BCC$\rightarrow$FCC is controlled by diffusion but at sufficiently high temperatures the transformation becomes massive and very rapid. Cementite dissolution may be controlled by substitutional diffusion and would be a rather slow process or by carbon diffusion and would then be quite rapid.

For approximate calculations one may use equations of the type Equation (19.50) to estimate the time needed for a certain degree of dissolution.

In practice one would not choose too high an austenitizing temperature due to rapid growth of the austenite grain size.

19.5.3 Martensite Hardening and Tempering

Classical hardening of steels is based on martensite formation. As martensite forms without any change in composition the BCC contains much more carbon than normally which yields a strong solution hardening effect. Moreover, due to the very fine internal structure of martensite even martensite in low-carbon steels have much higher hardness than BCC formed by diffusional process or a massive transformation. The hardening operation consists of austenitization, rapid quench to form martensite, followed by tempering. During tempering the steel is heated usually to temperatures lower than 450°C. Even though tempering lowers the hardness it is usually beneficial for the performance because the steel becomes less brittle. Tempering involves several phenomena like relaxation of the thermal stresses built up during quenching and the transformation stresses caused by the martensitic reaction. In addition, carbides precipitate and lower the carbon content of martensite. Tempering at high temperatures, above 600°C, leads to a rapid growth of cementite and complete homogenization of the martensitic structure and eventually a microstructure of spheroidal cementite in a BCC matrix. In steels with high contents of carbide-forming elements, such as W, Mo and V, annealing around 600°C leads to a precipitation of alloyed carbides that yield an increase in hardness. This is typical for alloyed tool steels.

19.5.4 Case Hardening

In many applications one wants a hard surface but a tough interior. In such cases carbon may be added to the steel surface at a temperature around 900°C in a carburizing process and the component is subsequently quenched to form martensite in the carbon-rich surface layer (see also Chapter 18). As an approximation one may use Equation (19.17) to predict the carbon concentration profile. However, it should be emphasized that in practical carburizing the reactions at the steel surface are usually very important and must be taken into account to make accurate predictions. Moreover, the operation usually involves several steps and Equation (19.17) does not strictly apply [27].

References

[1] Hillert, M., *Phase Equilibria, Phase diagrams and Phase Transformations Their Thermodynamic Basis.* Cambridge University, Cambridge, 1998.

[2] Saunders, N. Miodownik, A. P., *CALPHAD (calculation of phase diagrams): A Comprehensive Guide.* Pergamon materials series. Oxford, Elsevier Science, 1998.

[3] Kaufman, L. and Bernstein, H., *Computer Calculation of Phase Diagrams.* Academic Press, New York, 1970.

[4] Darken, L. S., Diffusion of carbon in austenite with a discontinuity in composition, *Trans. AIME*, 180, 1949, 430–438.

[5] Volmer, M. and Weber, A., Nucleus formation in supersaturated systems, *Z. Phys. Chem.*, 119, 1926, 277.

[6] Beeker, R. and Döring, W., Kinetische Vbehandlung der Keimbildung in übersättigten Dämpfen, *Ann. Physik.*, 24, 1935, 719–752.

[7] Turnbull, D., *Solid State Physics*, Vol. 3, p. 225. Academic Press, 1956.

[8] Kelton, K. F., Time-dependent nucleation in partitioning transformations, *Acta Materialia*, 48, 2000, 1967–1980.

[9] Spaepen, F., Deutsches Zentrum für Luft aund Raumfahrt, in *Proceedings Kurzschule Phasenübergänge in Schmelzen*, eds. Herlach, D. M. and Holland-Moritz, D., Köln 200. IB-333-01/1, pp. 45–78.

[10] Volmer, M., Über Keimbildung und Keimwirkung als Specialfälle der hetergenen Katalyse, *Z. Elektrochem.*, 35, 1929, 555.

[11] Cahn, J. W., The kinetics of grain boundary nucleated reactions, *Acta Metall.*, 4, 1956, 449.

[12] Clemm, P. J. and Fisher, J. C., The influence of grain boundaries on the nucleation of secondary phases, *Acta Metall.*, 3, 1955, 70–73.

[13] Nabarro, F. R. N., The strains produced by precipitation in alloys, *Pros. Phys. Soc. R. Soc.*, A175, 1940, 519.

[14] Zener, C., Kinetics of the decomposition of austenite, *Trans. AIME*, 167, 1946, 550.

[15] Hillert, M., The role of interfacial energy during solid state phase transformations, *Jernkont Ann.*, 141, 1957, 757.

[16] Ivantsov, G. P., *Dokl. Nauk SSSR*, 58, 1947, 567.

[17] Lifshitz, I. M. and Slyozov, V. V., The kinetics of precipitation from supersaturated solid solutions, *Chem. Solids*, 19, 1961, 35.

[18] Wagner, C., Theorie der Alterung von Niederschlägen durch Umlösen, *Z. Elektrochemie*, 65, 1961, 581.

[19] Hillert, M. and Borgenstam, A., Massive transformation in the Fe-Ni system, *Acta Mater.*, 48, 2000, 2765–2775.

[20] Lücke, K., Masing, G. and Nölting, P., *Z. Metallkde*, 64, 1956, 64–74.

[21] Cahn, J.W., The impurity-drag effect in grain boundary motion, *Acta Metall.*, 10, 1962, 789–798.

[22] Hillert, M. and Sundaman, B., Treatment of the solute drag on moving grain boundaries and phase interfaces in binary alloys, *Acta Metall.*, 24, 1976, 731–743.

[23] Hillert, M. and Sundman, B., A solute-drag treatment of the transition from diffusion-controlled to diffusionless solidification, *Acta Metall.*, 25, 1977, 11–18.

[24] Borisov, V. T., *Soviet Phys. Dokl.*, 7 1962, 50.

[25] Aziz, M. J., Model for solute redistribution during rapid solidification, *J. Appl. Phys.*, 53, 1982, 1158–1168.

[26] Krauss, G. and Marder, A. R., Morphology of martensite in iron alloys, *Metall. Trans.*, 2, 1971, 2342–2357.

[27] Sproge, L. and Agren J., Experimental and theoretical studies of gas consumption in the gas carburizing process, *J. Heat Treat.*, 6, 1988, 9–19.

20 Transformation Toughening

R. I. Todd and
M. P. S. Saran
University of Oxford

Abstract

The brittleness of ceramics limits the application of their more beneficial properties, such as high melting point and hardness. Transformation toughening is one of the most potent mechanisms for overcoming this problem. Its development in zirconia-based materials has been successful in spreading the use of engineering ceramics to products requiring high reliability, such as artificial hip joints, and which involve tensile stresses, as in shears for cutting carbon fiber. Controlled processing is critical to the operation of the transformation toughening mechanism. This chapter explains this by describing the relationship between the principles of the toughening mechanism, the microstructures required for its operation and the processing strategies that have been successful in the fabrication of tough materials.

20.1 Introduction

Ceramic materials offer many advantages for structural applications, including high hardness, stiffness, melting point, and wear resistance. Other properties, such as optical transparency and low electrical and thermal conductivity, contrast sharply with those of metals and so provide a strong incentive for the use of ceramics in applications where these characteristics are essential. Ceramics also have the obvious and serious disadvantage that they tend to be very brittle. The consequent unpredictable and catastrophic nature of failure limits their use, despite the desirable properties listed above.

Classical brittle fracture occurs when the concentrated stress at a single flaw in the material reaches a critical level, at which point a crack spreads through the material leading to failure. The macroscopic stress at which this happens (i.e., the strength) is determined by the size of the flaw and the resistance of the material to crack propagation (the toughness). One approach to coping with brittle fracture is to reduce the critical flaw size by improving the processing and surface preparation of components. This approach has been successful in increasing the strength of ceramics although it does not guard against damage during service.

The complementary approach of increasing the toughness of the material gives more comprehensive protection against brittle fracture. Several toughening mechanisms have been identified in ceramic materials, but only two are capable of giving significant improvements. One is crack interface bridging, in which intact or interlocking ligaments are left bridging the crack surfaces behind the crack tip as it moves through the material during the early stages of crack growth. This toughening mechanism can be very potent where fibers or whiskers bridge the crack and operates at both high and low temperatures. The mechanism also has the disadvantages that the toughening is anisotropic and the production of appropriate microstructures is difficult and expensive, with stringent requirements for fiber or whisker alignment (see Chapter 30 for more details).

The second effective toughening mechanism is transformation toughening, the subject of this chapter. This relies on the occurrence of a stress-induced displacive transformation in the concentrated stress field around the tip of a loaded crack that produces a volume expansion or a shear strain. The resulting stresses then exert closure forces on the crack as it subsequently grows through the transformed region. Like crack bridging, this mechanism is potent and relies on the loading of the crack area left behind the crack tip as it propagates through the material. In contrast to crack bridging, however, transformation toughening becomes ineffective at high temperature but is isotropic in its effect and materials processing can be relatively inexpensive.

Several ceramics exhibiting displacive transformations with the required characteristics have been identified,[1] but all practical applications of transformation toughening concern systems based on zirconia (ZrO_2). For this reason, only zirconia-based systems will be considered in this chapter. The control and optimization of various microstructures is the key challenge of processing zirconia-toughened materials successfully and the main aim of this chapter is to describe the principles involved and the practical processing routes that have been successful. This chapter ends with a description of the typical properties and applications of these materials.

20.2 Phases of Pure Zirconia and its Stabilization

Pure zirconia melts at 2850°C (one of its attractive features) and below this temperature exhibits three polymorphs at atmospheric pressure: cubic (c) between 2370 and 2850°C, tetragonal (t) in the range 1200 to 2370°C, and monoclinic (m) below 1200°C. These three solid polymorphs are the most important for the discussion of transformation toughening, though high pressures and doping can induce others. The t and m polymorphs can both be referred easily to the ideal fluorite structure of c-ZrO_2 with the addition

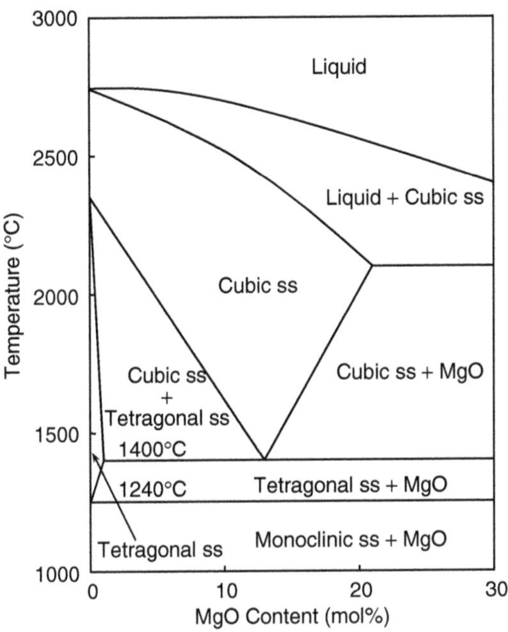

FIGURE 20.1 ZrO_2-MgO phase diagram. (Adapted from Grain, C.F., *J. Am. Ceram. Soc.*, 50, 288, 1967. With permission.)

of some distortion. The distortion is greatest for m-ZrO_2, in which the Zr^{4+} ions are coordinated by seven O^{2-} anions in contrast to the eightfold coordination of the higher temperature polymorphs. This is important in understanding the operation of the fluorite stabilizing dopants described below.

The $t \rightarrow m$ transformation is displacive and is accompanied by a dilatation of ~4% and a large shear strain of ~0.16. In practice, the temperature at which the transformation begins (the martensite start temperature, M_s, by analogy with martensitic transformation temperatures in steels) is ~950°C for coarse grained specimens. These large distortions render pure zirconia with conventional grain sizes (>1 μm) unusable for structural applications because the incompatibility between adjacent grains on cooling through the transformation after sintering causes such severe microcracking that the material loses all useful strength. Until recently this limited the exploitation of zirconia's many attractive properties.

Early attempts to solve this problem centered on the addition of other oxides that would "stabilize" the high temperature cubic phase. Most common were additions of CaO or MgO, the relevant phase diagram for which is shown in Figure 20.1. Both these oxides are sparingly soluble in ZrO_2 at room temperature, but show substantial solubility in the cubic phase at high temperature. With an appropriate amount of doping, the cubic phase undergoes eutectoid decomposition during equilibrium cooling in both systems (at 17 mol% CaO/1140°C or 13 mol% MgO/1400°C), but this transformation is so sluggish that the material can easily be cooled sufficiently quickly to retain c-ZrO_2 at room temperature. The damaging $t \rightarrow m$ transformation is thus avoided and such material is said to be "fully stabilized."

It was later found empirically that hypoeutectoid compositions, which contain some t-ZrO_2, have superior properties to the fully stabilized material. Such compositions are known as "partially stabilized zirconias" (PSZ). The initial emphasis was on improving the thermal shock resistance,[2] as the expansion caused by the $t \rightarrow m$ transformation during cooling compensated for the large thermal contraction of the c-zirconia matrix. Continued development showed that the best mechanical properties were obtained when the size of the t-ZrO_2 particles was very small and this was achieved initially by sintering at high temperature in the single-phase c-ZrO_2 phase field followed by controlled ageing in the $c + t$ field (Figure 20.1) to produce fine t-precipitates as shown in Figure 20.2. When the size of the t-precipitates formed is sufficiently small, they do not transform spontaneously to m-ZrO_2 during cooling and are

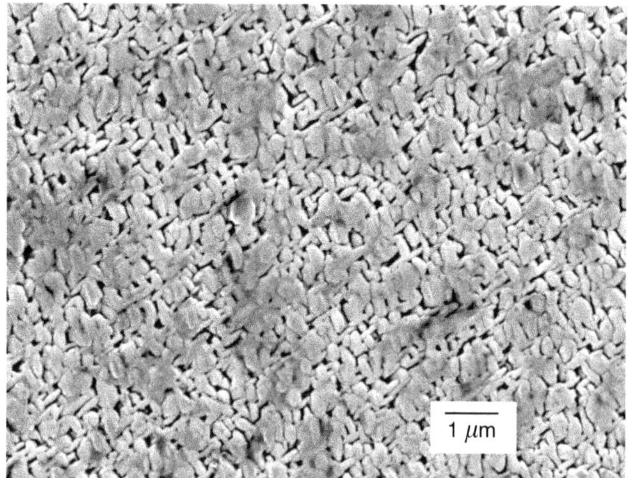

FIGURE 20.2 SEM image of a polished and thermally etched Mg-PSZ specimen showing fine *t*-precipitates inside a *c*-ZrO_2 grain.

retained metastably at room temperature. The improved properties observed are due to the ability of such precipitates to undergo stress-induced transformation in the vicinity of a loaded crack tip thus leading to transformation toughening, first identified by Garvie et al.[3]

20.3 General Strategy for Production of Metastable *t*-Particles

There is now a wide range of different zirconia-toughened materials, but PSZ illustrates three recurring themes regarding the production of metastable *t*-particles. The first is the addition of fluorite-stabilizing oxides. A number of such oxides has now been identified and several mechanisms seem to be involved in their success. Their stabilizing effect is usually attributable either to the dopant cation having a smaller charge than Zr^{4+} or to it having a different ionic radius, though it should be noted that several exceptions to these rules have been identified.[4] A lower charge on the cation leads to oxygen vacancies in order for charge neutrality to be maintained. The increased stability of the higher temperature polymorphs in the presence of oxygen vacancies has been demonstrated most clearly by Mommer et al.,[5] who showed that annealing monoclinic or tetragonal zirconia in a reducing atmosphere, which should increase the oxygen vacancy concentration, causes the $m \rightarrow t$ or $t \rightarrow c$ transformation to take place. Subsequent annealing in air showed these transformations to be reversible on restoration of the original oxygen vacancy concentration. This influence of oxygen vacancies can be rationalized in terms of the preference of the Zr^{4+} ion to be coordinated by seven O^{2-} ions as in the low temperature monoclinic phase, rather than eight, as in the high temperature phases. The addition of O^{2-} ion vacancies enables some of the Zr^{4+} ions to be sevenfold coordinated while simultaneously conforming to the higher symmetry structures with eightfold coordination owing to the occupation of some of the coordinating lattice sizes by oxygen vacancies.[6]

The size effect can be understood in terms of Pauling's rules. The cation/anion radius ratio for ZrO_2 is ≈ 0.6, which Pauling's first rule suggests should correspond to a coordination number of less than 8, eightfold coordination requiring a radius ratio of ≥ 0.732. The substitution of larger cations for Zr^{4+} should therefore favor the eight coordinated polymorphs, *c*-ZrO_2 and *t*-ZrO_2 over the seven coordinated *m* polymorph.

All the most common stabilizing agents possess either a smaller cation charge or a larger radius, or both: Ca^{2+} and Y^{3+} ions are both bigger than Zr^{4+} and have a lower charge, Mg^{2+} is similar in size to Zr^{4+} but has a lower charge and Ce^{4+} has the same charge but a larger radius than Zr^{4+}.

The second general defense against premature transformation of the t-particles during cooling illustrated by PSZ is the small size of the t-precipitates (Figure 20.2). Among suggestions to explain the size effect, the most plausible focus is on the ease of nucleation of the transformation.[7] One example is based on the assumption that each nucleation event leads quickly to the transformation of the whole particle from t- to m-ZrO_2. If the probability of a nucleation event is evenly distributed in volume or surface area, as appropriate, then small particles are statistically less likely to transform than large ones during cooling because they are less likely to contain a transformation nucleus.

A further reason to expect nucleation of the $t \rightarrow m$ transformation to be more difficult in small particles is that the scale of residual stress concentrations caused by thermal expansion mismatch between the t-ZrO_2 particles and their surroundings (either a different phase or differently orientated t-ZrO_2 particles) will be reduced.[7] The transformation involves both a dilatation and a shear strain, so stress concentrations that produce sympathetic elastic strains would be expected to help the transformation to nucleate owing to the consequent reduction in the strain energy produced. Any reduction in the extent of such stress concentrations will therefore assist in the retention of t-ZrO_2 to low temperatures.

The contribution of strain energy in nucleating the transformation is also the origin of the third general method of retaining metastable t-ZrO_2 particles below the equilibrium transformation temperature, namely, that the transformation should be constrained by a surrounding, initially nontransforming matrix, ideally with as high a stiffness as possible. Transformation of a t-ZrO_2 particle then creates a lot of elastic strain energy that reduces the total free energy gain on transformation and therefore inhibits its nucleation. In the example of PSZ, each of the t-precipitates shown in Figure 20.2 is constrained by the surrounding c-ZrO_2 matrix.

20.4 Transformation Toughened Zirconia Systems, Microstructures, and Nomenclature

Now that the principles behind the production of transformation toughened microstructures have been discussed, the most important microstructures used in practice will be described. Three distinct classes of material are in common use, and only three of the many possible stabilizers are now employed, namely MgO, Y_2O_3, and CeO_2, although CaO has some historical relevance.

20.4.1 Partially Stabilized Zirconia

The essential feature of PSZ is the dispersion of fine precipitates illustrated in Figure 20.2 for a commercial product containing magnesia as the stabilizing oxide (nomenclature Mg-PSZ). The figure shows only part of a cubic grain; owing to the high temperatures required for solution treatment in the single-phase cubic region, the cubic grain size is typically $>30\ \mu m$. The precipitate shape depends on the stabilizer. In Mg-PSZ, the t-ZrO_2 precipitates are oblate spheroids with typical aspect ratio $\sim$5 and are coherent with the c-ZrO_2 matrix. The rotation axes of the precipitates are the c axes of the tetragonal unit cell and these are aligned with the cubic axes of the matrix. There are therefore three possible precipitate variants in a given grain.

When the t-ZrO_2 precipitates become too large, through "overageing," or when they are stressed, they transform to m-ZrO_2. Each of the precipitates transforms in its entirety into several of the possible variants. These are thought to form sequentially to reduce the long-range strain field caused by the large shear strain of the $t \rightarrow m$ transformation.[8,9]

PSZ can also be produced with stabilization by CaO or Y_2O_3 (Ca-PSZ and Y-PSZ), but these are seldom used in practice due to lack of stability (CaO) or the very high temperatures $\sim$2000°C required for solution treatment and ageing (Y_2O_3). Commercial Mg-PSZs contain 8 to 10 mol% MgO, representing a balance between having a viable solution treatment temperature and a sufficiently high volume fraction of precipitates to give significant toughening.

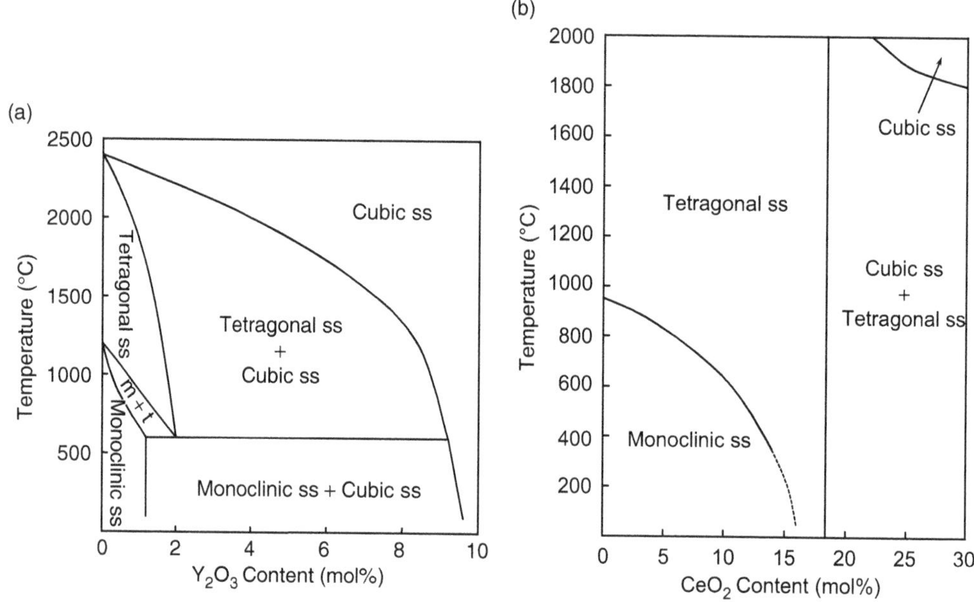

FIGURE 20.3 (a) ZrO_2-Y_2O_3 phase diagram. (Adapted from Scott, H.G., *J. Mater. Sci.*, 10, 1527, 1975. With permission.) (b) ZrO_2-CeO_2 phase diagram. (Adapted from Duwez, P. and Odell, F., *J. Am. Ceram. Soc.*, 33, 274, 1950. With permission.)

20.4.2 Tetragonal Zirconia Polycrystals

Stabilizers with substantial solubility in the t-phase can endow it with sufficient stability to be retained with particle sizes somewhat larger than the precipitates in Figure 20.2. This allows tetragonal zirconia polycrystals (TZP) to be produced by sintering in a fully or almost fully tetragonal field and cooling to room temperature without substantial transformation to m-ZrO_2. The resulting grain sizes are in the range 0.2 to 3 μm. The lower sintering temperatures required ($\sim\leq 1500$°C) and the absence of detailed heat treatment requirements offer obvious commercial advantages for TZP. Suitable stabilizers are Y_2O_3 (Y-TZP) and CeO_2 (Ce-TZP).

Y-TZP is produced with 2 to 3 mol% Y_2O_3. As the phase diagram in Figure 20.3a shows, some of these compositions contain a small proportion of cubic grains. These are usually larger than the tetragonal grains, and if they become very large can act as strength degrading flaws, but if kept fine are thought to play a role in increasing the strength, as will be discussed in Section 20.10.3.

CeO_2 is the only common stabilizer with the ability to produce thermodynamically stable t-ZrO_2 at room temperature (Figure 20.3b), for 16 to 18 mol% CeO_2. Typical Ce-TZP compositions contain $\sim$12 mol% CeO_2 and with metastable t-ZrO_2 grain sizes of 2 to 3 μm the M_s temperature is just below room temperature. The resulting ease of transformation leads to very wide $t \rightarrow m$ transformation zones (15 to 40 μm) around cracks compared with other transformation toughened materials.

20.4.3 Zirconia-Toughened Alumina

Dispersions of zirconia particles can be used to toughen other ceramics. Zirconia-toughened alumina (ZTA) is the most common example and allows the strength and toughness of alumina to be improved while retaining some of its advantages over zirconia (e.g., almost double the stiffness). ZTA is usually made by sintering mixtures of powders of alumina and zirconia (10 to 30 vol%) containing some stabilizing oxide. The typical size of the zirconia particles in ZTA is similar to the grain size of TZP, and often the same amounts of stabilizer are used as in TZP. It should be noted, however, that the high stiffness of alumina will constrain the $t \rightarrow m$ transformation more than in TZP, while the thermal expansion mismatch ($\sim 5 \times 10^{-6}$ K^{-1}) between the two ceramics during cooling after sintering causes tensile residual stresses in the zirconia which will help the transformation to occur. These conflicting effects cancel out to some extent.

20.5 Principles of Transformation Toughening

To optimize the processing of transformation toughened material it is essential to understand the toughening mechanism and how it relates to microstructure. In understanding the origin of transformation toughening, it is instructive to consider the propagation of a preexisting straight crack as it is gradually loaded in Mode I by the application of an external force. Initially, the crack tip is surrounded by untransformed material, which consists wholly or partly of metastable particles of t-ZrO_2 (Figure 20.4a). When a tensile load is applied, large stresses arise close to the crack tip and the t-ZrO_2 in the region over which a critical stress (assumed hydrostatic), σ_T, is exceeded transforms to make a frontal zone of transformation that expands as the stress intensity at the crack tip is increased (Figure 20.4b). This frontal zone is dilated compared with the surrounding, untransformed material resulting in residual stresses both within the frontal zone and in the untransformed material around it. The residual stress, σ_R, in the transformation zone is broadly compressive while the stress outside it has both tensile and compressive components and decays rapidly with distance from the particle. McMeeking and Evans[10] have shown that the effects of the positive and negative contributions cancel out so that the net stress intensity applied to the crack by the frontal zone is zero. Hence, at the point of crack extension, the externally applied stress intensity, K_∞ simply equals the local toughness of the material, T_L, and there is no transformation toughening.

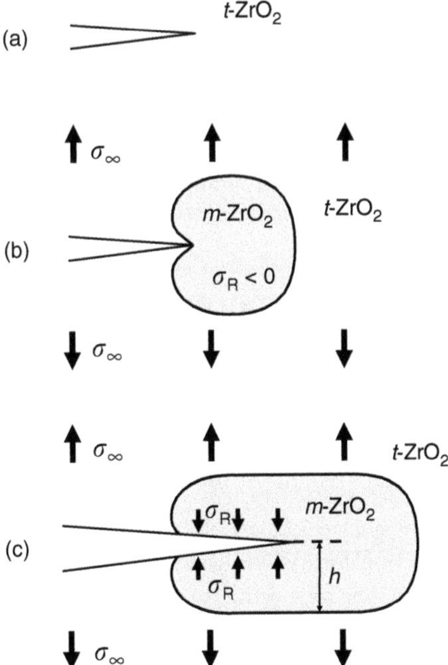

FIGURE 20.4 Schematic representation of events leading to transformation toughening. (a) Unstressed crack in matrix containing metastable t-ZrO_2 particles. (b) Stress-induced $t \rightarrow m$ transformation occurs ahead of crack to form frontal zone of transformation (σ_∞ —externally applied stress). (c) Crack grows into compressed transformation zone, which extends ahead of the crack.

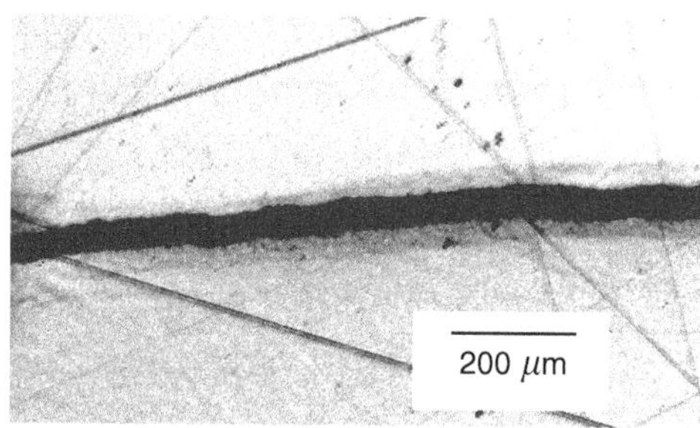

FIGURE 20.5 Nomarski interference contrast optical micrograph of the polished surfaces of a Ce-TZP after the passage of a crack across it, showing transformed "wake" around the crack surfaces.

As soon as the crack begins to grow the situation changes. The crack tip enters the frontal zone and the compressive stresses within it act to close the crack faces just behind the tip, producing a strongly negative internal contribution, K_T, to the total stress intensity (Figure 20.4c). The condition for further crack growth remains that the total stress intensity is equal to the local toughness of the material and thus the externally applied stress intensity required for crack growth increases and is given by:

$$K_{\infty c} = T_L - K_T \tag{20.1}$$

As K_T is negative, the macroscopically measured toughness, $K_{\infty c}$, is greater than T_L and the apparent toughness increases. This is the essence of the transformation toughening effect.

Continued crack growth has two important consequences. First, material continues to transform in the concentrated tensile stress field ahead of the crack tip as it propagates through the material. Second, as the transformation is irreversible, the transformed region, or "wake," left behind the moving crack tip exerts closure forces over a progressively increasing length of the crack faces behind the tip. This results in an increasingly negative K_T in Equation 20.1 and a correspondingly rising macroscopically measured toughness, $K_{\infty c}$. This increasing toughness with crack extension is known as *R*- or *T*-curve behavior. Eventually, the wake extends so far behind the crack tip that the most distant parts of it make little contribution and the toughness tends to a constant, plateau value, $K_{\infty c(p)}$. The dilatation in the wake causes uplift at the surface that is sufficiently pronounced in some materials for it to be seen through the optical microscope (Figure 20.5).

20.6 Simple Models for Transformation Toughening and Comparison with Experiment

Analytical models[10,11] for plateau toughness under simplifying assumptions described below give a toughening increment of:

$$\Delta K_c = -K_{T(p)} = \eta E f \Delta_T \sqrt{h}/(1-\nu) \tag{20.2}$$

where η is a dimensionless constant equal to 0.22, E is the effective Young's modulus, f the volume fraction of transforming particles, Δ_T the transformation dilatation, h the width of the transformed zone (see Figure 20.4c) and υ is Poisson's ratio. The zone width, h, follows simply from the near-field crack tip stress field as:[7,11]

$$h = \frac{\sqrt{3}(1+\nu)^2}{12\pi}\left(\frac{K_{\infty c(p)}}{\sigma_T}\right)^2. \tag{20.3}$$

Equation 20.2 and Equation 20.3 provide an excellent foundation for the discussion of transformation toughening, but do not provide a complete description of the process. One problem with the analysis is that it assumes the transformation to be complete inside the transformation zone and to be zero outside it. This is observed to be incorrect. Instead, the transformation zone is observed to be diffuse,[12,13] with a smaller fraction of transformed material toward the edge of the zone than near to the crack. This is known as a subcritical transformation zone and its consequence is to reduce the amount of toughening obtainable severely compared with Equation 20.2.[11] This only serves to exacerbate another problem with the equations, which is that the amount of toughening they predict is substantially less than that which is observed. The extent of the shortfall is as much as a factor of 4 with subcritical transformation.[8]

At least part of this discrepancy can be explained by revising some of the assumptions of the simple analysis of Equation 20.2 and Equation 20.3, notably that the nucleation of the transformation is sensitive only to the hydrostatic component of stress and only the hydrostatic component of the transformation strain contributes to the toughening. Considerable success in reconciling theory and experiment has been achieved by including more complex possibilities in models,[8] although it remains the case that no model can yet describe all the characteristics of the phenomenon.

20.7 Factors Influencing the Extent of Transformation Toughening

While Equation 20.2 and Equation 20.3 may oversimplify some aspects of transformation toughening, they remain a useful guide to the important variables that can be used to control transformation toughening through processing and variation of the composition. They also demonstrate how the extent of toughening depends on environmental factors such as temperature. The toughening increment increases with effective Young's modulus because a given dilatation then produces a bigger closure force. Little control over this parameter is possible for PSZ and TZP, but in ZTA, the higher stiffness of the alumina matrix has a marked influence on the achievable toughening.

Equation 20.2 also shows the importance of maximizing the volume fraction of transformable t-particles. This factor has already been alluded to in the range of practical compositions for Mg-PSZ (Section 20.4.1). The zirconia volume fraction in ZTA is also important, but in this case must be balanced with other factors such as the reduction in stiffness and hardness caused by the zirconia additions and the relevance of this to particular applications. In TZP, 100% of the material can transform in principle, but in practice the extent of transformation varies with distance from the crack as is determined by the transformability of the grains.

The remaining quantity of interest in Equation 20.2 and Equation 20.3 is the critical stress for transformation, σ_T. A low transformation stress leads to a wide transformation zone and high toughness. The effect of stabilizing oxides, particle size, residual stresses, and mechanical constraint on σ_T have already been discussed in Section 20.3 and similar arguments based on thermodynamics can be used to understand the influence of temperature on σ_T. The total free energy of transformation will generally be the sum of (1) the chemical free energy of the transformation, (2) the work done by an applied stress owing to the transformation strain, and (3) interfacial energies. The net reduction in free energy on transformation needs to be sufficient to drive the nucleation process. The chemical free energy release on transformation can be altered by doping as described in Section 20.2, but for any composition is also approximately proportional to the undercooling, ΔT, below the equilibrium transformation temperature. As the mechanical contribution is proportional to the applied stress, it follows that σ_T is high at the equilibrium $t \rightarrow m$ transformation temperature and decreases as the temperature is reduced beyond the equilibrium point. Equation 20.2 and Equation 20.3 show that this causes the transformed zone width and the toughness to increase as the temperature is reduced, and this is observed in practice, as shown in Figure 20.6. If the temperature is reduced too far, however, the chemical free energy becomes sufficient to nucleate the transformation without the need for a contribution from the applied stress and the $t \rightarrow m$ transformation begins to occur spontaneously (at the martensite start, or M_s temperature). This leads to a progressive reduction in toughness with further cooling as the capacity for the constrained, stress-induced transformation around the crack tip that is essential for toughening becomes exhausted. Thus, there is an optimum temperature for toughness, just above the M_s temperature. This simple argument also illustrates

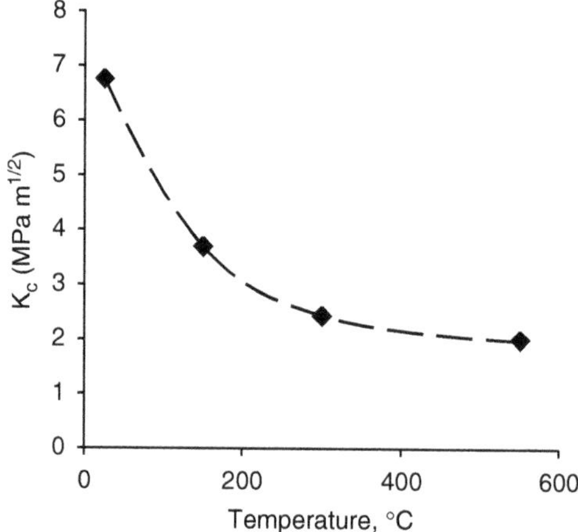

FIGURE 20.6 Toughness against testing temperature for a Ce-TZP. Transformation toughening is lost as the temperature approaches ~500°C, when t-ZrO_2 becomes stable.

an inherent problem with transformation toughening, namely the difficulty in producing a material with high toughness at high temperature. High temperature toughening requires a correspondingly high M_s temperature, but if the material is cooled below the design temperature, as must occur at least occasionally in practical applications, spontaneous and irreversible transformation of the t-ZrO_2 particles will occur, precluding subsequent toughening.

20.8 Other Toughening Mechanisms

Although the transformation toughening mechanism described above is the dominant toughening mechanism in most zirconia-based ceramics, other well-known mechanisms, such as crack deflection and crack interface bridging can also contribute. These two mechanisms arise in Mg-PSZ, for instance, as a result of the observed propensity of cracks to avoid transformed m-ZrO_2 particles[14] and make a relatively small contribution to the toughness of an optimally aged material. Two other toughening mechanisms that make bigger contributions to the total toughness are microcracking and ferroelastic switching and therefore deserve further comment. The former is of interest because it is linked directly to transformation toughening, and the latter is important because it can operate at high temperature.

20.8.1 Microcrack Toughening

The strains associated with the $t \rightarrow m$ transformation are so large that purely elastic accommodation in the material surrounding a transformed particle is unlikely. Additional accommodation is provided by microcracking under the action of the tensile stress components and this provides a further dilatation that can contribute to the toughness in the same way as primary transformation toughening itself. The reduction in stiffness of the cracked material ahead of the crack tip also reduces the stress intensity,[15] though both effects are partially offset by the fact that the microcracking also has an obviously deleterious effect on the inherent crack tip toughness, T_L. The contribution of microcracking to toughness is difficult to separate experimentally from the primary transformation effect and the degrading effect of the cracks on innate toughness is also difficult to assess. Notwithstanding these uncertainties, there are indications that the microcracking contribution to toughness in some zirconia toughened materials can be substantial.[16]

20.8.2 Ferroelastic Switching

If sufficient compressive stress is applied along the c-axis of t-zirconia, the c-axis can switch to one of the other cube-edge related directions in a ferroelastic transition.[17] It has been suggested that the resulting deformation can also be the source of closure stresses behind the crack tip leading to a toughening effect.[18] The ferroelastic transformation is most easy to study in the metastable t' phase that is formed by a displacive reaction on quenching ZrO_2 containing at least 3 mol% Y_2O_3 from the high temperature cubic regime (Figure 20.3a) rapidly to room temperature. This phase is crystallographically identical to the equilibrium t-phase, but contains more stabilizer and is reluctant to undergo the $t \rightarrow m$ transformation. Foitzik et al.[19] have estimated that a toughening contribution of 3 to 4 MPa $m^{1/2}$ can be obtained from ferroelastic switching in 3 mol% Y_2O_3-ZrO_2 heat treated to give the t' phase. The mechanism has the great potential benefit, compared with transformation toughening, that it is capable of operating at high temperature and is thought to be responsible for persistent observations of high toughness values at high temperature.[20]

20.9 The Strength of Transformation Toughened Materials

In materials exhibiting classical brittle behavior, the toughness, T, is a single-valued material constant. The condition for crack growth under a uniform stress, σ, is that the applied stress intensity, K_∞, exceeds T:

$$K_\infty = \psi\sigma\sqrt{c} > T \tag{20.4}$$

where c is the crack length and ψ is a dimensionless constant dependent on the shape of the crack. Thus, for materials containing flaws of similar shape, the strength is determined primarily by the size of the critical flaw and the toughness of the material.

On this basis, the experimental toughness values significantly in excess of 10 MPa $m^{1/2}$ displayed by transformation toughened materials might be expected to result in greatly increased strength compared with untoughened oxide ceramics, for which the innate toughness rarely exceeds 3 MPa $m^{1/2}$. There are indeed several reports of strengths well in excess of 1 GPa for Y-TZP and 2 GPa when up to 20% alumina was added to Y-TZP.[21] It is interesting to note, however, that of the commonly used transformation toughened materials, Y-TZP displays the smallest amount of toughening, with a toughness seldom significantly in excess of 10 MPa $m^{1/2}$; tougher varieties such as Ce-TZP and suitably aged Mg-PSZ usually have significantly lower strengths, typically 800 MPa or less. A more rigorous demonstration that strength does not simply scale with toughness in these materials can be obtained by plotting strength against toughness for a single system, in which the innate flaw size can be expected to remain approximately constant. All common zirconia-based ceramics show a *maximum* in such a plot, that is, the maximum strength occurs for an intermediate toughness value, often significantly less than that obtainable by optimum ageing. An example is shown in Figure 20.7 for Mg-PSZ.[22]

One reason for the departure from Equation 20.4 illustrated by Figure 20.7 is that the toughness is *not* single valued, but instead increases with crack length as was described in Section 20.5. An important consequence of this R-curve behavior is that, when determined by natural flaws, the strength depends mainly on the steepness of the initial part of the R-curve and does not depend on the plateau toughness toward which most of the previous work described in Section 20.6 and Section 20.7 was directed.[23] The observation of a maximum in strength–toughness plots can therefore be explained in terms of R-curve behavior if the R-curves for "tougher" (i.e., higher plateau toughness) materials have a shallower initial slope than those with less toughening. Early models for transformation toughening predict that this should be so,[10,23] and there is also experimental evidence for it.[24,25] Although precise comparisons are difficult to make with accuracy, Marshall and Swain[24] concluded that the effect was sufficiently strong to contribute significantly to the maximum in the strength–toughness relationship.

Another factor that may limit the strength that can be achieved in transformation toughened materials may be the transformation itself. Swain[22] noticed that very tough materials exhibited nonlinear stress–strain curves prior to failure in bending and that these were associated with the formation of deformation bands of transformed material $\sim$1 μm wide. Similar bands are observed around indentations as shown

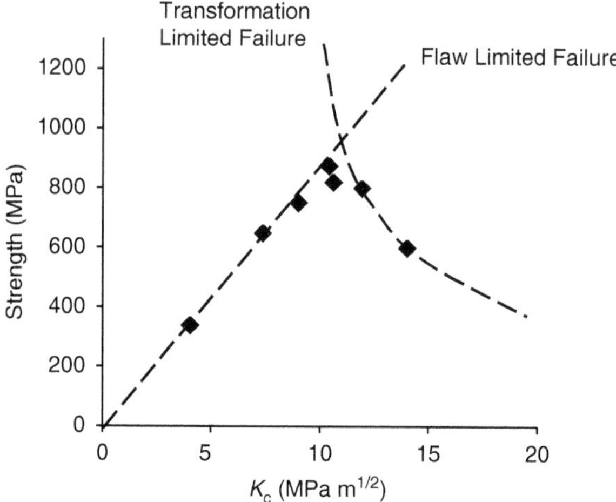

FIGURE 20.7 Strength against toughness for Mg-PSZ showing maximum strength at intermediate toughness. Also shown are lines corresponding to flaw-controlled failure and transformation limited failure. (Experimental data from Swain, M.V., *Acta Metall.*, 33, 2083, 1985.)

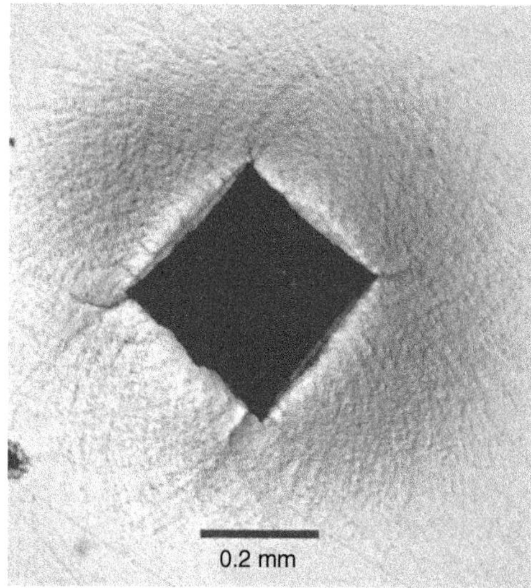

FIGURE 20.8 Vickers hardness indentation in the surface of a Ce-TZP specimen. Nomarski interference contrast optical micrograph. The stresses around the indentation have caused extensive $t \rightarrow m$ transformation, and the associated expansion has resulted in extensive surface uplift around the indentation. This reveals distinct bands of transformation, many grains in length, owing to the autocatalytic nature of the transformation process.

in Figure 20.8, and ahead of propagating cracks, and are thought to propagate by a cooperative "autocatalytic" process that is stimulated by the low critical stress for transformation that is also responsible for the high toughness (Equation 20.2 and Equation 20.3) of the materials that exhibit them. The bands contain considerable residual stresses[27] and are associated with microcracks that can act as critical flaws. Low toughness materials fail from preexisting flaws and the strength is proportional to the toughness (Equation 20.4) as shown in Figure 20.7. Swain[22] suggested that the lower strengths of very high toughness materials were a consequence of the formation of deformation bands at the (low) critical transformation stress, which act as large flaws and lead rapidly to failure. When the flaw-limited strength exceeds the transformation stress (inversely proportional to toughness increment according to Equation 20.2 and Equation 20.3) the strength becomes transformation limited and therefore falls with increasing toughness.

The influence of transformation bands on fracture initiation has also been cited as the reason that the addition of a nontransforming phase such as alumina often results in an increased strength, as in the work of Tsukuma et al.[21] mentioned earlier in this section. The alumina may block the development of long transformation bands and therefore reduce the residual stresses and crack size.[27]

The results reviewed in this section point the way for the use of processing to produce tailored microstructures suitable for particular applications. If high strength is important, for example, the balance of grain or precipitate size and stabilizer content needs to be kept below that for optimum plateau toughness. For reliability, and especially damage tolerance, the microstructure should be such that the t-ZrO_2 parts of the microstructure are on the point of transforming spontaneously to give maximum toughness. The strength may then be relatively low, but will also be well defined, and most importantly, the flaw size required to degrade the strength becomes very large compared with commonly encountered strength-controlling defects.[28] A material above the strength maximum such as a Mg-PSZ with a strength of 500 MPa and a toughness of 15 MPa $m^{1/2}$, for example, can sustain a semicircular surface flaw with a diameter of over 1 mm without loss of strength.

20.10 Zirconia Processing in Practice

Almost all structural zirconia is currently produced by sintering high quality powders that are commercially available. This section describes the production of these powders and their conversion to dense, strong components.

20.10.1 Powder Production

The principal sources for producing zirconia are Baddeleyite (ZrO_2) and zircon sand ($ZrSiO_4$). Baddeleyite is potentially the favored mineral owing to its low impurity content but currently accounts for <1% of total zirconia production owing to a lack of supply. Although zircon is a more abundant source, it has the disadvantage that its inherently high silica content must be removed to produce pure zirconia. This is achieved commercially by dissociating zircon using heat and chemical reactions. The methods described here are the most important for commercial production.

At sufficiently high temperatures, zircon will undergo thermal decomposition without the aid of chemical reactions in a plasma arc at temperatures above 6000°C. Zirconia solidifies in a cooler part of the furnace and forms dendrites. Cooling is too rapid to allow zircon to reform and the silica is left as a coating on the zirconia. It is removed by boiling with aqueous sodium hydroxide and the product is washed and milled to form a fine powder.

Zircon is also decomposed at lower temperatures by heating to between 2100 and 2250°C. The silica formed is liquid in this temperature range, and the zirconia is left as finely divided clusters with a range of sizes below 100 nm. The zirconia is purified using sulfuric acid, cleansed in a centrifuge, dried, milled with binders, and spray-dried to produce zirconia powders.

Most industrial production of zirconia powder favors chemical over purely thermal decomposition because of the lower temperatures involved and the ease of incorporating stabilizers into the processing route. The most widely used form of chemical dissociation begins with the separation of silica from zircon by a fusion reaction with sodium hydroxide at 600°C:

$$ZrSiO_4 + 4NaOH \rightarrow Na_2ZrO_3 + Na_2SiO_3 + 2H_2O$$

The sodium silicate is removed using water, which also hydrolyzes the zirconate to form a complex hydrated hydroxide that is then treated with either sulfuric or hydrochloric acid to produce a basic zirconium sulfate ($Zr_xO_{2x-y}(SO_4)_y$) or zirconyl chloride ($ZrOCl_2$) respectively. Ammonia is then added to precipitate zirconium hydroxide that is dried, calcined at typically 900°C, milled, and spray-dried to form fine zirconia powders. This route produces submicrometer powders with low agglomeration.

The chemical production routes provide the most desirable method to add stabilizers, because they can be coprecipitated to give atomic scale mixing and hence homogeneous powders with reproducible mechanical properties. In the basic route described above, a salt of the stabilizer (e.g., yttrium chloride) is added to the zirconyl chloride solution and coprecipitated by the ammonia. The drying stage is commonly accomplished by azeotropic distillation, which minimizes segregation of the stabilizer and reduces the tendency to form hard agglomerates.

Coprecipitation is not an option for the routes in which thermal decomposition alone is used to produce zirconia powder. In these cases, the stabilizer is dispersed as evenly as possible through the zirconia powder by mixing it with a solution of the nitrate of the stabilizer. When dried, this coats the powder and on calcination the nitrate decomposes to leave small particles of the stabilizing oxide on the surface of the zirconia particles. High temperature treatments are typically required for the stabilizer to diffuse into the zirconia particles and this leads to coarsening and agglomeration of the powder.

Although conventional thinking has been that a homogeneous distribution of stabilizer is desirable, recent research has suggested that a nonuniform distribution of stabilizer can enhance the toughness of Y-TZP.[29–31]

At present, thermal and chemical dissociation of zircon are the only techniques that allow large-scale production of zirconia at low cost. Other methods that have been used on a laboratory scale include fusion casting,[32] vapor-phase reaction,[33] hydrothermal precipitation[34] and sol-gel processing.[35] Though these techniques are capable of producing zirconia with properties as good or better than is made using the commercial powders currently available, the high costs required for production on an industrial scale mean that they do not provide financially viable alternatives.

20.10.2 Production of Mg-PSZ

Commercial fabrication of Mg-PSZ is from zirconia powders containing 8 to 10 mol% stabilizer plus sintering aids and/or scavengers to remove the silica that remains as an impurity in the powder. Silica is difficult to remove completely and if it is not controlled in this way reacts with MgO, preventing it from acting as a stabilizer. The process begins with conventional die-pressing to form green compacts (for more details see Chapter 24 and Chapter 25). These compacts may be cold isostatically pressed to improve sinterability and densification. The use of suitable binders allows green machining if required. The greens are sintered at between 1700 and 1800°C, in the cubic solid solution region of the phase diagram (see Figure 20.1), for about 2 h to produce a fully dense ceramic with a grain size of 30 to 60 μm.

At its simplest, the next stage is to age in the cubic + tetragonal region to produce a uniform dispersion of the all-important t-ZrO_2 precipitates within the cubic grains. There are two main ways of achieving this in practice. The easiest way to achieve reproducible results is first to cool the cubic ceramic rapidly to room temperature and then to reheat and age isothermally at a suitable temperature. For practical purposes, the "rapid cooling" needs to be ~500°C/h down to 1000°C, below which slower, furnace cooling will suffice because further diffusion-controlled development is slow. Cooling at 500°C/h through the $c + t$ region does result in some precipitation of t-ZrO_2, but the diameter of the lenticular precipitates formed is 30 to 60 nm, which is too small to give significant stress-induced transformation at room temperature.[8] Further ageing is therefore carried out just above the eutectoid temperature at ~1400°C to grow the t-ZrO_2 precipitates to a size giving the desired M_s temperature. Optimal ageing times for room temperature use are ~4 h, at which point the lenticular precipitates should be ~180 nm in diameter and ~40 nm thickness. Ageing for longer than this leads to further coarsening and an M_s above room temperature. The precipitates therefore transform spontaneously when cooled and toughening is lost.

The two-stage process described above is more time consuming and therefore expensive than the alternative possibility of forming the t-ZrO_2 precipitates during an isothermal ageing treatment imposed immediately after the sintering and solution treatment, without the intermediate steps of cooling to room temperature and reheating. There are some subtle differences between the precipitation behavior occurring at a given isothermal hold temperature with and without the intermediate excursion to room temperature, and it has been found that the optimum isothermal hold temperatures for use immediately after sintering are just below the eutectoid temperature. Hughan and Hannink[36] found that cooling at 500°C/h to 1340°C and holding for 2 h was capable of giving a strength of ~650 MPa. This compares favorably with the strength obtainable by the two-stage process above, but with the commercial advantage of a single firing schedule.

The toughest Mg-PSZ ceramics have been produced by using an additional "subeutectoid ageing" treatment for several hours at 1100°C after first ageing to just below the optimum level as above. Subeutectoid ageing has several effects, but the most important is thought to be the formation of the metastable δ-phase, $Mg_2Zr_5O_{12}$, at the interface between the preexisting t-ZrO_2 precipitates and the c-ZrO_2 matrix. This destabilizes the t-ZrO_2 precipitates, making them easier to transform under stress, by removing some of the stabilizing MgO from the precipitates and exerting a local stress on them by virtue of the lattice mismatch between the δ-phase and the t-ZrO_2.[25] Hannink et al.[25] found that the plateau toughness could be increased from ~8 MPa $m^{1/2}$ to almost 12 MPa $m^{1/2}$ by subeutectoid ageing at 1100°C for 8 h and similar results have been reported by Steinbrech[37] for a postfiring treatment of 90 min at the same temperature. Hughan and Hannink[36] noted nevertheless that subeutectoid ageing could not increase the strength beyond that of optimally single-fired Mg-PSZ.

20.10.3 Production of TZP and ZTA

While Mg-PSZ has an elaborate and sensitive fabrication methodology, polycrystalline TZPs are made through a more straightforward route in which the green body is sintered and cooled to form the desired microstructure consisting mainly of fine, equiaxed t-ZrO_2 grains of uniform size. Coprecipitated powders with particle sizes of 200 nm or less are used and the stabilizer content is matched with the relevant phase diagram. Typical stabilizer contents for Y-TZP are 2 to 3 mol% Y_2O_3, with 3 mol% being preferred because of its higher strength. This may be because the minority of nontransforming cubic grains that are formed with this composition can disrupt the formation of transformation bands in the same way as the alumina additions described in Section 20.9 though this has not been confirmed experimentally. Compositions for Ce-TZP are between 12 and 20 mol% CeO_2, with 12% being most common.

Powder processing to form green bodies is similar to that described above for Mg-PSZ. The fine powders used and the requirement to sinter in the t-phase field rather than the cubic field allow lower sintering temperatures than for PSZ. Y-TZP's are sintered at temperatures between 1300 and 1500°C for 1 to 2 h with resulting grains sizes of 0.2 to 3 μm. For Ce-TZP, firing is conducted at 1400 to 1500°C for 1 to 2 h and results in equiaxed grains of size 2 to 3 μm. Densities can be further improved by cold or hot isostatic pressing (HIPing), although the use of graphite heating elements and other high temperature parts should be avoided when HIPing Ce-TZP because the resulting reducing atmosphere reduces CeO_2 to Ce_2O_3.[8]

A further consideration in the practical processing of Y-TZP concerns its well-known susceptibility to hydrothermal ageing, that is, spontaneous surface transformation from t-ZrO_2 to m-ZrO_2 and consequently severe loss of strength when exposed to a combination of heat and moisture. Several approaches have been suggested to eradicate this problem, including the incorporation of CeO_2 [38] (note that Ce-TZP is less prone to this problem than Y-TZP) and reducing the grain size.[39] The efficacy of these and several other suggestions tend to rely fundamentally on increasing the stability of the t-ZrO_2, however, and this compromises the optimal mechanical properties. At present, industrial applications of polycrystalline TZP materials represent a balance between the high strength of Y-TZP and the superior stability against hydrothermal ageing and higher toughness of Ce-TZP.

Similar routes using mixed powders of zirconia and alumina are used to produce ZTA.

20.11 Properties and Applications of Transformation Toughened Zirconia

Prior to the discovery of transformation toughening, stabilized zirconia was mainly used in kilns and furnaces as thermal insulation. The ability to produce high density, stabilized zirconia has also led to the exploitation of other useful properties such as the ability to conduct oxygen ions when doped with Y_2O_3. This property of yttria stabilized zirconia forms the basis of the most common solid oxide fuel cell system and a range of oxygen-detecting sensors. The inherent properties of stabilized zirconia such as high melting point and relatively good thermal expansion matching to metals are also responsible for its use in thermal barrier coatings for protecting metallic components against the very high temperatures in gas turbines.

The majority of structural applications of zirconia are driven by its combination of high resistance to wear and chemical inertness, isotropic properties and versatile shape capability along with the high strength and reliability that has been the main theme of this article. In addition, oxide-based materials such as zirconia and ZTA are significantly cheaper than competitors such as sintered SiC and Si_3N_4 because the latter require high temperatures and controlled atmospheres during sintering. Table 20.1 gives a comparison of some properties of typical commercial ceramics. The obvious advantages of zirconia-toughened materials are the combinations of room temperature strength and toughness offered. Y-TZP is much stronger than competitor materials, while Ce-TZP has the greatest toughness. Mg-PSZ provides a good balance between strength and toughness and while it does not compete with Y-TZP in the room temperature properties shown, it is able to retain some toughening capability to higher temperatures. This is because m-ZrO_2 remains the thermodynamically stable phase to over 1000°C (Figure 20.1), whereas t-ZrO_2 becomes the stable form above ~500°C in Y-TZP and Ce-TZP, preventing the $t \rightarrow m$ transformation from toughening the material.

ZTA is attractive because it combines the exceptionally high strength and toughness of zirconia with the high stiffness and hardness of alumina. The presence of alumina also overcomes the poor thermal shock of zirconia to some extent because of its high thermal conductivity. While the high strength and relatively low stiffness of TZP and PSZ make them appear competitive in terms of the R_1 thermal shock parameter, which gives the maximum temperature reduction in rapid quenching so that the material can survive, the R_2 parameter, which describes the resistance to slower, more commonly encountered cooling

TABLE 20.1 Quoted Properties of Commercial Zirconia Ceramics and Some Competitor Materials

Properties	Units	Y-TZP[a]	Ce-TZP[b]	ZTA[c]	Mg-PSZ[d]	AL_2O_3[a]	Si_3N_4[a]	SiC[a]
Density	g/cm^3	6.05	6.20	4.10	5.70	3.92	3.25	3.12
Flexural strength, σ_f	MPa	1400	420	760	550	350	650	410
Compressive strength	MPa	2000	—	—	2000	2500	2000	2000
Modulus of elasticity, E	GPa	205	200	310	200	350	290	450
Poisson ratio, υ	—	0.30	—	—	0.30	—	0.24	0.17
Hardness	Hv	1350	900	1750	1100	1700	1500	2800
Fracture toughness	MPa $m^{1/2}$	10.0	15.0	6–12.0	8.4	4.5	8.0	4.0
Thermal expansion coefficient, α	$\times 10^{-6}$/ K	10.0	12.0	8.1	10.0	8.5	3.0	3.0
Thermal conductivity, k	W m^{-1}/K	2.0	2.0	23.0	2.5	28.0	25.0	100.0
Thermal shock parameter, R_1	K	478	123	212	193	82	568	252
Thermal shock parameter, R_2	kW/m	0.96	0.25	4.88	0.48	2.30	14.20	25.20

Thermal shock parameters: $R_1 = \sigma_f(1 - \upsilon)/E\alpha$, $R_2 = kR_1$.

Source:

[a] Dynamic-Ceramic Ltd.

[b] UltraHard Materials Ltd.

[c] Ceram Research Ltd.

[d] Morgan Advanced Ceramics.

FIGURE 20.9 Y-TZP pump components (by courtesy of Dynamic Ceramic Ltd.).

rates, shows these materials to perform poorly because of the low thermal conductivity of zirconia. The combination of high hardness, strength, and thermal shock resistance makes ZTA the material of choice for several demanding applications such as cutting tools. The use of more expensive SiC and Si_3N_4 is cost effective only for the most severe operating conditions requiring very high resistance to temperature, thermal shock or wear.

Some commercial Y-TZP precision components are shown in Figure 20.9. Toughened zirconia has found niches in many industries for its inertness and wear resistance, in applications ranging from seals, impellers, and bearings in manufacturing pumps for corrosive or abrasive products to valve seats and guides for turbochargers. Other wear-related applications include knives and scissors for cutting hard or tough materials like Kevlar, carbon fiber and metal sheet; applications in the paper industry, where hard TiO_2 particles used as whitener wear metal components rapidly; cams, dies and thread guides in textiles and metal wire production; ferrules and sleeves that connect and guide fiber optical cables. Its innate electrical and magnetic properties are used in precision insulators, ceramic substrates, and screwdrivers for electronic devices.

In recent years, zirconia has also played a major role in biomedical applications with the production of femoral heads for hip implants, knee and spinal joint replacements, endoscopic equipment, pace-maker covers, and dental restoratives.

As with other advanced materials, zirconia's properties can be tailored to its applications. For example, strength, toughness, corrosion resistance, and small grain size are emphasized in making knives and scissors, while chemical inertness, high toughness, wear resistance, and thermal shock properties are optimized in the production of pumps. The manufacture of dense zirconia, with small grain size and good surface finish, reduces friction in hip implants and fouling of seals, valve seats, and impellers.

20.12 Future Developments

A great deal of research has been done on transformation toughening since it was first identified by Garvie et al.,[3] and a detailed understanding of the behavior associated with the main microstructures described in this article has been established. As with any other area of research, some fine points in need of further elucidation remain, but these are unlikely to lead to major new developments. Instead, the current authors see three important areas for development in the immediate future.

The first is to apply transformation toughening to other microstructures. The excellent combination of properties offered by ZTA shows the potential of zirconia to improve already-competitive materials. There is no reason why transformation toughening should not be combined with other toughening mechanisms such as crack bridging to obtain even better properties. In a similar vein, laboratory studies of laminate structures that incorporate compressive residual stresses to hinder crack propagation are already starting to include toughened zirconia layers. The inherently high strength of the zirconia adds to the strengthening effect of the residual stresses.

The second area requiring development is the high temperature toughness of zirconia. The inherent problem with classical transformation toughening may be overcome by the ferroelastic switching described in Section 20.8.2 because it does not involve a phase change. Systematic studies of this effect in polycrystals are in the very early stages and there is a lot of scope for further exploration.

The final task for the future is to continue the commercialization of this exciting technology. Section 20.11 shows the success achieved to date in terms of the significant number of applications, as a similar section on any other new material might do. The fact that this list of successes is for a *structural ceramic* gives it added significance, however, because it shows that the natural reticence of designers to employ brittle materials, with their inherent unpredictability and catastrophic failure, is being broken down. There is good reason to hope that with successful experience will come confidence and that this will lead in turn to a rapid expansion in the range of applications benefiting from the advantages offered not only by zirconia but by other structural ceramics as well.

References

[1] Kriven, W.M., Possible alternative transformation tougheners to zirconia: crystallographic aspects, *J. Am. Ceram. Soc.*, 71, 1021, 1988.

[2] Heuer, A.H., Transformation toughening in ZrO_2 containing ceramics, *J. Am. Ceram. Soc.*, 70, 689, 1987.

[3] Garvie, R.C., Hannink, R.H., and Pascoe, R.T., Ceramic steel?, *Nature*, 258, 703, 1975.

[4] Shukla, S. and Seal, S., Mechanisms of room temperature metastable tetragonal phase stabilisation in zirconia, *Int. Mater. Rev.*, 50, 45, 2005.

[5] Mommer, N., Lee, T., and Gardner, J.A., Stability of monoclinic and tetragonal zirconia at low oxygen partial pressure, *J. Mater. Res.*, 15, 377, 2000.

[6] Kisi, E.H. and Howard, C.J., Crystal structures of zirconia phases and their inter-relation, *Key. Eng. Mater.*, 1, 153, 1998.

[7] Evans, A.G. and Cannon, R.M., Toughening of brittle solids by martensitic transformations, *Acta Metall.*, 34, 761, 1986.

[8] Hannink, R.J., Kelly, P.M., and Muddle, B.C., Transformation toughening in zirconia-containing ceramics, *J. Am. Ceram. Soc.*, 83, 461, 2000.

[9] Kelly, P.M. and Ball, C.J., Crystallography of stress-induced martensitic transformations in partially stabilized zirconia, *J. Am. Ceram. Soc.*, 69, 259, 1986.

[10] McMeeking, R.M. and Evans, A.G., Mechanics of transformation toughening in brittle materials, *J. Am. Ceram. Soc.*, 65, 242, 1982.

[11] Budiansky, B., Hutchinson, J., and Lambropoulos, J., Continuum theory of dilatant transformation toughening in ceramics, *Int. J. Solids Struct.*, 19, 337, 1983.

[12] Marshall, D.B., Shaw, M.C., Dauskardt, R.H., Ritchie, R.O., Ready, M.J., and Heuer, A.H., Crack-tip transformation zones in toughened alumina, *J. Am. Ceram. Soc.*, 73, 2659, 1990.

[13] Yu, C.S., Shetty, D.K., Shaw, M.C., and Marshall, D.B., Transformation zone shape effects on crack shielding in ceria-partially-stabilized zirconia (Ce-TZP)-alumina composites, *J. Am. Ceram. Soc.*, 75, 2991, 1992.

[14] Swain, M.V. and Hannink, R.H.J., R-curve behaviour in zirconia ceramics, in *Advances in Ceramics Vol. 12, Science and Technology of Zirconia II*, Claussen, N., Ruhle, M., and Heuer, A., Eds., American Ceramic Society, Columbus, OH, 1984, 225.

[15] Evans, A.G. and Faber, K.T., Crack growth resistance of microcracking brittle materials, *J. Am. Ceram. Soc.*, 67, 255, 1984.
[16] Ruhle, M., Evans, A.G., McMeeking, R.M., Charalambides, P.G., and Hutchinson, J.W., Microcrack toughening in alumina/zirconia, *Acta Metall.*, 35, 2701, 1987.
[17] Cain, M.G., Bennington, S.M., Lewis, M.H., and Hull, S., Study of ferroelastic transformation in zirconia by neutron diffraction, *Phil. Mag. B*, 69, 499, 1994.
[18] Virkar, A.V. and Matsumoto, R.L.K., Ferroelastic domain switching as a toughening mechanism in tetragonal zirconia, *J. Am. Ceram. Soc.*, 69, C224, 1986.
[19] Foitzik, A., Stadtwald-Klenke, M., and Ruhle, M., Ferroelasticity of t'-ZrO_2, *Z. Metallkd.*, 84, 397, 1993.
[20] Srinivasan, G.V., Jue, J.-F., Kuo, S.-Y., and Virkar, A.V., Ferroelastic domain switching in polydomain tetragonal zirconia single crystals, *J. Am. Ceram. Soc.*, 72, 2098, 1989.
[21] Tsukuma, K., Ueda, K., and Shimada, M., Strength and fracture toughness of hot-isostatic pressed Y_2O_3-stabilized ZrO_2-Al_2O_3 composites, *J. Am. Ceram. Soc.*, 68, C4, 1985.
[22] Swain, M.V., Inelastic deformation of Mg-PSZ and its significance for strength-toughness relationship of zirconia toughened ceramics, *Acta Metall.*, 33, 2083, 1985.
[23] Marshall, D.B., Strength characteristics of transformation-toughened zirconia, *J. Am. Ceram. Soc.*, 69, 173, 1986.
[24] Marshall, D.B. and Swain, M.V., Crack resistance curves in magnesia-partially-stabilized zirconia, *J. Am. Ceram. Soc.*, 71, 399, 1988.
[25] Hannink, R.H.J., Howard, C.J., Kisi, E.H., and Swain, M.V., Relationship between fracture toughness and phase assemblage in Mg-PSZ, *J. Am. Ceram. Soc.*, 77, 571, 1994.
[26] Sergo, V., Clarke, D.R., and Pompe, W., Deformation bands in ceria-stabilized tetragonal zirconia/alumina: I, measurement of internal stresses, *J. Am. Ceram. Soc.*, 78, 633, 1995.
[27] Becher, P.F. and Warwick, W.H., Fracture strength-fracture resistance response and damage resistance of sintered Al_2O_3-ZrO_2 (12 mol% CeO_2) composites, *J. Am. Ceram. Soc.*, 77, 2689, 1994.
[28] Stevens, R., *Introduction to Zirconia*, Magnesium Elecktron, Manchester, UK, 1986.
[29] Singh, R., Gill, C., Lawson, S., and Dransfield, G.P., Sintering, microstructure and mechanical properties of commercial Y-TZPs, *J. Mater. Sci.*, 31, 6055, 1996.
[30] Bowen, C.R., Tavernor, A.W., Luo, J., and Stevens, R., Microstructural design of sensor materials using the core-shell concept, *J. Eur. Ceram. Soc.*, 19, 149, 1999.
[31] Basu, B., Vleugels, J., and Van Der Biest, O., Microstructur-toughness-wear relationship of tetragonal zirconia ceramics, *J. Eur. Ceram. Soc.*, 24, 2031, 2004.
[32] Blackburn, S., Kerridge, C.A., and Senhenn, P.G., Electro-refined PSZ powder — An alternative zirconia for dense engineering components, in *Advances in Ceramics Vols. 24 A and B, Science and Technology of Zirconia III*, Somiya, S., Yamamoto, N., and Yanagida, H., Eds., American Ceramic Society, Westerville, OH, 1988, 193.
[33] Hori, S., Yoshimura, M., Somiya, S., and Takahashi, R., Al_2O_3-ZrO_2 ceramics prepared from CVD powders, in *Advances in Ceramics Vol. 12, Science and Technology of Zirconia II*, Claussen, N., Rühle, M., and Heuer, A., Eds., American Ceramic Society, Columbus, OH, 1984, 794.
[34] Hishinuma, K., Kumaki, T., Nakai, Z., Yoshimura, M., and Somiya, S., Characterisation of Y_2O_3-ZrO_2 powders synthesised under hydrothermal conditions, in *Advances in Ceramics Vols. 24 A and B, Science and Technology of Zirconia III*, Somiya, S., Yamamoto, N., and Yanagida, H., Eds., American Ceramic Society, Westerville, OH, 1988, 201.
[35] Tokudome, K. and Yamaguchi, T., Effect of solvation on the hydrolysis and properties of precipitates, in *Advances in Ceramics Vols. 24 A and B, Science and Technology of Zirconia III*, Somiya, S., Yamamoto, N., and Yanagida, H., Eds., American Ceramic Society, Westerville, OH, 1988, 1178.
[36] Hughan, R.R. and Hannink, R.H.J., Precipitation during controlled cooling of magnesia-partially-stabilized zirconia, *J. Am. Ceram. Soc.*, 69, 556, 1986.

[37] Steinbrech, R.W., Toughening mechanisms for ceramic materials, *J. Eur. Ceram. Soc.*, 10, 131, 1992.

[38] Sato, T., Ohtaki, A., Endo, T., and Shimada, M., Transformation of yttria-doped tetragonal-ZrO_2 polycrystals by annealing under controlled humidity conditions, *J. Am. Ceram. Soc.*, 68, C320, 1985.

[39] Winnubst, A.J.A. and Burggraaf, A.J., The aging behaviour of ultrafine-grained Y-TZP in hot water, in *Advances in Ceramics Vols. 24 A and B, Science and Technology of Zirconia III*, Somiya, S., Yamamoto, N., and Yanagida, H., Eds., American Ceramic Society, Westerville, OH, 1988, 39.

21

Bonding Processes

M. Powers
Agilent Technologies

S. Sen
University of California at Davis

T. Nguyentat
Aerojet

O. M. Knio and
T. P. Weihs
Reactive Nanotechnologies

Abstract

This chapter outlines four primary technological processes associated with the interfacial bonding of both similar and dissimilar advanced engineering materials; glass-to-metal sealing, ceramic-to-metal joining, diffusion bonding, and reactive joining. The first section will cover the principles and theoretical foundations for interfacial bonding including relative surface energy, wetting behavior and the Young Equation, sessile drop experiments, and the Ellingham diagram of the oxides. The following sections will provide overviews of the joining procedures and the basis for their application. In each case, references to the original work will be cited for further study and in-depth investigation. Case studies will be provided to define and highlight areas of application, as well as demonstrate the utility of a particular bonding process.

21.1 Introduction

From a historical perspective, the evolution of joining techniques based on chemical bonding and diffusion has been driven by the practical needs of industry. During the Second World War, the mass production of electronic equipment impelled requirements for hermetic vacuum tubes, lamps, and coaxial connectors dependent on glass-to-metal and ceramic-to-metal interfaces.[1,2] Later, advances in the aerospace industry drove viable progress in diffusion bonding procedures.[3] More recently, developments in sputter deposition processes and the emergence of nanotechnology have given rise to the nascent field of reactive joining with exothermic mutilayer foils.[4] Congruent with the progression of these bonding techniques has been the advancement in the fundamental theory and principles underlying the pragmatic approach, so that by now the scientific underpinning for the processes is well established. Contemporary developments in these fields owe significantly to the continuous expansion of their firm scientific foundation.

21.2 Principles of Interfacial Bonding

The fundamental thermodynamic driving force for the creation of an intimate atomic level interface between two mating surfaces is lowering the total free energy of the system. In general, this can be accomplished by the formation of chemical bonds at the interface and minimizing stress differentials and stress gradients in the interfacial zones. As an electronic structure across an interface is almost always the result of reactions that form equilibrium phases, a necessary condition for bonding is the generation of a stable chemical thermodynamic equilibrium. Stress differentials at interfaces can result from significant differences in the thermal expansion coefficients of mating components, epitaxial incompatibility between phases, surface defects or irregularities, and the presence of intermediate phases. These chemical and physical factors must be considered for the optimization of bonding processes used in engineering applications.

21.2.1 Surface Energy and Surface Tension

Given that atoms in the bulk of a solid crystalline material are arranged such that the internal energy of the system is minimized, the bonding forces between the atoms are essentially isotropic. This promotes long range atomic order and relatively simple close packed crystallographic structures in solid metals. Even in the bulk of liquid metals, ceramics, and glasses there is a propensity for short range ordering of atoms in the structure.

By contrast, the interatomic forces at the free surfaces of solid materials are asymmetric, which leads to a net force that pulls the atoms on the surface toward the bulk. As a result, work must be done against this force to add atoms to the surface and increase the surface area, because energy must be expended in displacing other atoms in the surface. Hence, the surface energy γ of a solid can be defined as the reversible work w_r necessary to increase the surface by a unit area:

$$dw_r = \gamma dA \tag{21.1}$$

This increase in free energy is also referred to as interfacial energy and has units of J/m^2 (energy per unit area). In crystalline solids, surfaces on planes with the densest atomic packing exhibit the lowest relative surface energy, so that surface energy in these materials is a function of crystallographic orientation.

Alternatively, the force required to incorporate atoms into the surface of a liquid by stretching of the surface is referred to as surface tension. Surface tension is defined as the force along a line of unit length perpendicular to the surface and is expressed in units of N/m (force per unit length). A dimensional analysis of the units for surface energy and surface tension suggests that they are a measure of the same quantity. This is true for liquids where the values are numerically equal, but not so for solids where the surface energy and surface tension are not exactly equivalent.[5] Nonetheless, the terms are often used interchangeably in the literature.

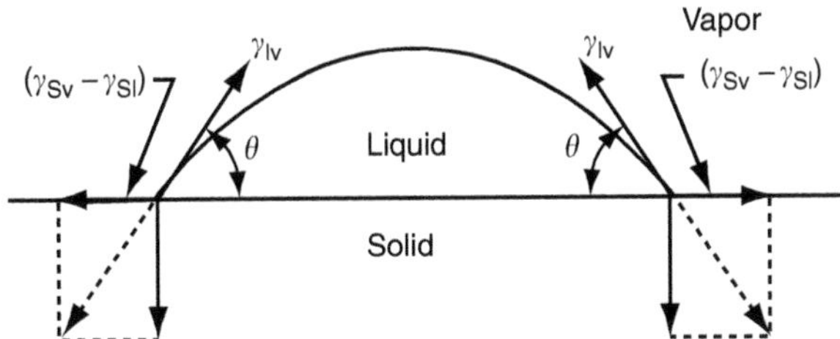

FIGURE 21.1 Equilibrium of forces on the periphery of a steady state sessile drop along a vertical plane through the center of the drop perpendicular to the interface for an acute or wetting contact angle formed when the surface energy of the solid (γ_{sv}) is greater than the surface energy of the liquid (γ_{lv}). (From Tomsia, A.P. and Pask, J.A., *Dent. Mater.*, 2, 10, 1986. With permission from the Academy of Dental Materials.)

By employing the concept of an idealized mathematical surface (with no associated volume) between two homogeneous phases, Gibbs showed that interfacial energy could be treated as an extensive excess quantity when compared with the sum of the surface energies of the separate phases. It follows that for chemically similar phases (i.e., liquid metal on solid metal or liquid oxide on solid oxide) and circumstances where there is chemical interaction between phases, the ensuing relative interfacial energy is comparatively low.

21.2.2 Wetting and the Young Equation

The degree to which a liquid phase will wet a solid it is in physical contact with, within a vapor–liquid–solid three phase system, is a direct consequence of the particular conditions of thermodynamic equilibrium. At the turn of the nineteenth century, Thomas Young balanced the horizontal components of the interfacial energies (surface tensions) at the point of contact in such a three phase system (Figure 21.1), and proposed that "for each combination of a solid and a fluid, there is an appropriate angle of contact between the surfaces of the fluid, exposed to the air, and to the solid."[6] This relationship is expressed in the now famous equation bearing Young's name as:

$$\gamma_{sv} - \gamma_{sl} = \gamma_{lv} \cos\theta \tag{21.2}$$

where γ_{sv} is the interfacial energy between the solid and vapor, γ_{sl} is the interfacial energy between the solid and liquid, γ_{lv} is the interfacial energy between the liquid and vapor, and θ is the liquid–solid contact angle measured inside the steady state sessile liquid drop.[7] Wetting is defined as the reduction of the solid–vapor interfacial energy (γ_{sv}) by the liquid, so that the driving force for wetting is the difference between the solid–vapor and solid–liquid interfacial energies ($\gamma_{sv} - \gamma_{sl}$) acting on the periphery of the liquid drop at the three phase contact line. The resisting force (γ_{lv}) is a result of the extension of the liquid surface, as the equilibrium geometry of a liquid drop is a sphere (i.e., the surface tension of a liquid tends to reduce its surface area to a minimum).

For the case depicted in Figure 21.1, $\gamma_{sv} > \gamma_{sl} > \gamma_{lv}$ so that the contact angle θ is acute and by definition, wetting of the solid by the liquid occurs. As the contact angle approaches zero, an extension of the liquid surface takes place with complete wetting (also defined as spreading) by the liquid. In practice, the driving force for wetting does not exceed γ_{lv} and spreading of the liquid drop does not occur, unless a chemical reaction takes place at the interface.[8] For the case where $\gamma_{sv} < \gamma_{sl} < \gamma_{lv}$, the contact angle is obtuse and wetting of the solid by the liquid is thermodynamically unfavorable. The condition where the contact angle $\theta = 90°$ can be defined as the boundary between wetting and nonwetting behavior of the liquid sessile drop.

Gibbs presented a more rigorous thermodynamic treatment of Young's construction, accounting for the effects of a gravitational field on the system.[9] It is now well established that the effect of chemical reactions on the relative interfacial energies of solid–liquid–vapor three phase systems, particularly at elevated temperatures where mass transfer effects can be significant, is such that the free energy of the reaction can enhance the driving force for wetting of the solid by the liquid.[10,11] This is true because mass transfer across the interfaces must result in a net decrease of the free energy of the system at any time, otherwise the reaction would not proceed. In effect, the magnitude of the relative interfacial energy between the solid and the liquid (γ_{sl}) is decreased by chemical reactions at the interface, so that the driving force for wetting of the solid and spreading by the liquid is improved.

It has also been found that the wetting behavior of a liquid on a solid can be influenced by the roughness of the surface microstructure. The derivation of Equation 21.2 assumes an ideally smooth and flat solid surface.[12,13] In real situations, the surface of a substrate is irregular and contains defects, voids, local deformation, and asperities.[14] Studies have demonstrated that marginal surface roughness can enhance wetting by reducing the liquid contact angle, broadening the three phase contact line and increasing the relative surface energy of the solid.[15–17] Conversely, when surface roughness is considerable it may not be possible for the liquid to fully penetrate and wet the surface irregularities.[18] More applications of the Young equation may be found in Chapter 8 and Chapter 30.

21.2.3 Sessile Drop Experiments

Although theoretical methods for the calculation of interfacial energies exist,[19,20] the generally accepted way to determine surface tensions and liquid–solid interfacial energies is by using an empirical approach. A variety of experimental techniques exist for the determination of the surface tension of a liquid, although, in most cases in the literature, the technique of choice has been the sessile drop method.[21] This technique provides a relatively straightforward way of accurately estimating surface tension of a liquid from the shape and the radii of curvature of a drop of the liquid resting on a horizontal solid surface, known as a sessile drop (Figure 21.1). The sessile drop method is ideally suited for measurements at high temperature, where true equilibrium conditions can be rapidly attained with minimal chance of reaction or contamination.[22]

The measurement of interfacial energy at a liquid–solid interface via the sessile drop method involves the following steps: (1) the formation of a drop of the liquid of interest on a smooth, horizontal surface of the solid, (2) the attainment of a relaxed shape of the drop that is in equilibrium under the combined effects of surface tension and gravity, and (3) measurement of the drop profile and calculation of surface tension. In the case of a glass-forming liquid, a common practice is to start with a cylinder of glass and increase the temperature sufficiently to allow for flow and deformation of the cylinder, which then forms a drop with an equilibrium sub-spherical shape. Sessile drops must be allowed to attain their equilibrium shape before measurement of the drop profile. The time τ required to attain a relaxed sessile drop shape depends on the shear viscosity η of the liquid, the surface tension γ, and the diameter of the drop D, according to the relation:[23]

$$\tau = D\eta/\gamma \tag{21.3}$$

It is critical to select an optimum drop size for an accurate application of the sessile drop method. If the drop size is large, then the gravity effect may be too high compared to the surface tension and the initial drop may break into smaller droplets. On the other hand, the deformation of the drop under gravity may be negligible for too small a drop size, which will introduce large error in the surface tension measurements.

Once the sessile drop is in equilibrium, a high-quality photograph of the drop can be taken with a camera and digitized to obtain the drop profile parameters needed for the calculation of γ. The calculation of γ is based on the relation between the shape factor β of the drop and its radius of curvature b at the apex, which can be written as:

$$\beta = b^2\rho g/\gamma \tag{21.4}$$

where ρ is the density difference between the liquid and the surrounding medium (usually air, inert gas, or vacuum) and g is the acceleration due to gravity.[22] The quantity β appears explicitly in the Laplace

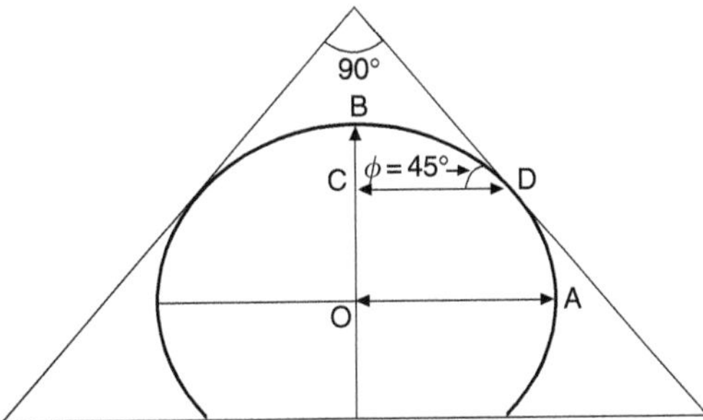

FIGURE 21.2 Typical profile of a sessile drop (bold curve) with an obtuse contact angle. The equatorial radius r_e and apex height h_e are equal to OA and OB, respectively. Tangents on either side of the profile at $\phi = 45°$ are according to the Dorsey construction (see text) where r_{45} and h_{45} are equal to CD and CB, respectively.

equation describing the equilibrium free surface of a sessile drop:

$$\frac{d^2z}{dx^2} + \left[1 + \left(\frac{dz}{dx}\right)^2\right]\frac{dz}{x\,dx} = (2 + \beta z)\left[1 + \left(\frac{dz}{dx}\right)^2\right]^{3/2} \tag{21.5}$$

Different numerical procedures are applied to obtain b, γ, and β from the experimentally measured drop profile, due to the lack of analytical solutions for these equations. One of these procedures involves computerized nonlinear fitting of the vertical (z) and horizontal (x) coordinates of the sessile drop free surface using Equation 21.4 and Equation 21.5. The parameters β and γ are optimized to obtain the best fit.[24] The other is a graphical procedure that utilizes the tables of Bashforth and Adams, which provide values of β as a function of the ratio of the equatorial radius r_e and the distance h_e between this diameter and the apex of the drop (Figure 21.2).[21,25] The measured values of r_e and h_e can be used to obtain β as well as values of r_e/b and h_e/b (and hence that of b) from these tables.[25] The surface tension γ can then be obtained using Equation 21.4, if the density of the liquid is known.

There are other, similar geometrical constructions for obtaining γ using the Bashforth–Adams tables. One such construction, proposed by Dorsey, is particularly notable for its simplicity. This method is based on the construction of two tangents on the two sides of the drop profile at $\phi = 45°$ (Figure 21.2). The corresponding radius r_{45} and height h_{45} are then used together with the equatorial radius $r_e(\phi = 90°)$ to obtain γ from the following equation:[26]

$$\gamma = \left(\frac{0.052}{f} - 0.12268 + 0.0481f\right) r_e^2 g\rho \tag{21.6}$$

where $f = [(r_{45} - h_{45})/r_e] - 0.41421$ and g and ρ have the same meanings as in Equation 21.4. An extension of Dorsey's method is based on similar tangent constructions at $\phi = 45°, 60°, 120°$, and $135°$.[22] The corresponding four β values are obtained from the tables of Bashforth and Adams and a weighted average of β is calculated giving stronger preference for those corresponding to higher ϕ values. This averaged β is then used to calculate γ using Equation 21.4.

It is clear from Equation 21.4 that the accuracy of the estimated value of γ depends critically upon the accuracy with which the density of the liquid is known at the temperature of interest.[21,22] The other factors that limit the accuracy of γ are errors introduced from measurements of the coordinates of the drop profile and the presence of surface contaminants or chemical reactions at the liquid–solid interface.

21.2.4 The Ellingham Diagram

The Ellingham diagram is one of the most useful thermodynamic tools for determining chemical reactions and equilibria germane to bonding processes. In 1944, Harold Ellingham published two diagrams that graphically depicted how the standard Gibbs free energy of formation changes with temperature for various oxide and sulfide reactions.[27] A few years later, Richardson and Jeffes published a revised version of Ellingham's diagram for the oxides that included a clever nomographic scale.[28] The scale allows estimation of the equilibrium partial pressure of oxygen for a given oxidation reaction at a particular temperature, or conversely the equilibrium temperature for the reaction at a specific oxygen partial pressure. The usefulness of the Ellingham diagram for metallurgical process applications is by now well established.

Under standard state conditions, the Gibbs free energy for a chemical reaction is given by:

$$\Delta G^\circ = \Delta H^\circ - T\Delta S^\circ \tag{21.7}$$

where ΔH° is the standard enthalpy change of the reaction (heat of reaction) and ΔS° is the standard change in entropy for the reaction. For the case of an oxidation reaction at equilibrium and standard state, the free energy of formation is also given by:

$$\Delta G^\circ = -RT \ln K = RT \text{ In } p_{O_2} \tag{21.8}$$

where K is the equilibrium constant for the reaction per mole of oxygen and p_{O_2} is the oxygen partial pressure. As the temperature dependence of ΔH° and ΔS° is modest when compared with ΔG°, over a temperature range where none of the reactants or products in the reaction undergoes a phase change, the value of ΔG° and therefore $RT \ln p_{O_2}$ is essentially linear with temperature. As such, a plot of ΔG° vs. temperature for these reactions results in a series of relatively parallel straight lines, whose intersections with the y-axis ($T = 0^\circ$) provide ΔH° and whose slopes are given by $-\Delta S^\circ$. As oxidation reactions generally involve a solid metal reacting with oxygen to form a solid oxide, most of the entropy contribution comes from the oxygen gas, so that the slope of the ΔG° plot is positive and the standard free energies become less negative as the temperature is increased (i.e., the oxide becomes thermodynamically less stable than the metal). By comparing the relative position of the ΔG° plots for selected oxides in the Ellingham diagram (Figure 21.3), the stability of the oxides (or oxygen affinity of the corresponding metals) as a function of temperature can be qualitatively compared. Because their standard free energy of formation is more negative, reactions that are lower on the figure have oxides of higher stability than those above them. This also means that a given metal can reduce the oxide of any metal with a ΔG° plot above it on the diagram.

The Ellingham diagram can also be used to determine the partial pressure of oxygen in equilibrium with a metal oxide at a particular temperature. The metal oxidation standard free energy plots in the diagram are based on one mole of oxygen reacting at a standard pressure of one atmosphere (when $\Delta G^\circ = 0$, $p_{O_2} = 1$). Richardson and Jeffes showed that a family of constant p_{O_2} lines can be constructed, which emanate from the origin of the Ellingham diagram ($\Delta G^\circ = 0$, $T = 0$) with slope $R \ln p_{O_2}$, each of which represent an effective $\Delta G^\circ = 0$ for a particular oxygen partial pressure. The intersection of a metal oxide ΔG° plot with one of these constant p_{O_2} lines indicates the equilibrium temperature for that particular value of p_{O_2}. In actual practice, the most common approach is to draw a straight line from the origin of the diagram (Point O in Figure 21.3) that intersects a selected oxidation reaction line at the temperature of interest and extends to the p_{O_2} nomographic scale outside the margin of the diagram. The intersection of this line with the scale indicates the partial pressure of oxygen in equilibrium with the oxide at that temperature. In other words, the metal will oxidize at this temperature if the partial pressure of oxygen in the ambient exceeds the equilibrium value determined, or the metal oxide will be reduced if the oxygen partial pressure is below this level. In an analogous fashion, the equilibrium partial pressure for H_2/H_2O can be determined by employing point H in Figure 21.3 and the equilibrium partial pressure for CO/CO_2 by using point C. The Ellingham diagram is not only useful for determining the relative stabilities of competing reactions, but also for gauging the effects of temperature and atmosphere on a particular oxidation reaction.

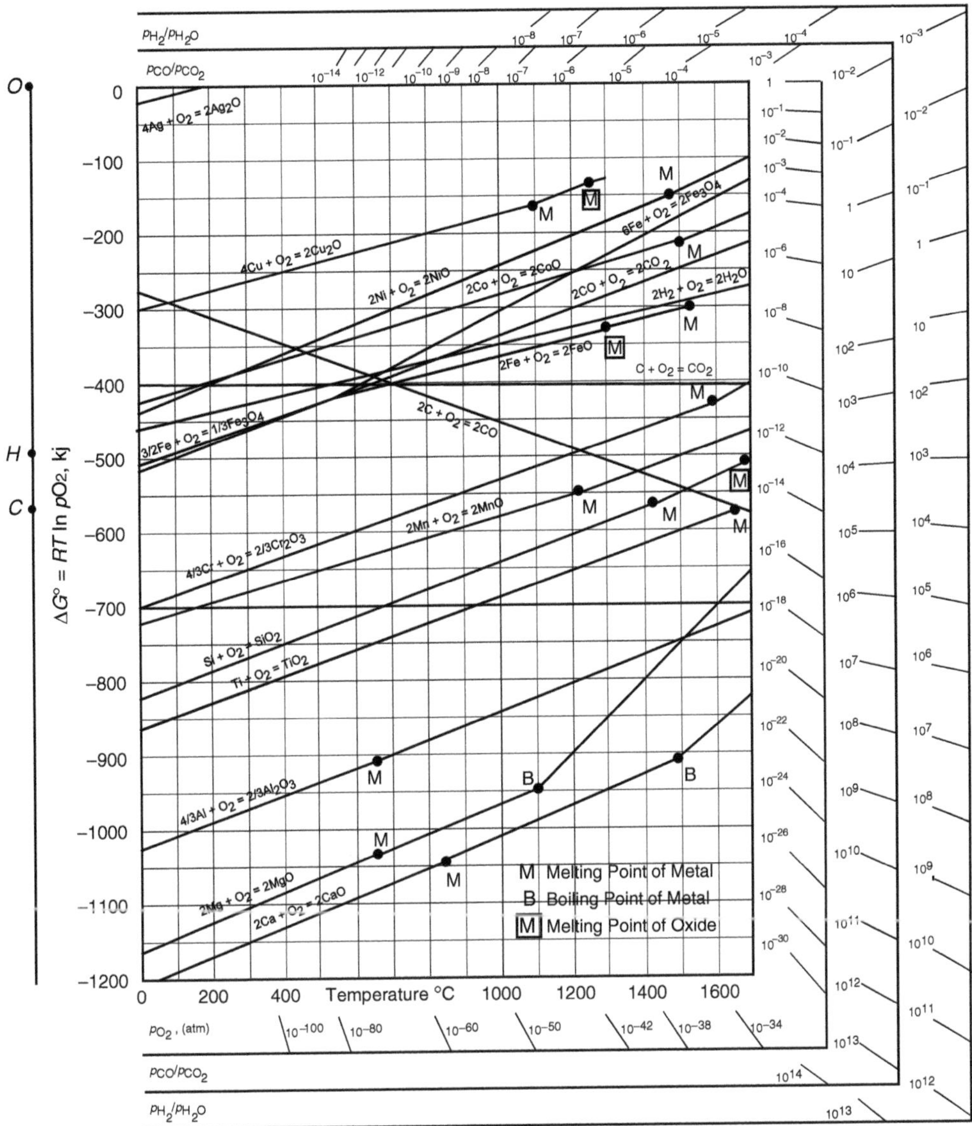

FIGURE 21.3 The Ellingham diagram for metallurgically important oxides. (From Gaskell, D.R., *Introduction to Metallurgical Thermodynamics*, 2nd ed., Hemisphere Publishing Corporation, New York, 1981. With permission from Taylor and Francis Group LLC.)

21.3 Glass-to-Metal Sealing

The distinct lack of thermodynamic affinity between glasses and metal alloys is the primary reason why the joining of these materials has historically involved considerable technical complexity. The basic requirements for a strong, hermetic glass-to-metal seal (GMS), whether based on a cylindrical bead design or fused integrally into a package style housing, are chemical bonding (i.e., electronic structure via atomic contact) and minimal stress differentials at the glass–metal interfaces, along with favorable stress gradients in the interfacial zones.[29] It is possible for two phases in intimate contact to form chemical bonds if they are at stable chemical thermodynamic equilibrium across their interface, whether or not the bulk phases are at equilibrium. Although methods have been established for modeling residual stress in glass seals and it is well known that stress differentials at glass-to-metal seal interfaces are minimized by matching the thermal expansion coefficients of the components,[30–33] the dependence of stress gradients in the interfacial zones on composition gradients that form during the fusing of the seals is not as well recognized.

21.3.1 Glass-to-Metal Seal Types

Although a number of classification schemes have been developed for glass-to-metal seals, such as those based on function or geometry,[34–41] the most direct and succinct is a classification based on the thermal expansion (i.e., residual stress) relationship of the glass seal components. As such, glass-to-metal seal types can be categorized as matched seals, compression seals, or what will be referred to here as hybrid seals.

For matched glass-to-metal seals, the component materials are selected so their thermal expansion characteristics are relatively close to each other throughout the sealing temperature range.[42] As a rule of thumb, this means the average coefficient of thermal expansion (CTE — expressed in units of cm/cm °C or simply /°C) for the glass and metal are within about 5 parts per million of each other. The strength of matched seals relies on chemical bonding developed during the sealing process between the glass and a metal oxide that has previously been grown on the metal components. An example of material selection for a matched glass-to-metal seal is borosilicate glass (CTE — 4.7×10^{-6}/°C) and ASTM F-15 alloy (Kovar®, CTE — 5.1×10^{-6}/°C) for a sleeve and pin configuration.

Compression glass-to-metal seals are designed such that the sleeve or housing material has a thermal expansion coefficient that is significantly higher than the glass and pin, particularly below the set point (often referred to as the softening point) temperature of the glass.[43] Typically, the difference in CTE between the glass and housing for compression seals is 5 to 10 parts per million, depending on seal geometry and material selection. During the sealing process, the housing contracts around the solidified glass and puts the glass under a compressive stress. This is advantageous because the compressive strength for most glasses is about an order of magnitude higher than the tensile strength, which imparts considerable mechanical stability to the glass seal assembly. In practice, the thermal expansion hierarchy for the components is designed such that the CTE of the housing is at least 5 ppm greater than the glass, and the glass CTE is slightly higher than the pin. This approach insures that the various glass–metal interfaces remain under compression when the glass seal is subjected to thermal cycling in service. An example of materials for a hierarchical compression seal is 316 stainless steel (CTE — 18.5×10^{-6}/°C) for the housing, soda-barium glass (CTE — 9.5×10^{-6}/°C) and ASTM F30 alloy (Alloy 52®, CTE — 9.3×10^{-6}/°C) for the pin.

The hybrid glass-to-metal seal combines the attributes of a compression seal and a matched seal in the same construction. The component materials are selected such that the hierarchy in their CTEs insures that the glass in the seal assembly will be under compression at room temperature. The metal alloy materials are selected so that it is possible to preoxidize the sleeve and pin prior to glass fusing and as such, realize chemical bonding at the metal–glass interfaces. An example of materials for a hybrid seal is AISI 1215 steel (CTE — 12×10^{-6}/°C) for the sleeve, borosilicate glass (CTE — 4.7×10^{-6}/°C) and ASTM F-15 alloy (CTE — 5.1×10^{-6}/°C) for the pin. For matched and hybrid glass-to-metal seal assemblies, tungsten (CTE — 4.6×10^{-6}/°C) and molybdenum (CTE — 5.5×10^{-6}/°C) are sometimes utilized for the center pin material.

21.3.2 Glass-to-Metal Sealing Process

The general process sequence for the fabrication of glass-to-metal seal assemblies involves a number of separate steps (Figure 21.4). First, the components (glass, center pin, and sleeve or housing) must be thoroughly cleaned (e.g., degreasing in a solvent followed by a light acid etch for metals or rinsing in mild solvents for glasses) and outgassed. If the metal components (sleeve and pin) are ferrous based, decarburization is performed to avoid gas formation (CO and CO_2 bubbles) at the glass–metal interface, during the glass fuse operation.[44] These bubbles are detrimental because they can lead to blistering and loss of adherence at the interface.[45] Decarburization can be achieved by exposing the parts to a moist hydrogen atmosphere with a dewpoint of about 20°C at 1000 to 1100°C for 7 to 15 min, depending on

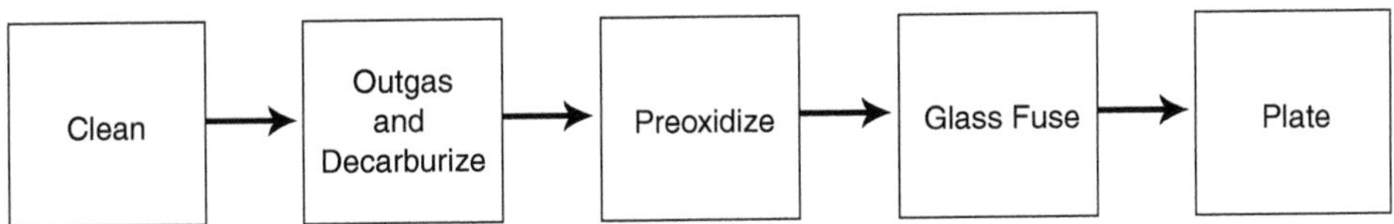

FIGURE 21.4 Generic process flow sequence for the fabrication of glass-to-metal seals.

the size of the part.[46] The key, particularly with Kovar®, is to select the decarburization temperature so that it is higher than the glass sealing temperature.

For the case of matched, hybrid, and sometimes the center pin for compression seals, the next step is to grow a contiguous intergranular oxide layer on the metallic components. The preoxidation procedure is critical to the ultimate reliability of glass seals that rely on chemical bonding, so process control is paramount. If the oxide is too thin, the strength of the glass-to-metal interface will be low due to superficial chemical bonding. Conversely, excessive oxide thickness may compromise the hermeticity and strength of the seal due to porosity in the oxide or bubbles that can form when the glass is fused. The preoxidation process is typically conducted in controlled atmosphere furnaces using a moist nitrogen environment (sometimes augmented with 0.25 to 0.5% hydrogen) at a dewpoint of 3 to 9°C and a set point temperature of 850 to 1050°C.[47–49]

The next step is the assembly of the GMS components into tooling fixtures for the glass sealing (glass fuse) procedure. Graphite is the most common material used for glass fuse tooling, especially for tooling that comes into direct contact with the glass, because graphite can be machined to very close tolerances and molten glass does not wet or stick to it (due to the interfacial energy relationship). The sealing process (Glass Fuse step in Figure 21.4) is performed in a continuous furnace under dry nitrogen (dewpoint < -20°C) or a lean exothermic gas atmosphere at temperatures ranging from 950 to 1000°C for 15 to 30 min, depending on the mass of the fixtures and oxide thickness on the parts. A dry nitrogen atmosphere is beneficial because it mitigates decomposition of the carbon tooling and possible unwanted oxidation of stainless steel compression seal components, which can occur in the presence of even small amounts of moisture. Ideally, matched and hybrid seals should emerge from the fuse furnace with minimal oxide remaining on their free surfaces, to facilitate the subsequent chemical cleaning necessary to remove the residual oxide prior to the electroplating step. In most cases, the fused glass-to-metal seals are electroplated, typically with several microns of gold over a diffusion barrier of nickel, to facilitate solder and lead termination processes that occur when the glass seals are incorporated into upper level assemblies.

The technological basis for achieving chemical bonding and favorable stress gradients in glass-to-metal seal assemblies is via controlled preoxidation of the metal alloy components.[50–56] If the metal oxide is of proper thickness (1 to 6 microns) and composition (low valence oxide), the molten glass will properly wet and dissolve the oxide when the GMS is fused at temperature. The glass at the oxide interface is quickly saturated because the solution rate of the oxide is faster than the diffusion rate of the dissolved oxide into the bulk glass.[57] This is the foundation for chemical bonding at the glass–oxide interface.

As the dissolved oxide diffuses further into the glass, a concentration gradient is formed that becomes more extended with time. The presence of the dissolved metal oxide in the molten glass alters the effective CTE of the glass, due to an increase in the O/Si ratio and the introduction of cations with varying degrees of covalency. The concentration gradients are then proportional to gradients in CTE, which generally results in more favorable stress gradients. Hence, the saturation of the glass with metal oxide near the interface and the subsequent diffusion of the oxide into the bulk of the glass results in chemical bonding and favorable stress gradients, as required. In effect, the metal oxide functions as a thermodynamic glue, so to speak, which enables bonding between the glass and metal.

21.3.3 Case Study: 200 GHz Coaxial Connector

The fundamental electrical performance of hermetic precision coaxial connectors in the mm-wave regime is critically dependent on the precision of the glass-to-metal seal transition geometry. If one considers the following equation, which provides a reasonable estimate of the maximum cutoff frequency for a coaxial transmission line,[58]

$$f_c = 7512[\varepsilon(d + D)]^{-1/2} \tag{21.9}$$

where ε is the dielectric constant in the transmission line, d is the outer diameter of the center conductor, and D is the inner diameter of the outer conductor, it is clear that f_c increases as the characteristic dimensions of the coaxial line decrease. Hence, as frequency ranges have pushed upward and concurrent glass seal geometries have become miniaturized, the resulting design requirements have stretched the limit of materials, metrology, and manufacturing process technology (Figure 21.5a).

FIGURE 21.5 (a) Photograph of glass-to-metal seals for high frequency electronic applications. The characteristic seal geometries decrease with increasing frequency and typically, the glass seals are designated by the size of the inner diameter of their coaxial outer conductor. Glass-to-metal seals shown here: 3.5 mm (26.5 GHz), 2.4 mm (50 GHz), 1.85 mm (67 GHz), and 1 mm (110 GHz). (b) Photograph of the 200 GHz coaxial connector (mated pair), also known as the "A200." (c) Optical micrograph of the 200 GHz connector in cross section showing the glass-to-metal seal. The nominal inner diameter of the sleeve (coaxial outer conductor) is 600 microns and the outer diameter of the center pin (coaxial inner conductor) is 254 microns. The pin steps down to a diameter of 89 microns inside the glass, to compensate for the discontinuity in dielectric constant between air and glass. (Courtesy of Agilent Technologies, Palo Alto, CA.)

The 200 GHz coaxial connector (Figure 21.5b) has a nominal inner diameter dimension for the sleeve (D in Equation 21.9) of 600 microns and nominal outer diameter dimension for the center conductor (d in Equation 21.9) of 89 microns, or about the diameter of an average human hair. For these miniature geometries, conventional machining tolerances represent a relatively large percentage of a particular component's size. And as the skin depth for signal transmission at 200 GHz is on the order of a quarter micron, the components in the glass-to-metal seal transition must exhibit microscopically smooth surface finishes.

The center conductor for the 200 GHz connector must support significant stress when it is mated with the center conductor of its matching connector launch assembly. The material selection for this component is tungsten, chosen for its high modulus of elasticity (400 GPa) and ability to be readily preoxidized (for chemical bonding with the glass). A 45° taper at the mating end of the center conductor is necessary to accommodate the connection procedure by establishing a wiping contact at the interface.[59] The center conductor also requires a step-down in diameter from 254 to 89 microns where the glass-to-metal seal is fused. Traditional approaches (machining and grinding) were investigated and found to be unworkable. A significant process breakthrough was achieved by turning the tungsten rod on a submersible lathe in a precision fine wire electrical discharge machining (EDM) unit.[60] This approach results in a precise, well-defined center conductor diameter step-down with reasonable surface finish and produces a taper on the mating end of the center pin that greatly facilitates the connector mating process.

The tiny glass bead for the 200 GHz glass-to-metal seal is fabricated from borosilicate glass in two halves, to aid in assembling the glass beads into the glass fuse tooling fixtures. Because of the design requirement for a well-defined and reproducible dielectric constant (driven by the high frequency application), the glass must be drawn into a tube with the correct outer and inner diameters, cut to length, cut in half longitudinally and finally polished on their flat surfaces, before final sealing to the metal. Fritted glass (ground and pressed with a binder) does not work for this application because it results in a myriad of tiny bubbles when the glass is sealed, leading to an ill-defined effective dielectric constant. The glass-to-metal seal assembly necessitates precision graphite tooling fixtures, to maintain the requisite geometry during the glass fuse process, and careful control of both the preoxidation and sealing furnace profiles. A cross section of the 200 GHz connector glass-to-metal seal is shown in Figure 21.5c. Considering the additional complications in fixturing, assembly, and metrology, it will be appreciated that fabrication of the glass-to-metal seal for the 200 GHz connector represents a formidable manufacturing challenge.

21.4 Ceramic-to-Metal Joining

As with glass-to-metal sealing, the basic requirements for strong, hermetic ceramic-to-metal joints are chemical bonding and minimal stress differentials at and near the interfacial junction. For both cases, these requirements are satisfied by the formation of a thermodynamically stable metal oxide and relatively close matching of the thermal expansion coefficients of the component materials (see Section 21.2). Researchers have investigated and modeled the residual stresses that can compromise the strength of ceramic-to-metal joints.[61,62] In general, the specific fabrication process steps used to achieve a strong, reliable hermetic seal are more involved for the case of ceramic-to-metal joining.

21.4.1 Mo–Mn Metallization Process

Although patents describing the metallization of electrical porcelains have been issued since the middle of the nineteenth century, the advent of modern vacuum tight ceramic-to-metal seals owes to the invention of refractory metal metallization by Pulfrich in the late 1930s.[63] His discovery led to the actual production of vacuum tubes during the Second World War. Later in the 1950s the rise of microwave electronics coupled with improvements in furnace technology catalyzed development of the moly-manganese process,[64,65] the traditional technique used for the fabrication of ceramic–metal assemblies.

The central feature of the Mo–Mn process is the metallization of ceramic components using a special "paint," mainly comprised of metal oxides and powders that are mixed with nitrocellulose lacquer and ground in a ceramic ball mill until a suspension having an average particle size of several microns is achieved. This is a time consuming process that can take up to 50 h to complete. The paint is applied to a ceramic substrate (layer thickness of about 25 microns) and fired under a wet N_2/H_2 atmosphere at a furnace temperature typically between 1400 and 1600°C. This sintering process produces reactions between metal oxides in the paint and the ceramic substrate, resulting in a chemically bonded metallization layer. The surface of the metallization layer is generally electroplated prior to final assembly with nickel or copper, which is subsequently sintered into the metallization. The plating serves to enhance wetting of the metallization by the braze alloy that will be used to produce the final assembly.

Finally, the metallized ceramic is brazed to a metal member in a continuous hydrogen furnace. This represents a conventional metal-to-metal joining operation, the technical details of which are well established. Although many different brazing alloys have been used for this step of the procedure, the prime consideration in all cases is to allow adequate time for the braze material to wet and flow while avoiding dissolution of the ceramic metallization.

21.4.2 Active Metal Brazing Process

The multitude of process steps for and associated expense of metallizing ceramics prior to ceramic-to-metal joining can be avoided by employing the technique of active metal brazing.[66–73] This process takes advantage of the thermodynamic affinity of certain high-temperature metals, primarily titanium and molybdenum, for oxygen. By alloying a more conventional braze filler material with one of these active metals, it is possible to produce a reliable, hermetic ceramic-to-metal interface using a single step brazing process.

Although active metal brazing alloys can be comminuted into powders and applied in lacquer carriers, the preferred approach is to prealloy the braze filler by arc melting and then die stamp annular preforms from a cold rolled thin foil of the alloy. The preforms are placed between the components to be joined and brazed in a vacuum furnace. When an active metal braze alloy is taken to temperature during a brazing operation, the oxygen affinity of the active metal constituent provides the driving force for its preferential diffusion to the braze filler–ceramic component interface. It is here that the oxidation reaction necessary for true chemical bonding takes place. The mechanism involves substitutional displacement of cations in the ceramic by the active metal component, in conjunction with an oxidation reaction between the active metal and oxygen in the ceramic. For example, a Ti doped braze filler in contact with sapphire (Al_2O_3) would react as follows:

$$3(\mathrm{Ti}) + \mathrm{Al_2O_3} \rightarrow 2(\mathrm{Al}) + 3\mathrm{TiO} \tag{21.10}$$

The dynamic equilibrium for this reaction is controlled by the local Ti and Al activities, while the reaction kinetics are controlled by Ti and Al diffusion. For most active metal braze alloys, the dissolution of ceramic cations in the braze melt is rather slow, so that wetting in the traditional sense (liquid melt transported by capillary action) does not occur.[74] For these alloys, wetting is manifested by the appearance of a grey-blue color at the interface, which is indicative of the formation of a tenacious and desirable oxide. If the ceramic component is transparent, the process provides its own built in metric, as the braze–ceramic interface can be visually inspected.

Generally, a relatively small percentage of the active constituent is preferred in active metal braze alloys (often <2 wt.%) to minimize or avoid the formation of brittle intermetallic phases in the brazement. As the active metal component readily getters oxygen, the process requires vacuum processing, so as to avoid oxidation reactions anywhere except the ceramic–metal interface (e.g., vacuum better than 5×10^{-5} torr, depending on the braze alloy).

The one step active metal brazing process is attractive not only because it promotes chemical bonding without the costly and time consuming premetallization of the ceramic component, but also because the active metal brazements offer excellent reliability for special applications involving resistance to high-temperature corrosion. It is important to note that this process can also be used to join metals and

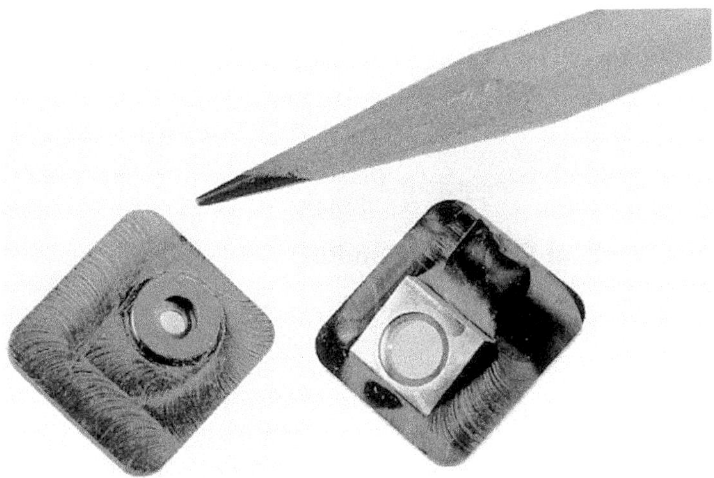

FIGURE 21.6 Photograph showing two examples of active metal brazed lid/sapphire window assembly designs. (From Powers, M., in *Joining of Advanced and Specialty Materials*, Singh, M., Indacochea, J.E., and Hauser, D., Eds., American Society for Metals International, Materials Park, 63, 1998. With permission from ASM International. Photograph courtesy of Agilent Technologies, Palo Alto, CA.)

nontraditional ceramics for which metallizations are not available, such as Si_3N_4, AlN, SiC, Y_2O_3, TiB_4, and graphite.

21.4.3 Case Study: Sapphire Laser Window

Measurement requirements used to characterize digital communications in the time domain have driven development of sophisticated optical communication analysis instruments. Integral to the success of one such optical telecommunications analyzer is a set of optical-to-electrical (O/E) plug-in modules that allow specific measurements at the various transmission rates commonly used in modern telecommunication systems.[75]

The heart of the O/E plug-in modules is a custom InP/InGaAs/InP p-i-n photodiode, which converts the incoming photons of light to a proportional electrical current. To protect this moisture sensitive photodiode, it is mounted (along with attendant amps, filters, and samplers) inside a hermetic hybrid microelectronic package assembly. A single-mode fiber is attached on the outside of the package assembly. The lid of the package has a sapphire window, which allows laser-generated light photons to pass from the fiber into the package and illuminate the photodiode.

The sapphire window is joined to an ASTM F-15 alloy (Kovar®) package lid using active metal brazing (Figure 21.6). This process eliminates the necessity of premetallization of the sapphire window by employing a quaternary active braze alloy of 63Ag–34.25Cu–1.75Ti–1Sn (in weight %), with a liquidus of 806°C. The hermetic ceramic-to-metal interface is developed *in situ* by taking the lid/window assembly to approximately 830°C in a vacuum of about 2×10^{-5} torr. By means of this relatively simple and robust joining process, the cost savings realized in fabrication of the lid/window assembly is about 70% when compared with the conventional Mo–Mn process approach.

21.5 Diffusion Bonding

Diffusion bonding is a method for joining metallic or nonmetallic materials to form metal-to-metal, metal-to-ceramic, or ceramic-to-ceramic joints. The diffusion bonding process is relatively straightforward for bonding of identical materials, but is significantly more challenging when dissimilar materials are employed. This bonding technique is based on the atomic diffusion of elements at the joining interface and, in principle, does not require an interlayer material that melts to form the bond. Diffusion bonding is a solid-state process that relies on atomic scale contact of the mating components. The process is generally accomplished by applying pressure to an assembly stack and then heating the assembly to approximately three-fourths of the absolute melting temperature of the lowest melting component, in a controlled atmosphere (usually inert or reducing). Joints formed in this manner are very stable and can usually be employed at relatively high service temperatures.

21.5.1 Fundamentals of Diffusion Bonding

The diffusion bonding process is based on the thermally activated transport of mass in the form of atomic migration through the lattice of a crystalline solid. Diffusion can be defined as the movement of atoms from a region of high concentration to a region of low concentration, which serves as the driving force for atomic movement. In general, the rate of diffusion is proportional to the gradient in concentration and is expressed mathematically by Fick's first law as:

$$J_x = -D(\partial c/\partial x) \tag{21.11}$$

where J_x is the flux of the diffusing species in the x direction, $(\partial c/\partial x)$ is the concentration gradient (which can change with time), and D is the diffusion coefficient, also known as the diffusivity. The transient nature of the concentration gradient is addressed in Fick's second law, which for the simple case where the diffusion coefficient is essentially independent of concentration can be expressed in a second order differential equation as:

$$\partial c_x/\partial t = D(\partial^2 c_x/\partial x^2) \tag{21.12}$$

The analytical solution to this equation depends on the geometry and boundary conditions for a particular situation and provides the means for determining salient diffusion parameters and resolving practical diffusion problems. A detailed treatment of diffusion theory is also presented in Chapter 18.

Migration of atoms in a crystalline solid can proceed by several mechanisms, such as the exchange of location between adjacent atoms, motion of atoms through interstitial sites, or motion of vacancies in the lattice structure of the material.[76] The exchange mechanism can occur if the solute and solvent atoms are relatively close to the same size. As this mechanism involves the interchange of atoms on sites in the lattice, there is an activation energy penalty due to distortion of the lattice that occurs during the exchange. If the solute atoms are significantly smaller than the solvent atoms, they can migrate via interstitial sites in the lattice with modest associated lattice distortion. The energy required for the interstitial mechanism depends on the crystal structure of the solvent material. For example, it is easier for solute atoms to move through a bcc material than through the corresponding fcc lattice (e.g., iron), as the bcc structure is more open. In the vacancy mechanism, an atom jumps to an adjacent unoccupied lattice site, leaving a lattice vacancy behind. Although there is some distortion of the lattice as the atom moves between its neighboring atoms to fill a vacancy, the associated energy is minimal. The vacancy mechanism is well established as the predominant means for diffusion in most metals and ionic compounds, due to the low activation energy required for atomic migration.

Diffusion is a kinetic process where temperature and the diffusivity of a particular material are important parameters. In general, the temperature dependence of the diffusion rate is manifested in terms of the diffusion coefficient D, which is defined as:

$$D = D_0 \exp(-Q/RT) \tag{21.13}$$

where the pre-exponential constant D_0 is a frequency factor that depends on the type of lattice and the oscillation frequency of the diffusing atom, Q is the activation energy per mole of diffusing species, R is the universal gas constant, and T is the temperature in Kelvin. This equation shows an Arrhenius relationship between the diffusion coefficient and temperature, so that for a given activation energy the diffusivity and resulting diffusion rate increase exponentially with increasing temperature. Consequently, elevated temperature (up to 0.9 of the melting temperature) is preferred for diffusion bonding because a stable and robust joint can be achieved in a relatively short time.

The activation energy for atomic diffusion at free surfaces, interfaces, and grain boundaries is low compared with that for bulk diffusion, due to looser bonding of the atoms and a higher oscillation frequency for the diffusing species.[77,78] This serves to enhance diffusion and facilitates diffusion bonding of components, assuming that intimate contact exists at the interface between them. The interfacial contact can be optimized by treatment of the bonding surfaces through a number of processes, such as mechanical machining, polishing, etching, cleaning, and coating.

Proper preparation of the component bonding surfaces for diffusion bonding is very important. The as-machined or formed surface finish is often not adequate for successful joining and usually requires some postprocessing to improve the smoothness of the surface. The surface finish of most metals and many ceramics can be enhanced by a variety of fabrication processes. Examples include media blasting with silica or silicon carbide, mechanical grinding using particle impregnated wheels, cloth or paper (often with silicon carbide or diamond for harder materials), lapping processes where the abrasive particles are free to rotate between the workpiece and a carrier substrate, and chemical polishing with agents that preferentially attack protrusions on the surface of the part. Mild etchants are sometimes used to remove surface oxides and in almost all cases, a cleaning process is required to remove organic contaminants left behind by the fabrication process.

The primary reason for coating some components is to create a clean, uniform, oxide free surface for the diffusion bonding process. A very thin coating layer of a common metal, such as nickel, gold, or silver, can be applied prior to bonding and normally dissolves into the substrate during the bonding process. Depending on the alloy system, the choice of coating material can affect the diffusion rate. Instead of a surface coating, it is sometimes possible to place a thin interlayer of nickel, gold, silver, platinum, or tantalum in the joint to improve the surface condition. This approach is particularly useful for diffusion bonding of refractory materials such as zirconia to alumina, or berryllia to copper.[79]

Material creep of the mating components is also an important process mechanism in diffusion bonding. Under high temperature and moderate applied loading, microscopic deformation through creep allows material to flow and fill voids or asperities in the bonding surfaces leading to full intimate contact between the components at the joint interface. This is generally accomplished by controlled uniaxial pressing of components that have been prepared with smooth flat surfaces. On the other hand, excessive loading pressure can lead to permanent macroscopic deformation, particularly when the bonding process temperature approaches the melting temperature of the components.

As diffusion bonding is driven by the thermally activated transport of atoms, the process can be effectively used to bond dissimilar materials that are difficult or impossible to weld. Examples include steel to copper alloys, steel to titanium alloys, and ceramics to other ceramics or to metals. In addition, diffusion bonding is a viable manufacturing process for the production of net shape and near net shape parts, because it does not cause significant macroscopic deformation when the process parameters are appropriately defined and controlled. Accordingly, surface condition, bonding temperature, and loading pressure are the primary factors that should be considered for optimization of the diffusion bonding process. Secondary factors such as component thermal conductivity, thermal expansion, and the bonding atmosphere of the chamber can also be important, particularly at high bonding temperatures where thermal effects and oxidation are accelerated.

21.5.2 Structural Integrity of Diffusion Bonded Joints

When properly utilized, diffusion bonding produces a monolithic structure. In general, the grain structure across the bond joint is similar to that of the bulk parent metal. However, depending on the degree of initial cold work in the components and hold time at the bonding temperature, recrystallization and grain growth may occur in the material during the diffusion bonding process. This can compromise certain material properties such as strength, ductility, and fatigue resistance, which in turn can impact the engineering design requirements of the assembly. If a surface coating or interlayer is employed at the joint interface, care must also be taken to avoid the possible formation of brittle intermetallic phases, which can degrade the strength and ductility of the bond, as well as adversely affect material properties such as electrical and thermal conductivity. As such, process variables including bonding temperature, hold time, and bond load, as well as bonding aids such as coating or interlayer metals, must be appropriately defined and selected for the materials to be bonded.

In some cases, particularly for diffusion bonding of dissimilar metals or alloys at high bonding temperatures, voids can form along the bond joint due to the Kirkendall effect.[80–82] This phenomenon occurs when one atomic species diffuses across the interface with a higher diffusion rate than another species diffusing in the opposite direction, and the diffusion mechanism is driven by the motion of vacancies.[80,82] As a result, vacancies tend to agglomerate in the side of the metal with faster moving atoms and form internal voids, which can adversely impact material properties such as strength, ductility, fatigue resistance, and fracture toughness. Kirkendall voids have been observed not only in diffusion bonded assemblies, but also in the solder joints of microelectronic assemblies.[80,81,83–85]

To mitigate internal voids in the bulk of substrate materials, consolidation of the diffusion bonded workpiece is sometimes necessary. In practice, this can be accomplished by a hot isostatic pressing (HIP) process, in which the workpiece is heated at elevated temperature in an inert atmosphere, such as argon, under relatively hydrostatic pressure for a specified hold time.[86] The HIP temperature is usually chosen to be within the solution anneal temperature range of the component material, so as to minimize the consolidation process residence time and avoid the formation of any undesired precipitates. The HIP process can also be used as a derivative diffusion bonding technique, where components are assembled into the geometry of the final product, encased into a form or mold, and then subjected to a diffusion bonding pressure/thermal cycle in an isostatic press.

Figure 21.7 shows the microstructure and various mechanical properties for the diffusion bonded joint between sheets of a typical 300 series austenitic stainless steel material. The grain structure across the joint interface is practically identical to that of the bulk material away from the interface. By comparing the ultimate tensile strength, yield strength, and ductility of the diffusion bonded sample with the corresponding bulk wrought alloy, it is clear that the properties of the diffusion bonded structure are comparable, and in some cases superior, to the parent stainless alloy. For this particular case, the ultimate tensile strength of the diffusion bonded sample is almost identical to that of the wrought material, both at ambient (24°C) and elevated service temperature (538°C). And although the yield strength for the joined specimen is lower (about 60% at elevated temperature), the ductility as a function of percent elongation is superior to the wrought sample, notably at the elevated service temperature. It is also noteworthy that the low-cycle fatigue behavior of the bonded joint corresponds to that of the bulk parent material.

21.5.3 Applications of Diffusion Bonding

Diffusion bonding has become a viable process for the fabrication of structural hardware and fluid/gas flow devices used in the aerospace and electronic industries. For example, thin plates of various metals and alloys can be diffusion bonded to produce cooling devices with extremely small and complicated flow channels. Assemblies fabricated in this manner have found applications in liquid rocket combustion chambers, missile sensor windows, injectors, heat exchangers, and cooled trailing edges.[87–91] The bonding concept is depicted in Figure 21.8 and involves a sequence of process steps where first the requisite apertures and channels are photo lithographically etched or laser cut into the plates, the plates are then arranged and stacked, and finally the plate stack is diffusion bonded to produce monolithic structures with internal flow channels.

Diffusion bonding has been employed for the fabrication of the copper liner for a liquid rocket combustion chamber, which requires extensive cooling due to the severe hot gas environment that results from the ignition of liquid oxygen in the chamber.[87,92,93] A similar process was used to fabricate a stainless steel window frame in the forebody of a land-based missile,[87] as shown in Figure 21.9. Cooling of

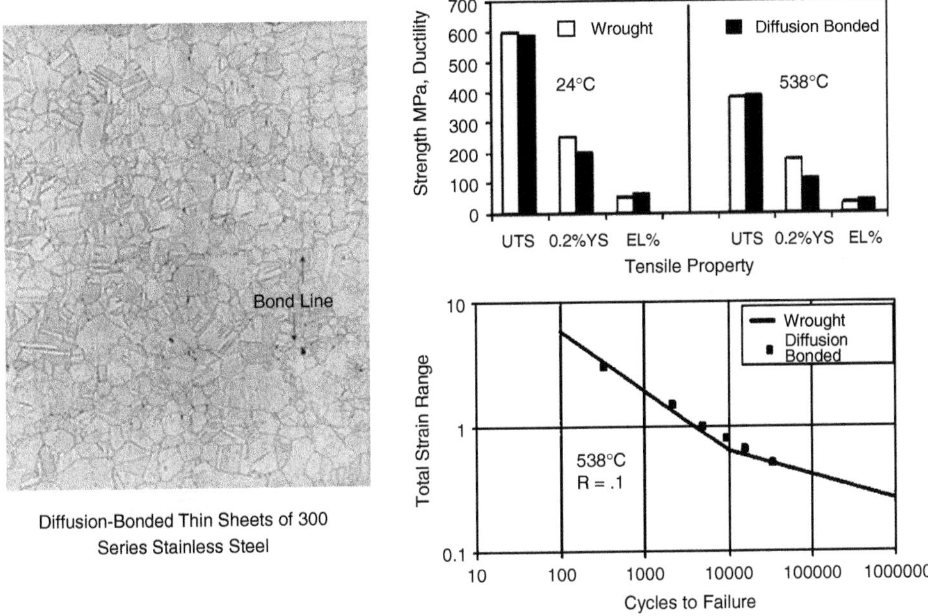

FIGURE 21.7 Left: Optical micrograph of typical microstructure in diffusion bonded 300 series stainless steel sheets. The location of the initial joint interface is only just discernable in some areas of the micrograph, owing to mild etching of the metallographic sample. Right: Cross bond ultimate tensile strength, yield strength, elongation and low-cycle fatigue life for diffusion bonded joint compared with properties of parent metal in wrought form. (Courtesy of Aerojet, Sacramento, CA.)

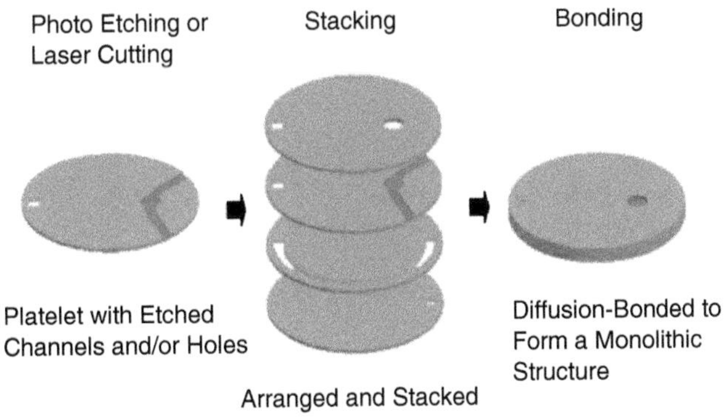

FIGURE 21.8 Schematic of the diffusion bonding process sequence to produce devices with internal flow channels.

the sapphire window is required to protect the electronic sensor underneath, due to severe hypersonic flight environments. Figure 21.9 also demonstrates several other applications of diffusion bonding, such as heat exchangers, regenerators, and cooled trailing edges.

Diffusion bonding has been demonstrated as an economical manufacturing process alternative for a number of products, such as horn arrays for antenna systems used in land and space surveillance (Figure 21.9). The production of such horn arrays through diffusion bonding of multiple etched platelets has proven to be simple and relatively low cost, compared with conventional techniques such as machining or electroforming of individual horns.[87,94,95]

In the past several decades, a variety of materials and diffusion bonding techniques have been developed for the production of commercial devices used in the aircraft, electronics, and automobile industries.

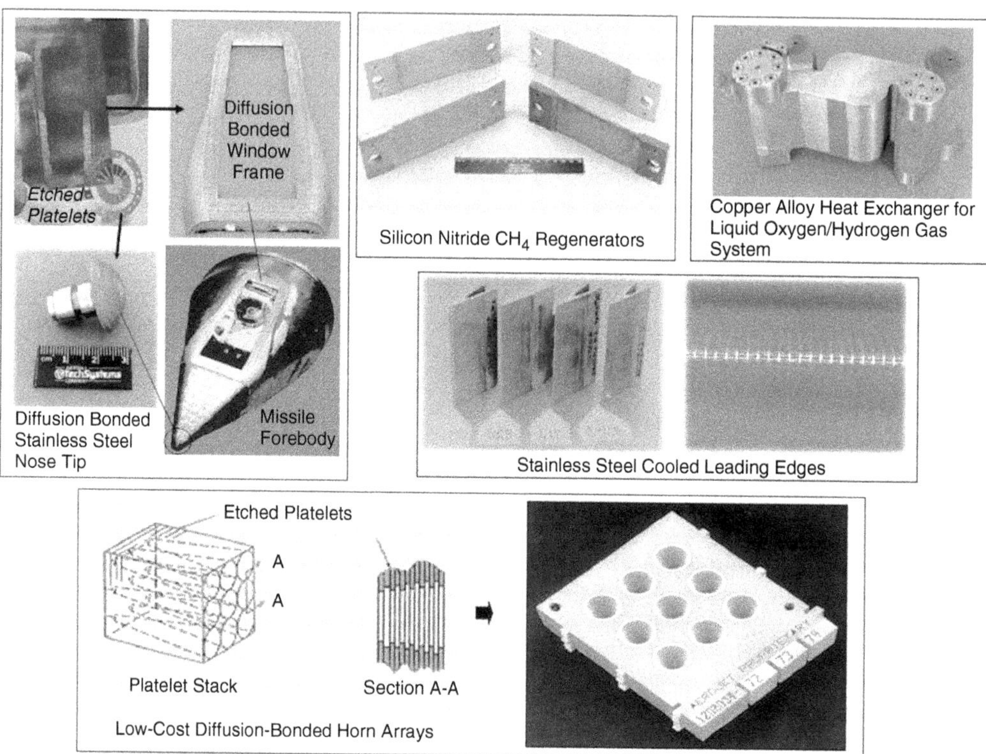

FIGURE 21.9 A variety of products fabricated via diffusion bonding. (Courtesy of Aerojet, Sacramento, CA.)

TABLE 21.1 Examples of Diffusion Bonding Materials in Aerospace Applications

Cu Alloys	*Ni Alloys*	*Co Alloys*
OFHC	Ni 200/201	Haynes® 188™
ZrCu	Inconel® 600/625	
Narloy-Z®	Inconel® 718	*Ceramics*
Glid Cop®	Haynes™ 214™	Si_3N_4
GrCop-84®		
		Dissimilar Systems
Mo Alloys	*Al Alloys*	Stainless Steel/ZrCu
Molybdenum	1100	Stainless Steel/Narloy-Z®
Mo–50Re	3000	Stainless Steel/Grcop-84®
	5000	Stainless Steel/Ni200
Fe Alloys	6061	A286/ZrCu
347		Ni/ZrCu
304/304L	*Ti Alloys*	Inconel® 600/ZrCu
316/316L	Titanium	Incoloy® 909/ZrCu
17-4PH	Ti–6Al–4V	Haynes® 188™/ZrCu
A286	Ti-15-3	Pt/Ni200
Incoloy909	Beta 21-S	Pt/MoRe

Examples include bonding of titanium alloy for turbofan disks, Hastelloy-x® alloy bonding for burner cans, niobium alloy bonding for heat shields, copper to berryllia bonding for electronic heat sink devices, zirconia to alumina bonding for oxygen analyzer sensors, and brake pad material to metal disk bonding for disk-brakes.[79,81] Table 21.1 presents a list of some of the engineering materials that have been successfully diffusion bonded and a number of other examples exist in the literature.[3,79,81,96–98]

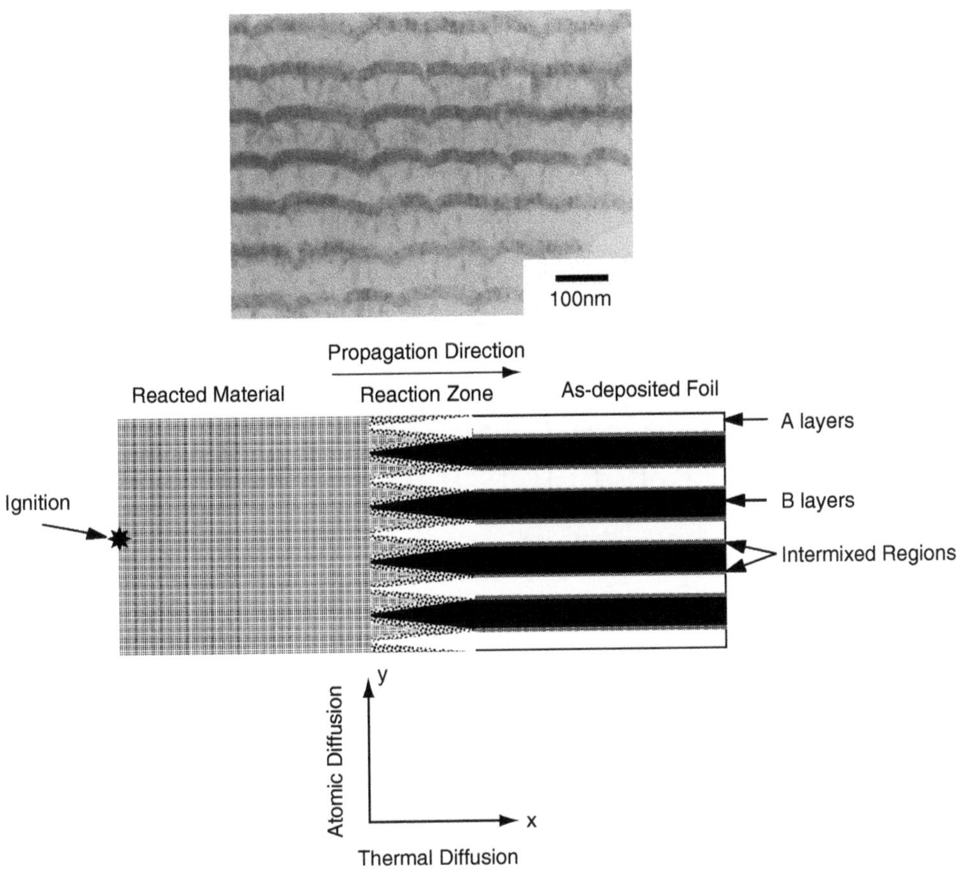

FIGURE 21.10 Top: Cross-sectional TEM micrograph of an as-deposited Ni/Al multilayer showing nanoscale layers. Bottom: Cross-sectional schematic of atomic and thermal diffusion during reaction propagation. Note that atoms diffuse normal to the layers while heat diffuses parallel to the layers. (From Gavens, A.J. et al., *J. Appl. Phys.*, 87, 1255, 2000. With permission from the American Institute of Physics.)

In summary, diffusion bonding is a practical method for the joining of advanced engineering materials. This bonding technique has been established as a straightforward precision process for the production of low cost, high-performance engineering devices, which are basic requirements for manufacturing technologies in the twenty-first century.

21.6 Reactive Joining

Reactive multilayer foils are a new class of nano-engineered materials that are typically fabricated by vapor depositing hundreds of nanoscale layers that alternate between elements with large negative heats of mixing, such as Ni and Al. An example is provided in Figure 21.10 top, which shows a cross-sectional transmission electron micrograph of a Ni/Al multilayer prior to ignition. These ignitable reactive foils support self-propagating exothermic reactions that travel at speeds ranging from 1 to 30 m/s and can be employed as local heat sources for joining of similar and dissimilar materials.[4,99–114]

21.6.1 Fundamentals of Reactive Joining

Self-propagating formation reactions in multilayer foils are driven by a reduction in chemical bond energy.[4,112] With a small thermal pulse, atoms diffuse normal to the layering (Figure 21.10 bottom) so

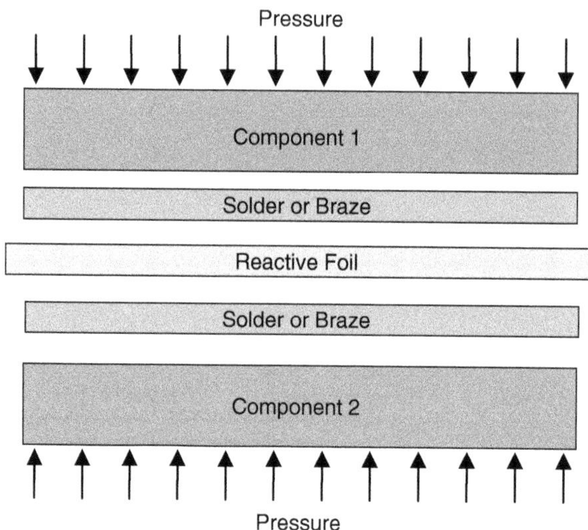

FIGURE 21.11 Schematic illustration of the geometric construction for the room temperature reactive joining process.

that A–A and B–B bonds are exchanged for A–B bonds. This local bond exchange produces a large quantity of heat that is conducted down the foil and facilitates further atomic mixing. The speed at which a reaction propagates along a multilayer foil depends on how rapidly the atoms diffuse normal to the layering and how rapidly heat is conducted along the foil.

Using a combination of empirical data and modeling, researchers have demonstrated that the velocities, heats, temperatures, and ignition thresholds of the reactions can be controlled.[4,115] This can be accomplished, for instance, by varying the thickness or composition of the reactive multilayers. Experimental observations have inspired the development of analytical and computational models for predicting reaction properties,[116–119] as well as computational models for predicting heat losses from reactive foils and the temperature evolution of neighboring materials when foils are ignited within an assembly.[120,121]

This scientific knowledge base has enabled the development of reactive joining, which relies on the use of reactive multilayer foils as local heat sources for room-temperature soldering or brazing.[122] As sketched in Figure 21.11, the fundamental methodology used in reactive joining is based on sandwiching a reactive multilayer foil between two solder or braze layers and the components to be joined, subjecting the resulting assembly to relatively low applied pressure, and then initiating the exothermic reaction by means of a moderate energy source. Heat released by the reaction results in melting of the solder or braze, which upon cooling bonds the components. This flux-free joining process is very rapid, typically completed in less than a second, and can be performed in air at room temperature. An additional feature of the reactive joining process stems from the localized nature of the heating, which naturally limits thermal exposure of the components and mitigates problems associated with differences in thermal expansion coefficients of the component materials. Both these problems are inherent in conventional furnace soldering or brazing processes. These features translate into key productivity, cost, or enabling technology advantages for several joining applications, such as component mounting onto printed circuit boards,[123] heat sink mounting onto chips,[124,125] hermetic sealing of MEMS and opto-electronic components,[126,127] metal–metal joining,[128–130] metal–ceramic bonding,[131] and the welding of metallic glass.[132]

21.6.2 Reactive Joining Design

Reactive joining technology has been the subject of intense development and validation in recent years. These efforts have shown that in general, there are several principal features that guide the design and development of a particular reactive joining process application. To highlight these features, it is essential to first identify two distinct reactive joining configurations. In the first geometry, which will be referred to as the preform configuration, the solder or braze utilized in the joint is applied as a freestanding sheet or preform. In the second geometry, called the prewet configuration, the solder or braze is applied to the component surfaces prior to the joining operation. A major design distinction between these two scenarios is that in the preform configuration, the reactive multilayer must provide sufficient heat to melt the entire solder or braze thickness, so that the faying surfaces of the components are wetted. By contrast, partial melting of the solder or braze layers is sufficient for the formation of strong bonds in the prewet configuration because the solder or braze is already chemically bonded to the components.

In addition to ensuring the necessary amount of melting of the solder or braze material, it has also been shown that the solder or braze must remain molten for a sufficiently long duration to allow adequate wetting of the component interfaces. In most cases, a melt duration of 0.5 ms or longer is sufficient for the formation of strong chemical bonds.[121,133] Both the amount and duration of melting depend on the properties of and factors related to the self-propagating reaction. These include reaction heat, reaction velocity, peak adiabatic temperature and overall foil thickness, the properties of the solder and braze material (heat capacity, thermal conductivity, density, heat of fusion, and melting temperature), as well as the properties of the components (density, heat capacity, and thermal conductivity). All of these parameters should be carefully considered in the design and development of reactive joining applications.

As alluded to earlier, the effect of thermophysical properties, the amount of melting within solder or braze layers during the reactive joining process and the thermal exposure of the components, can be computationally predicted using an efficient design model.[121] The model is based on the simulation of the energy conservation equation in its enthalpy form, a formulation that enables one to account for phase changes within the solder or braze layers in a straightforward manner.[119] When relevant, for example, as in reactive multilayer welding,[132] the phase changes within the components can be accounted for in the same fashion. The evolution of enthalpy (and consequently temperature) is obtained by integrating the energy conservation equation within the reactive foil, the solder or braze, and the components. Meanwhile, a simplified model of the self-propagating reaction is implemented, which is based on representing the latter in terms of a propagating front with prescribed velocity and heat release. The experimentally or computationally determined values of velocity and heat release are used for this purpose. This simplified representation is based on a detailed analysis of the effect of heat losses from the foil on the velocity of the self-propagating front.[120] Briefly, the latter indicates that for a sufficiently large foil thickness, particularly for the thickness range relevant to practical joining applications, heat loss from the foil to its surroundings has a small impact on the propagation velocity. Thus, the simplified representation of the self-propagating front enables us to overcome the otherwise insurmountable difficulty associated with the inherently broad spectrum of length scales that characterize reactive multilayer joining, which ranges from the nanoscale thickness of individual foil bilayers to the component length scale, typically on the order of several millimeters or larger. The validity of the predictive design model has been assessed in various applications, using transient temperature measurements during joining. Both infrared thermometry and the transient response of Si dies during joining have been used to verify design predictions.[132,134] These tests have demonstrated that the model accurately predicts the evolution of temperature within the component stack and as such, is a practical tool for reactive joining analysis and design.

Extensive evaluations of reactively soldered and brazed joints have shown that, due to cooling and densification induced by the reaction, cracking of the reactive multilayer foil occurs during the reactive joining process. These cracks are rapidly filled by molten solder or braze, the flow of which is enhanced by the pressure applied on the assembly during the reactive foil ignition process.[135–138] As a result, the reacted multilayer foil is embedded within a continuous solder or braze matrix and realization of this morphological structure is, in fact, essential to achieving optimal joint properties. To this end, optimization of the applied pressure during joining is also a key design parameter of the reactive joining process.

Finally, detailed evaluations have also demonstrated that when the parameters of the reactive joining process are suitably optimized, the resulting bonds exhibit mechanical properties that are similar to,[135–137] or in some cases better than,[138] those of conventionally soldered joints. In particular, these analyses indicate

FIGURE 21.12 Photograph of a printed circuit board with right angle connectors that were attached using reactive joining. (From Weihs, T.P. et al., in *Joining of Advanced and Specialty Materials VII*, Lienert, T.J., Weil, K.S., Zhou, Y.N., Smith, R.W., and Powers, M., Eds., American Society for Metals International, Materials Park, 75, 2005. With permission from ASM International. Photograph courtesy of Agilent Technologies, Palo Alto, CA.)

that failure modes of reactively soldered joints correspond to failures occurring within the solder. Thus, the mechanical properties of reactively soldered joints, including strength and resistance to thermal or mechanical fatigue, are in large part determined by the properties and behavior of the solder or braze alloy employed.

21.6.3 Case Study: Component Mounting

In this section, implementation of the reactive joining process using the preform configuration is illustrated in light of an exemplary component mounting application. Figure 21.12 shows a photograph of the component assembly, where gold plated right angle through-hole surface mountable connectors are joined to a printed circuit board (PCB) assembly by means of reactive joining. The configuration (see Figure 21.11) is fashioned by placing 25 μm thick freestanding eutectic Au–Sn solder performs on either side of an 80 μm thick annular shaped Ni–Al reactive foil preform. The resulting stack is positioned between the connector and gold plated connector pad on the PCB, a modest pressure is applied (1 to 2 MPa) and the reactive foil is ignited with a low voltage current source, thereby melting the solder.

The use of reactive multilayer soldering for this application leads to a number of inherent advantages. Foremost, localized heating enables the use of eutectic Au–Sn solder, an alloy with excellent mechanical properties and whose melting temperature (280°C) is on the order of, or slightly higher than, the glass transition temperature of typical PCB materials. Consequently, when the PCB is populated with electronic components in a subsequent solder reflow operation, the mechanical properties (strength, fatigue and creep resistance) of the board-connector joints are not compromised. The reactive joining process can be accomplished on a bench top (i.e., as opposed to in a conveyor furnace), it eliminates the need for flux and associated solvent cleaning steps, and it allows utilization of lead-free solder alloys with higher melting temperatures. In addition, due to the rapid nature of the reactive multilayer joining process, cost and productivity gains due to higher throughput can be achieved.

As detailed in the literature,[123,139] the reactive joining process has been optimized for this particular application via modeling and computational analysis. Specifically, a design study was performed to predict the thermal exposure of the components during joining and to assess its impact. An example of

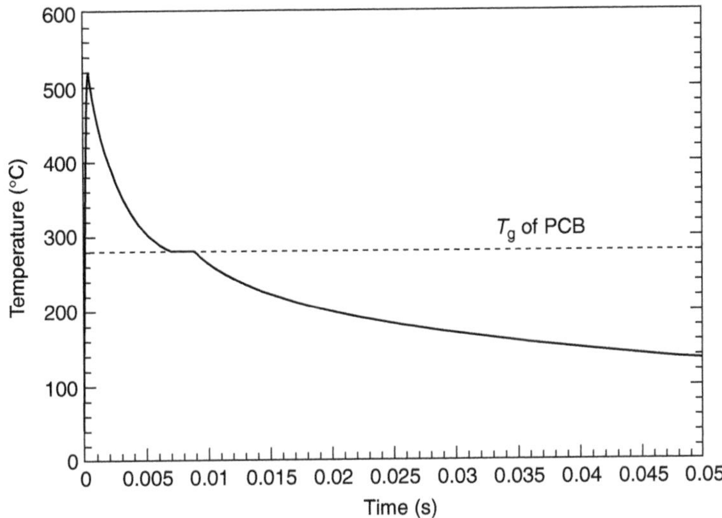

FIGURE 21.13 Temperature evolution at the surface of the PCB during reactive joining. (From Gavens, A.J., *J. Appl. Phys.*, 87, 1255, 2000. With permission.)

the design computations is presented in Figure 21.13, which shows the evolution of the temperature at the surface of the PCB during the reactive joining process. In this figure, the approximate glass transition temperature ($T_g \sim 270$ to 280°C) of the PCB material is depicted by a horizontal dotted line. It can be seen that the surface temperature of the PCB only exceeds T_g for about 8.5 ms. Subsequent experiments confirmed that short, localized thermal excursions above the glass transition temperature do not result in damage to the PCB. This is not necessarily the case for a furnace reflow process employing eutectic Au–Sn, where the PCB would be exposed to a temperature exceeding 300°C for several minutes, with the potential for significant degradation of the PCB material and construction.

Optimized connector–PCB joints have also been the subject of systematic evaluations for assessment of their quality, reliability, and ability to meet an over-torque specification. The evaluations included microstructural characterization using x-ray and optical microscopy, destructive and nondestructive torque testing, mechanical shock and vibration testing, as well as reliability testing.[123,139] Here, only the results of the mechanical strife and reliability testing are presented.

Torque strife testing to failure performed on reactively joined connector joints demonstrated that the strength of the bonds appreciably exceeded the torque specification of 1.7 Nm. Connector joints typically failed at torque values near or above 5.6 Nm. In almost all cases, the failures were manifested by either delamination of the connector pad metallization from the PCB or catastrophic failure of the connector housing, not inside the solder joint itself. This attests to the outstanding tensile and shear strength of eutectic Au–Sn solder, as well as the efficacy of the reactive joining process. To further assess the performance of the joints, nondestructive strife torque tests were repeated on samples after they were subjected to reliability testing, mechanical shock (1000 g acceleration in 0.5 ms), random vibration (50 to 2000 Hz with 7 g acceleration), and thermal cycle testing (−50 to 150°C at 20°C/min for 10 cycles). These experiments demonstrated that the joints maintained their mechanical integrity, even after the reliability testing. Accordingly, this component mounting application of reactive joining is an attractive alternative to conventional reflow soldering because it provides strong, durable joints and the potential for productivity enhancements.

21.6.4 Case Study: Large-Area Joining

This review of reactive joining technology concludes with an application of the reactive joining process to prewet configurations for large-area ceramic-to-metal bonding. Specifically, we describe the formation of large-area bonds between SiC armor tiles and Ti–6Al–4V backing plates. This armor mounting application is motivated by inherent limitations in standard joining methods; namely furnace soldering, furnace brazing, and epoxy mounting. The primary constraint for furnace soldering and brazing is a mismatch in CTEs between the metal alloy and the ceramic, which limits the size of SiC–Ti joints to relatively small areas (about 7 cm^2 for brazing). For larger joint areas, excessive contraction of the metal component relative to the ceramic tile occurs upon cooling from the reflow temperature, which results in cracking of the ceramic and delamination of the bond. As ceramic armor tiles typically measure approximately 100 cm^2, the default joining method is based on the use of structural epoxies. This approach is far from ideal because the mechanical strength of epoxy joints is fairly low, which limits the effectiveness of the ceramic armor for ballistic protection.

In the room temperature reactive joining process, limited heating of the components occurs because controlled amounts of heat are generated locally at the interface. Consequently, the thermal mismatch impediment that is inherent in furnace soldering or brazing, is naturally avoided. This enables the formation of large-area metallic bonds between the SiC tile and Ti-6-4 backing plate. Two prewet joint configurations were investigated for this application. One configuration involved predeposition of a 59Ag–27.25Cu–12.5In–1.25Ti active metal braze alloy (Incusil® ABA™) onto both the SiC and Ti-6-4 components. The braze material was applied as a paste with a layer thickness of about 50 microns, followed by a vacuum heat treatment above the melting temperature of the braze (805°C liquidus). The other prewet configuration was identical to the one just described, except that a layer of 65Sn–25Ag–10Sb eutectic solder alloy (233°C melting point) was deposited over the active metal braze layer. The multilayer reactive foil used for both configurations was fabricated by sputter depositing a 1 micron layer of the same active metal braze alloy onto the Al/Ni reactive foil structure *in situ*, without breaking vacuum.

As illustrated in Figure 21.14,[140] the resulting joints exhibit shear strengths that are significantly higher than those obtained using epoxy-based joining solutions. In addition, this evaluation established that the bonding process results in uniform strength across the entire joint interface. Finally, detailed proprietary evaluations of reactively joined SiC armor have demonstrated that the bonds meet thermal, mechanical, and environmental specifications for vehicle armor, and result in superior ballistic performance over epoxy-mounted armor.

The advantages of the reactive joining process outlined above — namely the limited thermal exposure of the components and mitigation of CTE mismatches — naturally extends to different material systems

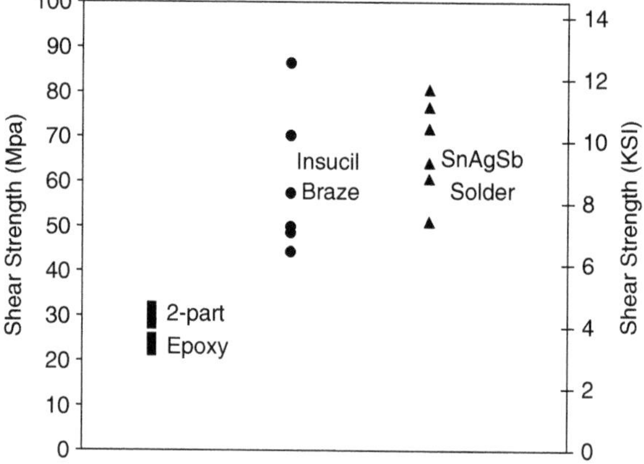

FIGURE 21.14 Left: Photograph of a large area, 10 cm × 10 cm, SiC tile joined to a Ti–6Al–4V plate using reactive joining. Right: Compressive shear strengths for SiC/Ti–6Al–4V joints obtained using (1) an Incusil® active metal braze melting layer and (2) a SnAgSb solder melting layer. Also shown for comparison are compressive strengths for epoxy joints. (From Doherty, K.J., in *Proceedings of the International Brazing and Soldering Conference*, San Diego, CA, February 16–19, 2003. Photograph courtesy of Reactive NanoTechnologies, Hunt Valley, MD.)

and other applications. These include large-area metal–metal joints for structural applications, such as sputter-target mounting. In all cases, the reactive joining process overcomes inherent limitations associated with furnace reflow processes, results in uniform bonds with minimal voiding, and provides an enabling approach for fabrication of strong large-area joints.

21.7 Conclusions

The evolution of bonding processes for joining of advanced materials has been driven not only by technological improvements in equipment and procedures, but by vital progress in the understanding of underlying scientific principles and the relationship between the materials science and the engineered application. The technologies of glass-to-metal sealing and ceramic-to-metal joining are by now well established and have found broad application in a number of different fields, from electronics to medical devices. Even so, significant developments continue to materialize, particularly in the areas of complex component assemblies and innovative materials. The field of diffusion bonding has grown from a specialized industrial technique for niche applications to a mature, established, and widely used process for the joining of both similar and dissimilar materials. By enabling a host of new joining applications through discontinuous innovation, the emergence of reactive joining with nanoscale exothermic foils is providing novel alternatives to traditional bonding process approaches and shows promise for increasing impact in the future.

This chapter is intended to provide the interested engineer, scientist, or student with an appreciation of the fundamental principles behind the bonding processes covered and how they relate to practical application of the technology. The information provided in this overview will help the reader understand the technical challenges associated with a particular process, aid in the selection of a bonding process for a specific application, and enable further investigation by detailed research of the literature and informed discussion with experts in the field.

Acknowledgments

This paper was inspired by and is dedicated to the memory of Joseph A. Pask (1913–2003), professor emeritus at the University of California, Berkeley and internationally respected leader in the field of modern ceramic science and engineering. We would like to thank Herbie Yu and Deb Yamaoka at Agilent Technologies for producing photographs and creating figures used herein. We would also like to thank David Gaskell at Purdue University for graciously providing the Ellingham diagram figure. We are grateful to Floyd Bishop, Don Estreich, Don Cropper, Matt Richter, Mike Whitener, Jon James, and Glenn Takahashi at Agilent Technologies, Tony Tomsia at the Lawrence Berkeley National Laboratory, Mike Hosking at Sandia National Laboratories, Ron Gronsky at U.C. Berkeley, Jim Shackelford at U.C. Davis, Toshi Oyama at Wesgo Metals and Ron Smith at Materials Resources International for input and stimulating collegial discussions. We thank Chau, Annie, and Michael Nguyentat for compiling the database used in the diffusion bonding section. We gratefully acknowledge the patience and support of our families during the writing of this paper.

References

[1] Monack, A.J., Theory and practice of glass-metal seals, *Glass Ind.*, 27, 389 (1946).

[2] Monack, A.J., Glass-to-metal seals in electronic components and applications, *Elec. Mfg.*, 39, 96 (1947).

[3] Derby, B., Diffusion bonding, in *Joining of Ceramics*, Nicholas, M.G., Ed., Institute of Ceramics/Chapman and Hall, London, 94 (1990).

[4] Weihs, T.P., Self-propagating reactions in multilayer materials, in *Handbook of Thin Film Process Technology, Vol. 2*, Glocker, D.A. and Shah, S.I., Eds., IOP Publishing, Bristol and Philadelphia, supplement to part F7, 1 (1995).
[5] Kingery, W.D., Bowen, H.K., and Uhlmann, D.R., Surfaces, interfaces and grain boundaries, in *Introduction to Ceramics*, John Wiley & Sons, New York, chap. 5 (1975).
[6] Young, T., An essay on the cohesion of fluids, *Phil. Trans. R. Soc.*, 95, 65 (1805).
[7] De Gennes, P.G., Wetting: Statics and dynamics, *Rev. Mod. Phys.*, 57, 827 (1985).
[8] Aksay, I.A., Hoge, C.E., and Pask, J.A., Wetting under chemical equilibrium conditions, *J. Am. Ceram. Soc.*, 12, (78), 1178 (1974).
[9] Gibbs, J.W., On the equilibrium of heterogeneous substances, *Trans. Conn. Acad.*, 3, 343 (1878).
[10] Pask, J.A. and Tomsia, A.P., Wetting, spreading and reactions at solid/liquid interfaces, in *Surfaces and Interfaces in Ceramic and Ceramic-Metal Systems*, Pask, J.A. and Evans, A.G., Eds., Plenum, New York, 441 (1981).
[11] Tomsia, A.P. and Pask, J.A., Chemical reactions and adherence at glass/metal interfaces: An analysis, *Dent. Mater.*, 2, 10 (1986).
[12] Roura, P. and Fort, J., Local thermodynamic derivation of Young's equation, *J. Colloid Interface Sci.*, 272, 420 (2004).
[13] Abraham, D.B. and Ko, L.F., Exact derivation of the modified Young equation for partial wetting, *Phys. Rev. Lett.*, 63, 275 (1989).
[14] Wenzel, R.N., Surface roughness and contact angle, *J. Phys. Colloid Chem.*, 53, 1466 (1949).
[15] Borgs, C., De Coninck, J., Kotecky, R., and Zinque, M., Does the roughness of the substrate enhance wetting, *Phys. Rev. Lett.*, 74, 2292 (1995).
[16] Parry, A.O., Swain, P.S., and Fox, J.A., Fluid adsorption at a non-planar wall: Roughness-induced first-order wetting, *J. Phys.: Condens. Matter*, 8, L659 (1996).
[17] Chow, T.S., Wetting of rough surfaces, *J. Phys.: Condens. Matter*, 10, L445 (1998).
[18] Shuttleoworth, R. and Bailey, G., The spreading of a liquid over a rough surface, *Discussions Faraday Soc.*, 3, 16 (1948).
[19] Lill, J.V. and Broughton, J.Q., Interfacial free energy calculations via virtual slip, *Model. Simul. Mater. Sci. Eng.*, 8, 345 (2000).
[20] Nikolopoulos, P., Agathopoulos, S., and Tsoga, A., A method for the calculation of interfacial energies in Al_2O_3 and ZrO_2/liquid-metal and liquid-alloy systems, *J. Mater. Sci.*, 29, 4393 (1994).
[21] Ellefson, B.S. and Taylor, N.W., Surface properties of fused salts and glasses: I, Sessile-drop method for determining surface tension and density of viscous liquids at high temperatures, *J. Am. Ceram. Soc.*, 21, 193 (1938); Ellefson, B.S. and Taylor, N.W., Surface properties of fused salts and glasses: II, Contact angle and work of adhesion on gold and platinum in various atmospheres, *J. Am. Ceram. Soc.*, 21, 205 (1938).
[22] Sangiorgi, R., Caracciolo, G., and Passerone, A., Factors limiting the accuracy of measurements of surface tension by the sessile drop method, *J. Mater. Sci.*, 17, 2895 (1982).
[23] Bagdassarov, N., Dorfman, A., and Dingwell, D.B., Effect of alkalis, phosphorus, and water on the surface tension of haplogranite melt, *Am. Mineral.* 85, 33 (2000).
[24] Maze, C. and Burnet, G., A non-linear regression method for calculating surface tension and contact angle from the shape of a sessile drop, *Surf. Sci.*, 13, 451 (1969).
[25] Bashforth, F. and Adams, J.C., *An Attempt to Test Theories of Capillarity*, Cambridge University Press, Cambridge (1883).
[26] Dorsey, N.E., New equation for determining surface tension from the form of a sessile drop or bubble, *J. Wash. Acad. Sci.*, 18, 505 (1928).
[27] Ellingham, H.J.T., Reducibility of oxides and sulphides in metallurgical processes, *J. Soc. Chem. Ind.*, 63, 125 (1944).
[28] Richardson, F.D. and Jeffes, J.H.E., The thermodynamics of substances of interest in iron and steel making from 0°C to 2400°C, *J. Iron Steel Ins.*, 160, 261 (1948).

[29] Pask, J.A., Glass-metal interfaces and bonding, in *Modern Aspects of the Vitreous State, Volume 3*, MacKenzie, J.D., Ed., Butterworths, London, chap. 1 (1964).
[30] Varshneya, A.K., Stresses in glass-to-metal seals, in *Treatise on Materials Science and Technology, Vol. 22*, Herman, H., Ed., Academic Press, San Diego, 241 (1982).
[31] Hull, A.W., Stresses in cylindrical glass metal seal with glass inside, *J. Appl. Phys.*, 17, 685 (1946).
[32] Martin, F.W., Stresses in glass-metal seals, *J. Am. Ceram. Soc.*, 33, 224, (1950).
[33] Hagy, H.E., Thermal expansion mismatch and stress in seals, *Electron. Packag. Prod.*, 18, 182 (1978).
[34] Miska, K.H., How to obtain reliable glass- and ceramic-to-metal seals, *Mater. Eng.*, 83, 32 (1976).
[35] Barber, D.R., Metal to glass electrode seals, *J. Sci. Instrum.*, 6, 138 (1929).
[36] Frary, F.C., Taylor, C.S., and Edwards, J.D., Glass-to-metal seals and joints, in *Laboratory Glass Blowing*, 2nd ed., Mcgraw-Hill, New York, 94 (1928).
[37] Gaylor, P.J., Glass-metal seals, *Glass Ind.*, 26, 509 (1945).
[38] Housekeeper, W.G., The art of sealing base metals through glass, *J. Trans. Am. Inst. Elec. Eng.*, 42, 954 (1923).
[39] Dalton, R.H., How to design glass-to-metal joints, *Prod. Eng.*, 36, 62 (1965).
[40] Scott, W.J., Glass-to-metal seal design, *J. Sci. Instrum.*, 23, 193 (1946).
[41] Matheson, R., Designing glass-to-metal seals, *Electron. Packag. Prod.*, 17, 197 (1977).
[42] Mairs, K.H., Fe-ni-Co alloys for glass-to-metal seals, *J. Metals*, 40, 460 (1952).
[43] Adam, H., Compressed glass-to-metal seals, *J. Soc. Glass Technol.*, 38, 285 (1954).
[44] Badger, A.E. and King, B.W., Remove carbon from iron and steel by heat treatment, *Ceram. Abstr.*, 18, 292 (1939).
[45] McLaren, H.D., Theory of blistering of cast-iron enamels, *J. Can. Ceram. Soc.*, 4, 54 (1935).
[46] Notis, M.R., Decarburization of an iron-nickel-cobalt glass sealing alloy, *J. Am. Ceram. Soc.*, 45, 412 (1962).
[47] Schmidt, J.J. and Carter, J.L., Using nitrogen based atmospheres for glass-to-metal sealing, *Metal Progress*, 128, 29 (1985).
[48] Yext, W.F. et al., Improved glass-to-metal sealing through furnace atmosphere composition control, in *Proceedings of the 33rd Electronic Components Conference*, IEEE, Orlando, 472 (1983).
[49] Buck, D.M., Ellison, T.L., and King, J.C.W., Nitrogen improves glass-to-metal seals, *Electron. Packag. Prod.*, 131, May (1984).
[50] Borom, M.P. and Pask, J.A., Role of "adherence oxides" in the development of chemical bonding at glass-metal interfaces, *J. Am. Ceram. Soc.*, 49, 1 (1966).
[51] Pask, J.A., New techniques in glass-to-metal sealing, *Proc. IRE.*, 36, 286 (1948).
[52] Tomsia, A.P. and Pask, J.A., Kinetics of iron-sodium disilicate reactions and wetting, *J. Am. Ceram. Soc.*, 64, 523 (1981).
[53] Abendroth, R.P., Oxide formation and adherence of an iron-cobalt-nickel glass sealing alloy, *Mater. Res. Standards*, 5, 459 (1965).
[54] Pask, J.A. and Fulrath, R.M., Fundamentals of glass-to-metal bonding: VIII; nature of wetting and adherence, *J. Am. Ceram. Soc.*, 45, 592 (1962).
[55] Hoge, C.E., Brennan, J.J., and Pask, J.A., Interfacial reactions and wetting behavior of glass-iron systems, *J. Am. Ceram. Soc.*, 56, 51 (1973).
[56] Brennan, J.J. and Pask, J.A., Effect of composition on glass-metal interface reactions and adherence, *J. Am. Ceram. Soc.*, 56, 58 (1973).
[57] Pask, J.A., From technology to the science of glass/metal and ceramic/metal sealing, *Ceramic Bull.*, 66, 1587 (1987).
[58] Maury, M.A., Microwave coaxial connector technology: A continuing evolution, *Microwave Journal 1990 State of the Art Reference*, 39, Sept. (1990).
[59] Botka, J. and Powers, M., 200 GHz broadband coaxial microwave connector, U.S. Patent Number 6,666,725 B2, December 23, 2003.

[60] Qu, J., Shih, A.J., and Scattergood, R.O., Development of the cylindrical wire electrical discharge machining process, part 1: Concept, design, and material removal rate, *Trans. ASME*, 124, 702 (2002).
[61] Levy, A., Thermal residual stresses in ceramic-to-metal joints, *J. Am. Ceram. Soc.*, 74, 2141 (1991).
[62] Charreyron, P.O., Patten, D.O., and Miller, B.J., Modeling of ceramic to metal brazed joints, *Ceram. Eng. Sci. Proc.*, 10, 1801 (1989).
[63] Pulfrich, H., Ceramic-to-metal seal, U.S. Patent 2,163,407, June 20 (1939).
[64] Nolte, H.J., Method of metallizing a ceramic member; Metallized ceramic, U.S. Patents 2,667,427 and 2,667,432, January 26, 1954.
[65] Chick, A.J. and Speck, L.S., Fabrication of metal to ceramic seals, U.S. Patent 2,708,787, May 24, 1955.
[66] Mizuhara, H., Huebel, E., and Oyama, T., High-reliability joining of ceramic to metal, *Ceram. Bull.*, 68, 1591 (1989).
[67] Mizuhara, H., Vacuum brazing ceramics to metals, *Adv. Mater. Process.*, 131, 53 (1987).
[68] Tanaka, T., Homma, H., and Morimoto, H., Joining of ceramics to metals, *Nippon Steel Tech. Rep.*, 37, 31 (1988).
[69] Stephens, J.J., Vianco, P.T., and Hosking, F.M., The active metal brazing of Tzm-mo and SiN ceramics, *JOM*, 48, 54 (1996).
[70] Nakahashi, M., Shirokane, M., and Takeda, H., Characterization of nitride ceramic-metal joints brazed with titanium containing alloys, in *Br. Ceram. Proc. No. 39*, Freer, R., Newsam, S., and Syers, G., Eds., Institute of Ceramics, Shelton, U.K., 25 (1986).
[71] Nicholas, M.G., Valentine, T.M., and Waite, M.J., The wetting of alumina by copper alloyed with titanium and other elements, *J. Mater. Sci.*, 15, 2197 (1980).
[72] Moorhead, A.J., Direct brazing of alumina ceramics, *Adv. Ceram. Mater.*, 2, 159 (1987).
[73] Lugscheider, E., Boretius, M., and Tillman, W., Interfacial reactions of active brazes with high-performance ceramics, in *Br. Ceram. Proc. No. 48*, Morrell R. and Partridge, G., Eds., Institute of Ceramics, Shelton, U.K., 139 (1991).
[74] Nicholas, M.G., Active metal brazing, in *Joining of Ceramics*, Nicholas, M.G., Ed., Institute of Ceramics/Chapman and Hall, London, 73 (1990).
[75] Miller, C.M. et al., Design of optical receiver modules for digital communications analysis, *Hewlett-packard J.*, 47, 22 (1996).
[76] Kazakov, N.F., *Diffusion Bonding of Materials*, 1st ed., Pergamon Press, New York, 274 (1985).
[77] Gust, W., Predel, B., and Nguyentat, T., Untersuchung der discontinuierlichen ausscheidung in polykristallinen Gold-nickel-Legierungen, *Zeit. Metall.*, 67, 110 (1976).
[78] Gust, W., Nguyentat, T., and Predel, B., Die discontinuierliche ausscheidung in nickelreichen Ni-cr-Mischkristallen, *Mater. Sci. Eng.*, 39, 15 (1979).
[79] Dunkerton, S.B., Diffusion bonding—An overview, in *Diffusion Bonding 2: Proceedings of the 2nd International Conference on Diffusion Bonding*, Stephenson, D.J., Ed., Elsevier Applied Science, New York, 1 (1991).
[80] Reed-hill, R.E. and Abbaschain, R., *Physical Metallurgy Principles*, Van Nostrand Reinhold, New York, 258 (1964).
[81] Nippes, E.F. et al., *Metals Handbook Vol. 6: Welding, Brazing and Soldering*, 9th ed., American Society for Metals International, Metals Park, OH, 677 (1983).
[82] Nakajima, H., The discovery and acceptance of the kirkendall effect: The result of a short research career, *JOM*, 49, 15 (1997).
[83] Zeng, K. et al., Kirkendall void formation in SnPb solder joints on bare Cu and its effect on joint reliability, *J. Appl. Phys.*, 97, 024508-1 (2005).
[84] Jeon, Y.D. et al., Studies of electroless nickel under bump metallurgy-solder interfacial reactions and their effects on flip chip solder joint reliability, *J. Electron. Mater.*, 31, 520 (2002).

[85] Fujikawa, H., Makiura, H., and Newcomb, S.B., Microstructural control of oxidation and corrosion resistant chromised nickel plated layers for Fe-ni-Cr alloys, *Mater. Sci. Technol.*, 18, 1347 (2002).

[86] Price, P.E. and Kohler, S.P., Hot isostatic pressing of metal powders, in *Metals Handbook Vol. 7: Powder Metallurgy*, 9th ed., American Society for Metals International, Metals Park, OH, 419 (1984).

[87] Nguyentat, T., Application of platelet diffusion bonding in aerospace technology, *Proceedings of the Sixth Japan International SAMPE Symposium*, Tanimoto, T. and Morii, T., Eds., Society for the Advancement of Material and Process Engineering — Japan Chapter, Tokyo, 321 (1999).

[88] Burkhardt, W.M. and Hayes, W.A., Formed platelet technology for low cost, long life combustion chamber, *Advanced Earth-to-Orbit Propulsion Technology*, NASA Conf. Pub. 3174, II, 190 (1992).

[89] Mueggenburg, H.H. et al., Platelet actively cooled thermal management devices, *Proceedings of the AIAA/SAE/ASME/ASEE 28th Joint Propulsion Conference*, Aiaa-1992-3127, American Institute of Aeronautics and Astronautics, Reston, VA, 23 (1992).

[90] Murphy, M., Schoenman, L., and Nguyentat, T., Development of a two-stage diffusion bonding process for titanium, Technical Publication App94-16r-2, Aerojet Propulsion Division, Sacramento, CA (1994).

[91] Franklin, J.E., Fabrication of ceramic fluidic devices for high performance applications, *Proceedings of the 4th Annual AIAA/BMDO Technology Readiness Conference*, AIAA Paper No. 05-b6, American Institute of Aeronautics and Astronautics, Reston, VA (1995).

[92] Hannum, N.P. and Price, H.G., Some effects of thermal-cycle-induced deformation in rocket thrust chambers, Nasa-tp-1834, (1981).

[93] Nguyentat, T., Gibson, V.A., and Horn, R.M., Nasa-z — A linear material for rocket combustion chambers, *Proceedings of the AIAA/SAE/ASME/ASEE 27th Joint Propulsion Conference*, Aiaa-91-2487, American Institute of Aeronautics and Astronautics, Reston, VA (1991).

[94] Haas, R.W. et al., Fabrication and performance of MMW and SMMW platelet horn arrays, *Int. J. Infrared Millimeter Waves*, 14, 2289 (1993).

[95] Clarricoats, P.J.B. and Oliver, A.D., *Corrugated Horns for Microwave Antennas: IEE Electromagnetic Waves Series*, Vol. 18, Peter Peregrinus Ltd., London (1984).

[96] Deschaux-beaume, F., Frety, N., and Colin, C., Diffusion bonding of si_3n_4-tin composite with nickel-based interlayers, *Metal. Mater. Trans. A*, 34A, 1627 (2003).

[97] Bailey, F.P. and Black, K.J.T., The effect of ambient atmosphere on the gold to aluminum solid state reaction bond, *J. Mater. Sci.*, 13, 1606 (1978).

[98] Suganuma, K. et al., Joining of Si_3N_4 to type 405 steel with soft metal interlayers, *Mater. Sci. Technol.*, 2, 1156 (1986).

[99] Barbee, T.W. and Weihs, T.P., Ignitable, heterogeneous, stratified structures for the propagation of an internal exothermic, chemical reaction along an expanding wavefront, U.S. Patent 5,538,795, July 23, 1996.

[100] Barbee, T.W. and Weihs, T.P., Method for fabricating an ignitable, heterogeneous, stratified structure, U.S. Patent 5,547,715, August 20, 1996.

[101] Floro, J.A., Propagation of explosive crystallization in thin rh-si multilayer films, *J. Vac. Sci. Technol. A*, 4, 631 (1986).

[102] Wickersham, C.E. and Poole, J.E., Explosive crystallization in zirconium/silicon multilayers, *J. Vac. Sci. Technol. A*, 6, 1699 (1988).

[103] Anselmi-tamburni, U. and Munir, Z.A., The propagation of a solid-state combustion wave in Ni-al foils, *J. Appl. Phys.*, 66, 5039 (1989).

[104] Ma, E. et al., Self-propagating explosive reactions in Al/Ni multilayer thin films, *Appl. Phys. Lett.*, 57, 1262 (1990).

[105] Clevenger, L.A., Thompson, C.V., and Tu, K.N., Explosive silicidation in nickel/amorphous-silicon multilayer thin films, *J. Appl. Phys.*, 67, 28 (1990).

[106] Bordeaux, F. and Yavari, A.R., Ultra rapid heating by spontaneous mixing reactions in metal-metal multilayer composites, *J. Mater. Res.*, 5, 1656 (1990).
[107] Dyer, T.S., Munir, Z.A., and Ruth, V., The combustion synthesis of multilayer NiAl systems, *Scripta Metal.*, 30, 1281 (1994).
[108] Weihs, T.P. et al., Self-propagating exothermic reactions in nanoscale multilayer materials, in *TMS Proceedings*, Orlando, FL, February 17–20, 1997.
[109] Van Heerden, D. et al., Metastable phase formation and microstructural evolution during self-propagating reactions in Al/Ni and Al/Monel multilayers, in *Mater. Res. Soc. Symp. Proc.*, Boston, MA, December 1 to 5, 1997.
[110] Reiss, M.E. et al., Self-propagating formation reactions in Nb/Si multilayers, *Mater. Sci. Eng. A.*, A261, 217 (1999).
[111] Myagkov, V.G. et al., Self-propagating high-temperature synthesis of nickel silicide in the nickel nitride plus silicon monoxide bilayer system, *Inorg. Mater.*, 35, 498 (1999).
[112] Gavens, A.J. et al., Effect of intermixing on self-propagating exothermic reactions in Al/Ni nanolaminate foils, *J. Appl. Phys.*, 87, 1255 (2000).
[113] Blobaum, K.J. et al., Deposition and characterization of a self-propagating CuOx/Al thermite reaction in a multilayer foil geometry, *J. Appl. Phys.*, 94, 2915 (2003).
[114] Weihs, T.P. et al., Freestanding reactive multilayer foils, U.S. Patent 6,736,942, May 18, 2004.
[115] Van Heerden, D. et al., Methods of controlling multilayer foil ignition, U.S. Patent Application 20050142495, June 30, 2005.
[116] Mann, A.B. et al., Predicting the characteristics of self-propagating exothermic reactions in multilayer foils, *J. Appl Phys.*, 82, 1178 (1997).
[117] Jayaraman, S. et al., A numerical study of unsteady self-propagating reactions in multilayer foils, *Proc. Combust. Inst.*, 27, 2459 (1998).
[118] Jayaraman, S. et al., Numerical predictions of oscillatory combustion in reactive multilayers, *J. Appl. Phys.*, 86, 800 (1999).
[119] Besnoin, E. et al., Effect of reactant and product melting on self-propagating reactions in multilayer foils, *J. Appl. Phys.*, 92, 5474 (2002).
[120] Jayaraman, S. et al., Numerical study of the effect of heat losses on self-propagating reactions in multilayer foils, *Combust. Flame*, 124, 178 (2001).
[121] Besnoin, E. et al., Method of controlling thermal waves in reactive multilayer joining and resulting product, U.S. Patent Application 20050136270, June 23, 2005.
[122] Makowiecki, D.M. and Bionta, R.M., Low temperature reactive bonding, U.S. Patent 5,381,944, January 17, 1995.
[123] Weihs, T.P. et al., Room temperature lead-free soldering of microelectronic components using a local heat source, in *Joining of Advanced and Specialty Materials VII*, Lienert, T.J., Weil, K.S., Zhou, Y.N., Smith, R.W., and Powers, M., Eds., ASM International, Materials Park, 75 (2005).
[124] Van Heerden, D. et al., A tenfold reduction in interface thermal resistance for heat sink mounting, *J. Microelectron. Electron. Packag.*, 1, 187 (2004).
[125] Subramanian, J.S. et al., Direct die attach with indium using a room temperature soldering process, in *Proceedings 37th IMAPS International Symposium on Microelectronics*, Long Beach, CA, November 14–18, 2004.
[126] Powers, M. et al., Room temperature hermetic sealing of microelectronic packages with nanoscale multilayer reactive foils, in *Proceedings Materials Science & Technology Conference and Exhibition*, Pittsburgh, PA, September 25–28, 2005.
[127] Van Heerden, D. et al., Hermetically sealed product and related methods of manufacture, U.S. Patent Application 20040200736, October 14, 2004.
[128] Wang, J. et al., Joining of stainless-steel specimens with nanostructured Al/Ni foils, *J. Appl. Phys.*, 95, 248 (2004).
[129] Wang, J. et al., Room-temperature soldering with nanostructured foils, *Appl. Phys. Lett.*, 83, 3987 (2003).

[130] Duckham, A. et al., Reactive nanostructured foil used as a heat source for joining titanium, *J. Appl. Phys.*, 96, 2336 (2004).
[131] Duckham, A. et al., Metallic bonding of ceramic armor using reactive multilayer foils, in *Proceedings of the 29th International Conference on Advanced Ceramics & Composites*, Cocoa Beach, Florida, January 23–28, 2005.
[132] Swiston, A.J. et al., Thermal and microstructural effects of welding metallic glasses by self-propagating reactions in multilayer foils, *Acta Mater.*, 53, 3713 (2005).
[133] Wang, J. et al., Effect of physical properties of components on reactive nanolayer joining, *J. Appl. Phys.*, 97, 114307 (2005).
[134] Pfahnl, A.C. et al., Heat sinks reactively soldered to ICs, in *Proceedings DesignCon East*, Boxborough, MA, April 6–7, 2004.
[135] Wang, J., Knio, O., and Weihs, T.P., Method of joining using reactive multilayer foils with enhanced control of molten joining materials, U.S. Patent Application 20050082343, April 21, 2005.
[136] Wang, J. et al., Investigating the effect of applied pressure on reactive multilayer foil joining, *Acta Mater.*, 52, 5265 (2004).
[137] Van Heerden, D. et al., Methods and device for controlling pressure in reactive multilayer joining and resulting product, U.S. Patent Application 20050121499, June 9, 2005.
[138] Wang, J. et al., Nanostructured soldered or brazed joints made with reactive multilayer foils, U.S. Patent Application 20050051607, March 10, 2005.
[139] Levin, J. et al., Room temperature soldering of connectors to PCB using reactive multilayer foils, in *Proceedings IMAPS 38th International Symposium on Microelectronics*, Philadelphia, PA, September 25–29, 2005.
[140] Doherty, K.J., Active soldering silicon carbide to Ti-6Al-4V, in *Proceedings of the International Brazing and Soldering Conference*, San Diego, CA, February 16–19, 2003.

22

Electrolytic Processes

Uwe Erb
University of Toronto

Abstract

This chapter deals with the synthesis of metals, alloys, and composite materials by electrolytic processes using aqueous solutions. The first part of the chapter covers the fundamentals of electrochemical reactions including electromotive force and driving force for electrolytic reactions, overpotential, reaction kinetics, and electrocrystallization. This is followed by several examples of processes in the three major application areas: surface finishing, direct manufacturing, and primary metal production. The final section deals with the application of electrolytic processes in the synthesis of various nanostructured materials such as bulk metal nanostructures, nanocrystalline particles, and template materials.

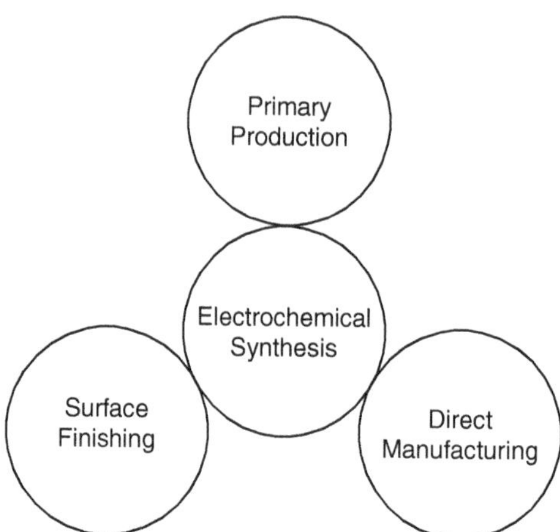

FIGURE 22.1 Main application areas of material processing by electrolytic methods.

22.1 Introduction

On the basis of the underlying fundamental chemical and physical principles, most material processing routes can be broadly categorized into one of five major groups. These are (1) solid-state processing (e.g., rolling, extrusion, severe plastic deformation, surface deformation); (2) liquid-phase processing (e.g., casting, rapid solidification, atomization); (3) vapor-phase processing (e.g., chemical or physical vapor deposition, molecular beam deposition); (4) chemical processing (e.g., precipitation, replacement); and (5) electrolytic processing (e.g., electrodeposition, calcination, electrowinning). Each processing route has unique features and is specifically designed to process distinct groups of materials in the most economical way. This section deals with electrolytic processing of materials which, in many respects, is similar to chemical processing in terms of the chemical reactions involved. However, the distinguishing feature of electrolytic processing is that it involves a well-defined interface at which charge transfer takes place. Most electrolytic reactions have clearly separated cathodic and anodic reactions. In processes such as electrodeposition or electrowinning an external power supply is required to drive the reaction, while in cases such as electroless deposition or galvanic conversion the electrons for metal ion reduction come either from a reducing agent added to the base electrolyte or from electrochemical displacement reactions.

The three major areas in which electrolytic processing is applied in industry are primary production of materials, surface finishing, and direct manufacturing (Figure 22.1). This chapter will first present some of the fundamental aspects of electrolytic processing by discussing issues such as electrode potential, charge transfer, and structure formation (Section 22.2). This will be followed by a description of specific examples from the three major application areas (Section 22.3 to Section 22.5). Finally, examples of applications of electrolytic processing in the synthesis of specific nanostructured materials will be presented (Section 22.6).

This chapter is mainly concerned with metallic materials such as pure metals, alloys, and metal–matrix composites. Electrolytic processing of other materials (e.g., semiconductors and ceramics) is mentioned whenever appropriate. However, a detailed description of electrolytic processing of these structures is beyond the scope of this review. Furthermore, this chapter deals mainly with electrolytic processes involving aqueous solutions. Electrolytic processing of materials using organic solutions, molten salts, or ionic liquids will only be briefly addressed.

22.2 Background

This section summarizes some of the important electrochemical principles of relevance to electrolytic processing of materials. For a more in-depth treatment of the subject, the reader is referred to the many excellent electrochemistry texts that have been published over the past several decades. (e.g., Reference 1 to Reference 10). These give full explanations with figures and derivations of the equations. Important terms often used in the literature which deal with electrolytic processing are given in italic print.

22.2.1 Electromotive Force and Driving Force for Electrolytic Reactions

When a metal is immersed in an aqueous electrolyte containing ions from the same metal there will be an exchange of ions between the electrolyte and the solid metal. After a certain period of time a *dynamic equilibrium* will be established at which the number of metal ions entering the solution from the solid is the same as the number of metal ions entering the solid from the solution:

$$M^{z+} + ze \leftrightarrow M \tag{22.1}$$

where M is the solid metal, M^{z+} is the charged ion, z is the number of electrons involved and e is the charge of the electron. Reaction from the left to right is called a *reduction reaction.* Conversely, the right to left reaction is referred to as *oxidation reaction.* At equilibrium the magnitudes of the currents associated with the anodic and cathodic reactions are the same, but in opposite directions. The current per unit area is referred to as the *exchange current density,* i_0. The consequence of this is the establishment of charge separation through an electrical double layer (*Helmholtz layer*) where the metal surface has a net negative charge (*inner Helmholtz plane*) while the aqueous solution adjacent to it has a positive charge (*outer Helmholtz plane*). Outside the Helmholtz layer is the *Gouy-Chapman layer*, a diffuse region of excess electric charge that can extend over considerable distances.

As in any other chemical reaction, the driving force for electrochemical reactions can be expressed in terms of the Gibbs energy of the system. Reactions proceed in the direction that would decrease the Gibbs energy. For processes that involve electron transfer, the driving force is directly related to the potential of the reaction of an electrolytic cell according to the equation describing electrochemical work:

$$\Delta G = -z \cdot F \cdot E \tag{22.2}$$

where z is the number of electrons taking part in the reaction, F is the Faraday constant (96,500 coulombs/mol) and E is the potential of the reaction.

The *standard potential* or *electromotive force* of a reaction can be measured using an electrochemical cell consisting of two half cells. In the simplest form the first half cell consists of a solid metal rod, M, immersed in a solution containing M^{z+} ions, usually at a concentration of 1 mol/L. This cell is connected to a second half cell, the *standard hydrogen electrode,* via an electrically conductive, but chemically inert salt bridge. The standard hydrogen electrode (standard half cell) consists of an inert platinum electrode in a 1 M solution of H^+ ions and is saturated with hydrogen gas bubbling through the solution at a pressure of 1 atmosphere and a temperature of 25°C. The shorthand to describe such an electrochemical cell is demonstrated for the case of a cell consisting of a rod of zinc immersed in a solution of $ZnCl_2$ (1 mol of Zn^{2+} ions) in cell 1, connected to the standard hydrogen electrode in cell 2:

$$\mathrm{Zn(s)} \mid \mathrm{Zn^{2+}} \vdots\vdots\, \mathrm{H^+(aq), H_2(g)} \mid \mathrm{Pt(s)}. \tag{22.3}$$

Here (s), (aq), and (g) describe the solid, aqueous, and gaseous states, respectively, and the two dotted lines indicate the salt bridge connecting the two half cells. When the two half cells are electrically connected via an external wiring circuit containing a volt meter, the potential difference between the two half cells can be measured in units of volt. Defining the standard potential of the hydrogen electrode as zero on the potential scale, the measured voltage against the Zn half cell is +0.76 V which is a measure of the electromotive force. Therefore, the half cell reactions in the two cells together with their standard potentials can be written as follows:

$$\text{Cell 1:} \quad Zn(s) \rightarrow Zn^{2+}(aq) + 2e \quad E_1^0 = 0.76V \tag{22.4}$$

$$\text{Cell 2:} \quad 2H^{+}(aq) + 2e \rightarrow H_2(g) \quad E_2^0 = 0V \tag{22.5}$$

The overall cell reaction is given as:

$$Zn(s) + 2H^{+} \rightarrow Zn(aq)^{2+} + H_2(g) \tag{22.6}$$

$$\Delta E = +0.76V + 0V = +0.76V \tag{22.7}$$

where ΔE is the standard potential of this electrochemical cell.

The cell reaction with a positive cell potential is consistent with the fact that the zinc metal dissolves spontaneously when immersed in an acid.

It should be pointed out that the *oxidation potential* of $E^0 = 0.76V$ for $Zn(s) \rightarrow Zn^{2+}(aq) + 2e$ is opposite in sign as for the *reduction reaction* $Zn^{2+}(aq) + 2e \rightarrow Zn(s)$ which is $E^0 = -0.76V$.

A relative ranking in terms of their electromotive force can be obtained by measuring cell reaction potentials for various elements relative to the standard hydrogen electrode, as described above for the case of zinc, and summarizing the results in a *table of standard potentials* at 25°C such as shown in Table 22.1. The values given in this table are for the reduction reactions. In this table the metals at the top

TABLE 22.1 Standard Electrode Potentials for Various Elements

Element	Electrode reaction	Standard potential, E (volts), 25°C	Element	Electrode reaction	Standard potential, E (volts), 25°C
Gold	$Au^{+} + e^{-} \rightarrow Au$	+1.68	**Iron**	$Fe^{2+} + 2e^{-} \rightarrow Fe$	−0.44
Gold	$Au^{3+} + 3e^{-} \rightarrow Au$	+1.50	Gallium	$Ga^{3+} + 3e^{-} \rightarrow Ga$	−0.53
Oxygen	$O_2 + 4H^{+} + 4e^{-} \rightarrow 2H_2O$	+1.23	Tantalum	$Ta^{3+} + 3e^{-} \rightarrow Ta$	−0.60
Platinum	$Pt^{2+} + 2e^{-} \rightarrow Pt$	+1.20	**Chromium**	$Cr^{3+} + 3e^{-} \rightarrow Cr$	−0.74
Iridium	$Ir^{3+} + 3e^{-} \rightarrow Ir$	+1.00	**Zinc**	$Zn^{2+} + 2e^{-} \rightarrow Zn$	−0.76
Palladium	$Pd^{2+} + 2e^{-} \rightarrow Pd$	+0.98	Chromium	$Cr^{2+} + 2e^{-} \rightarrow Cr$	−0.91
Mercury	$Hg^{2+} + 2e^{-} \rightarrow Hg$	+0.85	Niobium	$Nb^{3+} + 3e^{-} \rightarrow Nb$	−1.10
Silver	$Ag^{+} + e^{-} \rightarrow Ag$	+0.80	Vanadium	$V^{2+} + 2e^{-} \rightarrow V$	−1.13
Tellurium	$Te^{2+} + 2e^{-} \rightarrow Te$	+0.57	**Manganese**	$Mn^{2+} + 2e^{-} \rightarrow Mn$	−1.18
Copper	$Cu^{+} + e^{-} \rightarrow Cu$	+0.52	Zirconium	$Zr^{4+} + 4e^{-} \rightarrow Zr$	−1.53
Ruthenium	$Ru^{2+} + 2e^{-} \rightarrow Ru$	+0.45	Titanium	$Ti^{2+} + 2e^{-} \rightarrow Ti$	−1.63
Oxygen	$O_2 + 2H_2O + 4e^{-} \rightarrow 4OH^{-}$	+0.40	Aluminum	$Al^{3+} + 3e^{-} \rightarrow Al$	−1.66
Copper	$Cu^{2+} + 2e^{-} \rightarrow Cu$	+0.34	Hafnium	$Hf^{4+} + 4e^{-} \rightarrow Hf$	−1.70
Arsenic	$As^{3+} + 2e^{-} \rightarrow As$	+0.34	Uranium	$U^{3+} + 3e^{-} \rightarrow U$	−1.80
Rhenium	$Re^{3+} + 3e \rightarrow Re$	+0.30	Beryllium	$Be^{2+} + 2e^{-} \rightarrow Be$	−1.85
Bismuth	$Bi^{3+} + 3e^{-} \rightarrow Bi$	+0.20	Neodymium	$Nd^{2+} + 2e^{-} \rightarrow Nd$	−2.10
Antimony	$Sb^{3+} + 3e^{-} \rightarrow Sb$	+0.11	Cerium	$Ce^{3+} + 3e^{-} \rightarrow Ce$	−2.34
Tungsten	$W^{3+} + 3e^{-} \rightarrow W$	+0.10	Magnesium	$Mg^{2+} + 2e^{-} \rightarrow Mg$	−2.37
Hydrogen	$2H^{+} + 2e^{-} \rightarrow H_2$	0.00	Yttrium	$Y^{3+} + 3e^{-} \rightarrow Y$	−2.37
Lead	$Pb^{2+} + 2e^{-} \rightarrow Pb$	−0.13	Sodium	$Na^{+} + e^{-} \rightarrow Na$	−2.71
Tin	$Sn^{2+} + 2e^{-} \rightarrow Sn$	−0.14	Calcium	$Ca^{2+} + 2e^{-} \rightarrow Ca$	−2.87
Molybdenum	$Mo^{3+} + 3e^{-} \rightarrow Mo$	−0.20	Strontium	$Sr^{2+} + 2e^{-} \rightarrow Sr$	−2.93
Nickel	$Ni^{2+} + 2e^{-} \rightarrow Ni$	−0.25	Rubidium	$Rb^{+} + e^{-} \rightarrow Rb$	−2.93
Cobalt	$Co^{2+} + 2e^{-} \rightarrow Co$	−0.28	Potassium	$K^{+} + e^{-} \rightarrow K$	−2.93
Thallium	$Tl^{+} + e^{-} \rightarrow Tl$	−0.34	Arsenic	$As^{+} + e^{-} \rightarrow As$	−2.93
Indium	$In^{3+} + 3e^{-} \rightarrow In$	−0.34	Lithium	$Li^{+} + e^{-} \rightarrow Li$	−3.05
Cadmium	$Cd^{2+} + 2e^{-} \rightarrow Cd$	−0.40	Cesium	$Cs^{+} + e^{-} \rightarrow Cs$	−3.05

Note: Depending on the choice of cell diagram, the sign of the potential may be reversed in other references.

are chemically inert (e.g., gold, platinum, and palladium). Moving down in the table, the metals become more active (i.e., more susceptible to oxidation). Lithium and cesium show the highest tendencies for oxidation. Using the values given in Table 22.1, the cell potential of any pair of redox half reactions can be calculated. For example, for $Zn(s) + Cu^{2+} \rightarrow Zn^{2+} + Cu(s)$, the two half reactions are:

$$\text{reduction half reaction:} \quad Cu^{2+} + 2e \rightarrow Cu(s); \quad E^0 = +0.34V \tag{22.8}$$

$$\text{oxidation half reaction:} \quad Zn(s) \rightarrow Zn^{2+} + 2e; \quad E^0 = +0.76V \tag{22.9}$$

$$\text{which gives:} \quad \Delta E = (+0.34V) + (+0.76V) = +1.103V. \tag{22.10}$$

The table of standard potentials is of considerable technological importance in several areas. It can be used as a starting point in assessing the corrosion tendency of materials in the absence of corrosion inhibiting surface effects. It is also widely used as a first guide in the design of electrolytic materials processing methods to be discussed later in this chapter. However, there are a number of issues to be aware of when using tables of electromotive force values such as shown in Table 22.1. The first has to do with the convention regarding the sign of the potentials. Many European texts show the same values as shown in Table 22.1, but with the opposite sign. This was discussed in great detail by Bockris and Reddy.[6] While this issue is beyond the scope of this chapter, it should be pointed out that the sign of the potentials in Table 22.1 is consistent with the recommendation given by the International Union of Pure and Applied Chemistry (IUPAC). Second, the values listed in the table are only valid for standard conditions. There are a number of factors that can result in considerable changes in the potential values, such as the formation of complex ions, formation of adsorbed layers, and the like. Third, while the standard potentials give good information regarding the relative energy states, they provide no information with respect to the kinetics of processes (i.e., the rates of reactions).

22.2.2 Activity, Concentration, and Temperature Dependence; Pourbaix Diagrams

The potential of the M^{z+}/M electrode as a function of *metal ion activity* in solution, $a(M^{z+})$, and temperature, T, is given by the *Nernst equation*:

$$E = E^0 + \frac{RT}{zF} \ln a(M^{z+}) \tag{22.11}$$

where R, T, F, and z are the gas constant, absolute temperature, Faraday's constant, and number of electrons in the reaction, respectively. For many practical applications using dilute solutions the activity, $a(M^{z+})$, can be replaced with the *metal ion concentration*, $c(M^{z+})$, without introducing substantial errors. Thus, Equation 22.11 can be approximated as follows:

$$E = E^0 + \frac{RT}{zF} \ln c(M^{z+}). \tag{22.12}$$

At 298°K (25°C), the term (RT/F) has the value of 0.0257 V, giving:

$$E = E^0 + \frac{0.0257}{z} \ln c(M^{z+}) \tag{22.13}$$

or when converting the natural logarithm to decimal logarithm (factor of 2.303):

$$E = E^0 + \frac{0.0592}{z} \log c(M^{z+}). \tag{22.14}$$

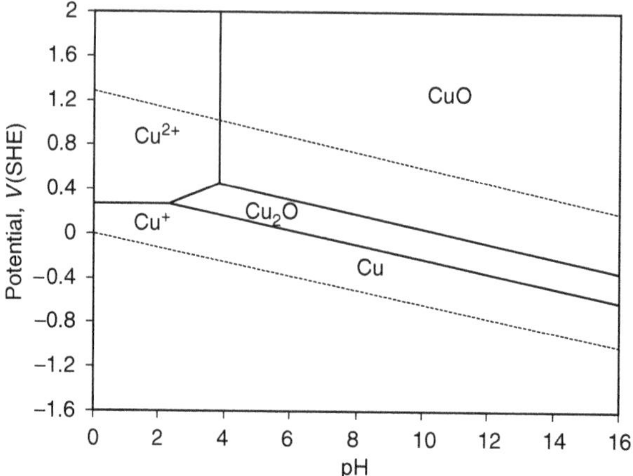

FIGURE 22.2 Pourbaix diagram for the water–copper system.

Using this equation it can be shown, for example, that the electrode potential at 25°C of a copper electrode immersed in aqueous copper sulfate solution ($z = 2$) changes from 0.34 V in a 1 *M* solution to 0.25 V in a 0.001 molar solution, that is, by $0.0592/z$ for each order of magnitude decrease in the ionic concentration.

When the Nernst equation is considered in conjunction with all possible electrochemical equilibria when a metal is in an aqueous solution containing its own ions, phase stability regions of various species can be plotted in potential — pH diagrams, also known as *Pourbaix diagrams.*[11,12] An example of a *Pourbaix diagram* for the water–copper system is shown in Figure 22.2. These diagrams are useful for electrolytic processes (as well as in corrosion studies) to identify all thermodynamically possible phases. However, again it is not possible to draw conclusions from such diagrams when it comes to reaction rates.

22.2.3 Activation Overpotential and Tafel Relationship

When an external potential is applied across the double layer, the electrode becomes polarized and current will flow either as anodic or cathodic current. The difference of the equilibrium potential, $\mathrm{E}(i = 0)$, and the potential of the same electrode as a result of flowing current, $\mathrm{E}(i)$, is usually referred to as the *activation overpotential* η_{a}:

$$\eta_{\mathrm{a}} = \mathrm{E}(i) - \mathrm{E}(i = 0). \tag{22.15}$$

The relationship between overpotential and current is usually expressed in the form of the *Tafel equations*:

$$\eta_{\mathrm{a}} = a \pm b \log^{i} \tag{22.16}$$

where the constants a and b for cathodic (−) and anodic (+) reactions are given as:

$$\text{anodic:} \quad a_{\mathrm{a}} = -(2.303RT/\beta zF) \log i_0 \tag{22.17}$$

$$b_{\mathrm{a}} = 2.303RT/\beta zF \tag{22.18}$$

$$\text{cathodic:} \quad a_{\mathrm{c}} = (2.303RT/(1 - \beta)zF) \log i_0 \tag{22.19}$$

$$b_{\mathrm{c}} = 2.303RT/(1 - \beta)zF. \tag{22.20}$$

In these equations i_0 is the exchange current density and β the geometrical transfer coefficient. This *activation overpotential* is a direct measure of the interference of the equilibrium conditions at the electrode. Larger exchange current densities require smaller overpotentials to produce a given net external anodic or cathodic current.

22.2.4 Limiting Current Density, Nernst Diffusion Layer, Reaction Kinetics

The Tafel equation describes the current–potential relationship for cases where charge transfer is the rate-determining step. However, for both anodic dissolution and cathodic deposition of metals the mobility of the ions in the solution presents a practical limit that results in a deviation from linear Tafel behavior, in an E-log i plot, at what is usually referred to as the *concentration overpotential.* Here, we only consider the case of cathodic deposition. If the deposition is carried out too fast, the ions M^{z+} from the bulk of the electrolyte cannot diffuse fast enough to the solid surface to become incorporated into the growing deposit. A current density limit will be reached at which the ions are deposited as soon as they arrive at the electrode. At this point the metal ion concentration at the electrode approaches zero and any further increase in current density will result in other electrochemical processes, such as the evolution of hydrogen bubbles. The *limiting current density*, i_L, for a system is given by:

$$i_{\mathrm{L}} = \frac{nFD}{\delta} c_{\infty} \tag{22.21}$$

where δ is the thickness of the so-called *Nernst diffusion layer*, D the bulk diffusion coefficient for M^{z+} ions in the solution and c_{∞} the concentration of M^{z+} ions in the bulk solution.

In Figure 22.3, the change in the metal ion concentration from c_{∞}, the bulk concentration in the electrolyte to zero at the electrode surface is schematically indicated. The thickness δ of the transition layer is typically on the order of micrometers. With Equation 22.21 the concentration overpotential, η_c, can be given as:

$$\eta_{\mathrm{c}} = \frac{2.303RT}{nF} \log\left(1 - \frac{i}{i_{\mathrm{L}}}\right) \tag{22.22}$$

where the current density i is a direct measure of the actual rate of the reduction process.

Both activation and concentration polarization occur at the same electrode. At low reaction rates (small current densities) activation polarization is predominant while higher reaction rates (high current densities) are concentration polarization controlled.

The combined effect is summarized in the *Butler–Volmer equation*:

$$\eta_{\mathrm{reduction}} = \alpha \log\left(\frac{i}{i_0}\right) + \frac{2.303RT}{i_0} \log\left(1 - \frac{i}{i_{\mathrm{L}}}\right) \tag{22.23}$$

where α is the expression $2.303RT/\beta zF$. This equation allows us to determine the kinetics of many reduction reactions from only three parameters: α, i_0 and i_L.

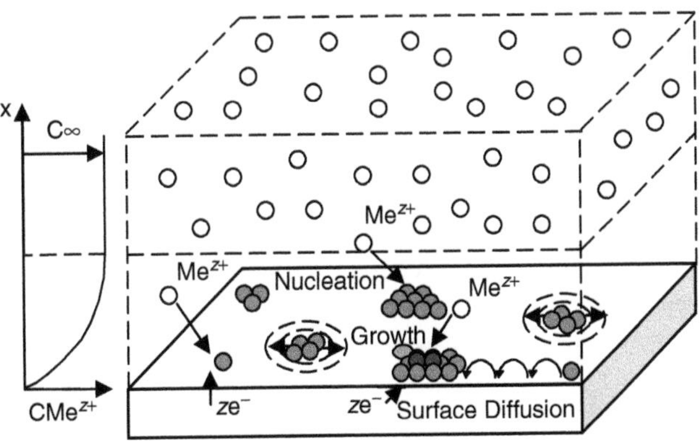

FIGURE 22.3 Schematic diagram showing various steps involved in electrodeposition.

22.2.5 Other Factors Contributing to Overpotential: Hydrogen Overvoltage

There are many other factors that can affect the overpotential such as adsorped species (*adsorption overpotential*), ionically conducting films or oxide films (*film or resistance overpotential*), formation of complexed ions (*complex ion overpotential*), ad-atom diffusion, and surface nucleation (*crystallization overpotential*).

One specific type of overpotential, sometimes referred to in short as the *overvoltage*, is the overpotential (required polarization) at which gas evolution appears at the electrode as a result of oxidation (e.g., oxygen evolution) or reduction (e.g., hydrogen evolution). In particular the *hydrogen overvoltage* is of great importance during electrodeposition processes (Section 22.3.1.1).

22.2.6 Electrocrystallization

When a metal is deposited from an aqueous solution onto a cathode, the hydrated ions in the solution must go through a series of complex steps before they can be incorporated as an atom into the growing electrodeposit. As pure metals always deposit in crystalline form (amorphous structures are only formed in certain alloy deposits), *electrocrystallization* is usually used as the term to describe this series of processes. Electrocrystallization can be considered to consist of two distinct stages.[5,6] The first stage, referred to as the *deposition stage*, is the mass transfer through the Nernst diffusion layer. Mass transfer is a function of the surface and bulk concentrations of ions, diffusion of the ions, and total overpotential. All of these are strongly dependent upon process parameters such as electrolyte composition, pH, temperature, agitation, and applied current density or potential. Fundamental mathematical expressions exist to describe these physical phenomena that give fairly good control during electrolytic processing.

The second stage of electrocrystallization is the actual *crystallization* process that involves the formation of adions on the cathode surface, adsorption and desorption reactions, surface diffusion of adatoms, as well as nucleation and growth of crystals. Some of these are shown in Figure 22.3 in very simplified form. To date, crystallization phenomena are less well understood than the deposition stage phenomena. They are usually described by using empirical relationships.

When deposition begins, the surface conditions of the cathode are very important not only in terms of cleanliness, adsorped layers, or oxide films, but also with respect to atomistic structure. All surfaces contain numerous defects such as steps, ledges, kinks, vacancies, dislocations, and intersecting grain boundaries. Processes such as adatom attachment and diffusion, as well as nucleation and growth are strongly influenced by these defects. Chapter 9 provides more details on surface growth.

Crystalline metal electrodeposits exhibit several growth forms including layers, blocks, pyramids, ridges, spiral growth forms, dendrites, powders, and whiskers.[13] These morphologies have been studied extensively and various models have been proposed to correlate specific growth forms with process parameters and substrate microstructure. For example, with increasing overvoltage the deposit structure changes from compact to powdery.

The internal microstructure evolution of the deposit in terms of grain size and shape is strongly dependent upon crystal growth *inhibition* that can be controlled, for example, by adding certain bath additives that reduce the surface mobility of adatoms or block certain sites for deposition. With increasing inhibition, the deposit structure changes from field oriented (FI) to basis oriented and reproduction type (BR), to twin transition types (TT), to field-oriented texture type (FT) and finally to unoriented dispersion type (UD):[14]

FI: Field-oriented crystal types such as whiskers, dendrites, or loose crystal powder.

BR: Basis reproduction type consisting of coherent deposits in which grain size and surface roughness increase with deposit thickness.

TT: Twin transition type showing high twin densities, usually in materials with low stacking fault energies.

FT: Field-oriented texture type showing coherent deposits with constant small grain size throughout the deposit.

UD: Unoriented dispersed type with very small grain size.

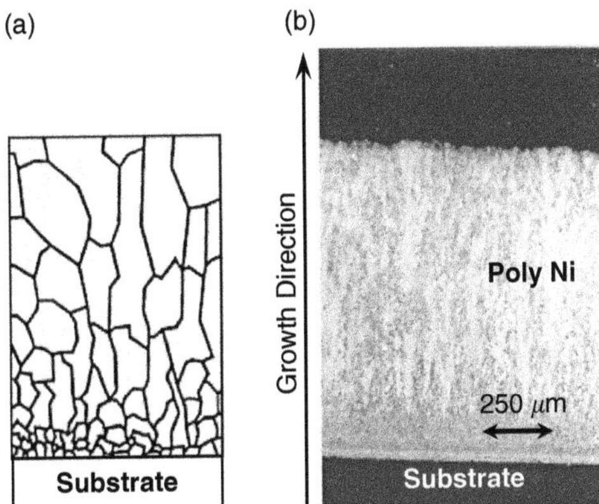

FIGURE 22.4 Cross-sectional structure of a typical Ni electrodeposit. (a) Schematic diagram and (b) thick deposit showing transition from equiaxed fine-grained to columnar textured structure.

With increasing deposit thickness considerable structural changes can be observed as a result of the competition between crystal nucleation and crystal growth. Under certain deposition conditions (e.g., BR growth conditions) the structure can change from initially fine grained, equiaxed structure to coarse grained columnar structure with extensive grain shape anisotropy and crystallographic texture. Figure 22.4 shows a schematic diagram of this transition as well as a cross-section through a nickel electrodeposit with such a change in the microstructure. It is generally accepted that this transition is a direct result of the growth competition between differently oriented grains, such that grains with lower surface energy grow faster than grains with high surface energy.[14]

Over the past decade, considerable progress has been made in the understanding of the early stages of electrodeposition, microstructural evolution with increasing deposit thickness and the formation of dendritic, densely branched, fractal, stringy, and needle-like morphologies.[15,16] Many studies in these areas apply powerful computer simulations (e.g., tight binding approximations, embedded atom models, Monte Carlo simulations) or experimental methods (e.g., *in situ* techniques such as scanning tunneling microscopy, small-angle neutron scattering or microwave reflectivity) to study these phenomena.

22.3 Surface Finishing

The broadest area of applications of electrolytic processes deals with various types of surface-finishing methods in which a coating is deposited or formed on the surface of finished or semifinished products. The reasons for applying coatings are to improve/modify surfaces in terms of appearance, corrosion resistance, wear resistance, oxidation resistance, heat and light reflectivity or absorption, electrical conductivity, magnetic properties, or to build up undersized or worn parts. The various process types include electrodeposition, electroless deposition, metal displacement, oxidation (anodizing) processes, and chemical treatments.

22.3.1 Electrodeposition

In electrodeposition the workpiece to be plated is made the cathode in an electroplating tank containing the metal ions in aqueous solution. The workpiece is electrically connected via a power supply with the anode that is usually made of the metal to be plated either in the form of a solid rod/sheet or, more commonly, pieces of the metal contained in an inert titanium basket. During electrodeposition, the anode dissolves slowly and replenishes the plating bath with metal ions as they are reduced on the cathode to form the coating. In some applications, dimensionally stable anodes are used (e.g., platinum or platinized titanium) that do not dissolve during the plating process.

TABLE 22.2 Watts Nickel and Sulfamate Nickel

	Watts Nickel	Sulfamate Nickel
Nickel sulfate, $NiSO_4 \cdot 6H_2O$ [g/l]	220–400	—
Nickel sulfamate, $Ni(SO_3NH_2)_2 \cdot 4H_2O$ [g/l]	—	300–450
Nickel chloride, $NiCl_2 \cdot 6H_2O$ [g/l]	30–60	0–30
Boric acid, H_3BO_3 [g/l]	30–45	30–45
Temperature (°C)	40–65	30–60
Cathode current density [mA/cm^2]	30–100	5–300
pH	2–4.5	3.5–5
Anode	Nickel	Nickel

22.3.1.1 Pure Metals

Electrodeposition of pure metals from aqueous solutions is the most widely used surface-finishing method. The basic process at the cathode follows the reduction reaction given in Equation 22.1, for example for the case of Ni:

$$Ni^{2+} + 2e \rightarrow Ni. \quad (22.24)$$

A large number of electroplating baths have been developed for nickel plating[14] which differ mainly in terms of metal salt types and concentrations, pH and bath additives used for very specific purposes. Two commonly used plating solutions for nickel are the Watts and the sulfamate baths, Table 22.2. Nickel sulfate/nickel sulfamate are the main sources of the nickel ions in the solutions. Nickel chloride also supplies some nickel ions, but its primary functions are to promote the dissolution of the anode nickel and to increase the electrical conductivity of the plating solution. Boric acid acts as a buffer in the solution. There are a number of other bath additives (e.g., saccharin, coumarin) that may be added for specific purposes such as stress relief or grain size refinement in the deposit. Many of these bath additives are proprietary chemicals supplied by many electroplating supply companies.

Potentially all metals with standard potentials more positive than the hydrogen reduction reaction (Table 22.1) could be electrodeposited from aqueous solutions. However, there are other factors that limit the number of electropositive metals that are routinely electrodeposited. These include the quality of the resulting deposits in terms of growth form, physical integrity and properties, the plating bath stability, ease and economics of operation, as well as environmental and health risk concerns when dealing with toxic chemicals.

On the other hand, there is also a relatively large number of metals with more negative potential than hydrogen that can still be plated, including nickel, cadmium, zinc, and chromium. The reason for this is the hydrogen overvoltage relative to the metal deposition overvoltage. The hydrogen overvoltage is a direct measure of the required activation energy for hydrogen reduction that varies greatly from metal to metal. The following example will illustrate that the metal zinc can be electroplated from an aqueous solution onto steel even though this would not be expected from its standard potential relative to hydrogen. By comparing only the two standard potentials in Table 22.1 for zinc (−0.76 V) and hydrogen (0 V) one would expect hydrogen evolution to occur instead. For zinc deposition, for example, from an acidic (pH = 1) 1 molar zinc sulfate solution, the reversible hydrogen potential as per Nernst equation is $E = -0.059\,\text{pH} \approx -0.06\text{V}$. On the other hand, the activation overvoltage for zinc deposition is very small and zinc can be plated at a potential very close to −0.76 V (Table 22.1). The hydrogen overvoltage on steel at pH = 1 increases from −0.56 V at a current density of 10 mA/cm^2 to −0.82 V at 100 mA/cm^2.[14] Therefore, at a current density of 100 mA/cm^2, the onset of hydrogen evolution is at a more negative potential than the potential required for zinc deposition, thus allowing for zinc deposition onto steel.

Nevertheless, with increasing negative potential it becomes more and more difficult to deposit metals from aqueous solutions. Chromium and manganese can still be plated. However, there are no practical aqueous solutions for electrolytic processing of elements such as Zr, Ti, Al, or other elements with even more negative standard potentials. For these metals electrolysis requires the use of molten salt or ionic

TABLE 22.3 Important Pure Metal Electrodeposits and Their Applications

Metal	Application	Typical thickness(μm)
Cadmium	Electrochemical protection of steel and cast iron	<25
Chromium	Hard chromium for enhanced wear properties on hydraulic shafts, piston rings, aircraft landing gears; rebuilding of mismachined or worn parts	100–250
Chromium	Decorative coatings over undercoatings such as nickel and copper–nickel	<5
Chromium	Black chromium for solar selective coatings	<5
Copper	Printed wiring boards; semiconductor interconnect technology; printing rolls for paper and textiles; build up of worn parts	10–200
Cobalt	Electroplated steel for batteries; magnetic coatings	1–20
Gold	Electronic components; wire bonding; high reflectivity surfaces; decorative finish; jewelry	0.5–10
Iron	Build up of worn parts; hard coatings on pistons	100–500
Lead	Corrosion protection of steel	10–100
Nickel	Corrosion resistant coatings; decorative coatings; magnetic coatings	10–200
Nickel	Black coatings for decorative purposes; nonreflective surfaces	1–1.5
Palladium	Interconnect products; electronic packaging; decorative coatings	0.5–5
Rhodium	Electronic components; decorative coatings	10–200
Silver	Decorative coatings; electronic coatings; medical-instrument coatings; jewelry	2–50
Tin	Corrosion protective coatings; solder applications	5–100
Zinc	Corrosion protection of iron and steel	5–20

liquid electrolytes (see Section 22.5.2). Taking all of these factors into consideration, only the elements given in bold in Table 22.1 are routinely processed using electrodeposition from aqueous solutions.

Table 22.3 lists several pure metal electrodeposits that are of considerable technological importance. The many different electroplating baths that have been developed for the various metals can be found in several books dealing with electrodeposited metals.[14–24] Depending on the application and type of coating, typical thicknesses range from <1 μm to hundreds of micrometers.

The physical, chemical and mechanical properties of electrodeposited metals depend strongly on grain size and grain shape, impurities codeposited with the metal, crystallographic texture, and internal stress. To a large extent, these microstructural characteristics can be controlled by the plating parameters such as bath composition, pH, and temperature as well as agitation of the electrolyte and current density used during deposition. Extensive tables showing electrodeposit properties are presented in several books on plating.[14,21,23]

22.3.1.2 Alloy Deposits

Commercial electrodeposition is not limited to pure metals. A large number of alloy deposits have been developed that contain two or more elements. Examples of important binary electrodeposits are given in Table 22.4. In alloy electrodeposition the bath formulations are more complex than for pure metal deposition. These baths contain several ionic species, and in many cases, complexing agents. For example, nickel–phosphorus alloy deposits can be electroplated from a solution containing nickel sulfate ($NiSO_4$), nickel carbonate ($NiCO_3$), phosphoric acid (H_3PO_4), and phosphorous acid (H_3PO_3).[19] There are several direct and indirect cathodic reactions during electrodeposition of Ni–P alloys:

$$\text{direct reactions:} \quad Ni^{2+} + 2e \rightarrow Ni \tag{22.25}$$

$$2H^+ + 2e \rightarrow H_2 \tag{22.26}$$

$$\text{indirect reactions:} \quad 12H^+ + 12e \rightarrow 12H \tag{22.27}$$

$$2H_3PO_3 + 12H \rightarrow 2PH_3 + 6H_2O \tag{22.28}$$

$$2PH_3 + 3Ni^{2+} \rightarrow 3Ni + 2P + 6H^+. \tag{22.29}$$

TABLE 22.4 Examples of Nickel, Cobalt, and Zinc Alloy Electrodeposits

Material	Application
Nickel–Chromium	Corrosion resistant coatings
Nickel–Cobalt	Decorative, mirror like deposits
Nickel–Copper	Compositionally modulated alloys
Nickel–Indium	Low friction and microelectronics coatings
Nickel–Iron	Soft magnetic coatings; decorative coatings
Nickel–Manganese	High ductility nickel coatings
Nickel–Phosphorus	Abrasion and corrosion resistant coatings
Nickel–Tungsten	High temperature oxidation/resistant coatings
Cobalt–Nickel	Magnetic coatings for storage devices
Cobalt–Molybdenum	High temperature oxidation resistant coatings
Cobalt–Phosphorus	Wear-resistant coatings
Cobalt–Tungsten	High temperature strength coatings
Zinc–Cobalt	Corrosion resistant coatings
Zinc–Nickel	Corrosion resistant coatings
Zinc–Tin	Coatings on fasteners for aluminum panels

Reaction 22.29 shows that phosphorus is reduced together with nickel, which explains the Ni–P alloy deposit formation. In this case, the phosphorus in the alloy deposit comes from the phosphorous acid in the plating bath. The phosphoric acid does not provide P for the alloy formation. Its primary role is to adjust the pH of the solution.

The composition of the electrodeposit may be quite different from the ratio of the relative concentrations of bath constituents. For example, for electrodeposited Zn–Ni electrodeposits, the ratio of the less noble metal (Zn) to the more noble metal (Ni) in the deposit is larger than in the bath. This effect is known as *anomalous codeposition.*[19]

The electrochemistry of alloy plating is much more complex than for simple metal deposition. The complexity is due to the presence of more than one ionic species in the solution, their mutual influence when it comes to double layer formation, electrochemical potentials, electrocrystallization, and deposit structure formation. Alloy deposition is beyond the scope of this chapter and the reader is referred to the references that treat alloy formation by describing simultaneous codeposition of several species for very specific alloy systems.[13,19,25]

22.3.1.3 Composite Deposits

Metal–matrix composites can be produced by adding second phase particles to the plating bath that are then codeposited with the metal matrix. An example is shown in Figure 22.5 which represents a nickel–silicon carbide composite consisting of a nickel matrix with about 5 vol.% SiC particles (average particle size: 0.3 μm). The coating was produced from a Watts type plating bath containing about 50 g/l of SiC powder. The purpose of the SiC particles is to increase the wear and scratch resistance of the nickel coating. Other hard particles that find application in wear-resistant coatings include alumina, titania, thoria, and diamond.[21] On the other hand, particles such as Teflon or molybdenum disulfide can be incorporated into the metal to make low friction surfaces. Several models have been proposed[26] that describe the particle incorporation in the deposits in terms of the various steps involved including initial weak adsorption of particles on the cathode surface, field-assisted adsorption, ionic double layer formation on the particles, and encapsulation in the metals matrix.

22.3.1.4 Semiconductors

Because of the very large negative reduction potential silicon cannot be electrodeposited from aqueous solutions. However, several compound semiconductors can be deposited from suitable aqueous electrolytes.[14,27] Examples of the main bath constituents for several II–VI and III–V compound semiconductors are given in Table 22.5.

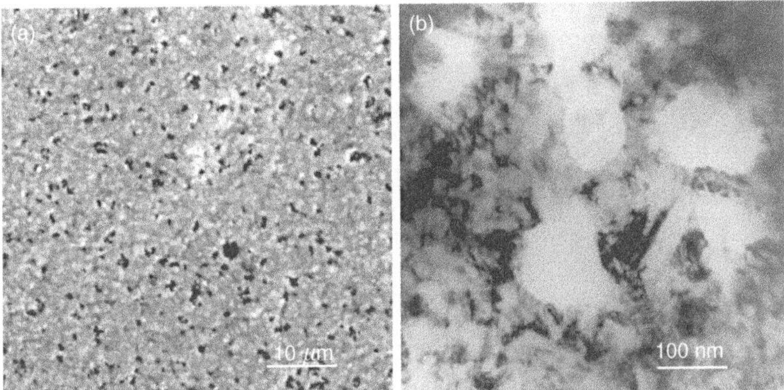

FIGURE 22.5 Ni–SiC composite electrodeposit: (a) scanning electron micrograph and (b) bright-field transmission electron micrograph.

TABLE 22.5 Main Bath Constituents for Semiconductor Deposition

Semiconductor	Bath Constituents
InAs	Indium chloride ($InCl_3$); Arsenic chloride ($AsCl_3$)
InSb	Indium chloride ($InCl_3$); Antimony chloride ($SbCl_3$)
PbS	Lead nitride ($Pb(NO_3)_2$); Sodium sulfite (Na_2SO_3)
CdTe	Cadmium sulfate ($CdSO_4$); Tellurium oxide (TeO_2)
CdSe	Cadmium chloride ($CdCl_2$); Selenium oxide (SeO_2)
ZnSe	Zinc sulfate ($ZnSO_4$); Selenic acid (H_2SeO_3)
GaAs	Arsenic oxide (As_2O_3); Gallium oxide (Ga_2O_3)

22.3.2 Electroless Deposition

Electroless deposition[14,28,29] was originally known as electrode-less deposition, describing the fact that no external power supply is connected to positive and negative electrodes. Also known in the literature as *autocatalytic reduction* or *chemical reduction*, the process still requires anodic and cathodic reactions. However, these occur on the same material, and the electrons required for these are supplied by a reducing agent in the plating bath according to the following general reactions:[14]

$$M^{z+} + Red \rightarrow M + Ox \tag{22.30}$$

where *Red* is the electron-providing reducing agent and Ox is the oxidation product of the reducing agent. The partial cathodic and anodic reactions are as follows:

$$\text{cathodic:} \quad M^{z+} + ze \rightarrow M \tag{22.31}$$

$$\text{anodic:} \quad Red \rightarrow Ox + n \cdot e \tag{22.32}$$

where n is the number of electrons produced during the oxidation of the reducing agent. The two half reactions occur on the same surface which could be either a metal substrate or a nonconductive substrate (e.g., glass, polymer) made active by first depositing catalytic nuclei such as tin or palladium. In other words, electroless deposition is an electrolytic method that can be used to deposit a metal coating onto nonconductors such as polymers or glass.

TABLE 22.6 Reducing Agents for Electroless Metal Deposition

Metal	Reducing Agents
Nickel	Sodium hypophosphite ($NaH_2PO_2 \cdot H_2O$)
Nickel	Sodium Borohydride ($NaBH_4$)
Nickel	Dimethylamine borane, DMAB ($(CH_3)_2$ $NHBH_3$)
Nickel	Hydrazine ($N_2H_4 \cdot H_2O$)
Copper	Formaldehyde (HCHO)
Gold	Potassium borohydride (KBH_4)
Gold	Dimethylamine borone, DMAB ($(CH_3)_2$ $NHBH_3$)
Palladium	Hydrazine ($N_2H_4 \cdot H_2O$)
Palladium	Hydroxylamine hydrochloride ($NH_2OH \cdot HCl$)

Some of the commonly used electroless metals that are commercially of considerable importance are nickel, copper, gold, and palladium. Table 22.6 shows several of the reducing agents used in their deposition.

The electrochemical reactions during electroless deposition are usually more complex than shown in Equation 22.30, and involve several intermediate steps. Without presenting all intermediate reactions, one important step during the deposition of electroless nickel using sodium hypophosphite as reducing agent is as follows:

$$H_2PO_2 + H \rightarrow P + H_2O + OH^-. \tag{22.33}$$

This reaction shows that phosphorus is codeposited with Ni, similar to the case of electrodeposition of Ni–P alloys (Section 22.3.1.2). This explains why electroless nickel can contain substantial amounts of phosphorus, up to 20 at.%, depending on the bath composition and temperature. A direct consequence of this is seen in the microstructural evolution of the deposits. For small phosphorus contents, the deposits are usually crystalline. With increasing phosphorus, the grain size continuously decreases, and for high phosphorus the deposits become amorphous.[14,28,29]

22.3.3 Displacement Deposition

Displacement deposition, also known as *immersion plating* or *galvanic plating* is similar to electroless deposition in that it does not require an external power supply. However, in contrast to electroless plating, no reducing agent is required for this process. In displacement deposition, the electrons for metal reduction come from the substrate itself. For example, a strip of zinc can be plated with copper from a copper sulfate solution at room temperature according to the reactions given in Equation 22.8 and Equation 22.9. In other words, the more electropositive metal copper replaces the more electronegative metal zinc on the zinc substrate. Of course, the thickness of galvanic Cu deposits is very limited, typically to a few micrometers, as this reaction needs direct contact between the zinc substrate and the copper sulfate solution. Once a closed layer of copper has been deposited the reaction will cease. Immersion plating solutions are commercially available[17] for a number of metals (e.g., Cu, Cd, Au, Pd, Pb, Pt, Rh, Ru, Ag) as well as several alloys (e.g., brass, bronze) in particular to plate substrate materials such as steel, aluminum, zinc, and copper. Of course, the choice of deposit/substrate combination is largely dictated by the standard potentials given in Table 22.1.

22.3.4 Chemical Conversion

From a processing point of view, chemical conversion techniques are borderline processes falling between chemical and electrochemical synthesis. In many cases, no clear electrified interface can be defined at which charge transfer occurs. Nevertheless, the most important chemical conversion processes are included here, because of (1) their current importance in surface finishing and (2) their as of yet unexplored future application potential in surface nanotechnology. To date, microstructural information on chemical conversion layers is very limited.

22.3.4.1 Antiquing of Brass, Copper, and Bronze

This process is a popular method in the manufacture of decorative hardware fixtures for many indoor and outdoor applications.[17] The surface color of copper, brass, and bronze articles is modified to various shades of brown to black or green, including verde green patina (verdigris) similar to natural patina on aged copper roofs, by an immersion technique. Typical solutions are based on polysulfide salts, selenium salts, or mild acids, and the color of the conversion layer comes from the compounds that form during the chemical reaction with the surface. Articles made of steel or zinc die casts are first electroplated with a thin layer of copper, brass, or bronze and then antiqued using the chemical conversion treatment.

22.3.4.2 Phosphating

Phosphate coatings are chemical conversion coatings applied to the surfaces of steels and aluminum alloys for articles such as nuts, bolts, and screws, as well as sheet steel for automotive and consumer product applications.[18] These coatings provide good corrosion and wear protection and, in many cases, are applied as a base for subsequent paint application. Using dilute solutions of phosphoric acid containing zinc, iron, or manganese salts, coating layers (typical thickness: 1 to 50 μm) of strongly adherent phosphate crystals are formed either during immersion or spray operations.

22.3.4.3 Chromate Conversion Coatings

Chromate conversion coatings are applied on various metals by chemical or electrochemical treatment with hexavalent chromium (chromic acid) or chromium salts (e.g., potassium chromate). During this treatment the surface of the metal is converted to a protective film that contains complex chromium compounds.[17,18] The thickness of the film is typically on the order of 1 μm and can vary in color from clear-bright, blue-bright and yellow to brown, olive, and black, depending on the substrate material. The main purpose of the chromate coating is to provide corrosion protection, hardness, and wear resistance and to serve as a base layer for subsequent paint application. Chromate coatings are routinely applied on aluminum, magnesium, and zinc alloys. More recently, trivalent chromium conversion layers have been developed, in view of the environmental and health risk concerns associated with the use of carcinogenic hexavalent chromium compounds.[17]

22.3.4.4 Blackening of Ferrous Metals

This conversion process uses various bath chemistries to produce attractive black finishes on ferrous metals that also provide moderate corrosion resistance.[17] In the hot alkaline nitrate black oxidizing solution, the black color comes from the black iron oxide (magnetite, Fe_3O_4) which is formed in boiling alkaline nitrate solution. A black zinc phosphating process has been developed in which steel, cast iron, or malleable iron is blackened before coating with a zinc–phosphate top layer.[17]

22.3.5 Anodizing

Anodizing of aluminum is a widely used surface-finishing process to improve the surface characteristics in terms of appearance (color), durability, and corrosion resistance of finished aluminum parts.[30,31] In this process, the natural protective aluminum oxide layer that forms in air to a thickness of only a few nanometers is artificially built up to considerable thickness (up to several tens of even hundreds of micrometers) by making the aluminum part the anode in a suitable electrolyte, connected via the power supply to a dimensionally stable (inert) cathode such as graphite or stainless steel. While anodic treatment in boric acid, borate, or nitrate baths produces only slightly thicker oxide films than air-exposure, sulfuric acid and chromic acid baths produce initially a thin oxide barrier, followed by the growth of a thicker,

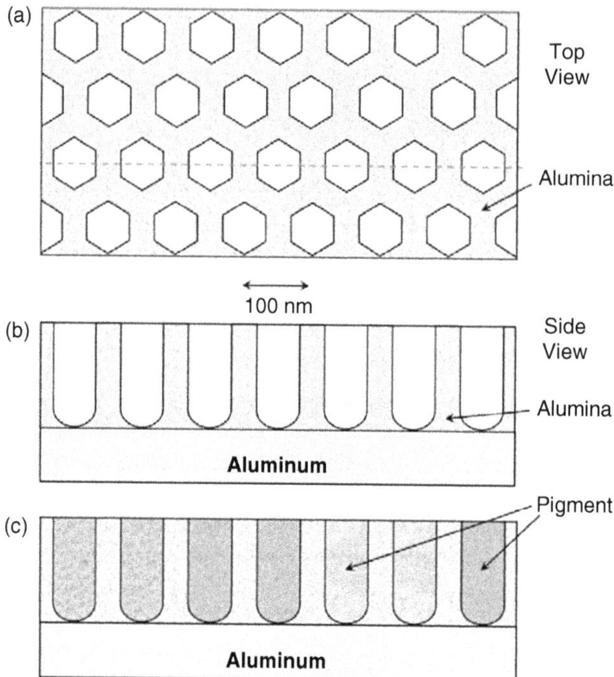

FIGURE 22.6 Nanoporous aluminum oxide produced by anodizing: (a) top view showing hexagonal pore structure, (b) cross-sectional view and (c) coloring by pigment addition.

porous layer on top of the initial layer. The reactions are as follows:

$$\text{cathodic reaction:} \quad 6H^+ + 6e \rightarrow 3H_2(g) \tag{22.34}$$

$$\text{anodic reactions:} \quad 2Al + 3O^{2-} \rightarrow Al_2O_3 + 6e \tag{22.35}$$

$$2Al^{3+} + 3H_2O \rightarrow Al_2O_3 + 6H^+. \tag{22.36}$$

It should be noted that there are two anodic reactions that produce Al_2O_3. The first Reaction 22.35 is at the Al/Al_2O_3 interface, while the second Reaction 20.36 takes place at the Al_2O_3/electrolyte interface, which requires the diffusion of Al^{3+} ions through the Al_2O_3 layer already formed on the base material.

Under suitable conditions this porous layer can be produced as irregular or regular arrangements of hexagonal columns with pores extending from the free surface down to the initial barrier layer. Figure 22.6a and Figure 22.6b show schematic diagrams of the top and side views of a regular oxide structure without giving details of the original barrier layer. For requirements such as scratch and corrosion resistance, the porous aluminum oxide is usually sealed, for example, by boiling in water, which hydrates and expands the oxide to close the pores. In color anodizing, the pores are impregnated prior to sealing with various dyeing compounds, mineral pigments, or microparticles formed from the microconstituents of the aluminum alloy during the anodizing process, to produce a wide variety of colors, in particular for architectural applications of aluminum (Figure 22.6c).

22.4 Direct Manufacturing Using Electrolytic Processing

Electrodeposition is not restricted to surface-finishing applications, but is also widely used to produce finished or semifinished products through direct manufacturing. Generally known as *electroforming*,[32–35]

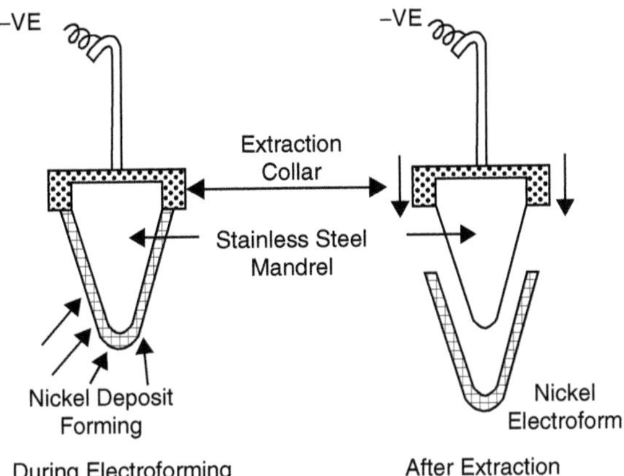

FIGURE 22.7 Net shape manufacturing of an electroformed product (courtesy of Nickel Development Institute, Toronto, Canada).

these processes are used to produce various product forms by electrodeposition onto suitably shaped cathodes, designed as temporary reusable mandrels during the deposition process (Figure 22.7). Following electrodeposition of the product, the mandrel is separated from the electroform by simple mechanical means or heating/cooling cycles. To allow for easy separation, mandrel materials such as titanium or stainless steel are commonly used that have a thin oxide layer on the surface to prevent strong adhesion of the electroformed product. Many metals have been used in electroforming, including nickel, copper, iron, gold, platinum, cobalt–nickel, and cobalt–tungsten alloys.

22.4.1 Large-Scale Manufacture

Direct net shape manufacturing by electroforming is applied in areas where other processing techniques are limited in their capabilities or simply too expensive. The most widely known example of electroformed nickel are the large printing rolls used as mandrels in textile printing in which the pattern is embedded in the surface of the electroformed nickel. Other examples in which the same principle was used are the stampers for records and CDs and the embossing tools for holographic images.

Sheet and foil products with thicknesses $<100\ \mu m$ can be produced in a continuous plating process onto a rotating drum that is partially or fully immersed as the cathode in a plating bath. The foil is mechanically stripped from the drum once the desired thickness is reached. The foil thickness is controlled by the current density and the residence time in the plating bath, given by the drum diameter and the rotational speed. One of the largest commercial applications of this type of foil product is the copper foil used as the metallic conductor in the manufacture of printed circuit board copper/epoxy laminates. Other foil applications include stock material for gaskets, pressure control membranes, hydrogen purification membranes, solar energy absorbers, and microfoils. If the rotating drum surface is modified to contain a surface pattern produced by using nonconductive paints, adhesives, or lithographic patterning, continuous metal mesh can be produced for application such as sieves, electric razor foils, printing screens, or centrifuge screens.

A large number of complex free-forms have been produced by electroforming on reusable complex-shaped cathodes. Examples include erosion shields for helicopter blades, thrust chambers for rocket engines, nozzles and precision reflectors and mirrors. Even very intricate art objects have been made by this process, known as *galvanoplasty*.[32]

22.4.2 Small-Scale Manufacture

In recent years, microsystems technology has advanced rapidly and microelectromechanical systems (MEMS) find increasing use in many automotive, aerospace, medical, defense, and biotechnology

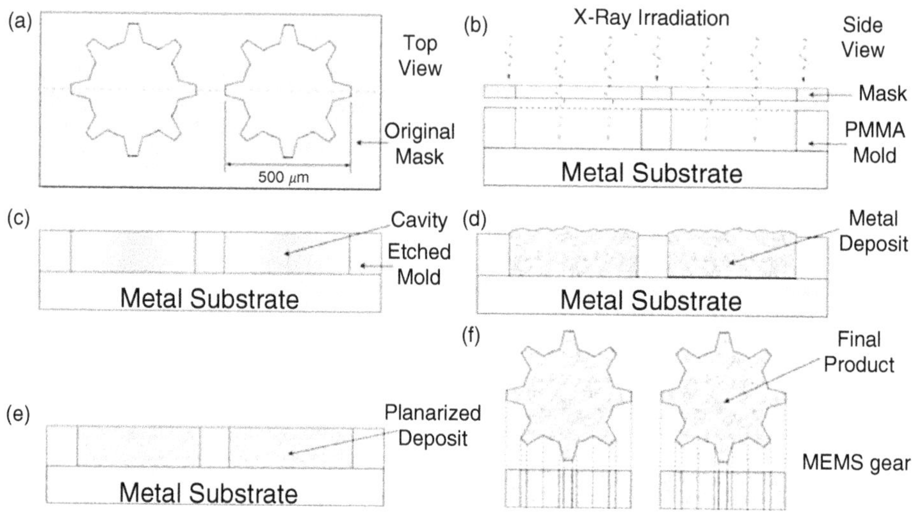

FIGURE 22.8 Various steps in the production of a MEMS gear by electrodeposition: (a) mask showing the shape of the gear, (b) irradiation of PMMA through mask, (c) etched mold with cavity, (d) mold cavity filled with electrodeposited metal, (e) planarization step and (f) final product removed from mold.

applications.[36,37] One specific area of MEMS manufacturing that deals with electrodeposition of small (<1 mm) metallic components is the process known as LIGA (German acronym for Lithographie [lithography], Galvanoformung [electroforming], and Abformung [molding]). (More details about such additive processes may be found in Chapter 26.) Figure 22.8 shows a series of schematic diagrams depicting the production of a MEMS gear via the LIGA process using polymethylmethacrylate (PMMA) as the mold material. Basically, LIGA involves two main subprocesses: preparation of a mold into which the metal is to be deposited and the actual deposition and removal of the final product. Typical metals that find applications in LIGA products are nickel, copper, gold, and nickel–iron alloys.

22.5 Primary Metal Production

While surface finishing is likely the area with the broadest range of electrolytic processes in terms of demonstrated technological flexibility, primary metal production is certainly the field where the largest tonnages of materials are handled. Electrolytic processes are of tremendous importance in the extraction and refining of metals for which pyrometallurgical processes are unfavorable in terms of free energy of reactions and economic considerations. Electrolytic reduction and refining is applied in numerous processing steps during the manufacture of many metals including nickel, copper, cobalt, lead, iron, aluminum, and magnesium. This chapter is limited to two of the most important processing routes: *electrowinning* and *electrorefining*, and *molten salt electrolysis*.

22.5.1 Electrowinning and Electrorefining

Electrowinning and electrorefining are two of the most important electrometallurgy processes in which huge quantities of metal ions are processed by electrolytic means, usually in very large electrolytic cells.[20] In electrowinning, metals are recovered from solutions derived from primary metal leaching and purification operations. The main purpose of electrorefining, on the other hand, is to electrolytically dissolve a metal in a solution and then redeposit it with higher purity than the starting material. The main differences between the two processes are the anode materials and the electrochemical reactions. In electrowinning the anode is the material to be refined, while electrowinning operations use dimensionally stable (e.g., inert lead) anodes. For this reason the anodic and cathodic reactions for both processes are quite different. For example, if the electrowinning electrolyte, coming from a sulfuric acid leaching of a metal oxide, is $MSO_{4(aq)}$ then the reactions are as follows:

$$\text{cathodic reaction:} \quad MSO_{4(aq)} + 2e \rightarrow M + SO_{4(aq)}^{-2} \tag{22.37}$$

$$\text{anodic reaction:} \quad H_2O \rightarrow 2H^+ + 1/2O_{2(g)} + 2e \tag{22.38}$$

In other words, the main reaction at the anode is the decomposition of water. On the other hand, during electrorefining the reactions are simply as follows:

$$\text{cathodic reaction:} \quad M^{z+} + ze \rightarrow M \tag{22.39}$$

$$\text{anodic reaction:} \quad M \rightarrow M^{z+} + ze \tag{22.40}$$

In both cases, the cathode materials are often starter sheets of the same material that is to be deposited. Assuming that there are no other cathodic reactions, the weight, W, of the deposit can be calculated using *Faraday's law*:

$$W = \frac{I \cdot t \cdot A}{z \cdot F} \tag{22.41}$$

where I is the current through the cell in amps, t the time in seconds and A the atomic weight of the deposited material.

22.5.2 Molten Salt Electrolysis

As pointed out in Section 22.3.1.1, reactive metals more electronegative than manganese cannot be processed electrolytically from aqueous solution. For this reason electrolytic methods for such metals require either water-free *organic electrolytes*[38] (not discussed in this chapter) or *molten salt electrolytes.* The latter are sometimes also referred to as *fused salt* or *high temperature inorganic melts.* Examples of materials that have been produced by this method include many refractories such as molybdenum, niobium, vanadium and tungsten, titanium, zirconium, magnesium, and aluminum, and several rare earth elements. While many of the fundamental concepts developed for aqueous electrolytic processes can be applied to molten salt electrolysis, considerable modifications and new considerations are required because the molten salt solvent is quite different from water as the solvent. In fact, the electrochemistry of fused salts is an entire research field on its own.[39,40]

Probably the most widely known molten salt process is the Hall–Héroult process for aluminum production introduced in 1866.[41] In this process, molten cryolite (Na_2AlF_6) is used as the solvent for alumina (Al_2O_3) which is usually chemically purified from bauxite in the Bayer process. Cryolite melts at about 940°C, and the Hall–Héroult process operates at temperatures between 960 and 1000°C. The anode material used is carbon and a typical electrolytic cell operates at 6 V. The cell is lined with carbon that acts as the cathode onto which Al is deposited. The overall cell reaction is:

$$2/3Al_2O_3 + C \rightarrow 4/3Al + CO_2. \tag{22.42}$$

During the process, the carbon from the anode is continuously consumed to produce CO_2. A typical cell produces about 300 kg of aluminum daily at an energy consumption of about 15 to 26 kWh/kg of aluminum. Aluminum with a purity of 99% or higher can be produced by this process. It is important to note that the fused salt electrolyte does not chemically react with the Al deposit.

22.6 Electrolytic Processing of Nanostructures

As a result of their outstanding physical, chemical, and mechanical properties, nanostructured materials have received considerable attention over the past 10 years. Since their first introduction in the early 1980s, several hundred synthesis methods for their production have been described in the literature.[42–52] Many of these methods are simple extensions of well-established production routes while others were specifically designed for various types of highly specialized nanomaterials. Many of these techniques are based on electrolytic processes.

In the following sections, examples will be presented for the synthesis of a variety of nanostructured materials by electrolytic processing including monolithic metals and alloys, structurally graded nanometals, nanocomposites, compositionally modulated alloy (CMA) nanostructures, and nanocrystalline oxide powders.

22.6.1 Monolithic Bulk Metal Nanostructures

Synthesis methods for the production of monolithic bulk electrodeposits with grain sizes <100 nm have been systematically developed since the early 1980s.[53–55] Initially, relatively simple electrolyte formulations and conventional direct current electrodeposition were employed to produce nanostructured deposits (e.g., Ni–P[56,57]), with particular attention to structures with relatively narrow grain size distributions and average grain sizes in the 5 to 50 nm range. Beginning in the late 1980s emphasis shifted toward the development of more complex alloy systems (e.g., Co–Fe, Ni–Fe, Ni–Mo, Co–Mo, Zn–Ni, Ni–Fe–Cr, etc.) and the use of pulsed current plating to promote crystal nucleation at higher achievable current densities.[58,59]

In the electrodeposition process there are two competing factors that govern the grain size of the final deposit. Figure 22.3 showed some of the steps involved in the electrocrystallization process. When a metallic ion is reduced it can either nucleate a new grain or join an existing grain and contribute to its growth. Nanostructure formation can be achieved by using operating parameters (e.g., bath composition, pH, temperature, current density, current on and off times during pulsed plating), which are conducive to massive nucleation throughout the entire electrodeposition process. It is important to avoid the transition from initially fine grained to coarse columnar structure often observed in conventional electrodeposition (Figure 22.4). In other words, nanostructured electrodeposits can be considered an extreme case of UD type deposits (Section 22.2.6). Their structure consists of nanosized grains throughout the entire thickness (Figure 22.9 and Figure 22.10).

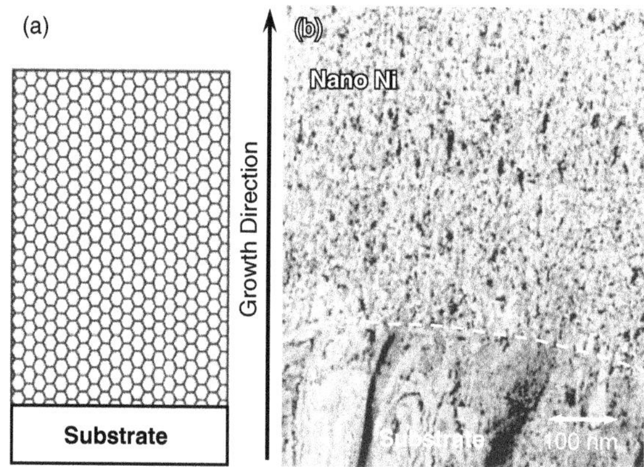

FIGURE 22.9 Cross-sectional structure of nanocrystalline nickel electrodeposit on bronze substrate: (a) schematic diagram and (b) transmission electron micrograph.

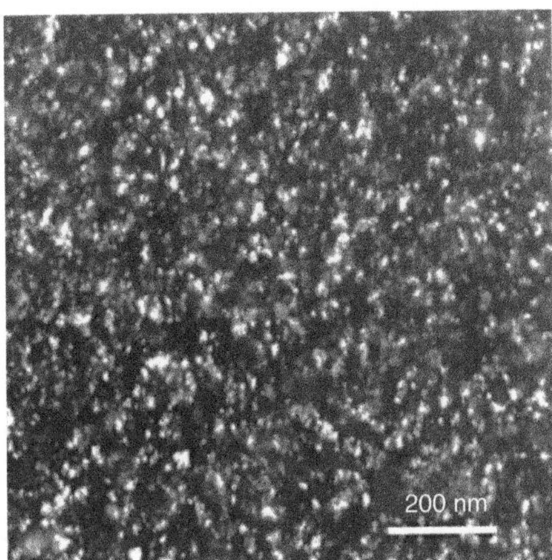

FIGURE 22.10 Darkfield electron micrograph parallel to the substrate of bulk nanocrystalline nickel (average grain size: 12 nm) in planar cross-section.

Electrodeposition processes have been developed for the synthesis of nanocrystalline deposits in many different shapes and forms.[54,55] These range from thin and thick corrosion and wear-resistant coatings to sheet and plate products for structural, magnetic, and electronic applications. Furthermore, the technology can be incorporated with little extra costs in conventional electroplating plants in processes such as rack plating, reel-to-reel plating, continuous strip plating, barrel plating, and brush plating.

One of the earliest industrial applications of this type of nanocrystalline deposits was the *in situ* electrosleeve nuclear reactor steam generator tubing repair technology.[60,61] This technology was the first large-scale structural application of nanomaterials in the world, specifically developed for Canadian and U.S. nuclear reactors since the early 1990s. In this process, steam generator tubes (e.g., Alloy 600 or 400) whose structural integrity was compromised by localized degradation phenomena (e.g., intergranular corrosion, pitting) were repaired by coating the inside of the tubes with a thick (1 mm) nanocrystalline Ni microalloy to restore a complete pressure boundary. The grain size of the materials was adjusted to be in the 50 to 100 nm range to give the required combination of strength, ductility, corrosion resistance, and thermal stability for this application. Figure 22.11 shows some of the tremendous property improvements that can be achieved when reducing the grain size in nickel electrodeposits from 10 μm to 10 nm.[59,62]

22.6.2 Structurally Graded Nanomaterials

While monolithic nanometals with relatively narrow grain size distributions have excellent properties in terms of hardness, tensile strength, and wear resistance, their tensile ductility is usually compromised regardless of the processing route.[59,63] In the past few years, it has been recognized that considerable ductility can be restored in these materials through broader or bimodal grain size distributions.[63] Recently, electroplating conditions have been developed for the synthesis of a variety of structurally graded nanometals (Figure 22.12), including bimodal distributions, alternate layers of different grain sizes, and grain size gradient deposits.[59,64] In such structures, a good compromise between strength and ductility can be achieved, whereby the smaller grains provide the strengthening in the metal while the larger grains allow for sufficient dislocation activity to result in reasonable ductility values. It should be noted that the deformation mechanisms in these materials are currently not completely understood.

22.6.3 Nanocomposites

Over the past several years, the concept of composite coatings (Section 22.3.1.3) has been extended to nanocrystalline materials.[54,55,59,65] Figure 22.13 shows several examples of submicrocrystalline/nanocrystalline second phase particles or fibers embedded in a nanocrystalline matrix. Typical examples of second phase

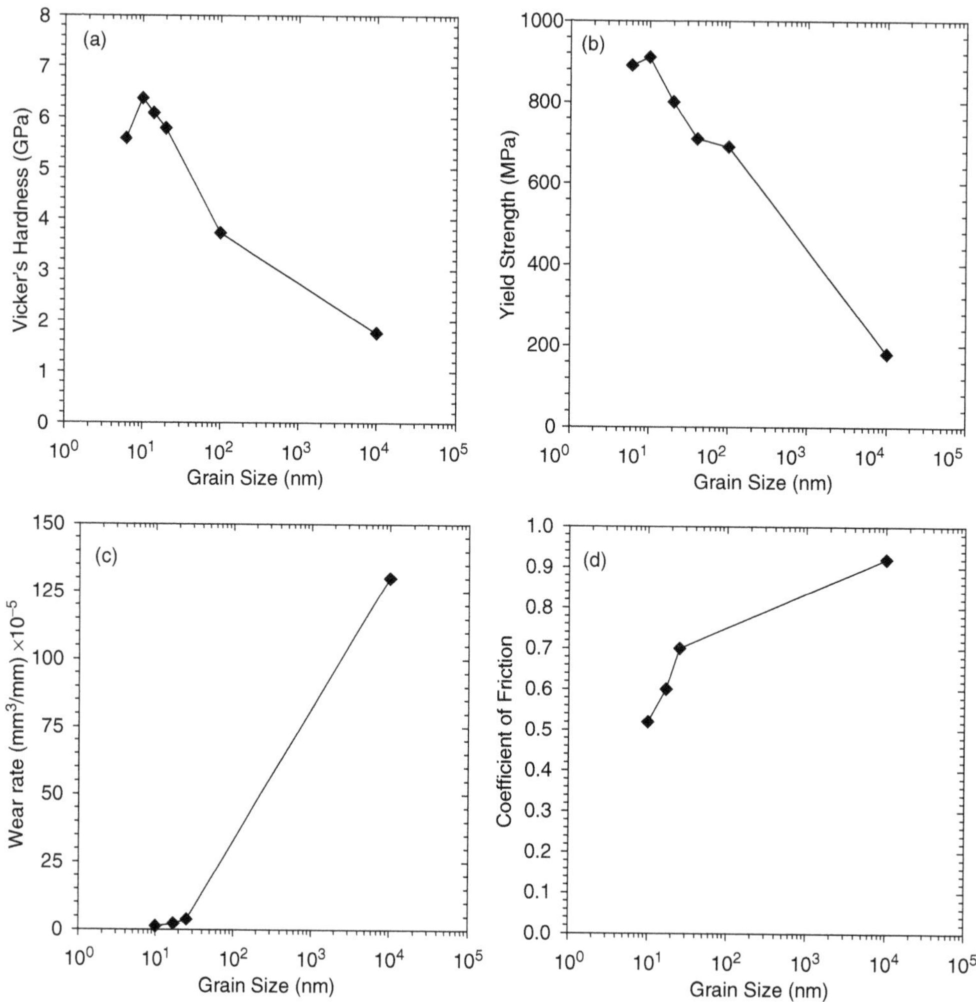

FIGURE 22.11 Effect of grain size in electrodeposited Ni on (a) Vickers hardness, (b) yield strength, (c) sliding wear rate and (d) coefficient of friction.

materials are Al_2O_3, SiC, or B_4C for improved hardness and wear resistance, or Teflon and MoS_2 for reduced coefficient of friction. Composite coatings can also be produced *in situ* by first codepositing an alloy as a supersaturated solid solution, followed by heat treatment to precipitate nanosized second phase particles in a nanocrystalline matrix.[62,66]

22.6.4 CMA Nanostructures

The idea of producing CMAs by electrodeposition (Figure 22.14) is not new. Some of the earliest work was apparently already done by A. Brenner in his 1939 Ph.D. thesis (as cited in Reference 67). Because of the unusual microstructure (i.e., individual layer thickness <10 nm), unexpected mechanical (e.g., strength, elastic constants), tribological and electromagnetic properties (e.g., giant magnetoresistance) have been reported for such materials, which has led to considerable research efforts since the early 1980s. The two main approaches to synthesize such materials are rotating substrate plating or potential-stepping deposition. In the rotating substrate method, the rotating cathode is situated between two physically separated baths and the thickness of each layer is determined by the plating rate in each bath and the rotation speed of the cathode. In the potential-stepping method, on the other hand, the cathode is placed

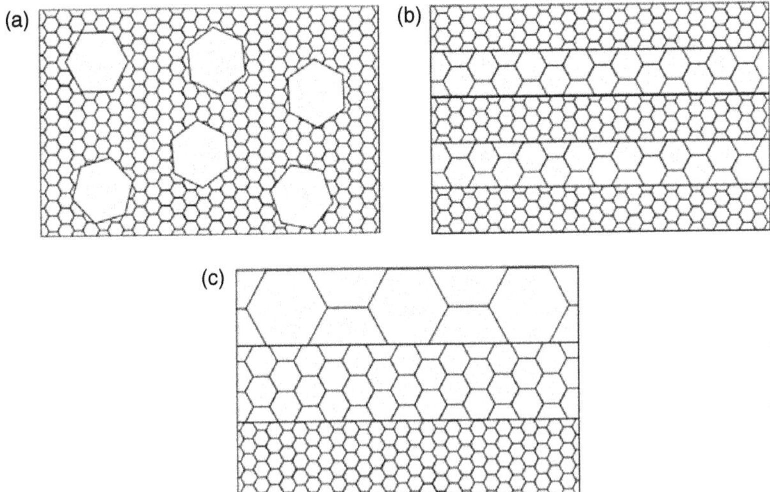

FIGURE 22.12 Schematic diagrams showing various types of structurally graded nanomaterials: (a) bimodal grain size distribution, (b) alternate layers with different grain sizes and (c) grain size gradient structure. (Courtesy of Integran Technologies, Toronto, Canada.)

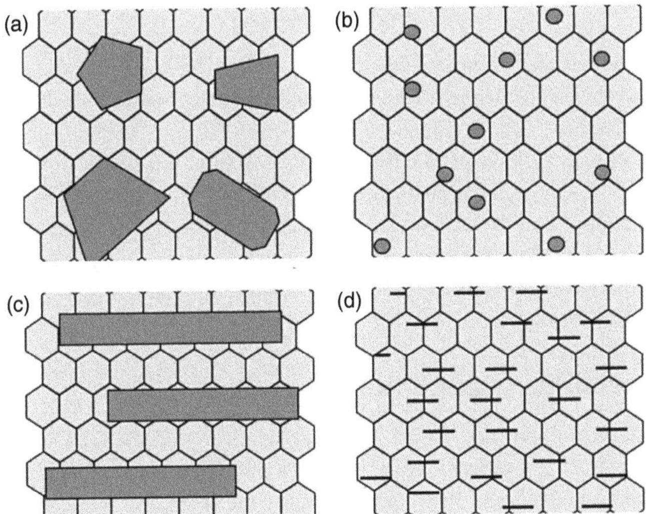

FIGURE 22.13 Examples of nanocomposite electrodeposits showing various configurations with different shapes and sizes of the reinforcing phase in a matrix of electrodeposited nanometal. (Courtesy of Integran Technologies, Toronto, Canada.)

into an electrolyte containing both species to be plated. By stepping the potential between predetermined values, the relative deposition rate of one species over that of the other for certain time intervals is controlled. Using these approaches, numerous compositionally modulated nanostructures have been produced, including Ag–Pd, Cu–Ni, Cu–Pb, and Ni–P[59,67] with modulation wavelength down to a few nanometers.

Even nanomodulated ceramic superlattices have been produced by the potential-stepping method. For example, layered structures consisting of $Tl_aPb_bO_c$ semiconducting oxides, with varying values for a, b, c in the alternating layers, were produced with modulation wavelengths between 6 and 13 nm.[68] These materials are of considerable interest for optical and electrical applications.

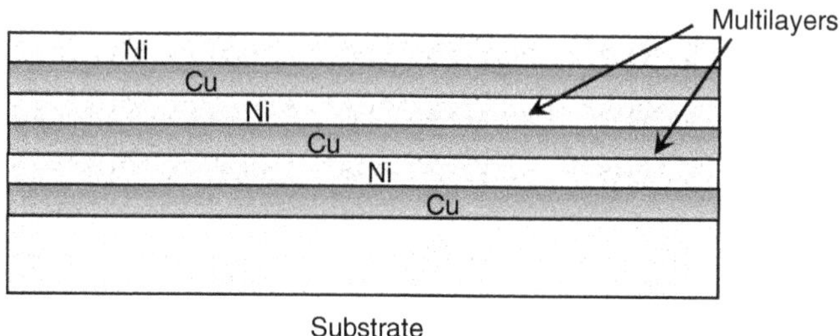

FIGURE 22.14 Schematic diagram showing a compositionally modulated material with alternating nanometer thick layers of Ni and Cu.

22.6.5 Nanocrystalline Particles

A number of electrolytic methods have been developed for the synthesis of various nanoparticles either on substrates or in colloidal solutions. For example, using the galvanic displacement approach (Section 22.3.3), very intricate structures of silver nanoparticles were recently grown on (111) and (100) oriented single crystals of germanium according to the following displacement reaction:[69]

$$Ge \leftrightarrow Ge^{4+} + 4e \tag{22.43}$$

$$4Ag^{+} + 4e \leftrightarrow 4Ag. \tag{22.44}$$

The hexagonal nanoparticles deposited in the form of vertical stacks on the substrate in what was described as nanoinukshuks, an analogy to the inukshuk structures made by the Inuit people in the Arctic using flat slabs of rock.

Another example of nanoparticle production using a combination of electrolytic and chemical reactions is the synthesis of β-CuI semiconductor nanocrystallites.[70] The three-step process involved (1) the electrodeposition of copper nanocrystals on a basal oriented single crystal of graphite from a Cu^{2+} containing solution, followed by (2) electrochemical oxidation of Cu to Cu_2O and (3) displacement of oxygen by iodine in an aqueous potassium iodine solution to obtain semiconducting β-CuI nanocrystallites with particle sizes up to about 20 nm.

22.6.6 Nanocrystalline Oxide Powders

Known as electrochemical deposition under oxidizing conditions (EDOC), one specific technique to produce nanostructured oxide powders is based on metal dissolution and deposition.[59,71,72] As shown in Figure 22.15, this method consists of three basic steps. At the sacrificial anode, metal is dissolved in an organic electrolyte (e.g., 2-propanol) containing a salt to increase the electrical conductivity of the bath. The second step is the cathodic reduction of metal ions to form metal clusters. These clusters are coated with a stabilizer (e.g., quarternary ammonia salts, betains, or ethoxylated fatty alcohols, not shown in Figure 22.15). In the final step, the colloidal metal particles are oxidized by introducing oxygen or air into the bath with simultaneous strong agitation or sonication. Subsequently, the oxide particles are filtered, washed with ethanol and dried. This method can produce simple oxides (e.g., ZnO, MgO, CuO, Fe_2O_3, SnO_2) or mixed oxides (e.g., $CoFe_3O_4$, In_2O_3/SnO_2) with high purity and particle size distributions much narrower than what can be achieved with other nano-oxide powder synthesis methods such as inert gas condensation, sol-gel, or flame pyrolysis. EDOC is a relatively inexpensive synthesis method and easy to scale up for large-scale production.

22.6.7 Template Nanotechnology by Anodizing

Over the past decade, the synthesis of highly ordered structures based on using ordered arrays of nanoporous honeycomb structures of anodic aluminum oxide (Section 22.3.5) has evolved rapidly as part of the

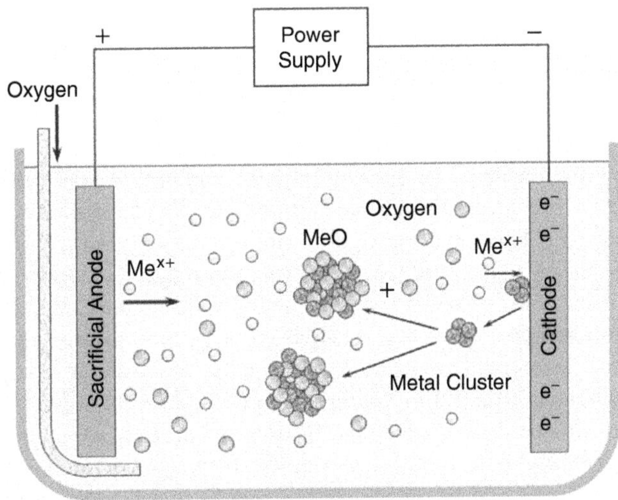

FIGURE 22.15 Schematic diagram showing an experimental setup for the production of nanocrystalline powders via EDOC. (Modified from Dierstein, A. et al., *Scripta Mater.*, 44, 2209, 2001.)

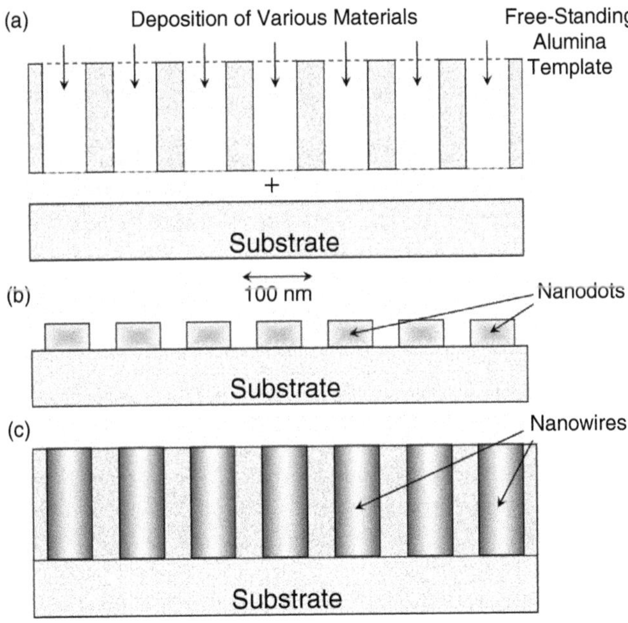

FIGURE 22.16 Use of nanoporous alumina template in nanodot and nanowire synthesis.

general area of template synthesis of nanostructures.[59,73–75] In these processes, the porous alumina films are extracted from previously anodized aluminum sheets (usually by dissolving the aluminum base) to serve as a template (Figure 22.16a) in further synthesis steps for a variety of nanostructures. For example, Figure 22.16b shows that deposition of other materials (e.g., metals, semiconductors) through such a template can be used to produce regular arrays of so-called quantum dots, which show very interesting electrical properties resulting from electronic state confinement in crystals with very small external sizes (<100 nm). If the pores in thick aluminum oxide layers are filled up completely (Figure 22.16c), this templating technique can be used to produce nanowires of varying diameters and lengths, which have been shown to display exceptional mechanical, electrical, optical, or chemical properties.

Besides nanowires, template synthesis has been used to produce nanocables with a radial metal/semiconductor junction.[76]

22.7 Conclusions

Electrochemical reactions provide the basis for an extensive number of electrolytic processes to synthesize materials in many different shapes and forms. Numerous processing routes have been developed over the past century for primary metal production, direct manufacturing processes, and surface-finishing applications. Today, many of the chemical and physical fundamentals of electrolytic processing are fairly well understood. However, numerous aspects of electrocrystallization require further study to achieve a better control of microstructure, and therefore properties, of the final material through appropriate choice of processing parameters such as electrolyte composition, pH, temperature, current density, and electrolyte agitation. While electrolytic processing is overall a relatively mature technology for conventional material synthesis, research over the past 20 years has shown that there are tremendous new opportunities for electrolytic processes in the general area of nanotechnology. Several nanostructured materials produced by electrolytic processes have already found application in industry, and many more processes are very close to commercialization. The outlook for electrolytic processing is excellent, and it is expected that many of the questions regarding electrocrystallization will be answered in the next few years through the extensive use of powerful computational as well as nanoscale chemical and microstructural analysis tools currently applied in the development of electrolytic nanomaterials processing technologies.

References

[1] Vetter, K.J., *Elektrochemische Kinetik*, Springer Verlag, Berlin, 1960.

[2] Potter, E.C., *Electrochemistry: Principles and Applications*, Cleaver Hume Press Ltd., London, 1961.

[3] West, J.M., *Electrodeposition and Corrosion Processes*, Van Nostrand, London, 1965.

[4] Raub, E. and Muller, K., *Fundamentals of Metal Deposition*, Elsevier, New York, 1967.

[5] Bockris, J.O.M. and Razumney, G.A., *Fundamental Aspects of Electrocrystallization*, Plenum Press, New York, 1967.

[6] Bockris, J.O.M. and Reddy, A.K.N., *Modern Electrochemistry*, Vols. 2, Plenum Press, New York, 1970.

[7] Bloom, H. and Gutmann, F., *Electrochemistry — The Past Thirty and the Next Thirty Years*, Plenum Press, New York, 1977.

[8] Bard, A.J. and Faulkner, L.R., *Electrochemical Methods*, Wiley & Sons, New York, 1980.

[9] Reiger, P.H., *Electrochemistry*, 2nd ed., Chapman and Hall, New York, 1993.

[10] Bockris, J.O.M. and Khan, S.U.M., *Surface Electrochemistry — A Molecular Level Approach*, Plenum Press, New York, 1993.

[11] Valensi, G., Van Muylder, J., and Pourbaix, M., *Atlas of Electrochemical Equilibria in Aqueous Solutions*, Pergamon Press, New York, 1966.

[12] Pourbaix, M., *Lectures on Electrochemical Corrosion*, Plenum Press, New York, 1973.

[13] Gorbunova, K.M. and Polukarov, Y.M., Electrodeposition of alloys, in *Advances in Electrochemistry and Electrochemical Engineering*, Vol. 5, Tobias, C.W., Ed., Interscience Publishers, Wiley & Sons, New York, 249, 1967.

[14] Schlesinger, M. and Paunovic, M., Eds., *Modern Electroplating*, John Wiley & Sons Inc., New York, 2000.

[15] Andricacos, P.C. et al., Eds., Electrochemical synthesis and modification of materials, *Mat. Res. Soc. Symp. Proc.*, 451, 1997.

[16] Merchant, H., Ed., *Defect Structure, Morphology and Properties of Deposits*, TMS, Warrendale, PA, 1–344, 1995.

[17] Stivison, D.S., Ed., *Metal Finishing —2004 Guidebook and Directory*, Elsevier, New York, 2004.

[18] Wood, W.G., Ed., *Metals Handbook* — 9th ed., Vol. 5: Surface Cleaning, Finishing and Coating, American Society for Metals, Metals Park, OH, 1982.

[19] Brenner, A., *Electrodeposition of Alloys: Principles and Practice*, Academic Press, New York, 1963.

[20] Boyer, H.E. and Gall, T.L., Eds., *Metals Handbook — Desk Edition*, American Society for Metals, Metals Park, OH, 1985.

[21] Safranek, W.H., *The Properties of Electrodeposited Metals and Alloys*, 2nd ed., American Electroplaters and Surface Finishers Society, Orlando, FL, 1986.

[22] Dennis, J.K. and Such, T.E., *Nickel and Chromium Plating*, John Wiley & Sons, New York, 1992.

[23] Dini, J.W., *Electrodeposition*, Noyes Publ., Park Ridge, NJ, 1993.

[24] Paunovic, M. and Schlesinger, M., *Fundamentals of Electrochemical Deposition*, John Wiley & Sons, New York, 1998.

[25] Andricacos, P.C. and Romankiw, L.T., Magnetically soft materials: Their properties and electrochemistry, in *Advances in Electrochemical Science and Engineering*, Vol., 3, Gerischer, H. and Tobias, C.W., Eds., VCH, Basel, Switzerland, 227, 1994.

[26] Celis, J.P., Roos, J.R., and Buelens, C., A mathematical model for the electrolytic codeposition of particles with a metallic matrix, *J. Electrochem. Soc.*, 134, 1402, 1987.

[27] Gomas, W.P. and Goossens, H.H., Electrochemistry of III–V compound semiconductors, in *Advances in Electrochemical Science and Engineering*, Vol., 3, Gerischer, H. and Tobias, C.W., Eds., VCH, Basel, Switzerland, 1, 1994.

[28] Okinaka, Y. and Osaka, T., Electroless deposition process: fundamentals and applications, in *Advances in Electrochemical Science and Engineering*, Vol., 3, Gerischer, H. and Tobias, C.W., Eds., VCH, Basel, Switzerland, 55, 1994.

[29] Mollory, G.O. and Hajdu, J.B., Eds., *Electroless Plating: Fundamentals and Applications*, American Electroplaters and Surface Finishers Society, Orlando, FL, 1990.

[30] Evans, U.R., *The Corrosion and Oxidation of Metals*, Edwards Arnold Publ. Ltd., London, 241, 1960.

[31] Wang, Y.M., Kuo, H.H., and Kia, S., Effect of alloy types on the gloss of anodized aluminum, *Plat. Surf. Fin.*, 91, 34, 2004.

[32] ASTM Special Technical Publication No. 318, *Electroforming — Applications, Uses and Properties of Electroformed Metals*, American Society for Testing and Materials, Philadelphia, PA, 1962.

[33] Watson, S.A., *Applications of Electroforming*, NDI Technical Series No. 10 054, Nickel Development Institute, Toronto, Ontario, 1989.

[34] Parkinson, R., *Electroforming — A Unique Metal Fabrication Process*, NDI Technical Series No. 10 085, Nickel Development Institute, Toronto, Ontario, 1998.

[35] Parkinson, R., *Nickel Plating and Electroforming*, NDI Technical Series No. 10 088, Nickel Development Institute, Toronto, Ontario, 2001.

[36] Lyshevski, S.E., *MEMS and NEMS*, CRC Press, Boca Raton, FL, 410, 2002.

[37] Lyshevski, S.E., *Nano- and Microelectromechanical Systems*, CRC Press, Boca Raton, FL, 1, 2001.

[38] Lehmkuhl, H., Mehler, K., and Landau, U., The principles and techniques of electrolytic aluminium deposition and dissolution in organoaluminum electrolytes, in *Advances in Electrochemical Science and Engineering*, Vol. 3, Gerischer, H. and Tobias, C.W., Eds., VCH, Basel, Switzerland, 163, 1994.

[39] Delimarskii, I.U.K. and Markov, B.F., *Electrochemistry in Fused Salts*, The Sigma Press, Washington, 1961.

[40] Mamantor, G. et al., Eds., *Molten Salts*, Proceedings Vol. 87-7, The Electrochemical Society Inc., Pennington, NJ, 1987.

[41] Sadoway, D.R., Inert anodes for the Hall-Héroult cell, *JOM*, 53, 34, 2001.

[42] Kear, B.H., Tsakalakos, T., and Siegel, R.W., Eds., Materials with ultrafine microstructures, *Nanostruct. Mater.*, 1, 1–106, 1992.

[43] Komarneni, S., Parker, J.C., and Thomas, G.J., Eds., Nanophase and nanocomposite materials, *Mater. Res. Soc. Symp. Proc.*, 286, 1–459, 1993.

[44] Yacaman, M.J., Tsakalakos, T., and Kear, B.H., First international conference on nanostructured materials, *Nanostruct. Mater.*, 3, 1–518, 1993.

[45] Schaefer, H.E. et al., Eds., Second international conference on nanostructured materials, *Nanostruct. Mater.*, 6, 1–1026, 1995.

[46] Suryanarayana, C., Singh, J., and Froes, F.H., Eds., *Processing and Properties of Nanocrystalline Materials*, TMS, Warrendale, PA, 1–496, 1996.

[47] Komarneni, S., Parker, J.C., Wollenberger, H.J., Eds., Nanophase and nanocomposite materials II, *Mater. Res. Soc. Symp. Proc.*, 457, 1–588, 1997.

[48] Trudeau, M.L. et al., Eds., Third international conference on nanostructured material, *Nanostruct. Mater.*, 9, 1–771, 1997.

[49] Ma, E. et al., Eds., *Chemistry and Physics of Nanostructures*, TMS, Warrendale, PA, 1–241, 1997.

[50] Muhammed, M. and Rao, K.V., Eds., Fourth international conference on nanostructured materials, *Nanostruct. Mater.*, 12, 1–1188, 1999.

[51] Inoue, A. et al., Eds., Fifth international conference on nanostructured materials, *Scripta Mater.*, 44, 1161–2372, 2001.

[52] Shaw, L.L., Suryanarayana, C., and Mishra, R.S., Eds., *Processing and Properties of Structural Nanomaterials*, TMS, Warrendale, PA, 1–222, 2003.

[53] Erb, U., Aust, K.T., and Palumbo, G., Electrodeposited nanocrystalline materials, in *Nanostructured Materials*, Koch, C.C., Ed., Noyes Publ./William Andrew Publ., Norwich, NY, 179, 2002.

[54] Palumbo, G., Gonzalez, F., Tomantschger, K., Erb, U., and Aust, K.T., Nanotechnology opportunities of electroplating industries, *Plat. Surf. Finish.*, 90, 36, 2003.

[55] Palumbo, G., McCrea, J., and Erb, U., Applications of electrodeposited nanostructures, in *Encyclopedia of Nanoscience and Nanotechnology*, Nalwa, H.S., Ed., American Scientific Publishers, Stevenson Ranch, CA, Vol. 1, 89, 2004.

[56] McMahon, G. and Erb, U., Structural transitions in electroplated Ni–P alloys, *J. Mater. Sci. Lett.*, 8, 885, 1989.

[57] McMahon, G. and Erb, U., Bulk amorphous and nanocrystalline Ni–P alloys by electroplating, *Microstruct. Sci.*, 17, 447, 1989.

[58] Erb, U. and El-Sherik, A., *Nanocrystalline Materials and Process of Producing the Same*, U.S. Patent No. 5,352,266, 1994.

[59] Cheung, C., Erb, U., and Palumbo, G., Nanostructure synthesis by electrodeposition, in *Processing and Fabrication of Advanced Materials*, Srivatsan, T.S. and Varin, R.A., Eds., ASM International, Materials Park, OH, 2004.

[60] Gonzalez, F. et al., Electrodeposited nanostructured nickel for *in-situ* nuclear steam generator repair, *Mater. Sci. Forum*, 225, 831, 1996.

[61] Palumbo, G. et al., *In-situ* steam generator repair using electrodeposited nanocrystalline nickel, *Nanostruct. Mater.*, 9, 736, 1997.

[62] Erb, U. et al., Grain size effects in nanocrystalline electrodeposits, in *Processing and Properties of Structural Nanomaterials*, Shaw, L.L. et al., Eds., TMS, Warrendale, PA, 109, 2003.

[63] Koch, C.C. and Scattergood, R.O., Grain size distribution and mechanical properties of nanostructure materials, in *Processing and Properties of Structural Nanomaterials*, Shaw, L.L. et al., Eds., TMS, Warrendale, PA, 45, 2003.

[64] Integran Technologies Inc., Toronto, unpublished data.

[65] Zimmerman, A.F. et al., Mechanical properties of nickel silicon carbide nanocomposites, *Mater. Sci. Eng.*, A328, 137, 2002.

[66] Erb, U., Palumbo, G., and Aust, K.T., Electrodeposited nanostructured films and coatings, in *Nanostructured Films and Coatings*, Chow, G.M. et al., Eds., NATO Science Series 3, Vol. 78, Kluwer Academic Publ., Dordrecht, The Netherlands, 11, 2000.

[67] Paunovic, M., Schlesinger, M., and Weil, R., Fundamental considerations, in *Modern Electroplating*, Schlesinger, M. and Paunovic, M., Eds., John Wiley and Sons, New York, 37, 2000.
[68] Switzer, J.A., Electrochemical architecture of ceramic nanocomposites, *Nanostruct. Mater.*, 1, 43, 1992.
[69] Aizawa, M. et al., Silver nano-inukshuks on germanium, *Nano Lett.*, 5, 815, 2005.
[70] Hsiao, G.S. et al., Hybrid electrochemical/chemical synthesis of supported luminescent semiconductor nanocrystallites with size selectivity: Copper (I) iodide, *J. Am. Chem. Soc.*, 119, 1439, 1997.
[71] Hempelmann, R. and Natter, H., *German Patent,* DE 198-40-841, 1998.
[72] Dierstein, A. et al., Electrochemical deposition under oxidizing conditions: EDOC, *Scripta Mater.*, 44, 2209, 2001.
[73] Al Malawi, D. et al., Electrochemical fabrication of metal and semiconductor nanowire arrays, in *Nanostructured Materials in Electrochemistry*, Searson, P.C. and Meyer, G.J., Eds., The Electrochemical Society, Inc., Pennington, NJ, 262, 1995.
[74] Masuda, H. and Fukuda, K., Ordered metal nanohole arrays made by a 2-step replication of honeycomb structures of anodic alumina, *Science*, 268, 1466, 1995.
[75] Sawitowski, T., Beyer, N., and Schulz, F., Bio-inspired antireflective surfaces by imprinting processes, in *The Nano-Micro Interface*, Fecht, H.J. and Werner, M., Eds., Wiley — VCH, Weinheim, Germany, 263, 2004.
[76] Ku, J.-R. et al. Fabrication of nanocables by electrochemical deposition inside metal nanotubes, *J. Am. Chem. Soc.*, 126, 15022, 2004.

V

Macroprocesses

23

Glass Processing

Alexis G. Clare
New York State College of Ceramics

Abstract

Glass is a unique material and is more a state of matter than it is a particular material. This chapter outlines the glass science behind processing and some of the unique features of modern glass processing. As there are a large number of processing modalities to cover, the chapter introduces the glass science behind the processing and suggests further reading at the end.

23.1 What is Glass?

In the manner that we talk about gases, solids, and liquids, glass is not a single material but rather a state of matter, a subset of the solid state. Even then we have to qualify that statement because while glass is essentially an elastic solid below its transformation region (known as the glass transition) and a liquid above it, the structure has all the attributes of a liquid except for the ability to flow on a reasonable time scale. The latter statement strictly means that a normal commercial glass at regular temperature is not going to flow on anything much shorter than a geological timescale. However, if you were able to take a photograph of the atoms showing their position, then you would see a structure that would be indistinguishable from that of a liquid, except that in a subsequent photograph of a liquid the atoms would have all moved, whereas in the glass they would be in much the same position as the first photograph. Essentially, a glass is an elastic solid without the structural periodicity and long-range order of a crystalline material.

If you have not gathered already, glass is quite an unusual solid in that it does not melt but "transitions:" from solid to liquid over a range of temperatures. More curious still is that the range over which the transition occurs is a function of the thermal history of the glass and also sensitive to the method used to measure it. Thus glass does not have a well-defined melting point as crystalline solids do. This is because the structure is cooling rate dependent and for most commercial glasses, a faster cooling rate would freeze in a larger volume structure than a slower cooling rate. Silica is a notorious exception to this trend.

Given that glass is basically a material with the attributes of a solid and the structure of a liquid, how does that happen and why are not all solids like this? The ASTM definition of a glass provides a large clue; "Glass is a substance that cooled from the melt without crystallizing." Thermodynamically most solids would crystallize, the inherent order resulting in a lower energy for the solid. However, thermodynamics still has to bow occasionally to kinetics and glasses can be formed from the melt by cooling rapidly enough such that the atoms do not have time to form an ordered structure. Cooling rapidly has many different meanings dependent upon the composition, for example, melted silica could be cooled relatively slowly and because the viscosity is so high it would not crystallize. On the other hand, to make a metallic glass requires very rapid cooling, sometimes more than thousands of degrees per minute.

So, given that glass would thermodynamically prefer to be crystalline, and given that extreme measures can be required to make a material form a glass, why bother? By far the most utilized properties of glasses, stemming from their inherently isotropic structure are their optical properties. With a few exceptions that will be introduced later, a majority of commercial glasses are highly transparent from the ultraviolet through to somewhere in the infrared. This is not just due to their electrically insulating nature, but also due to the fact that structural inhomogeneity generally occurs on such a small scale that light is not scattered as well as not absorbed. Add to that, that about 90% of commercial glasses are made from fairly innocuous oxides and of that 90% about 90% contain silica; then glass is also in general fairly chemically durable and can, in addition, be used in aqueous environments.

Glass is also very engineerable; it does not adhere as rigorously to stoichiometry as crystalline materials do in that you can incrementally change the properties of the glass almost continuously over a rather large range just by adding a property-modifying component. For example, you can easily make a fairly continuous range of properties by adding potassium oxide to silica. Note that you are adding the oxide thus maintaining stoichiometry to a certain extent but allowing a continuous variation of the alkali oxide content.

Having extolled the virtues of glass and given some idea that there is more to glass than sand soda, and potash, I do have to point out that while most glasses are oxide/silicate, there are some other exotic glasses for specific uses. Not only that, but glass can also be made into an incredible variety of shapes and forms for a far wider range of applications than most people would imagine. This chapter will address glass processing in general and then move to some of the more unusual processing technologies for some specialty applications.

23.2 Basic Glass Melting

23.2.1 Raw Materials

The raw materials chosen for making a glass from a batch are dependent upon the phase diagram of the system from which you are trying to make the glass and the viscosity temperature curve of the melt from which the glass will be cooled. For large commercial melts there is also the consideration of cost. Sometimes an intimate mix of two batch components can allow a low melting eutectic mixture that may save energy by allowing liquid formation at a lower temperature. A common mineral might contain two glassmaking components and the proper ratio of the two components can be achieved by topping up that mineral with one of the components. For oxide melts some of the components are introduced as the oxide, for example silica, whereas some are introduced as carbonates or as nitrates. It has been found that carbonates and nitrates decompose to the component oxides readily at glass melting temperatures and the release of gas on decomposition has two beneficial effects. First, the release of gases into bubbles helps to stir and mix the melt and in addition, the presence of bubbles helps sweep further dissolved gases out of the melt by forming larger bubbles that will actually float to the top of the melt. These processes are part of what is known as "fining" and are discussed further later.

Glass batches are quite simple to calculate taking into account any decomposition reactions that may occur. For example soda is often incorporated into the batch as sodium carbonate and it undergoes the decomposition reaction shown below:

$$Na_2CO_3 \iff Na_2O + CO_{2(g)}$$

The mole weight of Na_2CO_3 is 105.99 g/mol and the mole weight of Na_2O is 61.98 g/mol; therefore, once you have the weight of soda that you require in your melt if you simply multiply that weight by 1.71 then you can get the weight of sodium carbonate that you need to yield the required weight of soda. The same procedure is possible with minerals, too, for example, dolomite yields both calcia and magnesia. With a chemical formula of $CaCO_3{-}MgCO_3$ 1 kg of dolomite will yield 0.304 kg of CaO and 0.219 kg of MgO, the rest being evolved as carbon dioxide. In order not to overload one component clearly you can only use enough dolomite to give you the lesser of the two components as required for your batch. While most glassmaking books will give you gravimetric factors such as the 1.71 ratio for the sodium carbonate to soda, it is usually best to calculate the amount of raw material by writing out the reaction and calculating the yield in terms of moles and then converting to weight.

Most raw materials for commercial glassmaking are not terribly pure, it is very hard to eliminate iron from the raw materials in glassmaking, which is why commercial glasses often have a slight green or blue tint to them, particularly when you are looking through a thick part of the glass (e.g., a pane of window glass edge-on looks much greener than it does when looked at face-on). The other factor that will affect the efficiency and the homogeneity of melting is the size of the batch particles and the homogeneity of the batch mixing. In large commercial operations there are computer-controlled batch mixing and delivery into the furnace. Some industries even wet down batch to avoid losing fine powders. In recent years, there have been a number of attempts to improve melting by pretreating the batch, for example, obtaining a better intimacy of potential eutectics by pelletizing the batch under pressure. However, typically, any costs retrieved by using these methods are offset by the cost of applying the method.

23.2.2 Melting

Melting of glass can take place in a wide variety of situations. Researchers in the laboratory melt in crucibles typically in an electric furnace. Most large-scale commercial melting occurs in refractory tanks using gas-air firing or gas-oxygen firing. Crucible and refractory materials also vary very widely. For high temperature melting (up to 1600°C) the ideal crucible material is platinum or platinum/rhodium alloy. Since at 2006 prices the cost of a 150cc crucible is approximately $2500, you can imagine that most vast commercial melts are not done in platinum. One exception is the optical neodymium doped phosphate glasses that are made for the NOVA laser fusion project at Lawrence Livermore Laboratory. These glasses are made in a continuous melting process in a platinum-lined furnace. For high power laser applications no impurities can be tolerated (even small amounts of precipitated platinum) so care has to be taken not only to ensure that very pure raw materials are used but also to ensure that any platinum that is dissolved into the melt stays dissolved. Typical commercial melts are carried out in refractory lined tanks that last for a period of time and then have to be rebuilt. Glass melts are quite aggressive at about 1400°C and so are the vapors immediately above the melt. Many of the refractories are very high temperature ceramics containing materials such as chrome and zircon. The refractories are naturally porous but usually have a nonporous coating to slow down the corrosion processes by the glass melt.

Most commercial glass tanks have burners that lie just above the glass melt but recent studies have proposed a submerged burner system that could save considerable energy costs. The recent trend toward oxy-fuel firing in which the gas is mixed with oxygen has led to higher burner temperatures but at the same time, reaction between the fuel and oxygen has led to incorporation of more hydroxyl into the glass that sometimes is good and sometimes not. On the whole, electric melting is somewhat cleaner than combustion melting and some commercial melters are using a combination of both. The cost of electric melting has been rather prohibitive in the past but rising gas prices may make more electric melting more feasible. For melts that are below 1300°C it is possible to use refractory crucibles such as mullite, clay, or alumina. However, these crucibles are not easy to re-use, and typically result in contamination of the melt by the crucible materials. The advantages of platinum are that, apart from some phosphate melts the platinum does not significantly contaminate the glass and the crucibles can be cleaned and re-used relatively easily.

23.2.3 Fining

Most commercial glass melts have quite a shallow viscosity temperature curve meaning that if the temperature is raised slightly then viscosity does not decrease significantly. As a result trapped gas or a bubble has a difficult time rising to the top of the melt to release the gas into the atmosphere. Regardless of your reason for making glass, bubbles in the final solid are more often than not detrimental and so need to be avoided. In a regular glass a bubble would move up with a velocity given by a modification of Stokes' law for a solid sphere moving through a liquid. For a commercial soda lime silica glass where at about 1500°C a 10 μm diameter bubble would take over one and a half years to rise 1 m. This is clearly not acceptable and this process needs to be accelerated. The ideal situation would be a process that promotes the dissolution of gas as the temperature increases and the viscosity decreases, but that then would promote the solution of gas as the melt cools. The principle problem in oxide glasses is generally oxygen, that is, most of the undesirable bubbles that occur contain oxygen. In oxide glasses therefore, the control of oxygen release and dissolution can be achieved using the appropriate reduction–oxidation (REDOX) reaction. Ideally, one wants a reaction that promotes a reduced state at higher temperature, delivering oxygen that can help to sweep bubbles out of the melt when the melt is at low viscosity, and then promote an oxidized state at low temperature so that any oxygen remaining in bubbles as the melt viscosity increases, gets redissolved into the melt.

For many years both arsenic and antimony were chosen just for this purpose. The trivalent form of arsenic is stable at high temperatures while the pentavalent form is stable at low temperatures. As the temperature increases, the REDOX equilibrium favors the reduced state and oxygen is released into the melt as O_2. The oxygen gas helps bubbles rise through the melt by growing small bubbles into large bubbles and from Stokes' Law the large bubbles move faster through the melt. The action of the bubbles rising through the melt helps to stir the melt which, in turn, improves melt homogeneity. As the melt is cooled, the REDOX equilibrium favors the oxidized, pentavalent state so as the melt viscosity increases and bubbles can no longer rise, the oxygen redissolves to stabilize the pentavalent state. Ideally, at this point, if sufficient time has been allowed for fining, there should only be very small bubbles left in the melt and these are more likely to collapse than large bubbles, which also helps the oxygen to dissolve.

While arsenic and antimony have always worked very well as fining agents, their appearance on the list of environmentally regulated materials makes it more difficult to use them in glassmaking than it once was, despite the fact that they are typically present in very small quantities (usually less than one weight percent). More recently commercial glassmakers have tried to use both nitrate and sulfate fining. Instead of introducing alkali and alkaline earth elements as carbonates a small amount may be introduced as sulfate or nitrate. The decomposition temperatures of the nitrates and sulfates differ from those of the carbonates and nitrates in particular tend to lower the viscosity of the melt which is another aid to fining. However, there are also environmental issues with nitrate and sulfate emissions that can impact the application of this technique. Other redox equilibria are also being examined such as stannous/stannic and ferrous/ferric for fining, the latter being rather dependent on the level of coloration that might be tolerated in the glass.

23.2.4 Homogeneity and Phase Separation

One of the most appealing features of glass is that it can have optical homogeneity but this is not trivial to achieve. While glass does not typically have a microstructure in the way that other materials have microstructure, it can have inhomogeneity associated with it that can be on a scale rather smaller than the wavelength of light or, in some cases larger. We are going to look at what is meant by a homogeneous glass and then examine levels of inhomogeneity

A homogeneous glass is homogeneous on a length scale smaller than the wavelength of light. However, even an optically homogeneous glass such as a well-annealed silica gives rise to Rayleigh scattering that is due to inhomogeneity on the length scale of basic structural units. Typically, inhomogeneity on such a small scale does not limit any applications save one: optical fibers for telecommunications. In long-range telecommunications one is effectively passing light through very large thicknesses (km) of glass. In this case Rayleigh scattering which has a $1/\lambda^4$ dependence is the limiting factor of light lost from the fiber and indeed, telecommunications fiber can actually achieve that theoretical low loss limit (λ is wavelength). As we will see later in this chapter, the problem of impurities and other inhomogeneity has been solved in a rather unique fashion for optical fiber processing.

As mentioned previously, glass is unique in that the final structure and the amount of free volume in that structure are highly dependent upon the cooling rate of the glass. This can give rise to a very real problem when processing glass because with the exception of some exotic glasses the thermal conductivity of glasses and their viscous melts is not high (a typical soda lime silicate glass has a thermal conductivity of 1.7 W/m-K) and typical forming processes of glass can result in differential cooling across the glass piece. This differential cooling, then, causes different parts of the glass piece to have different volumes and one structure with more free volume adjacent to another with less, results in strain that can ultimately lead to failure of the glass piece. This problem is generally alleviated by annealing the glass. In annealing, the glass is either cooled or very slowly reheated to a temperature in which structural relaxations can occur on a reasonable time scale; in fact, the "Annealing Point" is defined as a temperature at which the viscosity is 10^{13} poise and where internal stresses built up during cooling are relieved within 15 min. The glass is held there for a time usually substantially longer than 15 min and then cooled at a rate slow enough to ensure that stresses are not reintroduced. The rate that is "slow enough" is not something that is easy to define. It depends on a number of variables not least of which is the shape of the article and this can become very complex for articles even as simple as a light bulb. Therefore, annealing schedules are often determined somewhat by trial, and hopefully not too many errors. It helps that one is able to visualize these stresses in the glass in a relatively simple way; if the glass piece is placed between two pieces of Polaroid that are crossed, the induced stresses have a tendency to rotate the plane of polarization after the first Polaroid such that a component of that light can pass through the second Polaroid. Thus, the stresses will then manifest themselves as bright fringes, sometimes colored as a result of the chromatic dispersion in the glass.

Even after annealing, however, one can sometimes see inhomogeneity in the glass. Sometimes this can be on quite a large scale where the inhomogeneity manifests as a distortion in the image seen through the glass. The reason for this, again, is structural. The refractive index of the glass depends on the structure and composition of the glass: ultimately the density and polarizability of the electrons in the structure. Hence, when there is structural or compositional inhomogeneity the light does not take a direct path through the glass but wavers resulting in image distortion. For good optical quality it is necessary to stir the glass when it is in the melt to ensure that there is no partitioning of composition or structure.

The tendency of a glass to phase separate due to thermodynamic driving forces is rather more the rule than it is the exception. This phase separation can take one of two forms: in the case of less stable glass forming compounds, crystallization can occur. Glassmakers generally refer to this as devitrification because it is usually deemed to be counterproductive except in some very important commercial materials called glass ceramics. Glass ceramics will be addressed separately later. Devitrification occurs via nucleation and growth stages where the nucleation can be either homogeneous or heterogeneous. (See Chapter 6 for more details on devitrification/crystallization.) The essence of good glassmaking is to ensure that during cooling one passes through the nucleation and growth stages fast enough that the kinetics of the melt prevent them from occurring. Unfortunately, the presence of heterogeneous nuclei can defeat this attempt. Heterogeneous nuclei may occur through contamination of the melt, or by inadequately melted batch. Crystals are particularly undesirable in glasses manufactured for their optical properties unless the crystals are small enough such that they do not significantly scatter the light. As we will see in glass ceramics, the composite nature of the glass ceramic can often lead to higher strength materials with superior thermal properties, very advantageous in certain applications.

The study and modeling of nucleation and crystallization from glass is a topic of great interest in the glass industry. Typically, these studies are carried out with thermal analysis or x-ray diffraction. Thermal analysis is usual by Differential Thermal Analysis (DTA) or Differential Scanning Cabrimetry (DSC) where either heat flow or temperature difference is measured as a function of temperature ramp against a standard material that does not undergo any phase transformations within the temperature range of interest (this is sometimes air and sometimes alumina). Endothermic or exothermic events are recorded thus crystallization exotherms can be observed. Isothermal studies can also be carried out to look at crystallization kinetics. Most of the work on nucleation and crystallization has been carried out by heat treatment followed by fast quenching and x-ray diffraction. The problem with this method is that it is not always easy to quench the glass fast enough to ensure that further nucleation or crystallization has not occurred during cooling, or indeed that it did not occur during heating of the glass. The nucleation rate is typically measured by first holding the glass above the liquidus to remove all nuclei and then cooling rapidly and holding at the nucleation temperature (which is typically lower than the crystallization temperature). The nuclei are too small to be seen; so they are counted by raising the temperature to the crystallization temperature allowing the nuclei to develop into crystals that can be seen in a microscope or measured in terms of quantitative x-ray diffraction. The study of devitrification has been somewhat improved by the use of hot stage microscopy (visible and electron) and hot stage x-ray diffraction. These eliminate the issues associated with quenching and, to a certain extent initial heating. The *in situ* x-ray diffraction uses a position sensitive detector so that about 15° 2θ can be examined simultaneously and therefore devitrification with phase identification can be observed in real time.

The other type of phase separation that can occur in glasses is liquid–liquid phase separation that manifests itself as the simultaneous existence of two glass compositions separated by an interface. This phase separation is either stable and occurs above the liquidus line in the phase diagram or it is metastable occurring below the liquidus line. The distribution of the two phases depends on a number of factors not least of which is the viscosity at which the phase separation occurs. Most of the characterization of liquid–liquid phase separation has occurred in metastable systems because the kinetics of the stable systems is very fast and therefore too difficult to control. In a metastable system the kinetics is such that phase separation occurs slowly, thus morphologies can be controlled and therefore characterized more consistently. In addition, if the phase separation is to be used to intentionally create a specified morphology, this is much easier to achieve in a metastable system. For example, phase separation is used in the Vycor® process for making high silica glass. Vycor is a Corning Incorporated composition that phase separates spinodally such that the two compositions form a continuous intertwining network. (See Chapter 4 for more details on spinodal decomposition and Vycor glasses.) One phase is a high alkali borate-rich phase while the other is very high silica. The less durable borate phase can be leached out and the remaining high-silica (about 96%) phase can be sintered to almost full density. As indicated there is more than one microstructure associated with the phase-separation process. Phase separation manifests itself through an immiscibility dome on the phase diagram (e.g., see Figure 4.3 and Figure 4.4 in Chapter 4). Within that immiscibility dome there is more than one microstructure. The previously mentioned spinodal microstructure is a central region of the dome known as the spinodal region. Outside of this region and yet still within the dome are the nucleation and growth microstructures. These are not to be confused with the nucleation and growth of crystalline phases although some of the processes are analogous, but what nucleates and grows are spherical droplets of one glass composition in another rather than the crystals. As the phase separation develops the chemical composition of the two phases remains unchanged but the relative volumes of the two phases change whereas in the spinodal region the structure is immediately intertwined and the two compositions are at first chemically similar but diverge rapidly.

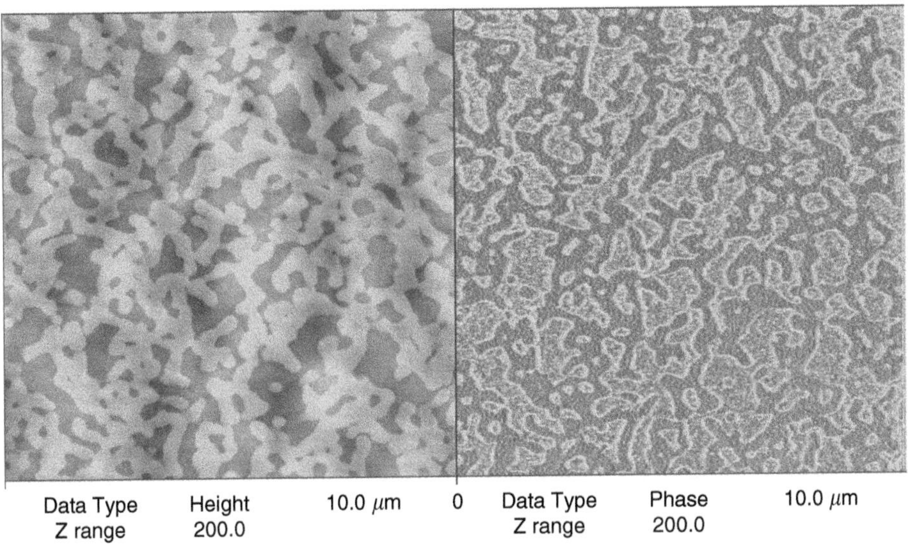

FIGURE 23.1 An atomic force microscopy (AFM) image of spinodal phase separation in a borosilicate glass. On the left is a topographical image of the glass after a light acid etch and on the right is a phase image of an unetched sample of the same glass.

Until recently the observation of phase separation has been rather difficult when it is on a scale not visible to the naked eye. Determination of the immiscibility dome has usually involved holding at a temperature well within the immiscibility dome to develop the phase separation such that it is visible to the naked eye in the form of light scattering giving the glass a milky appearance and then subsequently increasing the temperature until the milky appearance disappears. This is not very satisfactory as there can be quite a margin of error in the placement of the actual immiscibility interface. Observation of immiscibility on a smaller scale has always been rather tedious. It is rather difficult to see phase separation in the scanning electron microscope due to inadequate contrast between phases so it has usually been necessary to use the transmission electron microscope. At this point sample preparation becomes quite problematic because one either has to make a replica or ion beam thin the sample. Glasses are not very amenable to ion beam thinning because they are brittle and fracture catastrophically rather than thinning gradually. Nevertheless, studies have been completed with very nice TEM micrographs showing phase separation microstructure.

Recently, we have been using a combination of atomic force microscopy (AFM) of a lightly etched fracture surface to elucidate the morphology of phase separation. Figure 23.1 shows a sodium borosilicate glass that is known to undergo metastable immiscibility, having been heat treated in the spinodal decomposition region. Shown on the left is a topographical image after a light acid etch and on the right is a phase map that contrasts the two phases by contrasting the phase lag that is caused between the driven oscillations of the AFM tip and the response of that tip. The AFM has the added advantage of being able to carry out a more detailed analysis of both the size and to a certain extent the difference in chemical composition of the two phases. The difference in chemical composition is only qualitative in that it basically depends on the differential etch rate of the two phases.

23.3 Forming

In the previous description of basic glass melting a major step in between the melting and annealing was glossed over and that is forming. There are many different ways of forming the glass dependent on the application. Commercial glasses fall loosely into the following categories: flat, container, fiberglass, and specialty glass, the last being a sort of catch-all. This next section will describe the scientific basis behind these different forming modalities.

23.3.1 Flat Glass

Flat glass is quite a newcomer to glassmaking compared to container glass, as the principle reason for making flat glass was to see through it; production of flat glass really had to wait until the glassmakers had a reasonable ability to make homogeneous clear glass. Church windows were probably the first glass windows but there the objective was to provide the jewel like illustrations of biblical stories rather than visibility. In medieval times, glassmakers were able to make glass for windows by blowing either a cylinder, slicing it down the side and flattening it out while it was still hot, or by spinning a crown where the glass blower would blow a bubble, then turn the blowpipe on its side and spin it fast so that the glass splayed out and small panes could be cut out of the flat circle that had been formed. In both cases the thickness was not constant and therefore the image was not distortion free. Plate glass was made like lab melts by pouring onto a large area flat metal mold and then grinding and polishing to get good surfaces. It was not until after the industrial revolution and the appearance of mechanization that inexpensive flat glass could be made by rolling; however, it was not until 1902 when glass drawing to make sheet glass was patented. The drawing method used the adhesion of the molten glass to a metal "bait" to pull up a sheet for about half a meter where it was then draped over a roller. Similar methods were developed in France. None of these methods produced a truly flat glass as the roughness and evenness was really only as good as the roller. Plate glass was reserved for specialty applications such as store fronts and was still extremely expensive to make because of all the postprocessing steps. Flat glass processing was truly revolutionized in the 1950s when it was discovered by Sir Alastair Pilkington that instead of rolling the molten glass, it could be floated on a bath of molten tin. As dense liquids tend to have very flat surfaces and the top of a glass sheet, if it touched nothing else, would be smooth, this invention led to the smoothest flattest glass to date. Not only that but the thickness could be controlled by the volume and surface tension of the glass melt. A larger volume of glass channeled onto the molten tin gave a thicker glass. This results in a truly continuous process for flat glass, the only added complication being that while the glass is floating on the molten tin it must remain enclosed in a reducing atmosphere, otherwise the molten tin would oxidize. About 90% of all flat glass is produced this way and one can distinguish the tin side of the glass from the upper side by illuminating with a relatively low power UV source. Some of the tin from the molten metal bath diffuses into the glass surface and oxidizes to divalent tin. The divalent tin fluoresces under UV illumination. As the UV light will not pass through the glass only when the tin side of the glass faces the UV light does it fluoresce. However, not all compositions can undergo the float process.

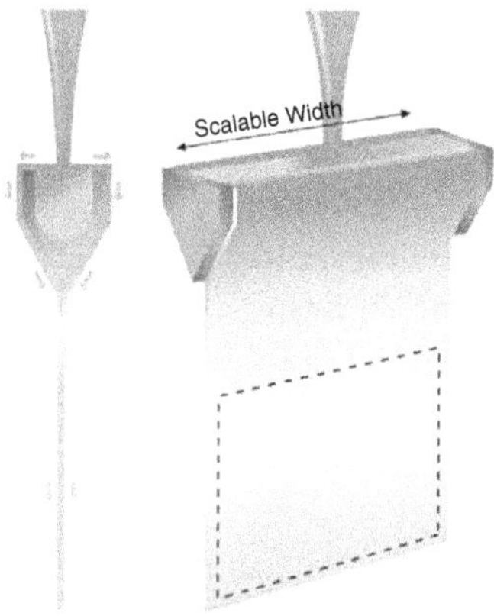

FIGURE 23.2 The fusion draw process for making thin flat glass. (Image Courtesy of Corning Incorporated.)

Active matrix liquid crystal displays (LCD) such as on a computer screen or some of the newer television screens require a glass substrate that does not contain alkali ions; otherwise these poison the active components in the LCD display. Unfortunately alkali-free glasses are not very amenable to the float process (although some compositions are now being developed that are able to be floated). An alternative method of making thin flat glass was needed and Corning, Inc. came up with the Fusion Draw process. In the Fusion Draw a trough of molten glass overflows from both sides and is drawn down creating exceptionally thin and smooth pieces of glass (Figure 23.2). The process is not limited by composition and glass sheet of about 1.8 m by 2.2 m can be drawn (http://www.corning.com/displaytechnologies/ww/en/discovery_center/fusion.aspx).

The category of "flat glass" does not always necessarily mean that the glass remains flat. Automotive glass is included in this category even though for modern cars it is far from flat and not completely transparent. The aerodynamic requirements of modern cars dictate curved surfaces for not only the windshield but also for the side windows. In general the windshield is a laminate that is sandwich layers of glass and a soft plastic to ensure that it does not shatter if it breaks. The side windows are tempered safety glass. Glass tempering uses the property of glass alluded to earlier where the volume depends upon the cooling rate. In tempered glass the glass is heated above the strain point and then the surface is rapidly cooled which freezes in a larger overall volume. However, the center of the glass cools more slowly which would tend to freeze in a smaller volume structure. Unfortunately the larger volume surface that was cooled quickly is now below the glass transition range (T_g) and is thus rigid. The rigid outer structure restricts the inner structure from forming the smaller volume and the competition that this sets up results in the outer surface being in compression. While the theoretical strength of glass is very high, the actual strength is much lower because surface flaws amplify the stresses at the crack tip resulting in catastrophic failure. However, if those flaws are placed under compression as they are in tempering, the cracks cannot propagate and the glass is substantially strengthened. To create the curved shapes of the windshields the glass is "slumped" over a mold of the correct shape for the car window. This can usually be carried out simultaneously with the tempering process. Tempered glass can be fractured particularly by impact from a sharp object but the fracture of tempered glass is different from normal glass because once the outer compressive layer is breached, the inner layer that is in competitive tension causes the glass to fly apart. However, the fractured pieces, unlike regular glass that produces sharp shards, are in fact more cubic, with blunter edges that are less likely to do damage. The problem with tempered glass is in order to achieve the temperature gradient needed to induce the competitive stresses, the glass has to be fairly thick which makes it quite heavy, so in applications where this is an issue alternative strengthening mechanisms are used.

Another way of strengthening flat glass that is used for aircraft windshields is chemical strengthening. Chemical strengthening again uses the principle of a compressive surface layer preventing crack propagation only this time the compressive surface layer comes from an ion exchange process. A glass which, for example, contains sodium ions is immersed in a molten salt bath of potassium nitrate. Potassium ions being larger than sodium ions exchange for the sodium ions in the glass surface but because the temperature of the molten salt is insufficient to allow structural relaxation, the potassium ions force themselves into the sodium sites which puts the surface under compression. The advantage of this method is that unlike tempering, it can be carried out on very thin pieces of glass. Strengths can be improved by almost a factor of ten dependent upon glass composition (Varshneya Private Communication).

23.3.2 Container Glass

Container glasses are typically soda lime silica glasses except for special applications such as pharmaceutical glasses. Pharmaceutical glasses are type I (borosilicate, chemically resistant), type II (dealkalized soda lime silicate glass), and type III (low alkali soda lime silicate glass). Most soda lime silicate bottles are resistant to aqueous and acidic solutions but are susceptible to basic solutions. In many cases color will be added to the glass for aesthetics or as a protective filter to stop light degradation of the product within. Most of the coloring agents used are transition metal oxides. For example, many containers have a brown color known as "amber" glass. The coloration of amber glass comes from the presence in the glass of both iron and sulfur. The sulfur is believed to replace an oxygen in the coordination polyhedron of the iron resulting in a charge transfer transition that gives rise to the amber coloration by absorbing the blue end of the spectrum.

Container glass forming is a rather automated process in which controlled volumes of molten glass are released from a furnace and dropped into a metal mold. The glass can either be pressed or blown into the mold to take the shape of the mold and to ensure a hollow center. Glasses are automatically annealed by being placed on a conveyor that processes them through a multizone furnace called a Lehr. Before going into the Lehr the bottles are sprayed with a lubricant coating to ensure that when they jostle they do not cause surface flaws that can make the container weak. A special form of container glass is the light bulb. Light bulbs are, of course, much thinner than regular containers and yet they have to be strong enough to support a vacuum. Light bulbs today are processed on a ribbon machine which is an ingenious device that sags a ribbon of glass through holes in a metal plate and blows them into molds that have a porous coating on the inside. The porous coating is wetted and the mold revolves rapidly as the glass sags, molded into shape by the layer of steam that forms as a result of the wetted porous mold. The revolution ensures that there is no seam present in the glass once the mold opens to release the light bulb. The steam surface of the mold ensures exceptional smoothness to the light bulb. Remarkably, this complex procedure still allows a ribbon machine today to make over 2000 light bulbs per minute.

There is some interest in strengthening containers in the same way as the flat glass by ion exchange. However, this would tend to be expensive as a postprocessing step for normal containers, although the benefits of chemically strengthened glass would be the reduced weight of the container. Special uses of glass are chemically strengthened, for example, the glass vials in emergency field inoculation equipment are strengthened to ensure their robustness in everyday use http://www.saxonglass.com/

23.3.3 Fiber Glass

Traditional fiberglass does not include optical fiber, which comes under specialty glass later. The commercial fiberglass category includes fibers for reinforcement and fibers for insulation. Glass fiber for reinforcement is usually a borosilicate glass composition and it is made from molten glass in platinum troughs being forced out through small apertures in the troughs and collected on a spinning drum. The process is continuous as long as the trough is continuously supplied with molten glass. This is achieved usually by a feed from a tank or by the addition of marbles of already melted and refined glass. The fibers range in thickness dependent on processing conditions but on the whole tend to be tens of microns in diameter. Glass fiber is extremely strong when it is first made because the surface is pristine. The theoretical strength of glass is much higher than that of steel. Unfortunately glass undergoes brittle fracture and thus is weakened by the presence of surface flaws that act as stress concentrators at the crack tip. In the absence of a coating or sizing fiberglass can weaken very quickly as surface flaws build up. Composites made from the fiber will have either long yarns or weaves when the strength required is directional or the glass fiber is sometimes chopped when the strength required is more uniform and multidirectional. The chopped fiber also ultimately lends itself better to fast processes such as injection molding or pressing.

Fiberglass is also formed into wool usually by a process similar to that used to make cotton candy. Glass melt from a tank is delivered to a spinning cup that has holes in it. The glass is forced out of the holes by centrifugal forces and the glass fiber is forced downward by blowers. As the fibers descend they are sprayed with a binder that helps to maintain the spring-like quality of the fiber. It is the springiness of the fiber that serves to trap the air thus making the glass wool such an excellent insulator. The fibers fall onto a backing material and pass through a binder curing oven.

The issues that have controlled fiberglass composition in more recent years have to do with its biocompatibility. Fibrous and friable materials such as asbestos have been shown in some cases to be a potential health hazard and glass fiber, because of its aspect ratio and size has come under the same scrutiny. In the 1990s it was generally felt that biopersistance of glass fiber (i.e., the amount of time the fiber remained in the lung) was directly related to their ability to cause inflammatory reactions. In 1994 Germany introduced a set of regulations requiring commercial fiber to have a KI index of above 40 where

$$KI = (c_{Na_2O} + c_{K_2O} + c_{B_2O_3} + c_{CaO} + c_{MgO} + c_{BaO}) - 2c_{Al_2O_3}$$

where c is the amount of oxide in the composition in wt.%. More recently, anomalies have been observed in that high alumina fiber that has KI index of less than 40 has not caused an inflammatory response in laboratory animals. Moreover, recent work by Technical Committee 26 of the International Commission on Glass has shown that high alumina fibers do in fact dissolve fairly readily in the lung, suggesting that *in vitro* experiments on biopersistance do not necessarily correlate absolutely to the potential health risks of glass fiber in the lung.

23.4 Specialty Glass

The category specialty glass has become very large in recent years as there have been an ever increasing number of applications for glass. The category still remains, however, because in terms of volumes the quantities in the specialty glass categories are substantially less than in the other three main categories. Because there are so many diverse types of glasses processed for different applications they will be grouped together in terms of their unique processing modes.

23.4.1 High Purity Glass

In the same way that regular water has a large number of impurities and even "pure water" has a stunning array of things other than water in it, any glass that is made from the melt is bound to contain large numbers of impurities even if it is made from extremely pure chemicals. Therefore for high purity glass you cannot pour from the melt. When is such high purity required? For example, when fabricating telecommunications optical fiber where light signals might have to be transmitted through kilometers of the glass before the signal can be reconditioned and amplified; even impurities on less than the parts per billion levels are unacceptable. Also when trying to fabricate lenses for high resolution lithography in which ultraviolet laser beams are used to map out intricate patterns on silicon wafers, small amounts of impurity not only block the intensity but the absorption can also damage the lens.

Simple processes for purifying water such as distillation can be adapted to make high purity glass, although not with the glass itself. The melting point of silica is too high to enable reasonable melt-quenched silica glass, so distillation above the boiling point is quite out of the question. However, there are some low melting temperature compounds of silicon that can be boiled, for example, silicon tetrachloride, which when reacted with high purity oxygen can form silica.

$$SiCl_{4(g)} + O_{2(g)} \iff SiO_{2(s)} + 2Cl_{2(g)}$$

The highly pure silica deposits as a low density soot that upon subsequent heat treatments is densified and consolidated into vitreous silica. The chlorine is rather useful in that it has a tendency to dry the silica. Although it may seem strange, most silica whether made by this route or by another route contains hydroxyl groups or Si–OH. Sometimes this can be favorable but for most applications needing high purity this is not favorable. In fact any presence of hydroxyl groups in optical fiber is very parasitic on the signals that are transmitted because even the overtones and combination bands of the hydroxyl vibrations, though very small, are enough such that over the long transmission distances they will cause substantial loss. The deposition can take place either on large, usually flat substrates (for planar waveguides or lenses) or in a cylindrical format either on the outside or the inside of a bait material for an optical fiber preform. In addition for applications such as planar waveguides or optical fiber preforms it is necessary to be able to dope the silica with other oxides. For wave guiding the principle of total internal reflection is used which requires that a core of material with a slightly higher refractive index than the outside (known as the cladding) is formed. Typically, this involves either incorporating into the core, an oxide of an element

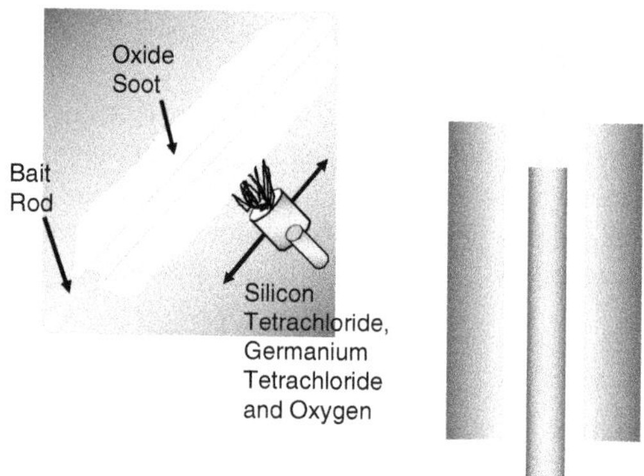

FIGURE 23.3 A schematic of one type of preform fabrication: an outside vapor deposition process. On the left is deposition of the silica soot followed by a consolidation into a solid preform on the right. The preform has the same core/clad structure that is required in the optical fiber.

that has higher polarizability than that of silicon, such as germanium, or incorporating into the cladding an oxide of an element that has a slightly lower polarizability than silicon such as boron. Many of the commercially available fibers have a germanium-doped core because one can deposit germania via the same approach as silica, that is, using a reaction of high purity germanium tetrachloride (Figure 23.3). Once the deposition has occurred, the low density soot has to be heated to consolidate into a glass, usually under a drying atmosphere. For optical fiber the large glass cylinder is a preform for the optical fiber which is drawn by locally heating the rod to a viscosity low enough to allow the preform to deform under tension while maintaining the core clad profile. The fiber is then coated with a polymer that is typically UV cured and then wound on a spool. The polymer protects the pristine fiber surface against flaws.

Sometimes, however, the dopant for the core of the fiber is somewhat more exotic, such as a rare earth oxide. Rare earth ions often have photoluminescent transitions that allow them to exhibit optical gain in certain hosts. As a telecommunications signal is transmitted over large distances, even in a very pure optical fiber, intensity of the signal is lost through unavoidable physical processes such as Rayleigh Scattering. Before the information is lost from the signal it has to be reconditioned and amplified. This can, of course, be carried out electronically but electronic signal amplification and conditioning is very slow and the advantage of optical fiber is that as we strive for the hundreds of gigabit and even terabit per second data rates this can only be achieved in an all-optical system. Hence optical amplification through stimulated emission is one key to the all optical process and is made possible by the inclusion of small quantities (less than a percent) of rare earth into the core of the fiber which is pumped with external light of wavelength shorter than the signal wavelength. Erbium is an excellent example having a transition at 1.55 microns which happens to be the point at which the aforementioned unavoidable losses in silica telecommunications fiber are at their lowest. The process by which the rare earth is introduced into the fiber is by solution doping. This is not as big a problem as it might seem because the quantities are rather small and any extra hydroxyl that might be introduced into the fiber is insignificant compared to the absorption that occurs due to the rare earth itself. Now it so happens that erbium coincidentally works with silica as the host glass, other rare earths are not so accommodating and if one requires amplification at other wavelength to obtain a broader variety of signal wavelengths then other rare earths have to be used in different glass hosts. Some of these will be discussed later in the exotic glass section.

An alternative way of making high purity glass is to use the sol gel method. Most of this work has been done in silica but there are many groups that are now extending this method to different glasses for new and exciting applications. The sol gel method is a chemical reaction method in which an organometallic compound is reacted to cause hydrolysis and then subsequently undergoes a condensation reaction. To facilitate the mixing of the organometallic and water, a common solvent is used, usually ethanol. An example of the reaction is shown below

$$Si(OR)_4 + 4H_2O \Longleftrightarrow Si(OH)_4 + 4ROH$$

$$Si(OH)_4 + Si(OH)_4 \Longleftrightarrow (OH)_3Si - O - Si(OH)_3 + H_2O$$

$$R = \text{alkyl group}$$

These reactions are catalyzed and the manner in which the catalysis occurs can substantially affect the final structure of the silica, for example, acid catalysis in the condensation step favors a more polymeric silica structure whereas a three-dimensional silica structure is obtained through base catalysis. As the gellation occurs the water is expelled and eventually a solid glass is obtained. There are some issues, however, in making a monolithic glass in this manner. The action of expelling water from the essentially porous structure builds up huge capillary pressures that have a tendency to crack the glass. In fact, if one requires a fully dense glass typically the sol gel glass has to be dried and the accelerated removal of the liquid can result in even more serious cracking. One way that this can be avoided is to carry out supercritical or hypercritical drying. This employs conditions above the critical points in the solid/liquid/vapor phase diagram of the liquid that is being extracted to ensure that there is no interface between the liquid and vapor that can lead to surface tension and capillary stresses. There are several other advantages to sol gel processing aside from being able to obtain high purity glasses. It is possible, with the sol gel method to make glasses from materials that would not otherwise form a glass; for example titanium dioxide can be made amorphous by this route. In addition there are often times when a porous material is needed and with sol gel processing one can often engineer controlled porosity. Lastly, since sol gel processing does not involve high temperatures (compared to traditional melt quenching), one can incorporate organic molecules such as dyes and even biological molecules. In the case of biological molecules, however, care needs to be taken to not heat the sample thus definitely obtaining a porous structure, and often biological molecules are more sensitive to the processing conditions of the glass and are not stable in the acidic or basic catalysis environments. Moreover, many biological molecules are adversely affected by the presence of large quantities of alcohol, therefore, instead of using alcohol as a common solvent, physical forms of mixing are used for example ultrasonic agitation to ensure large interfaces between the immiscible reactants.

23.4.2 Glass Ceramics

Although much of this chapter has been about avoiding devitrification, there is a class of materials known as glass ceramics in which devitrification is advantageous. Glass ceramics are typically processed in the way described for glasses in the previous sections but the composition typically contains a refractory nucleating agent which, upon postprocess heat treating, will allow dispersed and controlled devitrification (ceramming). This allows for the forming of interesting and useful shapes. However, the properties of some glasses are not always ideal. For example, it was already mentioned that most glasses undergo catastrophic failure due to small flaws that concentrate stresses. This type of brittle failure relies on a crack propagating unhindered through the material. Cracks can be blocked by inserting a material that provides and interface over which the energy of the crack can dissipate; this is the same type of principle as fiberglass reinforcement. In the case of a glass ceramic the interface occurs between the glass and the crystals. The exceptional fracture toughness of glass ceramics coupled with the ease of forming render them useful for applications such as computer hard drive substrates and even replacement teeth! In addition, apart from certain glass compositions such as Pyrex®, most glasses have a very poor thermal shock resistance. This is due to the fact that glasses have poor thermal conductivity and many compositions have high thermal expansion coefficients. Coupled with a low fracture toughness, the inability to disperse the heat evenly, the thermal stresses set up due to differential thermal expansion, causes most glasses to break if unevenly heated. This can be alleviated by devitrification because not only does the presence of the crystals increase the fracture toughness of the glass but also of the devitrifying phase that has a low or even negative coefficient of thermal expansion, then the over thermal expansion coefficient of the composite is low. The advantage of this is that the crystals do not have to be large and in fact Visionware® a Corning product that was invented in the 1960s is largely transparent and yet cookware made from this glass ceramic composition can be heated directly on a stovetop. Moreover, many electric stovetops, themselves are glass ceramic. Thermally stable glass ceramics are not only used for high temperatures but also for precision optical components where small changes in dimension cannot be tolerated. Typically since crystalline materials are somewhat denser than their chemical counterparts that are amorphous, devitrification is accompanied by a volume decrease.

23.4.3 Exotic Glass Processing

Exotic glasses really refer to those that are not common silicates, but it can also mean glasses that have a novel application. It has been stated earlier that one can in principle make a glass out of any material if it can be cooled fast enough. Indeed as technology has required, particularly in the photonic arena, new compositions not involving silica at all have evolved. There is a whole class of glasses made from fluoride compounds that can host certain rare earths and enable them to amplify light in a manner not possible in a silicate composition. Chalcogenide glasses made from arsenic, antimony, tellurium, selenium, and sulfur transmit light into the infrared and have large refractive indices for novel structures such as photonic crystal fibers. Nonoxide glasses are made in a similar way to the silicates described above but with a few more restrictions; typically water and, in some cases, oxygen have to be eliminated, thus melting usually involves an inert atmosphere glove box or even an evacuated, sealed silica ampoule. Even metals can be made into glass. Until recently metallic glasses were made by rapid quenching of a molten metal onto a chilled fast-rotating drum. They were interesting principally for their magnetic properties but their mechanical properties are of great interest, particularly now that they can be made in bulk with cooling rates as low as 1 K/s (www.memagazine.org/backissues/june98/features/metallic/metallic.html). Other oxide glasses such as phosphate and borate have some special uses such as the aforementioned laser fusion project and low temperature sealing glasses.

There are however some uses for glass that fall into the category of exotic even though they are silicate. In production now are silicate glasses with low silica content that will actually bond to animal bone and soft tissue. The prime composition invented by Larry Hench in the 1970s is called Bioglass® and is a commercial glass used to fill bony defects left behind by periodontal disease, fabricate middle ear implants, and even treat sensitive teeth. Another forming method that has recently been finding new application is the processing of glass microspheres. Glass microspheres are made by taking premelted glass frit and grinding it to a suitable particle size in the region of the sphere size required. The frit is fed into a flame working torch and the particles instantly form liquid droplets which, through surface tension effects, become spherical. If a blowing agent is included (something that rapidly forms a gas at the flame temperatures,) the spheres can be made hollow. Solid microspheres (about 20 microns in diameter) are commercially available as rare earth aluminosilicate glasses that are subsequently placed close to the core of a nuclear reactor. Under neutron bombardment the rare earth activates and becomes a beta radiation emitter but the half-life of the activated isotope is usually of the order of tens of days. Beta radiation has a short penetration depth in human tissue; therefore when the microspheres are injected in slurry into the hepatic artery they tend to accumulate in the high vascularity of any tumor present. Hence, the temporarily active microspheres can deliver a sizeable dose of radiation to a liver tumor without affecting the healthy surrounding tissue. Hollow microspheres are currently being examined as a possible storage medium for hydrogen fuel. Hydrogen can be loaded into the spheres at very high pressures because the spheres are not of a size that can support the fatal flaws that cause glass to be weak. In addition it has been found that the application of light to the glass can enable the passage of hydrogen into the glass and its release in a controlled manner.

23.5 Summary

This chapter has sought to explain some of the unique attributes of the material glass. It is many of those attributes that make glass such a useful material and many of those attributes that have presented the most challenges for consistent manufacture. While most glass melting is reasonably similar the forming modalities are what lead to the ability to apply glass to a wide range of applications. The key point to remember from this chapter is that all glass is not made equal, and that even glasses that have ostensibly the same composition can have quite different characteristics if they are processed in different ways.

Further Reading for Glass Science

Introduction to Glass Science and Technology by J.E. Shelby (2005) 2nd Edition, Royal Chemical Society.
Fundamentals of Inorganic Glasses by A.K. Varshneya (2005) 2nd Edition, Academic Press.
Glass Science by R. Doremus (1994) 2nd Edition, Academic Press.

Further Reading on Glass Processing

Engineered Materials Handbook, Section 6, Volume 4, (1991) ASM International.
Any of the Advances in Fusion and Processing of Glass Proceedings published by the American Ceramic Society (Ceramic Transactions) I is volume 29, II is volume 82 and II is volume 141.

Further Reading on Characterization

Experimental Techniques in Glass Science by Catherine Simmons and Osama El-Bayoumi Eds. (1994) American Ceramic Society.

24

Ceramic Processing

H. P. Buchkremer and
N. H. Menzler
Forschungszentrum Juelich, Institute for Materials and Processes in Energy Systems, IWV-1

Abstract

Wet ceramic processing technologies are described in this chapter, which is divided into two parts. The first part describes the manufacturing of components and substrates and the second part deals with coating techniques. The component manufacturing technologies covers techniques such as tape casting, powder injection molding, and the extrusion process. In the second part on coating technologies, the techniques described are layering by electrophoretic deposition, powder spraying, screen printing, and roller coating. The order in which the processes within the two sub-chapters are described follows as far as possible the increasing paste viscosity for the individual processes. It should be mentioned that no distinction is made between dense or porous components or layers. The difference is only noted in the description of examples.

24.1 Introduction

In this chapter, wet ceramic processing technologies are described. It is subdivided into two parts, the first dealing with the manufacturing of components and substrates, and the second describing coating techniques. The order of the processes within the subchapters follows as far as possible the increasing paste viscosity for the individual processes. It should be mentioned that no distinction is made between dense or porous components or layers. The difference is only noted in the description of examples.

24.2 Ceramic Processing

24.2.1 Component and Substrate Manufacturing Techniques

A component is taken to mean a three-dimensional part. Substrates can be described as three-dimensional with elongations in the x- and y-direction and with diminished thickness in comparison to the x- and

y-length. The ratio required determines the technology used. Tape casting is used if thin (50 to 800 μm) substrates of large x–y-dimensions are necessary. If thicker substrates are required they can either be processed by extrusion or by laminating or calendering of single tapes. Smaller substrates with more complex shapes are preferably manufactured by injection molding.

24.2.1.1 Tape Casting, Lamination, and Calendering

Tape casting is a wet chemical continuous shaping process (basic information on tape casting in Reference 1 to Reference 5). It is based on a water- or solvent-based slip (or slurry). A slip is a suspension composed of solid material in a solvent with the addition of mostly organic additives. During manufacturing the liquid slip is cast on a carrier tape made of polymers, metals, or textiles. Afterwards the tape is dried under controlled conditions, cut, possibly coated, debindered, and subsequently sintered.

Two-dimensional geometries with wet thicknesses of 200 to 1500 μm can be produced. Tape widths vary from 20 to 100 cm and the lengths of the casters are up to 25 m.

The main parameters affecting the thickness and quality of the green tape are the viscosity or rather the slip rheology, the interaction of the slip with the carrier tape (wetting behavior, adhesion), the adjustment of the doctor blade, the casting speed and the drying conditions (humidity, temperature, flow direction, and temperature of the drying air).

The use of organic additives enables the manufacturer to control various slip characteristics, to enhance slip processing and homogenization, and to fabricate qualitatively good products with high reproducibility. Major additives are binders, dispersants (deflocculants), plasticizers, softeners, de-foamers (de-airer), and separating agents. Despite these pressing agents, lubricants and fungicides can be added. All organic additives were only temporarily used and de-bindered during high temperature burning.

24.2.1.1.1 Slip Formation and Function of Organic Additives

The basic scheme of tape-casting slip manufacturing is presented in Figure 24.1. Important parameters during slip preparation are: complete dissolution of the organics in the solvent, homogeneous mixing of powders and organics to suppress separation, segregation, and agglomeration, careful de-airing to reduce solvent evaporation and, if necessary, a filtration step to cut off the maximum particle size of the powder or any impurities. The amount and type of organics depends on the powder to be cast, for example, the particle size, the particle size distribution, the specific surface area of the powder, the surface chemistry of the material (hydrophobic, hydrophilic), and on the solvent to be used. The main characteristics of the solvent are the viscosity, the vapor pressure under casting and drying conditions, and the hazardousness of the material. In most cases, more than one solvent is used. It is therefore necessary that the solvents should be an azeotropic mixture, which means that their evaporation behavior is comparable (with respect to evaporation conditions like temperature). The powder material can be characterized by the

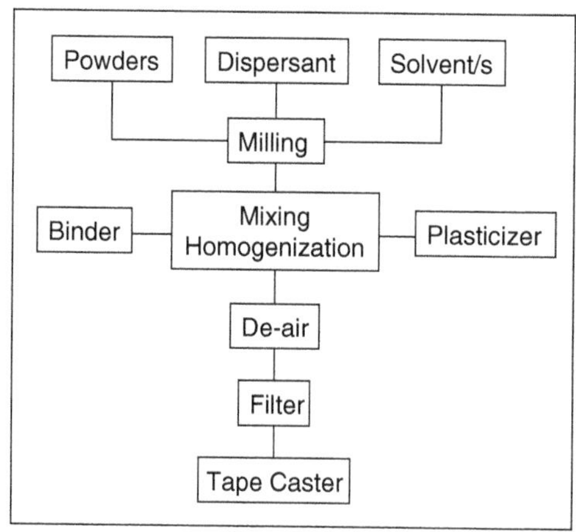

FIGURE 24.1 Basic scheme of tape cast slip formulation.

above-mentioned parameters and additionally by its shape (round, cylindrical, elongated) and its density. The particle sizes used depend on either the dispersability of the powder or on the tape thickness to be reached. A reference value is ten times the thickness of the final prepared tape to the particle size of the powder.

The functions of the various organics used are[6]:

- *Binder*: ensures stability, integrity, and flexibility of the green tape after drying
- *Plasticizer*: provides pliability and manageability
- *Dispersant*: reduces agglomeration and ensures homogeneous distribution of the powder particles
- *De-foamer*: reduces foaming during slip preparation, for example, by increasing surface tension
- *Separating agents*: prevent or reduce adherence of the cast tape on the support tape

24.2.1.1.2 Tape-Casting Process

In the tape-casting process the slip flows from a reservoir into an antechamber from which it spreads on the carrier tape. This is transported by an external transportation mechanism. The cast slip is drawn with the support tape under a wiper called a doctor blade. This doctor blade, the slip viscosity and the transportation speed govern the thickness and the width of the resulting tape. After moving under the doctor blade the tape is dried in a drying chamber under controlled conditions, for example, temperature, humidity, and air flow. This is the most critical step in the continuous casting process. If the drying speed is too low the tape is not dried sufficiently at the end of the line and therefore adheres too much either to the support or, if it is rolled up to itself. If the drying speed is too high the tape may crack due to the fact that the evaporation on the top side of the tape is faster than the transportation time for humidity from inside the tape to reach the surface. Industrial de-airing and conditioning systems and tape casters are common (e.g., made by Keko Equipment, Zuzemberk, Slovenia; HED International, Inc., Ringoes, NJ, USA; A.J. Carsten Comp. Ltd., Columbus, OH, USA; SAMA Maschinenbau, Weißenstadt, Germany and used by e.g., CeramTec, Plochingen, Germany and Kerafol, Eschenbach, Germany). Defects arising in the process are: nonuniform shrinkage in the x–y-direction due to textures caused by casting in one direction; cracking due to drying too quickly; adhesion to the carrier tape. The sides of the tape are cut off because at the outer part the tape has an uneven thickness.

After drying the tapes can be either unrolled, cut into stripes and unrolled, or punched into single plates. If holes are necessary they can even be made in the green state of the tape. The actual manufacturing process depends very much on the application of the tapes. They can be sintered to final density, coated in the green state, presintered and metallized, laminated, or calendered.

24.2.1.1.3 Laminating of Tapes

Lamination of single tapes is done either to increase the thickness of the samples or to create three-dimensional structures. Especially for the electronics industry laminated tapes with vias (holes within one tape layer) are needed to prepare LTCC substrates (*l*ow-*t*emperature *c*o-fired *c*eramics). A three-dimensional channel structure can be achieved by connecting the vias of the various layers. This third-dimension connection is necessary to enhance integration of the electronic parts. The greatest challenge during LTCC manufacturing is the relationship between the via positions after sintering (they must be of high precision) and their positions after cutting in the green state. That means that the shrinking behavior during the firing process must be controlled very well. Therefore, the characterization of shrinking behavior (esp. in the x–y-direction), uniformity of shrinkage (no textural effects), and the homogeneity of the firing equipment (homogeneous temperature distribution) are essential for this process. Figures 24.2 show schematically a cross section of an LTCC substrate and some photos of samples.

24.2.1.1.4 Example: Substrates for Solid Oxide Fuel Cells

The materials used for planar anode-supported solid oxide fuel cells (SOFCs) are nickel and yttria-stabilized zirconia (YSZ). The metallic nickel ensures electrical conductivity, catalyzes the formation of water and reforms higher hydrocarbons. The stabilized zirconia is necessary to match the coefficient of thermal expansion to the electrolyte and the stability and integrity of the cell during sintering and

(a)

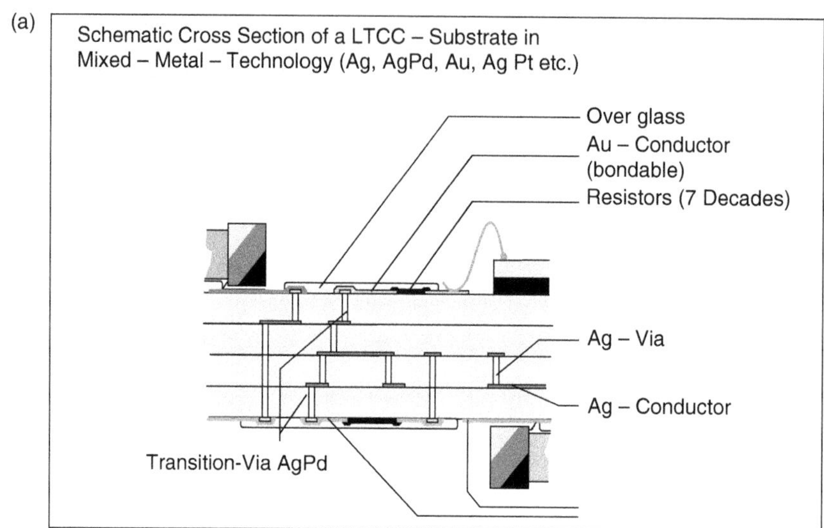

(b)

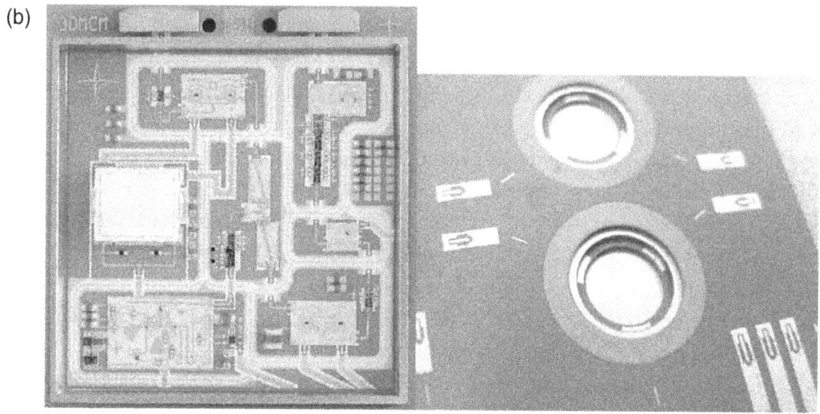

FIGURE 24.2 (a) Schematic cross section of an LTCC substrate (courtesy of Micro Systems Engineering GmbH, Berg, Germany); (b) Photographs of LTCC substrates (left-handed photo courtesy of EADS, Munich, Germany; right-handed photo courtesy of Micro Systems Engineering GmbH, Berg, Germany).

operation. Nickel oxide is used for manufacturing the anode substrate due to the fact that the subsequent sintering steps are run under oxidizing conditions. The first time an SOFC stack is heated the nickel oxide is reduced to the metal state.

For slip formulation the powders and a four-type solvent are used (e.g., toluene, isopropanol, methylethylketone, and ethanol). As the electrodes must be porous to ensure gas transport a space holder is incorporated. In the case of the substrate presented here graphite is used. As organic additives polvinylbutyral (binder), oleic acid (dispersant), benzylbutylphtalate, phtalac acid bis 2-ethylhexylester, and polyethylene glycol (plasticizers) are used. Additionally, cyclohexanone was added as an adhesive agent and sorbitan trioleate as a surface conditioner. Mixing and homogenization follows the classical route as described before. The basic characteristics of the materials and slip are: particle size distribution, BET surface, chemical composition, phase composition (YSZ), and viscosity. Typical particle sizes for both powders used are 0.2 to 6 μm. After debinding and sintering the resulting substrate is characterized with respect to electronic conductivity, permeation, mechanical stability, shrinkage, and cross sections (by optical microscopy and SEM). The thickness of the end fired tape varies between 300 and 800 μm, depending on the envisaged use. Figure 24.3 shows a typical cross-sectional view of a tape cast anode support.[7]

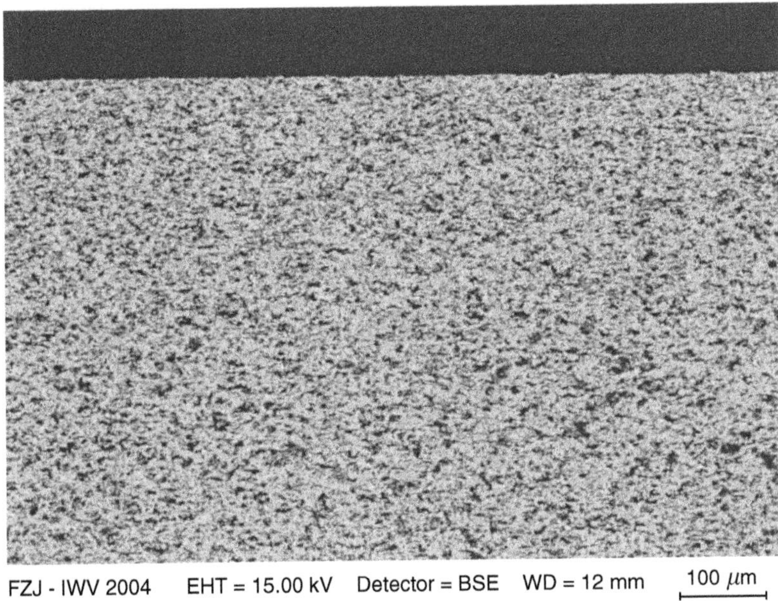

FIGURE 24.3 Cross section (SEM) of an anode support for SOFCs manufactured by tape casting (black: pores; gray: nickel oxide and zirconia) (courtesy of Forschungszentrum Jülich, Jülich, Germany).

24.2.1.2 Powder Injection Molding

The shape complexity of structural parts manufactured by conventional powder pressing techniques is often limited. To overcome this disadvantage powder injection molding was developed on the basis of the well-known plastic injection molding process. Similar to plastic injection molding, a molten polymer is mixed with a ceramic or metal powder and injected into a mold.[8–10] This allows the high-volume production of complex-shaped parts in near net shape geometry. Depending on the powders used in the process two different terms are used:

1. Ceramic Injection Molding (CIM)
2. Metal Injection Molding (MIM)

Sometimes the two definitions are combined into the generic term Powder Injection Molding (PIM). Due to the numerous similarities between the two techniques, in this chapter, the term powder injection molding is normally used.

Powder injection molding was developed in the 1980s of the last century in the United States.[11] (See Chapter 25 for more details on powder injection molding). Since that time a continuously growing interest in this process can be seen. At present, powder injection molding is a competitive manufacturing process for small precision components. A typical feature of this processing technique is its flexibility. Depending on the application, a wide variety of materials can be processed in complex shapes and also large- and small-volume manufacturing is possible.

Although powder injection molding allows the production of complex-shaped parts in a cost effective way to date there has been a limitation on size. This is mainly due to the exponentially growing debinding time for thick wall parts, which is directly related to the manufacturing costs.

24.2.1.2.1 PIM Process

As already mentioned, PIM overcomes the essential difficulties of powder pressing techniques by injecting a low-viscosity binder-powder mixture into a mold. The parts manufactured by powder injection molding therefore have characteristic properties, for example, high precision, near net shape, complex shapes, possibility of undercuts, and isotropic mechanical behavior.

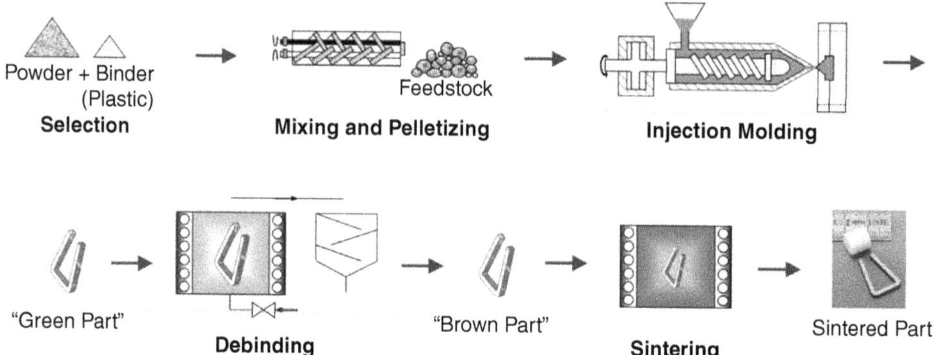

FIGURE 24.4 Powder injection molding processing steps.

The PIM process can be subdivided into the following steps (Figure 24.4):

- Binder/powder selection
- Mixing
- Molding
- Debinding
- Sintering

The PIM process is briefly described in the following.

Binder/Powder Selection

Binder: The binder[12,13] is added to the powder as a processing aid. It has to be optimized with regard to all the manufacturing steps and meet the following needs:

- Flow characteristics during injection
- Short debinding times
- No chemical reaction with the powder
- Low volume change during solidification/freezing
- Sufficient strength after shaping

Frequently used binders are polymers, for example, polyethylene, polyacetal, and wax. Most of them are thermoplastics but thermosetting plastics are also in use. Thermoplastics are widely used due to their good flow characteristics, at elevated temperatures. Most binders are multicomponent systems that allow the different binder properties to be tailored to an optimized binder-powder mixture with good flowability and no demixing during injection molding.

Powder: Almost any ceramic or metal powder can be processed by PIM, if they show sufficient sinterability. Metallic powders are coarser in most cases and can therefore be debinded faster than ceramic powders or parts with thicker walls can be manufactured cost effectively. Ceramic powders are injection molded at the present in particle sizes ranging from some 100 nm to about 50 μm. Coarser ceramic powders show low sinterability and require very high process temperatures or a liquid phase forming sinter additives. In general, metallic powders are more sensitive to contamination caused by binder components. Especially the widely used oxide ceramics are not attacked by typical PIM binders and can be processed easily.

Particle shape is important for some applications:

- Typically a high powder packing density[14] is required and this can be achieved more easily with spherical powders.
- Packing strength during debinding should be sufficient and this is easier to achieve with irregular-shaped powders.

The powders normally used have distributions of particles of different sizes. Typical properties of an optimized PIM powder are:

- Tailored particle size distribution (cost, packing, density)
- No interaction with the binder
- No agglomeration
- High reproducibility of powder manufacturing

Mixing

Powder and binder are mixed to form the "feedstock"[15] for the injection molding machine. Several types of mixing devices are available that differ in their effectiveness and basic principles. High shear rates are needed during the mixing procedure to coat the powder particles completely and destroy agglomerates in the feedstock. Widely used devices are z-blade and planetary mixers as well as extruders. The amount of binder should be as low as possible. This leads to less shrinkage during sintering and results in high-precision parts. Typically, the amount of binder in PIM feedstocks is in the range of 35 to 70 volume percent. Amounts of binder below a critical value lead to enhanced wear of the mixing and molding equipment and reduce part accuracy. As a first rough reference value, the binder should at least fill the free space between the particles. Depending on particle shape, surface wetting behavior, and binder properties, a lower or higher amount may be necessary under real conditions.[16] After finishing the mixing procedure the homogenized compound[17] is produced in suitable granules. These granules form the feedstock for the molding process.

Injection

The machines used for this step of the PIM process are in general modified plastic injection molding facilities. In particular, the tolerances between the screw and cylinder of the injection molding machines are adjusted to the particle size distribution of the powders to avoid clogging and destroying the unit. A typical PIM machine is made by Arburg, Germany.

The individual steps of an injection cycle are similar in plastic, metal, or ceramic molding. Granules or pellets of the feedstock are fed into the heated extruder part of the injection molding machine and there mixed again and the binder becomes molten resulting in a low-viscosity compound. This molten compound is usually injected into a heated mold where it rapidly consolidates. After opening the mold the shaped part is ejected and cools down to ambient temperature. Then the next cycle can be started.

Depending on the materials (binder, ceramic powder) the size of the part and different equipment parameters, a typical injection molding cycle[18,19] lasts from a few seconds to minutes and allows high-volume production when such machines are combined with robotics and handling systems. The produced bodies are called "green parts" because of their low mechanical properties in this state. Sufficient green strength of the workpiece is, however, necessary during handling and charging the parts into the extraction or debinding equipment.

An important indicator of the performance of a powder injection molding machine is its clamping force because this characterizes the maximum size of a processable part.

Debinding

Debinding means the removal of the binder from the green part[20,21] while maintaining the shape. During this process the part loses most of its green strength and becomes damageable. This fact requires careful control and minimum part handling and movement.

More than 90% of all produced PIM parts are debinded using two different processes:

1. *Thermal debinding*: This means heating the green part to cause the binder to be removed by decomposition or evaporation. In most cases decomposition and evaporation take place in parallel. Complete thermal debinding must be done with great care to avoid disruption of the green part. The gases and vapors have to leave the compact through the pores of the ceramic/metallic powder skeleton, which requires a binder decomposition/evaporation from the surface to the center of the

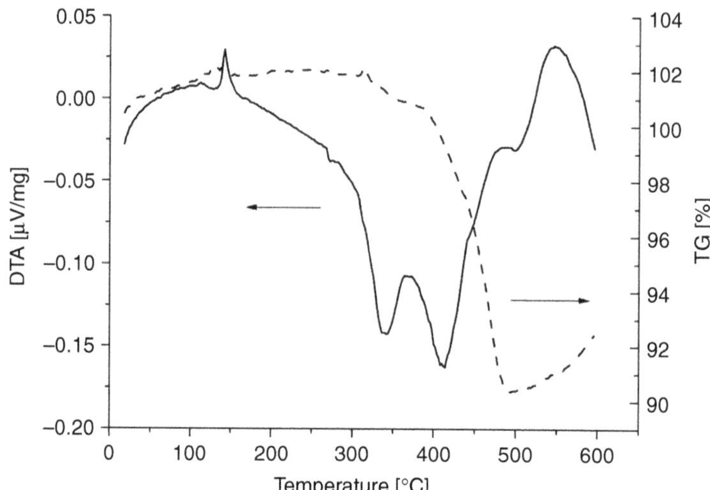

FIGURE 24.5 Results of differential thermal analysis (DTA) and thermal gravimetry (TG) of a two-component binder in a NiTi feedstock.

parts. Figure 24.5 shows such debinding behavior of a two-component binder — NiTi powder sample.

Thermal debinding is normally a time-consuming step and greatly dependent on the wall thickness of the green bodies. Catalytic debinding can drastically reduce the process time by using a gaseous catalyst, for example, nitric acid, and allows a continuous debinding-sintering line to be established, which for high volume production is in general more cost effective than batch production.

2. *Extraction debinding*: This debinding process is only possible with binders that are soluble in suitable solvents.[22,23] This can be done in a liquid or gaseous state of the solvents. Compared to thermal debinding the temperatures are low (<100°C) and the velocity is higher. Normally the binder is not completely removed during solvent extraction. This gives the debinded parts sufficient strength to feed them to the sintering equipment where the remaining binder is thermally removed.

The compacts after debinding are called "brown parts." From the "green" to the "brown" stage there is only a slight shrinkage detectable due to the fact that most of the binder was located in the free space between the powder particles. Brown parts are porous, show more or less zero strength, and must be carefully handled.

Sintering

Sintering is a thermal process where the powder particles grow together by solid-state diffusion processes, sometimes with additional liquid phase generation to support the transport of matter and enhance the wettability of the sintered powders. More details on sintering are given in Chapter 25.

Sintering of PIM parts is substantially the same as that of traditional powder parts, but in some cases the PIM parts are smaller in volume and weight. The sintering equipment is therefore similar to that for conventional parts. Depending on the materials to be sintered, furnaces permitting temperatures of up to 2000°C are necessary. Oxide ceramics can be sintered under air, but nonoxide ceramics and metals require controlled atmospheres,[24,25] sometimes in vacuum or inert gases.

Sintering is nearly always performed to produce parts with density having nearly theoretical value, except for applications where a degree of residual porosity is needed. This densification during sintering leads to a visible shrinkage of the parts and reaches values of about 17% linear. When the green part is produced with a homogeneous density the total shrinkage during sintering should be isotropic and result in high-precision parts with no tendency to warp. As already mentioned for debinding, sintering must

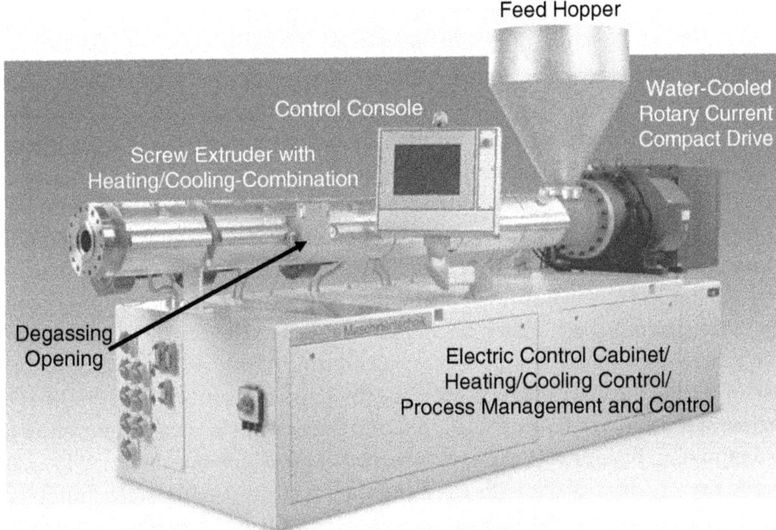

FIGURE 24.6 Industrial single screw extruder (courtesy of esde Maschinentechnik GmbH, Bad Oeynhausen, Germany).

also be very closely controlled to retain the shape of the part during shrinkage, which involves particle movements and rearrangement.

After sintering, the properties of PIM parts can be improved by different well-known methods if necessary. Typical postsintering processes are, for example, polishing, machining, hot isostatic pressing, and surface protection.

24.2.1.3 Extrusion

Extrusion is a traditional plastic forming method for manufacturing ceramic parts with identical dimensions over a great length. Commercial tiles, furnace tubes, bricks, pipes, catalyst supports, heat exchangers and the like are manufactured via extrusion. Compared to casting techniques like tape or slip casting, the major difference is the viscosity of the material to be shaped. Casting techniques are based on low-viscosity slips or suspensions with viscosities in the range of Pa s while extrusion masses have viscosities of 100 to 1000 Pa s, which is two to three orders of magnitude stiffer than for casting slips. Geometries producible by extrusion are three-dimensional parts with an elongation in the extrusion direction. It is a continuous process where the parts are cut directly after forming mostly by wire cutting. Extruded dimensions perpendicular to extrusion direction vary from millimeters to tens of centimeters.

24.2.1.3.1 Extrusion Process

Figure 24.6 shows a picture of an industrial single screw extrusion equipment. While the figure shows an extruder with only one screw, there are also machines with two screws rotating in parallel or in opposite directions. The advantage of two screws are more homogeneity, higher power input into the paste and by producing samples of large geometries a better distribution of screw power. This reduces screw damage and extrusion defects.

Normally water-based plastic raw material compounds with an amount of water between 15 and 20 wt.% are used. Because the drying step is a very critical and time-consuming stage, the amount of water in the compound is reduced by adding water-solulable binders like polyvinyl alcohol, methylcellulose, polyacrylamide, or polyethylenimine. Advantages of the extrusion process are continuous low-cost production, a high fabrication capacity and the ability to produce large shaped parts. Limitations of the process are the indispensable drying step, cost-intensive dies, the constraint of producing one geometry and a texturized microstructure due to one-dimensional pressing through the die. A major influence on the quality of the resulting parts is exerted by the raw material particle sizes and size distributions, the difference in particle sizes for compositions with more than one raw material (separation processes during mixing and pressing), the degree of agglomeration, the amount of solid fraction in the compound, the pressing speed, and the surface quality of the die. Another important factor is the rheological behavior of the compound. The preferred flow behavior for the material is shear-thinning or Bingham viscosity, which means a material with a yield stress and subsequently nearly Newtonian flow behavior.

The drying and debinding of the water and the binders has to be done slowly, carefully, and in the exact temperature ranges due to the evaporation of the organic parts. Normally, the temperature interval for the debinding process is between 300 and 1000°C. The heating rates or the dwelling times for debinding must be chosen with respect to the diffusion of the water from inside the compound to the surface and the transportation of the gaseous species from the organic binders to the surrounding atmosphere. Drying too fast leads to inherent tensions that may cause cracks during drying or subsequent sintering.

A critical factor during the extrusion process is the thickness to width ratio of the parts to be formed. While the maximum thickness of tape casting is limited, extrusion is limited by minimum thickness. For flat substrates a minimum of approx. 3 to 5 mm is needed. If three-dimensional parts like catalyst carriers are manufactured the thickness of the walls can be thinner (~1 mm). Thin-walled samples need careful adaptation of particle sizes, rheological behavior of the paste during pressing through the die and after pressing the parts must be stable enough to be freestanding.

Both dense and porous materials can be produced. By adding pore formers, pores of equivalent size are formed during debinding and sintering (only reduction due to material shrinkage). Due to the fact that extrusion is a one-dimensional pressing technique the risk of texturized microstructures is high. This must be suppressed by optimization of particle size distributions, paste formulation (e.g., adding organic aids), homogenization, and the pressing die. Modern applications of the extrusion process are the manufacturing of catalyst carriers (Figure 24.7a) and supports for solid oxide fuel cells (Figure 24.7b).

24.2.1.3.2 Extrusion Defects

Common extrusion defects are bending of samples, laminations, tearings, and segregation. Bending occurs mostly after drying or sintering of the parts. The reasons for bending may be inhomogeneous densities in the parts or nonuniform pressure distribution over the pressing die. Bending must be suppressed as the defects first occur after a subsequent manufacturing step and the material is in most cases fatigued. Lamination orientations are cracks within the parts that are either oriented in the extrusion or in the circumference perpendicular to the extrusion direction. Causes can be inhomogeneities in the pastes or incomplete re-knitting after pressing through complex-shaped dies. Tearings occur on the borders or surfaces of pressed parts. The reasons for those defects can be inhomogeneous drying, internal stresses due to contacts between the pressing paste and die or friction.

24.3 Coating Technologies

24.3.1 Electrophoretic Deposition

In contrast to the other shaping and coating technologies that operate without external influence, electrophoretic deposition (EPD) uses an external electrical field during coating. Suspensions used for EPD are based on water and the oxide powder to be coated; additionally organic additives are used for stabilization. A stable suspension is a prerequisite for successful coating via EPD. The viscosity of the suspension is nearly equivalent to that of the solvent used.

24.3.1.1 Influencing Parameters

Although EPD uses powder particles for coating, supported by an external electrical field, the particles must be polarized. Stability of the suspension ensures homogeneous coating with independent particles; therefore agglomeration defects are reduced. Besides homogeneity another fundamental requirement is a high electrophoretic mobility of the particles.[26] Additionally, viscosity of the suspension, the zeta

FIGURE 24.7 Applications of the extrusion process for the manufacturing of catalyst carriers (a) Catalyst carriers made of cordierite ceramic (Corning source: http://www.fh-koblenz.de/fachbereiche/fbwt/cdrom/produkte/filter/f11.htm) and (b) for porous support of a solid oxide fuel cell; left: support geometries, right: microstructure of a cathodic support (SEM of a fracture surface) (courtesy Siemens).

potential, and the shape of the particles determine the coating quality. By deposing particles on the substrate the concentration in the suspension becomes inhomogeneous and therefore the stability of the coating conditions changes. The packing of the individual particles starts when the electrophoretic and the electrostatic forces are higher than the Van der Waals barrier.[27,28]

Besides these suspension parameters, the electric parameters also define the layer quality. These include the applied current and voltage, deposition time, electrode material, and the changing of the parameters during deposition.

There are two contrary effects. On the one hand, the electrophoretic rate decreases with the intensity of the electric field and, on the other hand, by enhancing the electrical field the particle mobility increases.[29]

24.3.1.2 Basics of Electrophoretic Deposition

The particles within a suspension can be stabilized either by steric or electrostatic stabilization. (A basic description of the steric and electrostatic effects may be found in Chapter 2.) For steric stabilization normally organic molecules with a long chain length are used. One end of the organic chain adheres to the particle surface, the other one is solvated by the liquid. By means of this surface coating, the individual particles are prevented from adhering to each other and therefore from forming agglomerates. If electrostatic stabilization is used the particles are polarized with identical load and thus they repel each other. Agglomeration and sedimentation are minimized or prevented. If steric and electrostatic stabilization are combined electrosteric stabilization is created. EPD is based on the surface polarization of particles caused by an external electrical field.

If powder particles are incorporated into a liquid a potential difference arises. This charge formation is caused by one of the following mechanisms[30]:

- Adsorption or orientation of molecules at particle surface
- Selective adsorption of liquid ions on the powder
- Dissociation of ionic compounds on the particles
- Wear electricity (only relevant in dynamic processes)

The first two mechanisms are of major interest for stabilization in ceramic suspensions.

Particles with an electric charge in a liquid are surrounded by a layer with opposite electric charge liquid molecules. This surface layer is surrounded with a second layer of once again opposite charge. These two layers are called Helmholtz double layer.[30] This theoretical model was extended by Stern[31] by adding first a fixed adsorbed layer of counterions and a second surrounding area in which the concentration is drastically reduced with increasing distance from the particle. The potential difference between surface potential and potential at the boundary fixed layer/diffusive layer is called the zeta potential.

If a particle "coated" with such a double layer is moved within a liquid the outer diffuse layer is deformed and switched.

If charged particles are moved in a suspension under an electrical field there four different forces are found: the Coulomb force, the relaxation, the Stoke's wear, and the retardation. The Coulomb force is oriented parallel to the movement while the three others are oriented in the opposite direction. Based on these forces electrophoretic mobility can be defined. With this electrophoretic mobility, a relation for the mass of the coated powder can be calculated.[32] The mass of the coated powder depends on the electrode area, the particle concentration in the suspension, the electrical field strength, and the electrophoretic mobility.

24.3.1.3 Electrophoretical Deposition of Ceramics

First the experimental setup needs to be defined. There are three possibilities of coating a substrate (Figure 24.8). The direction of electrophoresis can be horizontal or vertical. In both cases, another external force becomes important: gravity. If the suspension is completely stabilized gravity can be neglected, but if the particles have a sedimentation tendency gravity becomes an influencing parameter. Gravity can influence layer thickness and quality. If the suspension is well stabilized, configuration (a) in Figure 24.8 is to be preferred. The setup is easy to establish but the fixation of the substrate on the electrode must be solved due to the need for good contact between sample and electrode. If the suspension tends to sedimentation coating against gravity is a good solution (configuration [b] in Figure 24.8). If coating against gravity is used normally only the finer particles are transported to the substrate. Again the substrate holder is not easy to realize. If configuration (c) is used gravity supports the coating and no sample holder needs to be developed. But such a device can only be realized if sedimentation is negligible. This emphasizes again

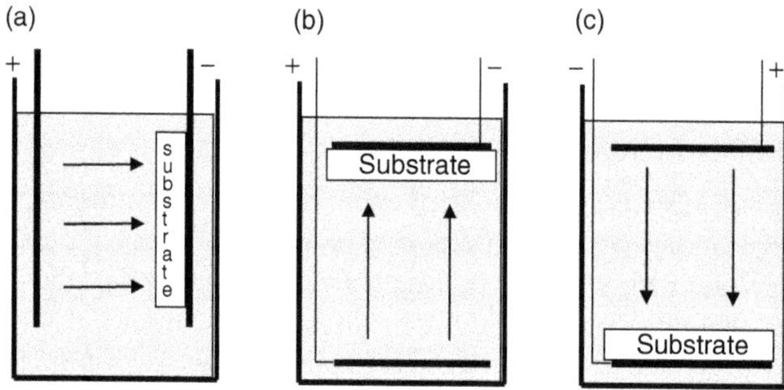

FIGURE 24.8 Possible electrode configurations for EPD; (a) horizontal coating, (b) and (c) vertical coating.

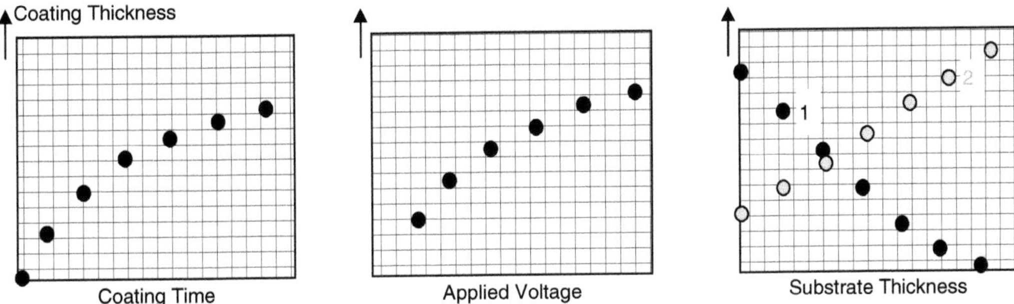

FIGURE 24.9 Schematic overview of influence of coating time, applied voltage, and substrate thickness on layer thickness (1: dependence if the relative permittivity of substrate ε_S is greater than that of the coated layer ε_L; 2: $\varepsilon_S < \varepsilon_L$).

the great necessity of a stable and homogeneous suspension, not only theoretically but also for practical reasons.

The thickness of the coating is influenced by the coating time, the voltage applied, and the substrate thickness. Figure 24.9 summarizes schematically the influencing parameters.

Reference 29 summarizes ceramic coatings for various applications made of alumina, alumina/zirconia multilayers, silicon carbide, alumina-metal, bioceramics, and semiconductors. Details can be found in the literature cited therein. The applied layer thicknesses vary from 450 nm up to 3 mm with deposition times of 30 s up to 30 min.

Due to the fact that electrophoresis applies individual particles to a substrate, the advantages of this technology are quite obvious. The substrates can be dense or porous, easily formed or have complex geometries, they are not limited by component size, and the coating times are short as illustrated by the data given above. The limitations of the technology are the particle size of the powder to be coated (big particles cannot be stabilized easily; influence of gravity), the opposite influence of the applied electrical field on particle mobility and electrophoretic deposition rate.

24.3.2 Wet Powder Spraying

In this subchapter, the wet powder spraying (WPS) process is described. This method uses suspensions to deposit layers of ceramic or metallic powders on top of structural parts. Wet powder spraying is used here as a collective term for some similar coating methods also familiar as "suspension spraying" or "slurry coating." The wet powder spraying technique was developed for the production of ceramic and metallic functional layers. It can be easily applied to manufacture porous layers, for example, multilayer arrangements used as graded filtering structures. More details on spray coating metals are given in Chapter 11.

24.3.2.1 Basic Principles

In the wet powder spraying process a suspension, consisting of a powder — carrier liquid-binder-mixture — is sprayed onto a substrate using a modified airbrush.[33] A spraying gun is used and is often mounted on a robotic system that allows a complex three-dimensional movement where flat, tubular, and three-dimensional parts can be coated. Sometimes the spraying gun is mounted in a fixed position and the part to be coated performs the movement, for example, rotation (Figure 24.10).

To obtain reproducible and high-quality coatings advanced controlling of the key parameters is necessary. The key parameters are the nozzle size of the gun, the distance between nozzle tip and the substrate surface, the viscosity and the feed rate of the suspension, the operating pressure and the operating speed between gun and structural part.

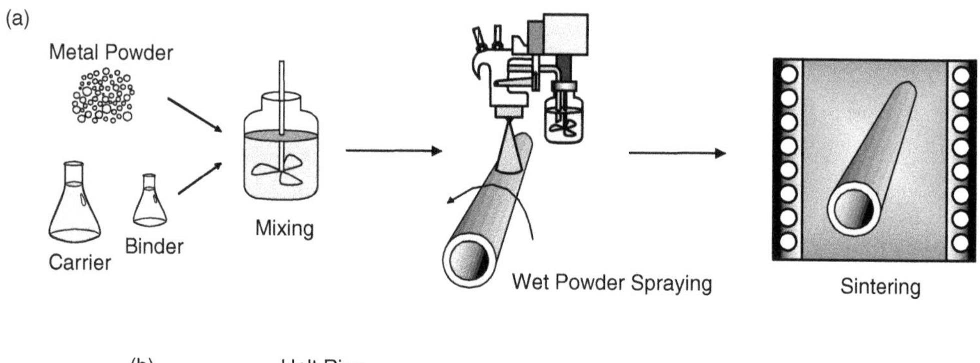

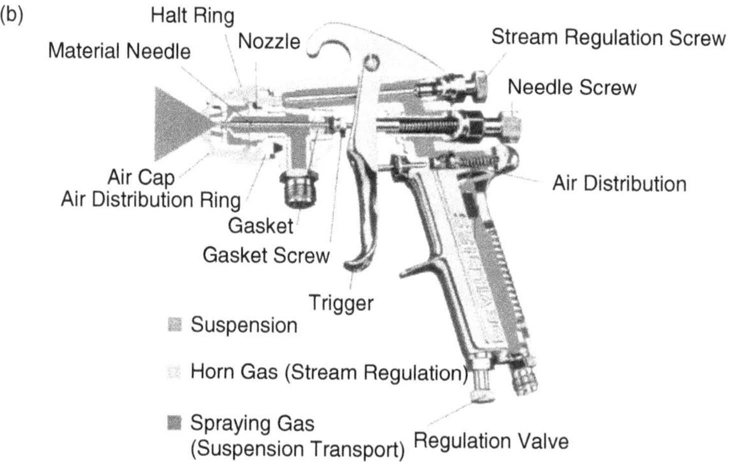

FIGURE 24.10 (a) Schematic diagram of the wet powder spraying method. (After Zhao, L., Manufacture and characterisation of composite graded filter membranes for microfiltration. Ph.D. thesis; Forschungszentrum Jülich reports JÜL-4079, 2003.) (courtesy of Forschungszentrum Jülich, Jülich, Germany); (b) Schematic drawing of a suspension spray gun (courtesy of DeVilbiss).

24.3.2.2 Spray Gun

The function of the spray gun (see Figure 24.10b) is the uniform distribution of the powder suspension onto the substrate surface. For this purpose, the suspension is atomized with the aid of pressurized air (gas). The droplets formed in the micrometer range and containing solid powder particles fly in the spraying direction and hit the part surface where they are further scattered.

To react flexibly to the different coating requirements the spray pattern of the gun can be modified by changing the ratio between horn and spraying air from a circular to elliptical shape. As in all spraying processes, wet powder spraying efficiency is also dependent on the overspray. Changing the spray pattern and the deposition rate are easily accessible parameters to influence the amount of overspray.

24.3.2.3 Suspension

Besides the spray parameters, the quality of the deposited layer is strongly influenced by the suspension properties.[34,35] The most important suspension properties are drying behavior and viscosity.

For an efficient deposition procedure the suspension requirements have to be optimized for each powder, also taking into account the surface properties of the substrate due to the suspension–surface interaction. Besides the carrier liquid, typical spraying suspensions contain a solid powder (ceramic or metallic), a dissolved binder and agents to influence viscosity, zeta potential, pH value, and drying behavior. Table 24.1 gives an example of a ceramic suspension used to deposit thin porous TiO_2 layers on top of a porous stainless steel support. The suspension is ethanol-based with the addition of terpineol

TABLE 24.1 Example of a Composition of TiO_2 Spraying Suspension

Composition	Weight Percent [%]
TiO_2	39.8
ethanol	37.1
terpineol	15.9
HAc	6.6
PEI	0.6

Courtesy of Forschungszentrum Jülich, Jülich, Germany.

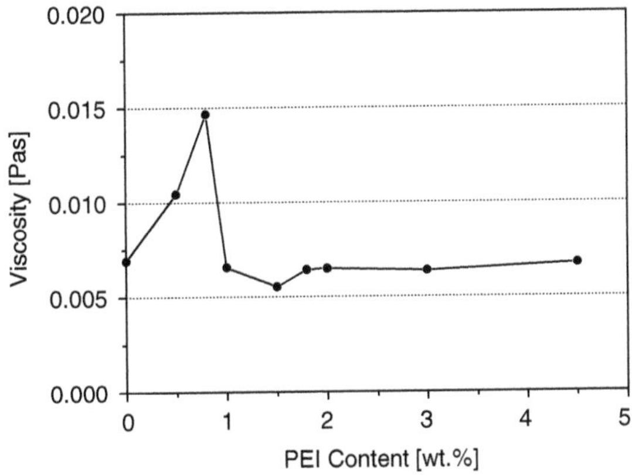

FIGURE 24.11 Influence of polyethylene imine content on the viscosity of a TiO_2 suspension (courtesy of Forschungszentrum Jülich, Jülich, Germany).

to extend the drying time of the sprayed layer. Acetic acid is added to shift the isoelectric point (zeta potential = 0) to useful pH value. Polyethylenimine acts as a binder to fix the TiO_2 particles in the layer and also as viscosity tuner (see Figure 24.11).

High-quality deposition results can be obtained by correctly dispersing powder particles in the liquid phase. Particles tend toward agglomeration due to the attractive forces — van der Waals, Coulomb, and physical friction — in the suspension. With decreasing particle size attractive forces increase, leading to enhanced agglomeration of particles.[35,36] The surface properties affect the stabilization of a suspension, which is attributed to a balance between the attractive and the repulsive forces. In general, suspensions can be dispersed and stabilized by electrostatic, steric, or electrosteric effects (see Section 24.3.1). Electrostatic stabilization is accomplished by generating a charge polarization on the particle surface. Steric stabilization is normally achieved by adsorption of polymeric additives that serve to form protective colloids. The presence of an adsorbed polyelectrolyte or polymer is required for electrosteric stabilization[36,37] forming double layer repulsion fields around the particles. The creation of a fully stabilized ceramic suspension requires a highly charged particle surface or an adsorbed polymer on the particle surface.[38]

Sometimes wet powder spraying suspensions cannot be produced in a fully stabilized manner. In this case, the particles tend to agglomerate and sedimentation takes place over time. Spraying of such unstabilized suspensions is done while stirring continuously to avoid agglomeration and sedimentation.

24.3.2.4 Spray Parameters

Spraying Distance: The spraying distance can be changed to optimize the deposition conditions. With increasing distance D the layer thickness, s. decreases ($s \sim 1/D^2$). The amount of overspray is also directly influenced by the spraying distance.

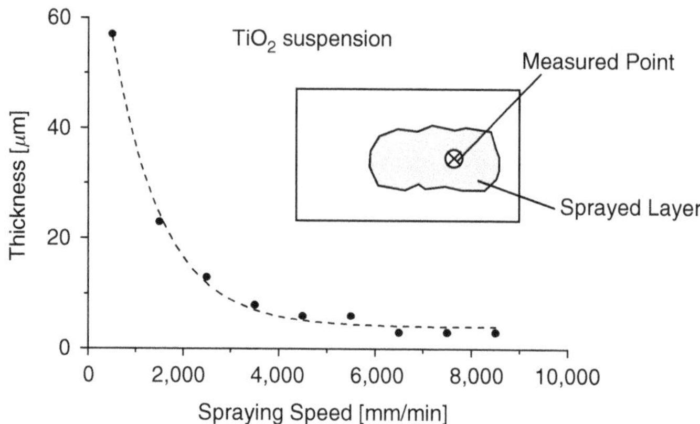

FIGURE 24.12 Relation between spraying speed and layer thickness (courtesy of Forschungszentrum Jülich, Jülich, Germany).

Spraying Speed: Similar to the spraying distance, also the spraying speed v can be chosen to change the layer thickness as $s \sim 1/v$. Figure 24.12 shows this relationship for a ceramic suspension. The spray gun was moved during spraying at a speed between 850 and 8500 mm/min resulting in decreasing layer thickness from 55 to about 5 μm exactly following the $s \sim 1/v$ relation.

Nozzle Size: The nozzle size has a considerable effect on the spraying efficiency and the result. Typical nozzle sizes vary in a wide range from about 0.1 mm to some millimeters, depending on the suspension viscosity, drying behavior, feed rate, substrate properties, and particle diameter. Under unfavorable conditions the spray nozzle can become clogged or operates in a pulsating unstable manner. To prevent plugging the nozzle diameter should be at least 10 to 20 times larger than the particle diameter.

Spraying pressure: The spraying pressure is directly related to the volume of gas passing the nozzle and dispersing the suspension in fine droplets. Finding the right spraying pressure is always a compromise between feed rate and overspray. The higher the pressure, the stronger the atomization will be and the smaller the zone with a homogeneous deposition thickness. Too large volumes of air can also cause powder agglomerates on the substrate surface. This is in particular the case if too much carrier liquid evaporates during the flight time. The particles then hit the surface in a "dry" state and tend to form agglomerates.

24.3.2.5 Powders

Wet powder spraying allows the processing of a wide variety of ceramic, metallic, and composite powders. In most cases, the only limitation is the sinterability and compatibility with the supporting material. The particle diameter can change from nanometer scale to about 100 μm.[39] Powder diameters between 200 nm and about 30 μm are state of the art. Depending on the application, spherical and irregularly shaped powders can be used and also mixtures of the two (see Figure 24.13). Often, deposition aims at a high green density of the layer to prevent cracking during sintering. For this reason, and also for reasons of cost, powders with a wide range of diameters are processed (see chapter on powder injection molding). The particle size distribution is an important parameter and only precise control can assure reproducible layer properties.

24.3.2.6 Sintering

In nearly all cases the sprayed layer has to be sintered to obtain sufficient mechanical properties or to achieve the intended function. Before the sintering procedure starts a drying step is necessary. During drying the solvent or the water evaporates and only the binder and additives remain in the porous powder layer. Sintering in most cases takes place in a solid-state below the melting temperature of the powder particles.[40,41] In a few applications also liquid phase sintering is used to support densification of the layer. More details on sintering may be found in Chapter 25.

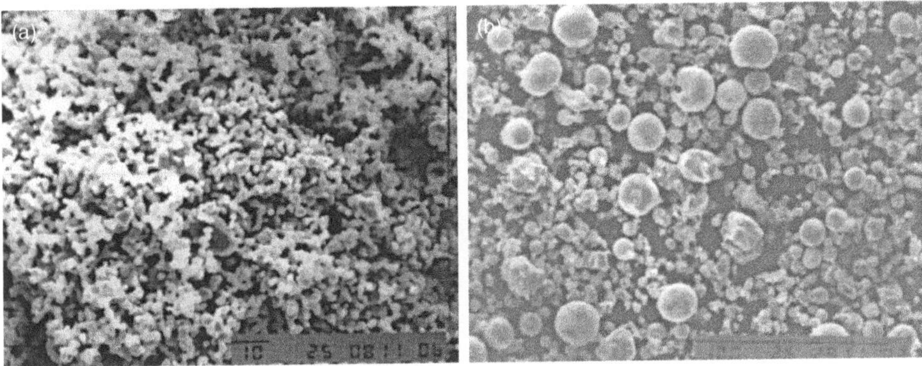

FIGURE 24.13 La–Sr–Mn oxide powders for solid oxide fuel cell cathodes (a) spray dried, (b) plasma atomized.

24.3.2.7 Applications

During the last few decades wet powder spraying has become a well-established processing method to manufacture layers on planar, tubular, and complex-shaped three-dimensional parts. Especially in tubular arrangements, it may be necessary to coat the inner wall. Although completely dense layers are achieved by this method (no open porosity; Figure 24.14a), the main applications are related to porous layers and structures.

There are different reasons for this tendency to produce porous structures, for example, shape and material flexibility as well as the possibility of coating inner surfaces to some degree. Wet powder spraying is a powerful means to produce planer and tubular filtering parts.[42–44] After spraying and sintering, the filter components are ready for use. Figure 24.14b shows such a porous layer formed of fine TiO_2 particles. Sinter necks are formed between the particles, but there is enough free space to act as a filter.

Depending on the particle size distribution and the spray parameters, wet powder spraying is able to process sharp pore size distributions of porous filtration layers. Figure 24.15 shows the average values of 20-μm-thick porous titanium oxide layers. Remarkable is the narrow pore size distribution and the excellent reproducibility.

24.3.3 Screen Printing

24.3.3.1 Introduction

Screen printing (SP) belongs to the group of widely used mechanical printing methods. These techniques are able to coat surfaces of different materials with paints or suspensions containing solid particles, for example, synthetics, metals, ceramics.[45,46] In this chapter, screen printing is described with respect to its relevance in industrial ceramic applications.

Printing is a very old technology. First steps in printing were taken around 2600 BC by the Sumerians. Also screen printing is well known and has been used for many centuries. Typical of screen printing is the pressing of the printing medium through the open structures of a screen. This technique was developed in the seventeenth century in Japan and China to print paintings onto paper and wooden surfaces.[47]

24.3.3.2 Basic Principles

The screen printing process enables the homogeneous and uniform deposition of a material in a layered structure onto a substrate. Therefore the material to be deposited is converted into a paste form with the right viscosity. This paste normally consists of organics, for example, polymers, solvents, and homogeneous distributed inorganic solid particles. During screen printing the paste is moved over the screen by a squeegee, whereas screen is pressed on the substrate and the paste is pressed through the meshes. To print only on desired areas open and masked screen arrays are used. The printed paste adheres to the surface

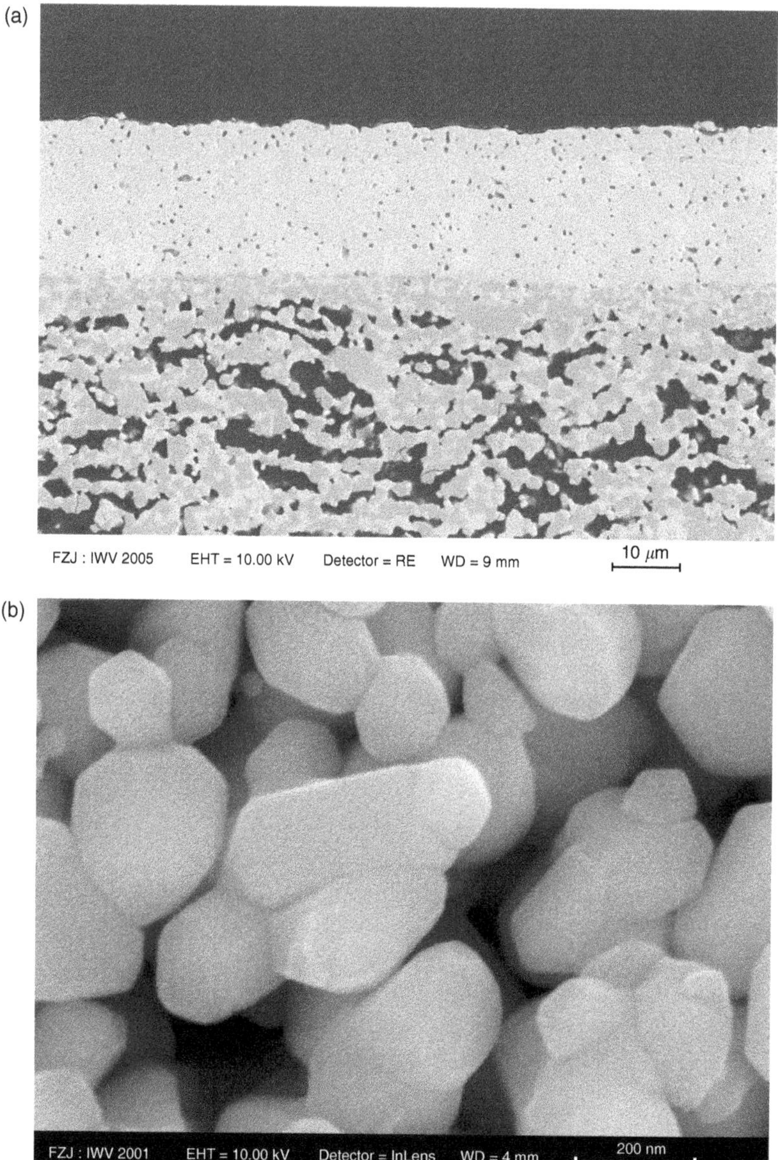

FIGURE 24.14 (a) Dense zirconia fuel cell electrolyte (no open porosity) layer manufactured by wet powder spraying and sintered at 1400°C (b): Surface structure of a porous TiO_2 powder layer sintered at 950°C (courtesy of Forschungszentrum Jülich, Jülich, Germany).

of the substrate. After removing the screen the paste forms an even and continuous film before fixing (Figure 24.16a).

Nowadays screens are manufactured from plastics or stainless steel and mounted in a metallic frame.[48] The screens are differentiated by the fabric fineness (mesh). The structuring of the screens into open and closed areas can be done by different methods. Polymer films are frequently used, which are laminated as light-sensitive emulsions and then light-hardened. The emulsion of unhardened areas is finally washed out.

Squeegees are manufactured from polymers of different degrees of hardness depending on the printing requirements. The quality of the screen-printed layers is influenced by a number of different values, for example, paste, screen, substrate, printing, posttreatment (Figure 24.16b).

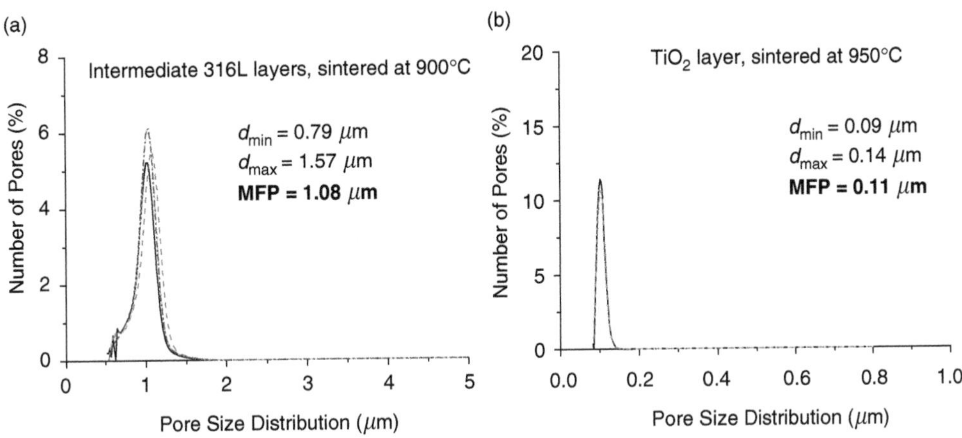

FIGURE 24.15 Pore size distributions in porous layers sintered at 950°C: (a) 316 L stainless steel, (b) TiO_2 (courtesy of Forschungszentrum Jülich, Jülich, Germany).

In most technical applications a precise layer thickness is important. Therefore, the squeegee pressure, the viscosity and the solids content of the paste have to be exactly controlled. The wet layer thickness can be calculated using the geometrical dimensions of the screen (cloth thickness mesh width, and wire diameter) but it is difficult to calculate the real layer thickness exactly, due to a number of influencing factors that are not well known, for example, line width, edge effects, polymer film thickness, and surface nature.

Screen-printed structures are in most cases functional layers, meaning that their technical and physical properties control the application potential. There are applications that require porous layer structures, for example, gas sensors and filtration membranes, and there is sometimes a need to produce dense layers, due to the better electrical and physical properties. Normally, the screen-printed layers are deposited on top of a nonshrinkable substrate. The layer can then shrink only in the vertical direction and complete densification becomes difficult. To support shrinkage during firing a shrinkable substrate can be taken into consideration (co-firing). It is also always helpful to start firing with a layer printed to a high green density, which means a lower shrinkage is necessary during firing resulting in a decreasing tendency to form cracks.

24.3.3.3 Screen Printing Machines

This chapter briefly describes the processes and the equipment necessary for the deposition of thick layers[49,50] onto substrates. In general, the printing process can be done with a stencil screen, which defines the printed pattern, and a squeegee, which forces the printing paste through the open areas of the screen onto the substrate surface. For simple printing requirements the process can be controlled by hand, but if accurate and reproducible deposition is needed then there are highly automated and controlled screen printing machines available to produce thousands of parts per hour. Typical machine features are:

- Screen mounting
- Squeegee and its control units
- Substrate holder
- Substrate moving mechanism
- Screen-substrate adjustment equipment

The *screen mounting* must hold the screen exactly in position during printing. Therefore, it has to be designed to resist the printing forces in lateral and vertical directions. At present, it is possible, if necessary for example for electronic devices, to maintain the position of the screen in the micrometer range (5 to 10 µm). To fulfill such precise positioning not only are the stiffness requirements high but air conditioning

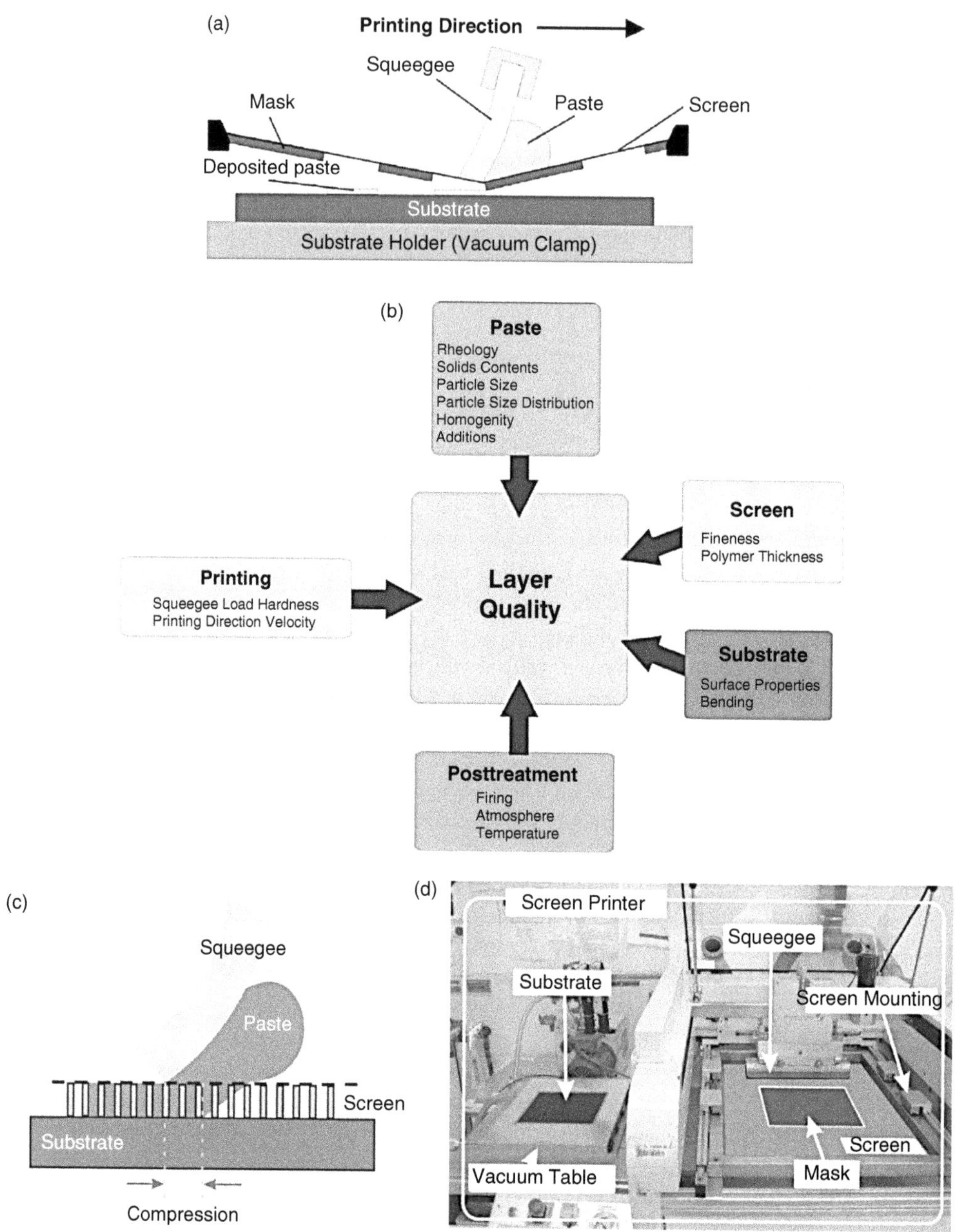

FIGURE 24.16 (a) Basic principles of the screen printing process. (b) Parameters influencing layer quality. (After Stolz, S., Siebdruck von elektrisch leitfähigen Keramiken zur Entwicklung heizbarer keramischer Mikrokomponenten. Ph.D. thesis Albert-Ludwigs-University of Freiburg. Forschungszentrum Karlsruhe, Scientific reports FZKA 6906, 2004; courtesy of Forschungszentrum Karlsruhe, Karlsruhe, Germany). (c) Schematic illustration of paste deposition by a squeegee. (d) Screen printing machine (EKRA type E1) (courtesy of EKRA, Germany).

and temperature control also play a critical role. In a highly automated process chain the screen mounting must allow fast positioning and replacement and also easy cleaning of the screen.

The elastic *squeegee* is moved with an inclined edge across the face of the screen, bringing it into contact with the substrate and pressing a quantity of paste through the screen over the pattern areas (Figure 24.16c).

The squeegee strongly influences the definition and thickness of the printed layer. The most important parameters for adjusting the squeegee to the printing requirements are composition, shape, and motion. Squeegees (doctor blades) are in most cases made from plastics or natural rubbers. They are produced over a wide range of hardness (shore hardness) and in different shapes. The selection of hardness, shape, design, attack angle, and holder design has up to now been an empirical process and demands a high degree of experience.

With respect to the printing result it is important to press the squeegee with constant force onto the screen during printing. The methods of generating the squeegee down pressure differ in accuracy and reproducibility and should be selected from the perspective of the application in question.

Last but not least the motion of the squeegee across the screen is an important printing parameter. This motion should be reproducible from print to print. Due to the ease of speed control hydraulic, electrohydraulic, or pneumatic hybrid systems are frequently used in advanced screen printers. With mechanical squeegee drives the speed is not so constant and it needs a lot of effort to achieve the results obtained with, for example, hydraulic systems.

The *substrate holder* is required to place the substrate precisely below the screen (see Figure 24.16d). During printing the substrate should not move and should maintain the position exactly. It is important that the substrate holder is adjusted to the accuracy of the screen positioning system. This also means the substrate holder should be able to assure reproducible substrate positioning in the micrometer range.

Advanced screen printing machines normally use vacuum tables to hold the substrates. The x–y-position is fixed by metallic stop positions mounted perpendicular to each other. A disadvantage of such guide rails is the fact that substrate edges are often neither ideally straight nor perpendicular to each other. Highly advanced precision printers are at present equipped with substrate positioning systems controlled by photodiodes or laser diodes. These systems identify the profile of the substrate or markers on the surface and level the substrate by computer control into an optimum position. Vacuum clamping of the substrates is not possible if the samples are too small or bent. In this case mechanical clamping is necessary.

24.3.3.4 Screens

The screen is one of the key components in the whole printing chain. It has two main functions to fulfill, the thickness control of the paste (ink) deposited onto the substrate and the definition of the printed pattern. During the printing action the squeegee presses the screen to contact the substrate. This leads to a sealing to prevent a lateral loss of paste into regions where printing is not required. Simultaneously, the squeegee forces the paste through the open areas of the screen and deposits it onto the substrate surface where it adheres after lifting (snap-off) the screen interrupts the printing process. Advanced screens are precisely manufactured components and in multilayer microelectronic devices assure an absolute accuracy of about 20 μm. To meet demands for such high accuracy there is a need for high-quality screens mounted on metallic or rigid composite frames.

At present, there are two main groups of screens available, the mesh type that is mainly used, and the foil type. In the latter, a thin metal foil is fixed on a metallic frame and etched to produce the stencil pattern. Etching allows precision structuring of such a foil screen. Typical advantages besides the manufacturing accuracy of such foil-type screens are their solvent resistance, long working life, and good cleaning behavior. But there are also limitations. The considerable stiffness and the low snap-off values mean that such screens are not useful for bent substrates and large-area printing. The mesh-type screens are widely used in advanced printing machines due to their flexibility in adjusting them to the applications. Typically the mesh is fabricated from plastics, for example, polyester, nylon, or stainless steel.

The plastic filaments of the screen are strong, flexible, and show good resistance to ink solvents and chemicals, but water adsorption leads to some swelling and there is also a pronounced temperature sensitivity. Stainless steel is, compared to plastic, stronger but displays a lower yield strength. Also swelling is not a problem and the chemical resistance and the temperature behavior are superior. This combination of properties leads to a longer lifetime and lower abrasion of stainless steel screens, but they are less resilient and therefore prone to more damaging of the screens during printing. Table 24.2 shows examples of screen mesh dimensions and theoretical paste (ink) volumes.

TABLE 24.2 Screen Mesh Dimensions and Theoretical Paste (Ink) Volume

Mesh Count Per inch	Wire Diameter, d (mm)	Mesh Opening, w (μm)	Theoretical Ink Volume, Vth (cm^3/cm^2)	Open Screen Area, a (%)
20	0.30	950	324	58
30	0.140	688	175	68
45	0.180	375	146	46
635	0.020	20	12	25
510	0.025	25	14	25
400	0.025	40	19	38
300	0.032	56	28	40
200	0.040	90	43	48
105	0.075	160	75	46
400	0.018	45	20	51
300	0.020	65	23	58
200	0.036	90	41	51

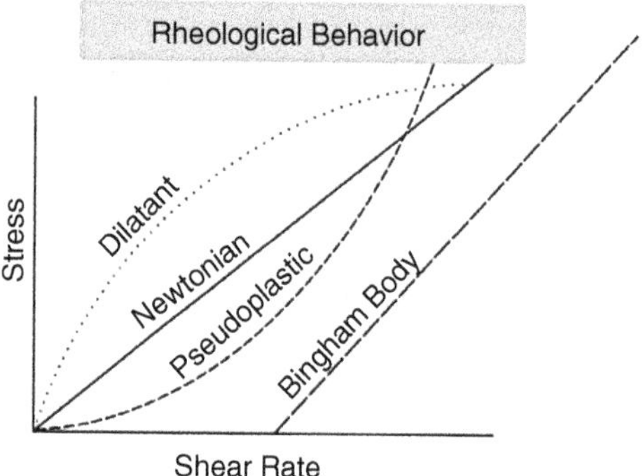

FIGURE 24.17 Schematic of the rheological behavior of ceramic suspensions.

At present, mesh wire diameters down to about 15 μm are in use with mesh counts per inch up to about 600 resulting in mesh widths in the 20 μm range. Such highly sophisticated screens guarantee a registration accuracy of less than 20 μm in microelectronics and thick film applications (e.g., from Koenen, Germany).

24.3.3.5 Screen Printing Pastes (Inks)

Screen printing pastes (inks) for ceramics processing are suspensions of fine particles (ceramics, metals) in organic vehicles (binders, solvents, additives). The viscosity of such pastes is normally adjusted in such a way that they are not able to flow through the screen meshes by the action of gravity without external pressure. On the other hand, the viscosity of the paste has to be so low that the squeegee is able to force it through the open meshes and distribute the paste homogeneously onto the substrate. After deposition the paste forms a smooth layer with a minimum of side flow and edge slumping. Additional requirements of pastes are long-term stability without decomposition and good drying behavior to avoid long drying times and crack formation. Some parameters influence the overall printing behavior of a paste and it is difficult to calculate the properties exactly in advance. In nearly all cases, the final paste composition is adjusted empirically. Knowledge of the flow behavior of pastes is indispensable and therefore viscosity and surface tension have to be described depending on the shear rate, velocity, and forces during printing.[51–54] In general, it is important to know which relationship between shear rate and applied stress can be found for a particular paste, for example Newtonian, dilatant, pseudoplastic, or Bingham types (Figure 24.17).

TABLE 24.3 Typical Commercially Available Chemicals for Screen Printing Pastes

Product	Manufacturer	Chemical Composition
Decoflux C 43	Zschimmer & Schwarz	Synth resin in terpineol
KD 2882	Zschimmer & Schwarz	Synth. polymers in org. solvents
Screen printing oil no. 221	W.C. Heraeus	Amongst others polymetacrylate in butyl glycolate
ESL 403	Electro Science Lab	Not specified

TABLE 24.4 Solvents for Paste Fabrication

Product	Manufacturer	Boiling Point (°C)	Vapor Pressure (mm Hg)	Dielectric Coefficient
Terpineol	Fluka	213–218	0.1 (38°C)	2,8
Diethylene glycol-monobutyl ether acetate	Fluka	245	0.04 (20°C)	unknown

TABLE 24.5 Binders for Paste Fabrication

Product	Manufacturer	Chemical Composition	Molecular Weight (g/mol)
Ethyl cellulose	Fluka	Cellulose ether	~60, 000
Paraloid B-66	Rohm und Haas	MMA[a]/BMA[b]-Copolymer	70,000
Pioloform BM 18	Wacker	Polyvinyl butyral	70,000–90,000

Molecular Weight: Manufacturer's Information.
[a]Methyl metacrylate.
[b]Butyl metacrylate

The viscosity (η) can be defined as the ratio between shear stress (τ) and shear rate ($\dot{s}$), $\eta = \tau/\dot{s}$. From Figure 24.17 it can be directly seen that the viscosity is only constant for a Newtonian liquid and also partly for the Bingham type. Frequently, ceramic pastes behave pseudo-plastically and therefore the viscosity varies over a wide range during screen printing. This behavior requires an empiric fine tuning of the paste composition. Normally, the ceramic particles are mixed into a commercially available liquid containing solvents, binders, dispersing agents, and rheology adjusting agents. A large variety of chemicals are used to produce screen printing pastes. Table 24.3 shows a small selection of such products, while Table 24.4 and 24.5 show solvents and binders for paste fabrication, respectively.

24.3.3.6 Applications

Due to the ability of screen printing technology to produce layers and structures in the micrometer range there is a continuous growth and new applications can be found. For decades screen printing has been established in the fast growing microelectronics market. Here a typical application is the printing of layered resistors, strip conductors, and dielectric structures. The resistor pastes contain small electrically conductive particles, for example, ruthenium oxide and glass particles, which together form finely distributed conducting particles in an insulating matrix. This allows the adjustment of the resistivity over some orders of magnitude. Strip conductors are printed from metal-containing pastes, which are covered by ceramic or glass pastes. These layered systems are fired in continuous or batch-type furnaces at temperatures from 800 to 900°C. The ongoing trend to miniaturize electronic components leads to requirements

TABLE 24.6 Screen-Printed Ceramics and Applications[55–58]

Material	Application
$Pb(Zr, Ti)O_3$	Piezo-electric sensors and transducers
$SrTiO_3$, SnO_2, $SmFeO_3$	Gas sensors
Cd S	Optical devices and sensors
ZrO_2, CeO_2	Solid oxide fuel cells (electrolyte)
$BaTiO_3$	Capacitors
$LaSrMnO_3$, LaSrCoFe oxide	Fuel cell (SOFC) cathodes
RuO_2 + glass	Piezo-resistive sensors and transducers
In–Sn oxide	Electric heaters

for thinner and more finely structured layers and this widens the range of applications for screen printing but forces the screen manufactures to move to finer and finer structures. Table 24.6 shows a variety of printed ceramic materials and the related applications.[55–58]

In the past decade, screen printing has also been extensively used to produce low-temperature co-fired ceramics (LTCC).[59] (see also Section 24.2.1.1) These three-dimensional structured devices are multilayer glass ceramic arrangements requiring high-precision screen printing and co-firing. During co-firing at low temperatures (900°C) the complete multilayer system is produced, in contrast to the normally used ceramic firing temperatures in the range of 1400°C. As an example, inductively heated chemical microreactors are made with screen-printed heating layers.[60]

24.3.4 Roller Coating

Roller or roll coating (RC)[61,62] is a coating technology usable mostly for nearly flat, even, continuous substrates. But also round geometries if continuous can be coated. Suspensions used for RC are comparable to those used for screen printing or slip casting. This means that they are composed of oxide powders, the solvent (usually an alcohol), and some organic additives for stabilization and adoption of rheological behavior. The simplest equipment is a hand roller made of polymeric or textile material. However, there are numerous industrial coating processes that can be included in the "roller coating group."

24.3.4.1 Industrial Coating Processes

Besides simple hand coating, various coating processes are possible for applying a ceramic layer on either a permanent support or substrate (ceramic, metal . . .) or on a nonpermanent support (plastic, glass . . .), which can either be dismantled before firing of the layer or which is burned out during heat treatment. Figure 24.18 displays some industrial coating processes known as roller coating.

The thickness of the layer formed by roller coating is influenced by the moving speed of the substrate, the contact pressure of the roller on the substrate, the viscosity and the amount of solid fraction of the slip. Besides classical roller coating, in which a roller transports the suspension to the substrate to be coated, there are a number of variants.

First, to produce noncontinuous layers the gravure coating principle should be mentioned; (See Figure 24.18e). Here cavities in the roller surface are filled with suspension while the elevations are not coated. This is ensured by a knife or a doctor blade that wipes the suspension off the elevations. Therefore only the filled cavities of the roller coat the surface of the substrate and subsequently a noncontinuous layer is formed. The thickness of the layer is determined by the deepness of the cavities and the amount of solid fraction within the suspension.

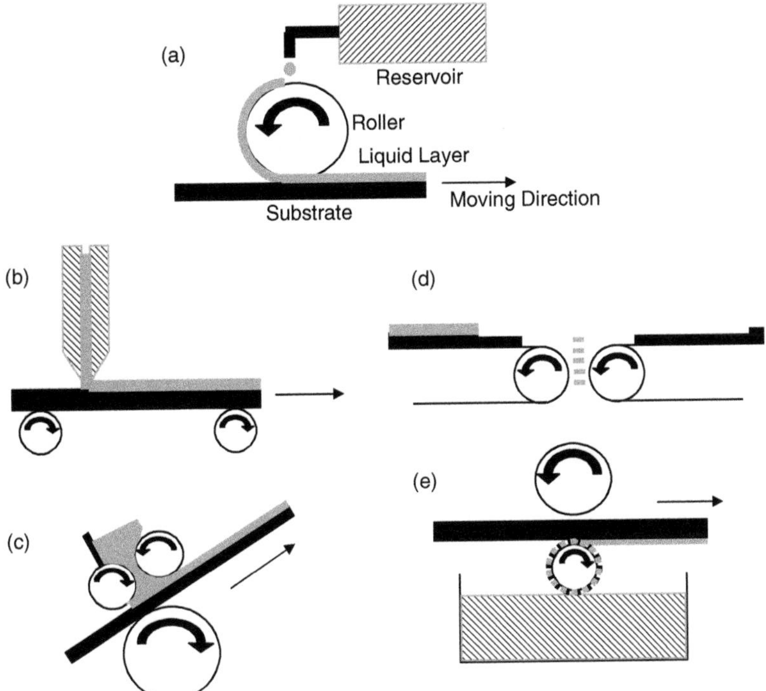

FIGURE 24.18 Schematic overview of possible industrial roll-coating techniques and comparable technologies; (a) typical/classical roller coating; (b) slot die (extrusion) coating; (c) reverse roll coating; (d) curtain coating; (e) gravure coating.

Another easy way of coating is the slot die coating technique (Figure 24.18b). Here the rollers are only the transportation support for the substrates to be coated. The suspension flows within a hollow cylinder in a vertical direction. The substrate is moved horizontally. By varying the viscosity of the suspension, the inner diameter of the cylinder and the moving speed of the substrate the thickness of the layer can be varied.

The reverse roll-coating technique (Figure 24.18c) is used when the suspension has a high viscosity that can be correlated to a high solid fraction in the solvent. The shear power must be quite high to coat such suspensions. This is ensured by using two rollers moving in reverse. Thus the suspension is constrained to move in the gap between the two rollers. Thicker layers can be obtained with suspensions of higher solid fractions.

The last technique is curtain coating (Figure 24.18d). Here the substrates are moved horizontally to the suspension that flows out of a reservoir located above the transport line. This technique is useful for coating smaller parts. Additionally, the thickness of the layers formed can be kept very thin due to the speed of the substrates and the low viscosity of the suspension. The suspension used can be composed quite easily because the requirements for the flow behavior are less stringent. Only sedimentation and agglomeration need to be avoided.

Especially with the technologies of roller coating, gravure coating, slot die coating, and reverse roll coating it is also possible to coat supports that are subsequently burnt out. Thus thin tapes or layers can be obtained after careful drying and sintering. Additionally, by adapting the slip parameters or by adding pore formers either dense or porous materials (coatings or tapes/layers) can be achieved.

While Figure 14.18a shows the classical roller coating technique with one roller there are numerous modifications. Figure 24.19 shows some examples.

Double coaters are used for coating both substrate sides in parallel with either the same material or with different materials. Single top coaters have the advantages of using suspensions with higher viscosities

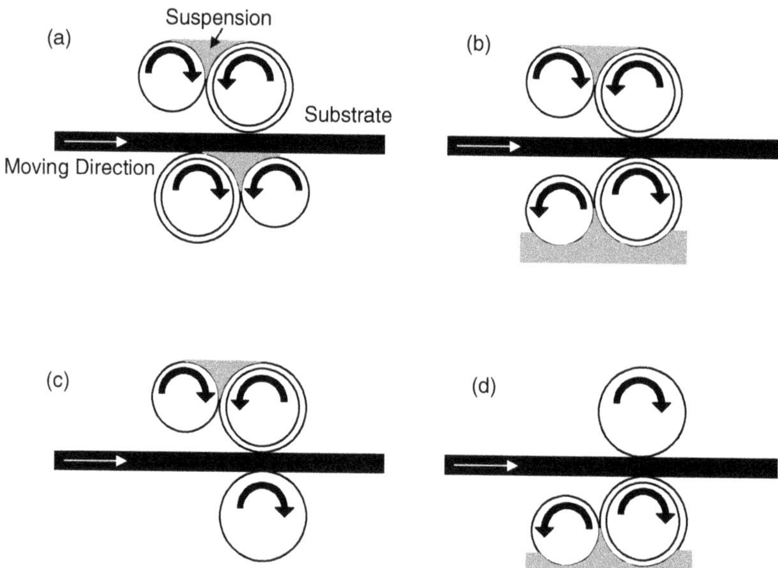

FIGURE 24.19 Schematic illustration of possible single and double coaters. (a) double coater model A; (b) double coater model B; (c) single top coater; (d) single bottom coater.

(support by gravity) and better homogenization (e.g., less sedimentation), while bottom coaters are sued for suspensions with less viscosity but quite good adhesion to the substrate in green state.

After coating careful and controlled drying and sintering are necessary to suppress layer defects like cracking or delamination.

24.4 Conclusions

Modern ceramic processing technologies comprise component and substrate manufacturing techniques, and coating technologies. The processing of components starts by using more classical methods like extrusion and goes via tape casting of two-dimensional large-area substrates and ends by fabrication of complex-shaped three-dimensional structural parts by powder injection molding. Common goals of these methods are cost reduction by enhancing the number of parts per time, near net shape manufacturing, high precision, and highly industrial automated fabrication. Besides conventional processing of structural parts more and more functionalities are integrated into the parts (e.g., lamination of structured tape casted foils for LTCC-ceramics). Similar trends as for structural parts can be seen for coatings, that is, thinner layers (sub-micro) finer microstructures for porous and dense layers and coating of complex-shaped components.

References

[1] Mistler, R.E., Twiname, E.R., *Tape Casting, Theory and Practice*, The American Ceramic Society, Westerville, OH, 2000.

[2] Richerson, D.W., *Modern Ceramic Engineering*, Marcel Dekker, Inc., New York, 1992.

[3] Reed, J.S., *Principles of Ceramics Processing*, John Wiley and Sons, Inc., New York, 1995.

[4] Schneider Jr., S.J., *Ceramics and Glasses. Engineered Materials Handbook, Vol. 4*, ASM International, USA, 1991.

[5] Kingery, W.D., *Introduction to Ceramics*, 2nd Edition. John Wiley and Sons, Inc., New York, 1976.

[6] Moreno, R., The role of slip additives in tape casting technology, I: Solvents and dispersants, *Am. Ceram. Soc. Bull.*, 71, 1521, 1992.
[7] Menzler, N.H., Zahid, M., Buchkremer, H.P., Foliengießen von Substraten für anodengestützte oxidkeramische Brennstoffzellen (SOFC), in *Technische Keramische Werkstoffe*, Kriegesmann J. (Ed.), Verlag Deutscher Wirtschaftsdienst, 2004, Chapter 3.4.6.1, 1–21.
[8] Beuers, J., Lang, E., Poniatowski, M., Pulvermetallurgischer Spritzguß — ein neues PM-Formgebungsverfahren eröffnet neue Anwendungsgebiete für die Pulvermetallurgie, *Metall*, 43, 963,1989.
[9] Bose, A., German, R.M., Potential of powder injection molding and hot isostatic pressing of Nickel Aluminide Matrix Composites, *Ind. Heating*, 55, 38, 1988.
[10] German, R.M., *Powder Injection Molding*, Metal Powder Industries Federation, Princeton NJ, 1990.
[11] Wich, R.E., Method and means for removing binder from a green body, U.S. Patent 4305756, 15 Dec. 1981.
[12] Onoda, G.Y., The rheology of organic binder solutions, in *Ceramic Processing Before Firing*, Onoda, G.Y. and Hench, L.L. (Eds.), John Wiley and Sons, New York, 1978, pp. 235–251.
[13] Chung, C.I., et al., Requirements of binder for powder injection molding, *Adv. Powder Metall.*, 3, 67, 1989.
[14] Sohn, H.Y., Moreland, C., The effect of particle size distribution on packing density, *Can. J. Chem. Eng.*, 46, 162, 1968.
[15] Erickson, A.R., Wiech, R.E., *Injection Molding in Metals Handbook*, 9th Edition, Vol. 7, 1984, pp. 495–500.
[16] German, R.M., Optimization of the powder-binder mixture for powder injection molding, *Adv. Powder Metall.*, 3, 51, 1989.
[17] Vinogradov, G.K., Malkin, A.Y., Rheological properties of polymer melts, *J. Polym. Sci.*, 2, 135, 1966.
[18] Hieber, C.A., Shen, S.F., A finite-element/finite-difference simulation of the injection molding filling process, *J. Non-Newtonian Fluid Mech.*, 7, 1–32, 1980.
[19] Bose, A., German, R.M., Injection molding of an intermetallic matrix composite, in *Modern Developments in Powder Metallurgy*, Vol. 18 ed. Gustafson, D.A, APMI, 1988, pp. 299–314.
[20] German, R.M., Theory of thermal debinding, *Int. J. Powder Metall.*, 23, 237–245, 1987.
[21] German, R.M., Debinding practice, in *Powder Injection Molding*, MPIF, Princeton, NJ, 1984, Chap. 13.
[22] Lin, S.T., German, R.M., Extraction debinding of injection molded parts by condensed solvent, *Powder Metall. Int.*, 21, 19–24, 1989.
[23] Johnson, A., Carlstrom, E., Hermannsson, L., Carlsson, R., Rate-controlled thermal extraction of organic binders from injection molded bodies, *Proc. Brit. Ceram. Soc.*, 33, 139–147, 1983.
[24] Brazenhall, M.V.J., Furnace atmospheres for sintering, *Metal Powder Report*, 45, 600–604, 1990.
[25] Krehl, M., Schulze, K., Petzow, G., The influence of gas atmospheres on the first stage sintering of high purity niobium powders, *Metall. Trans.*, 15A, 1111–1116, 1984.
[26] Sarkar, P., Nicholson, P.S., Electrophoretic deposition (EPD): Mechanisms, kinetics, and applications to ceramics, *J. Am. Ceram. Soc.*, 78, 1897–2002, 1996.
[27] Nicholson, P.S., Sarkar, P., The electrophoretic deposition of ceramics, in *Ceramic Transactions*, vol. 51. *Proc. 5th Int. Conf. on Ceram. Proc. Sci. and Technol.* (Ed. H. Hausner), Friedrichshafen, 11.-14.09.1994, Germany. The American Ceramic Society, Westerville, OH, 1995.
[28] Moreno, R., Ferrari, B., Advanced ceramics via EPD of aqueous slurries, *Am. Ceram. Soc. Bull.*, 79, 44–48, 2000.
[29] Hruschka, M., Entwicklung der "Elektrophoretischen Beschichtung" für die Herstellung dünner Elektrolytschichten. Ph.D. thesis, University of Bonn, Faculty of Mathematics and Natural Sciences, 1995. English translation: Development of electrophoretic deposition for manufacturing thin electrolyte layers (available in German only).
[30] Helmholtz, H., Studien über elektrische Grenzschichten, *Ann. Phys. Chem.*, 7, 337–382, 1879.

[31] Stern, O., Zur Theorie der elektrischen Doppelschicht, *Z. Elektrochem.*, 30, 508–516, 1924.

[32] Hamaker, H.C., Formation of a deposit by electrophoresis, *Trans. Faraday Soc.*, 36, 287–295, 1949.

[33] Zhao, L., Manufacture and characterisation of composite graded filter membranes for microfiltration. Ph.D. thesis; Forschungszentrum Jülich reports JÜL-4079. 2003.

[34] Baklouti, S., Pagnoux, C., Chartier, T., Baumard, J.F., Processing of aqueous α-Al_2O_3, α-SiO_2 and α-SiC suspensions with polyelectrolytes, *J. Eur. Ceram. Soc.*, 17, 1387, 1977.

[35] Terayama, H., Okumura, K., Sakai, K., et al., Aqueous dispersion behavior of drug particles by addition of surfactant and polymer, *Colloid. Surf. B*, 20, 73, 2001.

[36] Heijman, S.G.J., Stein, H.N., Electrostatic and sterical stabilization of TiO_2 dispersions, *Langmuir*, 11, 422, 1995.

[37] Van der Hoeven, Ph. C., Lykelma, J., Electrostatic stabilization in non-aqueous media, *Adv. Colloid Interface Sci.*, 42, 205, 1992.

[38] Morris, G.E., Skiner, W.A., Self, P.G., Smart, R.St.C., Surface chemistry and rheological behaviour of titania pigment suspensions, *Colloid. Surfaces A*, 155, 27, 1999.

[39] Vaßen, R., Stöver, D., Processing and properties of nanophase ceramics, *J. Mater. Process. Technol.*, 92–93, 77, 1999.

[40] Darcovich, K., Roussel, D., Toll, F.N., Sintering effects related to filtration properties of porous continuously gradient ceramic structures, *J. Membr. Sci.*, 183, 293, 2001.

[41] Stech, M., Reynders, P., Rödel, J., Constrained film sintering of nanocrystalline TiO_2, *J. Am. Ceram. Soc.*, 83, 1889, 2000.

[42] Zhao, L., Bram, M., Buchkremer, H.P., Stöver, D., Li, Z., Development of a graded porous metallic filter for microfiltration using the wet powder spraying technique, Proceedings of the PM 2001, Nice, France, 2001, p. 35.

[43] Prinz, D., Arnhold, V., Buchkremer, H.P., Kuhstoss, A., Neumann, P., Stöver, D., Graded high porous microfilters by powder metallurgical coating techniques, Materials Science Forum, Trans Tech Publications 308–311, Switzerland (1999), 59.

[44] Marshall, T.J., A relation between permeability and size distribution of pores, *J. Soil. Sci.*, 9, 1, 1958.

[45] Kosloff, A., *Ceramic Screen Printing*, The Signs of the Times Publishing Company, 3rd Edition, Cincinnati, OH, 1984.

[46] Holmes, P.J., Loasby, R.G., *Handbook of Thick Film Technology*, Electrochemical Publications Limited, Scotland, UK, 1976.

[47] Sergent, J.E., Harper, C.A., *Hybrid Microelectronics Handbook*, 2nd Edition, Mc Graw-Hill, Inc., New York, 1995.

[48] Stolz, S., Siebdruck von elektrisch leitfähigen Keramiken zur Entwicklung heizbarer keramischer Mikrokomponenten. Ph.D. thesis Albert-Ludwigs-University of Freiburg. Forschungszentrum Karlsruhe, Scientific reports FZKA 6906 (2004); English translation: Screen printing of electroconductive ceramics for the development of microheater devices (available in German only).

[49] Stojanovic, B.D., Foschini, C.R., Pavlovic, V.B., Pavlovic, V.M., Pejovic, V., Varela, J.A., Barium titanate screen-printed thick film, *Ceram. Int.*, 28, 293–298, 2002.

[50] Shanefield, D.J., Electronic thick film technology, in *Ceramic Films and Coatings*, J.B. Wachtman (Eds.), 284–302, Noyes Publ., Park Ridge, IL, 1993.

[51] Trease, R.E., Dietz, R.L., Rheology of pastes in thick-film printing, *Solid State Technol.*, 15, 39–43, 1972.

[52] Goodwin, J.W., The rheology of dispersions, *Colloid Sci.*, 2, 246–293, 1975.

[53] Gilleo, K., Rheology and surface chemistry for screen printing, *Screen Print.*, 128–132, 1989.

[54] Pugh, R.J., Bergström, L. (Eds.), *Surfactant and Colloid Chemistry in Ceramic Processing*, Marcel Dekker, New York, 1994.

[55] Carotta, M.C., Martinelli, G., Sadaoka, Y., Nunziante, P., Traversa, E., Gas-sensitive electrical properties of perovskite-type $SmFeO_3$ thick films, *Sens. Actuators B* 48, 270–276, 1998.

[56] Morten, B., Prudenziati, M., *Piezoresistive Thick-Film Sensors, Handbook of Sensors and Actuators, Volume 1: Thick Film Sensors*, Elsevier, Amsterdam, The Netherlands, 1994, pp. 189–209.
[57] Van Herle, J., Ihringer, R., Vasquez Cavieres, R., Constantin, L., Bucheli, O. Anode supported solid oxide fuel cells with screen-printed cathodes, *J. Eur. Ceram. Soc.*, 21, 1855–1859, 2001.
[58] Will, J., Mitterdorfer, A., Kleinlogel, C., Perednis, D., Gauckler, L.J., Fabrication of thin electrolytes for second-generation solid oxide fuel cells, *Solid State Ionics*, 131, 79–96, 2000.
[59] Kita, J., Dziedzic, A., Golonka, L.J., Bochenek, A., Properties of laser cut LTCC heaters, *Microelectron. Reliability*, 40, 1005–1010, 2000.
[60] Shaikh, A., Thick film pastes for AIN substrates, *Adv. Microelectron.*, 21, 18–21, 1994.
[61] http://www.tciinc.com/coating.html.
[62] http://www.uswebcon.com/slitting.htm.

25

Powder Processing

Randall M. German
Mississippi State University

Abstract

The fabrication of engineering components using powder technologies provides several advantages as there is minimal loss of material, melting can be avoided, and once a tool cavity is generated it can be replicated millions of times at fairly low cost. Consequently, pressing or molding followed by firing (sintering) is used to make many consumer and industrial items, such as automotive transmission gears, electronic capacitors, wrist watch cases, and light bulb filaments. This processing technology is shared by ceramics, polymers, metals, composites, and various ceramic-metal-polymer combinations, often delivering materials that are unattainable via alternative production routes. In the simplest form, powder technologies are similar to how students make vases in art class — the wet powder is first formed at low stresses and then fired to generate significant strength.

25.1 Introduction

Humans fabricate millions of automobiles, cellular telephones, computers, and other devices every year. These diverse products share a common interest in low-cost, net-shape, and high-volume production. Powder processing relies on the semi-fluid character of a powder to flow and fill a die at room temperature, taking on the die shape. Thus, unlike machining where there is wasted material and expense associated with mass removal, powder approaches simply form the needed mass of powder into the desired shape in a single step. Unlike casting, which is only applied to lower melting temperature metals, powder techniques are applicable to all materials, including diamonds, ceramics, and various compounds such as tungsten carbide. Indeed, many of the products formed using powders are not available as castings. In most applications, cost is an important factor, so the ability to fabricate complex shapes to final size has significant economic benefit. Further economic gains come from the fact that powder processes use automation with relatively low energy consumption, while exhibiting high material utilization. One widely employed option is to mix combinations of insoluble powders to form composites. Further, significant property customization is achieved by controlling the microstructure in terms of the relative sizes, shapes, and amounts of the phases, including controlled porosity for filters, bearings, and sound absorbers.

25.2 Conceptual Overview

Particles are discrete, small divisions of solids below 1 mm in size. They typically range from 0.1 to 200 μm. For reference, a human hair has a typical diameter near 100 μm, while the pigment used in paint is typically a particle in the 1 μm range. In engineering systems, it is common to mix various powder chemistries together to form composites. The next step is to shape the powder using pressure and polymers. Subsequently the powder is bonded using heat, in a process termed sintering. Soft powders are pressed to nearly full-density, but hard powders resist pressing and require high firing temperatures to induce sintering densification.

25.3 Shaping Options

Powder compacts are formed by four approaches:

- Use of slurries where the pores between particles are saturated with a polymer, such as in powder injection molding
- Use of particles or polymer-bonded granules where high die compaction pressures deform and bond the particles into a green body
- Use of simultaneous pressure and temperature to densify the powders, such as in hot isostatic pressing
- Use of freeform techniques to create the shape without tooling, such as from a computer-controlled laser beam (a topic covered in Chapter 26).

Polymer additions are an inherent part of shaping. Powder-polymer slurries are formed by options such as injection molding, tape casting, slip casting, or extrusion (see also Chapter 24). For some shapes it is most effective to rely on die compaction where the dry powder and polymer are pressed using upper and lower punches working in a die, similar to how an aspirin tablet is formed. Forming pressures can exceed 1 GPa, but more typically range from 350 to 500 MPa. To avoid tool damage the pressed shapes tend to have most of the complexity in the pressing orientation. The decision on what forming technique to apply depends on the component, material, and desired properties.

The powder-polymer feedstock is adjusted to match the forming process, ranging from dry granules to a stiff paste to a paint-type liquid. The powder and polymer are mixed in a ratio dictated by the desired rheology. Low pressure techniques such as slip and slurry casting require a solvent. After shaping, the polymer or binder is removed (debinding) and the powder sintered; these last two steps can be combined into a single thermal cycle. As significant shrinkage is associated with sintering densification, the final dimensions rely on uniform shaping. For this chapter, the primary powder shaping techniques are injection molding, die compaction, and hot isostatic pressing.

25.4 Shaping Processes

The decision on which forming process to apply to a powder is based on factors such as the component features, desired properties, annual production quantity, and component shape. The solutions range from simple cylinders to high complexity shapes, such as tea cups.

25.4.1 Powder Injection Molding

Powder injection molding builds on plastic injection molding using a feedstock that is high in particle content. The key production steps are outlined in Figure 25.1. The process begins by mixing selected powders and binders. The particles are small to aid in sintering densification, and often have average sizes below 20 μm with near spherical shapes. The binders are thermoplastic mixtures of waxes, polymers, oils,

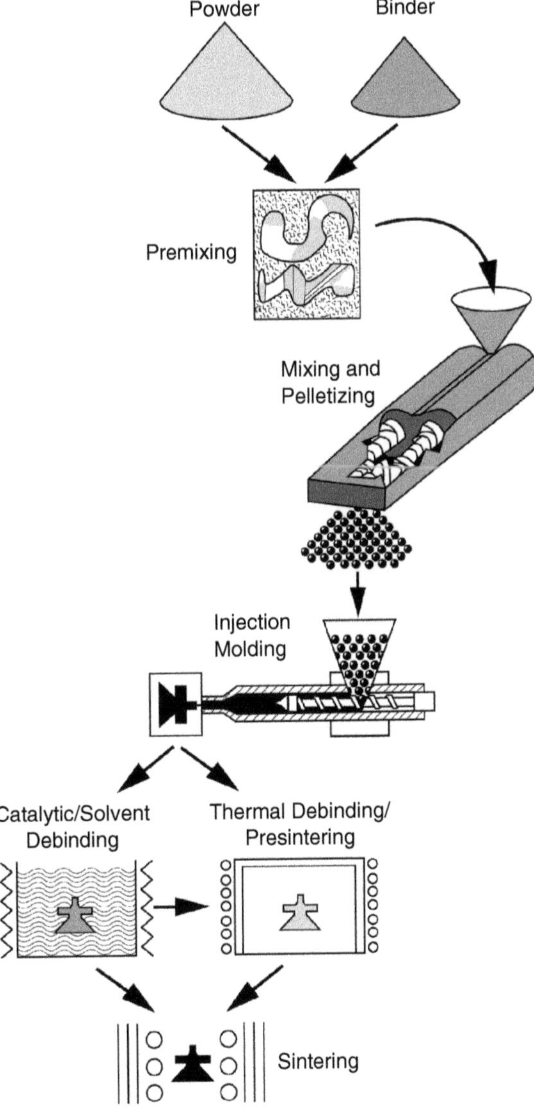

FIGURE 25.1 A schematic of the powder injection molding process, where a powder and polymer are mixed to form the feedstock, which is molded, and then the polymer (binder) is removed prior to sintering to near full density.

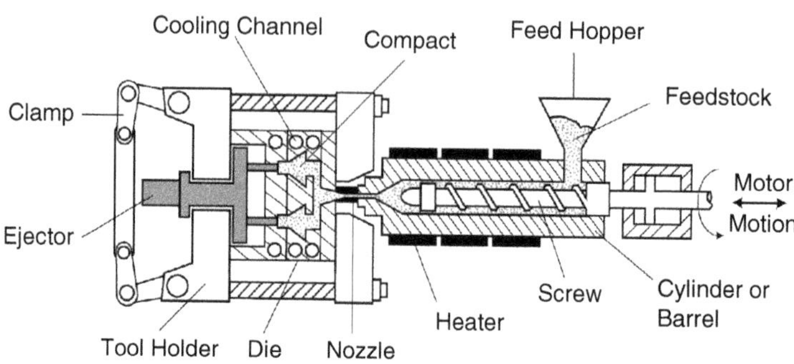

FIGURE 25.2 Cross-sectional view through a powder injection molding operation, showing the entry of the feedstock into the feed hopper and its heating and transport into a chilled die. The hot feedstock freezes to take on the die cavity shape, after which the die opens to eject the powder-polymer component, which is then debound and sintered.

lubricants, and surfactants (a favorite is 65 wt.% paraffin wax, 30 wt.% polypropylene, and 5 wt.% stearic acid). This binder is fully molten at about 150°C. The amount of binder tends to range near 40 vol.% of the mixture, depending on the powder characteristics; the weight fraction of binder can be 15% for low-density powders (such as alumina) and as low as 2% for high-density powders (such as tungsten, gold, and rhenium). When heated, this binder imparts viscous flow characteristics to the feedstock to allow filling of complex tool geometries. If the die is cold, then the binder freezes after injection, allowing the shaped particles to be ejected. Next, the binder is removed and the remaining porous powder structure sintered to near full density. The product may then be further densified, heat treated, or machined. The sintered compact has the shape and precision of an injection molded plastic, but is capable of performance levels unattainable with polymers.

The equipment used for shaping the compact is the same as used for plastic injection molding. A cross section through a typical molding operation is sketched in Figure 25.2. Most molding machines fill a die through a gate from a pressurized and heated barrel. A plunger or reciprocating screw generates the pressure needed to fill the die. The feedstock enters the barrel as cold granules from the loading hopper and is heated above the melting temperature of the binder as it progresses along the barrel. Because the feedstock is hot and the die is cold, filling must be accomplished in a split second to avoid premature freezing. Filling is performed by using the screw as a plunger. After filling the die, pressure is maintained on the feedstock during cooling to avoid shrinkage voids. After sufficient cooling, the compact is ejected and the cycle repeated.

Temperature and pressure provide the primary control parameters in molding. Progressive mold filling from the gate pushes air out of the die cavity through a vent. Vents are located at the last portion of the die to be filled and are opposite the gate location.

After molding, the binder is removed from the compact by a process termed debinding. There are several options including thermal, solvent, and capillary extraction methods. Thermal debinding is used most frequently, where the compact is heated slowly to 600°C in air to decompose the binder. The next step is sintering, which can be incorporated directly into the debinding cycle. As discussed later, sintering provides strong interparticle bonds and removes the void space by densification. Isotropic powder packing allows for predictable and uniform sintering shrinkage, so the original compact is oversized as appropriate for the final compact dimensions. Figure 25.3 is a photograph showing the size change from molding to sintering. Computer models exist to predict final size, density, shape, and even properties. These routines handle both the molding step (machine operation and tool design) and the shrinkage in sintering. After sintering, the compact has competitive strength and excellent microstructural homogeneity, leading to properties superior to those available with many other processing routes. For example, a simple 316L stainless steel has a tensile strength of 520 MPa and elongation to fracture of 50% when prepared by powder injection molding, but only half these values when prepared by traditional die compaction powder metallurgy.

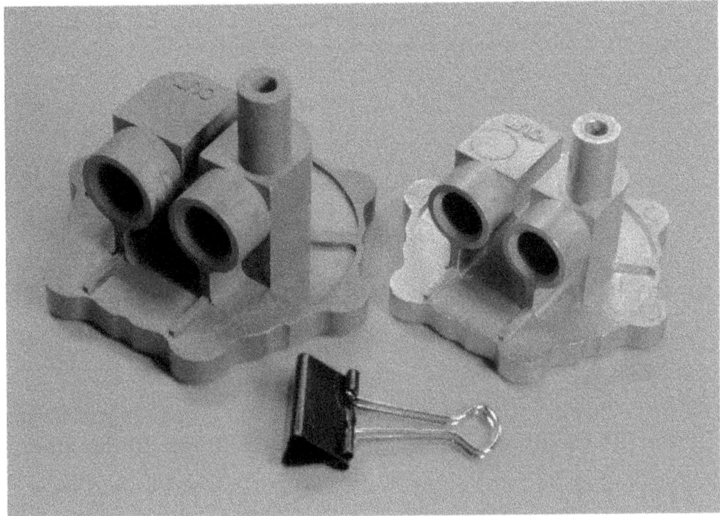

FIGURE 25.3 A stainless steel pump housing showing the shape complexity possible with powder injection molding and the size change between molding (left component) and sintering (right component).

TABLE 25.1 Nominal Powder Injection Molding Component Design Concepts

Restrictions
no inside closed cavities or undercuts on internal bores
must have round corners and draft angles for ejection
features generally must be larger than 0.1 mm
Desirable features
largest dimension below 125 mm
wall thickness less than 10 mm
flat surfaces to provide support
Allowed design features
shape complexity including blind holes, hexagonal holes, and holes at angles to each other
identification numbers, protrusions, studs, and surface textures on part
helical threads, internal threads, and external threads if needed

A primary concern in designing a tool set is component shrinkage. The shrinkage factor is the change in a dimension divided by the original dimension. As the target is the final component size, each dimension of the tool cavity is oversized to accommodate sintering shrinkage. One key advantage of powder injection molding is the ability to fabricate complex shapes that cannot be produced by alternative techniques, in production runs often reaching millions of units per year. Accordingly, complex tool designs are a necessary aspect of injection molding. As the molding pressures are low compared to the strength of the tool material, it is possible to have many inserts, slides, and features to increase the tool complexity.

The molding parameters are sensitive to the feedstock characteristics and mold design. Consequently, no single set of conditions is ideal. For example, the cycle time will vary from approximately 5 to 60 s, with a typical value of 20 s. Often, the mold fill rate is near 1.5 cm^3/s but can be as high as 75 cm^3/s, with peak pressures as high as 60 MPa and a feedstock temperature below 200°C. Depending on the feedstock, the mold temperature ranges from 30 to 140°C.

Although many complex geometries can be fabricated via injection molding, there are a few restrictions and desirable features, as listed in Table 25.1. Statistics show the typical powder injection molding (PIM) part is near 25 mm in maximum dimension and 8 g in median mass. There are larger components up to 15 kg in production by PIM, but these are specialized, lower production quantity structures such as ceramic pouring spouts used in steel processing. A constant and thin wall ensures rapid cooling and

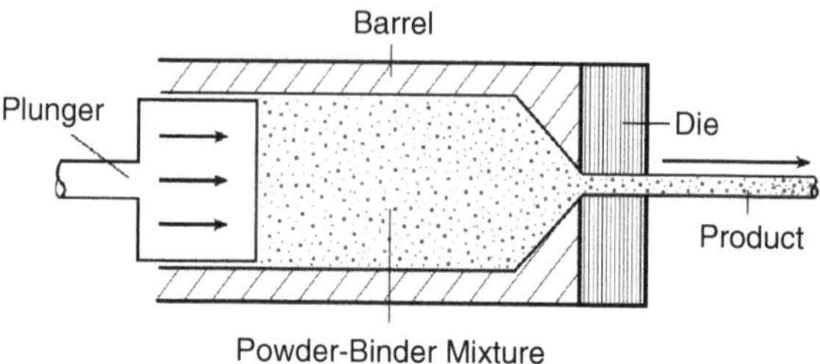

FIGURE 25.4 Powder-binder extrusion is used to form a product that is long and thin with a constant cross section, such as a rod, tube, honeycomb, or twist drill.

debinding. Best results occur when the thickness is not highly variable, as thickness determines how fast the equipment can run. From an economic view, many factors work against large structures:

1. The tooling is more expensive
2. The processing equipment is larger and more expensive
3. Cycle times are slower as heat transfer is slower in molding, debinding, and sintering

25.4.2 Other Shaping Options

Several related technologies are applied to powder shaping. They share the attribute of relying on a high polymer content for forming the particles. Extrusion is used to form long, thin shapes, such as rods, tubes, honeycombs, and twist drills. In many regards the feedstock is similar to that used in powder injection molding. As illustrated in Figure 25.4, the powder-binder mixture is placed in a heated cylindrical barrel and this mixture is compressed by a plunger. The shaping die is at the exit of the barrel. The angle of feedstock entry into the die is usually about 25° and the die length is usually two to four times the exit diameter. Flow depends on the pressure. Pressures are over 1 MPa, but this creates enormous forces on the dies for large components. Usually the feedstock is evacuated during mixing to avoid bubbles in the extrudate. Once the binder sets, it acts as a glue to hold the particles in place during handling and is subsequently removed as part of the sintering cycle.

The product shape depends on the extrusion die. In simple cases, the shape is a circular rod, but in other cases the tooling includes center holes, splines, or other features. Spiders are used to place cores in the die center, allowing for shapes such as honeycombs. This means the support is upstream from the die to minimize flow disruption. Further the output can be twisted to create the flutes on a twist drill. All of these products have constant cross sections and only need to be chopped to length after extrusion. An example product is the helical screw found in a twist drill. Other applications are in the forming capacitors, microelectronic substrates, porous tubes, welding rods, and automobile exhaust catalytic converter substrates. Since the 1940s, porous stainless steel filters have been fabricated using powder extrusion processes based on gelation of water soluble cellulose binders.

Solvents such as water are added to the feedstock to lower the viscosity sufficiently to allow pouring. That viscosity is similar to thick paint. In slip casting these slurries are poured into porous plaster of Paris molds. Once poured into the mold, the water is adsorbed by capillary action into the porous mold, resulting in a progressive increase in the solids loading and viscosity of the remaining slurry. A typical mold has pores of approximately 0.1 μm size. The slurry remains in the mold for several minutes and even several hours for thick components. This time is shortened by applying a vacuum to the outside of the mold, pressurizing the slurry, or subjecting the mold to centrifugal or rotational forces. The polymer eventually forms bonds between the particles to provide strength so that the dry shape can be handled after extraction from the mold. The green strength after drying might be 7 MPa. A typical binder for slip casting is a mixture of water, alginate, and surfactants. But water-containing solutions of cellulose, ammonium polymethyl methacrylate, or polyvinyl alcohol are also in use.

FIGURE 25.5 A slurry cast bronze statue formed using a wax-polymer binder, bronze powder, and rubber tooling. After the slurry was cast and cooled, the shape was placed in a furnace and slowly heated to burn out the binder and sinter the particles. This technology is ideal for smaller production quantities, such as encountered in the art field.

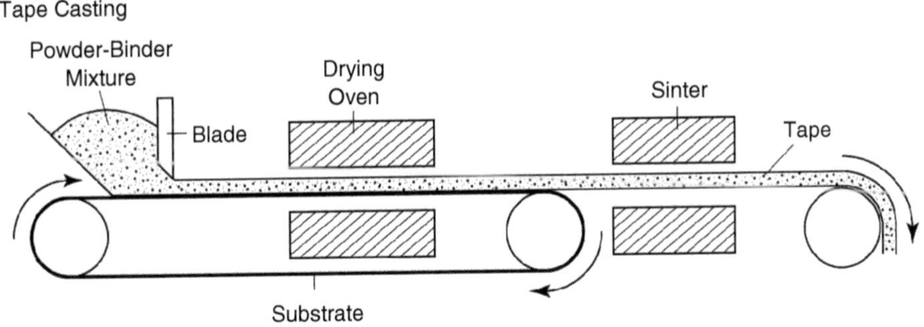

FIGURE 25.6 Tape casting is useful for forming powders into thin sheets as is appropriate for the semiconductor and electronics industry. The basic process is illustrated here, showing how the slurry of powder, binder, and solvent is spread onto a moving sheet, with subsequent drying of the solvent used to form a flexible sheet that is sintered.

A common application for slip casting is the fabrication of ceramic bathroom fixtures. It is also used to fabricate prototype shapes and even pilot tool sets. When the binder is a thermoplastic, then after mold filling the slurry is frozen. This variant, best termed slurry casting, is slow. Figure 25.5 shows an example component formed using bronze powder in a rubber tool with a thermoplastic binder. Slip and slurry casting are most effective for larger shapes, as there is little equipment cost and the tooling is relatively inexpensive.

Tape casting uses a similar low viscosity mixture of powder-solvent-polymer, but the forming step is via deposition of the mixture on a moving sheet. The process is shown schematically in Figure 25.6, where the powder-binder slurry is fed onto a moving plastic sheet that passes through a small gap. A doctor blade levels the slurry to a constant sheet thickness. Subsequently, the solvent is evaporated, leaving binder behind to bond the particles and provide handling strength. Binders for tape casting are similar to those discussed earlier. Water is a preferred solvent, but some binders require more toxic or hazardous solvents. Once the solvent has evaporated, the particles are held together by the remaining polymer into a thin, flexible sheet. Tape casting is used to form battery electrodes, brazing layers, microelectronic substrates, coatings, and thin foils.

25.5 Compaction Technologies

Powder compaction relies on pressure to deform and bond the particles, so typically much less polymer is required in shaping the powders. This is the most common means for shaping and densifying a powder. The particle size is large, as densification is attained by pressing the particles at high pressures. After compaction the component must have sufficient strength for handling (3 to 20 MPa), but full strength still requires sintering.

25.5.1 Die Compaction

One of the most common means of forming a shape from powders is by the application of pressure. As shown in Figure 25.7, the increase in density is initially rapid at low pressures, but the powder progressively resists densification as the pores between particles collapse. This graph shows the fractional density versus compaction pressure for a nonspherical copper powder. It also gives the corresponding green strength for each compact, which is important to handling the compacts prior to sintering. As the material hardness increases there is more resistance to densification in compaction.

A schematic view of die compaction is given in Figure 25.8. The powder starts at the apparent density with 4 to 6 contacting neighbors for each particle (coordination number). At this point the powder has no bonding strength. As pressure is applied, the particles rearrange, deform, and bond. Progressively more pressure is required to continue compaction as deformation hardens the particles. Eventually the

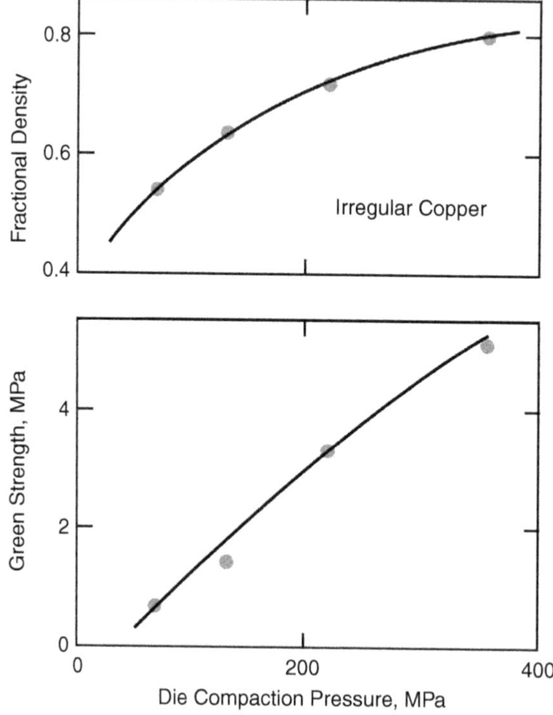

FIGURE 25.7 Compaction data for die pressed irregular (nonspherical) copper powder. The top plot shows the fractional density (based on 8.96 g/cm^3 theoretical density) versus applied compaction pressure and the bottom plot shows the corresponding green strength.

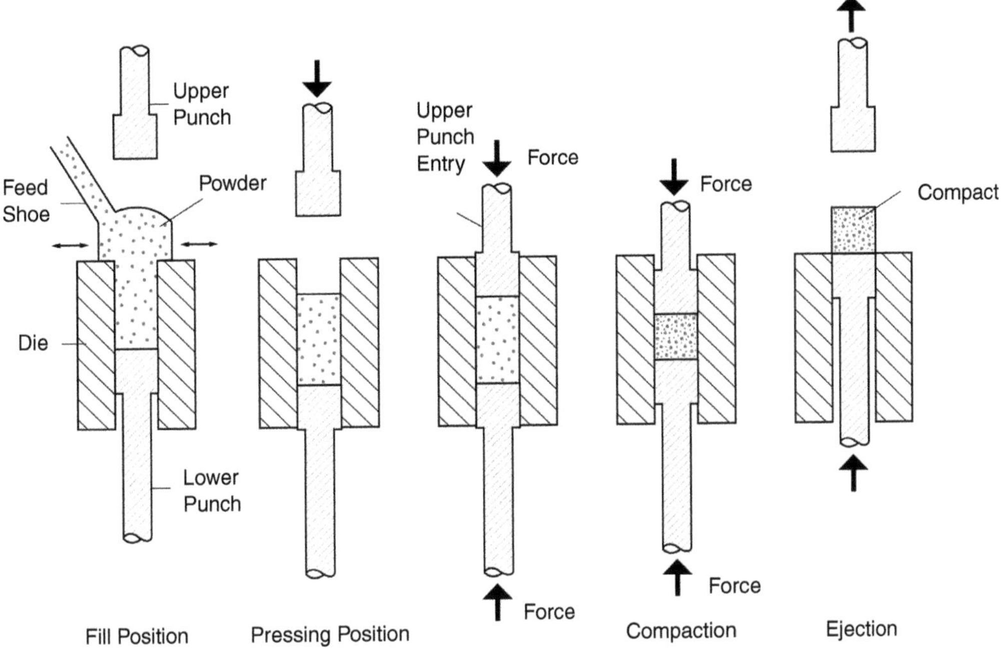

FIGURE 25.8 A schematic view of the events during die compaction, starting with filling the cavity from the powder feed shoe (upper punch is retracted out of the die and lower punch is in the fill position), moving to the pressing position with the lower punch moved down and the upper punch entering the die and pressing against the powder and lower punch. At the end of the pressing cycle the upper punch is retracted and the lower punch is used to eject the compacted powder.

TABLE 25.2 Compaction Pressure Ranges for Some Common Materials

Material	Median particle size, μm	Compaction pressure, MPa	Green density, % theoretical
Alumina	0.5	110–140	45–60
Aluminum	45	150–275	90–95
Brass	65	400–700	85–90
Bronze	85	200–300	85–90
Carbides	1.5	140–400	45–55
Ferrites	0.3	110–165	45–50
Iron	110	480–820	85–92
Stainless steel	65	700–840	85–88
Steel	90	500–820	88–92
Tantalum	1	70–140	30–40
Tungsten	2	240–500	55–60
Tungsten carbide	1	200–350	60–62

material hardens to a point of diminishing return. Depending on the powder and tool material, the peak compaction pressure ranges up to 1000 MPa. Table 25.2 lists a few common powders and typical compaction pressures. Very hard powders and very soft powders tend to be pressed at lower pressures.

Die compaction tooling is designed to apply pressure along a single axis (uniaxial). The progression of motions for such a tool set during pressing was illustrated in Figure 25.8. The die provides the cavity into which the powder is filled and pressed. The upper punch is retracted during powder filling. The lower punch position during powder entry is termed the fill position; it determines the amount of powder

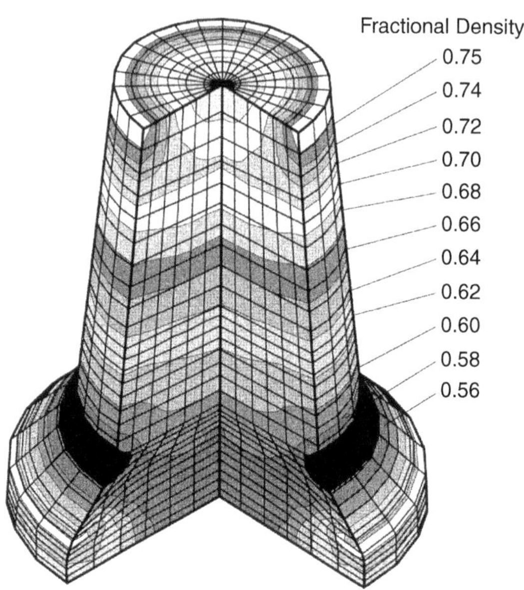

FIGURE 25.9 Computer simulated fractional density contours on a die compacted component, showing how steps and wall friction influence the density distribution. In turn the low-density regions shrink more in sintering, leading to distortion and loss of dimensional precision. For this reason, die compaction techniques are best applied to squat components.

placed in the die. The powder is dumped into the die cavity from an external feed shoe. Any variations in powder flow result in variations in the pressed parts mass and the sintered part size. During filling, the lower-punch position might change to help ensure uniform powder placement throughout the cavity. After filling, the lower-punch drops to a lower position to keep the powder near the center of the die during pressurization. Densification occurs by forced movement of the punches toward the compact center. Finally, the upper punch is removed and the lower punch is used to eject the compact and the cycle repeats. Although the punches are shown as simple geometries, actual shapes can be quite complex. Further, it is common to have core rods positioned within the punches to form holes in the compact. As shown by the density contours in Figure 25.9, the pressed green density can be a complicated function of position, making final dimensional specification difficult.

The compacted powder is termed a green body, so the density after pressing is the green density and the strength after pressing is the green strength. When pressure is transmitted from both the bottom and top punches, the process is termed double-action pressing. A similar effect is possible using a floating die, with the bottom punch held stationary and the die allowed to move while the upper punch applies pressure. When pressure is transmitted from only one punch the compact is less uniform and the process is termed single-action pressing.

After pressing, the green compact is mechanically locked into the die. The force required to push the compact out of the die is called the ejection force. Here the effect of a lubricant is most pronounced. Both the ejection force and die wear decrease with increasing lubrication. Many die compaction options exist, including hard and soft tooling, heating the die and powder, lubricating the powder or just the tooling, and changing the peak pressure, strain rate, number of repeat pressurizations, or dwell time of the pressure. However, there are important extremes in powder compaction — very hard particles that do not deform, and very soft particles that easily deform. Even at intermediate pressures softer particles (lead, tin, aluminum) exhibit higher green densities and can be pressed to near full density at 400 MPa or less. On the other hand, hard carbide, oxide, boride, or nitride powders do not deform in compaction. Hence, it is necessary to use a polymer as a binder to provide strength. The compaction pressures are relatively low and the pressed density is often below 60% of theoretical.

A major difficulty with die compaction traces to die-wall friction. This friction inhibits easy ejection, and more importantly causes density gradients in the green compact. Punch motion against the powder is similar to plowing snow: close to the punch the packing is dense, but far removed from the punch, the powder is unaffected. This pressure decay with distance is because the powder spreads load to the die wall, in the form of friction. As pressure determines green density, pressure gradients become density gradients.

Dimensional change during sintering is inversely proportional to the green density, so regions of low density undergo more size change on heating. Thus, differential green densities from die-wall friction result in warpage during sintering, giving poor sintered tolerances. Consequently, die compaction is best applied to shapes that minimize density gradients, which tend to be squat in the pressing direction. Density gradients are reduced when both the upper and lower punches move toward the compact center using double-action pressing. The level where the lowest density occurs in the compact is called the neutral plane, and in most instances the desire is for the neutral plane to be in the center of the compact.

As the compaction pressure and density increase, the die-wall friction also increases. As pressed density goes up, the wall pressures exceed 50% of the applied pressure. Thus, the higher the applied pressure, the higher the radial pressure and die-wall friction. Additionally, there is friction between the punches, core rods, and other parts of the tooling and compact. All of these contribute to density gradients in the green body.

Alloying increases strength, but at the same time decreases compressibility. In steels, carbon is desirable for sintered strength, but as an alloying addition in iron powder it causes a loss of compressibility. Hence in sintered steels the carbon (graphite) is usually mixed with iron for easy compaction, and dissolved into the iron to form steel during sintering. Thus, when compressibility is a concern, it is common to press a mixture of powders and use sintering to interdiffuse the particles to form the alloy.

25.5.2 Pressing Technology

Compaction involves decisions on the powder mixture, lubricant and lubrication technology, pressurization technology, tool design, and press. Friction between the die wall and the powder during pressing and ejection contribute to density gradients and tool wear. As the compaction pressure increases, ejection becomes more difficult and lubricants are required to minimize die wear. Polymer lubricants come as particles. They are mixed with the powder prior to pressing. The alternative is to apply a lubricant to the tooling between compaction strokes. For example, spray units induce a static charge on the lubricant particles to make them adhere to the die wall. However, die-wall lubrication slows the compaction cycle, because the tool-coating step must be performed prior to powder fill. Common lubricants are based on fatty acids, waxes, and metal stearates, so they tend to be fairly inexpensive. At low compaction pressures the lubricant reduces friction and improves compaction, but at high pressures the lubricant takes up volume and interferes with compaction. In practice, high-viscosity wax lubricants are best, especially when used in low concentrations (0.3 to 0.5 wt.%).

Most die compaction is performed in mechanical presses. The feed shoe places powder in the die cavity. That powder charge is transferred by punch motion to the die center prior to pressurization. As several events are occurring at once, small variations occur in the mass of powder in the cavity. Normally, this is in the ± 0.3 to 0.5% range. Accordingly, dimensions after sintering vary by about one-third this value, ± 0.1 to 0.2%. Uniformity is a problem with multiple-level parts, so one cure is to overfill the die and then push the excess powder out of the die prior to compaction. An alternative is to pull powder into the die cavity by placing the feed shoe over the tool cavity and then dropping the lower punch. Once filled, the powder mass is transferred to the pressing position to allow clearance for entry of the upper punch.

The pressing sequence involves feeding powder into the cavity, pressurizing the powder, and ejecting the compact out of the tooling. Pressures delivered from one direction cause large density and property gradients, especially as the height-to-diameter ratio increases. Double-acting pressurization delivers a more uniform stress to the powder, giving a more uniform product. The simplest case is for only one punch to move, with the die body and lower punch remaining fixed. For flat parts the upper punch is replaced by a fixed platen, a process called anvil pressing. Here, the lower-punch presses the powder against the flat plate covering the die. Anvil pressing is used for the production of simple shapes in high production volumes.

Most die compaction is performed with at least two pressing actions, such as the top and bottom punches moving toward the die cavity center. While punches move toward the die center during pressing, ejection corresponds to the lower-punch pushing to the top of the die body. A common variant is to have a fixed lower punch and a floating die. As pressure is applied to the upper punch, die-wall friction causes the die body to move against a spring. Although the lower punch is fixed, this still creates relative lower-punch motion toward the die center. When the die body moves down over a fixed lower punch for ejection, the process is termed withdrawal.

Uniform powder fill is difficult to achieve for features located near the compact center. For example, to form a flange located at the compact center requires a separate motion to shift the powder into proper location prior to pressurization. To perform this task, the upper punch is segmented and moved during powder transfer, prior to significant pressurization. If pressure is applied prior to powder transfer, then a crack will form where the powder segment is shifted.

Compaction presses come in many sizes. A common measure of a press is based on tonnage. This is the force available to compact a part, and is determined by the compaction pressure times cross-sectional area of the punch faces perpendicular to the pressing direction. Small components are fabricated on low tonnage presses and large components are fabricated on high tonnage presses. Presses can exceed 1000 tons or 9 MN. However, the bulk of powder compaction is performed on presses in the range of 0.5 to 2.2 MN.

A distinguishing factor in compaction presses is the number of part levels. Each distinct height step of more than 10% in the pressing direction requires a separate punch motion. These levels might be on either the upper or lower punch. Fill depth is another limitation. For thick parts, the fill height must be large, but this also increases the press size. Hence, the fill capability is important in selecting a press, especially for larger components.

The pressurization mechanisms vary between presses, and include hydraulic, mechanical, pneumatic, and hybrid systems. Hydraulic presses rely on a central pressure system and servo-valves to proportion hydraulic pressure to the pressing components. They are good at controlling pressure and are best suited to controlled-density pressing. Mechanical presses have a flywheel driven by an electric motor. Cams and levers take power from the flywheel to function the press. They are best at controlling green dimensions. Rotary presses follow the same motions, but use a spinning table with several dies that undergo the compaction cycle on each table rotation. When equipped with 50 tool sets, rotary presses can form 8000 compacts per minute. Anvil presses form simpler shapes, but provide faster pressing rates. In the small components typical to the electronics industry, the small presses rely on electric motors for coordination of the tool motions. Hybrids exist that select certain features from each press type. For example, a mechanical press might have pneumatic pressure buffers on the punches to control the peak pressure: the core technology is a mechanical press, but the upper punch has a pressure cylinder for controlling the final densification. Other designs rely on computer control systems to ensure uniform products.

In terms of press design, the partitioning of technology ranges from single-level presses that use anvils instead of an upper punch, to computer-controlled, multiple-level presses with integrated set-up systems for production of complicated multiple-level components. The computer numerically controlled compaction presses are the most sophisticated and rely on computer simulations to help properly design tooling and set-up operating parameters. They are applied to the most complex shapes.

The design of parts and tools for die compaction requires experience and common sense. The powder characteristics, expected properties, and dimensional tolerances of the compact are essential knowledge. Tooling is designed for long life and minimum wear. As specified tolerances on a component reflect the size after sintering, the green body must be more precise to allow for dimensional variation in sintering. Typical sintered tolerances are in the ± 0.2 to $\pm 0.5\%$ range, and in some instances are even tighter. Tooling tends to be 10-fold more precise.

Certain shapes and features are difficult to die compact. For example, punches with sharp edges and thin sections tend to buckle or bend and are easily damaged. Likewise, undercuts and slots or holes perpendicular to the pressing axis make ejection impossible. Such features are best added by machining after compaction or after sintering. Further, dimensional control is improved if the shape is rounded, smooth, tapered, and squat. Thus, tooling is kept as simple as possible along the axis of pressing. Most die compaction is applied to components with a projected area in the pressing direction of less than 65 cm^2. Very thick and very thin sections are problems, so generally section thickness is between 0.6 and 60 mm. Flat punches, or at least punch designs with less than 10 to 15% steps in component thickness are desirable. Features cannot be too close to each other, so close edge spacing between holes or holes and walls must be avoided.

25.5.3 Cold Isostatic Compaction

Although uniaxial die compaction is a dominant means to press a powder, by no means is it the only approach. Cold isostatic pressing (CIP) is preferred for complex shapes involving undercuts or large length-to-diameter ratios. In the CIP process, powder is sealed into flexible tooling. That powder-tool assembly is pressurized in a vessel filled with a fluid such as oil or water. Lubricants are rarely used. However, wax is added to hard materials to increase the green strength. Compaction pressures up to 1400 MPa are possible; however, cold isostatic pressing is usually performed at pressures below 420 MPa.

The tooling is rubber, latex, polyurethane, or polysilicone. These are selected because they can be dipped or cast around a master mold. If the tooling is soft, then problems arise with it becoming trapped in the pores between particles, so hard and stiff polymers are most popular. The rubber mold can be complex in shape, but must be oversized to compensate for powder compression. Usually, an external support is required to help hold the bag shape during powder loading and in some situations the bag is evacuated to remove air prior to compaction.

The two variants of cold isostatic pressing are the wet and dry bag techniques. With a wet bag, the filled and sealed mold is immersed in a fluid chamber that is pressurized by an external hydraulic system. For example, a hollow tube is formed by pressing the powder (inside a rubber envelope) against a solid core. After pressing, the wet rubber mold (or bag) is removed from the chamber and the compact extracted from the mold and pushed off the core.

The dry bag approach is favored in high-volume production, because the bag is built directly into the pressure cavity. The flexible bag deforms but is not ejected. Two end plugs allow powder loading and component unloading without removing the bag assembly.

Isostatic pressurization is useful for making large, thin, long, or homogeneous compacts that would be impossible to fabricate by die compaction. Isostatic pressures give a slight density advantage over die compaction as there is little die-wall friction. Also, density gradients are small in isostatically compacted components, allowing large sintering shrinkages without distortion. For this reason, it is widely applied to forming large shapes from cemented carbides, tool steels, stainless steels, and many ceramics.

25.6 Sintering

Sintering is the most complicated aspect of powder processing. It is also the most important aspect, as it converts the weak green compact into a high-performance component. Many events occur during the sintering cycle, including polymer burnout, particle bonding, dimensional change, and significant coarsening of the microstructure. After sintering the component must have the desired properties and proper size. About 70% of all sintering cycles involve the formation of a liquid phase, often by melting one of the powders. For example, a mixture of iron, copper, and graphite is used to form a steel (iron-carbon) with a solidified copper phase filling the interparticle voids.

At the atomic level, sintering induces contacting particles to bond together. It can occur at temperatures below the melting point by solid-state atomic transport events, but in many instances involves the formation of a liquid phase (for details on liquid phase sintering, see also Chapter 5, Section 5.2.2). The most important atomic events are surface diffusion (at lower temperatures) and grain-boundary diffusion (at

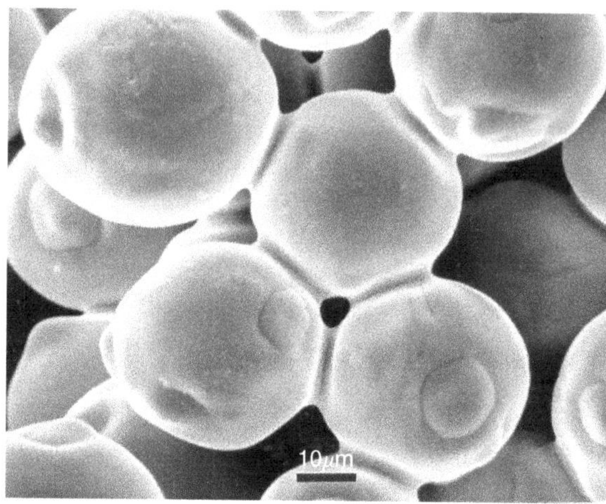

FIGURE 25.10 A scanning electron micrograph of the sinter necks formed by heating loose spherical nickel powder to approximately 1000°C for 30 min in vacuum. Initially the spheres were in point contact, but atomic diffusion induced by the high-temperature exposure resulted in bonding between the spheres. Higher sintering temperatures and longer sintering times result in nearly complete bonding and elimination of pores.

higher temperatures). Grain-boundary diffusion predominantly occurs at the contacts between particles. On a microstructural scale the bonding becomes evident as necks grow between touching particles. Figure 25.10 shows a scanning electron micrograph of the solid-state neck formation between sintering spheres with a grain boundary in the neck. Such neck growth increases the strength over the green strength and causes many beneficial property changes.

Inherently, sintering is due to the motion of atoms that occurs at high temperatures and the reduction in the surface energy associated with small particles. Powder fabrication puts energy into the material by creating surface area or surface energy. Then, in sintering, that energy is eliminated; surface energy per unit volume depends on the inverse of the particle size. Thus, smaller particles with high surface areas have more energy and sinter quickly. However, not all surface energy is available for sintering. For a crystalline solid, nearly every particle contact will evolve a grain boundary with an associated grain-boundary energy. So as neck growth removes surface energy, it adds grain-boundary energy. Obviously, this only occurs when the decrease in surface energy is greater than the increase in grain-boundary energy.

From a fundamental view, sintering is treated in terms of driving forces, mechanisms, and stages:

- Sintering driving forces describe the microscopic curvatures that cause bonding
- Sintering mechanisms describe the path of atomic motion in response to the driving forces
- Sintering stages describe the geometric progress resulting from the atomic motion; those stages in turn change the driving forces.

The driving forces or sintering stresses come from the curvatures in the microstructure. Sintering mechanisms are usually diffusion processes over the surfaces, along the grain boundaries, or through the crystalline lattice. The stages of sintering help describe the driving force and kinetics, and are used to mathematically model the process.

Sintering converts the green powder compact into a product with the desired size, shape, and quality. Figure 25.11 provides a sequence of five optical micrographs showing progressive particle bonding, pore elimination, and microstructure coarsening during sintering for a stainless steel powder (images are from quenched samples captured at 1000, 1100, 1200, 1300, and 1365°C during heating at 10°C/min). The geometric changes in sintering result from the increased atomic motion that accompanies heating. As temperature increases, atoms have more vibrational energy and jump to neighboring atomic sites with a frequency that approaches thousands and even millions of jumps per second. Thus, temperature

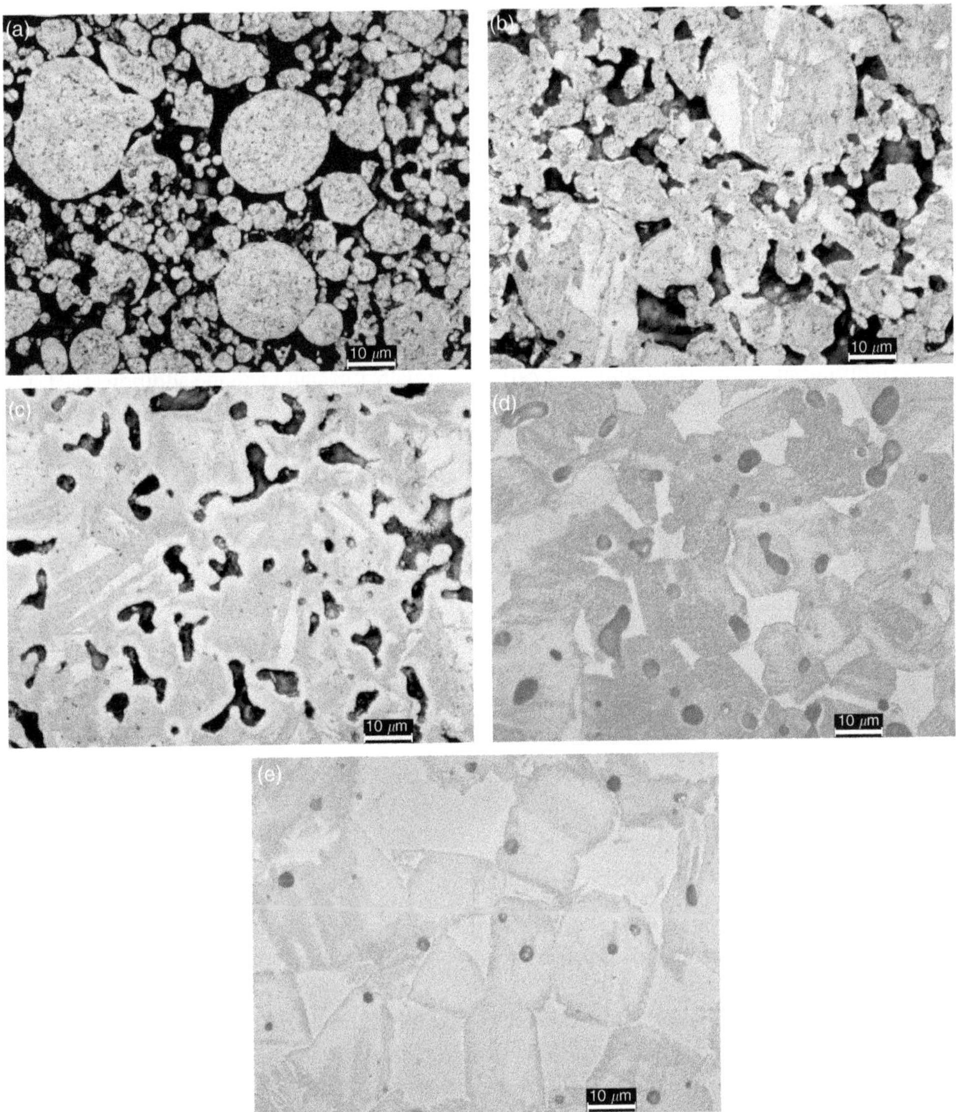

FIGURE 25.11 Five micrographs taken from a sequence samples showing progressive particle bonding, pore elimination, and microstructure coarsening during sintering for a stainless steel powder. Images a to e are from quenched samples captured at 1000, 1100, 1200, 1300, and 1365°C during heating at 10°C/min.

has a dominant effect on the rate of sintering. In defining a sintering cycle, the first consideration is what temperature is required to induce the desired level of sinter bonding. Along with particle bonding, sintering cycles have secondary objectives such as oxide reduction.

25.6.1 Sintering Atmospheres

The sintering atmosphere's most important function is to control the high-temperature chemical reactions. Further, the atmosphere helps in removing lubricants or binders, transferring heat to the compacts, extracting contaminants, and controlling the final chemistry such as the carbon and nitrogen levels.

The powder compact chemically interacts with the surrounding atmosphere during heating; in some cases the compact absorbs species from the atmosphere. Carburization and nitridation reactions are examples where the atmosphere adds mass to the component. The opposite reactions, such as oxide reduction, result in the transfer of species to the atmosphere. Thus, oxidation–reduction, carburization–decarburization, and similar reactions are an integral part of sintering.

The atmosphere composition depends on the material and desired chemical reactions. Oxide ceramics are often fired in air, but most high-performance materials require protective atmospheres; this is true even for high-performance oxides. Metals require protection from oxidation during sintering, and adsorbed oxygen and moisture must be removed from the green compact. A reducing atmosphere is often required for sintering metals. Even for metals there is no best atmosphere, as metals such as titanium and tantalum are very reactive with reducing gases like hydrogen, while stainless steels are very responsive to hydrogen. Besides hydrogen, reducing reactions can form carbon dioxide using carbon (graphite) or carbon monoxide. Similar to reduction, atmospheres are used to add or subtract species such as carbon or nitrogen during sintering. The addition of carbon is usually through methane additions (CH_4) while nitrogen is usually generated from molecular nitrogen N_2 or occasionally ammonia NH_3. These reactions are controlled with great accuracy, allowing final chemical specifications to be as tight as $\pm 0.05\%$.

Atmospheres for sintering include air, nitrogen, argon, oxygen, hydrogen, and various gas mixtures. Because of the need to control all chemical reactions, there is necessary concern with the impurity levels in the atmosphere. The concentration of various species such as oxygen, carbon monoxide, methane, carbon dioxide, and water vapor determine what happens during sintering. Several atmosphere conditions are possible:

- Oxidizing (carbon dioxide, water, or oxygen)
- Neutral (argon, helium, or vacuum)
- Reducing (hydrogen or carbon monoxide)
- Hydriding (hydrogen or ammonia)
- Dehydriding (vacuum or argon)
- Nitriding (nitrogen or ammonia)
- Carburizing (methane or propane)
- Decarburizing (carbon dioxide, water, or oxygen)

These are not exclusive. For example, it is possible to have a combination such as hydrogen and methane that is carburizing and reducing at the same time. More details about nitriding and carburizing may be found in Chapter 18.

Proper control and manipulation of the atmosphere provides an opportunity to change the degree of sintering and the material chemistry; however, the atmosphere also changes during sintering. This is because powder compacts carry contaminants into the furnace. In some situations the sintering atmosphere plays an active role in forming carbides, nitrides, oxides, or hydrides.

25.6.2 Polymer Burnout

An early task during heating to the sintering temperature is to extract the polymer used in shaping. If the polymer is carried into the high-temperature portion of the furnace, then the impurities will alter the compact chemistry, properties, and dimensions. During heating, the carbon level from the polymer progressively falls with higher temperatures. Most polymers thermally decompose into small molecules such as methane, ethane, butane, propane, carbon monoxide, carbon dioxide, or water. This decomposition starts as low as 150°C, usually peaks from 350 to 450°C, and is usually completed by 550°C. Some polymers (especially gelation binders such as cellulose) char into graphite, which is a contaminant if not removed using oxygen. The atmosphere has a role in polymer burnout as it provides reaction species for the polymers. Oxygen-containing species (O_2, CO_2, H_2O) help remove carbon polymer contaminants in the 600 to 800°C range. A flowing atmosphere sweeps away the burnout products.

Unfortunately, for some materials, carbon in the polymer will react with the powder and cannot be removed. For example, titanium reacts with binders and lubricants to form titanium carbide at temperatures over 350°C. Polymers with low decomposition temperatures are required in such situations. Besides carbides, improper burnout gives carbon contamination that might be accompanied by sooting, surface blemishes, or discoloration. Sooting happens with rapid heating and insufficient oxidant. The problem is more evident with higher furnace loadings, low atmosphere flow rates, and large or high-density components.

Slow heating is the best practice for removing a polymer. Further, the compact should reach at least 650°C before entering the high-heat portion of the furnace. A high atmosphere turnover (at least 10 times per hour) helps remove the burnout products. If there are no nitride or hydride reactions, a hydrogen-nitrogen atmosphere has generally proven most effective during polymer burnout. The nitrogen provides the flow while the hydrogen reacts with the decomposing polymer. Oxygen might be added to this atmosphere to assist in polymer burnout.

25.6.3 Sintering Furnaces

The sintering furnace controls the time-temperature path in the sintering cycle. Additionally, it holds the atmosphere, provides for removal of the lubricants and binders, and potentially performs post-sintering heat treatment. The furnace performs those functions in either a batch or continuous mode. The difference between a batch and continuous furnace depends on control of either the furnace temperature versus time or compact position versus time.

A batch furnace is loaded with the material to be sintered and then its temperature is cycled over several hours. As each cycle can be programmed differently, batch furnaces have a high level of flexibility. Further, vacuum sintering and pressure-assisted sintering are only performed in batch furnaces. Usually the furnace walls are kept cold via reflective heat shields around the working zone and water cooling on the outside. An alternative design is a hot wall where the heating elements are outside a retort that contains the sintering material and protective atmosphere. The work can be placed in the hot zone from the front, top, or bottom using various elevators, hoists, or lift-trucks.

A continuous furnace controls the component position versus time in a sequence of zones using a conveyor, such as a belt or pusher. Usually, the conveyor is a limitation in the furnace operating temperature. For lower temperatures the conveyor is fabricated from a wire mesh. Higher temperatures require a ceramic, graphite, or refractory metal belt or pusher. With refractory materials or graphite, maximum temperatures over 2000°C (3632°F) are possible. The higher the operating temperature, the lower the loading possible, so expense increases dramatically as sintering temperature climbs.

The first zones in a continuous furnace initiate polymer burnout and contaminant removal. Atmosphere composition and flow are adjusted in each zone to control reactions and to push contaminants away from the high temperature zone. After a transition region, the high heat region takes the component to the sintering temperature. The length of the high heat region is sized to ensure proper dwell time at the sintering temperature. Cooling takes place in the final zones where the compact is often subjected to a high gas flow. By selective placement of gas inlets and outlets, each furnace zone may contain a different atmosphere, allowing tailoring of the chemical reactions during sintering. Usually the furnace heating elements are external to the atmosphere, with heat radiating through a muffle.

25.6.4 Sintering Cycles

In practice, about 70% of all sintering involves a liquid phase. From a technical view, exceeding a minimum temperature is mandatory to form the liquid. Additionally, sintering cycles are designed to adjust product chemistry, homogenize mixed powders, eliminate porosity, and develop the desired microstructure.

Table 25.3 gives a few industrial sintering cycles. It includes information on the material, particle size, heating rate, peak temperature, hold time, and atmosphere. This summary illustrates the diversity of sintering cycles. Further, sintering cycles vary from site to site, in part reflecting differences in the material, powder sources, polymer addition, green body, furnace design, and final product specifications.

One difficulty in sintering comes from the different vapor pressures of materials. In some sintering cycles, one of the ingredients will evaporate. For example, in sintering brass, it is common to lose zinc,

TABLE 25.3 Example Production Sintering Cycles

Material	Particle size μm	Green density %	Atmosphere	Dew point °C	Heating rate °C/min	Peak temperature °C	Time min	Dimensional change %
Aluminum	60	90	N_2-H_2	−20	10	600	20	−2
Alumina Al_2O_3	0.4	58	H_2	−40	18	1830	60	−17
Bronze Cu-10Sn	75	70	N_2	−40	15	820	5	−2.2
Copper	32	56	H_2	−40	2	1045	120	−17
Fe-2Ni	5	64	H_2	−20	15	1250	60	−15
Fe-2Ni-0.5C	80	90	N_2-H_2	−40	5	1200	60	−0.04
Ni_3Al Intermetallic	20	70	H_2	−50	10	1340	60	−10
Silicon carbide	1	55	vacuum	—	10	2100	120	−17
Stainless steel 316L	44	62	H_2	−40	5	1385	90	−16
Steel Fe-2Cu-0.8C	100	88	N_2-H_2	−20	10	1120	30	0.0
Steel 4640	80	89	N_2-H_2	−40	5	1120	30	0.1
Tantalum	0.3	35	vacuum	—	15	1350	30	−3
Tantalum	2	62	vacuum	—	10	2400	120	−10
Titanium	50	75	vacuum	—	5	1400	120	−8
Tungsten	3	60	H_2	−10	70	2800	45	15
W-Cu	3	65	H_2	−40	10	1350	60	−14
W-7Ni-3Fe	3	40	H_2	18	5	1500	120	−25
WC-10Co	0.5	52	vacuum	—	8	1400	60	−19

leaving behind a copper-rich surface. Materials containing lead suffer the same problem, and sometimes compounds are the problem, as with the high vapor pressure of molybdenum oxide.

25.7 Full-Density Processes

Pores degrade the properties of materials. As a consequence, it is often necessary to ensure the total elimination of porosity to maximize final properties. This is especially true in high-performance components such as jet engine turbines, biomedical implants, and military systems. A wide variety of full-density powder processes have emerged that generally apply pressure during sintering. In general, these full-density powder techniques work because

- Most materials become softer as temperature increases
- Most materials become ductile as temperature increases
- At high temperatures most materials deform without hardening

As an example, think of common glass. At room temperature it is hard, brittle, and resists deformation; yet it is easily formed at low pressures when heated. In the same manner, elevated temperatures aid powder densification because most materials exhibit increased workability. Thermal softening at high temperatures greatly lowers the pressure required to densify the powder. Some information on sintering processes may be also found in Chapter 7.

Full-density techniques represent various combinations of pressure and temperature in the following clusters:

- Low-stress routes that operate at high temperatures, such as sintering, that are dominated by diffusion controlled processes
- Intermediate-stress routes that operate at intermediate temperatures via diffusional creep processes, such as hot isostatic pressing
- High-stress routes that operate at lower temperatures via plastic flow, such as powder forging
- Ultra-high-stress routes that attain full density at room temperature, such as explosive compaction

There are many additional factors, such as particle size, green density, and component size, that separate the processes, but this simple categorization helps sort out the key parameters for densification. Performance levels can be exceptional if control is gained over defects, impurities, microstructure, and product homogeneity. That is the primary justification for full-density product. Most engineering properties are improved by densification; indeed the full-density properties can exceed those possible via traditional metalworking routes. When densification is properly performed, there is no defect in the microstructure larger than the particle size, so microstructure control promotes significant performance gains when compared with other fabrication routes. Property gains can be seen in strength, hardness, ductility, fracture toughness, optical transmission, and wear resistance. Beyond property gains, property uniformity is another reason to use pressure-assisted sintering. For example, in the traditional fabrication of tool steels, residual inclusions tend to be aligned in the microstructure, resulting in anisotropic properties. When powders are used to form full-density tool steels, the properties are higher and isotropic. The strength changes with ingot diameter for both the transverse (perpendicular to rolling direction) and longitudinal (along rolling direction) orientations. On the other hand, the hot isostatic pressing gives higher and more uniform strength. Those attributes translate to higher-quality components. The lower a material's fracture toughness, the more sensitive it is to microstructure defects and residual pores, so property uniformity is very important to ceramics and low-ductility metals like tool steels.

25.7.1 Hot Pressing

Hot pressing is a manifestation of stress-enhanced densification. Similar to die compaction, hot pressing is performed in a rigid die using uniaxial pressurization as sketched in Figure 25.12. Graphite dies allow induction heating, but may contaminate the material. Other die materials include refractory metals such as molybdenum alloys, and ceramics such as alumina or silicon carbide. In those cases external heaters are required to heat both the die and compact.

During hot pressing the initial densification is by particle rearrangement and plastic flow at the particle contacts. Once the effective stress acting at the particle contacts falls below the *in situ* yield strength,

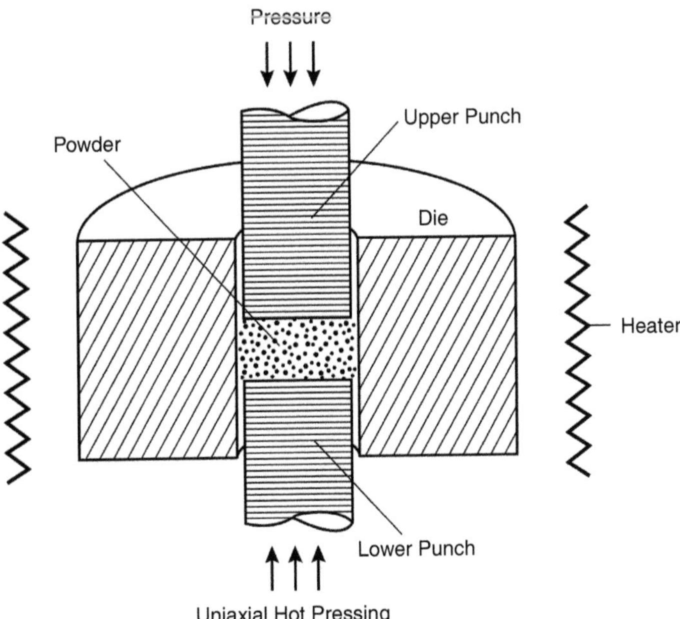

FIGURE 25.12 Uniaxial hot pressing relies on a graphite die and punches with external heating. The powder contained in the die is heated and pressed simultaneously; the heat softens the powder and the pressure squeezes the porosity out of the compact.

then further densification depends on diffusion rates. Temperature is a critical factor, and small grain sizes aid densification. Hot pressing cycles are slow when compared to die compaction, often with cycle times measured in hours because of the large thermal mass. Maximum temperatures depend on the die material and range up to 2200°C (4000°F) and maximum applied pressures are 50 MPa. Vacuum is often selected for the process environment to minimize contamination of the compact. Consequently, hot pressing equipment can be expensive. However, hot pressing is widely used to fabricate unique hard and brittle compositions. One large use is in the consolidation of diamond–metal composite cutting tools.

25.7.2 Spark Sintering

This technique, related to hot pressing, uses direct resistance heating through the punches, die, and powder bed during consolidation. Early demonstrations used a welding arc to sinter refractory metal powders with hydraulic loading to ensure electrical continuity. The electrical current induces self-heating, thereby eliminating the need for a furnace. Current densities range up to 300 A/cm^2. The peak temperature is limited by the available power, but 2700°C has been attained. To sustain electrical contact and to induce densification, the contacts are pressurized. Over the years many variants have arisen, leading to names such as spark plasma sintering, electro-consolidation, and field-activated sintering.

In one version of spark sintering the powder is contained in a punch and die assembly, typically constructed from graphite. Current is passed from the punches though the powder compact with a superimposed high-frequency alternating current. Pulsing the current helps remove contaminants, and very high frequencies and high current densities induce rapid densification. Pressures up to 100 MPa are applied as the compact is heated. The process times are short and are often measured in minutes. In production facilities with an automated rotary table for feeding new dies into the press, cycles of just 30 s have been demonstrated. One key advantage is that the component can have functional gradients when different powders are placed in different locations, yet still be consolidated in a single event.

25.7.3 Hot Isostatic Pressing

A popular means to compact powder to full density is hot isostatic pressing, often abbreviated HIP. Unlike hot pressing and spark sintering, which apply pressure along one axis, hot isostatic pressing applies pressure from all directions simultaneously. This gives less particle–particle shear. One consequence is that surface films on the particles can remain as prior particle boundary decorations that degrade properties, even when full density is achieved. Also, the compact surface can be contaminated by the container and needs to be removed after HIP by chemical dissolution, machining, or abrasion. This adds to the cost and makes HIP less of a net-shape technology; thus, it is often referred to as a near-net-shape process.

Figure 25.13 shows a schematic of the hot isostatic pressing sequence. To directly compact a powder requires encapsulation in a gas-tight container. The container may be fabricated from any material that is soft and deformable at the consolidation temperature, such as glass, steel, stainless steel, titanium, or tantalum. Prior to HIP, the container is filled with powder and heated under vacuum to remove volatile contaminants. After evacuation and degassing, the container is sealed. Failure to adequately degas the powder will lead to thermally induced porosity; after HIP when the component is exposed to a high temperature the collapsed gas-filled pores reform in what was a full-density compact. In the HIP step there is heating and pressurization, usually with the heating and pressure cycles being independently programmed to soften the container prior to powder densification. If a powder compact is sintered to less than 8% porosity, preferably less than 5%, then the pores are sealed and no container is required for densification, a process termed containerless HIP.

Powder consolidation takes place in a heated pressure vessel. High-pressure gas, such as argon or nitrogen, is used to transfer heat and pressure to the compact. Beyond initial particle rearrangement and plastic flow at the point contacts, most of the densification is by creep processes. Temperatures up to 2200°C and pressures up to 200 MPa are possible using HIP. Chambers come in various sizes up to 1.5 m

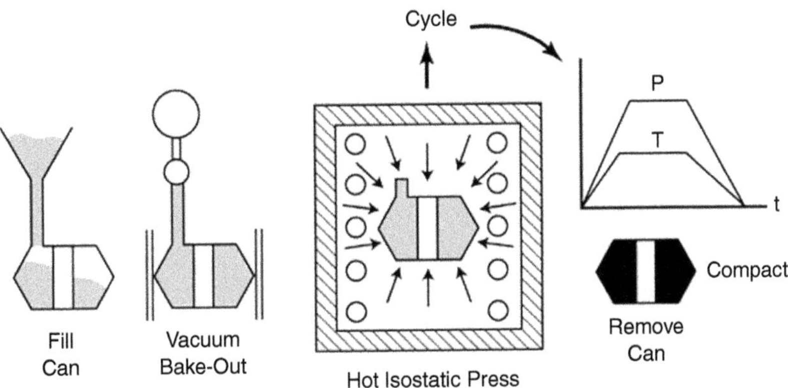

FIGURE 25.13 Hot isostatic pressing involves loading a powder into a preformed can, heating and evacuating the powder, sealing the can, and then simultaneous heating and pressurization to consolidate the powder into a dense body. The final step is removal of the can from the compact.

diameter and 2.5 m high. Cycle times run from 8 to 1 h, the latter being performed in repetitive operations such as in the production of tool steels.

The input material does not necessarily have to be fully alloyed. One success has been to reactively synthesize a material during the HIP cycle. Elemental powders are mixed and heated. During the heating, one powder might melt or react, generating a product phase that is simultaneously densified by pressure. For example the $TaAl_3$ intermetallic compound can be fabricated from a mixture of tantalum and aluminum powders. These reactive hot isostatic pressing (RHIP) cycles usually give more homogeneous products if pressurization is delayed until after the reaction.

There is much interest in faster HIP cycles. Approaches include adding HIP onto vacuum sintering cycles to employ containerless HIP for rapid densification at the end of sintering. Another variant used for large bodies is to heat the powder in the container outside the HIP. Then, as large bodies cool slowly, only a pressure cycle is employed to press the hot billet to full density. This is common for tool steel preforms that see subsequent forging or rolling. Another idea is to heat the powder compacts in a sealed pressure vessel and introduce a liquid that vaporizes to generate a pressure pulse. Liquid nitrogen will produce rapid pressurization when injected into a heated HIP container. The pressure pulses exceed the material yield strength, so densification occurs by plastic flow instead of creep. The lower pressure and longer time required with a conventional cycle results in lower final properties and more microstructure coarsening.

As already mentioned, because consolidation in HIP is hydrostatic and there is little shear on the particle surfaces, contaminated prior particle boundaries can exist even in full-density compacts. Such boundaries are weak and limit properties. Accordingly, for some applications it is necessary to subject the HIP compact to post-consolidation deformation. When properly executed, HIP provides a near-net-shape compact with full density and minimal microstructural defects or gradients. Thus, it is ideal for large, complex components intended for high-performance applications.

25.7.4 Pressure-Assisted Sintering

A closed pore condition is attainable with many materials through sintering to about 95% density. However, the rate of sintering densification slows with the elimination of porosity. One approach to enhanced sintering is to backfill the sintering furnace with pressurized gas to assist in the final elimination of pores. Initially, sintering is performed under vacuum to ensure a closed pore condition. After pore closure at about 95% density, the hot furnace is pressurized to squeeze out the remaining pores. Typical pressures are 5 to 15 MPa.

Obviously the furnace must be designed for both vacuum and pressure operation. In production the operating cost is about half that of vacuum sintering with a separate HIP treatment. By avoiding high temperatures and pressures the microstructure is more uniform. In processing WC-Co compositions with the higher pressures associated with HIP, the softer cobalt is preferentially pushed into the pores. In the sintered microstructure these appear as lakes of cobalt. The lower pressures in sinter-HIP treatments avoid this form of segregation and promote more uniform microstructures and properties.

A variant on the sinter-HIP process would be to remove the compacts from the vacuum sintering furnace for subsequent hot isostatic pressing in a separate cycle without encapsulation; a process also known as containerless hot isostatic pressing.

25.7.5 Powder Forging

Room-temperature deformation of a powder to full density requires pressures far higher than practical. At high temperatures, materials are softer and deform more easily. Hot pressing and powder forging are both uniaxial hot compaction techniques, but they differ in two major aspects. Hot pressing usually starts with a loose powder and hot forging usually starts with a sintered preform. Further, hot pressing has full lateral constraint, while powder forging starts without lateral constraint, causing the preform to bulge. This bulging phase is termed "upsetting" where the vertical force causes the compact to push against the wall. Bulging is possible as the sintered preform has some strength and will not crack if the radial strains are not too large.

Usually, the powder is pressed into a preform, sintered to 75 to 90% density, then hot forged to final size. In the forging step there are many variants in the lateral constraint and strain rate. If the preform weight is closely monitored, then forging provides a dense compact close to the final shape. When contrasted with wrought forging, powder forging operates with lower tooling costs, fewer steps, and less scrap. Die compaction and sintering provide the preform and significantly influence the success of the forging operation, but hot powder forging delivers the density and performance.

The behavior of the porous preform is a main concern in powder forging. Although the mass remains constant, the material volume decreases with pore collapse. The pore collapse is significantly different from that encountered under the hydrostatic conditions used in HIP.

Die-wall constraint is important in determining the actual stress and strain conditions. If properly performed, at the end of the forging stoke the powder totally fills the die and has achieved full density. Pores experience a higher shear during powder forging, leading to improved pore collapse and properties versus isostatic compaction. This difference in pore collapse contributes to improved bonding between the particles with hot forging. Both processes give total pore elimination when properly executed.

Lubricants have a large effect on forging. Without lubrication, the forged powder exhibits low-density regions because of drag on the punch or die wall. Also, friction causes circumferential tensile stresses as flow occurs in the die. The combination of poor lubrication and excessive strain without constraint will lead to cracking. Preform design is important to not exceeding the material limits. Major height strains (typically over 50%) ensure pore elimination and excellent particle bonding, but care is required to prevent buckling, cracking, or low-density regions. A fracture limit occurs with respect to total straining and a large initial height can result in buckling. Thus, there is also an upper limit to the height-to-diameter ratio of the preform. Also, the mass of the preform must be sufficient to fill the part shape at full density. All of these constraints are now captured in process design software to expedite preform and process design.

Temperature determines the stress necessary to achieve densification. Typical ferrous forging operations do not exceed 1200°C, and range closer to 800°C to extend tool life. Production rates are in the range of four forgings per minute. In forged steels, strength is significantly improved at the expense of ductility. More importantly, the fatigue properties are often superior to wrought products. Unfortunately, inclusions seriously degrade properties. To prevent inclusions, the preform is heated in a protective atmosphere and moved by a robot from the furnace to forging die in just a few seconds. Automotive connecting rods are one product forged from powder preforms. They are fabricated to 100% density, with less than 0.2 vol.% inclusions and less than 300 ppm oxygen. Forged connecting rods deliver yield strengths of 710 MPa, tensile strengths of 1000 MPa, fracture elongation of 15%, and a fatigue endurance limit of 400 MPa.

Other applications are found in truck and automobile transmission, engine, and brake systems. Although hot forging and preform design are challenging, the property gains from the process are very attractive.

A variant of powder forging is practiced in ceramics at very slow strain rates. Because of the limited plastic flow even for hot ceramics, the stresses are kept much lower and cycle times become long. Here the deformation looks like forging, as stress is applied along one axis, but diffusional creep is the controlling mechanism.

25.8 Summary

The conversion of powders into useful engineering components is a growing technology that is applicable to essentially all materials. Three processes give an idea of the forming technologies — die compaction followed by sintering, injection molding followed by sintering, and hot isostatic pressing. These techniques have been undergoing expanded use for the past few decades and continue to find new applications, material combinations, and property gains. Often the user community is unaware of the properties and benefits possible via powder techniques. More important, because powder techniques are still relatively new, there are opportunities for new discoveries. The opportunities are enormous, the time is short, the tasks ahead are large, but there is no shortage of growth areas.

Further Reading

Alman, D. E. and Newkirk, J. W. (eds.), *Powder Metallurgy Alloys and Particulate Materials for Industrial Applications*, The Materials, Metallurgical, and Minerals Society, Warrendale, PA, 2000.

Boothroyd, G., Dewhurst, P., and Knight, W., *Product Design for Manufacture and Assembly*, second edition, Marcel Dekker, New York, 2002.

Bose, A., *Advances in Particulate Materials*, Butterworth-Heinemann, Boston, MA, 1995.

Bose, A. and Eisen, W. B., *Hot Consolidation*, Metal Powder Industries Federation, Princeton, NJ, 2003.

Chiang, Y. M., Birnie, D., and Kingery, W. D., *Physical Ceramics*, John Wiley and Sons, New York, 1997.

German, R. M., *Sintering Theory and Practice*, John Wiley and Sons, New York, 1996.

German, R. M., *Powder Metallurgy of Iron and Steel*, John Wiley and Sons, New York, 1998.

German, R. M., *Powder Injection Molding — Design and Applications*, Innovative Material Solutions, State College, PA, 2003.

German, R. M. and Bose, A., *Injection Molding of Metals and Ceramics*, Metal Powder Industries Federation, Princeton, NJ, 1997.

James, P. J. (ed.), *Isostatic Pressing Technology*, Applied Science, London, UK, 1983.

Kubicki, B., *Sintered Machine Elements*, Ellis Horwood, New York, 1995.

Lee, P. W., Trudel, Y., Iacocca, R., German, R. M., Ferguson, B. L., Eisen, W. B., Moyer, K., Madan, D., and Sanderow, H. (eds.), *Powder Metal Technologies and Applications*, ASM Handbook volume 7, ASM International, Materials Park, OH, 1998.

Mostaghaci, H. (ed.), *Advanced Ceramic Materials*, Trans Tech Publications, Zurich, Switzerland, 1996.

Mutsuddy, B. C. and Ford, R. G., *Ceramic Injection Molding*, Chapman and Hall, London, UK, 1995.

Roberts, P. R. and Ferguson, B. L., Extrusion of metal powders, *Int. Mater. Rev.*, 36, 62, 1991.

Schneider, S., *Engineered Materials Handbook Volume 4 Ceramics and Glasses*, ASM International, Materials Park, OH, 1991.

Upadhyaya, G. S., *Sintered Metallic and Ceramic Materials Preparation, Properties and Applications*, John Wiley and Sons, New York, 2000.

26

Layer-Based Additive Manufacturing Technologies

Brent E. Stucker and
G. D. Janaki Ram
Utah State University

Abstract

Additive manufacturing involves the manufacture of complex-shaped parts layer-by-layer, in an additive fashion, directly from a computer-aided design model of an object. The introduction of additive manufacturing technologies has revolutionized product design, development, and manufacturing. The fundamental advantage of these technologies lies in their ability to rapidly produce complex-shaped parts without special tools, molds, or dies, in materials ranging from polymers to metals to ceramics, including multimaterial, functionally gradient and composite combinations as well. Virtually every industry, from consumer products to aerospace, has benefited from these technologies, resulting in better, faster, and cheaper products.

In this chapter we provide an overview of additive manufacturing technologies. We begin with a step-by-step description of a basic additive manufacturing process chain, followed by a useful classification of additive manufacturing technologies into categories. In the subsequent sections, additive manufacturing processes are discussed in detail within their respective categories. The emphasis is placed on process- and category-specific details rather than manufacturer- or machine-specific details. Since further advancements in the field are heavily dependent on an in-depth theoretical understanding of the underlying process principles, an entire section is dedicated to describe modeling efforts in additive manufacturing technologies. After introducing the technologies, various current and emerging applications of additive manufacturing technologies are discussed. The chapter ends with a discussion of future directions, which includes a few specific areas where more improvements must occur in order to attain a wider utilization of additive manufacturing technologies.

26.1 Introduction

26.1.1 Background

Additive manufacturing technologies are a relatively new group of technologies, developed since the late 1980s, wherein complex-shaped parts are built layer-by-layer in an additive fashion directly from a three-dimensional (3D) computer-aided design (CAD) model without using special tooling, molds, or dies. While these technologies were originally developed to produce prototypes, advancements in the field have made it possible to go beyond prototyping to direct manufacture of end-use components in a wide range of materials. A number of terms are in use to describe these technologies, including rapid prototyping (RP), additive manufacturing, and solid freeform fabrication.

More than 40 additive manufacturing processes have been developed and many more are under development. The basic approach for all these processes is the same, although specific process details can vary widely. The process begins with generating a 3D CAD model of the part to be built and slicing it into horizontal cross sections. These cross sections are systematically created and stacked in such a way that they form a 3D object. The introduction of these technologies has ushered in a new era in product design, development, and manufacturing. Knowledge of these technologies and their effective application is now vital for practicing manufacturing and design professionals.

26.1.2 History

Developments in several basic technologies in the 1960s to 1980s, most notably in the fields of lasers, materials, and computing technologies, including CAD and computer-aided manufacturing (CAM), set the stage for additive manufacturing. Charles Hull patented a breakthrough process which he coined "stereolithography" (SLA), in 1986 for automated manufacture of accurate plastic prototypes using an ultraviolet (UV) laser and photo-curable liquid polymers.[1] The process was commercialized soon thereafter. By the early 1990s, a number of new RP processes had been successfully commercialized, which included selective laser sintering (SLS), laminated object manufacturing (LOM), solid ground curing, and fused deposition modeling. Early materials were primarily polymeric and "RP" was coined as a generic term to describe the technologies. The use of RP parts as patterns for rubber molding, investment casting, and other pattern-based replications processes soon became a useful application for these technologies. In the mid-to-late 1990s, a major push was made in the area of generating tooling for injection molding and other mold-based mass-production processes, leading to development of a number of processes that involved using an additively manufactured part somewhere in the process chain of making a tool.

Since the introduction of additive manufacturing technologies, fabrication of end-use metallic components was a generally recognized need. In the late 1990s, laser cladding-based metal fabrication technologies became successfully commercialized. And, in parallel efforts, several more metal RP approaches were successfully commercialized, including laser and electron beam melting of powder beds and ultrasonic consolidation of metal foils. The market for additive manufacturing is now diverging into development aimed at creating low cost machines for rapidly producing prototypes (3D printers) and developments of higher-end machines that are capable of making functional parts accurately and repeatably (rapid manufacturing machines). This divergence will likely continue as these technologies mature.

26.1.3 Advantages of Additive Manufacturing

Additive manufacturing technologies offer numerous advantages for product design, development, and manufacture. Some of these advantages are highlighted below in each of the four major application areas of additive manufacturing.

26.1.3.1 Rapid Prototyping

The advantages of having a prototype prior to actual component manufacture are well known.[2] Prototypes help reduce the cost and time of product development by providing quick, easy, and relatively inexpensive feedback early in the design cycle. Additive manufacturing techniques have the following advantages over competing technologies for prototype fabrication: (1) few restrictions on form and shape, (2) rapid turnaround times for complex shapes, (3) reduced need for human interaction, (4) flexibility in terms of materials, and (5) efficient material utilization.

26.1.3.2 Rapid Tooling

Additive manufacturing technologies can produce various kinds of tooling (dies, molds, mold inserts, patterns, etc.) in a wide range of materials that can be directly used in conventional manufacturing operations. Some of the benefits of additive manufacturing as applied to rapid tooling are: (1) speed of fabrication, (2) ability to fabricate tools from multiple materials, and (3) use of conformal cooling channels.

26.1.3.3 Rapid Manufacturing

Additive manufacturing offers exciting possibilities for the manufacture of end-use components. When additive manufacturing techniques are used to manufacture parts, the need for tooling is eliminated, and along with it the typical restrictions of design for manufacture and the cost and time associated with creating the tooling. Some of the advantages of additive manufacturing as applied to rapid manufacturing include: (1) parts of virtually any geometrical complexity can be produced without tooling, (2) individually customized products can be produced based on digital input data, (3) designers can easily combine parts, leading to reduced assembly and inventory, (4) certain additive processes offer significant microstructural benefits (fine grain size, finer and uniformly distributed second phase particles, reduced segregation, etc.), leading to superior mechanical properties, (5) parts can be built in materials that are difficult to process using conventional methods, and (6) parts can be built as multimaterial and functionally graded material products.

26.1.3.4 Repair and Feature Addition

Additive manufacturing techniques facilitate accurate repair of defective or service-damaged components by adding precise amounts of material using very low and extremely localized heat inputs. Because the heat inputs are very low, the heat-affected zones (HAZs) are extremely narrow and repairing operations do not damage the bulk part. With additive manufacturing technologies it is possible to add features (e.g., a rib) to existing parts. This allows for economical fabrication of parts with significantly reduced manufacturing complexity and efficient material utilization.

26.2 Additive Manufacturing Process Chain

Most additive manufacturing processes employ the same basic process chain steps. These steps are: (1) CAD model construction, (2) STL file generation, (3) build file creation, (4) part construction, and (5) postprocessing.

26.2.1 CAD Modeling

Additive manufacturing requires 3D digital data as its input, typically a valid 3D "solid" model. 3D solid modeling is a field that is rapidly advancing, making it possible to generate highly complex models accurately and quickly. For an existing part for which no technical data is available, the data necessary for generating a CAD model can be created using reverse engineering methods.[3]

26.2.2 STL File Generation

Most additive manufacturing machines require an STL (STereoLithography) file. The file is a simple, neutral format designed such that most any CAD system can easily translate its 3D data into the STL format. STL files are a boundary representation solid model, made up of a triangular mesh, to approximate the bounding surfaces of the part, and a surface normal, a vector indicating the outward direction perpendicular to the triangle's surface. Most CAD packages allow a user-inputted maximum allowable deviation between the original model surface and the face of the triangle that represents the model surface. With a smaller deviation, model accuracy improves but the file size increases due to smaller, more numerous triangles. As a general rule, the maximum allowable deviation should be set to a level below the desired feature resolution.

26.2.3 Build File Creation

Prior to creating a build file, several steps may be required. Most additive manufacturing machines have software settings that take into account machine-dependent part shrinkage values. These values are applied to the STL file, and other geometry corrections are applied for machine/material specific distortion, and postprocessing procedures (such as finish machining). Other build machine-dependent set-up requirements may include: (1) part orientation, (2) support structure generation, (3) part placement, (4) slicing, and (5) process parameter selection.

Careful consideration of the build orientation is important to balance part quality and machine time. In most additive manufacturing processes the height of the part is the most significant determinant of build time. Each layer contributes to build time due to machine-specific operations such as platen repositioning, calibration, or material leveling. In most cases, upward-facing features are typically the most precise. Another important aspect of build orientation is that it will determine the amount and location of support structures. Thus, part orientation must consider build-time considerations, part quality, and supports.

Additive manufacturing processes often require support structures to prevent sagging or slumping of features during part build-up. Supports serve two purposes: (1) to rigidly anchor the part to the build platen, and (2) to support any overhanging features during part build-up. Certain processes, such as powder bed processes, do not need support structures, as these processes take advantage of the excess, unused material to contain and support the parts during the build process. Additive manufacturing systems are generally equipped with support generation software to automatically generate support structures. The goal is to minimize support structures, as they add to machine time, material cost and part finishing time, while ensuring satisfactory part build times and feature quality.

One of the unique aspects of additive manufacturing is that multiple unrelated parts can be built concurrently in one machine run. To fully capitalize on this capability there should be careful consideration of part placement and the overall build layout. These decisions can impact operational efficiency, throughput, build time, and part quality. It is generally desirable to arrange parts in such a way that the uppermost parts in a build end at a similar Z height, as it offers the greatest operational efficiency.

After the build layout is completed, the STL files for the part and supports are sliced into thin, horizontal cross sections or layers. Most systems offer a range of layer thicknesses between 0.05 and 0.5 mm. As parts are formed layer-by-layer, all parts exhibit some degree of "stair stepping" on features that are not parallel to the horizontal or vertical planes as a result of approximating the geometric features by horizontal layers. How a part is oriented thus determines which faces are subjected to the staircase effect. Stair stepping can be minimized by smaller layer thicknesses, but only at the expense of increased build time. After considering the above process steps, the build file is created with an appropriate choice of build parameters. Once the build file is created, part construction is typically fully automatic and unattended. Constructions times

TABLE 26.1 Classification of Major Additive Manufacturing Processes

Class	Commercial Examples
Photopolymerization processes	StereoLithography (SLA) Solid Ground Curing (SGC)
Extrusion processes	Fused Deposition Modeling (FDM) Fused Deposition of Ceramics (FDC)
Sheet lamination processes	Laminated Object Modeling (LOM)
Printing processes	3D Printing (3DP), Ballistic Particle Manufacturing (BPM), MultiJet Modeling (MJM)
Powder bed sintering/melting processes	Selective Laser Sintering (SLS), Direct Metal Laser Sintering (DMLS), Selective Laser Melting (SLM), Electron Beam Melting (EBM)
Metal deposition processes	Laser Engineered Net Shaping (LENS), Directed Light Fabrication (DLF), Direct Metal Deposition (DMD)
Hybrid and other processes	Shape Deposition Manufacturing (SDM), Segment Milling Manufacturing (SMM), Ultrasonic Consolidation (UC), Direct Write (DW) Electronics

can range anywhere from a few hours to a few days. Part construction details vary from process to process and will be discussed later. Similarly, the cleaning, postprocessing, and finishing methods employed are process, material, and application dependent and will be discussed later.

26.3 Classification and Process Details

There is a great diversity in additive manufacturing processes and there are many ways in which one can classify them. One of the better ways is to classify them according to the fundamental process by which the part is built, as shown in Table 26.1. In this section, the major additive manufacturing processes are discussed. The emphasis is laid on process-specific details rather than machine-specific details offered by various vendors.

26.3.1 Photopolymerization Processes

Photopolymerization processes involve polymerization of a photo-curable liquid resin, a polymer that solidifies as a result of electromagnetic irradiation. More details on polymerization processes may be found in Chapter 30. The vast majority of photopolymers used in commercial systems are curable in the UV range. There are a number of processes that have been developed based on photopolymerizable materials, with variations with respect to the type of laser or light, method of scanning or exposure, and other aspects. These processes can be broadly classified into point-wise and layer-wise curing processes.

26.3.1.1 Point-Wise Curing Processes

The most common terminology used for point-wise curing of photopolymers is stereolithography. In this process, parts are built point-by-point, line-by-line, and layer-by-layer on a platform in a vat of photo-curable liquid resin, as shown in Figure 26.1. Initially, a thin layer of liquid resin is formed on the top of the platform, typically employing some kind of recoating mechanism. A finely focused laser beam then draws the cross section of the layer on the surface of the liquid resin. The liquid resin is cured wherever the laser beam passes, forming the bottommost layer of the part, attached to the platform. The platform is then lowered by an elevation control system and a thin layer of liquid resin is formed (recoating) on top of the previously cured layer. The laser beam then scans the surface of the liquid resin according to the next slice cross section, forming the next layer of the part. This process is repeated sequentially until the entire part is built. Stereolithography requires support structures for building overhanging features. These supports are built simultaneously with the part, and are used to attach the part to the platform.

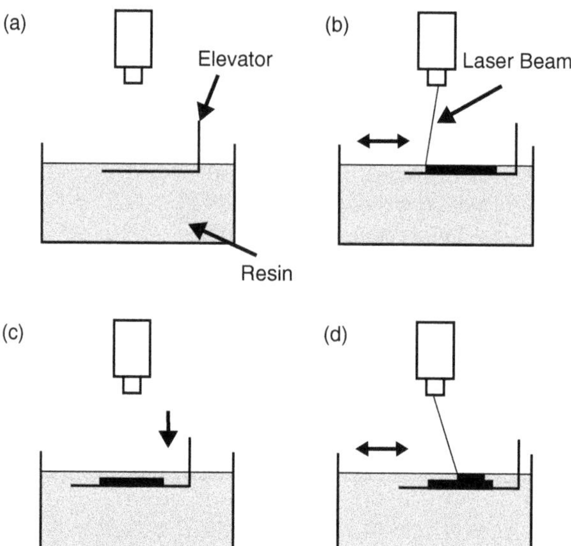

FIGURE 26.1 Schematic of the stereolithography process.

In these processes, a solid layer is generated as a series of overlapping trace solidifications called "voxels."[4] The voxels of one line overlap those of the adjacent line of a layer. Similarly, the voxels of one layer penetrate into the previous layer by a certain depth. Thus, the actual penetration depth (cure depth) of the laser must be slightly more than the layer thickness to ensure layer-to-layer bonding. Layer thickness is user defined and may vary from approximately 0.05 to 0.4 mm. The process takes place at room temperature without the need for special atmospheric preparation.

There are several companies around the world that manufacture prototyping machines based on point-wise photopolymerization technology. The most popular are "stereolithography" (SLA) machines manufactured and supplied by 3D Systems, USA. Most of the earlier models employ a He-Cd laser, whereas most recent models use higher power solid-state or semiconductor lasers. Some commercial systems embody certain process innovations that are noteworthy. For example, in E-Darts (Autostrade Co. Ltd., Japan) and COLAMM (Mitsui Zosen Corp., Japan) systems, the liquid resin is cured by laser light transmitted through a transparent window plate at the bottom of the resin chamber. Unlike many systems in the group, which build parts on a descending platform, these systems build parts on an ascending platform. Another interesting process innovation is the two laser beam process.[5] The process employs two laser beams located along different axes. In this method, the two lasers are focused to intersect at a particular point in the vat. The resin cures at this point of beam intersection. The benefits of this system include that a part can be formed starting from the inside and moving outwards, and that a movable support platform and recoating is not necessary, however the process is not commercially available.

A number of photo-curable resins are commercially available with varying physical, chemical, and mechanical properties to suit various application requirements. The desirable characteristics in a photo-curable resin are low viscosity, high photospeed, good surface wetting between the cured and liquid resin, low volumetric and linear shrinkage, low curling tendency, long shelf life, and good mechanical properties. Acrylate resins generally exhibit higher viscosity and higher curl distortion. Epoxies and vinylethers exhibit almost negligible curling tendency and superior mechanical properties, and are therefore considered more attractive for most prototyping applications.

Important process parameters in point-wise photopolymerization processes are laser power, scan speed, laser spot size, laser wavelength, hatch spacing, and layer thickness. To maintain accuracy and consistency during part building, the cure depth and the cured line width must be controlled, which are a function of laser power, spot size, and scan speed. A number of excellent studies are available describing parameter effects.[6,7] In brief, build parameters together with physical and chemical properties of the resin play a significant role in determining part accuracy and build time. The resin recoating process after completion of each layer is a critical step in photopolymerization processes. Application of a uniform layer of resin of a desired thickness is crucial for part dimensional accuracy and properties.

After part construction, a number of postprocessing steps occur. The parts from the vat are cleaned using suitable solvents to remove excess uncured liquid resin. It is important to remove parts from the vat within a controlled amount of time, as the infusion of liquid monomers into a semicured part results in "swelling." It is also typically necessary to cure parts using UV ovens at temperatures up to 175°C to increase the degree of cure, and thus the part strength, as parts are generally not fully polymerized during the build process. This postcuring processing results in shrinkage, which may lead to dimensional inaccuracies and distortion in the part.[8] The support structures are also manually removed either before or after the curing process. It is often important to postcure the parts quickly once the support structures are removed, to avoid green creep distortion, a phenomenon that causes delayed or latent curl distortion after supports have been removed. After curing, finishing operations like sanding, polishing, bead-blasting, or milling may be performed.

Photopolymerization processes are widely used and typically produce the best surface finishes of all additive manufacturing processes. Parts produced using these processes generally serve as concept models as well as prototypes for design verification and form and fit issues. Parts can be used as prototype tooling and low-volume production tooling in certain cases. End-use metal parts can be produced through investment casting and sand casting processes using patterns fabricated from these processes. The main limitation of these processes is that only photo-curable resins are usable for part construction. These materials are expensive and some exhibit bad odors and toxicity. Another disadvantage is the requirement of support structures for all overhanging features.

26.3.1.2 Layer-Wise Curing Processes

To overcome the speed limitations of photopolymer curing using a point energy source, a number of processes have been developed to cure entire layers of photopolymers at a time. Cubital Ltd., Israel, developed and commercialized a process called "solid ground curing" (SGC) based on this principle in 1991. In this process, a layer is generated by exposing a thin film of photo-curable resin to high-power UV light through a mask, which allows precise projection of the layer cross section onto the liquid resin. The process combines two subprocesses: generation of the mask and generation of the layer. From the slice data, the contour data are transferred to a glass plate by a technique similar to photocopying technology. When UV light is passed through this glass mask onto the liquid resin, the resin is polymerized wherever light is allowed to pass. The surrounding unexposed resin remains liquid, which is removed using a vacuum. The empty space in the layer is then filled with molten wax. The model surface is finally face-milled to provide a level, accurate layer thickness. A new layer of photopolymer is then applied to the uppermost surface and the process continues to build the next layer in the same manner.

As each layer is surface milled before generating a new layer, this process facilitates a high degree of accuracy in the Z direction. After part construction, the supporting wax is removed during postprocessing. In this process, part building times are very short and independent of the layer geometrical complexity. No postcuring is necessary as the parts are completely cured during the build process itself. This process never achieved widespread commercial success, due to the inherently complex and expensive nature of the process and machine. However, the benefits of the process have begun to be captured in other layer-wise processes by simplifying the procedure, as described below.

Envision Technologies GmbH, Germany, has commercialized prototyping machines based on layer-wise photopolymerization principles since 2001. The machine consists of a DLP (digital light processing technology) projector equipped with a mercury lamp located below a glass vat. A movable platform is positioned close to the bottom surface of the glass vat facing the DLP projector. The gap between the elevator plate and the bottom side of the vat defines the layer thickness. The DLP projector shines light in a pattern based on the layer information from the STL file, and the liquid resin is cured across the entire layer forming the first layer of the part rigidly attached to the elevator plate. A special coating is added to prevent the layer from adhering to the bottom surface of the glass vat. The elevator plate is moved upward by a layer thickness and fresh resin flows into the gap. The process is then repeated until the part is built completely layer-by-layer from top to bottom.

Objet Geometries Inc., Israel, is yet another company that manufactures prototyping machines based on a layer-wise photopolymerization principle. The machine contains a multinozzle print head that prints liquid polymer resin onto a platform according to the slice information. The entire layer is printed as the print head is swept across the platform. The printed liquid resin is cured by exposing it to UV light from two lamps located on the print head assembly. The printing and curing processes are repeated until the entire part is constructed. Support structures are also printed and cured simultaneously using a different photopolymer that is water-soluble. No postcuring is necessary. Similar technology has since been introduced by 3D Systems as part of their InVision line of 3D Printers.

26.3.1.3 Micropolymerization Processes

There are several processes developed exclusively for microfabrication applications using lamps, lasers, or x-rays as the energy source.[9,10] These processes, employing either point-wise or laser-wise curing, build complex-shaped parts that are typically less than 1 mm in size. They are referred to as microstereolithography (μSL) processes. One μSL process has been commercialized by MicroTEC GmbH, Germany. Their machines use a He-Cd laser and are capable of constructing small parts with layers as thin as 1 μm, with submicron precision and a feature definition of less than 10 μm. These processes have been successfully used for fabrication of parts for applications in microelectromechanical systems.[11] Fully dense micron-sized ceramic parts have also been successfully produced using microfabrication processes.[12] In this approach, nanoceramic powders are mixed with photopolymers and processed using a μSL process. Subsequently, the polymer is burned out and the porous ceramic part sintered to obtain a dense ceramic part.

26.3.2 Extrusion-Based Processes

Extrusion-based processes are currently the best selling processes for prototyping applications. Although originally developed for producing thermoplastic prototypes, these processes have also been shown to be suitable for producing functional parts in metals and ceramics. Commercialized systems in this group include: (1) fused deposition modeling (FDM) by Stratasys, Inc., USA, (2) multiphase jet solidification (MJS) by IPT GmbH, Germany, and (3) melted extrusion modeling (MEM) by Beijing Yinhua Co. Ltd., China.

Extrusion-based additive manufacturing processes fabricate parts by extruding molten thermoplastic material through a small nozzle to form a thin bead or "road" that is deposited in a predetermined pattern to complete each build layer. The FDM process uses an extrusion head, called the liquefier, which travels in the X and Y directions (Figure 26.2). Through this heated head passes a thin filament of build material (typically 1 mm dia) driven by rollers as needed. The build material, in a semimolten state, is extruded through a nozzle and deposited on the substrate as required by the part geometry. Upon deposition, the extrudate bonds to the previous layer and the adjacent road and solidifies quickly as a result of heat conduction. The deposited extrudate generally assumes an elliptical cross-section. The build chamber is maintained at a set temperature, depending on the type of build material, to keep the already deposited material at a warmer temperature so as to ensure proper inter-road and interlayer adhesion, which is key to part strength. Parts are built on a computer-controlled build platform, which is lowered as each layer is deposited, maintaining a constant distance between the extrusion tip and the working surface. The feed rate of the filament is used to control the amount of material extruded at any given time. Layer thicknesses are typically in the range of 0.1 to 0.5 mm. The extrudate width can range from 0.25 to 2.5 mm. Support structures are often generated using a separate extrusion head and are removed during postprocessing. In addition to thermoplastic filament-based extrusion, other extrusion designs have utilized liquid slurries and other precursor materials. The basic operating principle of these machines, however, is the same as the original FDM process.

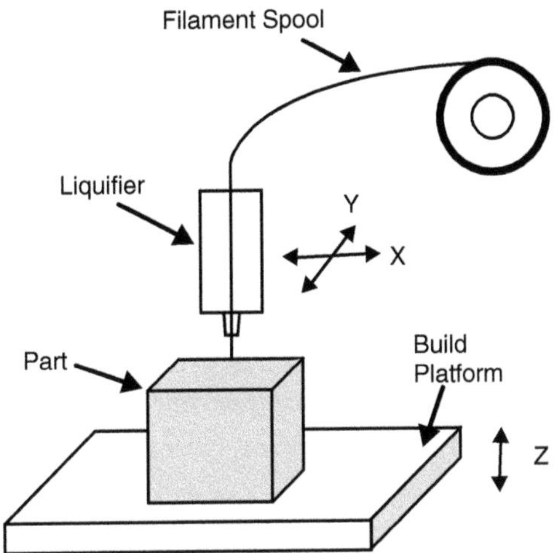

FIGURE 26.2 Schematic representation of the fusion deposition modeling (FDM) process.

A wide range of thermoplastic material can be used as build materials including investment grade waxes, polyolefins, polyamides, polyphenylsulfones, elastomers, nylon, and acrylonitrile butadiene styrene (ABS) to suit various application requirements. The most popular materials are ABS and polycarbonate. Polyphenylsulfones are used when high heat (up to 200°C) and chemical resistance, combined with good strength and rigidity, are required. Build materials are available in different colors and parts can be built in multicolors, if desired. Biocompatible materials such as poly(ϵ-caprolactone) have been used as well with extrusion-based processes for tissue engineering applications.[13]

Dense functional parts have been successfully produced in various metallic and ceramic materials using extrusion processes.[14,15] The extrudate consists of semimolten binder loaded with metallic or ceramic powder particles. The term "fused deposition of ceramics" (FDC) has been used extensively when using FDM to produce ceramic structures, including for scaffolding and orthopedic applications, such as β-tricalcium phosphate (TCP) scaffolds.[16] In general, after the part is constructed (forming a green part), the binder is removed from the part by dissolving it in a suitable solvent or by furnace processing to form the "brown part." The porous brown part is then sintered to attain a higher density, resulting in shrinkage. Alternatively, the green part can be infiltrated to obtain a dense part. The advantage of infiltration is that it results in little net shrinkage, thus more effectively retaining part dimensional accuracy.

Generally, separate extrusion heads, filament driving mechanisms, heating parameters, and nozzle diameters are employed for each type of build material. Important parameters that affect part quality are liquefier temperature, chamber temperature, standoff distance (distance between nozzle tip and working surface), filament feed rate, nozzle diameter, and deposition speed. It is important to carefully balance these parameters for satisfactory part build-up as shown by several investigators.[17,18]

Extrusion processes are quite suitable for producing 3D porous structures. By changing the direction of material deposition for consecutively deposited layers and the spacing between the material roads, parts with highly uniform internal honeycomb-like structures, controllable pore morphology, and complete pore interconnectivity with precisely controlled pore size and pore geometry can be produced. Zein et al.[13] and Kalita et al.[19] have successfully produced 3D interconnected controlled porosity scaffolds in bioabsorbable poly(ϵ-caprolactone) using the FDM process for bone graft applications. Similarly, 3D honeycomb ceramic structures with precisely controlled pore size, volume, and geometry can be produced using a sacrificial wax mold, as demonstrated by Bose et al.[20] In this approach, the FDM-created wax mold is infiltrated with a ceramic slurry, which is subsequently furnace processed to remove the mold material and sinter the ceramic structure.

Most parts produced by extrusion processes serve as concept models or form, fit, and function models. Patterns for investment casting can be produced using wax and ABS. Mechanical, thermal, and chemical properties of ABS parts produced by extrusion processes compare well with injection molded parts. The properties of ABS parts do not significantly change with time or environmental exposure and retain their dimensional and geometrical detail. Unlike many other processes, extrusion-based processes are nearly continuous and utilize time efficiently. The process proceeds from one layer to the next with little or no delay. Shrinkage of thermoplastics upon cooling is a concern with regard to part accuracy. Dimensional accuracy thus rests on the use of appropriate shrinkage compensation factors[21] and optimized deposition strategies.[22] The minimum feature size possible with extrusion processes is typically twice the road width.

The most obvious limitation of extrusion-based processes is surface finish. The characteristic elliptical nature of the extrudate results in a "wavy" surface finish. Smaller nozzle sizes and thinner layer thicknesses improve surface finish but make the building process slower. Build times are dependent on the material volume of the part and support structures, but are relatively independent of the part height. Further, nesting of parts or building multiple parts in a single build does not typically offer appreciable time saving.

26.3.3 Sheet Lamination Processes

One of the first commercialized additive manufacturing techniques was laminated object manufacturing. LOM involved layer-by-layer lamination of sheets of paper, cut using a laser, each sheet representing one cross-sectional layer of the part. A number of other processes have been developed based on sheet lamination, using other build materials and cutting strategies. The sheets can be either cut and then stacked or stacked and then cut. These processes can be categorized based on the mechanism employed to achieve bonding between layers: (1) gluing or adhesive bonding, or (2) welding or brazing.

26.3.3.1 Processes Based on Gluing or Adhesive Bonding of Sheet Materials

The most popular lamination build material has been paper with a thermoplastic adhesive on one side. A number of proprietary adhesive-backed paper and resin sheet materials are available. Sheet material thickness employed are in the range of 0.07 to 0.2 mm. Potentially, any sheet material that can be precisely cut using a laser or mechanical cutter and that can be bonded well using adhesives can be utilized for part construction. A further classification is possible within these processes. In one category, there are processes in which the laminate is bonded first to the substrate and is then formed ("bond-then-form" processes). In the second category, there are processes in which the laminate is formed first and then it is bonded to the substrate ("form-then-bond" processes).

26.3.3.1.1 Bond-then-Form Processes

In these processes, the building process typically consists of three steps in the following sequence: placing the laminate, bonding it to the substrate, and cutting it according to the slice contour. The majority of commercial systems in this group are LOM machines, introduced in 1991 by Helisys, Inc., USA (currently supplied by Cubic Technologies, USA), which use adhesive-backed laminates for part construction and employ a heated roller to produce bonding between the laminates. Initially, a double-sided adhesive tape is placed on a metallic build platform on to which a base is built. The building process begins by placing a laminate on the build platform. A heated roller is then passed across the laminate to thermally activate the adhesive and to produce bonding between the laminates. A CO_2 laser, controlled to cut to a depth of one layer thickness, cuts the cross-sectional outline. The unused material is left in place as a support material and is diced using a crosshatch pattern into small pieces called "tiles" or "cubes" for easy removal. After part construction, the crosshatched pieces of excess material are separated from the part using typical wood carving tools (decubing).

A similar system to LOM is paper lamination technology (PLT) by Kira Corp. Ltd., Japan. This system makes use of plain paper (no adhesive) as the build material. A laser printer is used to apply a resin powder on top of the previously deposited layer or substrate in a pattern that represents the current cross-sectional geometry. A new sheet of paper is then placed on it and is pressed against the previously deposited resin printed layer using a hot plate. As a result, the resin is melted and the two layers are adhered to each other.

A knife then cuts the deposited layer to the corresponding slice profile. Because the support material is not adhesively bonded, unlike in LOM, the removal process is easier.

Solidimension, Israel, developed a commercial prototyping system (SD300) in 1999 based on lamination of PVC plastic sheets using ink-jetted glue. The company has a distribution agreement with Graphtec Corp., Japan (in Asia and Europe under the name of "XD700") and 3D Systems, USA (under the name "InVision LD 3D Printer").[23]

Laminated manufacturing principles have also been successfully applied to fabrication of parts from ceramic and composite materials.[24,25] For example, Klosterman et al.[24] have fabricated SiC/SiC particulate composite parts from 250 μm thick ceramic tapes, prepared by tape casting using a powder mixture of SiC, graphite, carbon black, and a polymer binder. These tapes were used for part construction employing the standard LOM process. After part construction, the green part was debinded at 600°C and subsequently densified at 1600°C by reaction bonding involving infiltration of liquid silicon.

26.3.3.1.2 Form-then-Bond Processes

In form-then-bond processes, sheet material is cut to shape first and then bonded to the substrate. This approach is more popular for construction of parts in metallic or ceramic materials, although its use has primarily been at the research level. One commercial technology was the "Offset Fabbers" system introduced by Ennex Corp., USA.[26] In this process, a suitable sheet material with an adhesive backing is placed on a carrier and is cut to the outline of the desired cross-section using a plotting knife. Parting lines and outlines of support structures are also cut. The shaped laminate is then placed on top of the previously deposited layers and bonded to it. In another commercial process (Computer-Aided Manufacturing of Laminated Engineering of Materials), laminated manufacturing principles are applied to fabrication of ceramic parts. In this process, individual slices are laser cut from sheet stock of green ceramic tape. These slices are precisely stacked one over another. After assembly, the layers are laminated using warm isostatic pressing. The laminated green object is then fired to densify the object into a monolithic structure.

The form-then-bond approach facilitates construction of parts with internal features and channels that are not possible with a bond-then-form approach. Another advantage is that there is no danger of cutting into the previous layers, unlike in bond-then-form processes. Also, the time-consuming and potentially damage-causing decubing step is eliminated. However, these processes require external supports for building overhanging features, and some type of tooling or alignment system to ensure proper registry between layers.

26.3.3.1.3 Capabilities and Limitations

Laminated manufacturing processes have little shrinkage, residual stresses, and distortion problems. Fabricated parts from paper resemble plywood, a typical pattern-making material, and are amenable to common wood finishing operations. Parts produced by laminated manufacturing have been most successfully applied where traditionally wooden patterns have been used.[27] Other specific advantages include: (1) excellent part building speeds, especially for large parts, (2) nontoxic, stable, and easy to handle feedstock, and (3) low material, machine, and process costs. However, these processes have several limitations, including: (1) most parts require coating to prevent moisture absorption and excessive wear, (2) control of the parts' accuracy in the Z-dimension is difficult (due to a swelling effect, or Z growth), (3) mechanical and thermal properties of the parts are inhomogeneous due to the laminated structure, and (4) small part feature detail is poor.

26.3.3.2 Processes Based on Welding or Brazing of Sheet Materials

Complex-shaped 3D parts have been successfully fabricated from metallic sheets/foils employing ultrasonic welding (described in detail in Section 26.3.7.4), diffusion bonding, laser spot welding, and brazing techniques. For example, Yi et al.[28] fabricated 3D metallic parts using preshaped 1 mm thick steel sheets utilizing a diffusion bonding process. Himmer et al.[29] produced Al injection molding dies with intricate cooling channels using 0.1 mm thick low melting point Al alloy 4343 coated Al alloy 3003 sheets (total sheet thickness 2.5 mm). The sheets were initially laser cut to shape and were assembled precisely using mechanical fasteners. The sheets were then bonded together by heat treating the assembly in a furnace just above the coating material melting point. The bonded part was finally milled to precise part dimensions. More recently, Himmer et al.[30] have demonstrated satisfactory layer bonding using brazing and laser spot welding processes. Similarly, Wimpenny et al.[31] produced steel tooling with conformal cooling channels by brazing laser-cut-to-shape sheets.

26.3.4 Printing Processes

A number of additive manufacturing technologies have been developed using droplet printing. These processes can be classified into binder printing and direct printing processes. Binder printing processes involve selectively depositing fine droplets of a fluid binder into a powder bed, creating the shape of a cross-sectional slice by gluing the powder together. In direct printing processes, the material used for part building is directly printed through the print nozzle, creating the actual part.

26.3.4.1 Binder Printing

Massachusetts Institute of Technology, USA, faculty and students developed and patented the basic technology for binder printing processes. Their process was coined "three-dimensional printing" (3D Printing). In this process, a liquid binder is selectively deposited, sprayed, or printed in the form of fine droplets through a print head assembly onto a prelaid powder bed (Figure 26.3). Upon deposition the binder wets the powder particles and binds them together as well as to the previous layer. The inkjet head typically contains a large number of parallel nozzles, enabling the print head to cover 50 mm or more of the layer profile in a single pass. The print head passes over only those areas on the powder bed that contain the part. The areas on the powder bed that lay outside the part geometry remain loose and support the layers that will be printed subsequently. After forming a complete layer, the build platform is lowered by one layer thickness and a fresh layer of powder is laid. The binder is printed again according to the corresponding slice cross section. The entire process runs at or near room temperature and in ambient atmosphere. The binder accounts for approximately 10% of the part volume.

Basically, each binder droplet produces a small, roughly spherical agglomerate of particles called a "primitive." The part is constructed by stitching these primitives together with adjacent primitives between lines and layers. The process parameters optimized to achieve this condition include powder (shape, size, and distribution) and binder (viscosity, wetting) characteristics, powder bed density, binder droplet size, and delivery rate. Because the print head can generate a layer width of up to 50 mm in one pass, build time is not affected by the geometrical complexity, and is dominated by build height. Commonly used build materials are starch and plaster for generating concept models. Metallic and ceramic powders can also be used for generating tooling and patterns for various casting processes. Binder materials vary depending on the build material and the intended application.

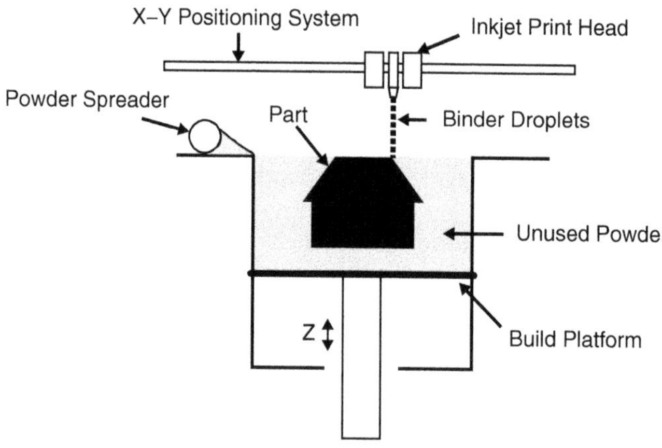

FIGURE 26.3 Schematic representation of the 3D printing process.

After part construction, the build platform is raised, loosely adhering powder is removed, and the part is typically further processed. Green parts generally exhibit a low mechanical strength and should be handled with care. Parts made from starch or plaster are often infiltrated with wax or epoxy, and sealed to prevent moisture absorption.

MIT has licensed their patented 3D printing technology to a number of organizations, which manufacture and supply systems for particular fields of use under various brand names. 3D Printing RP machines from Z Corp., USA, primarily use starch and plaster formulations for part construction. ProMetal systems from the Ex One Corp. Company, USA, are specially designed for construction of metal parts for tooling applications. These machines use steel powders, which typically are subsequently infiltrated with bronze after burning out the binder, to produce a finished component. Soligen Technologies, USA, is another company that manufactures 3D printing machines. Their direct shell production casting (DSPC) machines are specially made for fabrication of ceramic molds for metal casting. ZCorp and Ex One Corporation have also introduced ceramic mold materials for metal casting that work in their machines. Machines from Therics Inc., USA, are designed for manufacture of implantable drugs, solid oral tablets, and tissue engineering products by printing micro drops of binders, drugs, and other materials.

3D printers are typically simple to operate and are among the least expensive of all additive manufacturing systems. 3D printing processes are often five times faster than other additive manufacturing processes and enable generation of parts with complex color schemes, utilizing full-color printhead technologies. Multiple parts can be built simultaneously in one machine run. No support structures are required and there is little material waste, as the unused powder can be recycled after sieving.

Work is being done to produce functional metallic, ceramic, and composite parts using binder printing processes. For example, Moon et al.[32] have fabricated functionally graded SiC–Si composites using an inkjet printing process. The method involved fabrication of preforms by printing acetone-based furfuryl resin binder onto catalyst-coated carbon powder. The carbon preform was subsequently debinded and pressureless reaction infiltrated with liquid silicon to yield SiC–Si composite parts. Sun and coworkers[33] produced fully dense Ti_3SiC_2 parts using inkjet printing, cold isostatic pressing, and sintering processes. A significant process innovation is to print a reactive liquid to produce an *in situ* alloy. Cao et al.[34] have successfully fabricated Ti–Al intermetallic parts by printing fine droplets of liquid Al on to a Ti powder bed. The liquid Al and Ti exothermically react producing a piece of molten TiAl compound that immediately solidifies into a solid bead.

The most common limitation of these processes is poor feature definition and surface finish. The unavoidable spread of the liquid binder interferes with feature resolution. Also, considering the fragile nature of the green part, small features are easy to destroy during powder removal. In general, feature sizes down to 0.75 to 1 mm are achievable, which is more than adequate for a concept model. 3D printed parts generally exhibit a textured surface roughness commensurate with the powder particle size and droplet size used.

26.3.4.2 Direct Printing

Direct printing processes print the build material itself to form the part layer-by-layer. One of the most successful direct printing techniques, the ModelMaker, was commercialized in 1994 by Sanders Prototype Inc., USA (now Solidscape Inc.). Their systems use two inkjet-type print heads, one depositing a thermoplastic build material and the other depositing supporting wax. The build material is melted in the heated inkjet print head and is shot as microdroplets onto the model structure. After each layer is deposited, a cutter removes approximately 0.025 mm of the layer's top surface to provide a smooth even surface for the next layer. Those areas that are adjacent to the part are filled with wax by a second print head as a support material, which is removed during postprocessing. A high degree of precision is possible with this system and the parts exhibit excellent surface finish, but the build times are rather high. Parts produced are especially suitable for precision investment casting, and are widely used in the jewelry industry.[2]

One early process in this category was ballistic particle manufacturing (BPM), which was, however, not a commercial success. In this process, molten thermoplastic material was sprayed onto the substrate using a multiaxis nozzle. Upon impact the molten droplets solidified and welded to the previously deposited material. The unique multiaxis nozzle enabled printing from any direction, thus enabling printing normal to the surface, resulting in a smoother overall surface finish.

Other commercial systems, such as 3D Systems' MultiJet modeling techniques, consist of arrays of piezoelectric print heads arranged in a line. Small droplets of material are ejected through these print heads as required so that an entire layer is deposited in just one traversing run. Support structures in the form of thin needles or as a separate waxy material are generated simultaneously and are removed during postprocessing. Photopolymer-material systems based on this principle were discussed in Section 26.3.1.2.

A number of processes have been investigated for producing parts in various low melting point metals like tin, zinc, and lead.[35,36] The building process usually takes place inside a closed chamber under inert atmosphere to prevent oxidation of the part.

26.3.5 Powder Bed Sintering/Melting Processes

A schematic of a powder bed sintering process is shown in Figure 26.4. In this process, a thin layer of heat-fusible powder (typically 0.1 mm thick) is laid on the build platform using a counter-rotating roller. The process takes place inside an enclosed chamber filled with inert gas. The build chamber is maintained at an elevated temperature below the melting point of the powder material using a heater. This is to minimize the laser power requirements of the process and to prevent warping of the part as it is built.[37] A finely focused laser or electron beam is directed onto the powder bed and is moved in such a way that it heats and fuses the material to form the slice cross section. Surrounding powder remains loose and serves as support for subsequent layers. After completing a layer, the platform is lowered by one layer thickness and a new layer of powder is laid and leveled for creating the next layer of the part.

There are three general methods by which an energy source bonds material in a powder bed process: single-component direct method, two-component direct method, and two-component indirect method.[38]

In the single-component direct method a powdered material is directly sintered using a sufficiently high-power laser. The surface of the powder particles are melted upon exposure and are fused during cooling. In contrast to classical powder sintering processes (see Chapter 25), sintering for this case is not predominantly diffusion controlled. Instead, sintering occurs as a result of momentary thermal activation involving incipient melting and fusion of the powder particles.

The two-component direct method uses a mixture of two metal powders comprising a high melting point powder (the main structural material) and a low melting point material (acting as the binder). In this

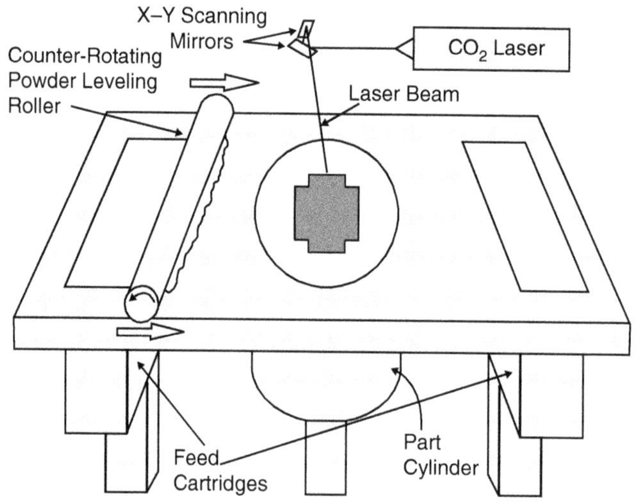

FIGURE 26.4 Schematic representation of a typical powder bed sintering process.

case, energy is so controlled so that it results in complete melting of only the low melting constituent, which flows through the pores between the solid high melting point powder particles, causing binding and densification upon solidification. The sintering mechanism in this case is similar to classical liquid phase sintering, and the end result is a part that is useable in its as-produced form without further thermal processing.

In the case of the two-component indirect approach, a polymer-coated metallic or ceramic powder is used for part construction. Alternatively, a dry mixture of metallic and polymer powders can also be used. During the building process, the polymer binder is melted and binds the particles together. The metallic powder particles remain largely unaffected by the heat of laser. The parts produced are generally porous. The green parts are subsequently debinded and infiltrated with a lower melting point metal to produce dense metallic parts.

26.3.5.1 Polymer-Melting-Based Powder Bed Processes

The basic technology for powder bed sintering processes was developed at the University of Texas at Austin, USA, in the late 1980s and early 1990s. The SLS process was originally developed for producing plastic prototypes using thermoplastic powdered materials. The principle was later extended to metallic and ceramic powders for fabrication of tooling and functional parts. Laser sintering machines that process polymer materials are commercially available from 3D Systems, USA, and EOS GmbH, Germany.

Laser sintering processes conceptually offer almost unlimited choices of build materials, including thermoplastics, metals, and ceramics. Thermoplastic materials were the first to be commercially developed for laser sintering processes because of their relatively low melting temperatures and low heat conductivities. Polyamide-based powders (such as DuraForm PA and DuraForm GF from 3D Systems) are used to create rigid and rugged plastic parts for functional applications. Polystyrene-based materials with a low residual ash content (such as PS 1500 and PS 2500 from EOS GmbH) are particularly suitable for making patterns for investment casting.

A number of proprietary metal powders are available from machine manufacturers for producing functional tools through the two-component indirect process. The most common powder combination is a stainless steel powder mixed with a combination thermoplastic/thermoset binder, which is subsequently infiltrated with bronze. Similarly, a number of commercially available binder-coated ceramic powders (zirconia- and silica-based) can be used for producing molds and cores for metal casting. Stucker et al.[39] have successfully employed the SLS process to produce complex-shaped ZrB_2–Cu composite EDM electrodes with a more homogeneous microstructure compared to a hot pressing route. The approach involved (1) fabrication of a green part from polymer coated ZrB_2 powder using the SLS process, (2) debinding and sintering of the ZrB_2, and (3) infiltration of the sintered, porous ZrB_2 with liquid copper. Similarly, Lu et al.[40] produced *in situ* TiC reinforced Cu electrodes using Cu, Ti, Ni, and C powder mixtures employing a laser sintering process.

Use of optimum process parameters is extremely important for producing satisfactory parts. Important process parameters include: (1) laser related parameters (laser power, spot size, pulse duration, pulse frequency), (2) scan related parameters (travel speed and scan spacing), (3) powder related parameters (particle shape, size, and distribution, powder bed density, layer thickness), and (4) temperature related parameters (powder bed, powder feeder, and atmospheric temperatures). It should be noted that most of these parameters are strongly interdependent and are mutually interacting. The required laser power, for instance, typically increases with the melting point of the material, but may vary depending upon the wavelength absorption characteristics of the material. The powder shape, size, and size distribution strongly influence the laser absorption characteristics and powder bed packing density. Powder bed temperature, laser power, scan speed, and scan spacing are to be carefully balanced with each other to maximize dimensional accuracy, surface finish, and mechanical properties. A number of studies focusing on process parameter effects and process parameter optimization in laser sintering processes are available.[41,42]

Laser sintering processes are versatile and can produce parts from materials as varied as polymers, metals, and ceramics. These processes do not require external supports, which saves time during part building and cleaning. Accuracy and surface finish of laser sintered parts are dependent on powder particle size, laser spot size, and layer thickness. Finer particle sizes produce smoother, more accurate parts, but are difficult to spread and handle.

26.3.5.2 Metal-Based Powder Bed Processes

There are several processes that have been developed to directly produce functional single-component metal parts. In contrast to direct polymer or two-component indirect metal methods, these direct metal methods use higher powered energy sources, tighter atmospheric control, and novel laser scanning strategies.

Selective laser-powder remelting (SLPR) is one such process (being commercialized primarily as the MCP Realizer) developed by Fraunhofer Institute for Laser Technology, Aachen, Germany. The process has been successfully applied to various metals and ceramics. This process, also known as selective laser melting (SLM), has achieved accuracy and surface roughness quality of less than 50 μm. Concept Laser GmbH manufactures prototyping machines based on the same basic concept, terming the process "Laser Cusing," a term derived from cladding and fusing. Their "M3 Linear" machine, commercially introduced in 2002, not only builds dense metallic parts, but also facilitates engraving, carving, and marking operations in the same machine.

A novel variation on laser-based powder bed processes is electron beam melting (EBM), which has been successfully applied to a variety of metals.[43] This process makes use of an electron beam as the heat source in place of a laser beam. This process was developed at Chalmers University of Technology, Sweden, and is commercialized by Arcam AB, Sweden. In this process, a focused electron beam scans across a thin layer of prelaid powder bed inside a vacuum chamber, causing localized melting and resolidification. Unique benefits of EBM vs. laser processing include higher energy conversion efficiencies and beam manipulation advantages, including more rapid focusing and defocusing of the beam, and more rapid beam movements. Disadvantages of EBM include the need for the processed material to be electrically conductive and the need for a vacuum atmosphere.

In addition to producing polymer-based laser sintering machines, EOS GmbH also produces a direct metal laser sintering (DMLS) machine that uses a two-component (liquid phase sintering based) direct metal process. Their commercially available systems have the largest install base of any metal powder bed system, with commercially available materials that mimic the properties of tool steel, stainless steel, and bronze materials. More recently, EOS has begun commercializing single-component metal powders for their DMLS machines as well.

A number of investigators have extensively studied metallic, ceramic, and composite functional parts formation using laser sintering processes.[44,45] Das and coworkers[46] employed a novel approach for fabricating fully dense parts using the SLS process. Initially, the SLS process was used to produce 3D parts with a gas impermeable skin, which encapsulates the porous internal volume, by carefully controlling the process parameters. These parts were then directly postprocessed to full density by containerless hot isostatic pressing (HIP). The SLS/HIP approach was successfully used to produce complex 3D parts in Inconel 625 and Ti–6Al–4V for aerospace applications.

26.3.6 Metal Deposition Processes

Metal deposition processes involve complete melting and deposition of metallic powders, using a focused high-power laser beam as the heat source, for construction of 3D parts. Research variants of this procedure include using an electron beam in place of the laser beam, or the use of a thin metal wire instead of metal powder as the build material. Various forms of this technology have been referred to as laser engineered net shaping (LENS), directed light fabrication (DLF), direct metal deposition (DMD), and others. Because these processes involve full melting and solidification of powdered material at every point, the resulting parts attain a high density during the build process itself in contrast to the porous partially sintered parts produced in many powder bed processes.

In these processes, a focused laser beam is directed onto a substrate situated on the build platform capable of computer-controlled motion (Figure 26.5). The laser generates a small molten pool on the substrate.

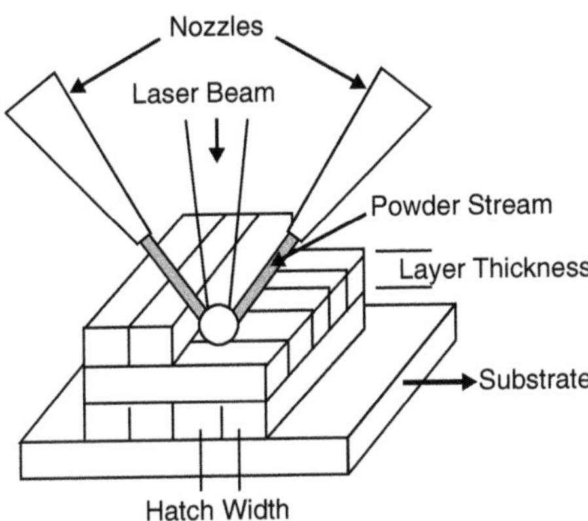

FIGURE 26.5 Schematic representation of a metal deposition process.

Powdered material is then injected into this pool, where it melts and solidifies rapidly as the laser beam moves away. In effect, a thin track of solidified metal welded to the layer below is formed along the line of laser scanning. A layer is generated by a number of consecutive overlapping tracks (similar in principle to FDM). The amount of track overlap is generally 25% of the track width and typical layer thicknesses employed are 0.25 to 0.5 mm. After each layer is formed, the laser head and powder delivery nozzle move one layer thickness away from the deposit keeping the same standoff distance throughout the build process. The deposition process occurs inside an enclosed chamber filled with inert gas, or in the presence of an inert gas shield. These processes require support structures for building overhanging features on the part. However, advanced machines with 5-axis capability can generate overhanging features without the need for support structures.[47]

The LENS process was originally developed by Sandia National Laboratories, USA, and was commercialized by Optomec Design Company, USA, in 1998, as the first commercially available metal deposition technology. AeroMet Inc., USA, also manufactured direct metal deposition machines targeted toward large aerospace structures until the business was closed in 2005. POM, USA, manufactures direct metal deposition machines with 5-axis capability (DMD machines) targeting the tool and die industry. Controlled metal buildup (CMB) was developed by RoderTec GmbH and Fraunhofer Institute for Production Engineering, where the build material is introduced in the form of a wire. After depositing a layer, it is shaped to the corresponding slice contour by a high-speed cutter. With these direct metal processes, practically any metallic material can be used for part construction. These processes are particularly suited for carrying out weld repairs and surface modifications.

Metal deposition processes aim to produce fully dense functional parts in metals and ceramics and are not meant for producing parts in plastic materials. These processes are not restricted to proprietary material formulations, as is typically the case with most other processes. Many metallic, intermetallic, ceramic, and composite materials in powder form have been successfully processed.[47–50] The powder size generally used ranges from 20 to 100 μm. Either prealloyed powders or suitably blended elemental powders can be used. Elemental powders can be delivered in precise amounts to the melt zone using separate feeders to generate various alloys and composite materials *in situ*.[51]

Several studies focusing on process parameter effects are available.[52,53] Important process parameters are laser focus height, scan spacing, powder feed rate, laser traverse speed, laser power, and laser spot size. Laser focus height is the height above the substrate at which the laser beam is focused to its maximum power density. Generally, depositions are conducted with the laser focus "buried" into the substrate about 1 mm to ensure adequate substrate melting and strong bonding. If it is desirable to minimize dilution, it is advantageous to focus the laser on or just above the substrate surface. The powder feed rate, laser power, and traverse speed are all interrelated and must be carefully balanced with each other. Sophisticated accessory equipment have been developed for accurately controlling the process by monitoring the melt pool dimensions and temperature.[54,55]

Direct metal deposition processes involve extremely high solidification cooling rates (10^3 to 10^5°C/sec) leading to several microstructural advantages.[54] These advantages include: (1) suppression of diffusion controlled solid-state phase transformations; (2) formation of supersaturated solutions and nonequilibrium phases; (3) formation of extremely fine, refined microstructures with reduced elemental segregation; and (4) formation of very fine second phase particles (inclusions, carbides, etc.). Laser deposited parts typically exhibit a layered microstructure with an extremely fine solidification substructure (Figure 26.6). The interface region generally shows no visible porosity and extremely thin HAZ, as can be seen, for example, on the microstructure at the interface region of a LENS deposited medical-grade CoCrMo (Figure 26.7). Some materials exhibit pronounced columnar grain structures aligned in the laser scan direction, while some materials exhibit fine equiaxed structures, and process parameters play a strong role in determining

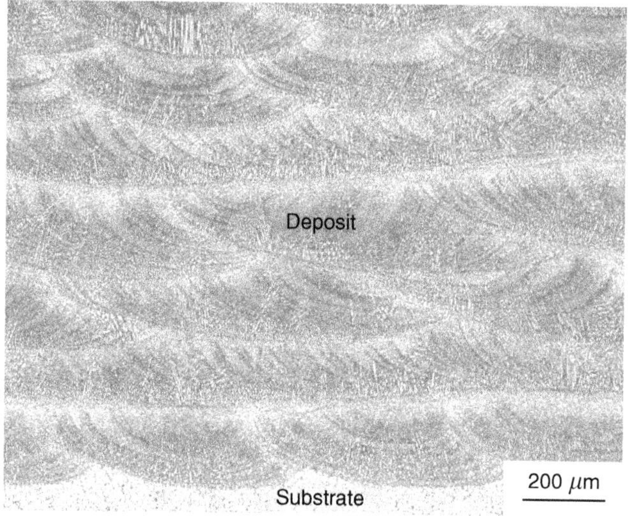

FIGURE 26.6 Microstructures of a CoCrMo LENS deposit showing characteristic layered structure and extremely fine solidification structure.

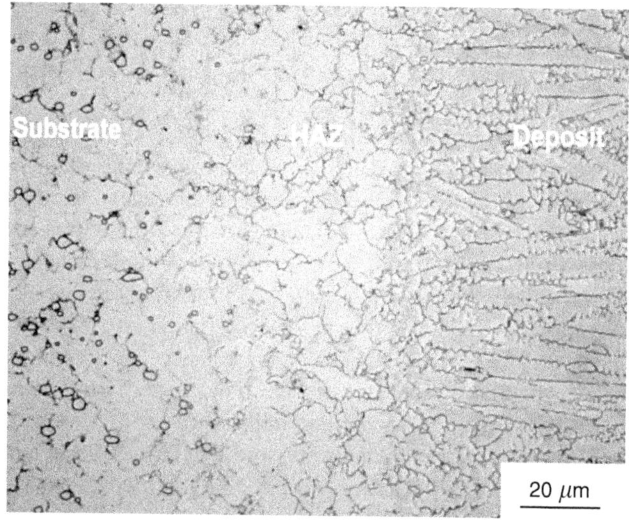

FIGURE 26.7 Microstructure at the substrate/deposit interface of a CoCrMo LENS deposit on a CoCrMo wrought substrate showing a sound interface with a narrow HAZ and epitaxial growth features. The particles seen on the substrate side are carbide particles.

the character of grain structure. Studies have shown that, with proper choice of processing parameters, it is possible to promote solidification in strongly columnar modes to produce directionally solidified parts and even single crystal parts using laser deposition processes.[56–58] The deposited material generally shows no visible porosity, although gas evolution during melting due to excessive moisture in the powder can cause large pores in the deposit. Parts generally show excellent layer-to-layer bonding, although lack-of-fusion defects can form at the layer interfaces when the parameters are not properly optimized.

Parts produced using these processes experience a complex thermal history in a manner very similar to multipass weld deposits. Variations in cooling rate may result as part construction is progressing due to heat build-up. Also, the heat energy introduced during the deposition of a layer can reheat previously deposited material into various phase fields. The thermal cycle experienced can be different at each point in the reheated material depending on its distance from the heat source leading to a variety of microstructures over a short distance. Reheating can also result in some degree of stress relieving due to annealing or tempering effects as well as cause precipitation of strengthening phases in age-hardenable systems. A number of publications describe microstructural evolution in laser deposited parts in Ti–6Al–4V, 304 stainless steel, and others.[48,59–62]

During laser deposition, residual stresses of significant magnitude can be generated as a result of solidification, which, in the worst case, can lead to cracking during part construction, especially when the deposited material is inherently brittle. Residual stresses pose a significant problem when dealing with metallurgically incompatible dissimilar material combinations. Formation of brittle intermetallic phases at interfaces in combination with residual stresses can lead to cracking. In such applications, use of a suitable interlayer can solve the cracking problem. For example, Kumar and Stucker have successfully deposited CoCrMo on a porous Ta substrate using the LENS process employing Zr as an interlayer, a combination that is otherwise quite prone to cracking.[63]

When compared to their wrought or cast equivalents, laser deposited parts generally exhibit superior yield and tensile strengths because of their relatively finer grain sizes.[47,64–66] With respect to tensile elongation, however, laser fabricated parts are typically inferior.[47,48] Parts exhibit better properties when the layers are oriented parallel to the stress axis (longitudinal) than when they are oriented perpendicular to it (transverse).[48] Wu et al.[65] reported superior creep resistance in a laser deposited Ti alloy compared to its wrought equivalent, which was attributed to a higher density of precipitate particles at the grain boundaries. Similarly, Xue and Islam[64] reported substantially higher stress rupture life in laser deposited superalloy IN 738 compared to its cast equivalent, which was attributed to uniform gamma prime precipitation and fine, uniform carbide distribution together with the directionally solidified columnar grain structure in the laser deposited material.

The main advantage of laser-powder deposition processes lies in their ability to produce fully dense parts without subsequent furnace processing. With these processes, materials that cannot be processed through conventional techniques can be formed in complex geometries with intricate internal cavities that cannot be machined directly. These processes can produce functionally graded materials and can facilitate microstructural design on a very fine scale.[67] For example, Banerjee et al.[68] produced a gradual transition in a LENS deposit composition and microstructure from binary Ti–8Al to Ti–8Al–20V within a length of 25 mm using elemental Ti, Al, and V powders. Liu and DuPont[50] successfully produced a functionally graded TiC/Ti composite using the LENS process. Similarly, Banerjee et al.[69] produced Ti–6Al–4V/TiB composite parts using the LENS process employing a blend of pure prealloyed Ti–6Al–4V and elemental B powders. These processes also have potential for depositing thin layers of select, dense corrosion and wear resistant metals on porous biomedical implants to improve their service performance (e.g., deposition of dense CoCrMo layers as a bearing surface onto porous Ta[63]). These processes are also capable of producing directionally solidified and single crystal turbine airfoils in superalloys for use in aircraft engines. Further, these processes can be utilized for effectively repairing and refurbishing defective and service-damaged components.[58]

The main limitations of laser deposition processes are poor resolution, surface finish, and build speed compared to other additive manufacturing technologies. However, compared to traditional welding deposition operations, these processes are superior in these respects.

26.3.7 Other Processes

Over the years dozens of approaches to additive manufacturing have been studied. Most commercialized technologies fall within the preceding categories, but there are a number of processes that have been researched and commercialized that are based on other additive approaches. Some of these other approaches are discussed below. This list, though not comprehensive, helps to illustrate the breadth of technology developments within additive manufacturing and the potential for years of technology developments to come.

26.3.7.1 Direct Write Electronics

"Direct Write" (DW) is the ability to write or print passive or active electronic components (conductors, insulators, batteries, capacitors, antennas, etc.) directly from a computer file without any tooling or masks. Many different DW technologies have been developed over the recent years, many following funding by the Defense Advanced Research Projects Agency (DARPA). These DARPA-funded projects included matrix assisted pulsed laser evaporation (MAPLE), nScrypt 3De, maskless mesoscale materials deposition (M^3D), and direct write thermal spray.

DW processes use some sort of 3D programmable dispensing or deposition head to accurately apply small amounts of material automatically to form circuitry or other useful mesoscopic devices. Using these technologies a variety of materials can be deposited (semiconductor and dielectric polymers, conductive metals, resistors, etc.) onto a variety of substrates (plastic, metal, ceramic, glass, and even cloth). With these processes it is possible to apply material traces as small as 10 μm in width even on curved or irregular surfaces. Integration of DW processes with other additive manufacturing processes makes it possible to produce parts with completely embedded electrical circuitry and other electronic devices with a huge application potential in electronics, aerospace, and satellite industries.[70]

A schematic of the MAPLE process is shown in Figure 26.8. In this process, a laser transparent quartz disk or polymer tape is coated on one side with a film (a few microns thick), which consists of a powdered material that is to be deposited and a polymer binder. The coated disk or tape is placed in close proximity and parallel to the substrate. A laser is focused onto the coated film. When a laser pulse strikes the coating, the polymer is evaporated and the powdered material is deposited on the substrate, firmly adhering to it. By appropriate control of the positions of both the ribbon and the substrate, complex patterns can be deposited. There is little to no heating of the substrate on which the material is transferred.

The M^3D process begins with atomization of a liquid molecular precursor or a colloidal suspension of metals, dielectric, ferrite, or resistor powder.[71] The aerosol stream is delivered to a deposition head using a carrier gas. The stream is then surrounded by a coaxial sheath air flow, and exits the chamber through an orifice directed at the substrate. This coaxial flow focuses the aerosol stream onto the substrate and

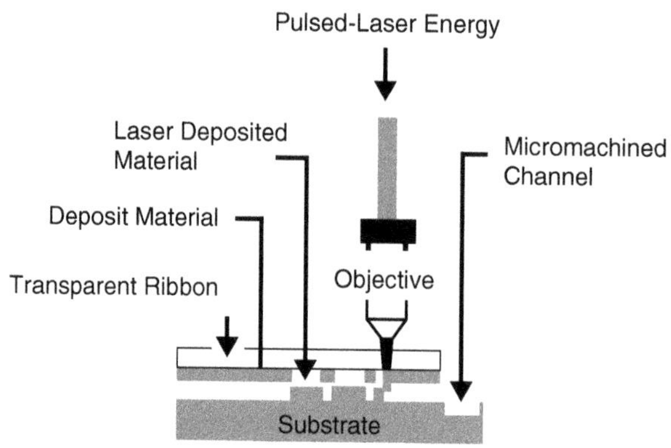

FIGURE 26.8 Schematic of the MAPLE direct write process.

allows for deposition of features with dimensions as small as 25 μm. The substrate is attached to a platen that moves under computer control, so that intricate geometries may be deposited. Either laser chemical decomposition or thermal treatment, often by a coaxial laser system, is used to process the deposit to the desired state.

Pen-based processes, such as Micropen and nScrypt 3De, are close competitors to laser-based DW processes. These processes use dispensing nozzles (50 μm to 2.5 mm in diameter) and a variety of metallic and ceramic slurries or "inks," to write intricate patterns. The electronic inks are heated at low temperatures to evaporate any fluid, leaving behind the dried metal or ceramic, and then fired to sinter the powders together. Two aspects of the pen-based processes make them extremely useful: (1) fine line traces can be deposited on nonplanar substrates, and (2) inks can be custom-designed for specific functional needs.

Thermal spray techniques for DW applications have been demonstrated by researchers at the State University of New York at Stony Brook.[72] With appropriate torch designs and powder particle size distribution, these processes were shown to result in satisfactory deposits (as thin as 5 μm) of a wide range of materials. Thermal spray techniques offer certain unique advantages in DW applications: (1) high writing speed, (2) flexibility with spray materials (metals, ceramics, polymers or any combinations of these materials and incorporation of mixed or graded layers), (3) useful material properties in the as-deposited state, and (4) very low thermal input during processing.

26.3.7.2 Segment Milling Manufacturing

There are a few processes that construct parts segment-by-segment, as opposed to layer-by-layer. These are collectively referred to as "segment milling manufacturing" (SMM) processes. In these processes a part is mathematically split into plate-like free partial bodies or segments rather than thin layers. Each segment is shaped to its 3D shape using a milling machine, often in combination with a deposition system. The shaped segments are then assembled to form a complete 3D part. Parts constructed using these processes are free from stair stepping and exhibit a smooth surface, an advantage when compared to many other additive processes. The accuracies achievable are distinctly higher than other layer-based additive processes and are dependent only on the accuracy of the milling machine. However, these processes are restricted to part geometries that can be divided. Examples of commercially available systems based on SMM include: (1) layer milling manufacturing (Zimmermann GmbH, Germany), (2) stratoconception (Charlyrobot, France), and (3) stratified object manufacturing (MEC GmbH, Germany).

26.3.7.3 Shape Deposition Manufacturing

Another process that incorporates concepts of additive plus subtractive technologies is shape deposition manufacturing (SDM), developed by researchers at Stanford University and Carnegie Mellon University of USA. The SDM acronym has come to encompass the generic method of adding material using a deposition system, machining away excess material, and continuing with addition and subtraction for fabricating 3D parts. In SDM of metals, molten metal droplets are typically deposited by one of three methods. The first is plasma-spray deposition, which involves the repeated spraying of molten metal, where individual spray droplets are of the order of 1 to 10 μm in diameter. The second method is termed microcasting, in which molten metal droplets of the order of 5 to 10 mm in diameter are deposited onto existing layers of the part using conventional arc welding processes.[73,74] The third uses a laser in place of the welding arc for more accurate deposition of the liquid metal.[75] In all these processes, after depositing each layer, a milling machine is used to shape the layer contour. Also, the top surface is face milled to produce a flat surface for the deposition of subsequent layers. These processes show considerable promise for producing high quality dense metallic parts, especially dies, quickly and economically. Although the accuracy and resolution achievable during the building process are generally quite poor, subsequent CNC machining can ensure adequate dimensional accuracy. Due to the additive plus subtractive nature of SDM, it has been used to demonstrate the ability to embed electronics and other functional components within a structure to build functioning encapsulated devices.[76]

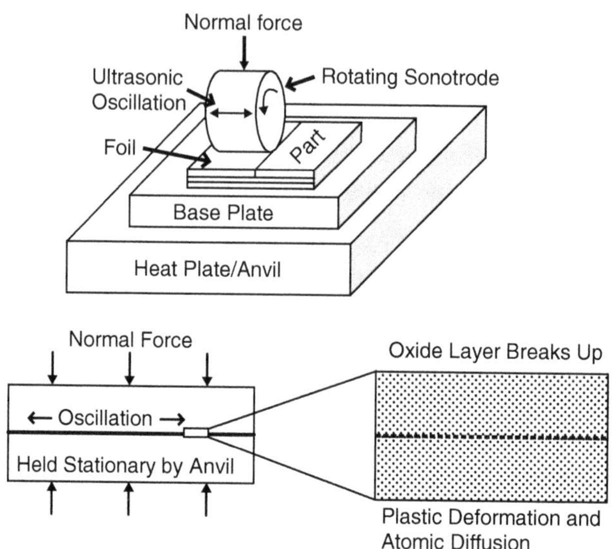

FIGURE 26.9 Schematic of the ultrasonic consolidation process.

26.3.7.4 Ultrasonic Consolidation

Ultrasonic Consolidation is a novel application of ultrasonic welding aimed at fabricating complex 3D structures from metal foils. The process is a marriage between sheet lamination processes and the "additive plus subtractive" nature of SDM processes. This technology was developed and commercialized by Solidica Inc., USA, in 2000.

Figure 26.9 illustrates the basic UC process. In this process a rotating ultrasonic sonotrode travels along the length of a thin metal foil (typically 100 μm thick and 25 mm wide) placed over the substrate. The thin foil is held closely in contact with the substrate by applying a contact pressure via the rotating sonotrode. The sonotrode oscillates transversely to the direction of welding at ultrasonic frequencies, while traveling over the metal foil. The combination of normal and oscillating shear forces results in generation of dynamic stresses at the interface between the two mating surfaces. The stresses produce plastic deformation at the interface, which breaks up the oxide film, producing relatively clean metal surfaces, across which plastic flow and atomic diffusion takes place, establishing a metallurgical bond. After depositing a strip of foil, another foil is deposited adjacent to it. This process repeats until a complete layer is placed. After placing a layer, a computer-controlled milling head shapes the layer to its slice contour. This milling can occur after each layer or, for certain geometries, after several layers have been deposited. Once the layer is shaped to its contour, the chips are blown away using compressed air and foil deposition starts for the next layer. The process takes place at room temperature, in an ambient atmosphere, or at slightly elevated temperatures of 93 upto 150°C using a heated platen. It is possible to simply insert prefabricated components (such as thermal management devices, sensors, computational devices, etc.) into machined cavities of the part under construction prior to encapsulation by subsequent material addition, thus creating structures with embedded functionality.

Many commercially available metal foils, including alloys of aluminum, titanium, magnesium, copper, and steel, can be used for part construction, although most of the work so far has been done on various aluminum alloys.[77,78] Amplitude of oscillation, contact pressure, temperature and weld speed are the most important parameters that must be carefully controlled to ensure proper bonding of the foil to the substrate.

It is worth mentioning that during ultrasonic consolidation processing, 100% bonding does not normally occur. Instead, metal-to-metal bonds are established at a number of points along the interface. One may find metal-to-metal bonded regions, oxide accumulated regions, and a few physical discontinuities

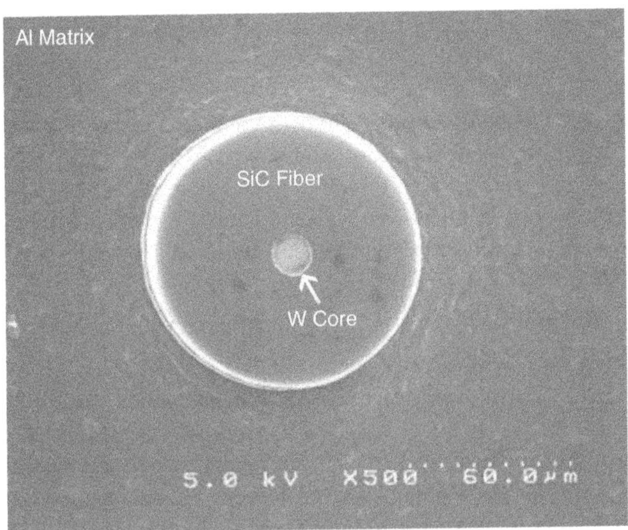

FIGURE 26.10 Microstructure illustrating sound metal flow around a SiC fiber embedded in Al alloy 3003 matrix using the UC process.

(no-contact regions) along the layer interface. These defects along the layer interfaces can be detrimental to part mechanical and corrosion performance. A parameter called "linear weld density" is generally used to represent the percentage of bonded area to the unbonded area along the interface.[77] Linear weld densities levels above 90% can be achieved using optimized parameters. When the surface roughness of a previously deposited layer is removed prior to deposition of a subsequent layer, linear weld densities of 100% have been achieved.

UC combines the advantages of additive and subtractive fabrication approaches, allowing complex 3D parts to be formed with high dimensional accuracy and surface finish, including objects with complex internal passageways, objects made up of multiple materials, and objects integrated with wiring, fiber optics, sensors, and instruments. Because the process does not involve melting, there is negligible shrinkage, residual stresses, and distortion in the finished parts. One unique aspect of UC is that highly localized plastic flow around embedded structures is possible, resulting in sound physical/mechanical bonding between the embedded material and matrix material. This is illustrated in Figure 26.10, an embedded SiC fiber that clearly reveals flow lines in the matrix material in a circular pattern around the SiC fiber evidencing extensive localized plastic flow. Kong et al. have also successfully embedded shape memory fibers in a 3003 Al alloy matrix using the UC process.[79] This ability to embed fibers within the matrix can be utilized in a number of ways, including manufacture of fiber-reinforced metal matrix composites with structural fibers for localized stiffening, optical fibers for communication and sensing, shape memory fibers for actuation, or wire meshes for planar or area stiffening.

26.4 Modeling

Modeling of the various additive manufacturing processes has been a major focus during the past five years. This is due to the fact that further advancements in the field are heavily dependent on an in-depth theoretical understanding of the underlying process principles. Manufacture of high quality functional parts requires careful process optimization and real-time process control, which critically depend on a quantitative understanding of the effects of process parameters on process characteristics and deposit characteristics.

Residual stress is one example of major modeling efforts that have occurred across additive manufacturing platforms. Cracking and dimensional inaccuracies arising from residual stresses are major concerns for many processes. An understanding of how residual stresses develop and how they lead to tolerance loss is a key issue. Modeling has been done for many additive manufacturing processes, including SLS,[80] LENS[48,81–84] and SDM processes.[85–87] For example, Klingbeil et al.[85] have developed a residual stress model for SDM processes. The results suggest that both the method of deposition and deposition path can have a substantial effect on the magnitude of residual stresses and warping. Substrate preheating and substrate insulation yield substantial payoffs for limiting residual stress-induced warping. The results also suggest that even small changes in maintaining 3D mechanical constraints can result in unacceptable warping and that mechanical constraint design should be based on deposition path.

26.4.1 Metal Deposition Modeling

One of the most active areas of process modeling for additive manufacturing has been in the laser-based metal deposition processes. Efforts have been made to understand thermal processing, the behavior of molten pools, solidification-related phenomena, and the role of process variables on material properties.[81,83,84] For example, Beuth and Klingbeil[81] in an analytical investigation discussed the role of process variables in laser-based direct fabrication. The authors showed how process variables relate to deposit characteristics by modeling the solidification process. A process map was constructed based on their model which predicted melt pool size, thermal gradients, microstructure, and residual stress build up in thin-walled structures with reasonable accuracy over a wide range of process parameters. Such an approach is useful, as it can allow precise correlation between process variables, solidification process and part microstructure and provide a means for process control.

Several research groups have reported closed-loop control optimization for deposition processes using a CCD camera[88] or a phototransistor[89] as the sensing device. Controlling the powder delivery system to achieve a stable or controllable powder supply[90] or to study the powder stream distribution and laser-powder interaction[91] is also an important area of research. Hu and Kovacevic[92] developed a closed-loop control system for heat input control in laser deposition processes where an optoelectronic sensor was used to sense the powder delivery rate in real time at a high sampling frequency. Such powder delivery rate sensors are particularly important when producing parts with multimaterial and graded compositions.

26.4.2 Powder Bed Sintering Modeling

Attempts have been made to understand the heating, melting, and solidification processes, and residual stresses and distortion in laser sintering processes.[93,94] Shiomi et al.[93] investigated using a pulsed laser beam, where the results suggested both the maximum temperature of the powder and the volume of the solidified part were affected by the peak power rather than the duration of the laser irradiation. Matsumoto and coworkers[94] predicted the residual stresses in laser sintering processes would result in deflection increase with increase in track length. Based on these observations, it was suggested that it would be beneficial to divide the scan area into small segments and then scan each segment with short tracks. This type of scan pattern has been utilized in various manufacturers' embodiments of powder bed melting/sintering processes, particularly EOS and Arcam machines.

26.4.3 Multimaterial Modeling

One of the most powerful aspects of additive manufacturing is its suitability for producing parts with multiple materials and functionally graded materials (FGMs). The ability to precisely control local composition opens up endless possibilities for component design. Despite these exciting possibilities, there is an absence of knowledge, methods, and tools in the area of computer representation and design of parts with local composition control. New CAD tools are needed to model graded compositions, and manufacturers need algorithms capable of converting these models into machine instructions for their fabrication.

Extensive research has gone into developing methods for design, representation, exchange, and processing of parts with graded compositions in the recent years.[95–97] Current approaches include volume meshing, cellular decompositions, tetrahedral meshing, and a parametric and feature-based methodology for the design of solids with local composition control.

26.5 Applications

Additive manufacturing technologies have been used in virtually every industry, from children's toys to aerospace components. The industry has grown into a billion-dollar industry and is currently on an upward trend.[23] Rapid prototyping has so far been the largest application class for these technologies. However, direct manufacturing applications are growing rapidly, which should eventually dominate the industry.

26.5.1 Rapid Prototyping

Additive manufacturing technologies have become an integral part of product design, development, and manufacture in recent years as the benefits of prototyping are increasingly being realized. Parts produced are being used as concept models for idea realization, presentation models for effective communication and customer feedback as well as fully functional prototypes for rigorous validation of form, fit, and functionality of design before committing to expensive and time-consuming traditional manufacturing operations. Successful examples of prototyped components reducing the cost and time associated with product development abound.[98] Automotive, aerospace, defense, medical, consumer electronics, telecommunications, sporting goods, toys, architecture, and virtually every other industry have success stories centered on rapid prototypes.

26.5.2 Rapid Tooling

Central to the theme of rapid tooling is the ability to produce long-lasting tooling with improved thermal and mechanical characteristics, incorporating shorter lead-times and with an overall cost savings. Additive manufacturing technologies are beginning to make headway as methods for rapid and economical manufacture of various tooling components such as: (1) dies and inserts for injection molding, die casting, and vacuum casting; (2) patterns, cores, and molds for sand casting; (3) sacrificial patterns and shells for investment casting; (4) dies for metal forming operations; and (5) many other tooling applications for series production of plastic and metallic functional parts.

26.5.3 Direct Manufacturing

The use of additive technologies for rapid manufacturing falls into two categories: indirect and direct. Indirect manufacturing involves manufacture of a pattern or a mold, as discussed previously, and has long been successfully adopted for manufacture of end-use components. Successful examples of this indirect manufacturing approach are plentiful.[99] Direct manufacturing methods, on the other hand, create final end-use components in a single operation. This is an area that is rapidly growing, with increasing numbers of manufacturers beginning to exploit the unique capabilities of additive processes for direct part production.

26.5.3.1 Aerospace and Automotive

The manufacturing environment in the aerospace industry is characterized by small production runs, high manufacturing complexity, and expensive materials, which ideally suits additive manufacturing technologies. Further, additive manufacturing technologies embody the characteristics of agile manufacturing, which has long been identified as a key enabler to world-class aerospace manufacturing. As a result, the aerospace industry has been one of the first to implement direct manufacturing applications. NASA has identified additive manufacturing as an enabling technology for future space missions.[100] The vision is to make use of additive manufacturing for large structures (including moon and Mars dwellings) and replacement parts as well as for repair of damaged components in space.

The automotive industry is another industry where additive manufacturing technologies hold great application potential. Automotive giants such as Daimler-Benz, Ford, Honda, General Motors, and Volkswagen, have already been using these techniques for manufacture of functional parts adopting both direct and indirect approaches.[101,102] Although direct manufacturing is primarily applied to custom or low-volume products at present, these technologies have potential for large-scale production of small, intricate components such as automotive electrical connectors.

26.5.3.2 Biomedical

Research into biomedical applications of additive manufacturing abound. These applications include manufacturing of medical devices, customized implants and prostheses, manufacture of controlled drug delivery systems, surgical planning, and tissue engineering.[103] Customized prosthetics and implants based on anatomical data from imaging systems such as computed tomography have been investigated. Additive manufacturing processes have also been employed to produce various artificial joints and load-bearing implants in biocompatible materials such as hydroxyapatite, titanium, CoCrMo, and others.

Tissue engineering is an important application area for additive manufacturing technologies.[104,105] Most tissue engineering strategies for creating functional replacement tissues or organs rely on a design that closely resembles the natural extracellular structure, a highly porous microstructure with interconnected pore networks that allow cell in-growth and reorganization, and special surface chemistry and material characteristics that are spatially variable, favor cellular attachment and proliferation, and exhibit controlled degradation. Additive manufacturing technologies offer several advantages over conventional techniques for producing scaffolds with these characteristics, and various processes and biocompatible materials have been demonstrated.[13,106,107]

26.5.4 Repair

Metal deposition processes have been successfully employed for repair of defective and service-damaged components.[108] This capability is particularly attractive for high value products such as turbine airfoils or large castings for construction equipment. The feasibility of repairing directionally solidified and single crystal superalloy aerofoils has been demonstrated by several investigators employing direct metal deposition processes.[56–58]

26.6 Future Directions

Despite the many benefits of additive manufacturing, less than 25% of the design, product development, and manufacturing communities are making use of it at present.[23] As with any new technology, industry's resistance to change and reluctance to work through the challenges of a developing technology are inhibiting adoption. Many experts believe that a general lack of awareness and understanding of the benefits and appropriate application of additive manufacturing technologies is hindering growth of the industry. Thus, education of the next generation of designers, manufacturers, and engineers is critical to the effective adoption of these technologies.

A few common drawbacks must be the focus of current and future developments to see sustained additive manufacturing industry growth. These include: (1) inability to meet tight dimensional accuracy and surface finish requirements, (2) lack of material options, (3) unsatisfactory material properties, (4) lack of consistency and repeatability in process and part characteristics, (5) high cost of equipment and materials, and (6) long part construction time for high-volume manufacture.

Notwithstanding the remarkable progress made over the years, many processes still fall short of industry expectations. Software approximation errors, unacceptable stair-stepping effects, and improper compensation for material shrinkage are still common sources of error in additive manufacturing processes.[109,110] Closed-loop process monitoring and control strategies are rare in the industry, which results in repeatability issues and a high dependence on operator skills. In addition, a quantitative understanding of the effects of process parameters, build style, support structures, and other factors on the magnitude of shrinkage, residual stresses, and distortion is rarely used to enhance product performance.

Broadening the range of materials that can be processed is critical for realizing widespread application of direct manufacturing. Greater benefits might be realized by developing application-specific or material-specific machines for direct manufacturing rather than all-purpose machines, as is the tradition with additive manufacturing companies.

Local composition control and microstructural design are two strong points of additive manufacturing that merit additional study. Further advancements in 3D CAD modeling and data representation methods are essential to fully realize the multimaterial capabilities of additive manufacturing technologies.

Finally, it should be noted that additive manufacturing technologies will likely never eliminate the need for conventional manufacturing technologies. High-volume manufacturing techniques like casting, forging, and injection molding will always have their place. Additive manufacturing should be treated as complementary techniques to conventional manufacturing techniques, and applied where appropriate. If properly applied, additive manufacturing has the potential to provide a giant leap forward in product design, development, and manufacturing.

References

[1] Hull, C., U.S. Patent No. 4575330, Apparatus for the production of three dimensional objects by stereolithography, March 11, 1986.

[2] Chua, C.K., Leong, K.F., and Lim, C.S., *Rapid Prototyping: Principles and Applications*, 2nd ed., World Scientific Publishing Co. Pvt. Ltd., Singapore, 2003.

[3] Giri, D., Jouaneh, M., and Stucker, B.E., Error sources in a 3D reverse engineering process, *Prec. Eng. J. Int. Soc. Prec. Eng. Nanotechnol.*, 28, 242, 2004.

[4] Gebhardt, A., *Rapid Prototyping*, Hanser Publishers, Germany, 2003.

[5] Burns, M., *Automated Fabrication: Improving Productivity in Manufacturing*, PTR Englewood Cliffs, NJ, Pretence Hall, 1993.

[6] Zhou, J.G., Herscovici, D., and Chen, C.C., Parametric process optimization to improve the accuracy of rapid prototyped stereolithography parts, *Int. J. Mach. Tools Manuf.*, 40, 363, 2000.

[7] Schaub, D.A. and Montgomery, D.C., Using experimental design to optimize the stereolithography process, *Qual. Eng.*, 9, 575, 1997.

[8] Jacobs, P.F., *Rapid Prototyping and Manufacturing: Fundamentals of Stereolithography*, 1st ed., Society of Manufacturing Engineering, Dearborn, MI, 1992.

[9] Yi, F., Wu, J., and Xian, D., LIGA technique for microstructure fabrication, *Microfab. Technol.*, 4, 1, 1993.

[10] Ikuta, K., Maruo, S., and Kojima, S., New microstereolithography for freely movable 3D micro structures — Super IH process for submicron resolution, in *Proceedings of the IEEE Micro Electro Mechanical Systems*, The Eleventh IEEE International Workshop on Micro Electro Mechanical Systems (MEMS '98), Heidelberg, Germany, January 25–29, 1998, 290.

[11] Carroza, M.C. et al., Piezoelectric-drive stereolithography fabricated micropump, *J. Micromech. Microeng.*, 5, 177, 1995.

[12] Varadan, V.K. and Varadan, V.V., Microstereolithography for fabrication of 3D polymeric and ceramic MEMS, in *Proceedings of the Conference on SPIE: MEMS Design, Fabrication, Characterization and Packaging*, Vol. 4407, Eds., U.F. Behringer and D.G. Uttamchandani, Edinburgh, UK, 2001, 147.

[13] Zein, I. et al., Fused deposition modeling of novel scaffold architectures for tissue engineering applications, *Biomaterials*, 23, 1169, 2002.

[14] Wu, G. et al., Solid freeform fabrication of metal components using fused deposition of metals, *Mater. Des.*, 23, 97, 2002.

[15] Grida, I. and Evans, J.R.G., Extrusion freeforming of ceramics through fine nozzles, *J. Eur. Ceram. Soc.*, 23, 629, 2003.

[16] Cornejo, I.A. et al., Development of bioceramic tissue scaffolds via fused deposition of ceramics, in *Proceedings of the Conference on Bioceramics: Materials and Applications III*, Ed., L. George, American Ceramic Society, Westerville, OH, 2000, 183.

[17] Anita, K., Arunachalam, S., and Radhakrishnan, P., Critical parameters influencing the quality of prototypes in fused deposition modeling, *J. Mater. Process. Technol.*, 118, 385, 2001.

[18] Lee, B.H., Abdullah, J., and Khan, Z.A., Optimization of rapid prototyping parameters for production of flexible ABS object, *J. Mater. Process. Technol.*, 169(1), 54, 2005.

[19] Kalita, S.J. et al., Development of controlled porosity polymer-ceramic composite scaffolds via fused deposition modeling, *Mater. Sci. Eng. C*, 23, 611, 2003.

[20] Bose, S., Suguira, S., and Bandyopadhyay, A., Processing of controlled porosity ceramic structures via fused deposition, *Scripta Mater.*, 41, 1009, 1999.

[21] Dao, Q. et al., Calculation of shrinkage compensation factors for rapid prototyping (FDM), *Comput. Appl. Eng. Edn.*, 1650, 186, 1999.

[22] Ziemian, C.W. and Crawn, P.M., Computer aided decision support for fused deposition modeling, *Rapid Prototyp. J.*, 7, 138, 2001.

[23] Wohlers, T., Wohlers Report 2005, Rapid Prototyping and Tooling State of the Industry Annual Worldwide Progress Report, Wohlers Associates, Inc., Fort Collins, CO, 2005.

[24] Klosterman, D. et al., Interfacial characteristics of composites fabricated by laminated object modeling, *Compos. A*, 29A, 1165, 1998.

[25] Zhang, Y. et al., Rapid prototyping and combustion synthesis of TiC/Ni functionally gradient materials, *Mater. Sci. Eng A*, A299, 218, 2001.

[26] Burns, M., Heyworth, K.J., and Thomas, C.L., Offset Fabbing, in *Proceedings of Solid Freeform Fabrication Symposium*, Eds., H. Marcus et al., Austin, TX, 1996.

[27] Mueller, B. and Kochan, D., Laminated object manufacturing for rapid tooling and patternmaking in foundry industry, *Comput. Ind.*, 39, 47, 1999.

[28] Yi, S. et al., Study of the key technologies of LOM for functional metal parts, *J. Mater. Process. Technol.*, 150, 175, 2004.

[29] Himmer, T., Nakagawa, T., and Anzai, M., Lamination of metal sheets, *Comput. Ind.*, 39, 27, 1999.

[30] Himmer, T. et al., Metal laminated tooling — A quick and flexible tooling concept, in *Proceedings of the Solid Freeform Fabrication Symposium*, Eds., D.L. Bourell et al., Austin, TX, 2004, 304.

[31] Wimpenny, D.I., Bryden, B., and Pashby, I.R., Rapid laminated tooling, *J. Mater. Process. Technol.*, 138, 214, 2003.

[32] Moon, J. et al., Fabrication of functionally graded reaction infiltrated SiC–Si composite by three-dimensional printing (3DP™) process, *Mater. Sci. Eng. A*, A298, 110, 2001.

[33] Sun, W. et al., Freeform fabrication of Ti_3SiC_2 powder-based structures — Part I: Integrated fabrication process, *J. Mater. Process. Technol.*, 127, 343, 2002.

[34] Cao, W.B. et al., Development of freeform fabrication method for Ti–Al–Ni, *Intermetallics*, 10, 879, 2002.

[35] Sachs, E. et al., Three-dimensional printing: Rapid tooling and prototyping directly from a CAD model, *Trans. ASME: J. Eng. Ind.*, 114, 481, 1992.

[36] Orme, M., Willis, K., and Cornie, J., The development of rapid prototyping of metallic components via ultra fine droplet deposition, in *Proceedings of the 5th Internation Conference on Rapid Prototyping*, Dayton, OH, 1994, 27.

[37] Stucker, B.E., The selective laser sintering process, in *LIA Handbook of Laser Materials Processing*, Eds., J.F. Ready, et al., Laser Institute of America & Magnolia Publishing, Inc., Orlando, FL, USA, 2001, 554.

[38] Kathuria, Y.P., Microstructuring by selective laser sintering of metallic powder, *Surf. Coat. Technol.*, 116–119, 643, 1999.

[39] Stucker, B.E. et al., U.S. Patent No. 5870663, Manufacture and use of ZrO_2/Cu composite electrodes, Feb. 9, 1999.

[40] Lu, L. et al., *In situ* formation of TiC composite using selective laser melting, *Mater. Res. Bull.*, 35, 1555, 2000.

[41] Hardro, P.J., Wang, J.H., and Stucker, B.E., Determining the parameter settings and capability of a rapid prototyping process, *Int. J. Ind. Eng.*, 6, 203, 1999.

[42] Chatterjee, A.N. et al., An experimental design approach to selective laser sintering of low carbon steel, *J. Mater. Process. Technol.*, 136, 151, 2003.

[43] Cormier, D. and Harrysson, O., Electron beam melting of gamma titanium aluminide, in *Proceedings of 16th Solid Freeform Fabrication Symposium*, Eds., D.L. Bourell et al., Austin, TX, 2005.

[44] Sercombe, T.B., Sintering of freeformed maraging steel with boron additions, *Mater. Sci. Eng. A*, A363, 242, 2003.

[45] Simchi, A. et al., On the development of direct metal laser sintering for rapid tooling, *J. Mater. Process. Technol.*, 141, 319, 2003.

[46] Das, S. et al., Processing of titanium net shapes by SLS/HIP, *Mater. Des.*, 20, 115, 1999.

[47] Lewis, G.K. and Schlienger, E., Practical considerations and capabilities for laser assisted direct metal deposition, *Mater. Des.*, 21, 417, 2000.

[48] Griffith, M.L. et al., Understanding the microstructures and properties of components fabricated by LENS, in *Proceedings of the Materials Research Society Symposium*, 625, 2000, 9.

[49] Brice, C.A. et al., Characterization of laser deposited niobium and molybdenum silicides, in *Proceedings of the Materials Research Society Symposium*, 625, 2000, 31.

[50] Liu, W. and DuPont, J.N., Fabrication of functionally graded TiC/Ti composites by Laser Engineered Net Shaping, *Scripta Mater.*, 48, 1337, 2003.

[51] Banerjee, R. et al., In-situ deposition of Ti-TiB composites, in *Proceedings of the Conference on Metal Powder Deposition for Rapid Manufacturing*, Eds., D. Keicher et al., San Antonio, Texas, USA, 2002, 263.

[52] Srivastava, D. et al., The optimization of process parameters and characterization of microstructure of direct laser fabricated TiAl alloy components, *Mater. Des.*, 21, 425, 2000.

[53] Smith, A.R. and Stucker, B.E., Laser deposition of pure titanium powder onto a cast CoCrMo substrate using laser engineered net shaping (LENS), in *Proceedings of the 2nd International Conference on Advanced Research in Virtual and Rapid Prototyping (VRAP 2005)*, Eds., P.J. Bartolo et al., Taylor & Francis (ISSN 1745-2759), Leira, Portugal, 2005, September.

[54] Hofmeister, W. et al., Solidification in direct metal deposition by LENS processing, *J. Metals*, September, 30, 2001.

[55] Hofmeister, W., Melt pool imaging for control of LENS process, in *Proceedings of the Conference on Metal Powder Deposition for Rapid Manufacturing*, Eds., D. Keicher et al., San Antonio, Texas, USA, 2002, 188.

[56] Gaumann, M. et al., Single crystal laser deposition of superalloys: Processing-microstructure maps, *Acta Mater.*, 49, 1051, 2001.

[57] Liu, W. and DuPont, J.N., Effects of melt-pool geometry on crystal growth and microstructure development in laser surface-melted superalloy single crystal: Mathematical modeling of single-crystal growth in a melt pool (Part 1), *Acta Mater.*, 52, 4833, 2004.

[58] Li, L., Repair of directionally solidified superalloy GTD-111 by laser-engineered net shaping, *J. Mater. Sci.*, in press.

[59] Kelly, S.M. et al., Microstructural study of laser formed Ti-6Al-4V, in *Proceedings of the Materials Research Society Symposium*, Eds., S.C. Danforth, D.B. Dimos, and F. Prinz, Materials Research Society, Pittsburgh, PA, Vol. 625, 2000, 3.

[60] Brooks, J.A., Headley, T.J., and Robino, C.V., Microstructures of laser deposited 304L austenitic stainless steel, in *Proceedings of the Materials Research Society Symposium*, Eds., S.C. Danforth, D.B. Dimos, and F. Prinz, Materials Research Society, Pittsburgh, PA, Vol. 625, 2000, 21.

[61] Banerjee, R. et al., Microstructural evolution in laser deposited Ni-25 at.% Mo alloy, *Mater. Sci. Eng.*, A347, 2003, 1.

[62] Stucker, B.E., Esplin, C., and Justin, D., An investigation of LENS-deposited medical-grade CoCrMo, in *Proceedings of the Solid Freeform Fabrication Symposium*, Eds., D.L. Bourell et al., Austin, TX, 2004.

[63] Kumar, S.S. and Stucker, B.E., Development of a Co-Cr-Mo to tantalum transition using LENS for orthopedic applications, in *Proceedings of the Solid Freeform Fabrication Symposium*, Eds., D.L. Bourell et al., Austin, TX, 2005.

[64] Xue, L. and Islam, M., Laser consolidation — A novel one-step manufacturing process from CAD models to net-shape functional components, in *Proceedings of the Conference on Metal Powder Deposition for Rapid Manufacturing*, Eds., D. Keicher et al., San Antonio, Texas, USA, 2002, 61.

[65] Wu, X. et al., Microstructure and mechanical properties of a laser fabricated Ti alloy, in *Proceedings of the Conference on Metal Powder Deposition for Rapid Manufacturing*, Eds., D. Keicher, et al., San Antonio, Texas, USA, 2002, 96.

[66] Keicher, D.M. and Smugeresky, J.E., The laser forming of metallic components using particulate materials, *J. Metals*, May 1997, 51–54.

[67] Ensz, M. et al., Critical issues for functionally graded material deposition by laser engineered net shaping, in *Proceedings of the Conference on Metal Powder Deposition for Rapid Prototyping*, Eds., D. Keicher et al., San Antonio, Texas, USA, 2002, 195.

[68] Banerjee, R. et al., Precipitation of grain boundary in a laser deposited compositionally graded Ti-8Al-xV alloy — An orientation microscopy study, *Acta Mater.*, 52, 377, 2004.

[69] Banerjee, R. et al., Direct laser deposition of in situ Ti-6Al-/4V-/TiB composites, *Mater. Sci. Eng. A*, A358, 343, 2003.

[70] Palmer, J.A. et al., Stereolithography: A basis for integrated meso manufacturing, in *Proceedings of the 16th Solid Freeform Fabrication Symposium*, Eds., D.L. Bourell et al., Austin, TX, 2005.

[71] Essien, M. and Renn, M.J., Development of mesoscale processes for direct write fabrication of electronic components, in *Proceedings of the Conference on Metal Powder Deposition for Rapid Manufacturing*, Eds., D. Keicher et al., San Antonio, Texas, USA, 2002, 209.

[72] Sampath, S. et al., Thermal spray techniques for fabrication of meso-electronics and sensors, in *Proceedings of the Materials Research Society Symposium*, Eds., S.C. Danforth, D.B. Dimos, and F. Prinz, Materials Research Society, Pittsburgh, PA, vol. 625, 2000, 181.

[73] Song, Y. et al., 3D welding and milling: Part I — A direct approach for freeform fabrication of metallic prototypes, *Int. J. Mach. Tools Manuf.*, 45, 1057, 2005.

[74] Zhang, Y. et al., Weld deposition-based rapid prototyping: a preliminary study, *J. Mater. Process. Technol.*, 135, 347, 2003.

[75] Fessler, J.R. et al., Laser deposition of metals for shape deposition manufacturing, in *Proceedings of the Solid Freeform Fabrication Symposium*, Eds., D.L. Bourell et al., Austin, TX, 1996, 117.

[76] Li, X. and Prinz, F., Metal embedded fiber bragg grating sensors in layered manufacturing, *J. Manuf. Sci. Eng.*, 125, 577, 2003.

[77] Kong, C.Y., Soar, R.C., and Dickens, P.M., Characterization of aluminium alloy 6061 for the ultrasonic consolidation process, *Mater. Sci. Eng. A*, A363, 99, 2003.

[78] White, D.R., Ultrasonic consolidation of aluminium tooling, *J. Adv. Mater. Process.*, 161, 64, 2003.

[79] Kong, C.Y., Soar, R.C., and Dickens, P.M., Ultrasonic consolidation for embedding SMA fibres within aluminium matrices, *Compos. Struct.*, 66, 421, 2004.

[80] Karapatis, N. et al., in *Proceedings of the Solid Freeform Fabrication Symposium*, Eds., H.L. Marcus et al., Austin, TX, 1998, 79.

[81] Beuth, J. and Klingbeil, N., The role of process variables in laser based direct metal solid freeform fabrication, *J. Metals*, Sept., 36, 2001.

[82] Rangaswamy, P. et al., Residual stresses in LENS® components using neutron diffraction and contour method, *Mater. Sci. Eng. A*, 399(1-2), 72, 2005.
[83] Dai, K. and Shaw, L., Thermal and mechanical finite element modeling of laser forming from metal and ceramic powders, *Acta Mater.*, 52, 69, 2004.
[84] Kahlen, F.J. and Kar, A., Residual stresses in laser deposited metal parts, *J. Laser Appl.*, 13, 60, 2001.
[85] Klingbeil, N.W. et al., Residual stress-induced warping in direct metal solid freeform fabrication, *Int. J. Mech. Sci.*, 44, 57, 2002.
[86] Nickel, A.H., Barnett, D.M., and Prinz, F.B., Thermal stresses and deposition patterns in layered manufacturing, *Mater. Sci. Eng.*, A317, 59, 2001.
[87] Beuth, J.L. and Narayan, S.H., Residual stress-driven delamination in deposited multi-layers, *Int. J. Solids Struct.*, 33, 65, 1996.
[88] Hofmeister, W.H., McCallum, D.O., and Knorovsky, G.A., Video monitoring and control of the LENS process, in *Proceedings of the AWS 9th International Conference on Computer Technology in Welding*, Detroit, MI, The American Welding Society, Sept. 28–30, 1999, 187.
[89] Mazumder, J. et al., Closed-loop direct metal deposition: Art to part, *Optics Laser Eng.*, 34, 397, 2000.
[90] Li, I. and Steen, W.M., Sensing, modeling and closed loop control of powder feeder for laser surface modification, in *Proceedings of ICALEO*, 1993, 965.
[91] Vetter, P.A. et al., Laser cladding: Relevant parameters for process control, *SPIE* 2207, 1994, 452.
[92] Hu, D. and Kovacevic, R., Sensing, modeling and control of laser-based additive manufacturing, *Int. J. Mach. Tools Manuf.*, 43, 2003, 51.
[93] Shiomi, M. et al., Finite element analysis of melting and solidifying processes in laser rapid prototyping of metallic powders, *Int. J. Mach. Tools Manuf.*, 39, 237, 1999.
[94] Matsumoto, M. et al., Finite element analysis of single layer forming on metallic powder bed in rapid prototyping by selective laser processing, *Int. J. Mach. Tools Manuf.*, 42, 61, 2002.
[95] Liu, H. et al., Methods for feature-based design of heterogeneous solids, *Comput. Aided Des.*, 36, 1141, 2004.
[96] Rajagopalan, S. et al., Representation of heterogeneous objects during design, processing and freeform-fabrication, *Mater. Des.*, 22, 185, 2001.
[97] Zhou, M.Y. et al., Modeling and processing of functionally grade materials for rapid prototyping, *J. Mater. Process. Technol.*, 146, 396, 2004.
[98] Grimm, T., *User's Guide to Rapid Prototyping*, Society of Manufacturing Engineers, Michigan, MI, USA, 2004.
[99] Hongjun, L. et al., A note on rapid manufacturing process of metallic parts based on SLS plastic prototype, *J. Mater. Process. Technol.*, 142, 710, 2003.
[100] Karen, M. et al., Solid freeform fabrication: An enabling technology for future space missions, in *Proceedings of the Conference on Metal Powder Deposition for Rapid Manufacturing*, Eds., D. Keicher et al., San Antonio, Texas, USA, 2002, 51.
[101] Rapid Prototyping Report, Volkswagen uses laminated object manufacturing to prototype complex gear box housing, 5(2), CAD/CAM Publishing Inc., 1995.
[102] Muller, H. and Sladojevic, J., Rapid tooling approaches for small lot production of sheet metal parts, *J. Mater. Process. Technol.*, 115, 97, 2001.
[103] Mohoney, D.P., Rapid prototyping in medicine, *Computer Graphics World*, 18, 42, 1995.
[104] Leong, K.F. et al., Solid freeform fabrication of three-dimensional scaffolds for engineering replacement tissues and organs, *Biomaterials*, 24, 2363, 2003.
[105] Sachlos, E. et al., Novel collagen scaffolds with predefined internal morphology made by solid freeform fabrication, *Biomaterials*, 24, 1487, 2003.
[106] Lam, C.X.F. et al., Scaffold development using 3D printing with a starch-based polymer, *Mater. Sci. Eng.*, C20, 49, 2002.

[107] Sachlos, E. et al., Novel collagen scaffolds with predefined internal morphology made by solid freeform fabrication, *Biomaterials*, 24, 1487, 2003.
[108] Capello, E., Colombo, D., and Previtali, B., Repairing of sintered tools using laser cladding by wire, *J. Mater. Process. Technol.*, 164–165, 990, 2005.
[109] Yan, X., A review of rapid prototyping technologies and systems, *Comput. Aided Des.*, 28, 307, 1996.
[110] Balsmeier, P., Rapid prototyping: State-of-the-art manufacturing, *Ind. Mgt.*, 39, 55, 1997.

27

Solidification Macroprocesses (Thermal — Mechanical Modeling of Stress, Distorsion and Hot-Tearing)

Michel Bellet
Ecole des Mines de Paris

Brian G. Thomas
University of Illinois at Urbana-Champaign

Abstract

Computational modeling of thermal–mechanical behavior has great potential for understanding the formation of defects during solidification, especially when combined with appropriate failure criteria. This chapter summarizes the constitutive equations that can be used to model the complex phenomena that govern thermal–mechanical behavior of solidification macroprocesses. Important issues in model development are discussed, including the numerical treatment of thermal–mechanical coupling, boundary conditions, solid, mushy, and liquid phases, contact, and methods to solve the highly nonlinear and coupled equations. The steps in model development are extended to the prediction of hot-tear crack formation and validation with numerical benchmarks and experiments. Finally, three different practical applications of thermal–mechanical models are presented. They range from distortion of complex-shaped foundry castings, taper optimization, and longitudinal crack formation during continuous casting of steel in the mold region, and the cyclic bulging and stresses that develop between support rolls during continuous casting of slabs.

27.1 Introduction

The application of numerical methods to the mechanical modeling in solidification analysis has received a continuously growing interest for the past 20 years. After having concentrated their efforts on the thermal and microstructural predictions, research teams have been more and more interested in the coupled thermomechanical analysis. These developments have been motivated by the efforts by the casting industry to increase the quality of final products while lowering the costs. Manufacturers are then interested in the development of new numerical tools able to model the thermomechanical response of castings during the processes. Accurate calculation of stress and distortions during casting is just the first step, however, as engineers are more interested in their practical consequences. These include residual stress and distortion, and defects such as segregation and the formation of cracks such as hot tears. As computing power and software tools advance, it is becoming increasingly possible to perform useful mechanical analysis of castings and their important related behaviors.

Despite the considerable progress in computational structural mechanics, and in the capabilities of intensive computing, the thermomechanical analysis of castings is still a challenge nowadays, for the following reasons:

- Many interacting physical phenomena are involved in stress–strain formation. Stress arises primarily from the mismatch of strains caused by the steep temperature gradients of solidification, and depends on the time- and microstructure-dependent inelastic flow of the material.
- Predicting distortions and residual stresses in cast products means being able to describe and calculate the history of the cast product and its environment on huge temperature intervals. This makes the mechanical problem highly nonlinear, involving liquid–solid interaction and quite complex constitutive equations. Also the identification of reliable values of the numerous parameters involved in those relations is a very difficult task.
- The coupling between the thermal and the mechanical problems is an additional difficulty. This coupling comes from the mechanical interaction between the casting and the mold components, through gap formation or the build-up of contact pressure, modifying locally the heat exchange. This adds some complexity to the nonlinear heat transfer resolution.
- As it does not make sense to perform such analyses without accounting for the presence of molds and their interaction with the castings, the problem to be solved is multidomain, often involving numerous deformable and interacting components.
- Shapes of cast parts are by essence complex, which first brings out frequent difficulties regarding the interface between CAD design and the mechanical solvers, and second demands great computational resources.
- In the case of solidification of semifinished products, by continuous casting, the computational demand is also outstanding, because of the characteristic dimensions to be considered: meters, tens of meters.

This chapter summarizes some of the issues and approaches in performing computational analyses of mechanical behavior, distortion, and hot tearing during solidification. The governing equations are presented first, followed by a brief description of the methods used to solve them. Finally, a few examples of recent applications in shape casting and continuous casting are introduced.

27.2 Constitutive Models for Metallic Alloys in Liquid, Mushy, and Solid State

The modeling of mechanical behavior requires solution of the equilibrium equations (relating force and stress), constitutive equations (relating stress and strain), and compatibility equations (relating strain and displacement). In casting analysis, the cast material may be in the liquid, mushy, or solid state. Therefore, mechanical modeling of casting processes has to consider constitutive models for each of these states.

27.2.1 Liquid State: Newtonian Model

Metallic alloys are generally considered as Newtonian fluids. Including thermal dilatation effects, the constitutive equation can be expressed as follows:

$$\dot{\boldsymbol{\varepsilon}} = \frac{1}{2\mu_l}\boldsymbol{s} - \frac{1}{3\rho}\frac{d\rho}{dt}I \tag{27.1}$$

in which the strain rate tensor $\dot{\boldsymbol{\varepsilon}}$ is split into a mechanical part, showing the linear relation between the strain rate tensor and the stress deviator $\boldsymbol{s}$, and a thermal part. In this equation, μ_l is the dynamic viscosity of the liquid alloy, ρ is the density, and $\boldsymbol{I}$ is the identity tensor. Taking the trace of this expression, $\mathrm{tr}\dot{\boldsymbol{\varepsilon}} = \nabla \cdot \boldsymbol{v}$, the mass conservation equation is recovered:

$$\frac{d\rho}{dt} + \rho\nabla \cdot \boldsymbol{v} = \frac{\partial\rho}{\partial t} + \nabla \cdot (\rho\boldsymbol{v}) = 0 \tag{27.2}$$

In casting processes, the liquid flow may be turbulent, even after mold filling. This may occur because of buoyancy forces or forced convection like in jets coming out of the nozzle outlets in continuous casting processes. The most accurate approach, direct numerical simulation, is generally not feasible for industrial processes, owing to their complex-shaped domains and high turbulence. To compute just the large-scale flow features, turbulence models are used, which increase the liquid viscosity according to different models of the small-scale phenomena. These models include the simple "mixing length" models, the two-equation models such as k-ε, and large eddy simulation (LES) models, which have been compared with each other and with measurements of continuous casting.[1–3]

27.2.2 Mushy State: Non-Newtonian Model

Metallic alloys in the mushy state are very complex two-phase liquid–solid media. Their mechanical response is highly dependent on the local microstructural evolution, which involves several complex physical phenomena. To overcome this difficulty, the mushy state may be considered in a first approach as a single continuum. The mushy material is then modeled as a non-Newtonian fluid, according to the following equations:

$$\begin{cases} \dot{\boldsymbol{\varepsilon}} = \dot{\boldsymbol{\varepsilon}}^{\mathrm{vp}} + \dot{\boldsymbol{\varepsilon}}^{\mathrm{th}} \\ \dot{\boldsymbol{\varepsilon}}^{\mathrm{vp}} = \dfrac{3}{2K}(\dot{\varepsilon}_{\mathrm{eq}})^{1-m}\boldsymbol{s} \\ \dot{\boldsymbol{\varepsilon}}^{\mathrm{th}} = -\dfrac{1}{3\rho}\dfrac{d\rho}{dt}\mathbf{I} \end{cases} \tag{27.3}$$

in which K is the viscoplastic consistency and m the strain rate sensitivity. Denoting $\sigma_{\mathrm{eq}} = \sqrt{3/2 s_{ij}s_{ij}}$ the von Mises equivalent stress scalar, and $\dot{\varepsilon}_{\mathrm{eq}} = \sqrt{2/3\dot{\varepsilon}_{ij}^{\mathrm{vp}}\dot{\varepsilon}_{ij}^{\mathrm{vp}}}$ the von Mises equivalent strain rate scalar, Equation 27.3 yields the well-known power law: $\sigma_{\mathrm{eq}} = K(\dot{\varepsilon}_{\mathrm{eq}})^m$. It can be noticed that the preceding Newtonian model is actually a particular case of the non-Newtonian one: Equation 27.1 can be derived from Equation 27.3 taking $m = 1$ and $K = 3\mu_l$. The solidification shrinkage is included in the third equation, as we can write in the solidification interval $\rho = g_s\rho_S + g_l\rho_L$ with ρ_S and ρ_L the densities at the solidus and liquidus temperatures, respectively. Hence, we have:

$$\mathrm{tr}\,\dot{\boldsymbol{\varepsilon}}^{\mathrm{th}} = -\frac{1}{\rho}\frac{d\rho}{dt} = -\frac{1}{\rho}(\rho_S - \rho_L)\frac{dg_s}{dt} \approx \frac{\rho_L - \rho_S}{\rho_L}\frac{dg_s}{dt} \tag{27.4}$$

27.2.3 Solid State: Elastic-Viscoplastic Models

In the solid state, metallic alloys can be modeled either as elastic-plastic or elastic-viscoplastic materials. In this latest class of models, one of the simpler is expressed as follows, but it should be mentioned that a lot of models of different complexity can be found in the literature:[4,5]

$$\dot{\boldsymbol{\varepsilon}} = \dot{\boldsymbol{\varepsilon}}^{\mathrm{el}} + \dot{\boldsymbol{\varepsilon}}^{\mathrm{in}} + \dot{\boldsymbol{\varepsilon}}^{\mathrm{th}} \tag{27.5a}$$

$$\dot{\boldsymbol{\varepsilon}}^{\mathrm{el}} = \frac{1+\nu}{E}\dot{\boldsymbol{\sigma}} - \frac{\nu}{E}\mathrm{tr}(\dot{\boldsymbol{\sigma}})\mathbf{I} + \dot{T}\frac{\partial}{\partial T}\left(\frac{1+\nu}{E}\right)\boldsymbol{\sigma} - \dot{T}\frac{\partial}{\partial T}\left(\frac{\nu}{E}\right)\mathrm{tr}(\boldsymbol{\sigma})\mathbf{I} \tag{27.5b}$$

$$\dot{\boldsymbol{\varepsilon}}^{\mathrm{in}} = \frac{3}{2\sigma_{\mathrm{eq}}}\left\langle\frac{\sigma_{\mathrm{eq}} - \sigma_0}{K}\right\rangle^{1/m}\mathbf{s} \tag{27.5c}$$

$$\dot{\boldsymbol{\varepsilon}}^{\mathrm{th}} = -\frac{1}{3\rho}\frac{d\rho}{dt}\mathbf{I} \tag{27.5d}$$

The strain rate tensor $\dot{\boldsymbol{\varepsilon}}$ is split into an elastic component, an inelastic (nonreversible) component, and a thermal component. Equation 27.5b yields the hypoelastic Hooke's law, where E is Young's modulus, ν the Poisson's coefficient, and $\dot{\boldsymbol{\sigma}}$ a time derivative of the stress tensor $\boldsymbol{\sigma}$. Equation 27.5c gives the relation between the inelastic strain rate tensor $\dot{\boldsymbol{\varepsilon}}^{\mathrm{in}}$ and the stress deviator, $\mathbf{s}$, in which σ_0 denotes the scalar static yield stress, below which no inelastic deformation occurs (the expression between brackets is reduced to zero when negative). In these equations, the temperature dependency of all the involved variables should be considered. The effect of strain hardening may appear in such a model by the increase of the static yield stress σ_0 and the plastic consistency K with the accumulated inelastic strain $\varepsilon_{\mathrm{eq}}$, or with another state variable that is representative of the material structure. The corresponding scalar equation relating stress and inelastic strain rate von Mises invariants is:

$$\sigma_{\mathrm{eq}} = \sigma_0 + K(\dot{\varepsilon}_{\mathrm{eq}})^m \tag{27.6}$$

Inserting this into Equation 27.5c simplifies it to:

$$\dot{\boldsymbol{\varepsilon}}^{\mathrm{in}} = \frac{3\dot{\varepsilon}_{\mathrm{eq}}}{2\sigma_{\mathrm{eq}}}\mathbf{s}, \quad \text{or, in incremental form,} \quad d\boldsymbol{\varepsilon}^{\mathrm{in}} = \frac{3d\varepsilon_{\mathrm{eq}}}{2\sigma_{\mathrm{eq}}}\mathbf{s}. \tag{27.7}$$

Although metallic alloys show a significant strain rate sensitivity at high temperature, they are often modeled in the literature using elastic-plastic models, neglecting this important effect. In this case, Equation 27.7 still holds, but the flow stress is then independent of the strain rate. It may depend on the accumulated plastic strain because of strain hardening.

27.2.4 Implementation Issues

As stresses and distortions are generated mainly in the solid phase, the mechanical modeling of a cast part may be restricted to its solidified region at any instant. However, this approach, often used in the literature, has several drawbacks. First, doing so, the volumetric shrinkage that affects the mushy zone cannot be taken into account. Second, the liquid and mushy regions may alter the distortion and stress of the solidified regions. Finally, the mechanical behavior of the mushy zone is of crucial importance to quality problems. Thus, it is preferable to consider the entire casting, including the mushy and liquid regions.

To avoid managing different constitutive equations for the different states, numerous authors have modeled the liquid and mushy state behavior simply by lowering the value of the Young modulus and taking Poisson coefficient close to one half, using then a single elastic-viscoplastic (or elastic-plastic) constitutive model for the entire casting.[6–13] However, this approach may suffer from numerical difficulty, and furthermore fails to account accurately for the significant thermal dilatation and shrinkage that affect the mushy and liquid regions.

To overcome these difficulties, other authors use a different constitutive equation for the different physical states of the alloy. The whole casting is modeled, and the constitutive equation is chosen according to the local state. One implementation[14] simply changes the constants in Equation 27.5 to model the different states. Using a physically reasonable (high) value for the elastic modulus, and setting $m = 1$, $\sigma_0 = 0$, and $K = 3\mu_l$ allows Equation 27.5 to approximate Equation 27.1 for a Newtonian fluid, as the generated elastic strains are very small. Another approach[15,16] uses a viscoplastic equation for the mushy and liquid states and an elastic-viscoplastic equation for the solid state.

In each of these previous mechanical models, the liquid, mushy, and solid zones are considered as a single continuum. The velocity of the liquid phase is not distinguished from the velocity of the solid phase, and the individual dendrites and grain boundaries are not resolved, so fluid feeding, porosity formation, and hot tearing are clearly oversimplified. In the context of stress–strain prediction, this approximation seems valid. To predict defect formation from first principles, however, might require a multiphase approach, involving the different phases, including liquid, solid, and even gas. Recent approaches, like those developed by Nicolli et al.[17] and Fachinotti et al.[18] take into account the mechanical interaction between a deformable solid skeleton and the liquid phase, the momentum transfer between the two phases being expressed by a Darcy law.

27.2.5 Example of Constitutive Equations

Material property data are needed for the specific alloy being modeled and in a form suitable for the constitutive equations just discussed. This presents a significant challenge for quantitative mechanical analysis, because measurements are not presented in this form, and only rarely supply enough information on the conditions to allow transformation to an alternate form. As an example, the following elastic-viscoplastic constitutive equation was developed for the austenite phase of steel by Kozlowski et al.[19] by fitting constant strain rate tensile tests from Wray[20,21] and constant-load creep tests from Suzuki et al.[22] to the form required in Equation 27.5 to Equation 27.7:

$$\dot{\varepsilon}_{eq} = f_{\%C}\langle\sigma_{eq} - \sigma_0\rangle^{1/m} \exp\left(-\frac{4.465 \times 10^4}{T}\right)$$

where

$$\begin{aligned} f_{\%C} &= 4.655 \times 10^4 + 7.14 \times 10(\%C) + 1.2 \times 10^4(\%C)^2 \\ \sigma_0 &= (130.5 - 5.128 \times 10^{-3}T)\varepsilon_{eq}^{f_2} \\ f_2 &= -0.6289 + 1.114 \times 10^{-3}T \\ 1/m &= 8.132 - 1.54 \times 10^{-3}T \\ &\text{with } T\ [\mathrm{K}],\ \sigma_{eq},\ \sigma_0\ [\mathrm{MPa}] \end{aligned} \tag{27.8}$$

This equation, and a similar one for delta-ferrite, have been implemented into the finite element codes CON2D[14] and THERCAST[23] and applied to investigate several problems involving mechanical behavior during continuous casting.

Elastic modulus is a crucial property that decreases with increasing temperature. It is difficult to measure at the high temperatures important to casting, owing to the susceptibility of the material to creep and thermal strain during a standard tensile test, which results in excessively low values. Higher values are obtained from high-strain-rate tests, such as ultrasonic measurements.[24] Elastic modulus measurements in steels near the solidus temperature range from ~1 GPa[25] to 44 GPa.[26] Typical modulus data by Mizukami et al.[27] include values ~10 GPa near the solidus and have been used in previous analyses.[28,29]

The density needed to compute thermal strain in Equation 27.1, Equation 27.4, or Equation 27.5d can be found from a weighted average of the values of the different solid and liquid phases, based on the local phase fractions. For the example of plain low carbon steel, the following equations were compiled[14] based on the solid data for ferrite (α), austenite (γ), and delta (δ) from Harste et al.[30,31] and the liquid (l) measurements from Jimbo and Cramb:[32]

$$\begin{aligned} \rho(\mathrm{kg/m^3}) &= \rho_\alpha f_\alpha + \rho_\gamma f_\gamma + \rho_\delta f_\delta + \rho_\mathrm{l} f_\mathrm{l} \\ \rho_\alpha &= 7881 - 0.324T(^\circ\mathrm{C}) - 3\times 10^{-5}T(^\circ\mathrm{C})^2 \\ \rho_\gamma &= \frac{100[8106 - 0.51T(^\circ\mathrm{C})]}{[100-(\%\mathrm{C})][1+0.008(\%\mathrm{C})]^3} \\ \rho_\delta &= \frac{100[8011 - 0.47T(^\circ\mathrm{C})]}{[100-(\%\mathrm{C})][1+0.013(\%\mathrm{C})]^3} \\ \rho_\mathrm{l} &= 7100 - 73(\%\mathrm{C}) - [0.8 - 0.09(\%\mathrm{C})][T(^\circ\mathrm{C}) - 1550] \end{aligned} \tag{27.9}$$

Specialized experiments to measure mechanical properties for use in computational models will be an important trend for future research in this field.

27.3 Model Development

27.3.1 Thermomechanical Coupling

Coupling between the thermal and mechanical analyses arises from several sources. First, regarding the mechanical problem, besides the strain rate due to thermal expansion and solidification shrinkage, the material parameters of the preceding constitutive equations strongly depend on temperature and phase fractions, as shown in the previous section. Second, in the heat transfer problem, the thermal exchange between the casting and the mold strongly depends on local conditions such as the contact pressure or the presence of a gap between them (as a result of thermal expansion and solidification shrinkage). This is explained in the next two paragraphs.

27.3.1.1 Air Gap Formation: Conductive–Radiative Modeling

In the presence of a gap between the casting and the mold, resulting from their relative deformation, the heat transfer results from concurrent conduction through the gas within the gap and from radiation. The exchanged thermal flux, q_gap, can then be written:

$$q_\mathrm{gap} = \frac{k_\mathrm{gas}}{g}(T_\mathrm{c} - T_\mathrm{m}) + \frac{\sigma(T_\mathrm{c}^4 - T_\mathrm{m}^4)}{(1/\varepsilon_\mathrm{c}) + (1/\varepsilon_\mathrm{m}) - 1} \tag{27.10}$$

with $k_\mathrm{gap} = (T)$ the thermal conductivity of the gas, g the gap thickness, T_c and T_m the local surface temperature of the casting and mold, respectively, ε_c and ε_m their gray-body emissivities, σ the Stefan–Boltzmann constant. It is to be noted that the conductive part of the flux can be written in more detail to take into account the presence of coating layers on the mold surface: conduction through a medium of thickness g_coat, of conductivity $k_\mathrm{coat}(T)$. It can be seen that the first term tends to infinity as the gap thickness tends to zero: this expresses a perfect contact condition, T_c and T_m tending toward a unique

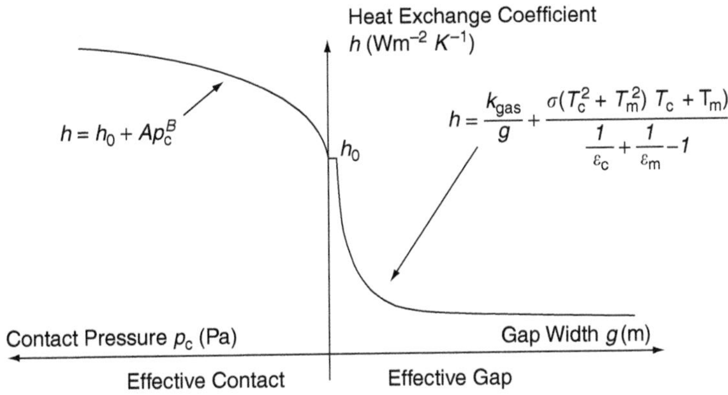

FIGURE 27.1 Modeling of the local heat transfer coefficient in the gap and effective contact situations.

interface temperature. The reality is somewhat different, showing always nonperfect contact conditions. Therefore, the conductive heat exchange coefficient $h_{\mathrm{cond}} = k_{\mathrm{gas}}/g$ should be limited by a finite value h_0, corresponding to the "no-gap" situation, and depends on the roughness of the casting surface. A recent benchmark exercise has demonstrated the significant impact of the consideration of gap formation on temperature prediction in shape casting.[33] Specific examples of these gap heat transfer laws are provided elsewhere for continuous casting with oil lubrication,[34] and continuous casting with mold flux.[35]

27.3.1.2 Effective Contact: Heat Transfer as a Function of Contact Pressure

In the case of an effective contact, the conductive heat flux increases with the contact pressure according to a power law.[36] Still denoting h_0 the heat exchange coefficient corresponding to no gap and no contact pressure, we can write, to ensure the continuity with the gap situation:

$$q_{\mathrm{contact}} = (h_0 + Ap_c^B)(T_c - T_m) \tag{27.11}$$

with p_c the contact pressure, A and B two parameters that depend on the materials, the presence of coating or lubricating agent, the surface roughness, and the temperature. The parameters and possibly the laws governing their evolution need to be determined experimentally.

Figure 27.1 is a graphic representation of the heat exchange coefficient in both cases. It should be noted that a smoothing method around the neutral position has been suggested by Laschet et al.[37]

27.3.2 Numerical Solution

We focus here on the resolution of the momentum conservation, from which distortions and stresses can be calculated. The energy conservation is not discussed in this chapter.

27.3.2.1 Momentum Conservation

At any time, and in any location of the solidifying material, whatever its current state (liquid, mushy, or solid) the momentum conservation is expressed by:

$$\nabla \cdot \boldsymbol{\sigma} + \rho \mathbf{g} - \rho \boldsymbol{\gamma} = \nabla \cdot \mathbf{s} - \nabla p + \rho \mathbf{g} - \rho \boldsymbol{\gamma} = 0 \tag{27.12}$$

where $\mathbf{g}$ denotes the gravity and $\boldsymbol{\gamma}$ the acceleration. The acceleration is actually noticeable only in liquid pools, when they are affected by convection.

The weak form of the preceding equation can be obtained. Keeping the velocity and pressure as primitive unknown variables, it is written as:[38]

$$\begin{cases} \forall \boldsymbol{v}^* \displaystyle\int_{\Omega} \boldsymbol{s}(\boldsymbol{v}) : \dot{\boldsymbol{\varepsilon}}^* \mathrm{d}V - \int_{\Omega} p \nabla \cdot \boldsymbol{v}^* \mathrm{d}V - \int_{\partial\Omega} \boldsymbol{T} \cdot \boldsymbol{v}^* \mathrm{d}S - \int_{\Omega} \rho \boldsymbol{g} \cdot \boldsymbol{v}^* \mathrm{d}V + \int_{\Omega} \rho \frac{\mathrm{d}\boldsymbol{v}}{\mathrm{d}t} \cdot \boldsymbol{v}^* \mathrm{d}V = 0 \\ \forall p^* \displaystyle\int_{\Omega} p^* \mathrm{tr}\, \dot{\boldsymbol{\varepsilon}}^{\mathrm{in}} \mathrm{d}V = 0 \end{cases} \tag{27.13}$$

where $\boldsymbol{T}$ is the external stress vector, $\boldsymbol{g}$ the gravity. In such a velocity-pressure formulation, the second equation is a weak form of the incompressibility of inelastic deformations.

27.3.2.2 Finite Element Formulation and Numerical Implementation

The thermal–mechanical modeling equations just presented must be solved numerically, owing to the complex shape of the casting process domain. Although finite-difference approaches are popular for fluid flow and solidification analysis to compute the temperature field, the finite element formulation is preferred for the mechanical analysis, owing to its historical advantages with unstructured meshes and accurate implicit solution of the resulting simultaneous algebraic equations.

Spatial discretization using finite elements, combined with time discretization using finite differences, yield a set of nonlinear equations $\boldsymbol{R}_{\mathrm{mech}}(\boldsymbol{V}, \boldsymbol{P}) = 0$, in which the unknowns are the velocity components and the pressure value at each node of the finite element mesh. In some formulations, the incompressibility condition is directly included in the momentum equation using a penalty technique to enforce it. This results in a velocity formulation $\boldsymbol{R}_{\mathrm{mech}}(\boldsymbol{V}) = 0$ in which the velocity field is the only unknown, but which is known to give rise to pressure locking problems when the incompressible inelastic strains become too large.[39] In the preceding formulations, nodal velocities can be replaced by nodal displacements, as they are linked by the time integration scheme.

Solving the previous global finite element nonlinear system $\boldsymbol{R}_{\mathrm{mech}} = 0$ can be a daunting task that is subject to significant convergence difficulties. Generally, it is solved using a full or modified Newton–Raphson method,[40] which iterates to minimize the residual error in Equation 27.13. These methods require local consistent tangent operators $\partial \mathbf{s}/\partial \dot{\boldsymbol{\varepsilon}}$ and $\partial p/\partial \dot{\boldsymbol{\varepsilon}}$ to form a global consistent stiffness matrix $\partial \boldsymbol{R}_{\mathrm{mech}}/\partial \boldsymbol{V}$.[39]

At the local level, when the constitutive equations involve strain rate dependency, then an algorithm is also required to integrate the constitutive equations to provide an estimate of $\mathbf{s}(\dot{\boldsymbol{\varepsilon}})$ and $p(\dot{\boldsymbol{\varepsilon}})$ and the previously mentioned tangent operators. When the constitutive equations are highly nonlinear, it is very useful to employ an implicit algorithm to provide better estimates of inelastic strain at the local level. Many methods have been developed[41–43] that require solving two or more ordinary differential equations at each local integration point. The bounded Newton–Raphson method developed by Lush et al.[42] and later improved upon by Zhu[43] was implemented into a user-subroutine in ABAQUS and found to greatly accelerate the solution.[44]

Alternatively, an operator splitting method can be used to march through time by alternating between the global and local levels without iteration at either level.[14,44]

27.3.2.3 Boundary Conditions: Modeling of Contact Conditions. Multidomain Approaches

At the interface between the solidifying material and the mold, a unilateral contact condition (i.e. including contact release) generally applies:

$$\begin{cases} \boldsymbol{\sigma n} \cdot \boldsymbol{n} \le 0 \\ g \ge 0 \\ (\boldsymbol{\sigma n} \cdot \boldsymbol{n}) g = 0 \end{cases} \tag{27.14}$$

where g is the local interface gap width (positive when air gap exists effectively) and $\boldsymbol{n}$ is the local outward unit normal to the part. The fulfillment of Equation 27.14 can be obtained by means of a penalty condition, which consists in applying a normal stress vector $\boldsymbol{T}$ proportional to the normal distance or the normal velocity difference (or a combination of both) via a penalty constant χ_p:

$$\boldsymbol{T} = \boldsymbol{\sigma n} = -\chi_p \langle (\boldsymbol{v} - \boldsymbol{v}_{\text{mold}}) \cdot \boldsymbol{n} \rangle \boldsymbol{n} \tag{27.15}$$

Different methods of local adaptation of the penalty coefficient χ_p have been developed, among which is the augmented Lagrangian method.[45] More complex and computationally expensive methods, such as the use of Lagrange multipliers may also be used.[46]

The possible tangential friction effects between part and mold can be taken into account by a friction law, such as a Coulomb model for instance. In this case, the previous stress vector has a tangential component, $\boldsymbol{T}_\tau$, given by:

$$\boldsymbol{T}_\tau = -\mu_{\text{f}} p_{\text{c}} \frac{1}{\|\boldsymbol{v} - \boldsymbol{v}_{\text{mold}}\|} (\boldsymbol{v} - \boldsymbol{v}_{\text{mold}}) \tag{27.16}$$

where $p_{\text{c}} = -\boldsymbol{\sigma}_n = \boldsymbol{\sigma n} \cdot \boldsymbol{n}$ is the contact pressure, and μ_{f} the friction coefficient.

The previous approach can be extended to the multidomain context to account for the deformation of mold components. The local stress vectors calculated by Equation 27.15 can be applied onto the surface of the mold, contributing then to its deformation. For most casting processes, the mechanical interaction between the cast product and the mold is sufficiently slow (i.e., its characteristic time remains significant with respect to the process time) to permit a staggered scheme within each time increment: the mechanical problem is successively solved in the cast product and in the different mold components. A global updating of the different configurations is then performed at the end of the time increment. This simple approach gives access to a prediction of the local air gap size g, or alternatively of the local contact pressure p_{c}, that are used in the expressions of the heat transfer coefficient, according to Equation 27.10 and Equation 27.11.[47]

27.3.3 Treatment of Regions in the Solid, Mushy, and Liquid States

27.3.3.1 Solidified Regions: Lagrangian Formulation

In casting processes, the solidified regions generally encounter small deformations. It is thus natural to embed the finite element domain into the material, with each node of the computational grid corresponding with the same solid particle during its displacement. The boundary of the mesh corresponds then to the surface of the casting. This method, called Lagrangian formulation, provides the best accuracy when computing the gap forming between the solidified material and the mold. It is also the more reliable and convenient method for time integration of highly nonlinear constitutive equations, such as elastic-(visco)-plastic laws presented in Section 27.2.3.

27.3.3.2 Mushy and Liquid Regions: ALE Modeling

When the mushy and liquid regions are modeled in the same domain as the solid (cf. discussion in Section 27.2.4), they are often subjected to large displacements and strains arising from solidification shrinkage, buoyancy, or forced convection. Similar difficulties are generated in casting processes such as squeeze casting, where the entire domain is highly deformed. In these cases, a Lagrangian formulation would demand frequent remeshings to avoid mesh degeneracy, which is both computationally costly, and detrimental to the accuracy of the modeling. It is then preferable to use a so-called arbitrary Lagrangian Eulerian formulation (ALE). In an Eulerian formulation, material moves through the computational grid, which remains stationary in the "laboratory" frame of reference. In the ALE formulation, the updating of the mesh is partially independent of the velocity of the material particles to maintain the quality of the computational grid. Several methods can be used, including the popular "barycentering" technique that keeps each node at the geometrical centroid of a set of its neighbors. This method involves significant extra complexity to account for the advection of material through the domain, and the state variables such as temperature and inelastic strain must be updated according to the relative velocity between the mesh and the particles. In doing this, some surface constraints must be enforced to ensure mass conservation, expressing that the fluxes of mesh velocity and of fluid particle velocity through the surface of the mesh should remain identical. A review on the ALE method in solidification modeling is available, together with some details on its application.[48]

27.3.3.3 Thermomechanical Coupling

Because of the interdependency between the thermal and mechanical analyses, as presented in the previous section, their coupling should be taken into account during the cooling process. In practice, the cooling time is decomposed into time increments, each increment requiring the solution of two problems: the energy conservation and the momentum conservation. With the highly nonlinear elastic-viscoplastic constitutive equations typical of solidifying metals, the incremental steps required for the mechanical analysis to converge are generally much smaller than those for the thermal analysis. Thus, these two analyses are generally performed in succession and only once per time increment. However, in the case of very rapid cooling, these solutions might be preferably performed together (including thermal and mechanical unknowns in a single set of nonlinear equations), or else separate but iteratively until convergence at each time increment, otherwise the time step has to be dramatically reduced.

27.3.4 Hot-Tearing Analysis

Hot tearing is one of the most important consequences of stress during solidification. Hot tearing is caused by a combination of tensile stress and metallurgical embrittlement. It occurs at temperatures near the solidus when strain concentrates within the interdendritic liquid films, causing separation of the dendrites and intergranular cracks at very small strains (on the order of 1%). This complex phenomenon depends on the ability of liquid to flow through the dendritic structure to feed the volumetric shrinkage, the strength of the surrounding dendritic skeleton, the grain size and shape, the nucleation of supersaturated gas into pores or crack surfaces, the segregation of solute impurities, and the formation of interfering solid precipitates. The subsequent refilling of hot tears with segregated liquid alloy can cause internal defects that are just as serious as exposed surface cracks. The hot tearing of aluminum alloys is reviewed elsewhere.[49] Hot-tearing phenomena are too complex and insufficiently understood to model in detail, so several different criteria have been developed to predict hot tears from the results of a thermal–mechanical analysis.

Casting conditions that produce faster solidification and alloys with wider freezing ranges are more prone to hot tears. Thus, many criteria are solely based on thermal analysis. That of Clyne and Davies[50] simply compares the local time spent between two critical solid fractions g_{s1} and g_{s2} (typically 0.9 and 0.99, respectively), with the total local solidification time (or a reference solidification time). The "hot cracking susceptibility" is defined as:

$$HCS_{\text{Clyne}} = \frac{t_{0.99} - t_{0.90}}{t_{0.90} - t_{0.40}} \tag{27.17}$$

Criteria based on classical mechanics often assume cracks will form when a critical stress is exceeded, and they are popular for predicting cracks at lower temperatures.[51–54] This critical stress depends greatly on the local temperature and strain rate. Its accuracy relies on measurements, such as the submerged split-chill tensile test for hot tearing.[55–57]

Measurements often correlate hot-tear formation with the accumulation of a critical level of mechanical strain while applying tensile loading within a critical solid fraction where liquid feeding is difficult. This has formed the basis for many hot-tearing criteria. That of Yamanaka et al.[58] accumulates inelastic deformation over a brittleness temperature range, which is defined, for example, as $g_s \in [0.85, 0.99]$ for a

Fe-0.15 wt.% C steel grade. The local condition for fracture initiation is then:

$$\sum_{g_{s1}}^{g_{s2}} \Delta\varepsilon_{in} \geq \varepsilon_{cr} \tag{27.18}$$

in which the critical strain ε_{cr} is 1.6% at a typical strain rate of $3 \times 10^{-4}\ s^{-1}$. Careful measurements during bending of solidifying steel ingots have revealed critical strains ranging from 1 to 3.8%.[58,59] The lowest values were found at high strain rate and in crack-sensitive grades (e.g., high-sulfur peritectic steel).[58] In aluminum rich Al–Cu alloys, critical strains were reported from 0.09 to 1.6% and were relatively independent of strain rate.[60] Tensile stress is also a requirement for hot tear formation.[58] The maximum tensile stress occurs just before the formation of a critical flaw.[60]

The critical strain decreases with increasing strain rate, presumably because less time is available for liquid feeding, and also decreases for alloys with wider freezing ranges. Won et al.[61] suggested the following empirical equation for the critical strain in steel, based on fitting measurements from many bend tests:

$$\varepsilon_{cr} = \frac{0.02821}{\dot{\varepsilon}^{0.3131}\,\Delta T_B^{0.8638}} \tag{27.19}$$

where $\dot{\varepsilon}$ is the strain rate and ΔT_B is the brittle temperature range, defined between the temperatures corresponding to solid fractions of 0.9 and 0.99.

More mechanistically based hot-tearing criteria include more of the local physical phenomena that give rise to hot tears. Feurer[62] and more recently Rappaz et al.[63] have proposed that hot tears form when the local interdendritic liquid feeding rate is not sufficient to balance the rate of tensile strain increase across the mushy zone. The criterion of Rappaz et al. predicts fracture when the strain rate exceeds a limit value that allows pore cavitation to separate the residual liquid film between the dendrites:

$$\dot{\varepsilon} \geq \frac{1}{R}\left[\frac{\lambda_2^2\,\|\nabla T\|}{180\mu_l}\frac{\rho_L}{\rho_S}(p_m - p_C) - v_T\frac{\rho_S - \rho_L}{\rho_S}H\right] \tag{27.20}$$

in which λ_2 is the secondary dendrite arm spacing, p_m is the local pressure in the liquid ahead of the mushy zone, p_C is the cavitation pressure, v_T is the velocity of the solidification front. The quantities R and H depend on the solidification path of the alloy:

$$R = \frac{1}{\|\nabla T\|}\int_{T_2}^{T_1}\frac{g_s^2 F(T)}{g_l^3}\,dT \quad H = \int_{T_2}^{T_1}\frac{g_s^2}{g_l^2}\,dT \quad F(T) = \frac{1}{\Delta T}\int_{T_2}^{T} g_s\,dT \tag{27.21}$$

where the integration limits are calibration parameters that also have physical meaning.[64] The upper limit T_1 may be the liquidus or the coherency temperature, while the lower limit T_2 typically is within the solid fraction range of 0.95 to 0.99.[65]

27.3.5 Model Validation

Model validation with both analytical solutions and experiments is a crucial step in any computational analysis and thermomechanical modeling is no exception. Weiner and Boley[6] derived an analytical solution for unidirectional solidification of an unconstrained plate with a unique solidification temperature, an elastic-perfectly plastic constitutive law and constant properties. The plate is subjected to sudden surface quench from a uniform initial temperature to a constant mold temperature.

This benchmark problem is ideal for the validation of computational thermal-stress models, as it can be solved with a one-dimensional mesh, as shown in Figure 27.2. Numerical predictions should match with arbitrary precision according to the mesh refinement. For example, the solidification stress analysis code, CON2D[14] and the commercial code ABAQUS were applied for the conditions in Table 27.1.[44] The

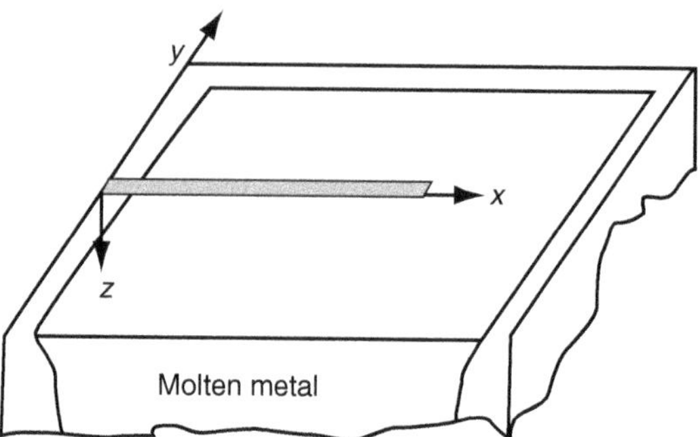

FIGURE 27.2 One-dimensional slice domain for modeling solidifying plate.

TABLE 27.1 Constants Used in Solidification Calculations

Conductivity, k [W m^{-1} K^{-1}]	33.0
Specific heat, C_p [J kg^{-1} K^{-1}]	661.0
Elastic modulus in solid, E_s [GPa]	40.0
Elastic modulus in liquid, E_l [GPa]	14.0
Thermal linear expansion coefficient [K^{-1}]	0.00002
Density, ρ [kg m^{-3}]	7500
Poisson's ratio, ν	0.3
Liquidus temperature, T_L [°C]	1494.45
Fusion temperature (analytical), T_F [°C]	1494.4
Solidus temperature, T_S [°C]	1494.35
Initial temperature, T_0 [°C]	1495.0
Latent heat, L [J kg^{-1} K^{-1}]	272000
Liquid viscosity, μ_l [MPa-s]	2.22×10^{-9}
Surface film coefficient, h [W m^{-2} K^{-1}]	250,000
Mold temperature, T_∞ [°C]	1000

solidification-temperature interval is spread over a small (0.1°C) artificial temperature range about the fusion temperature of 1494.4°C. The instantaneous surface quench is modeled with high convection, $q_{gap} = h(T - T_\infty)$, to lessen the instabilities caused by these extreme conditions. The elastic-perfectly plastic constitutive equation with $\sigma_0(\text{MPa}) = 0.03 + 20(1494.4°\text{C} - T)/494.4$ was transformed to a numerically challenging rate formulation with the form of Equation 27.5c, by setting $m = 1$ and $K = 6.67 \times 10^{-9}$ MPa-s.[44] This represents a limit case for this elastic-viscoplastic expression. To model the unconstrained plate with a single row of elements, a generalized plane strain condition was imposed in the y and z directions (parallel to the surface) by coupling the displacements of all nodes along the edges of the slice domain as shown in Figure 27.2.[44] The constant axial strain assumed under this condition is computed by satisfying a single extra scalar equation:

$$\int \sigma_{zz} dS = F_z \tag{27.22}$$

Figure 27.3 and Figure 27.4 compare the temperature and stress profiles in the plate two times. The temperature profile through the solidifying shell is almost linear. Because the interior cools relative to the fixed surface temperature, its shrinkage generates internal tensile stress, which induces compressive stress at the surface. With no applied external pressure, the average stress through the thickness must naturally equal zero, and stress must decrease to zero in the liquid. Stresses and strains in both transverse directions (y and z) are equal for this symmetrical problem. The close agreement demonstrates that both

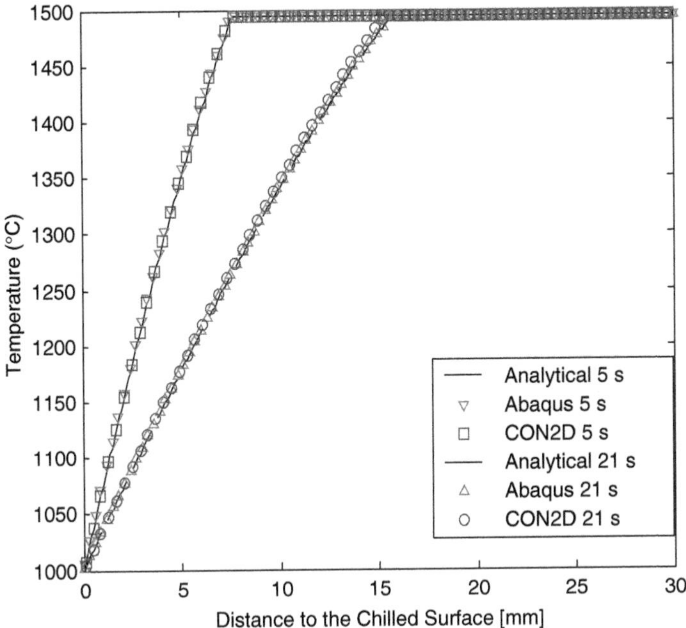

FIGURE 27.3 Temperatures through solidifying plate at different times comparing analytical solution and numerical predictions. (With permission from Koric, S. and Thomas, B.G., *Int. J. Num. Meths. Eng.*, (June) 2006, pp. 1955–1989.)

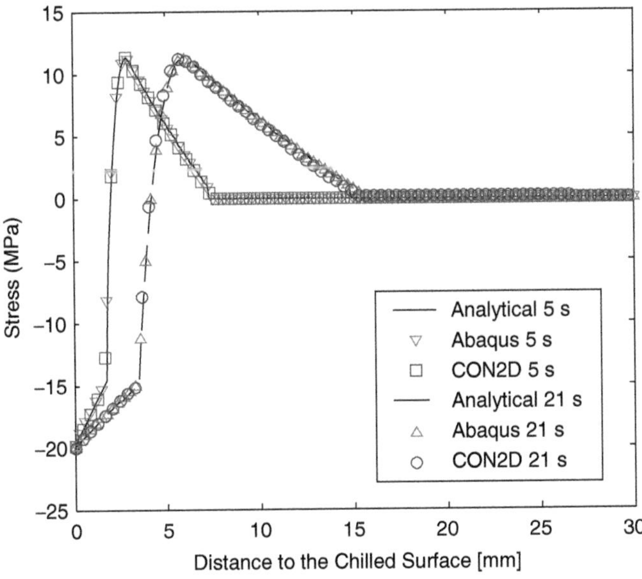

FIGURE 27.4 Transverse (Y and Z) stress through solidifying plate at different times comparing analytical solution and numerical predictions. (With permission from Koric, S. and Thomas, B.G., *Int. J. Num. Meths. Eng.*, in press, 2006.)

computational models are numerically consistent and have an acceptable mesh resolution. Comparison with experimental measurements is also required, to validate that the modeling assumptions and input data are reasonable. Plant experiments are especially important when validating predictions of more complex phenomena, such as hot tearing.

27.4 Applications

27.4.1 Application to Mold Casting

27.4.1.1 Braking Disks

The finite element software THERCAST for thermomechanical analysis of solidification[66] has been used in automotive industry to predict distortions of braking disks made of grey iron and cast in sand molds.[67] Particular attention has been paid to the interaction between the deformation of internal sand cores and the cast parts. This demands a global coupled thermomechanical simulation, as presented above. Figure 27.5 illustrates the discretization of the different domains involved in the calculation. The actual cooling scenario has been simulated: cooling in mold during 45 min, shake out and air cooling during 15 min. Figure 27.6 gives the temperature evolution at different points located in a horizontal cross section at mid-height in the disc, showing the influence of different physical phenomena: solidification after 2 min, solid state phase change after 20 min. The deformation of the core has been calculated, bringing out the thermal buckling of core blades. They are exposed to very high temperature, and their dilatation is too constrained, resulting in their deformation, as shown in Figure 27.7. This deformation causes a difference in thickness between the two braking tracks of the disk. Such a defect needs heavy and costly machining operations to get qualified parts. Instead, process simulation allows the manufacturer to test alternative geometries and process conditions to minimize the defect.

Similar thermomechanical calculations have been made in the case of plain disks, leading to comparisons with residual stress measurements by means of neutrons and x-ray diffraction.[68] As shown in Figure 27.8, calculations are consistent with measurements, the difference being less than 10 MPa.

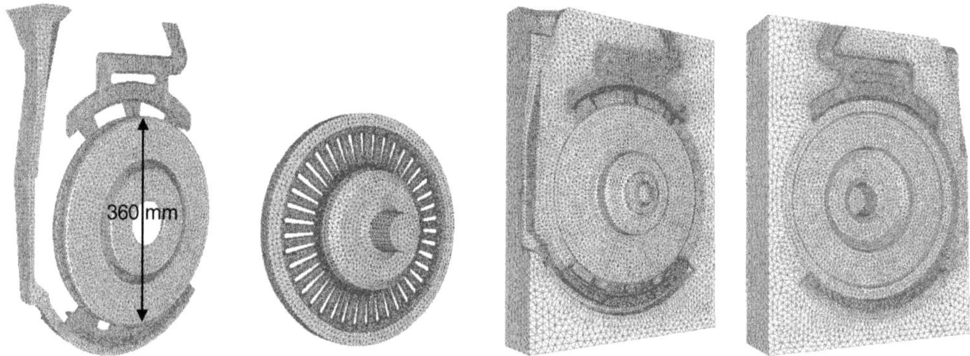

FIGURE 27.5 Finite element meshes of the different domains: part, core, and two half molds.

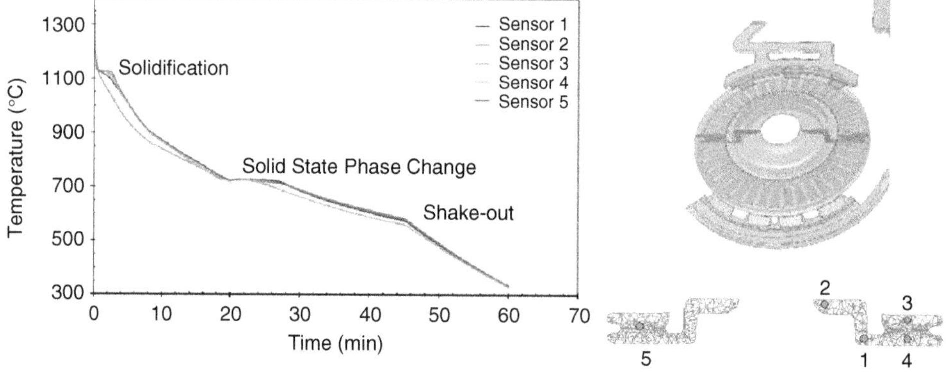

FIGURE 27.6 Temperature evolution in the part at different points located in the indicated section.

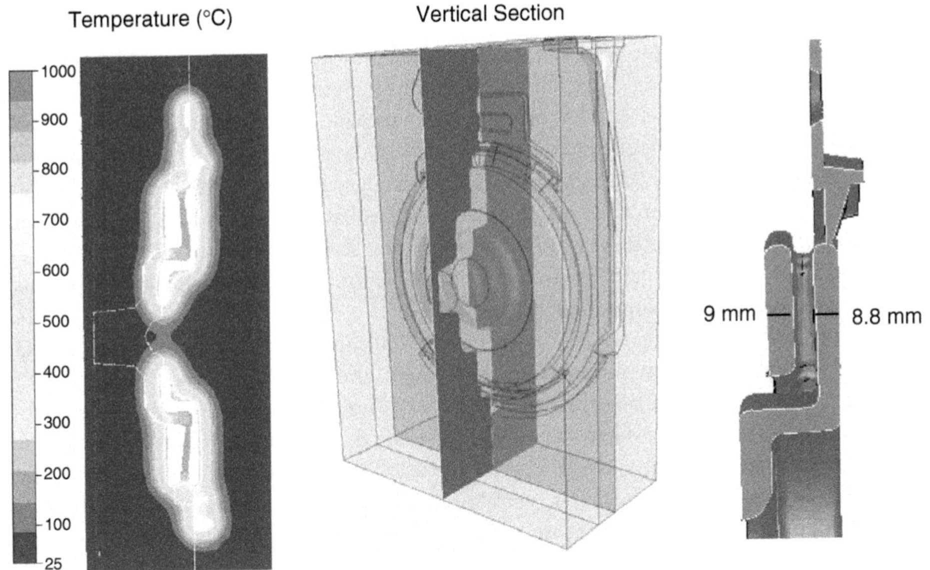

FIGURE 27.7 Deformation of core blades in a radial section, after a few seconds of cooling. On the left, displacements have been magnified by a factor 100. The temperature distribution is superimposed. On the right, the difference in thickness between the two braking tracks is shown.

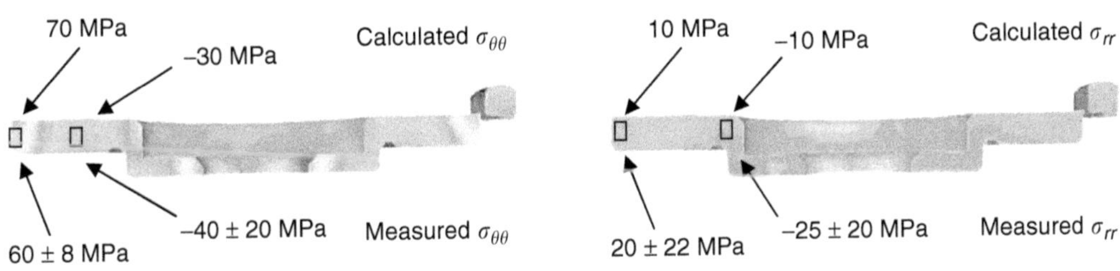

FIGURE 27.8 Residual hoop stresses (left) and radial stresses (right) in a radial section on as-cast plain disks made of grey iron. Top line: calculated values; bottom line: measured values.

27.4.2 Application to Continuous Casting of Steel, Primary Cooling in Mold

The two-dimensional transient finite element thermal–mechanical model, CON2D[14,44] has been applied to predict temperature, displacement, strain, stress, and hot tearing during primary cooling in continuous casting of steel. This Lagrangian model tracks a transverse slice through the strand as it moves downwards at the casting speed to reveal the entire three-dimensional stress state. The two-dimensional assumption produces reasonable temperature predictions because axial (z-direction) conduction is negligible relative to axial advection.[35] In-plane mechanical predictions should be reasonable because bulging effects are small and the undiscretized casting direction is modeled with the appropriate condition of generalized plain strain.

The mechanical properties of steel used with this model were discussed in Section 27.2.5 and Section 27.3.2 and in Table 27.1 and feature temperature-dependent elastic modulus and composition-dependent elastic-viscoplastic constitutive behavior and density. Two specific example applications with this model are presented next: (1) predicting the ideal taper of the mold to minimize gap formation during slab casting and (2) finding the critical casting speeds to avoid quality problems related to bulging below the mold.

27.4.2.1 Ideal Taper of a Slab Casting Mold

Thermal and mechanical behavior were simulated in a slice through the centerline of the wide face of the solidifying steel shell (Figure 27.2), as it moves down through the continuous casting mold. The total shrinkage strain predicted for this slice has been shown to provide an accurate and economical estimate of the ideal taper that should be applied to the narrow faces of the mold, to avoid formation of an air gap. The corner effects are reasonably small and were ignored in this analysis. The total heat flux (integrated from the heat flux profile) was forced to match an empirical equation that was obtained from a curve fit of many measurements under different conditions at a typical slab caster.[69] Computing an ideal taper is useful application of computational models, to minimize the problems associated with excessive air gap formation, or pressure on defect formation and productivity.

Sample results are given in Figure 27.9 for a typical 200-mm thick slab mold, based on standard conditions of 1000 mm width, 800 mm working mold length, and 1.5 m/min casting speed. These figures show the effect of steel grade on heat flux, shell thickness, surface temperature, and ideal narrow-face taper, all as a function of distance down the mold wall. Taper is presented as %/mold. A mold flux with 1215°C solidification temperature was assumed for the 0.13%C peritectic steel and a 1120°C mold flux for the low and high carbon steels, which is typical of industrial practice. The practice of adopting mold powders with high solidification temperature and low viscosity was proposed by Wolf to produce lower,

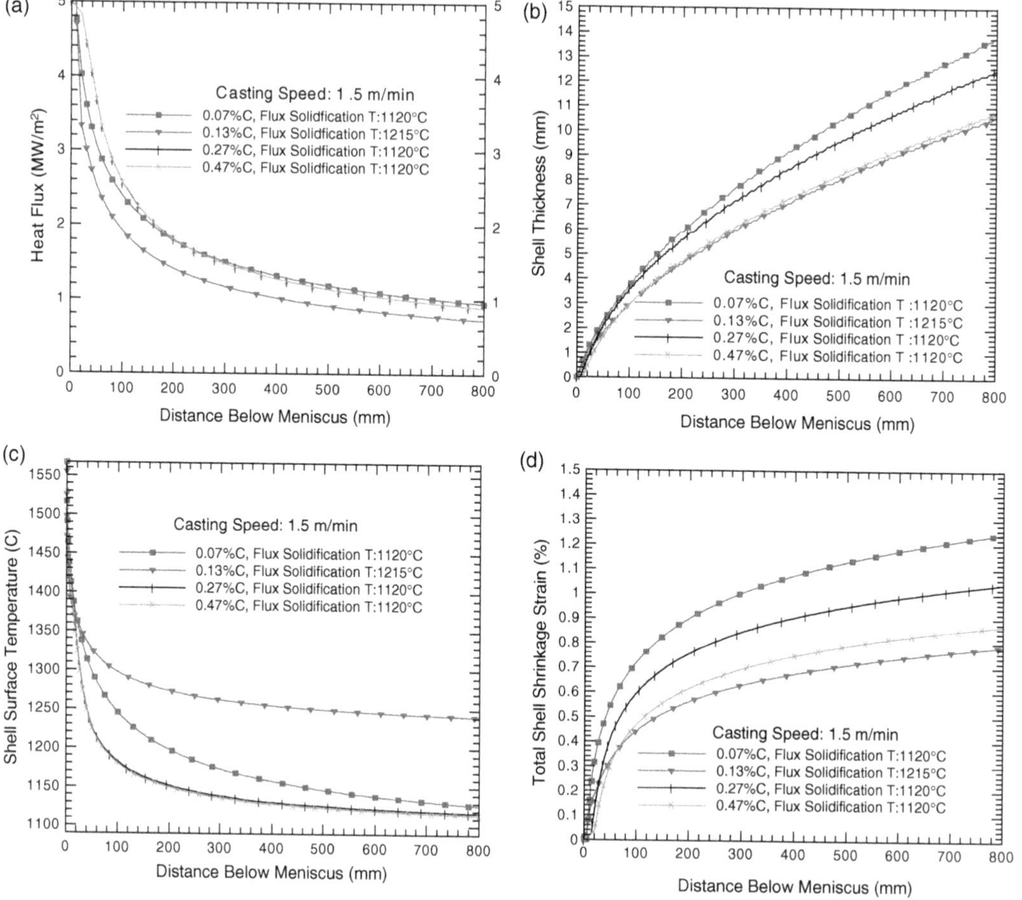

FIGURE 27.9 Effect of steel grade on heat flux (a), shell thickness (b), surface temperature (c), and ideal narrow-face taper (d) as a function of distance down the mold wall. (With permission from Thomas, B.G. and Ojeda, C., Ideal Taper Prediction for Slab Casting, ISSTech Steelmaking Conference, Indianapolis, IN, USA, April 27–30, 2003, 86, 396–308, 2003.)

but more uniform heat transfer rates to help avoid cracks in depression-sensitive grades, such as peritectic steels.[70] Slags with opposite properties are used for low and high carbon steels, to help avoid sticker problems.

The results show that the higher solidification-temperature mold flux, combined with the rougher shell surface produce a lower heat flux for the peritectic steel (Figure 27.9a). The lower heat flux produces a thinner shell (Figure 27.9b). It also produces a hotter shell surface temperature (Figure 27.9c). This effect appears to outweigh the importance of the extra shrinkage of the peritectic steels. Thus, peritectic steels experience less shrinkage and require less taper than either low or high carbon steels (Figure 27.9d).

The low carbon steel (0.07%C) experiences more inelastic strain than other grades, owing to its microstructure being mainly composed of soft, delta phase. The delta-ferrite phase creeps faster than the stronger austenite phase found in other steels. The extra creep generated in the compressive solid surface layer tends to allow the interior shrinkage to have more effect in this grade. The net effect is that low carbon steel experiences a thicker shell with more shell shrinkage and taper than for the other grades.

As for every case studied in this work, significantly more mold taper is needed just below the meniscus than near mold exit. Thermal shrinkage strain dominates the need for taper. However, before implementing new taper designs into an operating casting mold, these results must be modified to account for several phenomena that affect ideal taper. These include the thermal distortion of the narrow face, relative to its distortion at the meniscus, the expansion and thermal distortion of the wide face, the decrease in wide face perimeter due to the change in mold cavity dimensions down a funnel mold, and finally, the variation in thickness of the resolidified mold flux layers down the mold. More details are provided by Thomas and Ojeda,[71] but clearly, more work is needed before the model results can be safely used in practice.

27.4.2.2 Maximum Casting Speed to Avoid Hot Tears in Billet Casting

The model was next applied to predict the maximum casting speed allowable, while avoiding excessive bulging and hot tearing. Simulations start at the meniscus, 100 mm below the top of the mold, and extend through the 800-mm long mold and below, for a caster with no submould support.

The model domain is an L-shaped region of a two-dimension transverse section, shown in Figure 27.10. Removing the center portion of the section, which is always liquid, saves computational cost and allows a pressure boundary condition that avoids stability problems related to element "locking" that may occur with fixed-displacement boundary conditions.

The instantaneous heat flux, given in Equation 27.23, was based on fitting many plant measurements of total mold heat flux and differentiating.[28] It was assumed to be uniform around the perimeter of the billet surface to simulate ideal taper and perfect contact between the shell and mold. Below the mold, the billet surface temperature was kept constant at its circumferential profile at mold exit. This eliminates the effect of spray cooling practice imperfections on submold reheating or cooling and the associated complication for the stress/strain development. A typical plain carbon steel was studied (0.27%C, 1.52%Mn, 0.34%Si) with 1500.7°C liquidus temperature, and 1411.8°C solidus temperature:

$$q(\mathrm{MW/m^2}) = \begin{cases} 5 - 0.2444t(\mathrm{s}) & t \leq 1.0\ \mathrm{s} \\ 4.7556t(\mathrm{s})^{-0.504} & t > 1.0\ \mathrm{s} \end{cases} \tag{27.23}$$

Simulation results are presented here for one-quarter of a 120-mm square billet cast at speeds of 2.0 and 5.0 m/min. The latter is the critical speed at which hot-tear crack failure of the shell is just predicted to occur.

The temperature and stress distributions in a typical section through the wideface of the steel shell cast at 2.0 m/min are shown in Figure 27.11 — at four different times during cooling in the mold. Unlike the analytical solution in Figure 27.3, the surface temperature drops as time progresses. The corresponding stress distributions are qualitatively similar to the analytical solution (Figure 27.4). The stresses increase with time, however, as solidification progresses. The realistic constitutive equations produce a large region of tension near the solidification front. The magnitude of these stresses (and the corresponding strains) are not enough to cause hot tearing in the mold, however.

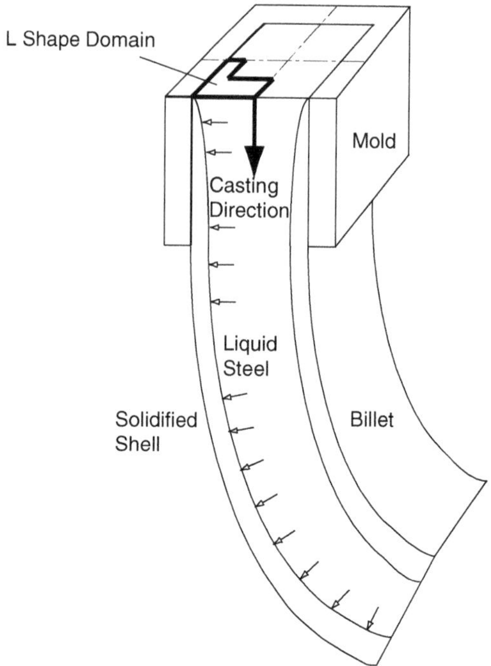

FIGURE 27.10 Model domain. (With permission from Li, C. and Thomas, B.G., *Modeling of Casting, Welding, and Advanced Solidification Processes*, Vol. X, Stefanescu, D., Warren, J., Jolly, M., and Krane, M., eds., TMS, Warrendale, PA (San Destin, FL, May 25–30, 2003), 2003, 385–392.)

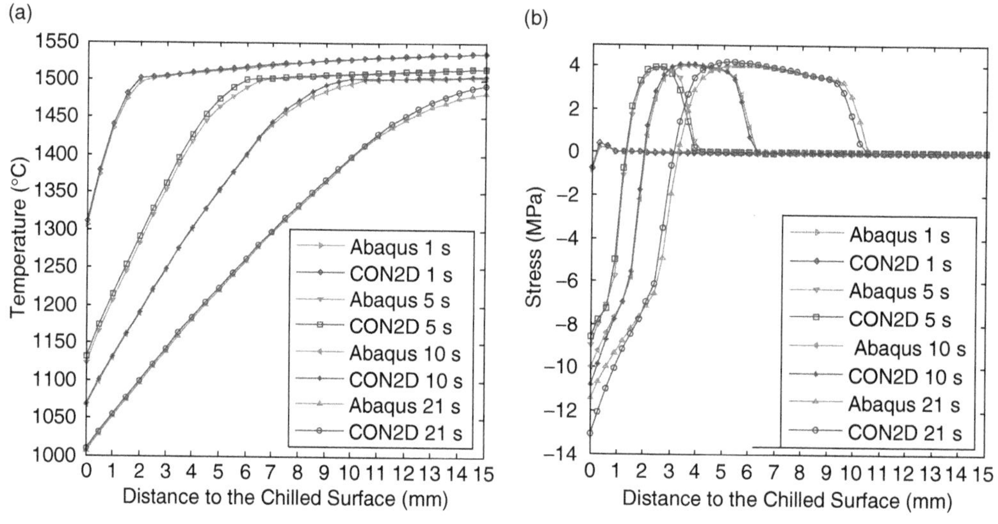

FIGURE 27.11 (a) Temperature distribution and (b) lateral (y and z) stress distribution along the solidifying slice in continuous casting mold. (With permission from Koric, S. and Thomas, B.G., *Int. J. Numer. Methods Eng.*, (June) 2006, pp. 1955–1989.)

Figure 27.12a shows the distorted temperature contours near the strand corner at 200 mm below the mold exit, for a casting speed of 5.0 m/min. The corner region is coldest, owing to two-dimensional cooling. The shell becomes hotter and thinner with increasing casting speed, owing to less time in the mold. This weakens the shell, allowing it to bulge more under the ferrostatic pressure below the mold.

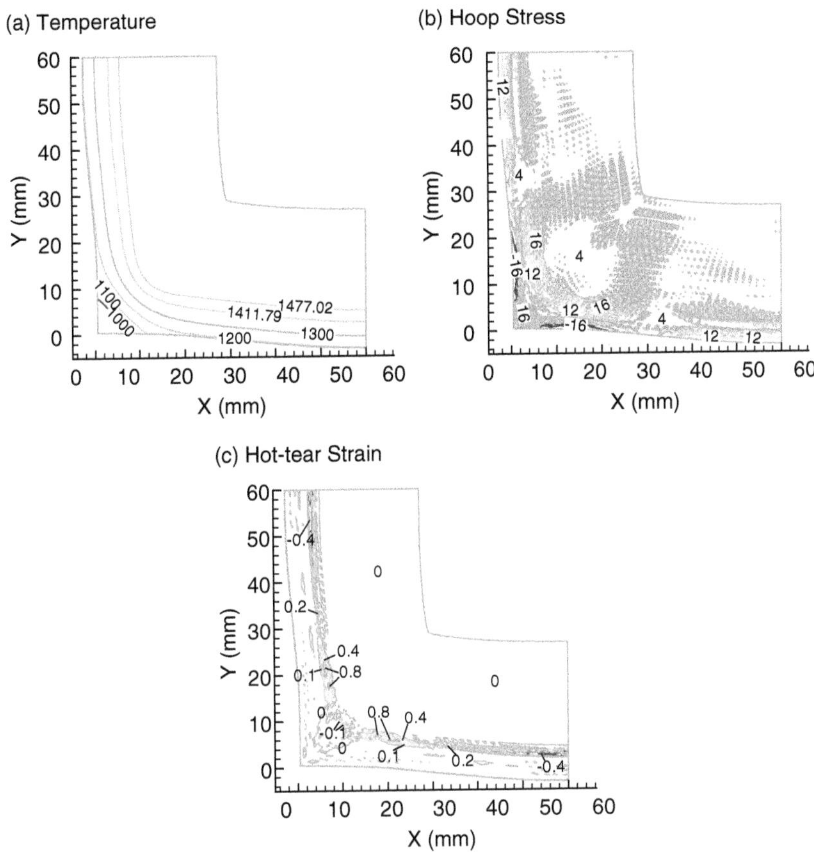

FIGURE 27.12 Distorted contours at 200 mm below mold exit. (With permission from Li, C. and Thomas, B.G., *Modeling of Casting, Welding, and Advanced Solidification Processes*, Vol. X, Stefanescu, D., Warren, J., Jolly, M. and Krane, M., eds., TMS, Warrendale, PA [San Destin, FL, May 25–30, 2003], 2003, 385–392.)

Figure 27.12b shows contours of "hoop" stress constructed by taking components in the x direction across the dendrites in the horizontal portion of the domain and the y direction in the vertical portion. High values appear at the off-corner subsurface region, due to a hinging effect that the ferrostatic pressure over the entire face exerts around the corner. This bends the shell around the corner and generates high subsurface tensile stress at the weak solidification front in the off-corner subsurface location. This tensile stress peak increases slightly and moves toward the surface at higher casting speed. Stress concentration is less and the surface hoop stress is compressive at the lower casting speed. This indicates no possibility of surface cracking. However, tensile surface hoop stress is generated below the mold at high speed in 27.12b at the face center due to excessive bulging. This tensile stress, and the accompanying hot-tear strain, might contribute to longitudinal cracks that penetrate the surface.

Hot tearing was predicted using the criterion in Equation 27.18 with the critical strain given in Equation 27.19, and a 90% temperature of 1459.9°C. Inelastic strain was accumulated for the component oriented normal to the dendrite growth direction, because that is the weakest direction and corresponds to the measurements used to obtain Equation 27.19. Figure 27.12c shows contours of hot-tear strain in the hoop direction. The highest values appear at the off-corner subsurface region in the hoop direction. Moreover, significantly higher values are found at higher casting speeds. For this particular example, hot-tear strain exceeds the threshold at 12 nodes, all located near the off-corner subsurface region. This is caused by the hinging mechanism around the corner. No nodes fail at the center surface, in spite of the high tensile stress there. The predicted hot-tearing region matches the location of off-corner longitudinal

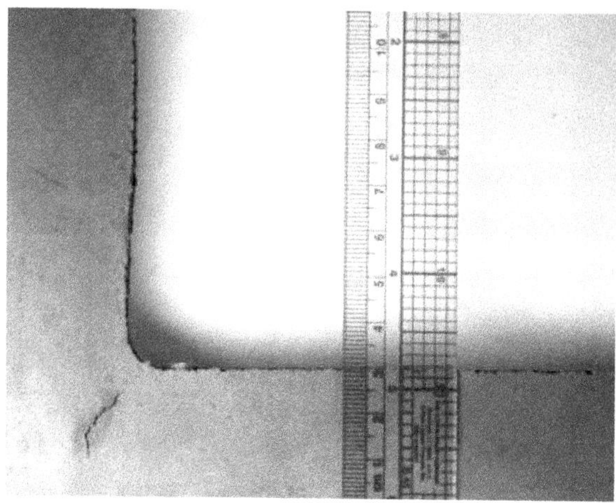

FIGURE 27.13 Off-corner internal crack in break-out shell from a 175-mm square bloom. (With permission from Li, C. and Thomas, B.G., *Modeling of Casting, Welding, and Advanced Solidification Processes*, Vol. X, Stefanescu, D., Warren, J., Jolly, M., and Krane, M., eds., TMS, Warrendale, PA (San Destin, FL, May 25–30, 2003), 2003, 385–392.)

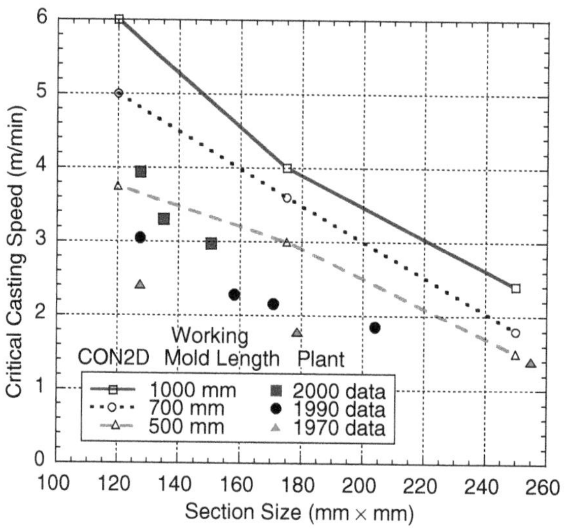

FIGURE 27.14 Comparison of critical casting speeds, based on hot-tearing criterion, and typical plant practice. (With permission from Li, C. and Thomas, B.G., *Modeling of Casting, Welding, and Advanced Solidification Processes*, Vol. X, Stefanescu, D., Warren, J., Jolly, M. and Krane, M., eds., TMS, Warrendale, PA [San Destin, FL, May 25–30, 2003], 2003, 385–392.)

cracks observed in sections through real solidifying shells, such as the one pictured in Figure 27.13. The bulged shape is also similar.

Results from many computations were used to find the critical speed to avoid hot tear cracks as a function of section size and working mold length, presented in Figure 27.14.[72] These predictions slightly exceed plant practice, which is generally chosen by empirical trial and error. This suggests that plant conditions such as mold taper are less than ideal, that other factors limit casting speed, or those speeds in practice could be increased. The qualitative trends are the same.

This quantitative model of hot tearing has enabled many useful insights into the continuous casting process. Larger section sizes are more susceptible to bending around the corner, so have a lower critical speed, resulting in less productivity increase than expected. The trend toward longer molds for the past three decades enables a higher casting speed without cracks by producing a thicker, stronger shell at mold exit.

27.4.3 Application to Continuous Casting of Steel, Secondary Cooling

Thermomechanical simulations are used by steelmakers to analyze stresses and strains all along the secondary cooling zone. One of their thrust is the prediction of the bulging of the solidified crust between the supporting rolls that is responsible for the tensile stress state in the mushy core, which in turn induces central macrosegregations.[73,74] Two- and three-dimensional finite element models have recently been developed, based on an original "global non steady-state" approach that provides results on the whole length of the caster. The approach implemented in THERCAST software is described in detail elsewhere.[23,75] The constitutive models are those presented in Section 27.2.5. The contact with supporting rolls is controlled by a penalty formulation derived from what has been presented in Section 27.3.2 using penalty coefficients attached to the different rolls and continuously adapted to control the numerical penetration of the strand. Figure 27.15 shows the results obtained on a vertical-curved machine (strand thickness 0.22 m, casting speed 0.9 m/min, material Fe-0.06 wt.% C) in a region located around 11 m below the meniscus. The pressure distribution reveals a double alternation of compressive and depressive zones. First, along the strand surface, the material is in a compressive state under rolls where the pressure reaches its maximum, 36 MPa. Conversely, it is in a depressive (tensile) state between rolls, where the pressure is minimum (−9 MPa).

Examination of the pressure state within the solid shell close to the solidification front (i.e., close to the solidus isotherm), reveals that the stress alternates. The steel is in a tensile state (negative pressure of about −2 MPa) when passing in front of rolls, while it is in a compressive state in between, the value of pressure being around 2 to 3 MPa. These results agree with previous structural analyses of the deformation of the

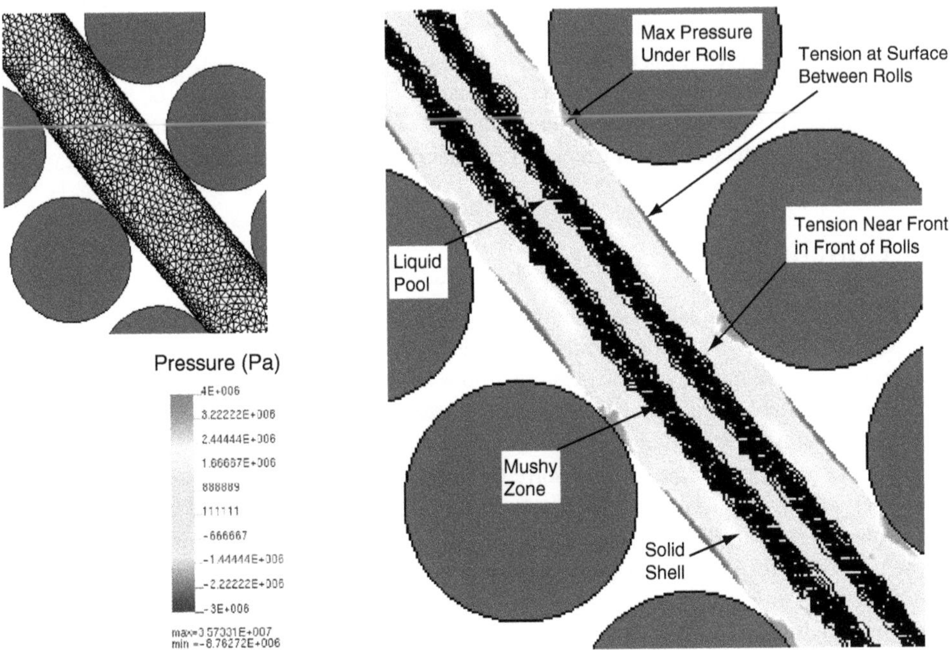

FIGURE 27.15 Illustration of the results of the calculations in the middle of the secondary cooling zone, at a metallurgical length of about 11 m. On the top left view, the finite element mesh can be seen, with a fine band of 20 mm. On the right view, the pressure distribution reveals compressive and depressive zones, the latter being close to the solidification front (the mushy zone is materialized by 20 lines separated by an interval $\Delta g_l = 0.05$). (From Bellet, M. and Heinrich, A., *ISIJ Int.* 44, 1686–1695, 2004. With permission from *ISIJ*.)

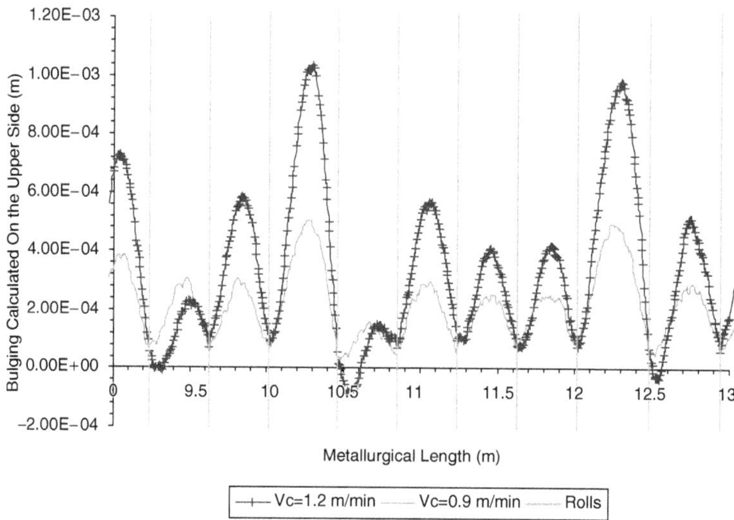

FIGURE 27.16 Slab bulging calculated at two different casting speed: 0.9 and 1.2 m/min. The slab bulging increases with the casting speed. (With permission after Triolet, N. and Bobadilla, M., *Proceedings of the MCWASP-XI, 11th International Conference on Modelling of Casting, Welding, and Advanced Solidification Processes*, Gandin, C.A. and Bellet, M., eds., The Minerals, Metals and Materials Society, Warrendale, PA, Opio, France, May 28, June 2, 2006, 753–760.)

solidified shell between rolls, such as those carried out in static conditions by Wünnenberg,[76] Miyazawa and Schwerdtfeger[73] or by Kajitani et al.[77] on limited slab sections moving downstream between rolls and submitted to the metallurgical pressure onto the solidification front. The influence of process parameters on the thermomechanical state of the strand can then be studied using such numerical models. An example is given by Figure 27.16, presenting the sensitivity of bulging to the casting speed. It can also be seen that bulging predictions are sensitive to the roll pitch, a larger pitch between two sets of rolls inducing an increased bulging. These numerical simulations can then be used to study possible modifications in the design of continuous casters, such as the replacement of large rolls by smaller ones to reduce the pitch and the associated bulging.[78]

27.5 Conclusions

Mechanical analysis of casting processes is growing in sophistication, accuracy, and phenomena incorporated. Quantitative predictions of temperature, deformation, strain, stress, and hot tearing in real casting processes are becoming possible. Computations are still hampered by the computational speed and limits of mesh resolution, especially for realistic three-dimensional geometries and defect analysis. Further developments are needed in fundamental models of defect formation over multiple length scales, and in efficient convergence algorithms to solve the equations. Fundamental measurements, including the quantification of interfacial conditions, material constitutive properties, and conditions for defect formation are also needed. However, solidification processes are growing in maturity and improvements gained by trial and error in the plant are becoming more expensive. As computing power and software tools continue to advance, advanced computational models will become even more important in the years to come. Future advances to casting processes will increasingly rely on advanced computational models such as the thermomechanical models discussed here.

Acknowledgments

The authors wish to thank the Continuous Casting Consortium and the National Center for Supercomputing Applications at the University of Illinois, the French Ministry of Industry, the French

Technical Center of Casting Industries (CTIF) and the companies Arcelor, Ascometal, Atlantic Industrie, Aubert et Duval, Erasteel, Industeel, and PSA Peugeot-Citroën, for support of this work.

References

[1] Thomas, B.G. et al., Comparison of four methods to evaluate fluid velocities in a continuous casting mold, *ISIJ Int.*, 41, 10, 1266, 2001.

[2] Yuan, Q., Zhao, B., Vanka, S.P., and Thomas, B.G., Study of computational issues in simulation of transient flow in continuous casting, *Steel Res. Int.*, 76, 1, Special Issue: Simulation of Fluid Flow in Metallurgy, 33–43, 2005.

[3] Yuan, Q. et al., Computational and experimental study of turbulent flow in a 0.4-scale water model of a continuous steel caster, *Metall. Mater. Trans.*, 35B, 5, 967–982, 2004.

[4] Estrin, Y., A versatile unified constitutive model based on dislocation density evolution, in *Constitutive modelling — Theory and Application*, MD-vol. 26/AMD-vol. 121, ASME, New York, 65–75, 1991.

[5] Agelet de Saracibar, C., Cervera, M., and Chiumenti, M., On the constitutive modeling of coupled thermomechanical phase-change problems, *Int. J. Plast.*, 17, 1565–1622, 2001.

[6] Weiner, J.H. and Boley, B.A., Elasto-plastic thermal stresses in a solidifying body, *J. Mech. Phys. Solids*, 11, 145–154, 1963.

[7] Smelser, R.E. and Richmond, O., Constitutive model effects on stresses and deformations in a solidifying circular cylinder, in *Modeling of Casting and Welding Processes IV*, Giamei, A.F. and Abbaschian, G.J., eds., Palm Coast, FL, The Minerals, Metals & Materials Society, Warrendale, PA, 313–328, 1988.

[8] Bellet, M. et al., Thermomechanics of the cooling stage in casting processes: 3D finite element analysis and experimental validation, *Metall. Trans. B*, 27, 81–100, 1996.

[9] Fjaer, H.G. and Mo, A., ALSPEN — A mathematical model for thermal stresses in DC-cast Al billets, *Metall. Trans.*, 21B, 6, 1049–1061, 1990.

[10] Wiese, J.W. and Dantzig, J.A., Modelling stress development during the solidification of grey iron castings, *Metall. Mater. Trans. A*, 21, 489–497, 1990.

[11] Inoue, T. and Ju, D.Y., Simulation of solidification and viscoplastic stresses during vertical semi-continuous direct chill casting of aluminium alloy, *Int. J. Plast.*, 8, 161–183, 1992.

[12] Moitra, A., Thomas, B.G., and Storkman, W., Thermo-mechanical model of steel shell behavior in the continuous casting mold, in *EPD Congress, Proceedings of TMS Annual Meeting*, Hager, J.P. ed., The Minerals, Metals, and Materials Society, Warrendale, PA, San Diego, CA, 547–577, 1992.

[13] Celentano, D., Oller, S., and Oñate, E., A finite element model for thermomechanical analysis in casting processes, *J. Phys.*, 3, 1171–1180, 1993.

[14] Li, C. and Thomas, B.G., Thermo-mechanical finite-element model of shell behavior in continuous casting of steel, *Metal. Materl. Trans. B*, 35B, 6, 1151–1172, 2004.

[15] Cruchaga, M.A., Celentano, D.J., and Lewis, R.W., Modeling fluid-solid thermomechanical interactions in casting processes, *Int. J. Numer. Methods Heat Fluid Flow*, 14, 167–186, 2004.

[16] Bellet, M., Jaouen, O., and Poitrault, I., An ALE-FEM approach to the thermomechanics of solidification processes with application to the prediction of pipe shrinkage, *Int. J. Numer. Methods Heat Fluid Flow*, 15, 120–142, 2005.

[17] Nicolli, L.C., Mo, A., and M'Hamdi, M., Modeling of macrosegregation caused by volumetric deformation in a coherent mushy zone, *Metall. Mater. Trans. A*, 36, 433–442, 2005.

[18] Fachinotti, V.D. et al., Two-phase thermo-mechanical and macrosegregation modelling of binary alloys solidification with emphasis on the secondary cooling stage of steel slab continuous casting, *Int. J. Numer. Methods Eng.*, 67, 1341–1384, 2006.

[19] Kozlowski, P. et al., Simple constitutive equations for steel at high temperature, *Metall. Trans. A*, 23A, 3, 903–918, 1992.

[20] Wray, P.J., Plastic deformation of delta-ferritic iron at intermediate strain rates, *Metall. Trans. A*, 7A, November, 1621–1627, 1976.

[21] Wray, P.J., Effect of carbon content on the plastic flow of plain carbon steels at elevated temperatures, *Metall. Trans. A*, 13A, 1, 125–134, 1982.

[22] Suzuki, T. et al., Creep properties of steel at continuous casting temperatures, *Ironmak. Steelmak*, 15, 2, 90–100, 1988.

[23] Costes, F., Heinrich, A., and Bellet, M., 3D thermomechanical simulation of the secondary cooling zone of steel continuous casting, in *Proceedings of the MCWASP X, 10th International Conference on Modeling of Casting, Welding and Advanced Solidification Processes*, Stefanescu, D.M., Warren, J.A., Jolly, M.R., and Krane, M.J.M., eds. The Minerals, Metals & Materials Society, Warrendale, Pennsylvania, USA, 393–400, 2003.

[24] Donsbach, D.L. and Moyer, M.W., Ultrasonic measurement of elastic constants at temperatures from 20 to 1100°C, in *Ultrasonic Materials Characterization, Special Publication 596*, Berger, H. and Linzer, M., eds., Nat. Bureau of Standards, 1980.

[25] Puhringer, O.M., Strand mechanics for continuous slab casting plants, *Stahl Eisen*, 96, 6, 279–284, 1976.

[26] Hub, D.R., Measurement of velocity and attenuation of sound in iron up to the melting point, *Proceedings of the IVth International Vong. Acoustics*, Copenhagen, paper 551, paper #551, 1962.

[27] Mizukami, H., Murakami, K., and Miyashita, Y., Mechanical properties of continuously cast zteels at high temperatures, *Tetsu-to-Hagane*, 63, 146, S652, 1977.

[28] Li, C. and Thomas, B.G., Maximum casting speed for continuous cast steel billets based on sub-mold bulging computation, in *Steelmaking Conference Proceedings*, 85, ISS, Warrendale, PA, Nashville, TN, March 10–13, 2002, 109–130, 2002.

[29] Li, C. and Thomas, B.G., Thermo-mechanical finite element model of bulging and hot tearing during continuous casting of steel billets, in *Modeling of Casting, Welding, and Advanced Solidification Processes*, X, Stefanescu, D., Warren, J., Jolly, M., and Krane, M., eds., TMS, Warrendale, PA, San Destin, FL, May 25–30, 2003, 385–392, 2003.

[30] Harste, K., Jablonka, A., and Schwerdtfeger, K., Shrinkage and formation of mechanical stresses during solidification of round steel strands, *4th International Confernce on Continuous Casting*, Centres de Recherches Metallurgiques and Verein Deutscher Eisenhuttenleute, Stahl und Eisen, Brussels, 633–644, 1988.

[31] Harste, K., Investigation of the shrinkage and the origin of mechanical tension during the solidification and successive cooling of cylindrical bars of Fe-C alloys, Ph.D. Dissertation, Technical University of Clausthal, 1989.

[32] Jimbo, I. and Cramb, A., The density of liquid iron-carbon alloys, *Metall. Trans. B*, 24B, 5–10, 1993.

[33] Kron, J. et al., Comparison of numerical simulation models for predicting temperature in solidification analysis with reference to air gap formation, *Int. J. Cast Metals Res.*, 17, 295–310, 2004.

[34] Kelly, J.E. et al., Initial development of thermal and stress fields in continuously cast steel billets, *Metall. Trans. A*, 19A, 10, 2589–2602, 1988.

[35] Meng, Y. and Thomas, B.G., Heat transfer and solidification model of continuous slab casting: CON1D, *Metal. Mater. Trans.*, 34B, 5, 685–705, 2003.

[36] Madhusudana, C.V. and Fletcher, C.V., Contact heat transfer — The last decade, *AIAA J.*, 24, 510–523, 1985.

[37] Laschet, G., Jakumeit, J., and Benke, S., Thermo-mechanical analysis of cast/mould interaction in casting processes, *Z. Metallkd*, 95, 1087–1096, 2004.

[38] Rappaz, M., Bellet, M., and Deville, M., Numerical modeling in materials science and engineering, in *Springer Series in Computational Mathematics*, Springer-Verlag, Berlin, 2003.

[39] Zienkiewicz, O.C. and Taylor, R.L., *The Finite Element Method*, 4th ed., McGraw Hill, New York, 1988.

[40] ABAQUS Theory Manual v6.0, Abaqus, Inc., Pawtucket, RI, 2004.

[41] Nemat-Nasser, S. and Li, Y.F., An explicit algorithm for large-strain, large-strain rate elastic-viscoplasticity, *Comp. Methods. Appl. Mech. Eng.*, 48, 205–219, 1992.

[42] Lush, A.M., Weber, G., and Anand, L., An implicit time-integration procedure for a set of internal variable constitutive equations for isotropic elastic-viscoplasticity, *Int. J. Plasticity*, 5, 521–549, 1989.

[43] Zhu, H., Coupled thermal-mechanical finite-element model with application to initial solidification, Thesis, University of Illinois, 1993.

[44] Koric, S. and Thomas, B.G., Efficient thermo-mechanical model for solidification processes, *Int. J. Numer. Methods. Eng.*, 66, 1955–1989, 2006.

[45] Glowinski, R. and Le Tallec, P., Augmented Lagrangian and operator-splitting methods in non-linear mechanics, *SIAM*, 9, 245, 1989.

[46] Wriggers, P. and Zavarise, G., On contact between three-dimensional beams undergoing large deflections, *Commun. Numer. Method. Eng.*, 13, 429–438, 1997.

[47] Jaouen, O. and Bellet, M., A numerical mechanical coupling algorithm for deformable bodies: application to part/mold interaction in casting process, in *Proceedings of 8th International Confernce on Modelling of Casting, Welding and Advanced Solidification Processes*, Thomas, B.G. and Beckermann, C., eds., The Minerals Metals and Materials Society, Warrendale, PA, San Diego (CA, USA), June 7–12, 1998, 739–746, 1998.

[48] Bellet, M. and Fachinotti, V.D., ALE method for solidification modelling, *Comput. Methods Appl. Mech. Eng.*, 193, 4355–4381, 2004.

[49] Eskin, D.G., Suyitno and Katgerman, Mechanical properties in the semi-solid state and hot tearing of aluminum alloys, *Prog. Mater. Sci.*, 49, 629–711, 2004.

[50] Clyne, T.W. and Davies, G.J., Comparison between experimental data and theoretical predictions relating to dependence of solidification cracking on composition, in *Solidification and Casting of Metals*, The Metals Society, London, 275–278, 1979.

[51] Kinoshita, K., Emi, T., and Kasai., Thermal elasto-plastic stress analysis of solidifying shell in continuous casting mold, *Tetsu-to-Hagane*, 65, 14, 2022–2031, 1979.

[52] Kristiansson, J.O., Thermal stresses in the early stage of solidification of steel, *J. Thermal Stresses*, 5, 315–330, 1982.

[53] Thomas, B.G., Samarasekera, I.V., and Brimacombe, J.K., Mathematical model of the thermal processing of steel ingots, Part II: Stress model, *Metall. Trans. B*, 18B, 1, 131–147, 1987.

[54] Okamura, K. and Kawashima, H., Calculation of bulging strain and its application to prediction of internal cracks in continuously cast slabs, in *Proceedings of the International Confernce on Comp. Ass. Mat. Design Proc. Simul., ISIJ*, Tokyo, 129–134, 1993.

[55] Ackermann, P., Kurz, W., and Heinemann, W., *In situ* tensile testing of solidifying aluminum and Al–Mg shells, *Mater. Sci. Eng.*, 75, 79–86, 1985.

[56] Bernhard, C., Hiebert, H., and Wolf, M.M., Simulation of shell strength properties by the SSCT test, *ISIJ Int. (Japan)*, 36, Suppl. Science and Technology of Steelmaking, S163–S166, 1996.

[57] Suzuki, M., Yu, C., and Emi, T., *In-Situ* measurement of tensile strength of solidifying steel shells to predict upper limit of casting speed in continuous caster with oscillating mold, *ISIJ Int., Iron Steel Inst. Jpn*, 37, 4, 375–382, 1997.

[58] Yamanaka, A. et al., Measurement of critical strain for solidification cracking, in *Modelling of Casting, Welding, and Advanced Solidification Processes — V*, Rappaz, M., Ozgu, M.R., and Mahin, K.W., eds., Davos, SW, TMS, Warrendale, PA, V, 279–284, 1990.

[59] Yamanaka, A., Nakajima, K., and Okamura, K., Critical strain for internal crack formation in continuous casting, *Ironmaking Steelmaking*, 22, 6, 508–512, 1995.

[60] Wisniewski, P. and Brody, H.D., Tensile behavior of solidifying aluminum alloys, in *Modelling of Casting, Welding, and Advanced Solidification Processes — V*, Rappaz, M., Ozgu, M.R. and Mahin, K.W., eds., Davos, SW, TMS, Warrendale, PA, V, 273–278, 1990.

[61] Won, Y.-M. et al., A new criterion for internal crack formation in continuously cast steels, *Metall. Mater. Trans. B*, 31B, 779–794, 2000.

[62] Feurer, U., Mathematisches modell der Warmrissneigung von binären aluminium legierungen, *Giessereiforschung*, 28, 75–80, 1976.

[63] Rappaz, M., Drezet, J.-M., and Gremaud, M., A new hot-tearing criterion, *Metall. Mater. Trans. A*, 30A, 2, 449–455, 1999.

[64] Drezet, J.M. and Rappaz, M., Prediction of hot tears in DC-Cast aluminum billets, in *Light Metals*, Anjier, J.L., ed. TMS, Warrendale, PA, 887–893, 2001.

[65] M'Hamdi, M. et al., The importance of viscoplastic strain rate in the formation of center cracks during the start-up phase of direct-chill cast aluminium extrusion ingots, *Metall. Mater. Trans. A*, 34, 1941–1952, 2003.

[66] Thercast. *presentation*. www.transvalor.com and www.scconsultants.com. 2006.

[67] Bellet, M., Aliaga, C., and Jaouen, O., Finite elements for a thermomechanical analysis of solidification processes, in *Modeling of Casting, Welding, and Advanced Solidification Processes IX*, Sahn, P.R., Hansen, P.N., Coley, J.G., eds., Shaker Verlag GmbH, Aachen, 10–17, 2000.

[68] David, S. and Auburtin, P., Numerical simulation of casting processes. Benefits of thermomechanical simulation in automotive industry, in *Conference Matériaux 2002, Tours, France, Proc.*, on CD, in French, Université Technologique de Belfort-Montbéliard, 5, 2002.

[69] Cicutti, C. et al., Mould thermal evaluation in a slab continuous casting machine, Steelmaking Conference Proceedings, 85, 97–107, 2002.

[70] Wolf, M.M., Continuous casting: Initial solidification and strand surface quality of peritectic steels, 9, Iron and Steel Society, Warrendale, PA, 1–111, 1997.

[71] Thomas, B.G. and Ojeda, C., Ideal taper prediction for slab casting, ISSTech Steelmaking Conference, Indianapolis, IN, USA, April 27–30, 2003, 86, 396–308, 2003.

[72] Li, C. and Thomas, B.G., Thermo-mechanical finite element model of shell behavior in continuous casting of steel, modeling of casting, welding and advanced solidification process X, San Destin, FL, May 25–30, 2003, TMS, 2003.

[73] Miyazawa, K. and Schwerdtfeger, K., Macrosegregation in continuously cast steel slabs: Preliminary theoretical investigation on the effect of steady state bulging, *Arch. Eisenhutten*, 52, 11, 415–422, 1981.

[74] Lesoult, G. and Sella, S., Analysis and prevention of centreline segregation during continuous casting of steel related to deformation of the solid phase, *Solid State Phenomena*, 3, 167–178, 1988.

[75] Bellet, M. and Heinrich, A., A two-dimensional finite element thermomechanical approach to a global stress–strain analysis of steel continuous casting, *ISIJ Int.*, 44, 1686–1695, 2004.

[76] Wünnenberg, K. and Huchingen, D., Strand bulging between supporting rollers during continuous slab casting, *Stahl und Eisen*, 98, 6, 254–259, 1978.

[77] Kajitani, T., Drezet, J.-M., and Rappaz, M., Numerical simulation of deformation-induced segregation in continuous casting of steel, *Metall. Mater. Trans. A*, 32, 1479–1491, 2001.

[78] Triolet, N. and Bobadilla, M., Mastering steel slab internal soundness and surface quality issues through thermomechanical modelling of continuous casting, in *Proceedings of the MCWASP-XI, 11th International Conference on Modelling of Casting, Welding, and Advanced Solidification Processes*, Gandin, C.A. and Bellet, M., eds., The Minerals, Metals and Materials Society, Warrendale, PA, Opio, France, May 28, June 2, 2006, 753–760, 2006.

VI

Multiscale Processes

28

Processing Nanoscale Structures to Macrocomposites

Hans J. Fecht
Forschungszentrum Karlsruhe, Institute of Nanotechnology and University of Ulm

G. Wilde
University of Münster, Institute of Materials Physics and Forschungszentrum Karlsruhe, Institute of Nanotechnology

Abstract

Nanocrystalline bulk materials, that is, dense and macroscopically extended materials with a grain size that is significantly smaller than 100 nm, have attracted considerable attention because of their modified and often improved properties. Although technologically as well as scientifically of increased interest, synthesizing significant amounts of continuous material with nanoscale grain size and without detrimental impurity concentrations still presents a mayor challenge. Additionally, nanocrystalline materials are particularly unstable against grain growth since they are far away from thermodynamic equilibrium and since they have short diffusion pathways. Thus, stabilization strategies, for example, based on a "nano-composite" approach, need to be designed and included into advanced processing methods. This chapter summarizes the current state of this emerging field by highlighting recent examples.

28.1 Introduction and Background

Nanostructured materials and composites as a new class of engineering materials with enhanced properties and structural length scales between 1 and 100 nm can be produced by a variety of different methods.[1]

TABLE 28.1 Effects of Nanomaterials and Applications Due to the Reduced Dimension

Effect of nanoscale	Applications
Higher surface to volume ratio, enhanced reactivity	Catalysis, solar cells, batteries, gas sensors
Lower percolation threshold	Conductivity of materials, sensors
Increased hardness/wear resistance with decreasing grain size	Hard coatings, tools, protection layers, sword making
Narrower bandgap with decreasing grain size	Opto-electronics
Higher resistivity with decreasing grain size	Electronics, passive components, sensors
Improved atomic transport kinetics	Batteries, hydrogen storage
Lower melting and sintering temperature	Processing of materials, low sintering materials
Improved reliability, fatigue	Electronic components, MEMS

Besides, the fabrication of clusters, thin films, and coatings from the gas or liquid phase, chemical methods such as sol-gel processes and electrodeposition are common methods of processing. As a versatile alternative however, mechanical methods have been developed that allow fabricating nanostructured materials in large quantities with a broad range of chemical compositions and atomic structures. These methods can be applied to powder samples, thin foils, and to the surface of bulk samples.

Some of the basic principles in improving the mechanical strength of a material have already been developed in early steel making. As an example, composites of iron and iron-carbides date back to 1500 B.C. with the Hittites in the Middle East having the monopoly at these times. Pattern forging was developed later on by smiths during the Iron Age and Merovingian, Roman, Anglo-Saxon, and Viking periods. As a prime example, Damascene steels consist of two different steel grades, welded in over a hundred layers. After etching, one of the steel grades turns dark and a beautiful pattern appears on the surface.[2]

As for the ancient sword manufacturing the development of strong, corrosion- and wear-resistant materials is still a primary goal of materials science and engineering these days. However, the potential of conventional methods of materials strengthening (e.g., cold working, solution hardening, formation of laminates, etc.) in the meanwhile has been almost completely exhausted.

Table 28.1 illustrates the effects and applications of the reduced dimensionality of nanomaterials with regard to different physical properties and the resulting applications. Furthermore, the addition of nanoparticles to an otherwise homogeneous material can lead to a change in the macroscopic material behavior. Thus, most material properties may be changed and engineered dramatically through the controlled size-selective synthesis and assembly of nanoscale building blocks.

For example with regard to mechanical metallurgy, we have shown recently that railroad steel undergoes significant changes in the microstructure and mechanical properties that finally account for progressive increase in noise and vibration levels during movement of high speed trains over such deformed rails after extended time periods. The microstructural changes (extreme deformation, work hardening, grain refinement to the nanometer scale, and wear/erosion) at the surface of the bulk components exposed to such severe plastic deformation processes are comparable to those observed during sword making of Damascus steels (Figure 28.1). The effect of etching in both cases produces very similar results and effects, such as "white etching layers" and the formation of corresponding geometrical patterns.

28.2 Nanoscale Effects

Physical and chemical property changes on a nanoscale are of great interest for science and technology these days. Most fundamental physical properties change dramatically when the characteristic length scale of a particular property coincides with the structural length scale of the nanostructure of a material. In the near future, this effect will allow tuning of the physical properties of a macroscopic material if the material consists of nanoscale building blocks with controlled size and composition.

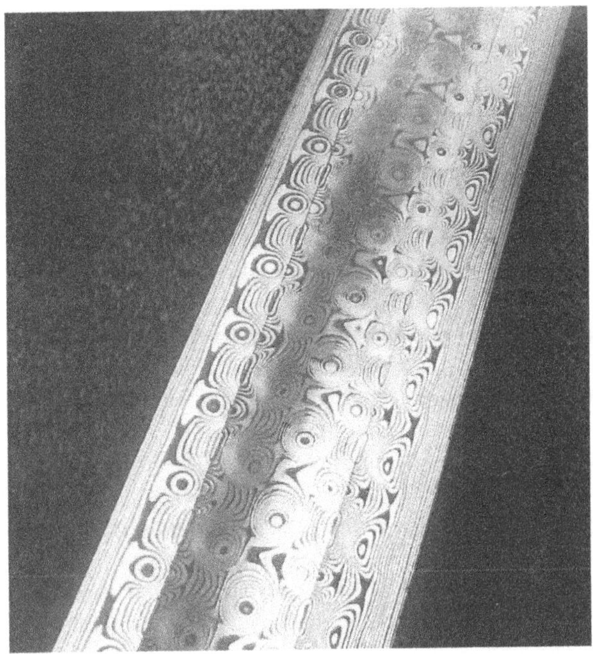

FIGURE 28.1 Damascene effect (from www.damasteel.se) showing a pattern that is named "**Muhammed's ladder.**"

Every property has a critical length scale, and if a nanoscale building block is made smaller than this critical length scale, the fundamental physics of that property changes. By altering the sizes of those building blocks, controlling their internal and surface chemistry, their atomic structure, and their assembly, it becomes therefore possible to engineer properties and functionalities in completely new ways. This fundamental behavior can generally be found for objects with different dimensionalities:

- 0-D (nanosized clusters)
- 1-D (nanowires)
- 2-D (thin-film-multilayers)
- 3-D (bulk nanostructures and composites)

when the length scale of a microstructure is on the order of a few nanometers.

In all these cases, interfaces and surfaces that separate the different particles, layers, and crystalline or noncrystalline domains from each other play the crucial role in controlling the properties and stability of nanostructures. Two effects are critical in this respect:

1. The atomic structure of the interface separating two domains and increasing the disorder in a nanostructure.
2. Finite size effects of the domains themselves.

These two structural aspects are in general inherently coupled with each other. Thus, it is necessary to develop a fundamental understanding of the correlation between a property and the characteristic length scale of a nanostructure to improve a particular property and develop specific applications, such as, for example, sensors, actuators, safety systems, and the like.

Of particular importance, for industrial applications are clusters (chemical industry for catalysis) and thin layers and coatings (micromechanical systems, energy systems, health & medicals) whereas nanowires are of more fundamental interest.

There are major unsolved questions and new phenomena observed in the field of interface controlled nanostructures when the size range is between the range corresponding to a cluster of atoms and a crystal, for example, in the mesoscopic range. To achieve the goals mentioned above a deeper understanding of fundamental issues becomes necessary. For example:

- Atomic diffusion through interfaces becomes an efficient mechanism of transport in a nanostructure at relatively low temperature in comparison with conventional matter. This fundamental effect can be used to considerably increase the sensitivity of a (gas) sensor or the kinetics of hydrogen diffusion in a hydrogen storage device.
- If the crystal size is smaller than the electron mean free path, boundary scattering dominates and, hence, the electronic conductivity as well as the temperature coefficient is found to decrease.
- Surface effects in magnetic materials do control the magnetic properties of thin layers or nanocomposites, leading to more efficient data storage devices and more sensitive magnetic sensors using effects, like GMR (giant magnetic resistance) and spintronics.
- The mechanical strength of engineering materials can be described by the familiar Hall–Petch relation where the strength increases as the square root of the grain size, and which is connected with dislocation pile-ups. For a typical grain size of 10 nm another strengthening mechanism associated with the stability of a dislocation loop becomes dominant. This mechanism can lead to the development of materials with superior strength and ductility, and thus improved performance and lifetime.
- Band gap changes in nanosized semiconductor particles and size-induced control of luminescence relaxation in oxide nanoparticles lead to changes in the optical properties and to interesting optoelectronic device applications.
- Thermodynamic phase equilibria are shifted or changed due to interface effects. This allows the generation of new materials being out of equilibrium in bulk and exhibiting properties not known up to now.
- Tribological properties are changed tremendously. For example, the interaction between an ultrasharp AFM-tip and a nanostructured sample surface is strongly influenced by size-dependent effects resulting in a change of frictional forces and energy dissipation mechanisms. This allows to reduce friction and wear in MEMS applications, microsystems, and microsurgical instruments and, thus increases their efficiency and lifetime and saves natural resources.

Furthermore, the addition of nanoparticles to an otherwise homogeneous material can lead to a change in the macroscopic material behavior. Most material properties may be changed and engineered dramatically through the controlled size-selective synthesis and assembly of nanoscale building blocks.

As such, it can be expected that the study of mechanically induced nanostructure formation and wear related phenomena, such as mechanical alloying, in the future not only opens new processing routes for a variety of advanced nanostructured materials but also improves the understanding of technologically relevant deformation processes from ancient sword making to modern materials engineering on a nanoscale level.

28.3 Solid-State Processing

28.3.1 Mechanical Alloying

In the 1970s, the method of mechanical attrition (MA) of powder particles followed by high temperature sintering was developed as an industrial process to successfully produce new alloys and phase mixtures (see also Chapter 13). For example, this powder metallurgical process allows the preparation of alloys and composites, which cannot be synthesized via conventional casting routes. This method can yield:[3]

1. Uniform dispersions of ceramic particles in a metallic matrix (superalloys) for use in gas turbines.
2. Alloys with different compositions than alloys processed from the liquid.
3. Alloys of metals with quite different melting points with the goal of improved strength and corrosion resistance.

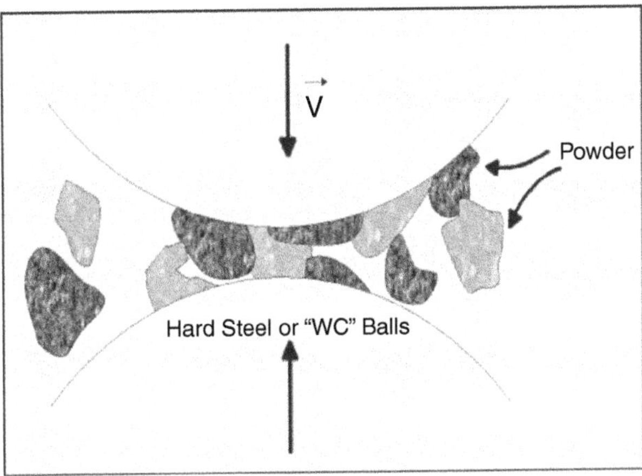

FIGURE 28.2 Principle of ball milling.

In the 1980s, the method of high-energy milling gained a lot of attention as a nonequilibrium solid-state process resulting in materials with nanoscale microstructures. The formation of nanocrystals or amorphous metallic materials has been observed in a broad range of alloys.[4–7] The amorphous phase formation occurs by intermixing of the atomic species on an atomic scale, thus softening and destabilizing the crystalline lattice[7] and driving the crystalline solid solution outside of its stability range against "melting" or amorphization.[8] This process is considered as a result of both mechanical alloying[9,10] and the incorporation of lattice defects into the crystal lattice.[11] More recent investigations demonstrate that the nanostructure formation also can occur for several unexpected cases, such as brittle ceramics, ceramic phase mixtures, polymer blends, and metal/ceramic nanocomposites.

During mechanical alloying or MA a large number of small powder particles of about 20 to 100 μm are placed together with hardened steel or tungsten carbide (WC) coated balls in a sealed container (Figure 28.2), which is shaken violently. Consequently, plastic deformation at high strain rates ($\sim 10^3$ to 10^4 s^{-1}) occurs within the particles and the average grain size of the powder particles can be reduced to few nanometers after extended milling. The temperature rise during this process is modest and is generally estimated to be <450 K. The collision time corresponds to typically 2 ms. More details on the mechanisms of mechanical alloying may be found in Chapter 13.

An alternate route to producing samples with high levels of both plastic deformation and interfacial area is cold rolling of layered elemental sheets. Each deformation cycle consisted of rolling the multilayer sandwich to a thickness of the order of 100 μm and subsequent folding (Figure 28.3). Here, the large increase in interfacial area is created internally with absolutely negligible contamination at ambient temperature.

28.3.1.1 Metals and Intermetallics

Upon ball milling, for most metallic elements and intermetallic compounds a refinement of the internal grain size is observed to typically 5 to 20 nm together with an increase of atomic level strains typically between 0.7 and 2.5% (intermetallics). The atomic disorder results in a decrease of about 20% of the enthalpy of fusion. The elemental processes leading to the grain size refinement include three basic stages as found by combined x-ray, electron and neutron diffraction/scattering analysis:

1. Initially, the deformation is localized in shear bands consisting of an array of dislocations at high density. Here, the dislocation cell size dimensions are basically a function of the acting shear stress σ resulting in an average cell size dimension of $L = 10\, G\, \mathbf{b}/\sigma$ with G being the shear modulus and $\mathbf{b}$ the magnitude of the Burgers vector.

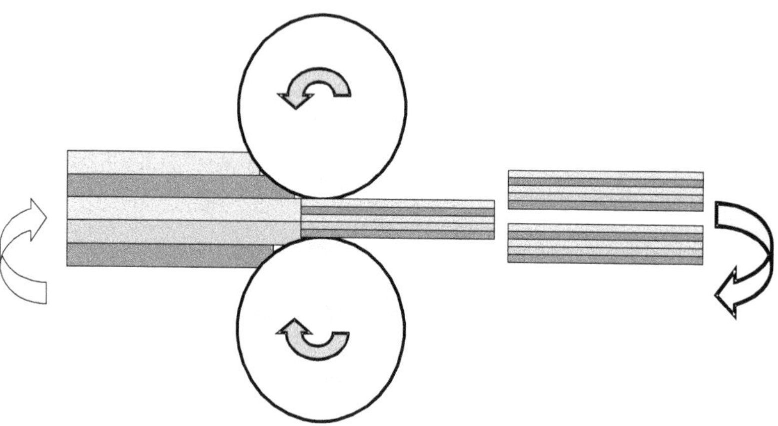

FIGURE 28.3 Principle of repeated cold rolling and folding.

2. At a certain strain level, these dislocations annihilate and recombine to small-angle grain boundaries separating the individual grains. The subgrains formed via this route are already in the nanometer size range.
3. The orientations of the single-crystalline grains with respect to their neighboring grains become completely random.

Extended solid solutions far beyond the thermodynamic equilibrium have generally been noted in the course of mechanical milling of alloys. In addition, for phase mixtures with negative enthalpies of mixing and large ($>15\%$) atomic size mismatch solid-state amorphization can occur. During this process long-range solute diffusion and solute partitioning are suppressed and therefore, highly metastable amorphous and nanocrystalline states become accessible.

28.3.1.2 Ceramics

Whereas ductile materials can be deformed as described above, it is surprising that nominally brittle materials, such as ceramics, also show a refined microstructure after extended MA. For example, ZrO_2 has been milled for up to 40 h resulting in a grain size reduction to several nanometers. At a grain size of less than about 12 nm a transition of the — for the bulk material — most stable monoclinic structure to a metastable orthorombic modification has been observed.[12] A simple estimate shows that below a critical grain size of about 30 nm the high temperature phase becomes more stable than the low temperature phase due to the defects incorporated.

Mechanical alloying does occur for ceramic powder mixtures, as well. For example, $(Fe,Cr)_2O_3$ solid solutions can be obtained from a Fe_2O_3/Cr_2O_3 powder mixture with an average grain size of about 20 nm.[13] Similar observations have been reported for ZrO_2/Y_2O_3 powder mixtures.[14] Also milling of ceramic materials can induce chemical processes. For example, extended periods of mechanical milling can lead to the deoxidation of oxides when appropriate materials are added. Mechano-chemical reactions have been reported for α-Fe_2O_3/Ti, Ag_2O/C[15] and CuO/Ca[16] mixtures.

28.3.1.3 Polymer Blends

Mechanical alloying of polymeric materials has been developed during the past decade, as well. Similar to metallic materials MA leads to an increase of the internal energy. For example, polyamide (PA), polyethylene (PE), acrylonitrile-butadiene-styrene (ABS), polypropylene (PP), and polystyrene (PS), have been investigated in detail.[17] To fracture the polymer chains, the milling process is conducted below the respective glass transition temperatures. As a result, the crystallinity of the powder material can be decreased considerably by mechanical milling. The corresponding storage of energy allows consolidation of the powder to bulk samples at lower temperatures than conventionally processed material. In addition, the achieved mechanical properties are considerably improved.

The milling process also allows mechanical alloying of polymer mixtures (PA/PE, PA/ABS),[18] as well as mechanical alloying of polymers with ceramic (PP/SiC) and metal powder (PS/Sn, PP/Al, PE/Cu[19]). By mechanical milling the normal compatibility restrictions for polymer formation from regular polymer science principles have been removed. This unique opportunity allows synthesizing new materials and materials combinations with enhanced properties that cannot be achieved by any other method.

28.3.1.4 Nanocomposites

Mechanical attrition is also a very versatile process to prepare nanocomposites. As this process is sensitive to contamination resulting from the milling environment atmospheric control can be used to purposely induce chemical reactions between the milled powders and their environment. By a proper choice of a reactive gas atmosphere (O_2, N_2, etc.) or a milling fluid (organic fluids) the metal powder can be intentionally modified by reactive milling to a nanocrystalline metal–ceramic composite[20] or fully reacted to a nanocrystalline ceramic (e.g. a metal nitride).[21]

The metal powders Ti, Fe, V, Zr, W, Hf, Ta, and Mo,[21,22] transform to a nanocrystalline nitride by high-energy ball milling under nitrogen gas flow. This solid-state interdiffusion reaction during reactive ball milling is triggered by fragmentation of the starting powder thus creating new surfaces. These freshly created surfaces react with the flowing nitrogen gas to form a nitride surface layer over the unreacted core particle. With further milling this reaction continues and a homogeneous nitride phase is formed and the unreacted core of metal disappears resulting in a nanostructured (often metastable) metal nitride with a grain size of typically 5 nm.

By ball milling in organic fluids as surfactants, which are sometimes used to prevent contamination by the milling tools, chemical reactions can be induced leading to the formation of fine carbides. For example, by milling Al (–Ti, –Zr or –Hf) alloys in hexane an average grain size of 9 nm can be achieved with carbon being dissolved in the matrix. During dynamic compaction at about 1300 K, grain growth occurs up to about 44 nm together with precipitation of ZrC particles, 7 nm in size. Such ultrafine-grained composites are expected to exhibit considerably improved strength and ductility.[23]

More recently, metallic glass/ceramic composites were obtained by mechanical alloying of multicomponent Zr-based elemental metallic powders together with SiC particles.[24,25] The distribution of SiC particles was uniform with sizes ranging from 1 μm down to values below 50 nm. It is further interesting to note that the SiC particles do not act as potent heterogeneous nucleation sites when the composite is heated to the crystallization temperature above the glass transition temperature. As such, mechanical alloying represents a convenient method to achieve dispersion-strengthened amorphous alloys with considerably improved strength and wear resistance by a powder metallurgical pathway.

28.3.2 Cold Rolling of Thin Sheets

An alternate route to producing samples with high levels of both plastic deformation and interfacial area is by cold rolling of layered elemental sheets that are folded between each deformation cycle (Figure 28.3). In this case, the large increase in interfacial area is created internally with absolutely negligible contamination. Similarly, in contrast to MA the uncertainty in the temperature during processing is removed as the sample is in firm contact with the massive rolls and deformation can be performed at a low strain rate to maintain ambient temperatures. This approach has been used to examine amorphous phase formation in several binary alloys such as Zr–Ni,[26] Cu–Er,[27] and Al–Pt[28] and also for the preparation of bulk Fe/Ag nanomultilayers with giant magnetoresistance.[29]

In some of the prior work on amorphous phase formation, deformation rates in excess of 1 s^{-1} were employed and some annealing was needed to complete the amorphization reaction. Recently, fully amorphous foils of a multicomponent $Zr_{65}Al_{7.5}Cu_{17.5}Ni_{10}$ alloy have been synthesized at ambient temperatures from a layered array of individual elemental sheets by repeated low-strain-rate (0.1 s^{-1}) cold rolling.[30] Figure 28.4a shows x-ray diffraction (XRD) patterns from the Zr–Al–Ni–Cu foils taken after

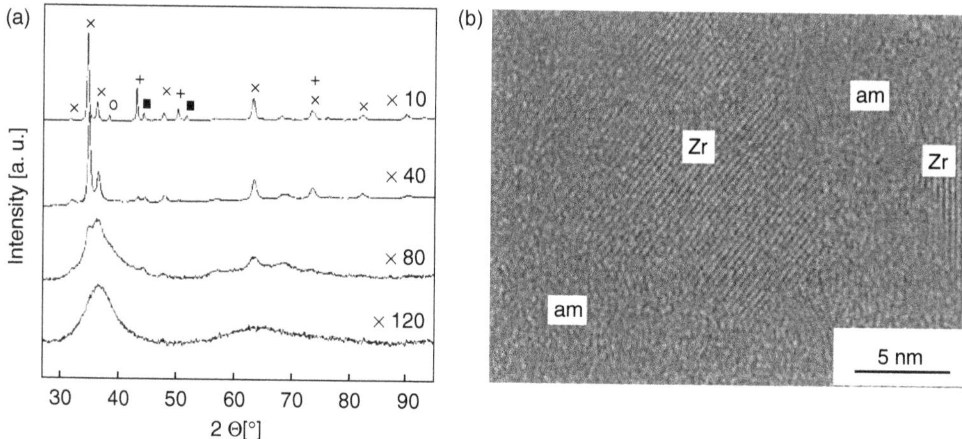

FIGURE 28.4 (a) XRD patterns of a Zr–Al–Ni–Cu alloy after cold rolling and folding for different numbers of successive cycles. The numbers indicate the number of repetitions that scales with the strain. × = Zr(hcp), o = Al(fcc), + = Cu(fcc), ■ = Ni(fcc). (b) HRTEM image of nanocrystalline/glass transition during cold rolling of thin foils of composition $Zr_{65}A_{7},5Cu_{17},5Ni_{10}$. (With permission from Sagel, A. et al., *Phil. Mag. Lett.*, 77, 109, 1998.)

10, 40, 80, and 120 deformation cycles. Each deformation cycle consisted of rolling the multilayer sandwich to a thickness of approximately 80 μm and subsequent folding. High-resolution transmission electron micrograph (HRTEM) analysis exhibits further evidence that a true amorphous phase has been formed as shown in Figure 28.4b.

The x-ray spectrum of the amorphous sample is basically identical to that of a metallic glass produced by liquid quenching with the same composition. Thermal analysis of the cold rolled amorphous sample with a DSC reveals a distinct glass transition at $T_g = 647$ K followed by a sharp exothermic crystallization peak at 745 K. It is worthy to note that very similar to amorphization reactions observed in mechanically alloyed Zr-based powder mixtures of similar composition the initial stage of cold rolling is characterized by the dissolution of solute into Zr along with a reduction in grain size to about 30 nm before the onset of the crystal-to-glass transition. The formation of similar amorphous phases from two inherently different initial states, that is, the solid and the liquid state, suggest that compositionally induced static disorder in a mechanically driven system can lead to the same final glass state, which is conventionally derived from freezing the dynamic disorder of a liquid to a glass.

However, special care needs to be taken in the case of vitrification reactions that develop under growth control[31] conditions. In this respect, marginal glass formers that require high quenching rates to prevent significant growth need to be distinguished from bulk glass-forming systems where it is often possible to completely avoid nucleation. The two different conditions that control the competition between vitrification and crystallization are schematically sketched in Figure 28.5. Calorimetry and microstructure analysis confirm that amorphous samples of $Al_{92}Sm_8$, a marginal glass-forming alloy, which have not been exposed to high temperatures in the liquid state before vitrification, exhibit a clear T_g signal.[32] The formation of a high number density of Al-nanocrystals, as observed during the heating of melt-quenched samples, does not occur for samples that were synthesized by a solid-state route, that is, by cold rolling.[31] This result shows clearly that homogeneous nucleation at temperatures at or below T_g does not cause the nanocrystallization, but that instead the nanocrystals develop by the growth of "quenched-in" nuclei.[33] Additionally, this result shows that the model of glass formation considering partitionless reactions is only valid in the limit of high cooling rates, R. The condition $R > R_{crit}$, with R_{crit}, the critical cooling rate for a complete avoidance of nucleation, has to be fulfilled to obtain similar amorphous states by MA or liquid undercooling.[34]

The present examples demonstrate that cold rolling offers an attractive alternative for the synthesis of nanostructured materials or multicomponent metallic glasses besides more traditional techniques such

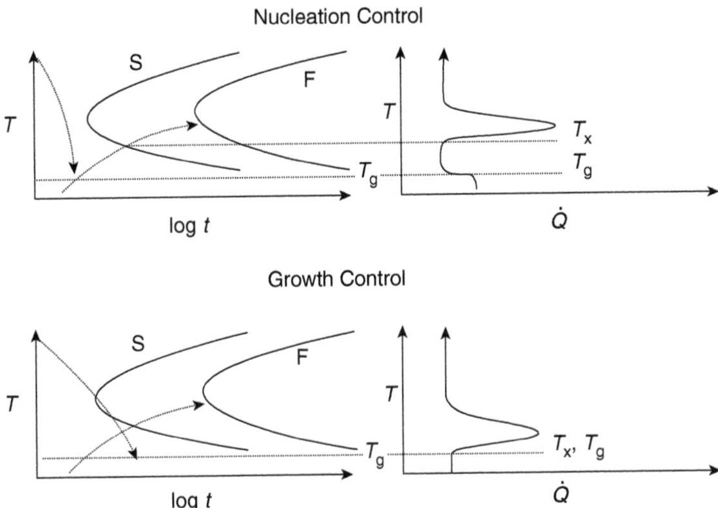

FIGURE 28.5 Glass formation kinetics: nucleation control vs. growth control. Quenching and reheating paths are shown on TTT diagrams (S-start; F-finish) and thermograms (dQ/dt: heat evolution rate). T_X marks the crystallization onset temperature during reheating after initial quenching and T_g is the glass transition temperature.

as MA or liquid undercooling. Due to the relatively simple experimental conditions, size restrictions do not limit the formation of bulk samples. Moreover, aside from the attractive option to synthesize bulk amorphous materials, this nonequilibrium processing route offers the additional option to obtain a nanocrystalline state or a composite microstructure with nanocrystals embedded in a residual amorphous matrix, for example, by careful thermal treatments of the as-deformed material. In addition to synthesizing partially or fully amorphous structures, the technique of repeated cold rolling and intermediate folding of thin sheets can also be used to produce nanocrystalline pure metals, alloys, or composite structures with extremely small grain sizes below 10 nm.[35–39] In this sense, the repeated rolling and folding method presents an attractive processing alternative to so-called severe plastic deformation techniques such as equal channel angular pressing or high-pressure torsion (HPT) straining, or to conventional powder attrition techniques as bulk quantities of massive materials with truly nanocrystalline microstructures can be obtained without the necessity of applying a high hydrostatic pressure. A similar procedure has previously been used to synthesize ultrafine-grained materials.[38] In this case, the rolling and folding procedure with an additional intermediate "brushing" operation has been denoted as accumulative roll bonding. In fact, cold rolling is a well-known process that is frequently applied in industrial processing of sheet metal. Yet, the deformation process used here differs from classical cold rolling as well as from accumulative roll bonding in the amount of strain (i.e., far larger than in both conventional processes) and in the absence of both, lubricating agents and intermediate brushing. Thus, due to friction at the roll surfaces, a shear strain component adds to the rolling strain that is active in conventional cold rolling processing.[40] Due to the continuous increase of strain during repeated cold rolling and folding of the multilayered sheet samples, the material is continuously energized and driven away from thermodynamic equilibrium. The microstructure response to this driven-system processing is given by the creation of dislocations that — if the creation rate exceeds the annihilation rate — lead to dislocation accumulation and finally to grain-boundary formation.

Pure Ni gives one example of nanostructure formation by repeated cold rolling and folding. The evolution of the texture during continued rolling deformation is indicated in Figure 28.6 together with the development of the layer thickness as a function of the number of folding and rolling (F&R) passes. The ratio of the (111)/(220) Bragg reflection intensity increased strongly up to 40 F&R passes (true strain, $\varepsilon_t = -28$). Further rolling up to 100 F&R passes ($\varepsilon_t = -70$) reduced the intensity ratio continuously down to a value close to 1. This result shows that a strong rolling texture developed in the early stages

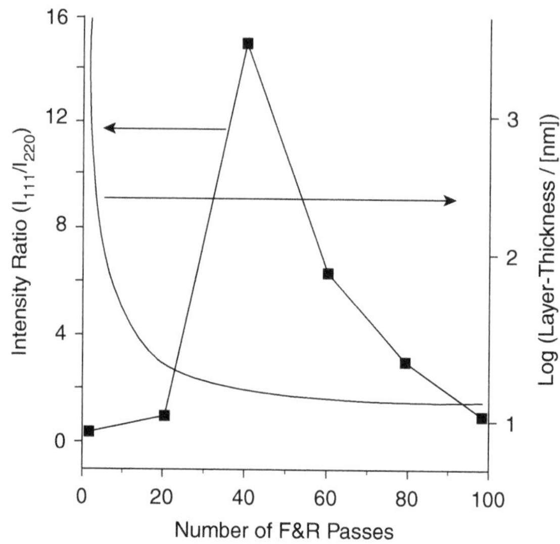

FIGURE 28.6 Ratio of the intensities I of the (111) and the (220) Bragg reflection of pure Ni (squares) vs. the number of F&R passes that scale linearly with ε_t. The lines represent guides to the eye. In addition, the layer thickness is also indicated schematically as a function of the number of passes. (With permission after Wilde, G., Rösner, H., and Dinda, G.P., *Adv. Eng. Mater.*, 7, 11, 2005.)

of rolling, which is indicative of a dominance of dislocation processes during that deformation stage.[41] At later stages of the overall deformation process, that is, at higher numbers of the F&R cycles (larger value of ε_t), further deformation proceeds at decreased grain sizes. Thus, it is anticipated that the local deformation mechanism is modified, and grain boundary sliding and grain rotation can become active or even dominant,[42] which leads to a continuous decrease of the texture with continued cold rolling. Thus, the intermediate stage needs to include a gradual transition from the deformation of individual layers that consist of strongly textured grains to a grain-size reduction of almost equiaxed grains with random orientations.

Additionally, the grain size distribution becomes narrower when the average grain size decreases down to the range of a few tens of nanometers. Dark-field transmission electron microscopy (TEM) micrographs that were taken at intermediate stages of the deformation process show larger regions with a similar orientation that are subdivided by bands of different orientation (Figure 28.7a). HRTEM analyses (Figure 28.7b) indicate that these band-like structures are regions of strain concentration where the crystal lattice is severely bent out of the diffraction condition. At even larger strain, the regions of common orientation disappear and the entire sample consists of nanoscale regions with different orientation with respect to the incident electron beam. After a short annealing treatment at rather low temperature to allow for some relaxation of the microstructure, the subsequently taken high-resolution electron microscopy (HREM) micrograph (Figure 28.7c) shows clearly that the microstructure consists of a dense array of nanocrystalline grains that are separated by a large fraction of high-angle grain boundaries. Additionally, Figure 28.7c displays the extent of microstructure refinement that has been achieved so far by the repeated cold rolling and folding process: clearly, continuous, massive materials with macroscopic dimensions and a microstructure that consists of individual grains with an average diameter below 10 nm can be synthesized.[35,36]

One issue with cold rolling is that bonding at the interfaces is often not perfect, if — for example — pure metals with melting points above 1500 K are deformed at room temperature and at low or moderate strain rates (compared to accumulative roll bonding) of the order of 1 s^{-1} without intermediate surface treatment. Moreover, cold rolling inherently produces foils or sheets, that is, materials that are less extended in one spatial direction. Thus, it is of interest to explore new processing pathways — as indicated

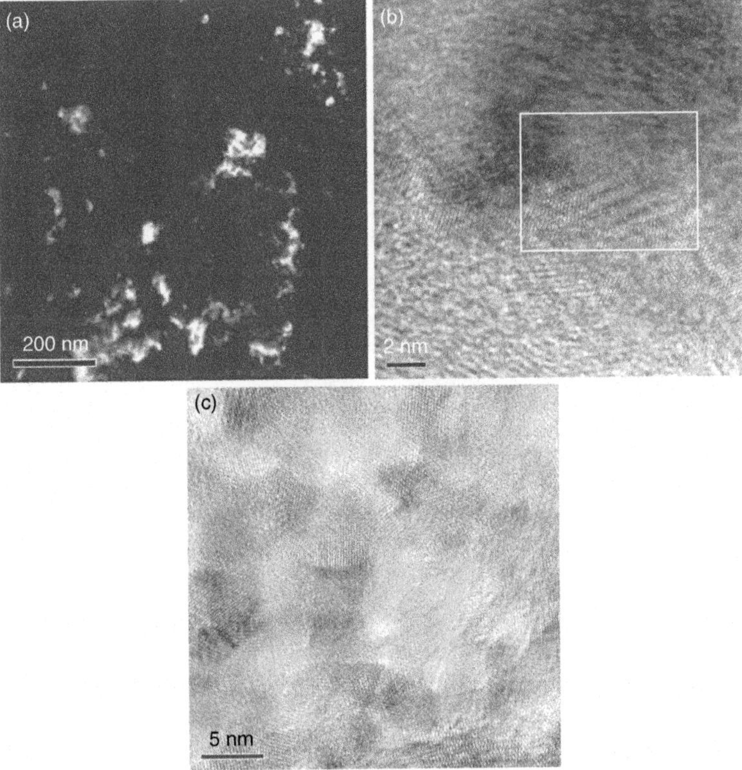

FIGURE 28.7 (a) Dark-field TEM micrograph of severely deformed Ni (80 passes). The bright regions show a substructure due to the high density of lattice defects. (b) HRTEM of the same deformation stage as in (a). The white frame indicates a region with high dislocation density. (c) HRTEM micrograph of a severely deformed Ni sample that has been thermally annealed for a short time at 350°C. Clearly, high-angle grain boundaries are visible that separate grains with an average diameter of about 10 nm. (With permission from Dinda, G.P., Rösner, H., and Wilde, G., *Scripta Mater.*, 52, 577, 2005.)

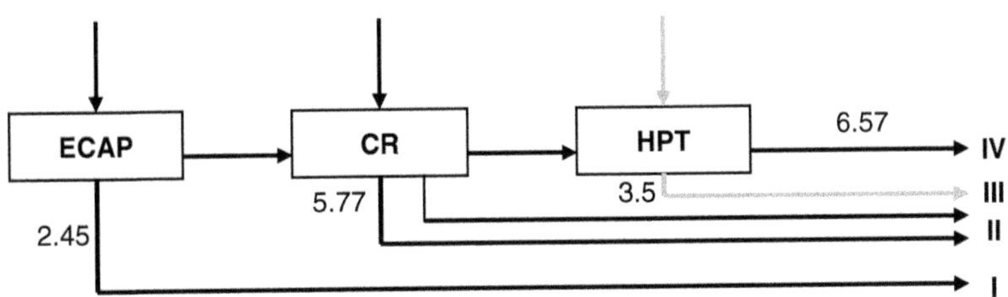

FIGURE 28.8 Summary of processing pathways that have been explored for Ni of similar purity (≥99.9%). The numbers indicate hardness values in GPa. ECAP: Equal Channel Angular Pressing, HPT: High-Pressure Torsion. (With permission from Wilde, G. et al., *Mater. Sci. Forum*, 425, 503–504, 2006.)

in Figure 28.8—that potentially allow for synthesizing massive nanostructured materials that are extended in all three spatial directions. It was observed that cold rolled material that was additionally processed by HPT at room temperature displayed full density and ductile behavior during deformation, as no cracks occurred during the torsion deformation (more details on severe plastic deformation such as HPT may be found in Chapter 13). Figure 28.9 shows (a) dark-field TEM overview and (b) high-resolution image of the microstructure of Ni after cold rolling ($\varepsilon = -80$) and subsequent HPT straining followed by annealing the sample at 300°C for a short time (heated from room temperature to 300°C by 20°C/min

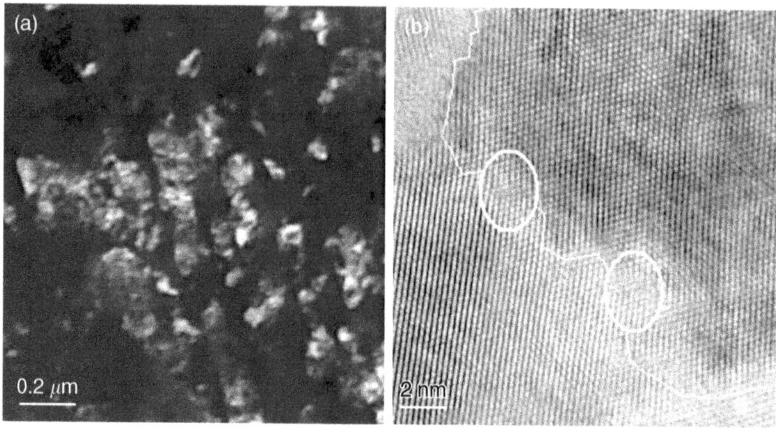

FIGURE 28.9 (a) Dark-field TEM image of Ni after cold rolling ($\varepsilon = -80$) and HPT, followed by short annealing at 300°C. (b) HRTEM micrograph of the same sample. The circle marks a dislocation situated at the grain boundary of two nanocrystalline grains. The white line indicates a grain boundary. (With permission from Wilde, G. et al., *Mater. Sci. Forum*, 425, 503–504, 2006.)

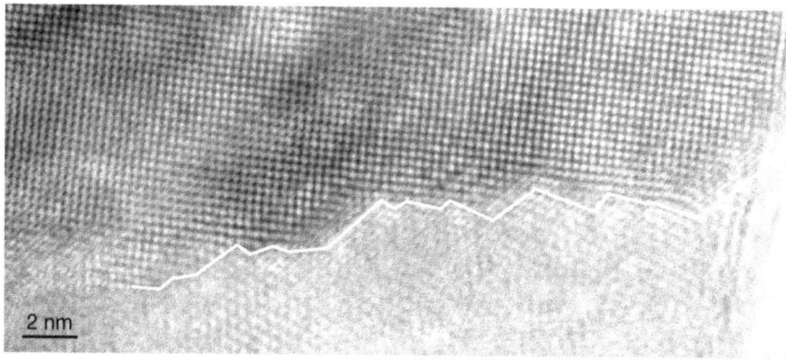

FIGURE 28.10 HRTEM image of Pd after 105 F&R passes by cold rolling. The white line indicates the position of the grain boundary between the upper grain (in ⟨100⟩ orientation) and the lower grain.

and subsequently cooled to 30°C by 150°C/min) to allow for some relaxation of the lattice defects to render the grain boundaries more observable.[43] It was also found that an initial ECAP treatment did not change the grain size obtained by cold rolling for up to 100 passes. However, texture analyses by XRD methods indicate that the remaining rolling texture after extended F&R decreased rapidly during HPT[35] and hardness measurements show a significant increase of the hardness after additional HPT.

The heat treatment did not result in flat grain boundaries, but instead the presence of grain boundaries with many kinks and facets is observed. The observation of such "rough" interfaces is not coupled to HPT processing, but to the application of a large strain. Clearly, the rough interface morphology is observed on as-deformed Pd-samples after cold rolling. Figure 28.10 shows the HREM image of a Pd sample after severe cold rolling to an equivalent strain of about $\varepsilon = -85$. In addition to the rough interfaces that resemble the "nonequilibrium" grain boundaries described by Valiev,[44] a very high density of lattice defects has been detected, including a high dislocation density at the grain boundaries and within the crystallites. These observations indicate that combining different nonequilibrium processing routes sequentially offer the potential for tuning the microstructure including the defect density and the defect distribution with obvious consequences concerning the performance and the stability of the materials.

Repeated cold rolling with intermediate folding presents a versatile and cost-effective alternative to conventional solid-state processing routes that can yield massive specimens of bulk nanocrystalline pure metals with grain sizes below 20 nm and, in the case of pure face-centered cubic (fcc) metals and binary

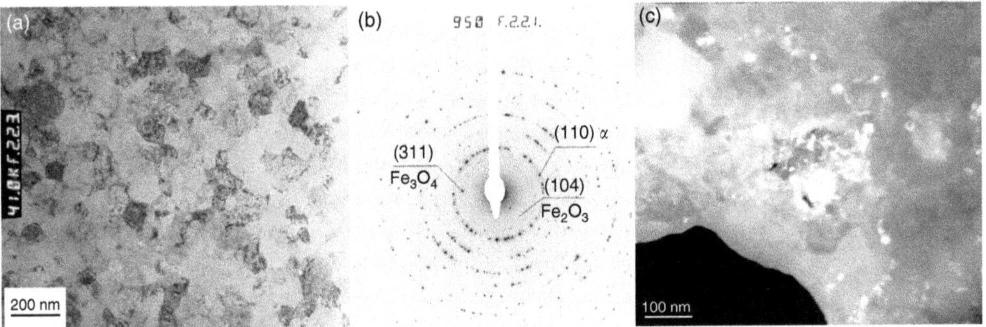

FIGURE 28.11 The microstructure of bulk nanocrystalline Fe-1 wt.% C compacts after HPT consolidation at 540°C (a) bright-field images, (b) selected area diffraction pattern, (c) dark field image in the (311) Fe_3O_4 reflection showing oxide particles distributed mainly along grain boundaries.

alloys such as $Ni_{50}Ti_{50}$,[34] can even produce microstructures with an average grain size below 10 nm. In addition, highly metastable or even nonequilibrium alloy phases are obtained under defined conditions. It also provides a new opportunity with enhanced versatility for synthesizing new materials and adjusting new microstructures by combined thermal and mechanical processing. Moreover, up scaling for continuous processing conditions to yield bulk quantities seems feasible and could provide a new perspective for the production of bulk nanocrystalline materials for advanced applications.

28.3.3 Powder Consolidation by HPT

HPT can be effectively applied for consolidation of nanocrystalline and amorphous powders prepared by ball milling, rapid quenching, and other techniques. The grain size in the resulting compacts is strictly dependent on the grain and particle size of the initial powders and on consolidation conditions. For example, after room temperature HPT consolidation of cryomilled Al-7.5 wt.% Mg powder, bulk nanocrystalline specimens with 97% relative density, and 30 nm grain size have been obtained.[45] HPT processing of powder blends of metal and metal oxides results in formation of nanocomposites consisting of a nanocrystalline matrix ($d \sim 50$ to 70 nm) and embedded oxide particles.[46,47] In some cases when the starting powder is too hard so that its room temperature consolidation is impossible, this consolidation can be performed at elevated temperatures and in vacuum to prevent oxidizing. For example, after HPT processing of ball-milled blends of powders of elemental Fe and C (carbon black) at 540°C, bulk specimens with composition of Fe-1 wt.% C and relative density of 96% have been obtained.[48] TEM investigations had shown that the microstructure of the compacts consists of nanocrystalline ferritic matrix with grain size of 70 nm and fine carbide and oxide particles distributed basically along the grain boundaries (Figure 28.11). Due to these particles, the grain growth was prevented at temperatures below 700°C.

Recently, HPT was applied also to produce bulk specimens out of very thin ribbons of amorphous $Fe_{77}Al_{2.14}Ga_{0.86}P_{8.4}C_5B_4Si_{2.6}$ and $Al_{88}Y_7Fe_5$ alloys obtained by rapid quenching.[49,50] It was shown that after HPT consolidation under a pressure of 6 GPa, fully dense samples can be obtained. Additionally, the torsion straining process induced significant changes in the metallic glass, which resulted in the formation of dispersed nanocrystallites inside the amorphous matrix.

Therefore, HPT deformation opens wide perspectives for consolidation of nanopowders and for synthesis of bulk nanostructured composites of various types (metal–ceramic, amorphous–crystalline) which cannot be produced using conventional processes.

28.3.4 Friction-Induced Surface Modifications

Many microscopic processes occurring during MA and mechanical alloying of powder particles exhibit common features with processes relevant in tribology and wear. For example, the effects of work hardening, material transfer, and erosion during wear situations result in similar microstructures of the wear surface as observed during MA.[51,52] In particular, during sliding wear, large plastic strains and strain gradients are created near the surface. Typical plastic shear strain rates can correspond here to several $10^3\ s^{-1}$.

Close to the surface of wear scars as well as in the wear debris of Cu, nanocrystalline structures have been observed by HREM with an average grain size of 4 to 5 nm.[53] Within the interiors of the grains no defects were observed suggesting that most of the defects are absorbed by the grain boundaries due to their proximity. However, this type of plastic deformation at high strain rates does not seem to be limited to metals and alloys,[54] but has been observed in ceramics[55] and diamond,[56] as well.

During sliding wear a special tribological layer develops on the surface of a sliding component being subjected to large plastic strains. This surface layer often is called the Beilby layer, which for a long time was thought to be amorphous because its microstructure could not be resolved with the instruments commonly used.[57] There are indeed some systems in which truly amorphous layers are produced by sliding,[58] but in most cases the subsurface layer with a thickness of several micrometers has a nanocrystalline structure. For example, during ultrasonic shot peening, Lu and coworkers observed the formation of nanocrystalline Fe-surfaces where the initially coarse-grained structure in the surface layer was refined into equiaxed ultrafine grains (about 10 nm) with random crystallographic orientation.[59,60]

As a further example of technical relevance, the development of high speed trains reaching velocities higher than 300 km/h is also a materials challenge concerning the mechanical integrity and safety required for railway tracks.[61,62] In particular, the interaction and slip between wheel and rail has been optimized and is controlled by sophisticated electronics whereas the materials for the rail have not been modified since two decades. In particular, on the steel surface (Fe-0.8 at.% C to 1.3 at.% Mn) where the local pressure typically exceeds 1.0 to 1.5 GPa solid-state transformations have been observed that are caused by friction-induced shear forces and which have strong similarities with MA of powder samples.

Corresponding XRD and TEM results indicate that the average grain size of the extremely deformed surface layer corresponds to about 20 nm whereas a gradient in grain size is observed further away from the surface reaching values up to 200 nm.[63] For example, Figure 28.12 exhibits TEM photographs of the nanocrystalline layer near the surface (a) in comparison with the initial pearlitic structure (b — microstructure on the left).

As a consequence hardness measurements have been performed using a nanoindenter at small loads as shown in Figure 28.12c. Steep hardness gradients have been found in cross section with a lateral resolution of a few micrometers, which are clearly correlated with the change in microstructure. As a result it is found that the hardness is increased from typical values for the pearlitic steel (S54) of ~2.5 to 13 GPa next to the surface.

This remarkable increase in hardness and mechanical strength of regions near the surface is clearly related to the fact that the continuous deformation process considerably decreases the average grain size. Similar results have been obtained for mechanically attrited α-Fe and α-Fe–C-powder processed in a ball-milling device. Here also an increase of hardness by a factor of 5 has been observed.[64] During MA of rail filings (identical composition as rail) a decrease in grain size to 7 nm after 50 h of milling time has been observed together with dissolution of the carbides. The same observation holds for the highly deformed rail surface, that is, a dissolution of carbides and supersaturation of the α-Fe with carbon. However, due to the powder milling experiments it is obvious that the main contribution to the hardness increase results from the grain size reduction and only partially from the highly strained martensite-like structure of the bcc. iron supersaturated with carbon.

Moreover, the wear resistance of the nanostructured areas is increased by a factor of two. Fretting wear measurements typical for the type of wear in wheel-rail contact reveal wear rates of $1.55 \times 10^{-5}\ mm^3\ m^{-1}$ for the nanostructured layer and $3.77 \times 10^{-5}\ mm^3 m^{-1}$ for the undeformed surface. As such, the large improvement of the mechanical properties clearly shows the importance of the formation of nanostructures for technologically relevant wear problems.

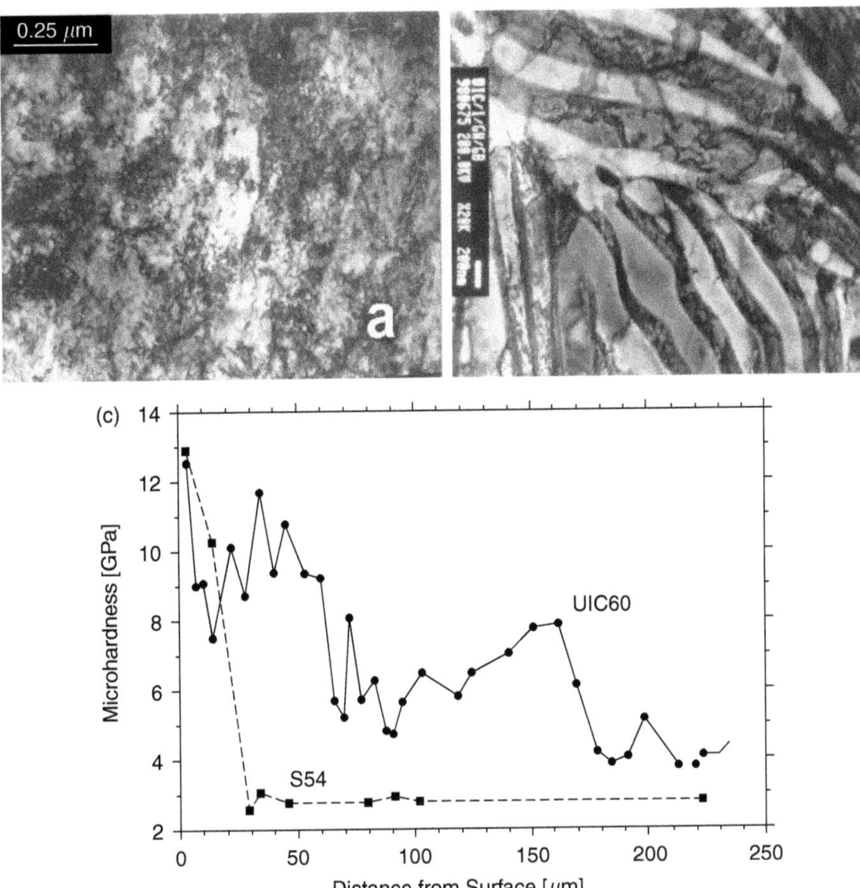

FIGURE 28.12 TEM bright-field images of the (a) nanostructured surface layer and (b) initial pearlitic steel sample. (c) Hardness vs. depth, that is, distance from surface for a high speed railway track (UIC60) with a nanostructured surface due to extreme wear conditions. S54 is pearlitic steel. (Courtesy of Ulm University.)

28.4 Stabilization of Ultra-Small Grained Composites

The minimization of the total interface free energy presents the driving force for coarsening of the microstructure, for example, by grain growth, that needs to be accounted for in general and that limits the processing options and the long-time stability of nanostructured materials — especially under thermal or mechanical load. This is a major challenge in any technological application of nanostructured materials involving elevated temperatures or mechanical stresses, because the driving force for grain growth is inversely proportional to the grain size. As a result, grain growth is often observed in nanostructured samples at rather low temperatures that can be hundreds of degrees lower than the onset of growth in conventional, polycrystalline materials.[65] One way to stabilize the nanoscale microstructures is by utilizing a composite approach with two phases that are both nanostructured and that are either mutually immiscible or that have very low interdiffusivities.

Different strategies have been reported to improve the stability of so-called *in situ* nanocomposites, that is, materials that do not require separate processing steps for creating a composite structure, thus leading to improved cost benefits. Those strategies for nanostructure stabilization use kinetic or thermodynamic approaches to prevent rapid boundary migration.[66] Both types of approaches can successfully prevent grain growth up to temperatures of about half the respective melting temperature, but the exponential dependence of the interface mobility on temperature limits the applicability range of kinetic mechanisms.

However, reducing or eliminating the thermodynamic driving force for growth, for example, by the method of segregation stabilization,[66–68] can lead to extraordinary stabilities even at high temperatures. A precondition for this method to become operative is that one component must have a high (negative) segregation enthalpy with respect to the second component. Recently, the feasibility of this approach was demonstrated for the alloy system Pd–Zr, in which Zr preferentially segregates to the grain boundaries, thus lowering both the mobility and the excess free energy of the internal interfaces.[69,70]

A specific situation that combines kinetic and thermodynamic stabilization aspects is given for nanostructured *in situ* composites that originate from glass-forming alloys via a so-called nanocrystallization reaction (for more details see Chapter 6). This method has been used to develop nanocrystalline materials in several interesting alloy categories including soft magnetic alloys,[71–73] hard magnetic alloys,[74,75] Mg-rich alloys with interesting hydrogen-storage capabilities,[76] and high strength aluminum alloys.[77] Characteristic of the nanocrystallization transformation is the occurrence of a primary crystallization at extremely high number densities. In most cases, an almost pure solid solution of the majority constituent crystallizes first. One consequence of the solute partitioning during primary crystallization is an enrichment of the amorphous matrix composition. If all of the solute rejected by the growing Al nanocrystal were distributed uniformly within the amorphous matrix, then the application of the parallel tangent construction reveals that the driving free energy for continued nanocrystal nucleation would decrease continuously with further reaction. In effect, the reduction of the thermodynamic driving force for nucleation, ΔG_v, would yield a reduction in nucleation rate even in the presence of a high heterogeneous nucleation site density. However, it has been established by atom probe field ion microscopy measurements that the large rare earth atom solutes do not distribute uniformly within the amorphous matrix, but tend to pile up at the interface between the nanocrystal and the matrix.[78,79] The nonuniform distribution will inhibit further transformation of existing nanocrystals, restrict the volume that is available for further nucleation and will also lead to a reduction in nucleation rate.

At the same time, this situation favors the formation of pronounced concentration gradients near the nanocrystal/glass interface that need to be accounted for in more complete analyses of the crystallization process. It has been shown theoretically[80] that the nucleation conditions for different phases can be markedly affected by the presence of a concentration gradient as well as by its steepness and curvature. This situation can be analyzed within the framework of nucleation in a concentration gradient as proposed by Desré.[80] After short time annealing (e.g., during heating the rapidly quenched material at a constant rate), the resulting concentration gradient is rather steep[81] and prevents the formation of any additional phase near the interface. Moreover, the width of spatial regions that have a composition suitable for the nucleation of an intermetallic phase can be smaller than the critical radius for nucleation. However, the diffusion fields are broadened if limited interdiffusion has occurred during a preannealing treatment. The resulting decrease of the concentration gradient involves two effects: the activation barrier for nucleation is decreased and the spatial regions of compositions that allow the nucleation of an additional phase with limited stability range are broadened. Thus, nucleation and growth of metastable phases are favored during the development from initial sharp gradients toward full diffusional equilibrium. Therefore, phases that are confined to narrow spatial ranges, that is, the interface region between the nanocrystals and the matrix (Figure 28.13), can develop in a controlled way. Such phases then present diffusion barriers and thus enhance the stability of the nanoscaled structure.[82] At the late stages of the interdiffusion process, the solute concentration along the diffusion path is not sufficient for the nucleation of metastable phases. Thus, the nucleation of the stable phase is again favored. These results indicate a new opportunity for the selection of a nucleating phase, which is important in boundary engineering or in any hetero-contact situation such as in microelectronics, as controlling the concentration gradient at the interface can modify the phase formation sequence to select a product phase with beneficial properties.

Based on the experimental results discussed above, it is apparent that the formation of the extreme nanocrystal number densities is inherently coupled to the presence of a high initial nucleation site density. Thus, the key strategy in enhancing the nanocrystal number density, and thus to improve both property performance *and* microstructure stability, is to promote the nucleation density of nanocrystals while minimizing the change of the amorphous matrix phase. One opportunity to modify the density of nuclei

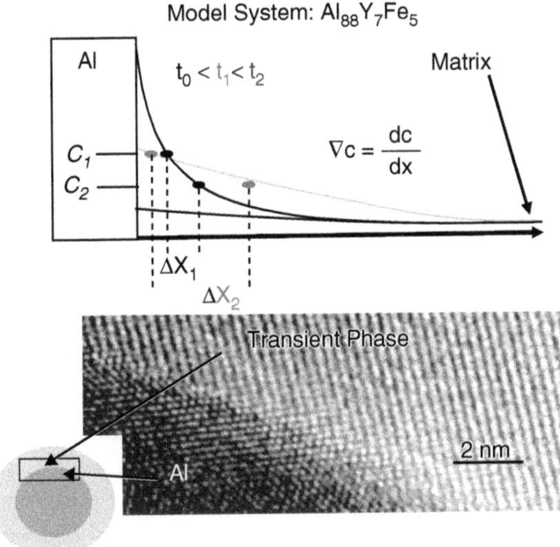

FIGURE 28.13 (Top) Schematic representation of the concentration profiles at the interface between an Al nanocrystal and the residual amorphous matrix at different annealing times, t. A decreasing steepness of the concentration gradient leads to an increase of the spatial ranges, ΔX, with a composition between C1 and C2. (Bottom) HRTEM micrograph of the interface region showing the presence of a metastable, ordered phase — a so-called transient phase — that only forms within a narrow window of concentration gradients. (With permission from Boucharat, N., Rösner, H., and Wilde, G., *Mater. Sci. Eng. C*, 23, 57, 2002.)

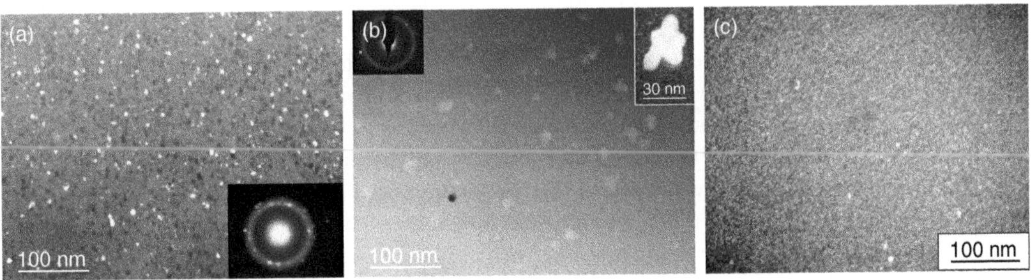

FIGURE 28.14 TEM images of an $Al_{88}Y_7Fe_5$ disk prepared by HPT. (a) Dark-field image and the corresponding SAED pattern. A high number density of Al-nanocrystals with a size of about 12 nm develops within the amorphous matrix. (b) Bright-field TEM image after annealing the as-quenched alloy at 245°C for 30 min. (With permission from Boucharat, N. et al., *Scripta Mater.*, 53, 823, 2005.) (c) Dark-field TEM image after cold rolling to an equivalent strain of $\varepsilon = -11.5$. The few white structures correspond to the nanocrystals that have formed during the rolling deformation. (With permission from Hebert, R.J. and Perepezko, J.H., *Mater. Sci. Eng. A*, 728, 375–377, 2004.)

in the undercooled melt is given by the kinetic balance that accounts for the formation of quenched-in nuclei itself. In this regime, small variations of the processing pathway such as the precipitation of an immiscible liquid component can lead to large variations of the resulting microstructure.

Alternative to changes of the processing pathway during melt quenching, the nanocrystal formation can be modified and enhanced by plastic deformation treatments at room temperature. Cold rolling of amorphous material induces the formation of nanocrystals and deformations to higher strain values increase the nanocrystal density even further, as indicated by quantitative TEM-microstructure analyses.[83] A typical microstructure obtained after HPT straining of an Al-rich amorphous alloy is presented in the dark-field TEM micrograph in Figure 28.14a, which reveals an extremely high number density of nanocrystals that seem to be homogeneously distributed within the amorphous matrix.[84] The number density thus obtained amounts to about 10^{22} m^{-3}. This has to be regarded as a lower limit as some nanocrystals have certainly been omitted from counting due to low contrast conditions or particle overlap. The results show clearly that the increase of the nanocrystal number density after the severe plastic deformation process is remarkable, as smaller values of 3 to 4.8×10^{21} m^{-3} and 1.0×10^{21} have been obtained after annealing[85,86] or after cold rolling to an equivalent strain of $\varepsilon = -11.5$,[87] respectively. For comparison, bright-field TEM images of the identical composition after thermal annealing or after cold rolling are indicated as in Figure 28.14b and Figure 28.14c.[84] XRD and TEM results clearly show that severe plastic deformation results in a considerably higher number density of Al nanocrystals than cold rolling or annealing at comparatively low temperatures. Moreover, the nanocrystals in the HPT sample are smaller compared to thermally generated nanocrystals and they are distributed homogeneously. These results indicate that either an extremely high density of shear bands has been induced by HPT, or that the application of the large shear strain has also modified the microscopic structure of the amorphous matrix not located in shear-band regions such that nanocrystal formation in the volume is favored even at ambient temperature. This scenario bears some relation to molecular dynamics simulation results that indicate collective rearrangements prior to shear-band formation and which are discussed in the framework of bond-orientational anisotropy and its relation to anelastic behavior.[88] Moreover, the results obtained by cold rolling also support the latter option. In any case, this result indicates that applying large shear strains presents a promising route for synthesizing homogeneous nanostructured Al-base composites with high stability of the microstructure that may not be obtainable by low strain — or strain-free methods.

28.5 Summary and Outlook

The solid-state processing methods of MA and mechanical alloying have been developed as a versatile alternative to other processing routes in preparing nanoscaled materials with a broad range of chemical composition and atomic structure. In this process, lattice defects are produced within the initially single-crystalline powder particles or within initially polycrystalline sheets during cold rolling. The internal refining process, with a reduction of the average grain size by a factor of 10^3 to 10^4, results from the creation and self-organization of dislocation cell networks and the subsequent formation of small-angle and high-angle grain boundaries within the powder particles during the mechanical deformation process. As a consequence, a change in the thermodynamic, mechanical, and chemical properties of these materials has been observed with the properties of nanophase materials becoming controlled by the grain size distribution and the specific atomic structure and cohesive energy of the grain or interphase boundaries. Such a transition from dislocation-controlled properties to grain-boundary-controlled properties is observed for nanocrystalline materials synthesized by other methods as well.[89]

One basic requirement for any application of these advanced materials is their stability under thermal — or mechanical load as well as their reliability characteristics. While a broad range of basic issues still need to be addressed concerning the coupling between stabilization strategies and materials properties, it seems clear that a composite approach that enables an effective diffusion control is suitable and most probably also necessary. In this respect — concerning the synthesis of nanostructured composites — solid-state processing methods seem especially suitable.

As such, MA offers interesting perspectives in preparing nanostructured powders with a number of different interface types in terms of structure (crystalline/crystalline, crystalline/amorphous) as well as atomic bonding (metal/metal, metal/semiconductor, metal/ceramic, etc.). Due to the broad range of possible atomic structures very different properties in comparison with conventional materials are obtained. For example, nanostructured particles prepared by MA can exhibit unusually high values in hardness,[90] enhanced hydrogen solubility,[91,92] enhanced catalytic properties,[93] magnetic spin-glass behavior,[94] and the like. This opens exciting possibilities for the preparation of advanced materials with particular grain- or interphase-boundary design.

New routes for nanostructure formation are presented by sequentially combining different nonequilibrium processing pathways that are based on rapid cooling or continuous strain energy input. The available permutations offer a wide range of options for tailoring the microstructure and the shape and quantity of the product nanostructure and — at the same time — present a wide field yet to be explored. First results on pure Ni indicate that the combination of cold rolling and HPT resulted in massive, nanocrystalline materials without texture and with the highest hardness observed for Ni.[93] Moreover, these strategies allow preparing stable, metastable, or even unstable product phases with properties that are improved or even completely unique.

Additionally, recent investigations by molecular dynamics simulations[95] together with experimental results such as the decrease of texturing during rolling, clearly indicate that the dominant mechanism that carries plastic deformation strongly depends on the characteristic size scale of the microstructure. Thus, it is expected that the study of MA and alloying processes in the future not only opens new processing routes for a variety of advanced nanostructured materials but also improves the understanding of technologically relevant deformation processes on a nanoscale level.

Acknowledgments

The continuous financial support by the Deutsche Forschungsgemeinschaft, the Helmholtz Association and the State of Baden-Wuertemberg is gratefully acknowledged. The authors would like to thank all the colleagues who have contributed to this topic over the past 10 years for the collaboration and stimulating discussions, in particular Drs. H. Gleiter (Karlsruhe, Germany), Y. Ivanisenko (Karlsruhe, Germany), W.L. Johnson (Caltech USA), J.H. Perepezko (UW-Madison, USA), and R. Valiev (Ufa, Russia).

References

[1] Fecht, H.-J., Formation of nanostructures by mechanical attrition, in *Nanomaterials: Synthesis, Properties and Applications*, Edelstein, A.S. and Cammarata, R.C., eds., Institute of Physics Publ., Bristol and Philadelphia, PA, 1996, p. 89.

[2] Srinivasan, S. and Ranganathan, S., *India's Legendary Wootz Steel — An Advanced Material of the Ancient World*, Tata Steel, India, 2004.

[3] Benjamin, J.S., Mechanical alloying, *Sci. Am.*, 234, 108, 1976.

[4] Koch, C.C. et al., Preparation of "amorphous" $Ni_{60}Nb_{40}$ by mechanical alloying, *Appl. Phys. Lett.*, 43, 1017,1983.

[5] Schwarz, R.B., Petrich, R.R., and Saw, C.K., The synthesis of amorphous Ni–Ti alloy powders by mechanical alloying, *J. Non-Cryst. Solids*, 76, 281, 1985.

[6] Hellstern, E. and Schultz, L., Amorphization of transition metal Zr alloys by mechanical alloying, *Appl. Phys. Lett.*, 48, 124, 1986.

[7] Ettl, C. and Samwer, K., Mechanical instabilities in extended solid solutions near the crystal-to-glass transition, *Mater. Sci. Eng. A*, 178, 245, 1994.

[8] Johnson, W.L., Thermodynamic and kinetic aspects of the crystal to glass transformation in metallic materials, *Prog. Mater. Sci.*, 30, 81, 1986.

[9] Fecht, H.-J. and Johnson, W.L., Entropy and enthalpy catastrophy as a stability limit for crystalline material, *Nature*, 334, 50, 1988.

[10] Shingu, P.H., ed., Mechanical alloying, *Mater. Sci. Forum*, 88–90, 1992.

[11] Fecht, H.-J., Defect-induced melting and solid-state amorphization, *Nature*, 356, 133, 1992.

[12] Qi, M. and Fecht, H.-J., Structural transition of zirconia during mechanical attrition, *Mater. Sci. Forum*, 187, 269–272, 1998.

[13] Michel, D., Mazerolles, L., and Chichery, E., Iron–chromium oxide solid solutions prepared by mechanical alloying, *Mater. Sci. Forum*, 99, 269–272, 1998.

[14] Tonejc, A.M. and Tonejc, A., Zirconia solid solutions ZrO_2-Y_2O_3 (CoO or Fe_2O_3) obtained by mechanical alloying, *Mater. Sci. Forum*, 497, 225–227, 1996.

[15] Tokumitsu, K. et al., Deoxidation of iron oxide by ball-milling, *Mater. Sci. Forum*, 181, 269–272, 1998.

[16] Schaffer, G.B. and Mc Cormick, P.G., Reduction of metal oxides by mechanical alloying, *Appl. Phys. Lett.*, 55, 45, 1989.

[17] Shaw, W.J.D., Current understanding of mechanically alloyed polymers, *Mater. Sci. Forum*, 19, 269–272, 1998.

[18] Pan, J. and Shaw, W.J.D., Effects of processing parameters on material properties of mechanically processed polyamide, *J. Appl. Polym. Sci.*, 56, 557, 1995.

[19] Karttunen, M. and Ruuskanen, P., The microstructure and electrical properties of mechanically-alloyed copper-polymer composites, *Mater. Sci. Forum*, 849, 269–272, 1998.

[20] Fu, Z., Fecht., H.J., and Johnson, W.L., Nanoscale phase separation induced by mechanical alloying in the iron–erbium–nitrogen system, *Mater. Res. Soc. Symp. Proc.*, 186, 169, 1991.

[21] Calka, A. and Williams, W.S., Synthesis of nitrites by mechanical alloying, *Mater. Sci. Forum*, 787, 88–90, 1992.

[22] El-Eskandarany, M.S. et al., Morphological and structural evolutions of nonequilibrium titanium nitride alloy powders produced by reactive ball milling, *J. Mater. Res.*, 7, 888, 1992.

[23] Schulson, E.M. and Barker, D.R., A brittle to ductile transition in NiAl of a critical grain size, *Scripta Metall.*, 17, 519, 1983.

[24] Lu, I.R., Preparation and thermal stability of ceramic/metallic glass composites prepared by mechanical alloying, *Mater. Lett.*, 35, 297, 1998.

[25] Moelle, C. et al., Formation of ceramic/metallic glass composite by mechanical alloying, *Mater. Sci. Forum*, 47, 269–272, 1998.

[26] Atzmon, M. et al., Formation and growth of amorphous phases by solid-state reaction in elemental composites prepared by cold working, *Appl. Phys. Lett.*, 45, 1052, 1984.

[27] Atzmon, M., Unruh, K.M., and Johnson, W.L., Formation and characterization of amorphous erbium-based alloys prepared by near-isothermal cold-rolling of elemental composites, *J. Appl. Phys.*, 58, 3865, 1985.

[28] Bourdeaux, F. and Yavari, A.R., Amorphization by solid-state reaction of crystalline aluminum and platinum multilayers prepared by cold rolling, *J. Appl. Phys.*, 67, 2385, 1990.

[29] Yasuna, K. et al., Bulk metallic multilayers produced by repeated press-rolling and their perpendicular magnetoresistance, *J. Appl. Phys.*, 82, 2435, 1997.

[30] Sagel, A. et al., Amorphization of Zr-Al-Ni-Cu during cold rolling of elemental foils at ambient temperatures, *Phil. Mag. Lett.*, 77, 109, 1998.

[31] Perepezko, J.H. and Wilde, G., Amorphization and alloy metastability in undercooled systems, *J. Non-Cryst. Solids*, 274, 271, 2000.

[32] Wilde, G., Sieber, H., and Perepezko, J.H., Glass formation versus nanocrystallization in an $Al_{92}Sm_8$ alloy, *Scripta Mater.*, 40, 779, 1999.

[33] Köster, U. and Meinhardt, J., Crystallization of highly undercooled metallic melts and metallic glasses around the glass transition temperature, *Mater. Sci. Eng. A*, 178, 271, 1994.

[34] Wilde, G., Sieber, H., and Perepezko, J.H., Glass formation in Al-rich Al-Sm alloys during solid state processing at ambient temperature, *J. Non-Cryst. Solids*, 621, 250–252, 1999.

[35] Dinda, G.P., Rösner, H., and Wilde, G., Synthesis of bulk nanostructured Ni, Ti and Zr by repeated cold-rolling, *Scripta Mater.*, 52, 577, 2005.

[36] Wilde, G., Rösner, H., and Dinda, G.P., Synthesis of bulk nanocrystalline materials by repeated cold-rolling, *Adv. Eng. Mater.*, 7, 11, 2005.

[37] Dinda, G.P., Rösner, H., and Wilde, G., Synthesis of ultrafine-grained alloys by repeated cold-rolling, in *Ultrafine-Grained Materials III*, Zhu, Y.T., Langdon, T.G., Valiev, R.Z., Semiatin, S.L., Shin, D.H., and Lowe, T.C., eds., The Minerals, Metals & Materials Society, 2004, 309.

[38] Saito, Y. et al., Ultra-fine grained bulk aluminum produced by accumulative roll-bonding (ARB) process, *Scripta Mater.*, 39, 1221, 1998.

[39] Dinda, G.P., Rösner, H., and Wilde, G., Crystal refinement by cold rolling in NiTi shape memory alloys, *Solid State Phenomena*, 55, 101–102, 2005.

[40] Huang, X. et al., Anisotropic compressive properties of porous copper produced by unidirectional solidification, *Mater. Sci. Eng. A*, 340, 265, 2003.
[41] Hughes, D.A. and Hansen, N., Microstructure and strength of nickel at large strains, *Acta Mater.*, 48, 2985, 2000.
[42] Hahn, H., Mondal, P., and Padmanabhan, K.A., Plastic deformation of nanocrystalline materials, *Nanostruct. Mater.*, 9, 603, 1997.
[43] Wilde, G. et al., New routes for synthesizing massive nanocrystalline materials, *Mater. Sci. Forum*, 425, 503–504, 2006.
[44] Valiev, R.Z., Nanostructuring of metals by severe plastic deformation for advanced properties, *Nat. Mater.*, 3, 511, 2004.
[45] Lee, Z. et al., Microstructure and microhardness of cryomilled bulk nanocrystalline Al–7.5%Mg alloy consolidated by high pressure torsion, *Scripta Mater.*, 51, 209, 2004.
[46] Stolyarov, V.V. et al., Processing nanocrystalline Ti and its nanocomposites from micrometer-sized Ti powder using high pressure torsion, *Mater. Sci .Eng. A*, 282, 78, 2000.
[47] Alexandrov, I.V. et al., Microstructures and properties of nanocomposites obtained through SPTS consolidation of powders, *Metal Trans. A*, 29A, 2253, 1998.
[48] Ivanisenko, Yu. et al., Nanocrystalline Fe-1 wt. %C compacts obtained by high pressure torsion of mechanically alloyed powder, *Ann. Chim. Sci. Mater.*, 27, 45, 2002.
[49] Sort, J. et al., Cold-consolidation of ball-milled Fe-based amorphous ribbons by high pressure torsion, *Scripta Mater.*, 50, 1221, 2004.
[50] Boucharat, N. et al., Nanocrystallization of amorphous $Al_{88}Y_7Fe_5$ alloy induced by plastic deformation, *Scripta Mater.*, 53, 823, 2005.
[51] Ivanisenko, Y.V. et al., Nanostructure and hardness of "white layer" on surface of rails, *Phys. Metals Metall.*, 83, 303, 1997.
[52] Fecht, H.-J., Nanostructure formation by mechanical attrition, *Nanostruct. Mater.*, 6, 33, 1995.
[53] Ganapathi, S.K. and Rigney, D.A., An HREM study of the nanocrystalline material produced by sliding wear processes, *Scripta Metall.*, 24, 1675, 1990.
[54] Doyle, F.D. and Aghan, R.L., Mechanism of metal removal in the polishing and fine grinding of hard metals, *Metall. Trans. B*, 6, 143, 1975.
[55] Mehrotra, P.K., Mechanisms of wear in ceramic materials, in *Proceedings of the International Conference on Wear of Materials*, Reston, VA, ASME, New York, 1983, p. 194.
[56] Humble, P. and Hannink, R.H.-J., Plastic deformation of diamond at room temperature, *Nature*, 273, 37, 1978.
[57] Beilby, G., *Aggregation and Flow of Solids*, Macmillan, London, 1921.
[58] Askenasy, P., Ph.D. thesis, California Institute of Technology, 1992.
[59] Tao, N.R. et al., Surface nanocrystallization of iron induced by ultrasonic shot peening, *Nanostruct. Mater.*, 11, 433, 1999.
[60] Lu, K., private correspondence (1999).
[61] Baumann, G., Fecht, H.-J., and Liebelt, S., Formation of white etching layers on rail treads, *Wear*, 191, 133, 1996.
[62] Baumann, G., Zhong, Y., and Fecht, H.-J., Comparison between nanophase formation during friction induced surface wear and mechanical attrition of a pearlitic steel, *Nanostruct. Mater.*, 7, 237, 1996.
[63] Bürkle, G., Mikrostruktur und mechanische Eigenschaften tribologisch beansprzchter nanoskaliger Eisen-Basis-Schichten, Thesis, Ulm University, 2003.
[64] Siegel, R.W. and Fougere, G.E., Mechanical properties of nanophase materials, in *Nanophase Materials*, Hadyipanayis, G.C. and Siegel, R.W., eds., Kluwer Academic Press, 1994, p. 233.
[65] Gertsman, V.Y. and Birringer, R., On the room-temperature grain growth in nanocrystalline copper, *Scripta Metall. Mater.*, 30, 577, 1994.

[66] Krill III, C.E. and Birringer, R., Grain-growth kinetics and thermal stability of nanocrystalline materials, in *Recrystallization and Grain Growth*, Gottstein, G. and Molodov, D.A., eds., Springer, Berlin, 2001, 205.

[67] Weissmüller, J., Alloy effects in nanostructures, *Nanostruct. Mater.*, 3, 261, 1993.

[68] Kirchheim, R., Grain coarsening inhibited by solute segregation, *Acta Mater.*, 50, 413, 2002.

[69] Srdic, V.V., Winterer, M., and Hahn, H., Sintering behavior of nanocrystalline zirconia doped with alumina prepared by chemical vapor synthesis, *J. Am. Ceram. Soc.*, 83, 1853, 2000.

[70] Weigand, H. et al., Interfacial free volumes and segregation effects in nanocrystalline $Pd_{85}Zr_{15}$ studied by positron annihilation, *Appl. Phys. Lett.*, 84, 3370, 2004.

[71] Koch, C.C., The synthesis and structure of nanocrystalline materials produced by mechanical attrition: A review, *Nanostruct. Mater.*, 2, 109, 1993.

[72] Suzuki, K. et al., High saturation magnetization and soft magnetic properties of bcc Fe-Zr-B alloys with ultrafine grain structure, *Mater. Trans. JIM*, 31, 743, 1990.

[73] Suzuki, K. et al., Low core losses of nanocrystalline Fe-M-B (M=Zr, Hf or Nb) alloys, *J. Appl. Phys.*, 74, 3316, 1993.

[74] Manaf, A., Buckley, R.A., and Davies, H.A., New nanocrystalline high remanence Nd-Fe-B alloys by rapid solidification, *J. Magn. Magn. Mater.* 128, 302, 1993.

[75] Davies, H.A. et al., Praseodymium and neodymium-based nanocrystalline hard magnetic alloys, *Mater. Res. Soc. Symp. Proc.*, 577, 27, 1999.

[76] Spassov, T., Rangelova, V., and Neykov, N., Nanocrystallization and hydrogen storage in rapidly solidified Mg–Ni–RE alloys, *J. Alloys Compd.*, 334, 219, 2002.

[77] Kim, Y.H. et al., Crystallization and high mechanical strength of Al-based amorphous alloys, *Mater. Trans. JIM*, 35, 293, 1994.

[78] Gloriant, T. et al., Nanostructured $Al_{88}Ni_4Sm_8$ alloys investigated by transmission electron and field-ion microscopies, *Mater. Sci. Eng.*, A, 304, 315, 2001.

[79] Hono, K. et al., Solute partitioning in partially crystallized Al-Ni-Ce(-Cu) metallic glasses, *Script. Mater.*, 32, 191, 1995.

[80] Hodaj, F. and Desré, P.J., Effect of a sharp gradient of concentration on nucleation of intermetallics at interfaces between polycrystalline layers, *Acta Mater.*, 44, 4485, 1996.

[81] Allen, D.R., Foley, J.C., and Perepezko, J.H., Nanocrystal development during primary crystallization of amorphous alloys, *Acta Mater.*, 46, 431, 1998.

[82] Boucharat, N., Rösner, H., and Wilde, G., Development of nanocrystals in amorphous Al-alloys, *Mater. Sci. Eng. C*, 23, 57, 2002.

[83] Hebert, R.J. et al., Dislocation formation during deformation induced synthesis of nanocrystals in amorphous and partially crystalline amorphous $Al_{88}Y_7Fe_5$ alloy, *Scripta Mater.*, 54, 25, 2006.

[84] Boucharat, N. et al., Nanocrystallization of amorphous $Al_{88}Y_7Fe_5$ alloy induced by plastic deformation, *Scripta Mater.*, 53, 823, 2005.

[85] Boucharat, N. et al., Devitrification of Al-based glass-forming alloys, *Mater. Sci. Eng. A*, 713, 375–377, 2004.

[86] Wu, R.I., Wilde, G., and Perepezko, J.H., Glass formation and primary nanocrystallization in Al-base metallic glasses, *Mater. Sci. Eng. A*, 301, 12, 2001.

[87] Hebert, R.J. and Perepezko, J.H., Effect of cold-rolling on the crystallization behavior of amorphous $Al_{88}Y_7Fe_5$ alloy, *Mater. Sci. Eng. A*, 728, 375–377, 2004.

[88] Tomida, T. and Egami, T., Molecular-dynamics study of structural anisotropy and anelasticity in metallic glasses, *Phys. Rev. B*, 48, 3048, 1993.

[89] Nieman, G.W., Weertman, J.R., and Siegel, R.W., Mechanical behavior of nanocrystalline Cu and Pd, *J. Mater. Res.*, 6, 1012, 1991.

[90] Kehrel, A., Moelle, C., and Fecht, H.-J., Mechanical properties of nanostructured niobium on a microscopic scale, in *NATO-Advanced Study Insitute on Nanophase Materials*, Hadjipanayis, G.C. and Siegel, R., eds., Kluwer Academic Publishers, 1994, 287.

[91] Moelle, C. and Fecht, H.-J., Thermodynamic properties and phase stability of nanocrystalline metals and hydrides, *Nanostruct. Mater.*, 3, 93, 1993.
[92] Ram, S. et al., Calorimetric study of the desorption of the interstitial hydrogen atoms in ferromagnetic Nd2Fe14BHx (x<=5) microcrystals, *Phys. Rev. B*, 56, 1, 1997.
[93] Zaluski, L. et al., Nanocrystalline hydrogen absorbing alloys, *Mater. Sci. Forum*, 853, 225–227, 1996.
[94] Zhou, G.F. and Bakker, H., Spin-glass behavior of amorphous Co_2Ge synthesized by mechanical milling, *Phys. Rev. Lett.*, 72, 2290, 1994.
[95] Yamakov, V. et al., Dislocation processes in the deformation of nanocrystalline aluminium by molecular-dynamics simulation, *Nat. Mater.*, 1, 45, 2002.

29

Thermomechanical Processing

John J. Jonas
McGill University

Matthew R. Barnett and
Peter D. Hodgson
Deakin University

Abstract

The softening mechanisms associated with deformation processing are described in detail. These include: dynamic recovery (DRV), static recrystallization (SRX), dynamic recrystallization (DRX), and metadynamic dynamic recrystallization (MDRX) or postdynamic recrystallization. The precipitation of second phases, such as the carbonitrides in microalloyed steels, is considered, as is the interaction between precipitation and recrystallization. The modeling of the softening processes is outlined together with that of the rolling load and grain size. A distinction is drawn between rough rolling, which involves full recrystallization between passes, and finish rolling, in which case no recrystallization takes place, leading to "pancaking" of the microstructure. The effect of such pancaking on phase transformation and grain refinement in complex materials such as steels is described, together with that of cooling rate through the transformation. Some attention is also paid to intercritical and warm rolling as well as to the production of ultrafine grained materials. Next, the effect of thermomechanical processing on mechanical properties is considered. Texture formation during rolling is characterized for both face-centered cubic (fcc) and body-centered cubic (bcc) crystal structures and recent advances in thermomechanical processing, both theory and practice, are depicted briefly. Finally, the processing of nonferrous metals, such as aluminum and titanium alloys, is considered.

29.1 Introduction

The resistance to flow (flow stress) of metals is almost an order of magnitude lower at elevated temperatures than at room temperature. Accordingly, most metals (with the notable exception of castings) are first reduced in cross section at "high" temperatures. Here, "high" refers to temperatures above half the melting point when expressed in Kelvin. Under these conditions, much less energy is expended in bringing about desired shape changes; advantage is also taken of the working of the metal to break up any brittle second phases that may be present and to reduce or even eliminate such undesirable alloying element and impurity segregation as may have been introduced during casting.

Given that the metal must in any event be "hot worked" during the first stages of manufacture, the possibility arises of using this stage of processing to carry out concurrent heat treatments that can simultaneously improve both the metallurgical structure as well as the properties. Such simultaneous working and control of the structure and properties is referred to as *Thermomechanical Processing* and is the subject of this chapter.

Thermomechanical processing first came into prominence in the mid-1970s as a technique for the production of high-strength plate for the fabrication of oil and gas pipelines. Such plate was required to have a fine grain size, as the latter controlled the toughness of the material, which was a particularly important property when the pipeline (or drilling platform) was to be used in near-arctic or low-temperature environments. The plate that was produced in this way generally contained about 0.04 to 0.05% Nb and was referred to as high-strength low-alloy (HSLA) steel. The "low alloy" designation arose because of the relatively low levels of added Nb. Despite the small Nb addition, the presence of this element is crucial to the success of this process and its role during rolling is therefore also one of the principal topics of this chapter.

Although thermomechanical processing is a term that is usually applied to *steel* rolling (in which case it is referred to as *controlled rolling*), the principles of this type of metallurgical treatment are now also being employed in the processing of other metals, such as the aluminum or titanium alloys. In the latter case, a phase change is involved, as in the case of steel, but this concerns a transition from a bcc to an hexagonal close packed (hcp) crystal lattice. Accordingly, reference will be made where appropriate to other alloy systems.

Whether in ferrous or nonferrous systems, two important principles are involved: the first is that the mechanism of recrystallization can be interrupted or controlled by the initiation of precipitation at a desired point of the process. When this is done appropriately, and a subsequent phase change is involved, considerable grain refinement can be achieved. Thus the interaction of recrystallization and precipitation (of carbonitrides in the case of steel) will be an important focus of this chapter. The second principle involves the initiation of precipitation *during* rolling, even when it does not prevent recrystallization. Here it produces strengthening without requiring the application of a separate heat treatment operation after rolling.

Before the influence of precipitation on processing and properties can be considered in detail, some more general features of softening during rolling must first be described.

29.2 Softening Mechanisms Associated with Rolling

29.2.1 Dynamic Recovery

This mechanism operates at all temperatures above absolute zero and is responsible for the decreasing slope of flow (stress–strain) curves at room temperature.[1,2] It involves the conversion of "forest" dislocations (which are highly effective obstacles to dislocation glide) into substructures that provide considerably less resistance to metal flow. This generally occurs by means of "cross-slip" at room temperature assisted by "climb" at higher temperatures. Increasing the temperature accelerates the effective rates of these softening (dislocation annihilation) mechanisms, so that less and less force need to be applied to the mobile dislocations to enable them to move.

At high enough strains, the effect of DRV is sufficient to overcome work hardening (the introduction of fresh forest dislocations), so that the slope of the flow curve is reduced to zero, leading to steady-state flow. Under the latter conditions, the effect of work hardening is exactly counterbalanced by that of DRV. An example of the influence of temperature on the flow curves of commercial purity aluminum is given in Figure 29.1a, while that of strain rate on the stress–strain curves of a Ti stabilized ultra-low carbon steel is presented in Figure 29.1b.

The attainment of steady-state flow is thus accompanied by the attainment of a steady-state substructure. At higher temperatures and lower strain rates this substructure is comprised of an arrangement of equiaxed subgrains. An example is given in Figure 29.2a for ferrite worked in torsion. Such a structure is commonly characterized by an average subgrain size and an average misorientation. The average subgrain size, d_{sg}, typically varies with the reciprocal of the steady-state flow stress, σ_{ss}, thus:

$$d_{sg} = 1/\sigma_{ss} \tag{29.1}$$

The evolution of the average misorientation is somewhat more complicated and is often seen to continue to increase with strain even in the "steady state." Subgrains are typically bounded by low angle boundaries characterized by misorientations less than 15°. A common range of average misorientation for warm worked ferrite or hot worked aluminum would be 1° to 3°.

29.2.2 Static Recrystallization

If a metal is unloaded after a certain amount of straining at an elevated temperature, it undergoes a type of softening known as SRX.[3] The latter involves a critical strain of up to 10%, below which no recrystallization occurs. This mechanism is characterized by the nucleation and growth of new, strain-free grains. As the boundaries of the new grains move, they sweep away the dislocations (work hardening) introduced by the prior deformation, thereby returning the metal to its initial fully soft (or annealed) state. The appearance of a sample undergoing SRX is illustrated in Figure 29.2b. The rate of this mechanism increases with temperature and with the amount of the prior strain. It also depends on the grain size of the starting material and to a lesser extent on the strain rate of the prior deformation. The grain size produced by SRX is a function of the applied strain and initial grain size in the manner illustrated in Figure 29.3.

The dependence of the fractional softening on time is depicted in Figure 29.4 for four typical steels.[4] It can be seen from Figure 29.4 (top) that the addition of V to a plain carbon (PC) steel has only a modest effect on the softening kinetics at 1000°C; that is, there is a small amount of observable retardation. By contrast, when Mo is added, which remains in solution at this temperature, there is significant retardation. This is entirely due to what is known as "solute drag;" the latter acts in part on the dislocations, thereby retarding nucleation, and in part on the boundaries that begin to move once recrystallization begins. Finally, there is the behavior of the Nb steel. At 1000°C, which is above the Nb(C,N) precipitation start temperature, the effects of solute drag can clearly be seen. By contrast, at 900°C (Figure 29.4 bottom), at which Nb(C,N) precipitation has begun, Andrade et al. showed that recrystallization (as evaluated in terms of the fractional softening) is completely arrested.[4]

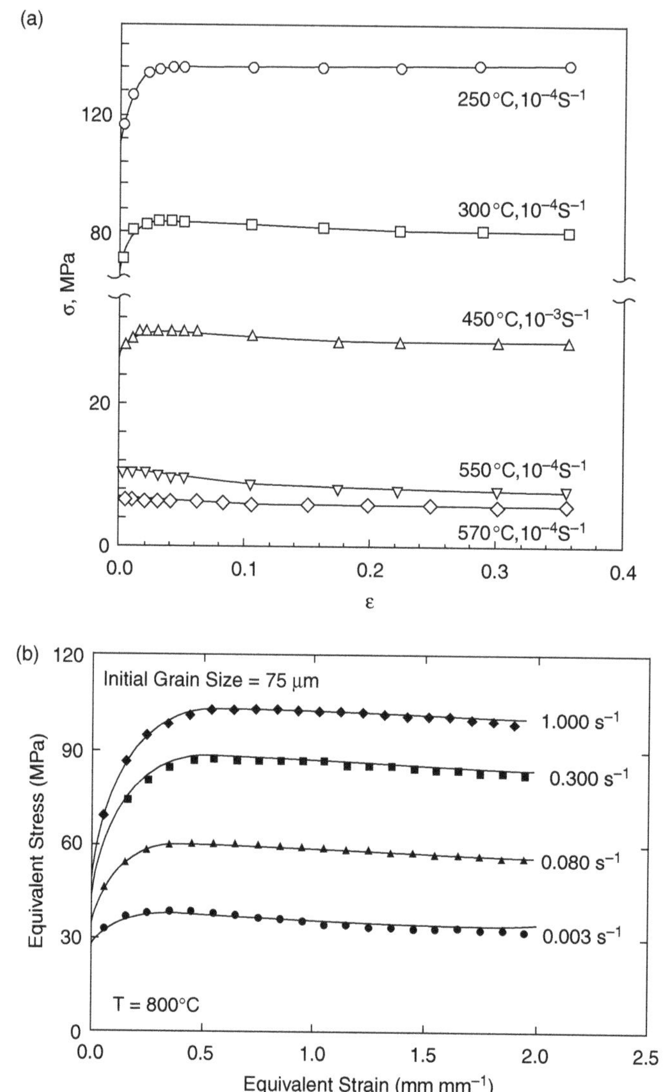

FIGURE 29.1 (a) Effect of temperature on the stress–strain curves of 5083 Al alloy (Al–4.7% Mg–1.6% Mn–0.2% Zr–0.18% Cr–0.1% Fe (in wt.%). (Used with permission from R. Kaibyshev, F. Musin, E. Avtokratova and Y. Motohashi; *Mater. Sci. Eng. A*; 2005; **392**; 373–379.); (b) Effect of strain rate on the stress–strain curves of alpha iron. (Used with permission from A. Oudin, M.R. Barnett and P.D. Hodgson; *Mater. Sci. Eng. A*; 2004; **367**; 282–294.)

This type of interaction between precipitation and recrystallization is at the core of thermomechanical processing and of controlled rolling, and will be considered in more detail below.

For modeling purposes, it is possible to fit equations to curves of this type, which can then be interpolated and even extrapolated to a limited degree. Such equations are of considerable use for modeling the separation force in rolling mills, which depends on the amount of softening achieved during the preceding pass. Equations have also been derived that specify the grain size produced by recrystallization. As will be seen in more detail below, the grain size equations are also highly useful for modeling purposes.[2]

29.2.2.1 The Recrystallization-Stop Temperature or T_{nr}

As can be seen from Table 29.1, typical plate mill interpass times are of the order of 20 s. Such long intervals are associated with the time required to reverse the direction of motion of a steel plate in a

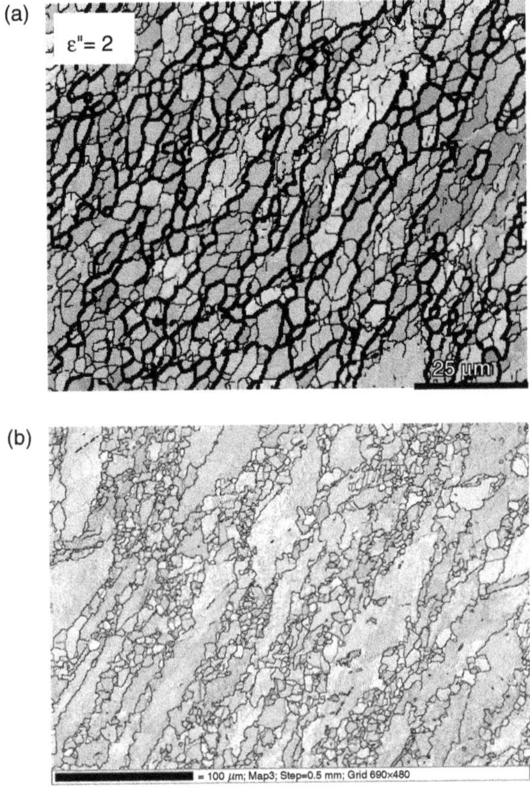

FIGURE 29.2 (a) EBSD (Electron Backscattered Diffraction) map of ferrite microstructure deformed in torsion at 1 s^{-1} and 765°C. The heavy black lines are grains, while the lighter lines are subgrains. (b) The appearance of a sample of hot worked stainless steel undergoing static recrystallization.

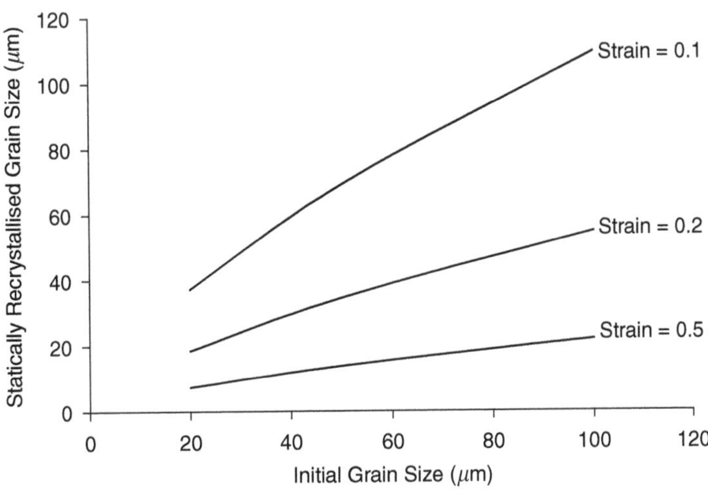

FIGURE 29.3 The effect of strain and initial grain size on the statically recrystallized grain size.

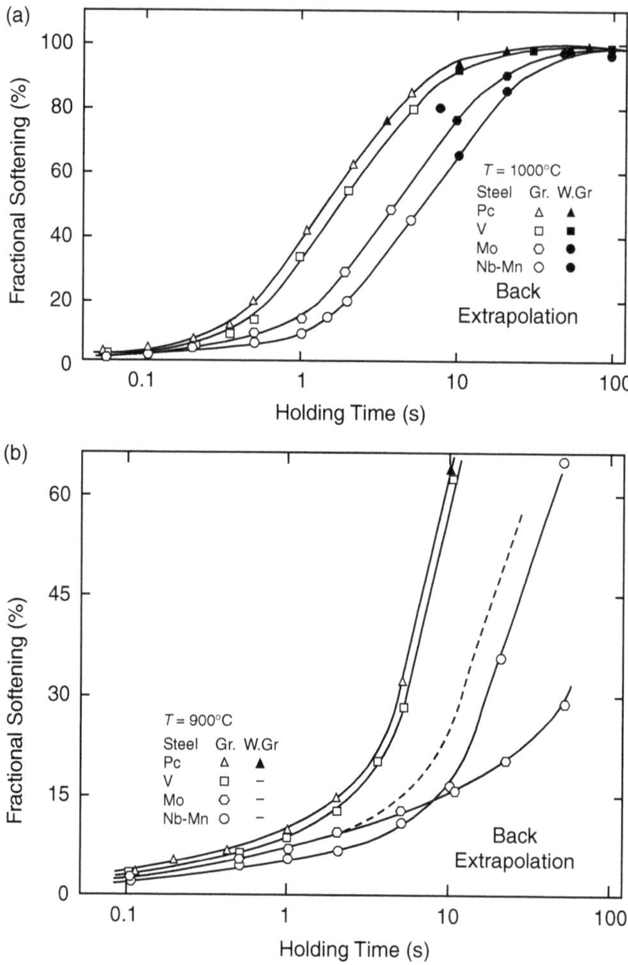

FIGURE 29.4 The dependence of fractional softening on time for four typical steels at 1000 and 900°C. (Used with permission from H.L. Andrade, M.G. Akben and J.J. Jonas; *Metall. Trans.*; 1983; **14A**; 1967.) Gr., W + Gr. refer to "graphite" and "tungsten". PC: plain carbon steel.

TABLE 29.1 Comparison of Hot Rolling Mills in Terms of Their Ranges of Strain Rate and Interpass Time

Mill Type	Strain Rate Range (s^{-1})	Interpass Time (s)
Reversing mills (plate, slabbing, roughing, etc.)	1–30	8–20
Hot strip mills	10–200	0.4–4
Wire rod mills	10–1000	0.015–1

reversing mill. At fairly high rolling temperatures (e.g., "roughing" temperatures), the material readily recrystallizes during the holding interval. However, as the material cools and the temperature unavoidably decreases during rolling, a stage of the schedule is reached when recrystallization is unable to go to completion, or even to produce substantial softening during the time available. This is often referred to as the "no-recrystallization" temperature or T_{nr}.[5] The prevention of recrystallization in the interpass interval is an essential component of the thermomechanical processing of microalloyed steels. In the PC

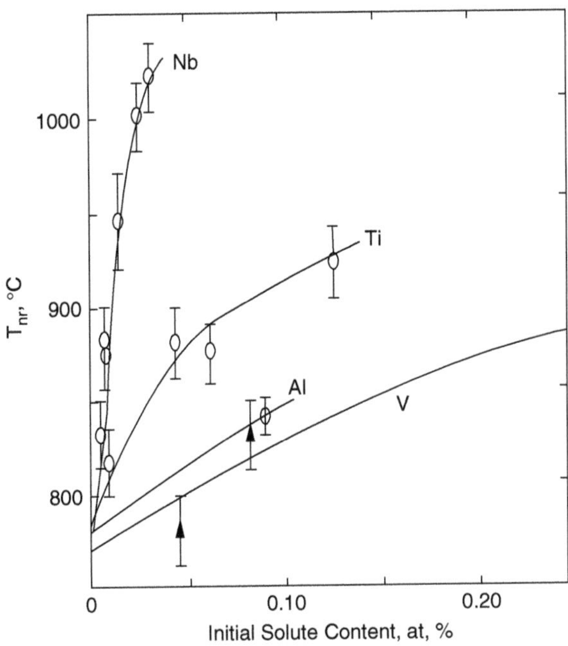

FIGURE 29.5 The influence of microalloy content on T_{nr}. (Used with permission from A.J. DeArdo; *Ironmak. Steelmak.*; 2001; **28**; 138–144.)

steels, which do not contain specific alloying additions that contribute to solute drag or to carbonitride precipitation, the workpiece continues to recrystallize while cooling until the upper critical temperature, A_{r3} (i.e., fcc to bcc transformation), is reached.

The transition from complete to incomplete recrystallization and then to no recrystallization at all does not take place at a single temperature, such as the melting point of a pure metal. Instead, it covers a range of up to 100°C. As the T_{nr} is of considerable importance in the design of rolling schedules, it will be taken up in more detail later. For the moment, it should be noted that the T_{nr} increases with the amount of Nb present and that it falls in the range 950 to 1050°C. Rolling carried out below this temperature is referred to as "finish rolling." The effects of other common microalloying elements on the T_{nr} are illustrated in Figure 29.5.

29.2.3 Dynamic Recrystallization

High stacking fault energy metals such as aluminum and α-iron (iron in its ferritic, bcc, or low-temperature form) only undergo softening by DRV during straining. By contrast, many other metals are subject to an additional dynamic softening mechanism known as dynamic recrystallization (DRX). In these cases, new grains are nucleated *during* deformation and the motion of their boundaries causes dislocation removal and therefore softening *during* straining. The metals and alloys that display this type of behavior are all the fcc metals except for aluminum. In aluminum, DRV is so effective that the driving force for DRX is not attained during the deformation. The metals that undergo DRX include the copper and nickel alloys, gold, silver, platinum, and especially iron (i.e., steel) in its high temperature (i.e., fcc or austenite) form. The hcp metals, for example, Mg, are also capable of undergoing DRX.

The additional softening mechanism has a marked effect on the flow curve, as illustrated in Figure 29.6 for steel.[6,7] Here, two types can be distinguished: (1) the single-peak and (2) the multiple-peak curves. It has been shown in detail[3] that the former type is associated with grain refinement during deformation, while the latter is a sign of grain coarsening. Only the grain refinement type of DRX occurs under most industrial conditions; this section is therefore concerned solely with that type.

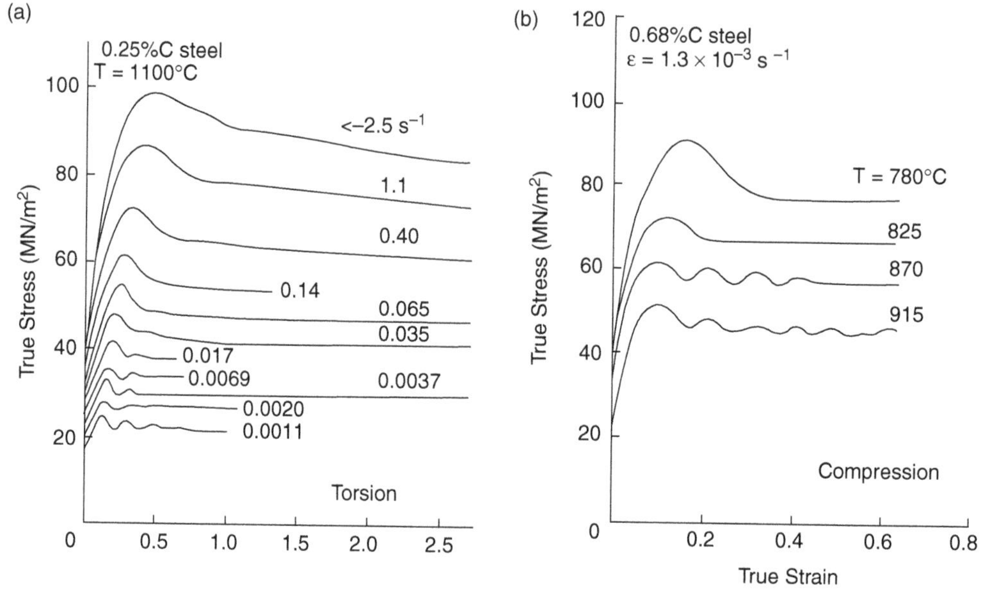

FIGURE 29.6 The effect of (a) strain rate and (b) temperature on the flow curve. (Used with permission from C.M. Sellars; *Deformation Processing and Structure*; ASM Materials Science Seminar; 1982; G. Krauss, Editor; p. 245 and R.A. Petkovic, M.J. Luton and J.J. Jonas; *Can. Metall. Quart.*; 1975; **14**; 137–145.)

DRX is initiated at strains of about one half the "peak strain," that is, the strain associated with the maximum in the flow stress. At the peak of the flow curve, the softening associated with DRX just balances the net strengthening produced by work hardening in association with DRV. After the peak, there is more softening than hardening until a steady state of flow is reached during which the softening attributable to DRX just matches the hardening (introduction of new dislocations) attributable to the combination of work hardening and DRV.

29.2.4 Metadynamic Recrystallization

The initiation of DRX during straining or rolling is of practical importance because it makes possible the occurrence of a particularly rapid type of softening after the completion of deformation. Such rapid "postdynamic" recrystallization takes place because the initial *coarse-grained* microstructure of metals that have been reheated for rolling has now been replaced by a *dynamic* microstructure that is much finer. As recrystallization is generally nucleated at grain boundaries, fine microstructures contain more nucleation sites and therefore recrystallize more rapidly than metals containing coarse-grained microstructures. The geometry of such DRX microstructures is independent of the strain.

The effect of strain on the time to 50% softening, t_{50}, is illustrated in Figure 29.7. It can be seen that the t_{50} for SRX decreases rapidly with strain (Figure 29.7a). Once the initial microstructure has been largely replaced by the dynamic microstructure, the t_{50} becomes strain independent. As can be seen from Figure 29.7a, the t_{50} for MDRX is much more sensitive to the strain rate of prior loading than that of SRX. Conversely, it is less temperature sensitive (Figure 29.7b).

29.3 Carbonitride Precipitation

The most important difference between the conventional processing of PC steels and that of the HSLA (microalloyed) steels involves the role of the Nb that has been added to the latter materials.[8,9] This has already been referred to briefly above. The Nb addition is put into solution at temperatures of about

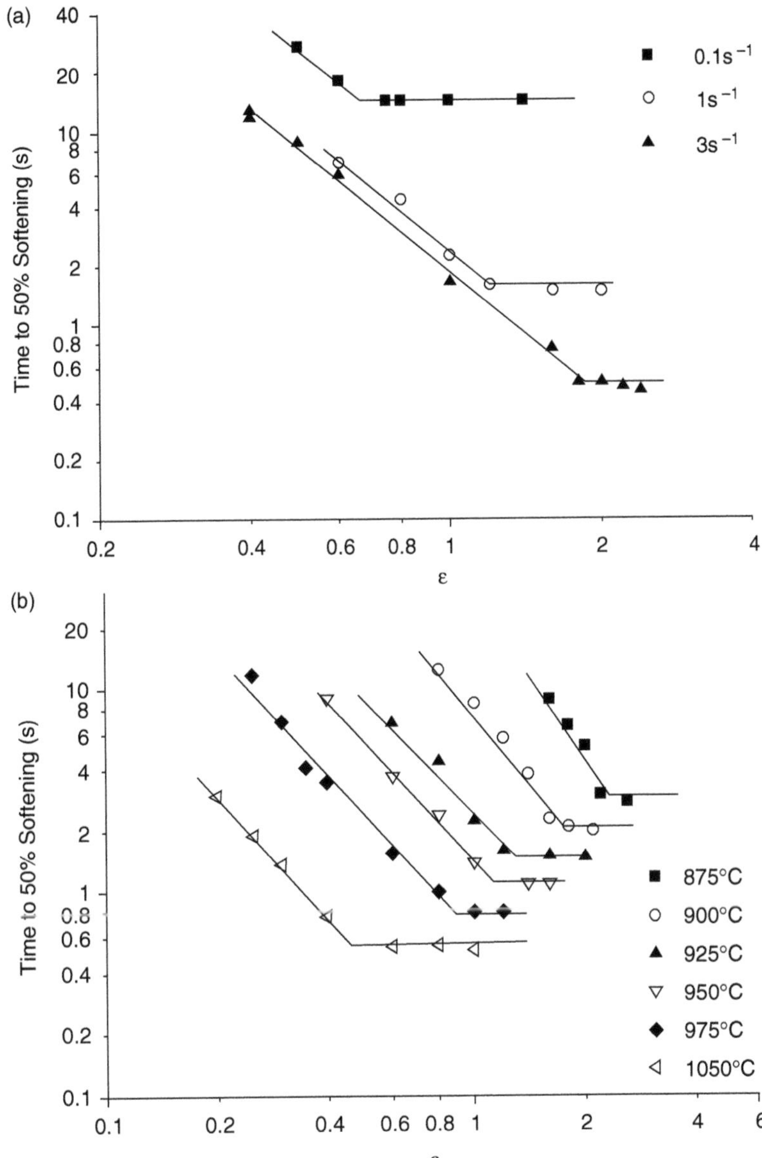

FIGURE 29.7 The effect of strain on the time to 50% softening, t_{50}.in HSLA steel. (a) The effect of strain rate on the time to 50% softening as a function of equivalent strain at $T = 925°C$. (b) The effect of temperature on time to 50% softening as a function of equivalent strain deformed at $\dot{\varepsilon} = 1\ s^{-1}$.

1200°C or higher. In the course of rolling, as the steel cools during its progress through the mill, the steel becomes supersaturated in Nb and precipitation occurs. The compound that forms is generally described as Nb(C,N) or niobium carbonitride. Its chemical composition is not exact or stoichiometric, as N can substitute for C in the lattice and some of the interstitial sites can also be taken up by vacancies. The solution temperatures of the various carbonitrides are listed in Table 29.2.[10]

Generally, nitrides are more insoluble than carbides.[10] Titanium is the most insoluble alloying element, while V is the most soluble. The kinetics of precipitation of Nb(C,N) depend of course on the amounts of Nb, C, and N in solution (i.e., on the "solubility product"). The rates observed in a typical HSLA steel are illustrated in Figure 29.8a.[8] Here it can be seen that the "nose" of the curve (i.e., the point of most rapid

TABLE 29.2 Calculated Equilibrium Carbonitride Solution Temperatures in Various Steels

Steel Type	Solution Temperatures, °C			
	T_{sol}	T_{range}	T_{ave}	$T_{average,sol}$
V	985	965	1025[a]	990
Nb	1050	1090	995[b]	1045
Nb–Mo	1050	1100	995[b]	1050
Nb–V	1050	1090	980[b]	1040
Nb–Mo–V	1050	1090	975[b]	1038

[a] No correction for effect of Mn on N activity.
[b] Corrected for effect of Mn on C activity.
Adapted from M.G. Akben, B. Bacroix and J.J. Jonas; *Acta Metall.*; 1983; 31; 161–174.

precipitation) is located at about 900°C and 10 s. Such a precipitation-start time (10 s) is shorter than the interpass time of plate mills, as indicated in Table 29.1.

The precipitation–time–temperature (PTT) curve of Figure 29.8 is frequently described as a C curve because of its characteristic shape. The upper rightward-trending branch indicates that the behavior at high temperatures is *nucleation* (driving force) controlled, becoming more rapid as the temperature is decreased below the solution temperature, at which the driving force is zero. Conversely, the lower rightward-trending branch signifies that, at relatively low temperatures, precipitate growth is *diffusion* controlled, eventually coming to a stop if the temperature (i.e., the diffusion rate) is low enough. The nose results from the intersection of these two opposite trends.

The kinetics depicted in Figure 29.8a represent what is known as "strain-induced" precipitation, which involves the formation of fine precipitates on the dislocations introduced during prior straining. Such precipitates are illustrated in Figure 29.8b, where they can be seen to be about <10 nm in diameter. If the material is undeformed, the relevant rates are much slower, for example, at least an order of magnitude slower. These particles generally form on grain boundaries and are much larger (50 to 100 nm). They do not play an important role in *thermomechanical processing*. Large (1 μm) TiN precipitates can, however, play a role in preventing grain coarsening during preheating as well as welding.

29.3.1 Interaction Between Precipitation and Recrystallization

The kinetics of carbonitride precipitation have already been considered above. Here it is of importance to compare the rates of these two mechanisms on the same time scale. This is done in Figure 29.9. Two features of this diagram are of particular importance. The first is that the RTT (recrystallization–time–temperature) plot only contains a downward-sloping (i.e., a diffusion controlled) arm. This is because the concept of a solution temperature (applicable to precipitation) above which there is no driving force does not apply here. Instead, the driving force is almost temperature independent. The second is that the difference in shape between the PTT and RTT curves leads to the situation where, at higher temperatures (where precipitation is nucleation controlled and quite slow), recrystallization precedes precipitation. Conversely, at lower temperatures, where precipitation is more rapid than the extrapolated rate of recrystallization, the latter is slowed considerably and may even be prevented from taking place.

The upper temperature region, which involves complete or near-complete softening between passes, coincides with what is known as "roughing," that is, the initial stages of rolling. Under these conditions, recrystallization takes place after every pass (or every second pass at lower temperatures). The lower temperature region, which involves the absence or near absence of softening, corresponds to the "finishing" stages of rolling. Under the latter conditions, no recrystallization occurs. The contrasting microstructural processes taking place within these two temperature ranges of rolling and their effects on the mechanical

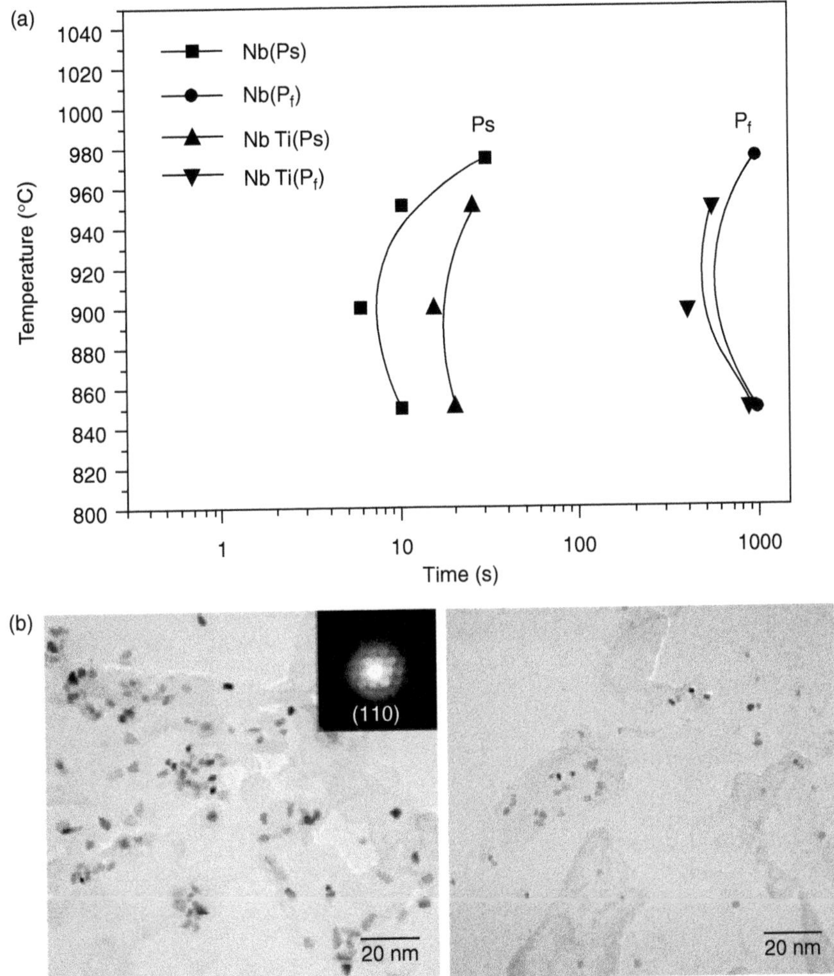

FIGURE 29.8 (a) PTT diagrams for the Nb and Ti–Nb grades, where P_s and P_f are the start and finish times for strain-induced precipitation, respectively. (b) Carbon extraction replicas exhibiting the NbC carbides formed in the (a) Nb grade and (b) Ti–Nb grade. (Used with permission from S.G. Hong, K.B. Kang, and C.G. Park; *Scripta Mater.*; 2002; 46; 163–168.)

properties of the final product are of considerable practical importance and will now be considered in turn. The relative kinetics of precipitation and of recrystallization for various rolling processes are compared in Table 29.3.

29.4 Modeling Rolling Load and Grain Size

As outlined above, the rolling load is sensitive to whether or not recrystallization occurred during the preceding pass. Accordingly, before the load can be modeled or predicted, something must be known about the recrystallization kinetics. The latter are usually determined experimentally by measuring the *fractional softening*, rather than the volume fraction of recrystallization. (Strictly speaking, the rolling load depends directly on the softening, and only partially on the fraction recrystallized, as the mechanism of static recovery also contributes to the measured softening.) An example of some fractional softening curves was presented above in Figure 29.4.

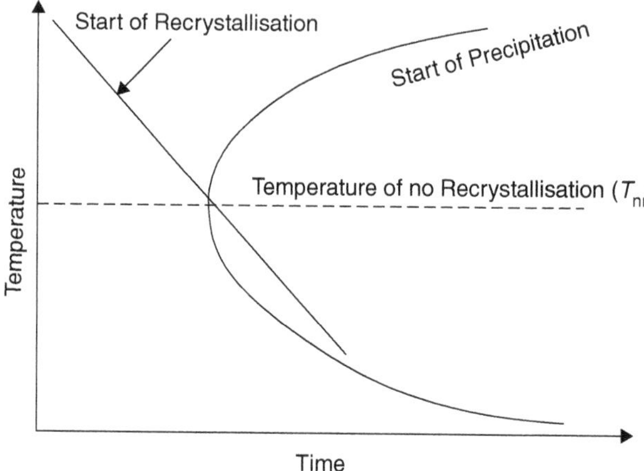

FIGURE 29.9 A schematic depiction of the variation in time required for the initiation of recrystallization and precipitation with changing temperature.

TABLE 29.3 The Relationship between Recrystallization and Precipitation for the Three Categories of Interpass Time

Interpass Time	Type of Process	T Range with Respect to T_{nr}	Role of Strain-Induced Precipitation (pptn)	Relation between pptn and Reaction
Long (plate)	Recrystallization Controlled Rolling (RCR)	Above	Absence required	SRX before pptn
Long (plate)	Conventional Controlled Rolling (CCR)	Below	Presence required	Pptn before SRX or DRX
Short (rod)	Dynamic DRCR+Metadynamic MDRCR	Below	Absence required	No SRX DRX before pptn
Intermediate (strip)	Various combinations	Above and below	Absence or presence	Various alternatives

The fractional softening is usually assessed by employing the technique of "interrupted mechanical testing" also known as "double-hit testing." According to this method, a sample is work hardened to a specific strain at a particular temperature and strain rate. The sample is then unloaded and held at temperature for a selected time before it is reloaded. The amount of softening occurring during the "delay" is then measured and normalized. This softening depends sensitively on the strain, strain rate, and temperature. It is also highly influenced by the chemistry of the steel, which is why these tests have to be carried out for each material of interest.

29.4.1 Kinetics of Fractional Softening

Once the data are available, they can be described in terms of the Avrami relation, which has the following form:

$$X = 1 - \exp(-0.693(t/t_{50})^n) \tag{29.2}$$

Here, X is fractional softening, t is the time, n is the time exponent, and t_{50} is a parameter that depends on the type of recrystallization taking place. For SRX:

$$t_{50} = A\varepsilon^{-p}\dot{\varepsilon}^{-q}d_0^r \exp(Q_{srx}/RT) \tag{29.3}$$

where ε and $\dot{\varepsilon}$ are the applied strain and strain rate, d_0 is the initial austenite grain size, Q_{srx} is the apparent activation energy for SRX (J/mol), R is the gas constant (8.31 J/mol.K), T is the deformation temperature (in K), and A, p, q, and r are material-dependent constants.

By contrast, for MDRX or postdynamic recrystallization:

$$t_{50} = B\dot{\varepsilon}^{-s}\exp(Q_{mdrx}/RT) \tag{29.4}$$

where Q_{mdrx} is the apparent activation energy for postdynamic (prestrain independent) recrystallization and B and s are material constants.

The important point to note here is that the SRX kinetics are strain and initial grain size dependent while these terms do not enter into the MDRX kinetics of Equation 29.4. Furthermore, the exponent s is larger than q signifying that MDRX is more rate dependent than SRX; conversely, Q_{srx} is greater than Q_{mdrx}, signifying that SRX is more temperature sensitive than MDRX.

Some typical t_{50} expressions and values for the material parameters in Equation 29.3 and Equation 29.4 for specific alloy elements and temperature ranges in PC, interstitial free, and HSLA steels may be found in Reference 11 to Reference 29.

29.4.2 Modeling of Rolling Load

The production of thermomechanically processed steels requires much tighter control over the rolling process if the various microstructural phenomena are to be controlled to obtain the desired final properties. Rolling in the full recrystallization region leads to much lower rolling loads than during controlled rolling, where the strain is accumulated from pass to pass. In the design of rolling schedules, therefore, it is necessary to be able to calculate the rolling loads taking into account the microstructural reactions that have a significant effect on the load.

A simple approach to calculate the rolling load, P, is to use an equation of the form:

$$P = k_m A_c Q_p \tag{29.5}$$

where k_m is the hot strength of the metal, A_c is the contact area between the metal and rolls and Q_p is a term that accounts for the geometry of the roll gap and the friction conditions. In flat rolling one example is:[30]

$$Q_p = \left(1 + \frac{1.6\mu\sqrt{R'\Delta h} - 1.2\Delta h}{h_{entry} - h_{exit}}\right) \tag{29.6}$$

where R' is the deformed roll radius, h_{entry} and h_{exit} are the strip entry and exit thicknesses, respectively, μ is the friction coefficient and $\Delta h = h_{entry} - h_{exit}$. In flat rolling $A_c = W\sqrt{R'\Delta h}$ with W being the strip or plate width. The hot strength of the metal is obtained using simple approaches based on the deformation parameters or more complex approaches based on the physical processes occurring during rolling. However, reasonably accurate rolling load predictions can be obtained from relatively simple models that combine an approach using the deformation parameters and the simple microstructure evolution equations above.

Many rolling mills use basic equations of the form:

$$k_m = k_0\varepsilon^x\dot{\varepsilon}^y\exp(Q_d/RT) \tag{29.7}$$

where Q_d is an apparent activation energy for deformation. One of the problems with this expression is that the basic relationship is based on a formulation more appropriate for creep than hot working. It is possible, therefore, to improve this basic equation by replacing the strain rate and temperature term by the hyperbolic sine law for hot working. However, in the following the above equation will be used. All of the additions outlined below can also be achieved with the more complex law.

Equation 29.7 is based only on the deformation parameters for a given rolling reduction and takes no account of the evolving microstructure. The processes that can affect the hot strength include the initial grain size, precipitation hardening, strain accumulation from a previous pass where there has not been complete recrystallization and transformation to ferrite between passes. For the majority of commercial hot rolling processes it is possible to ignore the grain size and precipitation hardening effects. Transformation during the rolling process is only used in a limited range of thermomechanical processes, such as warm rolling described later. The most important consideration, then, is to incorporate the effect of recrystallization between passes on the hot strength. This is achieved by introducing the concept of retained strain. If there is complete recrystallization between passes then the retained strain is zero, while for no recrystallization it is equivalent to the previous strain. For cases between these two extremes the retained strain is some fraction of the prior strain.

One simple approach is to assume that the fraction of retained work hardening (or retained "strain"), λ, is directly related to the amount of recrystallization between passes:

$$\lambda = 1 - X. \tag{29.8}$$

The amount of recrystallization, X, is calculated from Equation 29.2, with the appropriate t_{50} equation depending on the level of strain. Equation (29.7) is then modified to become:

$$k_{\mathrm{m}} = k_0(\varepsilon_i + \lambda\varepsilon_{i-1})^x \dot{\varepsilon}_i^y \exp(Q_{\mathrm{d}}/RT_i) \tag{29.9}$$

where the subscripts i and $i-1$ refer to the current and previous passes, respectively. In this way the retained work hardening can be accumulated over a number of passes.

At higher strain levels it is necessary to modify the strain function in Equation 29.7 and Equation 29.9 as the power law nature leads to significant overestimates of the stress; in some cases it is also necessary to include the effect of dynamic recrystallization.

29.4.3 Modeling of Austenite Grain Size

The austenite grain size can be modeled in a similar way to the rate of recrystallization. Again two quite distinct behaviors are seen for the static and post- or metadynamic regions. For static recrystallization the grain size has been modeled using:

$$d_{\mathrm{srx}} = D\varepsilon^{-0.5} d_0^{0.4} \exp(45,000/RT) \tag{29.10}$$

For the case when dynamic recrystallization occurs during deformation there are two possible recrystallized grain sizes:

$$d_{\mathrm{drx}} = 1.6 \times 10^4 Z^{-0.23} \tag{29.11}$$

$$d_{\mathrm{mdrx}} = 2.6 \times 10^4 Z^{-0.23} \tag{29.12}$$

where $Z = \varepsilon \exp(300,000/RT)$ and d_{drx} is the grain size after complete dynamic recrystallization and d_{mdrx} is the grain size after complete postdynamic recrystallization. In practice it is not possible to stop the postdynamic recrystallization process going to completion and so Equation 29.12 is more relevant. These equations and coefficients were developed for C–Mn and low-alloy steels. Similar equations have been developed for microalloyed steels.

After complete recrystallization there will be grain growth. This is modeled using equations of the form:

$$d^z = d^z_{\mathrm{recrys}} + k_{\mathrm{gg}}\, t \exp(Q_{\mathrm{gg}}/RT) \tag{29.13}$$

where d_{recrys} is the recrystallized grain size from either Equation 29.10 or Equation 29.12 (note it cannot be from Equation 29.11 as postdynamic recrystallization must be completed before grain growth can commence), t is the holding time and Q_{gg} is the apparent activation energy for grain growth. For classical, ideal grain growth the exponent z would be 2. However, much higher values are generally observed in steels and, for example, in C–Mn steels $z = 7$ with values for Q_{gg} and k_{gg} of 400 kJ/mol and 1.45×10^{27}, respectively. Higher values of the exponent z suggest a high level of impurity drag at the moving grain boundary; this effectively means that the rate of grain growth is quite low after the first few seconds. The addition of Ti to the steel at a stoichiometric ratio to N leads to very low levels of grain growth due to the pinning effect of TiN, which is sufficiently stable to withstand most reheating scenarios in industry. Other alloying additions can have more complex effects.

29.5 Rolling Processes

29.5.1 Rough Rolling

There are basically two kinds of roughing mills; one of these involves a single rolling stand, through which the workpiece is alternately passed in a forward and then a backward direction. This is known as a "reversing" mill. When larger tonnages of steel are involved, reversing roughing can be followed by reduction in a four-stand "continuous" roughing mill. In the latter case, the steel continues to move in the forward direction. In both types of mill, rolling begins at about 1200°C, once a slab or billet has emerged from the reheat furnace.

In part because of the relatively high temperatures, but also because of the reversal times and fairly low strain rates (velocities between rolling stands in the case of continuous mills), there is ample time for recrystallization between passes. This mechanism produces full softening of the steel and also successively reduces the grain size from about 200 to 300 μm after reheating to something in the range of 20 to 30 μm. Each pass distorts the grain shape inherited from the previous pass, but the grains then return to their equiaxed shapes as a result of every cycle of recrystallization. The effect of recrystallization on the microstructure during roughing is illustrated schematically in Figure 29.10.

29.5.2 Finish Rolling

Once the temperature has dropped below the T_{nr}, recrystallization no longer takes place and the grains become progressively more distorted with each additional reduction. The grains appear flattened under

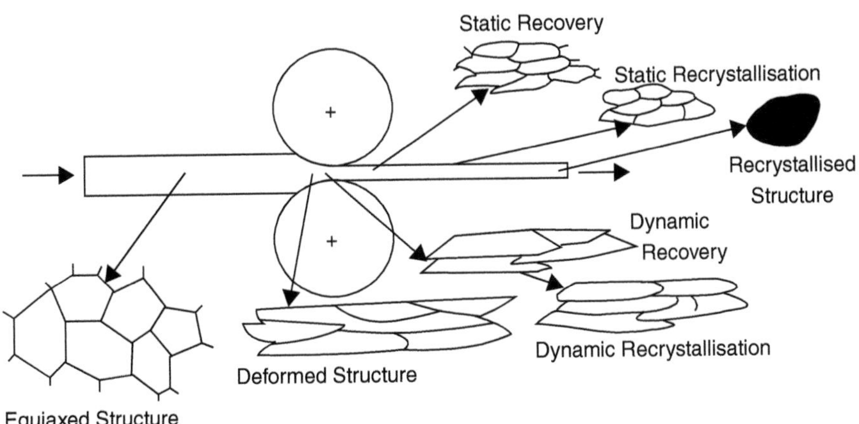

FIGURE 29.10 Schematic of restoration processes during roughing. (Used with permission from B.K. Panigrahi; *Bull. Mater. Sci.*; 2001; **24**; 361–371.)

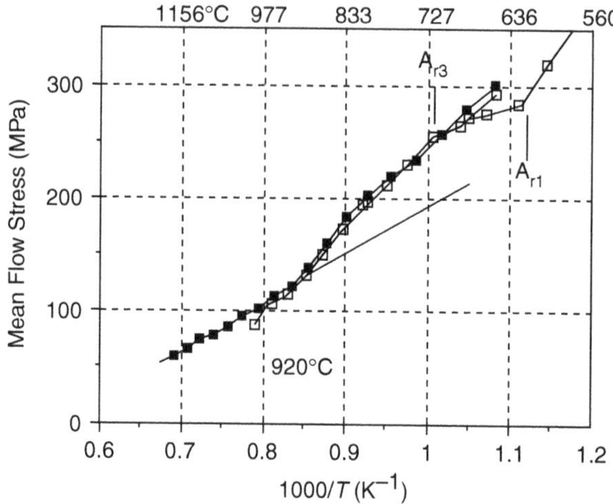

FIGURE 29.11 Stress vs. $1/T$ for 0.17% C, 0.036% Nb steel (Used with permission from T.M. Maccagno, J.J. Jonas, S. Yue, B.J. McGrady, R. Slobodian, and D. Deeks; *ISIJ Int.*; 1994; **34**; 917–922.)

the microscope, an appearance that is referred to as "pancaked." (Note that these "pancakes" are certainly not circular in the conventional way when flat products are being rolled, but become rather elongated; their width remains constant.) The pancaking produced by one or two passes in the absence of recrystallization is insufficient to achieve the goals of thermomechanical processing and it has now become well established that an accumulated strain of about one is required if adequate ferrite grain refinement is to be achieved during subsequent cooling. This is equivalent to about four or five successive passes of rolling without any recrystallization taking place between passes. The effect of pancaking on the microstructure is also depicted in Figure 29.10.

The rolling load during finishing is about four or five times higher (for a given width and reduction) than during roughing. This is partly because of the lower temperatures involved (1050–850°C instead of 1230–1030°C). However, the rolling load is also increased in this temperature range by what is referred to as the retained strain or accumulated work hardening (i.e., the lack of softening in the absence of recrystallization). The rolling loads in both roughing and finishing are represented schematically in Figure 29.11. The slow increase in load during roughing can be compared here with the more rapid increase from pass to pass during finishing. This diagram represents a rolling simulation carried out on a torsion testing machine.

29.6 The γ-to-α Transformation and Ferrite Grain Refinement

The main goal and principal effect of austenite pancaking is to produce significant grain refinement of the ferrite on subsequent transformation. Such refinement can be ascribed to several factors. First of all, flattening of the grains increases the total grain boundary surface area per unit volume. Because ferrite nucleation during the transformation of austenite begins at the gamma grain boundaries, an increase in boundary area increases the density of potential nucleation sites, thus contributing to the grain refinement.

A further factor leading to grain refinement is the retained work hardening in the austenite. The high dislocation density introduced by applying a pancaking strain of one greatly increases the density of nucleation sites per unit of grain boundary area. Straining not only raises the density of grain boundary nucleation sites, but also introduces "intragranular" sites, that is, sites located *within* the grains. These are situated on features of the deformation substructure known as shear bands and deformation bands.

As a result of the above factors, ferrite grain sizes of 4 to 6 μm can be readily achieved. In the absence of pancaking, that is, in transformation from *equiaxed* ferrite, the ferrite grain sizes fall in the range of

6 to 9 μm. Although these differences do not seem to be highly significant, they correspond to ASTM grain sizes of 11 to 12 in the case of thermomechanical processing and of 9 to 10 for conventional processing. As will be seen in the next section, such grain size differences lead to appreciable advantages when it comes to the fracture toughness and tensile properties.

29.6.1 Effect of Accelerated Cooling on Ferrite Grain Size

The ferrite grain size that forms on transformation depends on the actual temperature of transformation. The latter is not fixed (except under equilibrium conditions of very slow cooling rate) but is a function of cooling rate, decreasing as the cooling rate is increased. When the cooling rate is very high, martensite is formed, a microstructural component that is beyond the scope of this survey. At slightly slower cooling rates, bainite is formed, an important component of HSLA steels, but one that is not considered in detail here.

Turning to the ferrite grain size, this decreases as the cooling rate is increased into the range that leads to bainite formation. The reason is that rapid cooling suppresses the transformation, causing it to occur at lower and lower temperatures. Because the diffusivity decreases with the temperature decrease, and ferrite growth rates depend on diffusion, lower temperatures of transformation produce finer grain sizes. This important phenomenon has led to the introduction of many industrial processes for "accelerated cooling." Grain refinement occurs through the use of copious amounts of water, which has led to the frequent steel roller's remark that "water is the least expensive alloying element."

The combined effects of accelerated cooling rate ($\dot{T}$ in K/s) and controlled rolling (retained strain, ε_r) can be demonstrated by a simple equation that has been developed to predict the ferrite grain size, d_α, of C–Mn steels.

$$d_\alpha = \{(6.37CEQ - 0.4) + (24.2 - 59CEQ)\dot{T}^{-0.5} + 22(1 - \exp(0.015d_\gamma))\}\{1 - 0.45\sqrt{\varepsilon_r}\} \quad (29.14)$$

where CEQ is the carbon equivalent and d_γ is the austenite grain size before the controlled rolling phase. This equation shows that the composition of the steel has a direct effect on grain size as well as an indirect effect through the cooling rate sensitivity. Increasing the CEQ, cooling rate, and retained strain all decrease the grain size whereas decreasing the austenite grain size (usually through the addition of Ti to control grain growth) is required for a finer grain size.

29.6.2 Multiphase Steels

The emphasis in most of the discussion to date on thermomechanical processing of steels has focused on the refinement of the ferrite grain size as this provides the best approach to combining strength and toughness; key factors in structural steels. However, there are a range of other applications where properties such as formability are more important. This has become a topic of intense activity recently as automakers are now introducing a number of higher strength more formable steels; these are now termed Advanced High-Strength Steels (AHSS). The introduction of these new grades allows weight reduction in the car with improved safety (i.e., crash performance). The two major types of AHSS are dual phase (DP) and Transformation Induced Plasticity (TRIP) steels. The microstructure of the DP steel is ferrite and martensite, while the TRIP steel contains ferrite, bainite, and retained austenite. It is the transformation of the austenite to martensite during cold forming of the steel that gives the TRIP effect and provides high levels of work hardening and thereby strength and formability.

To produce these grades through thermomechanical processing without continuous annealing requires tight control of the run out table of the strip mill. For the DP steels the austenite to ferrite reaction takes place after rolling and then before the steel enters the coil water cooling is applied to both avoid the formation of pearlite and to cool the steel below the Ms (martensite start) temperature. Typically, coiling of these steels takes place around 250°C and so there are the competing needs to both form a substantial amount of ferrite and to be able to cool the steel to this temperature, without affecting productivity. The production of TRIP steels is even more complex. These steels contain a high level of Si and/or Al; for example, one family of TRIP steels is based on 0.2C-1.5Mn-1.5Si. The Si suppresses carbide formation, both at the higher temperatures where pearlite would form for this relatively high C content, and at the coiling temperature where a carbide free bainite forms. The initial austenite to ferrite reaction enriches the austenite in C. As for the DP steels the cooling avoids the formation of pearlite and in this case the steel is coiled at around 450°C where the bainite forms. As this is a carbide-free bainite there is further rejection of C into the remaining pools of austenite and these reach a C level where the austenite is stable to room temperature (i.e., the Ms is below room temperature).

29.6.3 Intercritical and Warm Rolling

Although the great bulk of finish rolling is carried out in the gamma (fcc or austenite) temperature range, that is, above the A_{r3} or upper critical temperature, it is also possible to finish roll at temperatures below the A_{r3} or even below the A_{r1} (the lower critical temperature). In the case of the latter, the material is ferritic, that is, in the bcc or ferritic phase. This is possible because ferrite below the A_{r1} (while cooler) is actually *softer* than austenite just above the A_{r3}. If no recrystallization occurs after such "warm" rolling, the work hardening introduced during these additional passes can be used to increase the yield strength of the material, when necessary. Excessive reductions below the A_{r1} (more than 2 or 3 passes) without recrystallization are to be avoided as they can lead to a susceptibility of rolled plate to a defect known as "laminations." The latter affect the fracture toughness.

There are a number of reasons why one might consider intentionally lowering the temperature of finish rolling into the ferritic reason. Chief among these is that it turns out to be a potent means of softening hot rolled low carbon strip product. For these grades, the typical grain size following hot rolling is 15 μm. For certain nonstructural forming operations a softer product is required and to achieve a significant drop in strength, ferritic or warm rolling can be employed to coarsen the final grain size to 30 μm. As will be shown further below, this change in grain size correlates to a drop in strength of approximately 50 MPa.

Another attraction with warm rolling is that it enables the rolling schedule designer to avoid unwittingly rolling in the intercritical region. This region, which corresponds to temperatures between the A_{r3} and the A_{r1}, is characterized by a flow stress that is extremely sensitive to temperature. If the strip edges cool during rolling such that they end up in the intercritical region, mill control becomes difficult due to the rapid load change. Under these conditions, the edges are actually softer than the strip centre and edge wave defects can result. When rolling thinner gauge (thickness) material the chances of over cooling during rolling are high and under these circumstances the engineers may choose to roll entirely in the ferritic region, thus avoiding the danger posed by overcooled edges.

29.6.4 Dynamic Strain-Induced Transformation and Ultrafine Ferrite

A recent advance in thermomechanical processing of steels has been the production of ultrafine (e.g., an average grain size of 1 μm or less) materials through a dynamic strain-induced transformation process. Here the austenite is deformed at temperatures significantly below the A_{e3} (the equilibrium austenite to ferrite transformation temperature) but above the A_{r3} (the continuous cooling transformation temperature). One example has been to use single pass rolling of a steel strip with a very large prior austenite grain size under conditions of high shear. The combined effect of the large austenite grain size that lowers the A_{r3} and the high cooling rate from the cold rolls leads to the formation of a highly refined equiaxed ferrite in the surface layers of the strip (Figure 29.12). Systematic studies have shown that the ferrite is definitely forming during the deformation and not after. It is possible that there is also a mix of transformation and then dynamic recrystallization of the ferrite that act as the refinement mechanisms. However, there are still a large number of unknowns at present related to the exact role of the concurrent deformation.

At present there are no commercial processes based on this concept but there are very large research activities all around the world, although the main activity is in Japan, China, and Korea. The main challenge from an industrial perspective is that very high levels of strain are required and this may have to

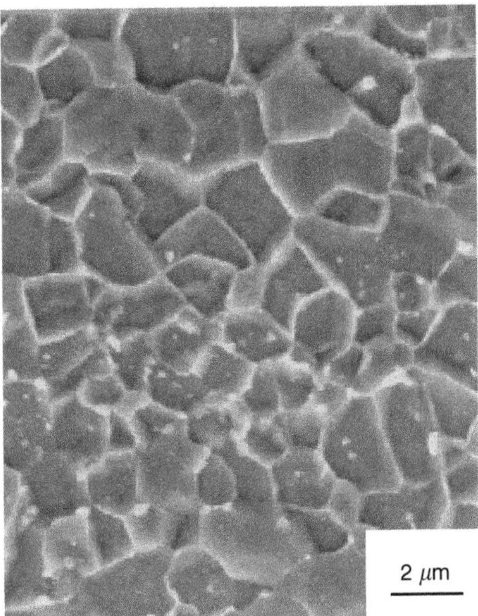

FIGURE 29.12 Ultrafine ferrite grains in surface region. Discrete carbides are also visible. (Used with permission from P.D. Hodgson, M.R. Hickson, and R.K. Gibbs; *Scripta Mater.*; 1999; **40**; 1179–1184.)

be in a single rolling pass. Also the ultrafine ferrite does not give a high level of ductility as the tensile and yield strengths are similar. It would seem that a better balance of properties is achieved with a grain size in the range 2 to 3 μm and also if the second phase can be changed as for the multiphase steels above.

29.7 Effect of Thermomechanical Processing on Mechanical Properties

The influence of grain refinement on the yield strength σ_y is best expressed in terms of the Hall–Petch equation:

$$\sigma_y = \sigma_0 + kd^{-1/2} \tag{29.15}$$

where σ_0 represents the yield strength of a single crystal and k is the grain boundary strengthening coefficient. The effect of controlled rolling on the yield strength is illustrated in Figure 29.13, where the marked influence of grain refinement can be readily seen. The yield stress (YS) and tensile strength (TS) of a simple ferrite + pearlite steel are also affected through precipitation hardening after transformation, during cooling to room temperature. Nb, Ti, and V can all provide added strengthening in this way and the choice of element depends on the application and processing route.

Industrially developed equations for the structure–property relationships of structural steels are:

$$YS = 62.6 + 26.1[Mn] + 60.2[Si] + 759[P] + 3286[N] + 19.7d_\alpha^{-0.5} \tag{29.16}$$

$$TS = 165 + 635[C] + 53.6[Mn] + 99.7[Si] + 652[P] + 3340[N] + 11d_\alpha^{-0.5} \tag{29.17}$$

Precipitation hardening equations are not as well developed but a common assumption is ~3000 MPa strengthening per wt.% Nb, while V is more complicated as it can form both carbides and nitrides with different strengthening capacity. One equation to describe the strengthening from V is:

$$\Delta YS, TS = 19 + 57\log\dot{T} + 700[V] + 7800[N]. \tag{29.18}$$

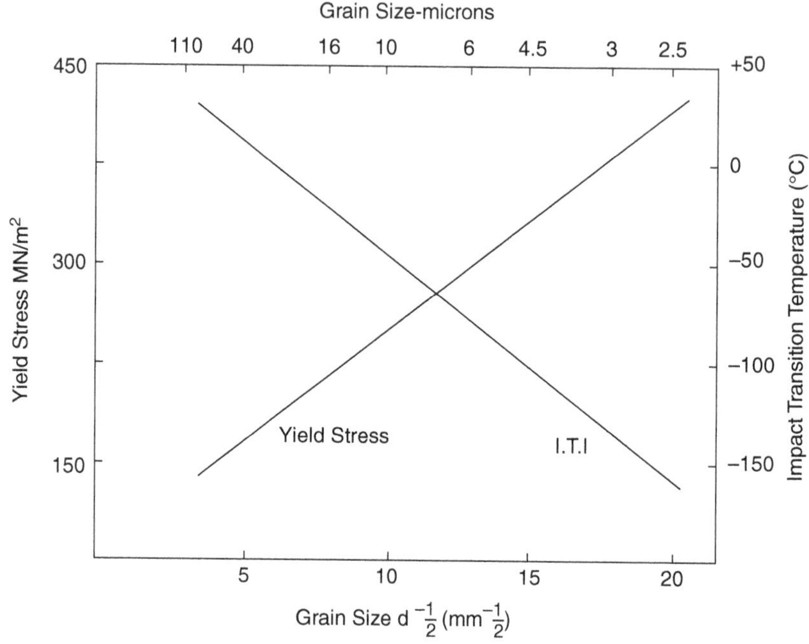

FIGURE 29.13 The effect of grain size on the yield strength and fracture toughness.

Examination of the above equations highlights some important concepts for thermomechanically processed steels. First, C only has a direct effect on the TS, with no effect on YS, although in reality increasing C does increase the YS by refining the ferrite grain size. Note that the effect of grain size refinement is much stronger for the YS than the TS, which is one of the reasons for the limited benefits from high levels of grain refinement as the YS/TS ratio approaches unity. The strong coupled effect of V and N has often meant that V is a preferred precipitation hardening element for steels produced by the electric arc route as these have higher residual N compared with the blast furnace route. In fact it is common in long products mills for bar and structural grades to add even more N to the melt to promote this strengthening.

Similar remarks can be made regarding the influence of grain refinement on the fracture toughness (Impact Transition Temperature), which is also depicted in Figure 29.13. Here the effect is even greater; steels that can be safely used at temperatures as low as −60°C can now be reliably produced by employing the principles of thermomechanical processing. These advances have made possible the construction of oil and gas pipelines across Siberia, Alaska, and the Canadian North. In a similar manner, the steels employed in North Atlantic and North Sea drilling platforms are produced in this way.

29.8 Texture Formation During Rolling

One of the disadvantages of controlled rolling is its effect on the texture and therefore on the anisotropy of mechanical properties. Recrystallized austenite contains what is known as the "cube" texture, which, on transformation is converted into the so-called "rotated cube," "Goss," and "rotated Goss" textures. Ferrite containing these three texture components has mechanical properties that are relatively isotropic; that is, the longitudinal and transverse yield strengths are approximately equal.

By contrast, the ferrite that forms from *pancaked* austenite is relatively anisotropic; that is, the transverse yield strength is higher than the longitudinal one. Such ferrite contains two principal texture components: the so-called "transformed copper" and "transformed brass." (The texture components formed during the plane strain rolling of fcc metals such as austenite are illustrated in Figure 29.14.) As long as appropriate

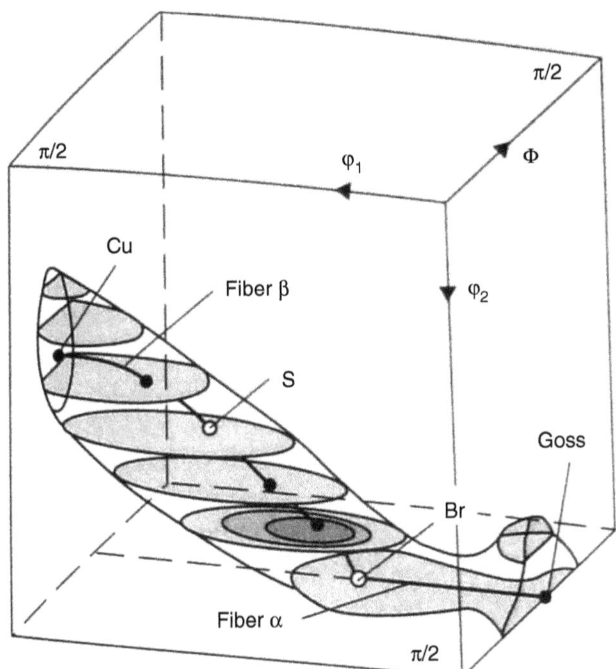

FIGURE 29.14 Three-dimensional view of Euler space with locations of some important ideal orientations and fibers (Bunge notation). (R.K. Ray, J.J. Jonas, and R.E. Rook; *Int. Mater. Rev.*; 1994; **39**; 129–172.)

measures are taken to compensate for the higher transverse yield strength of these steels, no particular problems need arise. A detailed discussion of these effects is beyond the scope of this chapter and for a fuller discussion see Reference 32.

29.8.1 Warm Rolling Textures

The textures developed during ferritic rolling follow the pattern of those produced during cold rolling, a much studied topic that is beyond the scope of this article. Suffice to say that during rolling two orientation fibers are established. One is characterized by the alignment of the ⟨111⟩ axes with the sheet normal. The other is typified by a ⟨110⟩ axis parallel to the rolling direction. The former is of considerable commercial significance as grains with this orientation impart superior deep drawability. The recrystallization of deformed ferrite is therefore often carried out in such a manner as to preserve these orientations while minimizing the ⟨110⟩ fiber.

Under certain conditions it is possible to achieve this feat for warm rolled material. However, in this case there are a number of additional difficulties, compared to cold rolled steels, that must be overcome. One of these is that the textures near the sheet surface often contain unfavorable texture components due to the shear strains generated in the roll bite.

This is not a problem in cold rolled alloys for two reasons; one is during cold rolling better lubrication is possible and the other is that the phase transformation following the austenite rolling that precedes cold rolling serves to diminish the retention of a strong surface texture.

The other important hurdle that must be overcome to achieve favorable recrystallization textures in warm rolled steels is that of the detrimental effect of solute carbon. At warm rolling temperatures considerable amounts of carbon can still be retained in solution (~200 ppm). This carbon interferes significantly with the generation of desired deformation structures. One elegant solution is to add Ti to the steel to combine with the solute carbon and form precipitates. These particles exert a favorable influence on recrystallization. But more importantly, they leave the matrix free of solute carbon. This allows for the development of favorable deformation inhomogeneities and thus to recrystallization textures that impart superior sheet deep drawability.

29.9 Recent Advances in Controlled Rolling

As indicated above, considerable additional grain refinement can be produced by employing the principles of accelerated cooling. The extent to which this can be used is often limited by geometric and other practical considerations. Methods have now been devised[33] to introduce rapid cooling immediately on exit from the roll bite. Although only applied to strip rolling for the moment, it now seems possible to produce steels with ferrite grain sizes of only 2 to 3 μm, a range that is equivalent to ASTM grain size numbers of 14 to 15. When applied to plate grades, such fine grain sizes will lead to steels that have acceptable toughness properties at temperatures as low as −80 to −90°C.

29.10 Processing of Nonferrous Metals

The reader will have noticed that this overview is focused primarily on steel rolling as this is where the very concept of thermomechanical processing originated. The control of recrystallization, which is at the very heart of thermomechanical processing, also proves to be very important in the hot working of nonferrous metals.

After steel, the most commonly rolled metal is aluminum. In this case, the hot rolling engineers need not concern themselves with DRX nor with a transformation to another phase during cooling. This difference means a significant variation in hot rolling strategy and a reduced potential for microstructure control. Understanding and manipulating SRX is consequently the key to the thermomechanical processing of aluminum. As mentioned in Section 29.8, the phase transformation that occurs following the conventional hot rolling of low carbon steel can weaken the texture, thus "smearing" out certain microstructure and crystallographic inhomogeneities. The absence of a phase transformation in aluminum means that greater care must be taken during hot rolling not to establish detrimental texture gradients through the strip thickness. These issues parallel those encountered in the rolling of steel in the ferritic region.

Recrystallization during the hot rolling of aluminum occurs between passes during rough rolling and in the coil following tandem hot or warm rolling. Depending on the process flow, the alloy and the product, recrystallization may or not be desired. When it is not desired, ensuring rapid cooling is important. Care must also be exercised to avoid arrays of coarse second phase particles. Under certain circumstances these can stimulate nucleation of recrystallization. The ability of aluminum to be formed into drink cans depends to a large degree on a favorable crystallographic texture and it turns out that this is achieved, at least in part, by avoiding recrystallization during hot rolling.

The properties of aluminum alloys are in many cases determined by the state of the second phase particles. Control of these during processing is therefore of vital importance. This is mostly accomplished by manipulating the *thermal* trace of the process with only a relatively minor role played by the deformation. However, in alloys where not all of the second phase particles are taken into solution during the homogenization step, the remaining particles can be broken up during rolling.

For aluminum alloys that are designed to be employed in an unrecrystallized state, controlling the deformation structure is obviously of prime importance. If the structure is hot worked, the subgrain size dictates a significant component of the strength. This is determined, as implied above, by the deformation conditions. In this regard the temperature and the strain rate are more important than the strain as the steady state is achieved at quite low strains during hot working. Lower temperatures and higher strain rates (rolling speeds) give finer subgrain sizes and therefore higher strength. The subgrain size also continues to evolve during recovery following deformation and this must be taken into account when modeling the properties of these alloys.

The discussion hitherto has focused on rolling. The other important thermomechanical process employed for a significant proportion of wrought aluminum is extrusion. Typically, this is carried out in one pass at high temperatures and involves strains as high as 10! Many extruded products are used in the as-deformed state and thus the considerations outlined in the previous paragraph come into play. In these cases it is imperative to avoid recrystallization, which has a tendency to occur in the subsurface regions. In this case too, the distribution, nature, and size of any second phase particles can be manipulated to retard the reaction and thus avoid it from occurring. An additional complication for extrusion is that particles can readily cause surface defects known as die lines. This places an important constraint on the control of the second phase.

Another metal of note that undergoes controlled thermomechanical processing is titanium. In this case, like steel, there is a solid state allotropic phase transformation that occurs within range of the working temperatures. Unlike steel, hot working in the "middle" of this phase transformation, that is, in the two-phase region, is quite common for titanium. The two phases in titanium differ from those in steel. The high temperature phase, β, in this case is bcc, as is ferrite in steel. The low temperature phase, α, is hcp in structure, like zirconium or magnesium, the latter of which will be discussed briefly further below.

The initial hot working stage in titanium processing is break-down forging and this occurs over a number of forging and reheating steps. These forging steps occur, alternatively, in the β and α–β regions. Such a procedure has been found to be optimal for developing a fine and homogeneous structure. This structure is ideally suited for subsequent hot working, which for plate and sheet products involves rolling. Deformation in these cases also frequently involves deformation in or through the two-phase field.

DRX is more readily initiated in the low temperature hcp α phase than in the bcc β phase. Dynamically recrystallized structures can be generated in the laboratory but commercial β phase Ti grades frequently display an elongated hot worked grain structure, which signifies the absence of β DRX. In one sense, the behavior of the two phases can be considered, for practical purposes, to be approximately opposite to that of the two steel phases. As mentioned above, in steel the high temperature γ phase is more inclined to exhibit DRX. Having said this it must be mentioned that explicit control of DRX does not feature in the common thermomechanical processing routes of the typical grades. However, control of SRX and the structures produced during the phase transformation are vitally important in this metal.

Iron is typically added to the commercial purity α alloys in small proportions. The iron is not soluble in the α phase and it leads to the formation of small regions of β phase that are stable even at room temperature. These islands of β are an intentional addition to the structure as they serve to retard grain growth during the recrystallization annealing treatments employed for these grades.

The production of α–β two-phase alloys, which have found considerable application in aerospace, relies on a careful marriage of theremomechanical processing and composition. A common processing route involves working in the α–β region to avoid grain coarsening followed by a β annealing treatment. Following annealing, accelerated cooling is carried out to generate martensite type structures with a characteristic lamellar spacing that is determined by the cooling rate. In another approach, which gives rise to a bimodal structure, α–β deformation is carried out to a sufficiently high degree as to ensure recrystallization of both phases during a subsequent annealing treatment in the two-phase region.

Turning now to magnesium, explicit control of the microstructure of magnesium alloys through thermomechanical processing is not widely practiced. Of course, processing conditions that give rise to excessive grain growth are generally avoided. In this regard an interesting technique of avoiding grain growth during homogenization has been developed. Homogenization is performed in a hydrogen atmosphere and during the heat treatment fine ZrH_2 particles are formed. These particles provide grain boundary pinning thus preventing grain growth.

As mentioned above, magnesium, which in most common alloys remains in its hcp state, readily undergoes DRX. Thus the structure following extrusion is typically comprised of an equiaxed grain structure, unlike aluminum. It turns out that this structure readily coarsens in the subsecond period before the extrusion reaches the conventional cooling stage. To overcome this growth, modified extrusion dies have been developed that incorporate water channels that feed fine jets directed to cool the extrusion immediately following exit from the die. Alloys in which particles are employed to retard this growth are currently under development.

29.11 Summary

The softening mechanisms associated with steel rolling (and with the rolling of other industrial metals) have been described. These include DRV, SRX, DRX, and MDRX or postdynamic recrystallization. The strain-induced precipitation of second phases is considered. In steels, these are principally the carbonitrides, which suppress recrystallization and make "controlled rolling" possible. In the nonferrous metals, such precipitation leads to in-process hardening and eliminates a postrolling heat treatment operation. Approaches to modeling the kinetics of softening are outlined and algebraic relations that describe these mechanisms are introduced. Some attention is also paid to grain size modeling.

Distinctions are drawn between "rough rolling," which generally involves recrystallization between passes, and "finish rolling," in which case recrystallization is suppressed (in the case of steels). The absence of recrystallization leads to "pancaking" of the microstructure, which has a considerable effect on the γ-to-α transformation and on ferrite grain refinement in steels. The effect of cooling rate through the transformation is also described. Some consideration is given to intercritical and warm rolling as well as to the production of ultrafine grained ferrite.

The processing of nonferrous metals, such as aluminum and titanium alloys, is considered briefly. This is followed by the effect of thermomechanical processing on the mechanical properties. The influence of plane strain rolling on the textures of fcc and bcc metals is described, together with the effects of the textures developed on the mechanical properties. Finally, some attention is paid to recent advances in the understanding and practice of thermomechanical processing.

Acknowledgments

The authors are indebted to the Natural Sciences and Engineering Research Council of Canada as well as the Australian Research Council for support of the investigations that led to the preparation of this review. They also acknowledge with gratitude the contributions of numerous graduate students and postdoctoral fellows to the research described here. One of the authors (J.J.J.) is grateful to Deakin University, Geelong, VIC, Australia for the grant of a Visiting Fellowship during which this article was prepared.

References

[1] R. Kaibyshev et al.; Deformation behavior of a modified 5083 aluminium alloy; *Mater. Sci. Eng. A*; 2005; **392**; 373.

[2] A. Oudin, M.R. Barnett, and P.D. Hodgson; Grain size effect on the warm deformation of a Ti–IF steel; *Mater. Sci. Eng. A*; 2004; **367**; 282–294.

[3] T. Sakai and J.J. Jonas; Dynamic recrystallization: mechanical and microstructural considerations; *Acta Metall.*; 1984; **32**; 189–209.

[4] H.L. Andrade, M.G. Akben, and J.J. Jonas; Effect of molybdenum, niobium and vanadium on static recovery and recrystallization on solute strengthening in microalloyed steels; *Metall. Trans.*; 1983; **14A**; 1967.

[5] A.J. DeArdo; Metallurgical basis for thermomechanical processing of microalloyed steels; *Ironmak. Steelmak.*; 2001; **28**; 138–144.

[6] C.M. Sellars; *Deformation Processing and Structure*; ASM Materials Science Seminar; 1982; G. Krauss, Editor; Metals Park, Ohio; 1984; p. 245.

[7] R.A. Petkovic, M.J. Luton, and J.J. Jonas; Recovery and recrystallization of carbon steel between intervals of hot working; *Can. Metall. Quart.*; 1975; **14**; 137–145.

[8] S.G. Hong, K.B. Kang, and C.G. Park; Strain-induced precipitation of NbC in Nb and Nb–Ti microalloyed HSLA steels; *Scripta Mater.*; 2002; **46**; 163–168.

[9] B.K. Panigrahi; Processing of low carbon steel plate and hot strip — An overview; *Bull. Mater. Sci.*; 2001; **24**; 361–371.

[10] M.G. Akben, B. Bacroix, and J.J. Jonas; Effect of vanadium and molybdenum addition on high temperature recovery, recrystallisation and precipitation behaviour of niobium-based microalloyed steels; *Acta Metall.*; 1983; **31**; 161–174.

[11] P.D. Hodgson and R.K. Gibbs; A mathematical model to predict the mechanical properties of hot rolled C-Mn and microalloyed steels; *ISIJ Int.*; 1992; **32**; 1329–1338.

[12] J.G. Williams, C.R. Killmore, and G.R. Harris; *International Conference on Physical Metallurgy of Thermomechanical Processing of Steels and Other Metals*; 1988; Iron and Steel Institute of Japan; Tokyo; 224.

[13] T. Siwecki; Modelling of microstructure evolution during recrystallisation controlled rolling; *ISIJ Int.*; 1992; **32**; 368–376.

[14] M. Militzer, E.B. Hawbolt, and T.R. Meadowcroft; Microstructural model for hot strip rolling of high-strength low-alloy steels; *Metall. Mater. Trans. A*; 2000; **31A**; 1247–1259.

[15] S. Cho, K. Kang, and J.J. Jonas; The dynamic, static and metadynamic recrystallisation of Nb-microalloyed steel; *ISIJ Int.*; 2001; **41**; 63–69.

[16] S. Cho, K. Kang, and J.J. Jonas; Mathematical modelling of the recrystallisation kinetics of Nb microalloyed steels; *ISIJ Int.*; 2001; **41**; 766–773.

[17] C.M. Sellars; *Hot Rolling and Forming Processes*; 1980; London; C.M. Sellars and J. Davies, Editors; The Metals Society; pp. 3–15.

[18] A.I. Fernandez, P. Uranga, B. Lopez, and J.M. Rodriguez-Ibabe; Static recrystallisation behaviour of a wide range of austenite grain sizes in microalloyed steels; *ISIJ Int.*; 2000; **40**; 893–901.

[19] P.D. Hodgson, S.H. Zahiri, and J.J. Whale; The static and metadynamic recrystallisation behaviour of an X60 Nb microalloyed steel; *ISIJ Int.*; 2004; **44**; 1224–1229.

[20] S.H. Zahiri, S.M. Byon, S. Kim, Y. Lee, and P.D. Hodgson; Static and metadynamic recrystallisation of interstitial free steels during hot deformation; *ISIJ Int.*; 2004; **44**; 1918–1923.

[21] A. Kirihata, F. Siciliano Jr., T.M. Maccagno, and J.J. Jonas; Mathematical modelling of mean flow stress during the hot strip rolling of multiply-alloyed medium carbon steels; *ISIJ Int.*; 1998; **38**; 187–195.

[22] P.D. Hodgson, D.C. Collins, and B. Perret; *The Use of Hot Torsion to Simulate the Thermomechanical Processing of Steel*; 7th International Symposium on Physical Simulation; 1997; NRIM; Tsukuba; Japan; pp. 219–229.

[23] O. Kwon; A technology for the prediction and control of microstructural changes and mechanical properties in steel; *ISIJ Int.*; 1992; **32**; 350–358.

[24] P. Pauskar and R. Shivpuri; *Integrated Microstructural–Phenomenological Approach to the Analysis of Roll Pass Design in Bar Rolling*; 40th MWSP; 1988; pp. 755–771.

[25] C. Roucoules, S. Yue, and J.J. Jonas; *Effect of Dynamic and Metadynamic Recrystallisation on Rolling Load and Microstructure*; 1st International Conference on Modelling of Metal Rolling Processes; 1993; Imperial College, London; pp. 165–179.

[26] P.D. Hodgson, J.J. Jonas, and S. Yue; *Strain Accumulation and Post-Dynamic Recrystallisation in C–Mn Steels*; International Conference of Grain Growth of Crystalline Materials; Materials Science Forum (Switzerland), Vol. 113–115; 1993; pp. 473–478.

[27] L.P. Karjalainen and J. Perttula; Characteristics of static and metadynamic recrystallisation and strain accumulation in hot deformed austenite as revealed by the stress relaxation method; *ISIJ Int.*; 1996; **36**; 729–736.

[28] D.Q. Bai, S. Yue, and J.J. Jonas; *Metadynamic Recrystallisation of Low Carbon Steels Containing Nb*; Thermomechanical Processing of Steel; 2000; Ottawa, ON; S. Yue and E. Essadiqi, Editors; MET SOC; pp. 669–683.

[29] P. Uranga, A.I. Fernandez, B. Lopez, and J.M. Rodriguez-Ibabe; Transition between static and metadynamic recrystallisation kinetics in coarse Nb microalloyed austenite; *Mater. Sci. Eng.*; 2003; **A345**; 319–327.

[30] T.M. Maccagno, J.J. Jonas, S. Yue, B.J. McGrady, R. Slobodian, and D. Deeks; Determination of recrystallization stop temperature from rolling mill logs and comparison with laboratory simulation results; *ISIJ Int.*; 1994; **34**; 917–922.

[31] P.D. Hodgson, M.R. Hickson, and R.K. Gibbs; Ultrafine ferrite in low carbon steel; *Scripta Mater.*; 1999; **40**; 1179–1184.

[32] R.K. Ray, J.J. Jonas, and R.E. Hook; Cold rolling and annealing textures in low carbon and extra low carbon steels; *Int. Mater. Rev.*; 1994; **39**; 129–172.

[33] C.P. Jongenburger, R. Koenis, M.R. van der Winden, and P.J. van der Wolk; Harvesting metallurgical knowledge for commercial yield; *Proc. 2nd Int. Conf. on Thermomechanical Processing of Steels*; Liege, Belgium, June 2004, ed. M. Lamberigts, pp. XIX–XXVI.

30

Multiscale Processing of Polymers and Nanocomposites

Carol Barry, Julie Chen, Joey Mead, and Daniel Schmidt
University of Massachusetts Lowell

Abstract

Recent discoveries in nanoscience have demonstrated the potential for novel functionality such as transparent chem-bio barrier films, flame-retardant and tough structural materials, and bio-specific tissue scaffolds. Although industrial processing of current commercial polymers occurs at very high rates and large volumes, many barriers to manufacturing of these novel materials remain. Yet, because of their carbon backbone and chain structure and the ability to modify this structure to facilitate directed self-assembly, polymers offer both an opportunity and a challenge for multiscale processing. While not comprehensive, this chapter discusses several key issues in achieving multiscale functionality through polymer and composites processing. Specific examples are provided of fundamental factors affecting dispersion of nanofillers, control of electrospun nanofiber morphology and patterning, layer instabilities in extrusion of multilayer films, and tooling materials for injection molding of nanofeatures.

Part 1 Multiscale Processing

30.1 Overview

With the explosion of discoveries of exciting new properties at the nanoscale comes the question of how to realize the potential of nanotechnology for commercial application. A key factor is the hierarchical manufacturing of products from the nano- to the micro- and the macroscale. For example, individual carbon nanotubes exhibit unique electrical, thermal, and mechanical properties, but to create a useful product, a manufacturing process must be created to integrate these nanotubes into a circuit or a structure without loss of the nanoscale properties. While much of the research in nanoscience and nanotechnology has been driven by the increasing demands of the semiconductor industry for smaller and smaller line widths and greater chip densities,[1] most of these efforts focus on evolution of the basic lithographic process. Multiscale processing is built into the layering of individual wires, components, and the like, building up to the chip level and then the board level. Polymers and polymer-based composites offer a new perspective of multiscale processing by introducing a very different type of processing because of the emphasis on high rates, large areas, and large volumes. For example, while lithography involves essentially "writing" of very precisely placed lines, polymer processes such as dispersion of fillers in composites, fiber spinning, extrusion of multilayer films, and molding of products all rely on some level of self-assembly of the polymer chains to create the final product. In many cases, this may mean a less precise, less ordered, yet still functional orientation and distribution of the nanoelement. In other cases, because of the biological, chemical, or physical driving force, the dimensions and placement can be very precisely controlled through self-assembly rather than direct manipulation (e.g., block copolymers).

While not comprehensive, this chapter discusses several key issues in achieving multiscale functionality through polymer and composites processing. In the first part, a brief introduction is provided to polymers, parameters of general interest in polymer processing, and specifically to block copolymers, which are of particular interest in patterning. The main focus of this chapter then turns to a discussion of fundamental factors affecting dispersion of nanofillers to form nanocomposites. In the subsequent sections, several different polymer processes are presented: external and internal "patterning" of electrospun nanofibers and discussion of modifications to commercial processes such as injection molding and extrusion to obtain nanoscale functionality.

30.2 Brief Introduction to Polymers and Multiscale Aspects

Polymers are materials comprised of a carbon backbone with varying complexity in terms of the size and interconnectedness of side chains. In general, as chain size, side chain bulkiness, and number of cross-links increase, the viscosity of the polymer melt increases, making it more difficult to process. For solvent-based processing, the ability to find a suitable solvent and the evaporation rate both affect processability. In many polymer processes, shear and extensional flows are instrumental in generating orientation of the polymer chains and thus crystallinity that then affects the mechanical, optical, and geometric shrinkage response of the material.

In composites, similar factors hold true, except that instead of the polymer chains, it is the filler or reinforcing element orientation that is of interest. For short fiber composites, the flow can orient the

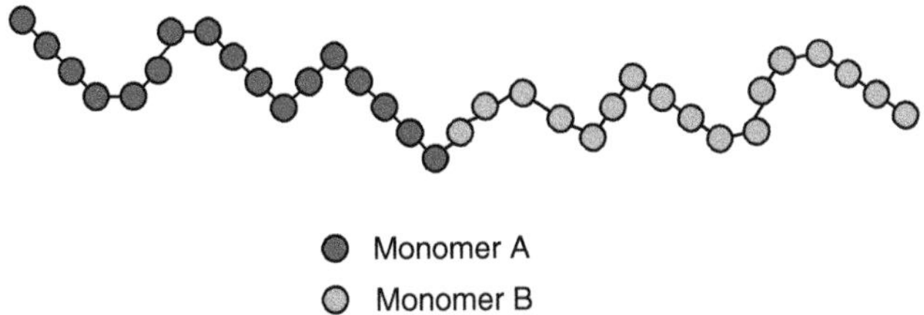

FIGURE 30.1 Block copolymer chain structure.

fibers, while in continuous fiber composites, the fiber orientation and general configuration affects the ability of the polymer to flow through the fiber perform.

As the size of the filler approaches the nanoscale, studies have shown that the filler itself (e.g., nanoclay) can affect the local orientation of the polymer chains to a degree sufficient to affect the overall properties of the material. In addition, relatively low loadings (e.g., greater than 10 vol.%) have been shown to result in significant increases in viscosity, making the polymer matrix very difficult to flow. For short, micron-size diameter fibers, the volume fraction can exceed 40% before viscosity increases affect processing.

30.2.1 Flexible Materials for Nanoscale Manufacturing

There are a variety of approaches to prepare polymeric structures with nanoscale "features" within them. One attractive method is to prepare micro or macroscale structures with embedded nanoscale features. For example, block copolymers, which are thermoplastic materials, contain nanoscale morphologies, but yet can be manufactured into micro and macroscale parts.

Block copolymers are of considerable interest because of their ability to self-assemble into a variety of useful morphologies.[2] Block copolymers are comprised of two (or more) different polymer chains covalently bonded together as depicted in Figure 30.1. Of particular interest is the phase separation of the two different portions of the chain, resulting in period structures with length scales in the range of 5 to 500 nm.[3]

The two-phase morphologies of block copolymers offer potential for applications such as filters, membranes, and high-density storage.[3] In addition, these morphologies can be used as flexible templates for assembly of nanodevices,[4] and the like, that are appropriately modified to "mate" with the block copolymer.[5,6] They have already been used to prepare ordered structures[7] incorporating nanorods,[8] nanoparticles,[9–12] and also as nanoreactors.[13]

Thin films of block copolymers can be prepared by spin-coating and dip-coating, while thicker structures can be formed using extrusion or injection molding processes.[3] The ability to obtain the desired morphology is a critical factor in the utilization of block copolymers for a given application. The specific morphology obtained (spheres, rods, lamella, and bicontinuous domains) is dependent on the relative volume fractions of both blocks, as well as the manufacturing process conditions, such as solvent choice and spin speed.[14] Interaction with the substrate surface can also play a large role in the morphology in very thin films. For example, poly(methyl methacrylate) (PMMA) preferentially wets silicon surfaces in polystyrene (PS)-PMMA block copolymers[14] affecting the resulting morphology. One issue for use of block copolymers is the uniformity of the morphology. Unguided the structures are not defect free over large areas (see Figure 30.2). Recently a number approaches for morphology control have appeared. For example, investigators[15,16] have used electric fields to control the morphology of diblock copolymers of PS and PMMA. The electric field was used to orient the cylindrical PS domains perpendicular to the surface. These structures were then used to prepare nanowire arrays through removal of the PMMA and electrodeposition into the porous structure. Electric fields have also been used to align the domains in the in-plane dimension.[14] The use of shearing forces is an attractive method to orient domains and has been

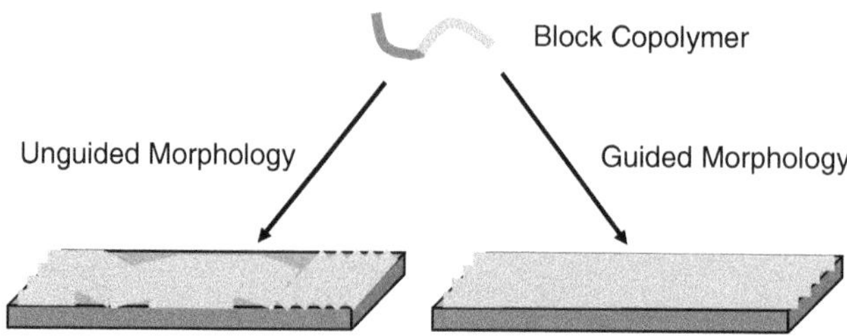

FIGURE 30.2 Block copolymer morphology with and without assembly control.

used successfully by a number of researchers.[17–19] Chemical functionalization is another approach to control morphology, where patterns of different chemical functionality in films of octadecyltricholorsilane on Si/SiO_2 have been successfully used to transfer the pattern into PS-PMMA diblock copolymers.[20]

Nanopatterned surfaces[21,22] have also been successfully used to control block copolymer morphology. Kim et al.[23] used a chemically modified surface to prepare defect-free nanopatterns over large areas. These approaches offer significant promise for control over the domain structure of block copolymers.

Polymeric materials lend themselves to a number of high rate-processing approaches including injection molding and extrusion. In the sections below we highlight methods to extend these techniques to prepare macro or microscale structures with nanoscale features.

30.3 Multiscale Processing of Nanocomposites: Dispersion in Filled Polymers

When a polymer is described as "filled" or we refer to a polymeric system as a composite, this implies that a second phase has been dispersed within a polymer matrix, so as to provide some set of properties distinct from those achievable with the polymer alone. While much of the emphasis in this area is on mechanical reinforcement, other, more "mundane" factors such as density and color or appearance often serve as important drivers for the inclusion of a dispersed phase. Regardless of the reasons for the introduction of this phase, one feature these systems generally share is the need for dispersion. Whether dispersion is achieved via physical mixing of neat (as in melt-blending) or diluted components (as in solvent-assisted blending), the creation of one component *in situ* (as when sol-gel chemistry is used to create ceramic particles within a solid polymer host), or the preparation of special materials where dispersion is "guaranteed" based on the molecular architectures employed (polymer-grafted particles), there are some issues that remain relevant throughout, when processing such materials and attempting to realize the desired structure and properties.

The most general statement we can make is that greater attractive interaction between elements in the dispersed phase will translate into the need for greater energy input to separate those elements. Even then, however, if the distance of separation is insufficient, they may still "feel" one another and be inclined to re-aggregate. Factors influencing this tendency for strong self-interaction in elements of the dispersed phase are listed in Table 30.1.

High levels of dispersion may be realized through appropriate processing in many cases, but to achieve the desired levels of dispersion with a minimum of effort, it behooves us to understand and control as many of the aforementioned factors as possible. With that in mind each of these factors will be discussed in detail. In the discussions that follow, the system being considered is, most generally, a polymer or polymer precursor (the continuous phase or matrix), to which solid particles (euphemistically referred to as elements of the dispersed phase, regardless of actual dispersion level) are somehow introduced. Likewise, unless otherwise specified, the matrix is assumed to be in the liquid state, either as a melt or a solution, consistent with the most common practices of melt and solution blending to create filled polymers.

TABLE 30.1 Factors Affecting the Level of Dispersion of Elements in the Dispersed Phase in a Filled Polymer System

Factor	Nature of influence
(a) Surface area (size/roughness)	Specific surface area increases as roughness increases and size decreases, increasing the tendency for self-interaction.
(b) Shape (topology/aspect ratio)	Shapes that pack efficiently mean more chances for larger areas of these elements to interact with one another, while higher aspect ratios increase the chances for contacts between elements.
(c) Mobility	The ability of the dispersed elements to move with respect to one another is required for a change in dispersion state
(d) Surface energy	A higher surface energy in elements of the dispersed phase indicates a greater tendency for self-interaction of those elements in the absence of a strongly interacting medium, and will discourage separation of those elements.
(e) Interfacial energy	Interactions between the dispersed phase and the continuous phase can overwhelm other factors if sufficiently strong, either forcing or preventing self-interaction of elements of the dispersed phase.

30.3.1 Dispersion: Effects of Surface Area

The tendency for smaller particles to aggregate more strongly than large particles of the same composition and structure is well known, and can be easily understood by means of a simple thought experiment. More details about aggregation may be found in Chapter 2. Consider a bag of glass marbles, all a few centimeters in diameter. They will not aggregate in any classical sense of the word, and will remain loose under all circumstances, regardless of how much compaction they experience (within reason; we must not break them!). Now consider what would happen if we were to reduce the diameter of our glass particles from centimeters to nanometers. As with any especially fine powder, it is obvious that agglomeration will be a much greater issue here than with our marbles. This is true in any medium, so long as a lack of mobility and the presence of strong interactions do not overwhelm this trend. Agglomeration was recognized as a significant problem almost 50 years ago, when fumed silica fillers (~5 to 50 nm "glass marbles") were developed to reinforce silicone rubber: "Fine particle size does not necessarily lead to good reinforcement. In practice, the situation is complicated by the fact that very finely divided fillers tend to agglomerate and are extremely difficult to disperse."[24]

Along these lines, it is also well known that, in the absence of aggregation, smaller particles generally tend to increase the viscosity of the filled polymer more than larger particles at any given concentration. Distributional effects also occur, especially at high concentrations. A broad particle size distribution will tend to give a reduced viscosity vs. a narrow distribution, due to simple geometric packing arguments (i.e., inter-particle contact will be inevitable at a lower concentration if the particles are all the same size than if they are different sizes). The latter may actually enhance aggregation, however, making the matter even more complex, as aggregates act as larger particles and increase the viscosity less than dispersed primary particles do. The Stokes–Einstein hydrodynamic approximation (Equation 30.1) gives us a useful means of understanding the relationship between viscosity and particle size:

$$d(H) = \frac{kT}{3\pi\eta D} \tag{30.1}$$

where $d(H)$ is the hydrodynamic diameter of the particle (just the hard-sphere diameter when dealing with a particle that is not swollen by the medium), k is Boltzmann's constant, T is the absolute temperature, η is the dynamic zero-shear viscosity of the dispersion, and D is the translational diffusion coefficient of the particle in the medium. While this is an approximation (most accurate for low concentrations of hard, impenetrable spheres in an incompressible fluid), it gives the important result that the viscosity is inversely related to the particle size, illustrating the viscosity issues inherent in the dispersion of fine particles in a polymer medium via physical mixing. The average particle size can in turn be related to the specific surface area of the particle chosen based on purely geometric considerations; the relation for spherical particles is given below (Equation 30.2), with a relation for cylindrical particles (Equation 30.3) included for comparison purposes:

$$SSA_{\text{sphere}} = \frac{SA_{\text{sphere}}}{m_{\text{sphere}}} = \frac{4\pi r^2}{\rho V_{\text{sphere}}} = \frac{4\pi r^2}{\rho \frac{4}{3}\pi r^3} = \frac{3}{\rho r} \tag{30.2}$$

$$SSA_{\text{cylinder}} = \frac{SA_{\text{cylinder}}}{m_{\text{cylinder}}} = \frac{2\pi r^2 + 2\pi rl}{\rho V_{\text{cylinder}}} = \frac{2\pi r^2 + 2\pi rl}{\rho \pi r^2 l} = \frac{2(r+l)}{\rho rl} = \frac{2}{\rho l} + \frac{2}{\rho r} \tag{30.3}$$

where *SSA* is specific surface area (generally measured in m^2/g), *SA* is surface area, *m* is mass, *V* is volume, *r* is radius, *l* is length, and ρ is the particle density. In the cylindrical case,* where we have a rod ($l > d$) or a disk ($d > l$), the general trend remains the same as in the simpler spherical case — the specific surface area is inversely proportional to the particle dimension. While it is not appropriate to attempt to apply the Stokes–Einstein equation to substantially nonspherical particles, the inverse relationship between dynamic zero-shear viscosity and particle size should also hold, qualitatively speaking. Thus, dynamic zero-shear viscosity is proportional to specific surface area — a reasonable conclusion considering that viscous drag will occur at all interfaces between the particles and the medium.

Here it should be noted that the above discussions refer to particles with smooth surfaces. What if they are not smooth? The rougher the particle surface, the greater the specific surface area, regardless of particle size, and the greater the potential for self-interaction as a result. Equation 30.2 and Equation 30.3 will underestimate the surface areas of roughened particles. Additionally, the use of the specific area to determine particle size will underestimate the particle size if the particles possess significant surface roughness. Surface roughness is likely to increase the viscosity as well, as the surface area is indeed higher and there is additional topography for the matrix to flow past, but the effects are likely to be small in most cases.

So, why bother using fine particles, if we have to deal with the twin problems of enhanced aggregation and increased viscosity? Increasing the particle size is an obvious means of avoiding such issues and reducing the amount of processing needed to produce the desired dispersion state. In fact, this is exactly what is done in many cases, and the use of larger particles does indeed save us substantial trouble (though surface roughness is generally much harder to control). Well before the age of nanotechnology, however, one of the basic arguments in favor of looking at such fine particles as reinforcing agents had already become clear, as can be seen in the excerpt from Fordham's 1961 text on silicones: "The factor common to all reinforcing fillers is high specific surface area, though whether this is the only — or even the principal — requirement has not yet been demonstrated with certainty."[25] Many years later, in the age of nanocomposites, there is still uncertainty, but many new and interesting forms of behavior have been noted in such systems,[26–31] making their disappearance (and therefore the need to understand the appropriate structure-processing-properties relations) even more important today.

30.3.2 Dispersion: Effects of Shape

First, particles whose shapes are conducive to very efficient packing prior to dispersion tend to be more difficult to disperse. Sheets will be much harder to shear apart than a collection of spherical particles of the same composition and volume, for instance, because the contact area will be much larger between parallel sheets than between spheres, while a bundle of parallel rods represents the intermediate case. Contact distances are necessary because the interaction forces expected between particles are short-range in nature, much weaker than covalent or ionic bonds to begin with, and inversely proportional to distance, regardless of type.

*Note that the two terms in the final expression represent the specific surface area due to the edges/sides and the specific surface area due to the (circular) faces of the cylinder, respectively; the two should be considered separately in systems where the edges and faces show distinct behaviors.

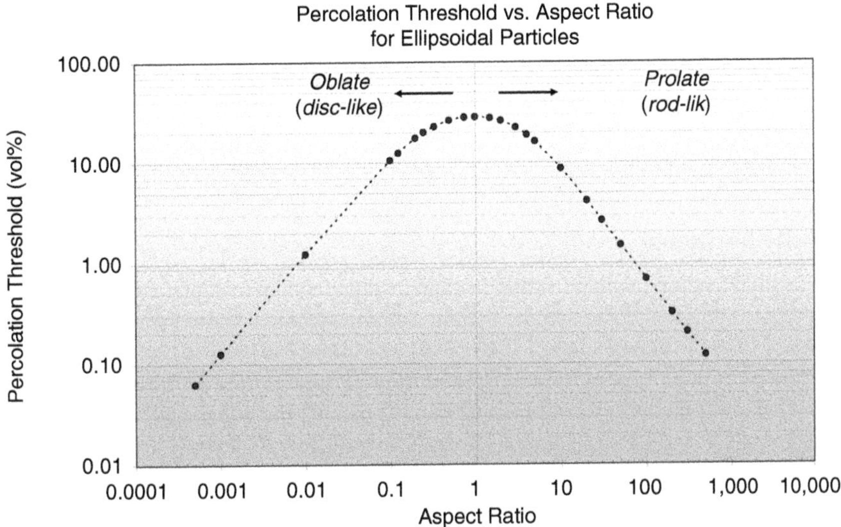

FIGURE 30.3 Percolation threshold vs. aspect ratio for ellipsoidal particles. (Data from Garboczi, E. J., et al., *Phys. Rev. E*, 52, 819, 1995. With permission.)

Additionally, aspect ratio, as measured by the length to diameter ratio of the elements of our dispersed phase, plays an important role as well. Specifically, percolation — the ability to trace a path from one side of the material to the other that passes solely through elements of the dispersed phase (i.e., particles) — becomes increasingly likely with increasing aspect ratio at any given particle concentration. This situation as it applies to filled polymers containing ellipsoidal particles of variable aspect ratios has been well-described, with the data reported for the volume fraction of particles present at percolation vs. the aspect ratio of the particles plotted in Figure 30.3.[32]

Percolation is relevant to dispersion in the same way as shape, in that it affects the degree to which inter-particle contacts are likely to occur given a certain volume of particles. While in contrast to the topological argument, percolation says nothing about packing efficiency — something more relevant when trying to break up an aggregate in the first place — it says something very significant about how readily we will be able to maintain dispersion at a given concentration and with a given aspect ratio. Specifically, above the percolation threshold we run out of room to add isolated particles. At this point, with inter-particle contact unavoidable, the degree of dispersion will decrease regardless of how the system is processed. The particles will no longer behave as isolated entities, and agglomeration will become increasingly likely as the concentration is increased above the percolation threshold. As many properties are detrimentally affected by percolation and agglomeration — with some significant exceptions where percolation is desirable (high electrical or thermal conductivity, for instance) — these arguments imply the existence of an optimal concentration for any given aspect ratio. While there is no sudden change in the viscosity at the percolation threshold, the more significant inter-particle percolation becomes, the greater the number of inter-particle interactions that must be overcome to achieve physical mixing.

Given that shape and aspect ratio can cause such problems, one may again wonder why we don't just stick with spherical particles. At a given volume fraction of particles, however, it is well known that higher aspect ratios generally give rise to more significant mechanical reinforcement, the result of more efficient load transfer. Likewise, sheet-like particles can produce enhanced barrier properties that would be difficult to achieve using other particle shapes.[33–37] Additionally, the tendency of high aspect ratio particles to align in applied fields (stress/shear/flow, electrical, magnetic, etc., dependent on particle characteristics) allows us to produce anisotropic materials, where the property enhancements are all concentrated along a particular axis or axes, rather than being averaged out over all directions. As before, then, it behooves us to understand the relationship between structure, processing, and properties.

30.3.3 Dispersion: Effects of Mobility

The mobility argument relates to the introductory remarks about systems where dispersion is "guaranteed." Specifically, there are certain situations where aggregation can be suppressed through restriction of particle mobility. In general, systems can be classified according to the level of particle mobility — fully restricted, partially restricted, or unrestricted — depending on their characteristics and the means by which they were obtained.

30.3.3.1 Fully Restricted

While it may not reflect the equilibrium structure the system might have liked to take at some point earlier in the processing, in this case the system is stuck in this new state and cannot get out of it by any means short of chemical or physical degradation. Two examples of this type shall be presented.

The first example is that of the polymer-grafted particle. If we consider a normal blend of an arbitrary type of particle with a polymer medium, dispersion is possible, but some agglomeration will probably occur somewhere in the medium at all but the lowest concentrations. Now consider the case where we bond individual polymer chains to the surface of our particles — making "hairy particles," if you will. So long as sufficient grafting occurs and the polymer–particle bonds remain intact, it will be physically impossible for the particles to agglomerate, because the grafted polymer will always be in the way, restricting the mobility of the particles if they get too close. Dispersion is "guaranteed" as a result, even though the breaking of that bond might otherwise allow the system to take a more "preferred" state with respect to particle agglomeration. If the right types of polymeric modifiers are used in the right amounts, such modified particles may behave as polymers themselves, displaying the ability to soften or melt, or, in extreme cases, to act as liquids at room temperature.[38] In such cases, the grafted particles may be used directly, with no further blending if in solid form, or chemically crosslinked to form networks composed solely of grafted particles, if in liquid form.

With that said, it may sometimes be necessary or desirable (to maintain transparency, reduce density or cost, etc.) to blend such modified particles into a polymer rather than using them by themselves. For this discussion we shall neglect the case where the particles are only lightly grafted; this will be treated later, in the section on interfacial energy (see Section 30.3.5). In addition to concentrating on particles with complete surface coverage, we will further limit our example by specifying that the polymer matrix is the same, chemically speaking, as the polymer grafted to our particles. We can then compare our control system A (unmodified particles plus polymer) with the *two* possibilities, B and C, we may observe in the grafted case:

1. We mix unmodified particles with polymer; both dispersion and agglomeration are possible.
2. We mix polymer-modified particles with polymer; the polymer chains at the particle surface are too short to mix with the matrix polymer; we have only increased our effective particle size as a result, and both dispersion and agglomeration are still possible.†
3. We mix polymer-modified particles with polymer; the polymer chains at the particle surface are long enough to mix and entangle with the matrix polymer, and dispersion is guaranteed.

These three situations are illustrated in Figure 30.4.

From the standpoint of a polymer chemist, a small silica particle with many polystyrene chains protruding from the surface looks a lot like a star polymer with many, many arms radiating from a central point and with each arm consisting of a silica "block" (toward the interior) and a polystyrene block (toward the exterior), or a spherical micelle of multiple diblock copolymer chains, with their matrix incompatible blocks inwards and their matrix-compatible blocks outwards. A physical analogy to reports in the literature of the phase behavior of exactly such blends of AB-type diblock copolymers with A-type homopolymer[39] indicates that the polymer grafts will probably have to have molecular weights similar to that of the polymer matrix to realize situation C rather than situation B. Likewise, the temperature dependence of such

†If the polymer grafts are allowed to consist of polymers other than the polymer matrix, case B is expected in all situations except where the two different polymers are actually known to mix (possible but very rare).

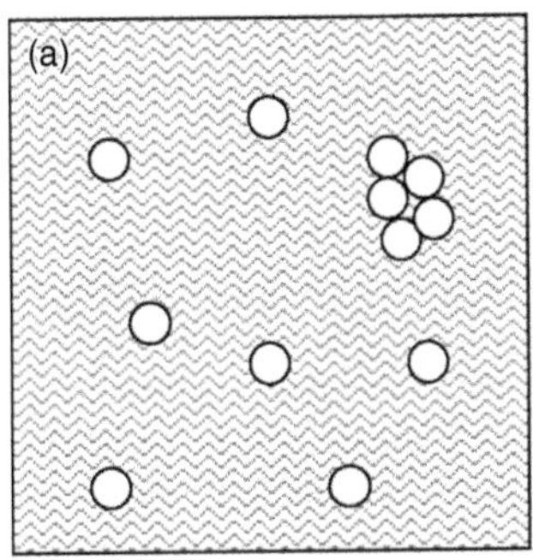

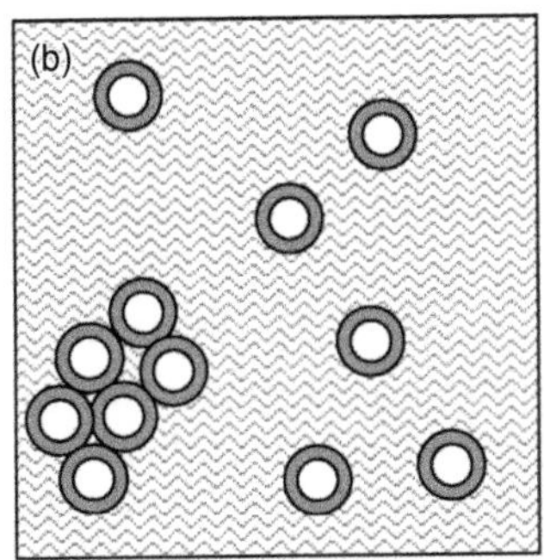

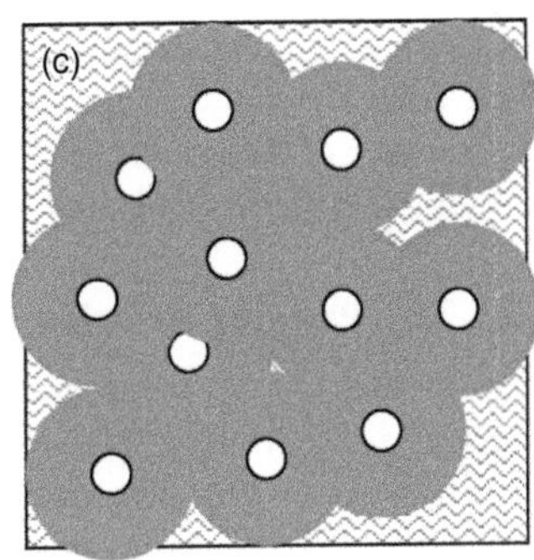

FIGURE 30.4 Three representative schematics of particle dispersion where: (a) the particles are unmodified and agglomeration is possible; (b) the particles are polymer-grafted, but the grafted polymer does not mix with the polymer matrix and agglomeration is possible; (c) the particles are polymer grafted, the grafted polymer *does* mix and entangle with the polymer matrix and dispersion is guaranteed.

phase behavior[40,41] makes it clear that it must be evaluated not only at room temperature but under all relevant process conditions as well.

We can also draw some conclusions about these systems from the standpoint of processing. Agglomeration to form larger particles (reducing viscosity) can occur in both cases A and B, and the opposing factors of greater particle size (lowering viscosity) and greater particle concentration by volume (increasing viscosity) in case B are likely to result in viscosities very similar to those observed in case A. Case C, however, is a different story; here, the grafted polymer chains are extended, mixed, and entangled with the matrix polymer, causing substantially more drag than in the noninteracting case. This combined with the higher dispersion levels expected in case C is likely to give rise to a higher viscosity than would be observed in either of the other cases.

The second example of the full restricted case is the polymer-entrapped particle. In this case, we disperse the particles to a very high degree in a liquid medium. This can be achieved by vigorous mechanical mixing or ultrasonication, the latter being one of the most effective means to disperse a variety of mineral systems[42–46] without damaging them,[47] though it can lead to degradation of both polymers[48–50] and other molecules potentially useful for particle surface modification.[51–55] Alternatively, such high levels of dispersion can be achieved via the polymerization and cross-linking process itself, in special instances where polymer formation and cross-linking between the particles is faster than elsewhere (usually caused by catalytic activity at their surface) and they are physically pushed apart as a result.[56–59]

Our particles will generally be inclined to settle and aggregate over time if we allow them. If we cross-link our medium much more rapidly than the particles are able to find each other and re-aggregate, however, we effectively trap them in the highly dispersed state. We can even attempt to trap them *during* the dispersion process (by cross-linking while ultrasonicating or physically mixing, for example), but this is not strictly necessary to achieve high levels of dispersion. Alternatively, *in situ* formation of our particles from molecular precursors may be carried out during network formation, though the end results are equivalent for the purposes of this example. Because they are trapped in a thermosetting network, one which cannot be melted or plastically deformed, the particles will never be allowed to proceed toward aggregation unless the polymer network is physically or chemically destroyed.

To be more explicit, this method will be effective so long as the distance between cross-links, or junction points in our network, is significantly less, even when the material is mechanically deformed, than the size of the particles in question. To put it another way, if the network is our fishing net and the particles are our fish, the holes in our net must not let the fish through, regardless of how the net is stretched. So long as this condition is met, dispersion is again "guaranteed," at least to the level that existed at the point where network formation occurred. The distance between cross-links is readily calculated, in turn, by knowledge of the structure and molecular weight of the polymer being crosslinked. Once the length of a single repeat in its extended conformation is estimated, it is trivial to calculate the number of repeats in the average chain and estimate the distance between cross-links. It should be noted here, however, that the trapping of these particles does not guarantee a strong interface between the particle and the matrix, meaning that properties enhancements are not necessarily guaranteed as a result.

30.3.3.2 Partially Restricted

The best example of *partially restricted* mobility is a situation analogous to the latter example given for the stable case, but involving a linear polymer rather than a network polymer / thermoset. Again, we disperse the particles to a very high degree in a liquid medium. In this case, however, our medium is a viscous polymer melt or solution — or, in special cases, a medium in which linear polymerization is occurring and forcing dispersion as a result.[60] The viscosities involved will suppress settling from occurring, in contrast to the previous case. In the end, then, the only way we can produce a solid polymer with particles dispersed at a greater level than they will tolerate, thermodynamically speaking, is by freezing the system *during* the dispersion process. Such situations are hard to realize, practically speaking, but have been demonstrated via the polymerization of solid polymer from gaseous monomer in the areas between the individual primary particles, so as to physically push them apart.[60] Alternatively, molecular precursors may be infiltrated into and reacted within a solid polymer matrix so as to again produce our particles *in situ*, though the end results are equivalent for the purposes of this example.

In the *fully restricted* situation, with the polymer particles trapped in the thermosetting resin, the particles might like to agglomerate but will never have the mobility to do so until the polymer surrounding them is destroyed. Now that we are dealing with a solid but *thermoplastic* matrix, however, with particles trapped in a similar manner, any additional melt or solution processing of the polymer will place the particles in a liquid medium once more, allowing them to find each other and self-associate to a thermodynamically stable level with only a minimal input of energy; hence, particle mobility is only *partially restricted*. This can lead to reduced dispersion levels and degraded properties unless high intensity processing is continued until the final step of the processing sequence. Likewise, even if the dispersion levels can be maintained in this fashion, if the interfacial strength is weak due to a lack of thermodynamic compatibility, the properties may be unaffected or may even degrade as a result.[60]

30.3.3.3 Unrestricted

This case describes the vast majority of filled polymers. Here, the particles are mixed with the polymer matrix as effectively as necessary for the application, with no attempt made to "trap" a greater degree of dispersion into the system than the particles will exhibit in the absence of processes designed to cause dispersion. That is, the particles are allowed to disperse freely according to the usual combination of processing and thermodynamics.

As we have discussed dispersion in a generally positive light up to now, the obvious question is why we do not always pursue systems where mobility is restricted and dispersion is guaranteed as a result. While it is possible to apply this technique in some systems (i.e., when network formation in the presence of the particles of choice is sufficient to realize the desired results), the difficulties in preparing particles grafted with high molecular weight polymer, combined with the properties enhancements achievable without going through so much trouble, often serve as substantial barriers to further application of this sort of approach.

30.3.4 Dispersion: Effects of Surface Energy

The surface energy (γ) of a solid surface is a very simple concept used to describe the energy necessary to create that surface in a material (i.e., by breaking bonds). Depending on the nature of the bonds that must be broken to create such a free surface, the surface energy (generally measured in units of energy per area) will be proportionally higher or lower. The work of adhesion, W_A, can also be defined in terms of surface energies. Specifically, for two identical surfaces of the same material with the same surface energy, the work of adhesion is defined as follows (Equation 30.4):

$$W_A = -2\gamma \tag{30.4}$$

It should be noted here that some texts prefer to refer to this as the work of cohesion, with work of adhesion reserved for contact between two *different* surfaces. In any case, the above result makes sense intuitively; as two surfaces are disappearing in this process, the energy used to create them is returned. The surface energy of a solid has been described using a variety of models, which most generally attempt to define it using various combinations of dispersive and polar components; none are judged to be accurate under all conditions.[61] With that said, the Fowkes approach,[62] most generally referred to as the Girifalco, Good, Fowkes, and Young combining rule, has been applied to polymer composites[63] and allows us to relate the work of adhesion to physical processes in a very simple manner by breaking things down into independent dispersive and polar components (Equation 30.5):

$$W_A = W_A^{LW} + W_A^{ab} \tag{30.5}$$

where W^{LW} refers to London / van der Waals (dispersive) interactions, and W^{ab} refers to so-called "acid–base" (polar) interactions. The aforementioned London / van der Waals dispersion forces are the weakest and are present between all materials (stronger in more polarizable species, weaker in less polarizable species), and drop off as $(\text{distance})^{-6}$, meaning that they are only active at very short range. The "acid–base" (polar) component, on the other hand, is a catch-all for the remaining intermolecular interactions involving interactions of (di)polar species (ion-dipole, ion-induced dipole, dipole-dipole, dipole-induced dipole, hydrogen bonding, the behavior of electron donors and acceptors, etc.), whose strengths span several orders of magnitude and drop off as $(\text{distance})^{-2}$ to $(\text{distance})^{-6}$, depending on their exact nature. As might be expected, the theoretical treatment of this term is less well established.

With that said, particles with high surface energies due to the presence of broken high-energy bonds or polar groups, for instance, will tend, in the absence of a strongly interacting medium, to aggregate much more strongly than those with few dangling bonds or polar groups. As such, aggregates of high surface energy particles will be more likely to form and harder to break up unless favorable interactions exist between the particles and the matrix. This is not only the expected result, but something that has been clearly demonstrated in the area of pharmaceutics,[64] where it is crucial to understand issues of particle dispersion to allow for ease of manufacture of many drugs.

In short, then, it appears that we are better off avoiding high surface energy fillers if we can avoid it. However, many fairly high surface energy materials (metals, oxide ceramics, etc.) have very useful properties as fillers; what is to be done? It is here that surface modifications come into play. Chemical modifiers can be directly bonded to the particle surface, as in the case of "hydrophobized", hexamethyldisilazane (HMDS) treated fumed silica, where methyl groups are covalently bound to the silica nanoparticles. The resulting drop in surface energy—thanks to the reduction in the polarity and polarizability of the surface achieved by replacing polar Si-OH and Si-O-Si groups with much less polar CH_3 groups—gives rise to a reduction in the surface energy, reducing the energy required to separate aggregates while otherwise maintaining the physical properties of the particles. Alternatively, nonbonding modifiers may be used instead; surface-active agents (surfactants), for instance, are of definite value here. Such molecules contain polar head-groups that interact strongly with the particle surface and nonpolar tails that cover the particle. The effect is the same as direct grafting, in that the exposed surfactant tails reduce the surface energy of the particle vs. its unmodified surface, and allow for easier separation of particle aggregates. In general, then, the ability to decorate the surface of any particle with nonpolar, low surface energy functionalities such as saturated hydrocarbons, fluorocarbons, silicones, or fluorosilicones (preferably lacking any unsaturated hydrocarbon substituents) is a recognized means of reducing the surface energy, and such methods, when practical, are frequently put to use to improve filler dispersion by allowing for easier de-aggregation. Fluorocarbon groups in particular are considered to produce the lowest surface energies possible, akin to the effects seen in Teflon (polytetrafluoroethylene, PTFE).

With that said, care must be taken. While enhanced dispersion will lead to increases in viscosity, a more subtle processing issue associated with such organic surface modification techniques is that they only function so long as the organic modifiers remain intact. If any step of the processing should degrade or destroy these modifiers, or cause them to detach from the particle surface and migrate elsewhere in the system, the advantages with respect to improved dispersion level will be lost, and the delocalization or degradation of modifier molecules may lead to plasticization, unintended reactions, coloring, outgassing, void formation, even corrosion of the processing equipment or occupational safety hazards, depending on the nature of the degradation products. On the other hand, there are specific instances where some amount of modifier degradation can actually *improve* dispersion[65] — though such situations are the exception rather than the rule.

The most common issue is that of thermal stability; if the modifiers used begin to thermally degrade under process conditions, it now becomes necessary to minimize processing times or even alter the process entirely to preserve the modifiers and ensure their effectiveness in improving dispersion. Modification strategies based on the direct covalent bonding of low surface energy groups to the particle surface tend to be fairly stable, thanks to the reasonable thermal stability of hydrocarbon, fluorocarbon, and silicone groups. The use of surfactant modifiers that interact with the particle surface via polar interactions, however, requires the presence of a polar head-group, many of which lack the thermal stability of the low surface energy functionalities to which they are attached. With that said, degradation due to other factors (shear, chemical reactions, etc.) is also possible, and should not be neglected. While the combination of shear and temperature has been shown to produce more degradation than temperature alone in nanoparticle-filled systems,[65] other factors are harder to predict, and must therefore be considered on a case by case basis.

30.3.5 Dispersion: Effects of Interfacial Energy

The elements of this discussion are quite similar to those of the previous discussion, with one important difference: Above, we are referring to the surface energy to create a free surface, which relates to the difficulty we will have in breaking up aggregates *in a noninteracting medium.* Clearly, however, it always costs energy to break up aggregates, and the surface energy is never zero because all materials experience dispersive interactions to some extent. The next step in improving dispersion, then, if we cannot make the particles dislike one another, is to address how much they like the matrix — that is, how well the matrix wets the particles. If interactions between the particle surface and the matrix are attractive in nature, such that some of the energy lost in separating the particles in an aggregate is gained in creating a new particle–matrix interface, dispersion will be enhanced as a result. In contrast to our previous treatment of work of adhesion, where we considered two identical surfaces coming together (with no interface formed after the fact), and related that to the difficulty in separating aggregated particles, here we must treat the case where we have two *different* surfaces coming together and forming an interface with its own interfacial energy. In this case, then, the work of adhesion is given as follows:

$$W_A = \gamma_{\text{Interface}} - \gamma_{\text{Matrix}} - \gamma_{\text{Particle}} \tag{30.6}$$

In other words, the work of adhesion is given by the energy required to create the interface, minus the energy returned to the system due to the fact that the free surfaces of the matrix and the particle have been consumed during the creation of the interface. We can show how these parameters relate to one another by considering the balance of forces when a droplet of the matrix material, in fluid form,[‡] comes into contact with the surface of an element of the dispersed phase, that is, one of our filler particles (Figure 30.5).

[‡]Note that we refer to the surface *energies* of solids, which can be defined but not measured (directly) and must be calculated based on one of a number of theoretical treatments, to distinguish them from the surface *tensions* of liquids, which can be both defined and measured directly.

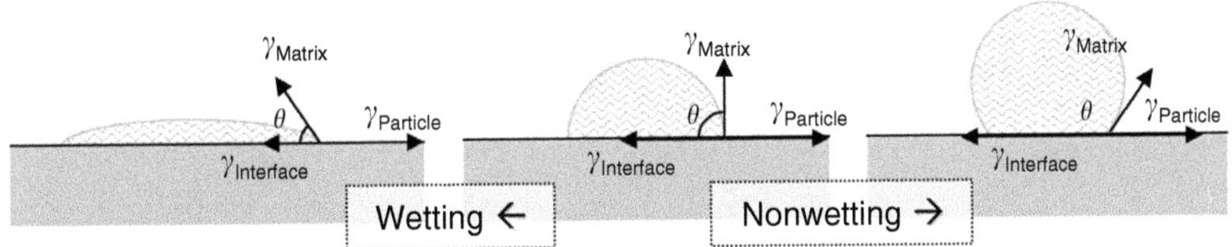

FIGURE 30.5 Schematic of the effects of the interfacial energy, $\gamma_{\text{Interface}}$, on the ability of a droplet of the continuous phase (polymer matrix) to wet the surface of an element of the dispersed phase (filler particle). Lengths of arrows correspond to magnitudes of surface energies/tensions involved, angles chosen satisfy Young's equation, and values of γ_{Matrix} and γ_{Particle} are the same in each case.

Young's equation (Equation 30.7) describes the force balance implied in Figure 30.5 in mathematical terms:

$$\gamma_{\text{Interface}} = \gamma_{\text{Particle}} + \gamma_{\text{Matrix}} \cos\theta \tag{30.7}$$

where the angle θ is the wetting angle (often referred to as the contact angle). As with all simple physical models, this force balance approach has its limits; here, we assume smooth surfaces, two phases and no impurities present in the system, and equilibrium conditions (i.e., no viscosity effects to slow or prevent wetting for instance). Still, it can point us in the right direction. More details about Young's equation may be found in Chapter 21. Note also that each of the surface energy terms mentioned here can be separated into the usual dispersive and polar components, so the same arguments mentioned in the section on the effects of surface energy apply.

With respect to $\gamma_{\text{Interface}}$ specifically, it should be clear that the surface energy will be low only if there are favorable interactions between the polymer and the matrix. Ironically, however, the same surface modifications likely to induce strongly favorable interactions between the particle and the matrix are likely to increase the surface energies of the individual phases as well. For instance, the introduction of polar groups at the particle surface to improve the interactions of the particles with a polar matrix (thus increasing $\gamma_{\text{Interface}}$) will also strengthen the interactions of the particles with one another (thus increasing γ_{Particle}), making particle aggregates more difficult to break up during initial processing and may not automatically give improved wetting as a result. Alternatively, the choice of very low surface energy particles will decrease our difficulties in de-aggregating them (via a decrease in γ_{Particle}), but may also decrease the attractive interactions between the particles and the matrix (via a decrease in $\gamma_{\text{Interface}}$), something that has significant consequences in terms of mechanical behavior due to the lack of load transfer. On the other hand, a weak interface can actually improve some mechanical properties as well, like toughness, because cracks will then propagate around the filler particles, rather than through them, and waste energy in doing so.

The above discussion should not be interpreted as indicating that this is a zero-sum game, however, and that any advantage we gain in decrease surface energy we lose in decreasing interfacial energy (or vice-versa). It is possible to optimize both factors to get the most out of the system, minimizing the difficulties associated with de-aggregation while maximizing the matrix–particle interactions to the extent we desire by again considering modification techniques akin to those described in the section on surface energy. In this case, however, we would choose modifiers that do not strongly interact with one another, but that *do* strongly interact with the matrix — giving rise to a sort of recognition at the molecular level. An extreme example of such a situation would be the introduction of a surface charge on all of our particles in opposition to the charges on groups present in the matrix. The particles will then repel each other, but will be strongly attracted to the matrix as a result. A similar technique might involve putting groups capable only of accepting hydrogen bonds on the surfaces of our particles, resulting in relatively weak inter-particle interactions, and relying on hydrogen bond donors in the matrix to give rise to strong interactions with the particles.

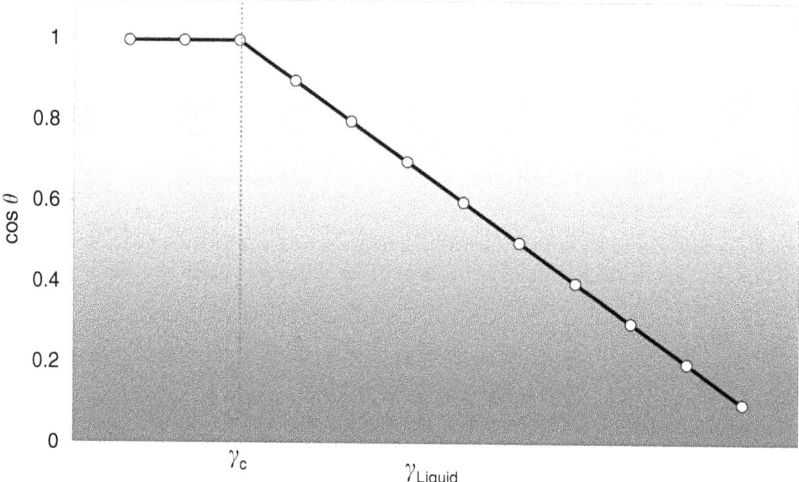

FIGURE 30.6 Cosine of the contact angle vs. surface tension of the liquid when a series of liquids with arbitrary surface tensions are placed on a surface and contact angles are recorded, according to the Zisman model (Zisman, *Adv. Chem. Ser.*, 43, 1, 1964.). Such plots may be used to determine the critical surface tension (γ_c) of a solid surface, that is, the maximum surface tension a liquid can have and still be able to fully wet the surface.

In addition to the above examples, an additional tool that can serve as a means for making intelligent choices about system composition so as to minimize the difficulties experienced during processing and maximize the chances of producing the desired structure and properties is the critical surface tension (γ_c) of a solid as first defined by Zisman.[66] This value is defined as the maximum surface tension a liquid can have and still be able to fully wet ($\theta = 0$) a solid surface. We can define this quantity using Young's equation, setting $\theta = 0$, and solving for $\gamma_c = \gamma_{Matrix}$ (Equation 30.8), as the matrix in this case is our wetting liquid in this case:

$$\gamma_c = \gamma_{Interface} - \gamma_{Particle} \tag{30.8}$$

Alternatively, if we expose our surface to a series of wetting liquids, each with a known surface tension, and monitor the contact angle each liquid makes with our solid surface, we can determine the value of the critical surface tension (γ_c) of that solid using the Zisman's relation (Equation 30.9):

$$\cos\theta = 1 + b(\gamma_c - \gamma_{Liquid}) \tag{30.9}$$

where b is a system-dependent constant and the equation is valid for values of θ between 0° and 90°. If we plot $\cos\theta$ vs. γ_{Liquid}, in other words, we expect to see the following result (Figure 30.6).

The utility of the critical surface tension is clear; our continuous phase (the matrix polymer) must have a surface tension less than the critical solid surface tension of the elements making up the dispersed phase (the filler particles) for strongly favorable interfacial interactions to exist. As the critical surface tensions of many bulk materials are either already known or readily determined by constructing a plot of the type shown in Figure 30.6, we can easily make rough, qualitative predictions of whether or not the interfacial energy of a given system will favor easy processing and the desired structure and properties. Likewise, because both the Zisman model for critical surface tension[66] and the Hilderbrand-Scatchard model for molecular solubility[67] assume that dispersive interactions are the most significant (a reasonable assumption for many polymers), there is a roughly linear relationship between the critical surface tension of a solid (γ_c) and the Hildebrand solubility parameter (δ). It must be emphasized, however, that as polar interactions become more significant (as would be expected in metals, ceramics, highly polar polymers, etc.), these models begin to break down.

In summary, while it is difficult to identify modification strategies that modify only the interfacial energy and not the surface energy (or vice-versa), it *is* possible, using theoretical considerations as guide where appropriate, to change one of the two more than the other so as to achieve the desired results. As with the modifications described in the section on surface energy, however, the stability and localization of any modifiers used to achieve changes in interfacial energy must be maintained or their effects may be lost and a host of additional problems may develop in addition to the loss of the desired structure / dispersion level — though again, in some exceptional cases modifier degradation may actually *improve* dispersion levels.[65] Likewise, if the change in interfacial energy ($\gamma_{\text{Interface}}$) due to the modifications is insufficient to make up for the tendency of the modified particles to aggregate, as quantified by the surface energy (γ_{Surface}) of the modified particles, the results will resemble those shown schematically in Figure 30.4b, where filler aggregation is still possible and perhaps even more favorable than in the unmodified case (Figure 30.4a). If the modifications succeed in substantially reducing the interfacial energy of the system, on the other hand, the effect is like that of soap on dirt in water — the desired level of dispersion (or, alternatively, emulsification, depending on the state of the phases) will be accessible with less processing and more control than would otherwise be the case, and the desired properties will be more readily achieved as a result.

30.4 Summary

These discussions show that there are a great many factors that combine with any processing method to influence the final structure, and consequently, properties, of any filled polymer or composite system. An attempt has been made to break these factors down into manageable categories, based on geometric (surface area, shape), physical (mobility), and physicochemical (surface and interfacial energy) concerns. As any approach to describing such a complex phenomenon in a limited space will be by definition incomplete, the focus here has been the identification of key parameters affecting processing, structure, and properties. This has served as a prelude to discussions of how to influence those parameters to achieve the desired level of dispersion (generally assumed to be high), and thus structure and properties, with the minimum energy input during processing. A summary of general relations and guidelines that have emerged as a result of these discussions is given below (Table 30.2).

There are caveats associated with every statement in Table 30.2, and there is no substitute for thorough consideration of both the system(s) of interest or for more detailed descriptions of the issues described here. With that said, it is hoped that this will serve as an effective introduction to the issues affecting dispersion in filled polymer systems.

Part 2 Multiscale Processing of Polymers — Fibers, Films, and Molded Structures

30.5 Overview

As with the introduction of any new technology, the implementation into commercial products of nanoscale elements and features will likely happen in stages. The first generation of products are fabricated by modification of existing manufacturing processes. For example, the addition of small quantities of nanoparticles and nanoclays in the manufacturing of coatings and polymer structures, respectively, has already led to products as diverse as sunscreen and automotive components. This chapter illustrates some of the relatively near-term challenges and potential approaches to controlling the structure and functionality at the nanoscale of micro- and macroscale polymer-based products. Examples include electrospinning of nanofibers, extrusion of multilayer films, and molding of nanofeatures in 3D structures.

TABLE 30.2 Practical Relations Between Factors Affecting Dispersion Levels and the Processing, Structure and Properties of Filled Polymer Systems

Factor	Relations
(a) Surface area (size / roughness)	• Specific surface area is proportional to particle roughness and inversely proportional to particle size. • Viscosity and is proportional to specific surface area. • Increased *SSA* improves many properties if dispersion can be realized, but makes dispersion much more difficult.
(b) Shape (topology / aspect ratio)	• Shapes capable of more efficient packing/inter-particle contact can make dispersion more difficult. • More extreme aspect ratios can improve many properties, but make inter-particle contact more likely and reduce the percolation threshold.
(c) Mobility	• Trapping particles in a dispersed state can be effective as a means of realizing some properties enhancements, though other properties will depend on the interfacial strength.
(d) Surface energy	• Surface energies should be minimized to encourage de-aggregation (though this may reduce the interfacial energy as a result, degrading some properties). • Modifier characteristics and stability must always be considered when processing modified fillers.
(e) Interfacial energy	• Interfacial energies should be tuned to maximize particle–matrix interactions (though this may increase the surface energy as a result, degrading dispersion levels). • Modifier characteristics and stability must always be considered when processing modified fillers.

30.6 Multiscale Processing of Nanofiber Assemblies: Control of Electrospun Fiber Alignment, Patterning, and Morphology

Commercial scale processes for manufacturing synthetic fibers, such as melt spinning and solvent (wet) spinning, have existed for over half a century. After the development of rayon and acetate in the early half of the 1900s, nylon and polyester fibers were developed in the mid-1900s by E.I. du Pont de Nemours and Company. These commercial processes are limited in their ability to fabricate fibers with submicron diameters, due to the increasing pressure with decreasing spinneret diameter. Electrospinning is one method that has been shown to produce fibers with nanoscale diameters for a wide variety of polymers.

Described as early as 1934 in a patent by Formhals,[68] electrospinning is a process where a high electric field is used to initiate and propel a fine jet of polymer solution from a source (e.g., a pipette or syringe) to a grounded or oppositely charged target. The jet initiation occurs when the electric field overcomes the surface tension in a Taylor cone formed at the tip of the pipette. Instability leads to a spiral whipping motion that further reduces the diameter of the jet. As the jet travels through the air, the solvent evaporates, increasing the solids content leading to fiber formation. At the extremes, high viscosity solutions require too high of an electric field to initiate spinning, while low viscosity solutions tend to break up into droplets. Figure 30.7a shows a schematic of a typical electrospinning setup with a point source on the left and a larger disk electrode at the source on the right. The process is driven by the electric fields, thus a larger source electrode approaches a parallel plate field, resulting in a smaller spread of the fiber deposition. This result is described in more detail by Bunyan et al.[69]

Under the basic configuration shown in Figure 30.7a, the electrospinning process creates a nonwoven mat of fibers as shown in Figure 30.7b, with fiber diameters ranging from hundreds of nanometers to tens of nanometers. Different process parameters values for electric field (applied voltage for a given source to target distance), polymer concentration, and solution conductivity can result in beaded fibers (fibers with sections of larger diameter)[70] and different fiber diameters. Significant studies of process conditions have been conducted for many polymers, including polyamides, polyacrylonitrile, polyethylene, polyethylene terephthalate, polypropylene, polybenzimidazole, and polycaprolactone. More details and references on process parameter effects can be found in Reference 71.

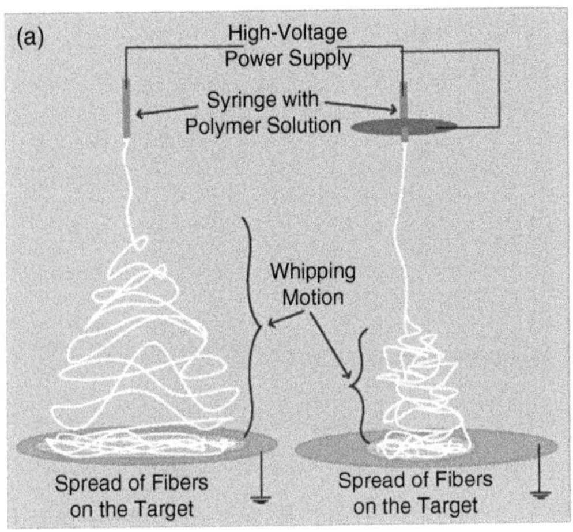

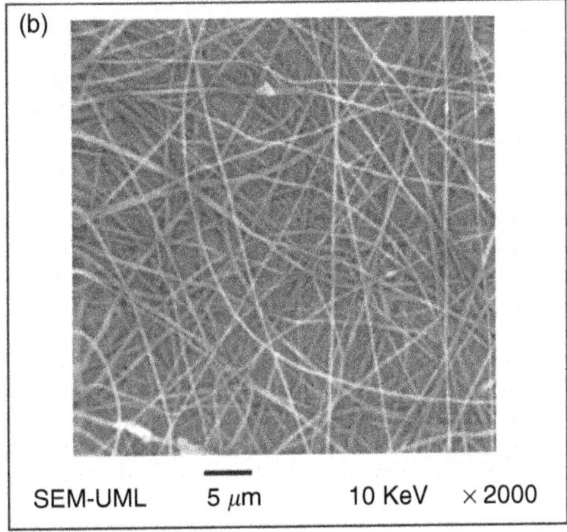

FIGURE 30.7 (a) Typical electrospinning setup and (b) SEM of electrospun nonwoven fiber mat.

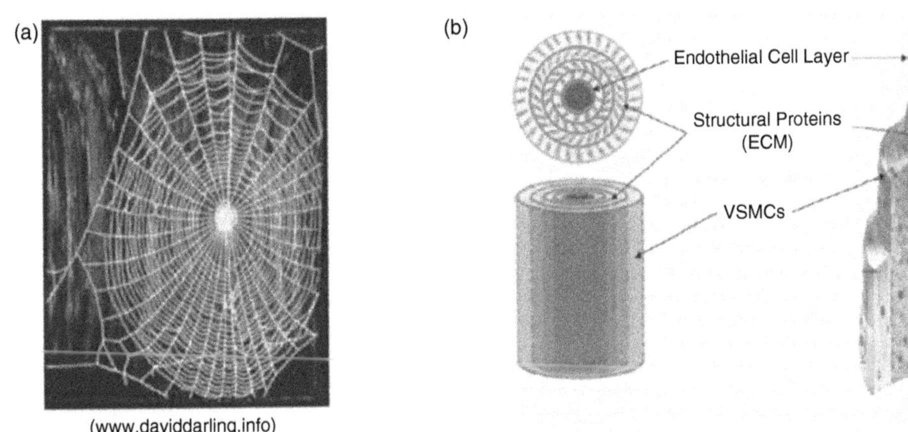

FIGURE 30.8 Examples of patterned and oriented fibers in nature (a) spider web; (b) collagen fibers in a blood vessel.

Multiscale processing of electrospun fibers ranges from self-assembly of core-sheath nanofibers, to dispersion of nanofillers in submicron diameter fibers, to alignment and patterning of the nanofibers. This section will discuss two main processing approaches, followed by a brief summary of some of the modeling approaches: (1) Alignment and Patterning of Nanofiber Assemblies; (2) Self-Assembly of Core-Sheath Fibers; and (3) Modeling of the Electrospinning Process.

30.6.1 Alignment and Patterning Using External Electric Fields

While a random mat of fibers can serve as a useful structure for applications such as filtration, there are many more applications that can be developed if control of orientation can be achieved. A multitude of examples exist in nature — for example, the combined radial and circumferential fibers of a spider web and the counter-rotating layers of helical collagen fibers in a blood vessel (Figure 30.8).

One approach to achieving simple alignment of the fibers is to spin onto a target that is moving horizontally at the same speed as the deposition of the fibers onto the target. Theron et al.[72] demonstrated spinning onto a rotating disk with a sharp metal edge. Figure 30.9 shows a schematic of such a setup, along with a scanning electron microscope (SEM) image of fibers collected at 500 rpm on a 10 cm diameter disk (translational speed of ~2.6 m/s).[73] Note that these fibers are still fairly large and still show some degree

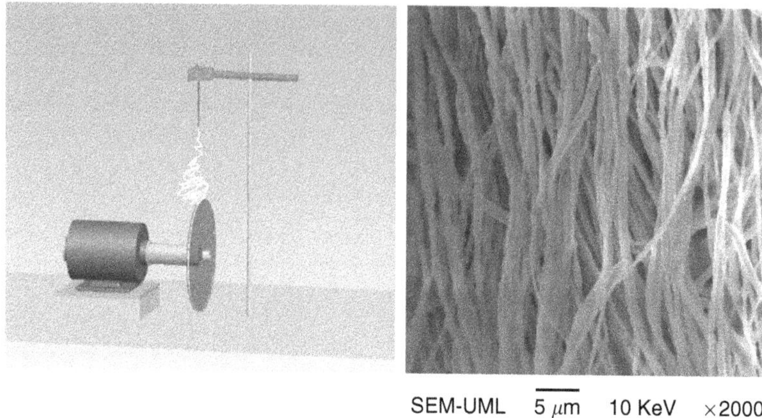

FIGURE 30.9 Electrospinning onto a rotating disc results in general alignment of the fibers. (After Bunyan, N., Control of Deposition and Orientation of Electrospun Fibers, MS thesis, University of Massachusetts Lowell, Department of Mechanical Engineering, 2003. Courtesy of University of Massachusetts, Lowell.)

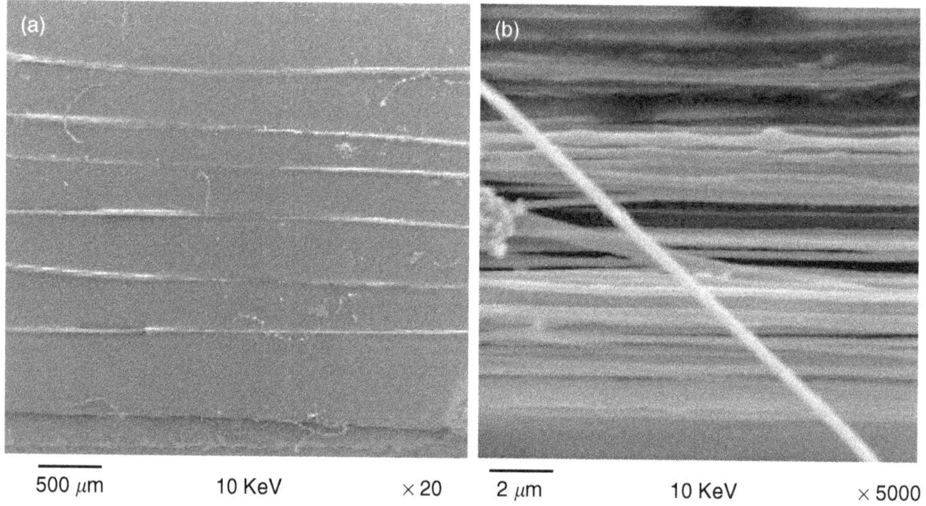

FIGURE 30.10 SEM images of (a) 6 steel wires (35 μm diameter) and (b) aligned electrospun fibers on one wire. (After Farboodmanesh, S., Mooskian, J., Lee, J., and Chen, J., University of Massachusetts Lowell Nanomanufacturing Center of Excellence 2006. Courtesy of University of Massachusetts Lowell.)

of misalignment. A second issue is the ability to collect such fibers in a continuous manner. Spinning onto the edge of a disk can result in a bundle of fibers or tow of length equal to the circumference of the disk. While the fiber is surmised to be continuously wound onto the disk, it is very difficult to unwind the fiber from the disk in one continuous length.

Other methods of achieving aligned electrospun fibers have been demonstrated by Dzenis[74] and Li, et al.,[75] using an insulating gap. This method creates very well aligned fibers across a short distance, effective for circuit-type "wiring," but not as effective for large distances and areas.

To address higher area applications, the edge of the disk was covered with six 35-μm stainless steel wires (see Figure 30.10a). The wires resulted in a focused electric field, drawing the electrospun fibers to its surface. The use of the small diameter wires led to highly aligned electrospun fibers (see Figure 30.10b), control of spacing of fiber deposition on a micro/macro scale via the wire spacing, and the ability to collect onto a continuous substrate.

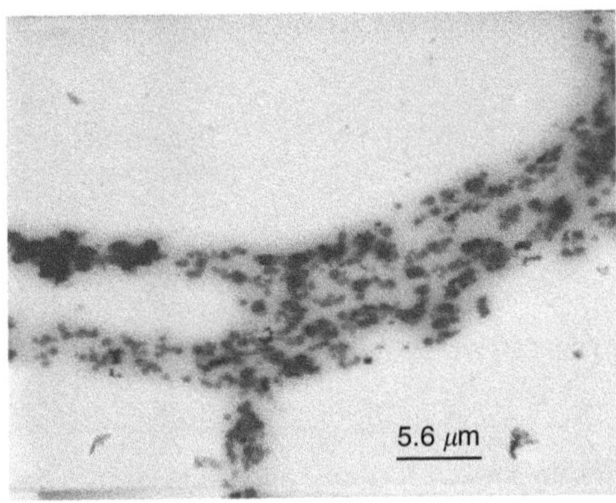

FIGURE 30.11 TEM image of uncured butyl rubber membranes at a 40 phr loading (N990). (After Viriyabanthorn, N., Stacer, R.G., and Sung, C., and Mead, J.L., in *Polymeric Nanofibers*, ed. D.H. Reneker and H. Fong, ACS Symposium Series 918, American Chemical Society, New York, 2006, Chapter 19. With permission.)

30.6.2 Internal Patterning (Core-Sheath Fibers)

The electrospinning method can also be used to prepare materials with nanoscale morphologies or nanoparticulate fillers. In the case of nanoparticulate fillers, the material behavior will be enhanced through the incorporation of the filler and the electrospinning process can be effective in dispersing the fillers. In the latter case, the earlier discussion on dispersion can be used to guide material development. Figure 30.11 shows an example of a TEM image of a carbon black filled electrospun fiber.[76] As seen in this figure, the mechanical compounding procedure used to incorporate the filler before electrospinning, seems to provide uniform distribution of carbon black throughout the fiber.

The effect of the filler on the electrospinning process and fiber structure, however, must be considered. For example, the effect of carbon black particle size, loading, and structure was found to have a significant effect on the morphology and mechanical properties of electrospun butyl rubber nonwoven mats.[76,77] Fiber diameter decreased with: increasing carbon black content, decreasing particle size, and increasing carbon black structure. The addition of carbon black filler was also found to decrease the number of beads. Density of the electrospun mats decreased with increasing carbon black content, smaller particle size, and higher structure. After correcting for the differences in density, decreasing carbon black particle size and increasing structure showed an increase in stress at break, ultimate elongation, and modulus of the electrospun membranes. Thus, although the base material properties were increased, the change in fiber structure and mat morphology may be a more important influence on the final properties of a structure.

Most electrospinning experiments are performed using constant viscosity solutions, however, nanoscale fillers will typically change the viscosity of the starting materials; thus, a constant viscosity solution will have a lower concentration of primary fiber material. In addition, the conductivity of the filler (in this case conductive carbon black) may also be a factor.

The preparation of novel fibers from mixtures of two or more polymers is an attractive method to prepare new and novel materials. Blends of two polymers and block copolymers have been extensively investigated.[78–85] Much of the work, however, has focused on the overall material properties, rather than a study of the internal structure of the nanofiber. Research on triblock copolymers of styrene-butadiene-styrene has revealed that the domain structure is elongated along the axis of the fiber,[86] while selective removal of one of the components from an electrospun nanofiber of a blend of polylactide and polyvinylpyrrolidone exhibited co-continuous phase morphology.[87] More recently, attention has turned to the development of controlled internal structures, such as depicted in Figure 30.12.

These types of structures could find application in nanowires prepared from a conducting polymer and an insulating polymer, or in fluidic channels from hollow fibers. There are a number of

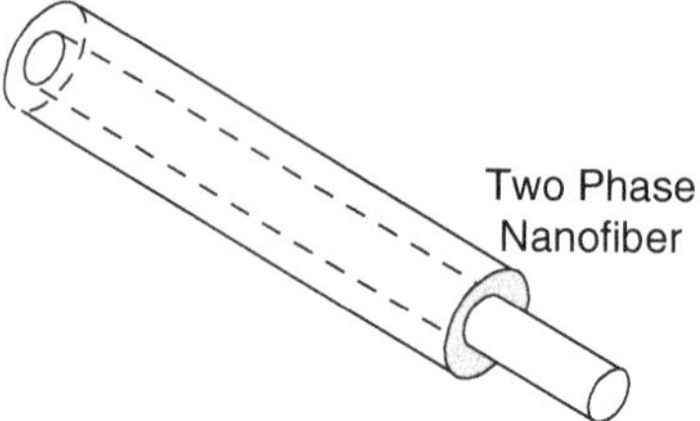

FIGURE 30.12 Schematic of unique core-sheath nanofiber structures.

potential approaches to fabricate these structures, including deposition of a second polymer onto an existing nanofiber, including *in situ* polymerization of polyaniline onto electrospun polymethyl methacrylate,[88] or chemical vapor deposition of poly (p-xylylene) (PPX) onto electrospun nanofibers of PLA.[89] Other approaches include co-electrospinning two polymer solutions with two coaxial spinnerets.[90] PEO/PDT, PLA/Pd, and PEO/PSU core-sheath structures have been electrospun using this approach.[90–92]

A fairly simple method to obtain these structures would be to electrospin directly from a solution of both polymers. In this case, the nanofiber morphology will depend on both thermodynamic and kinetic factors.[93] In the electrospinning process, the rapid solvent evaporation rates (milliseconds[94]) result in kinetic factors playing a much larger role in the structure formation. In addition to solubility parameters, viscosity and molecular weight will be important factors in structure development. Solution viscosity is well known to play a role in electrospun fiber geometry, however, the bulk viscosity of the two materials will become a dominant factor as the solvent evaporates.

In the case of electrospinning of polymer blends, the composition ratio of the two materials has been found to be a major factor in the morphology of the resulting nanofibers. In the case of polybutadiene (PB) and polycarbonate (PC) blends, it was determined that when the weight ratio of PB/PC blends was larger than 25/75, co-continuous structures with interconnected PB and PC nanolayers or strands were formed, oriented along the fiber axis.[95,96] When the weight ratio of PB/PC blends was smaller than 25/75, the morphology changed from co-continuous to core-sheath structures, with PB located in the interior of the fiber and PC located on the exterior. As the content of higher viscosity PB is increased, the overall solution viscosity will increase, thus providing greater resistance to the ability to phase separate and co-continuous structures are developed. As solvent evaporates and the solution becomes more concentrated, this effect will be magnified. This is further exemplified by changes in molecular weight. With increasing polymer molecular weight, the viscosity of the blend system increases, leading to changes in the morphology of the fiber as seen in Figure 30.13 for 25/75 weight ratio PB/PC blends. As the molecular weight of PC was increased from 21.9 to 27 kD the morphology changed from core-sheath to co-continuous structures.[96] The key role of kinetic factors in nanofiber structure is clearly illustrated in these results. Higher molecular weight material would be expected to exhibit better phase separation (larger domains) under equilibrium conditions.

Nanofibers with core-sheath structures have been prepared from blends of PS and polycarbonate,[97] PMMA and polycarbonate,[97] polyaniline and polycarbonate,[98] and polyaniline and polystyrene.[98]

30.6.3 Modeling of the Electrospinning Process

Modeling of the electrospinning process is still at a relatively basic level. Various combinations of models for the electrohydrodynamics, the charge transport, and the polymer dynamics have been developed. Hohman et al.[99,100] presented a detailed analysis of the jet initiation and the instability in the electrically forced fluid jets. Three different instabilities, namely the classical Rayleigh instability, electric field induced axisymmetric, and whipping instabilities, were addressed. Asymptotic approximations of electrohydrodynamic equations were developed. Hohman et al.'s model predicted both the jet initiation and

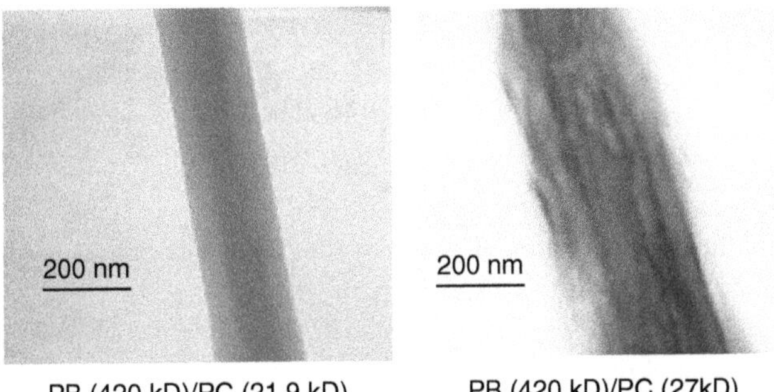

FIGURE 30.13 Effect of molecular weight on the nanofiber morphology for PB/PC blend ratios of 25/75. (After Wei, M., Kang, B., Sung, C., and Mead, J., in *Polymeric Nanofibers*, ed. D.H. Reneker and H. Fong, ACS Symposium Series 918, American Chemical Society, New York, 2006. With permission.)

the generation of droplets vs. fibers during the process as a function of parameters such as increasing electric field strength.

Reneker, et al. [101] analyzed and explained the bending instability using a mathematical model. The instabilities were modeled using viscoelastic dumbbells connected together with interactions following Coulomb's law. They predicted the three-dimensional path of the continuous jet both in the region where the instability grew slowly and in the region where the bending dominated the path of the jet. The model represented well the experimental data of jet paths that were obtained from high speed videographic observations. The theory developed by them showed the viscosity and surface tension as stabilizing forces acting on the charged jet at the pipette. Spivak et al.[102] considered the motion a viscous jet accelerated by an external electric field for modeling purposes. They also took into account the inertial, hydrostatic, viscous, electric, and surface tension forces.

As the deposition and patterning effort is dominated in great part by the electric fields, a charge transport model was developed. The geometry of the basic electrospinning setup, that is, the source (syringe and disk) and the target corresponds to a parallel plate geometry. With the source at positive potential and the target grounded, the ion-jet charge transport in electrospinning was modeled using the steady state charge distribution for two parallel plate electrodes. For the above space-charge system, the electrostatic potential variation in the space between the two plates due the charge distribution can be determined by the Poisson equation (Equation 30.10), which relates the electrostatic potential (V) in a dielectric to the charge density (q) via the permittivity (ε).

$$\Delta^2 V = -q/\varepsilon \tag{30.10}$$

The motion of polymer ions in the space between the source (disk) and the target can be described by the continuity equation, which states that the change in the total charge in a section is equal to the total current going out through its ends. The conservation law for current continuity for cylindrical symmetry under steady state conditions can be written in the form

$$\frac{\partial q(r,z)}{\partial t} + \nabla J(r,z) = 0 \tag{30.11}$$

where

$\boldsymbol{J}(r,z)$ = current density
$q(r,z)$ = space-charge density of the ions.

J_r and J_z are current density components that are given by

$$J_r(r,z) = \mu q(r,z)E_r(r,z) \quad (30.12)$$

$$J_z(r,z) = \mu q(r,z)E_z(r,z) \quad (30.13)$$

The two components of the current density are given by the product of charge mobility μ, charge density q, and the electric field components $E_r(r,z)$ and $E_z(r,z)$.

The boundary conditions for this set of equations include the voltages applied at the source (V_s) and the disk (V_d), the grounding of the target, and the relations arising from the cylindrical symmetry. The current density boundary conditions are specified as

$$J_r(r,0) = 0(E_r \text{ is negligibly small compared to } E_z) \quad (30.14a)$$

$$J_z(r,0) = \gamma E_z(r,0) \text{ for } 0 \leq r \leq R_s \quad (30.14b)$$

$$= 0 \text{ for } r > R_s$$

where γ = proportionality constant called the "injection parameter" $= \gamma \equiv \varepsilon\mu Vs/L^2$

The current density at the ion source (at $z = 0$) has the z-component $Jz(r,0)$ only, which can be assumed to be proportional to the field $Ez(r,0)$ at the source. The proportionality constant (γ) is introduced to specify the strength of the ion source. It has the units of conductivity, and combines the effects of process and material parameters, such as the flow rate, applied voltage, and polymer and solvent species and concentration. Starting with the initial condition that $q(r,z) = 0$ everywhere, the above set of coupled equations are solved by numerical iteration until the steady state is reached. The model can be used to predict the current density, J_z, at the target as a function of the radial location on the target.[69,73]

30.7 Extrusion and Injection Molding of Flexible Materials

This section discusses two common polymer processes — extrusion and injection molding — with a focus on how modifications to these processes can be made to enable nanoscale control of nanoscale features and elements.

30.7.1 Multilayer Extrusion

Nanolayered materials offer potential in a number of applications such as improved packaging for food, high strength materials, optical applications, and anticorrosion coatings.[103] In these applications the structure is designed to take advantage of the different properties of each layer — for example, a barrier layer combined with a scratch resistant outside layer. Layered systems have also shown unique behavior, including increased fracture toughness.[104,105]

Nanolayered composites can be formed layer by layer from solution.[106,107] This can be time consuming and often utilizes solutions that may pose environmental and disposal problems. Alternate methods to create multilayered materials utilize extrusion technologies.[108,109] These multilayer polymer films are commercially produced using feedblock or multi-manifold die technologies. Two or more extruders feed molten polymer to the feedblock or die where the melt stream is mechanically split to form the layers. Typical products have two to seven layers, however, films with over 2000 layers have been reported.[110] Schrenk and Wheatley[111] produced films with 769 layers using a feedblock system and four extruders. Baer et al.[108] created symmetric multilayer materials using a modification of the feedblock technology. Polymer combinations investigated include filled and unfilled polypropylene,[108] high-density polyethylene, low density polyethylene, linear low density polyethylene,[109] and polycarbonate.[112] Schrenk et al.[113] manufactured multilayer films for a wide variety of materials using two injection units fed into a feedblock system. An example of this approach is given in Figure 30.14.

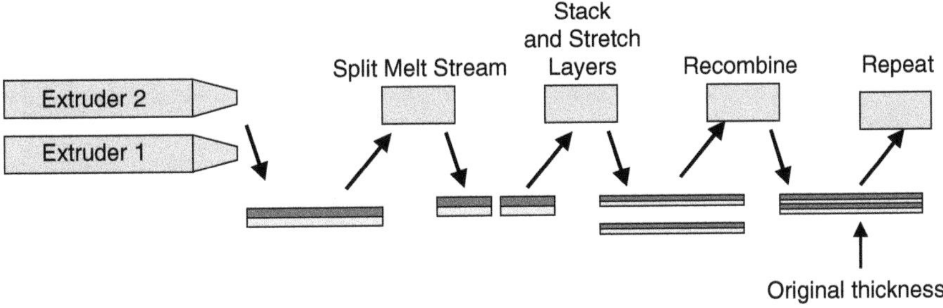

FIGURE 30.14 Schematic of multilayer extrusion system.

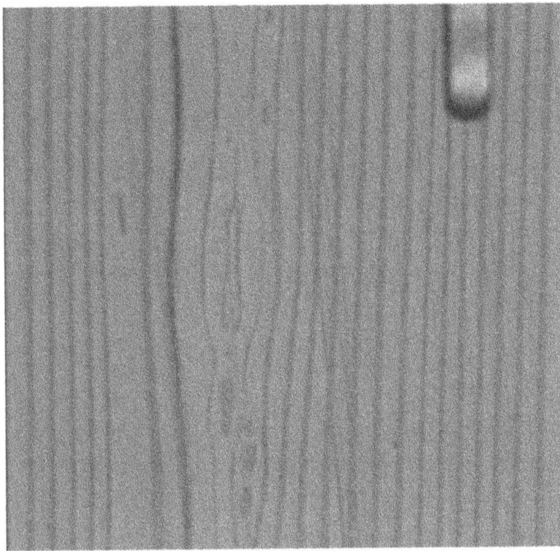

FIGURE 30.15 PP droplets as a discontinuous phase in the continuous phase of PC in a sample of 65 layers

In the multilayer extrusion process one of the key issues is stability of the layers. As the thickness of the layers decreases, individual layers may become discontinuous due to flow instabilities. One of the issues in producing nanometer thick layers is layer breakup into droplets.[114] Flow instabilities in conventional coextrusion are dependent on die geometry or design, asymmetry of the coextruded structure, and the elongational viscosity of the polymer melts.[115] Flow instabilities, as illustrated in Figure 30.15, lead to reduced properties and defects. Modeling has been extensively used to study the influence of polymer properties on flow instabilities in macro and microscale layers.[116–122] Layer thickness, and viscosity and elasticity ratios were found to be important factors. It has been suggested that flow instabilities occur when the interfacial shear stress reaches a critical value[123] and that decreasing the interfacial shear stress and reducing the elasticity differences between layers could decrease the flow instabilities.[124] Viscosity differences between materials may also produce defects (nonuniform layers) as a result of viscous encapsulation. In this case, the lower viscosity material will surround the higher viscosity resin.[125,126]

Flow instabilities are often investigated in fully developed flow, but in real extrusion systems, one must consider the effects of converging melt streams. This is particularly important in multilayer extrusion approaches where the melt stream is split many times to produce the layered structure. At the merge point of the two melt streams, material must accelerate from zero to its equilibrium velocity,[126] resulting in instabilities. Extensional flows play a role in these types of instability.[127]

30.7.2 Molded Polymer Nanostructures

There are a variety of processes that have been suggested for the manufacture of polymer-based devices, including microtransfer,[128] microcontact printing,[129] dip pen nanolithography,[130–132]

nanoimprint lithography,[133,134] hot embossing,[135–137] injection-compression molding,[138] and injection molding.[139–142] Some of these processes are also covered in Chapter 1 and Chapter 25. The injection molding process is used to produce approximately 30% of all the plastics components manufactured in the United States each year. Injection molding provides high-volume, three-dimensional net shape forming of polymeric components and is considered one of the most cost-effective manufacturing techniques available for producing large numbers of replicates. In the process discussed here, macro or microscale parts are molded of a single material containing nanoscale three-dimensional structures.

Nanoscale features are currently manufactured in digital versatile disks (DVDs). DVDs are typically injection or injection-compression molded in under 5 s[143–145] and contain depressions or pits that are 320 nm wide and 120 nm deep. Complete replication of the tooling projections is not required, as the "pits" and "lands" need only reflect light differently enough such that an opto-electronic detector and electronics can translate these feature differences into binary data. The machines, tooling, materials, and process were optimized for this one application and only a handful of materials, primarily bisphenol A polycarbonate, have been evaluated for compact disk and DVD molding.

Recently, research has begun to emerge in the rapid molding of nanofeatures — studying for example, the effect of processing conditions and material properties on nanoscale feature replication.[146,147] Tooling and mold parts were characterized using atomic force microscopy (AFM); two measurements, the depth ratio (relative depth of the molded features to the tooling depth) and the mean surface roughness, were developed to quantify feature replication. In these studies, injection velocity and processing pressures had little effect on replication, while the melt and mold temperatures were a factor in the ability to replicate nanoscale features from nickel tooling. Higher melt and mold temperatures typically produced greater depth ratios and smoother part surfaces.

The development of new tooling technologies is also required for the injection molding of nanoscale features. Table 30.3 provides the current tooling technologies. Conventional injection molds are machined from steel, aluminum, and beryllium-copper alloys,[148] however, CNC machining has a resolution of only 100 μm.[149] For smaller parts and features, toolmakers have turned to high precision electro wire discharge machining (WEDM); lithographic and etching techniques traditionally used in semiconductor fabrication — Lithographie Galvanoformung Abformung (LIGA); and laser ablation.[150] For more information on LIGA process, see Chapter 22 and Chapter 26. A version of micro-wire EDM uses 20-μm diameter wire to cut through a conductive workpiece, but currently is not a viable technique for nanoscale resolution.[151] Pulsed laser ablation deposits thin coatings under vacuum and low pressure background gases (He, Ar, H2, N2), of a wide range of materials (e.g., graphite, CVD diamond, copper and aluminum, ZnO and LiF, and various polymeric materials), on numerous substrates,[152] and can provide resolution of 1 μm in some materials.[153] With LIGA, high aspect ratio features as small as 0.5 μm in prepolymer-coated silicon are created by a high energy x-ray source; the pattern is later electroplated, usually with nickel. Although LIGA is a leading method for manufacturing of miniature metal parts, there are several drawbacks such

TABLE 30.3 Comparison of High-Resolution Tooling Techniques

Method	Max. resolution (μm)	Aspect Ratio	Material
CNC machining	100		
Micro milling	50–100	N/A	Steel
Micro wire EDM	1–50		
LIGA	0.5	$\leq$ 500	Nickel
Laser ablation	1–50	$\leq$ 600	metals, polymers
Electroforming	1	2.5	nickel alloys
Lithography			
UV	157	typically low, but up to 30[a]	Silicon or other semiconductors
EUV	13		
E-beam	<10		

[a] Ansari, K., van Kan, J.A., Bettiol, A.A., and Watt, F., *Appl. Phys. Lett.*, 85, 476–478, 2004.

TABLE 30.4 Comparison of Manufacturing Techniques for Nanoscale

	Hot embossing	Nanoimprint	Injection molding
Materials	Thermoplastics	Photoresist (PMMA, epoxy)	Thermoplastics or thermosets
Pressure (MPa)	>10	<0.1	>50
Temperature	>Tg	~ 25°C	150–400°C
Cycle time	1–10 min	2 min	3–14 s

as the soft nickel surface, which is subject to abrasion and wear, and the slow, multistep process.[154] DVD tooling, containing pits that are 100 to 400 nm wide and 140 nm deep,[155] are prepared using photolithography and electroforming of nickel. Lastly, ultraviolet (UV) and electron beam lithography of silicon wafers provides current resolution limits of 157 nm[156] and less than 10 nm,[157] respectively, however, the material used in these techniques is based on silicon rather than steel.

Silicon has several attractive factors as a tooling material, including high hardness, low linear thermal expansion, high thermal conductivity, and a flat surface compared to steel.[149,158] Silicon tooling has been employed for hot embossing, performed at temperatures near the softening temperature of a polymer under low pressures and heating times of 5 to 10 min.[159,160] Becker and Heim[158] hot embossed 8 μm features using silicon wafers, while Heyderman et al.[160,161] obtained features down to 50 nm using silicon and chromium-germanium coated silicon as the tooling. Recently, Grewell et al.[162,163] using infrared (IR), ultrasonic, and laser heating sources were able to reduce the heating time of hot embossing down to less than 10 s. Currently, 1 to 3 min cycle times are available in commercial hot embossing equipment.[164] Nanoimprint lithography, a process using a UV-transmittable quartz mold pressed into a low viscosity UV-curable prepolymer, produces a polymeric pattern with a typical cycle time of 2 min.[165] Shah et al.[166] used silicon wafers with feature sizes as small as 40 μm as tooling material for injection molding and Yu et al.[167] utilized silicon wafers as a substrate for making nickel mold inserts. D'Amore et al.[168] evaluated the injection-compression molding process using silicon wafers with V-shaped grooves of 1 μm opening width, and aspect ratios of 0.707. With modifications to the molding parameters and insert mounting, Yoon et al.[169] determined that silicon wafers were able to survive molding pressures of 50 MPa, however, gallium arsenide tooling failed at pressures as low as 5 MPa. Silicon tooling was found to survive over 3000 injection molding cycles without damage to the silicon surface.[170] Table 30.4 compares the differing techniques.

30.8 Concluding Remarks

The markets for polymeric materials have grown steadily since the significant commercial introduction of synthetic polymers after World War II. Properties such as low density, flexibility, tailorability, and biological compatibility have led to attempts to expand usage in products for structural, electronic, and biomedical applications. Introduction of nanoscale elements and fabrication of nanoscale geometries, can increase properties enough to achieve commercial viability in display technology, data storage, structural applications, and the like. Commercial viability also relies to a great extent on the creation of manufacturing processes to fabricate these products. Necessarily, early processes will be modifications of existing processes, such as the nanocomposites, injection molding, and extrusion described in this chapter. As we learn more about the response of these polymer materials at high rates in nanoscale geometries, completely new processes, akin to the shift from die-based fiber spinning to electric field controlled electrospinning, will be developed. In addition to the need for both experimental studies and analytical/computational models that address nanoscale and multiscale questions in manufacturing science, critical enabling technologies in rapid, on-line measurement and positioning will be needed to speed up the traditional, multi-decade technology transfer path.

References

[1] International Technology Roadmap for Semiconductors (ITRS) (http://public.itrs.net/).

[2] Hadjichristidis, N., Pispas, S., and Floudas, G., *Block Copolymers: Synthetic Strategies, Physical Properties, and Applications*, John Wiley & Sons, Hoboken, NJ, 2003.

[3] Hamley, I. M., Introduction to Block Copolymers, in *Developments in Block Copolymer Science and Technology*, ed. I.W. Hamley, John Wiley & Sons, Chicester, England, 2004, Chapter 1.

[4] McClelland, G. M., et al., Nanoscale patterning of magnetic islands by imprint lithography using a flexible mold, *Appl. Phys. Lett.*, 81, 1483, 2002.

[5] Kim, D. H., et al., On the replication of block copolymer templates by poly(dimethylsiloxane) elastomers, *Adv. Mater.*, 15, 811, 2003.

[6] Kim, Y. S., Lee, H. H., and Hammond, P. T., High density nanostructure transfer in soft molding using polyurethane acrylate molds and polyelectrolyte multilayers, *Nanotechnology*, 14, 1140, 2003.

[7] Maldovan, M., Carter, W. C., and Thomas, E. L., Three-dimensional dielectric network structures with large photonic band gaps, *Appl. Phys. Lett.*, 83, 5172, 2003.

[8] Chen, K. and Ma, Y., Ordering stripe structures of nanoscale rods in diblock copolymer scaffolds, *J. Chem. Phys.*, 116, 7783, 2002.

[9] Tokuhisa, H. and Hammond, P. T., Nonlithographic micro- and nanopatterning of TiO_2 using polymer stamped molecular templates, *Langmuir*, 20, 1436, 2004.

[10] Ali, H. A. et al., Properties of self-assembled ZnO nanostructures, *Solid-State Electron.*, 46, 1639, 2002.

[11] Clay, R. T. and Cohen, R. E., Synthesis of metal nanoclusters within microphase-separated diblock copolymers: ICP-AES analysis of metal ion uptake, *Supramol. Sci.*, 4, 113, 1997.

[12] Sohn, B. H. and Cohen, R. E., Electrical properties of block copolymers containing silver nanoclusters within oriented lamellar microdomains, *J. Appl. Polym. Sci.*, 65, 723, 1997.

[13] Liu, T., Burger, C., and Chu, B., Nanofabrication in polymer matrices, *Prog. Polym. Sci.*, 28, 5, 2003.

[14] Harrison, C., Dagata, J. A., and Adamson, D. H., Lithography with self-assembled block copolymer microdomains, in *Developments in Block Copolymer Science and Technology*, ed. I.W. Hamley, John Wiley & Sons, Chicester, England, 2004, Chapter 9.

[15] Schaffer, E., et al., Electrically induced structure formation and pattern transfer, *Nature*, 403, 874, 2000.

[16] Thurn-Albrecht, T., et al., Overcoming interfacial interactions with electric fields, *Macromolecules*, 33, 3250, 2000.

[17] Chen, Z. R., et al., Pathways to macroscale order in nanostructured block copolymers, *Science*, 277, 1248, 1997.

[18] Polis, D. L., et al., Nature of viscoelasticity in lamellar block copolymers: Contraction correlated to strain localization, *Phys. Rev. Lett.*, 83, 2861, 1999.

[19] Albalak, R. K. and Thomas, E. L., Microphase separation of block copolymer solutions in a flow field, *J. Polym. Sci. Part B: Polym. Phys.*, 31, 37, 1993.

[20] Yang, X. M., et al., Proximity X-ray lithography using self-assembled alkylsilixone films: Resolution and pattern transfer, *Langmuir*, 17, 228, 2001.

[21] Rockford, L., Mochrie, S. G. J., and Russell, T. P., Propagation of nanopatterned substrate templated ordering of block copolymers in thick films, *Macromolecules*, 34, 1487, 2001.

[22] Yang, X. M., et al., Proximity x-ray lithography using self-assembled alkylsilixone films: Resolution and pattern transfer, *Langmuir*, 17, 228, 2001.

[23] Kim, S. O., et al., Epitaxial self-assembly of block copolymers on lithographically defined nanopatterned substrates, *Nature*, 424, 411, 2003.

[24] Meals, R. N. and Lewis, F. M., *Silicones*. Reinhold Publishing Co., New York, 1959.

[25] Fordham, S., *Silicones*. Philosophical Library, Inc., New York, 1961.

[26] Giannelis, E. P., Krishnamoorti, R., and Manias, E., Polymer-silicate nanocomposites: Model systems for confined polymers and polymer brushes, *Adv. Polym. Sci.*, 138, 107, 1999.

[27] Giannelis, E. P., Organoceramic nanocomposites, in *Biomimetic Materials Chemistry*, 337, ed. S. Mann, Wiley VCH, New York, 1996.

[28] LeBaron, P. C., Wang, Z., and Pinnavaia, T. J., Polymer-layered silicate nanocomposites: An overview, *Appl. Clay Sci.*, 15, 11, 1999.

[29] Alexandre, M. and Dubois, P., Polymer-layered silicate nanocomposites: Preparation, properties and uses of a new class of materials, *Mater. Sci. Eng. R.*, 28, 1, 2000.

[30] Vaia, R. A. and Giannelis, E. P., Polymer nanocomposites: Status and opportunities, *MRS Bull.*, 26, 394, 2001.

[31] Schmidt, D., Shah, D., and Giannelis, E. P., New advances in polymer/layered silicate nanocomposites, *Curr. Opin. Solid State Mater. Sci.*, 6, 205, 2002.

[32] Garboczi, E. J., et al., Geometrical percolation threshold of overlapping ellipsoids, *Phys. Rev. E*, 52, 819, 1995.

[33] Nielsen, L. E., Models for the permeability of filled polymer systems, *J. Macromol. Sci. Chem.*, A1, 929, 1967.

[34] Mehta, B. S., DiBenedetto, A. T., and Kardos, J. L., Diffusion and permeation of gases in glass ribbon-reinforced plastics, *Int. J. Polym. Mater.*, 3, 269, 1975.

[35] Leddy, J., Characterizing flux through micro- and nanostructured composites, *Langmuir*, 15, 710, 1999.

[36] Fredrickson, G. H. and Bicerano, J., Barrier properties of oriented disk composites, *J. Chem. Phys.*, 110, 2181, 1999.

[37] Bharadwaj, R. K., Modeling the barrier properties of polymer-layered silicate nanocomposites, *Macromolecules*, 34, 9189, 2001.

[38] Bourlinos, A. B., et al., Surface-functionalized nanoparticles with liquid-like behavior, *Adv. Mater.*, 17, 234, 2005.

[39] Winey, K. I., Thomas, E. L., and Fetters, L. J., Isothermal morphology diagrams for binary blends of diblock copolymer and homopolymer, *Macromol.*, 25, 2645, 1992.

[40] Baek, D. M., Han, C. D., and Kim, J. K., Phase equilibria in mixtures of block copolymer and homopolymer, *Polymer*, 33, 4821, 1992.

[41] Vaidya, N. Y. and Han, C. D., Temperature-composition phase diagrams of binary blends of block copolymer and homopolymer, *Polymer*, 43, 3047, 2002.

[42] Karathanasis, A. D. and Hajek, B. F., Revised methods for rapid quantitative determination of minerals in soil clays, *Soil Sci. Soc. Am. J.*, 46, 419, 1982.

[43] Rasekh, H., Rose, K. W., and Worrall, W. E., The rheological stabilization of clay-water suspensions: A comparison of various methods, *Br. Ceram. Trans.*, 86, 132, 1987.

[44] Huang, Scott L., The influence of disaggregation methods on x-ray diffraction of clay minerals, *J. Sed. Pet.*, 59, 997, 1989.

[45] Lapides, I. and Heller-Kallai, L., Novel features of smectite settling, *Coll. Polym. Sci.*, 280, 554, 2002.

[46] Lu, S. C., Song, S. X., and Dai, Z. F., Dispersion of fine mineral particles in water, *Adv. Powder Technol.*, 3, 89, 1992.

[47] Piner, R. (Northwestern University, Evanston, IL, USA), personal communications 6/02-11/02.

[48] Basedow, A. M. and Ebert, K. H., Ultrasonic degradation of polymers in solution, *Adv. Polym. Sci.*, 22, 83, 1977.

[49] Price, G. J., The use of ultrasound for the controlled degradation of polymer solutions, *Adv. Sonochem.*, 1, 231, 1990.

[50] Price, G. J., Applications of high intensity ultrasound in polymer chemistry, in *Chemistry Under Extreme and Non-Classical Conditions*, Chapter 9, 381, ed. R. van Eldik and C. D. Hubbard, John Wiley & Sons, New York, 1997.

[51] Suzuki, Y., Warsito, Maezawa, A., and Uchida, S., Effects of frequency and aeration rate on ultrasonic oxidation of a surfactant, *Chem. Eng. Technol.*, 22, 507, 1999.

[52] Destaillats, H., Hung, H.-M., and Hoffmann, M. R., Degradation of alkylphenol ethoxylate surfactants in water with ultrasonic irradiation, *Environ. Sci. Technol.*, 34, 311, 2000.

[53] Pee, M. G. Y., Weavers, L. K., and Rathman, J. F., Sonochemical degradation of surfactants, *(Am. Chem. Soc.) Div. Environ. Chem. Prepr. Ext. Abs.*, 40, 645, 2000.

[54] Vinodgopal, K., Ashokkumar, M., and Grieser, F., Sonochemical degradation of a polydisperse nonylphenol ethoxylate in aqueous solution, *J. Phys. Chem. B*, 105, 3338, 2001.

[55] Yim, B., et al., Sonolysis of surfactants in aqueous solutions: An accumulation of solute in the interfacial region of the cavitation bubbles, *Ultrason. Sonochem.*, 9, 209, 2002.

[56] Messersmith, P. B. and Giannelis, E. P., Synthesis and characterization of layered silicate-epoxy nanocomposites, *Chem. Mater.*, 6, 1719, 1994.

[57] Lan, T. and Pinnavaia, T. J., Clay-reinforced epoxy nanocomposites, *Chem. Mater.*, 6, 2216, 1994.

[58] Lan, T., Kaviratna, P. D., and Pinnavaia, T. J., Synthesis, characterization and mechanical properties of epoxy-clay nanocomposites, *Polym. Mater. Sci. Eng.*, 71, 527, 1994.

[59] Pinnavaia, T. J., et al., Clay-reinforced epoxy nanocomposites: Synthesis, properties, and mechanism of formation, in *Nanotechnology ACS Symposium Series* 622, 250, ed. G.-M. Chow and K. E. Gonsalves, American Chemical Society, Washington, DC, 1996.

[60] Alexandre, M., et al., Polyethylene-layered silicate nanocomposites prepared by the polymerization-filling technique: Synthesis and mechanical properties, *Polymer*, 43, 2123, 2002.

[61] Asthana, R. and Sobczak, N., Wettability, Spreading, and interfacial phenomena in high-temperature coatings *J. Met. Electronic Ed. (JOM-e)* 52 (2000). (http://www.tms.org/pubs/journals/JOM/0001/Asthana/Asthana-0001.html)

[62] Fowkes, F. M., Role of acid-base interfacial bonding in adhesion, *J. Adhesion Sci. Technol.*, 1, 7, 1987.

[63] Zhandarov, S. and Mäder, E., Some reflections on the "self-consistent" approach to the calculation of acid-base parameters of solids and liquids from wetting data, *Comp. Interfac.*, 10, 41, 2003.

[64] Parsons, G. E., Buckton, G., and Chatham, S. M., The use of surface energy and polarity determinations to predict physical stability of nonpolar, nonaqueous suspensions, *Int. J. Pharm.*, 83, 163, 1982.

[65] VanderHart, D. L., Asano, A., and Gilman, J. W., Solid-state NMR investigation of paramagnetic nylon-6 clay nanocomposites. 2. Measurement of clay dispersion, crystal stratification, and stability of organic modifiers, *Chem. Mater.*, 13, 3796, 2001.

[66] Zisman, W. A., Relation of the equilibrium contact angle to liquid and solid constitution, *Adv. Chem. Ser.*, 43, 1, 1964.

[67] Hildebrand, J. H. and Scott, R. L., The solubility of non-electrolytes, *American Chemical Society Monograph* 17, Reinhold, New York, 1950.

[68] Formhals, A., Method and Apparatus for Production of Fibers, U.S. Patent No. 1,975,504 (filed 1934).

[69] Bunyan, N., et al., Electrostatic effects on electrospun fiber deposition and alignment, in *Polymeric Nanofibers*, ed. D. Reneker and H. Fong, ACS Symposium Series 918, American Chemical Society, 2006.

[70] Fong, H., Chun, I., and Reneker, D. H., Beaded nanofibers formed during electrospinning, *Polymer*, 40, 4585, 1999.

[71] Baker, A.-M., et al., Nanotechnology for polymers and plastics, in *Handbook of Plastics Technologies*, ed. C. Harper, McGraw-Hill, New York, 2006.

[72] Theron, A., Zussman, E., and Yarin, A. L., Electrostatic field-assisted alignment of electrospun fibers, *Nanotechnology* 12, 384–390, 2001.

[73] Bunyan, N., Control of Deposition and Orientation of Electrospun Fibers, MS thesis, University of Massachusetts Lowell, Department of Mechanical Engineering, 2003.

[74] Dzenis, Y., Spinning continuous fibers for nanotechnology, *Science*, 304, 1917, 25 June 2004.

[75] Li, D., et al., Collecting electrospun nanofibers with patterned electrodes, *Nano Lett.*, 5, 913, 2005.
[76] Viriyabanthorn, N., et al., Effect of carbon black loading on electrospun butyl rubber nonwoven mats, in *Polymeric Nanofibers*, ed. D. H. Reneker and H. Fong, ACS Symposium Series 918, American Chemical Society, New York, 2006.
[77] Threepopnatkul, P., et al., Fiber Structure and Mechanical Properties of Electrospun Butyl Rubber with Different Carbon Black Types, paper presented at the Fall 168th Technical Meeting of the Rubber Division, American Chemical Society, Pittsburgh, PA.
[78] Kim, J.-S. and Lee, D. S., Thermal properties of electrospun polyesters, *Polym. J.*, 32, 616, 2000.
[79] Drew, C., et al., Electrospun nanofibers of electronic and photonic polymer systems, *SPE Technical Papers*, 46, 1477, 2000.
[80] Park, W. H., et al., Effect of chitosan on morphology and conformation of electrospun silk fibroin nanofibers, *Polymer*, 45, 7151, 2004.
[81] Kahol, P. K. and Pinto, N. J., An EPR investigation of electrospun polyaniline-polyethylene oxide blends, *Synth. Met.*, 140, 269, 2004.
[82] MacDiarmid, A. G., et al., Electrostatically-generated nanofibers of electronic polymers, *Synth. Met.*, 119, 27, 2001.
[83] Ohkawa, K., et al., Electrospinning of chitosan, *Macromol. Rapid Commun.*, 25, 1600, 2004.
[84] Lee, K. H., et al., Mechanical behavior of electrospun fiber mats of poly(vinyl chloride)/polyurethane polyblends, *J. Polym. Sci.: Pt. B: Polym. Phys.*, 41, 1256, 2003.
[85] Norris, I. D., et al., Electrostatic fabrication of ultrafine conducting fibers: Polyaniline/polyethylene oxide blends, *Synth. Met.*, 114, 109, 2000.
[86] Fong, H. and Reneker, D. H., Elastomeric nanofibers of styrene-butadiene-styrene triblock copolymer, *J. Polym. Sci. B: Polym. Phys.*, 37, 3488, 1999.
[87] Bognitzki, M., et al., Preparation of fibers with nanoscaled morphologies: Electrospinning of polymer blends, *Polym. Eng. Sci.*, 41, 982, 2001.
[88] Dong, H., et al., Polyaniline/poly(methyl methacrylate) coaxial fibers: The fabrication and effects of the solution properties on the morphology of electrospun core fibers, *J. Polym. Sci.: Pt. B: Polym. Phys.*, 42, 3934, 2004.
[89] Bognitzki, M., et al., Polymer, metal, and hybrid nano- and mesotubes by coating degradable polymer template fibers (TUFT Process), *Adv. Mater.*, 12, 637, 2000.
[90] Sun, Z., et al., Compound core-shell polymer nanofibers by co-electrospinning, *Adv. Mater.*, 15, 1929, 2003.
[91] Li, D. and Xia, Y., Direct fabrication of composite and ceramic hollow nanofibers by electrospinning, *Nano Lett.*, 5, 933, 2004.
[92] Zhang, Y., et al., Preparation of core-shell structured PCL-r-gelatin bi-component nanofibers by coaxial electrospinning, *Chem. Mater.*, 16, 3406, 2004.
[93] Paul, D. R., Fibers from polymer blends, *in Polymer Blends*, Vol. 2., ed. D. R. Paul, and S. Newman, Academic Press, New York, 1978, Chapter 16.
[94] Fong, H. and Reneker, D. H., Elastomeric nanofibers of styrene-butadiene-styrene triblock copolymer, *J. Polym. Sci. B: Polym. Phys., 37*, 3488, 1999.
[95] Wei, M., Sung, C., and Mead, J., Phase morphology of electrospun nanofibers from polybutadiene (pb)/polycarbonate (PC) blends, *Polym. Preprints*, 44, 79, 2003.
[96] Wei, M., et al., Preparation of nanofibers with controlled phase morphology from electrospinning of polybutadiene (pb) /polycarbonate (pc) blends, in *Polymeric Nanofibers*, ed. D. H. Reneker and H. Fong, ACS Symposium Series 918, American Chemical Society, New York, 2006.
[97] Wei, M., et al., Phase morphology control of the electrospun nanofibers from the polymer blends, *2004 NSTI Nanotechnology Conference and Trade Show*, Boston, March, 2004.
[98] Wei, M., et al., Preparation of core-sheath nanofibers from conducting polymer blends, *Macromol. Rapid Commun.*, 26(16), 1127, 2005.

[99] Hohman, M. M., et al., Electrospinning and electrically forced jets. I. Stability theory, *Phys. Fluids*, 13, 2201, 2001.

[100] Hohman, M. M., et al., Electrospinning and electrically forced jets. II. Applications, *Phys Fluids*, 13, 2221, 2001.

[101] Reneker, D. H., et al., Bending instability of electrically charged liquid jets of polymer solutions in electrospinning, *J. Appl. Phys.* 87, 4531, 2000.

[102] Spivak, A. F., Dzenis, F., and Reneker, D. H., A model of steady state jet in the electrospinning process, *Mech Res. Commun.*, 27, 37, 2000.

[103] http://www.sciencenews.org/articles/20030809/bob9.asp, accessed November 29, 2005.

[104] Schrenk, W. J. and Alfrey, Jr. T., Some physical properties of multilayered films, *Polym. Eng. Sci.*, 9, 393, 1969.

[105] Baer, E., Jarus, D., and Hiltner, A., Microlayer coextrusion technology, *SPE Technical Papers*, 3947, 1999.

[106] Ferguson, G. and Kleinfeld, E., Multilayered nanostructures comprising alternating organic and inorganic ionic layers, U.S. Patent 5, 716, 709, 1998.

[107] Lesser, C., Gao, M., and Kirstein, S., Highly luminescent thin films from alternating deposition of CdTe nano-particles and polycations, *Mater. Sci. Eng.*, C 8–9, 159, 1999.

[108] Nazarenko, S., Hiltner, A., and Baer, E., Polymer microlayer structures with anisotropic conductivity, *J. Mater. Sci.*, 34, 1461, 1999.

[109] Nazarenko, S., et al., Creating layers of concentrated inorganic particles by interdiffusion of polyethylenes in microlayers, *J. Appl. Polym. Sci.*, 73, 2877, 1999.

[110] Wheatley, J. A. and Schrenk, W. J., Polymeric reflective materials (PRM), *SPE Technical Papers*, 39, 2864, 1993.

[111] Schrenk, W. and Wheatley, J., Coextruded elastomeric optical interference films, *SPE Technical Papers*, 34, 1703, 1988.

[112] Nazarenko, S., et al., Novel structures by layer multiplying coextrusion, *SPE Technical Papers*, 42, 1587, 1996.

[113] Schrenk, W. J., et al., Lamellar injection moulding process for multi-phase polymer systems, *SPE Technical Papers*, 39, 544, 1993.

[114] Ho, K., et al., Investigation of Interfacial Instabilities in Nanolayer Extrusion, SPE-62nd Ann. Tech. Conf. –Chicago, Vol. I, pp. 376, 2004.

[115] Ramanathan, R., et al., Wave pattern instability in multilayer coextrusion – an experimental investigation, *SPE Technical Papers*, 42(1), 224, 1996.

[116] Chen, K. P. and Zhang, Stability of the interface in co-extrusion flow of two viscoelastic fluids through a pipe, *J. Fluid Mech.*, 247, 489, 1993.

[117] Chen, K. P., Interfacial instability due to elastic stratification in concentric coextrusion of two viscoelastic fluids, *J. Non-Newtonian Fluid Mech.*, 40, 155, 1991.

[118] Anturkar, N. R., Papanastasiou, T. C., and Wilkes, J. O., Stability of coextrusion through converging dies, *J. Non-Newtonian Fluid Mech.*, 41, 1, 1991.

[119] Musarra, S. and Keunings, R., Co-current axisymmetric flow in complex geometries: numerical simulation, *J. Non-Newtonian Fluid Mech.*, 32, 253, 1989.

[120] Su, Y. Y. and Khomami, B., Interfacial stability of multilayer viscoelastic fluids in slit and converging channel die geometries, *J. Rheol.*, 36, 357, 1992.

[121] Wilson, G. M. and Khomami, B., An experimental investigation of interfacial instabilities in multilayer flow of viscoelastic fluids. 1. Incompatible polymer systems, *J. Non-Newtonian Fluid Mech.*, 45, 355, 1992.

[122] Van De Griend, R. and Denn, M. M., Co-current axisymmetric flow in complex geometries: Experiments, *J. Non-Newtonian Fluid Mech.*, 32, 229, 1989.

[123] Schrenk, W. J., et al., Interfacial flow instability in multilayer coextrusion, *Polym. Eng. Sci.*, 18, 620, 1978.

[124] Mavridis, H. and Shroff, R. N., Multilayer extrusion: Experiments and computer simulation, *Polym. Eng. Sci.*, 34, 559, 1994.
[125] Dooley, J., Hyun, K. P., and Hughes, K., An experimental study on the effect of polymer viscoelasticity on layer rearrangement in coextruded structures, *Polym. Eng. Sci.*, 38, 1060, 1998.
[126] Khomani, B., Interfacial stability and deformation of two stratified power law fluids in plane poiseuille flow, *J. Non-Newtonian Fluid Mech.*, 37, 19,1990.
[127] Khomani, B. and McHugh, A. J., Processing property interactions in poly(vinylidene fluoride). I. An analysis of melt stress history in an extensional flow geometry, *J. Appl. Polym. Sci.*, 36, 859, 1988.
[128] Zhao, X.-M., Xia, Y., and Whitesides, G. M., Fabrication of three-dimensional micro-structures: Microtransfer molding, *Adv. Mater.*, 8, 837, 1996.
[129] Reyes, D. R., et al., Micro total analysis systems. 1. Introduction, theory, and technology, *Anal. Chem.*, 74, 2623, 2002.
[130] Wang, X., et al., In *Linear Probe Arrays for Dip-pen Nanolithography*, International Conference on Micro & Nano Systems, Kunming, China, 2002; Kunming, China, 2002.
[131] Zhang, H., Chung, S.-W., and Mirkin, C. A., Fabrication of sub-50 nm Solid-state nanostructures. Based on dip-pen nanolithography, *Nano Lett.*, 3, 43, 2003.
[132] Yu, M. and Ivanisevic, A., Nanoscale surface patterning, Materials Research Society Symposium Proceedings, pp. Q8.19.1–Q8.19.5, 2003.
[133] Li, H. and Huck, S., Polymers in nanotechnology, *Curr. Opin. Solid State Mater. Sci.*, 6, 3, 2002.
[134] McAlpine, M. C., Friedman, R. S., and Lieber, C. M., Nanoimprint lithography for hybrid. Plastic electronics, *Nano Lett.*, 3, 443, 2003.
[135] Grass, B., et al., A new PMMA-microchip device for isotachophoresis with integrated conductivity detector, *Sens. Actuators B: Chem.*, 72, 249, 2001.
[136] Jaszewski, R. W., Gobrecht, J., and Smith, P., Hot embossing in polymers as a direct way to pattern resist, *Microelectron. Eng.*, 41–42, 575, 1998.
[137] Schift, H., et al., Nanoreplication in polymers using hot embossing and injection molding, *Microelectron. Eng.*, 53, 171, 2000.
[138] Liu, Y., et al., Microfabricated polycarbonate CE devices for DNA analysis, *Anal. Chem.*, 73, 4196, 2001.
[139] McCormick, et al., Microchannel electrophoretic separations of DNA in injection-molded plastic substrates, *Anal. Chem.*, 69, 2626, 1997.
[140] Edwards, T. L., et al., Rapid tooling using SU-8 for injection molding microfluidic components, *Proc. SPIE*, 4177, 75–82, 2000.
[141] Hulme, J. P., Fielden, P. R., and Goddard, N. J., Fabrication of a spectrophotometric absorbance flow cell using injection-molded plastic, *Anal. Chem.*, 76, 238, 2004.
[142] Kang, S., Replication technology for micro/nano optical components, *Jpn. J. Appl. Phys.*, 43, 5706, 2004.
[143] Schift, H., et al., Nanoreplication in polymers using hot embossing and injection molding, *Microelectron. Eng.*, 53, 171, 2000.
[144] Oshiro, T., Goto, T., and Ishibashi, J., Experimental study of DVD substrate quality by operating conditions in injection molding, *Annu. Tech. Conf.*, 55, 409, 1997.
[145] Sancoucy, M., private conversation, (2005).
[146] Srirojpinyo, C., et al., Interfacial effects in replication of nano-scale features, *Proc. Annu. Tech. Conf. Soc, Plastics Eng.*, 51, 754–758, 2005.
[147] Srirojpinyo, C., Yoon, S., Lee, J., Sung, C., Mead, J. L., and Barry, C. M. F., Effects of materials when injection molding nano-scale features, *Proc. Annu. Tech. Conf. Soc, Plastics Eng.*, 50, 743, 2004.
[148] Malloy, R. A., *Plastics Part Design for Injection Molding*, Hanser Gardner Publications, Cincinnati, OH, 1994.

[149] Becker, H. and Heim, U., Hot embossing as a method for the fabrication of polymer high aspect ratio structures, *Sens. Actuators*, 83, 130, 2000.
[150] Weber, L. and Ehrfeld, W., Molding of microstructures for high-tech-applications, *Proc. Annu. Tech. Conf. Soc, Plastics Eng.*, 930, 1998.
[151] Pham, D. T., et al., Micro-EDM-recent developments and research issues, *J. Mater. Process. Technol.*, 149, 50, 2004.
[152] http://www.chm.bris.ac.uk/pt/laser/ashfold/ablation.htm, last accessed November 22, 2005.
[153] Kim, J. T., et al., Fabrication of a micro-optical coupling structure by laser ablation, *J. Mater. Process. Technol.*, 146, 163–166 (2004).
[154] Hirata, Y., LIGA process — micromachining technique using synchrotron radiation lithography — and some industrial applications, *Nucl. Instrum. Methods Phys. Res. Section B: Beam Interactions Mater. Atoms*, 208, 21, 2003.
[155] Sharpless, G., *CD and DVD Disc Manufacturing*, Disctronics Manufacturing, 2002.
[156] Fay, B., Advanced optical lithography development, from UV to EUV, *Microelectron. Eng.*, 61–62, 11, 2002.
[157] Lehmann, F., et al., Fabrication of sub-10-nm Au-Pd structures using 30 keV electron beam lithography and lift-off, *Microelectron. Eng.*, 65, 327, 2003.
[158] Becker, H. and Heim, U., Silicon as tool material for polymer hot embossing, *IEEE International Conference on Micro Electro Mechanical Systems*, Orlando, 228, 1999.
[159] Lee, W., et al., Nanostructuring of a polymeric substrate with well-defined nanometer-scale topography and tailored surface wettability, *Langmuir*, 20, 7665, 2004.
[160] Heyderman, L. J., et al., Nanofabrication using hot embossing lithography and electroforming, *Microelectron. Eng.*, 57–58, 375, 2001.
[161] Chou, S. Y., Krauss, P. R., and Renstrom, P. J., Imprint of sub-25 nm vias and trenches in polymers, *Appl. Phys. Lett.*, 67, 3114, 1995.
[162] Grewell, D., et al., Feasibility of selected methods for embossing micro-features in thermoplastics, *Proc Annu. Tech. Conf. Soc. Plastics Eng.*, 1094, 2003.
[163] Grewell, D., Lu, C., Leeand, L. J., and Benatar, A., Infrared micro-embossing of thermoplastics, *Proc. Annu. Tech. Conf. Soc, Plastics Eng.*, 1231, 2004.
[164] http://www.jo-mt.de/downloads/Jenoptik_Froehling.pdf, last accessed November 22, 2005.
[165] Becker, H. and Locascio, L. E., Polymer microfluidic devices, *Talanta*, 56, 267–287, 2002.
[166] Shah, J., Su, Y.-C., and Lin, L., Implementation and analysis of polymeric microstructure replication by micro injection molding, *Proceedings of Micro-Electro-Mechanical Systems*, Nashville, TN, 1999, pp. 295–302.
[167] Yu, L., et al., Implementation and analysis of polymeric microstructure replication by micro injection molding, *Proc. Annual Tech. Conf. Soc, Plastics Eng.*, 85–789, 2001.
[168] D'Amore, A., et al., Concentration of information, *Kunststoffe Plast Eur.*, 4–7, 2004.
[169] Yoon, S., et al., The effects of tooling surfaces on injection molded nanofeatures, *Proc. Annu. Tech. Conf. Soc, Plastics Eng.*, 738–742, 2004.
[170] Yoon, S., et al., Evaluation of novel tooling for nanoscale injection molding, *SPIE International Symposia, Smart Structures & Materials/NDE*, San Diego, March 7–9, 107, 2005.

31

Multiscale Processes in Surface Deformation

Leon L. Shaw
University of Connecticut

Yuntian T. Zhu
Los Alamos National Laboratory

Abstract

Processes based on surface deformation are important methods for improving mechanical properties, especially fatigue and wear resistance. Examples of the surface deformation based methods include, but are not limited to, shot peening, surface mechanical attrition treatment, particle impact processing, hammer peening, laser shot peening, deep rolling, and roller-burnishing. In all of these methods multiscale processes, ranging from the introduction of macroscopic residual stresses and work hardening to the dislocation generation and rearrangement and to the formation of nanocrystalline grains, take place simultaneously. In this chapter processes relying on surface severe plastic deformation (S^2PD) through impacts of high-energy particles, shots and balls are reviewed. The topics discussed include processing parameters, development of deformation fields, local property alternation, evolution of surface roughness, formation of macroscopic residual stresses, microstructural evolution, and mechanical property enhancements. The perspectives of future developments are also discussed.

31.1 Introduction

Surface treatment processes can be broadly classified into two general groups, one being coating-based processes and the other noncoating-based processes. Noncoating-based processes can be further subdivided into two categories, that is surface deformation processes and non surface deformation processes. The former comprises shot peening (SP),[1–3] ultrasonic shot peening (USSP),[4,5] surface mechanical attrition treatment (SMAT),[6,7] high-energy shot peening (HESP),[8] surface nanocrystallization and hardening (SNH),[9,10] and particle impact processing (PIP).[11,12] The latter includes carburizing, nitriding, nitrocarburizing, plasma nitriding, ion implantation, laser surface treatment, induction heat treatment, and irradiation (See Chapters 10, 11, and 18). The property enhancements achieved from surface deformation processes are mainly due to plastic deformation, whereas the improvements achieved from non surface deformation processes are predominantly derived from composition changes, phase transformation, and structure modification. In this chapter, only surface deformation processes will be addressed (See Chapters 12–14 for bulk-deformation processes). The discussion will focus on five areas, that is, (1) overview of surface deformation processes, (2) fundamentals of surface deformation, (3) microstructural evolution during the process, (4) property enhancements, and (5) future developments. Each of these areas is described below.

31.2 Overview of Surface-Deformation Processes

Shot peening is a cold working process used widely by industry mainly to improve the fatigue life of metallic components.[1–3] In this process, a stream of spherical shots made of cast iron, steel, cut-wire, glasses, or ceramic beads is blasted against a workpiece. The shot size typically ranges from 0.25 to 1.0 mm,[13] although in some cases shots with diameter as small as 0.05 mm are also used.[12] The impact velocity of shots is high, typically ranging from 20 to 150 m/s.[13,14] Multiple and repeated impacts of the workpiece by these high-energy shots induce surface plastic deformation. When sufficient surface plastic deformation is produced, shot peening will lead to residual compressive stresses at the surface region,[2,3] a work-hardened surface layer,[8,15] and sometime a nano-grained surface layer.[12,14] These microstructural and stress-state changes can result in dramatic improvements in the fatigue strength of the workpiece.[2,3,16–18]

Recently, a new family of surface plastic deformation processes has been developed with the aim to create a nanocrystalline (nc) surface layer. These processes, pioneered by K. Lu and J. Lu,[19] rely on severe plastic deformation induced by impacts of high-energy balls (with sizes larger than shots) or by hammer peening, surface rolling, laser shock treatment, or machining to produce a nc surface layer. The common feature of these processes is surface severe plastic deformation (S^2PD). High-energy ball impact processes have received most of the attention because of their versatility in processing complex-shaped parts.[4–12,14,15] Several different names have been used for this group of processes, including ultrasonic shot peening,[4,5] surface mechanical attrition treatment,[6,7] high-energy shot peening,[8] surface nanocrystallization and hardening,[9,10] and particle impact processing.[11,12] Differences among these processes stem mainly from the approach via which the high velocity of balls is generated. In USSP the movement of balls is generated through collision between balls and a vibrating chamber driven by an ultrasonic generator. For SMAT, HESP, and SNH processes, the movement of balls is also generated through collision between balls and a vibrating chamber. However, the vibration of the chamber is driven by an electric motor. PIP, in contrast, is done through a high-pressure light-gas gun that accelerates particles to a desired impact speed. Different devices invented for generating high-speed balls offer a wide range of kinetic energies to produce various degrees of surface plastic deformation. Table 31.1 lists typical values for the sizes and speeds of balls and shots and their corresponding kinetic energies reported in the literature. It can be seen that PIP has the highest kinetic energy, followed by SNH, and then HESP. SP has the lowest kinetic energy that may overlap with the kinetic energy of USSP, depending on the diameter and velocity of balls and shots used. The kinetic energies of balls and shots are an important parameter in determining the degree of surface severe plastic deformation. As described in the next several sections, many studies, both experiments and

TABLE 31.1 Typical Parameters of Balls and Shots Used in SP and High-Energy Ball Impact Processes

Process	Diameter of balls or shots (mm)	Impact velocity (m/s)	Kinetic energy of balls or shots[a] (J)
SP[12,13]	0.25–1.0	20–150	9.2×10^{-6}–0.01
USSP[4,14]	0.4–3.0	<20	0.0001–0.02
HESP[b] [8]	4.0–8.0	2–3	<0.018
SNH[9,10,23]	4.0–8.0	5–15	0.0063–0.43
PIP[11,12]	4	120	1.88

[a]Kinetic energy is also dependent on the density of the ball and shot used, which is not listed here.
[b]SMAT has the same kinetic energy as HESP because similar equipment is used.

modeling, show unambiguously that the depth of plastic deformation zone is directly proportional to the kinetic energy of impacting balls and shots.[13,20–23]

31.3 Fundamentals of Surface Deformation

31.3.1 Processing Parameters

The main parameters of shot peening and high-energy ball impact processes are: (1) the radius of the shot or ball, (2) the impact velocity of the shot or ball, (3) the density of the shot or ball, (4) mechanical properties of the shot (or ball) and the workpiece, and (5) processing time. These processing parameters collectively affect the final microstructure, stress state, surface roughness, and properties of the processed workpiece. The major alternations induced by shot peening and high-energy ball impact processes to the surface region of the workpiece are: (1) introduction of residual compressive stresses, (2) formation of a work-hardened layer, (3) increased surface roughness, and (4) sometimes formation of a nc surface layer. Effects of the processing parameters on these alternations and the associated mechanisms are discussed in the following sections.

31.3.2 Development of Deformation Fields

Many studies of analytical and numerical approaches have been conducted to establish the deformation field of the workpiece induced by high-energy shots and balls.[13,21] The analytical approaches are mainly based on the Hertz theory of elastic contact between a sphere and a semi-infinite solid.[27] The normal pressure over the contact area upon impact and the stress field within the deformed solid can be calculated if the following two rules are allowed;

1. The stress field generated by elastic impact (with moderate velocities) is identical to that generated by elastic contact.[24]
2. The normal force upon impact is due to the deceleration of the moving ball.[25]

With these two rules, dynamic loading can be approximated by static contact, and the distribution of the normal pressure, $P(r)$, over a circular contact area of radius, a (Figure 31.1), can be calculated with the aid of[25]

$$P(r) = P_0 \frac{\sqrt{a^2 - r^2}}{a} \tag{31.1}$$

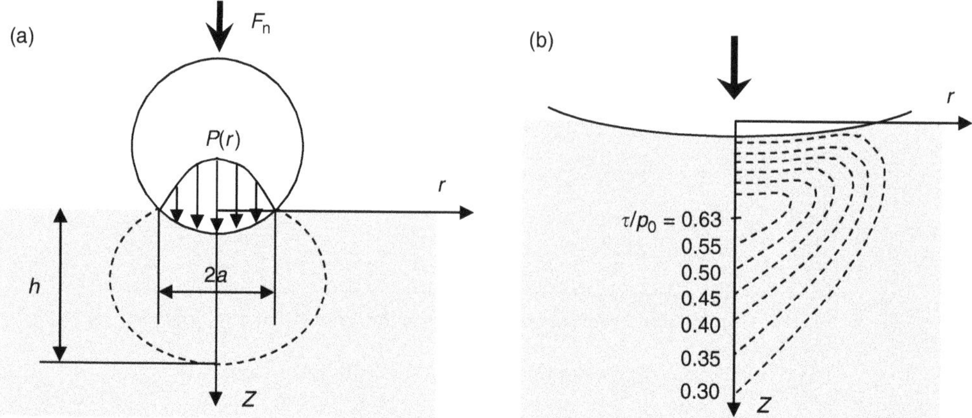

FIGURE 31.1 (a) Schematic of the distribution of Hertz pressure during ball impact; the area within the dash line represents the plastic zone in the deformed solid. (b) Schematic of the contours of the maximum shear stress within the deformed solid, normalized by the maximum pressure in the circle of contact.

where r is the distance from the center of the contact circle and P_0 is the maximum pressure in the circle of contact and given by[25]

$$P_0 = 1.145 E_H^{2/3} F_n^{1/3} R^{2/3} \left(\frac{1}{\pi}\right) \tag{31.2}$$

where R is the radius of the ball, E_H is the equivalent modulus related to Young's moduli and Poisson's ratios of the ball and solid materials, and F_n is the maximum normal force upon impact. The maximum normal force upon impact is related to the deceleration of the impacting ball, and can be calculated via[25]

$$F_n = \frac{2}{3} R^2 (2.5\pi\rho)^{3/5} v_b^{6/5} E_S^{2/5} \tag{31.3}$$

where v_b is the impact velocity of the ball, E_S is the elastic modulus of the solid, and ρ is the ball density. This maximum normal force can be treated as a concentrated load and used to compute the stress field within the impacted solid, as described by Shaw and DeSalvo.[28]

Figure 31.1a shows schematically the distribution of the normal pressure, $P(r)$, over a circular contact area and the plastic-elastic boundary within the impacted solid, while Figure 31.1b shows the contours of the maximum shear stress within the impacted solid normalized by the maximum pressure in the circle of contact. Based on the maximum shear Tresca yield criterion, plastic flow of the impacted solid will start at the point where $\tau/p_0 = 0.63$, and the plastic zone will expand following the contours defined in Figure 31.1b as the impact load increases. The depth of plastic zone, h, has been calculated for an elastic, perfectly plastic solid to be $1.816a$ where a is the radius of the contact circle,[28] and can be related to the parameters of shot peening and high-energy ball impact processes with the aid of[13]

$$\frac{h}{R} = 3\left(\frac{2}{3}\right)^{1/4} \left(\frac{\rho v_b}{\overline{p}}\right)^{1/4} \tag{31.4}$$

where $\overline{P}$ is the mean normal pressure in the circle of contact and equal to 2/3 P_0. R, ρ, and v_b have been defined before. The depth of plastic zone estimated from Equation 31.4 does not include work hardening, strain rate sensitivity, and thermal effects. Nevertheless, Equation 31.4 has been shown to be adequate in estimating the depth of plastic zone when compared with experiments.[13,28] Even more importantly, Equation 31.4 defines the relationship between the depth of plastic zone and the parameters of shot and

TABLE 31.2 Results of Finite Element Simulation of SNH and SP Processes

Process	Ball diameter (mm)	Ball density (g/cm^3)	Impact velocity (m/s)	Kinetic energy (J)	Depth of plastic zone (mm)	Maximum effective plastic strain
SNH	7.86	15	5	0.0476	1.6	0.18
SP	0.3	15	50	0.000265	0.15	0.00175
Hypo[a]	0.3	15	670	0.0476	1.2	0.315

[a] A hypothetical case that has the same kinetic energy as SNH, but with an extremely high impact velocity and a shot size equal to that in SP.

ball impact processes. As the mean normal pressure is proportional to the hardness of the impacted solid, Equation 31.4 reveals that the depth of plastic zone increases with shot (or ball) size, density, and velocity, and decreases with an increase in the hardness of the impacted solid. Based on these relationships, it can then be concluded that the depth of plastic zone increases with the kinetic energy of balls and shots.

The depth of plastic zone is one of the most important parameters because it determines the thickness of the work-hardened layer as well as the thickness of the surface layer within which residual compressive stresses are present. Finite element modeling has shown that the depth of plastic zone coincides with the thickness of the surface layer with residual compressive stresses.[20] Table 31.2 lists the depth of plastic zone for several shot- and ball-impact processes, computed via finite element modeling of dynamic impact of a WC/Co ball on a nickel-based C-2000® Hastelloy plate using LS-DYNA commercial code. The C- 2000 alloy is a single-phase material with a face-centered-cubic (fcc) crystal structure and a nominal chemical composition of (in weight percent): 23Cr, 16Mo, 1.6Cu, 0.01C, 0.08Si, and balance Ni. As expected, the depth of plastic zone increases with the kinetic energy of balls and shots. Furthermore, the finite element modeling also reveals that the depth of plastic zone increases with shot size for a given kinetic energy. Such a conclusion can also be obtained from the analytical formula by combining Equations 31.2, 31.3, and 31.4. Thus, high-energy ball impact processes can produce a deeper plastically deformed layer than shot peening even when balls and shots in these processes have the same kinetic energy.

Table 31.2 also lists the maximum effective plastic strain induced by impacts of balls and shots. The location of the maximum effective plastic strain is found not to be at the very impacted surface, but slightly below that surface. As expected, the maximum effective plastic strain increases with the kinetic energy of shots and balls. However, it is interesting to note that for a constant kinetic energy smaller shots generate larger effective plastic strain than large balls, even though the depth of plastic zone is smaller. Large plastic strain is necessary for the formation of nano-grains.[11] Thus, the finite element simulation suggests that for a given kinetic energy smaller balls are quicker in creating a nc surface layer than larger balls, whereas larger balls are more effective in creating a thicker nc surface layer than smaller balls.

31.3.3 Local Property Alternation

Plastic deformation during shot peening and high-energy ball impact processes will alter the surface property of the workpiece. Shown in Figure 31.2 are Vickers microhardness profiles of a nickel-based C-2000 Hastelloy plate treated with the SNH process. Corresponding to the largest plastic deformation and thus the highest work hardening, the region near the impacted surface exhibits the highest hardness. This high hardness derives mainly from work hardening because the nc surface layer is too thin ($\sim$50 μm for the sample processed for 180 min) to be measured with a Vickers indenter. It is also noted that the hardness decreases gradually as the location moves away from the impacted surface. Clearly, the decrease in hardness is due to the reduced plastic deformation as the position moves deeper into the solid. Commercially pure Ti processed using SP and HESP also exhibit similar phenomena.[8,15] However, the most interesting phenomenon to be noted in Figure 31.2 is the surface hardening saturation, that is, no additional hardening is obtained beyond 30-min processing. Such a phenomenon has also been observed with an Al-5052 alloy (not shown here). Surface hardening saturation has important implications in

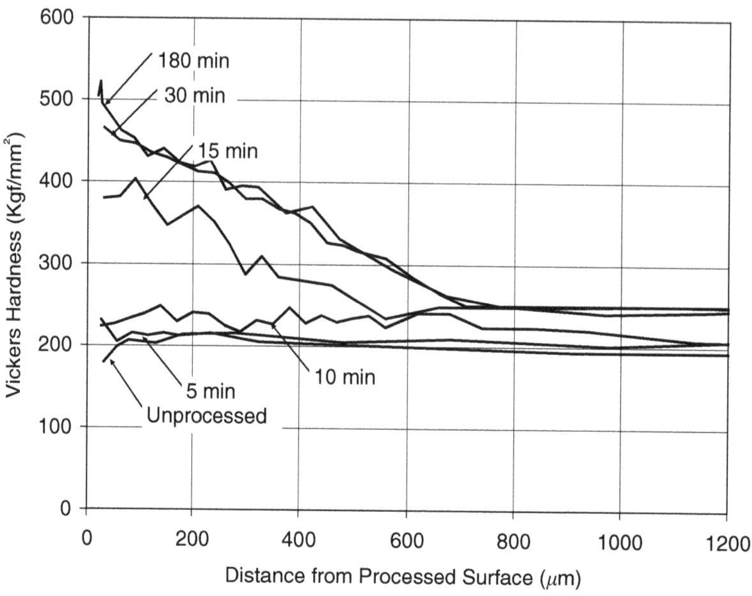

FIGURE 31.2 Vickers hardness profiles of nickel C-2000 specimens treated with the SNH process for various periods of time.

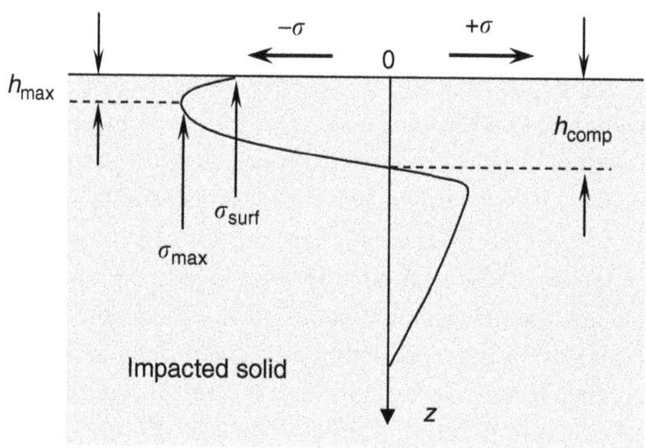

FIGURE 31.3 The typical residual stress profile produced by shot peening.

process optimization because prolonged processing could result in surface contamination of the workpiece by impacting balls.[29]

31.3.4 Residual Stresses

Plastic deformation on the top layer of the impacted surface not only causes the local property alternation, but also results in the formation of residual compressive stresses. Plastic deformation causes stretching of the top layer of the workpiece. Upon unloading, the elastically stressed sub-surface layers tend to recover their original dimensions, but the continuity of the material in both zones, the elastic and the plastic, does not allow this to occur. As a result, a residual compressive stress field followed by a tensile field is formed in the impacted workpiece. Figure 31.3 shows the typical residual stress profile generated via shot peening. The most important characteristics of the residual compressive stress field are (a) the depth of the maximum residual compressive stress, h_{max}, (b) the depth at which the residual stresses change sign,

h_{comp}, (c) the value of the surface residual stress, σ_{surf}, and (d) the value of the maximum compressive stress, σ_{max}. When the solid to be processed is very soft, σ_{max} can appear at the impacted surface rather than within the impacted solid.[30]

Both experiments and finite element modeling indicate that h_{comp} increases with the peening intensity and the kinetic energy of shots.[20,21,30] For a given kinetic energy, h_{comp} increases with shot size when the size of shots approaches 1 mm or larger.[21] However, h_{comp} becomes insensitive to the shot size when the shot size is below 0.7 mm while keeping the kinetic energy constant.[21] h_{max} also increases with the kinetic energy of shots, but is much less sensitive to the kinetic energy than h_{comp}. When keeping the kinetic energy constant, changing shot size or impact velocity has little influence on h_{max}. The values of σ_{surf} and σ_{max} are insensitive to the processing parameters (i.e., shot size, shot velocity, and shot kinetic energy). They are mainly determined by the mechanical characteristics of the impacted solid.[20,30] Although increasing the kinetic energy of shots increases the volume of the plastically deformed material and the effective plastic strain, σ_{surf} and σ_{max} remain unchanged.[21] Both σ_{surf} and σ_{max} increase with the yield strength of the impacted solid.[20,21,30,31] Furthermore, σ_{max} also exhibits strong correlation with the tensile strength of the impacted solid.[30] However, σ_{max} increases and σ_{surf} decreases with increases in the work-hardening rate of the impacted solid.[21]

The data presented above clearly indicate that h_{max}, σ_{surf}, and σ_{max} are insensitive to the processing parameters of shot peening and high-energy ball impact processes. Thus, when the kinetic energy of shots (or balls) and the impacted solid are fixed, large shots coupled with short processing times are preferred because such processing conditions will result in a thicker surface layer with residual compressive stresses and less surface contamination. Many studies[17] have revealed that fatigue life improves when a thicker surface layer with residual compressive stresses is present.

It should be mentioned that the magnitude of residual compressive stresses produced from surface deformation processes can be estimated using the Almen intensity, a widely known method and originally introduced by Almen.[32] The Almen intensity, determined from the residual arc height over a fixed length after the surface deformation process, is currently used by industry in the control of the shot peening process because of its simplicity. However, it should be cautious when using the Almen intensity to quantify the residual stresses because different residual stress profiles can result in the same Almen intensity.[21]

31.3.5 Evolution of Surface Roughness and Contamination

Surface roughness typically increases after shot peening and high-energy ball impacts.[9,23,29] Detailed studies[9,23] indicate that surface roughening in high-energy ball impact processes can be explained by the indentation process of impacting balls, and the surface roughness evolution can be divided into three stages: the roughness increase stage, the roughness decrease stage, and finally the steady-state stage. These three stages are related to different stages of the surface coverage by indents generated by impacting balls. The steady-state stage corresponds to the impact coverage of the entire surface multiple times, and the surface roughness at the steady-state increases with the shot size.

If cold adhesion between balls and the impacted solid takes place during processing, the subsequent separation of balls from the impacted solid will result in surface contamination and may lead to continued increases in surface roughness beyond the steady-state.[29] This will inevitably degrade the properties and performance of the impacted workpiece. Many studies have indeed revealed that optimization in the shot peening time is necessary; otherwise, the fatigue resistance of the impacted workpiece decreases rather than increases.[16]

31.3.6 Microstructure Changes

Plastic deformation induced by shot and ball impacts can dramatically change the microstructure at and near the surface region. Figure 31.4 shows the cross-sectional optical microstructure of a nickel C-2000 sample after SNH-processing for 180 min. As already mentioned, the C-2000 alloy is a single-phase

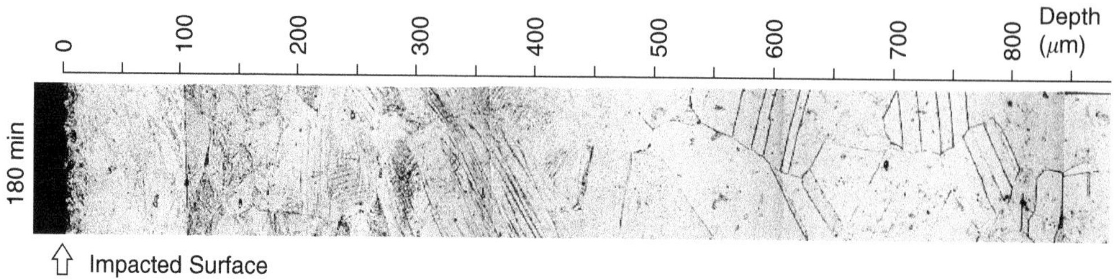

FIGURE 31.4 The cross-sectional optical microstructure of a nickel C-2000 sample after SNH-processing for 180 min.

material with a fcc crystal structure and a very low stacking fault energy (~1.22 mJ/m^2).[33] Note that many markings (i.e., straight and bent lines) are visible near the impacted surface region. These markings are confined within individual grains, and have been identified as deformation twins rather than deformation bands or shear bands, as revealed by the TEM analysis.[10] The density of deformation twin markings increases as the location approaches the impacted surface. However, right at the impacted surface the density of deformation twin markings are so high that individual twin markings can no longer be distinguishable. In fact, grain boundaries are also indistinguishable at the impacted surface region. TEM analyses reveal that this very surface region corresponds to a nc layer about 50 μm thick.[33] It is believed that deformation of the Ni C-2000 alloy, a low stacking-fault energy (SFE) material, during the SNH process is dominated by deformation twinning, and the surface nanocrystallization proceeds mainly with twin–twin intersections and subsequent interactions of the micro- and nano-twins with dislocations trapped within twins.[33]

Many studies[4–7] have shown that the dominating deformation mode and surface nanocrystallization mechanism in high-energy ball impact processes depend on the crystal structure and SFE of the impacted solid. Details of deformation mechanisms, microstructural evolution, and their dependency on material characteristics especially the crystal structure and SFE are described in the next section.

31.4 Microstructure Evolution

As shown in Section 31.3.6, surface deformation processes produce different grain refinement stages at different depths from the surface. This provides an opportunity to study the complete microstructural evolution in one sample. Furthermore, all of the surface deformation processes refine surface microstructures by plastic deformation. As such, their grain refinement processes should be similar although the depth of the refined grains may vary with the processing parameters associated with each technique.

Plastic deformation refines grains by accumulation and interaction of crystalline defects such as dislocations, twins, dislocation cells, grain boundaries, and subgrain boundaries. Therefore, it is expected that the deformation mode and mechanisms of a particular alloy affect the microstructural evolution. It is well known that crystalline structure and SFE are top two intrinsic material factors that affect the deformation mechanisms of metals and alloys. Specifically, face-centered-cubic (fcc) and body-centered cubic (bcc) systems have abundant dislocations slip systems to accommodate strains and therefore are less likely to deform by twinning. In contrast, twinning becomes necessary in hexagonal-close-packed (hcp) metals and alloys because they do not have the five independent slip systems required by the von Mises criterion for materials to deform continuously without void formation. SFE affects the generation and slip of partial dislocations associated with deformation twinning. Low SFE promotes the formation of deformation twins. In the following, the discussion on microstructural evolutions will be divided into three subsections: fcc metals and alloys, bcc metals and alloys, and hcp metals and alloys.

31.4.1 Microstructural Evolution in fcc Metals and Alloys

In this section, the microstrucural evolution of fcc metals with medium to high SFEs will be discussed first. This is then followed by discussion of fcc metals and alloys with low SFEs.

The microstrucural evolution of fcc metals with medium to high SFEs under large plastic deformation is well studied, especially in the early stage of grain refinement.[34–39] These metals deform primarily by dislocation slip. In the first stage of grain refinement, each grain is divided into many volume elements during plastic deformation[34–38] and there are differences in the number and selection of active slip systems among neighboring volume elements.[37,38] Each volume element deforms under a reduced number (<5) of slip systems, but a group of adjacent volumes act collectively to fulfill the von Mises criterion. Each volume element is usually subdivided into cells with dislocations forming cell boundaries. For this reason, the volume elements are referred to as cell blocks. Dislocations from neighboring cell blocks meet at their boundaries and interact to form cell block boundaries. This type of boundaries are called geometrically necessary boundaries as they are needed to accommodate the misorientation in neighboring cell blocks. The dislocation cell boundaries are called incidental boundaries as they are generated by statistical mutual trapping of glide dislocations.[38] The misorientations are very small across cell boundaries but are much larger across cell-block boundaries. The cell-block boundaries consist of dislocation walls as well as strings of small dislocation cells,[37,39] which are usually along the {111} slip planes.

In the second stage of grain refinement during a surface deformation process, with increasing surface deformation strains, the cell blocks become elongated and misorientations across cell and cell-block boundaries increases. Also the widths of cell blocks become smaller due to further division and elongation, and finally reduce to a point where the cell-block width equals the dislocations cell size, forming lamellar subgrains (see Figure 31.5).[5] Inside the lamellar subgrains are near equiaxed dislocation cells aligned along the longitudinal axis of the subgrains. In the third stage of grain refinement, misorientations across dislocations cell boundaries increase with further plastic strain so that they transform into subgrain boundaries. This effectively converts the lamellar subgrains into a string of equiaxed subgrains. In the fourth stage, the equiaxed subgrains further divide into smaller dislocation cells, which in turn convert into smaller subgrains as well as nanometer-sized grains, as their misorientations increases with increasing strain. Grain rotation may play a significant role in the formation of the nanometer-sized grains with high angle boundaries.[5]

Due to the intrinsic strain gradient, microstructures associated with the above four stages exist at different depths from the specimen surface. Note that for fcc metals with medium SFE such as

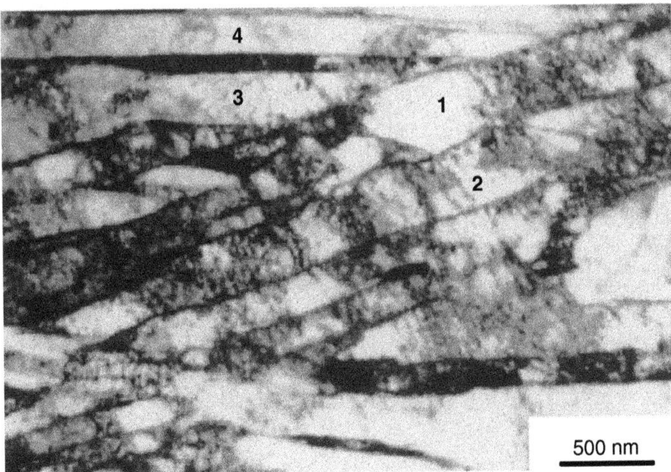

FIGURE 31.5 Lamellar subgrain structure at a depth of 60 mm from surface in a 7075 alloy specimen processed by ultrasonic shot peening. (Wu, X. et al., *Acta Mater.*, 50, 2075, 2002.) Dislocation cells are formed along the subgrains, which will transform into equiaxed subgrains upon further deformation.

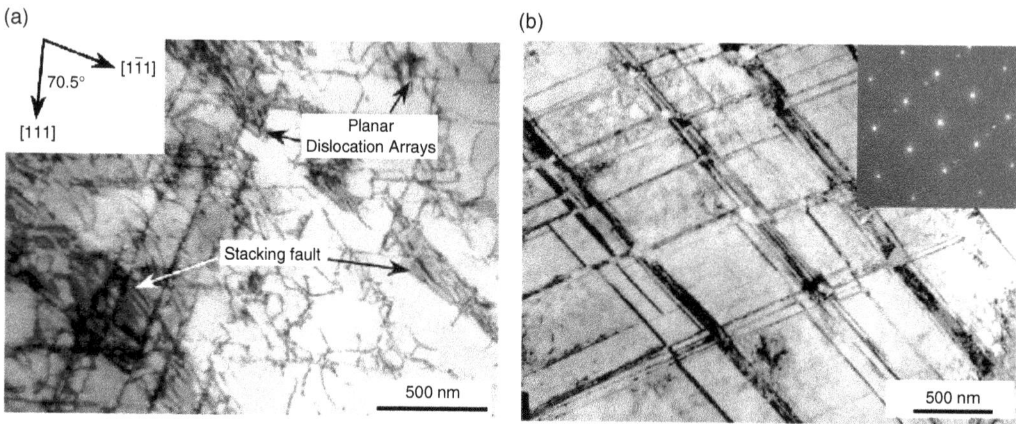

FIGURE 31.6 Microstructures formed in AISI 304 stainless steel processed by surface mechanical attrition treatment (SMAT). (a) Dislocation grids formed along {111} slip planes at a depth of 300 mm from the surface. (b) Twin grids formed at a depth of 300 mm. (Zhang, H.W. et al., *Acta Mater.*, 51, 1871, 2003.)

Cu, deformation twins may exist in nanograins smaller than 50 nm, because smaller grains promote deformation twinning.[40,41]

FCC metals and alloys with low SFE deform by both dislocation slip and twinning. In addition, lattice dislocations often dissociate into pairs of partial dislocations separated by wide stacking fault ribbons. These dislocations usually stay on one slip plane. It is hard for them to cross-slip into another plane, or to climb into a parallel plane to react with another dislocation because these two processes require the partials to first recombine into a lattice dislocation. Not surprisingly, these deformation characteristics were found to affect the grain refinement processes, as demonstrated in an AISI 304 stainless steels.[42]

In the 304 stainless steel processed by SMAT, abundant mechanical twins were observed in locations as deep as 300 μm from the hardened surface. Transmission electron microscopy (TEM) revealed dislocation arrays along {111} planes that form dislocation grids at a depth of 300 μm (see Figure 31.6a). At higher strains (closer to the surface at the depth of 150 μm), the dislocation grids transformed to twin grids (Figure 31.6b). With increasing strain (at a depth of 100 μm), the twins become thicker; twin grids become smaller and strain-induced martensitic transformation occurs at twin cross sections. At depths less than 40 μm, nanocrystalline grains with structures of both martensite and the initial fcc structures co-exist. In the final stage of nanostructure formation, grain boundary sliding, grain rotation, and recrystallization are believed to play major roles in the formation of high angle grain boundaries.[42]

The above grain refinement process, if the martensitic transformation is disregarded, should also be valid for fcc metals without martensitic transformation. This is indeed the case as found in a recent study on SNH processing of the very low SFE nickel C-2000 alloy (1.22 mJ/m^2).[33] Even at relative short processing times (60 min) and at locations far away from the impacted surface (600 μm), high densities of parallel micro- and nano-twins and twin intersections are generated. Inside these micro- and nano-twins many dislocations are trapped with the dislocation density as high as 1.5×10^{15} m^{-2}.[33] The twin intersections have divided the original coarse grains into submicron- and nano- sized regions. Subsequent interactions between twin boundaries and trapped dislocations further break down submicron-sized regions, result in curved twin boundaries with multiple steps, increase misorientations between neighboring nano-sized regions, and lead to the formation of diffuse, nonequilibrium, high-angle grain boundaries. Collectively, these various twin-dislocation interactions result in the final formation of polycrystalline equiaxed nanograins at the impacted surface.[33]

31.4.2 Microstructural Evolution in bcc Metals and Alloys

The microstructural evolution of bcc metals and alloys under a surface deformation processing is demonstrated in Fe processed by SMAT.[6] At a depth of 60 to 80 μm from the surface, parallel dense dislocation walls on {110}, {112}, and {123} slip planes were observed. With increasing strain (closer to the surface), these dense dislocation walls transform into sharper lamellar subgrains while transverse dislocation cell walls (dense dislocation walls) cut the subgrains into roughly equiaxed segments. This process repeats itself with higher strains, further dividing the grains and subgrains and finally leading to the formation of the nanostructures. In the process, the subgrain boundaries formed at early stages will later evolve into high angle grain boundaries, and dense dislocation walls will develop into subgrain boundaries. In the final stage of nanocrystallization, grain boundary sliding and grain rotation are also believed to play an important role in forming high angle boundaries. Interestingly, the nanocrystalline grains are found to form parallel arrays, indicating that they are formed from lamellae with nanoscale width. On the other hand, a microstructural evolution path that involves equiaxed subgrains, not the lamellar subgrains, was also found (see Figure 31.7).

31.4.3 Microstructural Evolution in hcp Metals and Alloys

The microstructural evolution of hcp metals and alloys during surface deformation processes is demonstrated in commercially pure Ti processed by SMAT.[7] At a depth of 300 μm where the material was subjected to relatively low strain, abundant deformation twins were formed, and dislocations were piled at twin boundaries or tangled. The activation of twinning is due to the fact that the hcp crystalline system does not have the five independent dislocation slip systems to satisfy the von Mises criterion. With increasing strain at a depth of 150 μm, lamellar structure and equiaxed subgrains of a few hundred nanometers was formed, similar to microstructures formed in Ti processed by equal channel angular pressing and cold rolling.[43] Some of the lamellar boundaries are sharp and straight, which could be from the initial twin boundaries.

With the strain increasing further at 60 to 70 μm, mostly equiaxed grains with sizes of 100 to 300 nm were formed, and no traces of twins could be observed. The grains were further refined to 50 to 250 nm at a depth of 15 to 30 μm. The lack of deformation twins at these small grain sizes suggest that other deformation mechanisms such as grain boundary sliding and grain rotation might have played roles in accommodating the applied strain so that twinning is no longer needed. These observations form a sharp contrast with fcc metals and alloys that tend to favor twinning when their grains are in the nanometer range.[40,41,44]

31.5 Property Enhancements

31.5.1 Fatigue Resistance

Shot peening improves fatigue resistance[2,3,16–18,45] so do high-energy ball impact processes.[10] The improvement in the case of shot peening is generally attributed to the presence of residual compressive stresses at the surface region. Furthermore, many studies[2,16,18,46] seem to support the notion that improvements in fatigue resistance are mainly due to the ability of residual compressive stresses in stopping the microcrack propagation and not in preventing fatigue crack initiation. This notion is consistent with the fact that a thicker surface layer with residual compressive stresses is more effective in improving the fatigue strength,[17] and larger shots, keeping other parameters constant, lead to better fatigue life.[46] Based on these results, it can be predicted that high-energy ball impact processes are likely to improve fatigue resistance more than shot peening because larger balls are used and a thicker surface layer with residual compressive stresses is created in high-energy ball impact processes (Table 31.2).

The roles of the nc surface layer and work-hardened region in improving fatigue resistance have not been well quantified yet. Nevertheless, a recent study[10] using finite element modeling to analyze the fatigue resistance improvement via the SNH process has shown that the work-hardened layer contributes more to the fatigue resistance improvement than residual compressive stresses. However, this may not be the case for shot peening because high-energy ball impact processes create a thicker work-hardened layer than

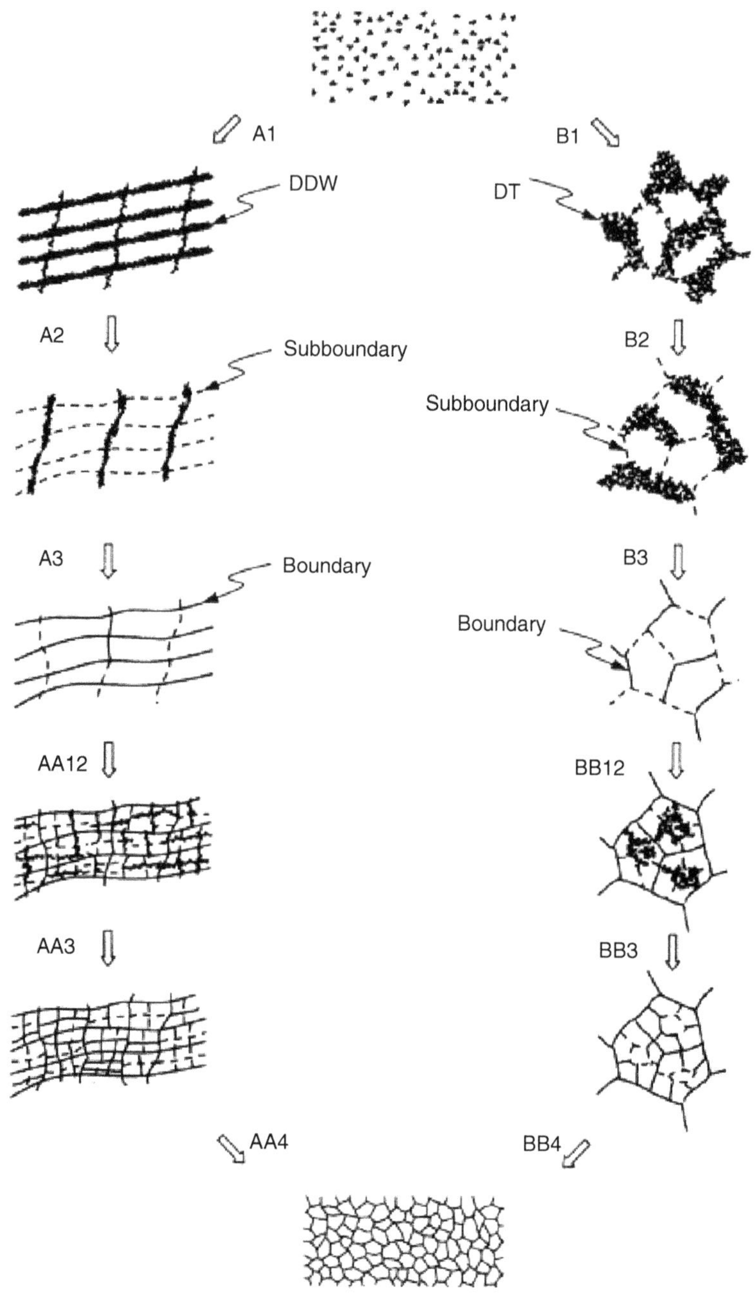

FIGURE 31.7 An illustration of two grain refinement processes in bcc Fe processed by SMAT. DDW means dense dislocation walls, DT — dislocation triangles. (Tao, N.R. et al., *Acta Mater.*, 50, 4603, 2002.)

shot peening. Clearly, more detailed studies are needed to define the individual contributions of residual compressive stresses, work-hardened layer, and nanograins at the surface.

As discussed in Section 31.3.5, surface roughness increases and surface contamination may occur during shot peening and high-energy ball impact processes.[9,29] The increased surface roughness and surface contamination could be harmful to fatigue resistance. As such, fatigue resistance improvements may not be observed if shot peening and high-energy ball impact processes are not performed properly.

This is the case when shot peening time exceeds the optimal duration.[16] "Over-processing" has also been observed with the nickel C-2000 alloy treated with the SNH process.[47]

31.5.2 Wear and Fretting Resistance

Both wear and fretting are surface-contact dominated processes, and thus it is expected that surface deformation processes can improve wear and fretting resistance if the processing conditions are properly selected. This is indeed the case, as reported by many studies.[48–53] Steels after shot peening have been shown to have better wear and fretting resistances than their counterparts without shot peening.[52,53] Ti alloys exhibit improvements in both fretting wear and fretting fatigue after shot peening.[49,50] The surface mechanical attrition treatment reduces the friction coefficient of steels against a diamond stylus, and improves the wear resistance of steels.[51]

All of the improvements via surface deformation processes listed above are related to one or all of the following mechanisms: (1) the presence of residual compressive stresses, (2) the decreased coefficient of friction, (3) the increased hardness, (4) the formation of a nc surface layer, and (5) the altered surface roughness. The function of compressive stresses is to close up microcracks at the surface induced by wear and fretting, and prevent them from propagation. Work hardening and a nanocrystalline surface layer can contribute to the increased hardness which increases abrasive wear resistance, decreases the friction coefficient, and may prevent adhesion with the antagonist. Surface roughness has dual functions; it could reduce wear and fretting resistance if a rough surface acts as potential stress raisers. However, a high degree of surface finish can increase the friction coefficient, and thus accentuate wear and fretting damage. The dual functions of surface roughness underscore the importance of process optimization because not all shot-peened solids exhibit improvements.[54]

31.5.3 Tensile Strength

As surface deformation processes only alter surface properties, it is expected that these processes cannot change the tensile strength of bulk materials significantly. However, when bulk materials are in a plate form (or a rod geometry), a 35% improvement in the tensile yield strength of mild steels with minimum degradation in ductility and toughness has been demonstrated using HESP.[55] A recent study[56] indicates that about 100% improvements in the tensile yield strength of the nickel C-2000 alloy can be achieved by SNH. Furthermore, it is noted that the ultimate tensile strength of the nickel C-2000 alloy is also increased by SNH, suggesting that the improvement in tensile strength is not due to work hardening alone, but also due to the contribution from the nc surface layer.

31.6 Future Developments

The latest development in high-energy ball impact processes, which have higher kinetic energy than shot peening, has offered new opportunities for surface deformation processes. Because of their high kinetic energies, high-energy ball impact processes can produce a thicker plastically deformed layer with residual compressive stresses and a thicker nc surface layer as well than shot peening. It is expected that these more pronounced microstructural and stress-state changes induced by high-energy ball impact processes can offer better improvements in fatigue, wear, and fretting resistances than shot peening can achieve. However, no direct comparisons have ever been conducted. Systematic studies in this area are anticipated in the near future.

Studies on separating individual contributions of residual compressive stresses, a work-hardened layer, and surface nanograins are scarce, and expected to be actively pursued in the near future. The understanding developed from such studies is essential for process optimization and full utilization of the potential of surface deformation processes.

Integration of high-energy ball impact processes with other surface treatments, such as carburizing and nitriding, can provide new avenues for strengthening materials. A recent work[57] has demonstrated that the nc surface of a pure iron plate created via SMAT can greatly enhance the nitriding reaction so that nitriding can be accomplished at 300°C rather than 500°C used in conventional nitriding processes. Such a trend in integrating several processing approaches into one to achieve synergy is expected to continue.

References

[1] Fuchs, H.O., *Mechanical Engineer's Handbook*, Wiley, New York, 1986.

[2] Batista, A.C. et al., Contact fatigue of automotive gears: Evolution and effects of residual stresses introduced by surface treatments, *Fatigue Fract. Eng. Mater. Struct.*, 23, 217, 2000.

[3] Liu, D. and He, J., Effect of shot peening factors on fretting fatigue resistance of titanium alloys, *Acta Metall. Sinica*, 37, 156, 2001.

[4] Tao, N.R. et al., Surface nanocrystallization of iron induced by ultrasonic shot peening, *Nanostruct. Mater.*, 11, 433, 1999.

[5] Wu, X. et al., Microstructure and evolution of mechanically induced ultrafine grain in surface layer of Al-alloy subjected to USSP, *Acta Mater.*, 50, 2075, 2002.

[6] Tao, N.R. et al., An investigation of surface nanocrystallization mechanism in Fe induced by surface mechanical attrition treatment, *Acta Mater.*, 50, 4603, 2002.

[7] Zhu, K.Y. et al., Nanostructure formation mechanism of α-titanium using SMAT, *Acta Mater.*, 52, 4101, 2004.

[8] Chen, C. et al., Surface nanostructures in commercial pure Ti induced by high energy shot peening, *Trans. Nonferrous Metals Soc. China*, 14, 215, 2004.

[9] Dai, K., Villegas, J., and Shaw, L., An analytical model of the surface roughness of an aluminum alloy treated with a surface nanocrystallization and hardening process, *Scripta Mater.*, 52, 259, 2004.

[10] Villegas, J. et al., Enhanced fatigue resistance of a nickel-based Hastelloy induced by a surface nanocrystallization and hardening process, *Phil. Mag. Lett.*, 85, 427, 2005.

[11] M. Umemoto, M., Todaka, K., and Tsuchiya, K., Formation of nanocrystalline structure in carbon steels by ball drop and particle impact techniques, *Mater. Sci. Eng.*, A375–A377, 899, 2004.

[12] Todaka, K. et al., Formation of nanocrystalline structure in steels by air blast shot peening and particle impact processing, *Mater. Sci. Forum*, 449–452, 1149, 2004.

[13] Al-Obaid, Y.F., Shot peening mechanics: Experimental and theoretical analysis, *Mech. Mater.*, 19, 251, 1995.

[14] Todaka, Y., Umemoto, M., and Tsuchiya, K., Comparison of nanocrystalline surface layer in steels formed by air blast and ultrasonic shot peening, *Mater. Trans.*, 45, 376, 2004.

[15] Ma, G. et al., Surface nanocrystallization of commercial pure titanium by shot peening, *Trans. Nonferrous Metals Soc. China*, 14, 204, 2004.

[16] Wagner, L., Mechanical surface treatments on titanium, aluminum and magnesium alloys, *Mater. Sci. Eng.*, A263, 210, 1999.

[17] Drechsler, A., Dorr, T., and Wagner, L., Mechanical surface treatments on Ti-10V-2Fe-3Al for improved fatigue resistance, *Mater. Sci. Eng.*, A243, 217, 1998.

[18] Song, P.S. and Wen, C.C., Crack closure and crack growth behavior in shot peened fatigued specimen, *Eng. Fract. Mech.*, 63, 295, 1999.

[19] Lu, K. and Lu, J., Surface nanocrystallization (SNC) of metallic materials — Presentation of the concept behind a new approach, *J. Mater. Sci. Technol.*, 15, 193, 1999.

[20] Meguid, S.A. et al., Three-dimensional dynamic finite element analysis of shot-peening induced residual stress, *Finite Elem. Anal. Design*, 31, 179, 1999.

[21] Guagliano, M., Relating Almen intensity to residual stresses induced by shot peening: A numerical approach, *J. Mater. Proc. Technol.*, 110, 277, 2001.

[22] Davies, R.M., The determination of static and dynamic yield stresses using a steel ball, *Proc. Royal Soc. London*, 197A, 416, 1949.

[23] Dai, K. et al., Finite element modeling of the surface roughness of 5052 Al alloy subjected to a surface severe plastic deformation process, *Acta Mater.*, 52, 5771, 2004.
[24] Love, A.E.H., *Mathematical Theory of Elasticity*, Cambridge University Press, 1934, 138–140.
[25] Davies, R.M., The determination of static and dynamic yield stresses using a steel ball, *Proc. R. Soc. London*, 197A, 416, 1949.
[26] Fathallah, R., Inglebert, G., and Castex, L., Prediction of plastic deformation and residual stresses induced in metallic parts by shot peening, *Mater. Sci. Technol.*, 14, 631, 1998.
[27] H. Hertz, in *On the Contact of Elastic Solids*, Jones and Schott, Eds., London Macmillan, 1896.
[28] Shaw, M.C. and DeSalvo, G.J., On the plastic flow beneath a blunt axisymmetric indenter, *Trans. ASME, J. Eng. Ind.*, 92, 480, 1970.
[29] Villegas, J., Dai, K., and Shaw, L., Surface roughness evolution in the surface nanocrystallization and hardening (SNH) process, in *Processing and Fabrication of Advanced Materials: XII*, T. Srivatsan and R. Varin, Eds., ASM International, Materials Park, OH, 2003, 358–372.
[30] Wang, S. et al., Compressive residual stress introduced by shot peening, *J. Mater. Proc. Technol.*, 73, 64, 1998.
[31] Meo, M. and Vignjevic, R., Finite element analysis of residual stress induced by shot peening process, *Adv. Eng. Software*, 34, 569, 2003.
[32] Almen, J. and Black, J.P.H., *Residual Stresses and Fatigue in Metals*, McGraw-Hill, Toronto, 1963.
[33] Villegas, J., *Investigation of the Effects of the Surface Nanocrystallization and Hardening (SNH) Process on Bulk Metallic Components*, Ph.D. thesis, University of Connecticut, 2005.
[34] Hansen, N., Cold deformation microstructures, *Mater. Sci. Technol.*, 6, 1039, 1990.
[35] Bay, B. et al., Evolution of fcc deformation structures in polyslip, *Acta Mater.*, 40, 205, 1992.
[36] Hughes, D.A. and Hansen, N., High angle boundaries formed by grain subdivision mechanisms, *Acta Mater.*, 45, 3871, 1997.
[37] Bay, B., Hansen, N., and Kuhlmann-Wilsdorf, D., Deformation structures in lightly rolled pure aluminium, *Mater. Sci. Eng.*, A113, 385, 1989.
[38] Kuhlmann-Wilsdorf, D. and Hansen, N., Geometrically necessary, incidental and subgrain boundaries, *Scripta Metall. Mater.*, 25, 1557, 1991.
[39] Huang, J.Y. et al., Microstructures and dislocation configurations in bulk nanostructured Cu processed by repetitive corrugation and straightening, *Acta Mater.*, 49, 1497, 2001.
[40] Liao, X.Z. et al., Deformation twinning in nanocrystalline copper at room temperature and low strain rate, *Appl. Phys. Lett.*, 84, 592, 2004.
[41] Liao, X.Z. et al., Grain size effect on the deformation mechanisms of nanostructured copper processed by high-pressure torsion, *J. Appl. Phys.*, 96, 636, 2004.
[42] Zhang, H.W. et al., Formation of nanostructured surface layer on AISI 304 stainless steel by means of surface mechanical attrition treatment, *Acta Mater.*, 51, 1871, 2003.
[43] Zhu, Y.T. et al., Nanostructures in Ti processed by severe plastic deformation, *J. Mater. Res.*, 18, 1908, 2003.
[44] Liao, X.Z. et al., Deformation twins in nanocrystalline Al, *Appl. Phys. Lett.*, 83, 5062, 2003.
[45] Zinn, W. and Scholtes, B., Mechanical surface treatments of lightweight materials — Effects on fatigue strength and near-surface microstructure, *J. Mater. Eng. Perf.*, 8, 145, 1999.
[46] Guagliano, M. and Vergani, L., An approach for prediction of fatigue strength of shot peened components, *Eng. Fract. Mech.*, 71, 501, 2004.
[47] Liaw, P.K., personal communication, 2005.
[48] Zhuang, W. and Wicks, B., Mechanical surface treatment technologies for gas turbine engine components, *ASME J. Eng. Gas Turbine Power*, 125, 1021, 2003.
[49] Lee, H., Jin, O., and Mall, S., Fretting fatigue behavior of shot-peened Ti-6Al-4V at room and elevated temperatures, *Fatigue Fract. Eng. Mater. Struct.*, 26, 767, 2003.
[50] Fu, Y. et al., Improvement in fretting wear and fatigue resistance of Ti-6Al-4V by application of several surface treatments and coatings, *Surf. Coat. Technol.*, 106, 193, 1998.

[51] Wang, Z.B. et al., Effect of surface nanocrystallization on friction and wear properties in low carbon steel, *Mater. Sci. Eng.*, 352, 144, 2003.
[52] Benrabah, A., Langlade, C., and Bannes, A.B., Residual stresses and fretting fatigue, *Wear,* 224, 267, 1999.
[53] Girish, D.V., Mayuram, M.M., and Krishnamurthy, S., Surface integrity studies on shot-peened thermal-treated En 24 steel spur gears, *Wear,* 193, 242, 1996.
[54] Fridrici, V., Fouvry, S., and Kapsa, Ph., Effect of shot peening on the fretting wear of Ti-6Al-4V, *Wear,* 250, 642, 2001.
[55] Liu, G. et al., Low carbon steel with nanostructured surface layer induced by high-energy shot peening, *Scripta Mater.*, 44, 1791, 2001.
[56] Shaw, L. and Liaw, P., unpublished data, 2005.
[57] Tong, W.P. et al., Nitriding iron at low temperatures, *Science,* 299, 687, 2003.

Index

Note: Page numbers in *italics* refer to illustrations.

C

D

J

K

L

M

T

Milton Keynes UK
Ingram Content Group UK Ltd.
UKHW052032071024
449327UK00027B/2523